Advances in Organic Chemistry
Methods and Results
VOLUME 9

Iminium Salts in Organic Chemistry
Part 1

Advisory Board

D. H. R. BARTON
Imperial College of Science and Technology, London, England

K. W. BENTLEY
Reckitt and Sons Ltd., Hull, England

A. J. BIRCH
The Australian National University, Canberra, Australia

CARL DJERASSI
Stanford University, California

A. ESCHENMOSER
Eidg. Technische Hochschule, Zürich, Switzerland

R. HUISGEN
Ludwig-Maximilians-Universität, Munich, Germany

E. C. KOOYMAN
Rijksuniversiteit, Leiden, The Netherlands

E. LEDERER
Institut de Chimie des Substances Naturelles, Gif-sur-Yvette, France

G. OURISSON
Université de Strasbourg, France

A. QUILICO
Instituto de Chimica Generale del Politecnico, Milan, Italy

R. A. RAPHAEL
University Chemical Laboratory, Cambridge, England

S. STALLBERG-STENHAGEN
University of Gothenberg, Sweden

G. STORK
Columbia University, New York

LORD TODD
University Chemical Laboratory, Cambridge, England

M. VISCONTINI
University of Zürich, Switzerland

A. WEISSBERGER
Eastman Kodak Company, Rochester, New York

K. WIESNER
University of New Brunswick, Fredericton, Canada

R. B. WOODWARD
Harvard University, Cambridge, Massachusetts

H. WYNBERG
University of Groningen, Holland

ADVANCES IN ORGANIC CHEMISTRY: Methods and Results
E. C. TAYLOR, *Editor*

Iminium Salts in Organic Chemistry
Part 1

Edited by

H. BÖHME

Universität Marburg/Lahn, Germany

H. G. VIEHE

Université de Louvain, Belgium

An Interscience® Publication

JOHN WILEY & SONS   New York • London • Sydney • Toronto

An Interscience® Publication

Copyright © 1976 by John Wiley & Sons, Inc.

All rights reserved. Published simultaneously in Canada.

No part of this book may be reproduced by any means, nor transmitted, nor translated into a machine language without the written permission of the publisher.

Library of Congress Cataloging in Publication Data:

Main entry under title:

Iminium salts in organic chemistry.

  (Advances in organic chemistry; v. 9, pt. 1)
  "An Interscience publication."
  Includes bibliographical references and indexes.
  1. Imines. I. Böhme, Horst, 1908– II. Viehe, Heinz Günter. III. Series.
QD251.A36 vol. 9, pt. 1 [QD401] 547'.008s [547'.04]
ISBN 0-471-90692-1   76-16155

Printed in the United States of America

10 9 8 7 6 5 4 3 2 1

Foreword

The exceptional role of the iminium grouping in many reactions that occur both in the laboratory and in nature has been recognized for a long time. Since then, however, organic chemistry has become such an extremely broad and diversified science that the enormous progress attained meanwhile in iminium chemistry, including new methods, reagents, ideas, new ways, and fields of application, may have escaped general attention. People engaged in this area have become aware that an urgent need exists for a book which not only gathers the vast amount of new material but reintegrates all of the recent achievements into a more general framework in terms of modern concepts of organic chemistry.

I strongly believe that the present work fulfills these requirements. Although I was engaged in the early discussions during the conception of this book I am now very impressed at seeing the final result. Both the editors and the authors have succeeded in creating a book from which, I am sure, the chemical community will profit for a long time.

Z. Arnold

Prague, Czechoslovakia,
February 1976

Series Editor's Note

Although most volumes in the *Advances in Organic Chemistry* series will continue to be multiauthored works presenting authoritative, critical, and timely discussions of new developments in synthetic and instrumental methodology, in line with the general objectives of the series as set forth in the Preface to previous volumes, the present volume, which will appear in two parts, marks a further expansion of the concept of *Advances*. The first departure from the normal format, as outlined above, will be found in Volume 7, which was a single-authored research monograph. The present volume is likewise devoted to a single topic, but is multiauthored and prepared under the general editorship of outside experts in the field. We hope that the rapidity of publication of the two types of research monographs in the *Advances* series will be attractive both to readers and to authors, and that the series as a whole will continue to present in a challenging, provocative, and stimulating manner new ideas, new techniques, and new methods that will become part of the classical repertoire of the practicing organic chemist.

EDWARD C. TAYLOR
Series Editor
Advances in Organic Chemistry

Preface

Research workers in nitrogen chemistry have felt the need for an adequate coverage of modern iminium salt chemistry. This book, we think, will satisfy this need.

Many discussions preceded the 1972 meeting in Marburg at which it was decided to "launch" this book. The project started with an encounter of H. G. Viehe with Z. Arnold in Prague, 1972, followed by others with L. Ghosez in Louvain, with H. Eilingsfeld, H. Pommer, and M. Pape in BASF-Ludwigshafen, with H. Bredereck in Stuttgart, with E. Küle and E. Grigat in Bayer-Leverkusen, and with C. Jutz in Munich. We feel honored and thank the authors for their extensive work and for their trust and confidence. To Prof. E. C. Taylor, the series editor, we address our repeated thanks for his masterly streamlining of this book.

May all the work serve well now!

H. Böhme
H. G. Viehe

Marburg, Germany
Louvain-la-Neuve, Belgium
August 1976

Contents

The Electronic Structure of Iminium Ions   1
    P. A. KOLLMAN

Structure Determination of Iminium Salts by Physical Methods   23
    R. MERÉNYI

Methyleniminium Salts   107
    H. BÖHME and M. HAAKE

The Vilsmeier-Haack-Arnold Acylations. C—C Bond-Forming Reactions of Chloromethyleniminium Ions   225
    C. JUTZ

Chemistry of Dichloromethyleniminium Salts (Phosgeniminium Salts)   343
    Z. JANOUSEK and H. G. VIEHE

α-Haloenamines and Keteniminium Salts   421
    L. GHOSEZ and J. MARCHAND-BRYNAERT

N-Heteroiminium Salts   533
    J. ELGUERO and C. MARZIN

Author Index   589

Subject Index   619

Advances in Organic Chemistry
Methods and Results
VOLUME 9

Iminium Salts in Organic Chemistry
Part 1

THE ELECTRONIC STRUCTURE OF IMINIUM IONS

By P. A. KOLLMAN, *Department of Pharmaceutical Chemistry, School of Pharmacy, University of California, San Francisco, California 94143*

CONTENTS

I. Introduction	1
II. Description of Methods	2
III. The Parent Iminium Ion	3
IV. Substituted Iminium Ions	8
A. Electronic Structure and Isomerization Energies	8
B. Rotational Barriers	13
C. "Stabilization" of Substituted Iminium Ions	14
D. Other Applications	18
References and Notes	19

I. Introduction

The iminium ion, $CH_2NH_2^+$, considered from the point of view of valence bond theory, contains contributions from the two resonance structures **1** and **2**:

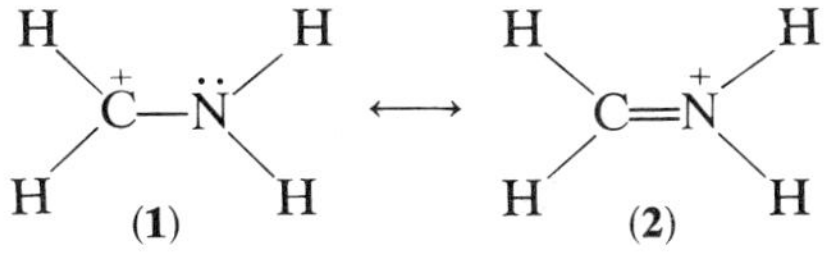

$$H_2\overset{+}{C}-\ddot{N}H_2 \quad (1) \quad \longleftrightarrow \quad H_2C=\overset{+}{N}H_2 \quad (2)$$

The ion is isoelectronic with ethylene and has many physical properties that are qualitatively similar to those of C_2H_4. From the point of view of the simple valence bond description above, however, the molecules differ somewhat, ethylene being predominately in resonance structure **4** and iminium containing significant contributions from structures **1** and **2**:

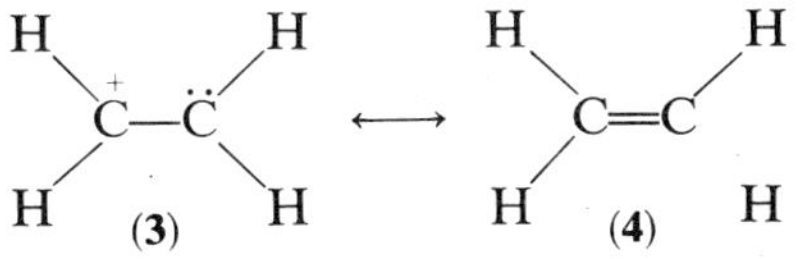

$$H_2\overset{+}{C}-\ddot{C}H_2 \quad (3) \quad \longleftrightarrow \quad H_2C=CH_2 \quad (4)$$

In any practical sense the chemistry of the two is very different; for example, a simple method of preparing alkenes is dehydration of alcohols, whereas iminium ions can be formed most simply by protonation of imines or by dissociation: $R_2NCH_2X \rightleftharpoons R_2N=CH_2^+ + X^-$.

In this chapter we examine the properties of iminium from the point of view of modern electronic structure theory. We address ourselves to two main questions. First, what do quantum-mechanical calculations predict for the structural and spectral properties of iminium? Second, what do these calculations predict for the relative stabilities, rotational barriers, and electronic structures of substituted iminium ions?

II. Description of Methods

All of the calculations described in this chapter solve the quantum-mechanical Schrödinger equation variationally, using an atomic orbital basis to represent molecular orbitals. The wave function is a single Slater determinant constructed from these molecular orbitals. This type of calculation is often referred to as LCAO-MO-SCF (Hartree-Fock): Linear Combination of Atomic Orbitals-Molecular Orbitals-determined via a Self-Consistent Field procedure. The optimum molecular energy determined that constrains the wave function to be a single determinant is called the Hartree-Fock energy; this is always greater than the exact molecular total energy. The difference between the exact (nonrelativistic) molecular energy and the Hartree-Fock energy is termed the correlation energy.

Here we consider semiempirical and nonempirical LCAO-MO-SCF solutions of the Schrödinger equation. There are many all-valence electron semiempirical methods, but the one discussed in this chapter is the CNDO/2 procedure described by Pople and Beveridge (1). This method was parameterized to reproduce charge distributions predicted from nonempirical calculations on simple molecules. Other methods, such as INDO (1), MINDO (2), and NDDO (3) are parameterized in a different fashion, include other terms in the Hamiltonian, and differ in the experimental properties they reproduce most successfully. It is a reasonable generalization that semiempirical molecular orbital methods are reliable in reproducing some molecular properties but are rarely predictive (2b).

The nonempirical calculations (often referred to as *ab initio*) are generally more reliable, but there is a great variation in their reliability, depending on the size of the atomic orbital basis set used to determine the molecular orbitals. In this chapter we consider mainly two types of basis sets, a single Slater (STO) to represent every atomic orbital, and a "double zeta" set, which uses two orbitals in the basis set per atomic orbital. Hehre, Pople, Ditchfield, and Stewart (4) have developed Gaussian representations of these two types of basis sets, and their STO-3G

(4a) and 431G (4b) bases are used for many of the calculations described below (5).

In general, for the molecules considered here, one would expect the CNDO/2 procedure to yield a reasonable representation of their electronic structures, and STO basis *ab initio* calculations to predict qualitatively correct molecular structures. The double zeta basis *ab initio* calculations allow one to predict with confidence electronic and molecular structures and the energetics of some reactions. Obviously the double zeta *ab initio* calculations are the most reliable, but they are also the most time consuming. For example, a CNDO/2 calculation on methyleniminium takes 0.2 sec, an STO-3G calculation 2 sec, and a 431G calculation 10 sec (all on a CDC 7600 computer). However, it must be emphasized that there are a number of interesting chemical properties which any Hartree-Fock calculation cannot adequately represent, such as dissociation energies, $\Delta E(HF \rightarrow H+F)$, and activation energies for chemical reactions $(H_2 + F \rightarrow HF + F)$. Calculations that include part of the "correlation energy" are capable of precisely representing some of these very important properties, but as yet only for quite small molecules (6).

III. The Parent Iminium Ion

It may be instructive to begin by comparing the basicity of imines and other heterocyclic compounds. The calculated proton affinity of methylenimine, CH_2NH, is 226 kcal/mole (7); similar calculations on formaldehyde [predicted proton affinity (PA) = 180 kcal/mole, compared with the experimental value of 161 kcal/mole] (8) lead to expectations that the experimental proton affinity of H_2CNH is ~207 kcal/mole. We expect from our calculations that CH_2NH will have a proton affinity similar to that of ammonia (experimental PA = 207 kcal/mole) and less than that of methylamine (PA = 216 kcal/mole) (9). Since a comparable inductive effect might be expected for a CH_2 and a CH_3 group, the difference between the proton affinities of methylamine, methylenimine, and HCN (PA = 180 kcal/mole) (10) follows the trend expected, decreased basicity paralleling the "%s character" of the nitrogen lone pair.

The calculated ground-state geometry of iminium appears to be quite similar to that of C_2H_4. The geometrical parameters for each are summarized in Table I. As one can see, the evidence is strong that the iminium ion contains a $C-N^+$ double bond ($R = 1.26$ Å) intermediate in length between the C–C double bond (1.32 Å) and the C–O double bond [for formaldehyde $R(C=O) = 1.21$ Å]. The more accurate and flexible 431G calculation (4b) predicts a somewhat smaller C–N distance than the minimal basis STO-3G calculation (4a), probably because the former can represent the C⋯N bond more accurately; for neutral hydrocarbons,

TABLE I

Geometries[a] for CH_2NH, $CH_2NH_2^+$, and CH_2CH_2

	$CH_2NH_2^+$		CH_2NH			CH_2CH_2	
	STO-3G	431G	431G	Experimental		STO-3G	Experimental (13)
$R(C-N)$	1.29	1.26	1.26	1.30	$R(C-C)$	1.31	1.34
$R(C-H)$	1.11	(1.11)	(1.09)	(1.09)	$R(C-H)$	(1.09)	1.09
$R(N-H)$	1.04	(1.04)	(1.00)				
$\theta(HCH)$	118	118	(118)	(118)[b]	$\theta(HCH)$	116	117
$\theta(HNH)$	116	115	(113)[c]				
$\nu(C-N)$[d]	1910	1980	1760		$\nu(C-C)$	1950	1620–1680 (14)

[a] Distances in angstroms, and angles in degrees; parameters in parentheses were not optimized.
[b] The two hydrogens were assumed to be of equivalent length and NCH to be the same for each; the microwave spectrum was consistent with slightly different structural parameters for the two hydrogens (see ref. 12).
[c] CNH angle determined by Lehn (11).
[d] Stretching frequencies calculated.

however, the minimal basis does very well in predicting geometrical parameters (6). X-ray structural evidence (15) on $(CH_3)_2\overset{+}{C}-N(CH_3)_2$ indicates a C—N bond length of 1.30 Å. Guanidinium, $(NH_2)_3C^+$, has a C—N bond length of 1.32 Å (16); a 431G calculation on guanidinium predicts a bond length of 1.30 Å (17). In the tetramethyl-substituted iminium ion, one might expect a longer C–N distance than in the unsubstituted compound because of methyl hyperconjugation with the carbonium ion center and C· · ·C repulsions. The comparison between the calculated and experimental guanidinium values supports a prediction of ~1.28 Å for the C–N bond length in iminium, CNH_4^+. The fact that the methyleniminium ion is predicted to have about the same bond length as methylenimine is further support for the "double-bonded" nature of CNH_4^+.

The predicted C–H and N–H bond lengths are in reasonable accord with what one expects for isoelectronic neutral species, although probably somewhat greater than the experimental values. The fact that the HCH angle in $CH_2NH_2^+$ is predicted to be 2° larger than the HCH angle in ethylene appears to be inconsistent with the prediction by a number of authors that AB_2 bond angles in similar systems can be predicted from electronegativity effects: the more electronegative the external atom, the smaller the BAB angle (18,19). However, if one looks at the Mulliken atomic populations in C_2H_4 and $CH_2NH_2^+$ (relatively insensitive to the HCH angle), one finds that $\overset{\delta-}{C}-\overset{\delta+}{H}$ polarity is greater in the iminium ion and thus the relative bond angles are consistent with the electronegativity picture.

III. THE PARENT IMINIUM ION

Rationalization of the relative HCH and HNH angles in iminium is not obvious via the same model; one would expect $\theta(HNH)$ to be greater than $\theta(HCH)$ if electronegativity effects were the key. It may be, however, that the greater π occupancy on the nitrogen half of the molecule shrinks the N—H bond hybrids to smaller angles,[†] but a more complete study of $CH_2NH_2^+$ and CH_2NH is needed for a better understanding of the bond angles found in these molecules.

As would be expected (6), the predicted stretching frequencies are uniformly higher than those found experimentally. The prediction that the stretching frequency of the C═N⁺ linkage in iminium is approximately equal to that of the ethylenic C═C and imine C═N is consistent with experimental observations on substituted (14,20) C═C, C═N, and C═N⁺ linkages, where $\nu(C═C) = 1620–1680$ cm⁻¹, $\nu(C═N) = 1640–1690$ cm⁻¹, and $\nu(C═N^+) = 1660–1690$ cm⁻¹.

The orbital energies and atomic populations for the three species C_2H_4, CNH_3, and CNH_4^+ are presented in Table II. The orbital energies are

TABLE II

Mulliken Populations and Orbital Energies of C_2H_4, CNH_3, and $CNH_4^{+\,a}$

Ethylene		Methyleniminium		Methylenimine	
Orbital energy, au	Atomic population	Orbital energy, au	Atomic population	Orbital energy, au	Atomic population
−11.2072	C: 6.34	−15.8854	C: 5.88	−15.5421	C: 6.08
−11.2056	C(π): 1.00	−11.5833	C(π): 0.53	−11.2660	C(π): 0.89
−1.0376	H: 0.83	−1.5696	N: 7.72	−1.2253	N: 7.55
−0.7821		−1.1686	N(π): 1.47	−0.8482	N(π): 1.11
−0.6388		−1.0313	H_N: 0.52	−0.6884	$H_{cis-N—H}$: 0.85
−0.5870		−0.9757	H_C: 0.68	−0.6118	H: 0.82
−0.4913		−0.8185		−0.4462	H_N: 0.70
−0.3772 (π)		−0.7907 (π)		−0.4147 (π)	
0.1875 (π*)		−0.1336 (π*)		0.1608 (π*)	
0.2651 (σ*)		−0.0168 (σ*)		0.2375 (σ*)	

[a] 431G basis set; minimum energy geometry.

[†] The extreme case of

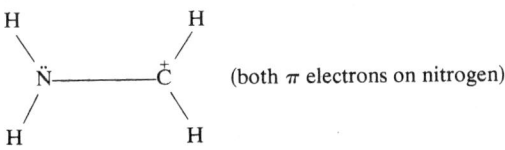 (both π electrons on nitrogen)

would lead to a predicted HCH angle of 120° (as in CH_3^+) and an HNH angle near 107° (NH_3).

similar in the ionic and neutral compounds, with the highest occupied and lowest unoccupied orbitals being of π symmetry. The orbital energies for CNH_4^+ are of much lower energy because of the positive charge. In CNH_4^+ the highest occupied and lowest empty orbitals are also of π symmetry. Comparing ethylene and methyleniminium, one finds a smaller π-π^* gap in ethylene, despite the fact that the iminium $\pi \rightarrow \pi^*$ transition appears to be at a longer wavelength (220–235 nm) (21) than the ethylene $\pi \rightarrow \pi^*$ (171 nm) (14). This is not very surprising, however, since experimentally one is observing the spectrum of the iminium ion in the vicinity of an anion, whereas the calculations described in Table II were done on the isolated cation. In view of the polarities predicted for the iminium ion (Table II), one might expect that the anion would be, on the average, nearer the NH_2 protons, and would destabilize the π ground state more than the π^* excited state (the ground-state π orbital contains 1.5 electrons on nitrogen and 0.5 on carbon; the π^*, the reverse polarity), and thus cause a red shift in the $\pi - \pi^*$ transition.

One of the curious features of methyleniminium is the fact that the Mulliken population on the nitrogen *increases* on protonation. Comparing methylenimine and methyleniminium, one finds that on protonation the total charge on the three protons in CNH_3 (2.37) is shared among the four in CNH_4^+ (2.40) and that the nitrogen and carbon lose only 0.03 electron on protonation. The C–N bond length causes us to conclude that the π bond is equally strong in the ion and in the imine, but far more ionic in CNH_4^+, where the nitrogen has 1.47 of the total of 2 π electrons. However, the Mulliken overlap population for the C–N bond is significantly smaller for the iminium ion than for the imine (0.66 versus 0.99). Thus the use of overlap populations to predict bond strength and length will not work with the two molecules CNH_3 and CNH_4^+. The very small populations on the hydrogens indicate that they should be quite far downfield in proton NMR, and this is what is observed (22).

The rotational barrier in the iminium ion is significant. With optimization of the parallel and perpendicular forms, one predicts a barrier of 71 kcal/mole for methyleniminium, which is significantly higher than the value predicted for methylenimine (57 kcal/mole) (11). The barrier for iminium appears to be lower than that of ethylene at the SCF level, but the important role of configuration interaction (*correlation energy*) in determining the barrier in C_2H_4 has been emphasized (23). On the basis of Buenker's calculations (23) one expects that the calculated values cited above are upper bounds for the actual barriers.

Why is the rotational barrier larger in $CH_2NH_2^+$ than in CH_2NH? In CH_2NH there is still a lone pair to stabilize the C^+ center in the perpendicular form, even though this "lone pair" is of a σ variety and

III. THE PARENT IMINIUM ION

much more tightly bound than the nitrogen π lone pair (thus the large barrier), the lone pair is more easily donated to the carbon than are the N–H bonding electrons of methyleniminium.

What makes iminium ions such relatively stable ions? We have compared (7) the stabilities of CH_2R^+, where R = H, CH_3, NH_2, and F, in an attempt to answer this question. Looking at the rotational barriers would give only a partial answer to this question, since for R = H and R = F there is obviously no (in the case of CH_3, very small) dependence of the energy on rotation of the R group. Comparing R = NH_2 and R = OH, it is clear that the π electrons of the nitrogen are more effective at stabilizing the carbonium ion center, since the rotational barriers of $CH_2NH_2^+$ and of CH_2NH are greater than the value for CH_2OH^+.

One way to compare the stabilization effect of the R group on the CH_2^+ carbonium ion fragment is to look at the energy for hydride transfer:

$$CH_3X + CH_2Y^+ \to CH_3Y + CH_2X^+ \qquad (1)$$

Since the heats of formation of a number of these species are known, one can determine ΔH for reaction 1 with different X and Y.

Similarly, one can carry out quantum-mechanical calculations on CH_3X and CH_2X^+ and compare the energy differences (ΔE) for the various substituent groups. The experimental and theoretically calculated differences for the various groups (relative to X = H) are presented in Table III. As one can see, the stabilizing influence on the carbonium ions follows the order X = NH_2 > OH > CH_3 > F.

It is also important to mention here that Radom et al. (24) have given extensive numerical support to the suggestion by Snyder (25) that the energies for reactions such as reaction 1 can be well described within the

TABLE III

Carbonium Ion Stabilization Energies

R	ΔH_{stab}[a]	ΔE_{stab}[b]	ΔE_{res}	ΔE_{induct}
R	0	0	0	0
CH_3	31–42	27	11	16
NH_2	96.5	89	66	23
OH	32–57	45	48	−3
F	4	−5	31	−36

[a] ΔH for the reaction $CH_2R^+ + CH_4 \to CH_3R + CH_3^+$, determined for heat of formation data for the above species; see ref. 10.

[b] ΔE calculated for the reaction of footnote a, evaluated from the total energies for the species.

Hartree-Fock framework; the agreement between theory and experiment in Table III is further support for this view. These types of reactions involve the same number of electron pairs in both reactants and products, and thus both reactants and products would be expected to have similar correlation energies (6).

We have further separated (7) the "resonance" and "inductive" effects on carbonium ion stability by carrying out SCF calculations with and without the $p\pi$ orbitals on the carbon to determine the "resonance" stabilization of these carbonium ions, and have attributed the remainder of the stability or instability to inductive effects, that is,

$$\Delta E_{stab}\{[E(CH_4) - E(CH_3^+)] - [E(CH_3X) - E(CH_2X^+)]\}$$
$$= \Delta E_{res}[E(CH_2X^+ \text{ with } p\pi \text{ orbital on C})$$
$$- E(CH_2X^+ \text{ without C } p\pi \text{ orbital})] + \Delta E_{induct} \quad (2)$$

These results are presented in Table III and clearly indicate that CH_3 and NH_2 groups are inductively stabilizing, OH is inductively neutral, and F is inductively destabilizing. The "inductive effect" for OH and NH_2 is somewhat surprising until one looks at the Mulliken populations and realizes that the *hydrogens* are playing an important role in "absorbing" the positive charge, thus compensating for the inductive withdrawing power of nitrogen and oxygen. Hence the iminium ion is stabilized in relation to the simplest carbonium ion, CH_3^+, by a substantial amount of resonance and inductive stabilization.

IV. Substituted Iminium Ions

A. ELECTRONIC STRUCTURE AND ISOMERIZATION ENERGIES

Since the iminium ion $CH_2NH_2^+$ has never been chemically isolated, one is naturally interested in the effects of various substituents on the electronic structure and properties of the parent iminium fragment. Here we consider the effects of substitution of CH_3, NH_2, OH, SH, F, Cl, —C=O, and —C=C groups on the iminium ion structure (26).

What is the effect of a methyl substituent on the iminium ion? We have examined two possibilities of dimethyl substitution and carried out electronic structure calculations at the CNDO/2 (1) and STO-3G *ab initio* (4a) level for two isomers **5** and **6**:

IV. SUBSTITUTED IMINIUM IONS

The difference in energy calculated for these two species is considerable, the C—CH$_3$ species being favored by 21 kcal/mole at the STO-3G level (the CNDO/2 calculations predict the N—CH$_3$ substituent to be more stable by 1 kcal/mole). The more trustworthy STO-3G results are consistent with one's intuition that the CH$_3$ groups play an important role in stabilizing the carbonium ion center, but comparison of the Mulliken populations for the CH$_3$-substituted species with the value for the parent species shows that the C—Me stabilizing effect comes, not from more net electrons in the carbonium carbon π orbital, but from an increase in the total nitrogen electron population (7.32) and in the nitrogen π population (1.52). In the parent compound these populations are N(total) = 7.28 and N(π) = 1.43; in the N–CH$_3$-substituted compound, N(total) = 7.16 and N(π) = 1.35. The carbon π populations are 0.57 in the parent and C—CH$_3$-substituted compounds and 0.66 in the N—CH$_3$-substituted compound. Our general working model is as follows: methyl groups attached to the positively charged carbon donate electrons into the C$^+$ π orbital and allow the nitrogen to retain more of its lone pair (1.52) than in the parent compound (1.43). In the N—CH$_3$ compound the interaction is a repulsive one between the C–H bond and the iminium N̈, forcing electrons from the nitrogen lone pair to the carbon, which ends up with 0.65 electron.

We have also examined the effects of difluorosubstitution on the iminium fragment and have considered the relative energies of the four difluorosubstituted compounds. The results are consistent with what one would predict: the CF$_2$ isomer **7**:

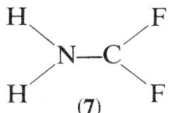

is the most stable, with the cis and trans 1,2-difluoro compounds of roughly equal energies and 45 kcal/mole less stable than the CF$_2$ isomer; the NF$_2$ isomer is an additional 34 kcal/mole higher in energy. The Mulliken populations on the C$^+$ for the CF$_2$ species indicate a large π population (0.77) but a quite small total population (5.33), indicating that these fluorosubstituted compounds are probably quite unstable (recall the inductive destabilizing effect of a fluoro group, discussed in Section III).

The fact that the cis and trans 1,2-isomers are very similar in energy is expected, considering the two 1,2-fluoro isomers of ethylene (27), where the cis is favored over the trans by ~0.3 kcal/mole. There is very little difference in the electronic structures of the two 1,2-difluoro isomers, and the carbon π population is 0.71, between the values for the CF$_2$ iminium ion (0.77) and the NF$_2$ isomer (0.63). The other populations are also close to "averages" of the values found for the CF$_2$ and NF$_2$ isomers.

One now inquires about the general effect of the substitution of π-donating groups such as NH_2, OH, F, SH, and Cl on the iminium ion fragment **8**:

$$\begin{array}{c} H \quad\quad\quad R \\ \diagdown\quad\quad\diagup \\ N\!-\!\overset{+}{C} \\ \diagup\quad\quad\diagdown \\ H \quad\quad\quad H \\ (8) \end{array}$$

There is a significant donation into the carbon π orbitals from these groups; the Mulliken populations on the C—N fragment for the various substitutions are present in Table IV. As one can see, these groups have a significant electron-donating effect on the carbon π orbital, but it is interesting that this effect is very similar for all the R's (0.10–0.13 electron).

How well does a double bond conjugate with the iminium fragment? We have examined the conjugation of C=C and C=O linkages with the $C\overset{+}{=}N$ fragment, carrying out the theoretical calculations on the isoelectric analogues of butadiene and acrolein. In these cases we found relatively little energy difference between species in which the double bond is attached to the nitrogen and those in which it is attached to carbon. Since complete geometry searches were not carried out, one should not overemphasize the absolute energy difference, but this difference is small. The results are summarized in Tables V and VI. The C=C-substituted iminium ions are similar to the Me-substituted compounds, carbon substituted ones being more stable than nitrogen-substituted by about 12 kcal/mole. This difference is *not* reflected in the $\overset{+}{C}$ π population, since the more stable compound has fewer (0.63) electrons in its π orbitals than does the less stable (0.65). Once again, the nitrogen π population is

TABLE IV

Mulliken Populations for the Substituted Iminium Ions

$$\begin{array}{c} H \quad\quad\quad R \\ \diagdown\quad\quad\diagup \\ N\!-\!\overset{+}{C} \\ \diagup\quad\quad\diagdown \\ H \quad\quad\quad H \end{array}$$

				R		
	H	NH_2	OH	F	SH	Cl
C:	5.70	5.66	5.68	5.63	5.90	5.79
C(π):	0.57	0.70	0.70	0.68	0.69	0.67
N:	7.28	7.34	7.31	7.30	7.32	7.27
N(π):	1.43	1.65	1.52	1.46	1.54	1.43

TABLE V

Mulliken Populations and Energies of $C_3NH_6^+$ Isomers

Compound	E(relative)	C_1(total)	(π)	C_2(total)	(π)	C_3(total)	(π)	N(total)	(π)
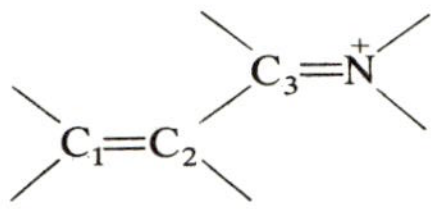 structure 1	0	6.01	0.74	6.08	1.12	5.76	0.63	7.31	1.52
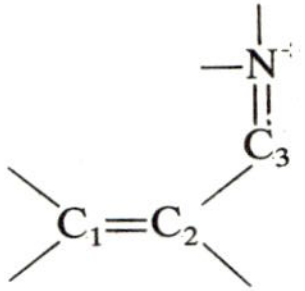 structure 2	6.2	6.02	0.78	6.07	1.09	5.76	0.62	7.31	1.52
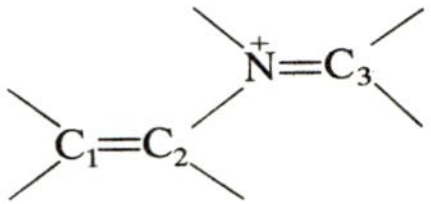 structure 3	11.9	6.06	0.85	5.96	1.13	5.83	0.65	7.22	1.37
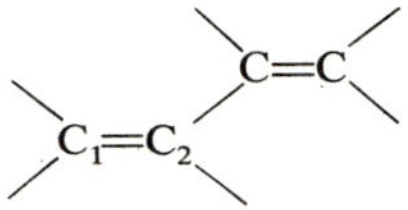 structure 4		6.13	1.00	6.06	1.00				
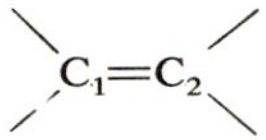 structure 5		6.12	1.00	6.12	1.00				
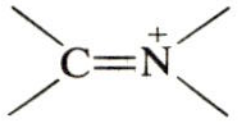 structure 6						5.80	0.57	7.28	1.43

TABLE VI
Mulliken Populations and Energies of $C_2NOH_4^+$ Isomers

Compound	E(relative)	C_1(total)	(π)	C_2(total)	(π)	$N(C_3)$(total)	(π)	O(total)	(π)
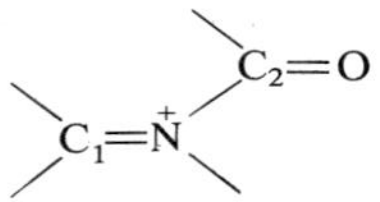 $C_1^+=N$, $C_2=O$	0	5.81	0.59	5.75	0.99	7.24	1.40	8.09	1.03
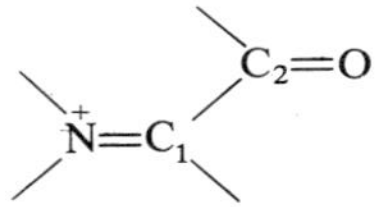 $N^+=C_1$, $C_2=O$	0.8	5.77	0.61	5.88	1.02	7.29	1.46	8.05	0.92
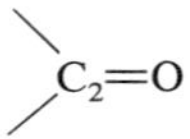 $C_3=C_1$, $C_2=O$				5.93	0.93			8.19	1.07
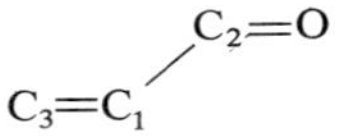		6.09	1.02	5.88	0.93	6.10	0.94	8.21	1.10
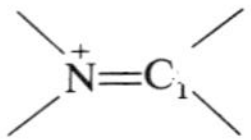 $N^+=C_1$		5.80	0.57			7.28	1.43		

greater for the more stable (1.52) than the less stable (1.37) compound. Hence, as we found in the Me substituted compounds, the greater stability can be rationalized in terms of the substituent (—C=C—) relieving the nitrogen of some of its electron-donating "responsibility."

However, —C=O substitution appears different, with the nitrogen-substituted isomer 0.8 kcal/mole more stable than the carbon-substituted, despite the greater nitrogen π population in $\overset{+}{N}$=C—C=O (1.46) than in C=$\overset{+}{N}$—C=O (1.40). One might rationalize the above facts because of the tendency for the $\overset{..}{N}$ lone pair to donate electrons to the C–O double bond via $\overset{+}{N}$=C$\diagdown^{O^-}$, whereas in the C—C=O case the carbonyl group must give up its electron to the iminium fragment. Looking at the electron populations in Table VI, one sees that in the C—C=O species π electrons are withdrawn from the carbonyl group and the C^+ gains, whereas in the N—C=O species there is little loss of π electrons by the —C=O group (in fact, some gain), but the nitrogen lone pair has to give up many of its π electrons to stabilize the C^+. It appears that, to withdraw electrons from the $\overset{..}{N}$ and C=O, one must pay comparable energetic prices and thus the two isomers are of nearly equal stability.

B. ROTATIONAL BARRIERS

One way to get a more quantitative measure of the resonance stabilization of carbonium ions is to look at rotational barriers (as mentioned previously); the rotational barriers for a number of substituted iminium ions are listed in Table VII (17). The absolute values for the barriers are exaggerated, as indicated by the results of more accurate calculations in parentheses, but the trends should be correct. In each case the NR_2 group was rotated perpendicularly to CR_2, so that the NH_2 resonance stabilization was lost. Looking at the barriers in $N(CH_3)_2$- and NF_2-substituted ions, one sees clearly that these groups had a greater resonance interaction with the CH_2 fragment than the parent compound, a finding consistent with their carbon π population being greater than that in iminium. Of the nitrogen-substituted species, only the N—C=O compound has a lower barrier than the iminium ion, perhaps because the $\overset{..}{N}$ is donating more of its electrons to C=O than C^+ in the planar species and thus less stabilization is lost on rotation. In terms of ability to donate π electrons to the C^+, one sees that the order of donating ability is $NH_2 > SH > OH > F > Cl$. It is interesting that apparently the SH group is a better donor than OH, but F is better than Cl; however, these calculations are not precise enough to be sure that this order is correct. Obviously, comparing the energies for the carbonium ion and neutral precursor and then

TABLE VII

Rotational Barriers in Substituted Iminium Ions

Compound	Barrier, kcal/mole
$\overset{+}{C}H_2-NH_2$	87.2 (70.5)
$\overset{+}{C}H_2-N(CH_3)_2$	92.1
$\overset{+}{C}(CH_3)_2-NH_2$	75.2
$\overset{+}{C}F_2-NH_2$	42.9
$\overset{+}{C}H_2-NF_2$	112
$\overset{+}{C}HOH-NH_2$	75.1
$\overset{+}{C}HSH-NH_2$	57.3
$\overset{+}{C}HF-NH_2$	77.2
$\overset{+}{C}HCl-NH_2$	83.3
$\overset{+}{C}HNH_2-NH_2$	39.1 (28.2)
$\overset{+}{C}(NH_2)_2-NH_2$	22.6 (14.1)
$\overset{+}{C}(C=C)H-NH_2$	67.4
$\overset{+}{C}(C=O)H-NH_2$	79.7
$\overset{+}{C}H_2-NH(C=C)$	95.7
$\overset{+}{C}H_2-NH(C=O)$	83.8
$\overset{+}{C}H(CH_3)-NH_2$	79.3

comparing with $\Delta E(CH_4-CH_3^+)$ is the best way to measure total "stabilization"; all that the rotational barrier results give us is some "feeling" for the resonance contribution.

C. "STABILIZATION" OF SUBSTITUTED IMINIUM IONS

We have examined the case of multiple NH_2 substitution on $CH_2NH_2^+$ in considerably greater detail (17) because the amino group is a very effective stabilizer of carbonium ions, guanidinium, $C(NH_2)_3^+$, being inert in boiling water. The results for the proton affinities of imine, amidine, and guanidine and the stabilization energies (relative to the methyl cation) of the iminium, amidinium, and guanidinium ions are presented in Table VIII.

Column 3 of this table, headed "Stabilization energy of RH^+," is the calculated energy for the reaction $XYZC^+ + CH_4 \rightarrow XYZCH + CH_3^+$. For $R = CH_2NH$ (X = H, Y = H, Z = NH_2) this is the energy for the reaction $CH_2NH_2^+ + CH_4 \rightarrow CH_2NH_2 + CH_3^+$, discussed previously and reported in Table III. Since the two studies (7,17) used somewhat different double zeta basis sets, it is encouraging that both predict the same energy for the reaction, 89 kcal/mole, in good agreement with the experimental value,

IV. SUBSTITUTED IMINIUM IONS

TABLE VIII

Proton Affinities, Stabilization Energies, and Rotational Barriers of Aminosubstituted Compounds (values in kcal/mole)

R	PA	Stabilization energy of RH^+	Rotational barrier in RH^+
CH_2NH	228	89[a]	70.5
$CH(NH_2)NH$	249	128	28.2
$C(NH_2)_2NH$	264	147	14.1

[a] Reference 7 and this study (431G) both lead to the same stabilization.

$\Delta H = 96.5$ kcal/mole. Note that the stabilization energy for amidinium ($R_1 = X$, $Y = Z = NH_2$) and guanidinium ($X = Y = Z = NH_2$) is even greater than that for iminium; the guanidinium ion is unusual because of its low acidity and great stability (stabilization energy calculated to be 147 kcal/mole).

As stated above, the rotational barrier results in Table VIII give only a qualitative picture of the "resonance stabilization" of these substituted iminium ions; for a better measure of their stabilization one must compare the calculated energies of the reaction $CZX-NHY^+ + CH_4 \rightarrow CH_3^+ + CHXZ-NHY$ for the different X, Y, and Z substituents in Table VII. These calculations were carried out with the STO-3G basis set and are presented in Table IX. Even though the STO-3G basis is not as reliable for reaction energies as is 431G, comparison of the iminium and amidinium stabilization energies (rows 1 and 5 of the table) indicates that STO-3G is probably capable of showing the correct trend in the stabilization energies. In relation to the parent iminium ion, substitution of NH_2, SH, $CH_2=CH$, CH_3, and $di(CH_3)_2$ at C^+ is stabilizing ($\Delta E > 99.2$ kcal/mole), and substitution of F, Cl, diFCl, diF_2, $diCl_2$, and $\underset{R}{C=O}$ (R = H, CH_3, OCH_3) is destabilizing ($\Delta E < 99.2$ kcal/mole). For monosubstitution, the effectiveness of stabilizing the carbonium ion center C^+ follows the order $NH_2 > C=C > SH > CH_3 > OH > F > Cl$. We expect the least reliable of the energies in Table IX to be those of halosubstituted compounds, because a single Slater representation of the atomic orbitals is poorest for the more electronegative elements, so that our prediction that SH is more stabilizing than OH and F is more stabilizing than Cl must remain tentative.

That substituting a $\underset{R}{\overset{O}{C}}$ group on the nitrogen "destabilizes" the

TABLE IX

STO-3G Calculations for $\Delta E(CXZ^+\text{---}NWY + CH_4 \rightarrow CH_3^+ + CHXZ\text{---}NWY)$

X	Z	Y	W	ΔE kcal/mole
H	H	H	H	99.2 (89)[a]
F	F	H	H	98.4
Cl	Cl	H	H	81.9
Cl	F	H	H	90.6
NH_2	H	H	H	136.1 (128)[a]
SH	H	H	H	123.6
OH	H	H	H	80.2
F	H	H	H	60.3
Cl	H	H	H	47.8
CH_2=CH	H	H	H	125.3
CH≡C	H	H	H	110.1
CH_3	H	H	H	114.1
CH_3	CH_3	H	H	127.5
H	H	HC=O	H	80.9
H	H	HC=O	HC=O	67.7
H	H	CH_3C=O	H	87.8
H	H	CH_3OCO	H	83.5
H	CH_2		H	91.9

[a] Values in parentheses were calculated using the 431G basis.

iminium ion (in relation to the neutral precursor) is consistent with the fact that N-o-haloalkyl amides and imides prefer to be neutral (22) rather than to form iminium ions. $R = CH_3$ stabilizes and $R = OCH_3$ destabilizes the iminium ion structure (in relation to $R = H$).

The alleniminium ion, CH_2=C=$\overset{+}{N}H_2$, is found to be slightly less stabilized in relation to ethylenimine $CH_2\text{---}CH_2$ (with N-H bridge), than the parent iminium ion (last row of Table IX).

Although Table IX lists compounds in which H^- is the reference anion, one can also study reactions using Cl^- or F^- as reference. To this end, we have carried out calculations on $CXYZ\text{---}NH_2$ and NH_2CXY^+, where X, Y, and Z include all combinations of H, F, and Cl. In fact, Table IX already includes all the examples in which $Z = H$, but we repeat these for comparison in Table X. As one can see from this table and Table IX, a single halosubstitution on the carbonium ion center is less favorable, relative to $CXY\text{---}NH_2^+$, than dihalosubstitution, no matter what the reference anion is. In each case F substitution is less destabilizing than Cl substitution.

IV. SUBSTITUTED IMINIUM IONS

An interesting experimental observation (22) is the following: in the equilibrium $RX \rightleftharpoons R^+ + X^-$, where R^+ is a representative iminium ion, the equilibrium favors the un-ionized compound for $X = F$ and the ionized compound for $X = Cl$. One could hope to predict the energetics of this reaction for $X = F$ and $X = Cl$ by carrying out molecular orbital calculations for RX, R^+, and X^-. Although reactions such as this, in which the number and type of chemical bonds differ in reactants and products, would not be expected to be well represented by single-determinant

TABLE X

STO-3G Calculated Relative Stabilities of Iminium Ions Formed via F^-, Cl^-, or H^- Abstraction

	F^- Abstraction		
Neutral	Cation	ΔE au	$\Delta(\Delta E)$,[a] kcal/mole
$CH_2F\text{—}NH_2$	$^+CH_2\text{—}NH_2$	98.24647	0
$CHF_2\text{—}NH_2$	$^+CHF\text{—}NH_2$	98.32208	47.4
$CHClF\text{—}NH_2$	$^+CHCl\text{—}NH_2$	98.33336	54.5
$CF_2Cl\text{—}NH_2$	$^+CFCl\text{—}NH_2$	98.27597	18.5
$CF_3\text{—}NH_2$	$^+CF_2\text{—}NH_2$	98.27166	15.8
$CFCl_2\text{—}NH_2$	$^+CCl_2\text{—}NH_2$	98.28273	22.7

	Cl^- Abstraction		
Neutral	Cation	ΔE, au	$\Delta(\Delta E)$,[b] kcal/mole
$CH_2Cl\text{—}NH_2$	$^+CH_2\text{—}NH_2$	454.79285	0
$CHFCl\text{—}NH_2$	$^+CHF\text{—}NH_2$	454.85977	41.9
$CHCl_2\text{—}NH_2$	$^+CHCl\text{—}NH_2$	454.86745	46.8
$CFCl_2\text{—}NH_2$	$^+CFCl\text{—}NH_2$	454.80300	6.4
$CF_2Cl\text{—}NH_2$	$^+CF_2\text{—}NH_2$	454.80118	5.2
$CCl_3\text{—}NH_2$	$^+CCl_2\text{—}NH_2$	454.80701	8.9

	H^- Abstraction		
Neutral	Cation	ΔE, au	$\Delta(\Delta E)$,[c] kcal/mole
$CH_3\text{—}NH_2$	$^+CH_2\text{—}NH_2$	0.79196	0
$CH_2F\text{—}NH_2$	$^+CHF\text{—}NH_2$	0.85411	39.0
$CH_2Cl\text{—}NH_2$	$^+CHCl\text{—}NH_2$	0.87424	51.6
$CHFCl\text{—}NH_2$	$^+CFCl\text{—}NH_2$	0.80577	8.7
$CHF_2\text{—}NH_2$	$^+CF_2\text{—}NH_2$	0.79329	0.8
$CHCl_2\text{—}NH_2$	$^+CCl_2\text{—}NH_2$	0.81959	17.3

[a] $\Delta E(\text{neutral} - \text{cation}) - \Delta E(CH_2FNH_2 - \overset{+}{C}H_2NH_2)$.
[b] $\Delta E(\text{neutral} - \text{cation}) - \Delta E(CH_2ClNH_2 - \overset{+}{C}H_2NH_2)$
[c] $\Delta E(\text{neutral} - \text{cation}) - \Delta E(CH_3NH_2 - \overset{+}{C}H_2NH_2)$

molecular orbital theory, we have calculated the energy for the above reaction with $R = CH_2NH_2$ and $X = H$, F, and Cl. The ΔE values calculated are 397, 397, and 196 kcal/mole for H, F, and Cl ionization. Although these energies are certain to be greatly exaggerated even in comparison to the experimental gas phase energies for the above reaction, the very large energy difference between the fluoride and the chloride is consistent with the experimentally observed differences in equilibrium constants.

D. OTHER APPLICATIONS

At the CNDO/2 level, we have examined the conformational properties of substituted creatine and creatine phosphate and have found the isomer stability for the isolated species to be dominated by the possibility of $N-H \cdots {}^-OOC$ intramolecular hydrogen bonding and, in the case of phosphocreatine, by the repulsion between PO_3^{2-} and the carboxyl group. The results of the calculations were consistent with the model for creatine substrate activity advanced by Kenyon and his co-workers (28).

In summary, one can conclude that electronic structure calculations can provide some insight into the factors that stabilize and destabilize carbonium ions. Some of the conclusions of this work are somewhat tentative (e.g., the relative resonance stabilizations of OH, SH, F, and Cl), but more precise calculations along the lines discussed above are clearly technically feasible, for example, a more precise examination of the relative stabilizing effects of OH, SH, F, and Cl. One has enough experience regarding the ability of accurate molecular orbital calculations to predict geometries and relative stabilities of known compounds to have some confidence in the results of such calculations on nonisolable intermediates. It is precisely in this area that such calculations can be most useful, since they are the most accurate method available for determining geometries and relative energies.

Further *ab initio* theoretical studies on the electronic spectra of iminium ions in the presence of anions is of interest, since these ions play a role in the chemistry of vision (29).

ACKNOWLEDGMENTS

I would like to thank the Academic Senate (University of California at San Francisco) and the NIH (GM-20564 and Career Development Award GM 70718) for partial support of this research and John McKelvey and Warren Hehre for the use of the CDC 7600 version of Gaussion 70. It is also a pleasure to acknowledge useful comments by H. Böhme, V. Buss, and P. von R. Schleyer.

REFERENCES AND NOTES

1. J. A. Pople and D. L. Beveridge, *Approximate Molecular Orbital Theory*, McGraw-Hill Book Company, New York, 1970.
2a. N. Bodor, M. J. S. Dewar, and D. H. Lo, *J. Amer. Chem. Soc.*, **94,** 5303 (1972) and references therein.
2b. See, however, R. C. Bingham, M. J. S. Dewar, and D. H. Lo, *J. Amer. Chem. Soc.*, **97,** 1285 (1975), and M. J. S. Dewar, *Science*, **187,** 1037 (1975) for another point of view.
3. R. Sustmann, J. E. Williams, M. J. S. Dewar, L. C. Allen, and P. von Schleyer, *J. Amer. Chem. Soc.*, **91,** 5350 (1969).
4a. W. J. Hehre, R. F. Stewart, and J. A. Pople, *J. Chem. Phys.*, **51,** 2651 (1969).
4b. R. Ditchfield, W. J. Hehre, and J. A. Pople, *J. Chem. Phys.*, **54,** 7241 (1971).
5. The calculations were carried out using Gaussian 70, a computer program developed in J. A. Pople's group at Carnegie Mellon University: I am grateful to J. McKelvey and W. Hehre for the use of the CDC 7600 version of Gaussian 70 (QCPE 236).
6. See H. F. Schaeffer, *The Electronic Structure of Atoms and Molecules: A Survey of Rigorous Quantum Mechanical Results*, Addison-Wesley, Reading, Mass., 1972, for an up-to-date view of the state of the art in regard to *ab initio* quantum-mechanical calculations applied to molecules.
7. P. A. Kollman, W. F. Trager, S. Rothenberg, and J. E. Williams, *J. Amer. Chem. Soc.*, **95,** 458 (1973).
8. A. C. Harrison, A. Ioko, and D. Van Raalte, *Can. J. Chem.*, **44,** 1625 (1966).
9. J. L. Beauchamp, *Ann. Rev. Phys. Chem.*, **22,** 527 (1971).
10. J. L. Franklin, J. G. Dillard, H. M. Rosenstock, J. T. Herron, K. Droxyl, and F. H. Field, *Ionization Potentials, Appearance Potentials and Heats of Formation of Gaseous Positive Ions*, NSRDS-NBS-26, U.S. Government Printing Office, Washington, D.C., 1969.
11. J. M. Lehn, *Theor. Chim. Acta*, **16,** 351 (1970).
12. K. V. L. N. Sastry and R. F. Curl, *J. Chem. Phys.*, **41,** 77 (1964).
13. L. S. Bartell, E. A. Roth, C. D. Hollowell, K. Kuchitsu, and J. E. Young, Jr., *J. Chem. Phys.*, **42,** 2683 (1965).
14. J. R. Dyer, *Applications of Absorption Spectroscopy of Organic Compounds*, Prentice-Hall, Englewood Cliffs, N.J., 1965.
15. L. M. Trefonas, R. L. Flurry, Jr., R. Majeste, E. A. Meyers, and R. F. Copeland, *J. Amer. Chem. Soc.*, **88,** 2145 (1966).
16. R. M. Curtis and R. A. Pasternak, *Acta Cryst.*, **8,** 675 (1955).
17. P. A. Kollman, J. McKelvey, and P. Gund, *J. Amer. Chem. Soc.*, **97,** 1640 (1975).
18. C. E. Mellish and J. W. Linnett, *Trans. Faraday Soc.*, **50,** 657 (1954); H. L. Bent, *Chem. Rev.*, **61,** 275 (1961).
19. P. Kollman, *J. Amer. Chem. Soc.*, **96,** 4363 (1974).
20. G. Opitz, H. Hellmann, and H. W. Schubert, *Liebigs Ann. Chem.*, **623,** 117 (1959).
21. S. Hünig and H. Hoch, *Fortschr. Chem. Forsch.*, **14,** 235 (1970).
22. N. J. Leonard and J. V. Paukstelis, *J. Org. Chem.*, **28,** 3021 (1963); H. Böhme and M. Haake, "Methyleniminium Salts," chapter in this book.
23. R. J. Buenker, *J. Chem. Phys.*, **48,** 1368 (1968).
24. L. Radom, W. J. Hehre, and J. A. Pople, *J. Amer. Chem. Soc.*, **93,** 289 (1971).
25. L. Snyder, *J. Chem. Phys.*, **46,** 3602 (1967).
26. For the iminium ions, the geometries used kept the structure as determined for the unsubstituted compound except that one or more hydrogens were replaced by

substituents. The particular geometries were as follows:
R = F or diF$_2$; R(C—F) optimized = 1.34 Å.
R(N—F) = 1.40 Å.
R = Cl or diCl$_2$; R(C—Cl) optimized = 1.67 Å.
R = OH; R(C—O) optimized = 1.34 Å; R(O—H) and θ(COH) taken from methanol structure; O–H bond trans to N–C double-bond.
R = SH; R(C—S) optimized = 1.70 Å; R(S—H) and θ(CSH) taken from CH$_3$SH structure; S–H bond trans to N–$\overset{+}{\text{C}}$ bond.
R = CH$_3$: R(N—C) and R(C—C) from ref. 15.

R = C(=O)—R'; structural parameters taken from appropriate amide structure with N-substitution; for C-substituted, structural parameters from acrolein.

R = —C=C$\diagdown$: C- and N-substituted structural parameters from butadiene compounds.

CH$_3$NH$_2$: experimental geometry.
CXYZNH$_2$: experimental geometry for methylamine except C—N line along C$_3$ axis; R(C—F) = 1.385 Å; R(C—Cl) = 1.782 Å; R(C—H) = 1.092 Å; most electronegative substituents trans to lone pair. X = SH, OH, C=C, and NH$_2$ geometries were taken from experimental data on appropriately substituted methane, with the methylamine part of structure left unchanged. The N—C(=O)(R) structures were taken from corresponding experimental structures of amides and esters, and the ethylenimine structure was obtained by experiment. The N(C=O)(C=O) compound was constructed using the amide structural parameters.

27. H. G. Viehe, *Chem. Ber.*, **93**, 1697 (1960).
28. G. Kenyon, G. E. Struve, P. A. Kollman, and T. I. Moder, "Conformational Analyses of Creatine and Phosphocreatine, a Steric Model for Phosphocreatine," *J. Amer. Chem. Soc.* (in press).
29. See *Acc. Chem. Res.*, **8** (March 1975), specifically, the article by B. Honig, A. Warshel, and M. Karplus, p. 92 of that issue.

Note Added in Proof

Since submitting this chapter we [P. Kollman, S. Nelson, and S. Rothenberg, "The Structure and Relative Energies of C$_2$H$_2$X$^+$ isomers (X=F, OH, NH$_2$, Cl, and SH)," *J. Amer. Chem. Soc.* (submitted)] have analyzed the relative stabilities of C$_2$H$_3$X$^+$ cations and found relative stabilization energies (at the 431G level) to be in the order NH$_2$ > SH > OH > Cl > F.

Our concluding paragraph was prophetic because very recently an article appeared on the nature of the protonated Schiff's base fragment in retinal [L. Salem and P. Bruckmann, *Nature*, **258,** 526 (1975)]. These authors found a charge transfer from the 7–11 part of retinal to the immonium end of the molecule when the molecule was rotated around the 11–12 bond in the $\pi-\pi^*$ excited state. This large change in polarity provides a beautiful mechanism by which the chromophore can interact with its protein environment and perhaps transmit the photon energy via a conformational charge in the protein rhodopsin.

STRUCTURE DETERMINATION OF IMINIUM SALTS BY PHYSICAL METHODS

By R. MERÉNYI, *Laboratoire de Chimie Organique, Université de Louvain, Place L. Pasteur, 1, B-1348-Louvain-la-Neuve, Belgium*

CONTENTS

I. Introduction	23
II. Nuclear Magnetic Resonance, Infrared, and Ultraviolet Tables I–XVIII	26
III. Infrared Spectroscopy	74
A. C=N Stretching Absorptions	74
B. NH Stretching and Binding Absorptions	75
C. Other Specific Absorptions	75
IV. Nuclear Magnetic Resonance Spectroscopy	76
A. Chemical Shifts	76
B. Coupling Constants	77
1. ^{1}H–^{1}H Couplings	77
2. ^{1}H–^{13}C Couplings	77
3. ^{1}H–N and ^{13}C–N Couplings	78
V. Barriers of Rotation of the C–N Bond (Table XIX)	78
VI. Ultraviolet and Visible Spectroscopy	84
VII. Mass Spectroscopy	86
VIII. Structure Determination by Diffraction Methods (Table XX)	89
IX. Other Physical Methods	94
References	96

I. Introduction

Most of the simple iminium salts are solids that hydrolyze easily on contact with moisture present in air, and decompose on melting or at even lower temperatures. Special care is needed for their investigation by physical methods, and thus literature data must be regarded with a critical eye. This point is also taken into consideration for this chapter and in the composition of the tables.

Depending on their structure, iminium compounds vary in reactivity; stabilization in this sense arises from aliphatic substitution (hyperconjugation and steric protection), and from conjugation with π systems and with lone-pair electrons. In practice, complex anions also have a stabilizing effect. There are iminium derivatives stable in aqueous solution, such as the amidinium salts, protonated or alkylated derivatives of *N*-heterocycles, amides, or thioamides (amidium and thioamidium salts), and their vinylogues as cyanines and protonated or alkylated merocyanines. This fact is also demonstrated by the increasing number of known, stable iminium compounds that are N—R_2, O—R, or S—R substituted.

Theoretical calculations (see the preceding chapter by Kollman or Ref. 1) and physical methods indicate that the positive charge in *simple iminium* salts is localized mainly on the nitrogen (mesomeric form **Ia**) with a minor contribution of the aminocarbonium form (**Ib**):

$$\left[\begin{array}{c} R \\ \\ R \end{array} \!\!\! C\!=\!\overset{+}{N} \!\!\! \begin{array}{c} R' \\ \\ R' \end{array} \longleftrightarrow \begin{array}{c} R \\ \\ R \end{array} \!\!\! \overset{+}{C}\!-\!N \!\!\! \begin{array}{c} R' \\ \\ R' \end{array} \right] X^-$$

(Ia) (Ib)

This situation is only slightly modified by aliphatic, halogen, and vinyl substitution and for the aromatic N-heterocycles. In the case of the parent iminium ion (R = R' = H) the calculated electron populations show the displacement of the positive charge onto the NH protons.

If a substituent Y in $\beta, \delta, \ldots$ position to the nitrogen allows the *delocalization* of a lone pair of electrons, the positive charge is distributed between N and Y according to their relative basicities. The contribution of $\alpha, \gamma, \ldots$ carbonium ions remains minor, with an alternation of π-electron densities ($C\pi$) on the carbons; generally, $C\pi < 1$ in $\alpha, \gamma, \ldots$, and $C\pi \geq 1$ in $\beta, \delta, \ldots$ position (2).

$$\left[Y \!\!\cdots\!\! (C \!\!\cdots\!\! C)_n \!\!\cdots\!\! N \right]^+ X^-$$

(**II**) $n = 0, 1, 2, 3, \ldots$;
Y = NR_2, OR, SR

A similar situation exists in molecules with a dipolar structure, such as amides, thioamides, amidines, and aminofulvenes with general structure **III**:

$$Z \!\!=\!\! (C \!\!-\!\! C)_n \!\!=\!\! N \;\equiv\; \overset{\delta-}{Z} \!\!\cdots\!\! (C \!\!\cdots\!\! C)_n \!\!\cdots\!\! \overset{\delta+}{N}$$

(**III**) $n = 0, 1, 2, 3, \ldots$;
Z = O, S, NR, ⌬

The charge distribution in **III**, as well as the C–N bond length, is

I. INTRODUCTION

practically the same in several cases, as in **II**. Nevertheless, only compounds of types **I** and **II** will be discussed in this review. Furthermore, for cyanines and cations of aromatic N-heterocycles, only representative references will be given from the abundant number of papers treating this subject.

In cases where anions are good nucleophiles, an equilibrium of the *ionic tautomer* (**I**) and the *covalent tautomer* (**IV**) is possible:

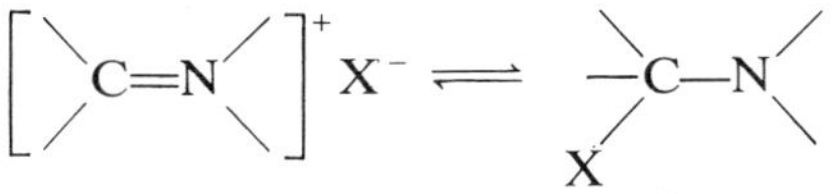

$$\left[\diagdown C = N \diagup \right]^+ X^- \rightleftharpoons \diagdown\underset{X}{C} - N \diagup$$

(**I**) (**IV**)

Spectroscopic evidence of this tautaumerism is given for FCH$_2$—N⟨O⟩ and [Cl$_2$C=NMe$_2$]$^+$Cl$^-$ (**43a**).* Nuclear magnetic resonance measurements show fast fluorine exchange for the first compound, indicative of ionization, but the preponderant presence of the covalent tautomer (**IV**) (3). Although NMR evidence is given for the iminium structure (**I**) of **43a** (4, 5), in the mass spectrum, [Cl$_3$C—NMe$_2$]$^+$ can be detected (6).

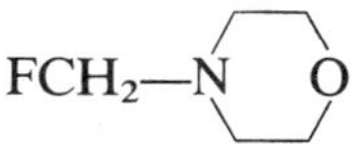

Halogen exchange is established for two other types of compounds; in both cases fast E–Z isomerization is shown:

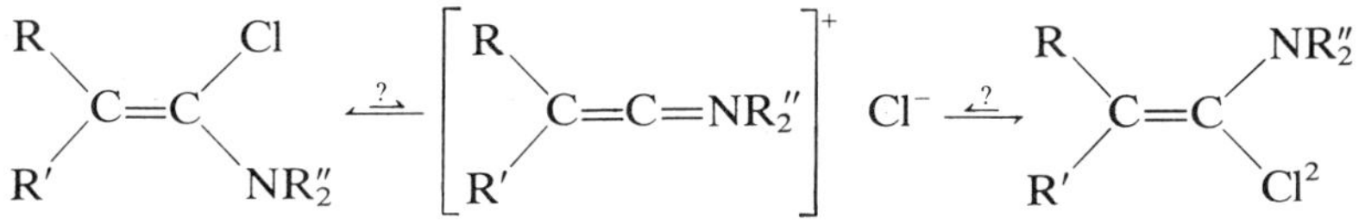

$$\underset{R'}{R}\diagdown C = C \diagup\underset{NR''_2}{Cl} \xrightarrow{?} \left[\underset{R'}{R}\diagdown C = C = NR''_2 \right]^+ Cl^- \xrightarrow{?} \underset{R'}{R}\diagdown C = C \diagup\underset{Cl^2}{NR''_2}$$

(see the chapter by Ghosez, p. 467) and

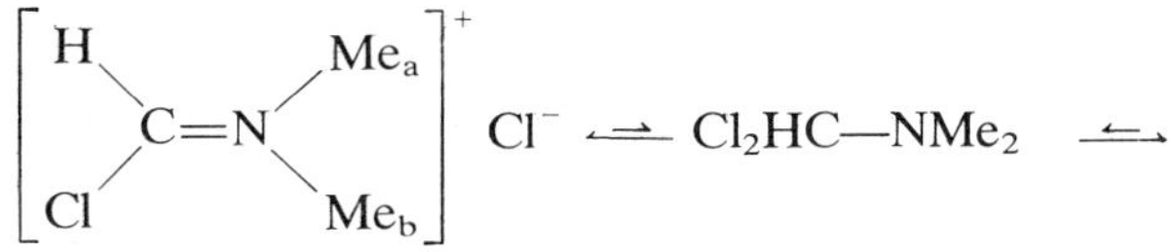

$$\left[\underset{Cl}{H}\diagdown C = N \diagup\underset{Me_b}{Me_a} \right]^+ Cl^- \rightleftharpoons Cl_2HC-NMe_2 \rightleftharpoons$$

(**24a**)

$$\left[\underset{Cl}{H}\diagdown C = N \diagup\underset{Me_a}{Me_b} \right]^+ Cl^-$$

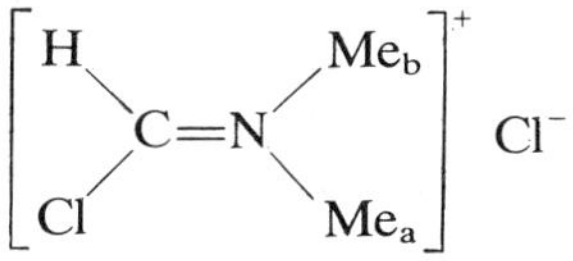

(**24a**)

---

*Throughout this chapter boldface arabic numbers refer to compounds listed in Tables I–XX.

(see Ref. 7). For the moment it is not clear whether the tautomers $\left[\begin{array}{c}R' \\ R\end{array}\!\!\!>\!\!C\!\!=\!\!C\!\!=\!\!NR_2''\right]^+ Cl^-$ for the first compound and $Cl_2HC\!-\!NMe_2$ for 24a are real intermediates or transition states only.* The fast exchange between 24a and dimethylformamide, its hydrolysis product, complicates the picture of the halogen exchange. These two processes probably give the substituent anion exchange in the case of the Vilsmeier-Haack-Arnold adduct (24d):

$$\left[\begin{array}{c}H\\ \\ Cl\end{array}\!\!\!>\!\!C\!\!=\!\!N\!\!<\right]^+ OPOCl_2^- \rightleftarrows \left[\begin{array}{c}H\\ \\ O\\ |\\ OPCl_2\end{array}\!\!\!>\!\!C\!\!=\!\!N\!\!<\right]^+ Cl^-$$

(a)   (b)

(24d)

Both tautomers (*a* and *b*) have been proposed as the structure of this compound (8, 9), and structure *b* has been singled out as the predominant one (10). Compound 24d would be an example of "tautomerism" between two ionic compounds in which two nucleophilic anions are in competition.

To introduce the discussion of the results of individual methods, the NMR, IR, and UV data for a great number of iminium compounds are given in extensive tabular form in Section II of this chapter. This systematic presentation of tables provides a general view of the spectroscopy of known compounds.

II. Nuclear Magnetic Resonance, Infrared, and Ultraviolet Tables I–XVIII

Tables I–X give the ^1H NMR and IR data of acyclic iminium compounds, that is, those in which $>\!\!C\!\!=\!\!N\!\!<$ is not part of a ring. In Table I are listed the simple iminium salts; in Table II, the double- or triple-bond conjugated ones; in Tables III–VII, the lone-pair conjugated compounds; and in Table VIII, their vinylogous derivatives. Tables IX and X contain data on cumulated and N-heterosubstituted iminium salts, and Tables XI

---

* Transformation of the anion by Lewis acids to a complex one (BF_4, $SbCl_5$, etc.) displaces the equilibrium to the iminium salt, as has been shown in the case of $[Me_2C\!=\!C\!=\!NMe_2]^+BF_4^-$ (261b; see the chapter by Ghosez, p. 468).

and XII data on cyclic compounds, that is, those in which >C=N<^+ is part of a ring. Tables XIII–XVII list the NMR coupling constants and the NMR of nuclei other than ^{1}H. Table XVIIIa gives UV data for some of the compounds listed in Tables I–XII, and Table XVIIIb for other compounds.

REMARKS, FOOTNOTES, AND ABBREVIATIONS FOR TABLES

All NMR chemical shifts are given with positive values at high frequency (low field); ^{1}H and ^{13}C chemical shifts are expressed in δ values, with reference to TMS as an internal standard.

For IR data the strongest skeletal vibration in the 1500–1750 cm^{-1} region is given: this corresponds for simple iminium salts to $\nu_{C=N}$ (Table I); in more complex systems it represents the antisymmetric stretching band for Y—C—N. As phase indication, "mull" refers to Nujol, Fluolube, Hostaflonoil, and similar mulls.

Footnotes for Tables I–XII

[a] Spectra of other derivatives also noted in the reference indicated.
[b] Spectra in other solvents also noted in the reference indicated.
[c] For a detailed list of bands see reference indicated.
[d] Coupled absorption modes: $\nu_{C=N}$–δ_{NH_2}.
[e] Raman data.
[f] Coupling constants given in Tables XIII–XVII.
[g] Inverse attribution (inverse configuration) is possible.
[h] External reference.
[i] Spectra of other nuclei given in Tables XIV–XVII.*
[k] Rotation barriers given in Table XIX.*
[l] Ultraviolet data given in Table XVIII.*
[m] For UV spectra see reference indicated.

Abbreviations

TFA	trifluoroacetic acid.
DMF	dimethylformamide.
DMSO	dimethyl sulfoxide.
Me	methyl.
Et	ethyl.
Ph	phenyl.

* The anion and the corresponding reference for these data may not always be the same.

TABLE I

Simple Iminium Compounds
For footnotes a–m see p. 27.

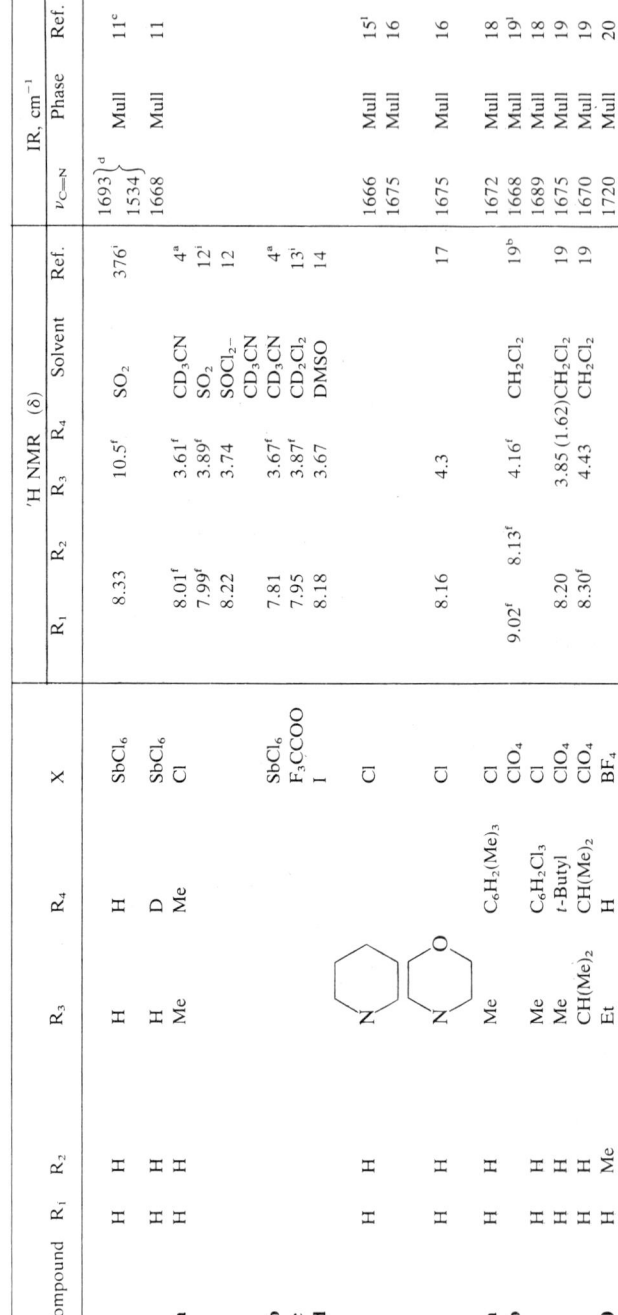

$$\left[\begin{array}{c} R_1 \\ C=N \\ R_2 \end{array}\begin{array}{c} R_3 \\ \\ R_4 \end{array}\right]^+ X^-$$

Compound	R_1	R_2	R_3	R_4	X	R_1	R_2	R_3	R_4	Solvent	Ref.	$\nu_{C=N}$	Phase	Ref.
1	H	H	H	H	SbCl$_6$	8.33		10.5^f		SO$_2$	376^j	1693^d 1534	Mull	11^e
2	H	H	H	D	SbCl$_6$							1668	Mull	11
3a	H	H	Me	Me	Cl	8.01^f 7.99^f 8.22		3.61^f 3.89^f 3.74		CD$_3$CN SO$_2$ SOCl$_2$–	4^a 12^i 12			
3b					SbCl$_6$	7.81		3.67^f		CD$_3$CN	4^a			
3c					F$_3$CCOO	7.95		3.87^f		CD$_2$Cl$_2$	13^i			
3d					I	8.18		3.67		DMSO	14			
4	H	H	(N-piperidine)		Cl							1666 1675	Mull Mull	15^l 16
5	H	H	(N-morpholine)		Cl	8.16		4.3			17	1675	Mull	16
6a	H	H	Me	C$_6$H$_2$(Me)$_3$	Cl	9.02^f	8.13^f	4.16^f		CH$_2$Cl$_2$	19^b	1672	Mull	18
6b	H	H	Me	C$_6$H$_2$Cl$_3$	ClO$_4$							1668	Mull	19^l
7	H	H	Me	t-Butyl	Cl							1689	Mull	18
8	H	H	Me	CH(Me)$_2$	ClO$_4$	8.20		3.85 (1.62)		CH$_2$Cl$_2$	19	1675	Mull	19
9	H	H	CH(Me)$_2$	H	ClO$_4$	8.30^f		4.43		CH$_2$Cl$_2$	19	1670	Mull	19
10	H	Me	Et		BF$_4$							1720	Mull	20

#					Anion				Solvent	Ref	IR	Medium	Ref
11	H	Me	n-Butyl[g]	Et[g]	SnCl$_6$						1682	KBr	15
12	H	Et	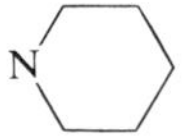		SnCl$_6$						1675	KBr	15
13	H	CH(Me)$_2$	Me	Me	BF$_4$	8.15[f]	2.9	3.4[f]	CDCl$_3$	21	1709		21
14	H	CH(Me)$_2$	 ⟨N⟩		ClO$_4$	8.35[f]	2.97	4.05 4.26[g]	CDCl$_3$	22	1697	Mull	22
15	H	CH(Et)$_2$	 ⟨N⟩		ClO$_4$	8.42[f]			CDCl$_3$	22	1690	Mull	22
16	H	C(Me)$_3$	 ⟨N⟩		ClO$_4$	8.27			CDCl$_3$	22	1682	Mull	22
17	H	CH$_2$Ph	Me	Me	ClO$_4$	8.15	3.96	3.4-3.7	CD$_3$CN	23[m]	1698		24
18	H	CHCl—CMe$_3$	Me	Me	ClO$_4$	7.59	4.0		DMSO	24[a]	1670		24
19	H	CHBr—CMe$_3$	Et	Et	ClO$_4$	7.84	4.02		DMSO	24[a]			
20	H	—P(=O)(OEt)$_2$	Me	Me	Cl	7.64[f]			CDCl$_3$	25			
21	Me	Me	H	H	SbF$_6$	2.73	2.82	9.53 3.66	SO$_2$–HSO$_3$F	26			
22	H	Cl	H	H	SbCl$_6$						1695 [d] / 1633	Mull	27[c]
23	H	Cl	D	D	SbCl$_6$						1633 [d] / 1610	Mull	27
24a	H	Cl	Me	Me	Cl	11.12		4.07 3.93	CDCl$_3$ SOCl$_2$	10 12[1]	1664	Mull	9
24b					SbCl$_6$						1681	CH$_2$Cl$_2$	28
24c	H	Cl	Me	Me	AlCl$_4$						1665	Mull	9
24d					OPOCl$_2$	10.17		3.93	CHCl$_3$	10	1667	Mull	9
25	H	Br	Me	Me	Br						1667	Mull	9
26	H	I	Me	Me	I						1628	Mull	9
27	Cl	Cl	H	H	SbCl$_6$						1659 [d] / 1568	Mull	29[c]
28	Cl	Cl	D	D	SbCl$_6$						1594	Mull	29

29

TABLE I (Cont.)

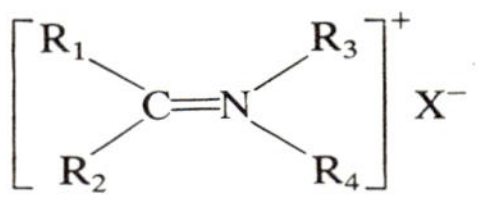

$$\left[\begin{array}{c}R_1\\R_2\end{array}C=N\begin{array}{c}R_3\\R_4\end{array}\right]^+ X^-$$

						¹H NMR (δ)					IR, cm⁻¹		
Compound	R₁	R₂	R₃	R₄	X	R₁	R₂	R₃	R₄	Solvent	Ref.	$\nu_{C=N}$	Phase	Ref.
29	Cl	Br	H	H	SbCl₆							1650 }ᵈ 1569	Mull	29
30	Cl	I	H	H	SbCl₆							1657 }ᵈ 1551	Mull	29
31	Cl	Cl	H	CCl₃	SbCl₆							1681	CH₂Cl₂	28
32	Cl	Cl	D	CCl₃	SbCl₆							1540	Mull	30
33a	Cl	Cl	H	Me	Cl							1520	Mull	30
33b					FSO₃			13.5	3.71ᶠ	SO₂–C₆D₆ HFSO₃	5ⁱ 5ⁱ			
34	Cl	Me	H	H	FSO₃							1685 }ᵈ 1592	Mull	31ᶜ
35	Cl	Me	D	D	FSO₃							1619	Mull	31
36	Br	Me	H	H	Br							1664 }ᵈ 1531	Mull	31ᵃ,ᶜ
37	Br	Me	D	D	Br							1626	Mull	31
38	I	Me	H	H	I							1637 }ᵈ 1503	Mull	31
39	I	Me	D	D	I							1603	Mull	31
40	Cl	CHMe₂	Me	Me	Cl	3.75	4.03	3.93ᵍ	CDCl₃	32			
41	Cl	CH₂CMe₃	Me	Me	Cl	3.25	4.0	3.96ᵍ	CDCl₃	32			
42	F	CHMe₂	Me	Me	BF₄	~3.4	3.53	3.46	CD₂Cl₂	33ᶠ			
43a	Cl	Cl	Me	Me	Cl		4.07	CD₃NO₂	34	1625	Mull	34	
							4.03	SOCl₂	5ⁱ				
							3.78ᶠ	SO₂	5ⁱ				
							3.85ᶠ	CD₃CN	5ⁱ				

30

No.	R	R'	N-substituent	Counterion			δ	Solvent	Ref	ν (cm⁻¹)	Medium	Ref
43b	Cl	Cl	Et	SbCl₆					5			
44	Cl	Cl	Me	Cl			3.85[f]	CD₃CN	35	1609	CHCl₃	35
							4.6	CDCl₃	5			
45	Cl	Cl	Me	FSO₃			3.90 4.77	CD₃CN	5			
							3.70 5.43	CDCl₃[a]	5[a]	1610	CH₂Cl₂	5
46	Cl	Cl	Me	Cl					5	1590	Mull	378
47	Cl	Br	Me	Cl			4.22 4.92	SO₂	12	1687	Mull	22[a]
48	Br	Br	Me	Br		2.47[f]	3.98[f]	SO₂	22[a]			
49	Me	Me	Me	ClO₄				CDCl₃				
50	Me	Me	(pyrrolidine)	ClO₄		2.52		CH₂Cl₂	22[a]	1690	Mull	22
51	Me	Et	(pyrrolidine)	ClO₄	2.47	2.77		CDCl₃	22	1680	Mull	22
52	Et	Et	(pyrrolidine)	ClO₄		2.78		CDCl₃	22	1665	Mull	22
53	=cyclopentylidene		(pyrrolidine)	ClO₄						1669	Mull	36
54a	=cyclohexylidene		(pyrrolidine)	ClO₄		2.78		SO₂	22	1670	Mull	22
										1665	Mull	37
										1669	Mull	36
54b	=cycloheptylidene		(pyrrolidine)	BF₄								
55	=cyclooctylidene		(pyrrolidine)	ClO₄		2.88		CDCl₃	22	1705	Mull	22
										1658	Mull	36
56	=cyclononylidene		(pyrrolidine)	ClO₄						1649	Mull	37

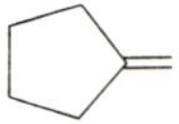

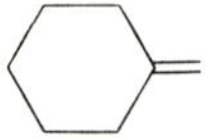

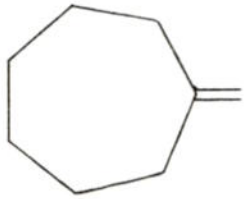

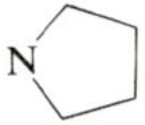

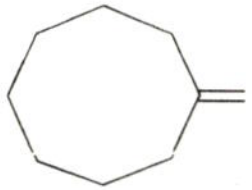

31

TABLE I (*Cont.*)

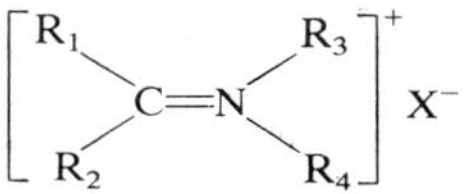

$$\begin{bmatrix} R_1 & & R_3 \\ & C=N & \\ R_2 & & R_4 \end{bmatrix}^+ X^-$$

Compound	R₁	R₂	R₃	R₄	X	¹H NMR (δ) R₁	R₂	R₃	R₄	Solvent	Ref.	IR, cm⁻¹ $\nu_{C=N}$	Phase	Ref.
57a	methylenecyclopentane		piperidine		ClO₄							1698	Mull	36
57b					SnCl₆							1644	KBr	15¹
57c					NO₃							1691	KBr	15¹
58	cyclohexylidene		morpholine		ClO₄	2.95				CH₂Cl₂	22	1640	Mull	22
59	methylenecyclopropane		Me	Me	SO₃F		2.26	3.71		CDCl₃	38			
60	methylenecyclobutane (Me₂)		Me	Me	BF₄			3.44	3.42	CDCl₃	39	1728	CHCl₃	39
61	H	Cl	Ph	Me	Cl									
62	Cl	Cl	Me	Ph	SO₃F			4.19		CD₃CN	5	1681	CH₂Cl₂	28
63	Me	CHMe₂	Me	Ph	ClO₄	2.60	2.75	3.83ᶠ		CD₃NO₂	40	1648	KBr	40
64	Me	CHMe₂	Ph	Me	ClO₄	2.18ᶠ		3.92ᶠ		CD₃NO₂	40			

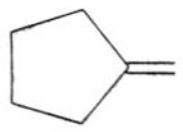

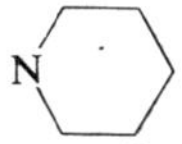

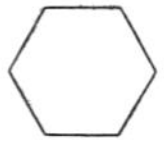

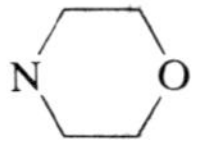

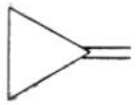

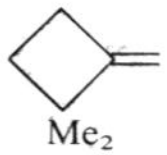

TABLE II.
Conjugated Iminium Compounds
For footnotes a–m see p. 27.

$$\left[\begin{array}{c}R_1 \\ \diagdown \\ C=N \\ \diagup \quad \diagdown \\ R_2 \quad\quad R_4\end{array}\right]^+ X^-$$

Compound	R_1	R_2	R_3	R_4	X	R_1	R_2	R_3	R_4	Solvent	Ref.	$\nu_{C=N}$	Phase	Ref.
								¹H NMR (δ)					IR, cm^{-1}	
65a	H	$H_2C=CH-$	Et	Et	$SnCl_6$							1646	KBr	41[a,l]
66a	H	$H_2C=CH-$			$SnCl_6$							1655	KBr	41[l]
67	H	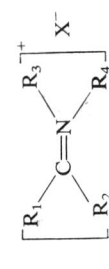	Me	Me	ClO_4	8.47	6.00	3.48	3.67[g]	$CDCl_3$	42[l]			
68a	H	Ph	Me	H	SO_3F	8.81	8.03	3.84[f]	10.0[f]	SO_2	26[a]			
68b					Cl	9.16[f]	8.38	3.73[f]	13.4	$CDCl_3$	43[l]			
							7.7							
68c					$(CO)_5CrBr$	9.44	7.93	3.90		Acetone	374[a,l]	1680	KBr	374[a,c]
							8.35							
69	H	Ph	Ph	H	SO_3F	9.15	7.85			SO_2	44[c]			
70a	H	Ph	Me	Me	ClO_4	8.8	7.6–8.0		3.93	TFA	23			
70b					BF_4	9.1[f]	7.4–7.9	3.87[f]	3.73[f]	CD_2Cl_2	45	1660	Mull	45
70c					F_3CCOO	8.91	7.3–7.8	3.99	3.92	TFA	45			
						9.01		4.01	3.95	$CDCl_3$	46			
70d					$(CO)_5CrBr$	9.57	7.88	4.08	4.13	Acetone	374[a,l]	1670	KBr	374[a,c]
							8.05							

33

TABLE II (Cont.)

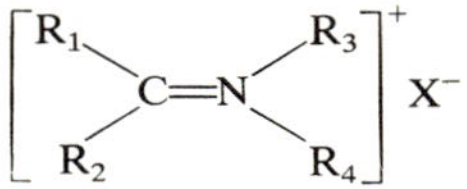

$$\left[\begin{array}{c} R_1 \\ R_2 \end{array} \!\!C\!\!=\!\!N\!\! \begin{array}{c} R_3 \\ R_4 \end{array} \right]^{+} X^{-}$$

Compound	R_1	R_2	R_3	R_4	X	R_1	R_2	R_3	R_4	Solvent	Ref.	$\nu_{C=N}$	Phase	Ref.
								\multicolumn{2}{c}{^{1}H NMR (δ)}			\multicolumn{3}{c}{IR, cm^{-1}}	
71	H	Ph-p-Me	Me	Me	ClO$_4$	8.11^c		3.9		Acetone	23			
72	H	Ph	CH$_2$Ph	CH$_2$Ph	BF$_4$	9.50		5.22	5.39	CDCl$_3$	47			
73	H	Ph-p-NO	Me	Me	F$_3$CCOO	9.27	8.2	3.95	4.10	CDCl$_3^j$	46			
74	H	Ph	 \multicolumn{2}{c}{pyrrolidine}	ClO$_4$	8.99^f				CDCl$_3$	22	1658	Mull	22
75	H	HC=CH–Ph						4.48	4.75^g	PhNO$_2$	373			
76	H	(methylfuryl)	\multicolumn{2}{c}{pyrrolidine}	ClO$_4$	8.71^f				CH$_2$Cl$_2$	22	1660	Mull	22
77a	H	(methylfuryl)	Me	Me	ClO$_4$	8.78^f				CDCl$_3$	22	1660	Mull	22
78	H	Ph	Ph	CH$_2$COOMe	F$_3$CCOO	8.60^c		3.73	3.82^g	CD$_3$CN	23^l			
79	H	Ph	H	COPh	SbF$_6$	9.3			(4.9)	C$_6$D$_6$	48^a			
80	H	Ph	H	COOEt	SbF$_6$	9.83^f				SO$_2$FSO$_3$H	49			
81	H	Ph	Br	Ph	Br	9.05^f				FSO$_3$H	49			
						9.15^g				TFA	50	1655	Mull	50^l
						9.45^g								
82	Cl	Ph	H	Me	Br							1618	Mull	51
83	Cl	Ph	D	Me	Br							1637	Mull	51
84	Br	Ph	H	Me	Br							1635	Mull	52

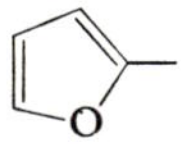

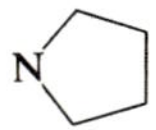

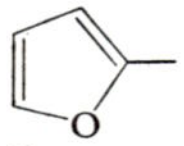

34

#	X	Structure	R1	R2	Y	δ	δ	δ	solvent	ref	ν (cm⁻¹)	medium	ref
85a	Cl	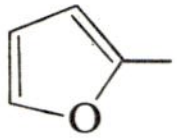 [2-methylfuran]	H	Me	Cl						1623	Mull	51
85b	Cl	[2-methylfuran]	D	Me	Br						1629	Mull	51
86a	Cl		Me	Me	Cl						1629	Mull	51
86b	Cl			Me	Br						1634	Mull	51
87a	Cl	Ph		Me	Cl						1634	CH$_2$Cl$_2$	28
87b		Ph			HCl$_2$						1642	Mull	51
87c					AlCl$_4$						1639	CH$_2$Cl$_2$	51
87d	Br		Me	Me	OPOCl$_2$ Br	7.8 / 7.97	4.04 / 4.21	3.86 / 3.97	CH$_2$Cl$_2$ / CDCl$_3$	375[a,i] / 52	1631 / 1640	Mull / KBr	51 / 52
88	Br	Ph	Me	Me	Br	7.72	4.05	3.87	TFA	52	1655 / 1600	Mull	52
89a	Br	Ph	[pyrrolidine]		Br$_3$	7.73	4.30	4.08	CD$_3$CN	52	1620	KBr	52
89b	Br	Ph	[piperidine]		Br	7.72	4.68	4.32	TFA	52	1615	Mull	52
90a	Br		Me	Me	Br$_3$	7.78 / 7.70	4.54 / 4.45	4.13 / 4.06	CD$_3$CN / CD$_3$CN	52 / 52	1620 / 1622	KBr / Mull	52 / 52
90b					BF$_4$								
90c													
91	Cl	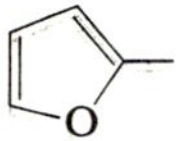 [2-methylfuran]	Me	Me	Cl						1621	Mull	51[a]
92	Cl	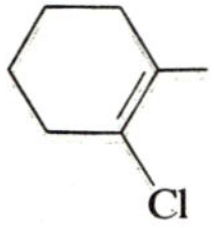 [2-methyl-1-chlorocyclohexene]	Me	Me	Cl		4.29[g]	4.20[g]	CDCl$_3$	53	1635	CHCl$_3$	54
93	Cl	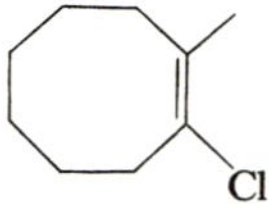 [2-methyl-1-chlorocyclooctene]	Me	Me	Cl		4.17[g]	4.09[g]	CDCl$_3$	53			

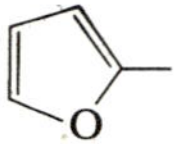

35

TABLE II (Cont.)

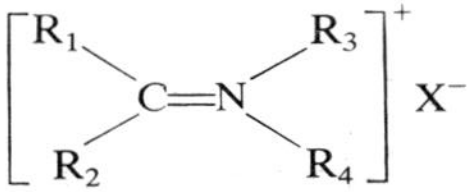

$$\left[\begin{array}{c} R_1 \quad R_3 \\ C=N \\ R_2 \quad R_4 \end{array}\right]^+ X^-$$

Compound	R_1	R_2	R_3	R_4	X	R_1	R_2	R_3	R_4	Solvent	Ref.	$\nu_{C=N}$	Phase	Ref.
94	Cl	Ph–C=CH / Cl	Me	Me	Cl			4.16		$CDCl_3$	54	1625	$CHCl_3$	54
95	Cl	Me–C=CPh / Cl	Me	Me	Cl			4.24		$CDCl_3$	54			
96	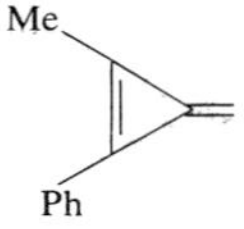 Me–△=CH₂–Ph		Me	Me	BF_4	2.80		3.49	3.61	CD_3NO_2	40[a,j]			
97	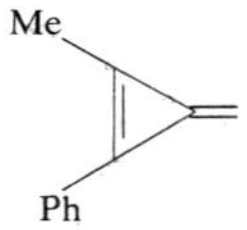 Me–△=CH₂–Ph		Me	Ph	BF_4	2.98		3.98[f]		CD_3NO_2	40[k]	1899 1599	KBr	40[l]
98	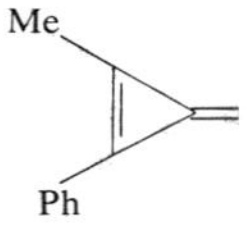 Me–△=CH₂–Ph		Ph	Me	BF_4	2.80			4.16	CD_3NO_2	40[k]			
99	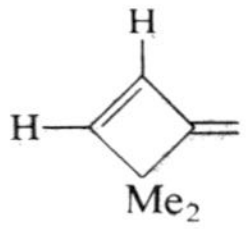 H–□=CH₂–Me₂ / H		Me	Me	BF_4			3.45	3.35[g]	$CDCl_3$	55	1720		55

#	Structure	R	R'	Anion	δ	δ	δ	Solvent	Ref	ν (cm⁻¹)	Medium	Ref
100	CMe₃, =CH₂, Me₂, H	Me	Me	BF₄	3.80		3.55[g]	CDCl₃	55[i]			
101	Me, =CH₂, Me, Me₂			ClO₄	3.60		3.42[g]	CDCl₃	55			
102	=CH₂ cyclohexene-Me	Me	CH₂Ph	ClO₄	3.63	6.87	5.18	CD₃NO₂	40[a,k]			
103	=CH₂ cycloheptatriene	H	Me	BF₄		7.65	3.18	CD₃CN	56[a,l]			
104	=CH₂ cycloheptatriene	H	Prop.	BF₄						1638	KBr	57[a,l]
105	=CH₂ cycloheptatriene	Et	Et	BF₄						1635	KBr	57[l]
106	=CH₂, Me, Ph, pyrrolidine			ClO₄		2.78		CH₂Cl₂	22	1658	Mull	22

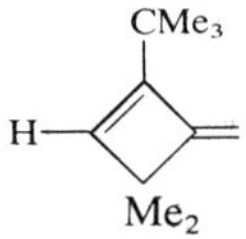

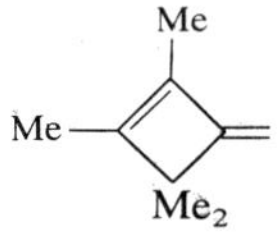

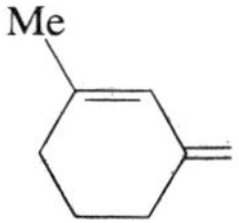

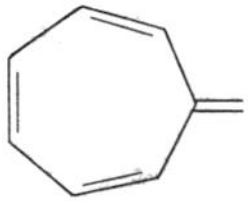

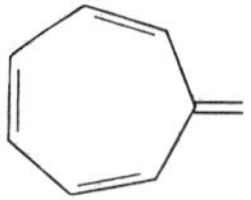

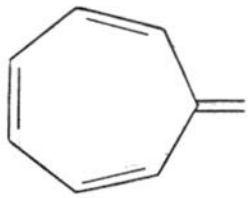

TABLE II (Cont.)

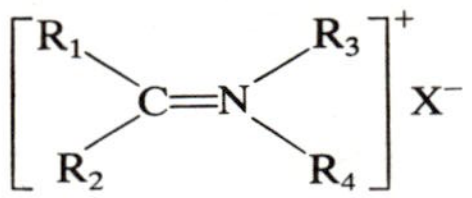

$$\begin{bmatrix} R_1 & R_3 \\ C=N \\ R_2 & R_4 \end{bmatrix}^+ X^-$$

Compound	R_1	R_2	R_3	R_4	X	¹H NMR (δ)				Solvent	Ref.	IR, cm⁻¹		
						R_1	R_2	R_3	R_4			$\nu_{C=N}$	Phase	Ref.
107	Me	PhHC=CH	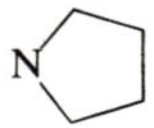 (pyrrolidine)		ClO₄	2.78				CH₂ClCN	22	1622	Mull	22
108	Ph	Ph	Me	H	SbF₆		8.0	3.79	9.7	SO₂ / HSO₃F	26			
109a	Ph	Ph	H	H	SbF₆			3.83	9.94	HSO₃F	58ᵃ			
									9.56	HSO₃F	58			

TABLE III

Amidinium Compounds
For footnotes a–m see p. 27.

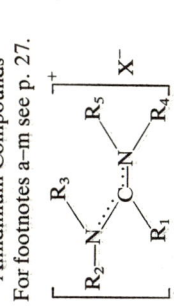

$$\left[R_2-N \overset{R_3}{\underset{R_1}{\cdots}} C \overset{R_5}{\underset{R_4}{\cdots}} N \right]^+ X^-$$

Compound	R_1	R_2	R_3	R_4	R_5	X	^{1}H NMR (δ)						IR, cm^{-1}		
							R_1	R_2	R_3	R_4	R_5	Solvent	Ref.	ν_a	Phase	Ref.
110	H	H	H	H	H	SnCl$_6$								1720 1715 }	Mull	59
111	H	H	H	H	CHCl$_2$	Cl								1705 }[d] 1601	Mull	60[a]
112	D	D	D	D	CDCl$_2$	Cl								1648	Mull	60
113	H	H	H	H	CHBr$_2$	Br								1689 }[d] 1604	Mull	60
114	D	D	D	D	CDBr$_2$	Br								1635	Mull	60
115	H	H	H	H	CHO	Br								1691 }[d] 1511	Mull	60
116	H	H	H	Me	Me	Cl	8.00	9.4	9.4	3.0[f]	3.2[f]	DMSO	61	1715		61[a]
117	H	H	n-Butyl	CHCl$_2$	n-Butyl	Cl	9.75	11.3	3.90	8.60	3.60	CDCl$_3$	62[a]			
118	H	H	Me	CHCl$_2$	Me	Cl	9.88[f]	11.2	3.33[f]	8.60	3.50	CDCl$_3$	62			
119a	H	Me	Me	Me	Me	PF$_6$	6.1	2.9	2.9	2.9		CD$_3$CN	63[a,j]			
							7.16	2.95	2.95	2.95			64	1720		64
119b	H	Et	Et	Et	Et	ClO$_4$	7.64	3.35	3.35	3.35		CDCl$_3$	65[a,j]			
120	H					ClO$_4$	7.72	3.62	3.62	3.62		CDCl$_3$	65			
121	H					ClO$_4$	7.98	3.89	3.89	3.89		CDCl$_3$	65[a,j]			

39

TABLE III (Cont.)

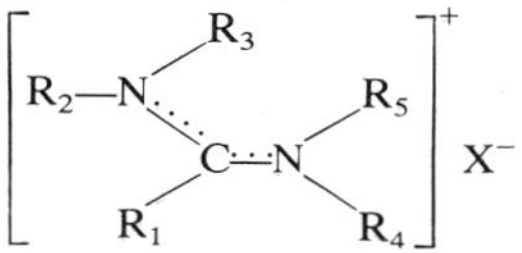

$$\left[\begin{array}{c} R_2-N \overset{R_3}{\underset{R_1}{\cdots}} C \overset{R_5}{\underset{R_4}{-N}} \end{array} \right]^+ X^-$$

Compound	R_1	R_2	R_3	R_4	R_5	X	R_1	R_2	R_3	R_4	R_5	Solvent	Ref.	ν_a	Phase	Ref.
							\multicolumn{6}{c}{^{1}H NMR (δ)}		\multicolumn{3}{c}{IR, cm^{-1}}
122	H	piperidine		piperidine		ClO_4	7.70 6.8	3.67			3.67	$CDCl_3$ SO_2	65 66			
123	H	Ph	Me	Ph	Me[g]	ClO_4								1652	KBr	67[l]
124	H	Ph	Ph	Me	Me	Cl	9.25	7.7		3.84	2.20[g]	$CDCl_3$	61	1684[d]	Mull	68[a]
125	Me	H	H	H	H	Cl $SbCl_6$ $SnCl_6$								1684[d] 1690[d] 1680[d]	Mull Mull Mull	69[a] 59[a,c] 70[a]
126	CCl_3	H	H	H	Me	Cl	2.24	9.18	8.45	3.14	3.06	DMSO	71[n]			
127	Me	H	H	Me	Me	Cl	2.38	9.73	8.88	3.18	3.16					
128	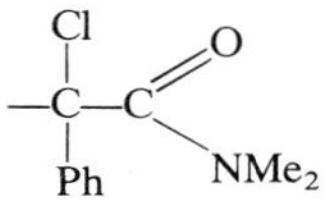 (Cl)(O=)C(Ph)(NMe$_2$)	H	H	Me	Me	Cl				3.29		DMSO	72[i]			
129	Me	H	Me	Me	H[g]	Cl		10.15	2.86[f]	2.94[f]	9.23	DMSO	73[a] 74,75[a,k] 74,75			
130	Me	Ph	Me	Ph	Me[g]	SO_4	2.45	7.6	2.98	3.14	7.1	14% H_2SO_4	74,75	1607	KBr	67[a]
131	Ph	Ph	Me	Ph	Me	ClO_4		7.37	3.32	7.37	3.32	D_2O	67	1597	KBr	67[a]
132	C≡CPh	Me	Me	CHMe$_2$	H	ClO_4	—	3.36	3.53	4.30	—	$CDCl_3$	76	1630	$CHCl_3$	76
133	C≡CPh	Me	Me	Me	Me	I		3.62		3.62		$CDCl_3$	76	1620	$CHCl_3$	76

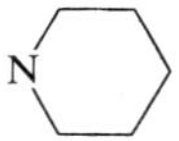

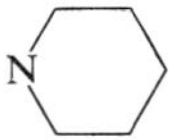

134 (t-But)(Me)C=C=C(Me)(t-But)							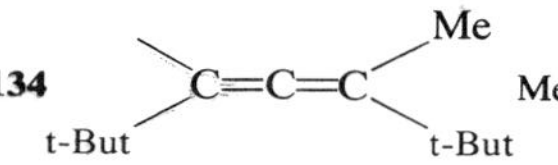 3.45	3.45	CDCl$_3$	77	1618 KBr 77[l]
135	Cl	H	H	H	H	Br					1680[d] Mull 59[c]
136	Br	H	H	H	H	SnCl$_6$					1678[d] Mull 59
137	Cl	Me[g]	H	Me	Me	SnCl$_6$	11.7	3.13	3.45[f]	CD$_3$CN	12[i]
138	Cl	H	Ph	Me	Me	Cl			3.66[f]	CDCl$_3$	12
139	Cl	Me	Me	Me	Me	Cl		3.62	3.62	CDCl$_3$	12[i] 1585 KBr 378
140	Cl	Me	Me	ClC=CH$_2$	Me	Cl		3.80	3.89	CDCl$_3$	54 1620 Film 78[a]
141 (Ph)(Ph)C=C(Me)(H) ClO$_4$	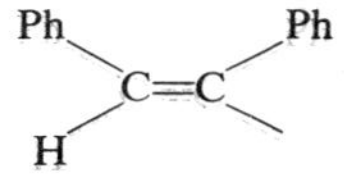			Me	Me	ClO$_4$	2.92	2.92	CDCl$_3$	79	

[a] Concentration varied from 0.06 to 9.85 mole%.

TABLE IV
Amidium Compounds (Protonated and Alkylated Amides)
For footnotes a–m see p. 27. See also chapter by Kautlehner, Volume 2.

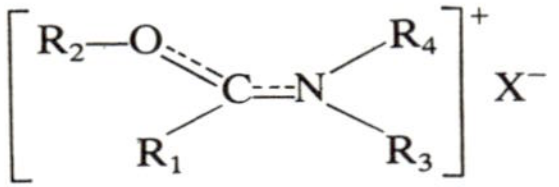

$$\left[R_2-O\underset{R_1}{\overset{R_4}{=}}C=N\underset{R_3}{\overset{}{}} \right]^+ X^-$$

Compound	R_1	R_2	R_3	R_4	X	^{1}H NMR (δ)					IR, cm^{-1}		
						R_1	R_2	R_3	R_4	Solvent	Ref.	$\nu_{C=N}$	Phase	Ref.
142a	H	H	H	H	SbCl$_6$	8.6	10.8	8.6^f	8.6^f	HFSO$_3$	81	1732	Mull	80
142b					FSO$_3$							1669 }	Mull	80
143	D	D	D	D	SbCl$_6$							1665 }		
144	H	H	Me	H	FSO$_3$	8.50		3.44^f	8.58^f	HFSO$_3$	82			
145a	H	H	Me	Me	Cl	8.65	15.4	3.2^f	3.4^g	CDCl$_3$	12			
145b					SO$_4$			4.09^f	4.16^f	H$_2$SO$_4$^{h}	37^k			
145c					FSO$_3$	8.38^f	9.98	3.43^f	3.53^f	HFSO$_4$^{h}	81			
145d					SbF$_6$	8.62	9.82^f	3.82^f	3.73^f	HFSO$_3$ } SO$_2$ClF }	83^a			
146	H	Ph	Ph	Me	SbF$_6$	8.78^f	10.49^f	7.16	3.88^f	HSO$_3$F } SO$_2$ }	83			
147	H	Me	H	H	SbCl$_6$							1712	Mull	80^{a,c}
148	D	Me	D	D	SbCl$_6$							1666	Mull	80
149	H	Me	Me	Me	MeSO$_4$	8.69		3.58	3.49	CDCl$_3$	84			
150	H	Et	Ph	Me	BF$_4$	8.48				CDCl$_3$	85^a			
151	H	Ph-p-Me	Me	Me	Cl			3.65	3.36	CD$_3$CN	86^a			
152a	Me	H	H	H	SbCl$_6$							1707	Mull	87
152b					ClO$_4$							1652	KBr	67
152c					FSO$_3$	2.67	10.72	8.24	8.36	HFSO$_3$	81^i			

#	R1	R2	R3	X				Solvent	Ref	IR	Medium	Ref
152d	Me	H		Cl						1718	Mull	88
153a	Me	D	Me	FSO$_3$	2.64		3.45	HFSO$_3$	81	1670	Mull	89d,c
153b	Me	D	Me	Cl						1662	Mull	89
154	Me	H	Me	Cl								
155	Me	H	▽N	SbF$_6$	2.96	9.06	5.00	HFSO$_3$/SO$_2$	90			
156a	Me	H	Me	Cl						1619^c	Mull	91
156b	Me	H	C$_6$H$_{11}$	Br						1616^c	Mull	91
157a	Me	D	Me	Cl						1616^c	Mull	91
157b	Me	D	C$_6$H$_{11}$	Br						1620^c	Mull	91
158	Me	Me	Me	ClO$_4$	2.62	4.35	3.50	3.39	TFA	23	1680	Mull	23
159	C≡CH	Et	H	BF$_4^-$	4.70	4.88	3.22^f	10.2	CD$_2$Cl$_2$	76^i	1665	CH$_2$Cl$_2$	76
160	C≡CPh	Et	Me	BF$_4^-$	7.7	4.90	3.48^f	3.73	CDCl$_3$	76^i			
161	HC=CH$_2$	Et	Me	ClO$_4$	6.2 / 6.5	4.56	3.39^f	3.46^g	CDCl$_3$	76			
162a	Ph	H	H	ClO$_4$							1654	KBr	92
162b				FSO$_3$									
163	Ph	H	Me^g	FSO$_3$		8.36	8.61^f	7.94^g	3.54^f	HFSO$_3$	93^l		
164	Ph	H	Et	ClO$_4$					HFSO$_3$	82^a	1677	KBr	92
165	Ph	Me	Et	ClO$_4$							1640	KBr	92
166	N$_3$	H	H	SbCl$_6$							1639	Mull	92
168	N$_3$	H	Me	SbCl$_6$							1670	Mull	94^a
169	F	H	H	SbF$_6$	i	11.07^f	8.48^f		HFSO$_3$/SO$_2$	95	1670	Mull	94^a
170	F	H	Me	SbF$_6$	i	11.5^f	3.65^f	8.37^f	SO$_2$	95			
171	F	H	Me	SbF$_6$	i		3.98	3.96	SO$_2$	95			
172	Cl	Ph	Me	Cl				3.87	CDCl$_3$	54	1665 / 1600	CHCl$_3$	54
173	H	COMe	Me	Cl	7.96	2.73		2.81	CDCl$_3$	84			
174	H	SOMe	Me	Cl	8.10	2.91		3.00	CDCl$_3$	84			

TABLE V
Thioamidium Compounds (Protonated and Alkylated Thioamides)
For footnotes a–m see p. 27.

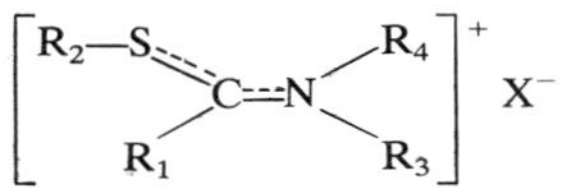

$$\left[\begin{array}{c} R_2-S \\ \diagdown \\ C=N \\ \diagup \quad \diagdown \\ R_1 \quad R_3 \end{array} \begin{array}{c} R_4 \\ \end{array} \right]^+ X^-$$

Compound	R_1	R_2	R_3	R_4	X	^{1}H NMR (δ) R_1	R_2	R_3	R_4	Solvent	Ref.	$\nu_{C=N}$	IR, cm^{-1} Phase	Ref.
175a	Me	H	H	H	SbCl$_6$							1548^d	Mull	96^l
176	Me	H	Me	Me	Cl							1623	KBr	97^c
177	Me	Me	H	H	SO$_4$							1610	KBr	97^{c,l}
178	Me	Me	Me	Me	I							1607	CHCl$_3$	98
179	Me	Me	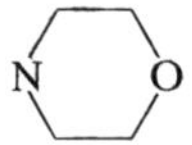 \multicolumn{2}{c}{N⌢O⌣}	I							1581	CHCl$_3$	98
180	Ph	Me	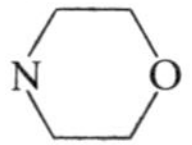 \multicolumn{2}{c}{N⌢O⌣}	I							1580	CHCl$_3$	98
181	Ph	Me	Me	Ph^g	I							1562, 1587^d, 1497	CHCl$_3$	98
182	Br	Me	H	H	Br							1545	Mull	99
183	Br	Me	D	D	Br									
184	Cl	Ph	Me	Me	Cl					3.96	CD$_3$NO$_2$	100		
185	Br	Ph	H	H	Br							1616^d, 1535	Mull	99

44

TABLE VI
Uronium, Thiouronium, and Guanidinium Compounds
For footnotes a–m see p. 27.

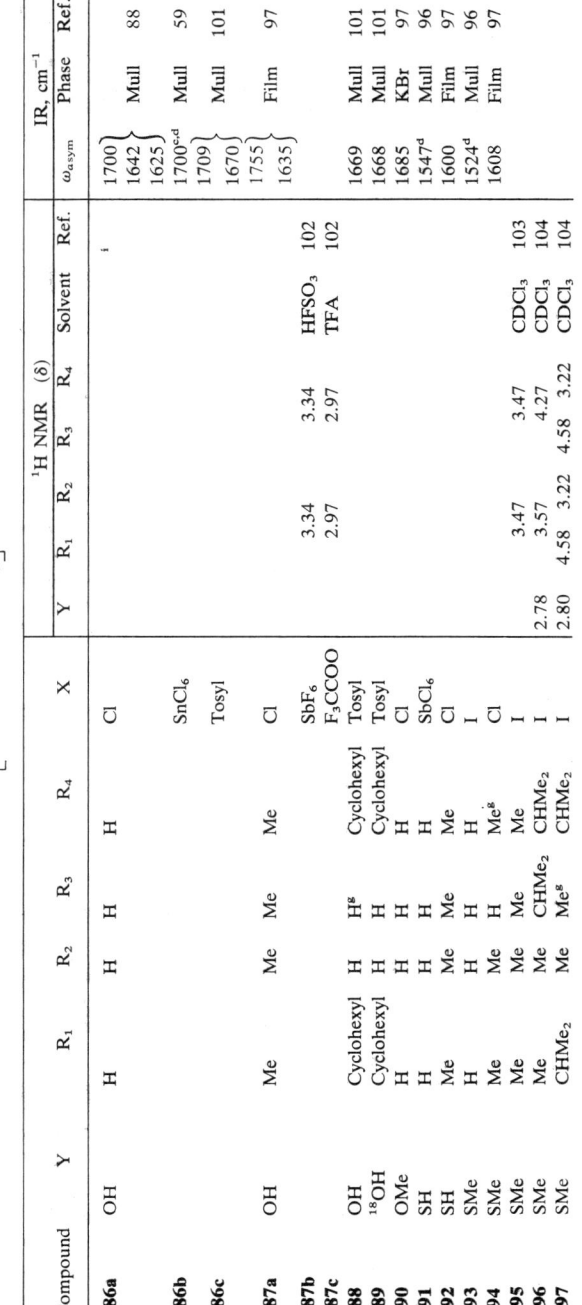

Compound	Y	R_1	R_2	R_3	R_4	X	1H NMR (δ) Y	R_1	R_2	R_3	R_4	Solvent	Ref.	IR, cm⁻¹ ω_{asym}	Phase	Ref.
186a	OH	H	H	H	H	Cl							i	1700 1642 1625	Mull	88
186b						SnCl₆								1700^c,d	Mull	59
186c						Tosyl								1709 1670	Mull	101
187a	OH	Me	Me	Me	Me	Cl								1755 1635	Film	97
187b						SbF₆		3.34		3.34		HFSO₃	102			
187c						F₃CCOO		2.97		2.97		TFA	102			
188	OH	Cyclohexyl	H	H^g	Cyclohexyl	Tosyl								1669	Mull	101
189	18OH	Cyclohexyl	H	H	Cyclohexyl	Tosyl								1668	Mull	101
190	OMe	H	H	H	H	Cl								1685	KBr	97
191	SH	Me	Me	Me	Me	SbCl₆								1547^d	Mull	96
192	SH	H	H	H	H	Cl								1600	Film	97
193	SMe	Me	Me	Me	Me^g	I								1524^d	Mull	96
194	SMe	Me	Me	H	Me	Cl								1608	Film	97
195	SMe	Me	Me	Me	Me	I	2.78	3.47		3.47		CDCl₃	103			
196	SMe	Me	Me	CHMe₂	CHMe₂	I		3.57		4.27		CDCl₃	104			
197	SMe	CHMe₂	Me	Me^g	CHMe₂	I	2.80	4.58	3.22	4.58	3.22	CDCl₃	104			

45

TABLE VI (Cont.)

$$\left[\begin{array}{c} R_2 \\ R_1 \end{array} N \cdots C \cdots N \begin{array}{c} R_4 \\ R_3 \end{array} \right]^+ X^-$$

Compound	Y	R_1	R_2	R_3	R_4	X	Y	R_1	R_2	R_3	R_4	Solvent	Ref.	ω_{asym}	Phase	Ref.
198	SMe	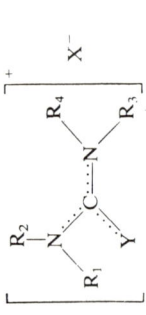		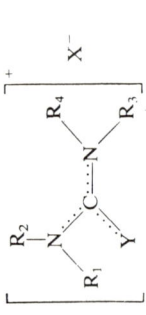		I	2.88		3.98	3.98		$CDCl_3$	104			
199	SPh-p-Me	Me	Me	Me	Me	ClO_4										
200a	NH_2	H	H	H	H	$SnCl_6$		3.27	3.27	3.27	3.27	$CDCl_3$	103	1670[c,d]	Mull	59
200b						$PtCl_6$								1667	Mull	96
201	ND_2	D	D	D	D	$PtCl_6$								1664	Mull	96
202	NHPh	Me	Me	Me	Me	F_3COO		3.16	3.00	2.98	2.50	TFA	105[k]	1587	Mull	96
203	NMe_2	Me	Me	Me	Me	Cl	2.94	2.94	2.94	2.94			106			
						PF_6	2.84	2.84	2.84	2.84			106			
Diprotonated Derivatives																
204	OH	NH_3^+	H	H	H	SbF_6	12.82	9.01	9.01		9.38	SO_2	107[a]			
205	OH	NH_3^+	H	H	Me[g]	SbF_6	12.55	9.01	9.40	3.74[f]		FSO_3H	107			
206	OH	$(NH_2Me)^+$	H	H	Me[g]	SbF_6	12.81	8.7(3.86)	9.50	3.82[f]		FSO_3H	107			
207	OMe	NH_3^+	H	H	H	SbF_6	4.73	9.05		9.47		FSO_3H	107			
208	NH_3^+	H	H	H	H	SbF_6	8.68	8.07	7.85	8.07	7.85	FSO_3H	107			
209	NH_3^+	H	H	H	Me	SbF_6	8.71	7.88	8.05	3.54		FSO_3H	107			
210	NH_3^+	H	H	Me	Me	SbF_6	8.63	7.54	3.69	3.62[f]		FSO_3H	107			

46

TABLE VII

Iminium Carbonates and Thiocarbonates
For footnotes a–m see p. 27.

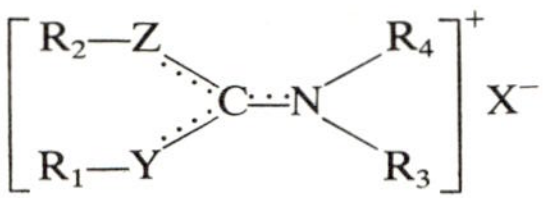

$$\left[\begin{array}{c}R_2-ZR_4 \\ \ddots C\cdots N \\ R_1-YR_3\end{array}\right]^+ X^-$$

Compound	R_1Y	R_2Z	R_3	R_4	X	R_1	R_2	R_3	R_4	Solvent	Ref.	$\nu_{C=N}$	Phase	Ref.
						¹H NMR (δ)						Ir, cm⁻¹		
211	HO	HO	H	H	SbF$_6$	10.1		7.45		SO$_2$	108[a]			
212	HO	HO	H	Me	SbF$_6$	9.91[f]		7.23	3.36[f]	HSO$_3$F	108			
213	HO	HO	Me	Me	SbF$_6$	9.30		3.38		HSO$_3$F	108			
214	HO	EtO	H	H	SbF$_6$	9.86[f]	4.86	7.40	7.35	HSO$_3$F	108			
215	HO	MeO	H	Et	SbF$_6$	9.68[f]	4.38	7.2	3.9	HSO$_3$F	108			
216	HO	MeO	Et	H	SbF$_6$	9.71	4.43	3.7	7.2	HSO$_3$F	108			
217	HO	EtO	Me	Me	SbF$_6$	9.18	4.80	3.46	3.31[g]	HSO$_3$F	108			
218	EtO	EtO	H	H	SbCl$_6$							1657	Mull	80[a,c]
219	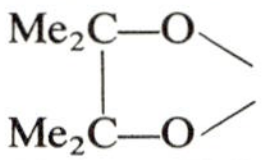 Me$_2$C—O	Me	Me	Cl	1.65		3.38		CDCl$_3$	109	1720	CHCl$_3$	109
220	Me$_2$C—O	PhO	Me	Me	Cl	7.25		3.76		CDCl$_3$	34	1710	CHCl$_3$	34
221	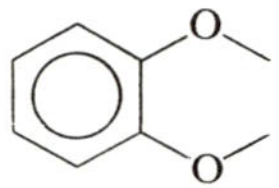 (catechol)	O	Me	Me	Cl							1692	CHCl$_3$	54
222	S—(CH$_2$)$_3$—S	Et	Et	Cl							1730		110
223	S—(CH$_2$)$_2$—S	H	H	Cl							1580	Mull	111[a,l]
224	S—(CH$_2$)$_2$—S	H	H	Cl							1570	Mull	111[a,l]
225	PhS	PhS	Me	Me	Cl			3.96		CD$_3$NO$_2$	100			

47

TABLE VIIIa

Vinylogues and Iminologues of Amidinium Compounds (Cyanines and Azacyanines)
For footnotes a–m see p. 27.

$$\left[\begin{array}{c} R_3' \\ N \\ R_2' \end{array} C \begin{array}{c} R_1' \\ Y \\ \end{array} C \begin{array}{c} R_3 \\ N \\ R_1 \end{array} R_2 \right]^+ X^-$$

Compound	R_1^n	R_2^n	R_3^n	Y	X	R_1	R_2	R_3	Y	Solvent	Ref.	$\nu_{C=N}$	Phase	Ref.
												IR, cm^{-1}		
												\multicolumn{3}{c}{¹H NMR (δ)}	
226	H	CH$_2$Ph	H	CH	ClO$_4$	7.90^f	4.52		5.57	CD$_3$CN-D$_2$O	112^m			
227	H	Me	Me	CH		7.69			5.37	DMSO	2^l			
						7.80^f	3.14	3.35	5.13^f	CDCl$_3$	113^a			
						7.68^f	3.15	3.35	5.13^f	C$_2$H$_2$Cl$_4$	114^{a,k}			
						7.49^f	3.03	3.23	5.19^f	D$_2$O	115^a			
228	H	\multicolumn{2}{c	}{(piperidine)}	CH	ClO$_4$	6.8			5.1	SO$_2$	66			
229	H	Me	Me	C(cyclohexenyl)	ClO$_4$	7.45	3.28	3.35^g		CDCl$_3$	116			
230	H	Me	Me	CPh	ClO$_4$	7.88	3.40	2.51		C$_2$H$_2$Cl$_4$	114^k			
231	H	Me	Me	CNMe$_2$	ClO$_4$		3.32		2.58	CDCl$_3$	117^l			
232	Ph	Me	Me	CH	ClO$_4$	7.42	2.9	3.54	5.74	C$_2$H$_2$Cl$_4$	114^k			
233	Br	H	H	CH	Br							1698^{c,d}	Mull	118
												1664		
234	Cl	Me	Me	CH	Cl		3.66^f		5.88	CDCl$_3$	119^{m,l}	1555	CHCl$_3$	54^l
235	Cl	Me	Me	CEt	Cl		3.57^f	3.28^g	2.7	CDCl$_3$	119^j			
236	Cl	Me	Me	CPh	Cl		3.58^f		7.4	CDCl$_3$	119^{m,l}			
237	Cl	Me	Me	CCl	Cl		3.68^f			CDCl$_3$	119^j			
238	Cl	Me	Me	CF	Cl		3.66^f		i	CDCl$_3$	120			
239	Cl	Et	Et	COPh	Cl		3.95			CDCl$_3$	72			

	R1	R2	R3								
240	H	Me	Me	(CH=CH=CH)	ClO4	7.76 7.68	3.10	3.23		DMSO CDCl3	2[k,l] 113	
						7.17	3.59	2.79		C2H2Cl4	114	
241	H	 ⌬N⌬		(CH=CH=CH)	ClO4	6.8[f]			5.33 7.0	SO2	66	
242	H	Me	Ph[g]	(CH=CH=CH)	ClO4	7.56[f]	3.54		5.1 6.8	CDCl3	113	
243a	H	Me	Me	(CH=CH)2=CH	ClO4	8.0	3.6	7.3	5.81[f] 7.70 8.1 6.4	CDCl3 DMSO	113 2[k,l] 114	
						7.68						
243b	H	Me	Me	(C=CH=NMe2)+	Br	7.16	2.7		β 5.3 γ 6.88 δ 5.69	CD3CN	113	
244	H	Me	Me	N	ClO4	8.18	3.14 3.55	3.35 3.70	g	CDCl3	121[l]	
245	H	Ph	Me	N	Cl	9.3	6.6	6.7[g]		CDCl3	61[l]	
246	H	Ph	Me	N	ClO4	8.1	6.9	3.2		SO2	66	
247	Cl	H	H	N	SbCl6							c 1587 1532 Mull 122[l]
248	Br	H	H	N	SbCl6							1595[c] Mull 123[a]
249	Cl	Me	Me	N	Cl		3.59[f]	3.64[g]		CDCl3	124[i]	1610 1674 CHCl3 123 54[l]
250	NMe2	Me	Me	N	SbCl6							1639 Mull 125[a]
251	Cl	H	H	N	SbCl6							1646[c] 1633 Mull 125
	NMe	Me	Me									
	Cl	D	D									
252	NMe2	Me	H[g]	CH	ClO4	2.77	2.97[f]	6.63	4.53	CDCl3	76	
	Ph	Et	Et									
253	NMe2	Me	Me	N	Cl		3.08	3.16		CDCl3	126	
	Cl											
254	NMe2	Me	Me	CH	Cl	3.02	3.02		3.60	CDCl3	119	1530 CHCl3 119
255	NMe2	Me	Me	N	BF4	2.96	2.96			CDCl3	126	

[n] R'1, R'2, and R'3 are given in the second line if different from R1, R2, and R3.
[g] The same as R1, R2 and R3.

49

TABLE VIIIb

Vinylogues of Amidium Compounds
For footnotes a–m see p. 27.

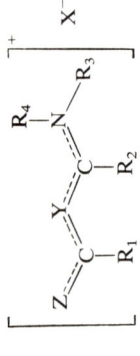

Compound	Z	R$_1$	Y	R$_2$	R$_3$	R$_4$	X				^{1}H NMR (δ)					IR, cm^{-1}	
								Z	R$_1$	Y	R$_2$	R$_3$	R$_4$	Solvent	Ref.	$\nu_{C=N}$	Phase	Ref.
257	OH	H	CH	H	Me	Me	ClO$_4$		7.77^f	5.79^f	7.82	3.17	3.35	H$_2$O	127			
258a	OEt	H	CH	H	Me	Me	Cl	4.43	8.48	6.23	9.15	3.54	3.72	CDCl$_3$	10			
258b							OPOCl$_2$	4.39	8.18	6.15	8.59	3.44	3.62	(CH$_2$Cl)$_2$	10			
259	OEt	H	CMe	H	Me	Me	OPOCl$_2$	4.45	8.60	2.07	8.22		3.62	(CH$_2$Cl)$_2$	10			

TABLE IX

Iminium Compounds with Cumulated Structures
For footnotes a–m see p. 27.

Compound	Cation	Anion	^{1}H NMR (δ)		Solvent	Ref.	IR, cm^{-1}		
							$\nu_{C=N}$	Phase	Ref.
261a	Me$_2$C=C=NMe$_2$	PF$_6$	1.98 (CMe)	3.52 (NMe)	CD$_3$CN	128			
261b		BF$_4$	2.03 (CMe)	3.57 (NMe)	CH$_2$Cl$_2$	39^j			
262	Cl$_2$C=N=CCl$_2$	SbCl$_6$					1855^n	Mull	30

^{n} ω_{asym}C—N—C.

TABLE X
N-Heterosubstituted Iminium Compounds
For footnotes a–m see p. 27.

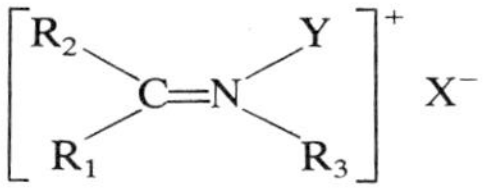

$$\left[\begin{matrix} R_2 \\ R_1 \end{matrix} C=N \begin{matrix} Y \\ R_3 \end{matrix} \right]^+ X^-$$

Compound	R_1	R_2	R_3	Y	X	R_1	R_2	R_3	Y	Solvent	Ref.	$\nu_{C=N}$	Phase	Ref.
263	Me	Me	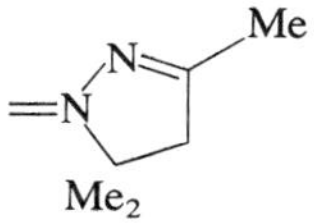 =N–N=C(Me)(CMe₂) (pyrazoline)		SnCl₆	2.70	2.61			DMSO	129			
264	H	n-Hexyl[g]	H	OH	Cl	8.14	2.77			CHCl₃	130[a]	1678	Mull	130[a]
265a	Me	Me	H	OH	Cl	2.28	2.19[g]			CHCl₃	130	1697	Mull	130
265b					SbF₆	2.26[f]	2.20[f,g]	12.0[f]	11.1	SO₂ / HFSO₃	131			
266	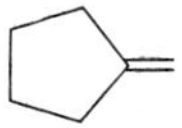 (cyclopentylidene)		H	OH	Cl	2.76				CHCl₃	130	1707	Mull	130
267	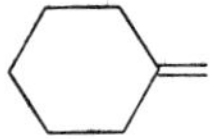 (cyclohexylidene)		H	OH	Cl	2.76				CHCl₃	132	1690	Mull	132
												1696	Mull	130
268	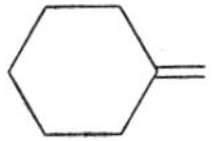 (cyclohexylidene)		D	OD	Cl							1656	Mull	130
269	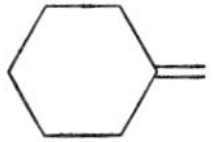 (cyclohexylidene)		H	OH₂⁺	Cl	2.86				Neat	132	1690	Film	132
270a	CMe₃	H	H	OH	SO₄	8.14[f]				H₂SO₄	133			
270b					F₃CCOO	8.08[f]				TFA	134[a]			

TABLE XI

Cyclic Iminium Compounds
For footnotes a–m see p. 27.

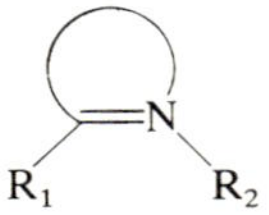

Compound	Cation	R_1	R_2	Anion	R_1	R_2	Solvent	Ref.	$\nu_{C=N}$	Phase	Ref.
						¹H NMR (δ)			IR, cm⁻¹		
271	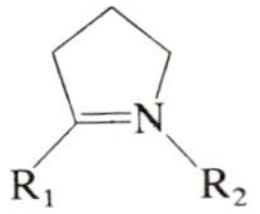	Me	Me	ClO₄	2.51	3.56	TFA	135[a,c]	1699	Mull	136[a]
272		n-Butyl	Me	ClO₄					1685	Mull	136
273		Ph	H	Cl					1653	Mull	137[a]
274a	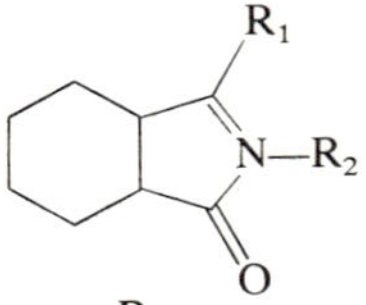	Ph	H	SbCl₆					1590	KBr	138[a,m]
274b				BF₄					1590	KBr	138
275a		Ph	Me	SbCl₆	8.0–8.4	3.92	CD₃NO₂	138[a,b]			
275b				BF₄	7.9–8.3	3.87	CD₃NO₂	138			
276	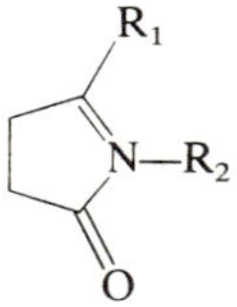	Me	Ph	SbCl₆	3.20	7.4–8.5	CD₃NO₂	138	1590	KBr	138[m]
277		Me	H	SbCl₆	3.10		CD₃NO₂	138[a,b]	1620	KBr	138[l]
278		Ph	Me	SbCl₆	7.7–8.4	3.97	CD₃NO₂	138			
279	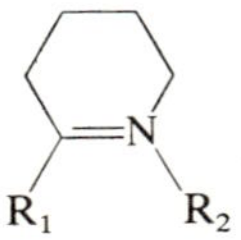	Me	Me	ClO₄	2.51	3.62[f]	TFA	135[a,c]	1686	Mull	136[a]
280		n-Butyl	Me	ClO₄					1690	Mull	139[a]
281		Et	Et	ClO₄					1684	Mull	136
									1671	Mull	139

52

#	Structure	Substituents		X			Solvent	Ref.	IR	Medium	Ref.
282	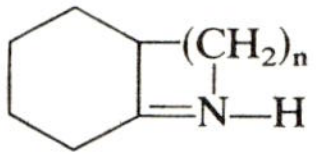 (CH₂)ₙ fused cyclohexane =N—H	n = 2		Cl					1684	CHCl₃	140
283		n = 3		Cl					1695	CHCl₃	140
284a	octahydroquinoline			ClO₄					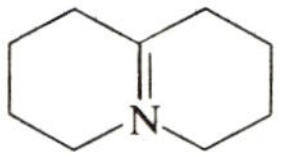 1696	Mull	141
284b				I					1685	KBr	141
284c				I₃					1688	Mull	142
									1686	Mull	142
285	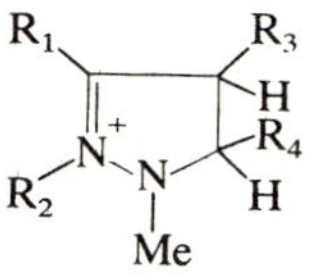 pyrazolinium ring with R₁,R₂,R₃,R₄,Me	$R_3 = Me, R_4 = Ph$	H	F_3CCOO	8.07	3.88	CDCl₃ / TFA	143			
286		$R_3 = R_4 = H$	Me	ClO₄	2.52	3.45	DMSO	144[a,c]			
287		$R_3 = R_4 = H$	Ph	Cl	2.43	7.7	CDCl₃	145[a]			
288		$R_3 = H, R_4 = Ph$	Me	Cl	8.86	3.30	DMSO	146[a,b]			
289		$R_3 = t\text{-}Bu, R_4 = Me$	CD₃	ClO₄		3.66[f]	CDCl₃	144			
290	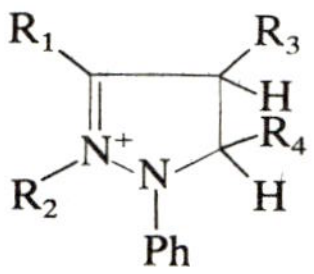 pyrazolinium ring with Ph	$R_3 = R_4 = H$	Me	Cl	2.82	3.52	CDCl₃	145[a,c]			
291		$R_3 = H, R_4 = Me$	Me	Cl	2.78	3.47	CDCl₃	145[l]			
292	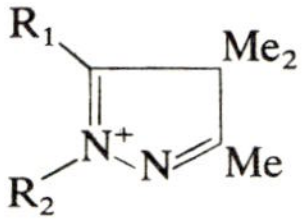 Me₂ pyrazolinium, Me	Me	Et	I	3.00	4.44	CDCl₃	147[l]			
293		Me	CH₂Ph	I	2.61	5.62	CDCl₃	147[l]			
294		Me	Ph	I	3.06	7.6–8.0	CDCl₃	147[l]			

TABLE XI (Cont.)

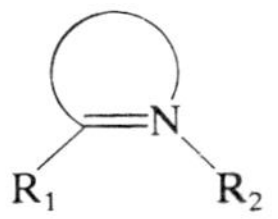

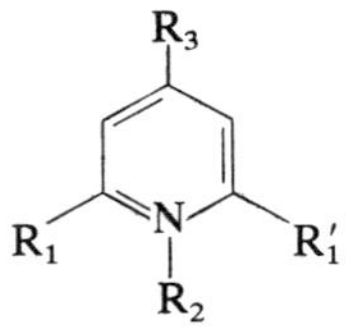

Compound	Cation		R_1	R_2	Anion	^{1}H NMR (δ)				IR, cm^{-1}		
						R_1	R_2	Solvent	Ref.	$\nu_{C=N}$	Phase	Ref.
295a	R_3=H		H	H	F$_3$CCOO	8.91^f	13.9^f	TFA	148	1636	Mull	149^a
295b					SbCl$_6$							
295c					BF$_4$	8.22		SO$_2$	150^a			
295d					Cl	8.81	16.22	CH$_2$Cl$_2$	151^{a,b,i}			
						8.66	15.70	CH$_3$CN	151			
						8.66	15.26	CH$_3$NO$_2$	151			
295e					Br	8.68	13.35	CH$_3$CN	151			
295f					I	8.65	9.50	CH$_3$CN	151			
296	R_3=H		H	Me	I		4.49^f	D$_2$O	152^i			
							4.40	D$_2$O	153^a			
						8.75	4.45	DMSO	12			
297	R_3=H		H	Et	I	9.10	4.57^f	D$_2$O	153^a			
298	R_3=Me		H	Me	I	8.73	4.23^f	D$_2$O	153			
299	R_3=COOMe		H	Me	I	8.50	4.47^f	D$_2$O	153			
						9.05						
300	R_3=COOH		H	Me	I	8.87	4.33^f	D$_2$O	153^a			
301	R_3=CF$_3$		H	Me	I	9.08	4.47^f	D$_2$O	153			
302a	R_3=H		H	COMe	Cl							
302b					SbF$_6$	8.17	(2.57)	SO$_2$ / FSO$_3$H	90^a	1619	Mull	91
303	R_3=H	$\{R_1=H, R_1'=F\}$	H	H	F$_3$CCOO				154^{a,i}			

		R_1	R_2				
304	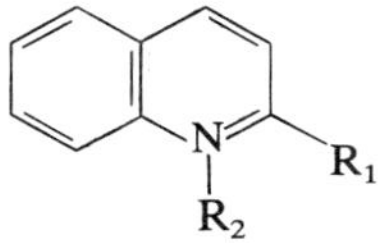	H	H	F$_3$CCOO	9.25[f]	TFA	12[i]
305	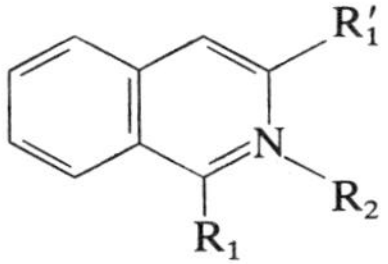	H	H	F$_3$CCOO	9.75[f]	TFA	12[i]
306		H	Me	I	10.13; R_1': 8.75 — 4.57	DMSO	79[i]
307	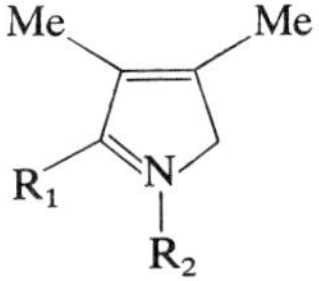	H	H	SO$_4$	8.80[f]	H$_2$SO$_4$	155[a,m]
308	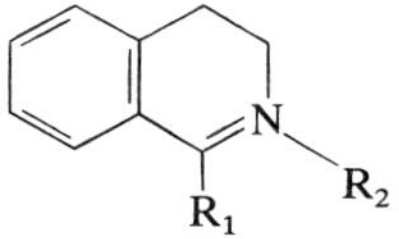	H	Me	I	9.45 — 3.80	DMSO	79

TABLE XII

Cyclic Amidinium, Thioamidium, Amidium, and Vinylogous Compounds
For footnotes a–m see p. 27.

Compound	Cation	Anion	R_1	R_2	R_3	R_4	Solvent	Ref.	$\nu_{C=N}$	Phase	Ref.
309a	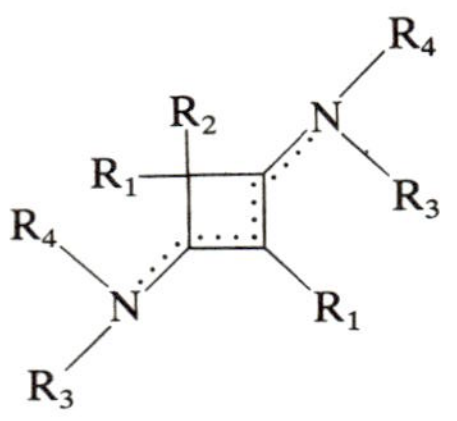 $R_1 = t$-But., $R_2 = H$, $R_4 = R_3 = Me$	Cl		4.30		3.37	CDCl$_3$	77	1692 1574	KBr	77[1]
309b		BF$_4$		3.96		3.26	CDCl$_3$	77	1694 1582	KBr	77[1]
310	$R_1 = $ OPh, $R_2 = H$, $R_4 = R_3 = Me$	Cl				3.47	CDCl$_3$	32			
311	$R_1 = $ Et, $R_2 = H$, $R_4 = R_3 = Me$	ClO$_4$			3.37	3.23^g	CDCl$_3$	120			
312	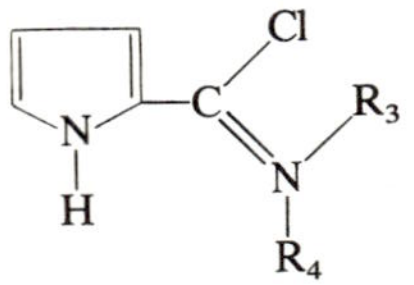 $R_3 = R_4 = Me$	Cl			3.99	3.83^g	CDCl$_3$	156			
313	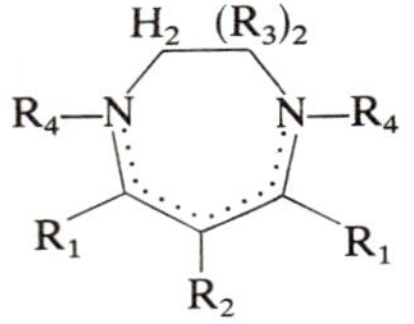 $R_1 = R_2 = R_3 = R_4 = H$		8.04^f	5.43^f			CD$_3$OD	112^m			
314	$R_1 = R_3 = Me$, $R_2 = R_4 = H$		2.29	5.14^f		9.0	(CD$_3$)$_2$CO	12			
315	$R_1 = Me$, $R_2 = R_3 = R_4 = H$					8.8		112^m			

#	Structure	Substituents	Anion				Solvent	mp	IR (cm⁻¹)		Ref.
316	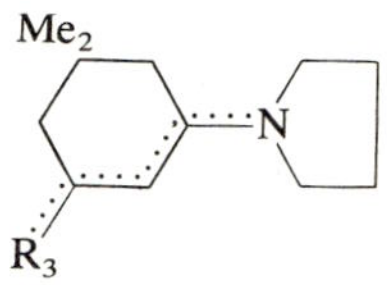	R_3 = NEt$_2$	I						1558	Mull	157
317		R_3 = SEt	I						1619[c] / 1568	Mull	157
318		R_3 = OEt	I						1626	Mull	157
319	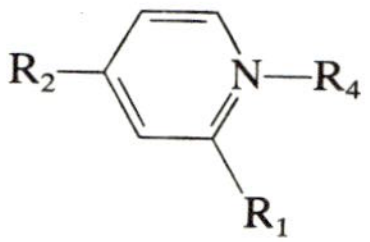	R_1 = OH, R_2 = R_4 = H	SbCl$_6$						1607 / 1631	Mull	149
320		R_1 = OMe, R_2 = R_4 = H	SbCl$_6$						1647 / 1638	Mull	149
321		R_1 = OH, R_2 = H, R_4 = Me	SbCl$_6$						1645	Mull	149
322		R_1 = OMe, R_2 = H, R_4 = Me	SbCl$_6$						1641	Mull	149
323		R_1 = R_4 = H, R_2 = OH	SbCl$_6$						1652	Mull	149
324a		R_1 = H, R_2 = OH, R_4 = Me	SbCl$_6$						1649	Mull	149
324b		R_1 = H, R_2 = NH$_2$, R_4 = Me	I	8.15	6.75	4.15[f]	D$_2$O	152[a]			
325			I			4.03[f]	D$_2$O	152			
326	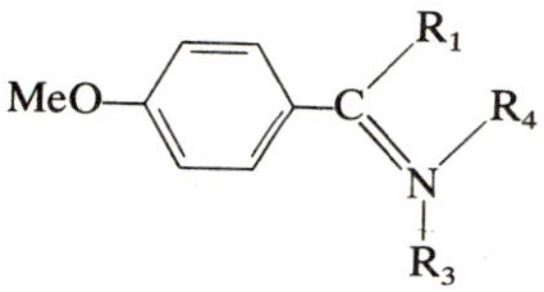 MeO-C$_6$H$_4$-C(R$_1$)=N-R$_3$ (R_4)	R_1 = R_2 = H, R_3 = R_4 = Me	ClO$_4$		3.4		TFA	23[e]			
327	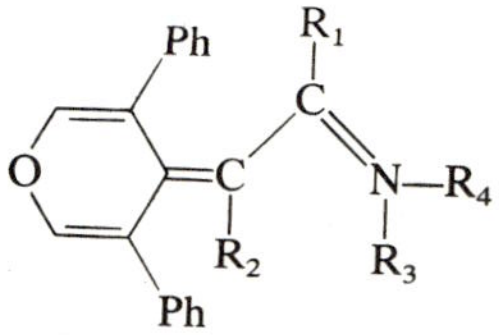 pyranone diimine	R_1 = R_2 = H, R_3 = R_4 = Me	ClO$_4$	8.85[f]	5.80	3.45	3.18[g]	DMSO	158[a,m]		

57

TABLE XII (*Cont.*)

Compound	Cation	Anion	¹H NMR (δ)						IR, cm⁻¹		
			R_1	R_2	R_3	R_4	Solvent	Ref.	$\nu_{C=N}$	Phase	Ref.
328	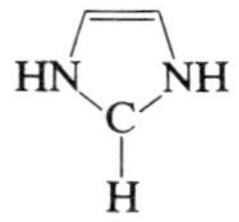 HN–C(H)–NH (imidazoline)	F_3CCOO	H_2: 8.78^f		$H_{4,5}$: 7.67		TFA	12^i			
329	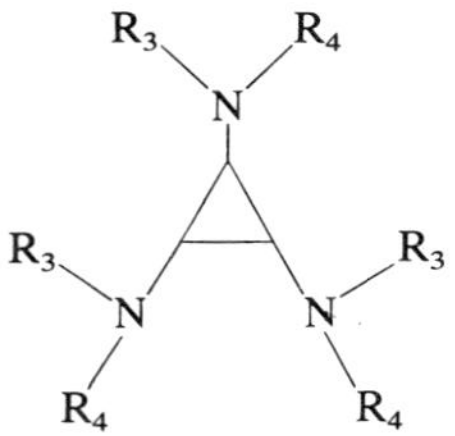 $R_3 = R_4 = Me$ cyclopropane tris-amino	ClO_4			3.16		$CDCl_3$	159^{a,l}			

58

TABLE XIII

$^1H-^1H$ NMR Coupling Constants of the Iminium Compounds[a]

Compound[b]	$^2J_{H,H\ C=N}$
6b	7.2 (CD$_2$Cl$_2$)
	7.6 (TFA)

Compound[b]	$^3J_{HC-NH}$
68a	4.5
68b	4.3
118	4.4
129	5.0
135	5.4
144	4.4
159	5.2
163	5.4
169	5.0
205	5.1
212	5.0
252	4.6

Compound[b]	$^3J_{HC=CH-N}$
226	12.0 (trans)
227	11.8 (trans)
240	11.0 (trans)
241	12.0 (trans)
257	11.5 (trans)
313	8.0 (cis)

Compound[b]	$^3J_{HO,H\ C=N}$
145c	4.7
145d	5.0
146	5.1

Compound[b]	$^4J_{HO,C=N,H}$
168	2.0
169	2.5
212	2.8
214	1.9
216	2.8

Compound[b]	$^3J_{HC=NH}$ cis	trans
68a		17.3
68b		16.5
79		17
80		17
118		(14)
144	(4.5)	
270a		14.4
307	4.8	
c	~6	~14

Compound[b]	$^4J_{HC=N-CH}$ cis	trans
3a	1.2	1.2
3b	1.25	1.25
6b	(1.3–1.4)	(0.8)
13		(1.9)
14		(2.0)
15		(2.1)
70b	1.0[d]	1.2[d]
74		(2.1)
75		(1.8)
76		(1.8)
144		(1.0)
145b	1.2	1.7
145c	0.7	1.1
145d	0.7	1.0
146		1.0
227	0.8	1.0

Compound[b]	$^4J_{HC-C=NH}$ cis	trans
265b	1.1	1.4

Compound[b]	$^5J_{HC-C=N-CH}$
49	(1.0)
50	(1.4)
289	0.7 (trans cyclic)

(*Continued overleaf*)

TABLE XIII (*Cont.*)

[a] Values given in hertz (cps); in parentheses if configuration has not been determined. For signs see p. 77.
[b] For formula, solvent, and reference, see Tables I–XII.
[c] Amidinium compounds: ref. 160.
[d] Configuration confirmed by Nuclear Overhauser Effect experiment.

^3J Coupling Constants in Protonated N-Heterocycles[a]

Compound	Name	$^3J_{HC=CH-N^+}$
295	Pyridinium	6.5 (5.5)
304	Quinolinium	5.7 (4.2)
305	Isoquinolinium	7.0 (6.0)

[a] TFA solution; coupling constants in hertz (cps); values of neutral compounds in parentheses: ref. 161.

TABLE XIV

^{14}N and ^{15}N NMR Data

a. Chemical Shifts[a]

Compound	Cation		Iminium (protonated compound)			Neutral compound		
		Anion	Chem. shift	Solvent		Chem. shift	Solvent	Ref.
1	$H_2C=NH_2$	$SbCl_6$	180	SO_2				376
3c	$H_2C=NMe_2$	F_3CCOO	199	CD_2Cl_2				12
68b	Ph–C(H)=N(H)(Me)	Cl	169	$CDCl_3$				43
145e	HO–C(H)=NMe$_2$	F_3CCOO	81	TFA		83	Neat	162
186a	H_2N–C(OH)=NH$_2$	Cl	66	cc HCl		56	H_2O	162
295a	$R_1=R_2=R_3=R_4=H$	F_3CCOO	169	TFA		292	Neat	163
295d		Cl	178	cc HCl		297	Neat	162
296	$R_2=R_3=R_4=H, R_1=Me$		173	cc HCl		286	Neat	164
			184[b]	CH_3OH		297[b]	Neat	165
		I	159	D_2O				152
319	$R_2=OH, R_1=R_3=R_4=H$	Cl	160	cc HCl		149	H_2O	162
330	$R_3=OH, R_1=R_2=R_4=H$	Cl	174	cc HCl		228	H_2O	162
323	$R_4=OH, R_1=R_2=R_3=H$	Cl	155	cc HCl		132	H_2O	162

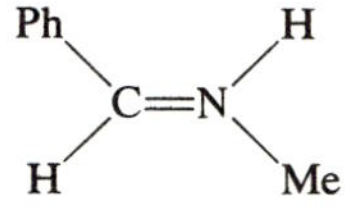

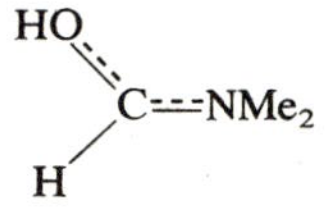

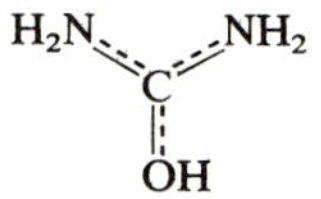

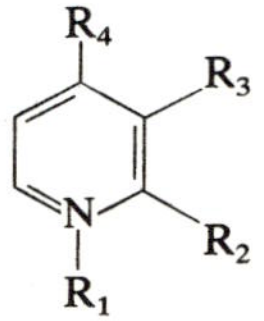

TABLE XIV (Cont.)

Compound	Cation	Iminium (protonated compound)			Neutral compound			Ref.
		Anion	Chem. shift	Solvent	Chem. shift		Solvent	
328a 328b	 HN⌐NH (pyrazolium)	F_3CCOO Cl	147 155	TFA cc HCl	186		Me_2CO	166 162
331	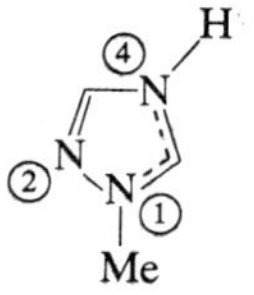 (4)N-N(1)-Me, (2), H (triazolium)	F_3CCOO	N_1: 151 N_2: 307 N_4: 151	TFA	N_1: 205 N_2: 306 N_4: 238		Me_2CO	166
332	 HN⌐S (thiazolium)	F_3CCOO	179	TFA	302		Me_2CO	166
304a 305	Quinolinium Isoquinolinium	Cl Cl	169 166	cc HCl cc HCl	183 186			164 164

[a] Parts per million relative to NH_4^+ ($CH_3NO_2 = 354$ ppm). High-frequency (low-field) shift is positive.
[b] ^{15}N.

b. ^{14}N–^{1}H (^{15}N–^{1}H) Coupling Constants[a]

Compound[b]	Solvent	Ref.	$^1J_{HN=}$	$^2J_{HC=N}$	$^2J_{H\overset{|}{C}-\overset{|}{N}=}$	$^3J_{H\overset{|}{C}-C=N}$	$^3J_{H\overset{|}{C}-\overset{|}{C}-\overset{|}{N}=}$
1	SO$_2$	376	65				
3a	CD$_3$CN	4			1.7		
3b	CD$_3$CN	4		2.0	1.7		
109b	SO$_2$	167	(−92.6)				
142b	HFSO$_3$	81	+62				
162b	HFSO$_3$	93	(95.5, 94.3)				
270a	H$_2$SO$_4$	133	(106)	(3.4)			
270b	TFA	134		(3.7)			
295a	TFA	148	70				
295d	CD$_3$OD	165		(−3.01)		(−3.98)	
295g	H$_2$SO$_4$	167a	(90.5)				
296	D$_2$O	169[c]		+1.0	0.4	+2.5	
297	D$_2$O	153					2.4
301	D$_2$O	153				3.0	
304	HFSO$_3$	170	(96)				
	D$_2$SO$_4$	171[c]		(−2.0)		(−4.5)	
306	H$_2$SO$_4$	170	(98)				

[a] Values in hertz (cps); ^{15}N–^{1}H couplings in parentheses: $J_{^{14}N-^1H} = -0.713 \cdot J_{^{15}N-^1H}$.
[b] For formulae, see Tables I–XII.
[c] Data for other derivatives also given.

TABLE XIV (Cont.)

c. ^{15}N–^{13}C NMR Coupling Constants[a]

Compound	Cation	Anion	$^1J_{^{15}N-C}$		$^2J_{^{15}N-C}$		$^3J_{^{15}N-C}$		Solvent	Ref.
295d	Pyridinium	Cl		12.0 (0.45)		2.1 (2.4)	C_4:	5.3 (3.6)	CD_3OD (neat)	165[b]
304b	Quinolinium	SO_4	C_2:	15.9 (1.4)	C_3:	1 (2.7)	C_4:	4.6 (3.5)	$H_2SO_4(CCl_4)$	172
			C_9:	13.8 (0.6)	C_{10}:	1 (2.1)	C_5:	0 (6)		
					C_8:	1 (9.3)	C_7:	2.7 (3.9)		

[a] Values in hertz (cps); $J_{^{14}N-^{13}C} = -0.713 \cdot J_{^{15}N-^{13}C}$; coupling constants of neutral compounds in parentheses.
[b] Data for other derivatives also given.

TABLE XV

^{13}C NMR Chemical Shifts of Iminium Compounds[a]
a. Acyclic Iminium Compounds

$$\begin{array}{c}\diagdown\\\diagup\end{array}C_1{=}\overset{+}{N}\begin{array}{c}{\diagup}C_3\\{\diagdown}C_2\end{array}$$

Compound[b]	$C_{1(\alpha)}$	C_2	C_3	C_β[c]	Solvent	Ref.
3a	168	49.7			SO$_2$–CD$_3$CN	12
33a	153.6	38.5			SO$_2$–C$_6$D$_6$	5
33b	164.1	38.7			SOCl$_2$–C$_6$D$_6$	5
43a	160.2		50.0		SOCl$_2$–C$_6$D$_6$	5
	168.0		49.7		SO$_2$–C$_6$D$_6$	5
68b	172.5	39.6			CDCl$_3$	43
68c	173.5	39.9			Acetone	374
70d	173.1	44.6	52.2		Acetone	374
100	182.4		42.9		CDCl$_3$	55
128	160.5	41.7	41.0		D$_2$O	72
137	155.8	33.7	43.3		CD$_3$CN	12
139	159	45			CDCl$_3$	173
153[b]	175.7	38.4	40.4		CDCl$_3$	76
159	157.2	30.7			CD$_2$Cl$_2$	76
160	156.4	39.1	43.0		CD$_2$Cl$_2$	76
227b	163.7	38.5	46.3	90.8	DMSO	174
234	159.4	42.5	47.7	89.8	CD$_3$CN	12
235	162.7		46.2	102.0	CDCl$_3$	12
236	161.8		46.8	102.9	CDCl$_3$	12
237	160.7	44.5	50.3	90.3	CDCl$_3$	12
240b	162.1	38.6	46.1	103.7		
				C$_\gamma$: 162.6	DMSO	174
249	152.2	43.5			CDCl$_3$	12
254b	169.1	40.8		72.4	CDCl$_3$	175
261b	215	44.1		88.6	CD$_2$Cl$_2$	32

[a] Values in δ (ppm relative to TMS); high-frequency (low-field) shifts are positive. For formulae see Tables I–XII.
[b] BF$_4^-$.
[c] For lone-pair delocalized systems, $\begin{array}{c}C_3\\C_2\end{array}{>}N({=}C_\alpha{=}C_\beta)_n{=}C_\alpha{=}N{<}\begin{array}{c}C_3'\\C_2'\end{array}$
$\quad\quad\quad\quad\quad\quad\quad\quad\quad\quad\quad\quad\quad\quad\quad(\gamma)$

TABLE XV (*Cont.*)

b. Protonated *N*-Heterocycles[a]

Compound	Heterocycle		C_α	Solvent	Ref.
333	Pyrazole		134.5 (134.1)	H_2O–HCl	176[b]
328a	Imidazole	C_2:	134.0 (135.7)	H_2O–HCl	176
328b			134.1 (135.8)	H_2O–H_2SO_4	177
334	L-Histidine Methyl ester	C_2:	135.1 (136.7)	H_2O–HCl	178
295a	Pyridine		141.2 (149.7)	TFA	179[b]
295d	Pyridine		141.9	H_2O–HCl	176
335	Pyrazine		142.5 (145.1)	H_2O–HCl	176
336	Pyrimidine	C_2:	151.7 (159.0)	H_2O–HCl	176
		C_4:	158.3 (157.0)		
337	Pyridazine		151.2 (152.3)	H_2O–HCl	176
304	Quinoline	C_2:	144.2 (149.5)	H_2O–acetone	180[b]
		C_{8a}:	136.7 (146.4)	HCl	
305	Isoquinoline	C_1:	145.9 (151.1)	H_2O–acetone	180
		C_3:	130.3 (141.0)	HCl	

[a] Chemical shift of the neutral compound given in parentheses.
[b] Data for other derivatives also given.

TABLE XVI
$^1H-^{13}C$ Coupling Constants[a]

Compound	$^1J_{=N-\overset{|}{C}H}$	$^1J_{H-C=N}$	^3J	Solvent	Ref.
3a	144.5	187		SOCl$_2$–CDCl$_3$	181
3c	145	188		CD$_2$Cl$_2$	181
9		186.2		TFA	19
24d	146	236		CH$_2$Cl$_2$	7
33a	146.5			SO$_2$	5
33b	146.5		7.4 (C=N—CH$_3$)	SOCl$_2$	5
43a	147			SOCl$_2$	5
43b	148			CD$_3$CN	5
48	147.5			SO$_2$	5
68b	143.2	179	6.1 (H_C=N_/C)	CDCl$_3$	43
97	142			CD$_3$NO$_2$	43
116	141			DMSO	61
137	143			CDCl$_3$	181
138	143			CDCl$_3$	181
145e[b]	146.5			TFA	181
153c[c]	142.5			CD$_2$Cl$_2$	181
159	144.0		4.3 (C=N—CH$_3$)	CD$_2$Cl$_2$	181
160	144.0			CD$_2$Cl$_2$	181
161	144.5			CDCl$_3$	76
235	143.1			CDCl$_3$	181
236	143.2			CDCl$_3$	181
237	143			CDCl$_3$	181
249	143.5			CDCl$_3$	181
261	147			CD$_2$Cl$_2$	32
296	145.5			D$_2$O	153
298	145			D$_2$O	153
299	146.5			D$_2$O	153
300	146			D$_2$O	153
301	149.5			D$_2$O	153
324b	143.5			D$_2$O	152
325	143.0			D$_2$O	152
328		219[d]		H$_2$O	178

[a] Values in hertz (cps); for formulae see Tables I–XII.
[b] F$_3$CCOO$^-$.
[c] BF$_4^-$.
[d] J_{C_2-H} of imidazolium; neutral compound: 209 Hz.

TABLE XVII

¹⁹F NMR Data

Compound[a]	Chemical shift[b]	$^3J_{FC=NH}$	Coupling constants ($^1H-^{19}F$)[c] $^3J_{HC\overset{|}{C}-N\atop F}$	$^4J_{FC=N\overset{|}{C}H}$	Solvent	Ref.
42	−31.5		27	(4.0) (2.4)	CD_2Cl_2	33
168	−27.8 (−12.7)	21.5 (trans)			SO_2–$HFSO_3$	95
169	−32.4 (−17.4)				$HFSO_3$	95
170	−37.0 (−24.9)			2.0 (cis and trans)	$HFSO_3$	95
338[d]	−79.4 (−60.7)		See reference		TFA	154[e]

[a] For formulae, see Tables I–XIII.
[b] Chemical shifts given in parts per million relative to CCl_3F; high-frequency (low-field) shift is positive. For protonated compounds, the chemical shift of the neutral compound is in parentheses.
[c] Values in hertz (cps); configuration has not been determined when values are in parentheses.
[d] 2-Fluoropyridinium.
[e] Data for other derivatives also given.

TABLE XVIII[a]

UV Spectra of Iminium Compounds

Compound[a]	λ_{max}, nm ($\epsilon \times 10^{-3}$)	Solvent	Ref.
4	219.5	Hexane	15[b]
	222.5	Dioxane	15
6b	263 (8.2), 289 (7.5)	CH_2Cl_2	19
17	232 (3.36), 302 (5.38)	MeCN	23
57a	223.5	Dioxane	15[b]
57c	222.5 (4.14)	MeCN	15
65a[c]	250 (56.4)	MeCN	41[b]
65b[d]	250 (10.7)	MeCN	41
66a[c]	260 (16.7)	MeCN	41
66b[d]	260 (2.25)	MeCN	41
67	390 (48.0)	MeCN	42
70d[d]	275 (6.9)	MeCN	182[b]
81	243 (29.6), 288 (5.1)	MeCN	50[b]
96	232 (7.7), 265 (7.9), 345 (2.1)	MeCN	40
97	244 (21.6)	MeCN	40
103	239, 324, 331	MeCN	56
104	239, 327, 333	EtOH	57
105	234, 328	EtOH	57
119a	224 (11.9)	H_2O	63[b]
119b	224	H_2O	65
120	223	H_2O	65
121	235	H_2O	65
122	230.5	H_2O	65
123	245 (21.0), 290 (3.05)	EtOH	183[b]
134	242 (10.2)	MeOH	77
162c	252.5 (13.8)	H_2SO_4	93
175b[e]	234 (14.4)	96% H_2SO_4	97
177	240 (9.9)	0.01 N H_2SO_4	97
223	258 (15.7)	MeOH	111[b]
224	242 (11.0), 268 (6.5)	MeOH	111
227	310 (54.7)	EtOH	184[b]
231	306 (28.2)		117
234	346	CH_2Cl_2	185[b]
235	288	CH_2Cl_2	185
236	278, 397	CH_2Cl_2	185
237	410	CH_2Cl_2	185
240	414 (120)	CH_2Cl_2	186[b]
243	515.5 (209)	CH_2Cl_2	186
244	326.5 (27.6)	MeCN	121
245	270 (40.0)		122
249	282 (32.0)	CH_2Cl_2	124
277	234 (3.6), 269 (8.5)	MeCN	138[b]
278	222 (8.3), 273 (11.2)	MeCN	138
291	254 (9.0)		145[b]
292	226 (9.25)	EtOH	147[b]

(*Continued overleaf*)

TABLE XVIII (*Cont.*)

Compound[a]	λ_{max}, nm ($\epsilon \times 10^{-3}$)	Solvent	Ref.
293	224 (12.8)	EtOH	147
294	270 (9.4)	EtOH	147
309a	227 (5.6), 290 (22.3)	MeOH	77
309b	221 (7.0), 290 (24.3)	MeOH	77
329	233 (16.6)		159

[a] For formulae, see Tables I–XII.
[b] Data for other derivatives also given.
[c] Anion: $SnCl_6^-$, strong absorption of the anion at ~230 nm.
[d] Anion: Cl^-.
[e] Anion: SO_4^-.

TABLE XVIIIb

UV Spectra of Iminium Compounds Not Listed in Tables I–XII

Compound	Cation	Anion	λ_{max}, nm ($\epsilon \times 10^{-3}$)	Solvent	Ref.
339	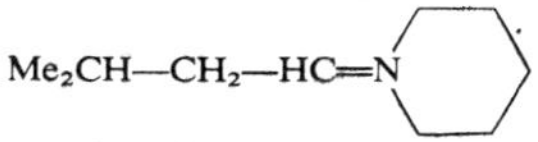 Me$_2$CH—CH$_2$—HC=N(piperidine)	Cl	222 (2.25)	CH$_3$CN	15[a]
340	R$_1$=R$_2$=R$_3$=H 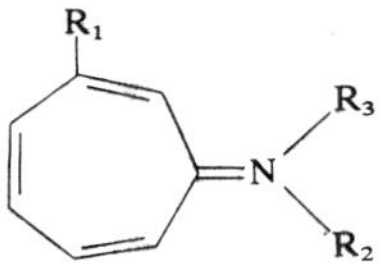	BF$_4$	234, 315, 324	CH$_3$CN	56
341	R$_1$=H; R$_2$, R$_3$=N(piperidine) (cycloheptatrienylidene with R$_1$, NR$_2$R$_3$) 	ClO$_4$	242.5 (25.0), 336.5 (20.8)	CH$_3$CN	187[a]
342	R$_1$=CMe$_3$, R$_2$=Me, R$_3$=Ph	SbCl$_6$	222 (10.5), 247 (16.05), 342 (10.63)	CH$_3$CN	40[a]
343	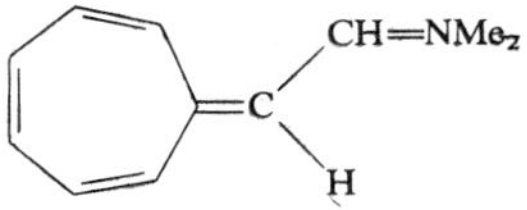 CH=NMe$_2$ (cycloheptatrienyl with C-H)	ClO$_4$	227 (14.8), 270 (7.3), 348 (9.3), 465 (18.6)	CH$_3$CN	187[a]
344	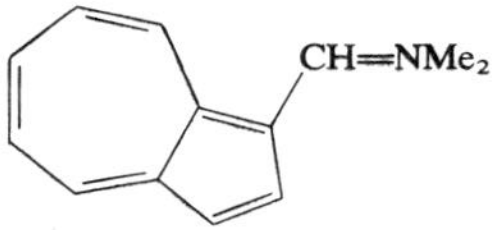 CH=NMe$_2$ (azulenyl)	ClO$_4$	276 (15.2), 323 (41.8), 402 (35.50), 485 (1.45)	CH$_3$CN	187
345	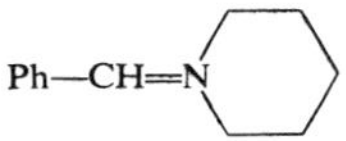 Ph—CH=N(piperidine)	Cl	275 (6.9)		182[a]

TABLE XVIIIb (Cont.)

Compound	Cation		Anion	λ_{max}, nm ($\epsilon \times 10^{-3}$)	Solvent	Ref.
346	Ph(HC=CN)$_n$—HC=NMe$_2$	$n = 1$	ClO$_4$	326 (29.5)	CH$_2$Cl$_2$	188[a]
347		$n = 2$	ClO$_4$	389 (42.6)	CH$_2$Cl$_2$	188
348		$n = 3$	ClO$_4$	436 (56.2)	CH$_2$Cl$_2$	188
349	HS=C(Me)=NHMe		SO$_4$	232 (7.08)	cc H$_2$SO$_4$	189[a]
350	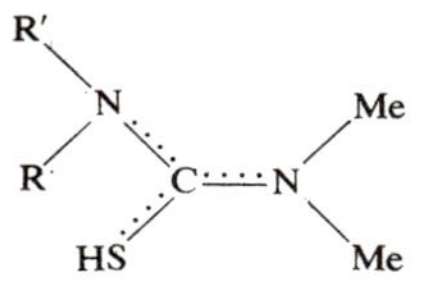 [structure]	R = R' = H	SO$_4$	End absorption only	cc H$_2$SO$_4$	189
351	[structure]	R = Me, R' = H[b]	SO$_4$	217 (10.5)	cc H$_2$SO$_4$	189
352		R = R' = Me	SO$_4$	230 (8.7), 246 (9.1)	cc H$_2$SO$_4$	189
353	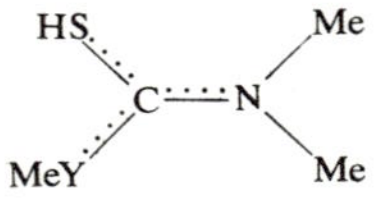 [structure]	Y = O	SO$_4$	221 (4.0)	cc H$_2$SO$_4$	189
354		Y = S	SO$_4$	250 (10.7)	cc H$_2$SO$_4$	189
355	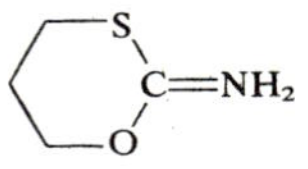 [structure]		Cl	233 (5.8)	MeOH	111
356	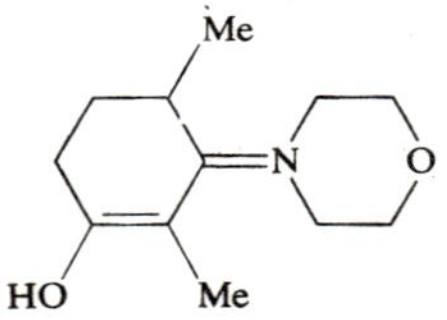 [structure]		ClO$_4$	317 (22.2)	EtOH	190[a]

72

#	Structure		Anion	Values	Solvent	Ref
357	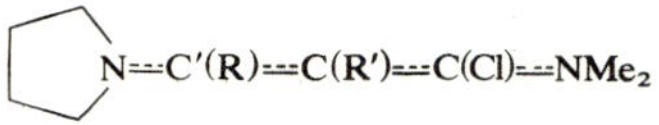 N≡C''(R)=C(R')=C(Cl)=NMe₂ (pyrrolidine)	R = H, R' = Ph	Cl	366	CH₂Cl₂	191[a]
358		R, R' = —(CH₂)₃—	Cl	370	CH₂Cl₂	191
359	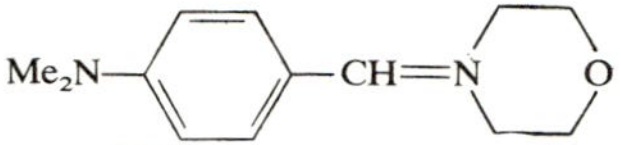 Me₂N–C₆H₄–CH=N-morpholine		Cl	398 (40)	MeCN	192
360	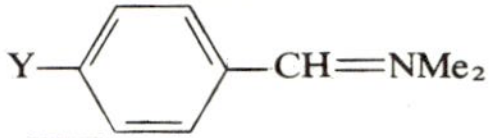 Y–C₆H₄–CH=NMe₂	Y = NMe₂	Cl	390 (46.8)	MeCN	182[a]
361		Y = OMe	Cl	320 (18.2)	MeCN	182
362		Y = SMe	Cl	355 (18.6)	MeCN	182
363	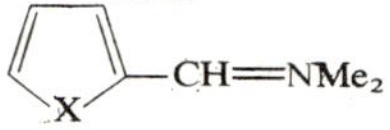 X–(thiophene)–CH=NMe₂	X = O	Cl	305 (21.4)	MeCN	182
364		X = S	Cl	282 (22.4), 315 (24.6)	MeCN	182
365		X = NMe	Cl	335 (28.2)	MeCN	182
366	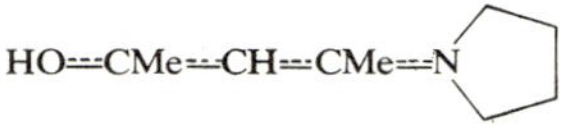 HO=CMe–CH=CMe=N(pyrrolidine)		ClO₄	302 (24.0)	EtOH	193[a]

[a] Data for other derivatives also given.
[b] Inverse attribution (inverse configuration) is possible.

III. Infrared Spectroscopy*

A. C=N STRETCHING ABSORPTIONS

The most characteristic IR absorption of simple iminium salts is the C=N stretching band at 1640–1700 cm^{-1}. The first iminium compounds identified by spectroscopic methods, the C-protonated enamines, as well as the cyclic† (136,139,194) or acyclic ones (15), show a strong absorption in this region. The C-protonation of an enamine that gives an iminium compound always leads to a shift of the strong band in this region toward higher frequencies. In enamines this band corresponds to $\nu_{C=C}$, and in the iminium compounds to $\nu_{C=N}$; the situation is analogous for dienamines (41). High-frequency shift is also observed on protonation of imines: $\nu_{C=N} \rightarrow \nu_{C=N^+}$ (11,140).

Generally, all unsubstituted and alkyl-substituted methyleniminium compounds absorb between 1640 and 1700 cm^{-1}. Higher frequencies are observed if BF$_4$ is the anion; however, this shift has not yet been explained. Halogen (Cl, Br, I) substitution on carbon gives values of 1590–1650 cm^{-1}. The lower frequency is due partially to the mass effect, and to the weakening of the double bond by electron donation from the halogen. This view is also supported by the high frequency of the C—X

stretching band: 688 cm^{-1} for $\left[\begin{array}{c} CH_3 \\ Cl \end{array} C=N \begin{array}{c} H \\ H \end{array} \right]^+$ (31).

For N-unsubstituted compounds there is a coupling effect between $\nu_{C=N}$ and δ_{NH_2}, which can be eliminated by deuteration.

On vinyl substitution the pure C=N stretching absorption can no longer be observed in the regions mentioned above. Nevertheless, there is always one or more strong bands (41,42,121,195).

Lone-pair conjugated compounds of type **II** (p. 24) show a symmetric and an antisymmetric Y—C—N stretching band, the latter being a strong band at 1670–1720 cm^{-1}. This means that for protonated amides (Y = OH) this band is generally at higher frequency than the "$\nu_{C=O}$" absorption of the corresponding amide (196). This fact first led to an erroneous conclusion of N-protonation. The symmetric stretching band occurs in the 1400–1550 cm^{-1} region.

Vinylogues of lone-pair conjugated iminium salts (cyanines, protonated or alkylated merocyanines, etc.) show, according to the asymmetry of the

---

* The spectral regions given in this section are relative to spectra in mull phase.

† $\diagup$C=N$^+\diagdown$ is part of a cycle.

molecule and to the length of the conjugated system, several strong bands in the 1500–1700 cm^{-1} region.

The C=N band of oximes shifts on protonation to higher frequencies (130,132).

B. NH STRETCHING AND BENDING ABSORPTIONS

The NH stretching and bending absorption bands must be observed in salts with complex anions; otherwise strong hydrogen bonding shifts appear.

The NH stretching band occurs in N-monosubstituted iminium salts at 1880–2200 cm^{-1} (30,70,137,140,195). The two NH$_2$ stretching bands can be observed at 3100–3200 and 3250–3300 cm^{-1}, respectively (11,27,31,70,197). Hydrogen bonding can shift them in chlorides down to 2500 cm^{-1} (31). For compounds of type **II** (Y = OR, NR$_2$) these bands occur at 3250–3500 cm^{-1} (69,87,123,125).

C. OTHER SPECIFIC ABSORPTIONS

The =CH$_2$ stretching in
$\begin{bmatrix} H & & CH_3 \\ & C{=}N & \\ H & & CH_3 \end{bmatrix}^+$
can be observed at 3125–3140 cm^{-1}, and a CH$_2$ bending vibration can also be assigned at 1420 cm^{-1} (198). In H$_2$C=NH$_2^+$SbCl$_6^+$ the bending frequency seems to be the same, whereas the CH$_2$ stretching is attributed to a band at 3002 cm^{-1} (11).

For
$\begin{bmatrix} Cl & & H \\ & C{=}N & \\ H & & H \end{bmatrix}^+$
these two bands appear at 3055 and 1352 cm^{-1} (27); in lone-pair conjugated compounds the CH stretching occurs at 3035 cm^{-1} (R = NH$_2$, R' = H) (59) and at 2960–3000 cm^{-1} (R = OH, O-alkyl, R' = H) (21).

The ClCCl ν_{asym} and ν_{sym} vibrations absorb at 882 and 634 cm^{-1}, respectively, in Cl$_2$C=NH$_2^+$SbCl$_6^-$ (29). A near-infrared study of the overtones of NH stretching vibrations was made on protonated amides (199,200).

Further vibrations are assigned in the following references: 8, 9, 11, 29–31, 59, 60, 62, 80, 87, 94, 96, 99, 118, 123, 125, 201–205.

Infrared data not listed in tables are given in the following references:
Simple iminium compounds: 206–217, 379.
Amidinium compounds: 202, 212, 218–222, 366.
Amidium, thioamidium, and uronium compounds: 223–232.
Cyclic iminium compounds: 203, 204, 233–236, 367.

IV. Nuclear Magnetic Resonance Spectroscopy[†]

In comparison to imines, the corresponding iminium compounds show, as main characteristics of the NMR spectra, a high-frequency[†] shift of all —CH proton signals, no change or low-frequency shift of α-carbons in CMR, and an important low-frequency shift of the nitrogen resonance. As far as coupling constants are concerned, the most characteristic changes appear in the $^1J_{CH}$ and $^1J_{CN}$ couplings, which become larger.

A. CHEMICAL SHIFTS

Neither PMR nor CMR spectra show chemical shifts for $RHC=NR_2'^+$ corresponding to a carbonium ion structure. For the isopropyl cation $\delta_{^1H} = 13$ ppm and $\delta_{^{13}C} = 317.5$ ppm (237) are observed, whereas the chemical shifts actually observed for $[RHC=NR_2']^+$ are $\delta_{^1H} = 7.5$–10.0 ppm and $\delta_{^{13}C} = 130$–180 ppm. These shifts correspond to a C–N double bond with less positive charge on the carbon than on the nitrogen, as is shown by theoretical calculations (see the chapter by Kollman) and infrared spectroscopy. The high-frequency proton shift is a consequence of the expected decrease in electron density at the α-carbon. Contrary to this high-frequency shift in PMR, a low-frequency shift of the α-carbon resonance is observed in ^{13}C spectra. This phenomenon can be explained by an important reduction of the paramagnetic contribution to the chemical shift on protonation of imines, corresponding to the loss of the low energy $n \to \pi^*$ transition (176,177). Thus the high-frequency shift expected for $C=N^+$ on the basis of electron density considerations is overcompensated for by this effect. The paramagnetic term of the screening constant is proportional to $1/\Delta E$, where ΔE is the average energy of excitation from the electronic ground state to the excited state, which is dominated in nitrogen compounds by the low-energy $n \to \pi^*$ and $n \to \sigma^*$ transitions. Thus for these compounds paramagnetic shifts are generally accompanied by bathochromic UV shifts (e.g., imines), and the lack of these transitions (higher ΔE) in the case of iminium compounds, by hypsochromic shifts.

The weight of the paramagnetic term relative to the diamagnetic term for different nuclei can be nicely shown with the pyridinium cations as an example (176). In proton resonance the electron density change determines the direction of the shift to high frequency for all the pyridine protons on protonation. However, the α-proton resonance shift (1.08 ppm) is smaller than the β- (1.71) or γ-proton shift (1.75 ppm).

[†] In this chapter chemical shifts are always expressed on the δ scale: high-frequency shift (=paramagnetic or low-field shift) is positive; low-frequency shift (=diamagnetic or high-field shift) is negative.

This finding, which is contrary to expectations based on electron density arguments, can be explained by a loss of paramagnetic contribution. For ^{13}C shifts the paramagnetic term is more important: there is a low-frequency shift (−8 ppm) for the α-carbons, and a high-frequency shift for β- (5) and γ-carbons (12 ppm). For ^{19}F chemical shifts the weight of the paramagnetic term is analogous to that for ^{13}C; correspondingly, 2-fluoropyridine on protonation shows a low-frequency shift (−19 ppm), and 3-fluoropyridine a high-frequency shift (16 ppm) (154).

The paramagnetic contribution is the most important in ^{14}N (^{15}N) NMR shifts. Pyridine shows a −113 ppm low-frequency shift on protonation, and analogous but smaller shifts are observed for other N-heterocycles or imines. The few known ^{14}N NMR data for simple iminium compounds are in the same domain as the chemical shifts of protonated N-heterocycles: 150–190 ppm relative to NH_4^+ as reference (−160 to −200 ppm relative to CH_3NO_2). ^{14}N (^{15}N) magnetic resonance could become excellent proof for the iminium function, as very few other functions (the terminal nitrogen of organic azides, R—O$\underline{N}$O, R_2N—$\underline{N}$O, and some isonitriles) give resonance lines in the same region. The site of protonation can be shown nicely by nitrogen resonance, for example, in the protonation on N_4 of 1,2,4-triazole (compound 331, Table XIVa).

B. COUPLING CONSTANTS

With few exceptions (some ^{15}N–^{1}H coupling) no sign determination has been made for coupling constants of iminium compounds. The signs of $^3J_{HC=NH}$ and $^2J_{H \atop H}{>}C=N$ coupling constants may be positive, as they were determined to be for other nitrogen derivatives (167,238,239).

1. ^{1}H–^{1}H *Couplings* (Table XIII)

The absolute values of these couplings are analogous to those for couplings in ethylene and allylic compounds. The only important differences occur for the geminal coupling, $^2J_{H \atop H}{>}C=N$, in methyleniminium compounds, which is +7 Hz, opposed to 1–2 Hz in ethylene derivatives, and for the allylic couplings, which have, for $^4J_{HC-C=NH}$ and $^4J_{HC=N-CH}$, $J_{trans} > J_{cis}$, whereas for allylic compounds the cis coupling generally has a higher absolute value.

2. ^{1}H–^{13}C *Couplings* (Table XVI)

The $^1J_{CH}$ couplings of N-methyl and N-methylene carbons are both characteristic for the iminium structure (178,181). In relation to the electronegativity of the more and more positively charged nitrogen or to other analogous parameters, the N–CH$_3$ ^1J coupling constant increases, following the sequence amine < amide < cyanine < amidinium < iminium, from 130 to 148 Hz. This NMR parameter is probably the most sensitive one for the iminium function. The $^1J_{HC=N}$ (methylene) coupling is of the order of 180–185 Hz in methylenammonium compounds and becomes higher with electronegative substituents such as chlorine (236 Hz for **24d**) and nitrogen [219 Hz for C$_2$ for imidazolium, in contrast to 209 Hz for imidazole (178)]. Very few data exist for J_{C-H} couplings with a separation of more than one bond.

3. ^{1}H–N and ^{13}C—N Couplings (Table XIVb and c)

^{14}N and ^{15}N coupling constants are related by the gyromagnetic ratios and have opposite signs: $J_{^{14}N-X} = -0.713 \cdot J_{^{15}N-X}$. $^1J_{H-N}$ coupling constants of iminium compounds are in the order of +60–70 (^{14}N) and −85 to −99 (^{15}N), respectively. The one-bond N–H coupling was related to the *s* character of this bond (167a,b). Thus, if measurable, it can characterize the iminium function. The same can be said of the one-bond ^{13}C–^{15}N coupling, but data are known only for pyridinium and quinolinium: 12–15 Hz versus 0.5–1.5 Hz in the neutral compounds. Some values of two- and three-bond N–H and N–C couplings are given in the tables.

Nuclear magnetic resonance data not listed in Tables I–XVII can be found in the following papers:

Simple, and double-bond conjugated iminium compounds: 210, 212, 214, 217, 240–249, 370–372, 374, 375, 380, 382, 395.

Lone-pair conjugated compounds: 212, 218, 221, 222, 227, 228, 230, 250–258, 368, 369, 383–386.

Vinylogues of lone-pair conjugated compounds: 245, 256, 259–263, 387.

Cyclic compounds: 168, 235, 264–273, 388, 389.

Other compounds: 131, 274.

V. Barriers of Rotation of the C–N Bond (Table XIX)

Rotation barriers can be measured using NMR spectroscopy (if they are no higher than ~25 kcal/mole) by observing the coalescence of signals of nuclei (R) in different environments:

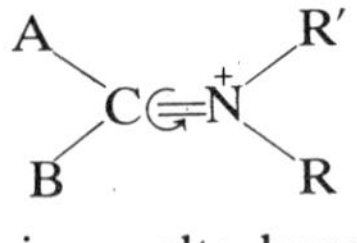

$$\begin{matrix} A & & R' \\ & \diagdown\!\!\!\diagup & \\ & C\!\!=\!\!\overset{+}{N} & \\ & \diagup\!\!\!\diagdown & \\ B & & R \end{matrix}$$

Simple, unconjugated iminium salts have a high barrier of rotation (calculated value: 70–90 kcal/mole; see the chapter by Kollman) which

TABLE XIX

Barriers of Rotation in Iminium Compounds

Compound	Cation	Anion	$\Delta G^{\ddagger}$ kcal/mole	Temp., °C	E_a, kcal/mole (log A)	Solvent	Ref.
96 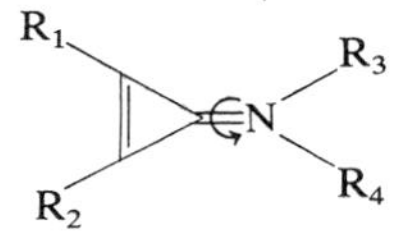	$R_1 = R_3 = R_4 = H, R_2 = Ph$	BF_4	25.4	+190		$PhNO_2$	279[a]
97	$R_1 = R_3 = Me, R_2 = R_4 = Ph$	BF_4	24.2[b]	+27.2		CD_3NO_2	279
98	$R_1 = R_4 = Me, R_2 = R_3 = Ph$	BF_4	24.0[b]	+27.2		CD_3NO_2	279
371 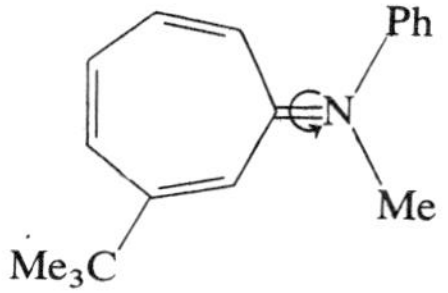		$SbCl_6$	16.0	+40		CD_3NO_2	280[a]
372	$R_1 = CD_3, R_2 = R^3 = Me,$ $R_4 = R_5 = H$	Cl NO_3			22.8 (13.5) 21.3 (12.7)	DMSO DMSO	283 283
373	$R_1 = R_5 = Ph, R_2 = R_3 = Me,$ $R_4 = H^d$	Cl	20.4[c]	+135	20.8 (13)	$PhNO_2$	286[a]
374	$R_1 = R_4 = Ph,^d R_2 = R_3 = R_5 = Me$	I	14.2 17.8	+1 +63		CH_2Cl_2 TFA	286 286
145b 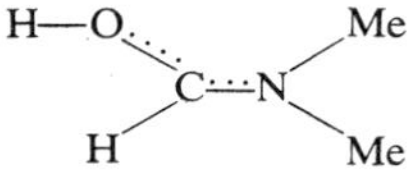		SO_4	20.0	+25		H_2SO_4	287
145c		F_3CCOO	20.7	+99	26.0 (16)	TFA	288

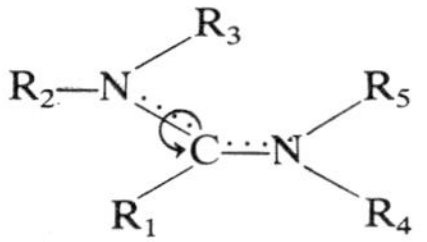

TABLE XIX (Cont.)

Compound	Cation	Anion	$\Delta G^{\ddagger}$ kcal/mole	Temp., °C	E_a, kcal/mole (log A)	Solvent	Ref.
202	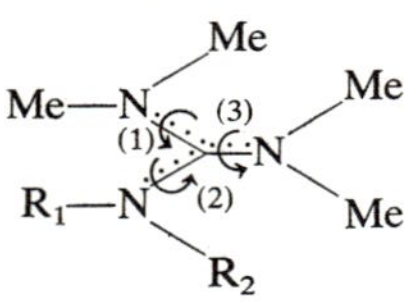 $R_1 = Ph, R_2 = H$	F_3CCOO	(1) 12.9 (2) 12.5 (3) 11.2	−25 −25 −48		$CDCl_3$–TFA	105[a]
375	$R_1 = Me, R_2 = H$ 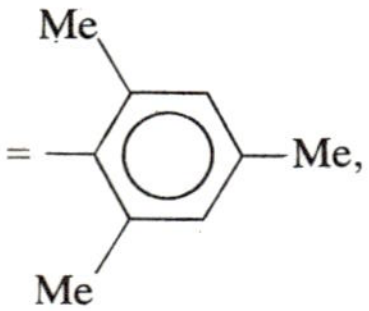	F_3CCOO	(1) 11.3 (2) 15.4 (3) 10.8	−51.5 +34 −53		$CDCl_3$–TFA	105
376	$R_1 = Me, R_2 = Me$ 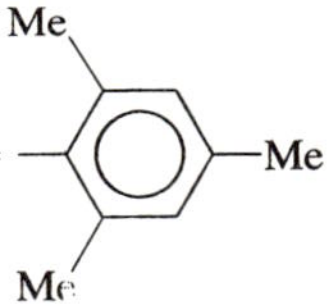	I	(2) 21.2	+147		$C_6H_3Cl_3$	103[a]
377	$R_1 = Me, R_2 = Ph$	I	(2) 15.0	+29		$CDCl_3$	103
227	$R_1 = R_2 = H$	ClO_4	21.5	+145.5	~17	Cl_2HC—$CHCl_2$ Ph_2Co	114[a] 115
230	$R_1 = H, R_2 = Ph$	ClO_4	18.1	+93		Cl_2HC—$CHCl_2$	114[a]
232	$R_1 = Ph, R_2 = H$	ClO_4	17.9	+93		Cl_2HC—$CHCl_2$	114

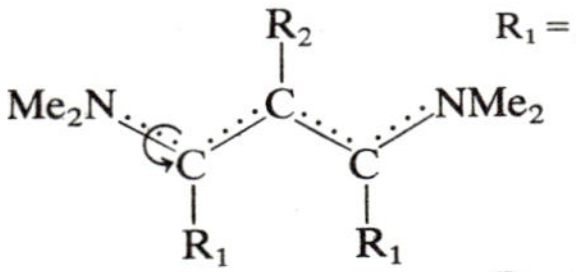

#	Structure					Ref
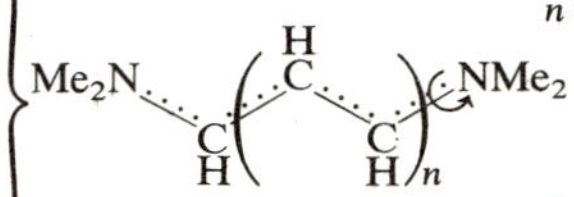 240	$Me_2N\cdots\overset{H}{\underset{H}{C}}-(\overset{H}{\underset{H}{C}})_n-NMe_2$, $n=2$		ClO_4	~10	Ph_2CO	115
243a	$n=3$		ClO_4	~7	CH_3CN, CCl_4	115
378	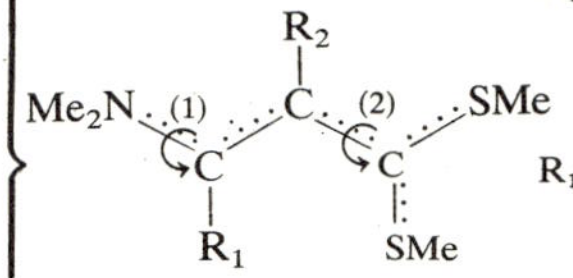 $Me_2N\cdots\overset{(1)}{\underset{R_1}{C}}-\overset{(2)}{\underset{SMe}{C}}\cdots SMe$, R_2	$R_1 = Ph, R_2 = H$	I	(1) 21.9 (2) 16.2	+147 +51	$Cl_2HC-CHCl_2$	262[b]
379		$R_1 = R_2 = H$	I	(1) 22.5 (2) 18.7	+150 +79	$Cl_2HC-CHCl_2$	262
380		$R_1 = H, R_2 = Me$	I	(1) 17.7 (2) 21.3	+115 +115	$Cl_2HC-CHCl_2$	262

[a] Data for other compounds also given.
[b] These compounds have been isolated: $k_{97\to98} = 15.6 \times 10^{-6}$, $k_{98\to97} = 20.3 \times 10^{-6}$ sec^{-1}, at 27.2°C.
[c] The corresponding amidine (no R_4) gives $\Delta G^\ddagger = 13.0$ kcal/mole at -16°C in CH_2Cl_2.
[d] Configuration can be inverse.

cannot be measured by this method. However, in some cases coalescence of signals is observed because of exchange processes, and the kinetic data of these can be determined. If the barrier of rotation is lowered by conjugation and at the same time exchange is possible, it is difficult to distinguish between the two processes. Two kind of exchanges are possible for iminium compounds.

1. N–H dissociation if the nitrogen is not disubstituted. The height of the barrier can be a function of dissociation of the iminium salt, inversion of the lone pair of the imine formed, or rotation of the C–N double bond in both:

$$\underset{B}{\overset{A}{\diagup}}C=\overset{+}{N}\underset{R}{\overset{H}{\diagdown}} \quad X^- \quad \underset{+HX}{\overset{-HX}{\rightleftharpoons}} \quad \underset{B}{\overset{A}{\diagup}}C=N\underset{R}{\diagdown} \quad \rightleftharpoons$$

$$\underset{B}{\overset{A}{\diagup}}C=N\underset{}{\overset{R}{\diagdown}} \quad \underset{-HX}{\overset{+HX}{\rightleftharpoons}} \quad \underset{B}{\overset{A}{\diagup}}C=\overset{+}{N}\underset{H}{\overset{R}{\diagdown}} \quad X^-$$

In lone-pair conjugated compounds, such as amidines (74,75), exchange can take place via the diprotonated amidine form in strong acids:

$$\left[\underset{R'}{\overset{R'}{\diagdown}}\overset{|}{\underset{}{\overset{+}{N}}}=\underset{\underset{A}{|}}{C}\underset{R}{\diagdown}\overset{+}{N}\underset{R}{\overset{H\;H}{\diagup}} \right]$$

If R = CH$_3$ its rate can be measured by the collapse of the doublet due to the coupling with NH. Exchange of the NH proton in guanidinium salts has also been studied (275).

2. Exchange of X if X and Y (= halogen, OAc) can be substituents or anions for the same compound:

$$\underset{X}{\overset{A}{\diagup}}C=\overset{+}{N}\underset{R}{\overset{R}{\diagdown}} \quad X^- \rightleftharpoons AX_2C-N\underset{R}{\overset{R}{\diagdown}}$$

or

$$\underset{X}{\overset{A}{\diagup}}\overset{+}{C=N}\underset{R}{\overset{R}{\diagdown}} \quad Y^- \rightleftharpoons AXYC-N\underset{R}{\overset{R}{\diagdown}} \rightleftharpoons \underset{Y}{\overset{A}{\diagup}}C=\overset{+}{N}\underset{R}{\overset{R}{\diagdown}} \quad X^-$$

V. BARRIERS OF ROTATION OF THE C–N BOND

This kind of exchange of X = Cl was studied in the cases of dimethylformamide chloride $\left(\begin{array}{c}\text{H}\\\text{Cl}\end{array}\!\!\!\!>\!\!\text{C}=\overset{+}{\text{N}}\!\!\!<\!\!\!\begin{array}{c}\text{Me}\\\text{Me}\end{array}\;\;\text{Cl}^-,\;\mathbf{24a}\right)$ and of the Vilsmeier-Haack-Arnold adducts: DMF + POCl$_3$ (COCl$_2$, SOCl$_2$) (7,276–278). An intermolecular exchange between DMF and **24a** could also be shown.

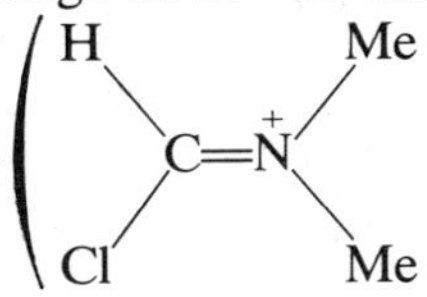

For both processes, 1 and 2, the dependence of kinetic data on solvent and concentration is characteristic.

If the iminium ion is conjugated, the barrier of rotation of the C–N bond has to be lower. The approximate value of the barrier of rotation of the iminium C–N bond conjugated with a double bond is 30–35 kcal/mole (40). Thus only if it is possible to form a cyclic cation with aromatic structure are the barriers low enough to be measured by NMR:

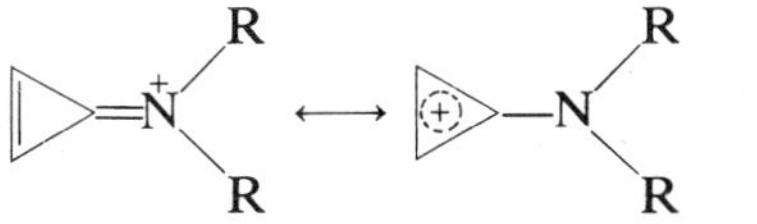

and

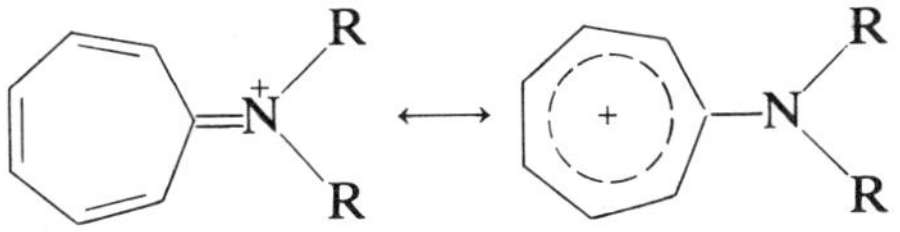

Even in this case the barriers are relatively high, showing that the positive charge is mostly on the nitrogen. Cyclopropeniminium salts (cyclopropylium stabilization of the transition state) give $\Delta G^{\ddagger}$'s in the order of 22–25 kcal/mole (**96–98**) (279,40), and cycloheptatrieniminium compounds (tropylium stabilization) 13–18 kcal/mole (**371**) (40,280).

Most of the kinetic data published on iminium compounds concern lone-pair conjugated systems:

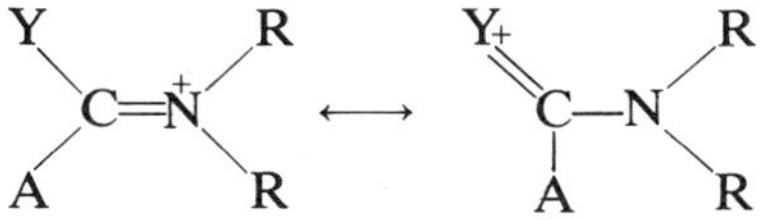

Even for this type of compounds (listed in Table XIX) there are few reliable data to lead to general rules. Nevertheless one can say that the presence of heteroatoms N, O, or S in β position to nitrogen (or in δ position in vinylogous systems) capable of supporting a positive charge lowers the barrier to the order of 20 kcal/mole (amidinium, amidium salts, and trimethine cyanines). If there is a third heterosubstituent to participate in the charge distribution, the barrier falls to 10–15 kcal/mole (guanidinium, uronium salts, iminium carbonates and thiocarbonates, etc.). The same order of diminution of the barrier can be observed for vinylogous systems: a supplementary C–C double bond lowers the barrier by ~3–7 kcal/mole, the effect of the first C=C group being more important than the second one.

It should be noted that the steric effect on barriers of rotation be as important as the electronic one, leading to a great variation in the values found with different substituents.

Compounds with polarized structure:

$$\underset{A}{\overset{X}{>}}C-N\underset{R}{\overset{R}{<}} \longleftrightarrow \underset{A}{\overset{X^{\delta-}}{>}}C=\overset{\delta+}{N}\underset{R}{\overset{R}{<}}$$

such as amides, amidines, and enamines, have been studied much more than iminium compounds with respect to rotation around the C–N bond (281). Protonation or alkylation of X leads to iminium salts, which should have a higher barrier of rotation for the C–N bond. Dimethylformamide shows practically no change on protonation (**145**). Amidinium salts certainly have a higher barrier of rotation than amidines (cf. **373**).

It would be interesting to have a systematic comparative study of these compounds. A comparative study of barriers of rotation and $^1J_{C-H}$ coupling constants of N–CH$_3$ for some amides, thioamides, and amidines has been made (282,283) Extension of this method to show the role of the steric factor or of exchange on the barriers would be interesting, since $^1J_{C-H}$ is sensitive only to electronic effects. Kinetic studies not included in Table XIX are given in refs. 73, 102, 127, 282, 284, 285, 373, 375.

VI. Ultraviolet and Visible Spectroscopy

Ultraviolet and visible spectral data are given in Table XVIII. No systematic studies of electronic spectra have been made for simple iminium compounds. The compound $[H_2C=N\text{(pyrrolidine)}]^+ Cl^-$ (**4**) absorbs at 219.5 nm in hexane, and there is a positive solvatochromic effect for this band in dioxane and tetrahydrofuran, as expected for a $\pi \rightarrow \pi^*$ transition. Very few changes can be observed on C-alkyl substitution of the methyleneiminium compound (**57, 340**). The ϵ values are of the order of 2–5000.

Conjugation with a double bond (**65**) or a phenyl group (**345**) leads to a bathochromic shift (250 and 275 nm, respectively), $\epsilon = 6$–10,000. In the literature data concerning compounds with the introduction of additional double bonds for the system

$$R\text{—}\langle\text{—}\rangle\text{—}(HC=CH)_n\text{—}HC=NMe_2^+ \ ClO_4^-$$

(346–348)

one can find a linear dependence of λ_{max} on n not only where R is an auxochromic (donor) group, such as —OR, —SR, or —NR$_2$, but also where R = H ($n = 0$–3). This type of linear dependence is expressed by the formula $\lambda_{max} = A \cdot n + B$, where A and B are constants for given substituents. This has been demonstrated for cyanine-type compounds, the electronic spectra of which have been discussed in general (289,290).

Whereas the bathochromic shift per double bond for compounds **346–348** is 53–55 nm, it is 103–105 nm in the case of the all-trans linear cyanine (see also ref. 291):

$$\left[Me_2N\overset{(CH}{\diagdown}\underset{CH)_n}{\diagup}\overset{CH}{\diagdown}\underset{}{\diagup}NMe_2 \right]^+ ClO_4^-$$

(227, 240, 243)

If cis double bonds are involved, for example by photoisomerization of penta- and trimethinecyanines (292) and for the cyclic all-cis form (293,294), a band with longer wavelength is observed (293). Also, for other compounds in which the cis configuration is fixed, new bands appear in the spectrum (63,112,116). Generally one observes a diminution of ϵ_{max} (377).

Protonation and alkylation of neutral species have been followed by UV spectroscopy by several authors. For the vinylogous amidines and amides (112,190,193,295), as for iminothiocarbonates (111) and oxazolines (296), the electronic spectra do not show important changes on protonation which leads to amidinium, amidium compounds, cyanines, etc.* Their polarized structure is one in which the lone pair of the nitrogen is still engaged in so far as the C–N bond has a real double-bond length (cf. Section VIII).† In the same manner the spectra of iminium compounds derived by C-protonation from simple enamines normally do not show any difference from those of the enamines or exhibit only a slight hypsochromic displacement:* −5 nm for compound **339** (15). For

---

* However, the spectrum can be modified if the configuration or conformation changes as a result of this process.

† The analogous situation (no important change of the chemical shift) can be observed in ^{13}C and ^{14}N (^{15}N) NMR for these compounds; namely, these chemical shifts are also related to the excited electronic states.

butadieneamines, however, this shift is more important: −10 to −20 nm (41), and the same order of shift is observed on protonation of indoles in the 3-position (195). However, general rules cannot be given because of the varying influence of substituents. A change in configuration or conformation can even lead to inverse shifts. In the case of thioamides and thiourea, where a low-frequency $n \to \pi^*$ band is observable, this band disappears on protonation and the strong absorption of 230–280 nm shifts to higher frequencies (97,189).

In some cases, if another chromophore is available for conjugation with the amide C=O, N-protonation is competitive with O-protonation and can be shown by UV, as in the case of benzamide (93).

The effect of substitution of chromophores and the steric disposition of these substituents can also be studied by UV spectroscopy. Thus phenyl substitution in α or β position in trimethinecyanines slightly shifts the absorption maximum; this effect is independent of the parasubstituent of the phenyl and can be considered as a steric effect only (184). This means that the benzene ring cannot be coplanar with the cyanine system (as is also shown by NMR). If there is no steric hindrance, several absorption bands corresponding to the different chromophoric systems can be observed; such is the case for γ-phenyl substitution on pentamethinecyanines or for branched cyanines with three of four NR_2 groups (42).

In the UV spectra of pyridine and other monocyclic azines in aqueous solution the $n \to \pi^*$ band shifts to shorter wavelengths because of hydrogen bonding. On protonation or alkylation this band disappears, and the $\pi \to \pi^*$ band exhibits a slight bathochromic shift (297). The UV maxima of α- and γ-pyridones show a hypsochromic shift on protonation and alkylation (−15 to −20 nm; ref. 297).

References not listed in tables are as follows:

Double-bond conjugated iminium compounds: 155, 195, 201, 280, 298–302.

Vinylogous amidinium compounds: 112, 116, 207, 259, 263, 291–294, 303–307.

Amidium and vinylogous compounds: 114, 196, 295, 308–310.

N-heterosubstituted compounds: 311.

Cyclic compounds: 267, 272, 312–315.

VII. Mass Spectroscopy

The iminium ion is one of the positive ions most often detected by fragmentation in the mass spectrometer. A review of these fragmentation products, however, is not the subject of this chapter. The presence and the relative stability of an iminium ion in the mass spectrometer as a particle rich in energy in high vacuum do not necessarily constitute proof of its stability in the solid state or in solution. Nevertheless it would be interesting to make a systematic and comparative study of iminium ions generated in the mass spectrometer and synthesized by chemical methods.

To obtain a mass spectrum of an organic salt by electron impact ionization* the compound must be transformed to a volatile product (316). For iminium salts there are three possibilities for obtaining a mass spectrum:

1. The thermal decomposition of the iminium salt gives a volatile product, and the mass spectrometer shows the spectrum of this secondary product. For protonated and "methylated" imines HX or MeX is thermally eliminated:

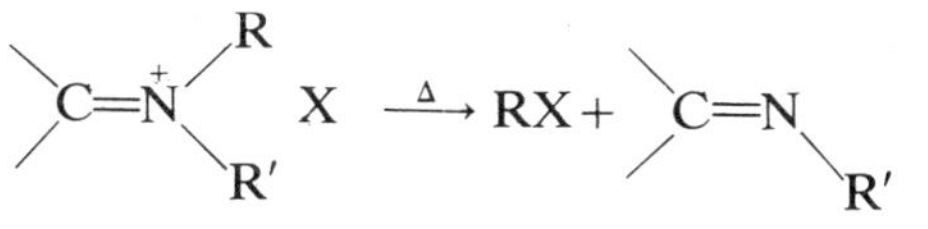

$$\diagdown C=\overset{+}{N}\diagup\overset{R}{\underset{R'}{\diagdown}} \quad X^- \xrightarrow{\Delta} RX + \diagdown C=N\diagdown R'$$

R = H, Me

In some special cases β elimination is possible:

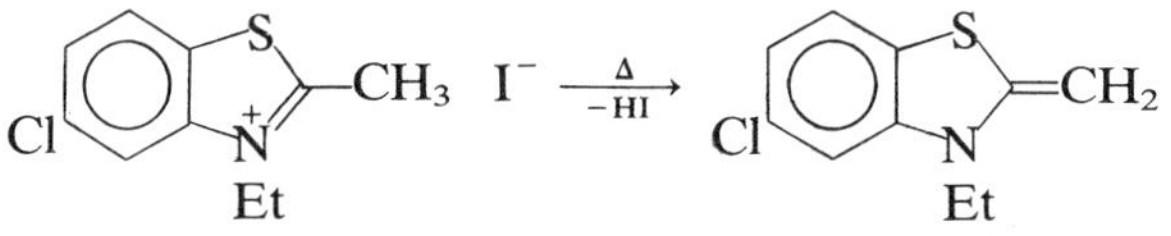

$$\text{(317)}$$

Neither the molecular ion nor the parent iminium ion appears in these spectra.

2. If there is an equilibrium between the ionic and the covalent forms of the compound (cf. p. 25), the latter can be volatilized:

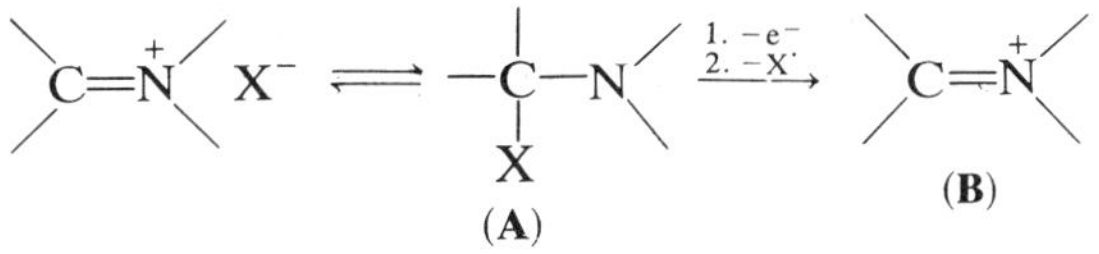

$$\diagdown C=\overset{+}{N}\diagup \quad X^- \rightleftharpoons -\underset{\underset{X}{|}}{\overset{|}{C}}-N\diagup \xrightarrow[2.\ -X^{\cdot}]{1.\ -e^-} \diagdown C=\overset{+}{N}\diagup$$

(A) \qquad\qquad (B)

The covalent tautomer gives a mass spectrum in which the parent iminium ion (B) is present, and the molecular ion formed from A can sometimes be observed.

3. The nucleophilic anion attacks the sp^3 carbon in α position to the nitrogen of a cyclic derivative, as shown for the von Braun degradation

---

* As yet, field ionization or field desorption techniques have not yielded better results than electron impact ionization.

(52):

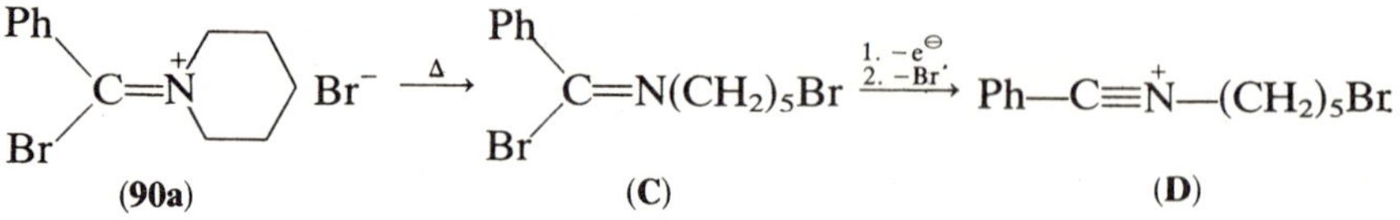

Ph\C(Br)=N⁺(piperidine) Br⁻ →^Δ Ph\C(Br)=N(CH₂)₅Br →^{1. -e⁻}_{2. -Br·} Ph—C≡N⁺—(CH₂)₅Br
(90a)   (C)   (D)

The molecular ion of the imine (**C**) cannot be distinguished from the radical ion of **A** formed by mechanism 2. The nitrilium ion (**D**) and the iminium ion (**B**) also have the same mass (318).

A mass spectroscopic study has been made of bromoiminium bromides (318):

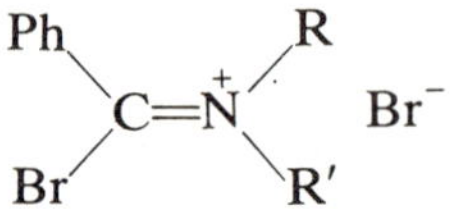

Ph\C(Br)=N⁺(R)(R') Br⁻

All three mechanisms have been shown for different substituents. If R = H, hydrogen bromide is liberated and the spectrum of the imine is observed. If R = R' = Me:

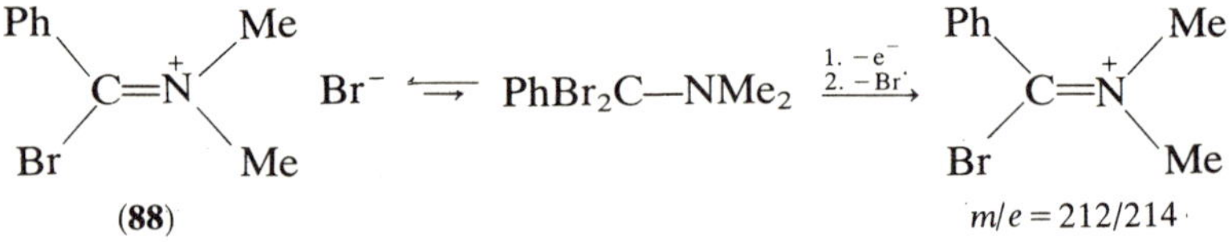

Ph\C(Br)=N⁺(Me)(Me) Br⁻ ⇌ PhBr₂C—NMe₂ →^{1. -e⁻}_{2. -Br·} Ph\C(Br)=N⁺(Me)(Me)
(88) $m/e = 212/214$

the iminium ion ($m/e = 212$), which can be formed only by fragmentation of PhBr₂C—NMe₂⁺ is present. The authors have evidence that there is simultaneous thermal decomposition, leading to the imine and MeBr. For compound **90a**, R, R' = —(CH₂)₅—, the process described under mechanism 3 is demonstrated.

Probably both mechanisms 1 and 2 play a role in the case of the mass spectrum of Cl₂C=N⁺Me₂Cl⁻, present almost exclusively in the ionic form in solution. In the spectrum even the molecular ion of the covalent tautomer Cl₃C—NMe₂⁺ ($m/e = 161$) can be detected, along with the iminium ion Cl₂C=N⁺Me₂ ($m/e = 126$) (5,6).

H₂C=N⁺Me₂Cl⁻ also gives a mass spectrum, probably via the covalent tautomer (mechanism 2). If BF₄ is the anion, the thermal degradation probably gives BF₃+F⁻. Then F⁻ attacks (following mechanism 2), yielding the volatile covalent tautomer, as observed for a 4-ethoxypyridinium derivative (319):

$$\left[\begin{array}{c}\text{OEt}\\ \text{Ph}\diagup\diagdown\text{Ph}\\ \text{Ph}\diagdown\diagup\text{Ph}\\ \text{N}\\ |\\ \text{Me}\end{array}\right]^{+} \text{BF}_4^{-} \xrightarrow{\Delta} \begin{array}{c}\text{OEt}\\ \text{Ph}\diagup\diagdown\text{Ph}\\ \text{Ph}\diagdown\diagup\text{Ph}\\ \text{F}\quad\text{N}\\ |\\ \text{Me}\end{array}$$

$$m/e = 461$$

These examples demonstrate that mass spectra can be useful to show the presence, even in minute quantity, of a covalent tautomer.

On the other hand, the preponderant presence of an iminium ion in the mass spectrum is not proof of the presence of the ionic tautomer in compounds for which the covalent tautomer is the preponderant or exclusive one, for example, $F_2HC-NMe_2$ and F_3C-NMe_2 (320,321).

Mass spectroscopic data are given in refs. 235, 236, 240, 322, 323, 382.

VIII. Structure Determination by Diffraction Methods (Table XX)

Only one simple iminium compound has been investigated by X-ray: N-dimethylisopropylideniminium perchlorate (**49,** Table XX) (197). The C–N bond length for this compound, 1.30 Å, corresponds to the length for a C–N double bond; it slightly exceeds that predicted for $H_2C=NH_2^+$ (1.28 Å; see the chapter by Kollman, p. 4).

For compounds in which the positive charge is delocalized on two or more heteroatoms and on carbons, one would predict longer and longer C–N bonds with increasing delocalization. This would correspond to the partial double-bond nature demonstrated by other methods (i.e., rotation barriers). In Table XX the C–N bond lengths are listed for some iminium compounds* of this kind, showing that the change in bond lengths corresponds more or less to the degree of charge delocalization.

As predicted by theoretical considerations (Kollman's chapter, p. 4), methyl substitution on the carbon would make the C–N bond longer (carbonium ion stabilization), but a methyl group on the nitrogen could have the opposite effect (iminium stabilization). Therefore the C–N bond in $Me_2N^+=C$ must be shorter than that in $H_2N^+=C$ (hydrogen bonding for the latter can even accentuate this difference). This phenomenon is shown in the asymmetric amidinium and bisguanide salts (see Table XX, compounds **128** and **383**).

---

* The following references report on diffraction methods not listed in Table XX: 250, 324–328, 390–392.

TABLE XX
C-N Double Bond Lengths in Iminium Compounds

Compound[a]	Cation		Anion	Bond length (σ), Å[b]		Ref.
49	$Me_2C=NMe_2$		ClO_4	1.302 (0.043)		197
100	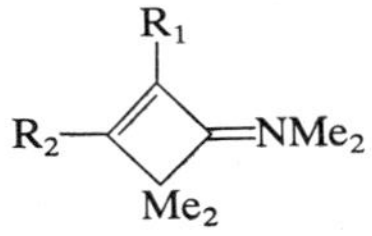 (structure with R_1, R_2, NMe_2, Me_2)	$R_1 = t$-butyl, $R_2 = H$	BF_4	1.287 (0.006)		329
101		$R_1 = R_2 = Me$	ClO_4	1.295 (0.007)		329
128	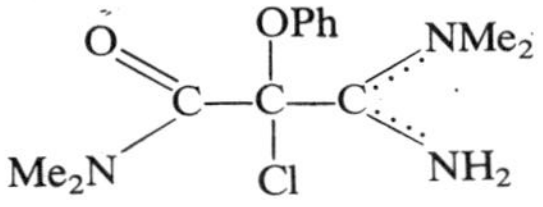 (structure: Me_2N, O, OPh, NMe_2, Cl, NH_2)		$Cl \cdot H_2O$	$CNMe_2$: 1.303 (0.005) CNH_2: 1.318 (0.005)		330
381	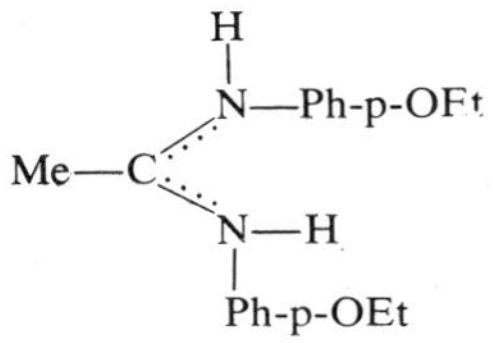 (structure: $Me-C$, $H-N-Ph$-p-OFt, $N-H$, Ph-p-OEt)			1.32		331
309	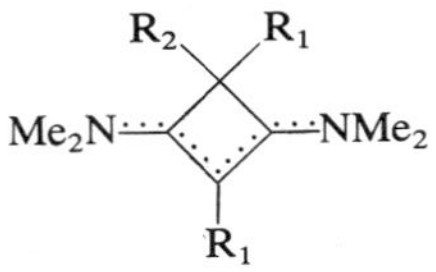 (structure with R_1, R_2, NMe_2, Me_2N)	$R_1 = t$-butyl, $R_2 = H$	Cl	1.314 (0.005)		332
311		$R_1 = Et$, $R_2 = H$	ClO_4	1.30		332

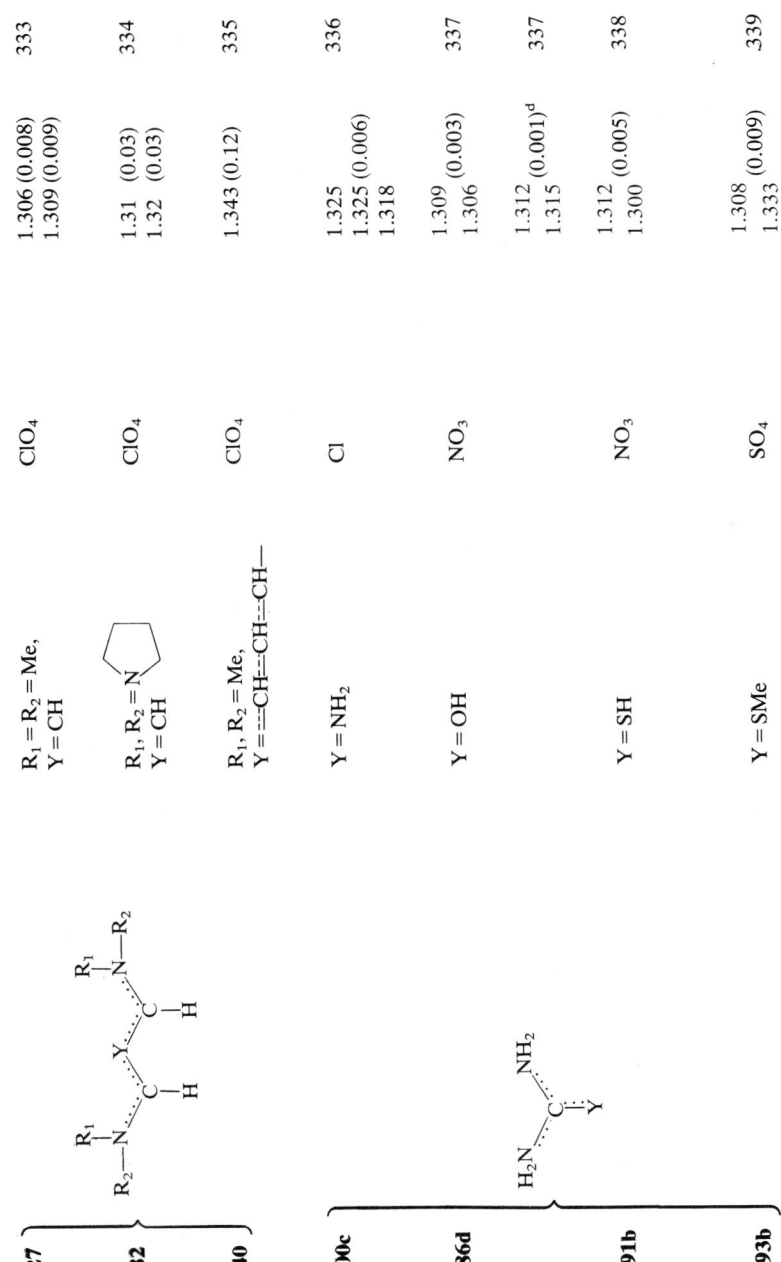

227	$R_1 = R_2 = Me$, $Y = CH$	ClO_4	1.306 (0.008) 1.309 (0.009)	333
382	$R_1, R_2 = N\langle\rangle$ $Y = CH$	ClO_4	1.31 (0.03) 1.32 (0.03)	334
240	$R_1, R_2 = Me$, $Y = -CH=CH=CH-$	ClO_4	1.343 (0.12)	335
200c	$Y = NH_2$	Cl	1.325 1.325 (0.006) 1.318	336
186d	$Y = OH$	NO_3	1.309 (0.003) 1.306	337
191b	$Y = SH$	NO_3	1.312 (0.001)[d] 1.315	337
193b	$Y = SMe$	SO_4	1.312 (0.005) 1.300	338
			1.308 (0.009) 1.333	339

91

TABLE XX (Cont.)

Compound[a]	Cation	Anion	Bond length (σ), Å[b]		Ref.
254b		ClO_4	C_1-N_1:	1.341 (0.007)	340
			C_1-N_2:	1.354 (0.007)	
			C_3-N_3:	1.350 (0.007)	
			C_3-N_4:	1.344 (0.007)	
383		Br	C_1-N_1:	1.376 (0.026)	341
			C_1-N_2:	1.349 (0.025)	
			C_2-N_4:	1.373 (0.033)	
			C_2-N_5:	1.321 (0.037)	
			C_2-N_3:	1.311 (0.028)	
			C_2-N_3:	1.313 (0.036)	
329	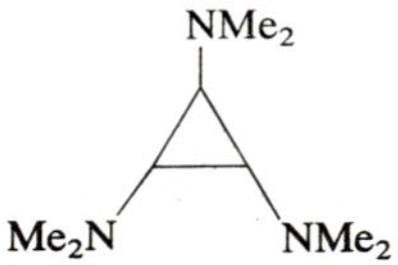	ClO_4		1.326 (0.007)	342
				1.337	
384	R = Me	ClO_4		1.29	343
385	R = Ph	ClO_4		1.31	344

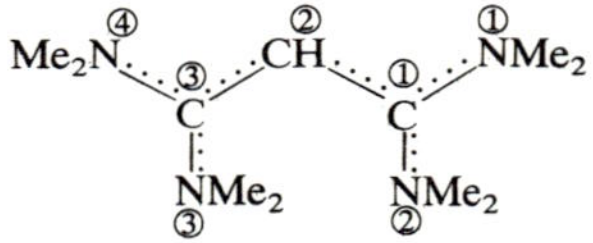

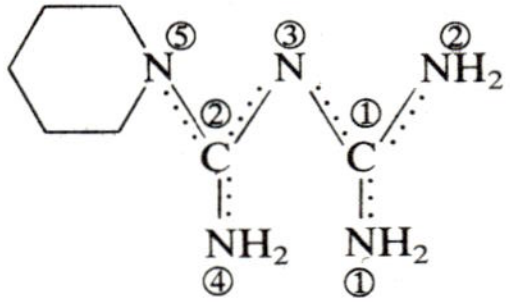

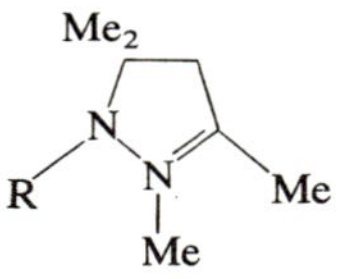

#	Structure	Substituents	Counterion	Bond length	Ref
386	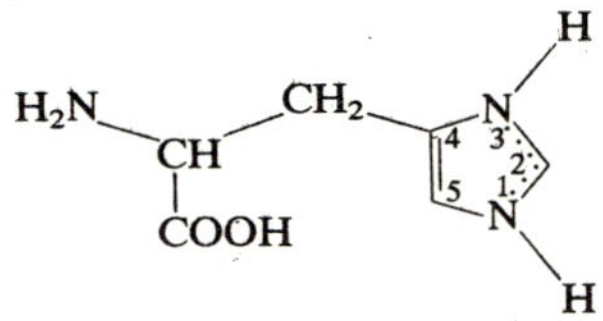 H₂N-CH(COOH)-CH₂-imidazole		Cl·H₂O	1.35 (1-2) 1.33 (2-3) 1.36 (1-5) 1.38 (3-4)	345
295d	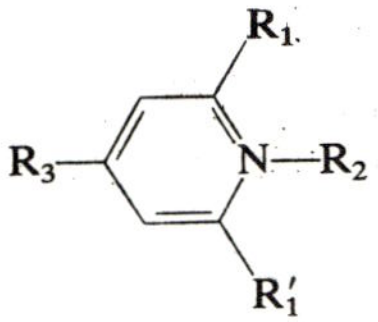 pyridinium with R₁, R₁', R₂, R₃	R₁ = R₁' = R₂ = R₃ = H	Cl	1.32	346
387	pyridinium with R₁, R₁', R₂, R₃	R₁ = R₁' = R₃ = Me, R₂ = Ph	ClO₄	1.366 (0.007) 1.358 (0.006)	347
388	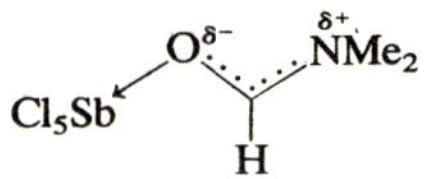 Cl₅Sb←O$^{\delta-}$=C(H)-NMe₂$^{\delta+}$			1.287 (0.012)	348
389	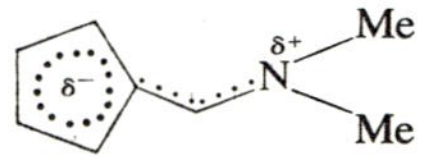 Me₂N$^{\delta+}$-CH₂-cyclopentadienyl$^{\delta-}$			1.331 (0.004)	49
390	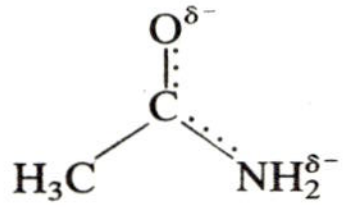 H₃C-C(O$^{\delta-}$)=NH₂$^{\delta-}$			1.351 (0.018) 1.317	349

[a] For spectroscopic data see Tables I–XII.
[b] Standard deviation in parentheses. Bond lengths are not thermally corrected.
[c] Bis-p-nitrophenyl phosphate ·H₂O.
[d] Neutron diffraction.

It is interesting to note that a number of compounds lacking a nominal charge, but polarized, like amides, amidines, and thioamides (especially their complexes with Lewis acids), or compounds with stabilized zwitterionic structures also show real C–N double-bond character. They are not discussed here, but some examples are included in Table XX for comparison.

IX. Other Physical Methods

X-ray photoelectron (*ESCA*) *spectra* of pyridinium and analogous compounds have been studied (350 and references therein, 351). The binding energies (BE) of the nitrogen $1s$ electrons that have been determined are generally inversely proportional to the electron densities on the nitrogen atom; higher BE values correspond to lower electron densities. Because samples are measured in the solid state, there is a correlation between ESCA and X-ray data. Although BE values for pyridinium and analogous compounds are between 397 and 402 eV,* the published values of different authors do not always correspond. The method is useful if a study is made with the same conditions for substituent and anion effects (350) and for compounds that can be studied only in the solid state (352).

Conductivity measurements have been made in SO_2 at low temperature (353,354) and in CH_3CN (355) to show the ionic nature of iminium compounds. The equivalent conductances in CH_3NO_2 for some simple iminium compounds are given in Table XXI and compared to the conductance of KI.

Conductometric titration was used to demonstrate the formation of a 1:1 complex from dimethylformamide and $PhOP(O)Cl_2$ (356).

Polarimetric measurements show a change in the molecular rotation of the enamines of ketosteroids on protonation (201).

Polarography was used to study the electrochemical reduction of

TABLE XXI
Equivalent Conductance, Extrapolated to Infinite Dilution, in CH_3NO_2 Solution at 0°C

Entry	Cation	Anion	Λ_∞	Ref.
3a	$H_2C=NMe_2$	Cl	141.0	6
3b		$SbCl_6$	128.6	6
			126.0	4
24a	$HClC=NMe_2$	Cl	51.3	6
24b		$SbCl_6$	79.2	6
43a	$Cl_2C=NMe_2$	Cl	28.7	6
43b		$SbCl_6$	74.3	6
KI			121.0	6

* Amines and ammonium compounds give BE values in the same range.

iminium salts (357) giving a radical, which can in some cases be observed by EPR. Preparative electrochemical reduction gives the dimeric diamine as the final product (393).

The half-wave potential determined by the polarographic study showed, in the case of polymethinium salts, a correlation with the calculated energies of the lowest unoccupied molecular orbital and with Hammett constants for substituted derivatives (358). Other polarographic studies on the same type of compounds (184) and on their cyclic derivatives (2,3-dihydrodiazepinium salts: ref. 359) have also been published. The Mannich intermediate was detected by polarography in mixtures of formaldehyde and dialkyl amines in water and in a water–acetonitrile solution of the corresponding aminals (360).

Some iminium salts of type $R_2C=\overset{+}{N}HR'X^-$ have been characterized by their *pressure of decomposition* (353).

Kinetic study of the formation of several iminium compounds and of reactions in which they are intermediates is the subject of numerous papers, from which the following examples may be mentioned.

Formation of the Vilsmeier reagents and the related exchange processes between dimethylformamide, its $POCl_3$, $COCl_2$, and $SOCl_2$ complexes, and N-dimethylchloromethyleneammonium chloride: 7, 277, 278.

Kinetics of the formylation reaction: 361, 362.

Mechanism and kinetics of enamine hydrolysis: 363.

Cleavage of alkyl dimethylaminobenzyl sulfides to N-dimethylbenzaliminium salts: 364.

Rates of hydrolysis of some iminium compounds: 182.

The Tscherniac-Einhorn reaction (condensations with hydroxymethyl amides): 365.

Nuclear quadrupole resonance (*NQR*) frequencies of ^{35}Cl have been determined for chloromethyleniminium compounds, and they are regarded as a confirmation of their iminium structure.* For **24d** NQR also confirms the $OPOCl_2^-$ anion (structure **a**, p. 26).

Compound	^{35}Cl NQR frequency (MHz)	Reference
24a	35.78	394
24d	36.31 (CCl)	394
	24.63 (PCl)	
87a	37.05	394
43a	39.74	394
	39.622	378
43c	39.398	378
	39.557	

*Cf. the NQR of α-chloroenamines, Ghosez', chapter, p. 467.

REFERENCES

1. P. A. Kollman, W. F. Trager, S. Rothenberg, and J. E. Williams, *J. Amer. Chem. Soc.*, **95,** 458 (1973).
2. R. Radeglia, E. Gey, K.-D. Nolte, and S. Dähne, *J. Prakt. Chem.*, **312,** 877 (1970).
3. H. Böhme and M. Hilp, *Chem. Ber.*, **103,** 104 (1970).
4. F. Knoll and U. Krumm, *Chem. Ber.*, **104,** 31 (1971).
5. J. Gorissen, Z. Janousek, B. Marin, C. Humblet, G. Duchêne, J. Motte-Collard, B. Stelender, R. Merényi, and H. G. Viehe, unpublished results.
6. B. U. Schlottmann, Dissertation, Marburg/Lahn, 1972.
7. G. J. Martin and S. Poignant, *J. Chem. Soc. Perkin* II, **1972,** 1964; **1974,** 642.
8. H. Bredereck, R. Gompper, K. Klemm, and H. Rempfer, *Chem. Ber.*, **92,** 837 (1959).
9. Z. Arnold and A. Holý, *Collect. Czech. Chem. Commun.*, **27,** 2886 (1962).
10. G. Martin and M. Martin, *Bull. Soc. Chim. Fr.*, **1963,** 1637.
11. J. Goubeau, E. Allenstein, and A. Schmidt, *Chem. Ber.*, **97,** 884 (1964).
12. R. Merényi, Louvain-la-Neuve, unpublished results.
13. A. Ahond, A. Cavé, Ch. Kan-Fan, H.-P. Husson, J. de Rostolan, and P. Potier, *J. Amer. Chem. Soc.*, **90,** 5622 (1968).
14. J. Schreiber, H. Maag, N. Hashimoto, and A. Eschenmoser, *Angew. Chem.*, **83,** 355 (1971); *Angew. Chem. Int. Ed.*, **10,** 330 (1971).
15. G. Opitz, H. Hellmann, and H. W. Schubert, *Liebigs Ann. Chem.*, **623,** 117 (1959).
16. H. Böhme and G. Lerche, *Chem. Ber.*, **100,** 2125 (1967).
17. H. Böhme and W. Stammberger, *Chem. Ber.*, **104,** 3354 (1971).
18. H. Böhme and D. Eichler, *Chem. Ber.*, **100,** 2131 (1967).
19. H. Volz and H-H. Kiltz, *Liebigs Ann. Chem.*, **752,** 86 (1971).
20. N. Wiberg and K. H. Schmid, *Angew. Chem.*, **76,** 381 (1964); *Angew. Chem. Int. Ed.*, **3,** 444 (1964).
21. R. Damico and C. D. Broaddus, *J. Org. Chem.*, **31,** 1607 (1966).
22. N. J. Leonard and J. V. Paukstelis, *J. Org. Chem.*, **28,** 3021 (1963).
23. N. J. Leonard and J. A. Klainer, *J. Heterocycl. Chem.*, **8,** 215 (1971).
24. P. Duhamel, L. Duhamel, and J. M. Poirier, *C. R. Acad. Soc. Paris*, **274C,** 411 (1972).
25. H. Gross and B. Costisella, *J. Prakt. Chem.*, **311,** 925 (1969).
26. G. A. Olah and P. Kreienbühl, *J. Amer. Chem. Soc.*, **89,** 4756 (1967).
27. E. Allenstein and A. Schmidt, *Chem. Ber.*, **97,** 1863 (1964).
28. H. H. Bosshard and H. Zollinger, *Helv. Chem. Acta*, **42,** 1659 (1959).
29. E. Allenstein and A. Schmidt, *Chem. Ber.*, **97,** 1286 (1964).
30. A. Schmidt, *Chem. Ber.*, **105,** 3050 (1972).
31. E. Allenstein and A. Schmidt, *Spectrochim. Acta*, **20,** 1451 (1964).
32. A.-M. Hesbain-Frisque and L. Ghosez, Louvain-la-Neuve, unpublished results.
33. A. Colens and L. Ghosez, Louvain-la-Neuve, unpublished results.
34. H. G. Viehe and Z. Janousek, *Angew. Chem.*, **83,** 614 (1971); *Angew. Chem. Int. Ed.*, **10,** 573 (1971).
35. J. Gorissen and H. G. Viehe, Louvain-la-Neuve, unpublished results.
36. W. Dörscheln, H. Tiefenthaler, H. Göth, P. Cerutti, and H. Schmid, *Helv. Chim. Acta*, **50,** 1759 (1967).
37. N. J. Leonard and K. Jann, *J. Amer. Chem. Soc.*, **82,** 6418 (1960).
38. E. Jongejan, W. J. M. van Tilborg, Ch. H. V. Dusseau, H. Steinberg, and Th. J. de Boer, *Tetrahedron Lett.*, **1972,** 2359.

39. J. Marchand-Brynaert, Ph.D. thesis, Université Catholique de Louvain, 1973.
40. A. Krebs, Habilitationsschrift, Heidelberg, 1971.
41. G. Opitz and W. Merz, *Liebigs Ann. Chem.*, **652,** 139 (1962).
42. Ch. Jutz, W. Müller, and E. Müller, *Chem. Ber.*, **99,** 2479 (1966).
43. M.-L. Hougardy and R. Merényi, Louvain-la-Neuve, unpublished results.
44. Y. Ito, S. Katsuragawa, M. Okano, and R. Oda, *Tetrahedron*, **23,** 2159 (1967).
45. T. R. Keenan and N. J. Leonard, *J. Amer. Chem. Soc.*, **93,** 6567 (1971).
46. P. A. Bather, J. R. Lindsay-Smith, and R. O. C. Norman, *J. Chem. Soc. C*, **1971,** 3060.
47. P. A. S. Smith and R. N. Loeppky, *J. Amer. Chem. Soc.*, **89,** 1147 (1967).
48. M. Vaultier, R. Danion-Bougot, D. Danion, J. Hamelin, and R. Carrié, *Tetrahedron Lett.*, **1973,** 2883.
49. H. L. Ammon, *Acta Cryst.*, **830,** 1731 (1974).
50. J. Singh Walia and P. Singh Walia, *Chem. Ind.*, **1969,** 135.
51. M. Grdinic and V. Hahn, *J. Org. Chem.*, **30,** 2381 (1965).
52. B. A. Phillips, G. Fodor, J. Gal, F. Letourneau, and J. J. Ryan, *Tetrahedron*, **29,** 3309 (1973).
53. H. G. Viehe, Z. Janousek, and M. A. Deffrenne, *Angew. Chem.*, **83,** 616 (1971); *Angew. Chem. Int. Ed.*, **10,** 575 (1971).
54. Z. Janousek, Ph.D. Thesis, Louvain, 1973.
55. C. Hornaert, A. M. Hesbain-Frisque, and L. Ghosez, Louvain-la-Neuve, unpublished results.
56. H. J. Dauben, Jr., and D. F. Rhoades, *J. Amer. Chem. Soc.*, **89,** 6764 (1967).
57. N. L. Bauld and Yong Sung Rim, *J. Amer. Chem. Soc.*, **89,** 6763 (1967).
58. R. van der Linde, J. W. Dornseiften, J. V. Veenland, and Th. de Boer, *Spectrochim. Acta*, **25A,** 375 (1969).
59. M. Kuhn and R. Mecke, *Chem. Ber.*, **94,** 3016 (1961).
60. E. Allenstein, A. Schmidt, and V. Beyl, *Chem. Ber.*, **99,** 431 (1966).
61. M. D. Scott and H. Spedding, *J. Chem. Soc. C*, **1968,** 1603.
62. W. Jentzsch, *Chem. Ber.*, **97,** 2755 (1964).
63. J. D. Wilson, Ch. F. Hobbs, and H. Weingarten, *J. Org. Chem.*, **35,** 1542 (1970).
64. N. Wiberg and J. W. Buchler, *Angew. Chem.*, **74,** 490 (1962); *Angew. Chem. Int. Ed.*, **1,** 406 (1962).
65. J. Ranftaand and S. Dähne, *Helv. Chim. Acta*, **47,** 1160 (1964).
66. G. Scheibe, W. Seiffert, H. Wengenmayr, and C. Jutz, *Ber. Bunsenges. Phys. Chem.*, **67,** 560 (1963).
67. C. Jutz and H. Amschler, *Chem. Ber.*, **96,** 2100 (1963).
68. M. Davies and A. E. Parsons, *Z. Phys. Chem.*, M.F., **20,** 34 (1959).
69. R. Mecke and W. Kutzelnigg, *Spectrochim. Acta*, **16,** 1216, 1225 (L960).
70. J. C. Grivas and A. Taurins, *Can. J. Chem.*, **37,** 1260 (1959).
71. E. Fluck and P. Meiser, *Angew. Chem.*, **83,** 721 (1971); *Angew. Chem. Int. Ed.*, **10,** 653 (1971).
72. T. L. Eggerichs and H. G. Viehe, Louvain-la-Neuve, unpublished results.
73. R. C. Neuman, Jr., G. S. Hammond, and T. J. Dougherty, *J. Amer. Chem. Soc.*, **84,** 1506 (1962).
74. G. S. Hammond and R. C. Neuman, Jr., *J. Phys. Chem.*, **67,** 1655 (1963).
75. G. S. Hammond and R. C. Neuman, Jr., *J. Phys. Chem.*, **67,** 1659 (1963).
76. J. S. Baum and H. G. Viehe, *J. Org. Chem.*, **41,** 183 (1976).
77. H. G. Viehe, R. Buijle, R. Fuks, R. Merényi, and J. M. F. Oth, *Angew. Chem.*, **79,** 53 (1967); *Angew. Chem. Int. Ed.*, **6,** 77 (1967).

78. E. Goffin, Ph.D. thesis, Louvain-la-Neuve, 1974.
79. R. Fuks, H. G. Viehe, and R. Merényi, Brussels and Louvain-la-Neuve, unpublished results.
80. E. Allenstein and A. Schmidt, Z. Anorg. Allg. Chem., **344,** 113 (1966).
81. R. J. Gillespie and T. Birchall, Can. J. Chem., **41,** 148 (1963).
82. T. Birchall and R. J. Gillespie, Can. J. Chem., **41,** 2642 (1963).
83. W. F. Reynolds, I. R. Peat, M. H. Freedman, and J. R. Lyerla, Jr., J. Amer. Chem. Soc., **95,** 328 (1973).
84. T. D. Smith, J. Chem. Soc. A, **1966,** 841.
85. H. G. Nordmann and F. Kröhnke, Angew. Chem., **81,** 1004 (1969); Angew. Chem. Int. Ed., **9,** 984 (1969).
86. V. A. Pattison, J. G. Colson, and R. L. K. Carr, J. Org. Chem., **33,** 1084 (1968).
87. W. Kutzelnigg and R. Mecke, Spectrochim. Acta, **18,** 549 (1962).
88. E. Spinner, Spectrochim. Acta, **15,** 95 (1959).
89. D. Cook, Can. J. Chem., **42,** 2721 (1964).
90. G. Olah and P. Szilágyi, J. Amer. Chem. Soc., **91,** 2949 (1969).
91. D. Cook, Can. J. Chem., **40,** 2362 (1962).
92. R. Gompper and P. Altreuther, Z. Anal. Chem., **170,** 205 (1959).
93. M. Liler, J. Chem. Soc. Perkin II, **1974,** 71.
94. W. Buder and A. Schmidt, Chem. Ber., **106,** 2877 (1973).
95. G. A. Olah, J. Nishimura, and P. Kreienbühl, J. Amer. Chem. Soc., **95,** 7672 (1973).
96. W. Kutzelnigg and R. Mecke, Spectrochim. Acta, **17,** 530 (1961).
97. M. J. Janssen, Spectrochim. Acta, **17,** 475 (1961).
98. J. D. S. Goulden, J. Chem. Soc., **1953,** 997.
99. E. Allenstein and P. Quis, Chem. Ber., **97,** 3162 (1964).
100. P. George and H. G. Viehe, Louvain-la-Neuve, unpublished results.
101. R. Stewart and L. J. Muenster, Can. J. Chem., **39,** 401 (1961).
102. G. Fraenkel and C. Franconi, J. Amer. Chem. Soc., **82,** 4478 (1960).
103. H. Kessler and D. Leibfritz, Tetrahedron Lett., **1969,** 397.
104. K. Hartke and G. Salamon, Chem. Ber., **103,** 133 (1970).
105. H. Kessler and D. Liebfritz, Tetrahedron, **25,** 5127 (1969).
106. C. F. Hobbs and H. Weingarten, J. Org. Chem., **36,** 2885 (1971).
107. G. A. Olah and A. M. White, J. Amer. Chem. Soc., **90,** 6087 (1968).
108. G. A. Olah and M. Calin, J. Amer. Chem. Soc., **90,** 401 (1968).
109. B. Leclef and H. G. Viehe, Louvain-la-Neuve, unpublished results.
110. H. Cross and J. Rusche, Angew. Chem., **76,** 534 (1964); Angew. Chem. Int. Ed., **3,** 511 (1964).
111. R. W. Addor, J. Org. Chem., **29,** 738 (1964).
112. E. Daltrozzo and K. Feldmann, Ber. Bunsenges. Phys. Chem., **72,** 1140 (1968).
113. S. Dähne and J. Ranft, Z. Phys. Chem. (Leipzig), **224,** 65 (1963).
114. M.-L. Filleux-Blanchard, D. le Botlan, A. Reliquet, and F. Reliquet-Clesse, Org. Magn. Resonance, **6,** 471 (1974).
115. G. Scheibe, C. Jutz, W. Seiffert, and D. Grosse, Angew. Chem., **76,** 270 (1964); Angew. Chem. Int. Ed., **3,** 306 (1964).
116. C. Jutz and W. Müller, Chem. Ber., **100,** 1536 (1967).
117. Z. Arnold, Collect. Czech. Chem. Commun., **38,** 1168 (1973).
118. E. Allenstein and P. Quis, Chem. Ber., **97,** 1857 (1964).
119. Z. Janousek and H. G. Viehe, Angew. Chem., **83,** 615 (1971); Angew. Chem. Int. Ed. **10,** 574 (1971).
120. M. Houtekie and L. Ghosez, Louvain-la-Neuve, unpublished results.

121. Z. Arnold, *Collect. Czech. Chem. Commun.*, **30,** 2125 (1965).
122. H. Gold, *Angew. Chem.*, **72,** 956 (1960).
123. E. Allenstein, *Z. Anorg. Allg. Chem.*, **322,** 276 (1963).
124. Z. Janousek and H. G. Viehe, *Angew. Chem.*, **85,** 90 (1973); *Angew. Chem. Int. Ed.*, **12,** 74 (1973).
125. E. Allenstein and P. Quis, *Chem. Ber.*, **96,** 2918 (1963).
126. B. Birdsell and F. Feeney, *J. Chem. Soc. Perkin* II, **1972,** 1643.
127. H. E. A. Kramer and R. Gompper, *Z. Physik. Chem.*, N.F., **43,** 349 (1964).
128. H. Weingarten, *J. Org. Chem.*, **35,** 3970 (1970).
129. J. Elguero and R. Jacquier, *Bull. Soc. Chim. Fr.*, **1965,** 2961.
130. H. Saito and K. Nukada, *J. Mol. Spectrosc.*, **18,** 1 (1965).
131. G. A. Olah and T. E. Kiovsky, *J. Amer. Chem. Soc.*, **90,** 4666 (1968).
132. H. Saito, K. Nukada, and M. Ohno, *Tetrahedron Lett.*, **1964,** 2124.
133. J. P. Kintzinger and J. M. Lehn, *Chem. Commun.*, **1967,** 660.
134. D. Crépaux and J. M. Lehn, *Mol. Phys.*, **14,** 547 (1968).
135. O. Cervinka, A. R. Katritzky, and F. J. Swinbourne, *Collect. Czech. Chem. Commun.*, **30,** 1736 (1965).
136. N. J. Leonard and V. W. Gash, *J. Amer. Chem. Soc.*, **76,** 2781 (1954).
137. M. C. Kloetzel, J. L. Pinkus, and R. M. Washburn, *J. Amer. Chem. Soc.*, **79,** 4222 (1957).
138. H.-D. Bartfeld and W. Flitsch, *Chem. Ber.*, **106,** 1423 (1973).
139. N. J. Leonard and F. P. Hauck, Jr., *J. Am. Chem. Soc.*, **79,** 5279 (1957).
140. B. Witkop, *J. Amer. Chem. Soc.*, **78,** 2873 (1956).
141. N. J. Leonard, A. S. Hay, R. W. Fulmer, and V. W. Gash, *J. Amer. Chem. Soc.*, **77,** 439 (1955).
142. N. J. Leonard and A. S. Hay, *J. Amer. Chem. Soc.*, **78,** 1984 (1956).
143. R. Jacquier, C. Pellier, C. Petrus, and F. Petrus, *Bull. Soc. Chim. Fr.*, **1971,** 646.
144. J. Elguero, R. Jacquier, and D. Tizané, *Tetrahedron*, **27,** 123 (1971).
145. J. L. Aubagnac, J. Elguero, and R. Jacquier, *Bull. Soc. Chim. Fr.*, **1969,** 3292.
146. J. L. Aubagnac, J. Elguero, and R. Jacquier, *Bull. Soc. Chim. Fr.*, **1971,** 3758.
147. J. Elguero, R. Jacquier, and D. Tizané, *Bull. Soc. Chim.*, **1968,** 3866.
148. I. C. Smith and W. C. Schneider, *Can. J. Chem.*, **39,** 1158 (1961).
149. C. L. Bell, J. Schoffner, and L. Bauer, *Chem. Ind.*, **1963,** 1435.
150. G. A. Olah, J. A. Olah, and N. A. Overchuk, *J. Org. Chem.*, **30,** 3373 (1965).
151. G. Kotowycz, T. Schaefer, and E. Bock, *Can. J. Chem.*, **42,** 2541 (1964).
152. F. W. Wehrli, W. Giger, and W. Simon, *Helv. Chem. Acta*, **54,** 229 (1971).
153. J.-F. Bielmann and H. Callot, *Bull. Soc. Chim. Fr.*, **1967,** 397.
154. W. A. Thomas and G. E. Griffin, *Org. Magn. Resonance*, **2,** 503 (1970).
155. Y. Chiang and E. B. Whipple, *J. Amer. Chem. Soc.*, **85,** 2763 (1963).
156. G. Duchène and H. G. Viehe, Louvain-la-Neuve, unpublished results.
157. N. J. Leonard and J. A. Adamcik, *J. Amer. Chem. Soc.*, **81,** 595 (1959).
158. R. Michelot and H. Khedija, *Tetrahedron*, **29,** 1031 (1973).
159. Z. Yoshida and Y. Tawara, *J. Amer. Chem. Soc.*, **73,** 2573 (1971).
160. K. M. Wellman and D. L. Harris, *Chem. Commun.*, **1967,** 256.
161. M. H. Palmer and B. Semple, *Chem. Ind.*, **1965,** 1766.
162. D. Herbison-Evans and R. E. Richards, *Mol. Phys.*, **8,** 19 (1964).
163. J. D. Baldeschwieler and E. W. Randall, *Proc. Chem. Soc.*, **1961,** 303.
164. M. Witanowski, *J. Amer. Chem. Soc.*, **90,** 5683 (1968).
165. R. L. Lichter and J. D. Roberts, *J. Amer. Chem. Soc.*, **93,** 5218 (1971).
166. H. Saitô, Y. Tanaka, and S. Nagata, *J. Amer. Chem. Soc.*, **95,** 324 (1973).

167a. G. Binsch, J. B. Lambert, B. W. Roberts, and J. D. Roberts, *J. Amer. Chem. Soc.*, **86**, 5564 (1964).
167b. A. J. R. Bourn and E. W. Randall, *Mol. Phys.*, **8**, 567 (1964).
168. E. Krakover and L. W. Reewes, *Spectrochim. Acta*, **20**, 71 (1964).
169. H. C. E. McFarlane and W. McFarlane, *Org. Magn. Resonance*, **4**, 161 (1972).
170. T. Axenrod, M. J. Wieder, G. Berti, and P. L. Barili, *J. Amer. Chem. Soc.*, **92**, 6066 (1970).
171. K. Tori, M. Ohtsuru, K. Aono, Y. Kawazoe, and M. Ohnishi, *J. Amer. Chem. Soc.*, **89**, 2765 (1967).
172. P. S. Pergosin, E. W. Randall, and A. I. White, *J. Chem. Soc. Perkin* II, **1972**, 1.
173. H.-O. Kalinowski and H. Kessler, *Org. Magn. Resonance*, **6**, 305 (1974).
174. R. Radeglia, G. Engelhardt, E. Lippmaa, T. Pehk, K.-D. Nolte, and S. Dähne, *Org. Magn. Resonance*, **4**, 571 (1972).
175. H.-U. Wagner, *Chem. Ber.*, **107**, 634 (1974).
176. R. J. Pugmire and D. M. Grant, *J. Amer. Chem. Soc.*, **90**, 697, 4232 (1968).
177. E. Lippmaa, M. Mägi, S. S. Novikov, L. I. Khmelnitski, A. S. Prihodko, O. V. Lebedev, and L. V. Epishina, *Org. Magn. Resonance*, **4**, 197 (1972).
178. M. W. Hunkapiller, S. H. Smallcombe, D. R. Whitaker, and J. H. Richards, *Biochemistry*, **12**, 4732 (1973).
179. A. Mathias and V. M. S. Gil, *Tetrahedron Lett.*, **1965**, 3163.
180. E. Breitmaier and K. H. Spohn, *Tetrahedron*, **29**, 1145 (1973).
181. R. Merényi, R. Verbruggen, and M. W. Baum, Louvain-la-Neuve, unpublished results.
182. H. Böhme and G. Auterhoff, *Chem. Ber.*, **104**, 2013 (1971).
183. D. H. Clemens, E. Y. Shropshire, and W. D. Emmons, *J. Org. Chem.*, **27**, 3664 (1962).
184. J. Kučera and Z. Arnold, *Collect. Czech. Chem. Commun.*, **32**, 1704 (1967).
185. G. de Voghel, T. L. Eggerichs, Z. Janousek, and H. G. Viehe, *J. Org. Chem.*, **39**, 1233 (1974).
186. H. E. Nikolajewski, S. Dähne, and B. Hirsch, *Chem. Ber.*, **100**, 2616 (1967).
187. C. Jutz, *Chem. Ber.*, **97**, 2050 (1964).
188. H. Hartmann, *J. Prakt. Chem.*, **312**, 1194 (1970).
189. M. J. Janssen, *Rec. Trav. Chim.*, **79**, 454 (1960).
190. J. R. Hargreaves, P. W. Hickmott, and B. J. Hopkins, *J. Chem. Soc. C*, **1968**, 2599.
191. H. G. Viehe, T. van Vyve, and Z. Janousek, *Angew. Chem.*, **84**, 991 (1972); *Angew. Chem. Int. Ed.*, **11**, 916 (1972).
192. H. Böhme and M. Haake, *Chem. Ber.*, **100**, 3609 (1967).
193. G. H. Alt and A. J. Speziale, *J. Org. Chem.*, **30**, 1407 (1965).
194. B. Witkop and J. B. Patrick, *J. Amer. Chem. Soc.*, **75**, 4474 (1953).
195. R. L. Hinman and E. B. Whipple, *J. Amer. Chem. Soc.*, **84**, 2534 (1962).
196. A. R. Katritzky and R. A. Y. Jones, *Chem. Ind.*, **1961**, 722.
197. L. M. Trefonas, R. L. Flurry, Jr., R. Majeste, E. A. Meyers, and R. F. Copeland, *J. Amer. Chem. Soc.*, **88**, 2145 (1966).
198. R. C. Neumann, Jr., and V. Jonas, *J. Phys. Chem.*, **75**, 3550 (1971).
199. S. Hanlon, S. F. Russo, and I. M. Klotz, *J. Amer. Chem. Soc.*, **85**, 2024 (1963).
200. S. Hanlon and I. M. Klotz, *Biochemistry*, **4**, 37 (1965).
201. J. L. Johnson, M. E. Herr, J. C. Babcock, A. E. Fonken, J. E. Stafford, and F. W. Heyl, *J. Amer. Chem. Soc.*, **78**, 430 (1956).
202. A. Schmidt, *Chem. Ber.*, **100**, 3725 (1967).
203. R. Foglizzo and A. Novak, *Spectrochim. Acta*, **26A**, 2281 (1970).

204. A.-M. Belloc and C. Garrigou-Lagrange, *Spectrochim. Acta*, **27A**, 1175 (1971).
205. D. Beierl and A. Schmidt, *Chem. Ber.*, **106**, 1637 (1973).
206. R. Kuhn and H. Schretzmann, *Chem. Ber.*, **90**, 557 (1957).
207. H. H. Bosshard, E. Jenny, and H. Zollinger, *Helv. Chem. Acta*, **44**, 1203 (1961).
208. O. Červinka and L. Hub, *Collect. Czech. Chem. Commun.*, **30**, 3111 (1965).
209. H. Gross, J. Gloede, and J. Freiberg, *Liebigs Ann. Chem.*, **702**, 68 (1967).
210. P. W. Hickmott and B. J. Hopkins, *J. Chem. Soc. C*, **1968**, 2918.
211. S. Hünig and H. Hock, *Fortschr. Chem. Forsch.*, **14**, 236 (1970).
212. H. Böhme, G. Auterhoff, and W. Höver, *Chem. Ber.*, **104**, 3350 (1971).
213. R. Mason and G. Rucci, *Chem. Commun.*, **1971**, 1132.
214. H. Ahlbrecht and M. Th. Reiner, *Tetrahedron Lett.*, **1971**, 4901.
215. R. Scheffold and E. Saladin, *Angew. Chem.*, **84**, 158 (1972); *Angew. Chem. Int. Ed.*, **11**, 229 (1972).
216. J. Marchand-Brynaert and L. Ghosez, *J. Amer. Chem. Soc.*, **94**, 2870 (1972).
217. A. Venot and G. Adrian, *Tetrahedron Lett.*, **1972**, 4663.
218. K. Hartke, F. Rossbach, and M. Radau, *Liebigs Ann. Chem.*, **762**, 167 (1972).
219. Z. Csürös, R. Soós, I. Bitter, and É. Kárpáti-Ádám, *Acta Chim. Acad. Sci. Hung.*, **73**, 239 (1972).
220. Z. Csürös, R. Soós, I. Bitter, and J. Pálinkás, *Acta Chim. Acad. Sci. Hung.*, **72**, 59 (1972).
221. S. Yanagida, T. Fujita, M. Ohoka, I. Katagiri, and S. Komori, *Bull. Chem. Soc. Jap.*, **46**, 292 (1973).
222. S. Yanagida, T. Fujita, M. Ohoka, I. Katagiri, M. Miyake, and S. Komori, *Bull. Chem. Soc. Jap.*, **46**, 303 (1973).
223. C. R. Redpath and J. A. S. Smith, *Trans. Faraday Soc.*, **58**, 462 (1962).
224. H. Bredereck, F. Effenberger, and G. Simchen, *Chem. Ber.*, **96**, 1350 (1963).
225. H. Bredereck, F. Effenberger, and H. P. Beyerlin, *Chem. Ber.*, **97**, 1834, 3076 (1964).
226. H. Eilingsfeld and L. Möbius, *Chem. Ber.*, **98**, 1292 (1965).
227. G. V. Boyd, *Chem. Commun.*, **1969**, 1147.
228. A. Pilotti, A. Reuterhäll, K. Torsell, and C.-G. Lindblad, *Acta Chem. Scand.*, **23**, 818 (1969).
229. P. Combelas, F. Cruege, J. Lascomb, C. Quivoron, and M. Rey-Lafon, *J. Chem. Phys.*, **66**, 668 (1969).
230. J. W. O. Tam and I. M. Klotz, *Spectrochim. Acta*, **29A**, 633 (1973).
231. P. Combelas, C. Garrigou-Lagrange, and J. Lascombe, *Biopolymers*, **12**, 611 (1973).
232. P. Combelas and C. Garrigou-Lagrange, *Spectrochim. Acta*, **30A**, 550 (1974).
233. N. J. Leonard, L. A. Miller, and P. D. Thomas, *J. Amer. Chem. Soc.*, **78**, 3463 (1956).
234. R. L. Pederson, J. L. Johnson, R. P. Holysz, and A. C. Ott, *J. Amer. Chem. Soc.*, **79**, 1115 (1957).
235. H. Böhme and K. H. Ahrens, *Tetrahedron Lett.*, **1971**, 149.
236. H. Newman and T. L. Fields, *Tetrahedron*, **28**, 4051 (1972).
237. G. A. Olah, *Angew. Chem.*, **85**, 183 (1973); *Angew. Chem. Int. Ed.*, **12**, 173 (1973).
238. E. W. Randall and D. G. Gillies, "Nitrogen Nuclear Magnetic Resonance," in *Progress in NMR Spectroscopy*, Vol. 6, J. W. Emsley, J. Feeney, and L. H. Sutcliff, Eds., Pergamon Press, Oxford, 1969.
239. D. Crépaux, J. M. Lehn, and R. R. Dean, *Mol. Phys.*, **16**, 225 (1969).
240. R. J. Harder and W. C. Smith, *J. Amer. Chem. Soc.*, **83**, 3422 (1961).
241. Z. Arnold, *Collect. Czech. Chem. Commun.*, **28**, 2047 (1963).

242. J. Elguero, R. Jacquier, and G. Tarrago, *Tetrahedron Lett.*, **1965**, 4719.
243. E. Elkik and P. Vaudescal, *C. R. Acad. Sci. Paris*, **264C**, 1779 (1967).
244. T. Moeller and A. H. Westlake, *J. Inorg. Nucl. Chem.*, **29**, 957 (1967).
245. A. F. McDonagh and H. E. Smith, *J. Org. Chem.*, **33**, 8 (1968).
246. L. Alais, P. Angibeaud, and R. Michelot, *C. R. Acad. Sci. Paris*, **269**, 150 (1969).
247. E. Elkik, *Bull. Soc. Chim. Fr.*, **1969**, 903.
248. E. Fluck and P. Meiser, *Chem. Ztg.*, **95**, 922 (1971).
249. G. R. Krow, C. Pyun, C. Leitz, J. Marakowski, and K. Ramey, *J. Org. Chem.*, **39**, 2449 (1974).
250. D. Herbison-Evans and R. E. Richards, *Trans. Faraday Soc.*, **58**, 845 (1962).
251. S. J. Kuhn and J. S. McIntyre, *Can. J. Chem.*, **43**, 995 (1965).
252. E. D. Becker, H. T. Miles, and R. B. Bradley, *J. Amer. Chem. Soc.*, **87**, 5575 (1965).
253. F. Vögtle, A. Mannschreck, and H. A. Staab, *Liebigs Ann. Chem.*, **708**, 51 (1967).
254. G. V. Boyd, *Chem. Commun.*, **1968**, 1410.
255. R. L. N. Harris, *Tetrahedron Lett.*, **1970**, 5217.
256. Z. Janousek, J. Collard, and H. G. Viehe, *Angew. Chem.*, **84**, 993 (1972); *Angew. Chem. Int. Ed.*, **11**, 917 (1972).
257. G. Ege and H. O. Frey, *Tetrahedron Lett.*, **1972**, 4217.
258. A. J. Weinheimer, E. K. Metzner, and M. L. Mole, Jr., *Tetrahedron*, **29**, 3135 (1973).
259. S. Dahne and J. Ranft, *Z. Phys. Chem. (Leipzig)*, **232**, 259 (1966).
260. C. Jutz and W. Müller, *Angew. Chem.*, **78**, 1059 (1966); *Angew. Chem. Int. Ed.*, **5**, 1042 (1966).
261. D. Lloyd, R. K. Mackie, H. McNab, and D. R. Marshall, *J. Chem. Soc. Perkin* II, **1973**, 1729.
262. D. le Botlan, M.-L. Filleux-Blanchard, G. le Coustumer, and Y. Mollier, *Org. Magn. Resonance*, **6**, 454 (1974).
263. I. H. Leubner, *Org. Magn. Resonance*, **6**, 253 (1974).
264. A. R. Katritzky and R. A. Y. Jones, *Chem. Ind.*, **1960**, 313.
265. A. R. Katritzky and R. E. Reavill, *Chem. Ind.*, **1963**, 753.
266. S. L. Johnson and K. A. Rumon, *J. Chem. Phys.*, **68**, 3149 (1964).
267. B. W. Roberts, J. B. Lambert, and J. D. Roberts, *J. Amer. Chem. Soc.*, **87**, 5439 (1965).
268. C. Toma and A. T. Balaban, *Tetrahedron Suppl.*, **7**, 9 (1966).
269. T. Goto and M. Isobe, *Tetrahedron Lett.*, **1968**, 1511.
270. D. A. Tomalia, N. D. Ojha, and B. P. Thill, *J. Org. Chem.*, **34**, 1400 (1969).
271. J. Epsztajn, E. Lunt, and A. R. Katritzky, *Tetrahedron*, **26**, 1665 (1970).
272. G. Pagani, *J. Chem. Soc. Perkin* II, **1973**, 1184.
273. C. S. Giam and J. L. Lyle, *J. Amer. Chem. Soc.*, **95**, 3235 (1973).
274. H. Böhme and M. Hilp, *Chem. Ber.*, **103**, 3930 (1970).
275. J. U. Lowe, Jr., R. D. Barefoot, and A. S. Tompa, *J. Org. Chem.*, **31**, 3315 (1966).
276. M. L. Filleux-Blanchard, M. T. Quemeneur, and G. J. Martin, *Chem. Commun.*, **1968**, 836.
277. G. J. Martin, S. Poignant, M. L. Filleux, and M. T. Quemeneur, *Tetrahedron Lett.*, **1970**, 5061.
278. G. J. Martin and S. Poignant, *J. Chem. Soc. Perkin* II, **1972**, 1964; **1974**, 642.
279. A. Krebs and J. Breckwoldt, *Tetrahedron Lett.*, **1969**, 3797.
280. A. Krebs, *Tetrahedron Lett.*, **1971**, 1901.
281. W. E. Stewart and T. H. Siddall, III, *Chem. Rev.*, **70**, 517 (1970).
282. R. C. Neuman, Jr., and L. B. Young, *J. Phys. Chem.*, **69**, 2570 (1965).
283. R. C. Neuman, Jr., and V. Jonas, *J. Phys. Chem.*, **75**, 3532 (1971).

284. H. E. A. Kramer and R. Gompper, *Tetrahedron Lett.*, **1963**, 969.
285. M.-L. Filleux-Blanchard, G. Le Coustumer, and Y. Mollier, *Bull. Soc. Chim. Fr.*, **1971**, 2607.
286. J. S. McKennis and P. A. S. Smith, *J. Org. Chem.*, **37**, 4173 (1972).
287. C. W. Fryer, F. Conti, and C. Franconi, *Ric. Sci. Rend. Sez.*, **A35**, 788 (1965).
288. F. Conti and W. von Philipsborn, *Helv. Chim. Acta*, **50**, 603 (1967).
289. G. W. Wheland, *Resonance in Organic Chemistry*, John Wiley and Sons, New York, 1955, p. 295.
290. A. Van Dormael, *Ind. Chim. Belg.*, **36**, 185, 267 (1971).
291. S. S. Malhotra and M. C. Whiting, *J. Chem. Soc.*, **1960**, 3812.
292. G. Scheibe, J. Heiss, and K. Feldmann, *Ber. Bunsenges. Phys. Chem.*, **70**, 52 (1966).
293. F. Dörr, J. Kotschy, and H. Kausen, *Ber. Bunsenges. Phys. Chem.*, **69**, 11 (1965).
294. K. Feldmann, E. Daltrozzo, and G. Scheibe, *Z. Naturforsch.*, **226**, 722 (1967).
295. G. H. Alt and A. J. Speziale, *J. Org. Chem.*, **29**, 794 (1964).
296. C. U. Pittman, Jr., S. P. McManus, and J. W. Larsen, *Chem. Rev.*, **72**, 357 (1972).
297. S. F. Mason, *J. Chem. Soc.*, **1959**, 1247, 1253.
298. S. Hünig and J. Utermann, *Chem. Ber.*, **88**, 1485 (1955).
299. E. B. Whipple, Y. Chiang, and R. L. Hinman, *J. Amer. Chem. Soc.*, **85**, 26 (1963).
300. Z. Arnold, *Collect. Czech. Chem. Commun.*, **30**, 2783 (1965).
301. C. Jutz and R. Heinicke, *Chem. Ber.*, **102**, 623 (1969).
302. V. Eisner and J. Kuthan, *Chem. Rev.*, **72**, 1 (1972).
303. G. F. Smith, *J. Chem. Soc.*, **1954**, 3842.
304. H. Fritz, A. Krekel, and H. Meyer, *Liebigs Ann. Chem.*, **664**, 188 (1963).
305. H.-R. Müller and M. Seefelder, *Liebigs Ann. Chem.*, **728**, 88 (1969).
306. M. Wakselman and E. Guibé-Jampel, *Tetrahedron Lett.*, **1970**, 4715.
307. Z. Arnold, J. Šauliová, and V. Krchňák, *Collect. Czech. Chem. Commun.*, **38**, 2263 (1973).
308. F. T. Edwards, H. S. Chang, K. Yates, and R. Stewart, *Can. J. Chem.*, **38**, 1518 (1960).
309. J. A. Van Allan, C. C. Petropoulos, and G. A. Reynolds, *J. Heterocycl. Chem.*, **6**, 803 (1969).
310. G. A. Reynolds and J. A. Van Allan, *J. Org. Chem.*, **34**, 2736 (1969).
311. P. Schiess and A. Grieder, *Tetrahedron Lett.*, **1969**, 2097.
312. A. R. Fersht and W. P. Jencks, *J. Amer. Chem. Soc.*, **91**, 2125 (1969).
313. R. R. Schmidt and E. Schlipf, *Chem. Ber.*, **103**, 3783 (1970).
314. G. W. Fischer, *Chem. Ber.*, **103**, 3470 (1970).
315. J. Elguero, R. Jacquier, and D. Tizané, *Tetrahedron*, **27**, 123 (1971).
316. H. J. Veith and M. Hesse, *Helv. Chim. Acta*, **52**, 2004 (1969).
317. P. R. Briggs, T. W. Shannon, and P. Vouros, *Org. Mass Spectrom.*, **5**, 545 (1971).
318. J. Gal, B. A. Phillips, and R. Smith, *Can. J. Chem.*, **51**, 132 (1973).
319. J. S. Baum, Ph.D. thesis, Rènsselaer Polytechnic Institute, Troy, N.Y., 1973.
320. F. S. Fawcett, C. W. Tullock, and D. D. Coffman, *J. Amer. Chem. Soc.*, **84**, 4275 (1962).
321. B. W. Tattershall, *J. Chem. Soc. A*, **1970**, 3263.
322. G. Dahms, A. Haas, and W. Klug, *Chem. Ber.*, **104**, 2732 (1971).
323. F. Bohlmann, H.-J. Müller, and D. Schumann, *Chem. Ber.*, **106**, 3026 (1973).
324. H. M. Sobell and K. Tomita, *Acta Cryst.*, **17**, 126 (1964).
325. S. W. Peterson and J. M. Williams, *J. Amer. Chem. Soc.*, **88**, 2866 (1966).
326. S. K. Porter and R. A. Jacobson, *J. Chem. Soc. A*, **1970**, 1356.
327. M. Sax, J. Pletcher, C. S. Yoo, and J. M. Stewart, *Acta Cryst.*, **B27**, 1635 (1971).

328. E. Oeser, *Chem. Ber.*, **107,** 627 (1974).
329. F. J. Chentli-Benchikha, J.-P. Declercq, G. Germain, and M. Van Meerssche, Louvain-la-Neuve, unpublished results.
330. J. Galloy, J.-P. Putzeys, G. Germain, J.-P. Declercq, and M. Van Meerssche, *Acta Cryst.,* **B30,** 2460 (1974).
331. J. W. O. Tam and I. M. Klotz, *J. Amer. Chem. Soc.*, **93,** 1313 (1971).
332. F. J. Chentli-Benchikha, A. Michel, J.-P. Declercq, G. Germain, and M. Van Meerssche, Louvain-la-Neuve, unpublished results.
333. B. W. Matthews, R. S. Stenkamp, and P. M. Colman, *Acta Cryst.*, **B29,** 449 (1973).
334. A. Zedler and S. Kulpe, *Z. Chem.*, **10,** 267 (1970).
335. F. J. Chentli-Benchikha, A. Michel, G. S. D. King, J.-P. Declercq, G. Germain, and M. Van Meerssche, Louvain-la-Neuve, unpublished results.
336. D. J. Haas, D. R. Harris, and H. H. Mills, *Acta Cryst.*, **19,** 676 (1965).
337a. S. Harkem and D. Feil, *Acta Cryst.*, **B25,** 589 (1969).
337b. J. E. Worsham and W. R. Busing, *Acta Cryst.*, **B25,** 572 (1969).
338. D. Freil and W. Song Loong, *Acta Cryst.*, **B24,** 1334 (1968).
339. C. H. Stam, *Acta Cryst.*, **15,** 317 (1962).
340. F. J. Chentli-Benchikha, J.-P. Putzeys, G. Germain, and M. Van Meerssche, Louvain-la-Neuve, unpublished results.
341. R. Handa and N. N. Saha, *Acta Cryst.*, **B29,** 554 (1973).
342. A. T. Ku and M. Sundaralingem, *J. Amer. Chem. Soc.*, **94,** 1688 (1972).
343. G. Lepicard, D. de Saint-Giniez, R. Jacquier, and C. Rérat, *C. R. Acad. Sci. Paris*, **267,** 1786 (1968).
344. J. L. Aubagnac, J. Elguero, B. Rérat, C. Rérat, and Y. Uesu, *C. R. Acad. Sci. Paris*, **274C,** 1192 (1972).
345. F. Donohue, L. R. Lavine, and F. S. Rollett, *Acta Cryst.*, **9,** 655 (1956).
346. C. Rérat, *Acta Cryst.*, **15,** 427 (1962).
347. A. Camerman, L. H. Jensen, and A. T. Balaban, *Acta Cryst.*, **B25,** 2623 (1969).
348. L. Brun and C.-I. Bränden, *Acta Cryst.*, **20,** 749 (1966).
349. W. C. Hamilton, *Acta Cryst.*, **18,** 866 (1965).
350. J. Jack and D. M. Hercules, *Anal. Chem.*, **43,** 729 (1971).
351. L. E. Cox, J. J. Jack, and D. M. Hercules, *J. Amer. Chem. Soc.*, **94,** 6575 (1972).
352. Gy. Gati, V. I. Nefedov, and J. W. Szalin, *Magy. Kém. Foly.*, **80,** 514 (1974).
353. F. Klages and W. Grill, *Liebigs Ann. Chem.*, **594,** 21 (1955).
354. F. Klages and G. Lukasczyk, *Chem. Ber.*, **96,** 2066 (1963).
355. G. J. Janz and S. S. Danyluk, *J. Amer. Chem. Soc.*, **81,** 3850 (1959).
356. F. Cramer, S. Rittner, W. Reinhard, and P. Desai, *Chem. Ber.*, **99,** 2252 (1966).
357. C. P. Andrieux and J.-M. Saveant, *Bull. Soc. Chim. Fr.*, **1968,** 4671.
358. A. Holy, J. Krupička, and Z. Arnold, *Collect. Czech. Chem. Commun.*, **30,** 4127 (1965).
359. H. P. Cleghorn, J. E. Gaskin, and D. Lloyd, *J. Chem. Soc. B*, **1971,** 1615.
360. M. Masui, K. Fujita, and H. Ohmori, *Chem. Commun.*, **1970,** 182.
361. P. Linda, G. Marino, and S. Santini, *Tetrahedron Lett.*, **1970,** 4223.
362. S. Alunni, P. Linda, G. Marino, S. Santini, and G. Savelli, *J. Chem. Soc. Perkin* II, **1972,** 2070.
363. P. Y. Sollenberger and R. B. Martin, *J. Amer. Chem. Soc.*, **92,** 4261 (1970).
364. W. M. Schubert and Yoshiaki Motoyama, *J. Amer. Chem. Soc.*, **87,** 5507 (1965).
365. H. E. Zaugg, R. W. de Net, J. E. Fraser, and A. M. Korte, *J. Org. Chem.*, **34,** 14 (1969).
366. O. D. Bonner, K. W. Bunzl, and G. B. Woolsey, *Spectrochim. Acta*, **22,** 1125 (1966).

367. J. Elguero, R. Gil, and R. Jacquier, *Spectrochim. Acta*, **23A**, 383 (1967).
368. G. Fraenkel and C. Nieman, *Proc. Nat. Acad. Sci. U.S.A.*, **44**, 688 (1958).
369. A. Berger, A. Loewenstein, and S. Meiboom, *J. Amer. Chem. Soc.*, **81**, 62 (1959).
370. E. Jongejan, H. Steinberg, and Th. J. de Boer, *Syn. Commun.*, **4**, 11 (1974).
371. G. Fodor, J. J. Ryan, and F. Letourneau, *J. Amer. Chem. Soc.*, **91**, 7768 (1969).
372. H. Volz, *Angew. Chem.*, **85**, 592 (1973); *Angew. Chem. Int. Ed.*, **12**, 586 (1973).
373. W. B. Jennings, S. Al-Showiman, and M. S. Tollay, *J. Chem. Soc. Perkin Trans.*, 2, **1975**, 1535.
374. E. O. Fischer, K. R. Schmid, W. Kalbfus, and C. G. Kreiter, *Chem. Ber.*, **106**, 3893 (1973).
375. M. L. Martin, G. Ricoleau, S. Poignant, and G. J. Martin, *J. Chem. Soc. Perkin Trans.*, 2, in press.
376. J. Galloy and R. Merényi, Louvain-la-Neuve, unpublished results.
377. G. Scheibe, H. J. Friedrich, and G. Hohlneicher, *Angew. Chem.*, **73**, 383 (1961).
378. V. P. Kukhar, V. I. Pasternak, M. I. Povolotskii, and N. G. Pavlenko, *Zh. Org. Khim.*, **10**, 499 (1974).
379. J. L. Thomas, *J. Amer. Chem. Soc.*, **97**, 5943 (1975).
380. D. S. Wulfman, L. N. McCullagh, and J. J. Ward, *J. Chem. Soc. D, Chem. Commun.*, **1970**, 220.
381. P. Dixneuf and R. Dabard, *J. Organometal. Chem.*, **37**, 167 (1972).
382. J. A. Deyrup and W. A. Szabo, *J. Org. Chem.*, **40**, 2048 (1975).
383. J. S. McKennis, Ph.D. thesis, University of Michigan, 1970; *Diss. Abstr. Int. B*, **31**, 4585 (1971).
384. S. R. de Lockerente, O. B. Nagy, and A. Bruylants, *Org. Magn. Res.*, **2**, 179 (1970).
385. H.-O. Kalinowski and H. Kessler, *Org. Magn. Res.*, **7**, 569 (1975).
386. G. J. Martin and N. Naulet, *Tetrahedron Lett.*, **1976**, 357.
387. W. Grahn and Ch. Reichardt, *Tetrahedron*, **32**, 125 (1976).
388. A. Katritzky, M. Kinus, and E. Lunt, *Org. Magn. Res.*, **7**, 569 (1975).
389. W. Giger and W. Simon, *Helv. Chim. Acta*, **53**, 1609 (1970).
390. C. Courseille, S. Geoffre, F. Leroy, and M. Hospital, *Cryst. Struct. Commun.*, **3**, 583 (1974).
391. E. Schaumann, A. Röhr, S. Sieveking, and W. Walter, *Angew. Chem.*, **87**, 486 (1975); *Angew. Chem. Int. Ed.*, **14**, 493 (1975).
392. G. Germain, J. P. Declercq, M. Van Meerssche, F. Hervens, and H. G. Viehe, *Bull. Soc. Chim. Belg.*, **84**, 1005 (1975).
393. C. P. Andrieux and J. M. Saveant, *J. Electroanl. Chem.* **26**, 223 (1970); **28**, 339 and 446 (1970).
394. G. Jugie, J. A. S. Smith, and G. J. Martin, *J. Chem. Soc. Perkin Trans.*, 2, **1975**, 925.
395. L. Alais, P. Angibeaud, R. Michelot, and B. Tchoubar, *Bull. Soc. Chim. Fr.*, **1970**, 539.

METHYLENIMINIUM SALTS

By H. BÖHME and M. HAAKE, *Pharmazeutisch-chemisches Institut, Philipps-Universität, D-355-Marburg/Lahn, Marbacher Weg, 6, Germany*

CONTENTS

I. Introduction	108
II. Synthesis and Preparation of Methyleniminium Salts	108
A. Reactions of Aldimines and Ketimines	108
1. Protonation	108
2. Alkylation	109
3. Acylation	110
4. Halogenation	112
B. Reactions of Enamines	112
1. Protonation	112
2. Alkylation	112
3. Acylation	114
4. Reactions with Halogens and Inorganic Halides	116
C. Cyanide Elimination from α-Aminonitriles	117
D. Condensation of Aldehydes or Ketones with Salts of Secondary Amines	118
E. Halogenation of N-α-Hydroxyalkyl Amides or Imides	119
F. Cleavage of Heterogeminals	121
1. Aminals	121
2. Dialkylalkoxymethyl- and Dialkylacyloxymethylamines	133
3. Dialkylalkylmercaptomethylamines	141
G. Hydride Abstraction from Tertiary Amines	142
H. Reactions of Amine Oxides and Nitrones	143
I. Fragmentation of Halomethylammonium Halides	145
J. Miscellaneous Methods of Preparation	146
K. Iminium Salts as Intermediates	149
III. Physical Properties of Methyleniminium Salts	150
A. Structure, Stability, and Solubility	150
B. Absorption in Infrared and Ultraviolet	151
C. Nuclear Magnetic Resonance	152
D. Mass Spectra	154
E. Miscellaneous Physical Properties	154
IV. Reactions of Methyleniminium Salts	154
A. Anion Exchange	155
B. Hydrolysis	156
C. Reactions with Hydrides	157
D. Deprotonation	157
E. C–O Bond Formation	162

F. C–S Bond Formation 164
G. C–N Bond Formation 169
H. C–P Bond Formation 175
I. C–C Bond Formation 179
   1. Reactions with Cyanides... 179
   2. Reactions with Active C–H Bonds................................. 180
   3. Reactions with Grignard Reagents and Other Organometallic Reagents .. 183
   4. Reactions with Multiple-Bond Systems 190
   5. Reactions with Aromatic Compounds 200
   6. Reactions with Heterocycles...................................... 203
   7. Reactions with Diazoalkanes 205
J. Miscellaneous Reactions... 206
   References 206
   Addendum... 217

I. Introduction

Iminium salts are derivatives of carbonyl compounds. The electronic distribution of the cation can be exemplified by resonance structures **A** and **B**:

$$\diagup_{\diagdown}C=\overset{+}{N}\diagup^{\diagdown} \longleftrightarrow \diagup_{\diagdown}\overset{+}{C}-N\diagup^{\diagdown}$$

(A)   (B)

This chapter considers mainly methyleniminium halides (MIH) and other isolable salts (MIS) derived from formaldehyde and secondary amines. Derivatives of other carbonyl compounds have already been discussed in several reviews (1–3). Therefore these derivatives are mentioned only insofar as their preparation and properties are of interest for comparison. The same is true also for derivatives from amides or imides (4–6).

II. Synthesis and Preparation of Methyleniminium Salts

A. REACTIONS OF ALDIMINES AND KETIMINES

1. *Protonation*

On protonation of imines methyleniminium, as well as alkyl- or arylmethyleniminium, salts are formed (1,2). The structures have been well established by NMR (7,8). Derivatives of aromatic aldehydes or ketones are described as stable salts, for example, **1**, which can be prepared from benzylidenemethylimine in chloroform with perchloric acid (9), or the hydrochloride **2** of benzophenonimine, which can be sublimed without decomposition at 230–250° (10). For protonation of methylenimine and acetaldehyde ethylimine see Section II J.

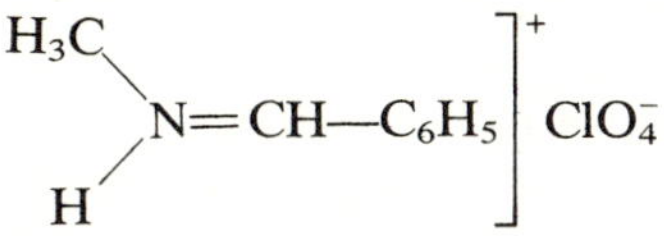

$$\left[\begin{array}{c}H_3C\\ \diagdown\\ N=CH-C_6H_5\\ \diagup\\ H\end{array}\right]^+ ClO_4^- \qquad \left[H_2N=C(C_6H_5)_2\right]^+ Cl^-$$

(1) 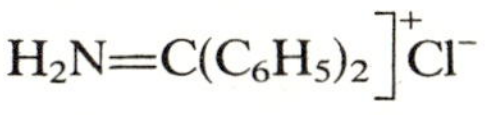   (2)

2. Alkylation

Alkylations with methyl iodide of imines derived from aromatic aldehydes and ketones have been known for many years. The first reaction seems to have been reported by Forster (11), who prepared the iodide **3** from benzylidene bornylamine and mentioned the instability of this salt in water. Such alkylations, which have been studied more intensively by Decker and Becker (12), can be easily performed for aliphatic amine derivatives, for example, dimethyldiphenylmethyleniminium iodide (**4**) (13), methylallylphenylmethyleniminium iodide (**5**) (14), and 3-methyl-2-butylidene-N-methyl-N-propyliminium iodide (**6**) (15). Because of

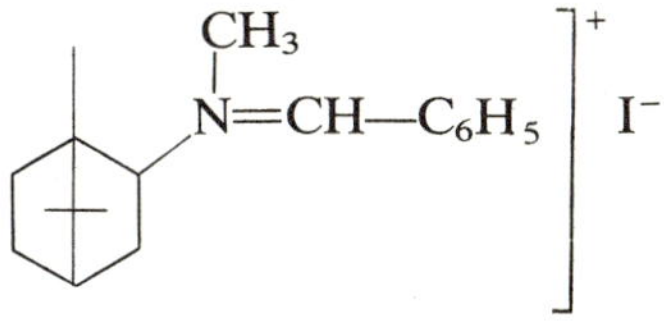

$$\left[\begin{array}{c}CH_3\\ |\\ N=CH-C_6H_5\end{array}\right]^+ I^- \qquad \left[(CH_3)_2N=C(C_6H_5)_2\right]^+ I^-$$

(3) 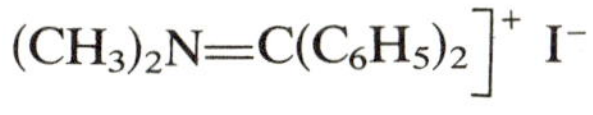   (4)

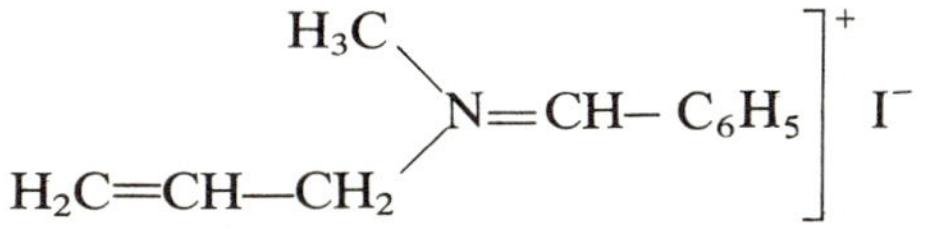

$$\left[\begin{array}{c}H_3C\\ \diagdown\\ N=CH-C_6H_5\\ \diagup\\ H_2C=CH-CH_2\end{array}\right]^+ I^-$$

(5)

$$\left[\begin{array}{c}H_3C \qquad CH_3\\ \diagdown \diagup\\ N=C\\ \diagup \diagdown\\ CH_3-CH_2-CH_2 \qquad CH(CH_3)_2\end{array}\right]^+ I^-$$

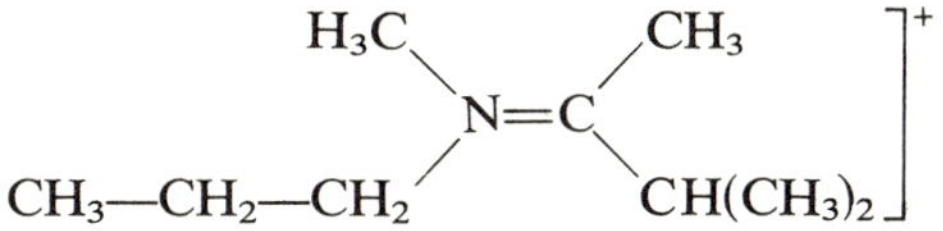

(6)

difficulties in the case of aromatic amine derivatives, the tertiary oxonium salts are more suitable alkylating reagents (16) and also have been used successfully for the alkylation of sterically hindered carbodiimides to salts **7a** (17a), as well as for monoquaternization of diaryl aldazines to salts of type **7b** (17b). Remarkable is the synthesis of aldehydes **10** via hydrolysis of perchlorates **9** that have been prepared from 3-aminopropenimides of type **8** with methyl iodide in the presence of perchloric acid (18).

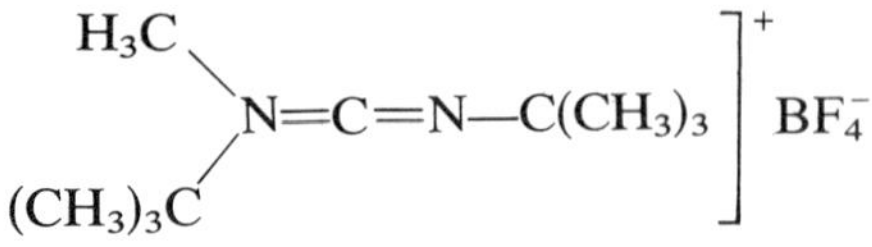

$$\left[\begin{array}{c}H_3C\\ \diagdown\\ N=C=N-C(CH_3)_3\\ \diagup\\ (CH_3)_3C\end{array}\right]^+ BF_4^-$$

(7a)

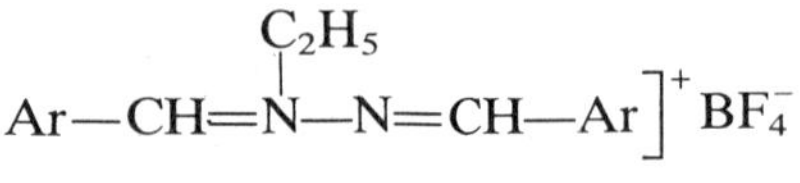

$$\left[Ar-CH=\underset{\underset{C_2H_5}{|}}{N}-N=CH-Ar\right]^+ BF_4^-$$

(7b)

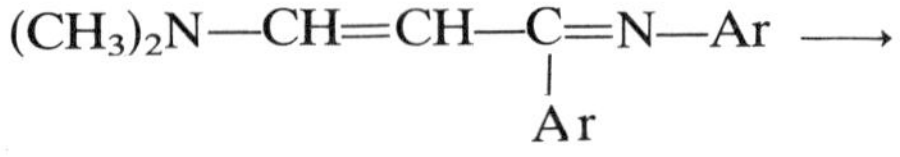

$$(CH_3)_2N-CH=CH-\underset{\underset{Ar}{|}}{C}=N-Ar \longrightarrow$$

(8)

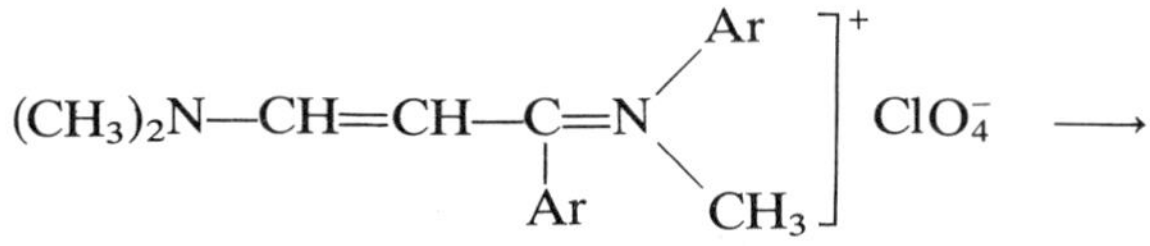

$$\left[(CH_3)_2N-CH=CH-\underset{\underset{Ar}{|}}{C}=N\diagup^{Ar}_{CH_3}\right]^+ ClO_4^- \longrightarrow$$

(9)

$$\underset{O}{\overset{H}{\diagdown}}C-CH=\underset{\underset{Ar}{|}}{C}-N\diagup^{Ar}_{CH_3}$$

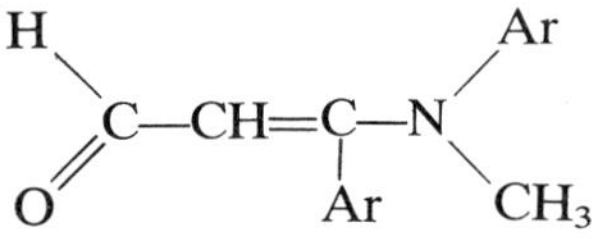

(10)

In connection with this, addition reactions of amide chlorides with imines should also be mentioned; these are discussed in chapter 12. An interesting example is the formation of **13** from benzalaniline (**11**) and dimethylchloromethyleniminium chloride (**12**) (19):

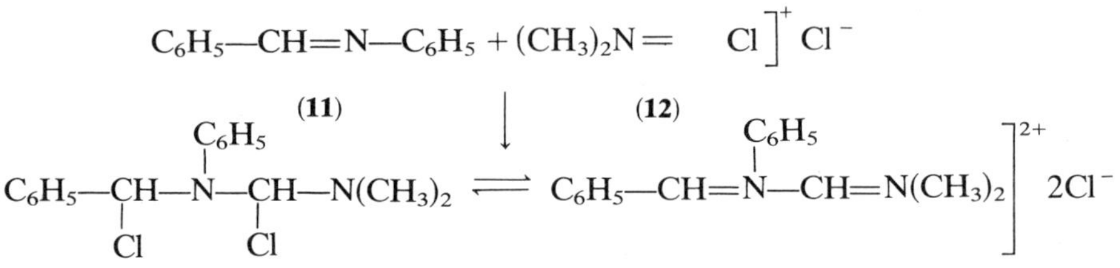

$$C_6H_5-CH=N-C_6H_5 + \left[(CH_3)_2N=Cl\right]^+ Cl^-$$

(11)   (12)

$$\underset{\underset{Cl}{|}}{C_6H_5-CH}-\underset{\underset{C_6H_5}{|}}{N}-\underset{\underset{Cl}{|}}{CH}-N(CH_3)_2 \rightleftharpoons \left[C_6H_5-CH=\underset{\underset{C_6H_5}{|}}{N}-CH=N(CH_3)_2\right]^{2+} 2Cl^-$$

(13)

3. Acylation

N-α-haloalkyl amides are formed by acid halide addition to aldimines. Generally these are described as compounds having covalent structure characteristics, although in reaction mixtures N-acyliminium ions may serve as intermediates (4–6,19a). For example, addition of benzoyl chloride to benzylidenemethylamine (**15**) (R = CH$_3$) yields **14** as a liquid that can be distilled and hydrolyzed to give equimolar amounts of hydrogen chloride, benzaldehyde, and N-methylbenzamide (20a). Analogous additions to formaldimines or ketimines have been performed with several types of acyl halides (20b–20d). With cyanoacetyl chloride and **15** (R = CH$_3$, C$_6$H$_5$) the analogous N-α-chlorobenzylamides **16** (21) are formed. By phosgene or chlorocarbonyl isocyanate addition to imines compounds **17** and **18** (22,23), and from dimethylketazine with ethyl chloroformate compound **19** were prepared (24). These compounds are of synthetic utility, especially for cyclization reactions (see Section IV-I-2 and IV-I-4-a). Cyclic imines and acid chlorides react in a similar way, for example, formation of compounds of type **20** from 3,3-dimethylindolenine by addition of benzoyl chloride (25,26) and other acyl chlorides (27).

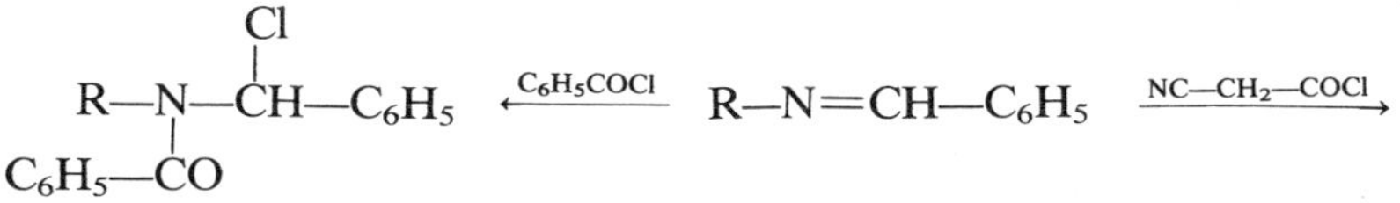

$$\underset{(14)}{\underset{C_6H_5-CO}{R-N-\overset{Cl}{\overset{|}{C}H}-C_6H_5}} \xleftarrow{C_6H_5COCl} \underset{(15)}{R-N=CH-C_6H_5} \xrightarrow{NC-CH_2-COCl}$$

$$\underset{(16)}{\underset{OC-CH_2-CN}{R-N-\overset{Cl}{\overset{|}{C}H}-C_6H_5}} \qquad \underset{(17)}{\underset{Cl-CO\ R''}{R-N-\overset{Cl}{\overset{|}{C}}-R'}} \qquad \underset{(18)}{\underset{OCN-CO}{R-N-\overset{Cl}{\overset{|}{C}H}-Ar}}$$

$$\underset{(19)}{\underset{OC-OC_2H_5}{(CH_3)_2C=N-N-\overset{Cl}{\overset{|}{C}}(CH_3)_2}}$$

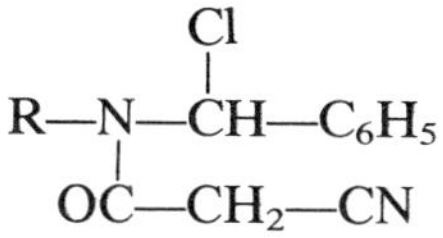

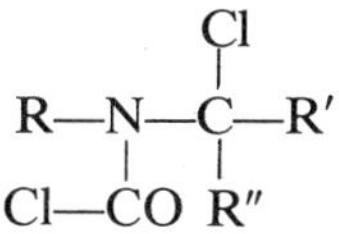

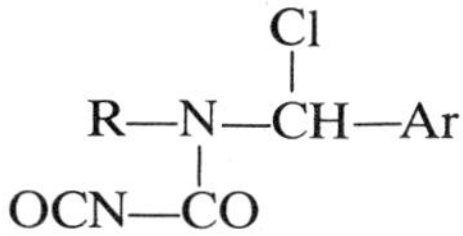

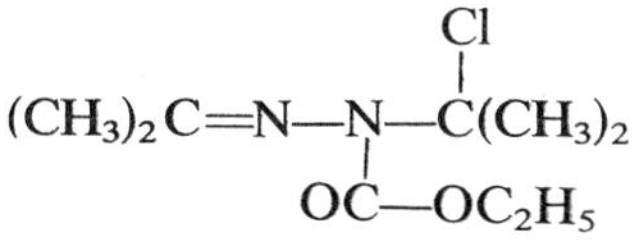

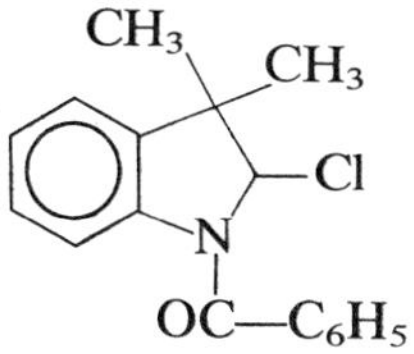

(**20**)

Analogous addition reactions of acid chlorides [e.g., acetyl chloride (28), oxalyl chloride (29), or phosgene (30)] to carbodiimides have also been reported.

4. Halogenation

Preparation of N-bromoiminium bromides (e.g., **21**) has been achieved by means of bromine addition to a cooled solution of an aldimine in carbon tetrachloride (31):

$$\left[\begin{array}{c} H_5C_6 \\ \diagdown \\ N{=}CH{-}C_6H_5 \\ \diagup \\ Br \end{array} \right]^+ Br^-$$

(**21**)

B. REACTION OF ENAMINES

Enamines are vinyl analogues of amines; they can be exemplified by resonance structures **22a** and **22b**. According to the ambident nucleophilic nature of an enamine, electrophiles may attack on nitrogen or on the β-carbon (32,33).

1. Protonation

Protonation of enamines with hydrogen halides (34–39), perchloric acid (40,41), or trifluoroacetic acid (39,42) is a well-known route to iminium salts. Tertiary enammonium salts **23a** can often be isolated as primary products, which undergo rearrangement to iminium salts **24a** more or less rapidly, depending on the structure of the enamine and the nature of the acid (32,43–45). Hydrogen chloride has also been added to N-vinylcarbazole and N-vinylphthalimide to give the N-α-chloroethylcarbazole (**25**) (46) and N-α-chloroethylphthalimide (**31**) (47). With enamides of type **28** hydrogen halides form, in a kinetically controlled reaction, N-α-haloalkylamides **27**, which rearrange to the thermodynamically more stable isomers **29** (48). β-Haloenamines (**30**) and hydrogen halides give, in a reversible reaction, β-haloiminium salts **31**. This reaction is of synthetic utility for the preparation of β-haloenamines because bromine addition to enamines (see Section II-B-4) also yields iminium salts **31** (49).

2. Alkylation

Analogously to protonation, alkylation of enamines can take place either on nitrogen or on the β-carbon, depending on the nature of the starting materials and the reaction conditions. In many cases primarily quaternary enammonium salts **23b** are formed, followed by subsequent rearrangement to C-alkylated iminium salts **24b**. The N-alkylated products can

$$\underset{(22a)}{\overset{|}{\underset{|}{N}}-\overset{|}{\underset{}{C}}=\overset{|}{\underset{}{C}}} \longleftrightarrow \underset{(22b)}{\overset{|}{\underset{|}{N}}=\overset{|}{\underset{}{C}}-\overset{|}{\underset{}{\bar{C}}}}$$

$$\Big\downarrow R^{\oplus}$$

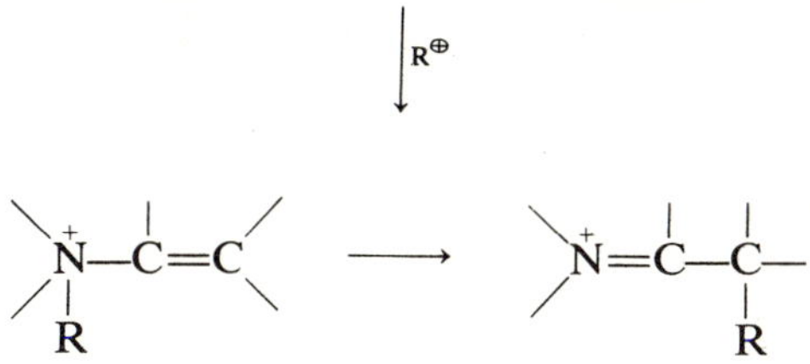

$$\underset{\underset{(23a): R=H}{(23b): R=alkyl}}{\overset{+}{\underset{R}{N}}-C=C} \longrightarrow \underset{(24)}{\overset{+}{\underset{}{N}}=C-\underset{R}{C}-}$$

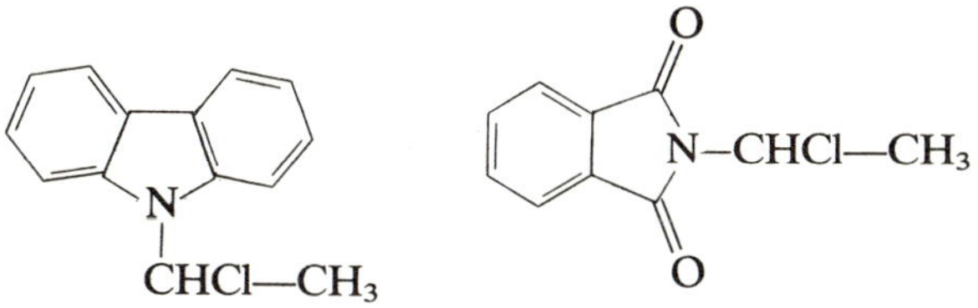

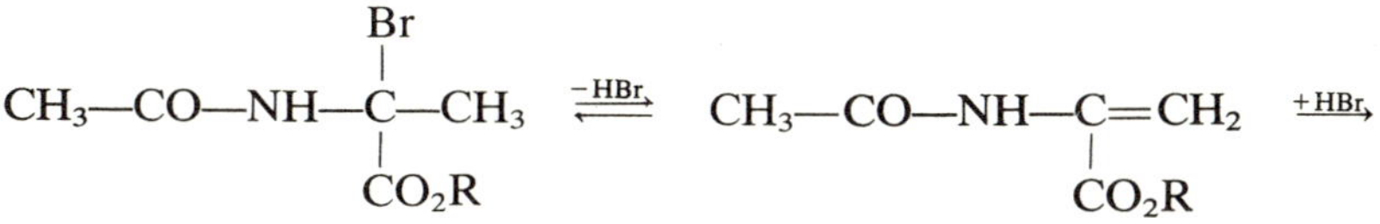

(25) carbazole-N-CHCl—CH$_3$

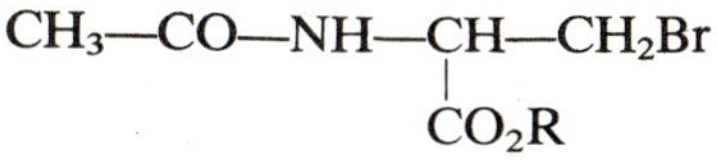

(26) phthalimide-N-CHCl—CH$_3$

$$\underset{(27)}{CH_3-CO-NH-\underset{CO_2R}{\overset{Br}{\underset{|}{C}}}-CH_3} \underset{}{\overset{-HBr,}{\rightleftarrows}} \underset{(28)}{CH_3-CO-NH-\underset{CO_2R}{\overset{}{\underset{|}{C}}}=CH_2} \xrightarrow{+HBr}$$

$$\underset{(29)}{CH_3-CO-NH-\underset{CO_2R}{\overset{}{\underset{|}{CH}}}-CH_2Br}$$

$$\underset{(30)}{R_2N-CR=CHal-R'} \underset{+N(C_2H_5)_3}{\overset{+HHal}{\rightleftarrows}} \underset{(31)}{[R_2N=CR-CHHal-R']^+ Hal^-}$$

undergo either intra- or intermolecular transfer of an alkyl group to carbon (50). C-alkylation appears to occur directly when N-alkylation is not possible for steric reasons or when solvents of high dielectric constant are used. Studies of theoretical considerations and of the reaction conditions that determine whether N- or C-alkylation will take place have shown that the facility of alkylation depends on the basicity of the enamine and the ease of formation of a trigonal atom in the transition state (1).

Alkylations of enamines are of synthetic utility since the reaction leads to the preparation of α-alkylated carbonyl compounds (51–54). Even the addition of carbon tetrachloride to enamine **32** yields iminium salt **33**, which can be hydrolized to give trichloropivalic aldehyde (**34**) (55,56). With trichloroacetic acid, enamines undergo several types of alkylation reactions. Participation of dichlorocarbene or trichloromethyl anion as well as formation and reaction of iminium salts are involved (45,57,58). With dimethylchloromethyleniminium chloride (**12**) enamine **35** yields compound **36** (19).

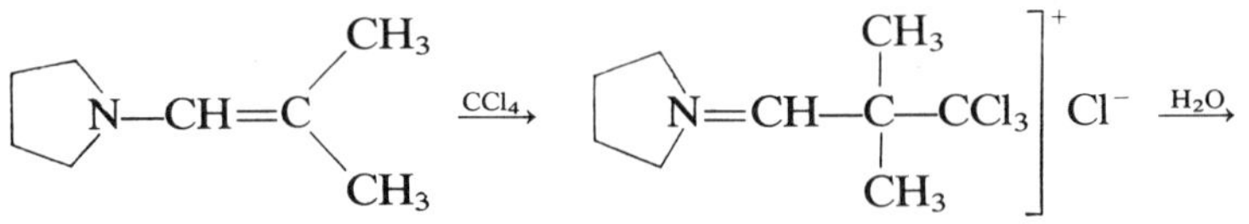

$$\text{(32)} \xrightarrow{CCl_4} \text{(33)} \xrightarrow{H_2O} \text{(34)}$$

$$R_2N-CH=C(CH_3)_2 \quad (35)$$

$$+$$

$$[(CH_3)_2N=CHCl]^+ \; Cl^- \quad (12)$$

$$\longrightarrow$$

$$R_2N-CH-\underset{\underset{Cl}{|}}{\overset{\overset{CH_3}{|}}{C}}-\underset{\underset{Cl}{|}}{\overset{\overset{CH_3}{|}}{CH}}-N(CH_3)_2$$

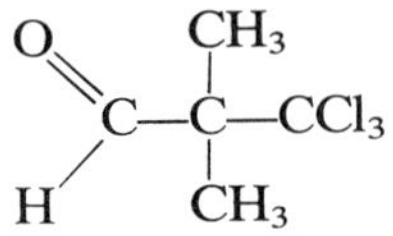

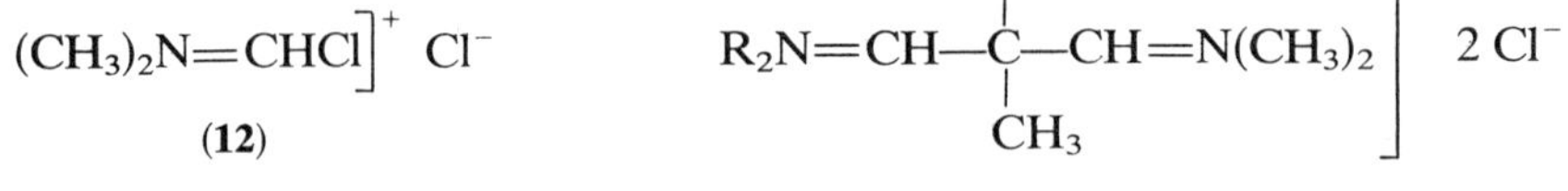

$$\Updownarrow$$

$$\left[R_2N=CH-\underset{\underset{CH_3}{|}}{\overset{\overset{CH_3}{|}}{C}}-CH=N(CH_3)_2\right]^{2+} \; 2\,Cl^-$$

(**36**)

3. Acylation

Acid chlorides that do not undergo ketene formation upon addition to enamines form acyl enamines (32,33,59). In contrast to the N-alkylated enamines, the N-acylated enamines are very unstable. Since they are acylating agents, β-C-acylated products are usually obtained in good yields. From acylation reactions of morpholine or piperidine derivatives

II. PREPARATION OF METHYLENIMINIUM SALTS

37 with 1 equivalent of benzoyl or acetyl chloride iminium salts **38** were isolated as intermediate products (56,58). The structure was elucidated from hydrolysis to **39** and hydrogen cyanide reaction to **40**, as well as lithium alanate reduction to **41**. For acetyl chloride addition to morpholine derivative **37** it has been shown that this acylation does not proceed via ketene formation with subsequent cycloaddition to a cyclobutanone intermediate and ring opening to **38** (60), although this often appears to be the mode of reaction with acid chlorides having an α-hydrogen (32,50).

$$R_2N-CH=C\begin{matrix}CH_3\\ \\CH_3\end{matrix} \xrightarrow{R'-COCl} \left[R_2N=CH-\underset{CH_3}{\overset{CH_3}{C}}-CO-R'\right]^+ Cl^-$$

(37) (38)

H_2O ↙ HCN ↓ $LiAlH_4$ ↘

$$\underset{H}{\overset{O}{C}}-\underset{CH_3}{\overset{CH_3}{C}}-CO-R' \qquad R_2N-\underset{CN}{\overset{H}{C}}-\underset{CH_3}{\overset{CH_3}{C}}-CO-R' \qquad R_2N-CH_2-\underset{CH_3}{\overset{CH_3}{C}}-\underset{OH}{\overset{}{CH}}-R'$$

(39) (40) (41)

$R_2 = $ ⟨cyclohexyl⟩ or ⟨morpholinyl⟩ ; $R' = C_6H_5$ or CH_3

Similarly, from enamines with α,β-unsaturated acyl chlorides the corresponding iminium salts were obtained; these intermediates are synthetically useful for the preparation of cyclohexane-1,3-diones (61), bicyclic or polycyclic β-diketones (62), and dialdehydes (63), as well as adamantane derivatives (64). The reactions are summarized and discussed in ref. 59. Enamine reactions with phosgene (65) and silicon tetrachloride (66) also were carried out to form iminium salts. Vicinal enamines **42** undergo the same type of reaction (65). With acetyl chloride, primarily iminium salt **43** is formed, deprotonation of which by a second endiamine molecule leads to an acylated endiamine (**44**), together with **45** (67):

$$R_2N-CH=CH-NR_2 \xrightarrow{CH_3COCl} \left[R_2N=CH-\underset{CO-CH_3}{\overset{}{CH}}-NR_2\right]^+Cl^- \xrightarrow[-HCl]{+42}$$

(42) (43)

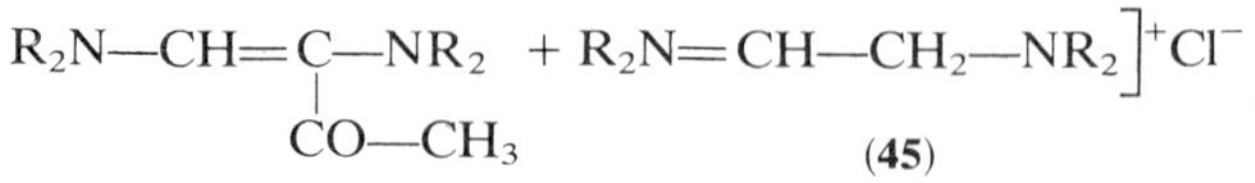

$$\text{R}_2\text{N-CH=C-NR}_2 \quad + \quad \left[\text{R}_2\text{N=CH-CH}_2\text{-NR}_2\right]^+ \text{Cl}^-$$
$$\underset{(44)}{\overset{|}{\text{CO-CH}_3}} \qquad\qquad (45)$$

4. Reactions with Halogens and Inorganic Halides

Bromine can be added to tertiary enamines to form salts of type **46** (68,69a–c). The structure was confirmed by hydrolysis and reaction with Grignard reagents (see Section IV-I-3). Analogous results were achieved with primary and secondary enamines (70). Bromine addition to 1,2-diaminoethenes yields 1,2-bisiminium bromides **47** (71,72), which are also accessible from cleavage reactions of glyoxal bisaminals (see Section II-F-1-c) (73,74):

$$\left[\text{R}_2\text{N=CH-CHBr-R'}\right]^+ \text{Br}^- \qquad \left[\text{R}_2\text{N=CH-CH=NR}_2\right]^{2+} 2\,\text{Br}^-$$
$$\qquad (46) \qquad\qquad\qquad\qquad (47)$$

Bromination of triaminoethenes gives salts of type **48** having an iminium and amidinium function (75); tetraaminoethenes afford bisamidinium salts **49** (76–78):

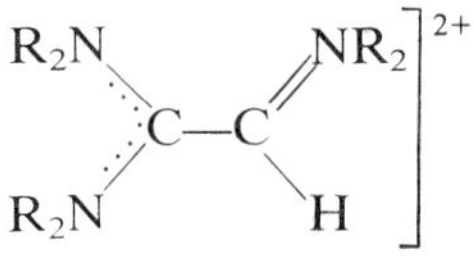

[Structures **(48)** and **(49)** showing C—C with R₂N substituents, 2 Br⁻ counterions]

Enamides also add bromine; for example, N-propenylphthalimide yields compound **50** (79), and N-styrylpyrrolidone gives dibromide **51** (80):

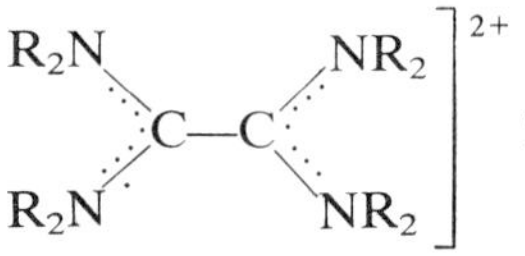

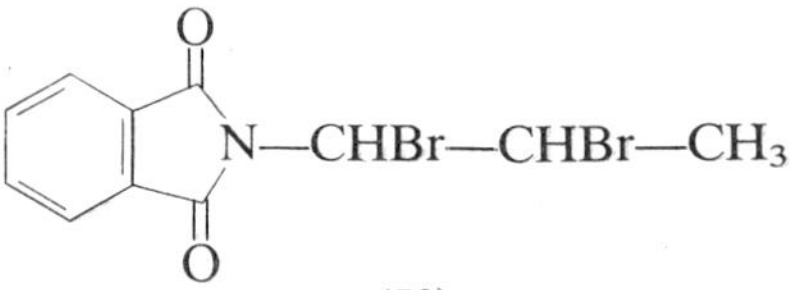

[Structure **(50)**: phthalimide-N—CHBr—CHBr—CH₃]
[Structure **(51)**: pyrrolidone-N—CHBr—CHBr—C₆H₅]

By nitrosyl chloride addition to enamines **52**, compounds of type **53**, which show both nitrosochloroamine and iminium salt nature (81), were prepared:

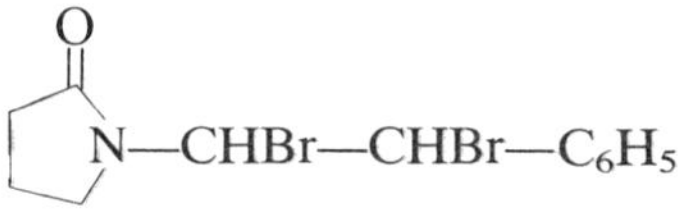

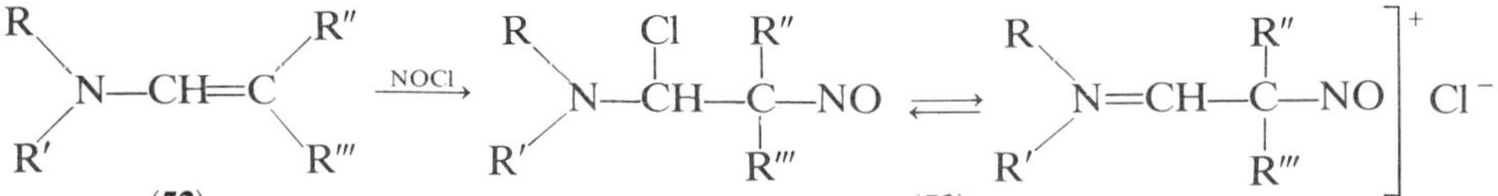

$$\underset{(52)}{\overset{R}{\underset{R'}{\diagdown}}\text{N-CH=C}\overset{R''}{\underset{R'''}{\diagup}}} \xrightarrow{\text{NOCl}} \underset{(53)}{\overset{R}{\underset{R'}{\diagdown}}\text{N-CH-}\overset{\overset{\text{Cl}}{|}}{\underset{\underset{R'''}{|}}{\text{C}}}\text{-NO} \rightleftharpoons \left[\overset{R}{\underset{R'}{\diagdown}}\text{N=CH-}\overset{R''}{\underset{R'''}{|}}\text{C-NO}\right]^+ \text{Cl}^-}$$

Halides of group IV elements of the periodic table react with ketene aminals **54** to form organometallic amidinium salts, for example, the crystalline germanium and tin derivatives **55a** and **55b** (66):

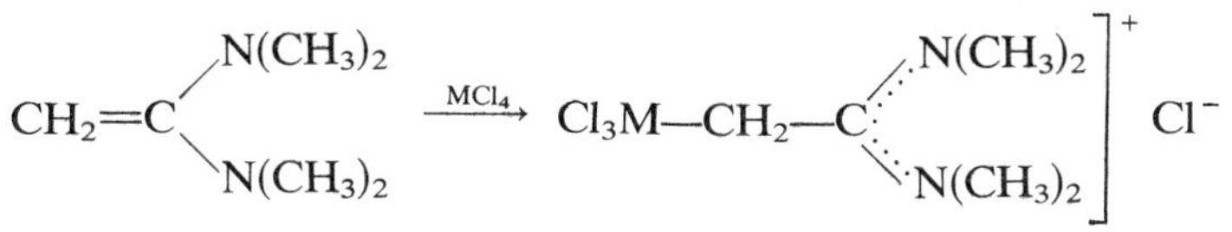

$$CH_2=C\begin{array}{c}N(CH_3)_2\\N(CH_3)_2\end{array} \xrightarrow{MCl_4} Cl_3M-CH_2-C\begin{array}{c}N(CH_3)_2\\N(CH_3)_2\end{array}\Bigg]^+ Cl^-$$

(54)   (55a) M = Ge
       (55b) M = Sn

Enamino ketones **56** undergo reaction with phosphorus pentachloride and other chlorides of acids, the anions of which also represent good leaving groups, to give chloroiminium salts of type **58** (82). The formation of these salts probably proceeds via intermediates **57** and **59**:

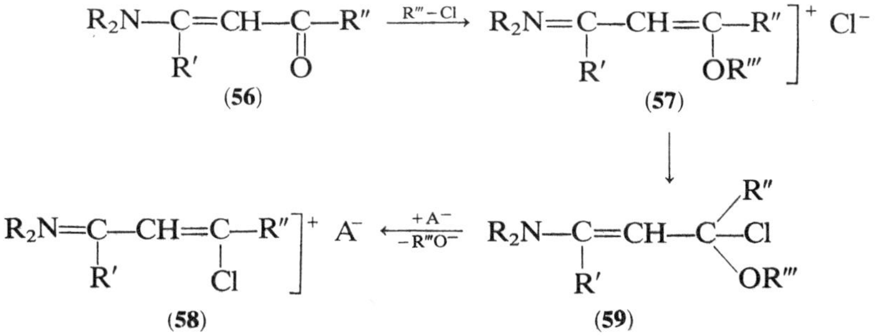

$$R_2N-\underset{R'}{C}=CH-\underset{O}{\overset{}{C}}-R'' \xrightarrow{R'''-Cl} \Bigg[R_2N=\underset{R'}{C}-CH=\underset{OR'''}{C}-R''\Bigg]^+ Cl^-$$

(56)   (57)

$$\Bigg[R_2N=\underset{R'}{C}-CH=\underset{Cl}{C}-R''\Bigg]^+ A^- \xleftarrow[-R'''O^-]{+A^-} R_2N-\underset{R'}{C}=CH-\underset{OR'''}{\overset{R''}{C}}-Cl$$

(58)   (59)

R'''—Cl = PCl$_5$, Cl$_3$CCOCl; A = Cl, ClO$_4$

C. CYANIDE ELIMINATION FROM α-AMINONITRILES

From α-aminonitriles **60** ternary iminium salts **61** have been prepared by means of silver nitrate in dry ether (83), for example, as the most simple symmetrically substituted iminium salt dimethylisopropyliden-iminium nitrate (**61**, R = R' = CH$_3$):

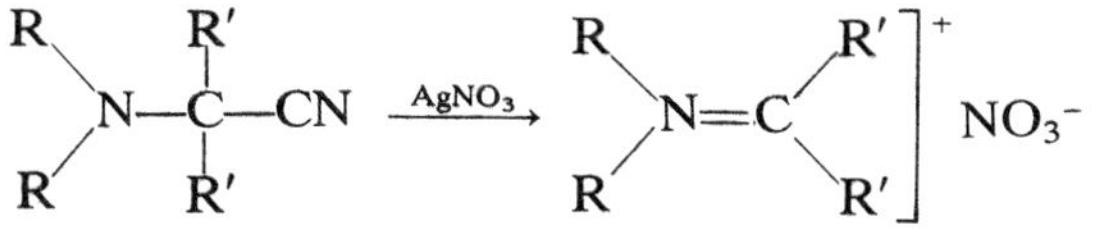

$$\begin{array}{c}R\\R\end{array}\!\!N-\underset{R'}{\overset{R'}{C}}-CN \xrightarrow{AgNO_3} \Bigg[\begin{array}{c}R\\R\end{array}\!\!N=C\begin{array}{c}R'\\R'\end{array}\Bigg]^+ NO_3^-$$

(60)   (61)

D. CONDENSATION OF ALDEHYDES OR KETONES WITH SALTS OF SECONDARY AMINES

Zincke and Würker (84) were the first to treat cinnamaldehyde with methylaniline and hydrogen chloride to form iminium salt **62**, which they characterized as a hexachloroplatinate. By an analogous procedure, using various aldehydes and ketones, a large number of iminium salts having complex anions have been prepared (85,86), for example, hexachlorostannates **63** as well as excellent yields of perchlorates (87), including 92% yield of dimethylisopropylideniminium perchlorate **64** as the initial link in this chain. The reactivities of the secondary amines, however, differ greatly. Reactions are easily achieved with dimethylamine and pyrrolidine, whereas reactions with morpholine, piperidine, and diethylamine are more difficult to achieve (88). With polyenealdehydes as starting materials, vinyl-analogous iminium salts **65** have been obtained (89), but it is necessary to add alkyl orthoformate in order to remove the water being formed. The tetraphenylborates and tetrafluoroborates (90–92) were also isolated. Through intramolecular condensation reactions of aminoketone perchlorates **66**, bicyclic iminium salts **67** have become known as interesting examples in connection with Bredt's rule (93).

Alkylation of diphenylcyclopropenone (**68**) with triethyloxonium fluoroborate gives salt **69**, which undergoes exchange of the ethoxy group with dimethylamine to form cyclopropylideniminium salt **70** (94,95). Analogous salts with alkylmercapto substituents have been described (96).

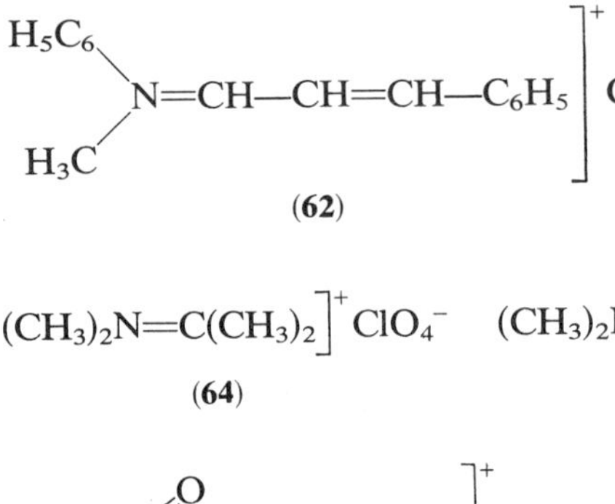

$\left[\begin{array}{c}H_5C_6 \\ \diagdown \\ H_3C \diagup \end{array} N=CH-CH=CH-C_6H_5\right]^+ Cl^-$ $\left[R_2N=C(CH_3)_2\right]^+_2 SnCl_6^{2-}$

(62)   (63)

$\left[(CH_3)_2N=C(CH_3)_2\right]^+ ClO_4^-$ $\left[(CH_3)_2N=CH-(CH=CH)_n-CH_3\right]^+ ClO_4^-$

(64)   (65) n = 1, 2, 3

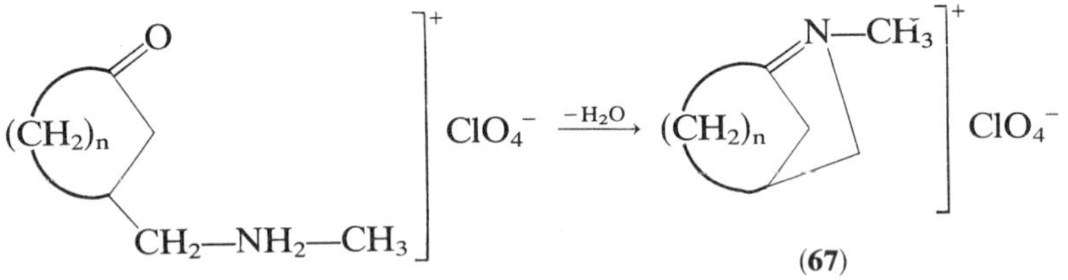

(66) ClO_4^- $\xrightarrow{-H_2O}$ (67) ClO_4^-

n = 4, 5, 9

II. PREPARATION OF METHYLENIMINIUM SALTS

$$\underset{(68)}{\underset{O}{\overset{H_5C_6 \diagdown \diagup C_6H_5}{\triangle}}} \xrightarrow{(C_2H_5)_3O]BF_4} \left[\underset{OC_2H_5}{\underset{(69)}{\overset{H_5C_6 \diagdown \diagup C_6H_5}{\triangle}}} \right]^+ BF_4^- \xrightarrow{(CH_3)_2NH}$$

$$\left[\underset{N(CH_3)_2}{\overset{H_5C_6 \diagdown \diagup C_6H_5}{\triangle}} \longleftrightarrow \underset{N(CH_3)_2}{\overset{H_5C_6 \diagdown \diagup C_6H_5}{\triangle}} \right]^+ BF_4^-$$
(70)

E. HALOGENATION OF N-α-HYDROXYALKYL AMIDES OR IMIDES

Dialkyl-N-α-hydroxyalkyl amines, which are formed by secondary amine addition to aldehydes, are relatively unstable compounds that can be isolated in sufficiently pure form only in the case of amines of low nucleophilicity (97). On the other hand, a large number of stable N-α-hydroxyalkyl amides and imides are known (4) that can be converted by means of inorganic acid or hydrogen halides to N-α-haloalkyl amides or imides. These are characterized by a covalent C–halogen bond (see Section III-A and ref. 5). Only a few examples of this type of compounds will be mentioned here because there already exist several excellent review articles (4–6). Among these compounds are derivatives from primary or secondary amides of aromatic, aliphatic, or heterocyclic carboxylic acids [e.g., **71** (5)], from lactams [e.g., **72**, **73** (98); **74**, **75** (99); **76** (100)], or from imides [e.g., **77** (101); **78** (102); **79**, **80**, **81** (103); **82** (104–106); **83** (107); **84** (108)]. Also N-α-haloalkylcarbamic acid chlorides, esters, and thiol esters have been described [e.g., **85** (109); **86** (110); **87** (111); **88** (112)], including N,N-bischloromethyl carbamates **89** (113). Frequently formaldehyde and benzaldehyde, as well as trichloro- or tribromoaldehydes, were used as oxo components, but phenyl-, furyl-, or thienylglyoxal, glyoxylic acid esters, and ninhydrin [e.g., **90** (114); **91** (115); **91a** (115a); **92** (116–120); **93** (121,122); **94** (123)] were also employed. With strong acids from α-hydroxyalkyl amides, N-acyliminium salts are formed as intermediates that play an important role in α-amido- or α-ureido-alkylation reactions (4–6,124). Cyclic acyliminium salts, (e.g., **95**, **96**) have been prepared from α-hydroxylactams by the action of strong protic or Lewis acids (125).

R—CO—NR′—CHR″—Hal

(71) R = H, alkyl, aryl
R′ = H, alkyl
R″ = H, alkyl, aryl

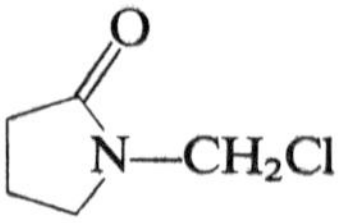

(72) pyrrolidinone-N—CH$_2$Cl

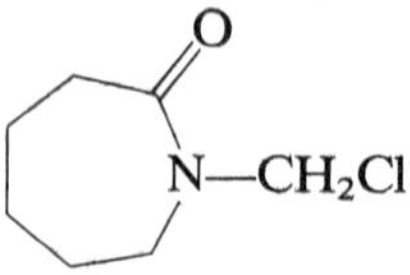

(73) caprolactam-N—CH$_2$Cl

(74) isoindolinone-N—CH$_2$Cl

(75) 3-phenyl-isoindolinone-N—CH$_2$Cl, C$_6$H$_5$

(76) ClCH$_2$—N(piperazinedione)N—CH$_2$Cl

(77) succinimide-N—CH$_2$Cl

(78) phthalimide-N—CH$_2$Cl

(79) tetrahydrophthalimide-N—CH$_2$Cl

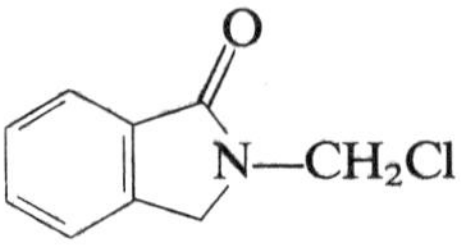

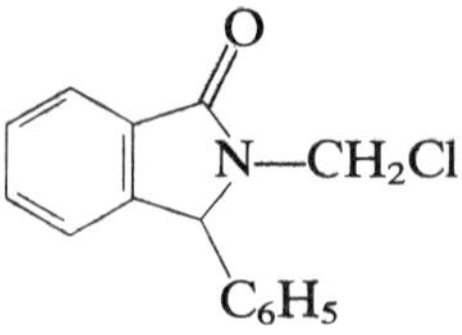

(80) norbornene-dicarboximide-N—CH$_2$Cl

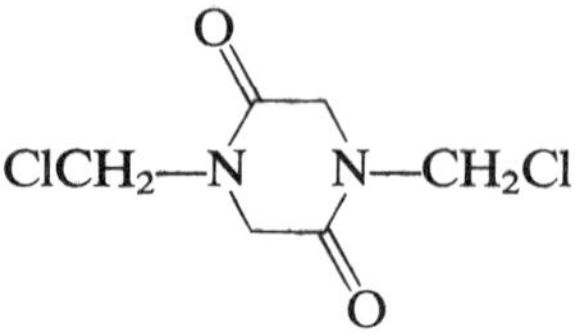

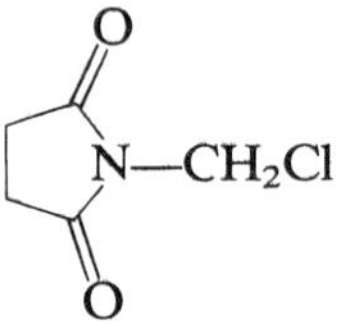

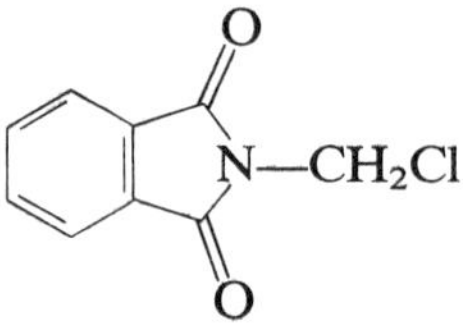

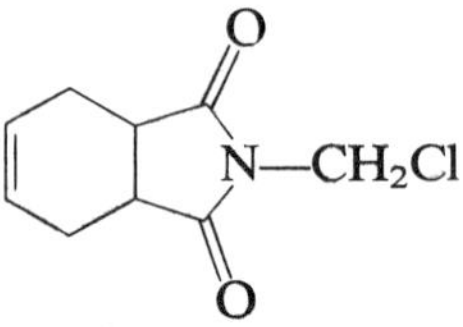

(81) naphthalimide-N—CH$_2$Cl

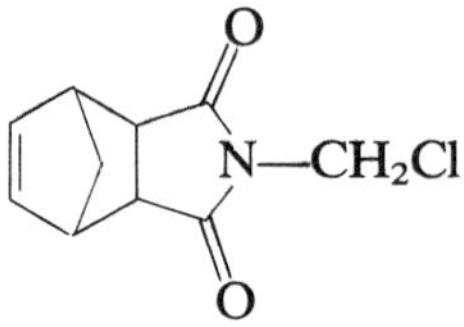

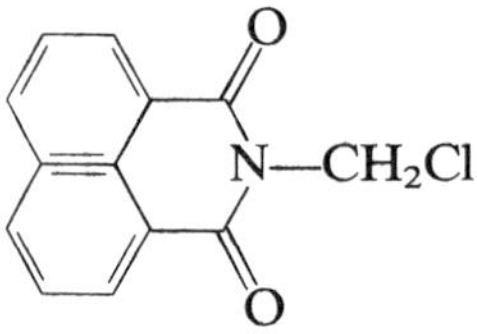

(82) isatin-N—CH$_2$Cl

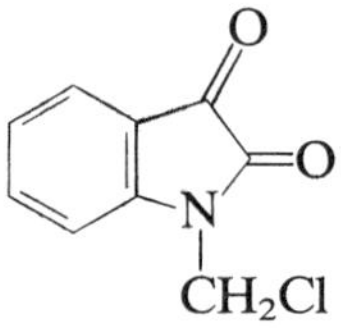

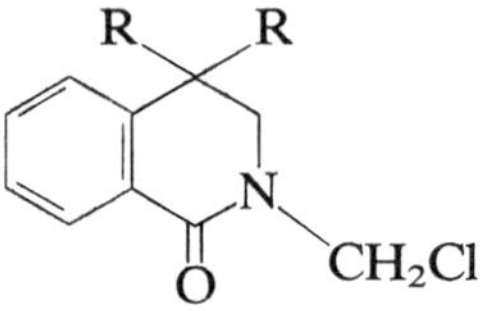

(83) 4,4-R,R-isoquinolinone-N—CH$_2$Cl

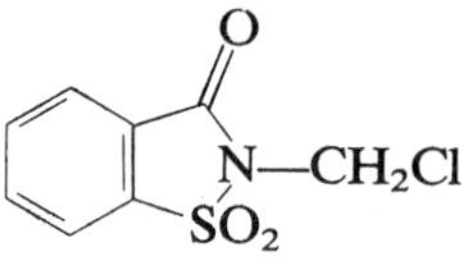

(84) saccharin-N—CH$_2$Cl (SO$_2$)

(85) Cl—CO—NH—CH$_2$Cl

(86) benzoxazolone-N—CH$_2$Cl

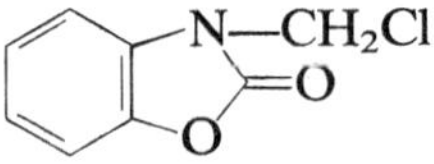

(87) RO—CO—N(R′)—CH$_2$Cl

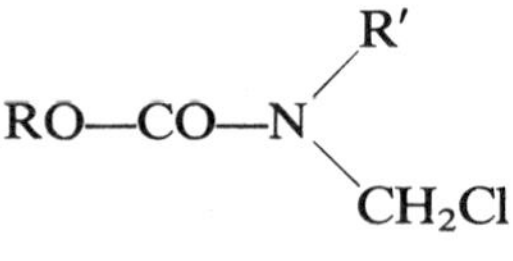

(88) RS—CO—N(R′)—CH$_2$Cl

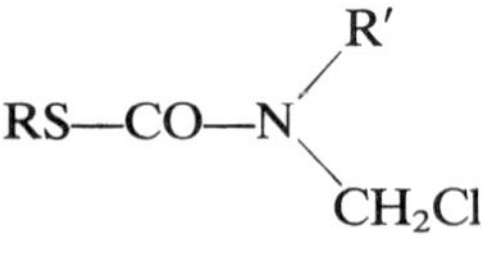

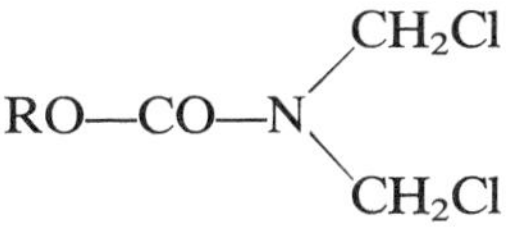

RO—CO—N(CH₂Cl)(CH₂Cl)

(89)

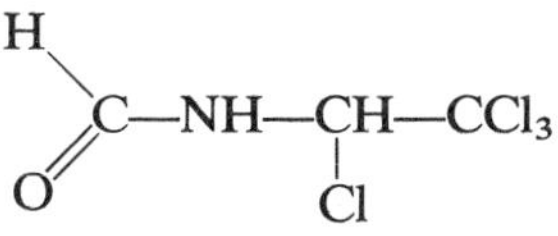

H\C(=O)—NH—CH(Cl)—CCl₃

(90)

(pyrrolidinone)N—CH(Cl)—CCl₃

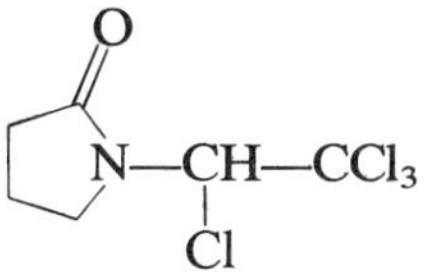

(91)

R—CHCl—CCl₂—CHCl—NH—CO—R′

(91a)

R = CH₃, C₆H₅; R′ = CH₃, C₆H₅, CH₃O

R—CO—CHCl—NH—CO—R

R = C₆H₅, C₄H₃O, C₄H₃S, alkyl—O

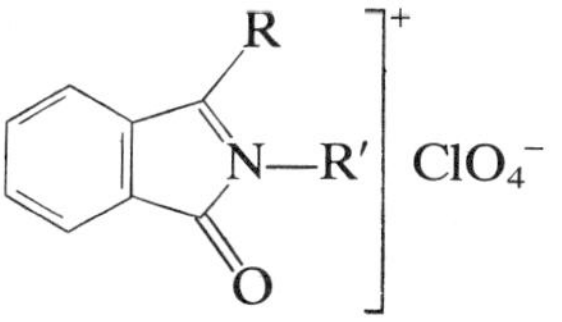

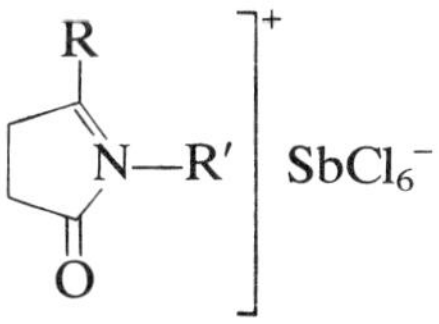

(92)

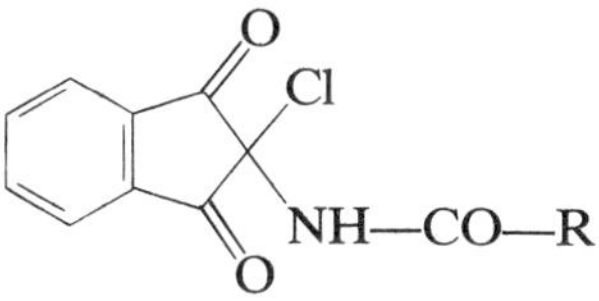

(93) 2-chloro-2-(NH—CO—R)-indane-1,3-dione

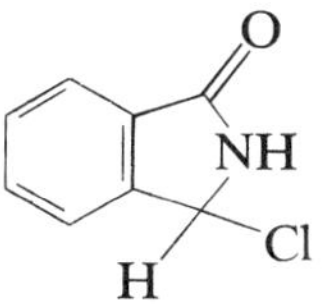

(94) 3-chloro-3H-isoindol-1(2H)-one (with H, Cl at C3, NH)

(95) [R, N—R′ isoindolone cation] ClO₄⁻

(96) [R, N—R′ pyrrolinone cation] SbCl₆⁻

F. CLEAVAGE OF HETEROGEMINALS

Heterogeminals are compounds, such as acetals, mercaptals, or aminals, having two heteroatoms on the same carbon, regardless of whether or not the two heteroatoms are the same and whether they carry hydrogen, alkyl, or acyl groups as further substituents.

1. Aminals

(a) Cleavage with Halogens or Cyanogen Bromide. By chlorination of aminals **97** at −50° *N*-chloroammonium salts of type **98** are formed. These undergo decomposition at room temperature to give *N*-chlorodialkylamines **99** and *N,N*-dialkylmethyleniminium chlorides (**100**) (126,127):

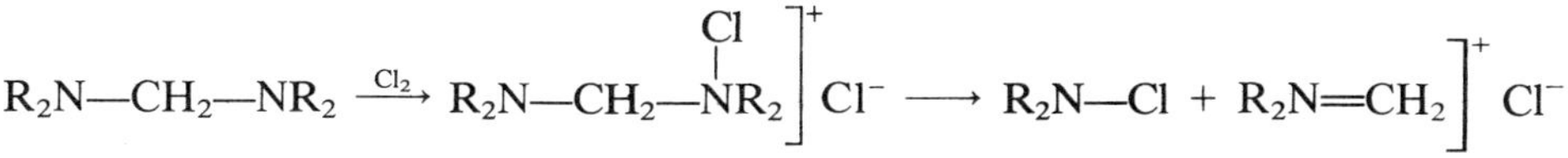

$$R_2N-CH_2-NR_2 \xrightarrow{Cl_2} [R_2N-CH_2-\overset{Cl}{\underset{}{N}}R_2]^+ Cl^- \longrightarrow R_2N-Cl + [R_2N=CH_2]^+ Cl^-$$

(97)   (98)   (99)   (100)

There is spectroscopic evidence for an intermediate analogous to **98** also in the bromination of 1,1-bis(dimethylamino)cyclopropane (**101**) at −65° in liquid sulfur dioxide, under which conditions N-bromodimethylamine and iminium salt **102** are formed (128):

$$(CH_3)_2N\underset{(CH_3)_2N}{\diagup}\!\!\bowtie \xrightarrow[-(CH_3)_2NBr]{+Br_2} \left[(CH_3)_2N\!\!=\!\!\triangleleft\right]^+ Br^-$$

(**101**)   (**102**)

Early in this century von Braun and Röver (129), by adding cyanogen bromide to tetramethylaminal **103**, isolated dimethylcyanamide (**104**) and a crystalline compound tentatively identified as a bisquaternary ammonium salt. Fifty years later this was recognized as N,N-dimethylmethyleniminium bromide (**105**) (126):

$$(CH_3)_2N\!-\!CH_2\!-\!N(CH_3)_2 \xrightarrow{BrCN} (CH_3)_2N\!-\!CN + \left[(CH_3)_2N\!\!=\!\!CH_2\right]^+ Br^-$$

(**103**)   (**104**)   (**105**)

(*b*) *Cleavage with Hydrogen Halides.* Protonation of aminals **97** by the addition of 1 equivalent of hydrogen halide in ether affords monotertiary salts **106** (130–132). From acetonitrile solutions of these salts bistertiary salts can also be isolated on further addition of hydrogen halide (133). However, if the aminal is added to an excess of hydrogen halide in an aprotic, polar solvent such as dimethylformamide or acetonitrile (which is able to dissolve the primarily formed salts), decomposition into dialkylammonium salts **107** and N,N-dialkylmethyleniminium halides **108** takes place (134). Although separation of the salts sometimes appears to be difficult, there are accessible iminium chlorides, bromides, or iodides with different substituents or substituents of the same kind on nitrogen [e.g., **109** (135)], as well as derivatives of benzaldehyde- or isobutyraldehydeaminals [e.g., **110** or **111**, respectively (36)]. Iminium salts of type **111** are also formed by protonation of the corresponding enamines (see Section II-B-1).

$$R_2N\!-\!CH_2\!-\!NR_2 \xrightarrow{HHal} \left[R_2N\!-\!CH_2\!-\!\overset{H}{\underset{|}{N}}R_2\right]^+ Hal^- \longrightarrow$$

(**97**)   (**106**)

$$\left[R_2NH_2\right]^+ Hal^- + \left[R_2N\!\!=\!\!CH_2\right]^+ Hal^-$$

                      (**107**)   (**108**)

II. PREPARATION OF METHYLENIMINIUM SALTS

$$\begin{bmatrix} H_3C \\ \diagdown \\ C_6H_5-CH_2 \diagup N=CH_2 \end{bmatrix}^+ Cl^- \quad [R_2N=CH-C_6H_5]^+ Cl^- \quad [R_2N=CH-CH(CH_3)_2]^+ Cl^-$$

(109) (110) (111)

With 1,3,5-trisubstituted hexahydrotriazines (112) or α-tripiperidene (114) as starting materials, cleavage reactions with hydrogen halides led to N-monoalkyliminium salts 113 and 115 (136), which are also accessible via protonation of the corresponding aldimines (see Section II-A-1).

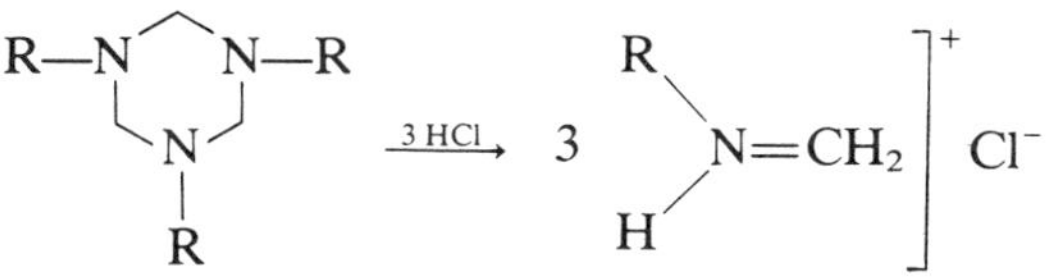

$$\text{(112)} \xrightarrow{3\,HCl} 3 \begin{bmatrix} R \\ \diagdown \\ H \diagup N=CH_2 \end{bmatrix}^+ Cl^-$$

(112) (113)

$$\text{(114)} \xrightarrow{3\,HCl} 3 \begin{bmatrix} \text{piperidinium} \end{bmatrix}^+ Cl^-$$

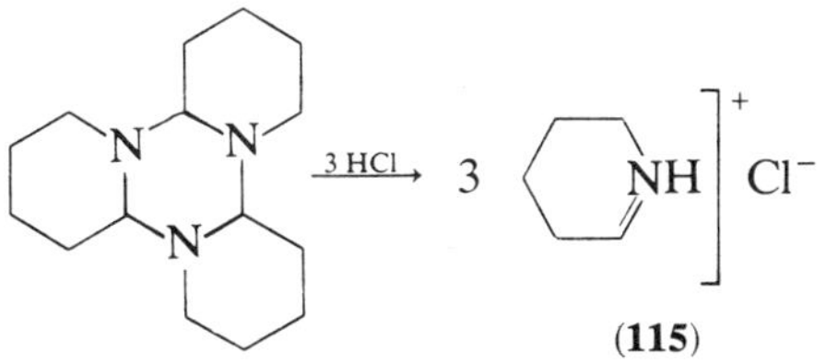

(115)

(114)

(c) *Cleavage with Acyl Halides, Alkyl Chlorocarbonates, Sulfonyl, Sulfinyl, or Sulfenyl Chlorides.* Cleavage of aminals with acyl halides or alkyl chlorocarbonates (137) is the most convenient method for the preparation of iminium salts 108. These reactions, which proceed exothermically and quantitatively, provide iminium salts that are almost analytically pure. Because of their insolubility in inert solvents like ether, these salts precipitate easily, while the second cleavage product (which is an acid dialkyl amide or dialkyl urethane) stays in solution. One can assume that, analogously to protonation of aminals, in the first step an addition product is formed which subsequently undergoes decomposition to an iminium salt (108) and the amide (116):

$$R_2N-CH_2-NR_2 \xrightarrow{R''-CO-Hal} \begin{bmatrix} R''-CO \\ | \\ R_2N-CH_2-NR_2 \end{bmatrix}^+ Hal^- \longrightarrow$$

(97)

$$R_2N=CH_2]^+ Hal^- + R''-CO-NR_2$$

(108) (116)

It appears to be difficult to propose a uniform scheme for the cleavage of unsymmetric aminals. Formation of an iminium salt such as **118** from aminal **117** is preferred if bulky substituents are present on one nitrogen (138,139). Exclusive formation of salt **121** from aminal **120** might be due to the superior stabilization of the iminium ion, compared to that of the corresponding phenyl analogue (see Section II-F-2-b) (140).

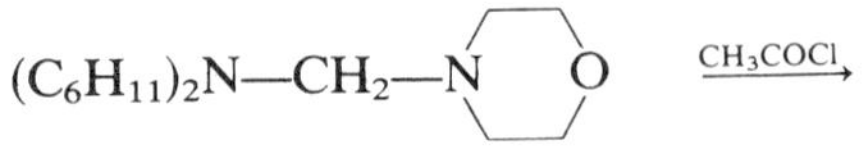

$(C_6H_{11})_2N—CH_2—N$ O
CH_3COCl

(**117**)

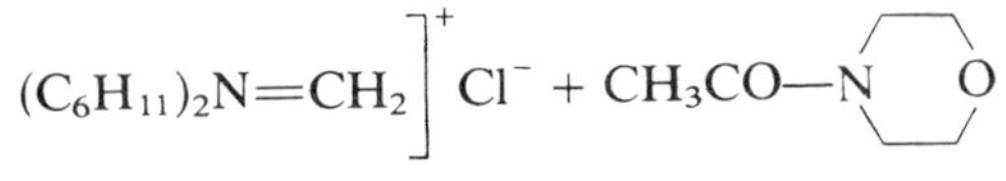

$(C_6H_{11})_2N{=}CH_2$ + Cl^- + $CH_3CO—N$ O

(**118**) (**119**)

Cl
H_3C
$N—CH_2—N$
Cl
Cl
C_6H_5COCl

(**120**)

$N{=}CH_2$ + Cl^- + Cl
Cl
CH_3
$N—CO—C_6H_5$
Cl

(**121**) (**122**)

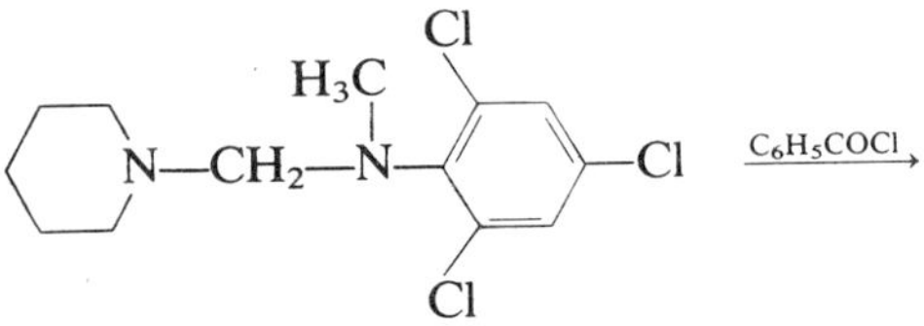

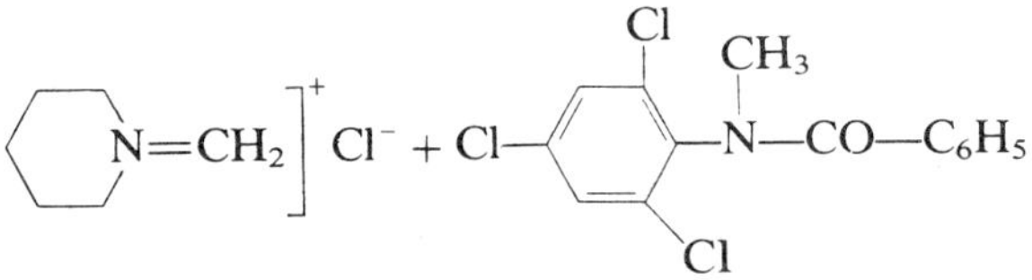

Unsymmetrical aminals from propylenimine and dicyclohexylamine or 2,6-dimethylpiperidine (e.g., **123**) also form, with acetyl chloride, the iminium salts of the more basic dialkyl amines (e.g., **124**), together with 1-acetyl-2-methylaziridine (**125**) (141):

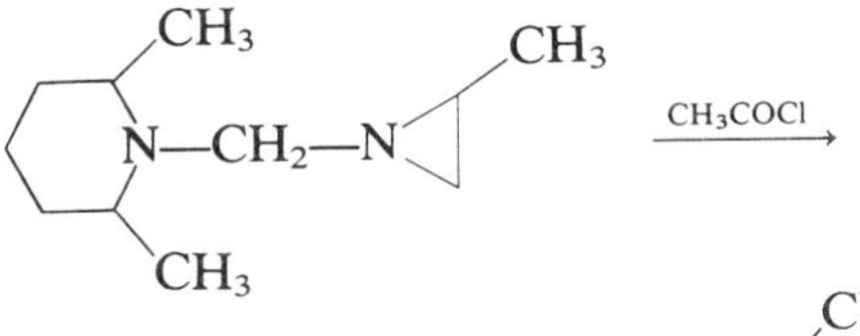

CH_3
CH_3
$N—CH_2—N$
CH_3
CH_3COCl

(**123**)

CH_3
+
$N{=}CH_2$
CH_3
Cl^- + $CH_3CO—N$
CH_3

(**124**) (**125**)

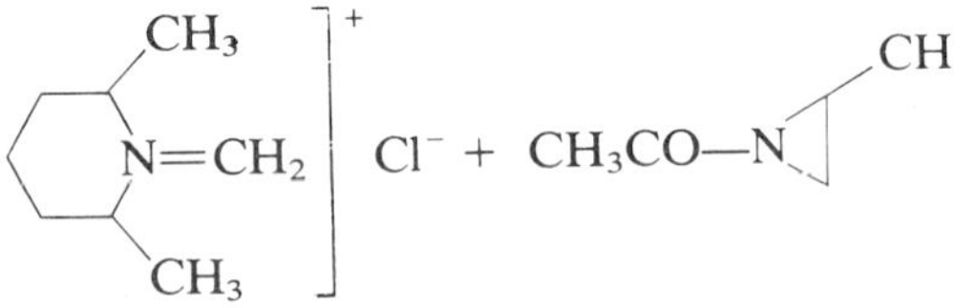

Perhaps for these cases one can assume a dissociation of the unsymmetrical aminal, partly into the iminium cation having the more basic nitrogen, and partly into a secondary amine anion (142). For unsymmetric acylamidomethyl- or diacylimidomethylamines **126**, **129**, and **131**, however, cleavage reactions with acetyl chloride have always afforded benzamidomethyl chloride (**127**), pyrrolidonomethyl chloride (**72**), and phthalimidomethyl chloride (**78**), combined with the formation of dialkyl acetamide (143,144). The same direction of cleavage has been found for N-morpholinomethylalkylnitramines **132**, which form N-chloromethylalkylnitramines **133** in addition to N-acetylmorpholine (**119**) (145). Obviously the structure of the cleavage products determines which of the competing pathways will be favored.

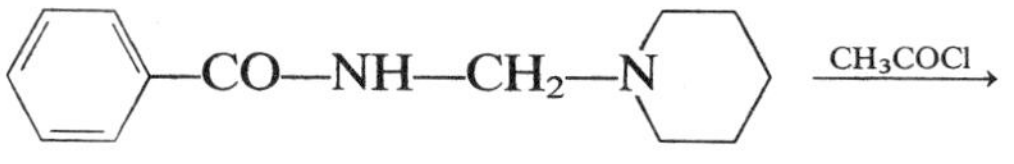

Ph—CO—NH—CH₂—N(piperidine) →[CH₃COCl] Ph—CO—NH—CH₂Cl + CH₃CO—N(piperidine)

(126)  → (127) + (128)

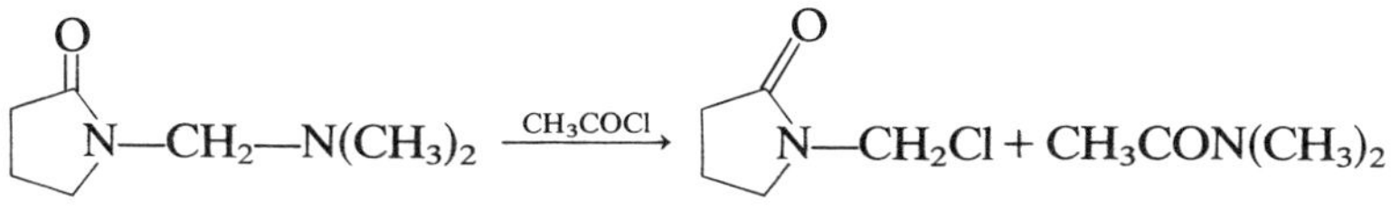

Pyrrolidinone-N—CH₂—N(CH₃)₂ →[CH₃COCl] Pyrrolidinone-N—CH₂Cl + CH₃CON(CH₃)₂

(129) → (72) + (130)

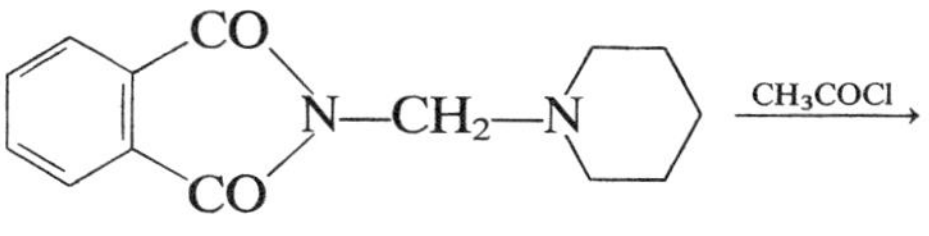

Phthalimide-N—CH₂—N(piperidine) →[CH₃COCl] Phthalimide-N—CH₂Cl + CH₃CO—N(piperidine)

(131) 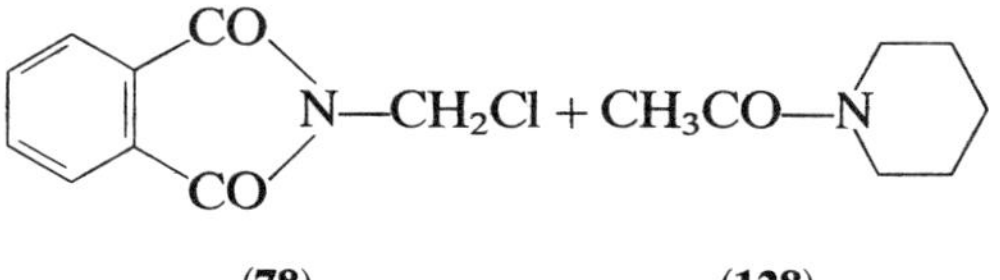 → (78) + (128)

R(O₂N)N—CH₂—N(morpholine) →[CH₃COCl] R(O₂N)N—CH₂Cl + CH₃CO—N(morpholine)

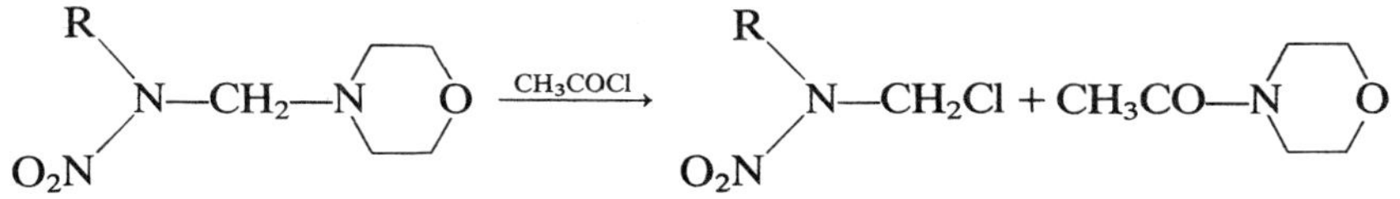

(132) → (133) + (119)

Iminium halides can be prepared as chlorides, bromides, or iodides, depending on the acid halide type. Also acid fluorides are able to cleave aminals. However, instead of salt-like iminium fluorides the fluoromethyl-dialkylamines **134** are formed, which are characterized by a covalent halogen–C bond (146) (see Section III-C). These compounds are similar in reactivity to iminium halides. Sometimes they are even more useful reactants because they can be dissolved in almost all organic solvents and thus react under homogeneous conditions. In the presence of boron trifluoride the cleavage of aminals with an acid fluoride affords dialkyl-methyleniminium tetrafluoroborates **135** (137):

$$R_2N-CH_2F \xleftarrow[-R'-CO-NR_2]{+R'-COF} R_2N-CH_2-NR_2 \xrightarrow[-R'-CO-NR_2]{+R'-COF+BF_3} \left[R_2N=CH_2\right]^+ BF_4^-$$

(134) (97) (135)

By acid halide cleavage iminium salts **136–144** are also accessible; the corresponding aminals are derived from unsubstituted or para-substituted benzaldehyde (137,147–149), as well as from furfural, thiophene carbaldehyde-(2), N-methylpyrrole carbaldehyde-(2) (148), pyridine carbaldehyde-(2) (150), hydratropaldehyde (151), or methyl- or phenyl-glyoxal (152). Finally bisaminals derived from glyoxal, terephthalaldehyde, or diphenyl biscarbaldehyde-(2,2′) (73,74) can also be prepared. It is also possible to have different substituents on nitrogen [e.g., **144** (153)].

$$\left[R_2N=CH-\underset{}{\bigcirc}-R'\right]^+ Hal^- \qquad \left[R_2N=CH-\underset{X}{\bigcirc}\right]^+ Hal^-$$

(**136a**) R′ = H
(**136b**) R′ = CH$_3$
(**136c**) R′ = (CH$_3$)$_2$CH
(**136d**) R′ = CH$_3$O
(**136e**) R′ = CH$_3$S
(**136f**) R′ = (CH$_3$)$_2$N

(**137a**) X = O
(**137b**) X = S
(**137c**) X = NCH$_3$

$$\left[R_2N=CH-\underset{N}{\bigcirc}\right]^+ Hal^- \qquad \left[R_2N=CH-CH\underset{C_6H_5}{\overset{CH_3}{\diagup}}\right]^+ Cl^-$$

(138) (139)

$$\left[R_2N=CH-CO-R'\right]^+ Hal^- \qquad \left[R_2N=CH-CH=NR_2\right]^{2+} 2\ Hal^-$$

(**140a**) R′ = CH$_3$
(**140b**) R′ = C$_6$H$_5$

(141)

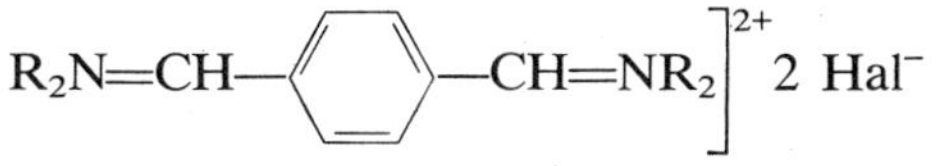

$$[R_2N=CH-\underset{}{\underset{}{\bigcirc}}-CH=NR_2]^{2+} \quad 2\,Hal^-$$

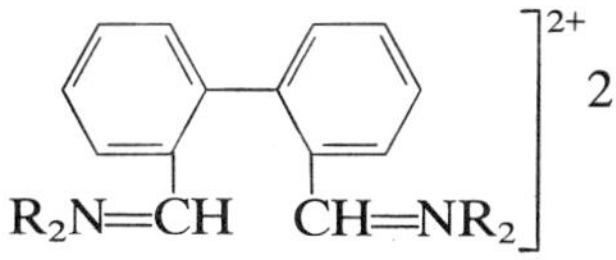

(142)

(143) [biphenyl with $R_2N=CH$ and $CH=NR_2$ substituents]$^{2+}$ 2 Hal$^-$

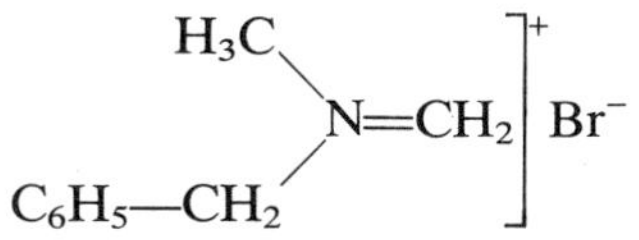

(144) $[\underset{C_6H_5-CH_2}{\overset{H_3C}{\diagdown}}N=CH_2]^+$ Br$^-$

Analogously, the corresponding hydroxylamine and hydrazine derivatives [e.g., **145–147** (97,142,151,154,155)] have been synthesized, as well as covalent trialkylfluoromethylhydrazines [e.g., **148** (156)]:

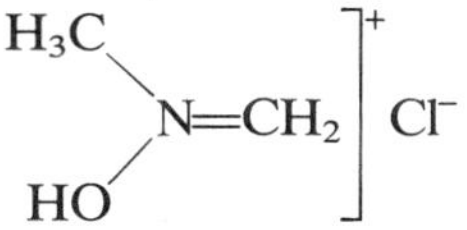

(145) $[\underset{HO}{\overset{H_3C}{\diagdown}}N=CH_2]^+$ Cl$^-$

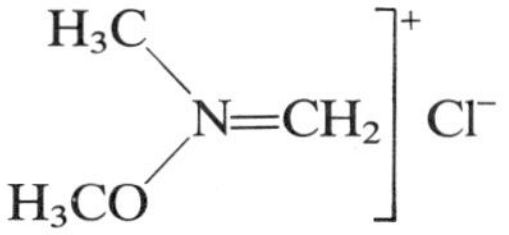

(146) $[\underset{H_3CO}{\overset{H_3C}{\diagdown}}N=CH_2]^+$ Cl$^-$

(147) $[\underset{H_3C}{\overset{R_2N}{\diagdown}}N=CH_2]^+$ Cl$^-$

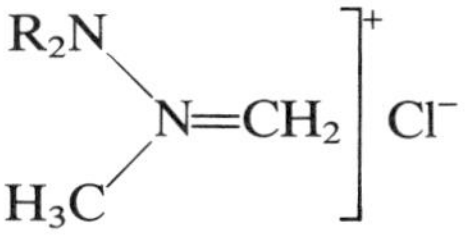

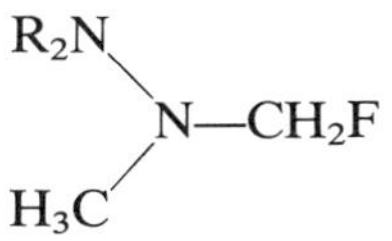

(148) $\underset{H_3C}{\overset{R_2N}{\diagdown}}N-CH_2F$

From cyclic aminals such as 1,3-dialkylimidazolidines **149** or 1,3,5-trialkylhexahydro-*sym*-triazines **112** iminium salts of types **150** (157) and **151** (158) were obtained. For **151**, however, covalent structure characteristics are predominant.

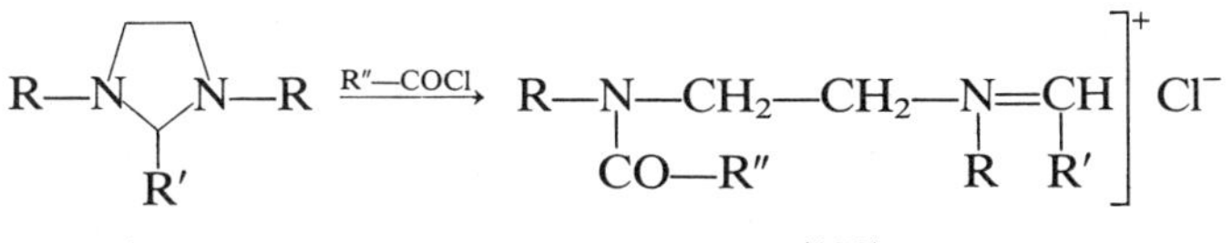

(149) [imidazolidine: R—N, N—R, R'] $\xrightarrow{R''-COCl}$ **(150)** $[\underset{CO-R''}{R-N}-CH_2-CH_2-\underset{R\ R'}{N=CH}]^+$ Cl$^-$

(112) [hexahydrotriazine with three R groups] $\xrightarrow{3\,R'-COCl}$ 3 $[\underset{R}{\overset{R'-CO}{\diagdown}}N=CH_2]^+$ Cl$^-$ $\rightleftarrows$ **(151)** $\underset{R}{\overset{R'-CO}{\diagdown}}N-CH_2Cl$

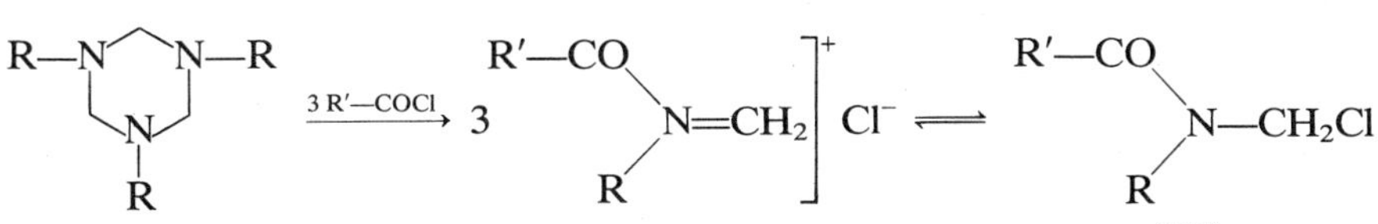

Cleavage of aminals **97** with sulfonic acid chlorides, for example, phenyl or tosyl chloride, to iminium chlorides **100** and sulfonamides **152** also proceeds with excellent yields (159,160):

$$R_2N\text{—}CH_2\text{—}NR_2 \xrightarrow{R'\text{—}SO_2Cl} [R_2N\text{=}CH_2]^+ Cl^- + R'\text{—}SO_2\text{—}NR_2$$
$$(97) \hspace{3cm} (100) \hspace{2cm} (152)$$

Sulfenic acid or sulfinic acid chlorides are also able to cleave aminals to form iminium halides, together with sulfenic or sulfinic acid amides, respectively (144,160).

(d) *Cleavage with Carbonic Anhydrides.* The only example appears to be the cleavage of aminal **103** with trichloroacetic anhydride, which gives a dimethylmethyleniminium trichloroacetate **153** (161). The other carbonic anhydrides that have been tried so far for cleavage of aminals **154** have led in all cases to the unpolar carbonic esters of α-dialkylamino-alkanols **155** (38) (see Section IV-E):

$$(CH_3)_2N\text{—}CH_2\text{—}N(CH_3)_2 \xrightarrow{(CCl_3CO)_2O}$$
$$(103)$$
$$[(CH_3)_2N\text{=}CH_2]^+ CCl_3CO_2^- + CCl_3\text{—}CO\text{—}N(CH_3)_2$$
$$(153)$$

$$R_2N\text{—}CHR^1\text{—}NR_2 \xrightarrow{(R^2\text{—}CO)_2O}$$
$$(154) \hspace{2cm} R_2N\text{—}CHR^1\text{—}O\text{—}CO\text{—}R^2 + R^2\text{—}CO\text{—}NR_2$$
$$(155)$$

(e) *Cleavage with Inorganic Acid Halides and Anhydrides.* In addition to acyl chlorides or alkyl chlorocarbonates phosgene has also been used for aminal cleavage reactions, in which iminium salts **100** and dialkylcarbamoyl chlorides **156** were formed (163). An interesting example is the formation of *N*-chloromethyl-*N*-phenylcarbamoyl chloride (**162**) and phenyl isocyanate from phosgene cleavage of *N,N'*-diphenyl-methylenediamine (**161**) (163). Analogous reactions have been achieved with different inorganic acid halides such as nitrosyl chloride, phosphorus trichloride, and thionyl or sulfuryl chloride; nitrosamines **157**, dialkyl-aminophosphorus dichlorides **158**, dialkylamidosulfinyl chlorides **159**, or dialkylamidosulfonyl chlorides **160** are formed as secondary products (162). In many cases the reactions are difficult to assess because these amide halides are also able to cleave the aminals **97**:

$$COCl_2 \xrightarrow[-100]{+97} R_2N-COCl$$
(156)

$$NOCl \xrightarrow[-100]{+97} R_2N-NO$$
(157)

$$PCl_3 \xrightarrow[-100]{+97} R_2N-PCl_2$$
(158)

$$SOCl_2 \xrightarrow[-100]{+97} R_2N-SOCl$$
(159)

$$SO_2Cl_2 \xrightarrow[-100]{+97} R_2N-SO_2Cl$$
(160)

$$C_6H_5-NH-CH_2-NH-C_6H_5 \xrightarrow{OCCl_2}$$
(161)

$$Cl-CH_2-N\begin{matrix}C_6H_5\\ \\CO-Cl\end{matrix} + C_6H_5-NCO$$
(162)

Nitrosyl perchlorate reacts like nitrosyl chloride (9) and cleaves aminals [e.g., benzaldehyde tetramethylaminal (**163**)] to give iminium perchlorates, such as dimethylbenzylideniminium perchlorate (**164**), which is also accessible from perchloric acid cleavage of the corresponding N,O-acetal (9) (see Section II-F-2-a):

$$(CH_3)_2N-\underset{\underset{C_6H_5}{|}}{CH}-N(CH_3)_2 \xrightarrow[-(CH_3)_2N-NO]{+ONClO_4} [(CH_3)_2N=CH-C_6H_5]^+ ClO_4^-$$
(163) **(164)**

With thionyl chloride, cleavage reactions also succeed for aminals of glyoxylic acid derivatives and yield iminium salts **165** and **166** (164), which are useful reagents for the preparation of amides and esters of N,N-dialkylated α-amino acids (165):

$$[R_2N=CH-CO_2CH_3]^+ Cl^- \qquad [R_2N=CH-CO-NR_2']^+ Cl^-$$
(165) **(166)**

By the action of sulfonic acid anhydrides or tetraalkyl pyrophosphates on aminals **97** iminium salts **167** and **169**, in addition to amides **168** and **170**, were obtained (166):

$$R_2N-CH_2-NR_2$$
$$(97)$$

$+R'-SO_2-O-SO_2-R'$ ↙   ↘ $+(C_2H_5O)_2PO-O-PO(OC_2H_5)_2$

$[R_2N=CH_2]^+ R'SO_3^- + R'-SO_2-NR_2$ $[R_2N=CH_2]^+ (C_2H_5O)_2PO_2^-$
$\quad\quad$ **(167)** $\quad\quad\quad\quad$ **(168)** $\quad\quad\quad\quad\quad$ **(169)**

$$+ (C_2H_5O)_2PO-NR_2$$
$$(170)$$

Reactions of aminals **97** with anhydrides based on two different acids (e.g., **171,172,174–176**) generally afford the iminium salts of the stronger acid (e.g., **167,173,169**), in addition to the amides derived from the lower acidic component (166):

$$R'-SO_2-O-CO-R'' \xrightarrow{+97} R_2N=CH_2]^+R'SO_3^- + R''-CO-NR_2$$
$\quad\quad$ **(171)** $\quad\quad\quad\quad\quad\quad\quad\quad\quad\quad$ **(167)**

$$R'-SO-O-CO-R'' \xrightarrow{+97} R_2N=CH_2]^+R'SO_2^- + R''-CO-NR_2$$
$\quad\quad$ **(172)** $\quad\quad\quad\quad\quad\quad\quad\quad\quad\quad$ **(173)**

$$R'-SO_2-O-SO-R'' \xrightarrow{+97} R_2N=CH_2]^+R'SO_3^- + R''-SO-NR_2$$
$\quad\quad$ **(174)** $\quad\quad\quad\quad\quad\quad\quad\quad\quad\quad$ **(167)**

$$R'-SO_2-O-PO(OC_2H_5)_2 \xrightarrow{+97} R_2N=CH_2]^+R'SO_3^-$$
$\quad\quad$ **(175)** $\quad\quad\quad\quad\quad\quad\quad\quad + (C_2H_5O)_2PO-NR_2$
$\quad\quad\quad\quad\quad\quad\quad\quad\quad\quad\quad\quad$ **(167)**

$$(C_2H_5O)_2PO-O-CO-R' \xrightarrow{+97} R_2N=CH_2]^+(C_2H_5O)_2PO_2^-$$
$\quad\quad$ **(176)** $\quad\quad\quad\quad\quad\quad\quad\quad\quad\quad + R'-CO-NR_2$
$\quad\quad\quad\quad\quad\quad\quad\quad\quad\quad\quad\quad$ **(169)**

In the case of phosphite derivatives secondary reactions take place (166). Treatment of aminals with tetraethyl pyrophosphite yields diethylphosphite amides and dialkylaminomethyldiethyl phosphonates **177**, the formation of which proceeds by a subsequent Michaelis-Arbusow rearrangement of the iminium salts initially formed. This reaction has also been performed starting from iminium halides and triethyl phosphite (162) (see Section IV-H).

$$R_2N-CH_2-NR_2 \xrightarrow[-(C_2H_5O)_2P-NR_2]{+(C_2H_5O)_2P-O-P(OC_2H_5)_2} R_2N=CH_2]^+(C_2H_5O)_2PO^-$$
(97)   (169)

$$\longrightarrow R_2N-CH_2-PO(OC_2H_5)_2$$
(177)

Dichlorosulfane is also able to cleave aminals. Iminium salts **100** and dialkylaminosulfur chlorides **178** are formed; these can cleave a further aminal molecule in the same way (144,160):

$$R_2N-CH_2-NR_2 \begin{matrix} \xrightarrow{SCl_2} R_2N=CH_2]^+ Cl^- + R_2N-SCl \\ (100) \quad\quad (178) \\ \xrightarrow{R_2N-SCl} R_2N=CH_2]^+ Cl^- + R_2N-S-NR_2 \\ (100) \quad\quad (179) \end{matrix}$$
(97)

Cyclic aminals undergo analogous reactions. Trimethylhexahydro-*sym*-triazine (**112**) is cleaved by phosphorus pentachloride in boiling methylene chloride to form methylbis(chloromethyl)amine (**180**) in good yields (167). From a mixture of hexamethylenetetramine and phosphorus pentachloride heated to 80–100°, one can isolate tris(chloromethyl)amine (**181**) (168,169). This compound was prepared earlier by another route (170). The reaction of hexamethylenetetramine with phosgene yields bis(chloromethyl)carbamoyl chloride (**182**) (171). All these compounds appear to possess covalent structures.

   $CH_3-N(CH_2Cl)_2$ $N(CH_2Cl)_3$ $Cl-CO-N(CH_2Cl)_2$

    (**180**)   (**181**)   (**182**)

The C–N bond cleavage of *N,N'*-diphenylmethylenediamine (**161**) to *N*-chloromethyl-*N*-phenylcarbamoyl chloride (**162**) or of 1,3,5-triphenylhexahydro-*sym*-triazine by means of phosgene (163) also yields covalent reaction products.

(*f*) *Cleavage with Alkyl or Aryl Halides.* Alkyl halides and aminals react in ether to form monoquaternary salts [e.g., **183** (132,172)], which can also be prepared from iminium salt **105** and tertiary amines (173). Salts of type **183** are also very reactive; they are easily hydrolyzed and undergo thermal cleavage to an iminium halide and tertiary amine in the first step (174). Further decomposition may be the reason for the failure of attempts to synthesize **105** from salts **183** by this procedure. The more reactive α-haloethers, however, cleave aminals to give iminium salts, for example, **185** and N,O-acetals, in fairly good yields (175).

On the basis of these results, which have been known for 20 years, it is not surprising that reactions of the aminal **103** with 1,4-dibromobutane or 1,5-dibromopentane form quaternary salts **187** (176). However, it is astonishing that the same study (176) resulted in failure to isolate a monoquaternary salt intermediate (**186**) or an iminium halide (**105**).

Other alkylating agents have also been used for aminal cleavage reactions. Spectroscopic evidence has shown that the formation of iminium salt **188** from 1,1-bis(dimethylamino)cyclopropane (**101**) and methyl fluorosulfonate proceeds via a monoquaternary salt with subsequent elimination of trimethylamine (109):

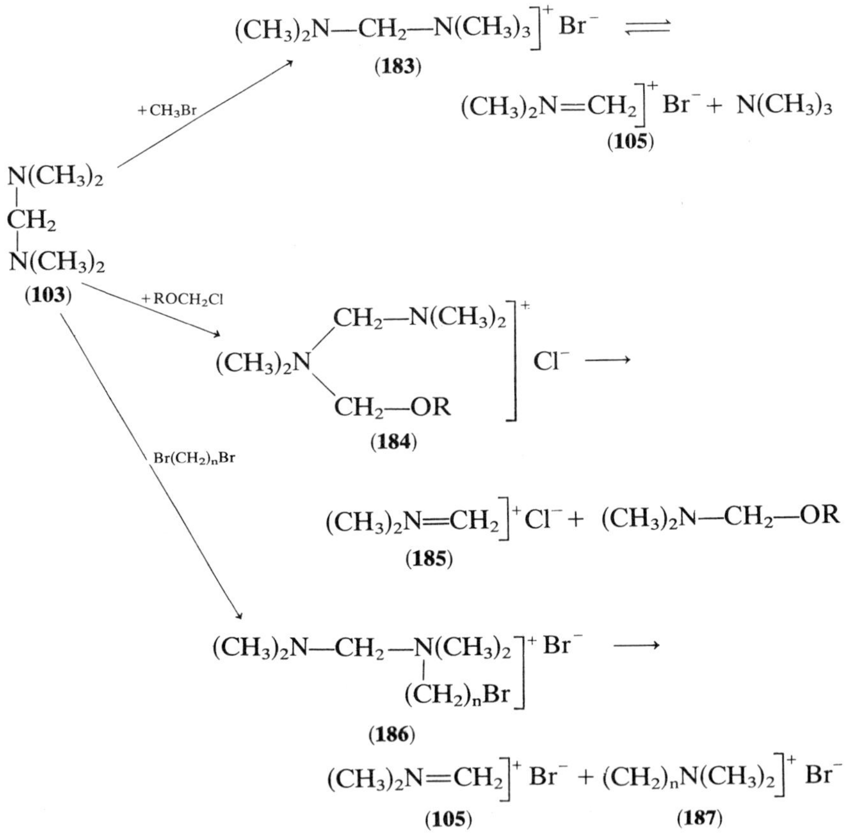

$$[(CH_3)_2N-CH_2-N(CH_3)_3]^+ Br^- \rightleftharpoons$$
(**183**)

$$[(CH_3)_2N=CH_2]^+ Br^- + N(CH_3)_3$$
(**105**)

$$\begin{array}{c} N(CH_3)_2 \\ | \\ CH_2 \\ | \\ N(CH_3)_2 \end{array}$$
(**103**)

+CH$_3$Br

+ROCH$_2$Cl

$$\left[(CH_3)_2N \begin{array}{c} CH_2-N(CH_3)_2 \\ \\ CH_2-OR \end{array} \right]^+ Cl^- \longrightarrow$$
(**184**)

Br(CH$_2$)$_n$Br

$$[(CH_3)_2N=CH_2]^+ Cl^- + (CH_3)_2N-CH_2-OR$$
(**185**)

$$\left[\begin{array}{c} (CH_3)_2N-CH_2-N(CH_3)_2 \\ | \\ (CH_2)_nBr \end{array} \right]^+ Br^- \longrightarrow$$
(**186**)

$$[(CH_3)_2N=CH_2]^+ Br^- + [(CH_2)_nN(CH_3)_2]^+ Br^-$$
(**105**) (**187**)

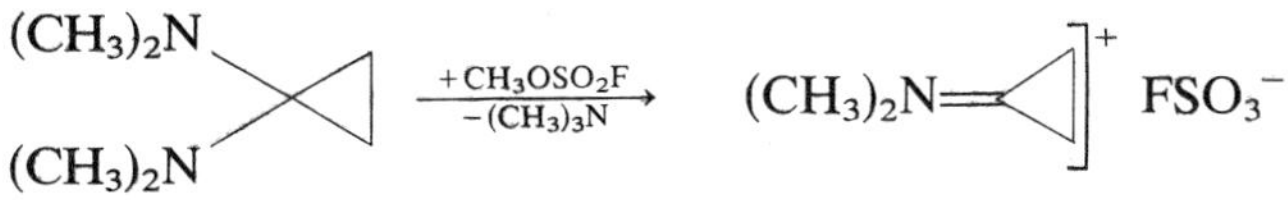

$(CH_3)_2N$\
$(CH_3)_2N$ ⟩⟨ $\xrightarrow[-(CH_3)_3N]{+CH_3OSO_2F}$ $(CH_3)_2N{=}\triangleleft\,]^+$ FSO_3^-

(101) (188)

Among aryl halides that are able to cleave aminals, the 1-halo-2,4-dinitrobenzenes have been used successfully. The reaction of 1-fluoro-2,4-dinitrobenzene (**189**) with bis(dimethylamino)methane (**103**) in nitrobenzene solution yields dimethylfluoromethylamine (**191**) in addition to 2,4-dinitro-*N,N*-dimethylaniline (**190**) in 90% yield (146). Also, with perfluorocyclobutene (**192**) aminal **103** is cleaved to form dimethylfluoromethylamine (**191**) in addition to **193** (177). For reactions of aminals with dihalomethanes see Section II-I.

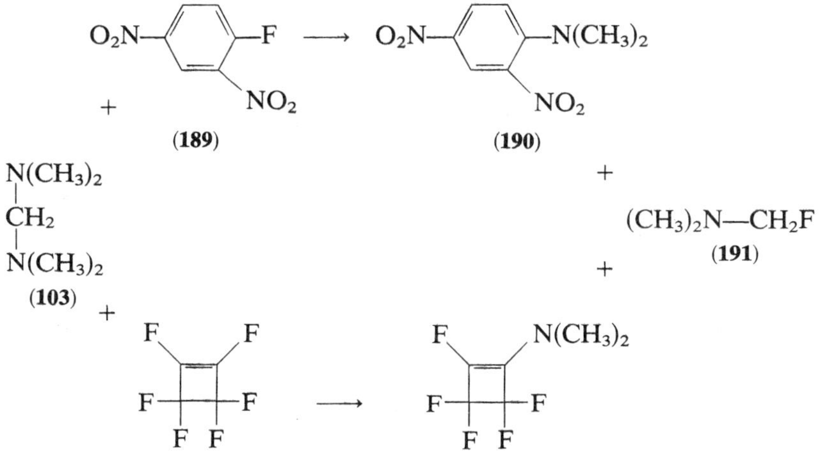

$O_2N{-}\langle\ \rangle{-}F \longrightarrow O_2N{-}\langle\ \rangle{-}N(CH_3)_2$
     NO_2 NO_2
(189)   (190)

+   +

$N(CH_3)_2$
|
CH_2 $(CH_3)_2N{-}CH_2F$
| (191)
$N(CH_3)_2$
(103)   +
      +
   F   F   F   N(CH_3)_2
    \ /   \ /
   F─┼─F   ⟶   F─┼─F
     │   │
    F F   F F
   (192)   (193)

2. Dialkylalkoxymethyl- and Dialkylacyloxymethylamines

(*a*) *Cleavage with Hydrogen Halides and Other Acids.* Equimolar amounts of hydrogen halides and dialkylalkoxymethylamines **194** in etheral solution yield crystalline ammonium salts **195** that are colorless, stable compounds under dry conditions. With an excess of hydrogen halides an oil precipitates, from which crystalline iminium salts **108** are obtained upon heating and drying at 80°C under vacuo. The yields are 80–90% of the salt, which can be purified by recrystallization from acetonitrile (178). This type of cleavage, which has been achieved with hydrogen chloride, bromide, or iodide, can also be applied to α-dialkylalkoxymethylamines with different nitrogen substituents and various alkyl or aryl groups on α-carbon, for example, **196–198** (134,178). An earlier reference (179) describes the isolation of an iminium chloride upon hydrolysis of diethyl aminomethyl isobutyl ether by means of aqueous hydrochloric acid. This result cannot be reproduced, however, and must be doubted because iminium salts are immediately and completely hydrolyzed in aqueous solution.

$$R_2N-CH_2-OR' \xrightarrow{HHal} [R_2\overset{H}{\overset{|}{N}}-CH_2-OR']^+ Hal^- \xrightarrow{-R'OH} [R_2N=CH_2]^+ Hal^-$$
$$(194) \qquad\qquad (195) \qquad\qquad\qquad (108)$$

$$\left[\begin{array}{c}C_6H_5-CH_2\\ \diagdown\\ H_3C\diagup N=CH_2\end{array}\right]^+ Cl^-$$
$$(196)$$

$$\left[\underset{\diagdown\diagup}{O}\overset{\diagup\diagdown}{N}=CH-C_6H_5\right]^+ Cl^- \qquad \left[\underset{\diagdown\diagup}{O}\overset{\diagup\diagdown}{N}=CH-CH(CH_3)_2\right]^+ Cl^-$$
$$(197) \qquad\qquad\qquad (198)$$

Stable ammonium salts have been obtained in ether with perchloric acid and equimolar amounts of dialkylalkoxymethylamines. For compounds having an aryl group attached to the α-carbon (e.g., **199**), the ammonium salt intermediate **200** decomposes spontaneously to give an iminium perchlorate (**164**) (9). This is probably due to the stabilizing effect of a phenyl group on the iminium carbon.

$$(CH_3)_2N-\overset{C_6H_5}{\underset{|}{CH}}-OC_4H_9 \xrightarrow{HClO_4} \left[(CH_3)_2\overset{H}{\underset{|}{N}}-\overset{C_6H_5}{\underset{|}{CH}}-OC_4H_9\right]^+ ClO_4^-$$
$$(199) \qquad\qquad\qquad (200)$$

$$\xrightarrow{-C_4H_9OH} [(CH_3)_2N=CH-C_6H_5]^+ ClO_4^-$$
$$(164)$$

Cyclic N,O-acetals, such as 2-dialkylaminodihydropyrans (**201**), react in the same way. Perchloric or trifluoroacetic acid effect ring opening with the formation of iminium salts **202** (180).

$$\underset{(201)}{\begin{array}{c}R^2\\ R^1 \diagup\diagdown\\ R_2N \diagdown O \diagup CH_3\end{array}} \xrightarrow{H^+} \left[R_2N=CH-\overset{R^1}{\underset{R^2}{\underset{|}{C}}}-CH_2-CH_2-CO-CH_3\right]^+$$
$$(202)$$

II. PREPARATION OF METHYLENIMINIUM SALTS

Cleavage by means of acids is easily achieved also for dialkylacyloxy-methylamines **203**, which are accessible from aminals **97** with acid anhydrides (38). Iminium halides, perchlorates, or trichloroacetates **204** can be prepared.

$$R_2N-CH_2-NR_2 \xrightarrow[-R'CONR_2]{+(R'CO)_2O} R_2N-CH_2-O-COR'$$
$$(97) \qquad\qquad\qquad\qquad (203)$$

$$\xrightarrow[-R'CO_2H]{+HX} [R_2N=CH_2]^+X^-$$
$$X = Cl, ClO_4, CCl_3CO_2$$
$$(204)$$

Analogously hydrogen halide addition to N-acetoxymethylnitramines **205** affords N-chloromethyl derivatives **206**. The fact that these can be distilled confirms a covalent rather than an iminium-salt type of structure (181).

$$\underset{NO_2}{R-N-CH_2O-COCH_3} \xrightarrow[-CH_3CO_2H]{+HCl} \underset{NO_2}{R-N-CH_2Cl}$$
$$(205) \qquad\qquad\qquad\qquad (206)$$

(b) *Cleavage with Acid Halides, Acid Anhydrides, and Cyanogen Bromide.* Cleavage of acetals with acyl halides, reported in 1937 by Post (182), proceeds similarly to the cleavage of aminals, but in consequence of low reaction rates it requires higher temperatures to form α-haloethers and carboxylic acid esters. This is easy to explain if the mechanism in both cases is based on a primary attack of an acyl cation, which of course proceeds preferentially and faster on the more nucleophilic nitrogen. Therefore one should expect that from acyl halide reactions with dialkyl-alkoxymethylamines **194** formation of N,N-dialkylamide (**116**) and α-haloether (**207**) will take place:

$$R_2N-CH_2-OR' \xrightarrow{R''-COHal} \left[\underset{}{R_2\overset{CO-R''}{\overset{|}{N}}-CH_2-OR'}\right]^+ Hal^- \longrightarrow$$

$$(194)$$

$$R''-CO-NR_2 + R'O-CH_2-Hal$$
$$(116) \qquad\qquad (207)$$

However, most of the results reported have demonstrated that the opposite mode of cleavage is preferred. The first example that seems to be known from the literature (183) describes the reaction of 1-methoxy-methyl-3,5-dinitro-1,3,5-triazacyclohexane (**208**) with acetyl chloride. In addition to formation of the *N*-chloromethyl compound **209** (yield 88%), which has been assigned a covalent structure, amide **210** is isolated only as a side product (yield 1%):

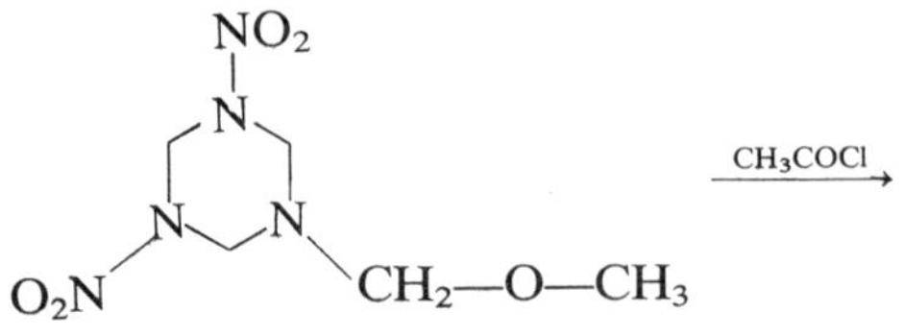

(**208**) $\xrightarrow{CH_3COCl}$

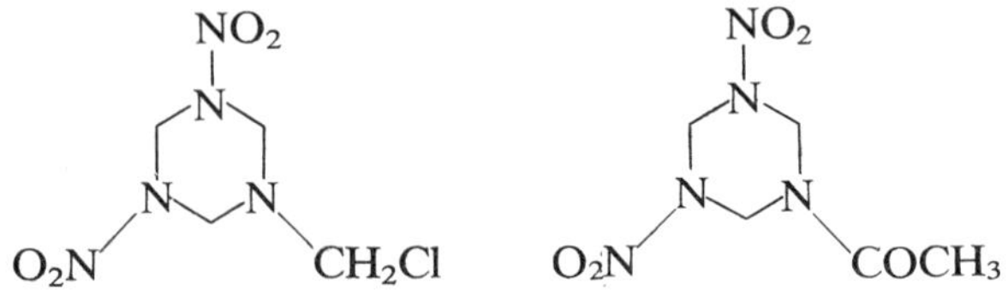

(**209**) (**210**)

From *n*-butyl-α-morpholinoisobutyl ether (**211**), on the addition of acetyl bromide in ether, iminium salt **212** has been prepared in 90% yield (137):

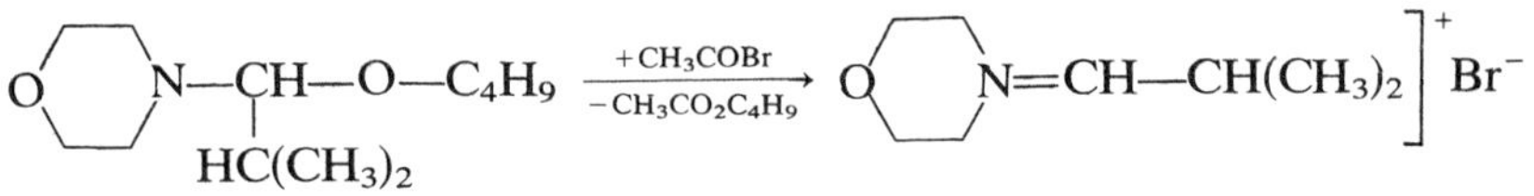

$$O\langle\rangle N-CH(HC(CH_3)_2)-O-C_4H_9 \xrightarrow[-CH_3CO_2C_4H_9]{+CH_3COBr} \left[O\langle\rangle N=CH-CH(CH_3)_2 \right]^+ Br^-$$

(**211**) (**212**)

Also ethoxymethylbis(β-chloroethyl)amine (**213**) and acetyl chloride in etheral solution give crystalline iminium salt **214** (184):

$$(ClCH_2-CH_2)_2N-CH_2-OC_2H_5 \rightarrow [(ClCH_2-CH_2)_2N=CH_2]^+Cl^-$$

(**213**) (**214**)

Analogous results are obtained from α-dialkylamino-α-methoxyacetic acid ester (**215a**) and amides (**215b**) (164) or dimethylaminomethoxymethanephosphonic acid esters (**215c**) (185), which have been cleaved by acetyl or thionyl chloride to give the corresponding iminium salts **216**:

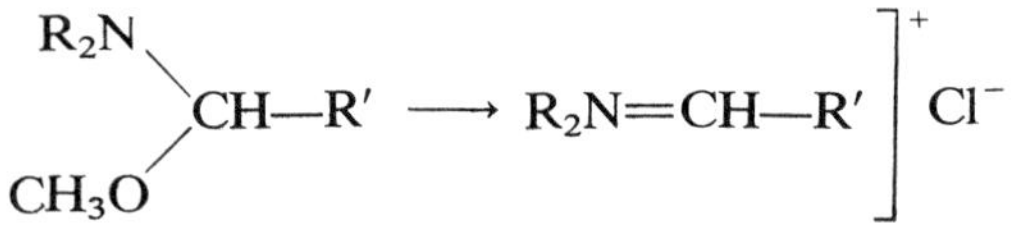

$$\begin{array}{c} R_2N \\ \diagdown \\ CH-R' \\ \diagup \\ CH_3O \end{array} \longrightarrow \left[R_2N=CH-R' \right]^+ Cl^-$$

(**215**) a R' = CO$_2$CH$_3$ (**216**)
b R' = CO—N$\langle\rangle$
c R' = PO(OCH$_3$)$_2$

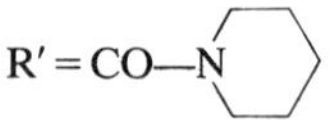

Iminium salts **167**, **169**, and **173** were prepared, together with esters of the cleaving acid component, in good yields from various N,O-acetals, for example from **194** with tetraalkyl pyrophosphates (**217**), with sulfonic-carboxylic acid anhydrides (**171**), with sulfinic-carboxylic acid anhydrides (**172**), with sulfonic-sulfinic acid anhydrides (**174**), with sulfonic-phosphorus acid dialkyl ester anhydrides (**175**), or with carboxylic-phosphorus acid dialkyl ester anhydrides (**176**). In all cases iminium salts of the stronger acid component, together with esters of the weak acid component of the cleaving anhydride, were formed (166).

$(C_2H_5O)_2PO—O—PO(OC_2H_5)_2$ + **194** ⟶

(**217**)
$$R_2N=CH_2]^+ (C_2H_5O)_2PO_2^- + (C_2H_5O)_2PO—OR'$$
(**169**)

$R''—SO_2—O—CO—R'''$ + **194** ⟶

(**171**)
$$R_2N=CH_2]^+ R''—SO_3^- + R'''—CO—OR'$$
(**167**)

$R''—SO—O—CO—R'''$ + **194** ⟶

(**172**)
$$R_2N=CH_2]^+ R''—SO_2^- + R'''—CO—OR'$$
(**173**)

$R''—SO_2—O—SO—R'''$ + **194** ⟶

(**174**)
$$R_2N=CH_2]^+ R''—SO_3^- + R'''—SO—OR'$$
(**167**)

$R''—SO_2—O—PO(OC_2H_5)_2$ + **194** ⟶

(**175**)
$$R_2N=CH_2]^+ R''—SO_3^- + (C_2H_5O)_2PO—OR'$$
(**167**)

$(C_2H_5O)_2PO—O—CO—R''$ + **194** ⟶

(**176**)
$$R_2N=CH_2]^+ (C_2H_5O)_2PO_2^- + R''—CO—OR'$$
(**169**)

Phthalimidomethyl ethyl ether (**218**) and acetyl chloride form *N*-chloromethylphthalimide (**78**) and ethyl acetate (186):

$$\text{Phth-N-CH}_2\text{-O-C}_2\text{H}_5 \xrightarrow{\text{CH}_3\text{COCl}} \text{Phth-N-CH}_2\text{Cl} + \text{CH}_3\text{CO}_2\text{C}_2\text{H}_5$$

(**218**)   (**78**)

Similarly bis(methoxymethyl) derivatives **219a** of 5-ethyl-5-phenyl-barbituric acid have been cleaved by acyl halides in the presence of Lewis acids to give the bifunctional bis(halomethyl) derivatives **219b** (186a):

$$\text{CH}_3\text{O-CH}_2\text{-N}\underset{\text{(219a)}}{\overset{\begin{array}{c}\text{C}_6\text{H}_5 \quad \text{C}_2\text{H}_5\end{array}}{\diagup\!\!\diagdown}}\text{N-CH}_2\text{-OCH}_3 \xrightarrow[\text{Sn(Hal)}_4]{\text{RCOHal}} \text{Hal-CH}_2\text{-N}\underset{\text{(219b)}}{\overset{\begin{array}{c}\text{C}_6\text{H}_5 \quad \text{C}_2\text{H}_5\end{array}}{\diagup\!\!\diagdown}}\text{N-CH}_2\text{-Hal}$$

Cleavage of alkoxymethylnitramines **220** with acetyl chloride in the presence of zinc chloride gives the covalent *N*-chloromethyl derivatives **206**, also in high yields (181):

$$\underset{\text{NO}_2}{\text{R-N-CH}_2\text{-OR}'} \xrightarrow[\text{ZnCl}_2]{\text{CH}_3\text{COCl}} \underset{\text{NO}_2}{\text{R-N-CH}_2\text{Cl}} + \text{CH}_3\text{CO-OR}'$$

(**220**)   (**206**)

Alkylaryliminium salts (e.g., **222a** or **222b**) are accessible by means of acetyl chloride cleavage of the corresponding methylmethoxymethylarylamines **221a** or **221b** (140):

$$\text{R-Ar(R,R)-N(CH}_3)\text{-CH}_2\text{-O-CH}_3 \xrightarrow[-\text{CH}_3\text{CO}_2\text{CH}_3]{+\text{CH}_3\text{COCl}} [\text{R-Ar(R,R)-N(CH}_3)=\text{CH}_2]^+ \text{Cl}^-$$

(**221**)   a: R = Cl   b: R = CH$_3$ (**222**)

Analogous reactions have been performed for the preparation of *N*-halomethyl-*sym*-triazines (187). Cleavage products of diethyl aminomethyl butyl ether (**223**) with catechol-phosphorus acid chloride (**224**) at 20° are iminium salt **225** and butylcatechol-phosphite (**226**), both of which undergo a Michaelis-Arbusow type of reaction with the formation of a

II. PREPARATION OF METHYLENIMINIUM SALTS

phosphonic acid ester at higher temperatures (see Section IV-H) (188):

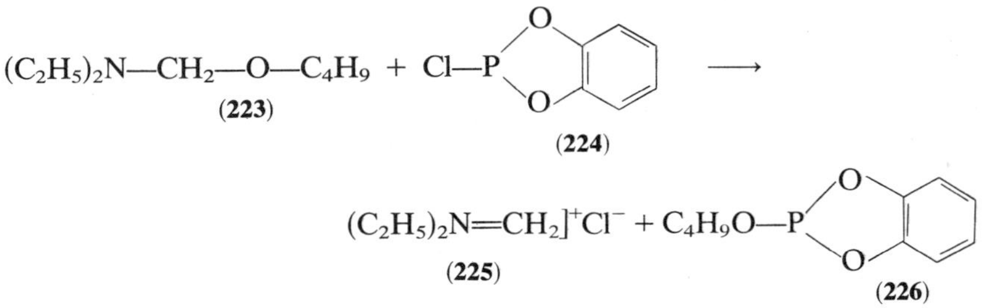

$(C_2H_5)_2N-CH_2-O-C_4H_9$ + Cl—P(catechol) ⟶
(**223**)   (**224**)

$[(C_2H_5)_2N=CH_2]^+Cl^-$ + C_4H_9O-P(catechol)
(**225**)   (**226**)

On the other hand, piperidinomethyl-n-butyl ether (**229**), upon acetyl chloride addition in ether, gave an α-haloether (**228**), together with acetylpiperidine (**128**) (56). Iminium salt **121**, together with n-butyl acetate (**227**), was isolated only in traces. However, on reaction of **229** with phosgene in ether, in addition to piperidine-N-carboxylic acid chloride, iminium salt **121** and α-haloether **228** were obtained in equimolar amounts (162):

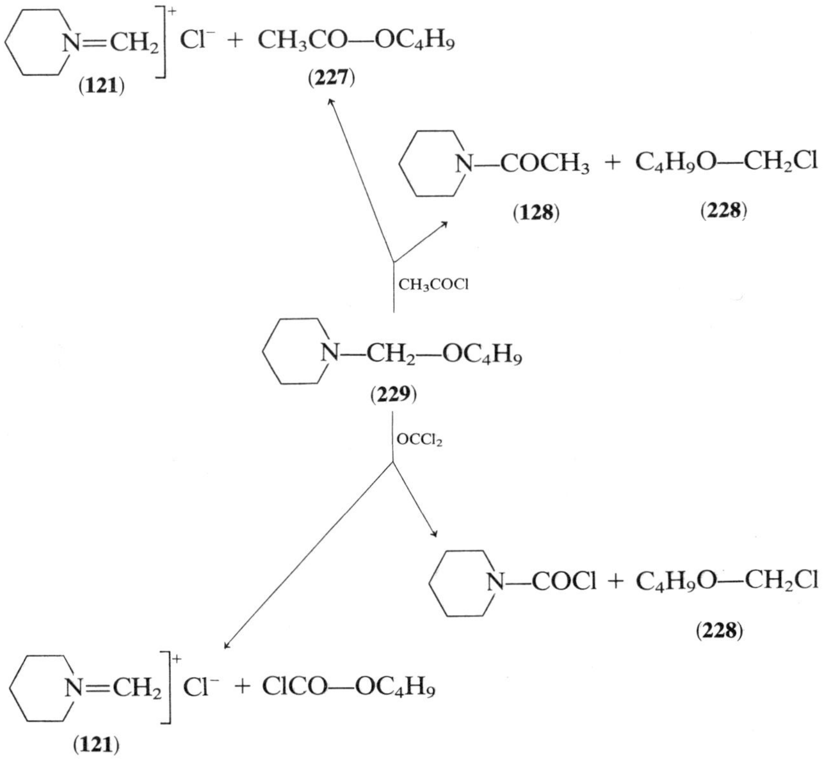

$[\text{piperidine}{-}N{=}CH_2]^+ Cl^-$ + $CH_3CO-OC_4H_9$
(**121**)   (**227**)

piperidine-N—COCH$_3$ + $C_4H_9O-CH_2Cl$
(**128**)   (**228**)

| CH$_3$COCl

piperidine-N—CH$_2$—OC$_4$H$_9$
(**229**)

| OCCl$_2$

piperidine-N—COCl + $C_4H_9O-CH_2Cl$
                                (**228**)

$[\text{piperidine}{-}N{=}CH_2]^+ Cl^-$ + $ClCO-OC_4H_9$
(**121**)

In the patent literature (189) ring opening of 3-alkyloxazolidines **230** with acyl halides has been reported to yield α-haloethers **231**; this has also been confirmed by other investigations (157):

$$CH_3N\text{—}O \xrightarrow{CH_3COCl} CH_3\text{—}N\text{—}CH_2\text{—}CH_2\text{—}O\text{—}CH_2Cl$$
$$\qquad\qquad\qquad\qquad\qquad\quad |$$
$$\qquad\qquad\qquad\qquad\quad CH_3\text{—}CO$$

(230) (231)

Also described have been reactions of cyanogen bromide with dialkylamino ethers **232**, which give dialkylcyanamide (**233**) and α-haloethers **234** in the first step. This is followed by a cleavage of further N,O-acetal **232** by α-haloethers **234**. As final products acetals **235**, as well as iminium salts **236**, are formed (190):

$$R_2N\text{—}CH(C_6H_5)\text{—}OR' \xrightarrow{BrCN} R_2N\text{—}CN + R'O\text{—}CH(C_6H_5)\text{—}Br$$

(232) (233) (234)

$$\mathbf{232 + 234} \longrightarrow C_6H_5\text{—}CH(OR')_2 + [R_2N\text{=}CH\text{—}C_6H_5]^+ Br^-$$

(235) (236)

The examples mentioned above do not permit one to formulate a uniform concept for the action of acyl halides on dialkylamino ethers. Systematic experiments have shown (191) that the nature of the starting N,O-acetals or acid halides, as well as the reaction conditions, especially the type of solvent, will influence the nature of the products that are preferentially formed. In addition, complications arise from the fact that α-haloethers act, with aminals, similarly to acid halides (see Section II-F-1-e).

From dialkylacyloxymethylamines, for example, **237** (see Section II-F-1-d), with acyl chlorides iminium chlorides **100** were obtained in high yields (38). This reaction is again favored because of the possibility of formation of carbiminium ions. Therefore with trichloroacetyl fluoride the iminium trichloroacetate **238**, in addition to acetyl fluoride, is formed instead of a covalent α-fluoroamine

$$R_2N\text{—}CH_2\text{—}O\text{—}CO\text{—}CH_3$$
(237)

CH₃COCl ↙   ↘ CCl₃COF

$[R_2N\text{=}CH_2]^+Cl^- + (CH_3CO)_2O \qquad\qquad [R_2N\text{=}CH_2]^+CCl_3CO_2^- + CH_3COF$

(100) (238)

Analogously to alkoxymethylnitramines **220**, acyloxymethylnitramines **239** are also cleaved by acetyl chloride in the presence of aluminum chloride to give the covalent N-chloromethyl derivatives **206** (192):

$$\underset{NO_2}{R-N-CH_2-O-CO-CH_3} \xrightarrow{\underset{AlCl_3}{CH_3COCl}} \underset{NO_2}{R-N-CH_2Cl}$$

(239)   (206)

3. Dialkylalkylmercaptomethylamines

Hydrogen halide addition to dialkylmercaptomethylamines **241** (see Section IV-F) leads to hydrohalides **240**, which are thermally stable salts that can be sublimed in high vacuum (193,194).

(a) *Cleavage with Acyl Halides and Halogens.* Like aminals and acetals, mercaptals are also cleaved by acyl halides to give α-halosulfides and thiolic acid esters (195). From dialkylaminomethylalkyl or dialkylaminomethylaryl sulfides **241** in etheral solution iminium salts **100**, together with alkyl thiolates **244**, have been prepared in excellent yields by the action of acyl halides (194). For these reactions, as well as for cleavage reactions of **241** with bromine or iodine, which give iminium halides **245** and disulfides **246** (196), one could assume an equilibrium between the intact N,S-acetal **241** and the dissociated form, consisting of the iminium ion **242** and the mercaptide ion **243**:

$$\left[\underset{R_2N-CH_2-SR'}{H}\right]^+ Cl^- \xleftarrow{+HCl} R_2N-CH_2-SR' \rightleftharpoons \left[R_2N=CH_2\right]^+ + R'S^-$$

(240)   (241)   (242)   (243)

$$+R''-COCl \swarrow \qquad\qquad \searrow +I_2$$

$$\left[R_2N=CH_2\right]^+ Cl^- + R''-CO-SR' \qquad 2\left[R_2N=CH_2\right]^+ I^- + R'SSR'$$

(100)   (244)   (245)   (246)

From kinetic hydrolysis studies of ethyl-α-dimethylaminobenzyl sulfide (**247**) it was established (197) that unimolecular carbon–sulfur heterolysis is the rate-determining step in the first stage with the formation of iminium and mercaptide ions:

$$(CH_3)_2N-\underset{C_6H_5}{CH}-SC_2H_5 \longrightarrow \left[(CH_3)_2N=CH-C_6H_5\right]^+ + C_2H_5S^-$$

(247)

The results have been supported by cleavage experiments involving N,S-acetals (e.g., **248**) with Grignard reagents, which gave exclusively tertiary amines (198):

$$R_2N-CH_2-SC_6H_5 \xrightarrow{R'MgCl} R_2N-CH_2-R' + C_6H_5SMgCl$$
(248)

Attempts to cleave N-alkylmercaptomethylamides with acyl halides failed, even upon heating at 60° for many days (186).

G. HYDRIDE ABSTRACTION FROM TERTIARY AMINES

This method was introduced by Meerwein, who obtained iminium salts **249** from reactions of diazonium fluoroborates with tribenzylamine (199):

$$[Ar-N\equiv N]^+ BF_4^- + (C_6H_5-CH_2)_3N \longrightarrow$$

$$[(C_6H_5-CH_2)_2N=CH-C_6H_5]^+ BF_4^- + ArH + N_2$$
(249)

Analogous hydride abstractions from tertiary amines have been performed successfully with trityl fluoroborate, bromide (200), perchlorate, and hexachloroantimonate or with dianisylmethyl perchlorate (201); the corresponding iminium salts (**250**) were obtained:

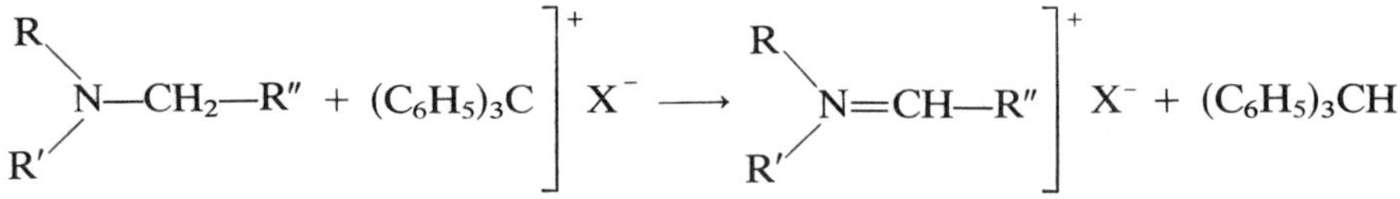

$$\left[\begin{array}{c}R\\ \diagdown\\ N-CH_2-R''\\ \diagup\\ R'\end{array}\right]^+ + (C_6H_5)_3C \quad X^- \longrightarrow \left[\begin{array}{c}R\\ \diagdown\\ N=CH-R''\\ \diagup\\ R'\end{array}\right]^+ X^- + (C_6H_5)_3CH$$

(250a) X = BF$_4$
(250b) X = Br
(250c) X = ClO$_4$
(250d) X = SbCl$_6$

This method is especially useful for the preparation of iminium salts derived from sterically hindered tertiary amines, provided that the iminium ion has a lower hydride acceptor strength than the starting carbenium ion. In addition to salt **249** (199,200), there have also been described such salts as **251** (199), **252**, and **253** (201).

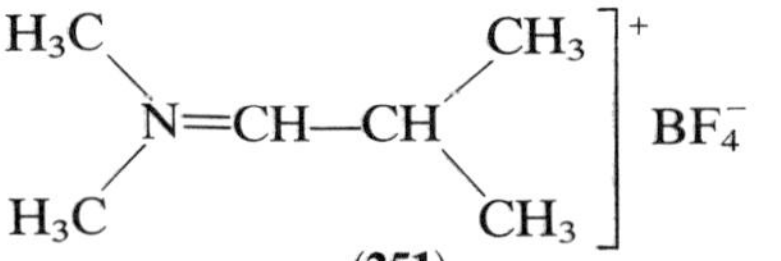

$$\left[\begin{array}{c}H_3C\\ \diagdown\\ N=CH-CH\\ \diagup\\ H_3C\end{array}\begin{array}{c}CH_3\\ \diagdown\\ \\ \diagup\\ CH_3\end{array}\right]^+ BF_4^- \qquad \left[\begin{array}{c}(CH_3)_2CH\\ \diagdown\\ N=CH_2\\ \diagup\\ (CH_3)_2CH\end{array}\right]^+ ClO_4^-$$

(251) 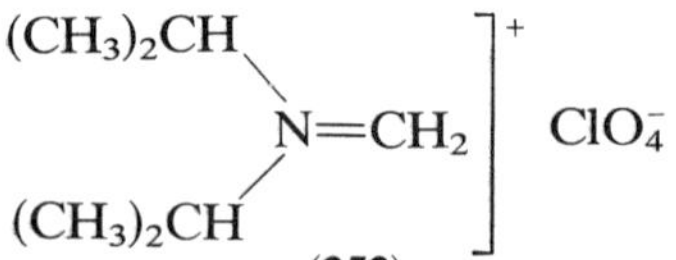 (252)

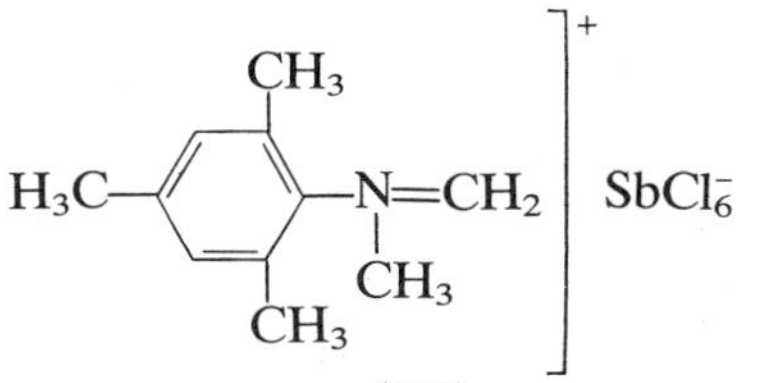

$$[\text{H}_3\text{C}-\underset{\underset{\text{CH}_3}{|}}{\overset{\overset{\text{CH}_3}{|}}{\text{C}_6\text{H}_2}}-\underset{\underset{\text{CH}_3}{|}}{\text{N}}=\text{CH}_2]^+ \text{SbCl}_6^-$$

(253)

Hydride abstraction reactions also lead to iminium salts of nonbenzoid aromatic systems, for example, salts **255**, from tropylidene derivatives **254** (202,203):

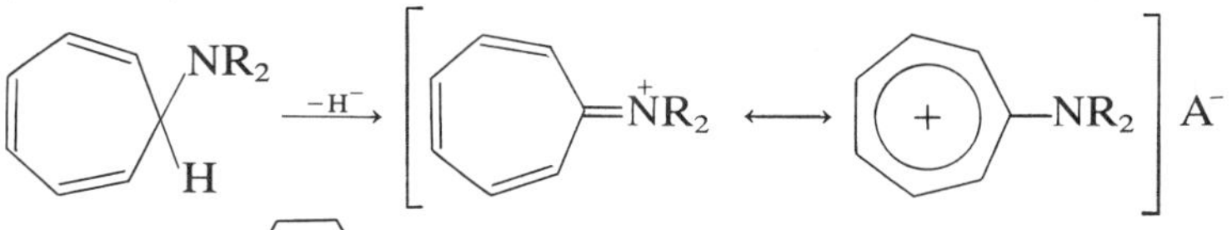

(254) a $R_2 = $ cyclohexyl , $A = ClO_4$ (255)

b $R = C_2H_5$, $A = BF_4$

The first salt of this type, N,N-pentamethylene-2,4,6-cycloheptatrienylideniminium perchlorate (**255a**), was prepared by bromination of 7-piperidinotropylidene (**254a**) followed by anion exchange with perchloric acid (204). Subsequently, for the purpose of hydride abstraction from **254**, for example, trityl fluoroborate was used to form **255b** (202,203).

Mercuric acetate dehydrogenation reactions (see Section II-K) of tertiary amines (e.g., **256**) can be considered formally to be hydride abstractions because of the formation of iminium ion intermediates (e.g., **258**). The mechanism of this oxidative method, which has been used especially for the generation of cyclic carbiminium ions, has been postulated to proceed through proton abstraction from an initially formed mercurated complex (e.g., **257**) with the amine nitrogen (40):

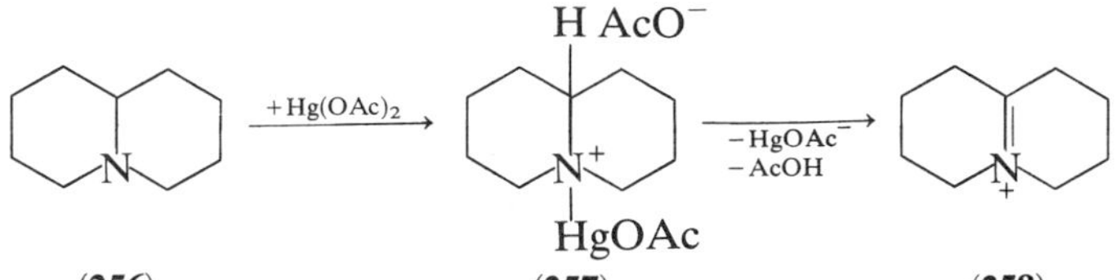

(256)   (257)   (258)

H. REACTIONS OF AMINE OXIDES AND NITRONES

From a mixture of trimethylamine oxide (**261**) and acetic anhydride in chloroform or methylene chloride solution, dimethylaminomethyl acetate

(259) can be isolated (205,206). This has the properties of a covalent compound and is also accessible via acetic anhydride cleavage of bis-(dimethylamino)methane (38). However, trifluoroacetic anhydride forms a crystalline iminium trifluoroacetate (260) (207). Sulfur dioxide addition to trimethylamine oxide also affords a crystalline product, which has been assigned the structure of an iminium salt (262) (208,209). Analogous derivatives are described for dimethylarylmethylamine oxides (210). Trimethylamine oxide, treated with 1 equivalent of acetyl chloride in the presence of triethylamine, gives initially 259 and triethylammonium chloride. On further addition of acetyl chloride to the homogeneous methylene chloride solution, precipitation of the iminium chloride 185 takes place (211).

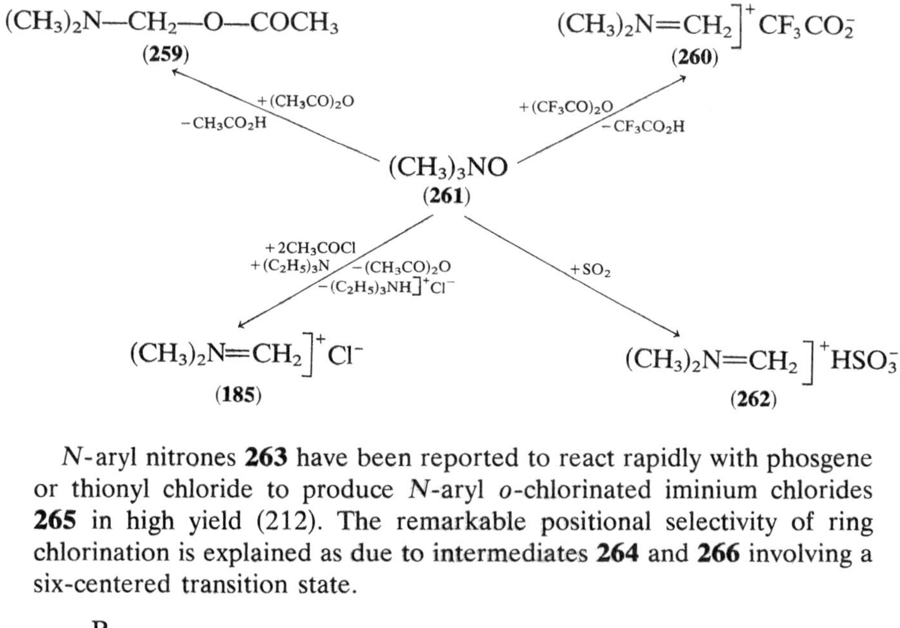

$(CH_3)_2N-CH_2-O-COCH_3$
(259)

$[(CH_3)_2N=CH_2]^+ CF_3CO_2^-$
(260)

$+(CH_3CO)_2O$
$-CH_3CO_2H$

$+(CF_3CO)_2O$
$-CF_3CO_2H$

$(CH_3)_3NO$
(261)

$+2CH_3COCl$
$+(C_2H_5)_3N$
$-(CH_3CO)_2O$
$-[(C_2H_5)_3NH]^+Cl^-$

$+SO_2$

$[(CH_3)_2N=CH_2]^+ Cl^-$
(185)

$[(CH_3)_2N=CH_2]^+ HSO_3^-$
(262)

N-aryl nitrones 263 have been reported to react rapidly with phosgene or thionyl chloride to produce N-aryl o-chlorinated iminium chlorides 265 in high yield (212). The remarkable positional selectivity of ring chlorination is explained as due to intermediates 264 and 266 involving a six-centered transition state.

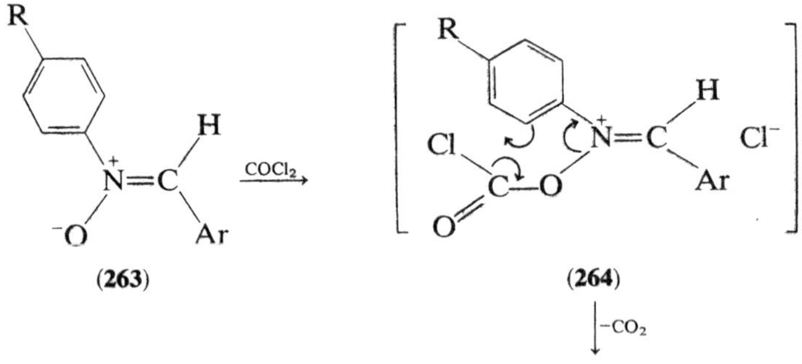

(263)

(264)

$\downarrow -CO_2$

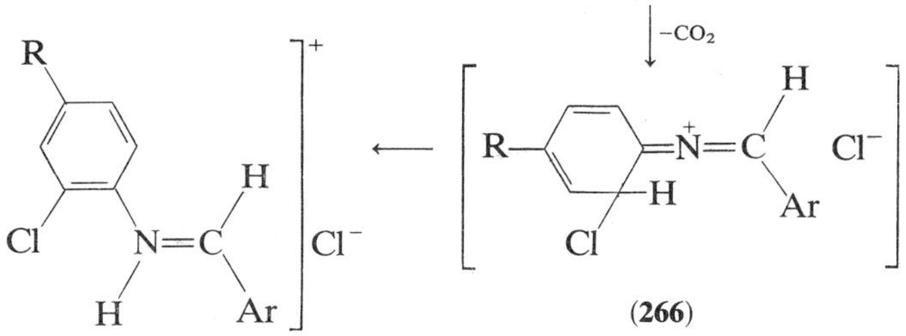

[Structures (265) and (266) with arrow labeled $-CO_2$]

(265) → (266)

I. FRAGMENTATION OF HALOMETHYLAMMONIUM HALIDES

If trimethylbromomethylammonium bromide (**267b**), which can be prepared from trimethylamine and methylene bromide (213), is heated above 160°, fragmentation into methyl bromide and dimethylmethyleniminium bromide (**105b**) can be observed (133). Analogous fragmentation reactions have been reported for **267a** (214) and **267c** (213), which gave the corresponding chloride **105a** (215) and iodide **105c** (216), respectively:

$$[(CH_3)_3N-CH_2Hal]^+Hal^- \rightarrow [(CH_3)_2N=CH_2]^+Hal^- + CH_3Hal$$
$$(267) \qquad\qquad (105)$$

a Hal = Cl
b Hal = Br
c Hal = I

Also, monoquaternary salts of aminals (132,172) undergo reversible dissociation into an iminium salt and a tertiary amine, sometimes even at low temperatures (173) (see Section II-F-1-f). An interesting example from a practical point of view is salt **268**. This can be prepared by methylene bromide addition to bis(dimethylamino)methane (**103**) and decomposes at room temperature to form two molecules of dimethylmethyleniminium bromide (**105b**) (215). The reaction of **103** with fluoroiodomethane proceeds exothermically and yields dimethylmethyleniminium iodide (**105c**), together with dimethylfluoromethylamine (**191**) (217). These can easily be separated because of the excellent solubility of **191** in unpolar solvents. From a mixture of methylene bromide and butyl-α-dimethylaminobenzyl ether the iminium bromide **105b** rather than the primarily formed ammonium salt **269** was isolated (215). Also catechol-dichloromethylene ether and triethylamine yield iminium salt **270** with elimination of ethyl chloride (218).

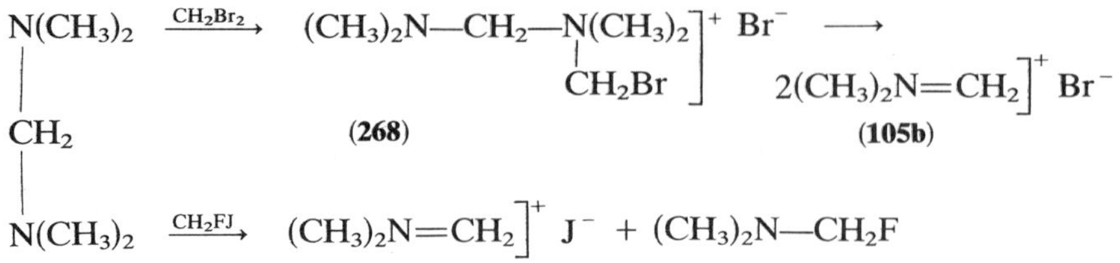

$$\underset{\underset{CH_2}{|}}{N(CH_3)_2} \xrightarrow{CH_2Br_2} \underset{(268)}{\left[(CH_3)_2N-\underset{\underset{CH_2Br}{|}}{CH_2}-N(CH_3)_2\right]^+ Br^-} \longrightarrow \underset{(105b)}{2\left[(CH_3)_2N=CH_2\right]^+ Br^-}$$

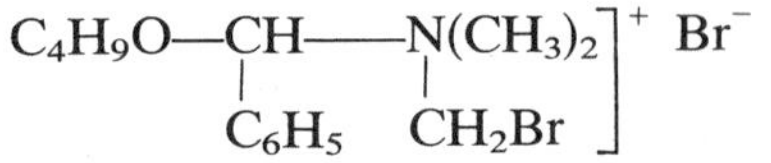

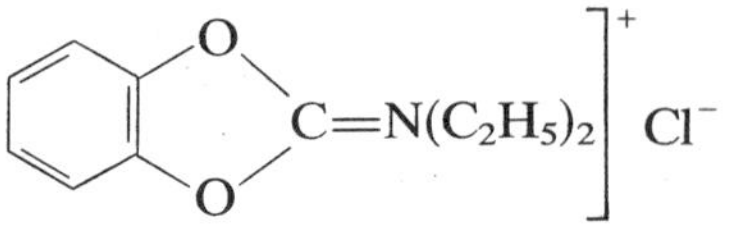

$$\underset{(103)}{N(CH_3)_2} \xrightarrow{CH_2FJ} \underset{(105c)}{\left[(CH_3)_2N=CH_2\right]^+ J^-} + \underset{(191)}{(CH_3)_2N-CH_2F}$$

$$\underset{(269)}{\left[\underset{\underset{C_6H_5}{|}}{C_4H_9O-CH}-\underset{\underset{CH_2Br}{|}}{N(CH_3)_2}\right]^+ Br^-} \qquad \underset{(270)}{\left[\text{benzodioxole}=C=N(C_2H_5)_2\right]^+ Cl^-}$$

J. MISCELLANEOUS METHODS OF PREPARATION

In special cases iminium salts are accessible from alkyl azides. From methyl azide an antimony(V)chloride adduct (**271**) was formed which could be cleaved in the presence of hydrogen chloride to give **272**, the hexachloroantimonate of the protonated methylenimine (219):

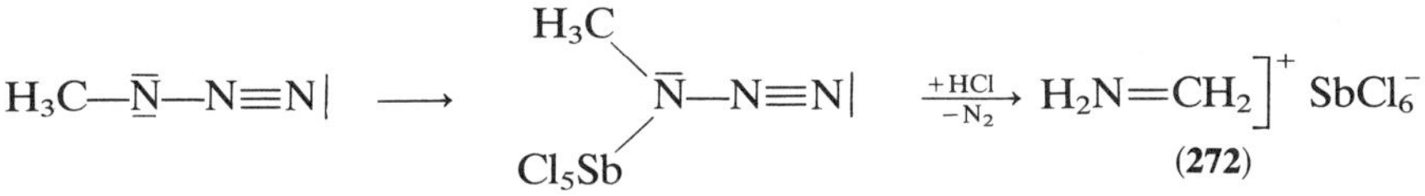

$$H_3C-\bar{N}-N\equiv N| \longrightarrow \underset{\underset{Cl_5Sb}{}}{\overset{H_3C}{\diagdown}}\!\!\!\!\!\!\!\!\!N-N\equiv N| \xrightarrow[-N_2]{+HCl} \underset{(272)}{\left[H_2N=CH_2\right]^+ SbCl_6^-}$$
(**271**)

The intermediate from ethyl azide addition to triethyloxonium fluoroborate collapses under elimination of nitrogen and hydride transfer to **273**, the tetrafluoroborate of the protonated acetaldehyde ethylimine (90):

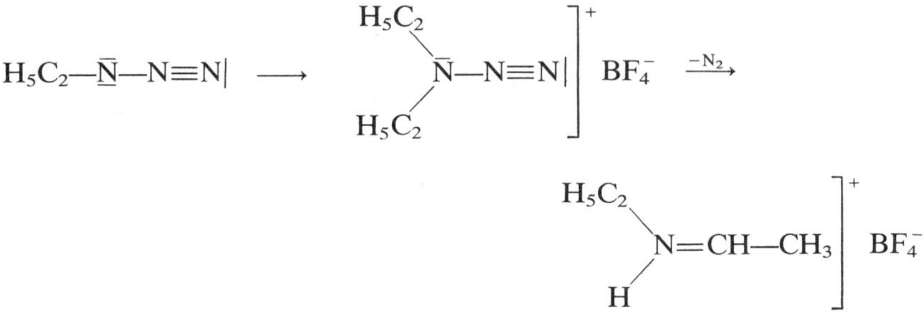

$$H_5C_2-\bar{N}-N\equiv N| \longrightarrow \left[\underset{\underset{H_5C_2}{}}{\overset{H_5C_2}{\diagdown}}\!\!\!\!\!\!\!\!\!N-N\equiv N|\right]^+ BF_4^- \xrightarrow{-N_2}$$

$$\underset{(273)}{\left[\underset{\underset{H}{}}{\overset{H_5C_2}{\diagdown}}\!\!\!\!\!\!\!\!\!N=CH-CH_3\right]^+ BF_4^-}$$

The decomposition of aziridinones (**274**), which takes place with decarbonylation upon protonation with hydrogen chloride in ether within a few minutes, yields crystalline aldiminium chlorides (**275**), which are easily hydrolyzed in contact with moisture (220):

$$\underset{(\mathbf{274})}{\underset{\underset{O}{\overset{\|}{C}}}{R-N-CH-R'}} \xrightarrow{HCl} \underset{(\mathbf{275})}{\left[\underset{H}{\overset{R}{\diagdown}} N = CH-R' \right]^+ Cl^-} + CO$$

The preparation of iminium salts directly by action of halogens on tertiary amines has not been described, although their formation as intermediates has been suggested (221). The reaction of trichloromethanesulfenyl chloride with triethylamine (222), as well as vanadium(IV) chloride reduction with trimethylamine (223), is presumed to proceed via methyleniminium salts. Related to this is the formation of iminium chloride **105**, together with **227**, in the trimethylamine reaction with N-halohexamethyldisilazane (**276**) (224):

$$\underset{(\mathbf{276})}{(CH_3)_3Si-\underset{Hal}{\overset{|}{N}}-Si(CH_3)_3} \xrightarrow{N(CH_3)_3}$$

$$\underset{(\mathbf{105})}{\left[(CH_3)_2N=CH_2 \right]^+ Cl^-} + \underset{(\mathbf{277})}{(CH_3)_2Si-NH-Si(CH_3)_3}$$

All these reactions may possibly be characterized by a common step of nucleophilic attack of the tertiary amine nitrogen on the positive halogen.

Finally, high-temperature chlorination and photochemical chlorination, which yield isocyanide dichlorides (225) and α-perchloroamines (226), respectively, should be mentioned as significant methods.

Chlorination of N-alkyl amides under photochemical conditions (UV irradiation at 40–130°) leads to acylimide dichlorides, for example, **280** (227) from **278** via **279** as the intermediate product, which can also be obtained by other routes (228). Chlorination of N-methyl-N-phenylcarbamoyl chloride (**281**) occurs upon irradiation with UV light in carbon tetrachloride to give the N-chloromethyl-N-phenylcarbamoyl chloride (**162**), which can also be synthesized by aminal cleavage reactions of **161** (see Section II-F-1-e) (163). The reaction of chlorine with N-methylphthalimide at 160–170° gives N-chloromethylphthalimide (**78**) quantitatively (229). With N-bromosuccinimide N-ethylphthalimide is substituted in the α-position. By subsequent elimination of hydrogen halide and bromine addition N-1,2-dibromoethylphthalimide (**282**) is formed (230); this is also accessible from N-β-bromoethyl- or N-vinylphthalimide with bromine (231) (see Section II-B-4). Chlorination of N-acylpiperidine also proceeds via substitution in the α position; by several hydrogen halide elimination and chlorine addition steps compound **283** is formed (232).

CCl$_3$—CO—NH—CH$_3$ ⟶ CCl$_3$—CO—NH—CH$_2$Cl ⟶

(**278**)   (**279**)

CCl$_3$—CO—N=CCl$_2$

(**280**)

Cl—CO—N(C$_6$H$_5$)—CH$_3$ ⟶ Cl—CO—N(C$_6$H$_5$)—CH$_2$Cl ⟵

(**281**)   (**162**)

C$_6$H$_5$—NH—CH$_2$—NH—C$_6$H$_5$

(**161**)

(**282**) phthalimido-N—CHBr—CH$_2$Br

(**283**) 2,2,3-trichloro-1-acetylpiperidine derivative

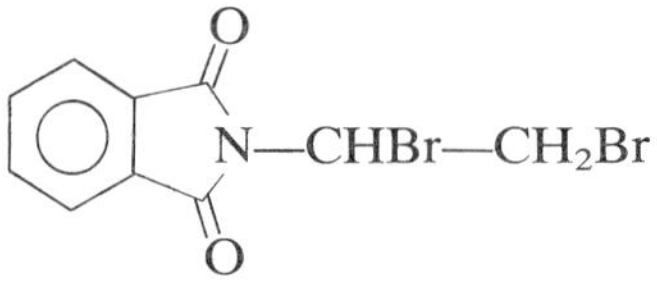

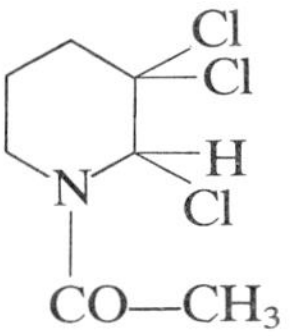

N-halomethyl amides or imides have also been prepared by fragmentation reactions. Decarbonylation of phthalimidoacetyl chloride takes place at high temperatures to give N-chloromethylphthalimide (**78**) (233). From reactions of N-haloamides or N-haloimides (e.g., **284**) with diazomethane, compounds of type **286** have been obtained (234,235). It has been suggested that methylene insertion into the N–Cl linkage proceeds through an ion-pair intermediate (**285**), starting with abstraction of the positive halogen by diazomethane; an alternative pathway through a halocarbene could be ruled out experimentally (235).

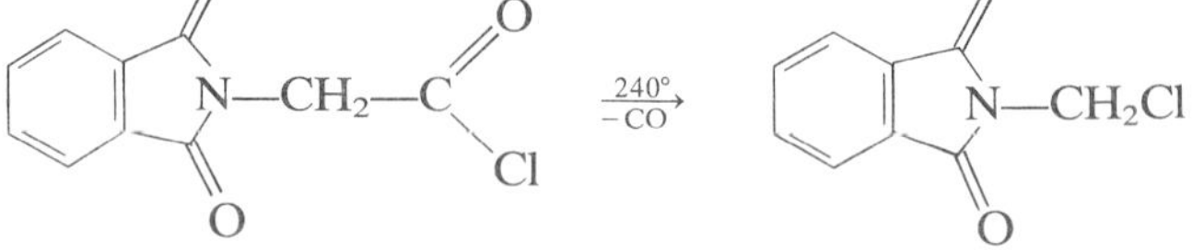

phthalimido-N—CH$_2$—C(=O)Cl $\xrightarrow[-CO]{240°}$ phthalimido-N—CH$_2$Cl

(**78**)

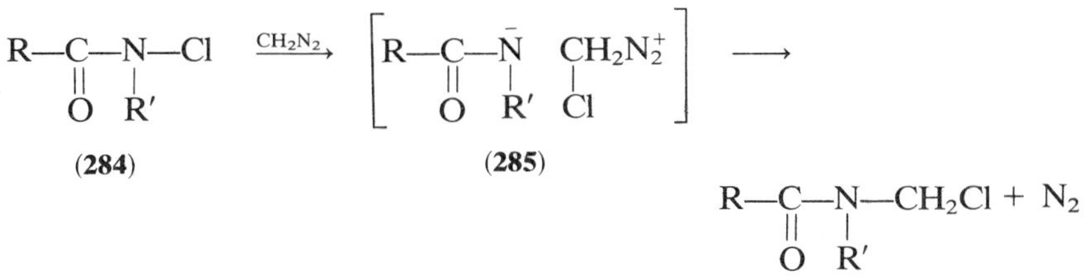

$$\underset{(284)}{\underset{\underset{R'}{|}}{\overset{\overset{O}{\|}}{R-C-N-Cl}}} \xrightarrow{CH_2N_2} \left[\underset{(285)}{\underset{\underset{R'}{|}\ \underset{Cl}{|}}{\overset{\overset{O}{\|}}{R-C-N}\ CH_2N_2^+}} \right] \longrightarrow$$

$$\underset{(286)}{\underset{\underset{R'}{|}}{\overset{\overset{O}{\|}}{R-C-N-CH_2Cl}}} + N_2$$

Finally, isolation of an iminium salt (288), in addition to the amide (289), has also been reported from Beckmann rearrangement experiments of benzoylferrocenoxime (287) with benzenesulfonyl chloride (236):

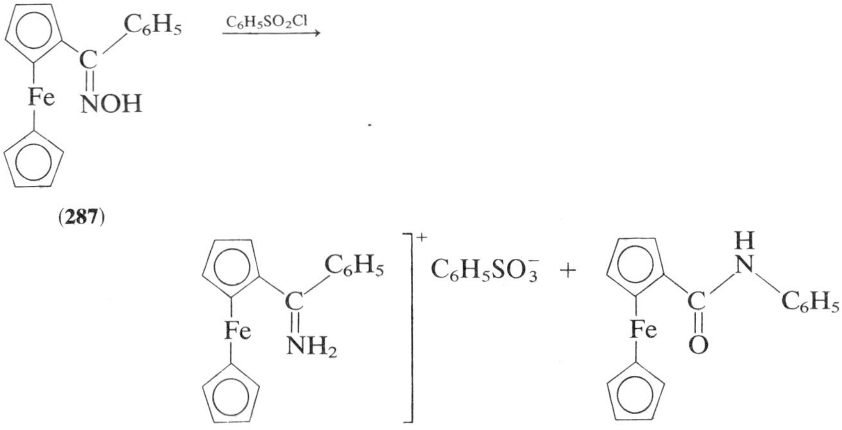

[Structure of 287: ferrocene with C(=NOH)C₆H₅ substituent]

(287)

$\xrightarrow{C_6H_5SO_2Cl}$

[Structure 288: ferrocene with C(=NH₂⁺)C₆H₅, C₆H₅SO₃⁻] + [Structure 289: ferrocene with C(=O)NH-C₆H₅]

(288) (289)

K. IMINIUM SALTS AS INTERMEDIATES

It was postulated that MIS serve as reactive intermediates in various reactions, even before their preparation had been achieved. Some well-known reactions are the Amadori rearrangement (237), the Eschweiler-Clarke methylation (238), the Leuckart-Wallach reaction (239), the Mannich condensation (240–242), and the Strecker synthesis (88). Iminium salts also appear to be intermediates in the Polonovski reaction (243,244) and in Ugi's four-component condensation (245,246), as well as in heterolytic fragmentation reactions [e.g., γ-aminoalkyl halides or certain α-aminoketoxime esters (247)]. Other examples are the diazonium salt decomposition in dimethylaniline (248), the decarbonylation of α-dialkylamino acid chlorides (249), the von Braun amide degradation (250), and the action of triethylamine on trichloroacetyl chloride (251). The generation of iminium intermediates has also been proposed (a) for oxidation reactions of tertiary amines with mercury(II) and other metal acetates (40,252–254), as well as with many other oxidants such as bromine (221), chlorodioxide or hypochlorites (255–257), nitrous acid or nitroxyl fluoroborate (258), and trichloromethanesulfenyl chloride (222), and (b) for solvolysis reactions of enamines (259) or aziridines (260,261), 1-chloroaziridines (262), and 3-chloro-1-azirines (263), as well as (c) for cleavage reactions of N-tertiary α-amino alcohols (264), α-amino ketones (265,266), bistertiary diamines (267), etc., with lead(IV) or mercury(II)acetate. Finally, iminium intermediates have been proposed in the reaction of dialkyl aminomethyl phenyl sulfides or dialkylformamides with Grignard reagents (197,268), in the amine catalysis of β-ketole dehydration (269), in the anodic dealkylation of aliphatic amines in acetonitrile (270), in the photochemical demethylation of tertiary amines (271), and in the preparation of cyclic enamines from tertiary amides by means of dialkylaluminum hydride reduction (272).

III. Physical Properties of Methyleniminium Salts

A. STRUCTURE, STABILITY, AND SOLUBILITY

In the first paper concerned with the preparation and properties of α-haloamines (127) a carbiminium salt type of structure (e.g., **290a**) was assigned to these compounds, in accordance with the chemical and physical properties of the chloro and bromo compounds. However, an equilibrium with the covalent form (e.g., **290b**) could not be ruled out.

$$(CH_3)_2N=CH_2 \leftrightarrow (CH_3)_2N-CH_2]^+Cl^- \rightleftharpoons (CH_3)_2N-CH_2Cl$$

(**290a**)   (**290b**)

The structure parameters of methyleniminium halides (MIH) have not been investigated to the same extent as for the planary dimethylisopropylideniminium ion (**64**). Here the bond distances (C=N, 1.30_2 Å; C—CH$_3$ and N—CH$_3$, 1.51_3 Å) and bond angles (H$_3$C—N—CH$_3$ and H$_3$C—C—CH$_3$, 125.4°; H$_3$C—C=N and H$_3$C—N=C, 117.3°) are known from X-ray-analysis (273). Contrary to the iminium salt character of α-haloamines, the corresponding N-α-haloalkyl amides and imides are usually characterized by covalent structural formulas (e.g., **291a**). Although in reactions N-acyliminium ions (e.g., **291b**) may possibly serve as intermediates (4–6), the reactivity is much decreased.

Most of the dialkylmethyleniminium chlorides, bromides, and iodides with small alkyl groups on nitrogen have melting points in the range of 100–150°; melting should be carried out in sealed tubes because of the ease of hydrolysis. Dimethylmethyleniminium chloride, bromide, and

$$\left[\begin{array}{cc} \underset{\underset{\overset{|}{O}}{\overset{\|}{R-C-\bar{N}-CH_2Cl}}}{R'} & \underset{\underset{\overset{|}{O}}{\overset{\|}{R-C-\bar{N}-CH_2}}}{R'} \\ \updownarrow \rightleftharpoons \updownarrow \\ \underset{\underset{\overline{|O|}}{\overset{|}{R-C=\overset{+}{N}-CH_2Cl}}}{R'} & \underset{\underset{\overset{|}{O}}{\overset{\|}{R-C-N=CH_2}}}{R'} \\ \textbf{(291a)} & \textbf{(291b)} \end{array}\right]^+ Cl^-$$

iodide can be sublimed *in vacuo* without decomposition. The solubility of DMIH's is dependent on the nature of the N-alkyl groups. Usually they are not soluble in nonpolar solvents. To a small extent they can be dissolved in dry aprotic, polar solvents such as acetonitrile, dimethylformamide, or nitromethane. Larger N-alkyl substituents enhance the solubility in methylene chloride, chloroform, tetrahydrofuran, or dioxane. Fluoromethyldialkylamines **134** (146) are low-boiling liquids and are characterized by a covalent halogen–C bond. Therefore they are soluble in almost all organic solvents, including ether, pentane, and carbon disulfide. Dimethyldifluoromethylamine (274) and dimethyltrifluoromethylamine (275,276) (see chapter 12 and 5) are also distillable liquids:

$$R_2N-CH_2F$$

(**134a**) $R = CH_3$

(**134b**) $R_2 = $ ⌬

B. ABSORPTION IN INFRARED AND ULTRAVIOLET

Methyleniminium salts, as well as tri- and tetra-alkylated iminium salts (1,87), show typical IR absorption spectral bands at 1660–1690 cm^{-1} (in Nujol) for the C–N double bond (37,277); the C–H absorption of the methylene group appears at 3110–3150 cm^{-1} (201,216). The C=N absorption of the corresponding dialkylchloromethyleniminium chlorides (see chapter 12) is in the same range, but for dialkyldichloromethyleniminium chlorides (see chapter 5) this absorption is shifted to lower frequencies. Depending on the alkyl substituents, absorption bands at 1590–1650 cm^{-1} have been observed (278).

In the IR spectra of fluoromethyldialkylamines **134** no absorption for a C–N double bond is present (146). A broad band at 835 cm^{-1} can probably be assigned to an extremely low-frequency absorption of the C–F bond.

TABLE I
UV Data on Aryl- and Heteroaromatic-Substituted Iminium Chlorides (Solvent: CH_3CN)

[Structures (292) and (293)]

Compound	X	R	λ_{max}, nm	log ϵ_{max}	Ref.
292a	CH_2	H	275	3.84	148
292b	CH_2	OCH_3	320	4.26	148
292c	CH_2	SCH_3	355	4.27	148
292d	CH_2	$N(CH_3)_2$	390	4.67	148
292e	O	$N(CH_3)_2$	398	4.70	149
293a	O	—	305	4.33	148
293b	S	—	282	4.35	148
			315	4.39	
293c	NCH_3	—	335	4.45	148

The UV spectra of iminium salts are characterized by $\lambda_{max} = 220\text{--}235$ nm (33); a bathochromic shift results from substitution of the methylene hydrogens by aromatic or heterocyclic groups. This has been demonstrated for compounds **292** and **293** (Table I), which are accessible from aminal cleavage (see Section II-F-1-c) (148,149) and are suitable subjects for photometric studies of hydrolysis (see Section IV-B).

C. NUCLEAR MAGNETIC RESONANCE

The ^{1}H NMR spectra of MIH have been measured in several deuterated solvents such as acetonitrile, chloroform, nitromethane, dimethylsulfoxide, and trifluoroacetic acid. Low shielding of the methylene protons, the signals of which usually appear around $\tau = 2$ ppm as a broadened singlet (87,216,279–282), confirms the important contribution of the iminium resonance structure in these compounds. Depending on the nature of the anion X of **294**, the signals have been found to be more or less resolved multiplets in the sequence $X = SbCl_4 \rightarrow AlCl_4 \rightarrow SbCl_6$ (282). For chloromethyleniminium salts the position of the single proton is far downfield [$\tau = -1.08$ (279) or $\tau = 0.85$ ppm (283)]. Methyl proton signals of dimethylmethyleniminium salts appear as a broadened singlet in the range of $\tau = 6$ ppm (279); these were likewise found to be multiplets in the case of complex anions (282). A broad singlet is also characteristic of the methylene protons of fluoromethyldialkylamines **134** (146). In contrast to the situation for MIS, the chemical shifts of both the methylene protons ($\tau = 5.05$ ppm for **134a**, $\tau = 5.13$ ppm for **134b** in carbon tetrachloride) and the methyl protons (singlet at $\tau = 7.48$ ppm for **134a** in carbon tetrachloride) indicate a covalent C–halogen bond. This is also confirmed by the change of the methylene proton singlet into a doublet ($J = 60$ Hz) at low temperatures. The ^{19}F NMR spectra of N-fluoromethylmorpholine (**134b**) at room temperature also show a singlet that changes to a triplet at low temperatures. From these results fluoride exchange has been suggested (217); this seems more likely than a temperature-dependent equilibrium between the covalent α-fluoroamine and an ionic iminium fluoride. As in the case of fluoromethyldialkylamines, the methylene proton signal of covalent dimethyl aminomethyl methyl ether (**295**) (191) appears at $\tau = 6.1$ ppm.

The structures of the dialkyl aminomethyl esters **296** are either covalent or ionic, depending on the nature of the acyl group. The derivatives of acetic and chloroacetic acid are characterized by a covalent O–C bond, while those of trichloro- or trifluoroacetic acid are iminium salts, as indicated by C=O IR-absorption (1715 cm^{-1} for **296a**, 1650 cm^{-1} for **296b**), as well as by NMR data (see Table II).

$$(CH_3)_2N{=}CH_2]^+X^- \qquad (CH_3)_2N{-}CH_2{-}OCH_3$$
$$(294) \qquad\qquad\qquad (295)$$
$$(CH_3)_2N{-}CH_2{-}O{-}CO{-}R' \qquad (CH_3)_2N{=}CH_2]^+R'CO_2^-$$
$$(296a) \qquad\qquad\qquad (296b)$$

Nuclear magnetic resonance studies of iminium salts have also been reported for protonated aldimines (7,8) as well as for cyclopropylideniminium salts of type **70** (95). Here the barrier of rotation around the C–N double bond which usually in iminium salts exceeds 30 kcal/mol (95), was found to be decreased to 22–25 kcal/mol due to charge delocalization in the cyclopropene ring.

TABLE II
NMR Data of Dimethyl Aminomethyl Esters **296**

296	R'	NMR (τ, ppm)		Solvent	Ref.
		CH$_3$	CH$_2$		
a	CH$_3$	7.6	5.1	CD$_3$CN	206
a	CH$_2$Cl	7.3	5.2	CCl$_3$D	206
b	CCl$_3$	6.6	2.4	CD$_3$CN	206
b	CF$_3$	6.1	2.1		207

D. MASS SPECTRA

The typical fragment of highest intensity in mass spectra of tetramethyldiaminomethane (**103**) and many other trimethylamine derivatives, $(CH_3)_2N$—CH_2—R, is the dimethylmethyleniminium ion (284). This is also the base peak of iminium chloride **290a** (278) ($m/e = 58$, $I_{rel} = 100\%$). However, although with much less intensity, the molecular ion peak ($m/e = 93/95$, $I_{rel} = 15\%$), as well as the CH_2Cl fragment ($m/e = 49/51$, $I_{rel} = 5\%$), are also present. This is an interesting finding, for it indicates the existence of dimethylmethyleniminium chloride (**290a**), possibly together with some covalent dimethylchloromethylamine (**290b**), at least under the conditions of the mass spectrometer. Molecular ion peaks ($I_{rel} = 20\%$) have also been observed for dimethylchloromethyleniminium chloride (see chapter 12) and especially for dimethyldichloromethyleniminium chloride (see chapter 5), in addition to the trichloromethyl cation (278).

E. MISCELLANEOUS PHYSICAL PROPERTIES

From conductometric studies performed with dimethylmethyleniminium chloride (**290**) (278) and the hexachloroantimonate **294** (282) it became evident that the chloride is dissociated to a high extent.

By polarographic methods the existence of iminium ions was also demonstrated in aqueous solutions (285).

IV. Reactions of Methyleniminium Salts

As a result of their carbiminium salt nature, the MIS are powerful electrophiles that can easily attack all types of strong and weak nucleophiles, including heteroatoms with lone pairs of electrons, carbanions, or electron-rich multiple bonds, as well as aromatic systems. Although in most cases the reactions of the MIS result in aminomethylation of the nucleophilic component, deprotonation of the carbiminium ion can also occur. The synthetic utility of the MIS as typical "Mannich reagents" has been demonstrated also for reactions in which the usual Mannich condensation proceeds with low yields or fails completely.

Although systematic studies are not yet available, the reactivity of MIS appears to be influenced both by the basicity of the parent amine component and by the type of anion. The basicity appears to have opposite effects on the stability and the reactivity of MIS; that is, basicity-enhancing substituents on nitrogen that stabilize the carbiminium ion weaken its electrophilic potential. The aminomethylation of dimethylaniline, for example, can be achieved only once with N,N-dimethylmethyleniminium chloride (**105a**), which is a fairly stable MIS. With the

IV. REACTIONS OF METHYLENIMINIUM SALTS 155

unstable N-methyl-N-arylmethyleniminium chlorides (**222**) or the N-methylenemorpholinium chloride (**361**), however, both of which represent MIS of weak basic amine components, bisubstitution of the aromatic ring has been achieved (see Section IV-I-5). In addition, the N-chloromethylacylamides (**71**), as well as the N-fluoromethyldialkylamines (**134**), are relatively less stable but very reactive electrophiles. The reactivity effect of the leaving anion seems to follow the order of the basicity, e.g., $F > Cl > ClO_4$. Furthermore, sterical requirements must be considered for iminium salts having additional substituents on iminium carbon (e.g., **64**), which often may be responsible for reactions of these salts that proceed slowly with low yields, or with formation of undesirable side products if more vigorous reaction conditions are employed.

A. ANION EXCHANGE

Anion exchange with the easily accessible MIH **100** is the simplest procedure by which many of the other salts can be prepared. For example, in acetonitrile or acetonitrile–nitromethane, perchlorates **297** (127), tetrafluoroborates **298**, and tosylates **299** (9), were obtained by means of the corresponding silver salts. The preparation of **299** has also been achieved from **100** by heating with methyl tosylate with elimination of methyl chloride (9). Furthermore there should be mentioned reactions of iminium chlorides with Lewis acids, which were carried out in 1,2-dichloroethane or methylene chloride, yielding tetrachloroantimonates **300**, tetrachloroaluminates **301**, and hexachloroantimonates **302**, respectively (281,282). Complex salts with copper(I) chloride were also described (286).

$$[R_2N=CH_2]^+ ClO_4^- \quad [R_2N=CH_2]^+ BF_4^- \quad [R_2N=CH_2]^+ CH_3-C_6H_4-SO_3^-$$
$$(297) \qquad\qquad (298) \qquad\qquad\qquad (299)$$

+AgClO$_4$ ↖   ↑+AgBF$_4$ ↗+AgOSO$_2$—C$_6$H$_4$—CH$_3$ or CH$_3$O—SO$_2$—C$_6$H$_4$—CH$_3$

$$[R_2N=CH_2]^+ Cl^-$$
$$(100)$$

↙+SbCl$_3$ ↓+AlCl$_3$ ↘+SbCl$_5$

$$[R_2N=CH_2]^+ SbCl_4^- \quad [R_2N=CH_2]^+ AlCl_4^- \quad [R_2N=CH_2]^+ SbCl_6^-$$
$$(300) \qquad\qquad (301) \qquad\qquad (302)$$

From tris(chloromethyl)amine (**181**) and methylbis(chloromethyl)amine (**180**) with Lewis acids iminium salts **303**, **304**, and **305** were isolated (169). The N-α-haloalkyl amide **14** gave, upon tin(IV) chloride addition, a hexachlorostannate (**306**) (287); from 3-chlorophthalimidine with antimony(V) chloride iminium salt **307** was formed (123),

$$[(ClCH_2)_2N=CH_2]^+ SbCl_6^-$$
(**303**)

$$[(ClCH_2)_2N=CH_2]_2^+ SnCl_6^{2-}$$
(**304**)

$$\left[\begin{array}{c}ClCH_2\\ \\ CH_3\end{array}\!\!N=CH_2\right]^+ SbCl_6^-$$
(**305**)

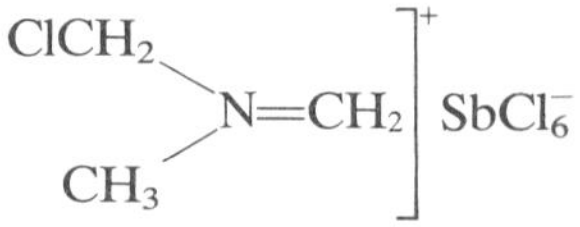

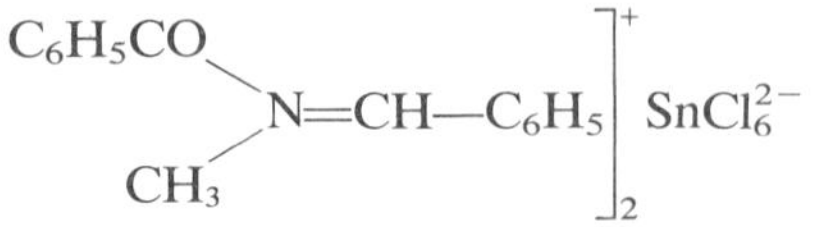

$$\left[\begin{array}{c}C_6H_5CO\\ \\ CH_3\end{array}\!\!N=CH-C_6H_5\right]_2^+ SnCl_6^{2-}$$
(**306**)

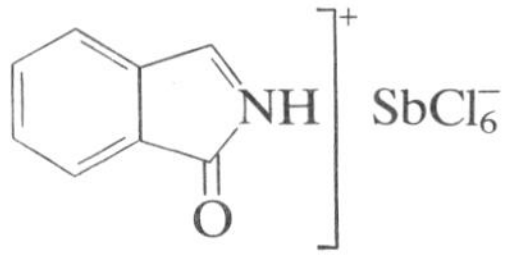

[isoindolin-1-one NH]$^+$ SbCl$_6^-$
(**307**)

The iminium nitrate **308** can be prepared from dimethylaminomethyl acetate (**295**) with silver nitrate in acetonitrile (166):

$$(CH_3)_2N-CH_2-O-COCH_3 \xrightarrow[-AgOAc]{+AgNO_3} (CH_3)_2N=CH_2]^+NO_3^-$$

(**295**)   (**308**)

B. HYDROLYSIS

In contact with water, iminium salts are hydrolyzed more or less rapidly with formation of the carbonyl component and a secondary ammonium salt. Therefore, for the preparation of iminium salts, moisture must be excluded, and usually handling under dry atmosphere in sealed equipment is especially recommended for MIS.

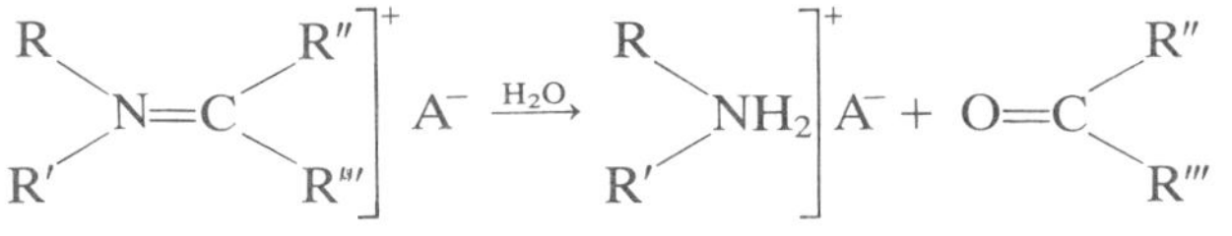

$$\left[\begin{array}{c}R\\ \\ R'\end{array}\!\!N=C\!\!\begin{array}{c}R''\\ \\ R'''\end{array}\right]^+ A^- \xrightarrow{H_2O} \left[\begin{array}{c}R\\ \\ R'\end{array}\!\!NH_2\right]^+ A^- + O=C\!\!\begin{array}{c}R''\\ \\ R'''\end{array}$$

The ease of hydrolysis provides a simple way for the analysis of the salts. By means of dissolution in water the anion, as well as the carbonyl component, can be determined, for example, formaldehyde from MIS either by the volumetric oxime method (288), the photometric chromotropic acid method (289), or the gravimetric dimedone method (290).

In special cases it has proved possible to study the rates of hydrolysis. The UV-absorption maxima of iminium salts bearing aromatic or heterocyclic substituents on iminium carbon (e.g., **292** or **293**) (see Section II-F-1-c) usually appear (25–60 nm) at longer wavelengths than the maxima of the corresponding aldehydes formed upon hydrolysis. This offers a possibility for spectroscopic studies of the hydrolysis rates. In acetonitrile–water, for example, the group of absorption curves is characterized by several isosbestic points; at constant water concentration the rates of hydrolysis for several iminium salts have been compared (148,149).

Hydrolysis studies of tertiary enamines in aqueous solutions of strong acids have indicated that iminium ion intermediates are involved (291). Dimethylchloromethylen- and dichloromethyleniminium salts (see chapter 12 and 5) are also very easily hydrolyzed, as are the covalent fluoromethyl amines (**134**). On the other hand, the α-polychloroamines, which are obtainable in a high-temperature chlorination or photochlorination reactions of tertiary amines, formamide chlorides, or thioformamides (225,226), are attacked by water very slowly at room temperature.

C. REACTIONS WITH HYDRIDES

Reduction of iminium salts, for example, with complex hydrides, yields tertiary amines (1,2). This reaction has also been studied with respect to the stereochemistry of the products obtained (292,293). However, hydride addition has not as yet been applied to MIS in order to form methyldialkylamines. By means of Grignard reagents (e.g., isopropyl-, *tert*-butyl-, cyclohexyl-, or arylmagnesium halides), but also upon phenyllithium addition, the iminium salts derived from cyclohexanone, aromatic aldehydes, or benzophenone can be reduced to give tertiary amines (147,294,295–297) (see Section IV-I-3). Hydride transfer has also been observed in the case of MIS reactions with alkali salts of formic acid (see Section IV-D), as well as for the coupling reaction of salt **309** with dimethylaniline (see also Section IV-I-5). From the originally formed phenyl-analogous aminal **310**, in a second step hydride abstraction by **309** leads to dimethyl-*tert*-butylamine (**311**) and iminium salt **312**, which yields *p*-dimethylaminobenzaldehyde upon hydrolysis (201).

D. DEPROTONATION

Deprotonation of MIH on the iminium carbon has been achieved in methylene chloride or acetonitrile by the action of triethylamine, which gives the hydrochloride almost quantitatively (282,298). The intermediate **314**, as a nucleophile, adds to further **313** with formation of salt **316**, deprotonation of which leads to the diaminoethylene (**315**). All attempts to capture intermediate **314** with typical carbene-trapping reagents, such

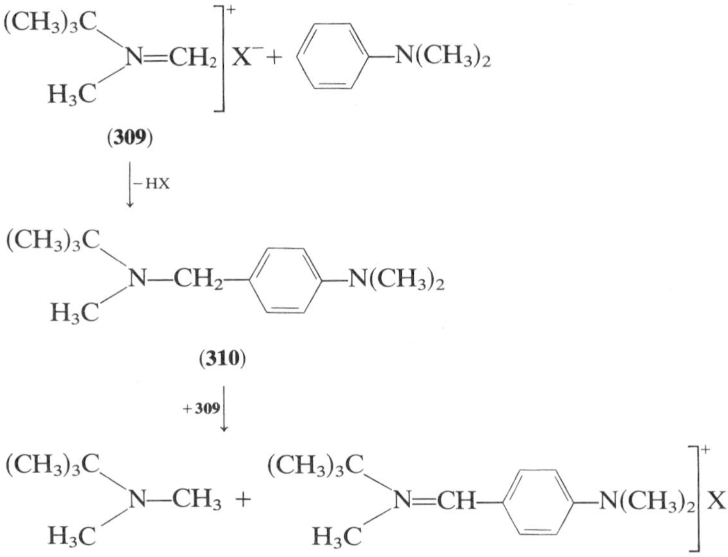

$$[(CH_3)_3C(H_3C)N=CH_2]^+ X^- + \langle\text{C}_6H_4\rangle-N(CH_3)_2$$

(309)

$\downarrow -HX$

$$(CH_3)_3C(H_3C)N-CH_2-\langle\text{C}_6H_4\rangle-N(CH_3)_2$$

(310)

$\downarrow +309$

$$(CH_3)_3C(H_3C)N-CH_3 \;+\; [(CH_3)_3C(H_3C)N=CH-\langle\text{C}_6H_4\rangle-N(CH_3)_2]^+ X^-$$

(311)   **(312)**

as triphenylphosphine, tetracyanoethylene, or cyclohexene, have failed, probably because of low electrophilicity as compared to iminium ions **313**.

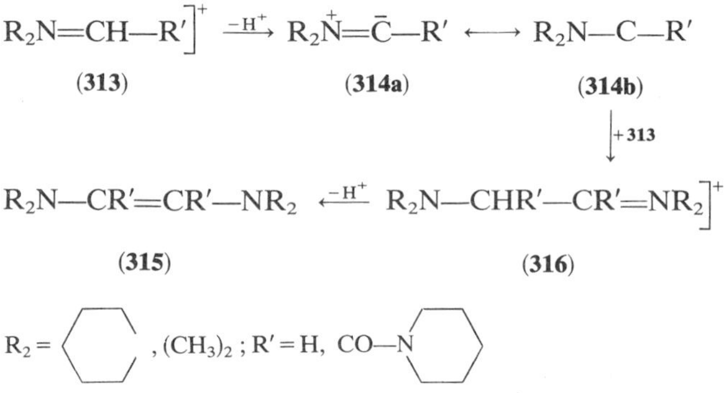

$$[R_2N=CH-R']^+ \xrightarrow{-H^+} R_2\overset{+}{N}=\bar{C}-R' \longleftrightarrow R_2N-\ddot{C}-R'$$

(313)   **(314a)**   **(314b)**

$\downarrow +313$

$$R_2N-CR'=CR'-NR_2 \xleftarrow{-H^+} [R_2N-CHR'-CR'=NR_2]^+$$

(315)   **(316)**

$R_2 = \langle\text{C}_5H_{10}\rangle, (CH_3)_2 ; R' = H, CO-N\langle\text{C}_5H_{10}\rangle$

In the case of iminium salts having C–H bonds in the α position to the iminium carbon, by reaction with base a deprotonation to enamines takes place (for the reverse reaction see Section II-B-1). The pyrroline derivative **317**, for example, yields with potassium hydroxide, 1,2,5-trimethyl-Δ^2-pyrroline **(318)** (299), pyrazolinium salt deprotonations afford 3-pyrazolines (300) (see chapter 7), and from iminium salt **319** enamine **320**

(301) is formed:

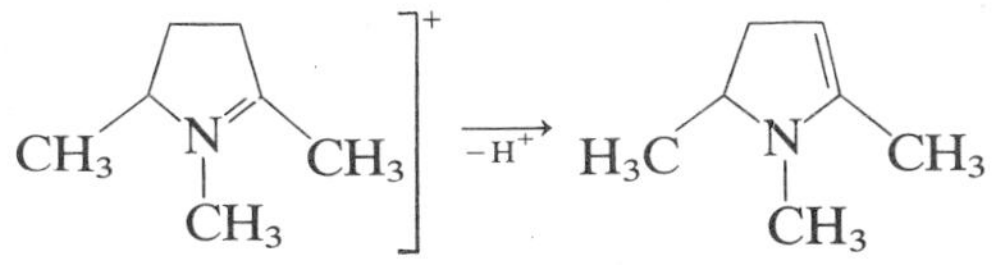

(317) → [−H⁺] → (318)

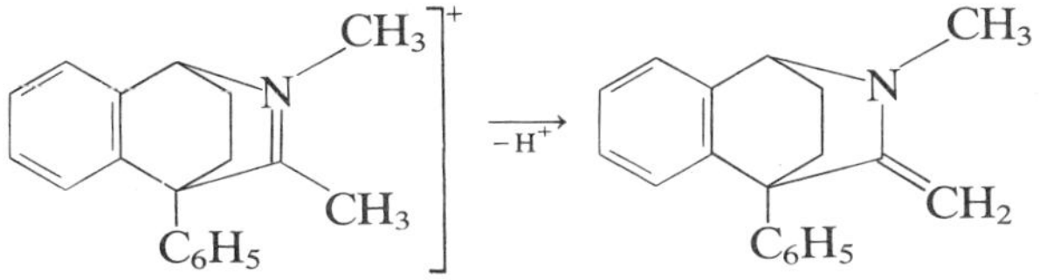

(319) → [−H⁺] → (320)

Similarly, deprotonation of acetophenone derivatives **321** by means of phenyllithium led to **322** (15); from iminium salts derived from methylbenzyl or methylisobutyl ketone (e.g., **323** and **325**) enamines **324** and **326** were obtained (15).

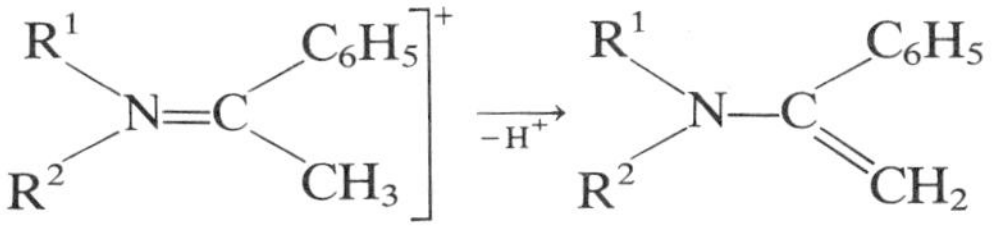

(321) → [−H⁺] → (322)

a: $R^1 = C_6H_5$; $R^2 = CH_3$

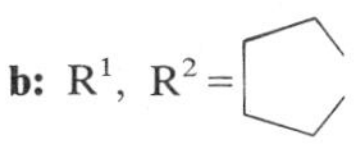

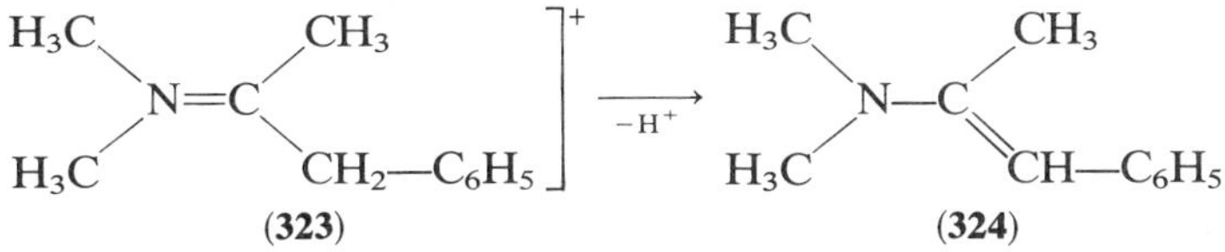

b: $R^1, R^2 =$ ⬠

(323) → [−H⁺] → (324)

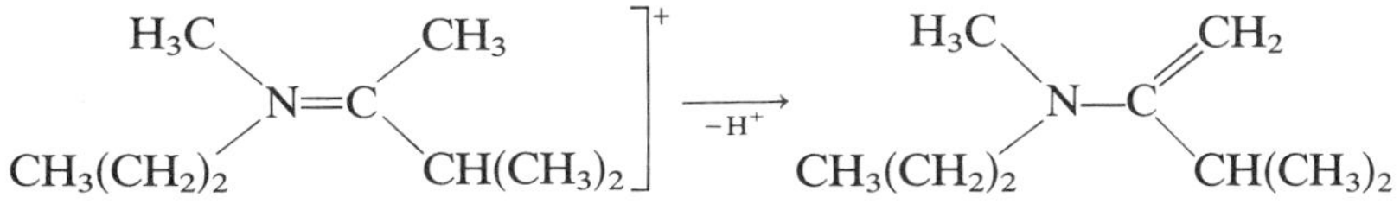

(325) → [−H⁺] → (326)

From derivatives of hydratropaldehyde (e.g., **139**) with triethylamine the E and Z isomers (**327,328**) are formed in the same ratio as from the thermal cleavage reaction of the corresponding aminal (**329**) (151):

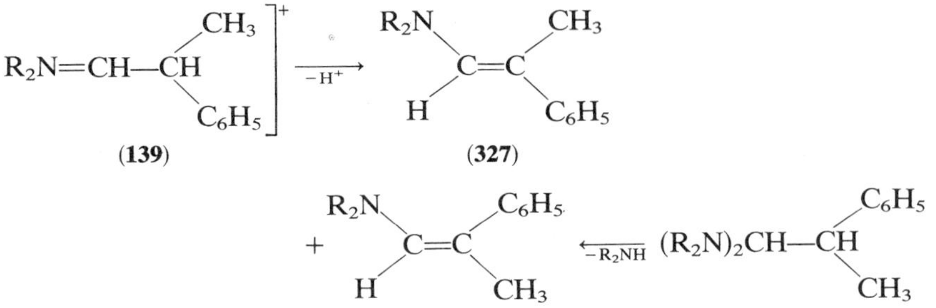

$$R_2N=CH-CH\begin{matrix}CH_3\\ \\C_6H_5\end{matrix}\Bigg]^+ \xrightarrow{-H^+} \begin{matrix}R_2N\\ \\H\end{matrix}C=C\begin{matrix}CH_3\\ \\C_6H_5\end{matrix}$$

(**139**)   (**327**)

$$+ \begin{matrix}R_2N\\ \\H\end{matrix}C=C\begin{matrix}C_6H_5\\ \\CH_3\end{matrix} \xleftarrow{-R_2NH} (R_2N)_2CH-CH\begin{matrix}C_6H_5\\ \\CH_3\end{matrix}$$

(**328**)   (**329**)

The mobility of hydrogen atoms in the α position to the iminium carbon also has been demonstrated on the basis of H–D exchange in refluxing D_2O (300,302), as well as aldol condensation reactions of iminium salts (e.g., **330**) which have been applied for the synthesis of polymethines, for example, **331** (303,304):

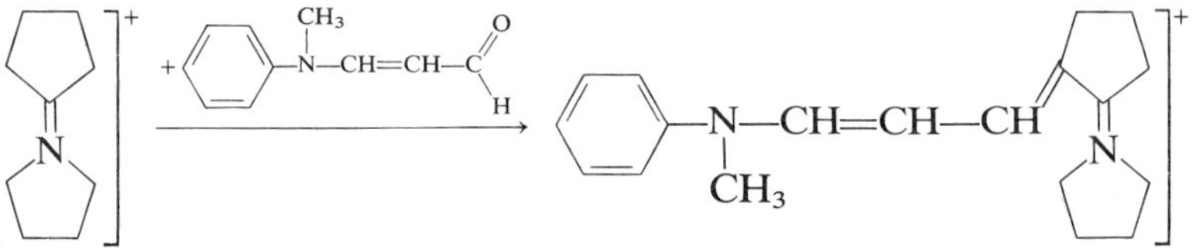

(**330**)   (**331**)

The deprotonation of a methyl group attached to iminium carbon may also be considered in connection with a surprising reaction course, which has been found for protonation experiments of dialkylaminocroton acid esters (**334**). The initially formed salts **332** or **333** (compare II-B-1) presumably in the first step under elimination of alcohol lead to a resonance-stabilized ketene **335**. This adds a second molecule **332** to give an intermediate, which undergoes ring closure to **336**. From this by loss of dialkylammonium salt the final derivative **337** of the 6-aminosalicylic acid can be formed (39,305a,305b).

The deprotonation of iminium salts from *vic*-endiamines is mentioned in Section II-B-3.

Furthermore, deprotonation has also been achieved for C–H bonds adjacent to iminium nitrogen. The action of arylmagnesium halides or phenyllithium on dimethyldiphenylmethyleniminium iodide (**4**), for example, leads to formation of the ylid (**338**). As a nucleophile, this adds more **4** to give an iminium salt (**340**), which upon hydrolysis leads to a

IV. REACTIONS OF METHYLENIMINIUM SALTS

secondary amine (**339**) (295):

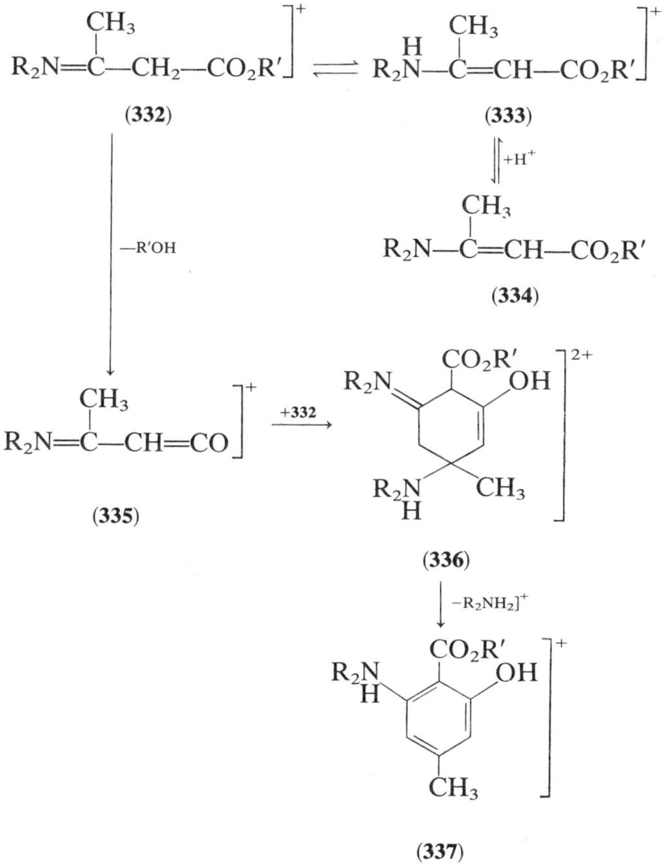

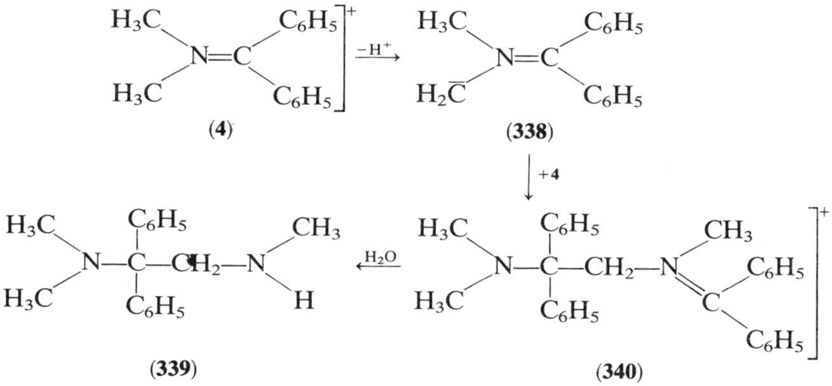

$$[R_2N=\overset{CH_3}{\underset{}{C}}-CH_2-CO_2R']^+ \rightleftharpoons [R_2\overset{H}{\underset{}{N}}-\overset{CH_3}{\underset{}{C}}=CH-CO_2R']^+$$

(**332**)   (**333**)

$\Updownarrow +H^+$

$$R_2N-\overset{CH_3}{\underset{}{C}}=CH-CO_2R'$$

(**334**)

$\downarrow -R'OH$

$$[R_2N=\overset{CH_3}{\underset{}{C}}-CH=CO]^+ \xrightarrow{+332} \text{(336)}$$

(**335**)

(**336**) = cyclohexene ring with $R_2N=$, CO_2R', OH, $R_2\overset{H}{N}$, CH_3 substituents, $2+$ charge

$\downarrow -[R_2NH_2]^+$

(**337**) = benzene ring with $R_2\overset{H}{N}$, CO_2R', OH, CH_3 substituents, $+$ charge

$$[\underset{H_3C}{\overset{H_3C}{>}}N=C\underset{C_6H_5}{\overset{C_6H_5}{<}}]^+ \xrightarrow{-H^+} \underset{H_2\overset{-}{C}}{\overset{H_3C}{>}}N=C\underset{C_6H_5}{\overset{C_6H_5}{<}}$$

(**4**)   (**338**)

$\downarrow +4$

$$\underset{H_3C}{\overset{H_3C}{>}}N-\overset{C_6H_5}{\underset{C_6H_5}{C}}-CH_2-\overset{H}{\underset{}{N}}\overset{CH_3}{\underset{}{}} \xleftarrow{-H_2O} [\underset{H_3C}{\overset{H_3C}{>}}N-\overset{C_6H_5}{\underset{C_6H_5}{C}}-CH_2-N\underset{\overset{\|}{C}\underset{C_6H_5}{\overset{C_6H_5}{}}}{\overset{CH_3}{}}]^+$$

(**339**)   (**340**)

E. C–O BOND FORMATION

Reactions of MIH **100** with alcohols in the presence of tertiary amines as proton acceptors yield dialkyl aminomethyl ethers **194** (196). These are also accessible from the condensation of secondary amines with formaldehyde and alcohol (306) or via α-haloethers **341** with 2 moles of secondary amine (307). With phenols, however, aminomethylation of the aromatic ring is preferred (see Section IV-I-5).

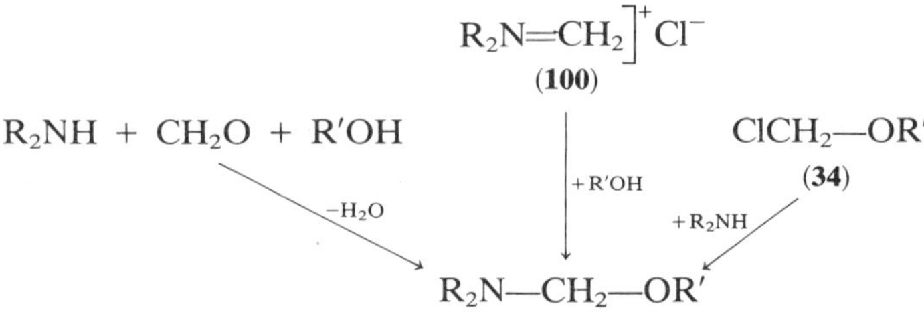

$$[R_2N=CH_2]^+ Cl^-$$
(100)

$$R_2NH + CH_2O + R'OH \xrightarrow{-H_2O} \underset{(194)}{R_2N-CH_2-OR'} \xleftarrow{+R_2NH} \underset{(34)}{ClCH_2-OR'}$$

(middle arrow: +R'OH)

Tris(chloromethyl)amine (**181**) and sodium methylate react to give tris-(methoxymethyl)amine (**342**) (169); with sodium trimethyl silanolate and trimethylsilyl methanolate, compounds **343** and **344** are obtained (308):

$$N(CH_2-O-CH_3)_3 \qquad N[CH_2-O-Si(CH_3)_3]_3$$
(342) (343)

$$N[CH_2-O-CH_2-Si(CH_3)_3]_3$$
(344)

On iminium carbon substituted iminium salts, as well as N-chloromethyl amides, imides, or nitramines, undergo analogous reactions with alcohols or phenols, for example, to form **345** (309), **346** (20), **347** (229), **348** (310), **349** (311), **350** (145), and **351** (116):

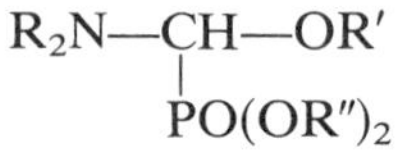

$$\underset{PO(OR'')_2}{R_2N-CH-OR'}$$
(345)

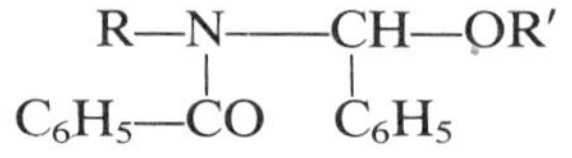

$$\underset{C_6H_5-CO \;\; C_6H_5}{R-N-CH-OR'}$$
(346)

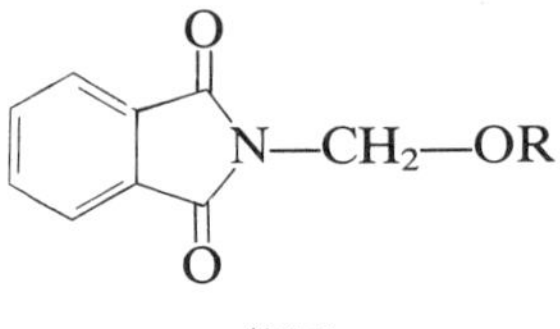

Phthalimide-N–CH$_2$–OR
(347)

$$C_6H_5-CO-NH-CH_2-OR$$
(348)

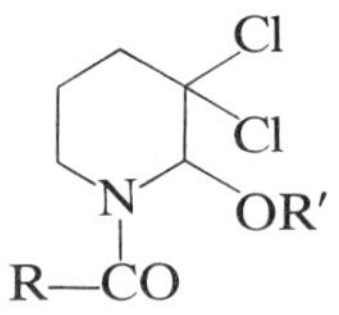

(piperidine with Cl, Cl, OR', N–CO–R)
(349)

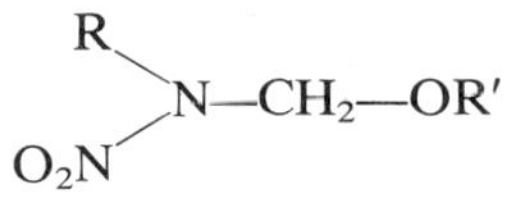

$$\underset{O_2N}{R}\!\!>\!\!N-CH_2-OR'$$
(350)

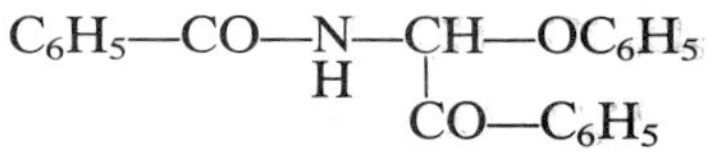

$$C_6H_5-CO-\underset{H}{N}-\underset{|}{CH}-OC_6H_5$$
$$CO-C_6H_5$$
(351)

Reactions of β-haloiminium salts **352** with an excess of alcoholate were reported to give α-aminoacetals **353** via aziridinium ion intermediates (312):

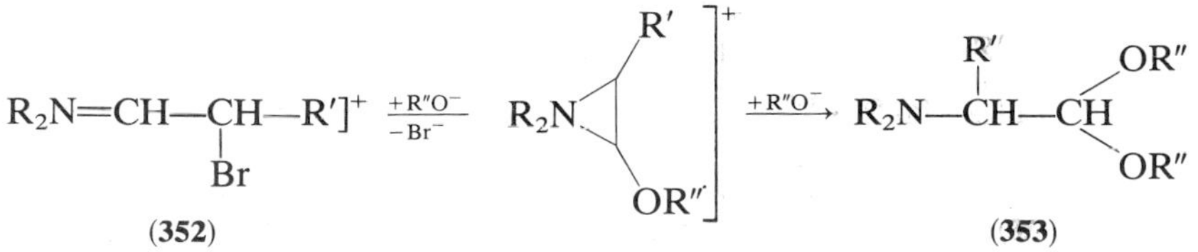

$$R_2N=CH-\underset{Br}{\underset{|}{CH}}-R']^+ \xrightarrow[-Br^-]{+R''O^-} R_2N\underset{OR''}{\overset{R'}{\triangleleft}}\Bigg]^+ \xrightarrow{+R''O^-} R_2N-CH-\underset{OR''}{\overset{R''}{\underset{|}{CH}}}{}^{OR''}$$

(352)   (353)

With β-haloalcohols (e.g., ethylenechlorohydrin) and MIH the dialkyl-(β-chloroethoxymethyl)amines **355** were synthesized; these undergo ring closure to oxazolidinium salts **356** (313). These and analogous products were also obtained from reactions of iminium salts with oxiranes [e.g., ethylene oxide, epichlorohydrin, and styrene oxide (313)]. Similar reactions were performed with γ-haloalcohols or oxetanes; intramolecular cyclization of the dialkyl-(γ-chloropropoxymethyl)amines **357** led to tetrahydrooxazinium salts **358** (313).

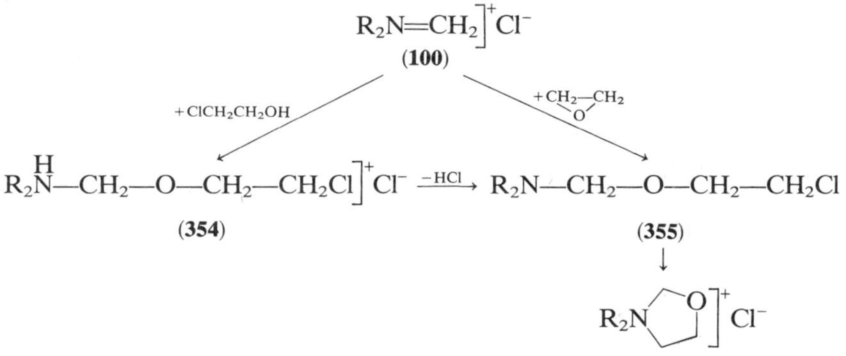

$$R_2N=CH_2]^+Cl^-$$
(100)

$$R_2\overset{H}{N}-CH_2-O-CH_2-CH_2Cl]^+Cl^- \xrightarrow{-HCl} R_2N-CH_2-O-CH_2-CH_2Cl$$
(354)   (355)

$$R_2N\overset{O}{\underset{}{\diagdown}}\Bigg]^+ Cl^-$$
(356)

$$R_2N-CH_2-O-CH_2-CH_2-CH_2Cl \longrightarrow R_2N\overset{O}{\underset{}{\diagdown}}\Bigg]^+ Cl^-$$
(357)   (358)

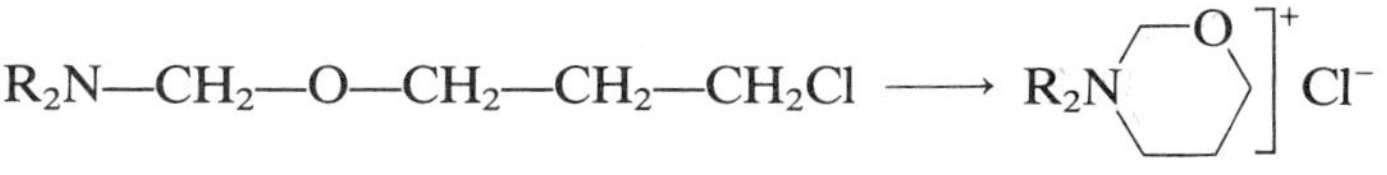

Alkali salts of carboxylic acids are good nucleophiles that add to MIH with the formation of the corresponding dialkyl aminomethyl esters **360** (206). These are also accessible via cleavage of aminals **97** by means of carbonic anhydrides (see Section II-F-1-d), as well as from reactions of these compounds with amine oxides **359** (Polonovski reaction) (205,206,243) or from secondary amine oxidation with diacyl peroxides (314). Attempts to react MIH with sodium formate to prepare formoxymethyldialkylamines failed. However, decarboxylation and hydride transfer, which occurred with the formation of tertiary amines in the case of N-methylenemorpholinium chloride (**361**), for example, led to N-methylmorpholine (**362**) (206).

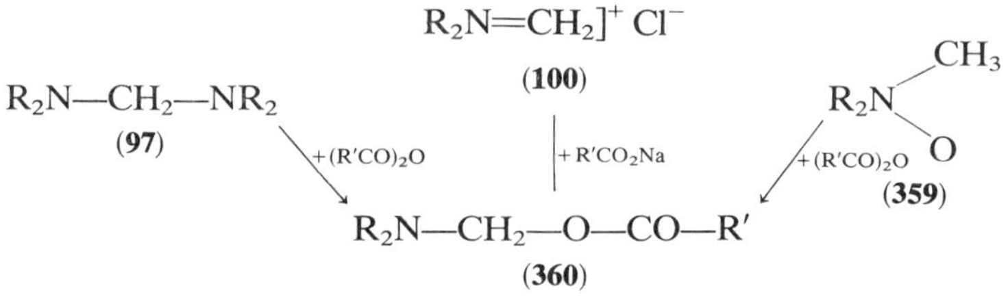

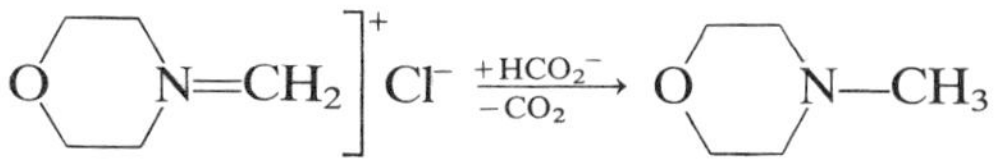

$$R_2N=CH_2]^+ \, Cl^-$$
(**100**)

$R_2N-CH_2-NR_2$ (**97**)

$R_2N\begin{smallmatrix}CH_3\\ \diagdown\\ \diagup\\ O\end{smallmatrix}$ (**359**)

$+(R'CO)_2O$ $+R'CO_2Na$ $+(R'CO)_2O$

$R_2N-CH_2-O-CO-R'$
(**360**)

$\left[O\begin{smallmatrix}\diagup\diagdown\\ \diagdown\diagup\end{smallmatrix}N=CH_2\right]^+ Cl^- \xrightarrow[-CO_2]{+HCO_2^-} O\begin{smallmatrix}\diagup\diagdown\\ \diagdown\diagup\end{smallmatrix}N-CH_3$

(**361**)   (**362**)

From N-acylchloromethyl amides or imides and alkali salts of carboxylic acids the corresponding N-acyloxymethyl amides or imides were prepared, for example, **363** (103), **364** (99), and **365** (116). The action of sodium acetate on tris(chloromethyl)amine (**181**) yields tris(acetoxymethyl)amine (**366**) (169).

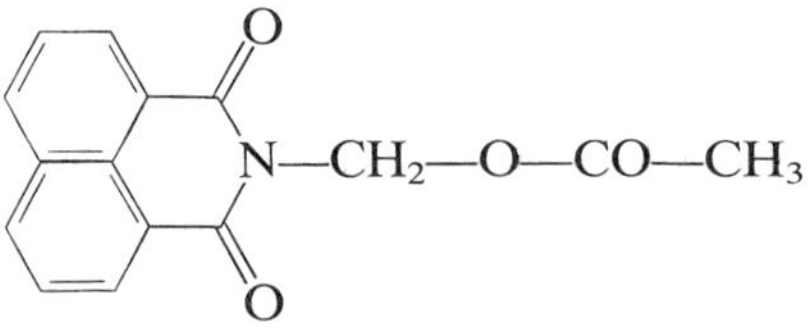

(**363**) naphthalimide-N—CH$_2$—O—CO—CH$_3$

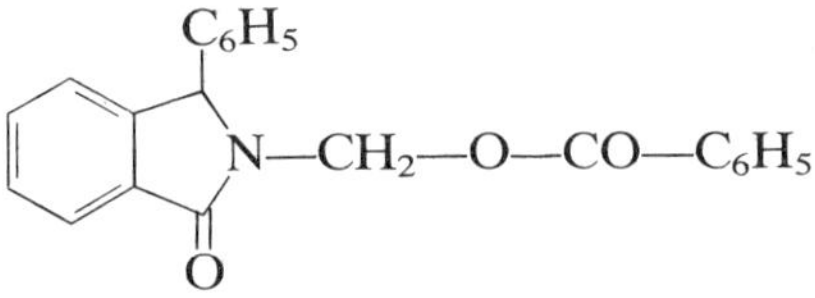

(**364**) 3-phenyl-isoindolin-1-one N—CH$_2$—O—CO—C$_6$H$_5$

$C_6H_5-CO-NH-\underset{\underset{CO-C_6H_5}{|}}{CH}-O-CO-C_6H_5$
(**365**)

$N(CH_2-O-CO-CH_3)_3$
(**366**)

F. C–S BOND FORMATION

With alkali thiophenolates in methanol (315), with mercaptans, or with thiophenols in the presence of tertiary amines as proton acceptors (196), the MIS afford dialkyl aminomethyl thioethers **241**. These have also been obtained from condensation of secondary amines with formaldehyde and mercaptans or thiophenols (306,316,317).

$$R_2N=CH_2]^+Cl^- + R'SH \xrightarrow{-HCl} R_2N-CH_2-SR'$$

(100)   (241)

Monoalkylmethylenimium chlorides **113** and mercaptans or thioacetic acid serve as starting materials for C–S bond formation in hydrochlorides **367** and **368** (136). N,S-acetals of types **369** (309) and **370** (318) are the reaction products of mercaptans with phosphonoiminium chlorides **216** and N-chloromethylnitramines **133**, respectively. Tris(chloromethyl)-amine (**181**) and ethanethiol react with the substitution of two halogens to form **371** (169); with alkali triphenylthiosilanolate, however, **372** has been obtained (308).

$$R-NH_2-CH_2-S-R']^+Cl^- \qquad R-NH_2-CH_2-S-CO-CH_3]^+Cl^-$$

(367)   (368)

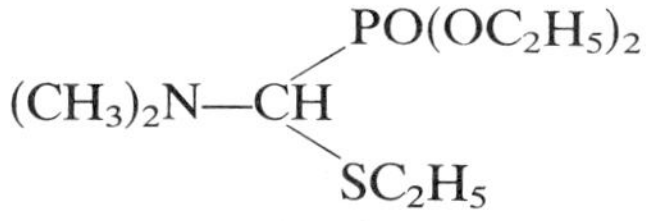

$$(CH_3)_2N-CH\begin{matrix}PO(OC_2H_5)_2\\SC_2H_5\end{matrix} \qquad \begin{matrix}R\\O_2N\end{matrix}\rangle N-CH_2-SR'$$

(369) 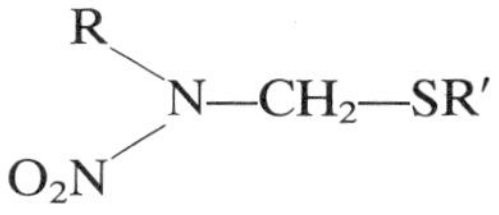   (370)

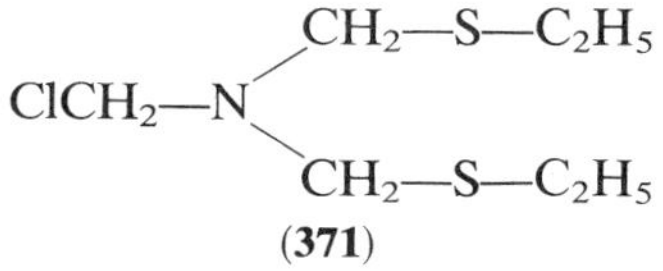

$$ClCH_2-N\begin{matrix}CH_2-S-C_2H_5\\CH_2-S-C_2H_5\end{matrix} \qquad N[CH_2-S-Si(C_6H_5)_3]_3$$

(371)   (372)

Iminium salt reactions with β-chloroethylmercaptan in acetonitrile afford hydrochlorides of dialkyl-(β-chloroethylmercaptomethyl)amines **373**. The free bases **375** undergo ring closure in methylene chloride at room temperature to give dialkylthiazolidinium chlorides **376** (319). These are also formed in acetonitrile from MIS reactions with ethylene sulfides, probably via episulfonium salt intermediates **374**. With peracetic acid **376** were oxidized to sulfones **377**, while with tertiary oxonium ions alkylations to sulfonium-ammonium salts (e.g., **380**) were achieved (320). By analogous reactions of **100** with γ-chloropropylmercaptan dialkyl-(γ-chloropropylmercaptomethyl)amines **378** and their cyclization products,

N,N-dialkyltetrahydro-1,3-thiazinium chlorides **379**, were prepared (319).

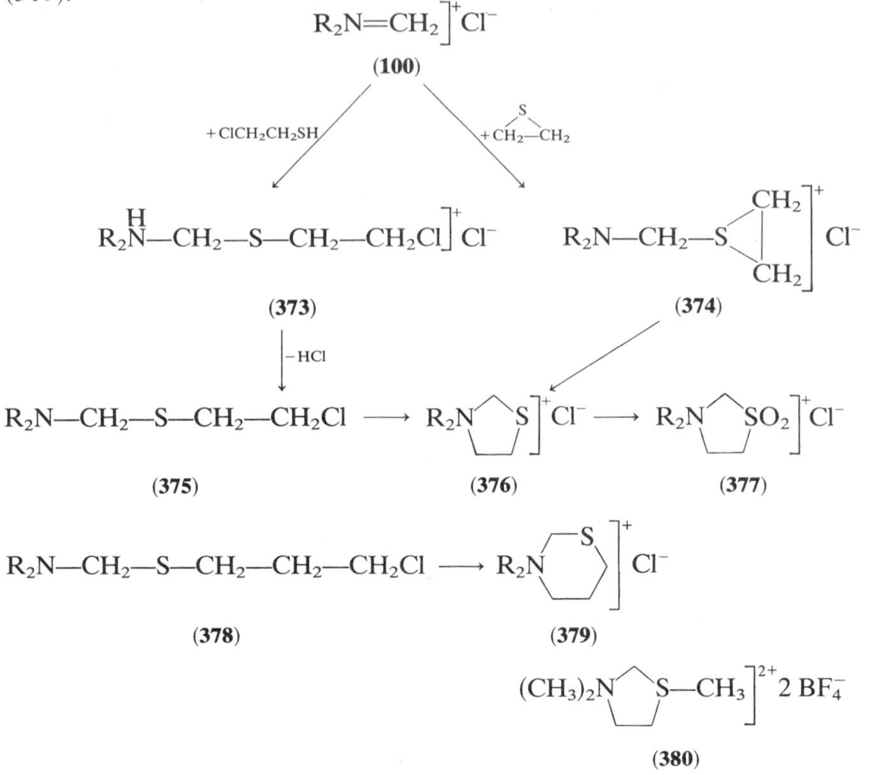

$$[R_2N{=}CH_2]^+Cl^-$$
(100)

+ClCH$_2$CH$_2$SH → $[R_2\overset{H}{N}{-}CH_2{-}S{-}CH_2{-}CH_2Cl]^+ Cl^-$
(373)

+CH$_2$–CH$_2$ (S) → $[R_2N{-}CH_2{-}S\overset{CH_2}{\underset{CH_2}{\big|}}]^+ Cl^-$
(374)

(373) $\xrightarrow{-HCl}$ R$_2$N—CH$_2$—S—CH$_2$—CH$_2$Cl → $[R_2N{\frown}S]^+ Cl^-$ → $[R_2N{\frown}SO_2]^+ Cl^-$
(375)   **(376)**   **(377)**

R$_2$N—CH$_2$—S—CH$_2$—CH$_2$—CH$_2$Cl → $[R_2N{\diagup}^{S}{\diagdown}]^+ Cl^-$
(378)   **(379)**

$[(CH_3)_2N{\frown}S{-}CH_3]^{2+}\ 2\,BF_4^-$
(380)

Treatment of MIH with alkali salts of *O*-alkyl thiocarbonates, xanthogenic acid, or monoalkyl trithiocarbonates afforded *S*-dialkylaminomethyl-*O*-alkyl thiocarbonates and dialkylaminomethylalkyl dithio- and trithiocarbonates **381–383** (321). Analogous reactions of MIH with alkali dithiocarbamates led to dialkylaminomethyl dithiocarbamates **384** (322), which were also obtained by the reaction of carbon disulfide with aminals (323,324).

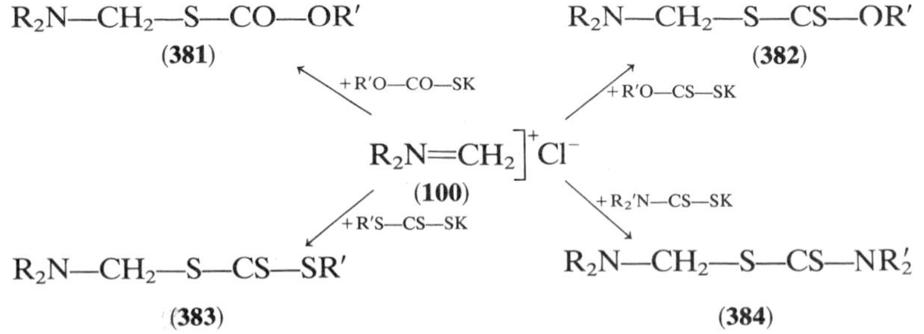

R$_2$N—CH$_2$—S—CO—OR'   R$_2$N—CH$_2$—S—CS—OR'
(381)   **(382)**

+R'O—CO—SK ↖   ↗ +R'O—CS—SK

$[R_2N{=}CH_2]^+Cl^-$
(100)

+R'S—CS—SK ↙   ↘ +R$_2$'N—CS—SK

R$_2$N—CH$_2$—S—CS—SR'   R$_2$N—CH$_2$—S—CS—NR$_2'$
(383)   **(384)**

Furthermore, C–S bond formation has also been carried out with N-acylhaloalkyl amides and imides. With mercaptans and thiophenols α-acylamidoalkyl sulfides, for example, **385** (186), **386** (186), **387** (186,325), **388** (106), **389** (311), and **390** (116) are formed. The thiolysis of N-chloromethyl carbamates (**87** and **89**) leads to the corresponding thiols (**391** and **392**) (111–113). With alkali sulfinates the products are sulfones such as **393** (103), **394** (99), **395** (106), and **396** (311). Sodium sulfite yields sulfonic acids, for example, **397** (326).

$C_6H_5-CO-NH-CH_2-SR$

(385)

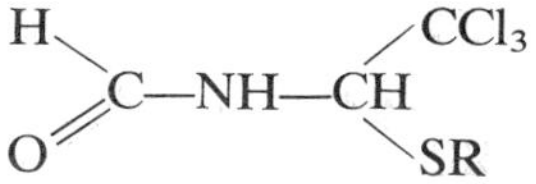

$\underset{O}{\overset{H}{>}}C-NH-CH\underset{SR}{\overset{CCl_3}{<}}$

(386)

(387) phthalimide-N-CH$_2$-SR

(388) isatin-N-CH$_2$-SR

(389) 2,3,3-trichloropiperidine with SR and N-CO-R

$C_6H_5-CO-NH-CH(S-C_6H_5)-CO-C_6H_5$

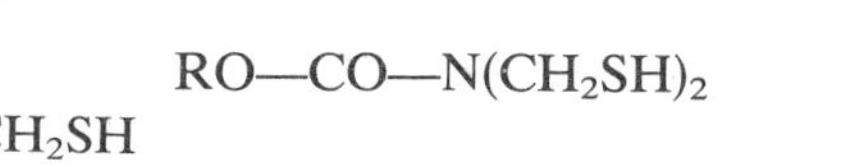

(390)

$RO-CO-N\underset{CH_2SH}{\overset{R'}{<}}$

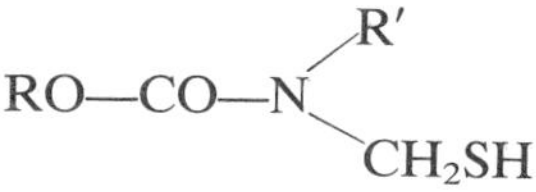

(391)

$RO-CO-N(CH_2SH)_2$

(392)

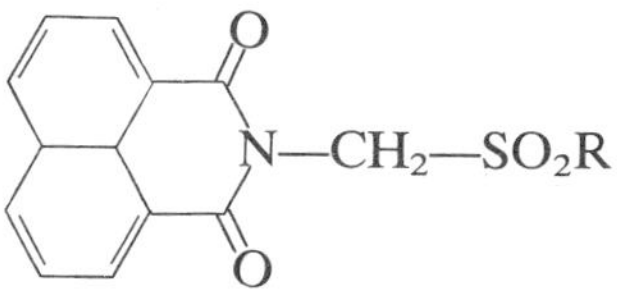

(393) naphthalimide-N-CH$_2$-SO$_2$R

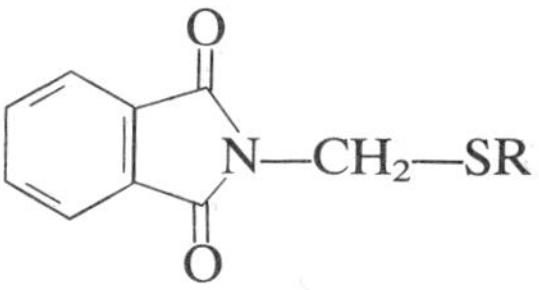

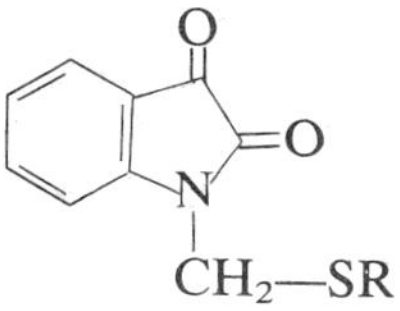

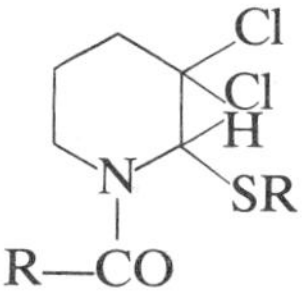

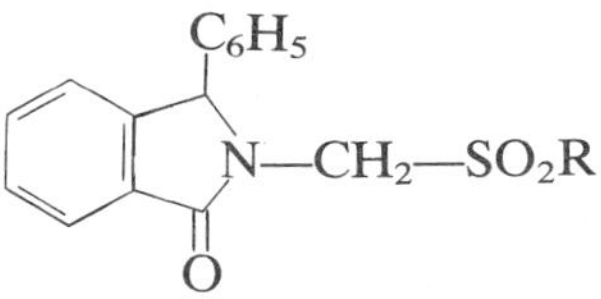

(394) 3-phenylisoindolin-1-one-N-CH$_2$-SO$_2$R

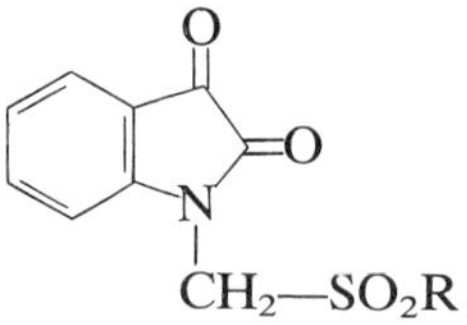

(395) 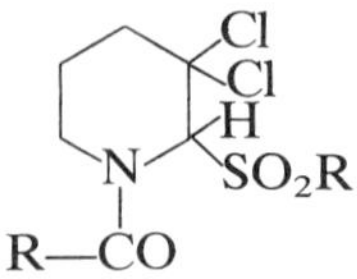 (396) 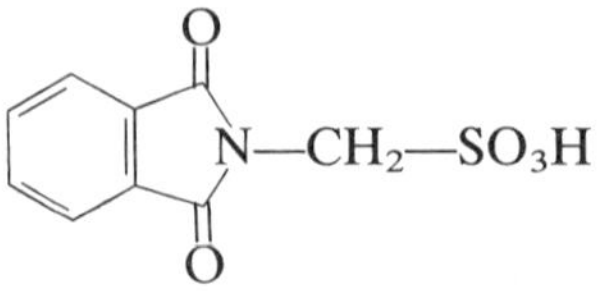 (397)

Condensation of the phenylglyoxal derivative ω-chloro-ω-acylaminoacetophenone (92) with thioamides or o-aminothiophenol provides an interesting route to the thiazole and benzothiazine heterocycles (398, 399) (120):

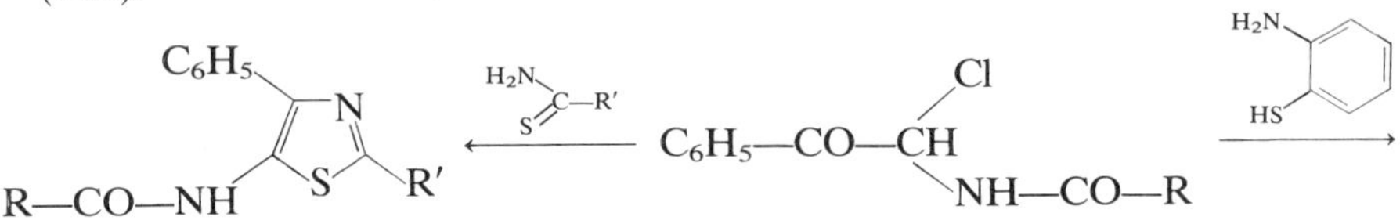

(398)   (92)

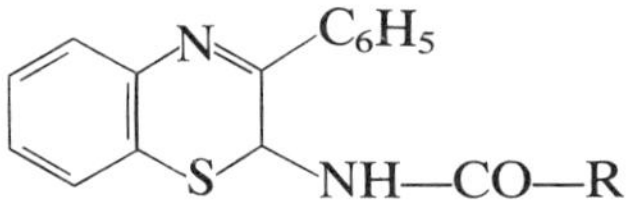

(399)

Analogously to the formation of 381–384, the reaction of N-acylchloromethylimides (e.g., 78, 82) with alkali salts of O-alkyl thiocarbonates, xanthogenic acid, or monoalkyl trithiocarbonates yields derivatives 400a–400c (321) and 401a (106), respectively. Depending on the reaction conditions, the thiocyanate ion attacks an electrophilic carbon at either the sulfur or the nitrogen side (see Section IV-G). Thus heating N-bromomethylphthalimide with potassium thiocyanate in acetone yields thiocyanate 400d (327). The analogue 401b, however, could only be isolated from N-chloromethylisatin with a silver thiocyanate suspension in acetone or acetonitrile at low temperatures (328).

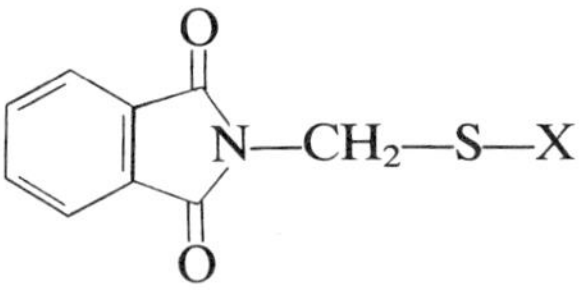

(400a) X = CO—OR'
(400b) X = CS—OR'
(400c) X = CS—SR'
(400d) X = CN

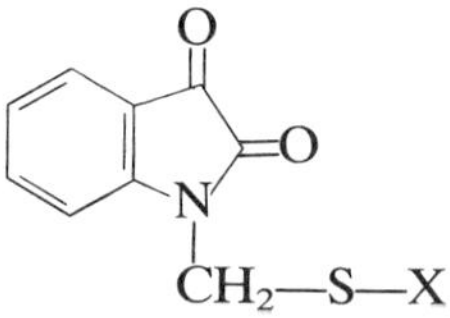

(401a) X = CS—NR$_2$
(401b) X = CN

G. C–N BOND FORMATION

Reactions of iminium salts with primary or secondary amines are of interest only in special cases for the preparation of aminals. Thus from MIH and 2,4,6-trichloro- and 2,4,6-trimethyl-*N*-methylaniline in acetonitrile solution the hydrochlorides of unsymmetrical aminals **402** were prepared (140). With cyanamide the bishydrochlorides of bis-(dialkylaminomethyl)cyanamides **403** are formed; these give the free bases upon treatment with base. These compounds are also accessible from condensation of the components in the presence of triethylamine, or from MIH with bissodium cyanamide (329).

Nitramines and iminium salts, as well as *N*-chloromethylnitramines, were starting materials for aminals of types **404** (330) and **405** (331). Hydroxylamines yielded *N*-hydroxyaminals **406** (332). The reaction of phosphonoiminium salts with secondary amines led to aminals **407** (332). Several intermediates are involved in the preparation of aminoalkylated aminals of type **408** from α-haloiminium salts **352** with alkali amides as reactive nucleophiles (312). Similarly, with MIH the corresponding derivatives of amides, imides (333), and azaanalogous sulfones (334) have been synthesized (e.g., **409, 410, 411**). The aminomethylation of tetracycline with MIH is of synthetic utility for the preparation of water-soluble antibiotics of type **412** [Reverin® (335)].

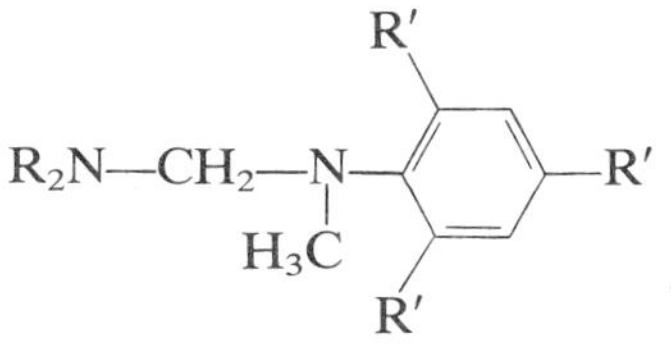

(**402**) R′ = CH$_3$, Cl

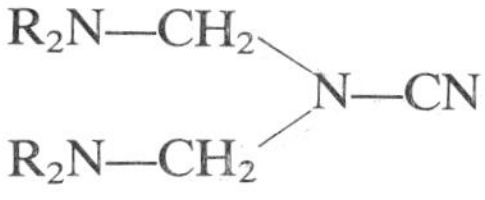

(**403**)

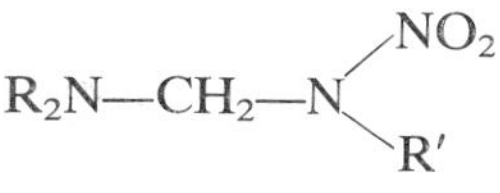

(**404**)

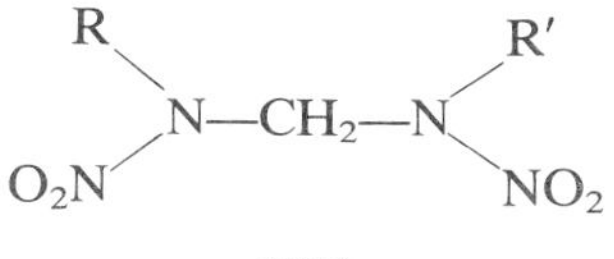

(**405**)

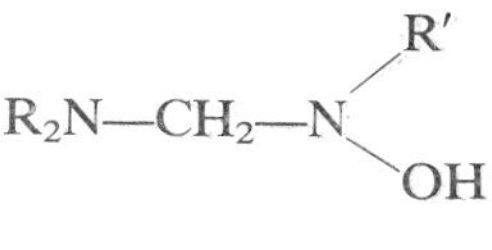

(**406**)

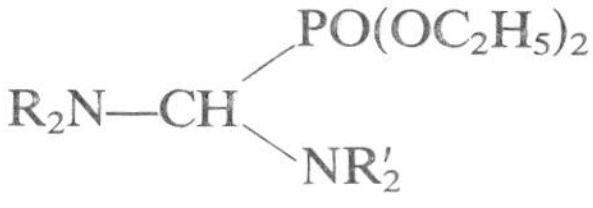

(**407**)

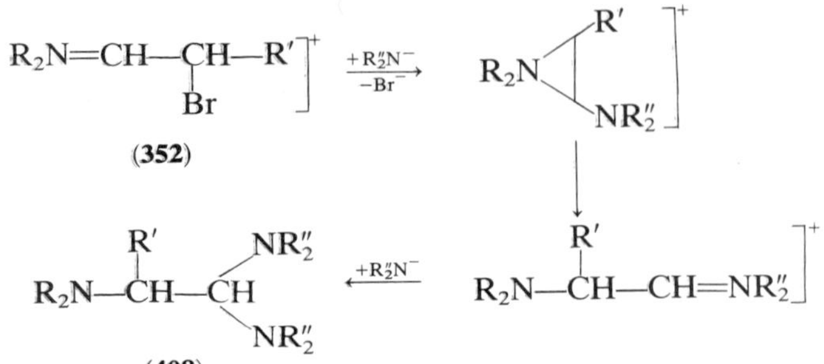

$$R_2N=CH-CH-R' \overset{+R_2''N^-}{\underset{-Br^-}{\longrightarrow}} \left[R_2N \underset{NR_2''}{\overset{R'}{\triangle}} \right]^+$$
$$\overset{Br}{}$$
(352)

$$R_2N-\underset{}{\overset{R'}{CH}}-CH\underset{NR_2''}{\overset{NR_2''}{\diagdown}} \overset{+R_2''N^-}{\longleftarrow} \left[R_2N-\overset{R'}{\underset{}{CH}}-CH=NR_2'' \right]^+$$

(408)

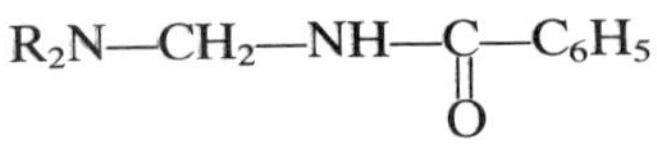

$$R_2N-CH_2-NH-\underset{O}{\overset{}{C}}-C_6H_5$$

(409)

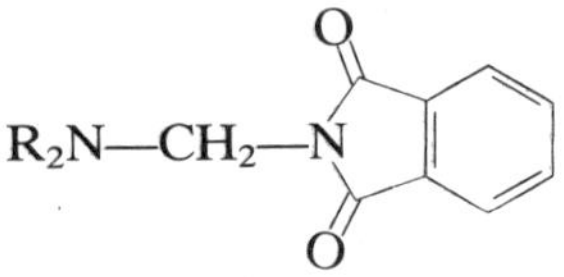

R₂N—CH₂—N (phthalimide)

(410)

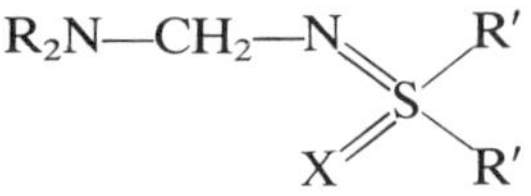

$$R_2N-CH_2-N\underset{X}{\overset{S}{\diagup\diagdown}}\underset{R'}{\overset{R'}{}}$$

(411)

X = O, NH

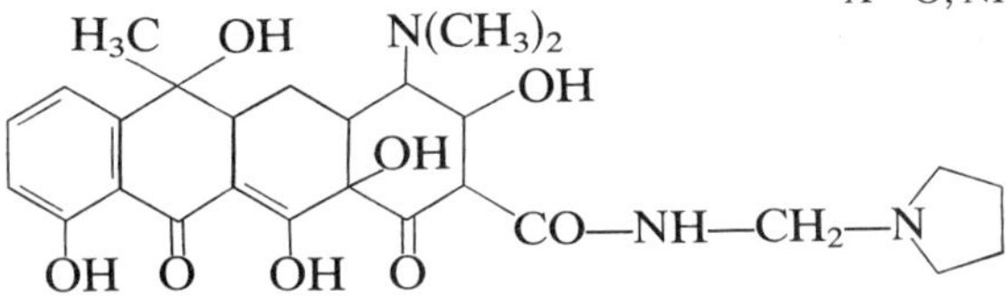

(412) [tetracycline-like structure with CO—NH—CH₂—N(pyrrolidine)]

Reactions of MIH with tertiary amines have been studied in detail. In aprotic, polar solvents monoquaternary salts of aminals (173) are obtained. Often these are also accessible by alkyl halide addition to aminals (132,172). Salts of type **183** are highly reactive electrophiles, which undergo hydrolysis and thermal decomposition very readily (174) (see Section II-F-1-f). In both cases the first step seems to be dissociation into an iminium salt and the tertiary amine. The reverse reaction, as mentioned above, can be applied also for the preparation of monoquaternary salts of unsymmetrical aminals [e.g., **413–416** (173,336)]. In the case of pyridine as a weak tertiary amine ($pK_a = 5.29$) the state of equilibrium allows isolation of the thermally unstable salts (**417**) only at low temperatures, because at room temperature dissociation back into pyridine and the iminium salt is energetically favored (173,337). Increasing stability is

IV. REACTIONS OF METHYLENIMINIUM SALTS 171

observed for analogous salts of the more basic 4-methylpyridine (pK_a = 6.11). In the case of 2-methylpyridine, however, steric hindrance to quaternization on nitrogen by means of an iminium salt favors a Mannich-type aminomethylation reaction on the 2-methylcarbon (see Section IV-I-6).

Aziridines are also quaternized by MIH; thus from N-butylaziridine and N-methylenepiperidinium chloride the monoquaternary salt **418** (R = n-C_4H_9) was prepared (338). However, in the case of N-β-phenylethylaziridine, instead of the expected salt **418** (R = C_6H_5—CH_2—CH_2) the rearranged product **419** was isolated. Similarly, bisquaternary or tertquaternary imidazolinium salts **421** were obtained from reactions of MIH with dialkyl- or monoalkyl-β-chloroethylamines. The intermediate monoquaternary salts **420** could be isolated at low temperatures (338). From 1-chloro-3-dimethylaminopropane with dimethylmethyleniminium chloride the bisquaternary hexahydropyrimidinium salt **423** is formed, via monoquaternary salt **422**, in low yields. Therefore for the preparation of this type of salts it appears to be more useful to carry out the alkylation of 1,3-dimethylhexahydropyrimidines directly with alkyl iodides in acetonitrile, or with trialkyloxonium salts in dichloroethane or nitromethane (339).

$(CH_3)_2N=CH_2]^+$ Hal⁻ $\xrightleftharpoons{(CH_3)_3N}$ $(CH_3)_2N$—CH_2—$N(CH_3)_3]^+$ Hal⁻ $\xleftarrow{CH_3Hal}$

(**105**) (**183**)

$(CH_3)_2N$—CH_2—$N(CH_3)_2$

(**103**)

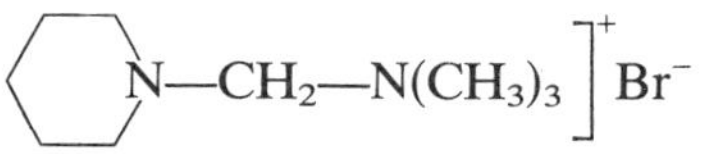

[⟨N⟩—CH_2—$N(CH_3)_3$]⁺ Br⁻   [$(CH_3)_2N$—CH_2—N(CH₃)⟨⟩]⁺ Br⁻

(**413**) 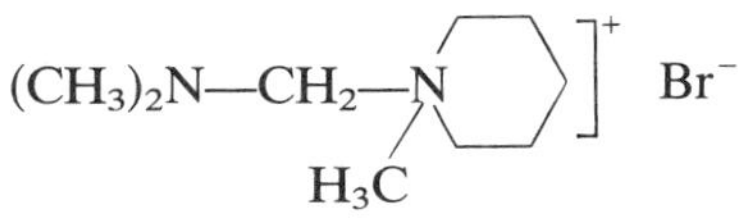   (**414**)

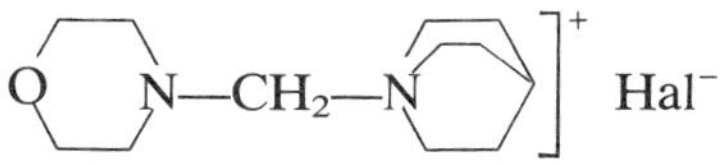

[O⟨N⟩—CH_2—N⟨⟩]⁺ Hal⁻   [R_2N—CH_2—N⟨N-N⟩]⁺ Hal⁻

(**415**) 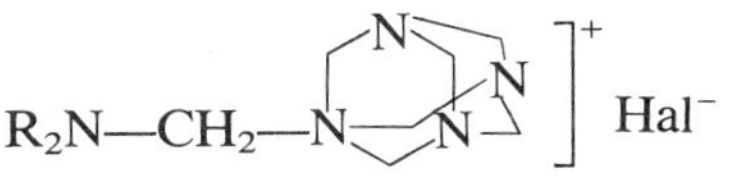   (**416**)

[R_2N—CH_2—N⟨⟩]⁺ Hal⁻   [⟨N⟩—CH_2—N△—R]⁺ Cl⁻

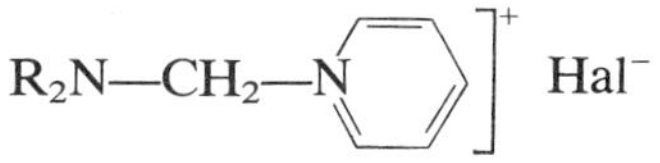

(**417**) 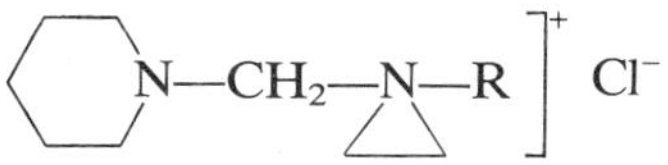   (**418**)

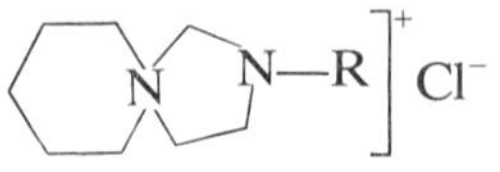

(419) [structure]

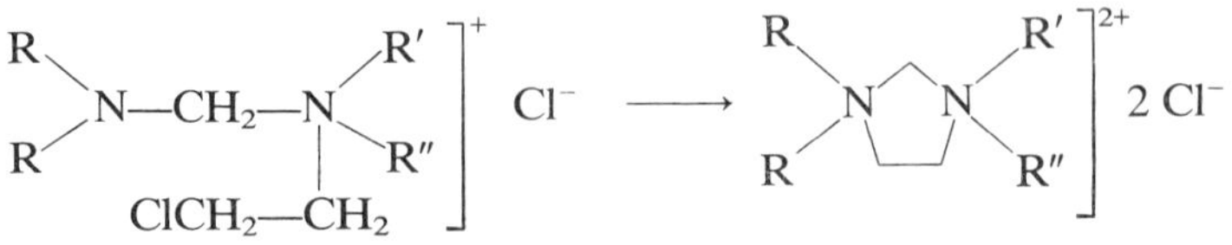

(420) → **(421)**

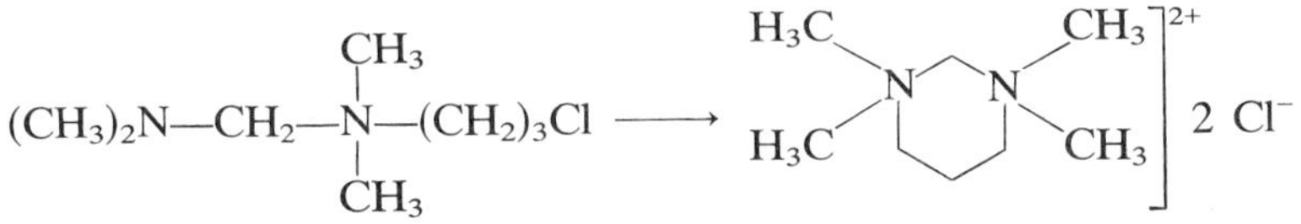

(422) → **(423)**

Heterocyclic compounds have also been synthesized from MIH and β-substituted hydrazines. For instance, *N*-methyl-*N*-β-chloroethylhydrazine (**424**) forms a hydrochloride (**425**), deprotonation of which under ring closure affords the triazinium salt **427** (340). The same synthetic route with chloroacetic *N'*,*N'*-dimethylhydrazide (**426**) via salt **428** and subsequent deprotonation leads to the imidazolidinium chloride **429** (340):

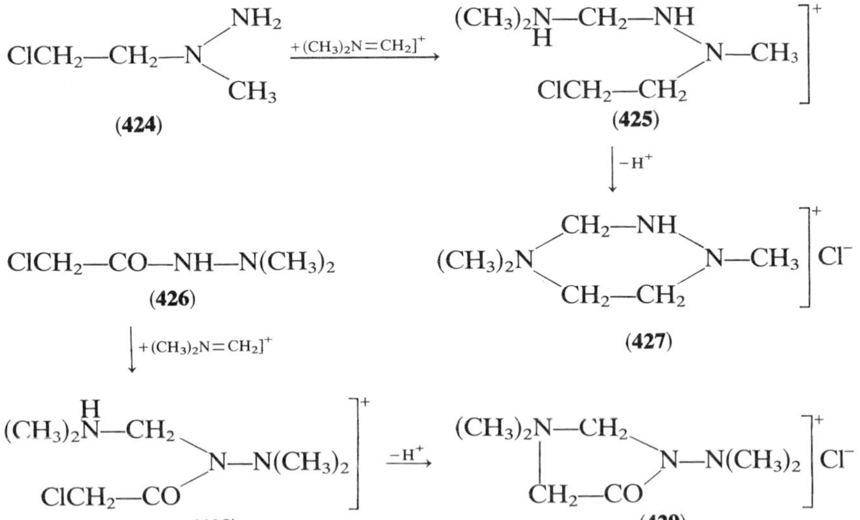

(**424**) $\xrightarrow{+(CH_3)_2N=CH_2]^+}$ (**425**) $\xrightarrow{-H^+}$ (**427**)

(**426**) $\xrightarrow{+(CH_3)_2N=CH_2]^+}$ (**428**) $\xrightarrow{-H^+}$ (**429**)

Aminals have also been used as nucleophilic components in MIH reactions, but because of their instability the monoquaternary salts could not be isolated. When **103** was added at −20° to a suspension of **361** in acetonitrile, a clear solution was obtained, indicating that reaction took place. Upon addition of ether a mixture of 80% **185** and 20% **361** was precipitated. This result leads to the conclusion that in solution dissociation equilibria of the intermediate salt **430** with mixtures of **103**+**361**, as well as of **431**+**185**, are present (175):

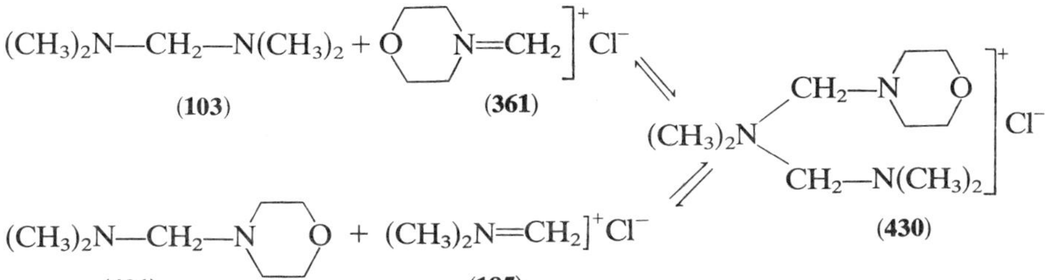

$(CH_3)_2N—CH_2—N(CH_3)_2$ + $\left[O\overset{\frown}{\underset{\smile}{N}}=CH_2 \right]^+ Cl^-$
(**103**)   (**361**)

$\left[(CH_3)_2N\underset{\diagdown}{\diagup} \begin{matrix} CH_2—N\overset{\frown}{\underset{\smile}{\,}}O \\ CH_2—N(CH_3)_2 \end{matrix} \right]^+ Cl^-$
(**430**)

$(CH_3)_2N—CH_2—N\overset{\frown}{\underset{\smile}{\,}}O$ + $[(CH_3)_2N=CH_2]^+ Cl^-$
(**431**)   (**185**)

Reactions of triethylamine with the iminium chloride (**185**) effect deprotonation of the MIH on the iminium carbon (see Section IV-D).

By means of sodium azide in aqueous solution or with a suspension of silver azide in methylene chloride the MIH have been converted to dialkylazidomethylamines **432** (341). These are distillable liquids; in contact with water they are decomposed to give a dialkylammonium azide and formaldehyde. Their reactions with phenylacetylene yield aminomethylphenyl-1,2,3-triazoles (**433**, **435**), while alkylation with methyl iodide affords quaternary salts **434**.

$R_2N—CH_2—N_3$
(**432**)

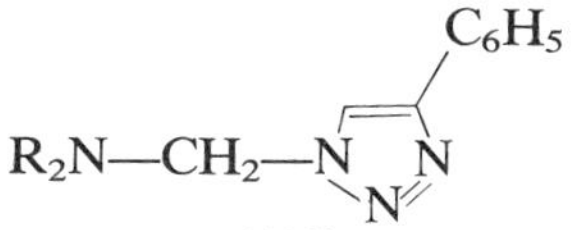

$R_2N—CH_2—N\underset{\diagdown N\diagup}{\overset{\frown}{N}}\!\!\overset{C_6H_5}{\underset{}{N}}$
(**433**)

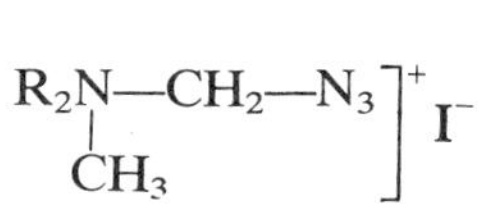

$\left[\begin{matrix} R_2N—CH_2—N_3 \\ | \\ CH_3 \end{matrix} \right]^+ I^-$
(**434**)

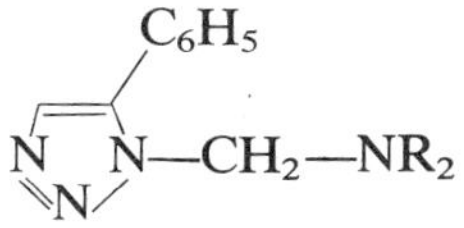

$\underset{\diagdown N\diagup}{N\overset{\frown}{\,\,}N}\!\!\overset{C_6H_5}{\underset{}{}}—CH_2—NR_2$
(**435**)

With amines the N-chloromethylamides or N-chloromethylimides undergo reactions analogous to those observed with MIH. From trichloroacetamido- or benzamidomethyl chloride, for example, with aniline, piperidine, or pyridine compounds such as **436**, **437**, and **438** (228,310), and with bis-β-chloroethylamine condensation products **439** (342) were prepared. N-Halomethylphthalimide with aniline or pyridine gave **440**

R—CO—NH—CH$_2$—NH—C$_6$H$_5$

(436)

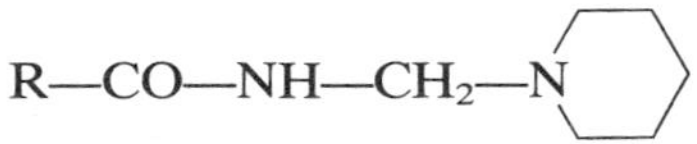

R—CO—NH—CH$_2$—N(piperidine)

(437)

R—CO—NH—CH$_2$—N$^+$(pyridinium)

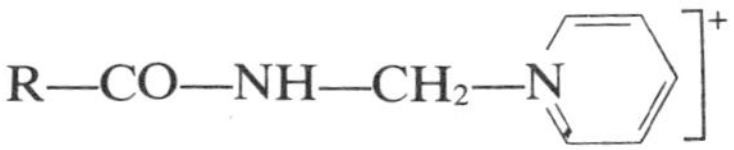

(438)

R—CO—NH—CH$_2$—N(CH$_2$CH$_2$Cl)$_2$

(439)

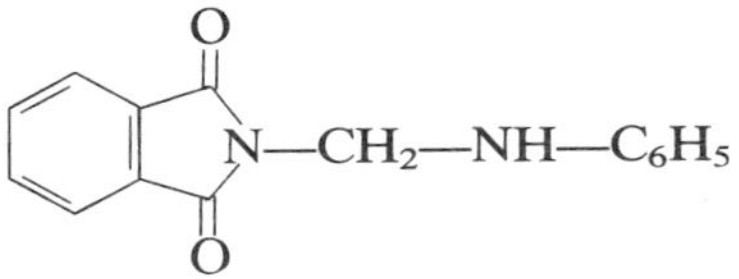

(phthalimide)—N—CH$_2$—NH—C$_6$H$_5$

(440)

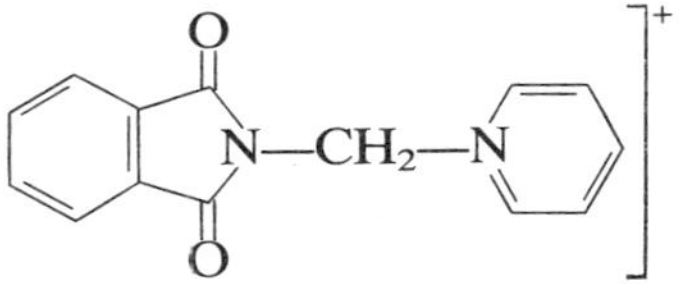

[pyridinium—N—CH$_2$—N(phthalimide)]$^+$

(441)

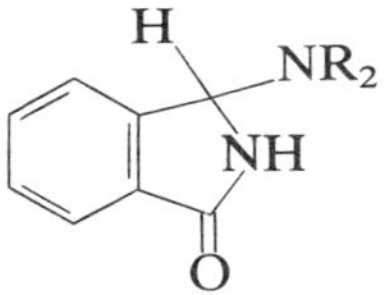

(isoindolinone)—CH(NR$_2$)(NH)—H

(442)

(C$_6$H$_5$—CO—NH—CH$_2$)$_3$N

(443)

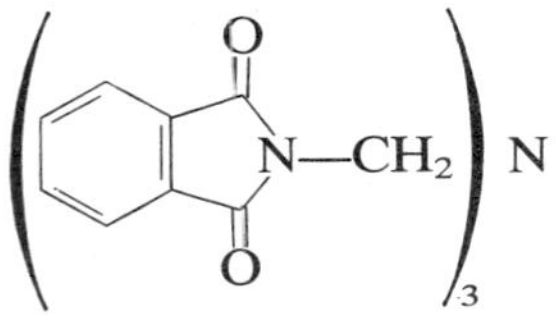

(phthalimide—N—CH$_2$—)$_3$N

(444)

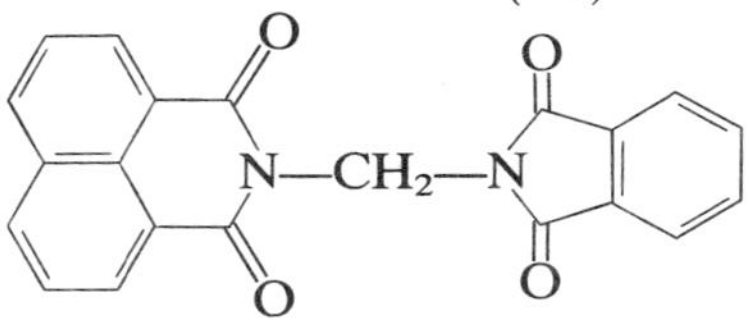

(naphthalimide)—N—CH$_2$—N(phthalimide)

(445)

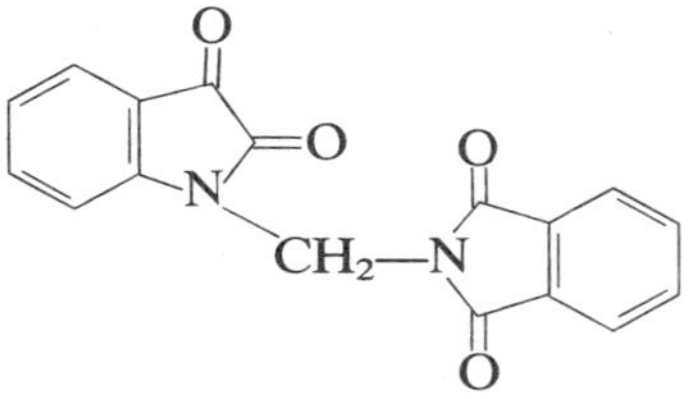

(phthalimide)—N—CH$_2$—N(phthalimide)

(446)

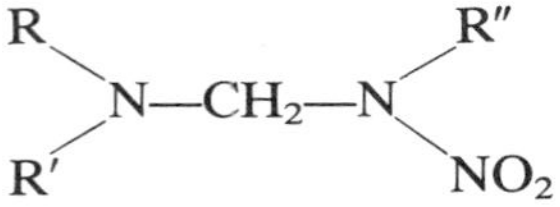

R—N(CH$_2$—N(R')(R''))—NO$_2$...

R\N—CH$_2$—N(R'')—NO$_2$
R'/

(447)

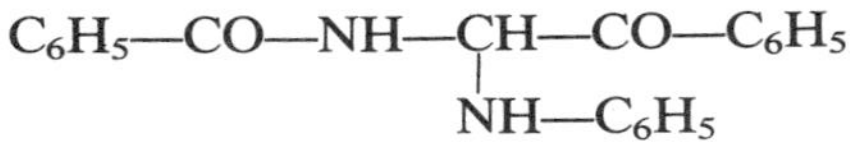

C$_6$H$_5$—CO—NH—CH(NH—C$_6$H$_5$)—CO—C$_6$H$_5$

(448)

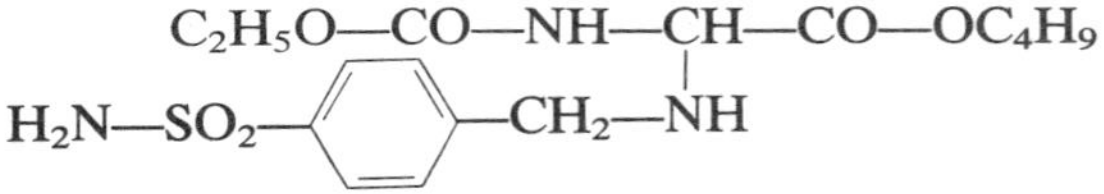

C$_2$H$_5$O—CO—NH—CH(CH$_2$—NH—C$_6$H$_4$—SO$_2$—NH$_2$)—CO—OC$_4$H$_9$

(449)

174

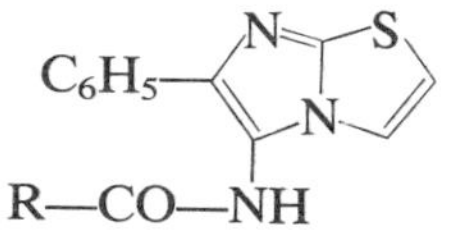

IV. REACTIONS OF METHYLENIMINIUM SALTS 175

(450) C₆H₅-[imidazothiazole]-R—CO—NH

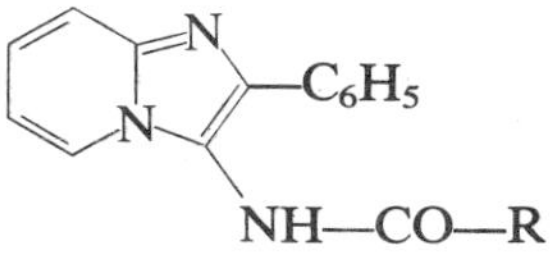

(451) [imidazopyridine]-C₆H₅, NH—CO—R

(343) or **441** (344), respectively. Reactions of 3-chlorophthalimidine (**94**) with secondary amines led to **442** (123). Ammonia reacted with **127** and **78** to give trisubstituted derivatives **443** (345) and **444** (344). With imides methylene bisimides such as **445** (103) and **446** (106,346) were obtained. The products of secondary amine reactions with N-chloromethylnitramines are aminals of type **447** (145). With derivatives of phenylglyoxal or glyoxylic acid esters **92** compounds **448** and **449** have been synthesized (116,117) and, 2-aminothiazole or 2-aminopyridine yields heterocycle **450** or **451**, respectively (120).

Depending on the reaction conditions, thiocyanates (see also Section IV-F) and N-chloromethylamides or N-chloromethylimides have also been converted to isothiocyanates. Thus N-chloromethylphthalimide (**78**) and potassium thiocyanate, with a catalytic amount of sodium iodide in dimethylformamide (347), give derivative **452**, which is also formed in an eutectic melt of potassium and sodium thiocyanate (106) or with trimethylsilyl isothiocyanate (348). Isothiocyanates **453** and **454** could be prepared from the corresponding N-chloromethylamides by heating with potassium or ammonium thiocyanate in acetone or dimethylformamide (328,346,349). With silver cyanate and N-chloromethylamides in boiling benzene isocyanates **455** were prepared (350).

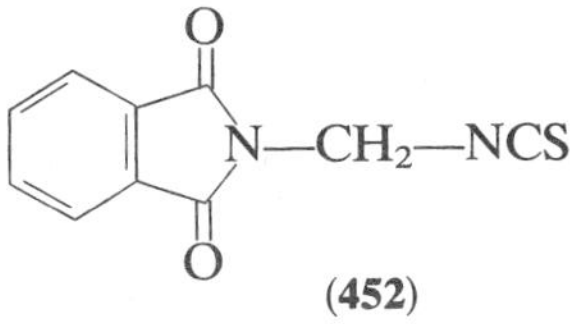

(452) phthalimide-N—CH₂—NCS

(453) R—CO—NH—CH₂—NCS

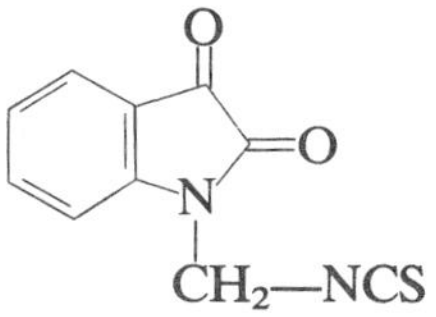

(454) [indolin-2-one]-CH₂—NCS

(455) R—CO—NH—CH₂—NCO

H. C–P BOND FORMATION

Reactions of dimethylmethyleniminium chloride with lithium diphenyl (351) or dimethylphosphide (352) gave the dimethylaminomethylphosphines **456**, which can be distilled and are inflammable upon contact with air. The same type of compounds can also be obtained by condensation of dialkyl amines with formaldehyde and dialkyl phosphines (352a). Triphenylphosphine and MIH in methylene chloride form an equilibrium with phosphonium salts **457**, which can be precipitated at low temperatures by the addition of ether (337). Because of the state of equilibrium at room temperature they decompose more or less readily into the thermodynamically favored starting components; only the morpholine derivative was found to be fairly stable. In water, however, they are all rapidly hydrolyzed to give triphenylphosphine and formaldehyde in quantitative yield. Thermally more stable are the corresponding salts of trialkyl phosphines; however, the stability depends on the nature of the iminium salt. Triethyl-(4-dimethylamino-α-morpholinobenzyl)phosphonium chloride (**458**), which could be obtained as a colorless salt from a concentrated ice-cooled solution of the components in acetonitrile, rapidly dissociates at higher temperatures into the yellow iminium salt **292e** and triethylphosphine. In UV-absorption studies the dissociation was shown to be almost complete in dilute solutions at room temperature. In contrast, the triethylphosphonium salt **459** is remarkable stable (m.p. 140°) (337); in aqueous solution no decomposition occurs, and conversion into a pikryl sulfonate has been achieved. Although no hydrolysis takes place in acidic solution, with an excess of base triethylphosphine and formaldehyde are immediately formed. The tributyldimethylaminomethylphosphonium chloride **460** (282) has also been described; this, upon the action of potassium tert-butoxide or methyllithium, affords the same reaction products as are obtained in the deprotonation reaction of dimethylmethyleniminium chloride (see Section IV-D); no ylide is formed:

$(CH_3)_2N\text{—}CH_2\text{—}PR_2$
(**456**)

$[R_2N\text{—}CH_2\text{—}P(C_6H_5)_3]^+ \text{Hal}^-$
(**457**)

[morpholino-N=CH—C$_6$H$_4$—N(CH$_3$)$_2$]$^+$ Cl$^-$ + (C$_2$H$_5$)$_3$P ⇌
(**292e**)

[morpholino-N—CH(P(C$_2$H$_5$)$_3$)—C$_6$H$_4$—N(CH$_3$)$_2$]$^+$ Cl$^-$
(**458**)

[morpholino-N—CH$_2$—P(C$_2$H$_5$)$_3$]$^+$ Br$^-$
(**459**)

$(CH_3)_2N\text{—}CH_2\text{—}P(C_4H_9)_3]^+Cl^-$
(**460**)

With trialkyl phosphites **461** MIH **100** undergo a Michaelis-Arbusov type of reaction. Elimination of an alkyl halide leads to formation of dialkylaminomethyldialkyl phosphonates **462** (162,353), which are also obtained by condensation of secondary amines with formaldehyde and dialkyl phosphites (354) or by the reaction of dialkyl aminomethyl ethers with dialkyl chlorophosphites (188). Aromatic or heteroaromatic substituted derivatives of type **463** can be converted by means of sodium hydride into reactive carbanions. These are strong nucleophiles and have been used for reactions with various carbonyl reagents to form enamines **465**, which are substituted in the α and β positions by aromatic or heteroaromatic groups and are converted to the corresponding carbonyl compounds **464** upon hydrolysis (150). With butyllithium as base, compounds of type **462** have been shown to undergo an analogous reaction sequence, by means of which ketones are transformed into α-substituted aldehydes (355).

$$R_2N=CH_2]^+ \; Cl^- + P(OR')_3 \longrightarrow R_2N-CH_2-PO(OR')_2 + R'Cl$$
$$\quad\quad (100) \quad\quad\quad\quad (461) \quad\quad\quad\quad\quad (462)$$

$$R_2N-\underset{R''}{\underset{|}{CH}}-\underset{\diagdown O \diagup}{\overset{||}{P}}(OR')_2 \xrightarrow{-H^+} \left[R_2N-\underset{R''}{\underset{|}{\bar{C}}}-\underset{\diagdown O \diagup}{\overset{||}{P}}(OR')_2 \longleftrightarrow R_2N-\underset{R''}{\underset{|}{C}}=\underset{|\underline{O}|}{\overset{|}{P}}(OR')_2 \right]^-$$

(463)

$$\Big\downarrow \begin{array}{l} +R'''COR'''' \\ -O_2P(OR')_2^- \end{array}$$

$$R''-\underset{O}{\overset{||}{C}}-CH \underset{R'''}{\overset{R''''}{\diagup}} \xleftarrow{H_2O} R_2N-\underset{R''}{\underset{|}{C}}=\underset{R'''}{\underset{|}{C}}-R''''$$

$$\quad\quad (464) \quad\quad\quad\quad\quad (465)$$

The formation of C–P bonds has also been achieved with N-chloromethylamides and N-chloromethylimides; for example, triphenylphosphine gave the phosphonium salts **466** and **467** (356). From phthalimido-, succinimido-, and o-sulfobenzimidomethyl halides and sodium diphenyl phosphide the corresponding imidomethylated phosphines **468–470** were prepared (357). In an analogous fashion, the corresponding arsines are also accessible. With trialkyl phosphites (358,359) or phosphinous acid esters (360) reactions of the Michaelis-Arbusov type take place with the formation of α-amidoalkylphosphonic or phosphinous acid esters (e.g., **471–473**), and also of α-amidoalkyldiaryl phosphinoxides **474**. N-Tetrachloroethylbenzamide and triethyl phosphite or diphenylphosphinic acid esters yield derivatives **475**, which undergo HCl elimination to give the enamides **476** (359,359a).

[R—CO—NH—CH$_2$—P(C$_6$H$_5$)$_3$]$^+$ Cl$^-$

[Phthalimido-N—CH$_2$—P(C$_6$H$_5$)$_3$]$^+$ Br$^-$

(466) R = CF$_3$, CCl$_3$, C$_6$H$_5$ (467)

Phthalimido-N—CH$_2$—P(C$_6$H$_5$)$_2$

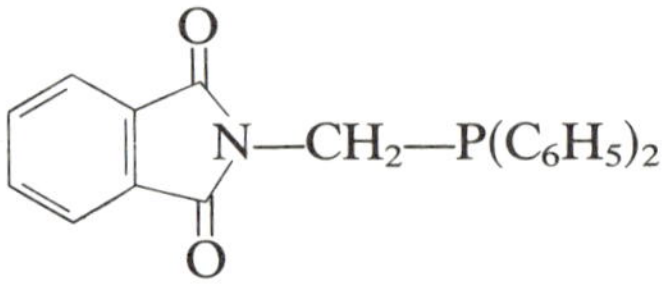

(468)

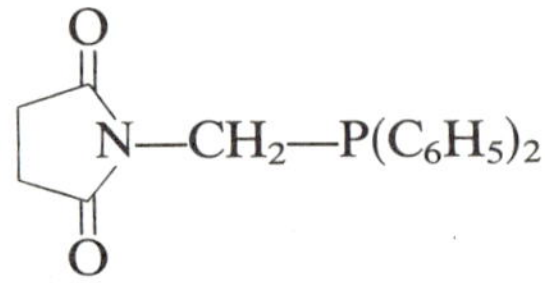

Succinimido-N—CH$_2$—P(C$_6$H$_5$)$_2$

(469)

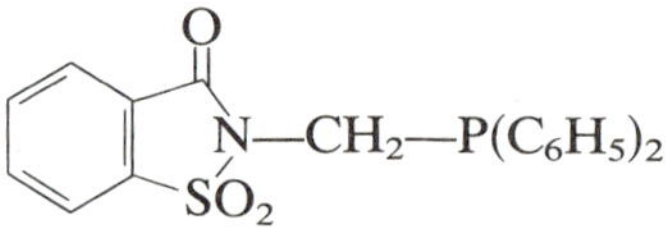

Saccharin-N—CH$_2$—P(C$_6$H$_5$)$_2$

(470)

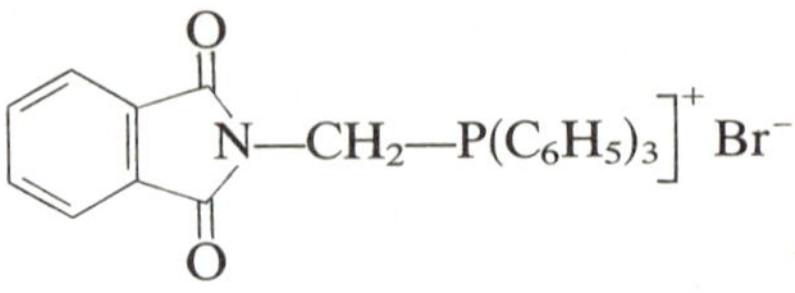

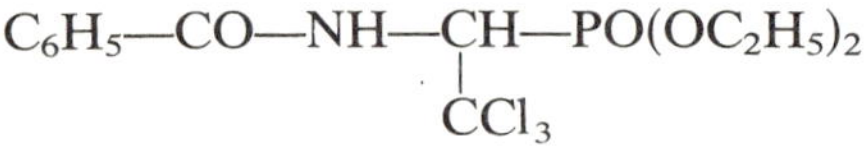

C$_6$H$_5$—CO—NH—CH(CCl$_3$)—PO(OC$_2$H$_5$)$_2$

(471)

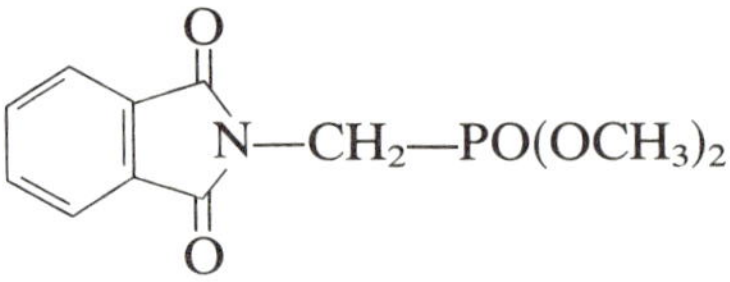

Phthalimido-N—CH$_2$—PO(OCH$_3$)$_2$

(472)

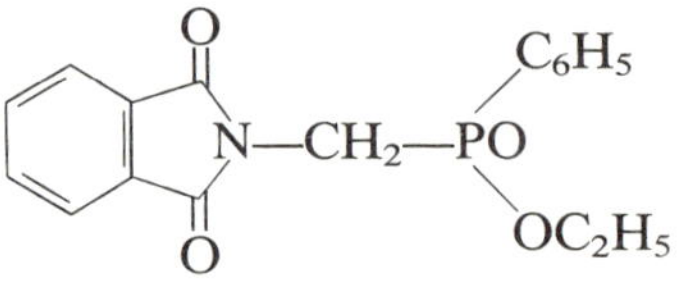

Phthalimido-N—CH$_2$—PO(C$_6$H$_5$)(OC$_2$H$_5$)

(473)

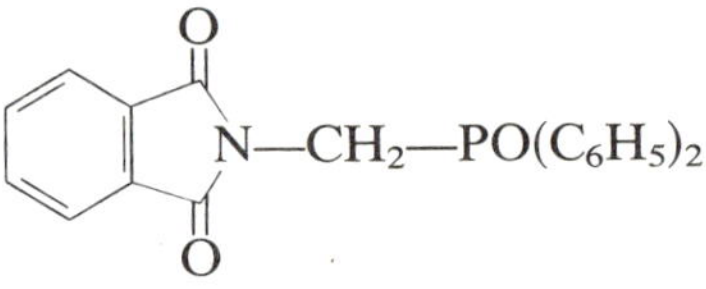

Phthalimido-N—CH$_2$—PO(C$_6$H$_5$)$_2$

(474)

C$_6$H$_5$—CO—NH—CH(CCl$_3$)—POR$_2$ $\xrightarrow{-HCl}$ C$_6$H$_5$—CO—NH—C(=CCl$_2$)—POR$_2$

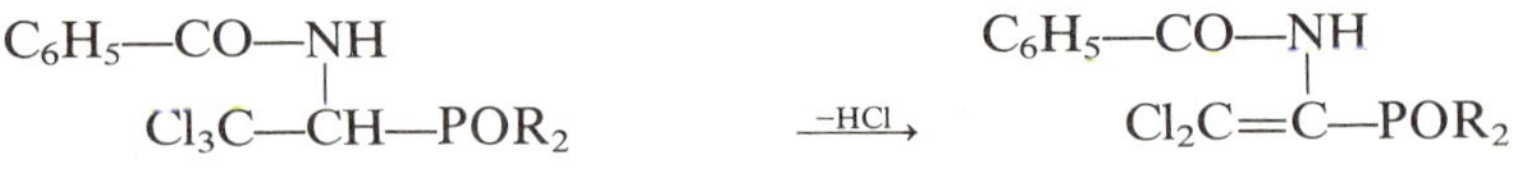

(475) (476)

R = C$_2$H$_5$O, C$_6$H$_5$

I. C–C BOND FORMATION

1. *Reactions with Cyanides*

Solvolysis of MIH with hydrogen cyanide is an exothermic reaction that yields the hydrochlorides of dialkylaminoacetonitriles **477** (127). This reaction has often been used for the characterization of iminium salts, for example, in the case of cleavage reactions of imidazolidines with acyl chlorides (157). The nitriles obtained from hydrochlorides **477** with base are usually stable, covalent compounds. They have also been isolated from MIS reactions in aqueous alkali cyanide solutions (13,315). Iminium salts **110** and **111**, which are cleavage products of benzaldehyde or isobutyraldehyde aminals, and also the piperidene hydrochloride (**115**) can be dissolved in hydrogen cyanide to form **478–480** quantitatively. They have also been prepared exothermally from aminals themselves in hydrogen cyanide solution (36). The cyano group of α-aminonitriles can enter into a number of other reactions, including elimination to iminium salts (see Section II-C).

$$[R_2N=CH_2]^+ \text{ Hal}^- \xrightarrow{\text{HCN}} [R_2\overset{H}{N}-CH_2-CN]^+ \text{ Hal}^-$$
$$(108) \qquad\qquad\qquad (477)$$

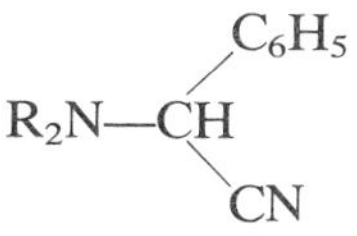

$$R_2N-\underset{CN}{\overset{C_6H_5}{CH}} \qquad R_2N-\underset{CN}{\overset{CH(CH_3)_2}{CH}} \qquad \underset{CN}{\underset{|}{\bigcirc}}{NH}$$

$$(478) \qquad\qquad (479) \qquad\qquad (480)$$

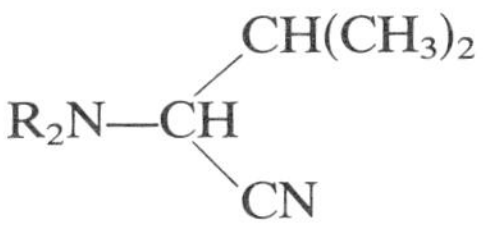

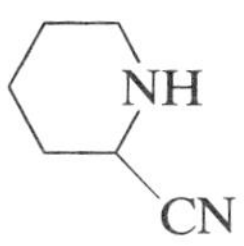

In the case of α-haloalkyl amides potassium cyanide reactions have already been reviewed for compounds of type **481**, which give the dehydrochlorinated products **482** of the initially formed α-cyanoalkyl amides (5). From **483** as precursors nitriles **485** were synthesized by means of hydrogen cyanide addition to the corresponding N-acylimine intermediates **484** (5,361,362):

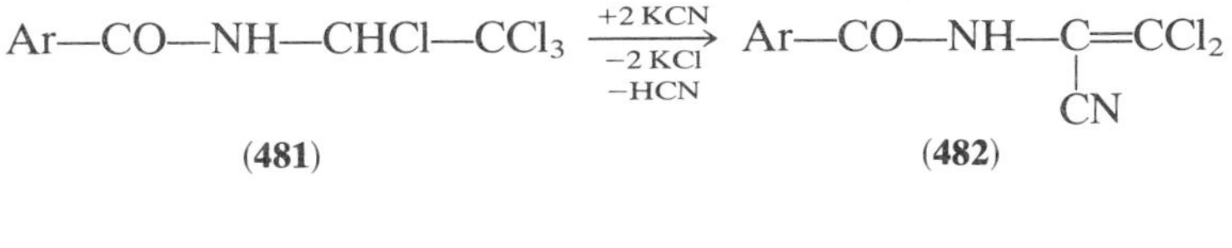

$$\text{Ar}-\text{CO}-\text{NH}-\text{CHCl}-\text{CCl}_3 \xrightarrow[\substack{-2\text{KCl} \\ -\text{HCN}}]{+2\text{KCN}} \text{Ar}-\text{CO}-\text{NH}-\underset{CN}{\overset{|}{C}}=CCl_2$$
$$(481) \qquad\qquad\qquad\qquad (482)$$

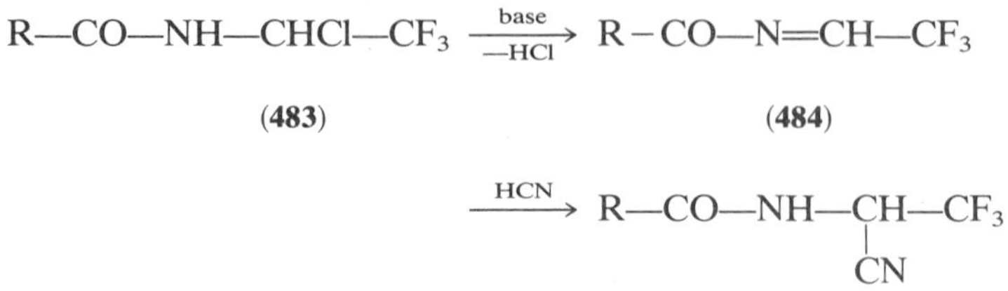

$$\text{R—CO—NH—CHCl—CF}_3 \xrightarrow[-\text{HCl}]{\text{base}} \text{R—CO—N=CH—CF}_3$$

(483) (484)

$$\xrightarrow{\text{HCN}} \text{R—CO—NH—CH(CN)—CF}_3$$

(485)

2. Reactions with Active C–H Bonds

Many reactions of MIH with tertiary CH-acidic compounds or with alkali salts of these have been described. As indicated by structures **486–492**, they include the use of monosubstituted β-dicarbonyl compounds such as monoalkyl- or monoarylmalonic esters (363), methanetricarboxylic acid esters (364), α-alkylacetoacetic esters (363), alkyl- or arylindanediones-1,3 (140,363), 1,1-bis(alkylsulfonyl)ethanes, and halobis(alkylsulfonyl)methanes (365).

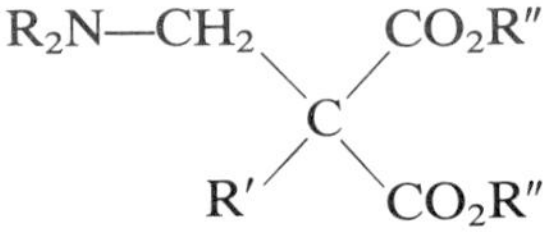

(486) $R_2N\text{—}CH_2\text{—}C(R')(CO_2R'')_2$

(487) $R_2N\text{—}CH_2\text{—}C(CO_2R')_3$

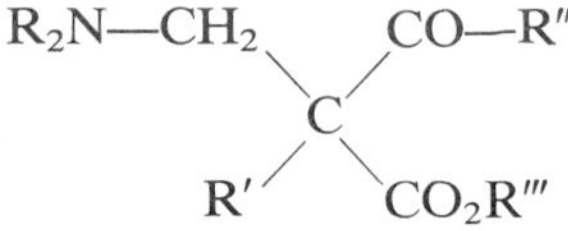

(488) $R_2N\text{—}CH_2\text{—}C(R')(CO\text{—}R'')(CO_2R''')$

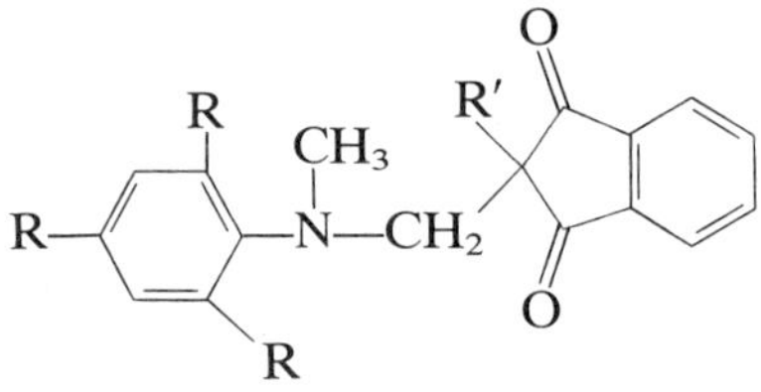

(489) arylindanedione with $R_2N\text{—}CH_2$— substituent (CH₃, N—CH₂, aryl ring with R substituents)

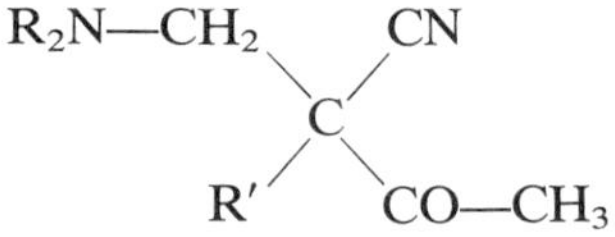

(490) $R_2N\text{—}CH_2\text{—}C(R')(CN)(CO\text{—}CH_3)$

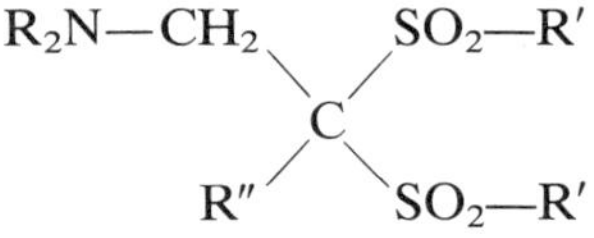

(491) $R_2N\text{—}CH_2\text{—}C(R'')(SO_2\text{—}R')_2$

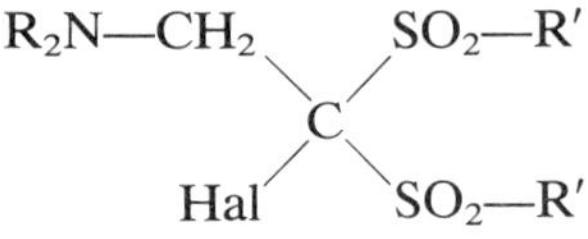

(492) $R_2N\text{—}CH_2\text{—}C(\text{Hal})(SO_2\text{—}R')_2$

Secondary CH-acidic methylene compounds also react with MIH, primarily with the formation of aminomethylation products, for example, **493** from cyclohexanone and **494** from dimedone (165). In many cases, however, the monosubstituted product undergoes a secondary amine 1,2-elimination to an olefinic derivative, which can be isolated

IV. REACTIONS OF METHYLENIMINIUM SALTS

when suitable substituents are present (e.g., in **495**) (366). Usually in a subsequent Michael addition of a second molecule of the CH-acidic component the methylene bisderivative is formed as the final product.

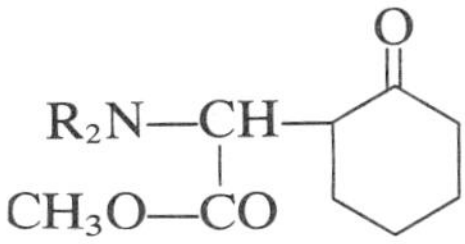

<pre>
        O   O
        ‖   ‖
R₂N—CH—⟨cyclohexanone⟩   R₂N—CH—⟨ ⟩—CH₃   Cl   Cl
   | | ‖   \   /
CH₃O—CO   CH₃O—CO   ⟨ ⟩—CH₃   ⟨cyclopentadiene⟩
                                        ‖   Cl / \ Cl
                                        O   C
                                                     / \
                                                    H   R

     (493)   (494)   (495)
</pre>

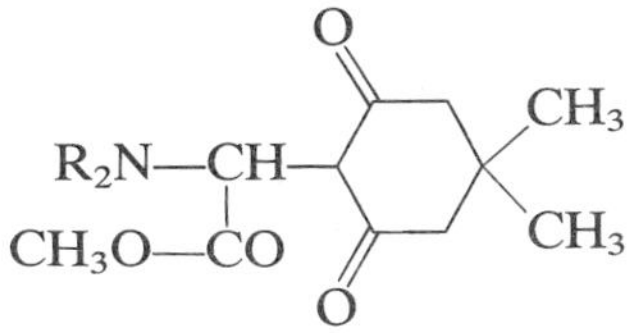

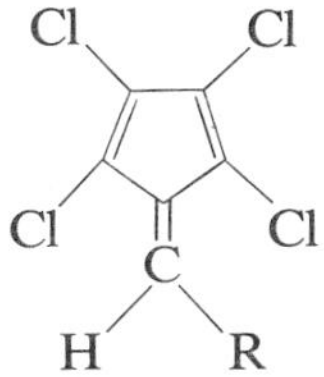

Aminomethylation by means of MIH has been performed also on CH-acidic methyl groups to form, for example, derivatives of alkyl aryl ketones, chalcones, or nitromethane, as indicated by structures **496–500** (165,201,363):

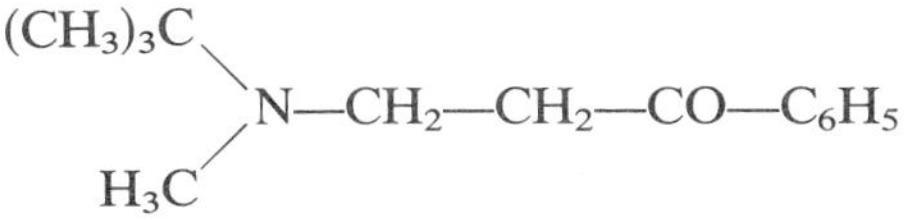

$(CH_3)_3C$
$\quad\quad\quad\diagdown$
$\quad\quad\quad\quad N-CH_2-CH_2-CO-C_6H_5 \quad\quad R_2N-CH-CH_2-CO-Ar$
$\quad\quad\quad\diagup \quad\quad\quad\quad\quad\quad\quad\quad\quad\quad\quad\quad\quad\quad\quad |$
$H_3C \quad\quad\quad\quad\quad\quad\quad\quad\quad\quad\quad\quad\quad\quad\quad\quad (R'O)_2PO$

$\quad\quad\quad$ (496) 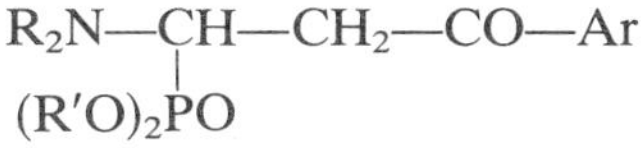 $\quad\quad\quad\quad\quad\quad\quad\quad\quad\quad\quad\quad\quad\quad$ (497)

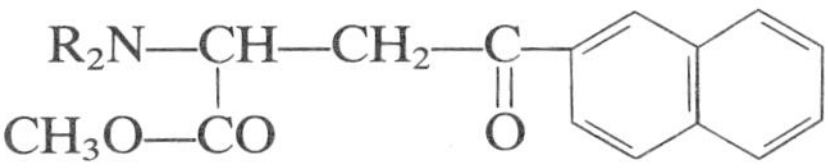

$\quad\quad\quad\quad R_2N-CH-CH_2-C-[naphthyl]$
$\quad\quad\quad\quad\quad\quad | \quad\quad\quad\quad\quad\quad ‖$
$\quad\quad\quad\quad CH_3O-CO \quad\quad\quad\quad O$

$\quad\quad\quad\quad\quad\quad\quad$ (498)

$R_2N-CH-CH_2-CO-CH=CH-C_6H_5 \quad\quad R_2N-CH_2-CH_2-NO_2$
$\quad\quad\quad\quad |$
$\langle piperidine \rangle$
$\quad N-CO$

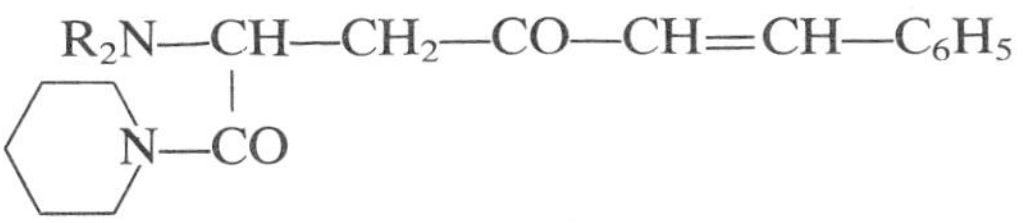

$\quad\quad\quad$ (499) 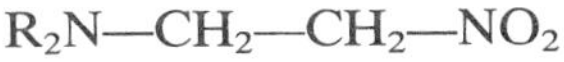 $\quad\quad\quad\quad\quad\quad\quad\quad\quad\quad\quad\quad\quad\quad$ (500)

The reaction products summarized in structures **486–500** are often not accessible by the usual aminomethylation methods, such as the Mannich condensation. For corresponding reactions of vinyl-analogous formamidinium salts see chapter 4 and 12.

For amidomethylation reactions of CH-acidic compounds, such as β-diketones and methylene-active esters, the N-halomethylamides or N-halomethylimides are the reagents of choice (98,99,103,117,365,367).

Most of the reactions reported in this category have already been reviewed (4–6). Under basic conditions, amidoalkylation reactions with N-haloalkylated secondary carbonamides, for example, the N-acyl-α-chloroglycine derivatives **501**, appear to proceed by an elimination-addition mechanism via acyl imines (e.g., **502**), forming compounds of type **503** (368). Of interest also are some novel ring-closure reactions that have been observed for suitably substituted N-α-haloalkyl amides. Thus from acylation products of aldimines with cyanoacetyl chloride (**16**; see Section II-A-3) the β-lactams **504** are formed (369) by elimination of HCl. Aldimines and malonyl chloride yield **505** (370) and the anisidine derivative **506** reacts with phenoxyacetyl chloride with formation of **507** (371). From dimethylketazine and ethyl chloroformate a pyrazoline derivative (**508**) has been obtained (24) via addition product **19**.

R—CO—CHCl—NH—CO—R' $\xrightarrow{-HCl}$

(**501**)

R—CO—CH=N—CO—R' $\xrightarrow{+HY}$

(**502**)

R—CO—CHY—NH—CO—R'

(**503**)

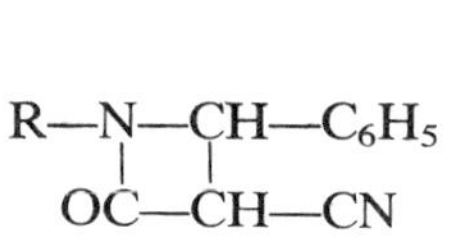

$$\begin{array}{c} R-N-CH-C_6H_5 \\ |\quad\;\;| \\ OC-CH-CN \end{array}$$

(**504**)

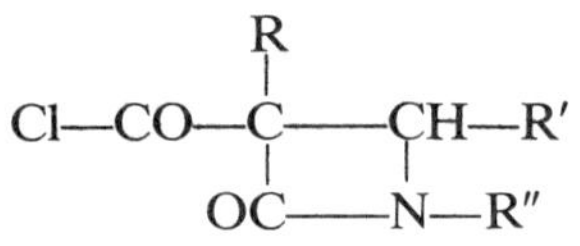

$$\begin{array}{c} \quad\;\; R \\ \quad\;\; | \\ Cl-CO-C-CH-R' \\ \quad\;\;|\quad\;\;\;\;| \\ \quad\;\; OC-N-R'' \end{array}$$

(**505**)

$$\begin{array}{c} H_5C_6 \quad CO_2CH_3 \\ \diagdown\;\;\diagup \\ C \\ \| \\ N-\bigcirc-OCH_3 \end{array} \longrightarrow \begin{array}{c} C_6H_5O \quad CO_2CH_3 \\ \diagdown\quad | \\ CH-C-C_6H_5 \\ |\quad\quad | \\ OC-N-\bigcirc-OCH_3 \end{array}$$

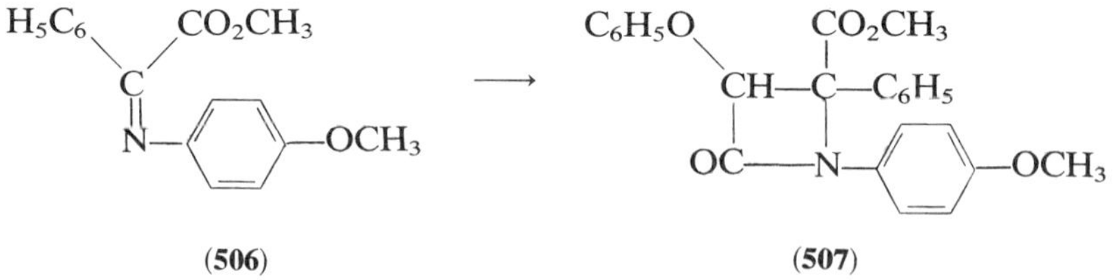

(**506**) (**507**)

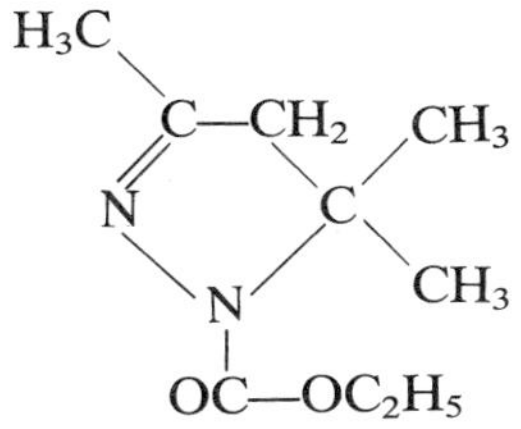

(508)

3. Reactions with Grignard and Other Organometallic Reagents

With Grignard reagents and MIH tertiary amines can be prepared. Dimethylmethyleniminium bromide, for example, reacts with phenyl- or methylmagnesium bromide to produce amines of type **509** (127); from methyl-*tert*-butyl and methylmesitylmethyleniminium perchlorate compounds **510** and **511** have been obtained (201). With allylmagnesium bromide, γ,δ-unsaturated amines (e.g., **512** or **513**) were synthesized; to these, hydrogen chloride can be added to form γ-chloroamines [e.g., **514** (147)]. Tris(chloromethyl)- and methylbis(chloromethyl)amines **181** and **180** are converted with phenylmagnesium bromide to tribenzyl- and methyldibenzylamine, respectively (169).

$(CH_3)_2N-CH_2-R$

(509)

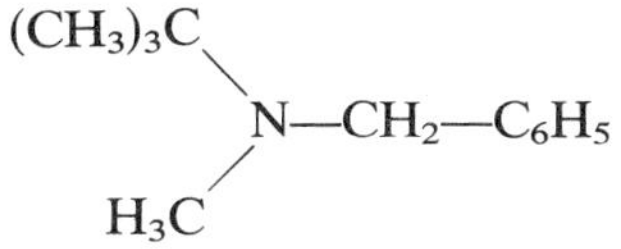

(510)

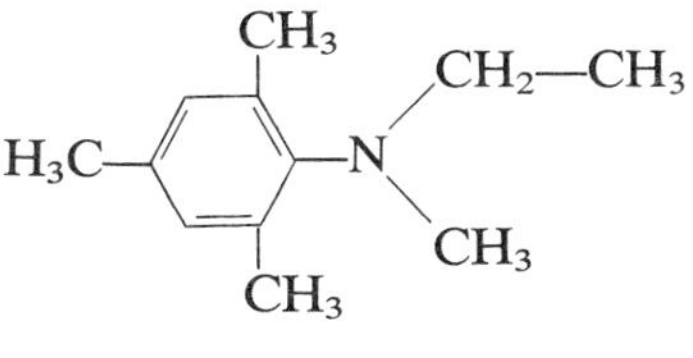

(511)

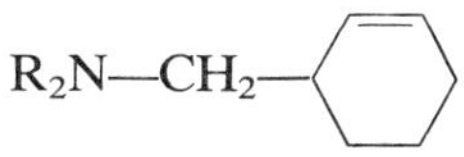

(512)

$R_2N-CH_2-CH_2-CH=CH_2 \xrightarrow{HCl} R_2N-CH_2-CH_2-CHCl-CH_3$

(513) (514)

Iminium salts derived from aliphatic, aromatic, or heteroaromatic aldehydes, as well as from aromatic or alicyclic ketones, undergo analogous reactions. Thus N-propylidenepiperidinium chloride (**515**) and benzylmagnesium chloride give 1-piperidino-1-phenylbutane (**516**) (372), and the iminium perchlorates **517** react with p-methoxy- or p-dimethylaminophenylmagnesium bromide to give tertiary amines of type **518** (295):

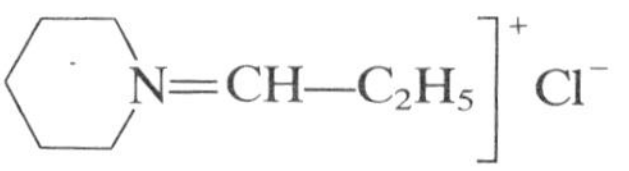

(**515**) 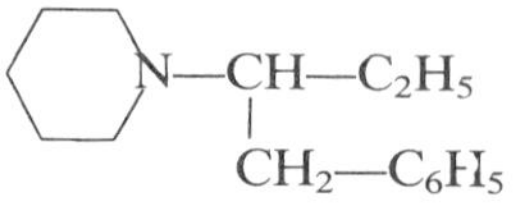   (**516**)

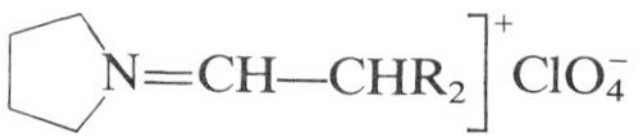

(**517**) 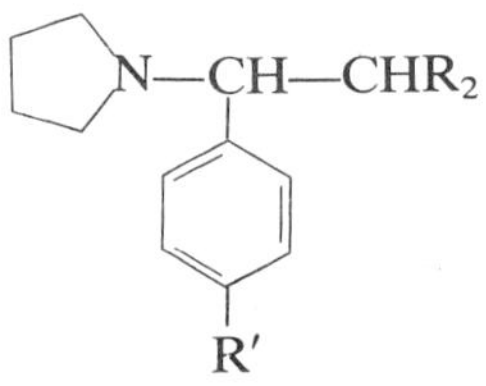   (**518**)

a   R = CH$_3$, R' = CH$_3$O
b   R = CH$_3$, R' = (CH$_3$)$_2$N
c   R = C$_2$H$_5$, R' = CH$_3$O
d   R = C$_2$H$_5$, R' = (CH$_3$)$_2$N

Reactions of dialkylarylmethyleniminium chlorides with aryl- or benzylmagnesium bromide give products of types **519** (147) and **520** (15). The yellow iminium chloride **292e** (see Section III-B) and the bisiminium salts **142** (see Section II-F-1-c) react with methylmagnesium bromide to give tertiary amines **521** and **522** (149,75). Arylmagnesium bromide additions to iminium salt **524**, which is a derivative of furfural, lead to amines of types **523** and **525** (15).

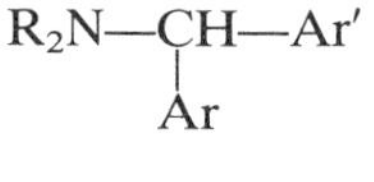

R$_2$N—CH—Ar'   R$_2$N—CH—CH$_2$—C$_6$H$_5$
       | |
       Ar   Ar

(**519**) 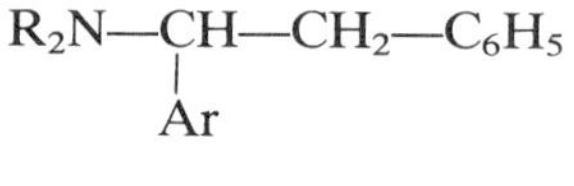   (**520**)

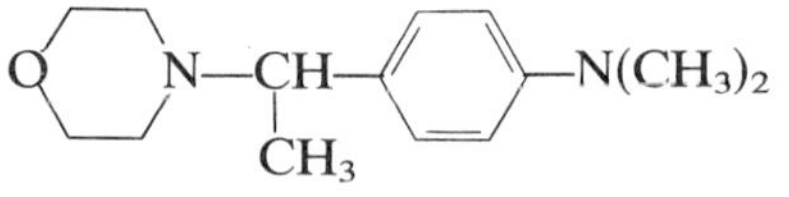

(**521**)

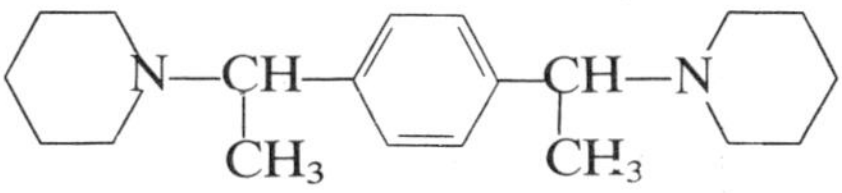

(522)

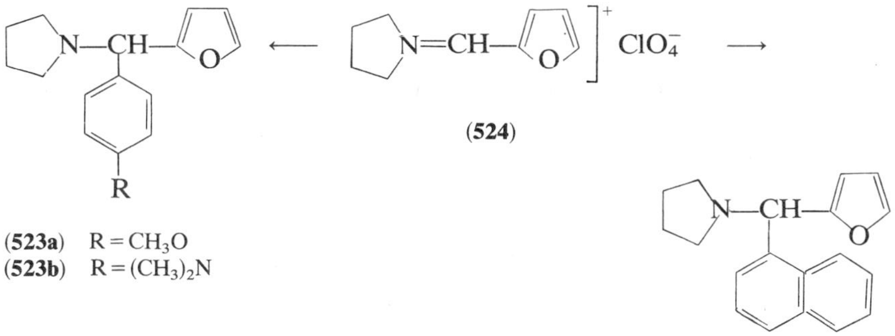

(524)

(523a) R = CH₃O
(523b) R = (CH₃)₂N

(525)

The first examples of the reaction of iminium salts derived from ketones, such as **4**, **526**, and **527**, are those with benzylmagnesium chloride to give the tertiary amines **528** (13), **529** (253), and **530** (315). Iminium salts **531** and **533**, derivatives of cyclohexanone and cyclopentanone, respectively, react with arylmagnesium bromides to produce the corresponding amines **532** and **534** in moderate yields (15).

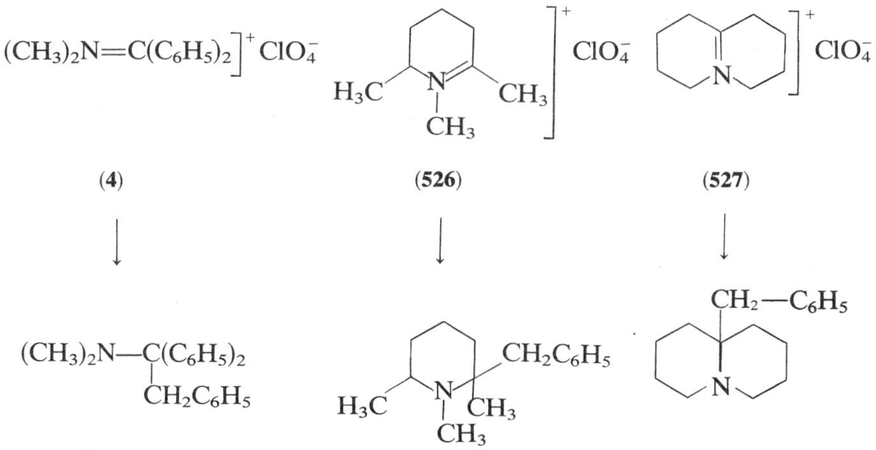

(4)   (526)   (527)

↓   ↓   ↓

(528)   (529)   (530)

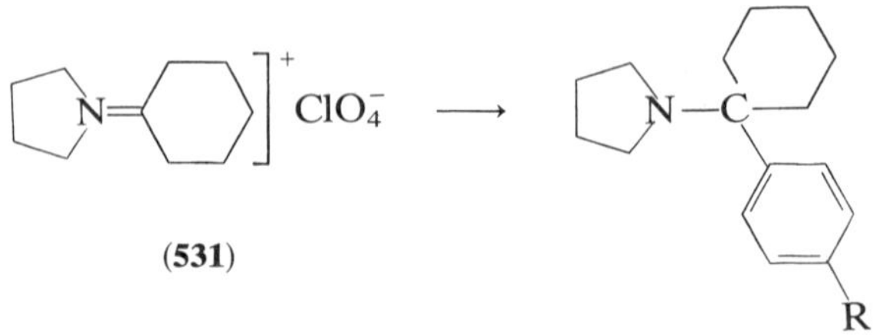

(531) → (532)

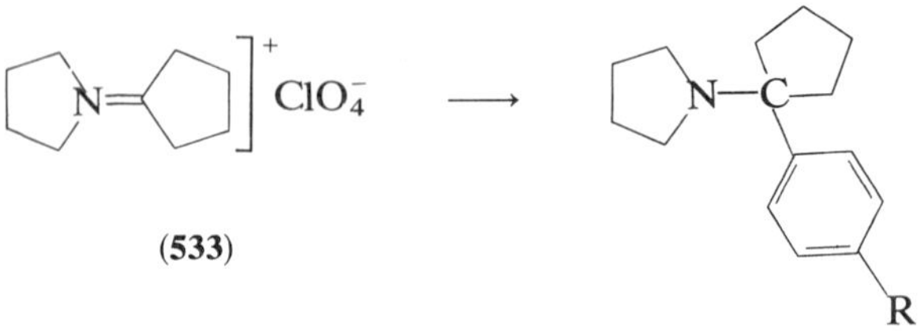

(533) → (534)

a R=CH₃,
b R=CH₃O

Occasionally in the case of Grignard reagents having bulky groups such as isopropyl-, *tert*-butyl-, cyclohexyl-, or certain arylmagnesium halides, a hydride transfer to the iminium salts derived from cyclohexanone (294), aromatic aldehydes (147), or benzophenone (15) was observed to be the preferred reaction (see Section IV-C). The two competing pathways have been examined further concerning the stereochemistry of the reaction components (294).

Iminium salts of type **46**, which can be obtained by bromine addition to enamines (see Section II-B-4), react with alkylmagnesium bromides to give branched β-bromoalkylamines **535**. On further addition of Grignard reagent tertiary amines **537** are formed (68) via anion exchange to **536** and subsequent hydrolysis:

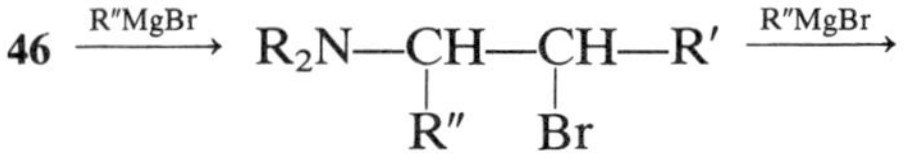

$$46 \xrightarrow{R''MgBr} \underset{(535)}{R_2N-\underset{R''}{CH}-\underset{Br}{CH}-R'} \xrightarrow{R''MgBr}$$

$$\underset{(536)}{R_2N-\underset{R''}{CH}-\underset{MgBr}{CH}-R'} \xrightarrow{H_2O} \underset{(537)}{R_2N-\underset{R''}{CH}-CH_2R'}$$

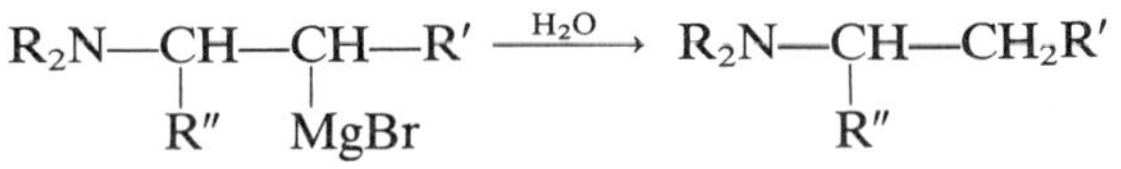

IV. REACTIONS OF METHYLENIMINIUM SALTS

The α-chlorine atom in **481** could also be replaced with Grignard reagents to give amides of type **538** (6). For optimal yields the reactions require inverse addition of 2 equivalents of the organometallic reagent again implying the intermediacy of an N-acyl imine (see Section IV-I-1).

$$\text{Ar—CO—NH—CH(Cl)—CCl}_3 \xrightarrow{\text{RMgBr}} \text{Ar—CO—NH—CH(R)—CCl}_3$$

(481)   (538)

Other organometallic reagents behave similarly. Phenyllithium adds to piperidene hydrochloride (**115**) with the formation of 2-phenylpiperidine (**539**) (36), and N-methylenemorpholinium bromide (**361**) and *tert*-butyllithium give N-neopentylmorpholine (**540**) (297). Iminium salts **197** and **198**, prepared from N,O-acetals of benzaldehyde or isobutryaldehyde, have been converted with phenyllithium to benzylamines **541** and **542** (178). From iminium salts derived from other aliphatic or heteroaromatic aldehydes and phenyl-, methyl-, or *n*-butyllithium, tertiary amines of types **543** and **544** were synthesized (15,373). Similarly the dehydrochinazolidinium perchlorate (**527**) and lithium-α-picoline gave amine **545** in 65% yield (315). Starting from tetraalkyliminium perchlorates (e.g., of type **64**), the reaction with butyllithium at −60° affords only about 7% of the adduct **546** or **547**. Under the same conditions derivatives of cyclohexanone or cyclopentanone, (e.g., iminium salts **531** and **533**) are transformed by reaction with butyllithium to amines **548** in 45% and **549** in 14% yield. Obviously in these cases a deprotonation in α position to iminium carbon takes place as a competing reaction, which leads to the formation of enamines (see Section IV-D). Finally, with butyllithium iminium salt **140b**, a cleavage product of ω,ω-bismorpholinoacetophenone, has been converted successfully to α-morpholinocaprophenone (**550**) (152).

(539) 2-phenylpiperidine

(540) N-neopentylmorpholine: O(CH$_2$CH$_2$)$_2$N—CH$_2$—C(CH$_3$)$_3$

(541) O(CH$_2$CH$_2$)$_2$N—CH(C$_6$H$_5$)—C$_6$H$_5$

(542) O(CH$_2$CH$_2$)$_2$N—CH(C$_6$H$_5$)—CH(CH$_3$)$_2$

(543) pyrrolidine-N—CH(R)—CH(CH$_3$)$_2$, R = CH$_3$, C$_4$H$_9$, C$_6$H$_5$

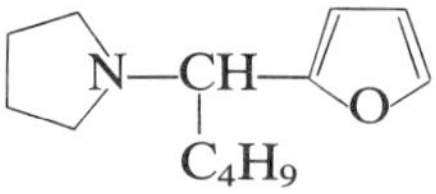

(544) 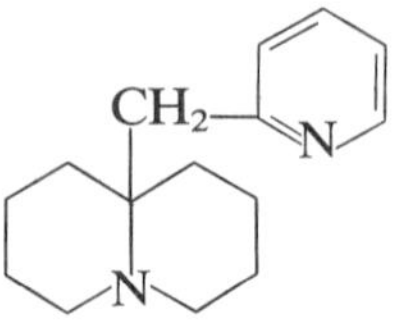 (545) 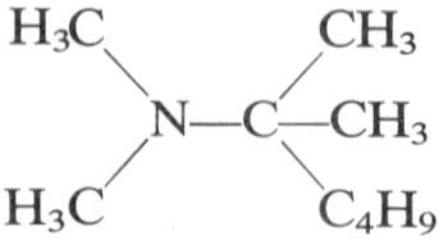 (546)

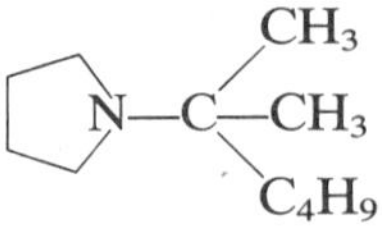

(547) 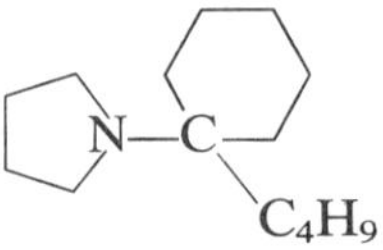 (548)

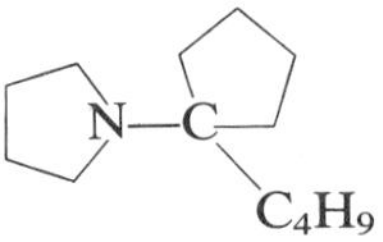

(549) 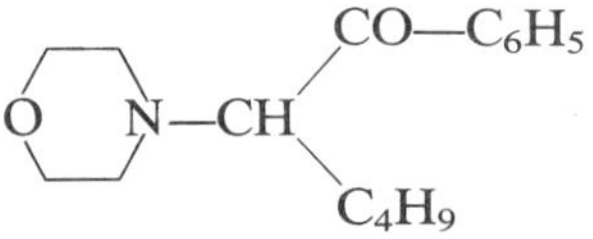 (550)

With lithium compounds of phenol ethers and MIH the corresponding dialkyl aminomethyl phenol ethers were obtained (374), for example, **551–554**, which are usually not directly accessible by aminomethylation of the phenol ethers with MIH (see Section IV-I-5). Analogously, but only in small yields, iminium salts derived from other aldehydes or from cyclohexanone (e.g., **517, 531**) have been converted to amines **555** and **556**.

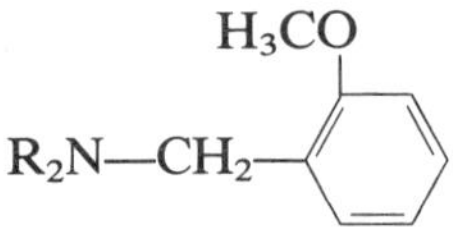

(551) 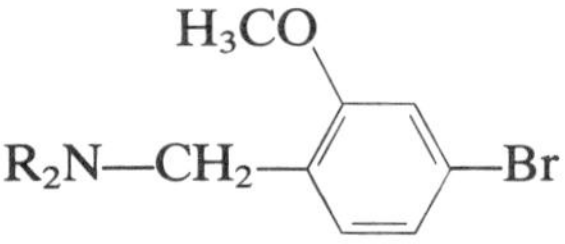 (552) 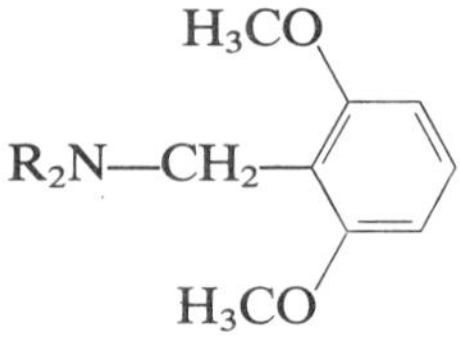 (553)

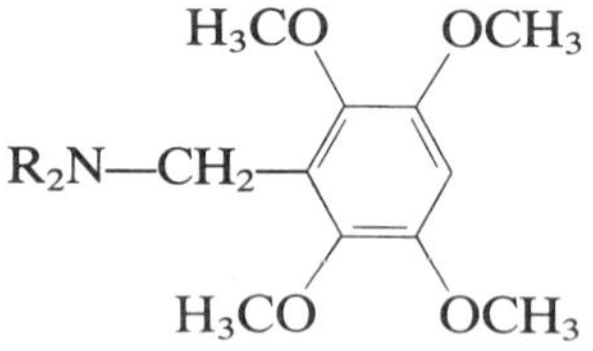

(554) 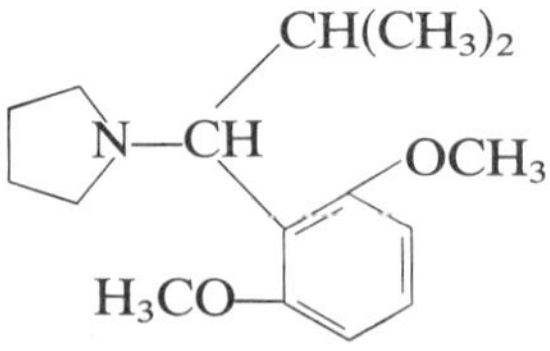 (555)

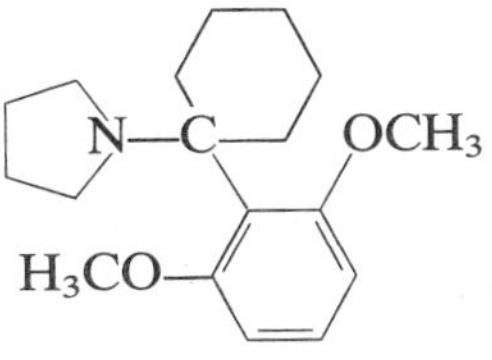

IV. REACTIONS OF METHYLENIMINIUM SALTS

(556)

Furthermore, reactions with organolithium compounds bearing heteroatoms such as sulfur, oxygen, silicon, or halogens on the anionic carbon have also been reported. Synthesis of α-aminoaldehydes **558** has been achieved via hydrolysis of the aminomercaptals **557**, which are accessible from lithium 1,3-dithiane (375) and MIH (376). Similarly the synthesis of sulfoxides of aminomercaptals **559** has been described (377). 1-Cyanoisochromane, isothiochromane, and homoisochromane also give lithium compounds by reaction with phenyllithium. The addition of MIH to these affords the corresponding aminomethylation products **560**, **561** (378), and **562** (379). Finally, synthesis of the tertiary amine **563**, starting from tris(chloromethyl)amine (**181**) and trimethylsilylmethyllithium (308), has been reported.

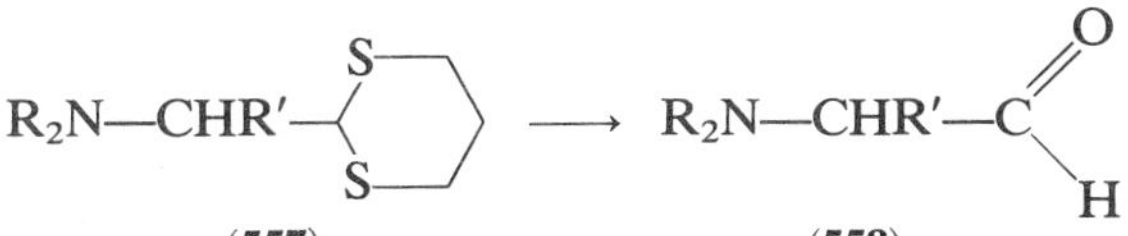

R_2N—CHR'—[dithiane] ⟶ R_2N—CHR'—C(=O)H

(557)   (558)

R_2N—CHR'—CH(S—CH$_3$)(SO—CH$_3$)

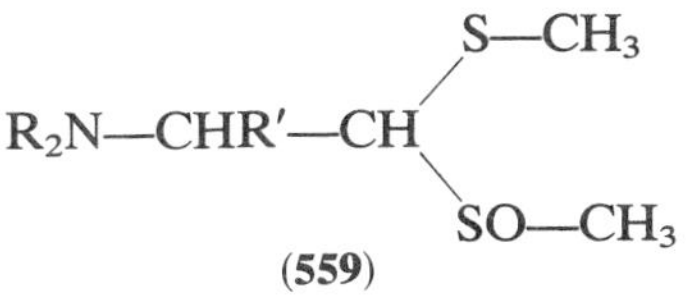

(559)

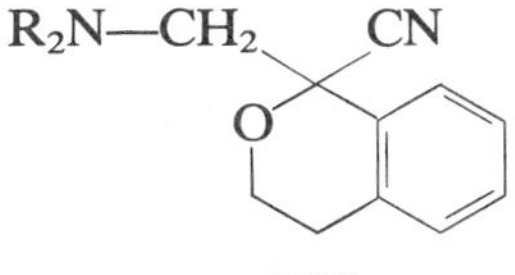

(560)

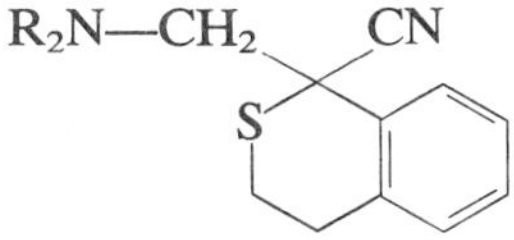

(561)

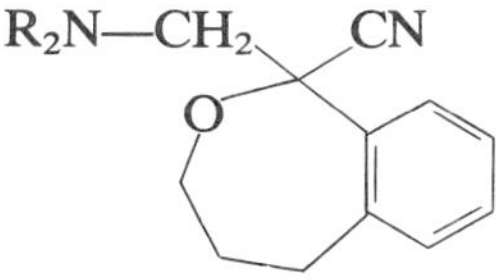

(562)

$N[CH_2—CH_2—Si(CH_3)_3]_3$

(563)

Methyleniminium halides have also been treated with organometallic reagents, prepared from 1-chloro-2,2-diphenylethenes and butyllithium at low temperatures (380), to obtain the dialkyl-(2-chloro-3,3-diphenylallyl)amines **564** (381). With trichloromethyllithium the dialkyl-(β,β,β-trichloroethyl)amines **565**, and with triphenylmethyllithium the dialkyl-(β,β,β-triphenylethyl)amines **566** were synthesized (382):

R_2N—CH_2—CCl=$C(C_6H_5)_2$ R_2N—CH_2—CCl_3

(564)   (565)

R_2N—CH_2—$C(C_6H_5)_3$

(566)

α-Chloroalkyllithium and alkyldilithium sulfones, sulfoxides, and sulfonamides also have been utilized in reactions with 1 or 2 equivalents of MIH (383). The products, such as **567a, 567c, 568, 569a**, and **569c**, can enter into various further reactions, for example, alkylation of **569a** to **569b**, or β-elimination of a secondary amine to **570**.

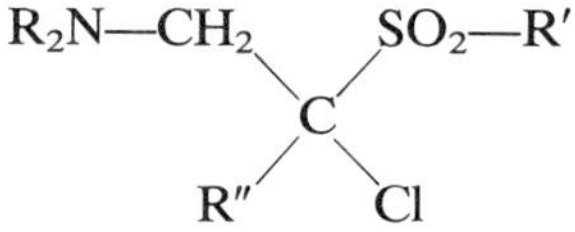

(567) 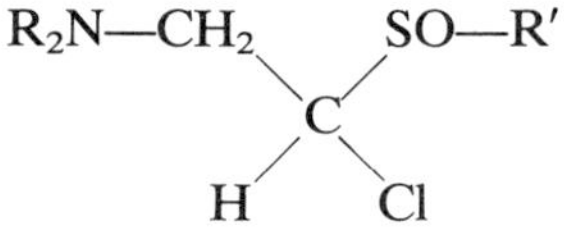 (568)

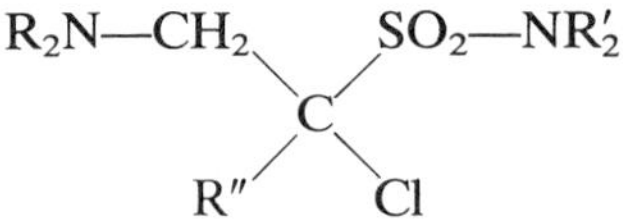

(569) 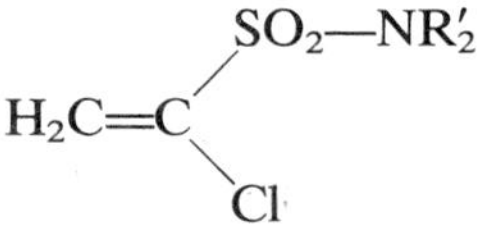 (570)

a: R″ = H; **b:** R″ = CH_3; **c:** R″ = R_2N—CH_2

In some cases alkylaluminum reagents were also used (384). From N-methylenepiperidinium chloride (**121**) and ethylaluminum sesquichloride or tri-n-hexylaluminum the tertiary amines **571a** and **571b** were prepared in good yields. Analogous reactions of iminium salts of type **216b** led to α-piperidinoalkane acid piperidides **572a–572e**, preparation of which by means of Grignard reagents failed.

4. Reactions with Multiple-Bond Systems

(a) *Olefins.* In contrast to cyclohexene, 2-methylbutene-1, or styrene, the aminomethylation of α-methylstyrene (**573**) by means of MIH takes

IV. REACTIONS OF METHYLENIMINIUM SALTS

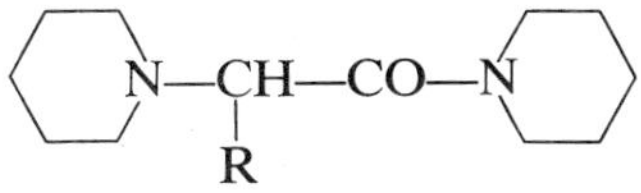

(571) ⟨N⟩—CH$_2$—R

(572) ⟨N⟩—CH(R)—CO—⟨N⟩

a: R = C$_2$H$_5$
b: R = n-C$_6$H$_{13}$
c: R = n-C$_8$H$_{17}$
d: R = n-C$_{10}$H$_{21}$
e: R = n-C$_{12}$H$_{25}$

place (but slowly). The attack proceeds on the vinylmethylene carbon, forming a carbonium ion (574), stabilization of which by proton elimination in the 2- or 4-position leads to a mixture of Hofmann and Saytzeff products 575 and 576 (385). Reactions of MIH could be achieved more rapidly, and also in higher yields, with 1,1-diphenylethylene and its derivatives having electron-donating substitutents in the paraposition (e.g., a p-methoxy- or p-dimethylamino group). Thus compounds of type 577 and also twofold-aminomethylated products of type 578 were obtained (385).

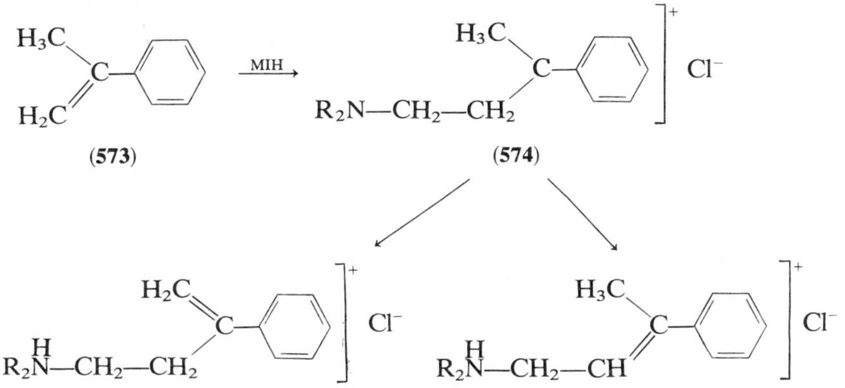

(573) H$_3$C–C(=CH$_2$)–C$_6$H$_5$ →[MIH] (574) [H$_3$C–C(C$_6$H$_5$)–CH$_2$–CH$_2$–NR$_2$]$^+$ Cl$^-$

(575) [H$_2$C=C(C$_6$H$_5$)–CH$_2$–CH$_2$–N(H)R$_2$]$^+$ Cl$^-$

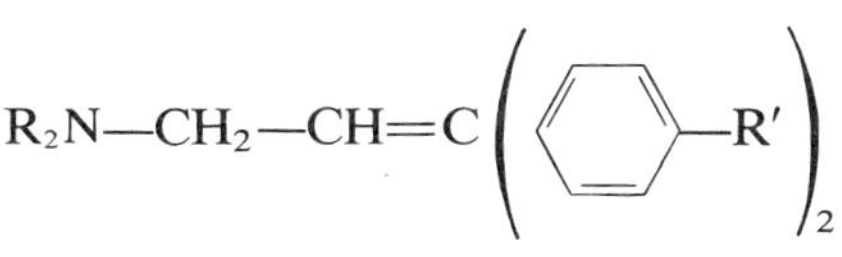

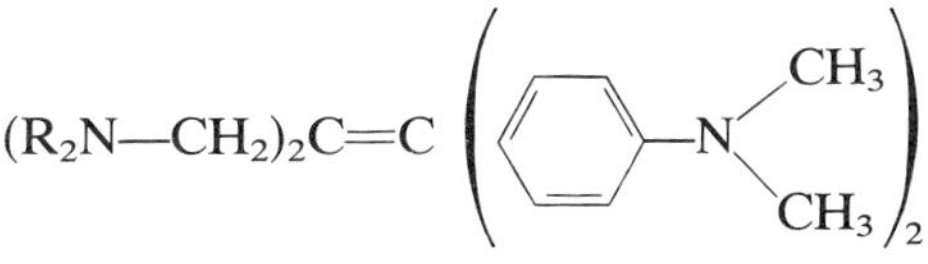

(576) [H$_3$C–C(C$_6$H$_5$)=CH–CH$_2$–N(H)R$_2$]$^+$ Cl$^-$

(577) R$_2$N—CH$_2$—CH=C(C$_6$H$_4$—R')$_2$

(578) (R$_2$N—CH$_2$)$_2$C=C(C$_6$H$_4$—N(CH$_3$)$_2$)$_2$

Similarly with p-methoxystyrene, anethol, isoeugenol methyl ether, isosafrol, and 4-methyl- and 2,4,6-trimethylstyrene derivatives 579a–d and 580a–b were prepared (385). In MIH reactions with 4-methoxy- and 4-dimethylaminocinnamic acid, aminomethylation and decarboxylation take place simultaneously with the formation of hydrochlorides of allyl amines **579a** and **579e**. Cinnamic acid itself does not react, but *p*-dimethylaminocinnamic ester and aldehyde yield **581a** and **581f**, respectively (386).

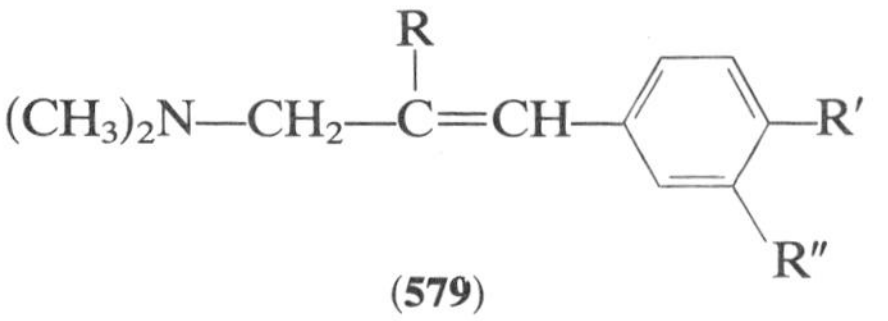

$$(CH_3)_2N-CH_2-\underset{|}{\overset{R}{C}}=CH-\underset{R''}{\overset{}{C_6H_3}}-R'$$

(**579**)

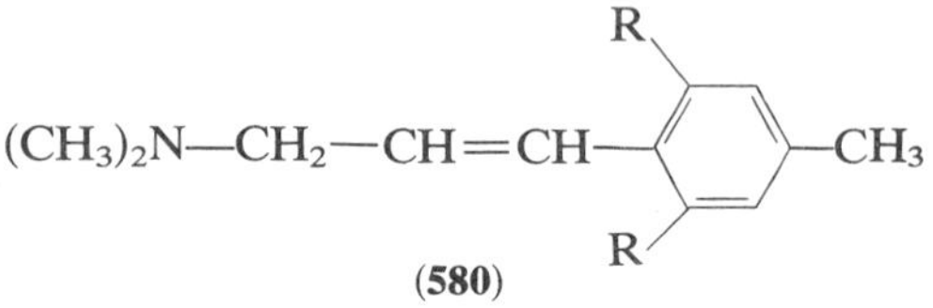

$$(CH_3)_2N-CH_2-CH=CH-\underset{R'}{\overset{R}{C_6H_3}}-CH_3$$

(**580**)

$$(CH_3)_2N-CH_2\underset{R'-\overset{O}{C}}{\overset{}{\diagdown}}C=CH-C_6H_4-N(CH_3)_2$$

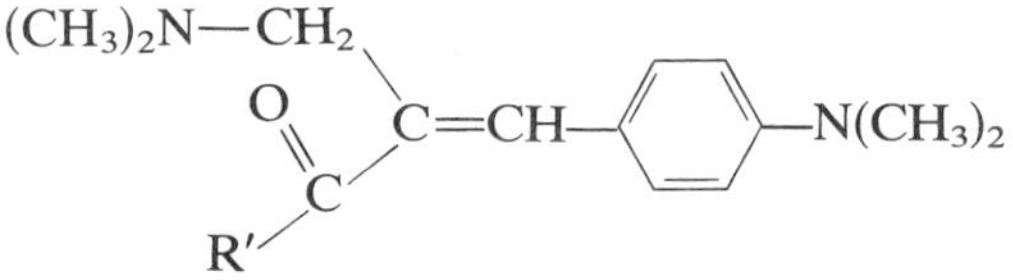

(**581**)

a R = R″ = H, R′ = CH₃O
b R = CH₃, R′ = CH₃O, R″ = H
c R = CH₃, R′ = R″ = CH₃O
d R = CH₃, R′ + R″ = O—CH₂O
e R = R″ = H, R′ = (CH₃)₂N
f R′ = H

Amidoalkylation of monoolefins or conjugated dienes by means of *N*-acyliminium ions has been examined over the last 15 years in considerable detail. These reactions, which have recently been reviewed (6), also include amidoalkylation of acetylenes and ketenes. The work has demonstrated that *N*-acyliminium ions generated from compounds of type **582** can give any one or several of the three products **583–585**, depending on the reaction conditions and the type of reactants. For example, *N*-hydroxymethylamides **582** (X = OH) and olefins in glacial acetic acid at 5–15° with an equivalent of sulfuric or sulfonic acid undergo cycloaddition to **583**, deprotonation products of which are 4*H*-5,6-dihydro-1,3-oxazines, which can be isolated and solvolyzed to products of type **584** (387,388). These are also accessible directly from **582** and olefins in a mixture of acetic acid, acetic anhydride, and a sulfonic acid at 80°. In trifluoroacetic acid at 80° or strong sulfuric acid (>70%) at 30°, however, unsaturated amides **585** are obtained (70–95% yields) (389).

N-acyliminium ions **586**, generated, for example, from *N*-chloromethylamides or *N*-chloromethylimides **582** (X = Cl) with stannic

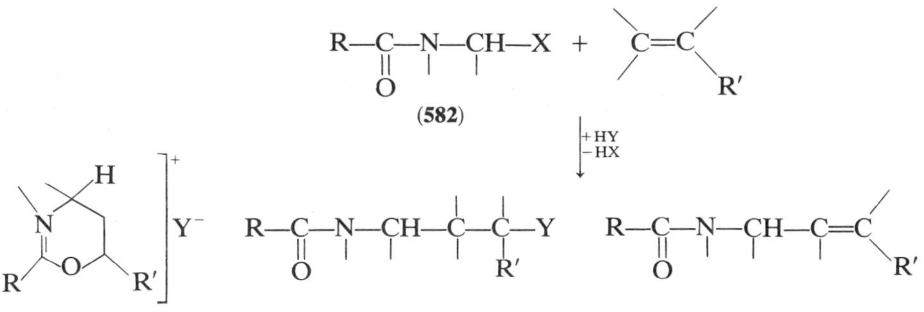

IV. REACTIONS OF METHYLENIMINIUM SALTS 193

$$R-\underset{\underset{O}{\|}}{C}-\underset{|}{N}-\underset{|}{CH}-X \ + \ \underset{R'}{\overset{}{C}}=\underset{R'}{\overset{}{C}}$$

(582)

$\Big\downarrow \begin{smallmatrix}+HY\\-HX\end{smallmatrix}$

(583)   (584)   (585)

chloride, were found to add both regiospecifically and cis stereospecifically to unsymmetrical olefins, giving 4H-dihydro-1,3-oxazinium salts of type **583** (6,390). With 1,3-dienes the corresponding 6-vinyl derivatives of **583** are formed (387,390); N-thioacyliminium ions and olefins give 4H-5,6-dihydro-1,3-thiazinium salts (390). 1,4-Cycloaddition of the cation **586** appears to occur in a concerted manner, involving a six-membered cyclic transition state **587**. With more electron-rich olefins (e.g., α-methylstyrene) in addition to the concerted reaction a simultaneous two-step reaction course via a carbocation (**589**) and reversible bond opening between the oxygen and C_6 in the cycloadduct **588** are proposed (390,391). Both mechanisms have been discussed in more detail in connection with diastereogenic reactions of chiral amidomethylium ions of type **586** to form diastereomeric oxazines **588** (391), as well as in a summary of cationic polar cycloadditions (392).

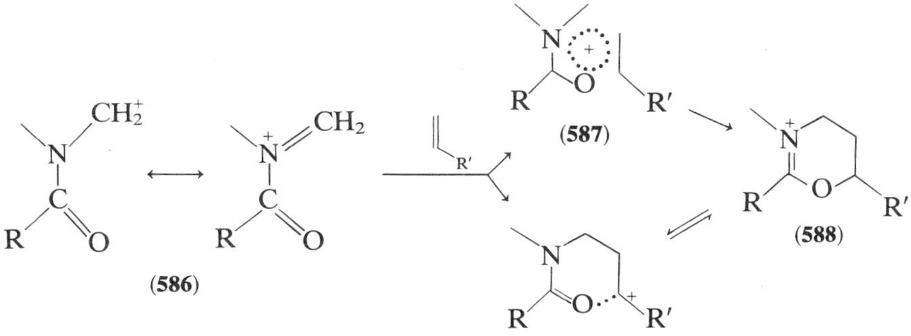

(586)   (587)   (588)   (589)

Oxazinium ions **588** can also be generated directly from three-component reactions of amides, formaldehyde, and olefins in glacial acetic acid–sulfuric acid, or from acyl chlorides, N-alkyl aldimines, and olefins in the presence of stannic chloride (390). They are intermediates in reactions of 1,1-dichloroolefins with N-acyliminium precursors of type

582 in 95% sulfuric acid (Tscherniac-Einhorn synthesis reagent), which upon hydrolysis yield β-amidocarboxylic acids **590** in excellent yields (393). With vinyl chloride or 1,2-dichloroethylene, under similar conditions β-amidoaldehydes (e.g., **591**) are obtained.

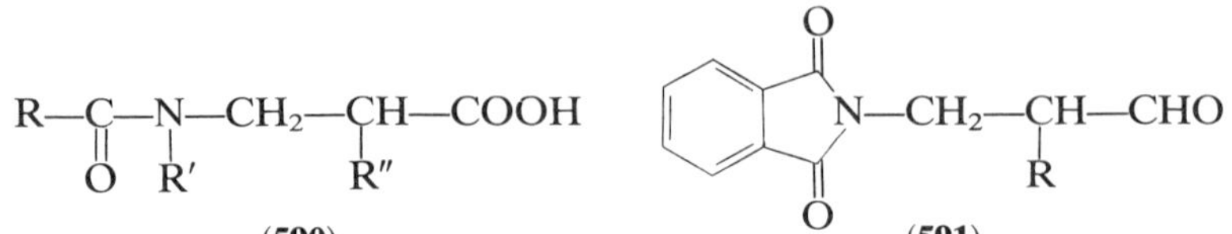

$$R-\underset{\underset{O}{\|}}{C}-\underset{\underset{R'}{|}}{N}-CH_2-\underset{\underset{R''}{|}}{CH}-COOH \qquad \text{(phthalimido)}N-CH_2-\underset{\underset{R}{|}}{CH}-CHO$$

(590)　　　　　　　　　　　(591)

Formation of the oxazinium hexachloroantimonates **593** and **594**, which were isolated from phthalimidomethyl hexachloroantimonate (**592**) upon reaction with the two isomers of 1,2-dichloroethylene in carbon tetrachloride, has been reported to proceed with more than 90% steric purity. Hydrolysis of **593** and **594** gave aldehyde **591** (R = Cl) (393).

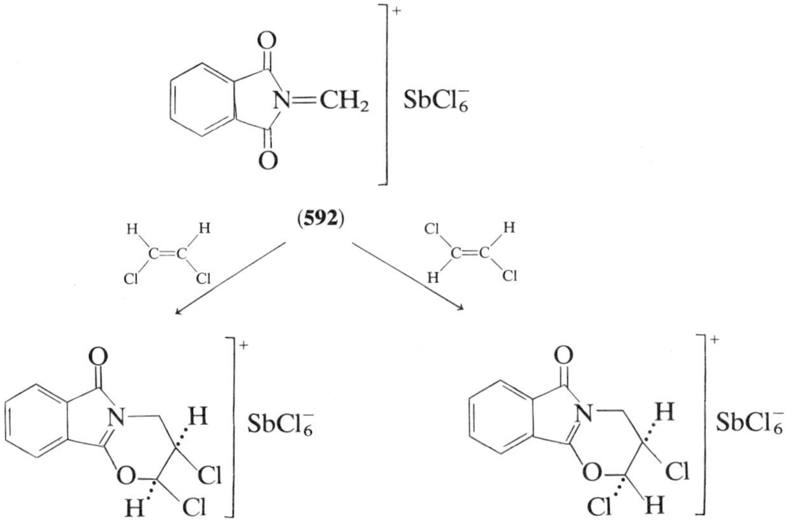

(592)

(593)　　　　　　　　　　　(594)

(b) *1,3-Dienes.* As dienophiles, MIH can be added to 2,3-dimethylbutadiene (**595**) in methylene chloride to give quaternary Δ^3-piperidenium salts **597** in a Diels-Alder type of reaction (394). Analogous results have been reported in the case of isoprene (395). For these reactions polarization by the positive charge on nitrogen seems to be an important factor, since the same type of reaction obviously has not yet been described for imines. This view is supported by the diene reaction of the cyclic iminium perchlorate **596** to give **598** (396). Diene reactions of the free base of **596** have failed.

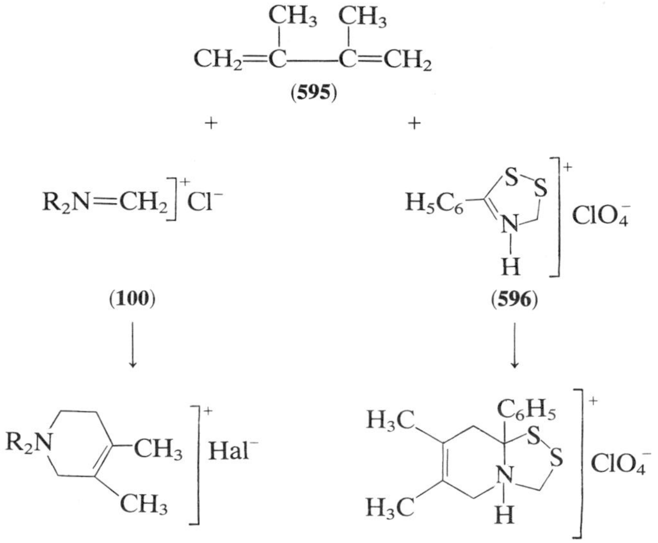

$$CH_2=\underset{CH_3}{\underset{|}{C}}-\underset{CH_3}{\underset{|}{C}}=CH_2$$
(**595**)

+ +

$[R_2N=CH_2]^+ Cl^-$

(**100**)

$\begin{bmatrix} H_5C_6\overset{S-S}{\underset{N}{\bigvee}} \\ H \end{bmatrix}^+ ClO_4^-$

(**596**)

↓ ↓

$\begin{bmatrix} R_2N\overset{}{\underset{CH_3}{\bigvee}}-CH_3 \end{bmatrix}^+ Hal^-$

(**597**)

$\begin{bmatrix} H_3C\underset{H_3C}{\overset{C_6H_5}{\underset{N}{\bigvee}}}\overset{S-S}{\underset{H}{\bigvee}} \end{bmatrix}^+ ClO_4^-$

(**598**)

(c) *Enol Ethers.* Methyleniminium halides and α,β-unsaturated ethers **599** yield α-haloethers **600**, which rearrange easily to the hydrochlorides of aminomethyl vinyl ethers **601** (313). With ethyl vinyl ether, dihydro-4H-pyran or 2-methyl-4,5-dihydrofuran as starting materials, derivatives **602**, **603**, and **604** were prepared. Analogous reactions of other alkoxyolefins have also been achieved with ketene acetals and allene ethers (397). In the case of methylisobutenyl ether the secondary reaction cannot take place, and isolation of the α-haloether **606** is possible. Hydrolysis of this leads to dialkylaminopivalic aldehyde **605**, while in methanol the acetal **607** is formed (313).

$CH_2=CH-OR' \xrightarrow{MIH} R_2N-CH_2-CH_2-CHCl-OR' \longrightarrow$
(**599**) (**600**)

$[R_2\overset{H}{N}-CH_2-CH=CH-OR']^+ Cl^-$
(**601**)

$R_2N-CH_2-CH=CH-OC_2H_5$ $R_2N-CH_2-\langle \rangle-O$ $R_2N-CH_2-\langle \rangle O$

(**602**)   (**603**) 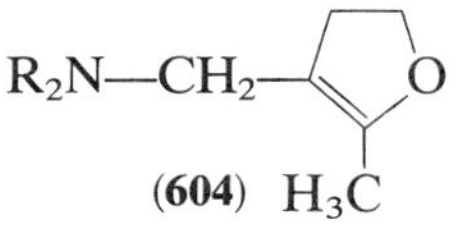   (**604**) H_3C

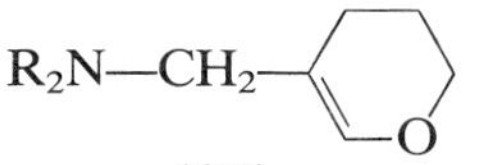

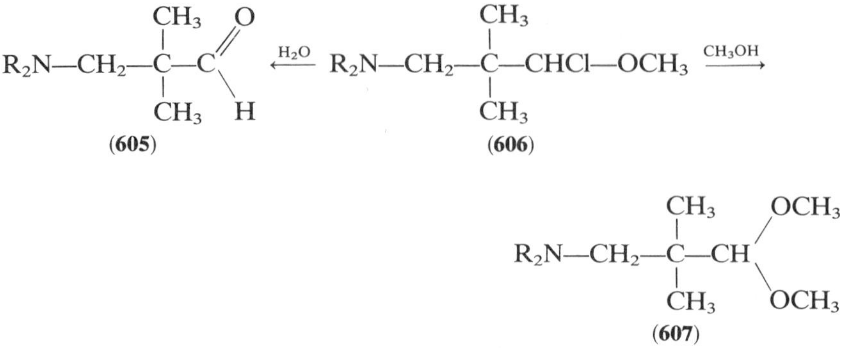

$$\underset{(605)}{R_2N-CH_2-\underset{\underset{CH_3}{|}}{\overset{\overset{CH_3}{|}}{C}}-\overset{O}{\underset{H}{C}}} \xleftarrow{H_2O} \underset{(606)}{R_2N-CH_2-\underset{\underset{CH_3}{|}}{\overset{\overset{CH_3}{|}}{C}}-CHCl-OCH_3} \xrightarrow{CH_3OH}$$

$$\underset{(607)}{R_2N-CH_2-\underset{\underset{CH_3}{|}}{\overset{\overset{CH_3}{|}}{C}}-CH\underset{OCH_3}{\overset{OCH_3}{<}}}$$

Reactions of MIH with enol borinates, which can be obtained from diazoketones with trialkyl boranes, are of synthetic utility, especially with respect to regiospecific aminomethylation of ketones. Under the usual conditions of a Mannich reaction, the electrophilic attack of the iminium ion proceeds on the enol form. Therefore unsymmetrical ketones, although preferentially aminomethylated on the more substituted α-carbon, always give a mixture of the two isomers (241). With enol borinates and iminium salts in tetrahydrofuran–dimethyl sulfoxide, however, usually one product is obtained; for example, from **608** via **609** product **610** is formed in excellent yield without any other isomers (398):

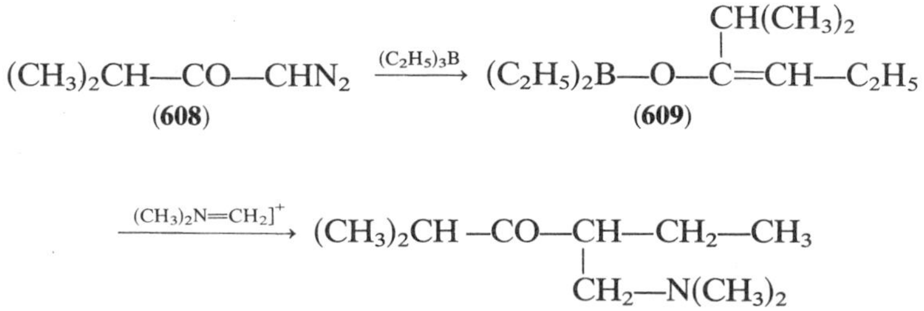

$$\underset{(608)}{(CH_3)_2CH-CO-CHN_2} \xrightarrow{(C_2H_5)_3B} \underset{(609)}{(C_2H_5)_2B-O-\underset{}{\overset{\overset{CH(CH_3)_2}{|}}{C}}=CH-C_2H_5}$$

$$\xrightarrow{(CH_3)_2N=CH_2]^+} \underset{(610)}{(CH_3)_2CH-CO-\underset{\underset{CH_2-N(CH_3)_2}{|}}{CH}-CH_2-CH_3}$$

Presumably iminium ion intermediates are also involved in reactions of dialkyl aminomethyl ethers or aminals with bis(trifluoromethyl)ketene (**611**) to give adducts **612** (399):

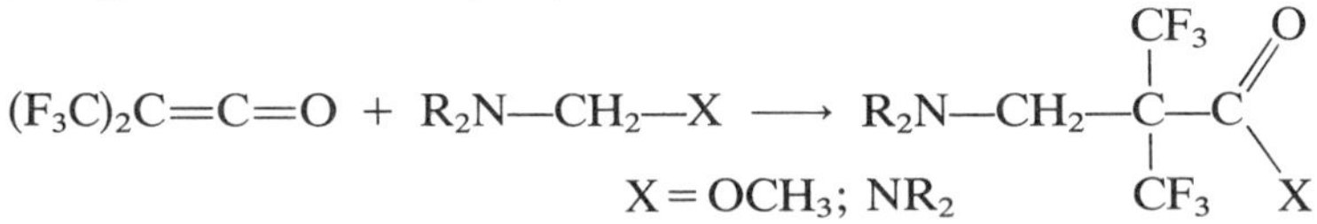

$$\underset{(611)}{(F_3C)_2C=C=O} + R_2N-CH_2-X \longrightarrow \underset{(612)}{R_2N-CH_2-\underset{\underset{CF_3}{|}}{\overset{\overset{CF_3}{|}}{C}}-\overset{O}{\underset{X}{C}}}$$

$$X = OCH_3; NR_2$$

(d) *Enamines.* With enamines MIH have been reacted to form iminium salts that, upon hydrolysis, yield dialkylaminopivalic aldehydes (400).

The possibility that the adducts can isomerize was demonstrated for N-methylenepiperidinium chloride (**121**) addition to 1-morpholinoisobutene (**613**), as well as for the addition of N-methylenemorpholinium chloride (**361**) to 1-piperidinoisobutene (**614**). In both cases the same ratio of a mixture of iminium salts **615** and **616** was obtained in acetonitrile. This was concluded from chromatographic analysis of the hydrolysis products, which showed mainly morpholinopivalic aldehyde (**618**) in addition to small amounts of piperidinopivalic aldehyde (**617**). These results indicate that iminium salts **615** and **616** are convertible via hydride transfer in an equilibrium, which is located mainly at **616** as the iminium derivative of the more basic piperidine (400).

The treatment of 1,1,2-triaminoethenes **619** with MIH in methylene chloride has led to amidinium salts **620**, which can be hydrolyzed to give 2,3-dialkylaminopropionic acid dialkylamides **621** (75):

$$R'_2N-CH=C\begin{matrix}NR'_2\\NR'_2\end{matrix} \xrightarrow{MIH} \left[\begin{matrix}R_2N-CH_2\\R'_2N\end{matrix}CH-C\begin{matrix}NR'_2\\NR'_2\end{matrix}\right]^+ Cl^- \xrightarrow{H_2O}$$

(**619**)   (**620**)

$$\begin{matrix}R_2N-CH_2\\R'_2N\end{matrix}CH-C\begin{matrix}O\\NR'_2\end{matrix}$$

(**621**)

For certain α-chloroalkyl derivatives of secondary amides (e.g., **622**) reactions with enamines in the presence of base (triethylamine or an excess of enamine) were reported to proceed through N-acyl imines as reactive intermediates (e.g., **623**), which add to the enamine with the formation of 1,4-cycloadducts (e.g., **625**) or β-amidoaldehydes (e.g., **624**) upon subsequent hydrolysis (6):

$$R-\underset{O}{\underset{\|}{C}}-NH-\underset{Cl}{\underset{|}{CH}}-CCl_3 \xrightarrow[-HCl]{base} R-\underset{O}{\underset{\|}{C}}-N=CH-CCl_3$$

(**622**)   (**623**) $\downarrow +(CH_3)_2C=CH-N(CH_3)_2$

$$R-\underset{O}{\underset{\|}{C}}-NH-\underset{CCl_3}{\underset{|}{CH}}-\underset{CH_3}{\overset{CH_3}{\underset{|}{C}}}-CHO \xleftarrow{H_2O}$$

(**624**)   (**625**)

[Structure **625**: six-membered ring with N, O, substituents H, CCl₃, CH₃, CH₃, H, N(CH₃)₂, R]

(e) *Ynamines.* Interesting reaction products were obtained from MIH additions to ynamines. For example, from iminium salts **64** and ynamine **626** at room temperature, presumably via the 2,2-cycloaddition intermediate **627**, acrylamidinium salts **628** have been synthesized (401):

$(CH_3)_2N=C(CH_3)_2]^+$
(**64**)

$$+ \longrightarrow (CH_3)_2N-\underset{|}{\overset{(CH_3)_2N-C(CH_3)_2}{C}}=C-C_6H_5 \Big]^+ \longrightarrow$$
(**627**)

$(CH_3)_2N-C\equiv C-C_6H_5$
(**626**)

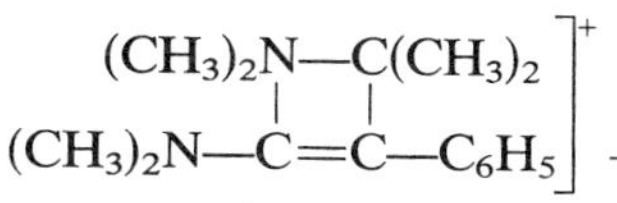

$$\begin{bmatrix}(CH_3)_2N\\ \diagdown\\ (CH_3)_2N\diagup\end{bmatrix}C-\underset{\underset{C_6H_5}{|}}{C}=C(CH_3)_2\Big]^+$$
(**628**)

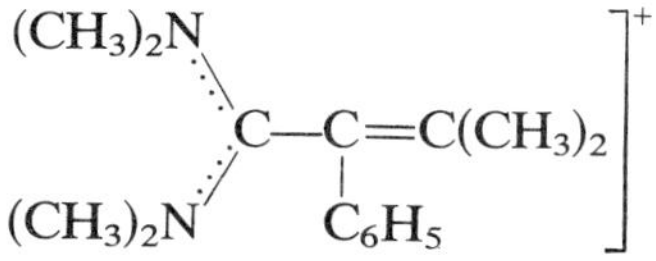

(f) *Isonitriles.* With isonitriles MIH react in accordance with the mechanism of Ugi's four-component condensation (245,246). Thus from β-phenylethylisonitrile and MIH **185** an adduct very sensitive to hydrolysis was formed. Since no assignment could be obtained by spectroscopic examination, the structure can be either an imide chloride (**629**) or a nitrilium chloride (**630**). Both structures are consistent with secondary reaction products; hydrolysis gives the amide **631**, and hydrazoic acid addition yields the tetrazole derivative **632** (402).

$(CH_3)_2N-CH_2-CCl=N-CH_2-CH_2-C_6H_5$
(**629**)

$(CH_3)_2N-CH_2-C\equiv N-CH_2-CH_2-C_6H_5]^+\ Cl^-$
(**630**)

$(CH_3)_2\overset{H}{N}-CH_2-CO-NH-CH_2-CH_2-C_6H_5]^+\ Cl^-$
(**631**)

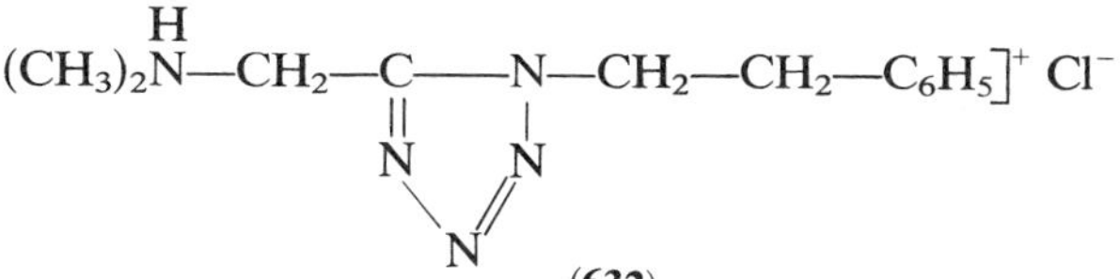

$(CH_3)_2\overset{H}{N}-CH_2-\underset{\underset{N}{\overset{\|}{N}}\diagdown_{N}\diagup}{\overset{}{C}}-N-CH_2-CH_2-C_6H_5]^+\ Cl^-$
(**632**)

(g) *Phosphine Alkylenes.* From reactions of alkylidene phosphoranes **633** with MIH the aminomethylation products **634** were obtained; these undergo a Wittig reaction with carbonyl reagents to give the substituted dialkylallylamines **635** (403). With, as starting materials, dimethylphenyl-methyleniminium chloride (**136a**) and phosphine alkylenes of type **636** having a methylene group in β-position to the phosphorus, the phenyl-substituted allenes **638** can be synthesized (404). In the first stage a phosphonium salt (**637**) is formed; deprotonation of this by action of (**636**), followed by intramolecular elimination of amine and phosphine from **639**, leads to **638**.

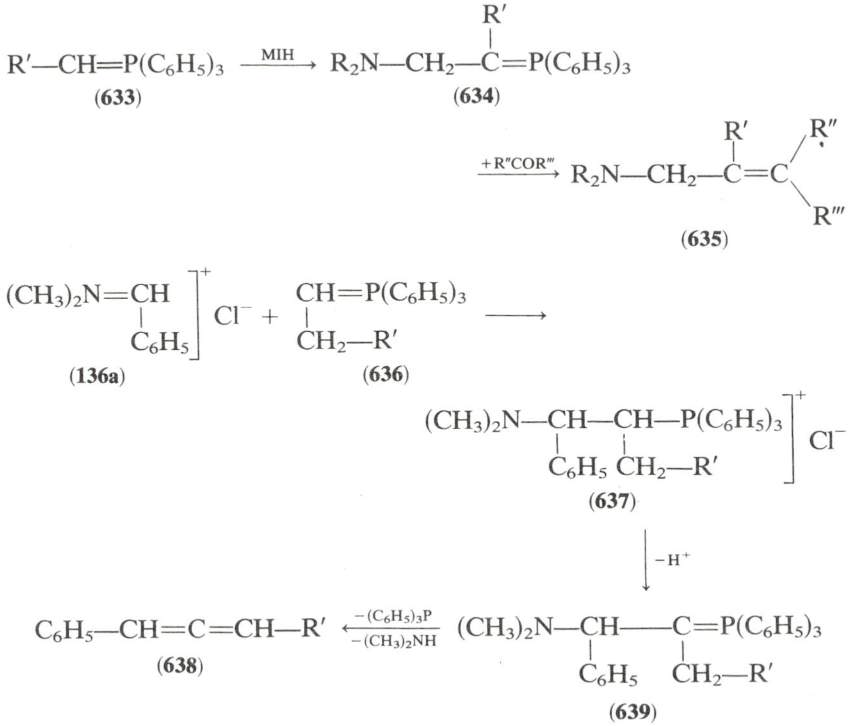

$$R'-CH=P(C_6H_5)_3 \xrightarrow{\text{MIH}} R_2N-CH_2-\underset{\underset{R'}{|}}{C}=P(C_6H_5)_3$$
(633) (634)

$$\xrightarrow{+R''COR'''} R_2N-CH_2-\underset{\underset{R'}{|}}{C}=C\underset{R'''}{\overset{R''}{<}}$$
(635)

$$\left[\begin{array}{c} (CH_3)_2N=CH \\ | \\ C_6H_5 \end{array} \right]^+ Cl^- + \begin{array}{c} CH=P(C_6H_5)_3 \\ | \\ CH_2-R' \end{array} \longrightarrow$$
(136a) (636)

$$\left[\begin{array}{c} (CH_3)_2N-CH-CH-P(C_6H_5)_3 \\ | | \\ C_6H_5 CH_2-R' \end{array} \right]^+ Cl^-$$
(637)

$$\downarrow -H^+$$

$$C_6H_5-CH=C=CH-R' \xleftarrow[-(CH_3)_2NH]{-(C_6H_5)_3P} (CH_3)_2N-\underset{\underset{C_6H_5}{|}}{CH}-\underset{\underset{CH_2-R'}{|}}{C}=P(C_6H_5)_3$$
(638) (639)

5. Reactions with Aromatic Compounds

Methyleniminium halides have been used for aminomethylation mainly of arenes whose nucleophilicity is enhanced by hydroxy, alkoxy, or dimethylamino groups. For example, substitution of the aromatic ring has been achieved with phenolates in dioxane (174,405), with phenols in ether in the presence of triethylamine, and without a proton acceptor in methylene chloride (165) or acetonitrile (385). Some typical examples are compounds **640–646**, which were obtained with phenol (174,196), methyl salicylate (174), *p*-hydroxybenzoic esters (405), eugenol or isoeugenol (385), or dihydroxy-α,β-diethylstilbenes (406).

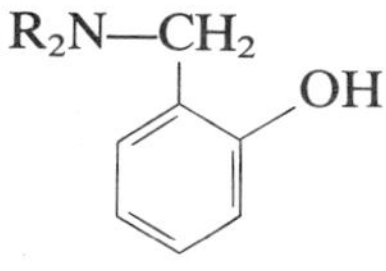

IV. REACTIONS OF METHYLENIMINIUM SALTS

(640) R₂N—CH₂ — C₆H₄ — OH (ortho)

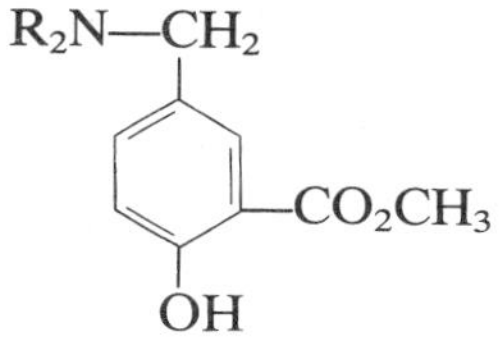

(641) R₂N—CH₂ substituted benzene with —CO₂CH₃ and —OH

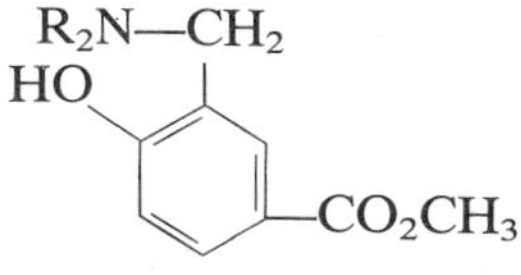

(642) R₂N—CH₂ substituted benzene with HO— and —CO₂CH₃

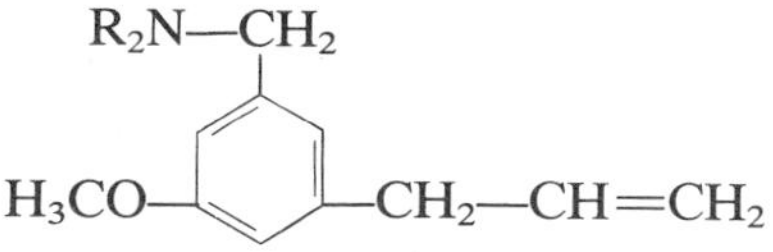

(643) R₂N—CH₂ substituted benzene with H₃CO— and —CH₂—CH=CH₂

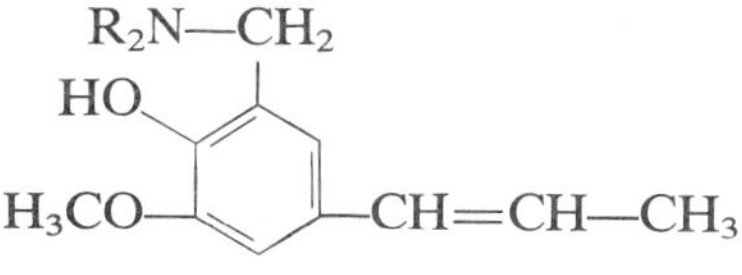

(644) R₂N—CH₂ substituted benzene with HO—, H₃CO—, and —CH=CH—CH₃

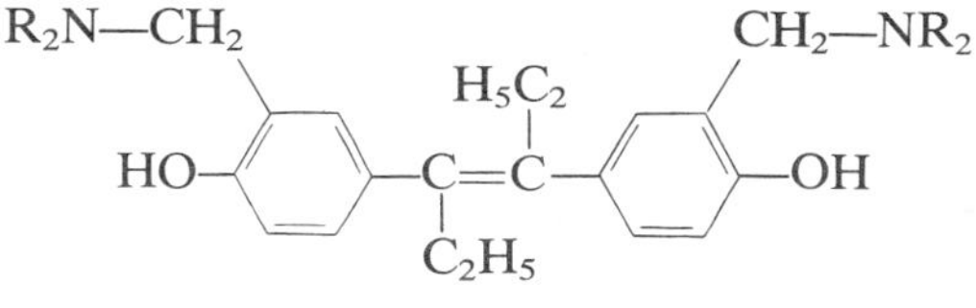

(645) R₂N—CH₂—Ar—C(C₂H₅)=C(C₂H₅)—Ar—CH₂—NR₂ with HO— and —OH substituents

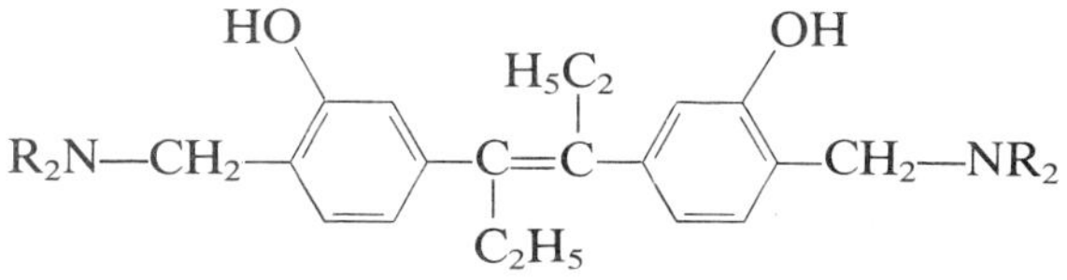

(646) R₂N—CH₂—Ar—C(C₂H₅)=C(C₂H₅)—Ar—CH₂—NR₂ with HO— and OH— substituents

Monoalkylmethyleniminium chlorides **113** (136), which can be prepared by means of hydrogen chloride cleavage of 1,3,5-trisubstituted hexahydrotriazines, preferentially attack the para position of phenol or phenol derivatives to form the hydrochlorides of secondary aminomethyl phenols **647** (407). Hydroquinones yield the benzoxazine hydrochlorides (**648**) (408).

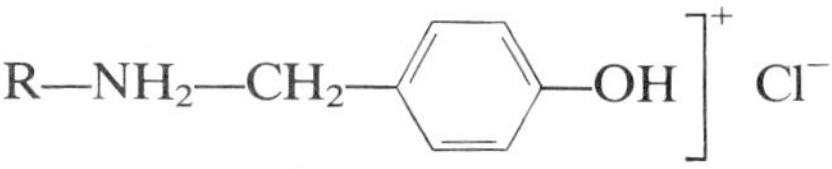

(647) [R—NH₂—CH₂—C₆H₄—OH]⁺ Cl⁻

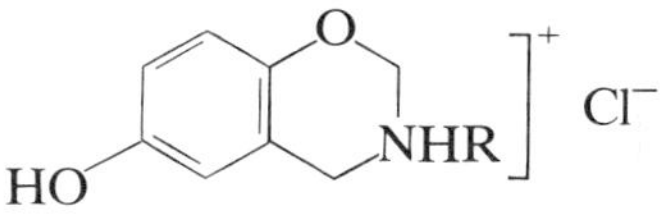

(648) benzoxazine with HO— and NHR, Cl⁻

Usually difficulties arise for aromatic substitution by means of iminium salts derived from aldehydes other than formaldehyde or from ketones, although in the case of iminium salts of type **216** aminoalkylations to compounds of type **649** or **650** could be easily performed (165):

(649): R₂N—CH(—CO₂CH₃) attached to 2-hydroxynaphthalen-1-yl

(650): R₂N—CH(—CO—N(piperidine)) attached to 2-hydroxynaphthalen-1-yl

Aminomethylation of phenol ethers also appears to be difficult. For example, reactions of anisol with MIH failed even upon heating or in the presence of Friedel-Crafts catalysts. Two methoxy groups in the 1,3-position, however, increase the nucleophilicity of C_4 sufficiently to permit the formation of **651** and **652** from resorcinol dimethyl ether and phloroglucinol trimethyl ether (405), respectively. In other cases the aminomethylation was successfully carried out using lithium reagents of the phenol ethers (see Section IV-I-3).

(651): R₂N—CH₂—(2,4-dimethoxyphenyl)

(652): R₂N—CH₂—(2,4,6-trimethoxyphenyl)

Reactions with dialkyl anilines or hydrochlorides of these are easy to perform in acetonitrile. The dialkyl-(*p*-dialkylaminobenzyl)amines **653** are phenyl-analogous aminals that can enter into other reactions, such as cleavage with acyl halides (409) or formation of monoquaternary salts **654** with additional MIH (281). The preparation of methylene-group-substituted aminoalkyl derivatives usually failed with the exception of compounds **655**, which are reaction products of iminium salts of type **216** (165). In the case of para-substituted aromatic reactants, the aminomethylation takes place in the ortho position [e.g., **656** (281)]. With iminium halides of weak basic amines such as morpholine or methyltrichlorophenylamine (e.g., **361** or **222a**) used as starting materials, twofold aminomethylation has also been achieved [e.g., to **657** (281) and **658** (410)].

Many reactions involving the α-amidoalkylation of arenes by means of N-acyliminium-type salts have been reported, and this work has been reviewed in detail (4–6). Here reactions of cycliminium cations can also be included, formation of which was achieved by acyl halide addition to quinoline, isoquinoline, or derivatives of these. For example, with dialkyl

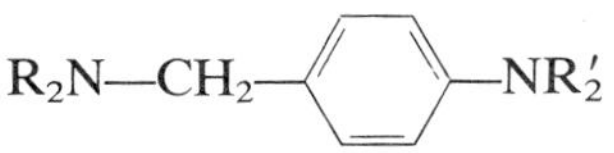

[Structures (653)–(658) shown]

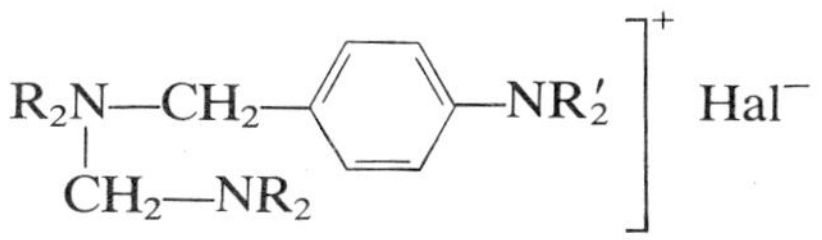

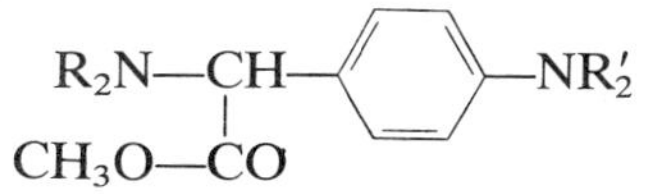

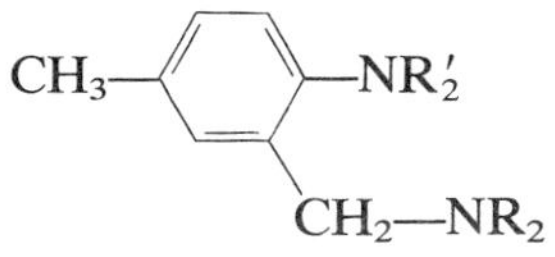

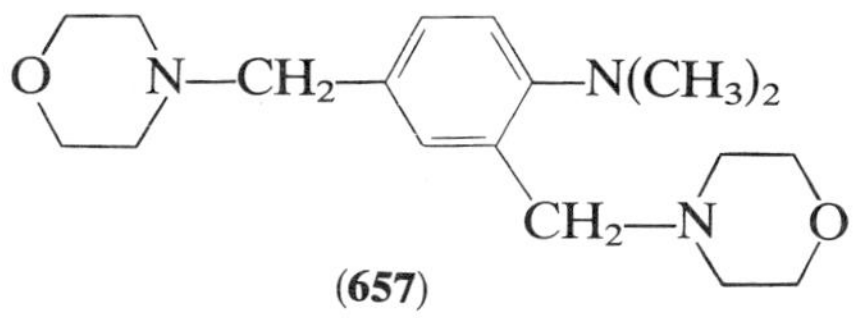

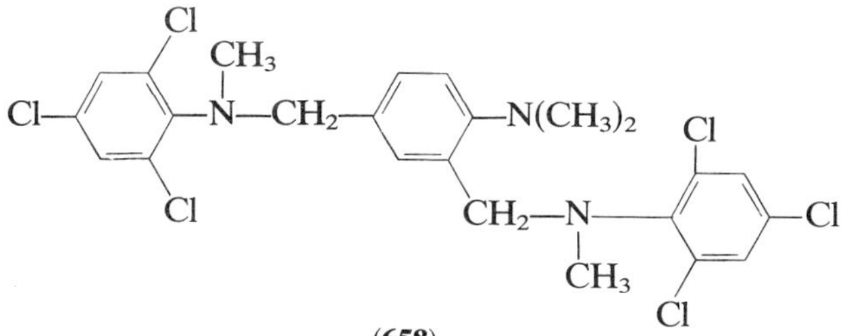

anilines or dihydroindoles the *p*-substitution products **659** and **660** were obtained (411–412):

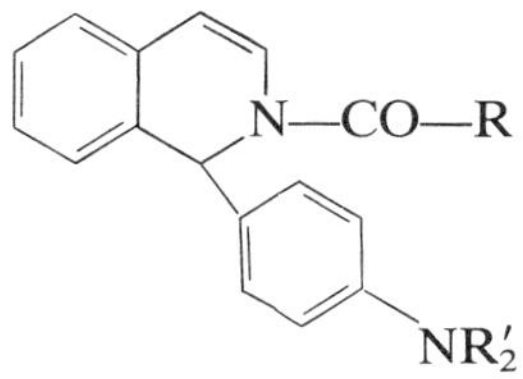

[Structures (659) and (660) shown]

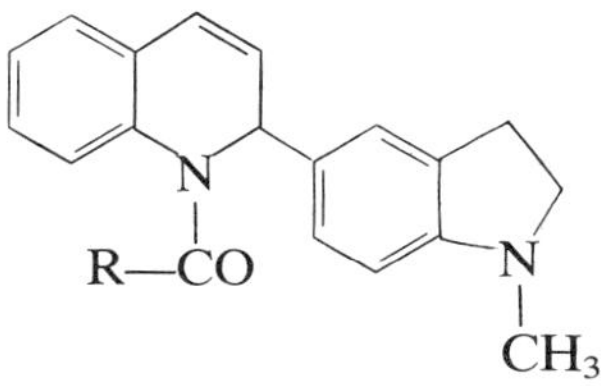

6. Reactions with Heterocycles

Methyleniminium halides have also been used for the aminomethylation of heterocyclic compounds. Reaction products **661–667** represent a selection of those which are obtainable from furan or 2-methylfuran (413), 1-*p*-tolylpyrrole or indole (165), 2-methylpyridine (173), 1-phenyl-3,4-tetramethylenepyrazolone-(5) (414), 2,3-dimethyl-1-phenyl-pyrazolone-(5), or 2-acetylthiazolidine (165) as starting materials. Also, aminomethylation of racemic dicyanocobalt(III) heptamethylcorrine by means of MIH in methylene chloride has been reported (216).

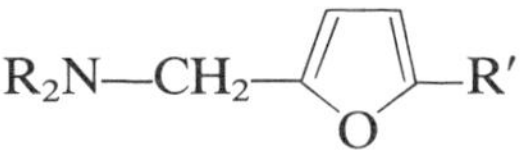

(**661**) R₂N—CH₂—(furan)—R'
R'=H, CH₃

(**662**) piperidine-N—CO—CH(NR₂)—pyrrole(N—C₆H₄—CH₃)

(**663**) piperidine-N—CO—CH(NR₂)—indole

(**664**) R₂N—CH₂—CH₂—pyridine

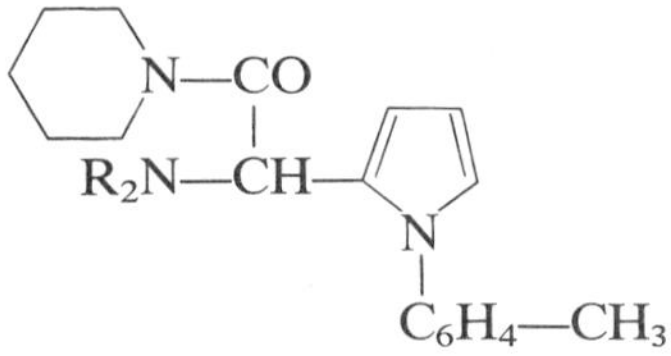

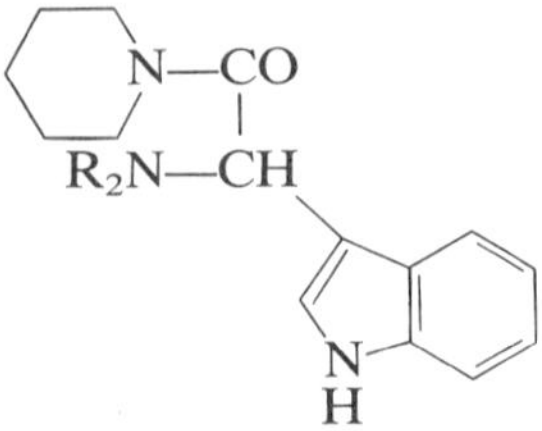

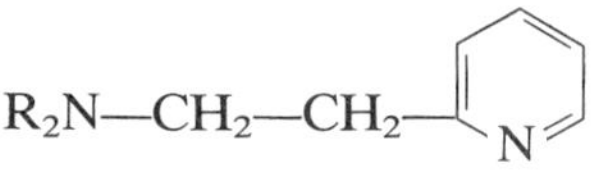

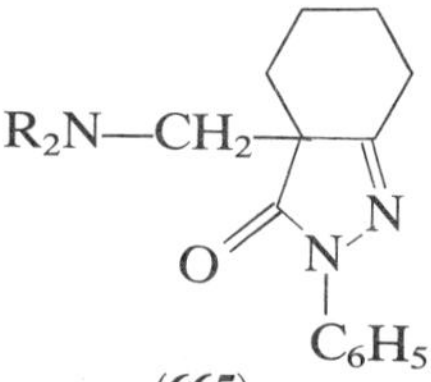

(**665**) R₂N—CH₂— (pyrazolone with cyclohexane, N—C₆H₅)

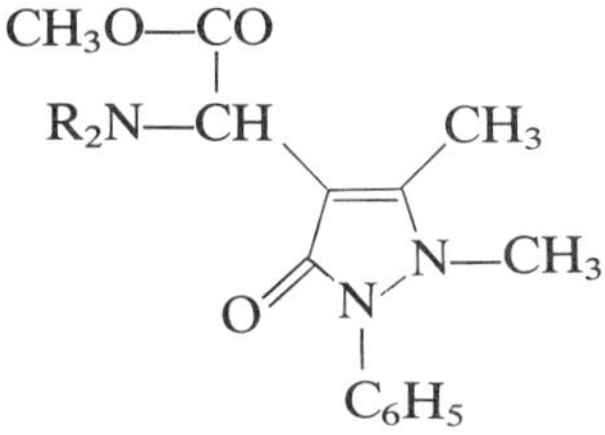

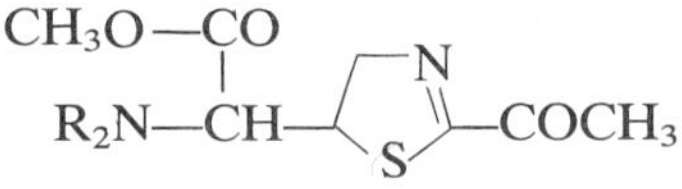

(**666**) CH₃O—CO—CH(NR₂)— (pyrazolone with CH₃, N—CH₃, N—C₆H₅)

(**667**) CH₃O—CO—CH(NR₂)— (thiazoline)—COCH₃

Likewise amidoalkylation of pyrrole with N-acylcycliminium salts is known, for example, formation of a one- or twofold-substituted product such as **668** or **669** (416):

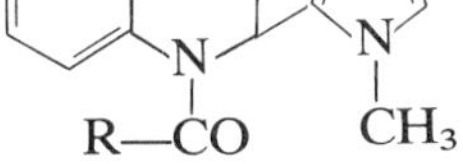

(**668**) dihydroquinoline(N—CO—R)—pyrrole(N—CH₃)

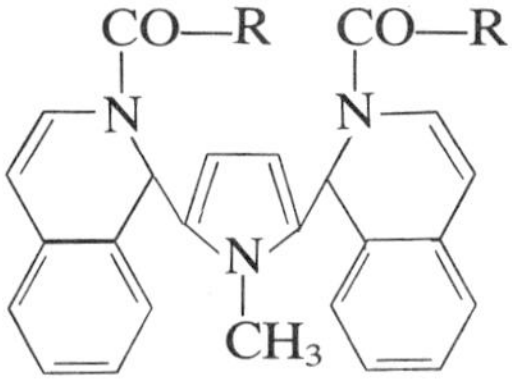

(**669**) bis(dihydroisoquinoline N—CO—R) linked by N-methylpyrrole

Adducts from Schiff's bases and acyl halides (see Section II-A-3) undergo analogous reactions; for example, the amidoalkylation of indole by means of **670** to **671** has been described (416):

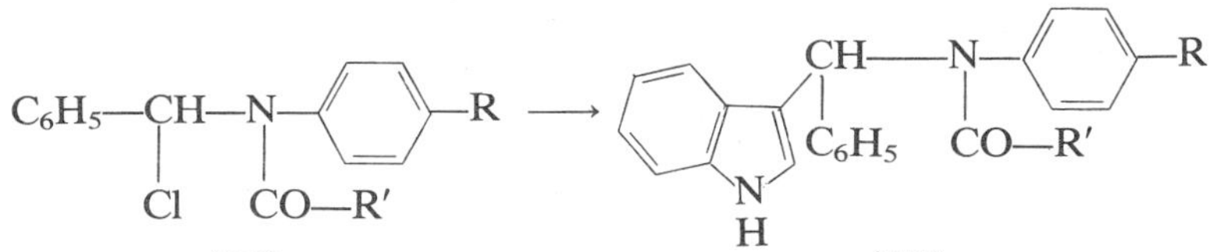

$$C_6H_5-\underset{\underset{Cl}{|}}{CH}-\underset{\underset{CO-R'}{|}}{N}-\underset{}{\bigcirc}-R \longrightarrow$$

[indole with substituents: CH—N—C₆H₄—R, C₆H₅, CO—R']

(**670**)   (**671**)

7. Reactions with Diazoalkanes

Addition of MIH to diazoalkanes (e.g., diazomethane or diazoacetic ester) with elimination of nitrogen leads to the formation of β-haloamines **674** (43,174). The mechanism of this reaction presumably involves an aziridinium salt (**673**) as an intermediate; this has also been reported for diazoalkane reactions with tetraalkyliminium perchlorates (417). 2-Aryl-substituted aziridinium fluoroborates were isolated from reactions of N,N-dialkylarylideniminium fluoroborate suspensions in tetrahydrofuran with diazomethane (418,419).

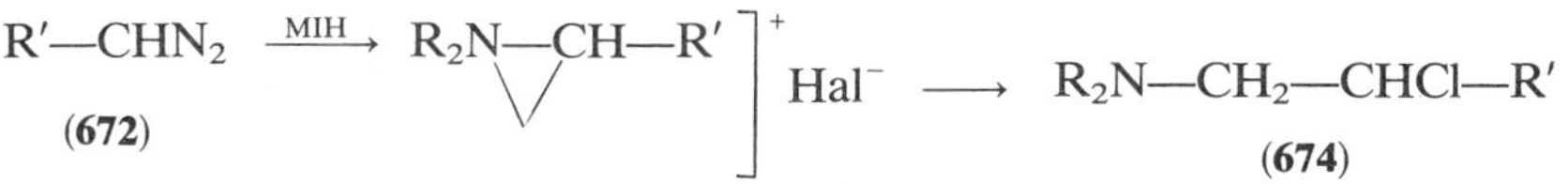

$$R'-CHN_2 \xrightarrow{MIH} \left[R_2N-\underset{\triangle}{CH-R'} \right]^+ Hal^- \longrightarrow R_2N-CH_2-CHCl-R'$$

(**672**)   (**673**)   (**674**)

$R' = H, CO_2C_2H_5$

By means of n-butyllithium the deprotonation of aziridinium salts **676**, prepared by diazomethane addition to iminium salts **675**, has been reported to proceed with the formation of exomethylene compounds **678** and pyrrole via the ylid intermediate **677**. This reaction has been utilized in a new synthetic approach to steroids with an exomethylene function (420).

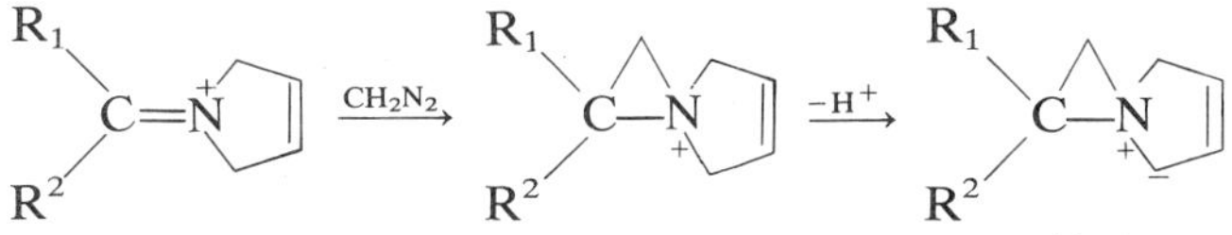

$$\underset{R^2}{\overset{R_1}{\diagdown}}C=\overset{+}{N}\diagup \xrightarrow{CH_2N_2} \underset{R^2}{\overset{R_1}{\diagdown}}\underset{\triangle}{C-\overset{+}{N}}\diagup \xrightarrow{-H^+} \underset{R^2}{\overset{R_1}{\diagdown}}\underset{\triangle}{C-\overset{+}{N}}^-$$

(**675**)   (**676**)   (**677**)

$$677 \longrightarrow \underset{R^2}{\overset{R_1}{\diagdown}}C=CH_2 + HN\diagup$$

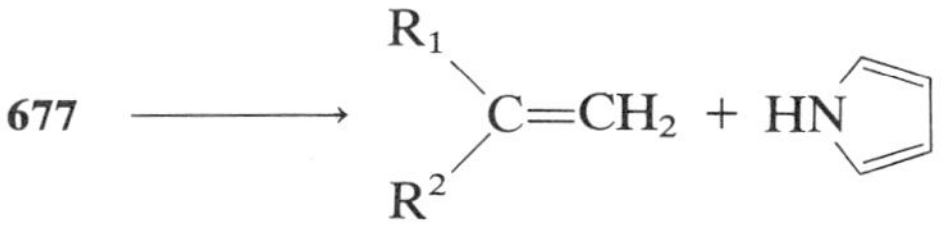

(**678**)

J. MISCELLANEOUS REACTIONS

If mixtures of dialkylarylmethyleniminium chlorides **679** and magnesium powder are heated in tetrahydrofuran, the symmetrical 1,2-dialkylamino-1,2-diarylethanes **680** can be isolated in up to 50% yield (147). Possibly this is a radical reaction on the surface of the metal. This view is supported by the fact that polarographic studies have indicated the possibility of radical dimerization of iminium salts, for example, of **681** to the diaminoethane **682**, which was identified as the hydrochloride (421).

$$[R_2N{=}CH{-}Ar]^+ Cl^- \xrightarrow[-MgCl_2]{+Mg} R_2N{-}\underset{Ar}{\overset{Ar}{CH}}{-}\underset{}{\overset{}{CH}}{-}NR_2$$
(**679**) (**680**)

$$\left[\underset{}{N}{=}C(CH_3)_2\right]^+ \longrightarrow \underset{}{N}{-}\underset{CH_3}{\overset{CH_3}{C}}{-}\underset{CH_3}{\overset{CH_3}{C}}{-}N$$
(**681**) (**682**)

In photochemical reactions the addition of methanol to the double bond of iminium salts **683** was achieved with the formation of C-hydroxymethylation products **684**. Analogous reactions were reported with formamide (422).

$$[R_2N{=}CR'_2]^+ \xrightarrow[-H^+]{+CH_3OH} R_2N{-}\underset{R'}{\overset{R'}{C}}{-}CH_2{-}OH$$
(**683**) (**684**)

REFERENCES

1. J. V. Paukstelis, in A. G. Cook, *Enamines*, Marcel Dekker, New York, 1969, p. 169.
2. P. A. S. Smith, *The Chemistry of Open-Chain Organic Nitrogen Compounds*, Vol. I, W. A. Benjamin, New York, 1965, p. 291.
3. S. Patai, *The Chemistry of the Carbon-Nitrogen Double Bond*, Interscience, New York, 1970, p. 63 (S. Dayagi and Y. Degani); p. 255 (K. Harada); p. 299 (J.-P. Anselme).
4. H. Hellmann, in W. Foerst, *Neuere Methoden der Präparativen Organischen Chemie*, Vol. II, Verlag Chemie, Weinheim/Bergstr., 1960, p. 190.
5. H. E. Zaugg and W. B. Martin, *Org. React.*, **14**, 52 (1965).
6. H. E. Zaugg, *Synthesis*, 1970, 49.
7. G. A. Olah and P. Kreisenbühl, *J. Amer. Chem. Soc.*, **89**, 4756 (1967).
8. G. R. Krow, C. Pyun, C. Leitz, and J. Marakowski, *J. Org. Chem.*, **39**, 2449 (1974).

9. H. Böhme and E. Köhler, *Sitzber. Ges. Beförd. Ges. Naturw. Marburg*, **83/84,** 535 (1961); *C. A.,* **59,** 11416 (1963).
10. A. Lachmann, *J. Amer. Chem. Soc.,* **46,** 1477 (1924).
11. M. O. Forster, *J. Chem. Soc.,* **75,** 934 (1899).
12. H. Decker and P. Becker, *Liebigs Ann. Chem.,* **395,** 362 (1913).
13. C. R. Hauser and D. Lednicer, *J. Org. Chem.,* **24,** 46 (1959).
14. A. L. Morrison and H. Rinderknecht, *J. Chem. Soc.,* 1950, 1478.
15. H. Böhme and P. Plappert, *Chem. Ber.,* **108,** 3574 (1975).
16. S. Hünig and J. Utermann, *Chem. Ber.,* **88,** 1485 (1955).
17a. K. Hartke, F. Rossbach, and M. Radau, *Liebigs Ann. Chem.,* **762,** 167 (1972).
17b. S. S. Mathur and H. Suschitzky, *Tetrahedron Lett.,* **1975,** 785.
18. A. Reliquet and F. Reliquet-Clesse, *C. R. Acad. Sci. Paris,* **276C,** 429 (1973).
19. Y. Ito, S. Katsuragawa, M. Okano, and R. Oda, *Tetrahedron,* **23,** 2159 (1967).
19a. G. Uray and E. Ziegler, *Z. Naturforsch.* **30b,** 245 (1975).
20a. H. Böhme and K. Hartke, *Chem. Ber.,* **96,** 600 (1963).
20b. J. P. Chupp, J. F. Olin, and H. K. Landwehr, *J. Org. Chem.,* **34,** 1192 (1969).
20c. K. W. Ratts and J. P. Chupp, *J. Org. Chem.,* **39,** 3745 (1974).
20d. V. U. Ahmad, A. Basha, and A. W. Rahman, *Z. Naturforsch.,* **30b,** 128 (1975).
21. H. Böhme, S. Ebel, and K. Hartke, *Chem. Ber.,* **98,** 1463 (1965).
22. H. Kiefer, *Synthesis,* **1972,** 39.
23. H. Hagemann and K. Ley, *Angew. Chem.,* **84,** 1062 (1972).
24. H. Böhme and S. Ebel, *Pharmazie,* **20,** 296 (1965); *C. A.,* **64,** 5066 (1966).
25. H. Leuchs, G. Wulkow, and H. Gerland, *Ber. Dtsch. Chem. Ges.,* **65,** 1586 (1932).
26. H. Leuchs and A. Schlötzer, *Ber. Dtsch. Chem. Ges.,* **67,** 1572 (1934).
27. K. Takayama, K. Harano, and T. Taguchi, *Yakugaku Zasshi,* **94,** 540 (1974); *Chem. Inform.,* **1974,** 40–259, 40–260.
28. K. Hartke and J. Bartulin, *Angew. Chem.,* **74,** 214 (1962).
29. H. D. Stachel, *Angew. Chem.,* **71,** 246 (1951).
30. H. Ulrich and A. A. R. Sayigh, *J. Org. Chem.,* **28,** 1427 (1963).
31. J. S. Walia and P. S. Walia, *Chem. Ind. (London),* **1969,** 135.
32. G. H. Alt, in A. G. Cook, *Enamines,* Marcel Dekker, New York, 1969, p. 116.
33. S. Hünig and H. Hoch, *Fortschr. Chem. Forsch.,* **14,** 235 (1970).
34. E. E. P. Hamilton and R. Robinson, *J. Chem. Soc.,* **109,** 1029 (1916).
35. H. Böhme, *Angew. Chem.,* **68,** 224 (1956).
36. H. Böhme, H. Ellenberg, O.-E. Herboth, and W. Lehners, *Chem. Ber.,* **92,** 1608 (1959).
37. G. Opitz, H. Hellmann, and H. W. Schubert, *Liebigs Ann. Chem.,* **623,** 117 (1959).
38. H. Böhme, H. J. Bohn, E. Köhler, and J. Roehr, *Liebigs Ann. Chem.,* **664,** 130 (1963).
39. J. Grätzel von Grätz, Dissertation Marburg, 1976.
40. N. J. Leonard, A. S. Hay, R. W. Fulmer, and V. W. Gash, *J. Amer. Chem. Soc.,* **77,** 439 (1955).
41. U. Edlund, *Acta Chem. Scand.,* **27,** 4027 (1973).
42. L. Alais, P. Angibeaud, and R. Michelot, *C. R. Acad. Sci. Paris,* **269c,** 150 (1969).
43. G. Opitz and A. Griesinger, *Liebigs Ann. Chem.,* **665,** 101 (1963).
44. H. E. A. Kramer, *Liebigs Ann. Chem.,* **696,** 15 (1966).
45. L. Alais, R. Michelot, and B. Tschoubar, *C. R. Acad. Sci. Paris,* **273c,** 261 (1971).
46. J. Pielichowski, *Rozniki Chem.,* **40,** 1765 (1966); *C. A.,* **66,** 75868 (1967).
47. K. Kata, *Kogyo Kagaku Zasshi,* **59,** 1006 (1956); *C. A.,* **52,** 10002 (1958).
48. A. L. Love and R. K. Olsen, *J. Org. Chem.,* **37,** 3431 (1972).

49. L. Duhamel, P. Duhamel, and J.-M. Poirier, *Tetrahedron Lett.*, **1973**, 4237.
50. H. I. House, *Modern Synthetic Reactions*, W. A. Benjamin, Menlo Park, Calif., 1972, p. 570.
51. G. Stork, R. Terrel, and J. Szmuszkovicz, *J. Amer. Chem. Soc.*, **76**, 2029 (1954).
52. G. Stork and H. K. Landesman, *J. Amer. Chem. Soc.*, **78**, 5128 (1956).
53. G. Stork, A. Brizzolara, H. Landesman, J. Szmuszkovicz, and R. Terrel, *J. Amer. Chem. Soc.*, **85**, 207 (1963).
54. J. Szmuszkovicz, *Advan. Org. Chem.*, **4**, 1 (1963).
55. E. Elkik and P. Vaudescal, *C. R. Acad. Sci. Paris*, **264C**, 1779 (1967).
56. K. Osmers, Dissertation, Marburg/Lahn, 1968.
57. A. Lukasiewiz and J. Lesinka, *Tetrahedron*, **21**, 3247 (1965); **24**, 7 (1968).
58. G. H. Alt and A. J. Speziale, *J. Org. Chem.*, **31**, 1340 (1966).
59. P. W. Hickmott, *Chem. Ind. (London)*, **1974**, 731.
60. T. Inukai and R. Yoshizawa, *J. Org. Chem.*, **32**, 404 (1967).
61. J. R. Hargreaves, P. W. Hickmott, and B. J. Hopkins, *J. Chem. Soc. C*, **1968**, 2599.
62. N. F. Firrel, P. W. Hickmott, and B. J. Hopkins, *J. Chem. Soc. C*, **1970**, 1477.
63. P. W. Hickmott and B. J. Hopkins, *J. Chem. Soc. C*, **1968**, 2918.
64. P. W. Hickmott, H. Suschitzky, and R. Urbain, *J. Chem. Soc. Perkin I*, **1973**, 2063.
65. A. Halleux and H. G. Viehe, *J. Chem. Soc. C*, **1970**, 881.
66. H. Weingarten and J. S. Wager, *J. Chem. Soc. Chem. Commun.*, **1970**, 854.
67. L. Duhamel, P. Duhamel, and G. Plé, *Tetrahedron Lett.*, **1974**, 47.
68. R. L. Pederson, J. L. Johnson, R. P. Holysz, and A. C. Ott, *J. Amer. Chem. Soc.*, **79**, 1115 (1957).
69a. A. Kirrmann, E. Elkik, and P. Vaudescal, *C. R. Acad. Sci. Paris*, **262**, 1268 (1966).
69b. L. Paul, E. Schuster, and G. Hilgetag, *Chem. Ber.*, **100**, 1087 (1967).
69c. R. Carlson and C. Rapp, *Acta Chem. Scand.*, **28**, 1060 (1974), **29**, 634 (1975).
69d. J. Pielichowski and J. Kyziol, *Mh. Chem.*, **105**, 1306 (1974).
70. H. Ahlbrecht and M. T. Reiner, *Tetrahedron Lett.*, **1971**, 4901.
71. A. Halleux and H. G. Viehe, *J. Chem. Soc. C*, **1968**, 1726.
72. L. Duhamel, P. Duhamel, and G. Plé, *C. R. Acad. Sci. Paris*, **751C** (1970).
73. H. Böhme, G. Auterhoff, and W. Höver, *Chem. Ber.*, **104**, 3350 (1971).
74. L. Duhamel, P. Duhamel, and G. Plé, *Tetrahedron Lett.*, **1972**, 85.
75. H. Böhme and W. Höver, *Liebigs Ann. Chem.*, **748**, 59 (1971).
76. R. L. Pruett, J. T. Barr, K. E. Rapp, C. T. Bahner, J. D. Gibson, and R. H. Lafferty, Jr., *J. Amer. Chem. Soc.*, **72**, 3646 (1950).
77. N. Wiberg and J. W. Buchler, *Angew. Chem.*, **74**, 490 (1962).
78. O. Tsuge, K. Yanagi, and M. Horie, *Bull. Chem. Soc. Jap.*, **44**, 2171 (1971).
79. S. Gabriel, *Ber. Dtsch. Chem. Ges.*, **44**, 1905 (1911).
80. H. Böhme and G. Berg, *Chem. Ber.*, **99**, 2127 (1966).
81. D. M. Kunovskaya and K. O. Ogloblin, *Zh. Org. Khim.*, **10**, 972 (1974); *Chem. Inform.* **1974**, 34–133.
82. G. H. Alt and A. J. Speziale, *J. Org. Chem.*, **29**, 794 (1964).
83. H. G. Reiber and T. D. Stewart, *J. Amer. Chem. Soc.*, **62**, 3026 (1940).
84. T. Zincke and W. Würker, *Liebigs Ann. Chem.*, **338**, 107 (1905).
85. M. Lamchen, W. Pugh, and A. M. Stephen, *J. Chem. Soc.*, **1954**, 4418.
86. R. Kuhn and H. Schretzmann, *Chem. Ber.*, **90**, 557 (1957).
87. N. J. Leonard and J. V. Paukstelis, *J. Org. Chem.*, **28**, 3021 (1963).
88. J. W. Stanley, J. G. Beasley, and I. W. Mathison, *J. Org. Chem.*, **37**, 3746 (1972).
89. H. E. Nikolajewski, S. Dähne, and B. Hirsch, *Z. Chem.*, **8**, 63 (1968).
90. N. Wiberg and K. H. Schmid, *Angew. Chem.*, **76**, 381 (1964).

91. N. M. Libman, *Zh. Org. Khim.*, **3,** 1235 (1967); *C. A.,* **67,** 90342 (1967).
92. T. R. Keenan and N. J. Leonard, *J. Amer. Chem. Soc.,* **93,** 6567 (1971).
93. H. Newman and T. L. Fields, *Tetrahedron,* **28,** 4051 (1972).
94. R. Breslow, T. Eicher, A. Krebs, R. A. Peterson, and J. Posner, *J. Amer. Chem. Soc.,* **87,** 1320 (1965).
95. A. Krebs and J. Breckwoldt, *Tetrahedron Lett.,* **1969,** 3797.
96. Z. Yoshida, S. Yoneda, T. Miyamoto, and S. Miki, *Tetrahedron Lett.,* **1974,** 813.
97. G. Zinner, W. Kliegel, and W. Ritter, *Chem. Ber.,* **99,** 1285 (1966).
98. H. Böhme, G. Driesen, and D. Schünemann, *Arch. Pharm. (Weinheim),* **294,** 344 (1961).
99. H. Böhme and G. Meyer, *Pharmazie,* **25,** 283 (1970).
100. E. Cherbuliez and G. Feer, *Helv. Chim. Acta,* **5,** 678 (1922).
101. E. Cherbuliez and G. Sulzer, *Helv. Chim. Acta,* **8,** 567 (1925).
102. S. Gabriel, *Ber. Dtsch. Chem. Ges.,* **41,** 242 (1908).
103. H. Böhme and G. Seitz, *Arch. Pharm. (Weinheim),* **299,** 695 (1966).
104. H. Böhme and H.-H. Otto, *Arch. Pharm. (Weinheim),* **300,** 922 (1967).
105. F. Knotz, *Sci. Pharm.,* **38,** 227 (1970); *C. A.,* **74,** 125548 (1971).
106. H. Böhme and H. Schwartz, *Arch. Pharm. (Weinheim),* **306,** 684 (1973).
107. H. Böhme and G. Meyer, *Arch. Pharm. (Weinheim),* **303,** 514 (1970).
108. H. Böhme and F. Eiden, *Arch. Pharm. (Weinheim),* **292,** 642 (1959).
109. F. W. Hoover, H. B. Stevenson, and H. S. Rothrock, *J. Org. Chem.,* **28,** 1825 (1963).
110. H. Zinner, H. Herbig, and H. Wigert, *Chem. Ber.,* **89,** 2131 (1956).
111. W. Ritter, Ger. Pat. 2,040,175; *C. A.,* **76,** 112939p (1972).
112. Farbenfabriken Bayer AG (G. Jäger, C. Metzger, W. Ritter, and R. Wegler), Ger. Pat. 2,119,518; *C. A.,* **78,** 58109u (1973).
113. W. Ritter, private communication.
114. F. Feist, *Ber. Dtsch. Chem. Ges.,* **47,** 1180 (1914).
115. H. Böhme, F. Eiden, and D. Schünemann, *Arch. Pharm. (Weinheim),* **294,** 307 (1961).
115a. B. S. Drach and G. N. Miskevich, *Zh. Org. Khim.* **11,** 316 (1975); *Chem. Inform.* **1975,** 18–223.
116. D. Matthies, *Arch. Pharm. (Weinheim),* **301,** 867 (1968).
117. D. Matthies, *Pharmazie,* **25,** 522 (1970); *C. A.,* **74,** 13397 (1971).
118. D. Matthies, *Synthesis,* **1972,** 380.
119. B. S. Drach, I. Y. Dolgushina, and A. V. Kirsanov, *Zh. Org. Khim,* **9,** 414 (1973); *Chem. Inform.,* **1973,** 20–299.
120. B. S. Drach, I. Y. Dolgushina, and A. D. Sinitsa, *Khim. Geterosikl. Soedin,* **1974,** 928; *Chem. Inform.,* **1974,** 44–127.
121. D. Matthies and K. Hain, *Synthesis,* **1973,** 154.
122. D. Ben-Ishai and Z. Inbal, *J. Org. Chem.,* **38,** 2251 (1973).
123. R. R. Schmidt and E. Schlipf, *Chem. Ber.,* **103,** 3783 (1970).
124. H. Petersen, *Synthesis,* **1973,** 243.
125. H. D. Bartfeld and W. Flitsch, *Chem. Ber.,* **106,** 1423 (1973).
126. Farbwerke Hoechst AG (H. Böhme and E. Mundlos), DBP 951, 269; *C. A.,* **53,** 22022 (1959).
127. H. Böhme, E. Mundlos, and O.-E. Herboth, *Chem. Ber.,* **90,** 2003 (1957).
128. E. Jongejan, W. J. M. van Tilborg, C. H. V. Dusseau, H. Steinberg, and T. J. de Boer, *Tetrahedron Lett.,* **1972,** 2359.
129. J. von Braun and E. Röver, *Ber. Dtsch. Chem. Ges.,* **36,** 1196 (1903).
130. E. Schmidt and P. Köhler, *Arch. Pharm. (Weinheim),* **240,** 231 (1902).

131. W. C. Hunt and E. C. Wagner, *J. Org. Chem.*, **16,** 1792 (1951).
132. H. Böhme and N. Kreutzkamp, *Sitzber. Ges. Beförd. Ges. Naturw. Marburg,* **76,** 3 (1953); *C. A.,* **49,** 3182 (1953); *Naturwissenschaften,* **40,** 340 (1953); *C. A.,* **49,** 2433 (1955).
133. H. Böhme, M. Dähne, W. Lehners, and E. Ritter, *Liebigs Ann. Chem.*, **723,** 34 (1969).
134. H. Böhme, W. Lehners, and G. Keitzer, *Chem. Ber.*, **91,** 340 (1958).
135. Farbwerke Hoechst AG (H. Böhme), DBP 1,081,465; *C. A.,* **56,** 9932 (1962).
136. D. D. Reynolds and B. C. Cossar, *J. Heterocycl. Chem.*, **8,** 597 (1971).
137. H. Böhme and K. Hartke, *Chem. Ber.*, **93,** 1305 (1960).
138. G. Zinner and W. Kliegel, *Chem. Ber.*, **98,** 4036 (1965).
139. G. Zinner and W. Kliegel, *Chem. Ber.*, **100,** 2515 (1967).
140. H. Böhme and D. Eichler, *Chem. Ber.*, **100,** 2131 (1967).
141. G. Zinner and W. Kilwing, *Chem. Ztg.*, **97,** 156 (1973).
142. G. Zinner, W. Kliegel, W. Ritter, and H. Böhlke, *Chem. Ber.*, **99,** 1678 (1966).
143. H. Böhme, K. Hartke, and A. Müller, *Chem. Ber.*, **96,** 595 (1963).
144. W. Pasche, Dissertation, Marburg/Lahn, 1968.
145. B. Unterhalt and D. Thamer, *Synthesis,* **1973,** 302.
146. H. Böhme and M. Hilp, *Chem. Ber.*, **103,** 104 (1970).
147. H. Böhme, M. Haake, and G. Auterhoff, *Arch. Pharm. (Weinheim),* **305,** 10 (1972).
148. H. Böhme and G. Auterhoff, *Chem. Ber.*, **104,** 2013 (1971).
149. H. Böhme and M. Haake, *Chem. Ber.*, **100,** 3609 (1967).
150. H. Böhme, M. Haake, and G. Auterhoff, *Arch Pharm. (Weinheim),* **305,** 88 (1972).
151. L. Duhamel, P. Duhamel, and P. Siret, *Tetrahedron Lett.,* **1972,** 3607.
152. H. Böhme and Y. S. Sadanandam, *Arch. Pharm. (Weinheim),* **306,** 227 (1973).
153. Farbwerke Hoechst AG (H. Böhme and K. Hartke), DBP 1,110,652; *C. A.,* **56,** 3330 (1962).
154. H. Ulrich and A. A. R. Saigh, *J. Chem. Soc.,* **1963,** 1098.
155. A. E. Milgrom, L. P. Levitan, V. A. Malii, Y. K. Sakharov, and V. S. Udalova, USSR Pat. 230,174; *C. A.,* **71,** 12600 (1969).
156. H. Böhme and M. Hilp, *Chem. Ber.*, **103,** 3930 (1970).
157. H. Böhme and K. Osmers, *Chem. Ber.*, **105,** 2237 (1972).
158. Farbenfabriken Bayer AG (H. Kritzler, K. Wagner, and H. Holtschmidt), DBP 1,153,756; C. **1964,** 33–2216.
159. K.-H. Meyer-Dulheuer, Dissertation, Marburg/Lahn, 1964.
160. E. Kulow, Diplom.-Arbeit, Marburg/Lahn, 1974.
161. H. Böhme and P. Backhaus, *Liebigs Ann. Chem.*, **1975,** 1790.
162. H. Böhme, L. Koch, and E. Köhler, *Chem. Ber.*, **95,** 1849 (1962).
163. H. Ulrich, R. Richter, P. J. Whitman, and A. A. R. Sayigh, *J. Org. Chem.*, **39,** 2987 (1974).
164. H. Gross, J. Gloede, and J. Freiberg, *Liebigs Ann. Chem.*, **702,** 68 (1967).
165. J. Gloede, J. Freiberg, W. Bürger, G. Ollmann, and H. Gross, *Arch. Pharm. (Weinheim),* **302,** 354 (1969).
166. H. Böhme and K.-H. Meyer-Dulheuer, *Liebigs Ann. Chem.*, **688,** 78 (1965).
167. E. Fluck and P. Meiser, *Chem. Ztg.*, **95,** 922 (1971).
168. E. Fluck and P. Meiser, *Angew. Chem.*, **83,** 721 (1971).
169. E. Fluck and P. Meiser, *Chem. Ber.*, **106,** 69 (1973).
170. T. Moeller and A. H. Westlake, *J. Inorg. Nucl. Chem.*, **29,** 957 (1967).
171. Farbenfabriken Bayer AG (H. Holtschmidt and K. Wagner), DBP 1,132,118; *C. A.,* **58,** 3323 (1963).
172. H. Böhme and W. Lehners, *Liebigs Ann. Chem.*, **595,** 169 (1955).

173. H. Böhme and M. Haake, *Liebigs Ann. Chem.*, **705,** 147 (1967).
174. H. Böhme, E. Mundlos, W. Lehners, and O.-E. Herboth, *Chem. Ber.*, **90,** 2008 (1957).
175. H. Böhme and W. Höver, *Chem. Ber.*, **103,** 3918 (1970).
176. R. M. Ottenbrite and G. R. Myers, *Can. J. Chem.*, **51,** 3631 (1973).
177. Z. E. Samojlova and R. G. Kostjanovskij, *Izv. Akad. Nauk. SSSR, Ser. Khim.*, **1970,** 1030; *C. A.*, **73,** 66352 (1970).
178. H. Böhme and H. Ellenberg, *Chem. Ber.*, **92,** 2976 (1959).
179. T. D. Stewart and W. E. Bradley, *J. Amer. Chem. Soc.*, **54,** 4172 (1932).
180. A. Venot and G. Adrian, *Tetrahedron Lett.*, **1972,** 4663.
181. J. Majer and J. Denkstein, *Collect. Czech. Chem. Commun.*, **31,** 2547 (1966).
182. H. W. Post, *J. Org. Chem.*, **1,** 231 (1937).
183. K. W. Dunning and W. J. Dunning, *J. Chem. Soc.*, **1950,** 2925.
184. H. Böhme and H. Orth, *Arch. Pharm. (Weinheim)*, **300,** 148 (1967).
185. H. Gross and B. Costisella, *J. Prakt. Chem.*, **311,** 925 (1969).
186. H. Böhme and A. Müller, *Arch. Pharm. (Weinheim)*, **296,** 54 (1963).
186a. J. A. Vida and W. R. Wilber, *J. Med. Chem.*, **16,** 602 (1973).
187. American Cyanamid Company (M. T. Beachem and J. C. Oppelt), U.S. Pat. 3,317,529; *C. A.*, **67,** 116896 (1967).
188. B. E. Ivanov and S. S. Krokhina, *Izv. Akad. Nauk. SSSR, Ser. Khim.*, **1967,** 2782; *C. A.*, **69,** 67494 (1968).
189. Badische Anilin- & Soda-Fabrik AG (E. Plötz), DBP 961,804; C. **1957,** 12307.
190. D. Martin and W. Weise, *Chem. Ber.*, **99,** 3367 (1966).
191. H. Böhme and A. Sickmüller, unpublished results.
192. A. G. Korepin. R. G. Gafurov, and L. T. Eremenko, *Izv. Akad. Nauk. SSSR. Ser. Khim.*, **1974,** 474; *Chem. Inform.*, **1974,** 24–207.
193. H. Ellenberg, Dissertation, Marburg/Lahn, 1958.
194. H. Böhme and G. Lerche, *Chem. Ber.*, **100,** 2125 (1967).
195. H. Böhme and J. Roehr, *Liebigs Ann. Chem.*, **648,** 21 (1961).
196. H. Böhme and K. Hartke, *Chem. Ber.*, **96,** 604 (1963).
197. W. M. Schubert and Y. Motoyama, *J. Amer. Chem. Soc.*, **87,** 5507 (1965).
198. I. E. Pollak, A. D. Trifunac, and G. F. Grillot, *J. Org. Chem.*, **32,** 272 (1967).
199. H. Meerwein, H. Allendörfer, P. Beekmann, F. Kunert, H. Morschel, F. Pawellek, and K. Wunderlich, *Angew. Chem.*, **70,** 211 (1958).
200. R. Damico and C. D. Broaddus, *J. Org. Chem.*, **31,** 1607 (1966).
201. H. Volz and H.-H. Kiltz, *Liebigs Ann. Chem.*, **752,** 86 (1971).
202. N. L. Bauld and Y. S. Rim, *J. Amer. Chem. Soc.*, **89,** 6763 (1967).
203. H. J. Dauben, Jr., and D. F. Rhoades, *J. Amer. Chem. Soc.*, **89,** 6764 (1967).
204. C. Jutz, *Chem. Ber.*, **97,** 2050 (1964).
205. R. N. Renaud and L. C. Leitch, *Can. J. Chem.*, **46,** 385 (1968).
206. H. Böhme and P. Backhaus, *Liebigs Ann. Chem.*, **1975,** 1790.
207. A. Ahond, A. Cavé, C. Kan-Fan, H. P. Husson, J. de Rostolan, and P. Potier, *J. Amer. Chem. Soc.*, **90,** 5622 (1968).
208. A. B. Burg, *J. Amer. Chem. Soc.*, **65,** 1629 (1943).
209. H. Z. Lecher and W. B. Hardy, *J. Amer. Chem. Soc.*, **70,** 3789 (1948).
210. P. A. Bather, J. R. Linday-Smith, and R. O. C. Norman, *J. Chem. Soc. C,* **1971,** 3060.
211. P. Backhaus, Dissertation, Marburg/Lahn, 1974.
212. D. Liotta, A. D. Baker, S. Goldstein, N. L. Goldman, F. Weinstein-Lanse, D. Felsen-Reingold, and R. Engel, *J. Org. Chem.*, **39,** 2718 (1974).
213. E. Schmidt and F. M. Litterscheid, *Liebigs Ann. Chem.*, **337,** 37 (1904).

214. H. Böhme and E. Boll, *Chem. Ber.*, **90,** 2013 (1957).
215. H. Böhme, M. Hilp, L. Koch, and E. Ritter, *Chem. Ber.*, **104,** 2018 (1971).
216. J. Schreiber, H. Maag, N. Hashimoto, and A. Eschenmoser, *Angew. Chem.*, **83,** 355 (1971).
217. M. Hilp, Dissertation, Marburg/Lahn, 1969.
218. H. Gross and J. Rusche, *Angew. Chem.*, **76,** 534 (1964).
219. J. Goubeau, E. Allenstein, and A. Schmidt, *Chem. Ber.*, **97,** 884 (1964).
220. K. Bott, *Tetrahedron Lett.*, **1968,** 3323.
221. H. Böhme and W. Krause, *Chem. Ber.*, **84,** 170 (1951).
222. A. Senning and P. Kelly, *Acta Chem. Scand.*, **26,** 2877 (1972).
223. R. Kiesel and E. P. Schram, *Inorg. Chem.*, **12,** 1090 (1974).
224. N. Wiberg, F. Raschig, and K. H. Schmid, *J. Organomet. Chem.*, **10,** 29 (1967).
225. H. Holtschmidt, *Angew. Chem.*, **74,** 848 (1962).
226. K. Grohe, E. Klauke, H. Holtschmidt, and H. Heitzer, *Liebigs Ann. Chem.*, **730,** 140 (1969).
227. K. Grohe, E. Degener, H. Holtschmidt, and H. Heitzer, *Liebigs Ann. Chem.*, **730,** 133 (1969).
228. H. Böhme and F. Eiden, *Arch. Pharm. (Weinheim)*, **289,** 677 (1956).
229. F. Sachs, *Ber. Dtsch. Chem. Ges.*, **31,** 1225 (1898).
230. H. E. Zaugg, *J. Amer. Chem. Soc.*, **76,** 5818 (1954).
231. M. Backstez, *Ber. Dtsch. Chem. Ges.*, **46,** 3087 (1914).
232. H. Böhme and H. Dehmel, *Arch. Pharm. (Weinheim)*, **304,** 397 (1971).
233. S. Gabriel, *Ber. Dtsch. Chem. Ges.*, **41,** 242 (1908).
234. R. A. Corral and O. O. Orazi, *Tetrahedron Lett.*, **1964,** 1693.
235. O. O. Orazi, R. A. Corral, and H. Schuttenberg, *Tetrahedron Lett.*, **1969,** 2639; *J. Chem. Soc. Perkin* I, **1974,** 2087.
236. H. Patin, *Tetrahedron Lett.*, **1974,** 2893; H. Patin and D. Mourot, *C. R. Acad. Sci., Sér. C.* **281,** 737 (1975).
237. F. Weygand, *Ber. Dtsch. Chem. Ges.*, **73,** 1259 (1940).
238. A. C. Cope, E. Ciganek, L. J. Fleckenstein, and M. A. P. Meisinger, *J. Amer. Chem. Soc.*, **82,** 4651 (1960).
239. V. Franzen, *Chem. Ztg.*, **80,** 779 (1956).
240. H. Hellman and G. Opitz, α-*Aminoalkylierung*, Verlag Chemie, Weinheim/Bergstr., 1960, p. 6.
241. H. O. House, *Modern Synthetic Reactions*, 2nd ed., W. A. Benjamin, Menlo Park, Calif., 1972, p. 654.
242. M. Tramontini, *Synthesis*, **1973,** 703.
243. R. Huisgen and W. Kolbeck, *Tetrahedron Lett.*, **1965,** 783.
244. H. Volz and L. Ruchti, *Liebigs Ann. Chem.*, **763,** 184 (1972).
245. I. Ugi and C. Steinbrückner, *Chem. Ber.*, **94,** 734 (1961).
246. I. Ugi, *Isonitrile Chemistry*, Academic Press, New York, 1971, p. 145.
247. C. A. Grob, *Angew. Chem.*, **81,** 543 (1969).
248. H. Suschitzky and C. F. Sellers, *Tetrahedron Lett.*, **1969,** 1105.
249. V. I. Maksimov, *Tetrahedron*, **21,** 687 (1965).
250. G. Fodor, J. J. Ryan, and F. Letourneau, *J. Amer. Chem. Soc.*, **91,** 7768 (1969).
251. R. R. Fraser and R. B. Swingle, *Tetrahedron*, **25,** 3469 (1969).
252. N. J. Leonard, L. A. Miller, and P. D. Thomas, *J. Amer. Chem. Soc.*, **78,** 3463 (1956).
253. N. J. Leonard and F. P. Hauck, Jr., *J. Amer. Chem. Soc.*, **79,** 5279 (1957).
254. B. Rindone and C. Scolastico, *Tetrahedron Lett.*, **1974,** 3379.

255. D. H. Rosenblatt, L. A. Hull, D. C. de Luca, G. T. Davis, R. C. Weglein, and H. K. R. Williams, *J. Amer. Chem. Soc.*, **89,** 1158 (1967).
256. W. H. Dennis, Jr., L. A. Hull, and D. H. Rosenblatt, *J. Org. Chem.*, **32,** 3783 (1967).
257. G. W. Gribble and R. B. Nelson, *J. Org. Chem.*, **38,** 2831 (1973).
258. P. A. S. Smith and R. A. Loeppky, *J. Amer. Chem. Soc.*, **89,** 1147 (1967).
259. M. Neuenschwander and A. Niederhauser, *Chimia*, **27,** 379 (1973).
260. M. Vaultier, R. Danion-Bougot, D. Danion, J. Hamelin, and R. Carrié, *Tetrahedron Lett.*, **1973,** 2883.
261. F. Texier and R. Carrié, *Bull. Soc. Chim. Fr.*, **1973,** 3437.
262. P. G. Gassman and D. K. Dygos, *J. Amer. Chem. Soc.*, **91,** 1543 (1971).
263. J. Ciabattoni and M. Cabell, Jr., *J. Amer. Chem. Soc.*, **93,** 1482 (1971).
264. H. J. Roth, *Arch. Pharm. (Weinheim)*, **294,** 427 (1961).
265. H. Möhrle and D. Schittenhelm, *Chem. Ber.*, **104,** 2475 (1971).
266. H. Möhrle, W. Haug, and E. Federolf, *Arch. Pharm. (Weinheim)*, **306,** 44 (1973).
267. H. Möhrle and S. Mayer, *Arch Pharm. (Weinheim)*, **302,** 481 (1969).
268. C. Hansson and B. Wickberg, *J. Org. Chem.*, **38,** 3074 (1973).
269. D. J. Hupe, M. C. R. Kendall, and T. A. Spencer, *J. Amer. Chem. Soc.*, **95,** 2271 (1973).
270. S. D. Ross, *Tetrahedron Lett.*, **1973,** 1237.
271. D. Herlem, Y. Hubert-Brierre, and F. Khuong-Huu, *Tetrahedron Lett.*, **1973,** 4173.
272. F. Bohlmann, H. J. Müller, and D. Schumann, *Chem. Ber.*, **106,** 3026 (1973).
273. L. M. Trefonas, R. L. Flurry, Jr., R. Majeste, E. A. Meyers, and R. F. Copeland, *J. Amer. Chem. Soc.*, **88,** 2145 (1966).
274. Z. Arnold, *Collect. Czech. Chem. Commun.*, **28,** 2047 (1963).
275. R. J. Harder and W. C. Smith, *J. Amer. Chem. Soc.*, **83,** 3422 (1961).
276. B. M. Tattershall, *J. Chem. Soc. A*, **1970,** 3263.
277. K. Hartke, Dissertation, Marburg/Lahn, 1959.
278. B. U. Schlottmann, Dissertation, Marburg/Lahn, 1972.
279. G. Martin and M. Martin, *Bull. Soc. Chim. Fr.*, **1963,** 1637.
280. A. F. McDonagh and H. E. Smith, *J. Org. Chem.*, **33,** 8 (1968).
281. U. Bomke, Dissertation, Marburg/Lahn, 1969.
282. F. Knoll and U. Krumm, *Chem. Ber.*, **104,** 31 (1971).
283. B. U. Schlottmann, unpublished results.
284. L. P. Vatlina, V. M. Vlasvo, S. A. Polyakov, and V. V. Takhistov, *Zh. Org. Khim.*, **8,** 459 (1972); *Chem. Inform.*, **1972,** 27–097.
285. M. Masul, K. Fujita, and H. Ohmori, *J. Chem. Soc. Chem. Commun.*, **1970,** 182.
286. R. Mason and G. Rucci, *J. Chem. Soc. Chem. Commun.*, **1971,** 1132.
287. R. R. Schmidt, *Chem. Ber.*, **103,** 3242 (1970).
288. W. Krause, *Pharm. Zentralhalle Dtschl.*, **90,** 218 (1951).
289. E. Bremanis, *Z. Anal. Chem.*, **130,** 44 (1949).
290. *Methoden der Organische Chemie* (Houben-Weyl), 4th ed., Vol. II, Georg Thieme Verlag, Stuttgart, 1953, p. 456.
291. W. Mass, M. J. Janssen, E. J. Stamhuis, and H. Wynberg, *J. Org. Chem.*, **32,** 1111 (1967).
292. O. Cervinka, *Collect. Czech. Chem. Commun.*, **30,** 2403 (1965).
293. D. Cabaret, G. Chauvière, and Z. Welvart, *Tetrahedron Lett.*, **1966,** 4109.
294. D. Cabaret, G. Chauvière, and Z. Welvart, *Tetrahedron Lett.*, **1968,** 549.
295. H. Böhme and P. Plappert, *Chem. Ber.*, **108,** 2827 (1975).
296. D. Cabaret, G. Chauvière, and Z. Welvart, *Bull. Soc. Chim. Fr.*, **1969,** 4457.
297. M. Haake, Dissertation, Marburg/Lahn, 1966.

298. H. Gross, J. Gloede, and D. Kunath, *Tetrahedron Lett.*, **1967,** 4089.
299. O. Cervinka, *Collect. Czech. Chem. Commun.*, **25,** 1183 (1960).
300. J. L. Aubagnac, J. Elguero, and R. Jacquier, *Bull. Soc. Chim. Fr.*, **1969,** 3306.
301. G. N. Walker and D. Alkalay, *J. Org. Chem.*, **32,** 2213 (1967).
302. J. L. Aubagnac, J. Elguero, and R. Jacquier, *Bull. Soc. Chim. Fr.*, **1969,** 3316.
303. H. E. Nikolajewski, S. Dähne, D. Leupold, and B. Hirsch, *Tetrahedron*, **24,** 6685 (1968).
304. C. Jutz and E. Schweiger, *Synthesis*, **1974,** 193.
305a. I. A. Zaitsev, M. M. Shestaeva and V. A. Zagorevskii, *Zh. Org. Khim.*, **2,** 1769 (1966); *C. A.*, **66,** 55194j (1967).
305b. H. Böhme and F. Martin, unpublished results.
306. C. M. McLeod and G. M. Robinson, *J. Chem. Soc.*, **1921,** 1470.
307. A. F. Childs, L. J. Goldsworthy, G. F. Harding, F. E. King, A. W. Nineham, W. L. Norris, S. G. P. Plant, B. Selton, and A. L. L. Tompsett, *J. Chem. Soc.*, **1948,** 2174.
308. W. Fink, *Helv. Chim. Acta*, **56,** 1117 (1973).
309. H. Gross and B. Costisella, *Liebigs Ann. Chem.*, **750,** 44 (1971).
310. H. Böhme, R. Broese, A. Dick, F. Eiden, and D. Schünemann, *Chem. Ber.*, **92,** 1599 (1959).
311. H. Böhme and H. Dehmel, *Arch. Pharm. (Weinheim)*, **304,** 403 (1971).
312. L. Duhamel, P. Duhamel, C. Collet, A. Haider, and J. M. Poirier, *Tetrahedron Lett.*, **1972,** 4743.
313. H. Böhme and P. Wagner, *Chem. Ber.*, **102,** 2651 (1969).
314. R. Huisgen and F. Bayerlein, *Liebigs Ann. Chem.*, **630,** 138 (1960).
315. N. J. Leonard and A. S. Hay, *J. Amer. Chem. Soc.*, **78,** 1984 (1956).
316. R. R. Renshaw and D. E. Searle, *J. Amer. Chem. Soc.*, **59,** 2056 (1937).
317. G. F. Grillot, H. R. Felton, B. R. Garret, H. Greenberg, R. Green, R. Clementi, and M. Moskowitz, *J. Amer. Chem. Soc.*, **76,** 3969 (1954).
318. D. Thamer and B. Unterhalt, *Synthesis*, **1973,** 303.
319. H. Böhme and G. Dähler, *Chem. Ber.*, **103,** 3058 (1970).
320. H. Böhme, G. Dähler, and W. Krack, *Liebigs Ann. Chem.*, **1973,** 1686.
321. H. Böhme and H.-H. Otto, *Arch Pharm. (Weinheim)*, **300,** 647 (1967).
322. N. Kreutzkamp and H. Y. Oei, *Arch. Pharm. (Weinheim)*, **299,** 906 (1966).
323. A. Ehrenberg, *J. Prakt. Chem.* [2], **36,** 117 (1887).
324. R. A. Donia, J. A. Shotton, L. O. Bentz, and G. E. P. Smith, Jr., *J. Org. Chem.*, **14,** 952 (1949).
325. C. P. Lo, *J. Org. Chem.*, **26,** 3591 (1961).
326. J. E. Balaban, *J. Chem. Soc.*, **1926,** 568.
327. G. W. Pucher and T. E. Johnson, *J. Amer. Chem. Soc.*, **44,** 822 (1922).
328. H. Böhme and H. Schwartz, *Arch. Pharm. (Weinheim)*, **307,** 775 (1974).
329. H. Böhme and W. Stammberger, *Chem. Ber.*, **104,** 3354 (1971).
330. B. Unterhalt and D. Thamer, *Synthesis*, **1973,** 676.
331. J. Denkstein and V. Kadeřábek, *Collect. Czech. Chem. Commun.*, **31,** 2904 (1966).
332. G. Zinner, N.-P. Lüpke, and U. Dybowski, *Arch. Pharm. (Weinheim)*, **305,** 64 (1972).
333. H. Böhme and F. Martin, unpublished results.
334. M. Haake, unpublished results.
335. Farbwerke Hoechst AG (F. Lindner, W. Siedel, and A. Söder), DBP 1063,598; *C. A.*, **55,** 16507 (1961).
336. H. Böhme and M. Haake, *Arch. Pharm. (Weinheim)*, **300,** 682 (1967).
337. H. Böhme and M. Haake, *Chem. Ber.*, **105,** 2233 (1972).

338. H. Böhme and H. Orth, *Chem. Ber.*, **99**, 2842 (1966).
339. H. Böhme and M. Dähne, *Liebigs Ann. Chem.*, **723**, 41 (1969).
340. H. Böhme and F. Martin, *Chem. Ber.*, **106**, 3540 (1973).
341. H. Böhme and D. Morf, *Chem. Ber.*, **91**, 660 (1958).
342. H. Schönenberger, L. Binde, and A. Adam, *Arch. Pharm. (Weinheim)*, **306**, 64 (1973).
343. F. Sachs, *Ber. Dtsch. Chem. Ges.*, **31**, 3230 (1898).
344. F. B. Kipping and F. G. Mann, *J. Chem. Soc.*, **1927**, 528.
345. A. Müller, Dissertation, Marburg/Lahn, 1961.
346. F. Knotz, *Sci. Pharm.*, **41**, (1973); *C. A.*, **79**, 92152h (1973).
347. E. Profft, H. Teubner, and W. Weuffen, *Arch. Exp. Veterinärmed.*, **21**, 225 (1967); *C. A.*, **68**, 19819 (1968).
348. Farbenfabriken Bayer AG (G. Zumach, B. Anders, F. Grewe, E. Kühle, and H. Kaspers) South African Pat. 6,802,546; *C. A.*, **71**, 12858 (1969).
349. F. Knotz, *Sci. Pharm.*, **38**, 26 (1970); *C. A.*, **73**, 7318 (1970).
350. B. S. Drach, E. P. Sviridov, and A. V. Kirsanov, *Zh. Org. Khim.*, **8**, 1825 (1972); *Chem. Inform.*, **1973**, 1-191.
351. A. M. Aguiar, K. C. Hansen, and J. T. Mague, *J. Org. Chem.*, **32**, 2383 (1967).
352. K. L. Lundberg, R. J. Rowatt, and N. E. Miller, *Inorg. Chem.*, **8**, 1336 (1969).
352a. K. Issleib, M. Lischewski, and A. Zschunke, *Z. Chem.*, **14**, 243 (1974).
353. H. Gross, G. Engelhardt, J. Freiberg, W. Bürger, and B. Costisella, *Liebigs Ann. Chem.*, **707**, 35 (1967).
354. E. K. Fields, *J. Amer. Chem. Soc.*, **74**, 1528 (1952).
355. S. F. Martin and R. Gompper, *J. Org. Chem.*, **39**, 2814 (1974).
356. B. S. Drach, E. P. Sviridov, and A. V. Kirsanov, *Zh. Obshch. Khim.*, **42**, 953 (1972); *Chem. Inform.*, **1972**, 31-367.
357. A. Tzschach and K. Kellner, *J. Prakt. Chem.*, **316**, 851 (1974).
358. D. Seyferth and R. S. Marmor, *Tetrahedron Lett.*, **1970**, 2493.
359. B. S. Drach and E. P. Sviridov, *Zh. Obshch. Khim.*, **43**, 1648 (1973); *Chem. Inform.*, **1973**, 46-315.
359a. B. S. Drach, E. P. Sviridov and A. V. Kirsanov, *Zh. Obshch. Khim.* **45**, 12 (1975); *Chem. Inform.* **1975**, 15-312.
360. I. C. Popoff, L. K. Huber, B. P. Block, P. D. Morton, and R. P. Riordan, *J. Org. Chem.*, **28**, 2898 (1963).
361. F. Weygand, W. Steglich, and F. Fraunberger, *Angew. Chem.*, **79**, 822 (1967).
362. F. Weygand, W. Steglich, I. Lengyel, A. Fraunberger. A. Meierhofer, and W. Oettmeier, *Chem. Ber.*, **99**, 1944 (1966).
363. H. Böhme, E. Mundlos, and G. Keitzer, *Chem. Ber.*, **91**, 656 (1958).
364. H. Böhme and L. Häfner, *Chem. Ber.*, **99**, 281 (1966).
365. H. Böhme, H. Russmann, and M. Junga, *Arch. Pharm. (Weinheim)*, **305**, 924 (1972).
366. F.-G. Fick and K. Hartke, *Tetrahedron Lett.*, **1974**, 3105.
367. H. Böhme, R. Broese, and F. Eiden, *Chem. Ber.*, **92**, 1258 (1959).
368. D. Matthies, *Arch. Pharm. (Weinheim)*, **307**, 801 (1974).
369. E. Ziegler and T. Wimmer, *Chem. Ber.*, **99**, 130 (1966).
370. A. K. Bose, J. C. Kapur, B. Dayal, and M. S. Manhas, *Tetrahedron Lett.*, **1973**, 3797; *J. Org. Chem.*, **39**, 312 (1974).
371. A. K. Bose, M. Tsai, J. C. Kapur, and M. S. Manhas, *Tetrahedron*, **29**, 2355 (1973).
372. G. Opitz and A. Griesinger, *Liebigs Ann. Chem.*, **665**, 101 (1963).
373. G. Opitz, A. Griesinger, and H. W. Schubert, *Liebigs Ann. Chem.*, **665**, 91 (1963).
374. H. Böhme and U. Bomke, *Arch. Pharm. (Weinheim)*, **303**, 779 (1970).

375. D. Seebach, *Synthesis*, **1969**, 17.
376. P. Duhamel, L. Duhamel, and N. Mancelle, *Tetrahedron Lett.*, **1972**, 2991.
377. L. Duhamel, P. Duhamel, and N. Mancelle, *Bull, Soc. Chim. Fr.*, **1974**, 331.
378. H. Böhme and F. Ziegler, *Arch. Pharm. (Weinheim)*, **307**, 287 (1974).
379. H. Böhme and V. Hitzel, *Arch. Pharm. (Weinheim)*, **306**, 948 (1973).
380. G. Köbrich, *Angew. Chem.*, **79**, 15 (1967).
381. H. Böhme and W. Stammberger, *Arch. Pharm (Weinheim)*, **305**, 383 (1972).
382. H. Böhme and W. Stammberger, *Arch. Pharm. (Weinheim)*, **305**, 397 (1972).
383. H. Böhme and W. Stammberger, *Liebigs Ann. Chem.*, **754**, 56 (1971).
384. H. Reinheckel, H. Gross, K. Haage, and G. Sonnek, *Chem. Ber.*, **101**, 1736 (1968).
385. H. Böhme and W. Fresenius, *Arch. Pharm. (Weinheim)*, **305**, 601 (1972).
386. H. Böhme and W. Fresenius, *Arch. Pharm. (Weinheim)*, **305**, 610 (1972).
387. W. Seeliger and W. Diepers, *Liebigs Ann. Chem.*, **697**, 171 (1966).
388. W. Seeliger, E. Aufderhaar, W. Diepers, R. Feinauer, R. Nehring, W. Thier, and H. Hellmann, *Angew. Chem.*, **78**, 913 (1966).
389. Union Oil Company of California (W. D. Schaeffer), U.S. Pat. 3,190,882; *C. A.*, 9821 (1965).
390. R. R. Schmidt, *Angew. Chem.*, **85**, 235 (1973).
391. R. R. Schmidt and A. R. Hoffmann, *Chem. Ber.*, **107**, 78 (1974).
392. C. K. Bradsher, in A. R. Katritzky and A. J. Boulton, *Advan. Heterocycl. Chem.*, **16**, 289 (1974).
393. K. Bott, *Chem. Ber.*, **106**, 2513 (1973).
394. H. Böhme, K. Hartke, and A. Müller, *Chem. Ber.*, **96**, 607 (1963).
395. A. T. Babayan, G. T. Martirosyan, and D. V. Grigoryan, *Zh. Org. Khim.*, **1968**, 984; *C. A.*, **69**, 43320 (1968).
396. H. Böhme and K. H. Ahrens, *Tetrahedron Lett.*, **1971**, 149; *Arch. Pharm. (Weinheim)*, **307**, 828 (1974).
397. M. Mühlstädt and H. Luther, *Z. Chem.*, **10**, 349 (1970).
398. J. Hooz and J. N. Bridson, *J. Amer. Chem. Soc.*, **95**, 602 (1973).
399. M. V. Urushadze, E. G. Abduganiev, Z. E. Samoilova, Kh. Khafizov, E. M. Rokhlin, and R. G. Kostyanovskii, *Izv. Akad. Nauk SSR, Ser. Khim.*, **1973**, 176; *Chem. Inform.*, **1973**, 23–175.
400. H. Böhme, K. Osmers, and P. Wagner, *Tetrahedron Lett.*, **1972**, 2785.
401. R. Fuks, G. S. D. King, and H. G. Viehe, *Angew. Chem.*, **81**, 702 (1969).
402. W. Stammberger, Dissertation, Marburg/Lahn, 1971.
403. H. J. Bestmann, *Fortschr. Chem. Forsch.*, **20**, 1 (1971).
404. H. J. Bestmann, *Angew. Chem.*, **77**, 651 (1965).
405. H. Böhme and D. Eichler, *Arch. Pharm. (Weinheim)*, **300**, 679 (1967).
406. H. Schönenberger, D. Adam, G. Alonso, and A. Adam, *Arch. Pharm. (Weinheim)*, **305**, 300 (1972).
407. D. R. Reynolds and B. C. Cossar, *J. Heterocycl. Chem.*, **8**, 605 (1971).
408. D. D. Reynolds and B. C. Cossar, *J. Heterocycl. Chem.*, **8**, 611 (1971).
409. H. Böhme and K. Hartke, *Chem. Ber.*, **93**, 1310 (1960).
410. D. Eichler, Dissertation, Marburg/Lahn, 1965.
411. A. K. Sheinkman, A. N. Prilepskaya, and A. N. Kost, *Khim. Geterosikl. Soedin*, **1970**, 1515; *Chem. Inform.*, **1971**, 16–383.
412. A. K. Sheinkman, A. K. Tokarev, and S. N. Baranov, *Khim. Geterosikl. Soedin*, **1971**, 82; *Chem. Inform.*, **1971**, 22–261.
413. H. Böhme and A. Müller, *Arch. Pharm. (Weinheim)*, **296**, 65 (1963).
414. H. Böhme and A. Müller, *Pharm. Zentralhalle Deutschl.*, **101**, 615 (1962).

415. A. K. Sheinkman and A. A. Deikalo, *Khim. Geterosikl. Soedin*, **1971**, 1654; *Chem. Inform.*, **1972**, 13–313.
416. A. K. Sheinkman and A. P. Kucherenko, *Khim. Geterosikl. Soedin*, **1973**, 1432; *Chem. Inform.*, **1974**, 5–246.
417. N. J. Leonard and K. Jann, *J. Amer. Chem. Soc.*, **82**, 6418 (1960).
418. T. R. Keenan and N. J. Leonard, *J. Amer. Chem. Soc.*, **93**, 6567 (1971).
419. M. B. Kass, A. P. Borsetti, and D. R. Crist, *J. Amer. Chem. Soc.*, **95**, 959 (1973).
420. Y. Hata and M. Watanabe, *J. Amer. Chem. Soc.*, **95**, 8500 (1973).
421. C. P. Andrieux and J.-M. Saveant, *Bull. Soc. Chim. Fr.*, **1968**, 4671.
422. W. Dörscheln, H. Tiefenthaler, H. Göth, P. Cerutti, and H. Schmid, *Helv. Chim. Acta*, **50**, 1759 (1967).

ADDENDUM

To Section II-A-2

Iminium salts were also prepared by alkylation of ketimines or aldimines with methyl fluorosulfonate and methyl triflate (423).

To Section II-A-3

Addition of nitrosyl chloride to imines yields covalent N-chloromethyl nitrosamines (424).

To Section II-B-1

Reactions of 1,2-diamino ethylenes (**42**) with hydrogen chloride afford ammonium-iminium dichlorides (**685**) which permit the synthesis of novel polyfunctional compounds (425).

$$R_2N\text{—}CH\!=\!CH\text{—}NR_2 \xrightarrow{2\,HCl} \left[R_2N\!=\!CH\text{—}CH_2\text{—}NHR_2\right]^{2+} 2Cl^-$$
$$\quad\quad(42)\quad\quad\quad\quad\quad\quad\quad\quad\quad\quad(685)$$

To Section II-E

By passing hydrogen chloride into dioxane solutions of primary nitramines (**686**) in the presence of paraformaldehyde the N-chloromethyl nitramines (**133**) are formed (426).

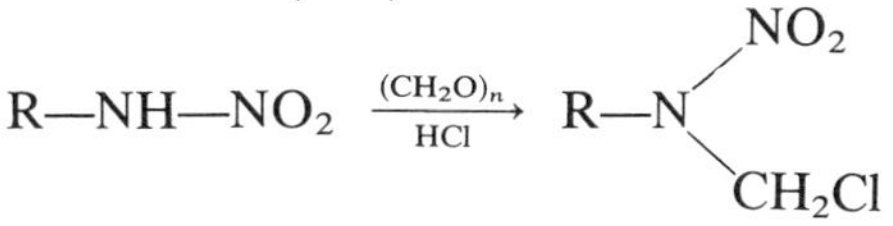

$$R\text{—}NH\text{—}NO_2 \xrightarrow[HCl]{(CH_2O)_n} R\text{—}N\!\!\begin{array}{c}\nearrow NO_2\\ \searrow CH_2Cl\end{array}$$
$$\quad(686)\quad\quad\quad\quad\quad(133)$$

To Section II-F-2-b

By cleavage of the α-ethers of dimethylnitrosamine with PCl_3 the corresponding α-chloromethyl-methylnitrosamine is accessible (424).

To Section II-J

N-Methyl-methyleniminium hexachloroantimonate (**687**) has been obtained with excellent yield upon treatment of methyl azide with nitrosyl hexachloroantimonate. Analogously ethyl or isopropyl azide via **688** afford approximately 45% of the ethyliden- or isopropylideniminium hexachloroantimonate of type **689** (427).

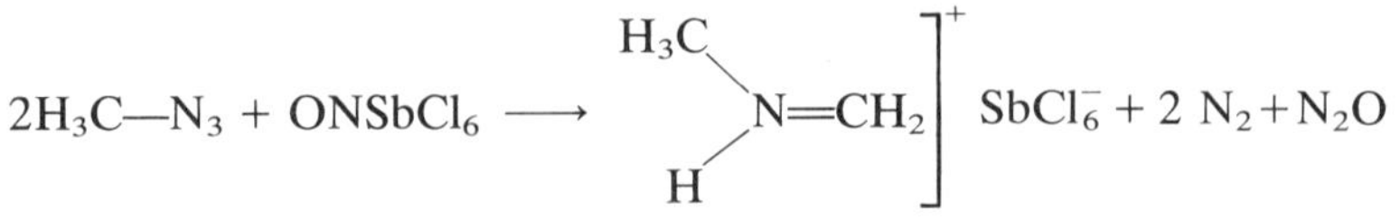

$$2H_3C-N_3 + ONSbCl_6 \longrightarrow \left[\begin{array}{c} H_3C \\ \\ H \end{array} N=CH_2 \right]^+ SbCl_6^- + 2\,N_2 + N_2O$$

(**687**)

$$\begin{array}{c} R \\ \\ R' \end{array} CH-\bar{N}-N\equiv N| + HSbCl_6 \longrightarrow$$

$$\left[\begin{array}{c} R \\ \\ R' \end{array} CH-\underset{H}{N}-N\equiv N| \right]^+ SbCl_6^- \xrightarrow{-N_2} \left[\begin{array}{c} R \\ \\ R' \end{array} C=NH_2 \right]^+ SbCl_6^-$$

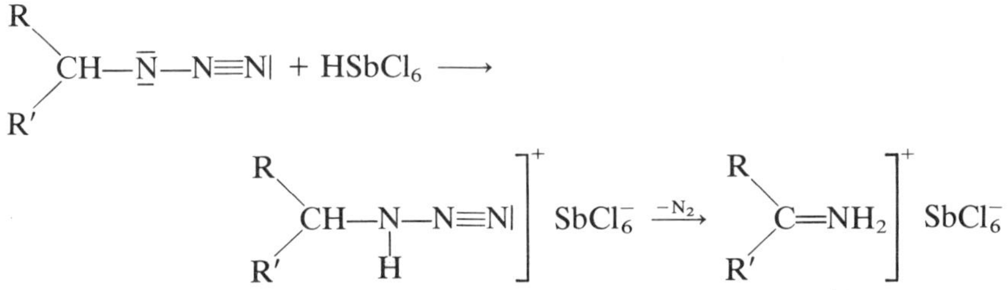

(**688**)   (**689**)

Oxidation of enamines with tetrabromomethane via radical cations yields stable iminium pentabromocarbonates, which can be converted by anion exchange into the corresponding perchlorates (428).

The oxidation of 1,2-diamino ethylenes (42) with quinone leads to diiminium salts (**690**) (429).

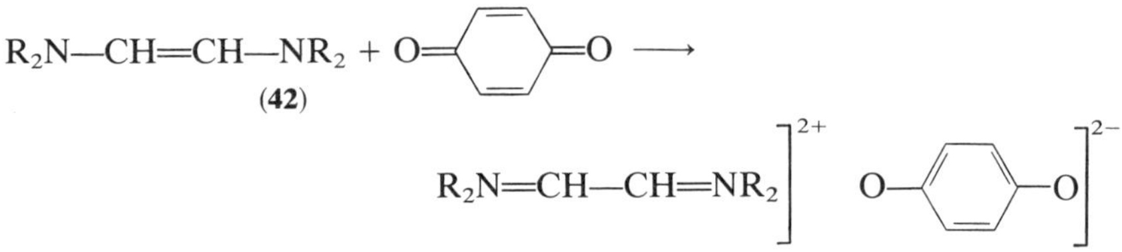

$$R_2N-CH=CH-NR_2 + O=\!\!\!\bigcirc\!\!\!=O \longrightarrow$$
(**42**)

$$\left[R_2N=CH-CH=NR_2 \right]^{2+} \left[O-\!\!\!\bigcirc\!\!\!-O \right]^{2-}$$

(**690**)

In case of acetamidoindandione derivatives the β-elimination is not possible, therefore with O_2SCl_2 or Br_2 only α-halogenated compounds of type **93** are formed (430).

To Section III-A

In case of bicyclic α-haloamines a double-bonded bridgehead carbon would characterize the iminium halide structure, thus in accordance with

Bredt's rule only a covalent halogen-carbon bond has been established (for examples see ref. 431).

To Section IV-C

The reduction of iminium salts to the corresponding ammonium salts can also be carried out with 1,4-dihydropyridine derivatives. As model substances of NADH-mediated enzymatic reduction of the C=N-linkage in biological processes these were shown to effect stereospecific reduction in case of steroidal iminium systems (432).

To Section IV-D

Proton abstraction from ketiminium salts with bases yield aziridines via azomethine ylides. The methyl fluorosulfonate (**691**) react with bis(trimethylsilyl)amide to give 1-*tert*-butyl-2,2diphenylaziridine (**629**) (423).

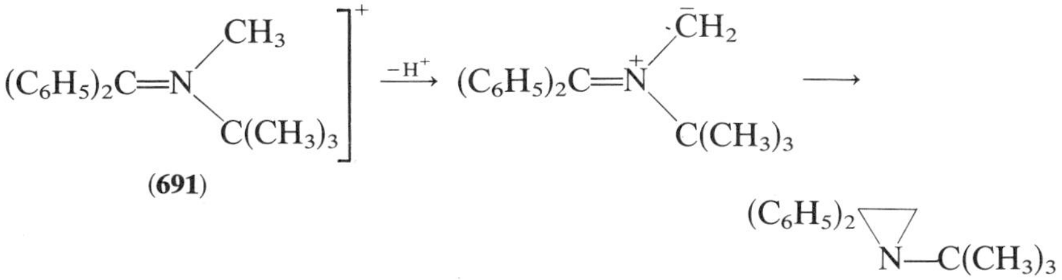

$$(C_6H_5)_2C=N\begin{matrix}CH_3\\C(CH_3)_3\end{matrix}\Bigg]^+ \xrightarrow{-H^+} (C_6H_5)_2C=\overset{+}{N}\begin{matrix}CH_2^-\\C(CH_3)_3\end{matrix} \longrightarrow$$

(**691**)

$$(C_6H_5)_2\overset{\triangle}{\underset{N-C(CH_3)_3}{}}$$

(**692**)

Related aldiminium salts (**693**) afford products, which are apparently derived from initial loss of the aldiminium vinyl proton, for example, aminomethylaziridines (**694**) and 1,2-diaminostilbenes (**695**) (423). *N*-Acyliminium salts are also deprotonated by triethylamine. The adduct **696** prepared from *N*-isopropylidenaniline and acetyl chloride (see Section II-A-3) leads to *N*-isopropenylacetanilide (**697**) in addition to the 1,3-oxazetidine (**707**) (see Section IV-J) (433).

To Section IV-E

The potassium salt of *N*-hydroxyphthalimide (**698a**) with *N,N*-dibenzylmethyleniminium chloride yields compound **698b**. Upon heating, this compound undergoes ring expansion to **699b** which can be hydrolyzed to **699a**. With methyleniminium chlorides (**121**) and (**361**) the silver salt of **699a** affords derivatives **699c** and **699d**. *N,N*-Dibenzylmethyleniminium chloride leads to **699b** in addition to **700b**. Upon heating **700b** rearranges to isomer **699b** (434).

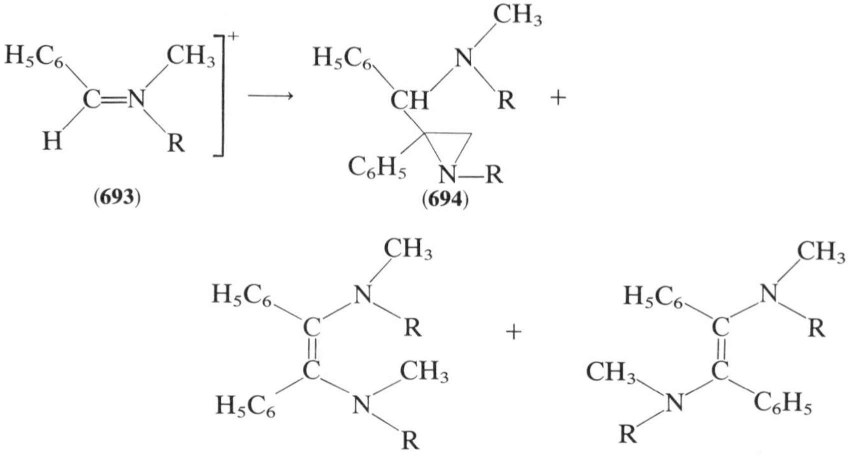

[Structures 693 → 694 + 695 + (structure with CH3/C6H5)]

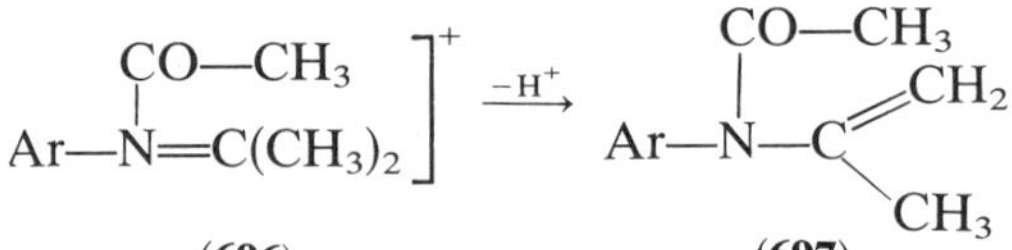

[Structures 696 → 697 with -H+]

[Structures 698, 699, 700]

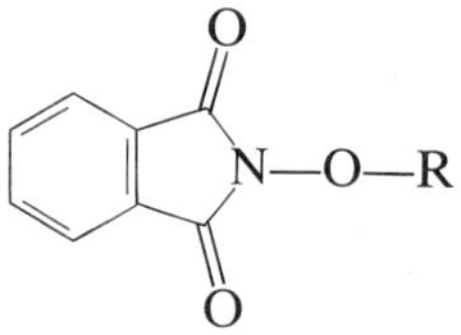

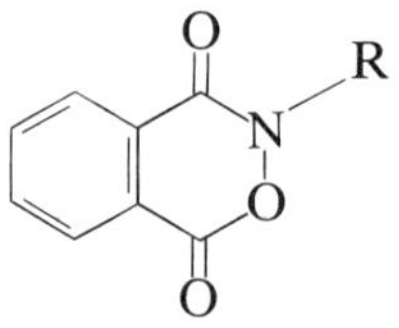

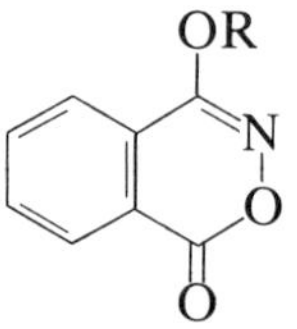

a: R = H;   b: R = CH$_2$N(CH$_2$—C$_6$H$_5$)$_2$;   c: R = CH$_2$—N◯;   d: R = CH$_2$—N◯O

To Section IV-G

Novel heterocycles **703** having azaanalogous sulfone structure are accessible from reactions of sulfodiimides (**701**) with the bifunctional N-α-chlorobenzyl N-methylcarbamide chloride (**702**) (435).

To Section IV-H

As potential sources for iminium salts (**705**) aziridines (**704**) have also been converted with trimethyl phosphite in the presence of acids to the corresponding α-aminophosphonic esters (**706**) (436).

(701) (702) (703)

(704) (705) (706)

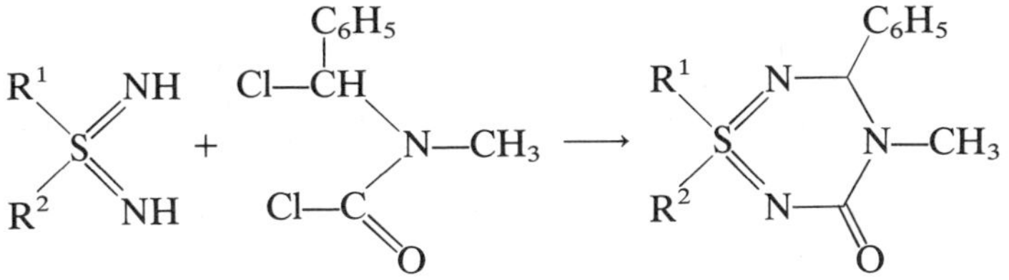

To Section IV-I-2

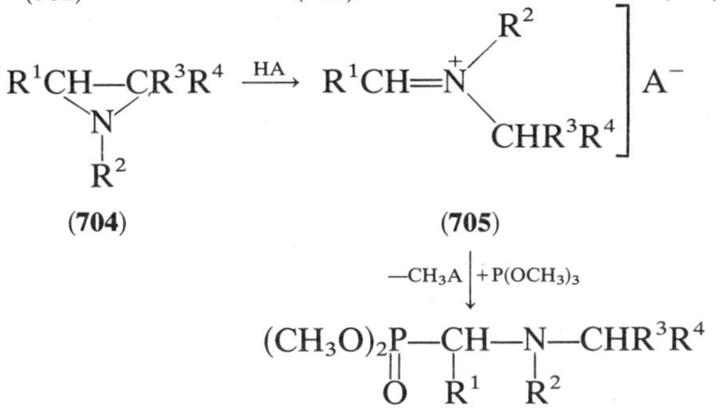

The aminomethylation by means of methyleniminium salts which proceeds regioselectively in case of unsymmetrical ketones (441) can also be achieved with aldehydes and sterically hindered ketones (207,442).

To Section IV-I-4-d

N,N-Tetramethylenhydrazones of aliphatic aldehydes as aza-analogous enamines react with methyleniminium salts under aminomethylation on the azomethin-carbon (443).

To Section IV-J

By addition of acetylchloride to *N*-isopropylidenaniline (see Section II-A-3) the *N*-acetyliminium salt (**696**) is formed which undergoes ring closure and yields the 1,3-oxazetidine (**707**) in addition to *N*-isopropenylacetanilide (**697**) as the final products (see Section IV-D) (433).

(707)

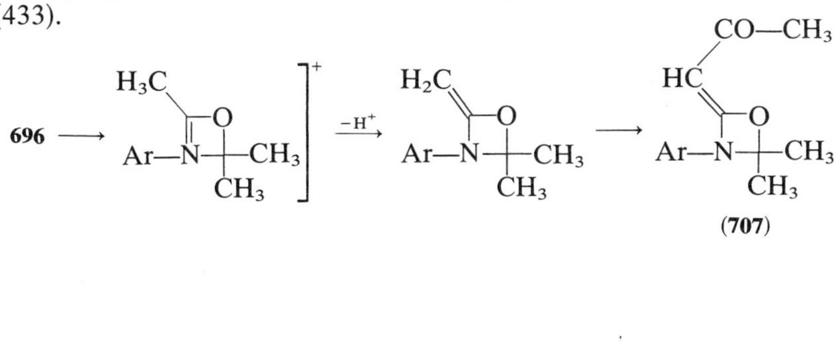

Acylchlorides readily undergo addition to 3,3-dimethyl-2-phenylazirine. The products, for example, **708** (see Section II-A-3), can be converted to other functionalized *N*-acylaziridines and oxazolines (**709**) (437).

$$\begin{array}{cc}
\text{R-CO} & \\
| & \\
\text{R' N CH}_3 & \text{R''} \quad \text{O} \quad \text{N} \\
\diagdown \diagup & \diagdown \diagup \diagdown \\
\text{H}_5\text{C}_6 \quad \text{CH}_3 & \text{H}_5\text{C}_6 \quad \text{CH}_3 \\
(\mathbf{708}) & (\mathbf{709})
\end{array}$$

R' = Cl, AcO, N$_3$; R'' = CH$_3$O

Sulfonyl isocyanates and isothiocyanates react with enamines to yield adducts which represent stable iminium dipoles (**710**) in the crystalline state. In solution these form an equilibrium with the β-lactame (**711**) or thietane (**712**) respectively (438–440).

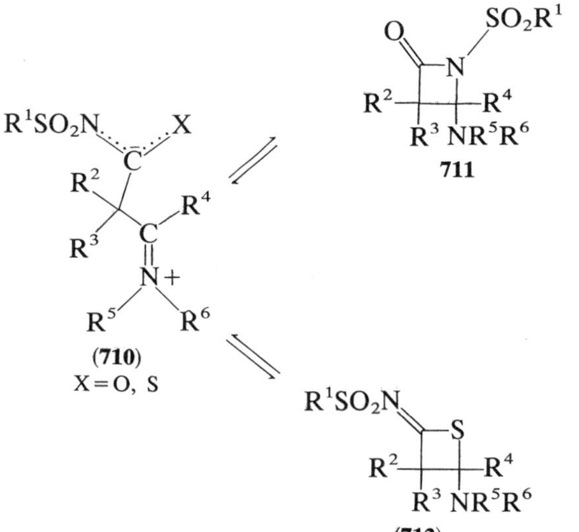

$$\begin{array}{c}
\text{R}^1\text{SO}_2\text{N} \cdots \text{X} \\
\diagdown \text{C} \diagup \\
\text{R}^2 \quad \quad \text{R}^4 \\
\diagdown \text{C} \diagup \\
\text{R}^3 \quad \| \\
\text{N}+ \\
\diagup \diagdown \\
\text{R}^5 \quad \text{R}^6 \\
(\mathbf{710}) \\
\text{X} = \text{O, S}
\end{array}
\quad \rightleftharpoons \quad
\begin{array}{c}
\text{O} \quad \text{SO}_2\text{R}^1 \\
\diagdown \diagup \\
\text{N} \\
\text{R}^2 \text{---} \text{R}^4 \\
\text{R}^3 \quad \text{NR}^5\text{R}^6 \\
\mathbf{711}
\end{array}$$

$$\begin{array}{c}
\text{R}^1\text{SO}_2\text{N} \quad \quad \text{S} \\
\diagdown \diagup \\
\text{R}^2 \text{---} \text{R}^4 \\
\text{R}^3 \quad \text{NR}^5\text{R}^6 \\
(\mathbf{712})
\end{array}$$

REFERENCES

423. J. A. Deyrup and W. A. Szabo, *J. Org. Chem.* **40,** 2048 (1975).
424. M. Wiessler, *Tetrahedron Lett.* **1975,** 2575.
425. G. Plé, *Bull. Soc. Chim. Fr.* **1975,** 2212.
426. A. L. Fridman, L. O. Konshina, and S. A. Petukhov, *Zh. Org. Khim.* **11,** 1187 (1975), *Chem. Inform.* **1975,** 37–148.
427. P. Volgnandt and A. Schmidt, *Z. Naturforsch.* **30b,** 295 (1975).

428. F. Effenberger, W. Podszun, W. W. Schoeller, W. G. Seufert, and W.-D. Stohrer, *Chem. Ber.* **109,** 306 (1976).
429. L. Duhamel and G. Plé, *C. R. Acad. Sci. Ser. C.* **280,** 779 (1975).
430. L. S. Geita, I. E. Dalberga, and A. K. Grinvalde, *Zh. Org. Khim.* **11,** 803 (1975); *Chem. Inform.* **1975,** 29–226.
431. H. Bochow and W. Schneider, *Chem. Ber.* **108,** 3475 (1975).
432. U. K. Pandit, R. A. Gase, F. R. M. Cabré, and M. J. de Nie-Sarink, *J. C. S. Chem. Commun.* **1975,** 211.
433. H. Iwamura, M. Tsuchimoto, and S. Nishimura, *Tetrahedron Lett.* **1975,** 1405.
434. G. Zinner and V. Ruthe, *Liebigs Ann. Chem.* **1975,** 2006.
435. M. Haake, *The Chemistry of S,S-Diorgano-Sulfodiimdes*, in A. Senning, *Topics in Sulfur Chemistry*, Vol. 1, Georg Thieme Publishers, Stuttgart, 1976.
436. M. Vaultier, R. Danion-Bouget, D. Danion, J. Hamelin, and R. Carrié, *C. R. Acad. Sci. Ser. C.* **280,** 213 (1975).
437. A. Hassner, S. S. Burke, and J. Cheng-fan I, *J. Amer. Chem. Soc.* **97,** 4692 (1975).
438. E. Schaumann, S. Sieveking, and W. Walter, *Tetrahedron Lett.* **1974,** 209.
439. E. Schaumann, S. Sieveking, and W. Walter, *Tetrahedron.* **30,** 4147 (1974).
440. E. Schaumann, A. Röhr, S. Sieveking, and W. Walter, *Angew. Chem.* **87,** 486 (1975).
441. Y. Jasor, M.-J. Luche, M. Gaudry, and A. Marquet, *J.C.S. Chem. Commun.* **1974,** 253.
442. G. Kinast and L.-F. Tietze, *Angew. Chem.* **88,** 261 (1976).
443. R. Brehme and H. E. Nikolajewski, *Tetrahedron* **32,** 731 (1976).

THE VILSMEIER-HAACK-ARNOLD ACYLATIONS. C—C BOND-FORMING REACTIONS OF CHLOROMETHYLENIMINIUM IONS

C. Jutz, *Organisch-Chemisches Institut, Technische Universität, Arcisstrasse 21, D-8000 München, Germany*

CONTENTS

I. Introduction	226
A. History of the Chemistry of Chloromethyleniminium Salts	226
B. Comparison of Chloromethyleniminium Chloride with Related Derivatives of Formic Acid	227
II. C—C Bond-Forming Reactions of the Chloromethyleniminium Salts and Related Compounds	229
A. General	229
1. Electrophilic Potential of Halomethyleniminium Salts	232
2. C—H Substitutions of the Nucleus in Aromatic Hydrocarbons	234
3. Substitutions of the Nucleus in Heteroaromatic Compounds	237
B. Reactions of Chloromethyleniminium Salts with Olefinic Double Bonds	243
1. Substitution of Styrenes and Aryl Polyenes	244
2. Substitution of Aliphatic Olefins	250
C. Reactions of Chloromethyleniminium Salts with Enamines, Dienamines, and Trienamines	258
1. Enamines	258
2. Dienamines	263
3. Trienamines	266
D. Reactions of Chloromethyleniminium Salts with Enol Ethers, Acetals, and Ketals	266
1. Vinyl Ethers and Acetals	267
2. Enol Ethers and Ketals	269
3. Vinylogous Enol Ethers and Acetals	271
4. Cyclization during Formylation of Acetals	274
E. Reactions of Chloromethyleniminium Salts with Methyl and Methylene Ketones	274
1. Monosubstitutions of Ketones	276
2. Reactions of Chloromethyleniminium Salts with Polyenals and Polyenones	285
3. Polysubstitutions of Ketones by Chloromethyleniminium Salts	288
4. Cyclizations by Action of Chloromethyleniminium Salts on Ketones	294
5. Bromo- and Iodoformylations of Ketones	297
6. Chloroacroleins and Chlorovinyl Ketones by Other Methods	298
F. Reactions of Chloromethyleniminium Salts with Carbonamides Possessing an α-Methylene Group	301
1. Lactams	303

         2. Pyrrolinones and Phthalimidines 304
         3. Pyrazolones 305
         4. Hydroxypyrimidines 307
     G. Reactions of Chloromethyleniminium Salts with Activated Methylene and
        Methyl Groups 308
         1. Methyl Groups in Polymethinium Salts 309
         2. Methyl Groups in Cycloiminium Salts and Related Compounds 309
         3. Methyl Groups in Pyrylium Salts 309
         4. Alkyltropylium Salts 311
         5. Methyl Groups in Electron-Deficient Heteroaromatics 312
         6. Cyclizations Involving Formylation of Activated Methyl Groups 314
         7. Methyl Groups in Nitro Compounds 316
         8. Carboxylic Acid Derivatives, Nitriles 316
     H. Reactions of Chloromethyleniminium Salts with Carboxylic Acids 318
         1. Substituted Acetic Acids 318
         2. Malonic Acids... 320
     I. Reactions of Chloromethyleniminium Salts with Hydrazones and Azines.... 321
     J. Reactions of Chloromethyleniminium Salts with Aliphatic Diazo Compounds 324
  III. Cyclizations with Chloromethyleniminium Salts........................... 324
  IV. Vilsmeier-Haack-Arnold Related Acylations 328
     A. Vinylogous Formylations 330
     References 333

I. Introduction

A. HISTORY OF THE CHEMISTRY OF CHLOROMETHYLENIMINIUM SALTS

That formanilide is a formylating agent in the presence of $POCl_3$ was discovered by Dimroth and Zoeppritz as early as 1902 (1). Good yields of aldehydes, however, were obtained only with resorcinol; the reaction failed with N,N-dialkylanilines. The use of disubstituted formamides such as methylformanilide (MFA) and dimethylformamide (DMF) instead of formanilide led to the successful "transfer of the formyl group from the nitrogen atom of an amide to the carbon atom of a substrate" by Vilsmeier (2,3). Vilsmeier even recognized that the reactive species was a 1:1 complex of MFA or DMF and $POCl_3$. Later Witzinger (4) commented on the electrophilic nature of this complex and related this so-called Vilsmeier-Haack reaction to other electrophilic aromatic substitutions, for example, Friedel-Crafts acylations. The use of the new reagent, however, was still confined largely to reactive aromatics and heteroaromatics. It was largely due to the investigations of Arnold (5-32) during the last 18 years that the very reactive, electrophilic species, in most cases chloromethylenedimethyliminium chloride, was shown to effect a smooth substitution of a great number of pure aliphatic substrates, often in complex multistep reactions. This "acylation" method

I. INTRODUCTION

should therefore preferably be referred to as the Vilsmeier-Haack-Arnold acylation. The expression "formylation" is also misleading, since it tends to obscure the wide-ranging and valuable synthetic potential of this versatile agent.

In all cases the prime product of substitution, of course, is not the aldehyde itself but an iminium salt, which often is also isolable. Further reaction of this salt (e.g., in the work-up of reaction mixtures) is under the control of the chemist, a fact that greatly enhances the scope of the reaction. The Vilsmeier-Haack-Arnold acylation method has therefore become an extremely valuable synthetic tool. Not only can aldehyde groups be introduced into various aromatic and heteroaromatic rings, but also many substitutive cyclizations and condensations are now both feasible and synthetically valuable, bringing otherwise only difficultly accessible or inaccessible compounds within reach of the reaction.

Several reviews have been written on this subject (33–39); some (33–37) contain detailed lists of compounds synthesized. In this survey of reactions of chloromethyleniminium salts, therefore, known facts will be repeated only if they are necessary for an understanding of the subject under discussion.

B. COMPARISON OF CHLOROMETHYLENIMINIUM CHLORIDE WITH RELATED DERIVATIVES OF FORMIC ACID

To compare some different halogen-containing derivatives of formic acid and their electrophilic properties, it seems opportune to write the following examples as *pro forma* equilibria:

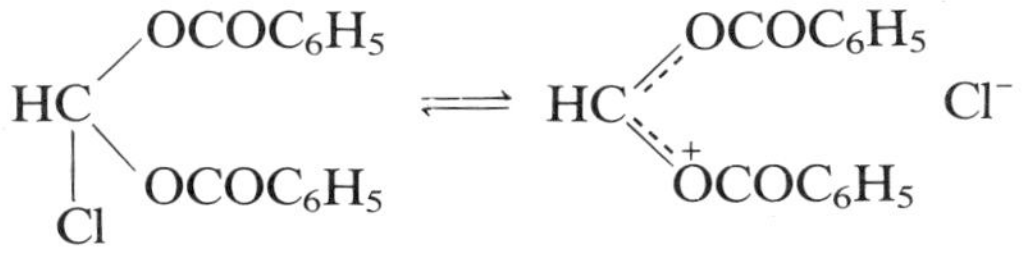

$$HC\begin{pmatrix}OCOC_6H_5\\Cl\\OCOC_6H_5\end{pmatrix} \rightleftharpoons HC\begin{pmatrix}OCOC_6H_5\\\overset{+}{O}COC_6H_5\end{pmatrix} Cl^-$$

(1)

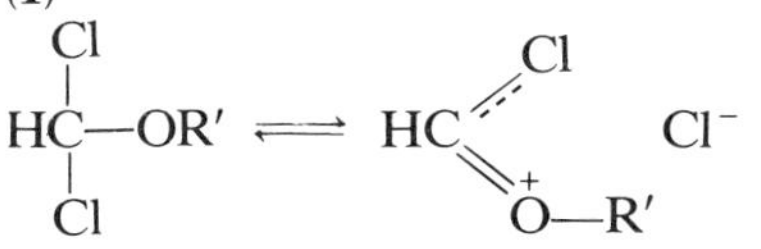

$$HC\begin{pmatrix}Cl\\OR'\\Cl\end{pmatrix} \rightleftharpoons HC\begin{pmatrix}Cl\\\overset{+}{O}-R'\end{pmatrix} Cl^-$$

(2)

$$HC\begin{pmatrix}Cl\\N(R)_2\\Cl\end{pmatrix} \rightleftharpoons HC\begin{pmatrix}Cl\\\overset{+}{N}R_2\end{pmatrix} Cl^-$$

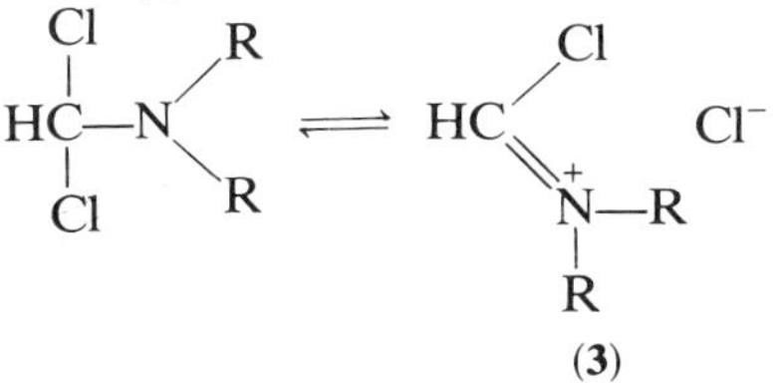

(3)

$$\begin{array}{c}\text{O—R}'\\|\\\text{HC—N}\\|\\\text{Cl}\end{array}\begin{array}{c}\text{R}\\\\\text{R}\end{array}\quad\rightleftharpoons\quad \text{HC}\begin{array}{c}\diagup\text{O—R}'\\\\\diagdown\overset{+}{\text{N}}\text{—R}\\|\\\text{R}\end{array}\quad\text{Cl}^{-}$$

(4)

$$\begin{array}{c}\text{R}\\|\\\text{N—R}\\|\\\text{HC—N—R}\\|\quad|\\\text{Cl}\quad\text{R}\end{array}\quad\rightleftharpoons\quad \text{HC}\begin{array}{c}\diagup\overset{|}{\underset{|}{\text{N—R}}}\\\\\diagdown\overset{+}{\text{N}}\text{—R}\\|\\\text{R}\end{array}\quad\text{Cl}^{-}$$

(5)

R' = alkyl, aryl, —CO—aryl; R = alkyl (CH$_3$).

The chloromethylene dibenzoate (**1**) is a stable derivative of the parent compound of this family, formyl chloride. Formyl chloride itself has a half-life of 1 hr at −60°, decomposing to hydrogen chloride and carbon monoxide (40), and can be regarded as the formylating agent in the known Gattermann-Koch aldehyde synthesis. Compound **1** can alkylate aromatics (e.g., mesitylene, anisole) only in the presence of dry aluminum chloride (41). Likewise the dichloromethyl alkyl ether **2** (R' = ethyl, *n*-butyl) is an effective alkylating agent, and many aromatic and heteroaromatic aldehydes can be prepared with it. Benzene, which is moderately reactive, gives benzaldehyde in 80% (42) yield. Compound **2**, however, alkylates only in the presence of a Friedel-Crafts catalyst; TiCl$_4$ is particularly suitable. In **1** and **2** dissociation to a reactive cation can therefore be enforced only by complexing the nucleophilic chloride ion. The cations thus formed are, as expected, strong electrophiles.

In the following examples the covalently bonded tertiary amines **3** (dichloromethylamines), **4** (alkoxy- or aryloxychloromethylamines), and **5** [bis(diamino)chloromethanes] are unknown and exist in their ionized salt forms as chloromethyleniminium chlorides **3**, alkoxy- or aryloxy-methyleniminium chlorides **4**, and tetrasubstituted formamidinium chlorides **5**. In the series of **1–5** the chloromethyleniminium chlorides **3** are the first members that dissociate freely without added Friedel-Crafts catalysts, and a dynamic equilibrium with the covalent form must be taken into consideration.

The exceptional role of **3** is also evident by comparison with **4** and **5**. As expected, **3** possesses a very high electrophilic potential and is therefore extremely sensitive to moisture and protic solvents. The formamidinium salts **5** are weak electrophiles that react only with strong bases (e.g., carbanions) by addition and elimination of a secondary amine to give aminomethylene compounds. Similarly salts of type **4** react with bases and reactive methylene compounds in the presence of added base. Only if one substitutes the R' on the oxygen of **4** by strong electron-attracting residues can the cation reach moderate to strong electrophilicity. Thus cyanuric chloride forms with DMF in the cold a white, crystalline adduct with proposed structure **6** (43,44). With **6** as electrophile at room temperature the following aldehydes were obtained: from dimethylaniline, *p*-dimethylaminobenzaldehyde (88%); from 1-dimethylaminonaphthalene, 4-dimethylamino-1-naphthaldehyde (47%); from pyrrole, pyrrol-2-carboxaldehyde (65%); from indole, indol-3-carboxaldehyde (31%); from 1,3-dimethoxybenzene, 2,4-dimethoxybenzaldehyde (36%); and from thiophene, thiophene-2-carboxaldehyde (5%). With cyclopentadiene, 1,1-diphenylethylene, anthracene, and 2-methoxynaphthalene **6** failed to give aldehydes (45):

$$Me_2\overset{+}{N}{=}CH{-}O{-}\underset{\underset{Cl}{N{=}N}}{C_3N_3}{-}O{-}CH{=}\overset{+}{N}Me_2 \quad 2\,Cl^-$$

(**6**)

From the available experimental facts it can clearly be established that, among the ionized derivatives of formyl chloride, the salts of type **3**, the so-called Vilsmeier-Haack-Arnold complexes, are without doubt the most reactive. Thus many compounds that do not react with **6** are substituted smoothly by **3**.

II. C–C Bond-Forming Reactions of Chloromethyleniminium Salts and Related Compounds

A. GENERAL

Many papers on the structure and reactivity of the so-called Vilsmeier-Haack-Arnold complexes have been published (2–4,18,46–59), as well as some erroneous opinions and speculations without experimental verification. Our present knowledge about the formation and structure of these complexes can be summarized in the following scheme:

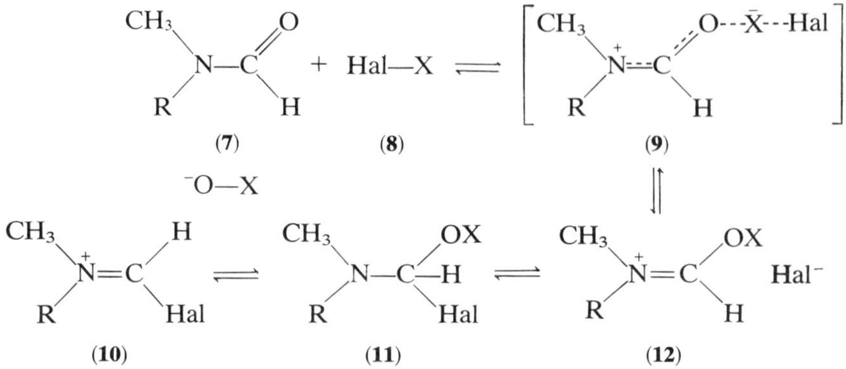

$$\underset{R}{\overset{CH_3}{\diagdown}}N-\underset{H}{\overset{O}{\overset{\|}{C}}} + Hal-X \rightleftharpoons \left[\underset{R}{\overset{CH_3}{\diagdown}}\overset{+}{N}\cdots C\underset{H}{\overset{O\cdots X\cdots Hal}{\diagup}}\right]$$

(7)   (8)   (9)

$$^-O-X$$

$$\underset{R}{\overset{CH_3}{\diagdown}}\overset{+}{N}=C\underset{Hal}{\overset{H}{\diagup}} \rightleftharpoons \underset{R}{\overset{CH_3}{\diagdown}}N-\underset{Hal}{\overset{OX}{\overset{|}{C}-H}} \rightleftharpoons \underset{R}{\overset{CH_3}{\diagdown}}\overset{+}{N}=C\underset{H}{\overset{OX}{\diagup}} \quad Hal^-$$

(10)   (11)   (12)

$R = CH_3, C_6H_5$; $X = OPCl_2, OSCl, Ar-CO, Ar-O-CO$, and others

At first the formamide **7** and the acid halide **8** react in a second-order acid-base association, passing through **9** (which is not an intermediate, but rather an energy-rich transition state) to the salt or salt-like, tight ion pair **12**. The formation of iminium ion **12** should be a reversible process, and the rate of the reaction may be influenced by the basicity of **7**, the Lewis acid property of halide **8**, and the polarity of the solvent used. In the following equilibria either **10**, **11**, or **12** is favored, depending on the nucleophilicity and leaving tendency of OX or halide as anions. Thus phosphorous oxychloride forms with dimethylformamide (DMF) or methylformanilide (MFA) salts of type **10** (18,52–54), but with thionyl chloride the relatively stable ion of type **12** is obtained (60). Finally the basic fluoride ion, being a poor leaving group, is mainly covalently bonded in difluoromethyldimethylamine (23,61) and represents a case of type **11**. When one fluorine ion is complexed by addition of boron trifluoride, a stable tetrafluoroborate salt **13** results immediately:

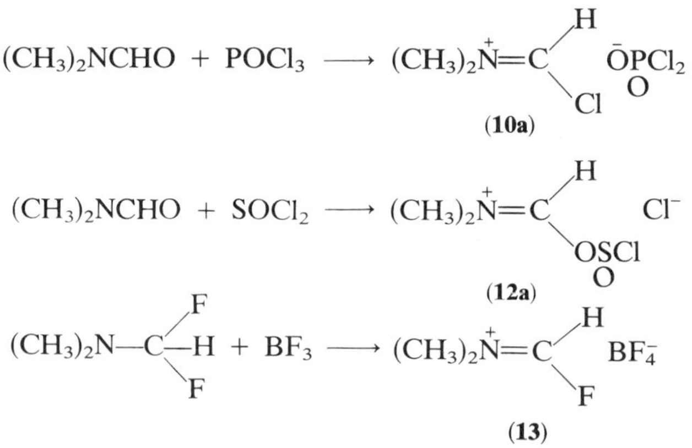

$$(CH_3)_2NCHO + POCl_3 \longrightarrow (CH_3)_2\overset{+}{N}=C\underset{Cl}{\overset{H}{\diagup}}\quad \bar{O}PCl_2\!=\!O$$

(10a)

$$(CH_3)_2NCHO + SOCl_2 \longrightarrow (CH_3)_2\overset{+}{N}=C\underset{OSCl=O}{\overset{H}{\diagup}}\quad Cl^-$$

(12a)

$$(CH_3)_2N-\underset{F}{\overset{F}{\overset{|}{C}-H}} + BF_3 \longrightarrow (CH_3)_2\overset{+}{N}=C\underset{F}{\overset{H}{\diagup}}\quad BF_4^-$$

(13)

II. C—C BOND-FORMING REACTIONS

We can therefore conclude that the basicity of the anions is graduated as follows: $F^- \gg O_2SCl^- > Cl^- > O_2PCl_2^- > BF_4^-$.

The immediate evolution of carbon dioxide in the reaction of DMF with phosgene to give chloromethylenedimethyliminium chloride (**3a**) (49) and the rapid exchange of the chloride in **3a** by iodide or bromide (17,18) and also by fluoride (23) can be seen as evidence of the mobile equilibria between **10**, **11**, and **12**. In the first case, the exchanging chloroformate ion undergoes fragmentation to carbon dioxide and chloride ion:

$$(CH_3)_2\overset{+}{N}=CH-O-C-Cl \longrightarrow (CH_3)_2\overset{+}{N}=CH-Cl \; Cl^- + CO_2$$
$$\phantom{(CH_3)_2\overset{+}{N}=CH-}Cl^- \; O (\mathbf{3a})$$

In an analogous manner the action of oxalyl chloride on DMF leads to **3a**. By fragmentation of the intermediate chlorocarbonyl formate ion, carbon dioxide and carbon monoxide are evolved:

$$Cl-C-C-O^- \longrightarrow Cl^- + CO + CO_2$$
$$O \; O$$

In another equilibrium, which is especially important when DMF is used as solvent, the chloromethylenedimethyliminium ion in **3a** and **10a** attacks the weakly nucleophilic DMF. In the cation **14** thus formed, an allylic-like migration of the chlorine takes place (53,54,59):

$$DMF + \mathbf{3a} \rightleftharpoons (CH_3)_2\overset{+}{N}\text{=-=}CH\text{-=-}O-\underset{\underset{Cl}{|}}{CH}-N(CH_3)_2 \; Cl^-$$

$$\Updownarrow$$

$$\mathbf{3a} + DMF \rightleftharpoons (CH_3)_2N-\underset{\underset{Cl}{|}}{CH}-O\text{-=-}CH\text{-=-}\overset{+}{N}(CH_3)_2 \; Cl^-$$
$$ (\mathbf{14})$$

This equilibrium, which weakens the electrophilic potential of **3a**, can be suppressed by proton-donating solvents and also by the presence of hydrogen chloride in the reaction mixture. Protonation on the oxygen atom of the DMF is then the competing reaction. With the more nucleophilic 3-dimethylaminoacrolein (**15**) **3a** reacts analogously by exchange to give the very insoluble 3-chloroallylidenedimethyliminium chloride (**16**), a vinylogue of **3a** (62):

$$\mathbf{3a} + (CH_3)_2N\diagup\!\!\!\diagdown C(H)=O \xrightarrow{\text{in CHCl}_3} DMF + (CH_3)_2\overset{+}{N}\diagup\!\!\!\diagdown CH=CHCl \quad Cl^-$$

(15)   (16)

1. Electrophilic Potential of Halomethyleniminium Salts

As shown in the preceding section, the formation of halomethyleniminium salts (**10** or **12**) depends on the basicity of **7** and the Lewis acid strength of halide **8**. However, the same substituent effects that enhance the basicity of **7** also stabilize the ions **10** or **12** which result, and thus weaken their electrophilic nature. On the other hand, iminium salts with electron-attracting groups on nitrogen tend to dissociate to their starting components. In the preparation of acid chlorides from acids by thionyl chloride in the presence of DMF, the initially formed salt **12a** acts as the effective dehydrating and transforming agent (47). Thus 2-naphthalenesulfonic acid gives a 97–100% yield of the sulfonyl chloride with thionyl chloride in the presence of DMF. With methylformanilide instead of DMF, the yield decreases to 24.5%, and with diphenylformamide no sulfonyl chloride is formed. These findings are in accord with the fact that thionyl chloride is a weak Lewis acid. With the relatively basic DMF an iminium salt of type **12** (here **12a**) was readily formed, but with the more weakly nucleophilic MFA the formation of the corresponding iminium salt proceeded incompletely. No such salt could be formed with the much less basic diphenylformamide.

Carbonyl chloride (phosgene) seems to react as a somewhat stronger Lewis acid than thionyl chloride. This is suggested, in particular, by the fragmentation step in the formation of the chloromethyleniminium chloride **3a** from DMF and carbonyl chloride, an irreversible process.

Amides with electron-attracting groups, for example, *N*-dimethyl(2-cyano-3-dimethylamino)acrylamide (**17**), fail to react with carbonyl chloride, whereas the stronger Lewis acid, phosphorus oxychloride, readily transforms **17** into the expected dimethyl(1-chloro-2-cyano-3-dimethylaminoallyliden)iminium salt **18** (63):

$$(CH_3)_2N-CH=C(CN)-CON(CH_3)_2 + POCl_3 \longrightarrow$$

(17)

$$(CH_3)_2N-CH=C(CN)-C(Cl)=\overset{+}{N}(CH_3)_2 \quad PO_2Cl_2^-$$

(18)

To demonstrate the formylating potential of chloromethyleniminium salts, Dallacker and Eschelbach (64) prepared the $POCl_3$ complexes of various para-substituted N-methylformanilides: p-Y—C_6H_4—$N(CH_3)CH=O$, where Y = H, CH_3, iso-C_3H_7, tert-C_4H_9, CH_3O, Cl, NO_2. The 3,4-methylenedioxythioanisole (**19**) was converted to the corresponding 6-methylmercapto-3,4-methylenedioxybenzaldehyde (**20**) at 50°C and at 80°C:

$$\text{(19)} \xrightarrow[\text{2.) } H_2O, NaOAc]{\substack{\text{1.)} \quad \underset{CH_3}{\overset{R}{>}}NCHO + POCl_3 \\ \text{in dichlorobenzene}}} \text{(20)}$$

At 50°C DMF–POCl$_3$ (the salt **10a**) afforded only a 28% yield of **20**, which increased to 75% at 80°C. In the same way MFA–POCl$_3$ (Y = H) gave an 86% yield at 50°C and 79% at 80°C. The slight decrease in yield at elevated temperatures is caused by self-formylation of the para-unsubstituted MFA (2,3). In contrast to the expectation that MFA–POCl$_3$ with Y = NO$_2$ should possess the strongest electrophilic properties, the poorest yield (50% of **20**) was obtained in the experiment with methyl-p-nitrophenylformamide and POCl$_3$ at 50°C; moreover, at higher temperatures extensive decomposition occurred. With the complex from methyl-p-methoxyphenylformamide (Y = OCH$_3$) the best results of the series (93% of **20** at 50° and 80°C) were achieved, even though the methoxy group is a strong electron-donating group. It may be concluded from these findings that in the case of N-methyl-p-nitrophenylformamide and POCl$_3$ the formation of the corresponding iminium salt is incomplete.

It has been claimed that the salt from dimethylthioformamide and POCl$_3$ is superior in formylating activity to both DMF–POCl$_3$ (**10a**) and MFA–POCl$_3$ (65). It is difficult to understand, however, why this salt, containing the same chloromethylenedimethyliminium ion as **10a** (from DMF–POCl$_3$), should be a stronger electrophile. Perhaps it is important that these salts can exist in certain solvents as tight ion pairs, depending on the identity of the anion (see Section II-A). Likewise **3a** and **10a** possess the identical cation. But here the stronger formylating power of DMF–POCl$_3$ (**10a**) over **3a** can be explained easily; **3a** is only slightly soluble in the solvents employed (chlorinated hydrocarbons: e.g., dichloroethane, dichloromethane, chloroform, and o-dichlorobenzene), and thus the true concentration of the formylating species in the heterogeneous mixture is small in comparison to homogeneous solutions of **10a**. In practice, DMF–POCl$_3$ is often used without solvent.

The chloromethylenedimethyliminium ion as the moderately soluble chloride **3a** or the dichlorophosphate **10a** is most frequently employed in the Vilsmeier-Haack-Arnold reactions for other reasons besides the availability of DMF. Usually **3a** is obtained from stoichiometric proportions of DMF and phosgene. For laboratory use it is more convenient to drop a measured volume of oxalyl chloride into a cooled, stirred chloroform solution of DMF. For clean reactions and simple work-up, use of **3a** seems to be especially advantageous. Its hydrolytic decomposition frees only 2 equivalents of acid per mole, whereas **10a** generates 6 equivalents of acid. Undoubtedly, **10a**, generated from DMF and phosphoryl chloride, is, as mentioned, the more efficient agent, as long as a great excess of DMF is avoided. Use of DMF as a solvent, by shifting the equilibrium toward **14**, substantially weakens the electrophilic power of **3a** and **10a**. Of course DMF can be replaced by either formylpiperidine or formylpyrrolidine, and phosphoryl chloride may be replaced by phosphoryl bromide. The higher electrophilic potential of the iminium salt from methylformanilide and phosphoryl chloride, the original reagent of Vilsmeier, is partially compensated for by the steric demand of this agent. Another disadvantage is the thermal sensitivity of the complex. Long reaction times at temperatures above 80°C are accompanied by decreased yields and extensive decomposition.

2. C–H Substitutions of the Nucleus in Aromatic Hydrocarbons

Electrophilic substitution of aromatic and heteroaromatic compounds by **3a** or **10a** usually proceeds as a second-order reaction (55–57). The mechanism is pictured schematically for the substitution of *N*-dimethylaniline (**21**) and **3a**:

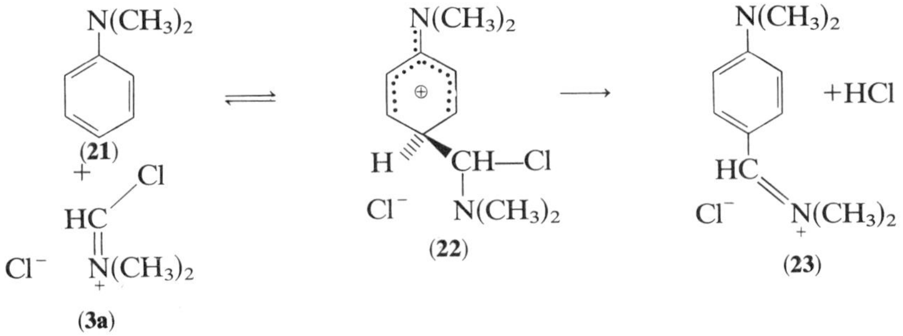

Undoubtedly the rate-determining step is the formation of the energy-rich Wheland σ complex **22**. Loss of a proton and elimination of the chloride to give the resonance-stabilized iminium salt **23** is a very fast and probably concerted process.

It should be emphasized that in all Vilsmeier-Haack-Arnold substitutions such iminium salts are the final products in the original reaction mixture. Indeed, for a long time this fact was ignored, and, surprisingly, it has even been neglected in more recent publications. Therefore formulations such as **23′**, the unionized form of **23**, are untenable and incorrect:

II. C—C BOND-FORMING REACTIONS

$$\underset{(23')}{\underset{N(CH_3)_2}{\overset{N(CH_3)_2}{\text{C}_6H_4-\overset{|}{\underset{|}{C}}H-Cl}}}$$

Solvolysis of the resulting intermediate iminium salts (e.g., **23**) with dilute aqueous alkali affords the desired aldehyde, reaction with hydrogen sulfide gives the thioaldehyde (65), and, finally, reduction with sodium borohydride yields a tertiary amine.

As moderately strong electrophiles of moderate steric bulk and of high selectivity, **3a** and **10a** substitute benzene derivatives preferentially para to a donating group. In the series of benzenoid hydrocarbons, benzene, hydrindene, naphthalene and methylnaphthalenes, phenanthrene, dibenzo(*a*,*h*)anthracene, and chrysene are unreactive, whereas acenaphthene is alkylated in the 5 position. A good survey of Vilsmeier-Haack-Arnold formylations of hydrocarbons is given in Refs. 34–37. The favored site of substitution in other polycyclic aromatic hydrocarbons can be determined with good reliability by means of the cationic localization energy, L_r^+, calculated by the simple HMO method (66) for individual positions.

With **3a** or **10a** and MFA–POCl$_3$, monosubstitution is the rule. The electrostatic repulsion of the charge in the resulting iminium ion and its deactivating effect by electron attraction prevent further attack of the cationic agent. However, under very strong reaction conditions (16 hr, 65°C), with a large excess of **10a**, *N*-dimethylaniline (**21a**) and *N*,*N*-3,5-tetramethylaniline (**21b**) are converted to 4-dimethylaminoisophthaldialdehyde (**24a**) (m.p. = 72°C, 14%) and to 4-dimethylamino-2,6-dimethylisophthaldialdehyde (**24b**) (m.p. = 79°C, 60–80%):

R⟶⟵R ⟨benzene ring⟩ N(CH$_3$)$_2$	CHO R⟶⟵R ⟨benzene ring⟩ CHO N(CH$_3$)$_2$
(**21a**) R = H	(**24a**) R = H
(**21b**) R = CH$_3$	(**24b**) R = CH$_3$

Reagents: 1. **10a**; 2. H$_2$O, NaOH

These are obtained in addition to the expected monoformylated compounds 4-dimethylaminobenzaldehyde and 4-dimethylamino-2,6-dimethylbenzaldehyde (67).

A second known case of disubstitution is represented by azulene (**25**).

With an excess of **10a** at 90–95°C **25** yields a yellow dication **26** and, after hydrolysis, azulene-1,3-dialdehyde (**27**) (43%) (68):

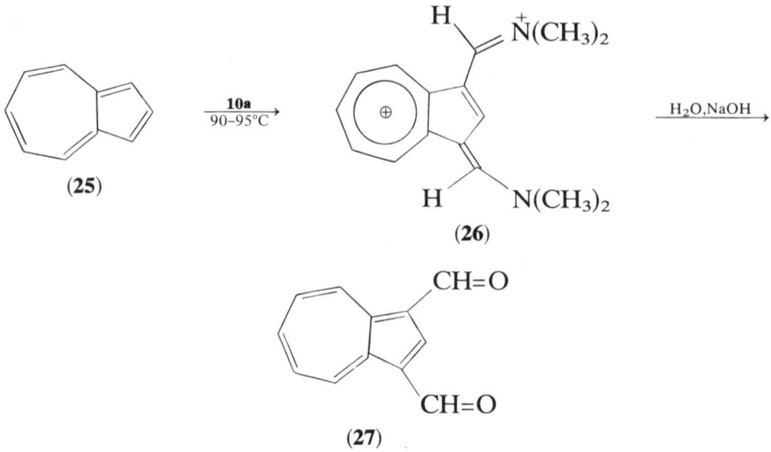

The dication **26** can be considered a combination of the tropylium ion with a pentamethinium salt.

The presence of the chloromethyleniminium ion and an aromatic ring, susceptible to smooth intramolecular electrophilic substitution, leads to cyclization (69,70). One example of this intramolecular type of Vilsmeier-Haack-Arnold reaction has been described (70):

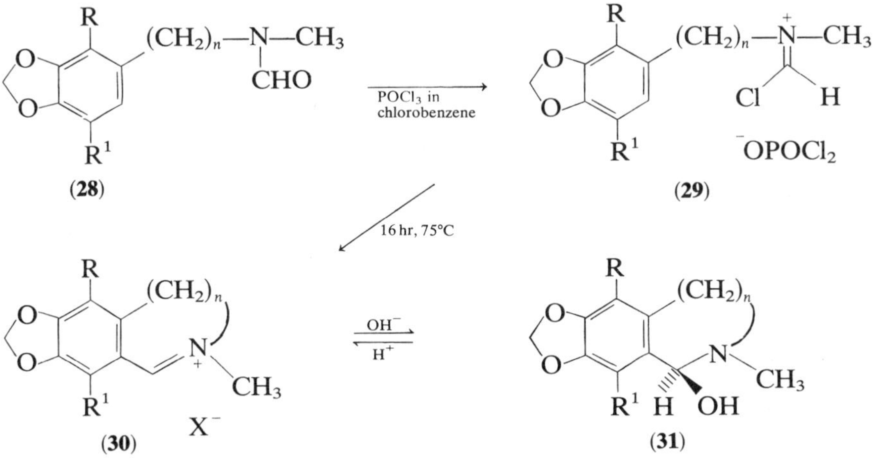

With formamides **28**, $n = 2$ and 3, cyclization of the intermediate chloroiminium salt **29** occurs in good yields. For isolation, the salts **30** were converted by alkali to the stable hemiaminals: the 1-hydroxy-1,2,3,4-tetrahydroisoquinolines (**31**, $n = 2$) and 1-hydroxy-2,3,4,5-

TABLE I

31	($n=2$)	R	R^1	Yield, %	m.p. °C	($n=3$)	Yield, %
a		H (hydrastinine)	H	66	112		82
b		H (cotarnine)	OCH$_3$	67	126		
c		OCH$_3$ (isocotarnine)	H	94	101		
d		OCH$_3$	OCH$_3$	50	171–173[a]		

[a] As **30d** chloride.

tetrahydro-1H-2-benzazepine (**31a**, $n=3$). The corresponding isoindolenium salt **30a**, $n=1$, could not be obtained from **28a**, $n=1$. Attempted cyclization of **28**, $n>3$, yielded only polymeric products. The formation of **30**, $n=2$, from **28**, $n=2$, can be interpreted as a special case of the well-known Bischler–Napieralski reaction.

3. Substitutions of the Nucleus in Heteroaromatic Compounds

Susceptible to electrophilic attack, the electron-rich five-membered aromatic heterocycles (furans, thiophenes, selenophenes, and pyrroles) are usually substituted in the 2- or 5-position by **3a**, **10a**, or MFA–POCl$_3$. 1-Alkyl or 1-aryl pyrazoles react in the 4-position. Benzo(b)thiophenes and indoles yield, as expected, preferentially the 3-formyl, and benzo(b)furans the 2-formyl, derivatives, provided that they are not highly substituted (see literature cited in refs. 33–37).

In the indole series, the resulting iminium salts **34** exhibit a particularly high stability. They lose a proton in cold base, affording enamines **35**, which are hydrolyzed to aldehydes **36** by boiling water (71). Indole, 2-methyl-, and 2-phenylindole (**32**) possess such high nucleophilicity that even the interaction of the weakly electrophilic acyloxymethyleniminium halides prepared from DMF and benzoyl chloride or acetyl bromide (58,72) provides a convenient, mild route to the corresponding aldehydes **36** (73). 2-Methoxycarbonyl- and 4-cyanoindole, however, do not react with this mild reagent **33**. The formylation by decarboxylation of a substituted dipyrrylmethane-2-carboxylic acid with benzoyloxymethylenedimethyliminium chloride has been reported (74).

(*a*) *Furans.* 3-Ethoxycarbonyl-, 3-*p*-methoxybenzoyl-, and 3-cyanofurans with alkyl or aryl substitutuents in position 2 give with **10a**, after hydrolysis, the corresponding 5-formyl derivatives (15–66%) (75).

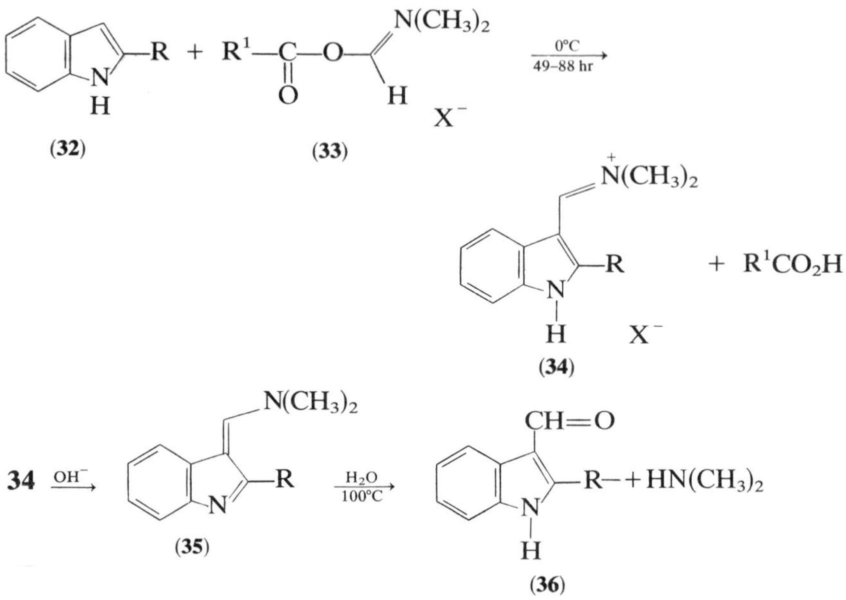

(b) *Thiophenes.* Formylation of 3-phenylthiophene with **10a** leads to a mixture of positional isomers, 4-phenylthiophene-2-carboxaldehyde (96%) and 3-phenylthiophene-2-carboxaldehyde (6%) (76).

The electron-donating effect of the alkylseleno group was demonstrated in the following experiment: from 2-methylselenothiophene with **10a**, 5-methylselenothiophene-2-carboxaldehyde (85%) was obtained as expected; 2-methylseleno-5-methylthiophene led to 2-methylseleno-5-methylthiophene-3-carboxaldehyde (26%) (77). The greater susceptibility to formylation of selenophene as compared to thiophene was demonstrated in experiments with MFA–POCl$_3$. A 1:1:1 mixture of thiophene, selenophene, and the MFA–POCl$_3$ complex gave the two heterocyclic aldehydes, selenophene-2-carboxaldehyde and thiophene-2-carboxaldehyde in an 83:17 ratio. Similarly, with 2-thienyl-2-selenienylmethane, after reaction and work-up, the mixture of the two isomeric aldehydes contained 80% 2-thienyl-2-(5-formylselenienyl)methane and 20% 2-(5-formylthienyl)-2-selenienylmethane (78).

A dealkylation takes place in the reaction of **10a** or MFA–POCl$_3$ above 50–70°C with 2-methoxy-5-methylthiophene (37). Alkaline hydrolysis after reaction at 20°C converts the iminium salt **38** to the methoxyaldehyde **40**. At 50–70°C **38** undergoes dealkylation to the enaminoketone **39**, hydrolysis of which gives, finally, the hydroxyaldehyde **41** (79):

R = CH$_3$, C$_6$H$_5$

(c) *Pyrroles.* As would be expected, 2,5-dimethylpyrroles with or without substituents on nitrogen (H, C_6H_5, $C_6H_5CH_2$) react with **10a** to give pyrrol-3-carboxaldehydes (80). More interestingly, even α-unsubstituted pyrroles react to some extent in the 3 position; the amount of β attack depends on the steric demand and the electronic effects of the nitrogen substituent. Strongly electron-withdrawing groups on nitrogen (CH_3CO-, C_6H_5CO-, $EtO-CO-$) reduce the overall susceptibility of the pyrrole nucleus to electrophilic substitution and lead to the exclusive formation of the corresponding 2-formyl derivatives (in 61, 74, and 54% yield, respectively) (81):

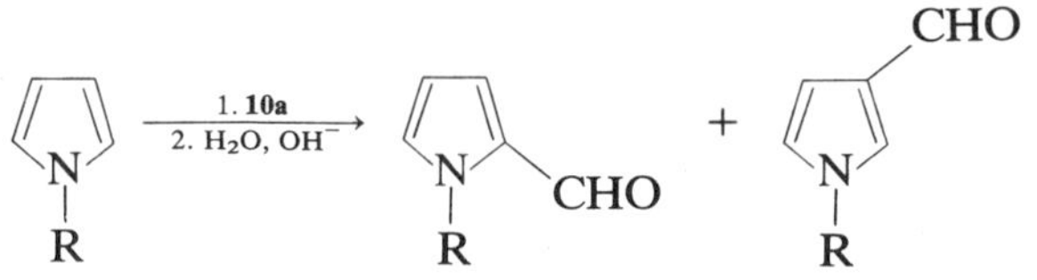

Nearly the same reactivity is shown by the two β-pyrrole positions (in rings A and B) and the methine carbon atoms in the dimethyl ester of the deuteroporphyrin-IX-Cu(II) **44**. A mixture of isomers of **45** and **46** is obtained after reaction with **10a** (82):

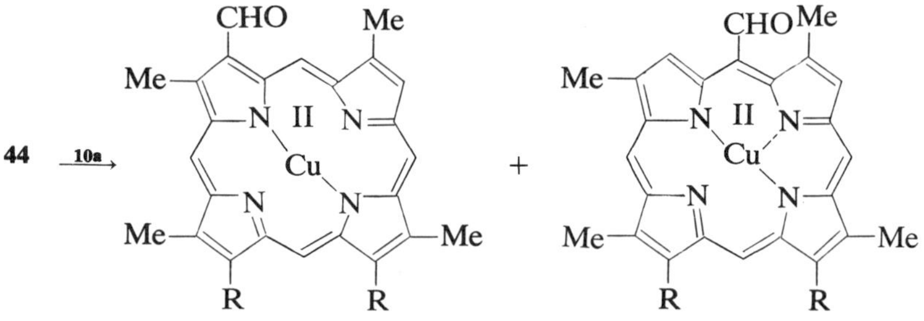

(**45**) R = $CH_2CH_2CO_2H$ (**46**) + 2 ISOMERS

TABLE II

Formylation of 1-*R*-Pyrroles by **10a**

R	Yield, %	42:43	R	Yield, %	42:43
CH_3	89	∞	$C_6H_5CH_2$	89	6.2:1
C_2H_5	85	11.5:1	$4\,CH_3OC_6H_4$	93	7.0:1
i-C_3H_7	79	1.9:1	C_6H_5	93	9.0:1
t-C_4H_9	69	1:14	$4\,NO_2C_6H_4$	88	7.3:1
			$2.6\,Me_2C_6H_3$	71	6.3:1

II. C—C BOND-FORMING REACTIONS

(d) *Benzo(b)furans.* In benzofurans interaction with **10a** leads as expected to 2-formyl derivatives, or to 3-formyl derivatives if the 2-position is occupied by alkyl substituents, whereas 2,3-dialkylbenzofurans are unreactive (83). If a methoxy group is present in position 4, 5, or 7, however, substitution of the 2-methylbenzofurans takes place in the benzene nucleus. In contrast, the more reactive 2-position of the furan ring in 3-methylbenzofurans is attacked by **10a** even if one or two methoxy groups are present on the benzene ring. This proves that the 3-position in benzofurans is much less susceptible to electrophilic attack than the 2-position (84).

(e) *Indoles.* As mentioned, indoles are preferentially substituted by **10a** in position 3 of the heterocyclic ring, even if a methoxy group is present in the 5-position of the benzene ring. The indole system may also be viewed as an enamine in which the nitrogen and β position are bridged by the phenyl ring. Moderate yields of formylation product **48** could be obtained from indole **47a**, an example of substitution in the benzene ring. Indole **47b** was unreactive (85).

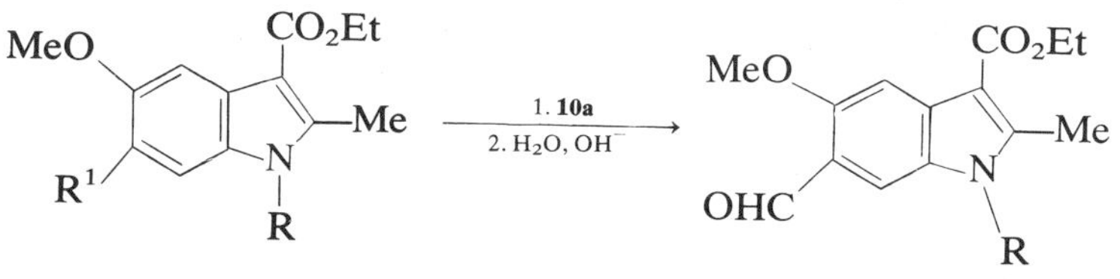

(47a) $R^1 = H$
(47b) $R^1 = Me$

(48) 10–27% yield

($R = H, Me, C_6H_5, C_6H_5CH_2$—)

(f) *Other Condensed Heterocycles.* 3-Thioformylindolizines (**51**), obtained by alkylation of the indolizines **49** with **10a**, followed by solvolysis of the stable intermediate iminium salts **50** with hydrogen sulfide in water, have been described (86):

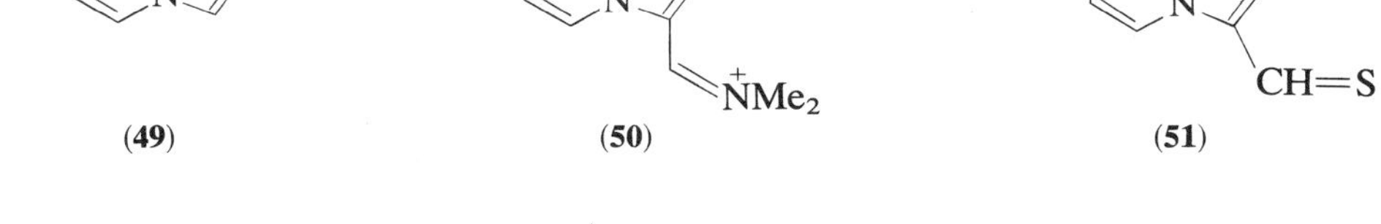

(49)   (50)   (51)

The thieno- and seleno(3,2-d)pyrazoles (**52**) are substituted by **10a** and lead to the formyl derivatives **53** (87):

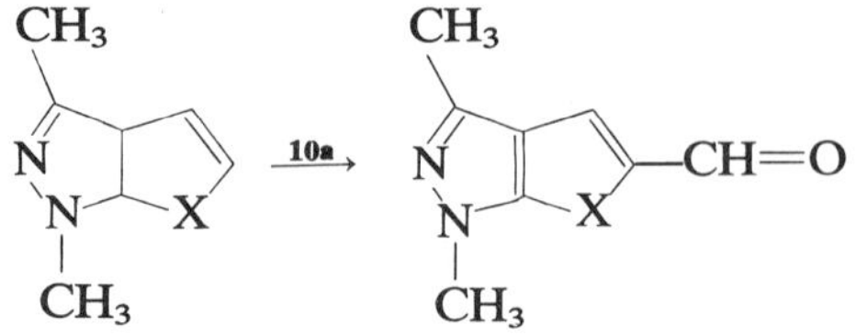

(**52a**) X = S (**53a**) 86%
(**52b**) X = Se (**53b**) 88%

In the 7-azaindole **54** the formyl group was introduced in the 3-position, as expected, by treatment with **10a**, yielding **55** (88):

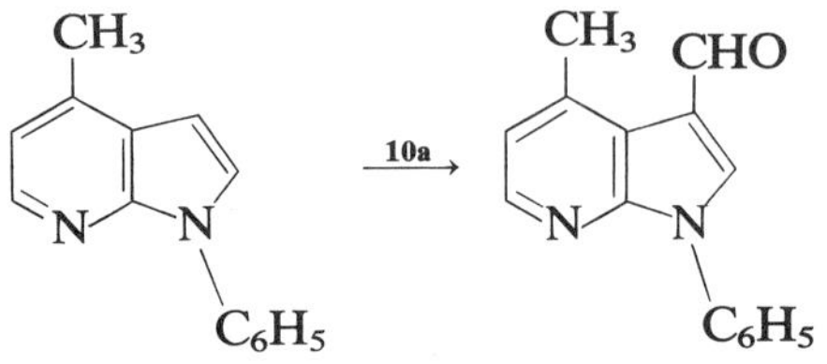

(**54**) (**55**)

The more interesting 6a-thiathiophthene (trithiapentalene) system **56** has been the subject of several investigations, especially in acylation reactions with **10a**. The parent compound **56a** could only be converted to the aldehyde **57a** in moderate yields (21–23%) by the more efficient dimethylthioformamide–POCl$_3$ salt (DMSF–POCl$_3$). Likewise **57b** was obtained from **56b** (89). The 2-phenyl-6a-thiathiophthene (**56c**) yields 62% of **57c** with DMSF–POCl$_3$, whereas with **10a** only 4% and with MFA–POCl$_3$ less than 1% of **57c** could be isolated (65). The introduction of a formyl group in the diaryl-6a-thiathiophthenes **56d** succeeded in 50–73% yield. The site of substitution was determined by degradation experiments and by synthesis (90).

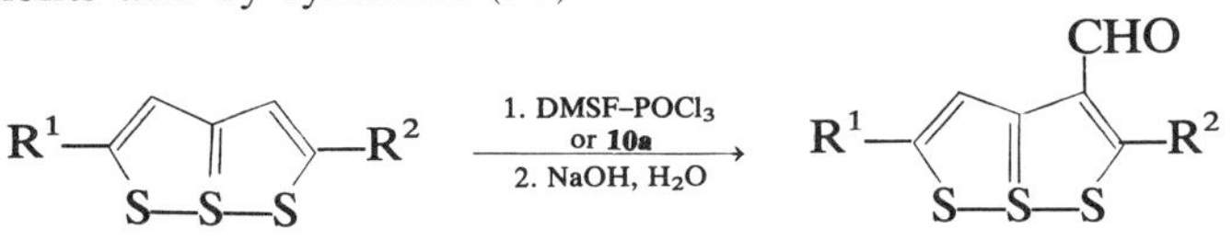

(**56**) (**57**)

    a R^1, R^2 = H, D;
    b R^1 = t-C$_4$H$_9$, R^2 = H
    c R^1 = C$_6$H$_5$, R^2 = H;
    d R^1 = C$_6$H$_5$, R^2 = 4-CH$_3$C$_6$H$_4$, 4-CH$_3$OC$_6$H$_4$

In contrast to the behavior of **56a–d** the 2,5-dimethyl-6a-thiathiophthene **58** was attacked at the methyl group, not the heteroaromatic nucleus, and yielded, after basification, the enamine **59** (89):

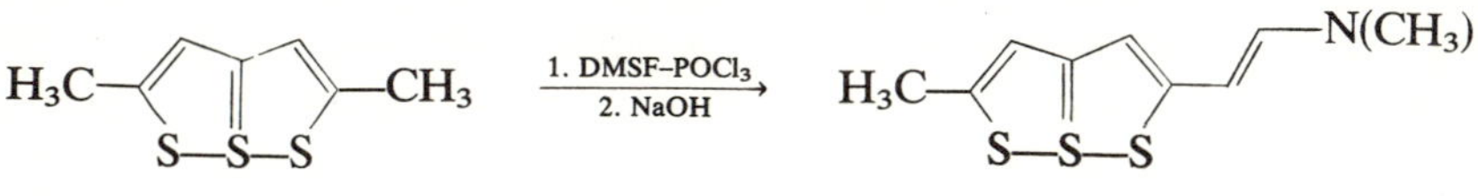

$$H_3C-\underset{S-S-S}{\underset{|}{\diagdown}}-CH_3 \xrightarrow[\text{2. NaOH}]{\text{1. DMSF-POCl}_3} H_3C-\underset{S-S-S}{\underset{|}{\diagdown}}-CH=CH-N(CH_3)_2$$

(58)   (59)

B. REACTIONS OF CHLOROMETHYLENIMINIUM SALTS WITH OLEFINIC DOUBLE BONDS

Because of investigations of Witzinger (4) it became evident that olefinic double bonds are also susceptible to electrophilic substitution by **3a**, **10a**, or MFA–POCl$_3$. Since styrene and 1,1-diphenylethylene extended aromatic systems were the only olefins reacted, Witzinger did not recognize the full importance of his findings. As will be shown later, the broad spectrum of reactions of chloromethyleniminium salts with various aliphatic compounds can be regarded as a substitution on a previously formed or *in situ* generated C—C double bond. We illustrate the mechanism schematically for the substitution of styrene (**60**):

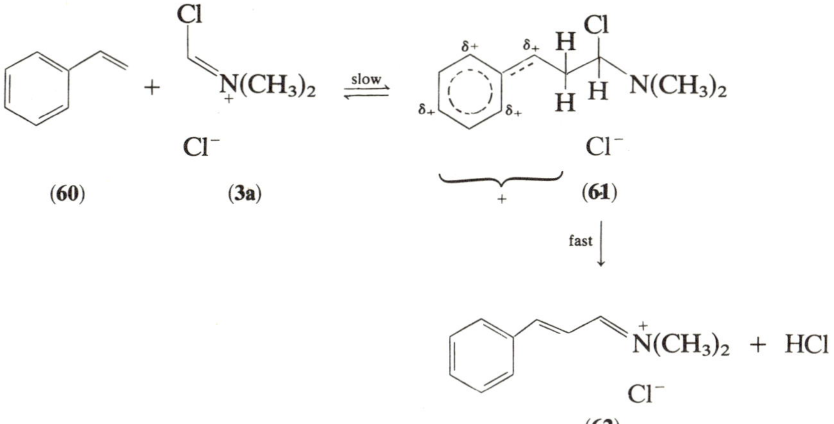

$$\text{Ph-CH=CH}_2 + \underset{Cl^-}{\overset{Cl}{\underset{+}{\diagdown}}{N(CH_3)_2}} \underset{\text{slow}}{\rightleftharpoons} [\text{complex } \mathbf{61}] \xrightarrow{\text{fast}} \text{Ph-CH=CH-}\overset{+}{N}(CH_3)_2 \, Cl^- + HCl$$

(60)   (3a)   (61)   (62)

Electrophilic attack of **3a** on styrene (**60**) leads in a slow, rate-determining step to the energy-rich carbonium ion **61**; here the stabilized benzylic system is comparable to the Wheland σ complex in aromatic substitution (see Section II-A-2, formula **22**). The activation energy of this reaction should correspond to the energy of **61** or, more generally, the carbonium ion formed from the olefin. The ease and the site of substitution in an unsaturated hydrocarbon (neglecting steric influences) may therefore be estimated by simple HMO calculations of the corresponding cationic localization energies, L_r^+ (66).

Worth mentioning is the fast elimination of hydrogen chloride and ionization of **61** and similar cations to yield stabilized iminium salts of type **62**. This may be the most important and fundamental difference between chloromethyleniminium ions and other electrophilic agents in their reaction with olefinic compounds. These agents also attack the C—C double bond generating a primary cation; however, the reaction is usually completed by a subsequent nucleophilic addition.

1. Substitution of Styrenes and Aryl Polyenes

Numerous cinnamaldehydes have been prepared by treating substituted styrenes (**63**) with **10a** in DMF as solvent at 55–60°C (method A) or with **10a** in 1,2-dichloroethane (method B) (91). Generally, better yields were obtained by method A, and, as expected, electron-releasing parasubstituents increased the yield by stabilizing the intermediate carbonium ion:

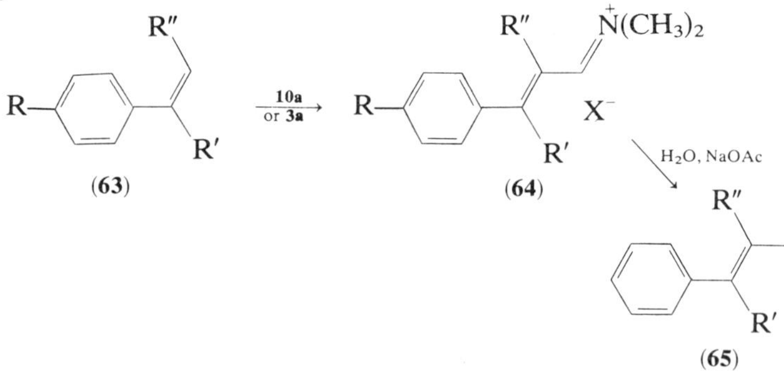

TABLE III
Formylation of Styrenes

65	R	R'	R"	Method A, %	Method B, %	
a	H	H	H	41	38	
b	CH_3	H	H		46	
c	H	CH_3	H	52	37	75% with **3a** in chloroform (93)
d	CH_3	CH_3	H	62		
e	i-C_3H_7	CH_3	H		34	
f	OCH_3	H	H			96% with excess of **10a** in benzene (92)
g	OCH_3	H	CH_3	68	54	

II. C—C BOND-FORMING REACTIONS

In some instances the formed, sometimes colored, iminium salt **64** separates in crystalline form from the reaction mixture. Likewise, the 5-arylpent-2,4-dien-1-als **67** are prepared in high yields from the corresponding 1-arylbuta-1,3-dienes **66** (92):

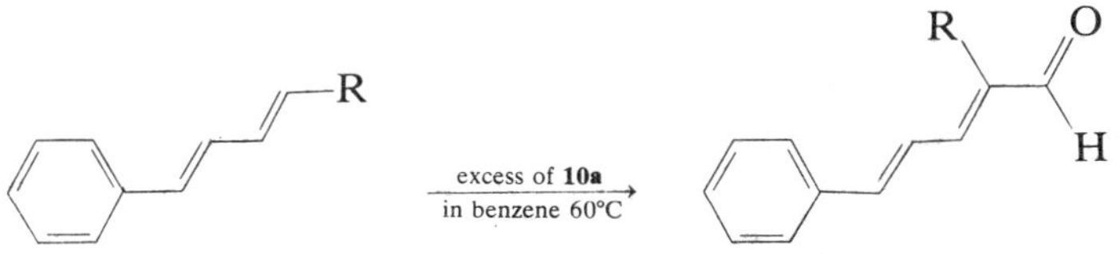

(**66a**) R = H
(**66b**) R = CH$_3$

(**67a**) 92%
(**67b**) 91%

Styrenes and other conjugated aryl polyenes, which may be either inaccessible or sensitive to acid-catalyzed polymerization, can be replaced by the corresponding carbinols **68**. Under the dehydrating effect of an excess of **10a** the olefins formed *in situ* (**69**) are immediately substituted and may be isolated as perchlorate salts (**70**) (94):

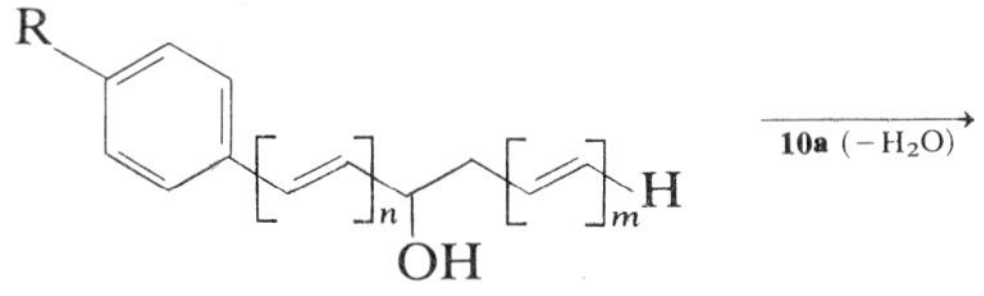

(**68**)

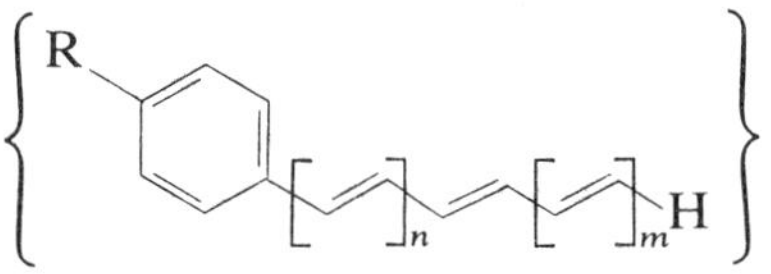

(**69**)

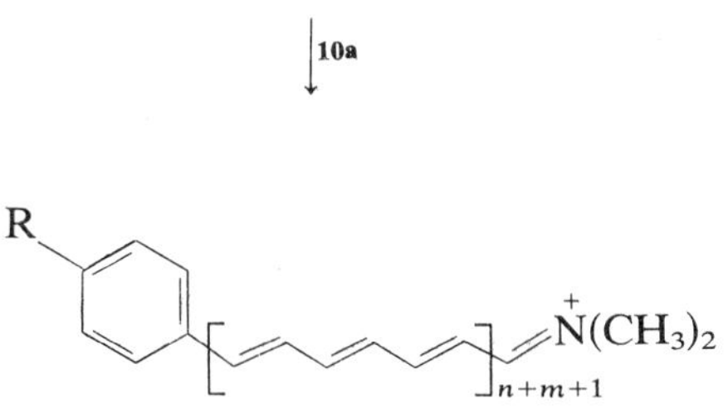

(**70**)   X^e

(ClO$_4^-$)

TABLE IV

Preparation of ω-Arylpolyeniminium Perchlorates

70	R	n	m	Yield, %
a	H	0	0	52
b	CH₃O	0	0	76
c	CH₃S	0	0	60
d	H	1	0	35
e	CH₃O	1	0	62
f	CH₃S	0	1	70
g	H	2	1	45

Iron(III)hemin (**71**) is readily transformed by **10a** to the dialdehyde **72**: Additional attack on the methine groups is not observed (95).

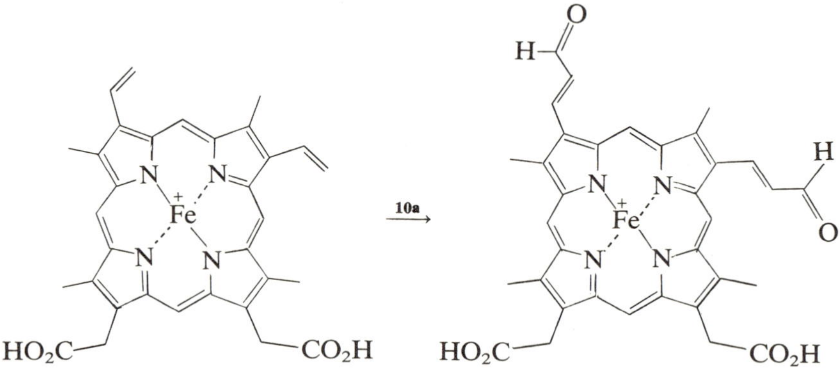

(71) → 10a → (72)

The 3,4-methylenedioxystyrenes **73** react in the same manner as **63** to afford a mixture of the stereoisomeric iminium salts **74** (Z form) and **75** (E form). Since the methylenedioxy group also activates the 5-position of the benzene ring by electron donation, **74** is susceptible to an intramolecular electrophilic substitution. In **74** and **75** the energy barrier for rotation about the double bond is strongly diminished by resonance stabilization. If slightly greater thermodynamic stability of **75** (E) over **74** (Z) is assumed, elevated temperatures and especially bulky substituents R' should shift the equilibrium in favor of **74** (Z), which undergoes cyclization to the aminoindenes **76**. In agreement with this interpretation, Witiak et al. (96) provide the following experimental data: when the reaction was run at low temperatures, only the aldehydes **77** (E form) and the starting olefin could be isolated, whereas heating the reaction mixture for 3 hr at 95–100°C made possible the isolation of **76**. Alkaline hydrolysis of **76** affords, via the isomeric enamines, the indanones **78**.

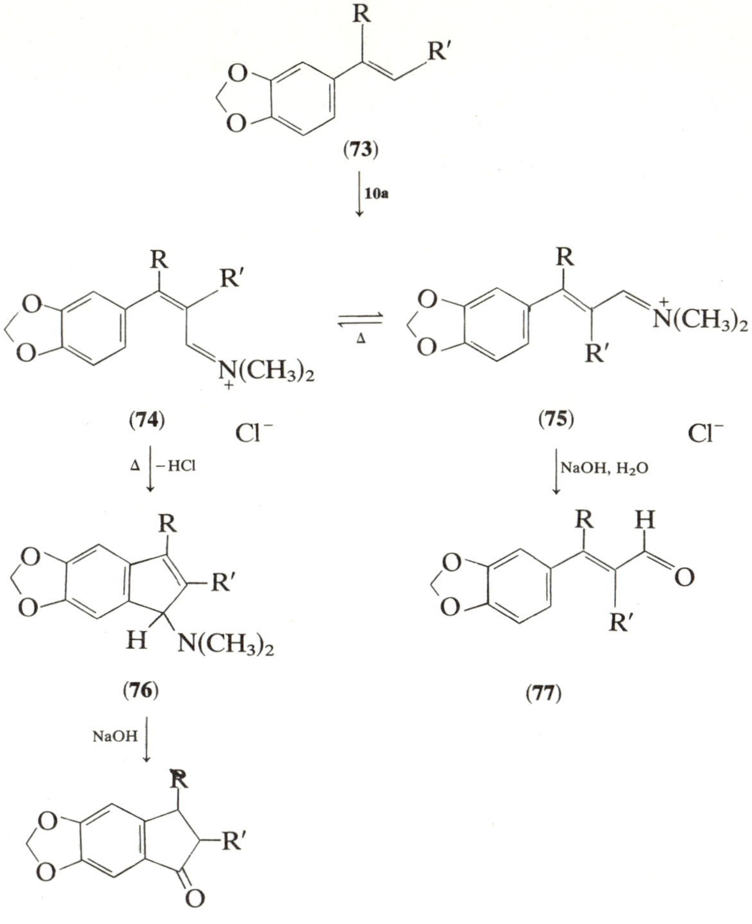

TABLE V

Aminoindenes by Formylation of 3,4-Methylendioxy Styrenes (**73**)

	R	R'	**76**, %	**77**, %	(At low °C, %)
a	H	CH$_3$	47	23	(48)
b	H	n-C$_3$H$_7$	71	—	(48)
c	H	n-C$_4$H$_9$	70	—	
d	H	C$_6$H$_5$	25	—	
e	C$_6$H$_5$	H	0	70	

Stilbene itself is unreactive toward **10a**. Activation by the dimethylamino group, however, allows substitution of the double bond. From 4-dimethylaminostilbene (**79**) and equimolar amounts of **10a** in DMF at 40–60°C, 4-dimethylamino-α-phenylcinnamaldehyde (**80**) was obtained after hydrolysis. An excess of **10a** further substituted **79** in the benzene ring, and work-up of the reaction mixture led to the dialdehyde **81** (97). In contrast, 2,4-dimethoxystilbene was attacked exclusively in the benzene ring by **10a**, and 4,6-dimethoxy-3-stilbenecarboxaldehyde was isolated (39%).

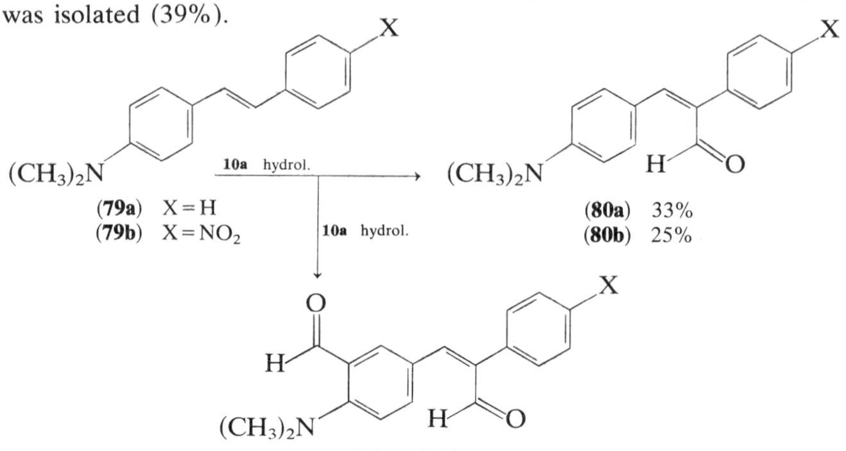

(**79a**) X = H
(**79b**) X = NO$_2$

(**80a**) 33%
(**80b**) 25%

(**81a**) 60%

Under forcing conditions (tenfold excess of **10a**, 10 hr at 70–95°C) the higher vinylogues of stilbene also react, treatment of α,ω-diphenylpolyenes **82a–c** with **10a** yields, along with starting hydrocarbons, the iminium salts **83a–c**, which form the conjugated aldehydes **84a–c** after hydrolysis:

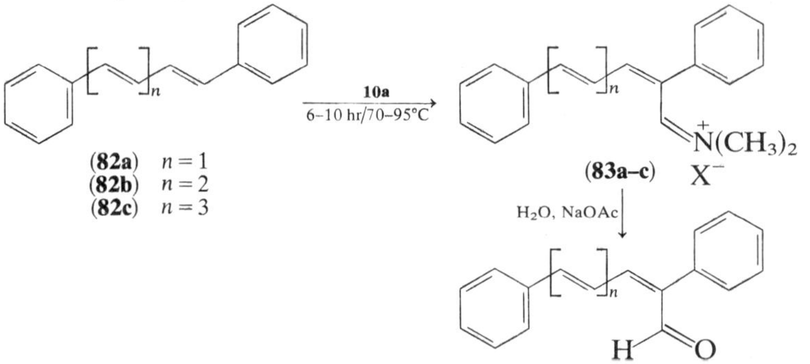

(**82a**) n = 1
(**82b**) n = 2
(**82c**) n = 3

(**83a–c**) X$^-$

(**84a**) 25%
(**84b**) 65%
(**84c**) 49%

From 1,4-diphenylbuta-1,3-diene (**82a**), 2,5-diphenylpenta-2,4-dien-1-al (**84a**) is formed, along with the deep red 4-dimethylamino-6-hydroxy-3,5-diphenylfulvene-2-carboxaldehyde (**88a**). The amount of the amphoteric **88a** increases, at the expense of **84a**, with longer reaction times and higher temperatures. The first step in the formation of **88a** is probably intramolecular, electrophilic substitution of the iminium ion **83a** to give **85**. This is followed by proton abstraction to yield the aminocyclopentadiene derivative **86**, which undergoes further substitution by **10a** to afford the diiminium salt **87**. Finally, hydrolysis of **87** leads to the fulvene derivative **88**. The introduction of donor substituents (e.g., methoxy groups) in the para positions of the benzene rings of **82a** should facilitate intramolecular ring closure of **83a** to **85**. With 1,4-bis(p-methoxyphenyl)-1,3-butadiene, the corresponding fulvene derivative **88b** was obtained in 82% yield (92) as the sole product:

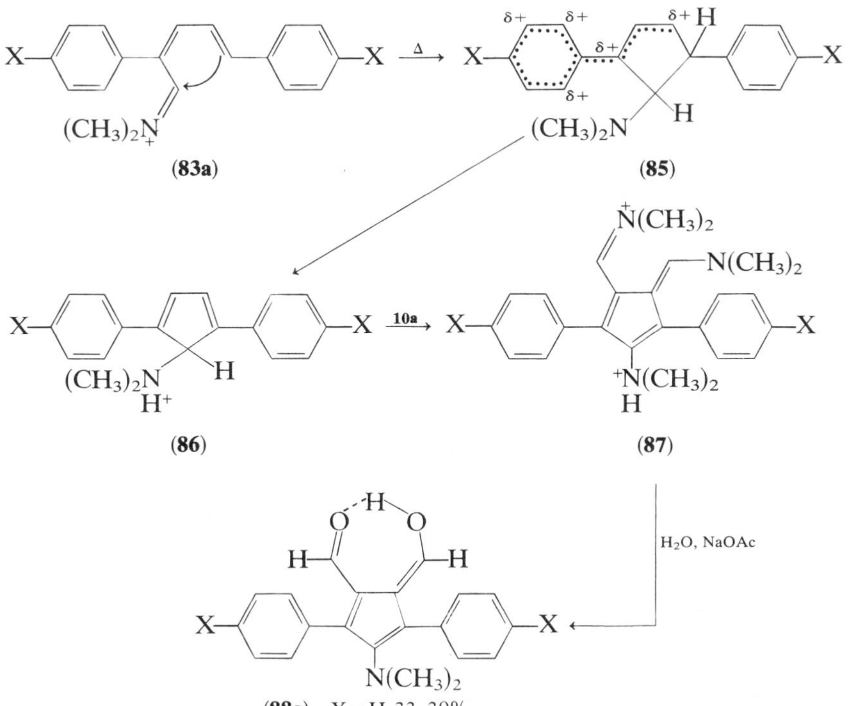

(**88a**) X = H 33–39%
(**88b**) X = OCH$_3$ 82%

Some complex polysubstitutions have been observed in the styrene series. With an excess of **3a** (or **10a**) the formylation of 2-phenylpropene

(**63c**) does not stop at the stage of the iminium salt **64c**. Rather, the dieneamine **89**, formed by proton abstraction of **64c**, undergoes substitution to the pentamethinium salt **90**, which upon further attack at the free β-enamine position yields the diperchlorate **91** in 98% yield (93). Such salts as **91** are valuable intermediates in synthesis; for example, **91** was transformed to the 4-phenylpyridine-3-carboxaldehyde (**92**) by boiling with an aqueous solution of ammonium chloride:

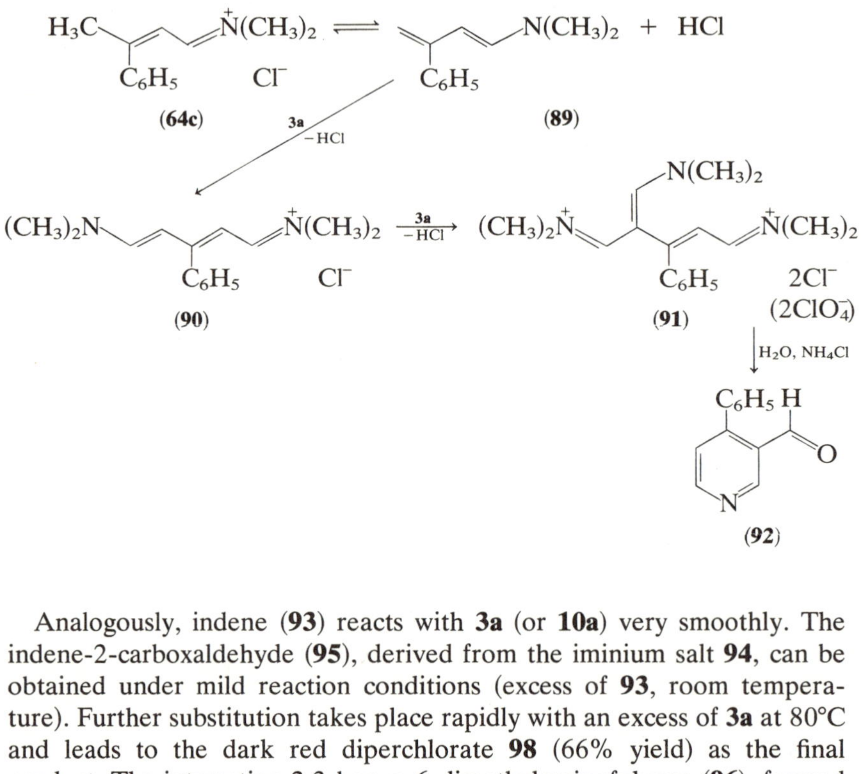

Analogously, indene (**93**) reacts with **3a** (or **10a**) very smoothly. The indene-2-carboxaldehyde (**95**), derived from the iminium salt **94**, can be obtained under mild reaction conditions (excess of **93**, room temperature). Further substitution takes place rapidly with an excess of **3a** at 80°C and leads to the dark red diperchlorate **98** (66% yield) as the final product. The interesting 2,3-benzo-6-dimethylaminofulvene (**96**), formed by proton abstraction from **94**, is assumed to be the intermediate. This further substitution is due to the high electron density of the 1- and 3-positions of the indene ring in **96** (26).

2. *Substitution of Aliphatic Olefins*

Purely aliphatic olefins lacking strong polarizing substituents may be attacked by the chloromethyleniminium ion according the scheme **60** → **62** (see Section II-B-1). Since in such monoolefins the carbonium ion should be of very high energy, only β-alkyl-substituted ethylenes with little steric bulk (e.g., exomethylene groups) are susceptible to reaction with **3a** or **10a**. For the same reason dienes and trienes show much higher reactivity. However, in such compounds cationic polymerization can compete with substitution. Polysubstitution is the rule, and in only a few instances can the reaction be stopped at a stable intermediate. This is so because the first step is rate determining and needs, in general, harsher conditions than the following steps. Isobutene (**99**) represents the simplest olefin that has been alkylated with **3a**. Reaction in dichloroethane at 60°C affords the yellow salt **105**, isolated in 73% yield as the triperchlorate (93). The reaction occurs by a sequence of several proton abstractions and substitutions which are depicted by structures **100–105**.

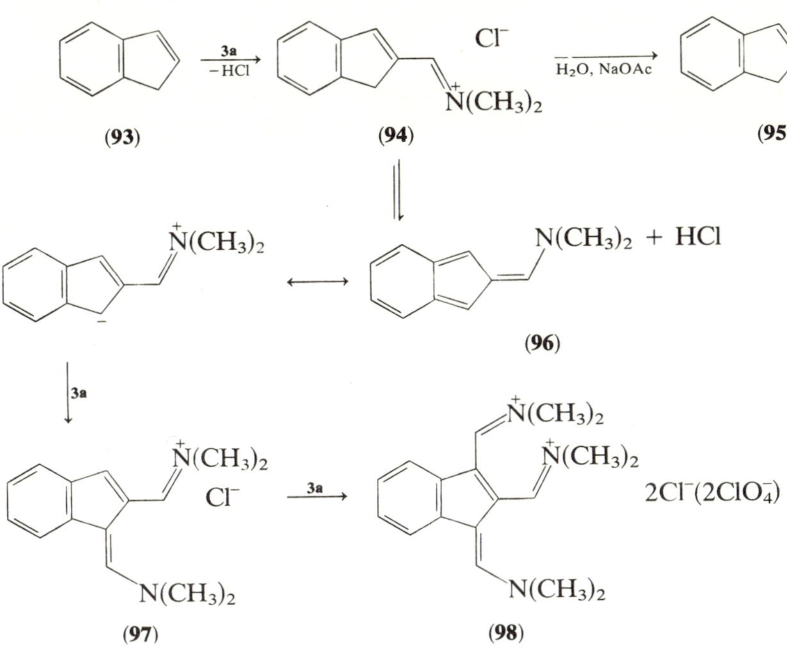

Compound **105** could also be obtained by subjecting independently prepared pentamethinium salt **102** to the same reaction conditions (22). Boiling in aqueous ammonium chloride transformed **105** into 2,7-naphthyridine-4-carboxaldehyde (**106**) (22,93).

Under very similar conditions, methylenecyclohexane (**107**) affords the trimethinium perchlorate **111** in low yields (98). Generated by the postulated equilibria, the dienamine **109** undergoes substitution by **3a** at the more basic β-enamine carbon atom.

The alkylation of 2-methylenebornane (**112**) allows isolation of each formylation product by variation of the reaction conditions. In chloroform at 30–60°C **3a** substitutes **112** to give the monoiminium salt **113**, obtained in 45% yield as the perchlorate salt. With the more reactive **10a**, at 90°C the trisubstitution product **115** can be isolated in 30% yield as the diperchlorate. It appears that formation of the intermediate dienamine **114** is impeded because of the strained endocyclic double bond. To simplify structural assignments, **115** was transformed to pyrido(4,3:2′,3′)bornen-4-carboxaldehyde (**116**) in 93% yield (98).

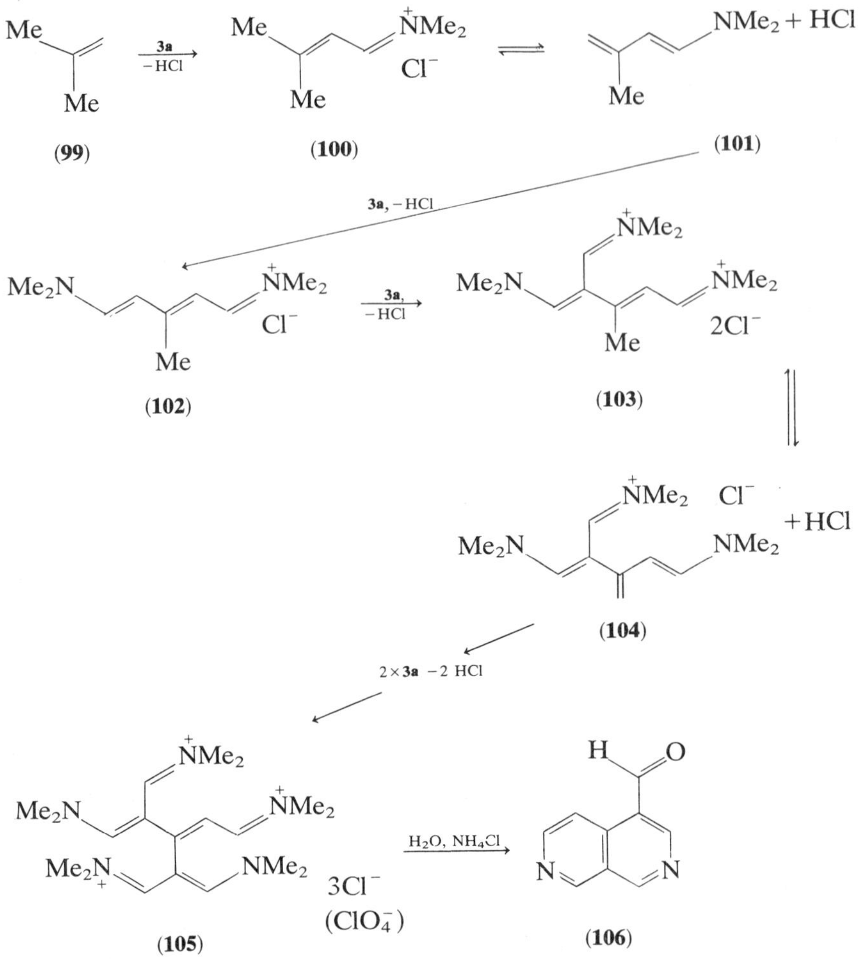

Proton abstraction to an enamine is precluded during the reaction of camphene (**117**) with **10a**, and thus the final product was the iminium salt **118**, which could be obtained as the perchlorate salt in 69% yield. Also

Me = CH$_3$

present in the reaction mixture was isobornyl chloride (25%), formed from released hydrogen chloride reacting with **117** (98).

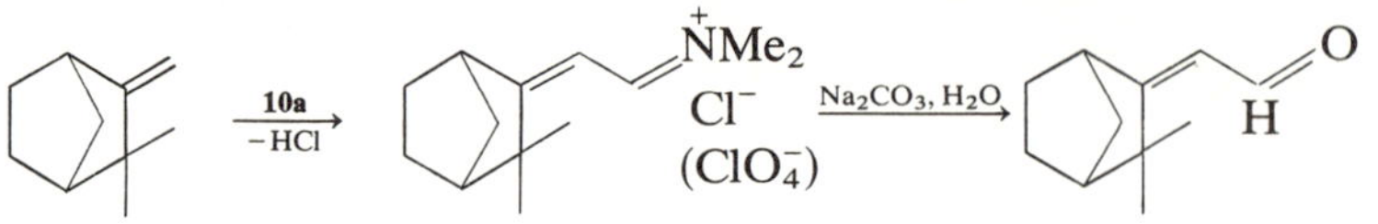

(**117**) →[10a, −HCl] (**118**) [with $\overset{+}{N}Me_2$, Cl^- (ClO_4^-)] →[Na_2CO_3, H_2O] (**119**) (81%)

Monosubstitution was also observed with **10a** on 17-methylene-5α-androstan-3β-ol acetate (**120a**) (14 days, room temperature), yielding **120b**, which could be isolated in moderate yield after work-up. The related 3-methylene-5α-androstanol acetate (**121**) proved to be surprisingly unreactive under all conditions used (99).

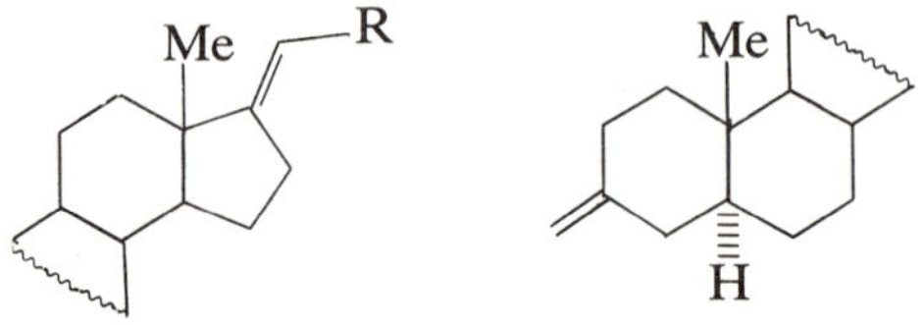

(**120a**) R = H
(**120b**) R = CH=O

(**121**)

Until recently, open-chain 1,3-dienes (1,3-butadienes) had not been investigated as to their behavior toward chloromethyleniminium ion. Whereas endocyclic olefinic double bonds are unreactive, the corresponding 1,3-dienes should be susceptible to substitution by **3a** and/or **10a**. Under vigorous conditions (24 hr, 80–90°C) the 3,5-androstadien-17β-ol acetate (**122**) reacted slowly with **10a**, exclusively at position 3, and yielded, after hydrolysis, **123** (20%), as well as the isomeric aldehyde **124** (15%) (99):

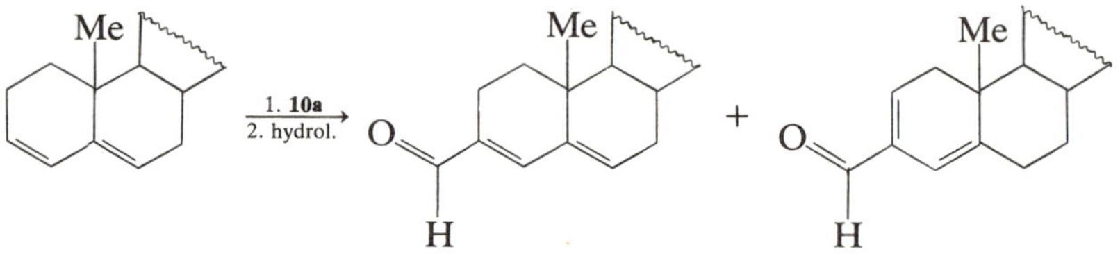

(**122**) →[1. 10a, 2. hydrol.] (**123**) (20%) + (**124**) (15%)

Increased reactivity was observed when an additional methyl group was introduced in position 6. Thus 6-methyl-17α-(1-propynyl)-3,5-androstadien-17β-ol acetate readily undergoes formylation at room temperature.

The reaction of 3-methyl-3,5-androstadien-17β-ol acetate (**125**) with **10a** is rather surprising. Reaction at room temperature for 24 hr yielded the 3-formylmethylene compound **128** as the sole product, even if the reaction mixture was later heated, whereas at 80°C (1 hr) the 6-formyl derivative **126** was obtained as the major product. Under the acidic conditions of the reaction apparently an equilibrium exists between **125** and the energetically less stable 4-en-3-methylene **127**. At room temperature only **127** reacts with **10a** and so displaces the equilibrium in its favor. However, at higher temperatures the energy barrier is overcome and attack on position 6 in **125** competes successfully with the slow equilibrium between **125** and **127** (99):

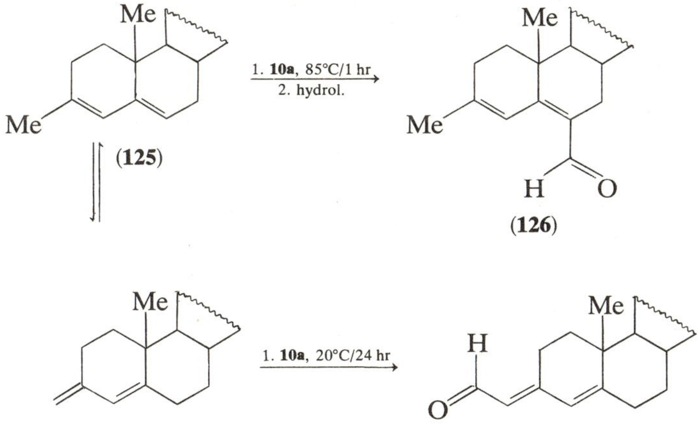

Cyclopentadiene (**129**), the simplest cyclic 1,3-diene, reacts smoothly with **3a** or **10a** at −10°C and more vigorously at room temperature (10,100). A triformylation occurs via **131**, formed by proton abstraction, and continues through **132** to **133**. Although the 6-dimethylaminofulvene (**131**) is a stable compound, it could not be prepared in this way. Formed in the rate-determining step, **131** reacts as a strong nucleophile with **3a** or **10a**, much faster than does **129**. Moreover, **132** and its hydrolysis products, **134** and **135**, are more easily synthesized by starting from otherwise prepared **131** (101). Depending on hydrolytic conditions during the work-up and subsequent treatment, the various, synthetically valuable aminoaldehydes and polyaldehydes **134–138** can be prepared.

As expected, linear 1,3,5-triene systems react smoothly with **3a** or **10a**. From 2,4,6-androstatriene-17β-ol benzoate (**139**) the corresponding 2-formyl derivative **140a** was obtained in 66% yield by action of **10a** in DMF (2 hr, 70°C). The nearly equivalent position 7 of the triene is more sterically hindered and is not attacked (127). The related 2-formyl-6-methyl-2,4,6-pregnatrien-17α-ol-20-one acetate (**140b**) was isolated in 20% yield after reaction of the carbinol **141** with **10a** (2 hr, 100°C). Presumably dehydration of the alcohol and rearrangement to an intermediate 6-methyl-2,4,6-triene occurs before formylation (99). 1,1,2,3,5,6-Hexamethyl-4-methylene-2,5-cyclohexadiene (**142**), a remarkably basic hydrocarbon (**10a**) with a cross-conjugated 2-vinyl-1,3-butadiene system, undergoes very rapid acylation by **10a**, even below

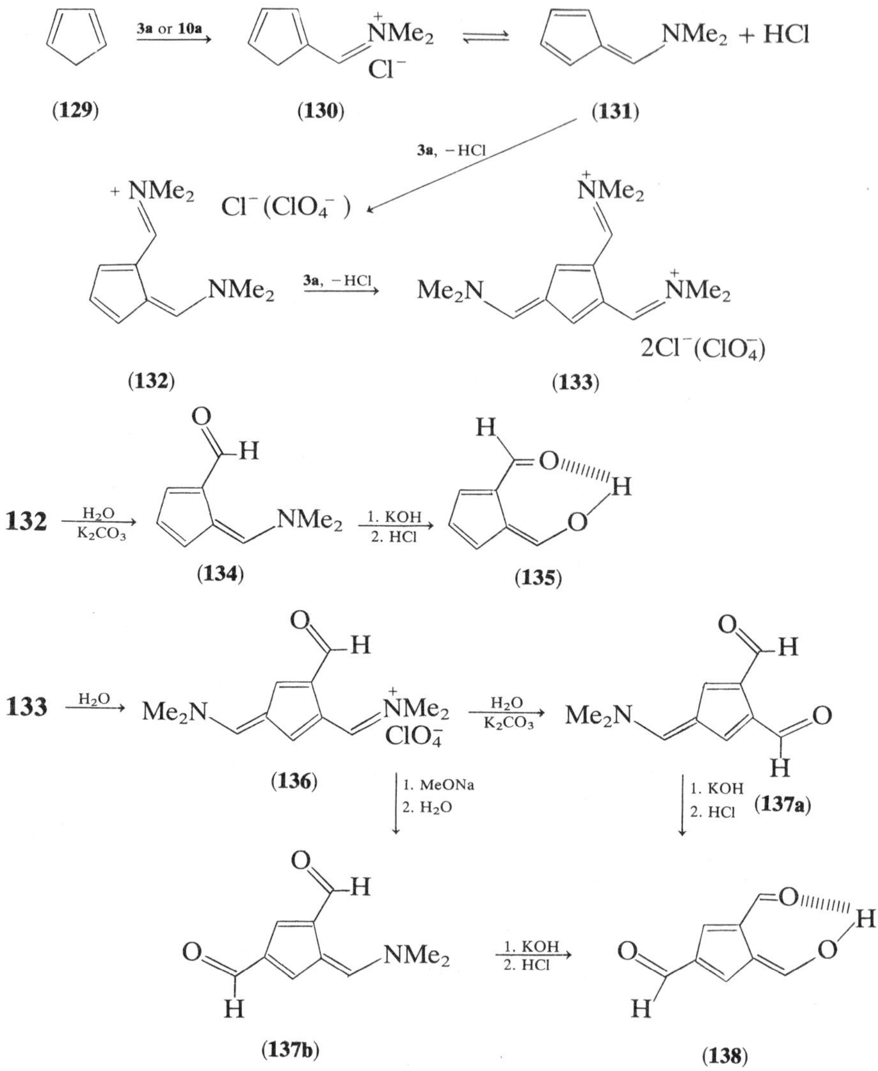

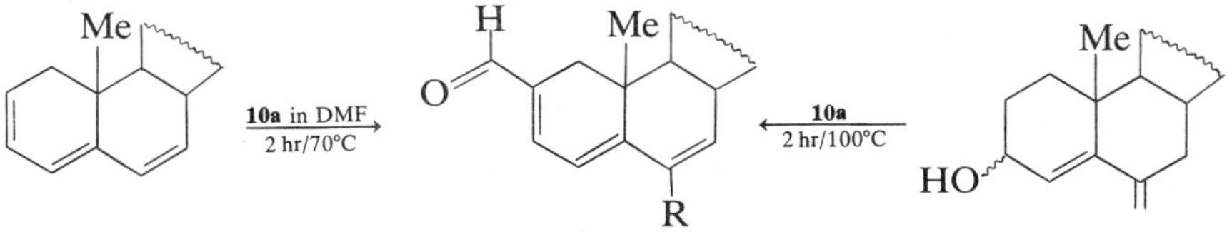

II. C—C BOND-FORMING REACTIONS

(139) → **10a** in DMF, 2 hr/70°C → (140a) R=H; (140b) R=CH$_3$ ← **10a**, 2 hr/100°C ← (141)

0°C, to give the yellow iminium salt **143** (92%). Likewise, 1,1-dimethyl-4-methylene-1,4-dihydronaphthalene and 9,9-dimethyl-10-methylene-9,10-dihydroanthracene, which are related to styrene and 1,1-diphenylethylene, respectively, need much stronger reaction conditions (60–80°, 2 hr) and afford salts **144** and **145** (98).

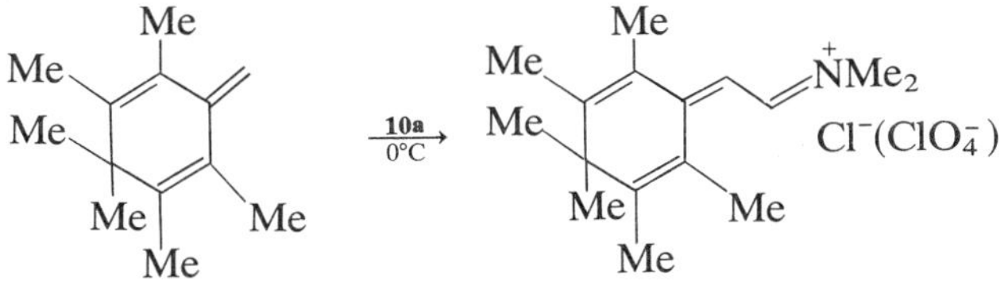

(142) → **10a**, 0°C → (143) 92%

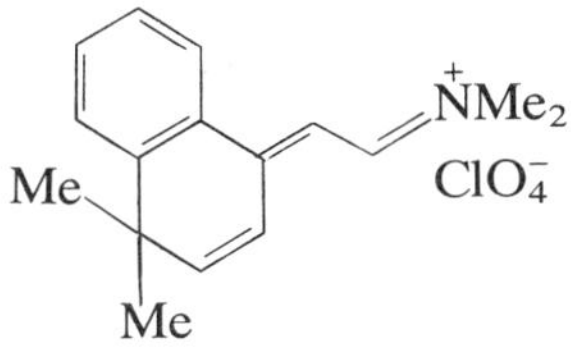

(144) 94% 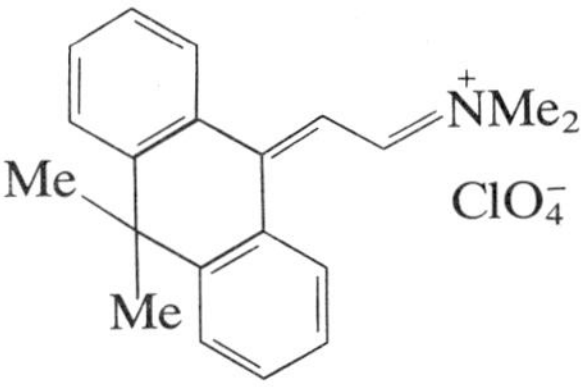   (145) 74%

A cross-conjugated system is also present in 6,6-diphenylfulvene (**146a**) and 6-(4-dimethylaminophenyl)fulvene (**146b**). These two compounds containing the 1-aryl-1,3,5-hexatriene moiety (**146b** is also a phenylogue of **131**) yield, after alkylation with **10a** and hydrolytic workup, red aldehydes **147a** and **147b** (100):

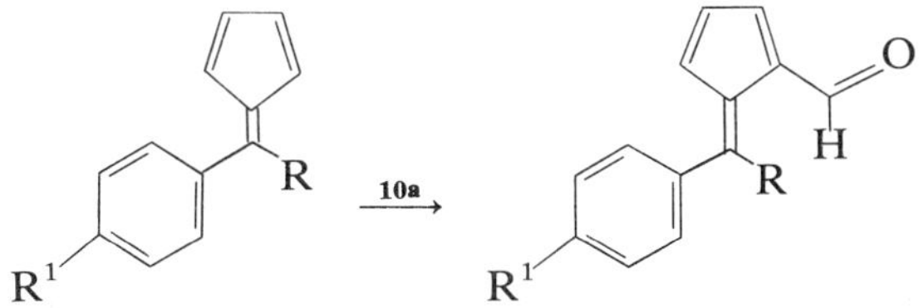

(146a) R = C$_6$H$_5$, R' = H
(146b) R = H, R' = NMe$_2$

→ **10a** → (147)

In a similar manner the hemicyanine salt **149** could be obtained from a heterologous sesquifulvalene, the 4-(1-indenylidene)-1,4-dihydropyridine **148**, which is related to **131** (102):

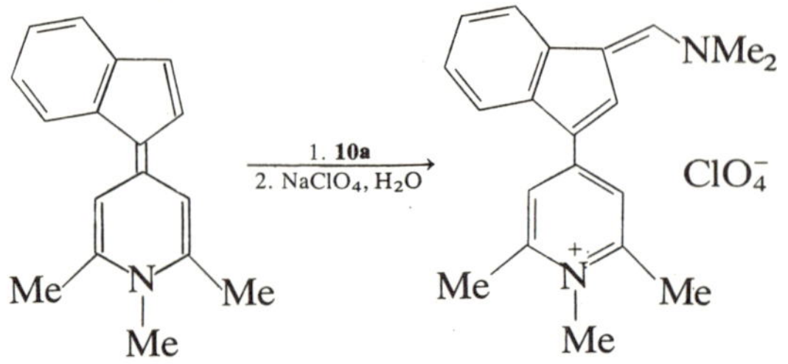

(**148**) → [1. **10a**; 2. NaClO₄, H₂O] → (**149**) 71%

C. REACTIONS OF CHLOROMETHYLENIMINIUM SALTS WITH ENAMINES, DIENAMINES, AND TRIENAMINES

1. *Enamines*

The smooth formylation of 1,3,3-trimethyl-2-methylenindoline (**150**) (34,36,103) and of 1-phenyl-1-(1-methyl-2-phenylindolyl-3)-ethylene (**152**) (33–36,104) by MFA–POCl$_3$ or **10a** has been known for some time. Although the corresponding aldehydes, **151** and **153**, were isolated in high yields, the close relationship of **152** and, particularly, of **150** to

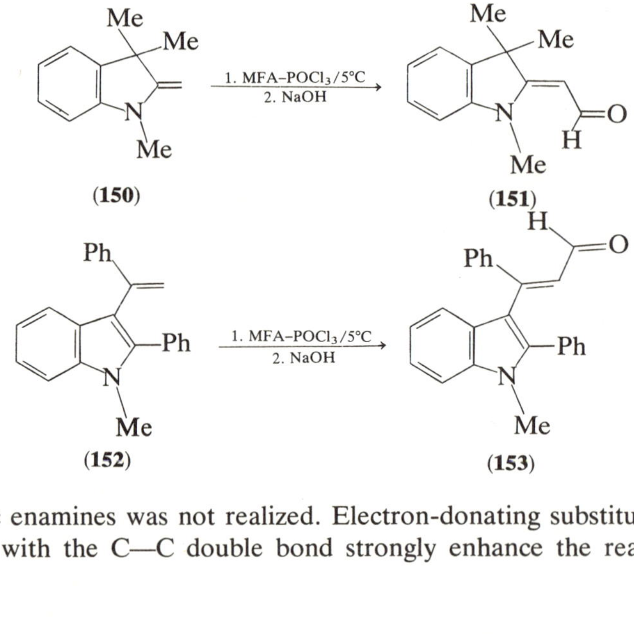

(**150**) → [1. MFA–POCl$_3$/5°C; 2. NaOH] → (**151**)

(**152**) → [1. MFA–POCl$_3$/5°C; 2. NaOH] → (**153**)

aliphatic enamines was not realized. Electron-donating substituents conjugated with the C—C double bond strongly enhance the reactivity of

II. C—C BOND-FORMING REACTIONS

vinyl ethers and enamines to electrophiles, especially **3a** or **10a**. As mentioned in Section II-B-2, an intermediate enamine (or dienamine) formed by deprotonation of the generated iminium ion is the cause in some cases for further rapid substitution.

Aliphatic enamines can be frequently and advantageously utilized in preparing many different 1,3-dicarbonyl compounds or their derivatives. Thus 1-morpholino-1-cycloalkenes **154** and also 1-morpholino-3,4-dihydronaphthalene (**156**) (easily prepared by azeotropic distillation of a mixture of the corresponding ketone and morpholine in toluene) react with **10a** to yield the ketoaldehydes **155** and **157**, existing, most probably, in the hydroxymethylene form (105):

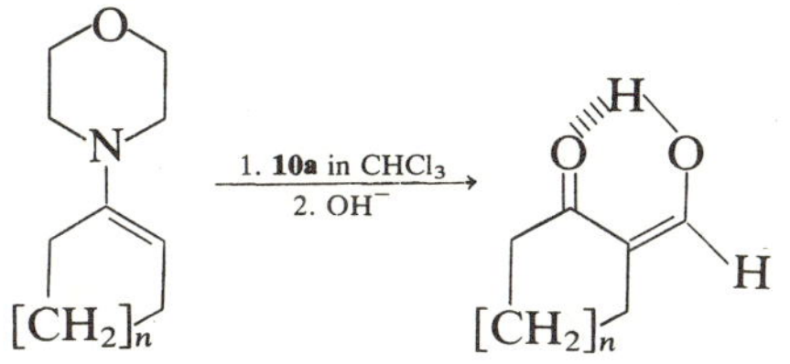

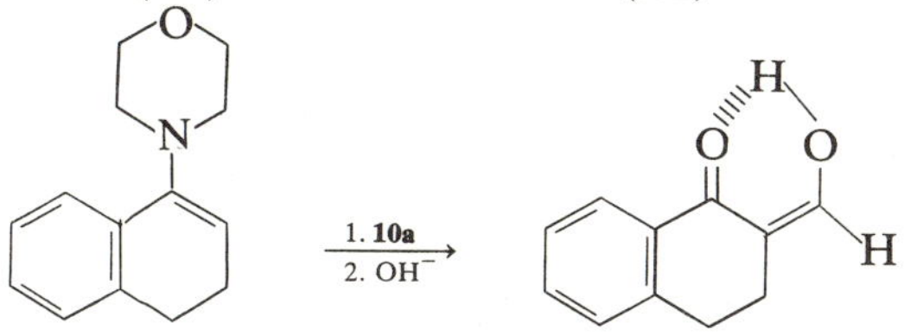

The condensation of an enamine or dienamine with **10a** leads initially to an isolable, resonance-stabilized, polymethinium salt, a vinylogue of the formamidinium ion. Further attack by **10a** is, however, possible and depends on structure and work-up. Control is easily effected by choosing the appropriate conditions.

Trimethinium salts can be generated only if the starting enamine contains a replaceable hydrogen at the β position. However, only a limited number of such simple enamines are easily prepared. As derivatives of malondialdehyde, the trimethinium salts yield, upon hydrolysis with saturated aqueous potassium carbonate at 60–70°C, β-dialkylaminoacroleins. Further action of sodium or potassium hydroxide produces the salts of substituted malondialdehydes.

A series of dimethylaminomalonaldehyde derivatives **159a–161a** (29), as well as the piperidino and morpholino analogues **160b** and **161b** (106), has been prepared from the highly reactive 1,2-bis(dialkyamino)ethylenes

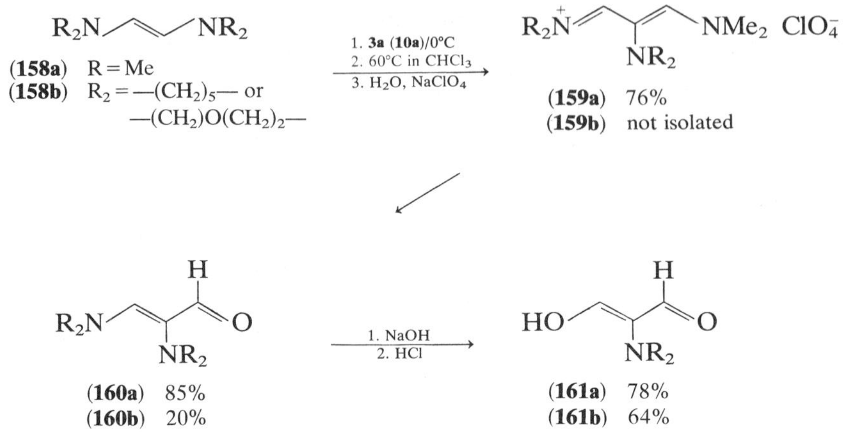

$R_2N\diagup\!\!\!\diagdown NR_2$

(158a) R = Me
(158b) $R_2 = -(CH_2)_5-$ or $-(CH_2)O(CH_2)_2-$

1. 3a (10a)/0°C
2. 60°C in CHCl$_3$
3. H$_2$O, NaClO$_4$

$R_2\overset{+}{N}\diagup\!\!\!\diagdown NMe_2$ ClO$_4^-$
         NR_2

(159a) 76%
(159b) not isolated

$R_2N\diagup\!\!\!\diagdown{}^H\!\!\diagdown O$
        NR_2

(160a) 85%
(160b) 20%

1. NaOH
2. HCl

$HO\diagup\!\!\!\diagdown{}^H\!\!\diagdown O$
        NR_2

(161a) 78%
(161b) 64%

158a and **158b**. The similar 2-dimethyl aminovinyl ethers **163**, formed *in situ* by the action of **3a** or **10a** on dimethylaminoacetaldehyde acetals **162** (see also Section II-D), are attacked by the chloromethyleniminium ion on the more highly electron-rich carbon atom in β position to the dimethylamino group, and 2-alkoxytrimethinium salts **164** are obtained (29):

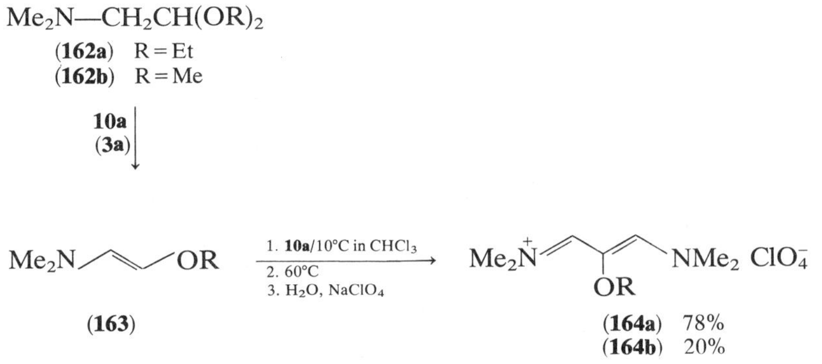

$Me_2N-CH_2CH(OR)_2$
  (162a) R = Et
  (162b) R = Me

  10a
  (3a) ↓

$Me_2N\diagup\!\!\!\diagdown OR$

(163)

1. 10a/10°C in CHCl$_3$
2. 60°C
3. H$_2$O, NaClO$_4$

$Me_2\overset{+}{N}\diagup\!\!\!\diagdown NMe_2$ ClO$_4^-$
         OR

(164a) 78%
(164b) 20%

1,4-Bis(dimethylamino)-1,3-butadiene (**165**) reacts like two isolated enamines, not like a dienamine. A double alkylation by **3a** or the related chloromethylenepentamethyleniminium chloride (from N-formylpiperidine and phosgene) takes place, yielding salt **166**, which, upon stepwise hydrolysis, gives **167** and the symmetrical tetraformylethane (**168**) (19).

Enamines with strong electron-attracting substituents are still susceptible to substitution by **10a**, provided that a position β to the nitrogen atom is

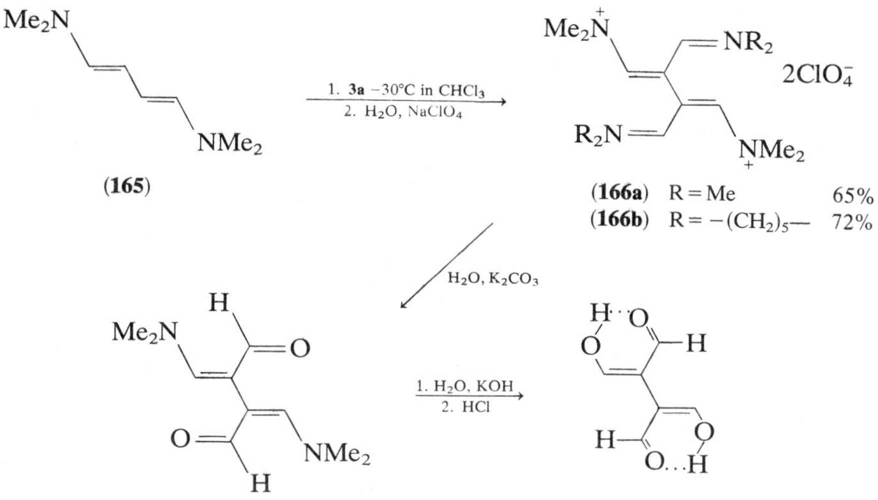

(165)

(166a) R = Me 65%
(166b) R = –(CH₂)₅— 72%

(167) (168)

free. The β-dimethylaminoacrylonitrile **169** affords the 2-cyanotrimethinium perchlorate **170** (107):

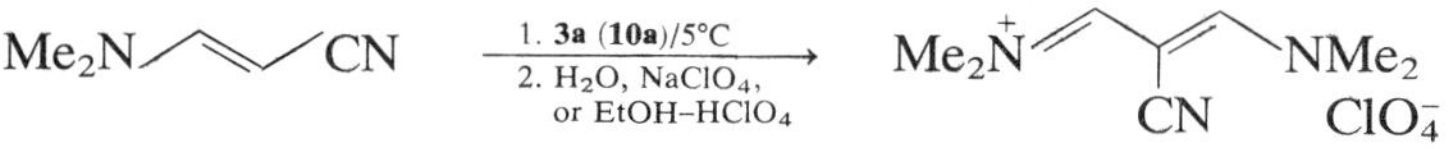

(169) (170) 40–75%

3-Methylenephthalimidines **172** behave like cyclic enamides and undergo a very smooth substitution reaction on the exocyclic carbon atom. As precursors of **172**, the more easily obtainable 2-acetylbenzamides (present as cyclic tautomers) **171** may also be employed, provided that an excess of **3a** is used to consume one molecule of water. The 3-formylmethylene derivatives **173** readily condense with several active methylene compounds (including **172**, in acetic acid–$POCl_3$), forming cyanine dyes (108):

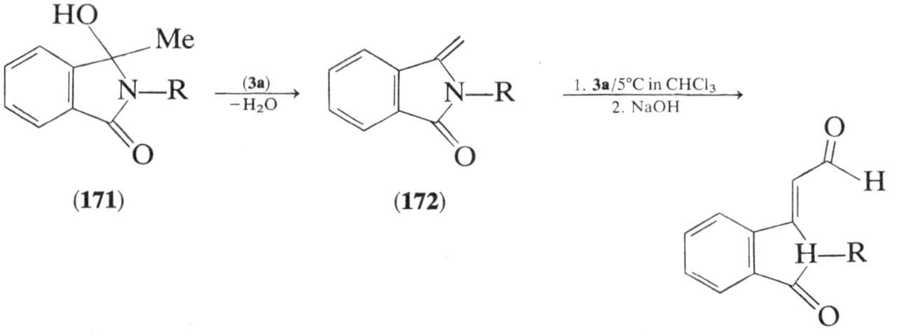

(171) (172)

(173)

TABLE VI
3-Formylmethylenephthalimidines

173	R	Yield, %	173	R	Yield, %
a	Me	99	e	$CH_2C_6H_5$	81
b	Et	80	f	C_6H_5	96
c	n-Pr	84	g	$p\text{-ClC}_6H_4$	94
d	c-Hex	82	h	$p\text{-NO}_2C_6H_4$	81

Like **172**, 5,5 dimethyl-4-methylene-1,3-oxazolidin-2-one was transformed by **10a** to the corresponding 4-formylmethylene derivative **174** (109):

(**174a**) R = Me
(**174b**) R = Et

Even unsubstituted positions β to the nitrogen atom in trimethinium and pentamethinium salts, with their extensive delocalization of electrons and alternating charge densities, are attacked by strong electrophiles in analogy to the parent enamines. Some trimethinium salts can be nitrated and brominated like aromatic compounds (107) in typical electrophilic substitution reactions.

Similarly, several trimethinium perchlorates (**175a–e**) may be acylated to the diperchlorates **176** by heating with an excess of **10a** in DMF (1–12 hr). Hydrolysis with NaOH affords the acyl malondialdehydes **177a** (11), **177b**, and **177c** (21). The salt **176d** was not isolated; however, hydrolysis with aqueous potassium carbonate gave the dimethylamide of 2-formyl-3-dimethylaminoacrylic acid (**178e**) in 76% yield (16). The corresponding 2-acyl-3-dimethylaminoacroleins **178b** and **178c** were obtained by the action of dimethylamine on the copper chelates of **177b** and **177c** (21). Compound **178a** was also formed from **177a** by treatment with dimethylcarbamoyl chloride in pyridine at 50–60°C. Careful hydrolysis of **176a–c** by aqueous potassium carbonate should also lead to **178a–c**, but this has been successfully tested only with **176a** (16). Diperchlorate **176a** is also generated by the action of a large excess of **10a** on malonic acid (16), acetaldehyde, vinyl ether, or 3-dimethylaminoacrolein (15). Consequently, triformylmethane has become a readily available compound.

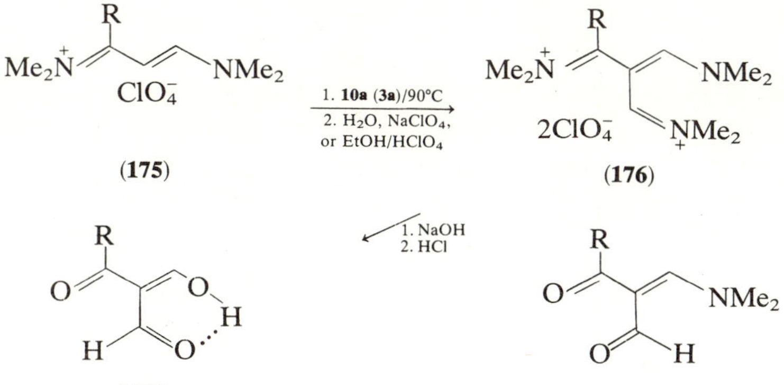

II. C—C BOND-FORMING REACTIONS

(175) → 1. **10a** (3a)/90°C; 2. H₂O, NaClO₄, or EtOH/HClO₄ → (176)

(176) → 1. NaOH; 2. HCl → (177) and (178)

Me = CH₃; Bu = C₄H₉; Ph = C₆H₅; Et = C₂H₅

TABLE VII

Formylation Products of Trimethinium Salts

Compound	R	%	Ref.
176a	H	86	110
176b	t-Bu	90	21
176c	C₆H₅	91	21
176d	Cl	—	16
176e	NMe₂	76	111
177a	H	84	
177b	t-Bu	80	
177c	Ph	85	

Similarly the 1-(*p*-dimethylaminophenyl)trimethinium perchlorate **179** is alkylated by **10a** at 90°C at position 2 of the methine chain; simultaneously an attack on the activated benzene nucleus occurs. After alkaline work-up, the acylmalondialdehyde **180a** is isolated. The related chloro derivative **181**, a phenylogous polymethinium salt formed by the action of **3a** on *p*-dimethylaminoacetophenone, undergoes only monosubstitution.

After work-up with ice water, two different dialdehydes, **182** and **183**, were obtained. Treatment of **183** with NaOH led to **180b** (21).

2. Dienamines

The highly reactive 1-dimethylamino-1,3-butadiene (**184**) undergoes disubstitution with **3a** in chloroform at 0°C, and after hydrolysis of the

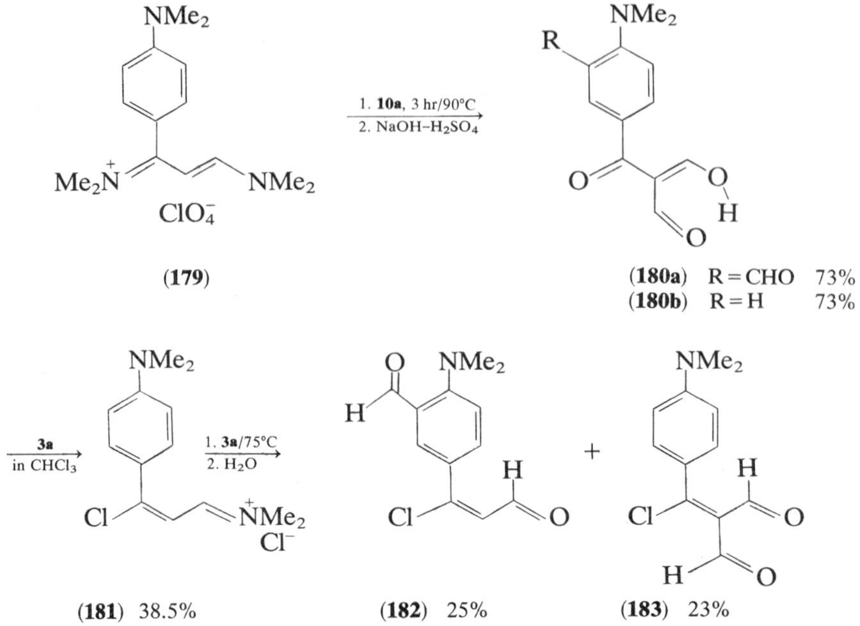

THE VILSMEIER-HAACK-ARNOLD ACYLATIONS

intermediate salt **185** (not isolated) with aqueous potassium carbonate a homogeneous dialdehyde, 4-dimethylaminomethylene-2-pentene-1,5-dial (**186**), was obtained in 35% yield. Better results (65% yield) were achieved by using 1,3-bis(dimethylamino)-1-butene (**187**), a precursor of **184**, which is easily prepared from crotonaldehyde and dimethylamine.

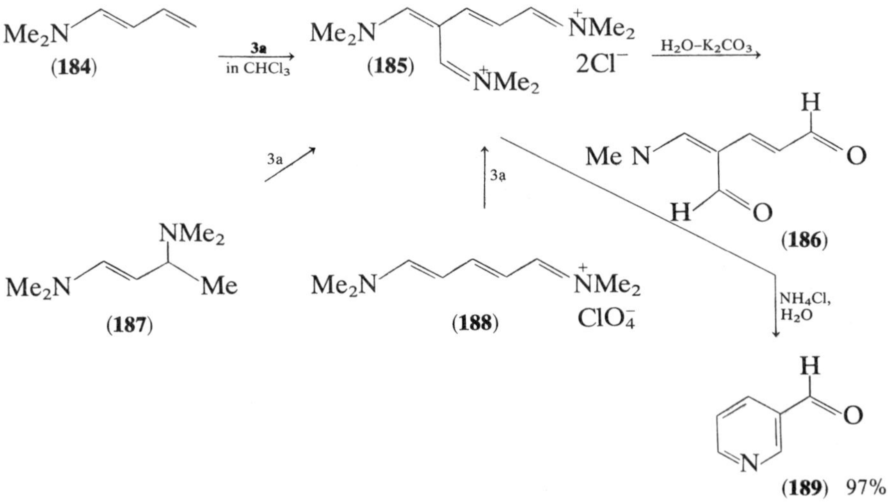

The direct use of the pentamethinium salt **188**, which is probably an intermediate in the reaction of **184** and **187** with **3a**, affords **186** in 79% yield. Cyclization of **185** by boiling with aqueous ammonium chloride represents a valuable route from crotonaldehyde to nicotinic aldehyde (**189**), a simple three-step process, in 40% overall yield (9,112).

Analogously to **184**, 3-N-pyrrolidino-3,5-androstadien-17β-ol propionate (**190**) is disubstituted by **10a** in trichloroethylene at 50°C (2 hr). Hydrolytic work-up with aqueous sodium acetate and diluted NaOH affords the pyrrolidinodialdehyde **191** (36%), which cyclizes with ammonia in methanol to the pyridine **192** (69%), or with nitromethane and sodium methoxide in methanol to the nitrobenzene derivative **193** (86%) (113). As enamines, **191**, **192**, and **193** undergo facile hydrolysis to the corresponding ketones (113):

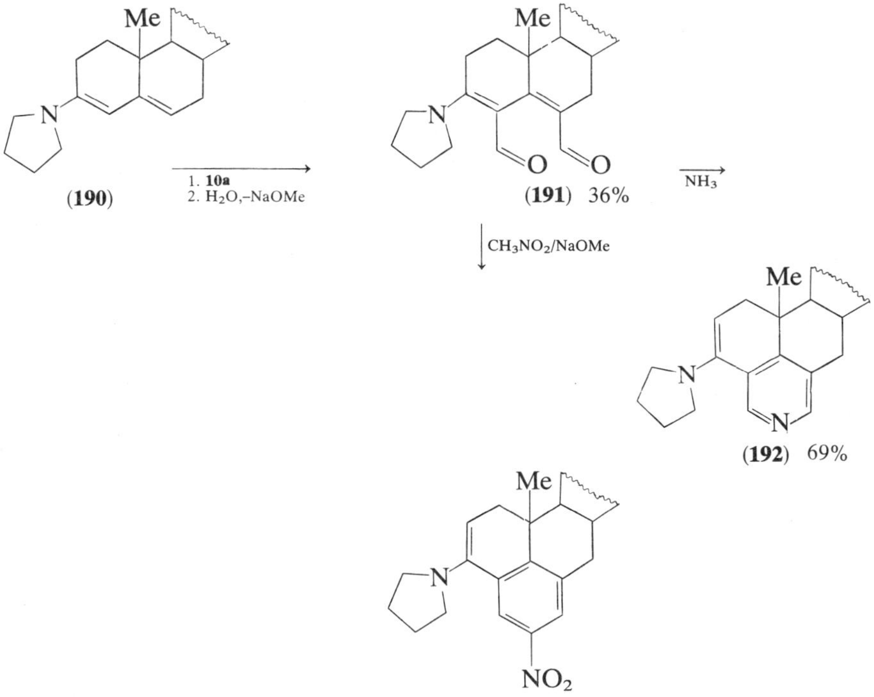

1-Dimethylaminomethylenindene (**194**) also contains a dienamine moiety and yields, with **3a** or **10a** in chloroform at 0–5°C, the pentamethinium salt **195** (61%) (26). No further alkylation with **3a** takes place because both positions β to nitrogen atoms are substituted.

(194) → [1. 3a/0–5°C in CHCl₃; 2. H₂O–NaClO₄] → **(195)** 61% (ClO₄⁻ salt with =NMe₂⁺ and =CHNMe₂ substituents)

3. Trienamines

The chemistry of formylation of cyclopentadiene, mentioned in Section II-B-2, reveals 6-dimethylaminofulvene (**131**) to be a highly reactive, transient intermediate. The monosubstitution of **131**, a cross-conjugated trienamine, takes place very rapidly with **3a**, even at −60°C, and leads in 98% yield to **132** (100).

D. REACTIONS OF CHLOROMETHYLENIMINIUM SALTS WITH ENOL ETHERS, ACETALS, AND KETALS

Compared with enamines, the corresponding enol ethers are considerably lower in nucleophilicity. Therefore polysubstitution by **3a** or **10a** is not normally observed. Open-chain enol ethers with a free β position are generally more easily accessible and much more stable than the corresponding enamines. Moreover, instead of the enol ethers, the related, even more easily available acetals can be successfully employed in formylation reactions. Indeed, the yields of reactions employing acetals are generally higher than those using enol ethers, since the latter compounds are also susceptible to cationic polymerization.

The formation of enol ethers *in situ* by the action of **3a** or **10a** on acetals or ketals by elimination of one molecule of alcohol was discovered when compound **199** was unexpectedly isolated after the reaction of the ethylene ketal of a 3-keto-5-ene steroid **196** with **3a** was followed by mild hydrolytic work-up. Thus **3a** attacks one of the ketal oxygen atoms, followed by ring opening and elimination of hydrogen chloride. The resulting enol ether **197** is then acylated by another molecule of **3a** to the bisiminium salt **198**, whose hydrolysis affords **199** (39,114)

The reaction of **3a** or **10a** with an enol ether, acetal, or ketal terminates with the formation of a 3-alkoxyallylideniminium salt (e.g., **202** or **207**), usually not isolated. The identity and ratio of the reaction products depend strongly on the mode of work-up. In most formylation experiments of this sort it is advisable first to quench the reaction mixture with ice and then to hydrolyze the iminium salt with cold aqueous potassium

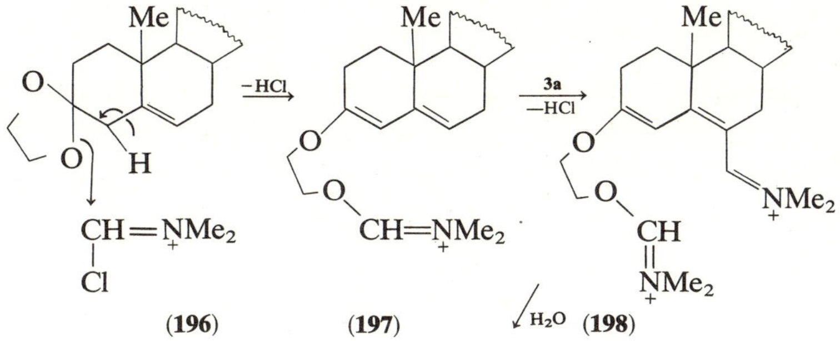

II. C—C BOND-FORMING REACTIONS

(196) (197) (198)

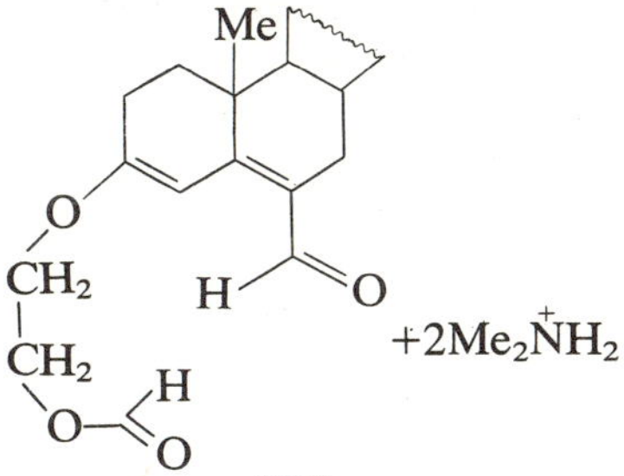

(199)

carbonate. In this way a β-dimethylaminoacrolein (203,210) or a β-dimethyl aminovinylketone (211) results.

As a route to numerous 1,3-dicarbonyl compounds and their derivatives, the Vilsmeier-Haack-Arnold formylation of enol ethers, acetals, and ketals represents a versatile, valuable supplement to the formylation of enamines.

1. Vinyl Ethers and Acetals

It was Arnold (5) who first observed formylations of acetals. A cooled suspension of **3a** (2.5 moles) in dichloroethane reacts in a slightly exothermic manner with substituted acetaldehyde acetals (**200**) (1.0 mole) or vinyl ethers (**201,202**). After basic work-up and heating with aqueous potassium carbonate, the β-dimethylaminoacroleins **203** are the main products. These serve as the most suitable starting materials for the preparation of various derivatives of malondialdehydes (e.g., trimethinium salts). Vinyl ethers **201** were treated with **10a** under nearly identical conditions (115) and were subsequently used in an investigation of the structure of the complex **10a** by HNMR techniques (50).

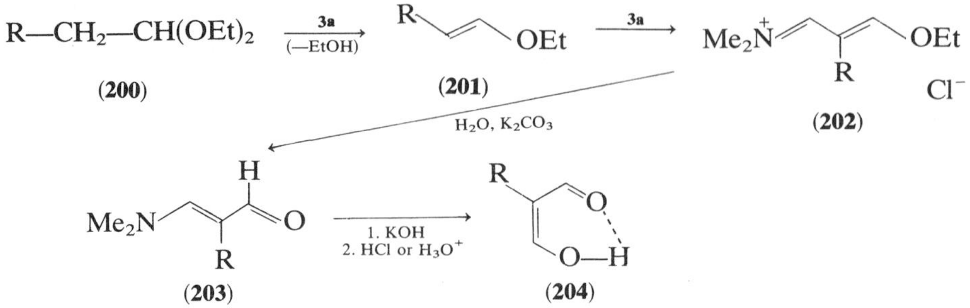

R—CH$_2$—CH(OEt)$_2$ $\xrightarrow[\text{(−EtOH)}]{\text{3a}}$ R⧸=⧹OEt $\xrightarrow{\text{3a}}$ Me$_2$N$^+$=⧸—⧹OEt Cl$^-$
(200)   (201)   R
                  H$_2$O, K$_2$CO$_3$ (202)

Me$_2$N—⧸=⧹—CHO $\xrightarrow[\text{2. HCl or H}_3\text{O}^+]{\text{1. KOH}}$ enol form (204)
     R
(203)

TABLE VIII

Preparation of β-Dimethylaminoacroleins from Acetals and Enol Ethers

203	R	Yield from **200**, % (5)	Yield from **201**, % (5)	Yield from **201** + **3a**, % (115)
a	H	69	78	57
b	Me	81		68
c	Et	75		77
d	n-Prop	70		
e	i-Prop	48		
f	n-Bu	60		
g	n-C$_5$H$_{11}$	89		
h	Ph—CH$_2$	70		
i	Ph	87		

Sometimes the slightly soluble, crystalline dianil salts are useful for identification and isolation. These may be prepared from a reaction mixture of **202** and aniline by addition of sodium perchlorate. The valuable trimethinium salts **205** can be obtained (22) from **203** by the application of Bredereck's method (116) for the preparation of amidinium salts:

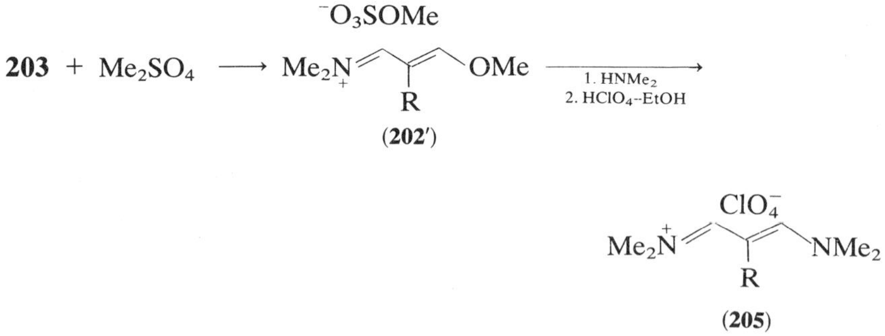

203 + Me$_2$SO$_4$ ⟶ Me$_2$N$^+$=⧸(R)—⧹OMe $^-$O$_3$SOMe $\xrightarrow[\text{2. HClO}_4\text{–EtOH}]{\text{1. HNMe}_2}$
                 (202′)

Me$_2$N$^+$=⧸—⧹NMe$_2$ ClO$_4^-$
      R
    (205)

In order to synthesize novel malondialdehydes with bulky substituents, R = cyclobutyl, cyclopentyl, and cyclohexyl, as well as adamantyl-1 and adamantyl-2 (117), the required vinyl ethers **202** were prepared by a Wittig reaction of the corresponding aldehydes, R—CHO, with methoxymethylenetriphenylphosphorane. Analogously, 2-methoxy- and 2-ethoxymalondialdehydes were obtained by formylation of 1,2-dimethoxy- or 1,2-diethoxyethylene (**206**) with MFA–POCl$_3$ (118). The intermediates of these reactions were the corresponding β-methylanilino-acroleins **207**, acid hydrolysis of which affords malondialdehydes **204k, l**, representing ethers of triose reductone:

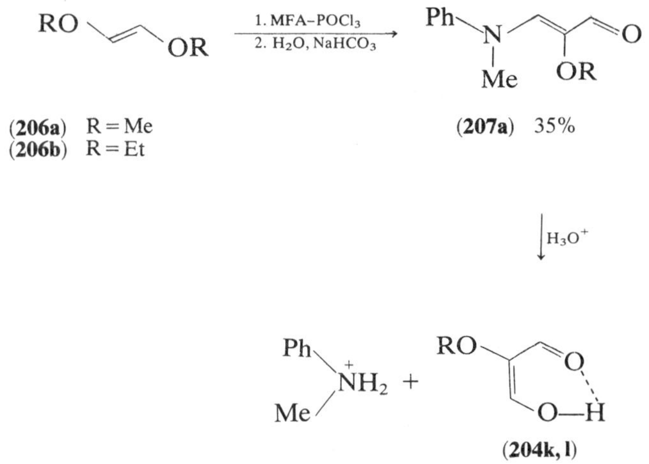

(**206a**) R = Me
(**206b**) R = Et

(**207a**) 35%

(**204k, l**)

Surely, alkoxymalondialdehyde and its derivatives may now be better prepared by the action of **3a** or **10a** on alkoxyacetaldehyde acetals.

2. Enol Ethers and Ketals

Ketals **205** react like acetals with **3a** in dichloroethane at 40°C. Arnold (6) isolated a mixture of easily separable products, the composition of which depends on the structure of the starting ketal and the mode of work-up (i.e., alkaline or not alkaline treatment of the resulting reaction mixture). As shown in the following reaction scheme, all the formylation products (**208–211**) are obtained by different transformations of the first-formed iminium salt (**207**). Also several interconversions between the individual compounds are possible. Dimethylamine effects a slow but irreversible transformation of the sterically crowded β-dimethylamino-acroleins **210** to the thermodynamically more stable β-dimethyl amino-vinylketones **211**.

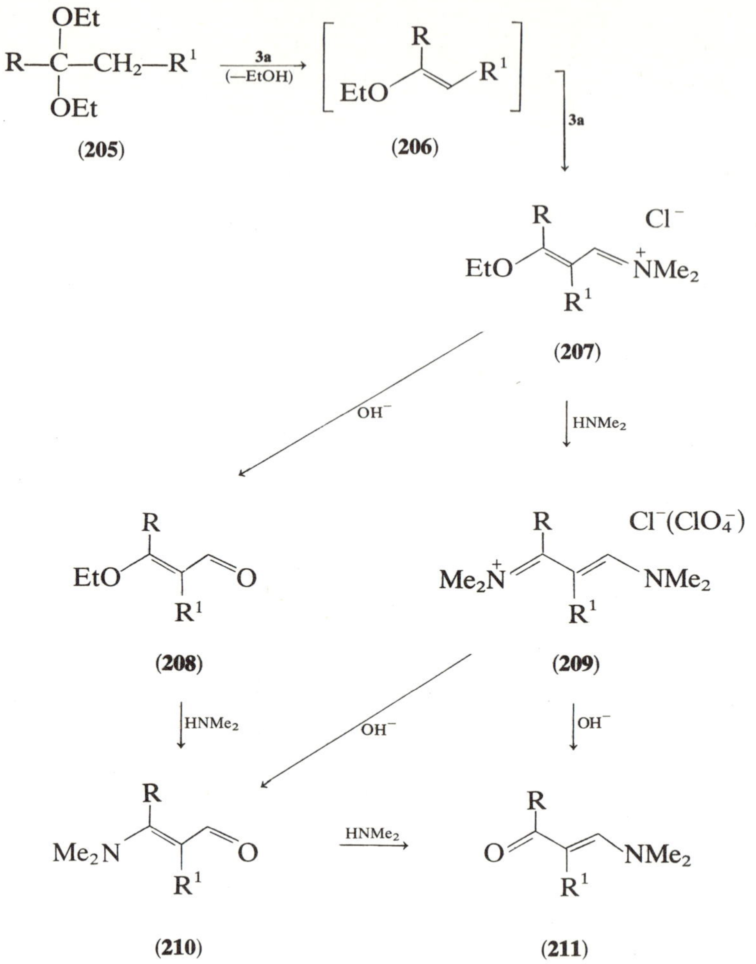

TABLE XI

Stereoisomerism of 3-Chloroacroleins

	R	R¹	208	209	210	211	Mode of work-up
a	Ph	H		44.6	25.7		H_2O–$NaOAc$/K_2CO_3
b	Ph	Me	92.1				H_2O/K_2CO_3
c	Me	H		56	ca. 2.3	2.3	H_2O/K_2CO_3
d	t-Bu	H	82			5	H_2O/K_2CO_3
e	—$(CH_2)_3$—			ca 35	13	H_2O/K_2CO_3
f	—$(CH_2)_4$—		59				H_2O/K_2CO_3

Several enol ethers of 14-hydroxydihydrodeoxycodeinone (**212**) were treated with **10a** in dichloroethane at 60–70°C, followed by hydrolytic work-up with buffered aqueous solutions (pH 8–9.5) to yield the corresponding 7-carboxaldehydes **213**. Under these conditions simultaneous formylation of the 14-hydroxy group to the tertiary formate ester **213a** occurs; this is easily removed by hydrolysis to afford **213b**. In the reaction of the 6-methoxy compound (R = Me), the by-products, 14-chloro-6-methoxy-$\Delta^{6,7}$-dihydrodeoxycodeine and 6-chloro-14-hydroxy-$\Delta^{6,7}$-dihydrodeoxycodeine-7-carboxaldehyde, were also isolated (119). Compound **213** was converted to the oxime, semicarbazone, hydrazone, and anil, each of which could be cyclized to the corresponding oxazole, pyrazole, and quinoline derivatives.

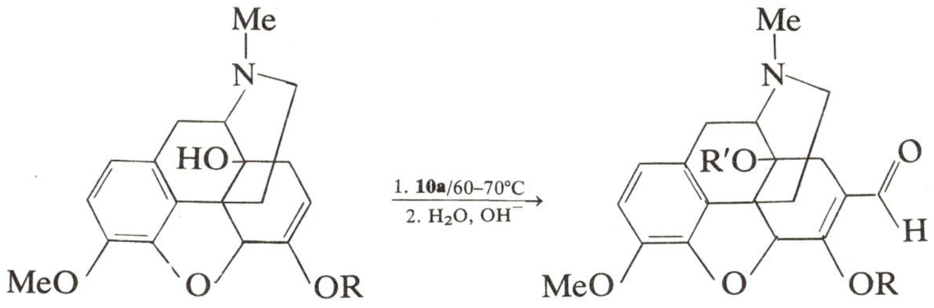

(**212**) R = Me, Et, n-Bu

(**213a**) R' = CHO
(**213b**) R' = H 50–64%

1-Methyl-4-alkoxy-2-pyridones (**214**) contain an enol ether moiety and may be converted by **10a** to the 3-carboxaldehydes **215** in 67–73% yields (120):

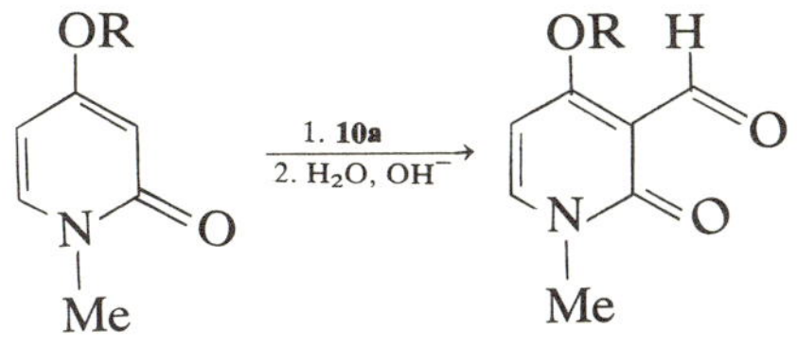

(**214**)

(**215**) 67–73%

3. Vinylogous Enol Ethers and Acetals

In the same way that acetaldehyde diethylacetal (**200a** = **216**, $n = 0$) forms ethyl vinyl ether (**201a** = **218**, $n = 0$) *in situ* by the action of **3a** or **10a**, either crotonaldehyde diethylacetal (**216a**) or 1,1,3-triethoxybutane (**217a**) can be used as a precursor of 1-ethoxy-1,3-butadiene (**218a**).

Compounds **216a**, **217a**, and **218a** react smoothly with **3a** in chloroform or 1,2-dichloroethane at 0°C to afford the iminium salt **219a**, hydrolysis of which with aqueous potassium carbonate gives the aminoaldehyde **220a** and the pentamethinium salt **221a**, derivatives of glutacondialdehyde (9). Compounds **220a** and **221a** are readily interconvertible in the usual manner.

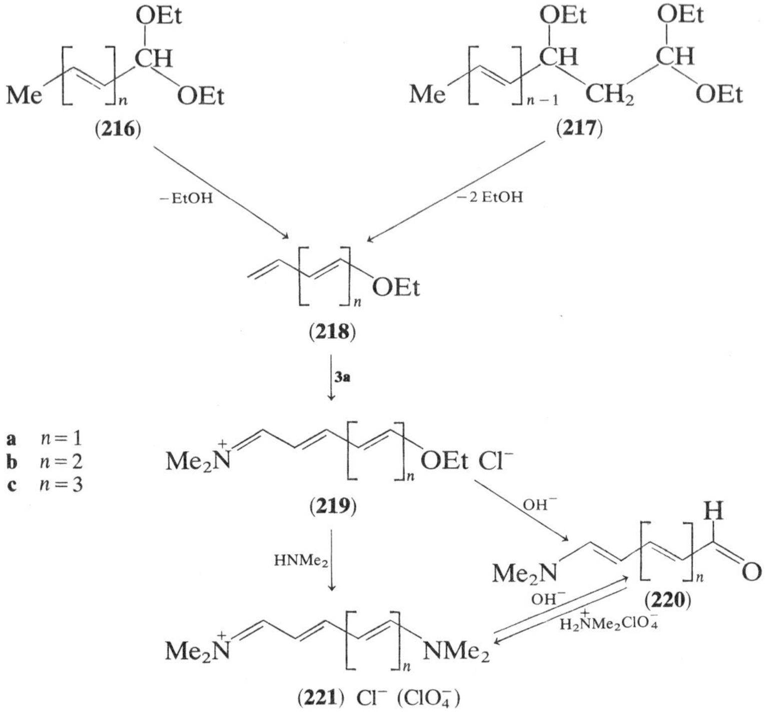

a $n = 1$
b $n = 2$
c $n = 3$

The next higher vinylogue, the heptamethinium perchlorate **221b**, may be prepared from sorbaldehyde diethylacetal (**216b**). Addition of 1,1,3-triethoxy-4-hexene (**217b**) to **3a** in DMF at −40°C, followed by the usual work-up, made possible the isolation of **221b** in an overall yield of 82% (24). 3-Ethoxy-4-hexenal and 1,1,3,5-tetraethoxyhexane were also formylated by **10a** in DMF, yielding respectively **220b** and **221b** (40–60%) (121).

Even the nonamethinium perchlorate **221c** has been synthesized. With **3a** in DMF at −20°C, 1,1,3-triethoxy-2,6-octadiene (**217c**) affords a 76.6% yield of the aforementioned compound; 2,4,6-octatrienal diethylacetal (**216c**), a 57% yield (24). 5-Ethoxy-2,6-octadienal and 1,1,3,5,7-pentaethoxyoctane reacts with **10a** in DMF to give **220c** and **221c** in fair yield (121).

Compounds of type **217** are easily prepared by the condensation of acetals **216** with ethyl vinyl ether in the presence of boron trifluoride etherate (122).

Substituted 5-dimethylamino-2,4-pentadienals **223** were obtained in 42–50% yield by the addition of the 1-ethoxydienes **222** (0.05 mole) to a mixture of **10a** (0.05 mole) and DMF (0.05 mole) in dichloroethane at 0–15°C ($\frac{1}{2}$ hr), followed by heating to 55–60°C ($\frac{1}{2}$ hr) and work-up with aqueous potassium carbonate (123):

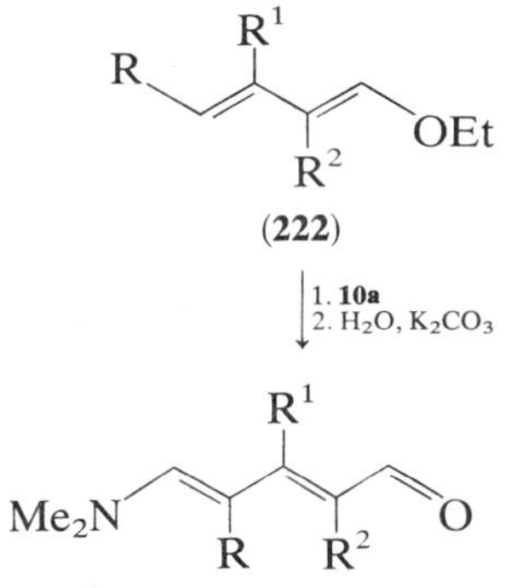

(**222**) → (**223**)

a R, R^1, R^2 = H
b R, R^1 = H; R^2 = Me
c R, R^2 = H; R^1 = Me
d R^1, R^2 = H; R = Me
e R = H; R^1, R^2 = Me

Of course the ethoxydienes **222a**, **222b**, and **222e** may be used to prepare the corresponding aminodienals **223a**, **223b**, and **223e** (42, 45, and 48% yield). However, unexpectedly, no aminodienal could be prepared from **222c**, and **223b** (50%) was obtained from **222d** (123).

3-Alkoxy- (usually 3-methoxy-) 3,5-dienes **224**, prepared from various

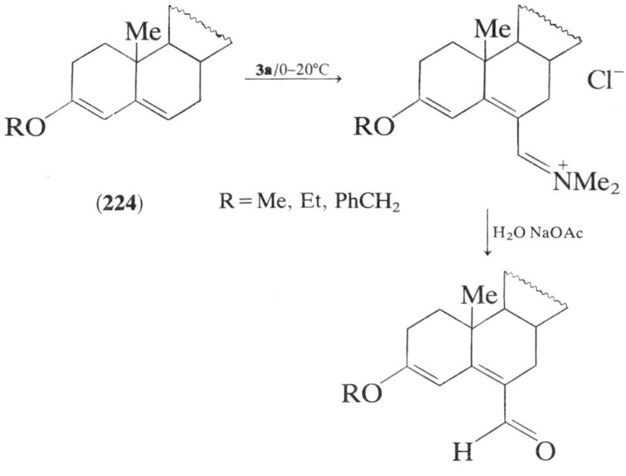

(**224**) R = Me, Et, $PhCH_2$

(**225**)

steroidal 4-en-3-ketones, were treated with **3a** (molar ratio 1:1–1:3) in dichloroethane at 0–20°C (1–2 hr) (114). The desired 3-alkoxy-6-formyl-3,5-diene steroids **225** (approaching 90% yields in the most favorable cases) were isolated after hydrolysis by aqueous sodium acetate. There is little effect from **3a** on most functional groups except for epoxy and some hydroxyl groups (formation of chlorohydrins and formate esters) (114).

4. Cyclization during Formylation of Acetals

A surprising cyclization occurs during the formylation of 2-ethyl-2-hexenaldiethylacetal (**226**) with an excess of **3a** (molar ratio 1:5) in dichloroethane at 50–60°C (3 hr). Decomposition of the reaction mixture by ice and potassium carbonate liberates 5-dimethylamino-4-ethoxy-1-ethyl-3-ethylidenecyclopentene (**229**) in 55% yield. Perhaps an intramolecular substitution of iminium salt **228** may have occurred (15).

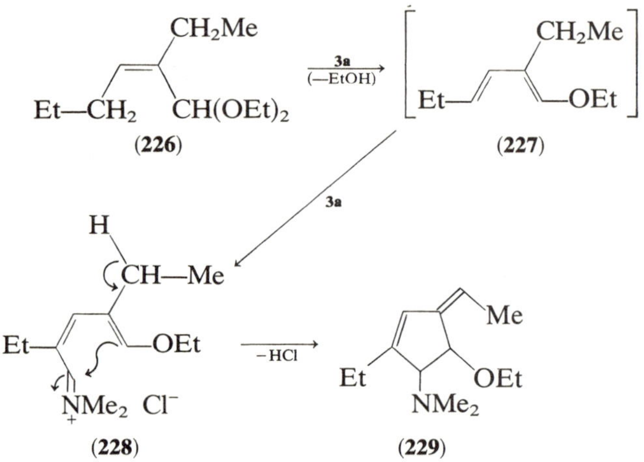

E. REACTIONS OF CHLOROMETHYLENIMINIUM SALTS WITH METHYL AND METHYLENE KETONES

The smooth reaction of methyl and methylene ketones with **3a** or **10a**, first discovered by Arnold and his coworkers in 1958 (8,124), normally yields a substituted 3-chloroacrolein. Thus an aldehyde group is introduced, and, in general, the oxygen function is replaced by chlorine. The earliest example of this type of alkylation, often called "chloroformylation," is the conversion of anthrone (**230**) to 10-chloro-9-anthracenecarboxaldehyde (**232**) by MFA–POCl$_3$ at 10°C (24 hr) in excellent yield (125). One observes an intense red-violet reaction due to the iminium ion **231**. The color fades upon decomposition with aqueous sodium acetate.

II. C—C BOND-FORMING REACTIONS

(230) → [MFA-POCl₃, 10°C, 24 hr.] → (231) → [H₂O, NaOAc] → (232) 98% + HN(Me)Ph

Since its discovery, the reaction of **3a** and **10a** with ketones has been studied extensively and has found broad application in synthesis. Moreover, reagents and starting materials are readily available and inexpensive, and the 3-chloroacroleins thus synthesized show extreme versatility. As derivatives of β-ketoaldehydes, they represent vinylogues of acid halides; therefore the chlorine atom may be readily displaced by nucleophiles. Closely related to β-chlorovinyl ketones, the 3-chloroacroleins also find application in numerous heterocyclic ring syntheses, for example, in pyrazoles, triazoles, pyridines, pyrimidines, pyrylium salts, and many other systems (126).

Mechanistic Considerations. Although "chloroformylation" has been widely used, no clear conception of its mechanistic course has been developed.

Arnold (8) suggested that the ketone enolizes before reaction with **3a** or **10a**; this can explain the fact that only sufficiently nucleophilic olefins are substituted by the reagent. In the chloroformylation of acetophenone (**233**) he has shown that α-chlorostyrene (**235**), once suspected of being an intermediate, does not react with **3a** or **10a**. In the chloroformylation of some steroidal dienones, chlorotrienes are obtained (127–130) in addition to the expected chloroformyltrienes. Since these chlorotrienes may also be converted by **10a** to the same chloroformyltrienes, Laurent and Wiechert (128) have concluded that the chlorotrienes may be intermediates in the chloroformylation of the dienones. However, since structurally similar steroidal trienes can be formylated as smoothly or even more readily than the mentioned chlorotrienes by **10a** (see Section II-B-2), such conclusions seem to be untenable.

Dimethyl aminovinyl ketones, expected by direct formylation of the enol form of the ketones, could indeed be isolated in certain cases (129), provided that the reaction was performed at low temperatures.

For all practical preparations, reagents **3a** and **10a** are always employed in excess, usually in molar ratios of 4:1 or 5:1 to ketone, with or without a solvent. In chloroformylations one frequently observes an induction period after which the markedly exothermic reaction sets in.

The following reaction scheme may be proposed as a mechanism of the chloroformylation of acetophenone (**233**). The reaction begins with an electrophilic attack of **3a** on the carbonyl oxygen of **233**, the only weakly basic center in the substrate, slowly forming **234** and a molecule of hydrogen chloride. Aryloxymethylenedimethyliminium chlorides, closely related to **234**, formed from arylchloroformates and DMF (131), are well known and possess considerable electrophilic potential. Further substitution of **234** by a second molecule of **3a** to give the dication **237** seems therefore very improbable. In a side reaction the weakly nucleophilic chloride ion may displace DMF from **234**, forming α-chlorostyrene (**235**). A key role in this reaction course may be played by the generation of hydrogen chloride in the conversion of **233** to **234**. It catalyzes the enolization equilibrium between **233** and **233'**. The latter undergoes a rapid substitution by **3a**, with further evolution of hydrogen chloride, to afford the β-dimethyl aminovinyl ketone (**236**) in its O-protonated form. Moreover, **234** may formylate **233'** to **236**. With the increasing concentration of hydrogen chloride, the rate of enolization should be enhanced, and consequently the reaction should also be accelerated in an autocatalytical mode. As shown for 3-dimethylaminoacrolein (**15**), free β-dimethyl aminovinyl ketones (**236**) also react with **3a** (Section II-A), yielding the labile bisiminium chloride **237**, which collapses very readily to the stable chloroacrolein iminium salt (**238**). Salt **238** can be isolated as its perchlorate in 98% yield (15). The usual work-up with aqueous sodium acetate then affords 3-chloroacrolein (**239**).

1. *Monosubstitutions of Ketones*

A considerable number of methyl ketones, as well as acyclic and cyclic methylene ketones **240**, have been converted via the iminium ions **241** to 3-chloroacroleins **242**. In practical preparations the ketone is added gradually with cooling to the reagent, which is always employed in excess in molar ratios ranging from 1:2.5 to 1:5. Although **10a** (with a small excess of DMF) can also be used without solvent, in order to keep the considerably exothermic and sometimes violent reaction under better control, especially in large-scale preparations, the use of a solvent, such as dichloroethane (DCE) or trichloroethylene (TCE), is advisable. After the initial reaction has ceased, the mixture is further heated for a period of time and then quenched by ice and neutralized by cold aqueous sodium

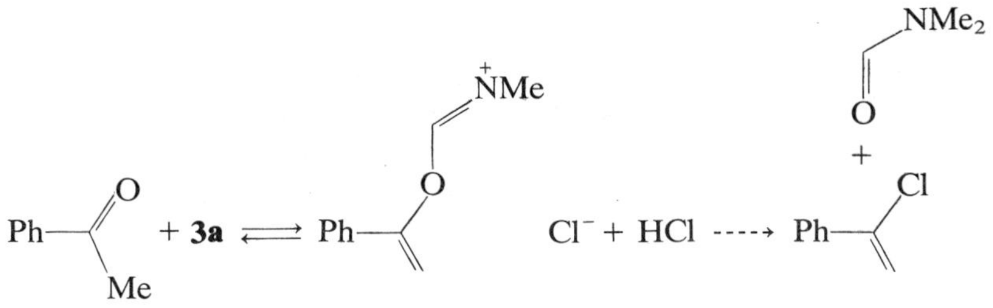

(233) Ph–C(=O)–Me + **3a** ⇌ (234) Ph–C(O–CH=N⁺Me)=CH₂

O=CH–NMe₂ + Ph–C(Cl)=CH₂ (235)
Cl⁻ + HCl ····→

(233) Ph–C(=O)–Me ⇌ (HCl) (233′) Ph–C(OH)=CH₂ → **3a** (also **234**)

Ph–C(OH)=CH–CH=N⁺Me₂ + HCl ⇌ Ph–C(=O)–CH=CH–NMe₂ (236) + 2HCl

↓ 3a

3a + 234 --?--→ (237) Ph–C(O–CH=N⁺Me₂)=CH–CH=N⁺Me₂ 2Cl⁻ —Δ→ (238) Ph–C(Cl)=CH–CH=N⁺Me₂ Cl⁻ + O=CH–NMe₂

(236) ←—H₂O low Temp.— (237)

(238) ↓

(239) Ph–C(Cl)=CH–CHO

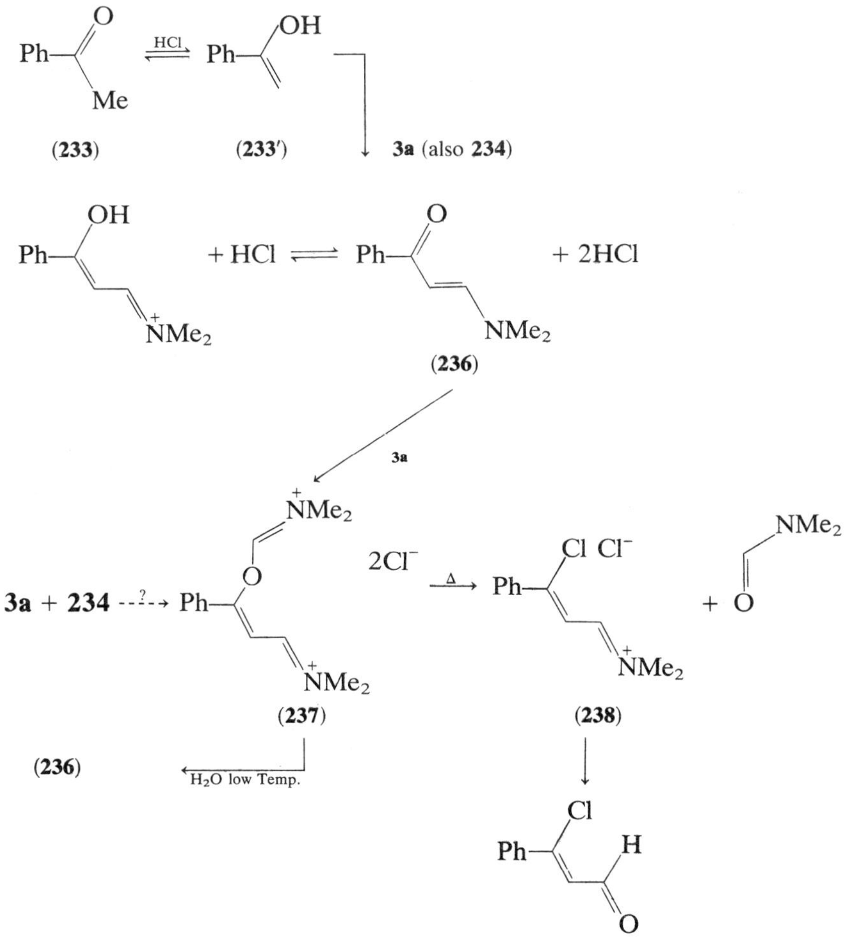

acetate or sodium carbonate. Basic conditions during the work-up must be carefully avoided. The low molecular weight aliphatic 3-chloroacroleins (**242**) are colorless, distillable liquids that are pungent lachrymators and are quite unstable. Within a few hours spontaneous, sometime violent, decomposition occurs, catalyzed by traces of alkali. The aromatic compounds show much greater stability.

TABLE X

3-Chloroacroleins **242**

242	R^1	R^2	Reaction time, hr	Reaction temp., °C	Agent	Solvent	Yield, %	Ref.
a	Me	H	0.5	20–40	**10a**	No	39	8
			3–4	60	**10a**	No	32	131
b	Me	Me ⎫	0.6	35–40	**10a**	No	67	8
	Et	H ⎭	3–4	60	**10a**	No	78	131
c	Et	Me	0.5	35–40	**10a**	No	77	8
d	CHMe$_2$	H	3–4	60	**10a**	No	14	131
e	t-Bu	H	3–4	50–55	**3a**	DCE	80	8
			1.5	70	**3a**	DCE	80	
			3–4	60	**10a**	No	51	131
f	Me	CHMe$_2$ ⎫	3–4	60	**10a**	No	20	131, 137
	Me$_2$CHCH$_2$	H ⎭						
g	Me	Et	3–4	60	**10a**	No	59	131
h	Me	n-Bu	3–4	60	**10a**	No	61	131
i	Me	—(CH$_2$)$_5$Me	3–4	60	**10a**	No	65	131
j	—(CH$_2$)$_2$—		3	50–55	**10a**	No	33	15
k	—(CH$_2$)$_3$—		0.5	35–40	**10a**	No	66	8
			3	55–60	**10a**	TCE	82	134
l	—(CH$_2$)$_4$—		0.3	30–40	**10a**	No	54	8
			3	55–60	**10a**	TCE	83	134
m	—(CH$_2$)$_5$—		0.7	35–40	**10a**	No	65	8
			2–3	55–60	**3a**	TCE	88	134
n	—(CH$_2$)$_6$—		0.8	35–40	**10a**	No	63	8
			3	55–60	**10a**	TCE	77	134
o	Ph	H	2	40–45	**10a**	No	47	8
			3	60–70	**3a**	DCE	98	15
p	4—Cl—C$_6$H$_4$	H	3	40	**10a**	TCE	30	136
q	4—Br—C$_6$H$_4$	H	3	40	**10a**	TCE	24	136
r	4—Ph—C$_6$H$_4$	H	3	60	**10a**	TCE	36	136
s	Ph	Ph	3	80	**10a**	TCE	60	136, 138
t	Ph	Me	3	40–45	**10a**	No	91	8
			5	50–55	**3a**	DCE	60	8
u	4—MeO—C$_6$H$_4$	H	3	40	**10a**	TCE	24	136
v	4—NO$_2$—C$_6$H$_4$	H	3–4	60	**10a**	No	71	131
w	3—NO$_2$—C$_6$H$_4$	Me	3–4	60	**10a**	No	43	131
x	3,4—(MeO)$_2$C$_6$H$_3$	H	3–4	60	**10a**	No	45	131, 133
y	3,4—(MeO)$_2$C$_6$H$_3$	Me	3–4	60	**10a**	No	56	131

(**240**) (**241**) (**242**)

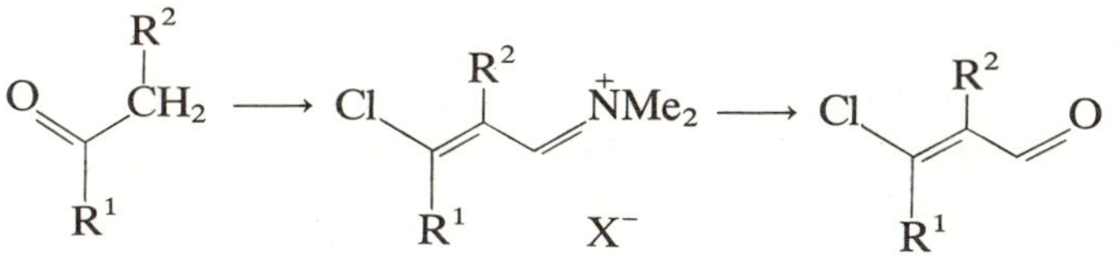

It should be pointed out that, in ketones with a second methylene group adjacent to the carbonyl group, not only may further substitution by **3a** or **10a** occur under stronger reaction conditions, but also two monosubstituted structural isomers (**242**) may be formed if the two methylene groups are not equivalent. Usually the formation of a 2-alkyl **242** by attack on the methylene group predominates. However, 4-methyl-2-pentanone (**240f**) gives, by reason of the steric bulk of the isopropyl group, a mixture in which the 3-chloro-5-methyl-2-hexenal is the main product (137). Moreover, cis-trans stereoisomerism also exists in compounds of type **242** (137,138,152).

(**240**) (**242b**) 78% yield E+Z-form (**242b′**) traces

(**240f**) (**242f**) E-form: 10% Z-form: 1% (**242f′**) E-form: 17% Z-form: 72%

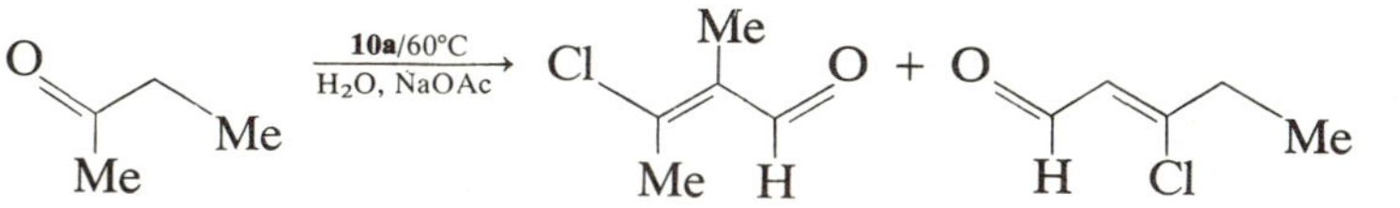

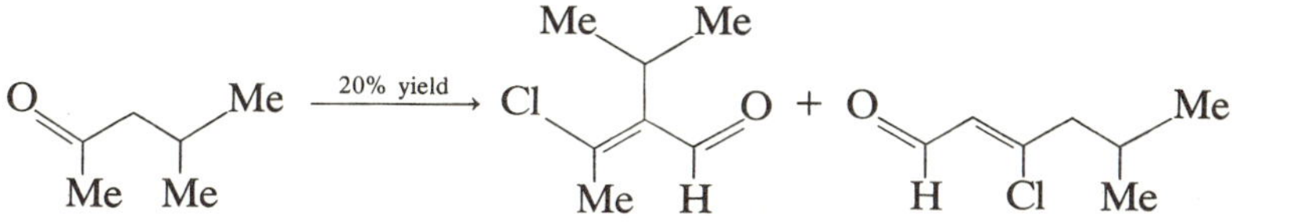

Structural assignments have been based on ^{1}H NMR spectra, including solvent shift studies (137,152). In some instances the 2,4-dinitrophenylhydrazones of the Z and E forms could be chromatographically separated (152). The two isomeric 3-chloro-2,3-diphenylacroleins (**242s**) may be differentiated by their solubilities and melting points (138). During chloroformylation of 1-acetonaphthone (**243a**), migration of the acetyl group takes place and one obtains the same 3-chloro-3-(naphthyl-2)-acrolein (**244b**) as is obtained from 2-acetonaphthone (**243b**) (131).

TABLE XI

Stereoisomerism of 3-Chloroacroleins

242	R^1	R^2	Yield, %	Z:E
b	Me	Me	77	31:69
c	Et	Me	58	25:75
	Me	CO_2Et	44	59:41
	Me	CO_2Me	50	100:00
s	Ph	Ph	60	17:83
t	Ph	Me	92	7:93
	Me	Ph	56	58:42

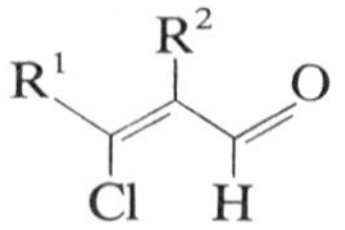

(242) Z-form

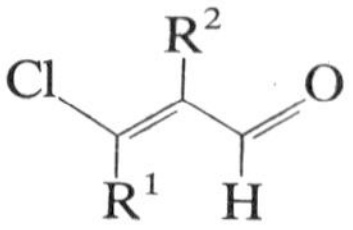

(242) E-form

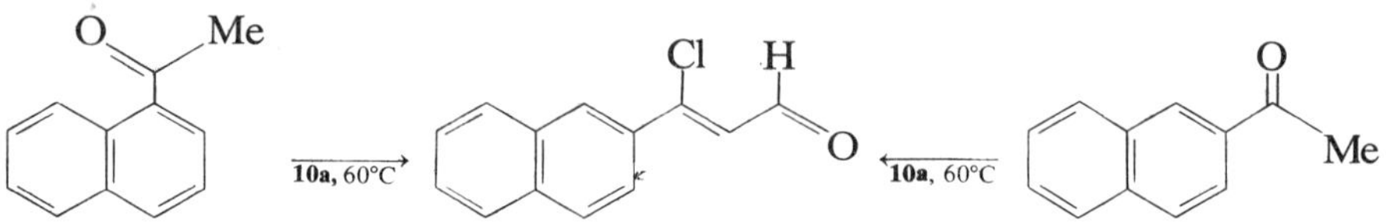

(243a) → [10a, 60°C] → (244b) 56% ← [10a, 60°C] ← (243b)

Except for 2-acetylthiophene (**243d**), conversion of the aromatic ketones listed in Table XII proceeds, depending on the reaction conditions, in satisfactory yield to the corresponding chloroacroleins. The thiophene ring shows considerable sensitivity toward the acidic reaction conditions, and much polymerization results. Isolation of the intermediate iminium salts of type **241** as their perchlorates is successfully effected by quenching the reaction mixture with ethanol and adding ethanolic perchloric acid (to avoid hydrolysis).

The considerable synthetic value of the choroacroleins has already been mentioned. However, one important conversion will be cited here: the fragmentation reaction. Chloroacroleins (and their iminium salts), provided that they are unsubstituted in position 2 (thus derived from methyl ketones) and possess enolizable carbonyl groups, undergo a smooth fragmentation by alkali to the corresponding ethynyl compounds **245** in good to excellent yields (131):

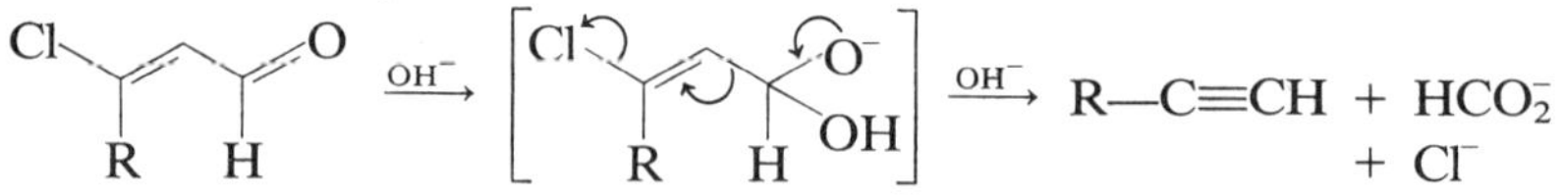

(242, 244) $\xrightarrow{OH^-}$ [intermediate] $\xrightarrow{OH^-}$ R—C≡CH + HCO_2^- + Cl^- (245)

TABLE XII

3-Chloroacroleins 244

244	Starting ketone 243	Product (yield)	Reaction time, hr	Reaction temp., °C	Agent	Solvent	Ref.
c	3-Acetyl-acenaphthene	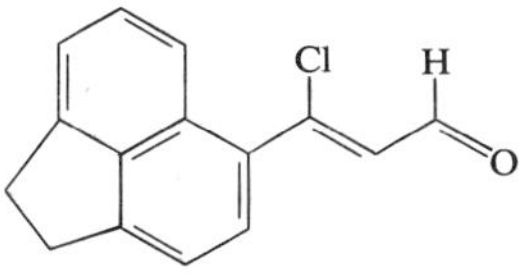 (71%)			10a	DMF	140
d	2-Acetyl-thiophene	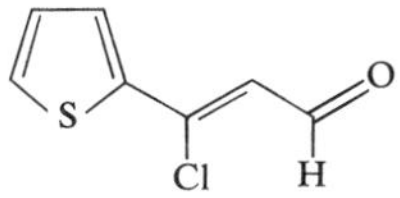 (11%)	3–4	60	10a	No	131
e	4-Acetyl-antipyrine	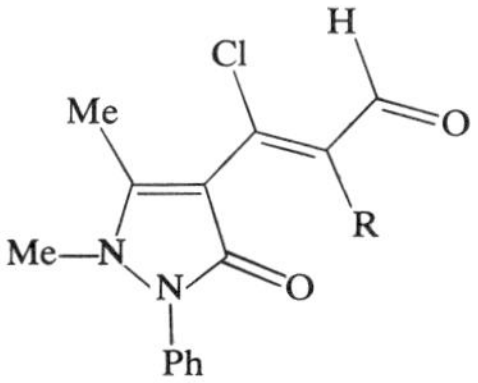 e R=H 73%	8	20	10a	DMF	131
f	4-Propionyl-antipyrine	f R=Me 75%	2	50	10a	DMF	131
g	1-Acetyl-1'-chloroferrocene	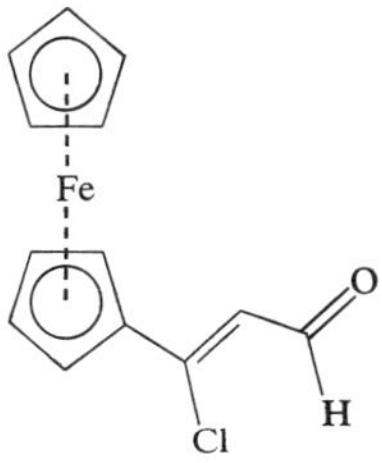			10a	DMF	141
h	1-Acenaphthenone	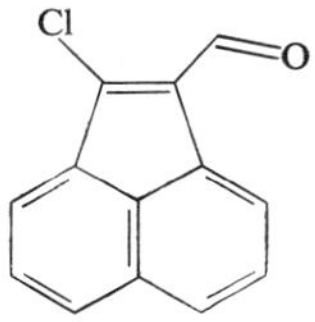 (80%)	3	50	10a	TCE	136

TABLE XII (Continued)

244	Starting ketone 243	Product (yield)		Reaction time, hr	Reaction temp., °C	Agent	Solvent	Ref.
i	1-Tetralone			3	55–60	10a	TCE	134
j	4-Chromanone			3	35	10a	TCE	136
		i X=CH$_2$	77%					
		j X=O	36%					
k	4-Thiochromanone	k X=S	68%	22	22	10a	DMF	136
l	Benzsuberone			3	55–60	10a	TCE	134
m	5-Tetrahydro-benzoxepinone			38	22	10a	DMF	136
		l X=CH$_2$	75%					
		m X=O	70%					
n	5-Tetrahydro-benzthiepinone	n X=S	54%	42	22	10a	DMF	136

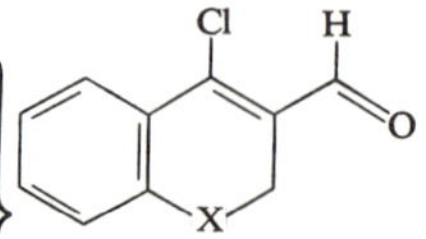

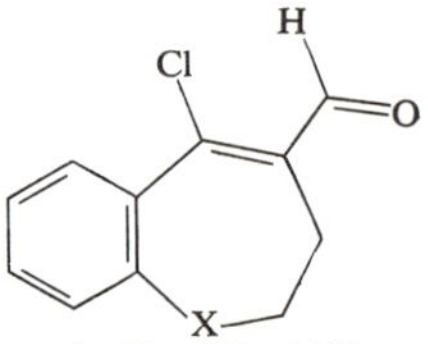

The reaction has been applied to the following compounds: **242e**, **242n**, **242o–242r**, **242v**, **242x**, **244b–244e**, and **244g**. It is preferently effected by dripping a dioxane solution of the chloroacrolein into a boiling mixture of sodium hydroxide in aqueous dioxane (131,140,141). Starting from acetylaromatics, readily available by Friedel-Crafts acylation, this method represents a valuable new approach to substituted acetylenes.

Ziegenbein (134) had previously observed the reverse transformation: the addition of **10a** to phenylacetylene (**245o**), forming the corresponding chlorocinnamaldehyde (**242o**), which was converted by heating with formamide to 4-phenylpyrimidine:

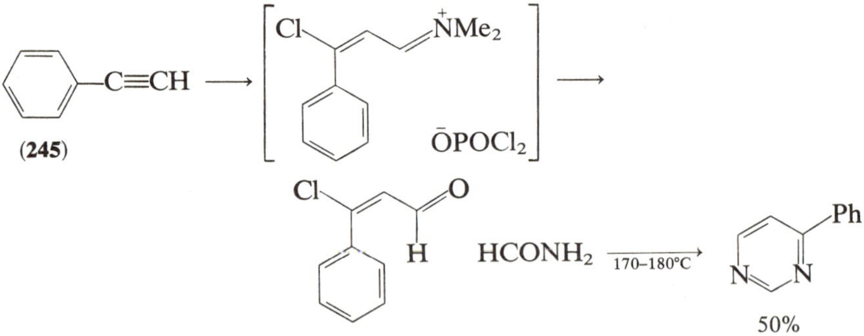

Obviously, functionalized ketones (e.g., **246**) can also be converted to chloroacroleins (**247**) in fair yields. That attack occurs exclusively on the methylene group was evident from the structure of pyrazole **248a**, formed from **247a** and phenylhydrazine (135):

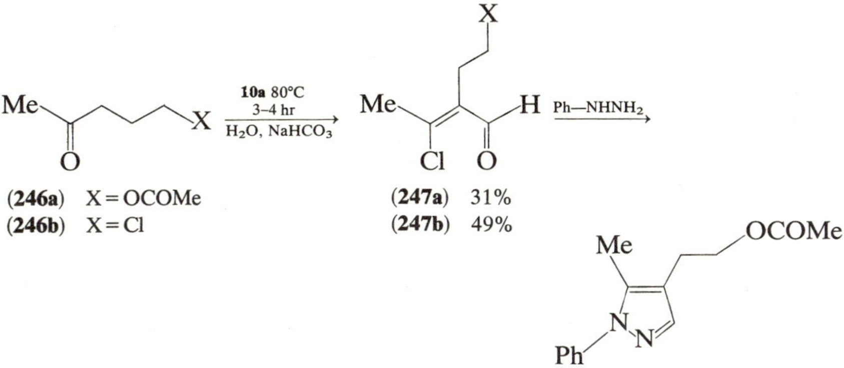

(**246a**) X = OCOMe
(**246b**) X = Cl

(**247a**) 31%
(**247b**) 49%

(**248a**)

The classical Vilsmeier-Haack-reagent, MFA–POCl$_3$, has been used in a chloroformylation of 6,7-dihydro-5H-benzo(b)thiophene-4-one (**249**). The four formylated products **250–253** were identified after work-up (139).

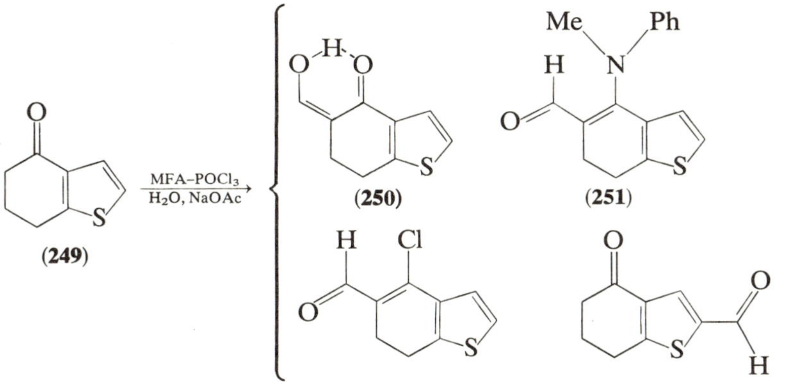

(**249**)

(**250**)

(**251**)

(**252**)

(**253**)

Chloroformylation of Steroidal Ketones. Reaction of **10a** with a 17-ketosteroid, the 3β-acetoxyandrost-5-en-17-one (**254**), affords 3β-acetoxy-17-chloro-16-formylandrosta-5,16-diene (**255a**) (129,146), with

3β-acetoxy-17-chloroandrosta-5,16-diene (**255b**) as a by-product (146):

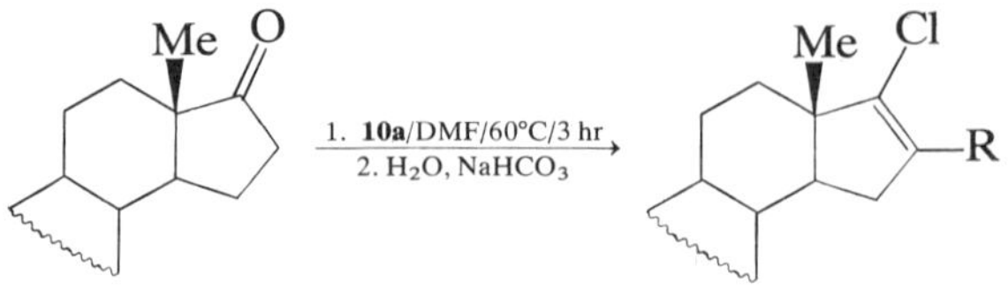

(**254**) → 1. 10a/DMF/60°C/3 hr; 2. H₂O, NaHCO₃

(**255a**) R=CHO 69%
(**255b**) R=H

The carbonyl of 3-ketosteroids is flanked by two methylene groups. Which of the two undergoes substitution by **10a** seems to depend solely on the stereochemistry of the A-B ring junction. 5α-Steroids **256**, for example, 17:-acetoxy-5α-androstan-3-one (146), 17β-acetoxy-17α-methyl-5α-androstan-3-one (129), and 5α-pregnan-3,20-dione, (147), give the corresponding 3-chloro-2-formyl compounds (**257**) in 22–27% yield. Remarkably, the acetyl group in 5α-pregnan-3,20-dione was unaffected by the reaction. The unusual use of acetyl chloride as solvent in the chloroformylation reaction has been described by some French workers (147,148). Under mild conditions (20°C), 17β-acetoxy-2-dimethylaminomethylene-17α-methyl-5α-androstan-3-one (**258**) could be obtained from the parent ketone (129):

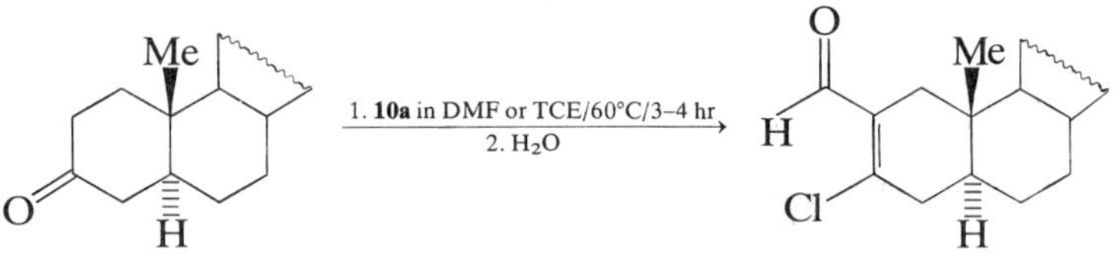

(**256**) → 1. 10a in DMF or TCE/60°C/3–4 hr; 2. H₂O → (**257**) 22–27% (129, 146, 147)

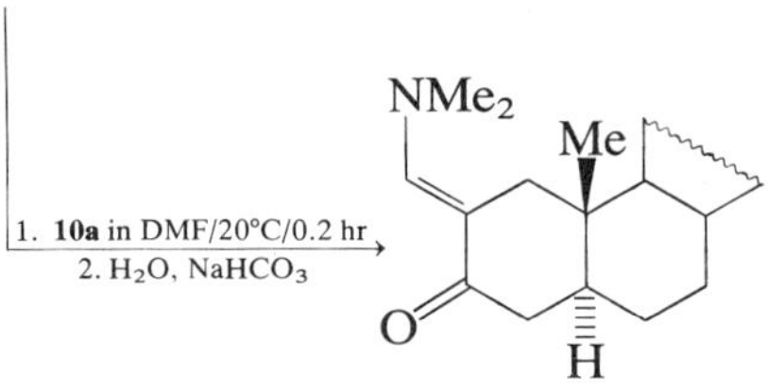

1. 10a in DMF/20°C/0.2 hr; 2. H₂O, NaHCO₃ → (**258**) 34%

Likewise, 17β-acetoxy-4,4-dimethylandrost-5-en-3-one yields 17β-acetoxy-3-chloro-2-formyl-4,4-dimethylandrosta-2,5-diene in 62% yield upon treatment with **10a** in DMF for 4 hr at 50–60°C, followed by work-up of the mixture with aqueous sodium hydrogen carbonate (129).

Conversely, 17β-acetoxy-5β-androstan-3-one (**259**) was converted to

17β-acetoxy-3-chloro-4-formyl-5β-androst-3-ene (**260**) in 20% yield (146,148):

(**259**) → 1. **10a** in TCE or AcCl/3 hr/70°C, 2. H₂O, NaOAc → (**260**) 20%

2. Reactions of Chloromethyleniminium Salts with Polyenals and Polyenones

A valuable extension of the scope of chloroformylations has been realized by the use of vinylogues of methyl and methylene ketones. Like acetone, acetaldehyde should react with **10a** to yield 3-chloroallylidenedimethyliminium chloride (**16**) ($n = 0$). However, this conversion does not occur. Under the reaction conditions an acid-catalyzed polycondensation takes place. The vinylogous acetaldehydes—crotonaldehyde (**261a**), sorbicaldehyde (**261b**), octa-2,4,6-trienal (**261c**), and deca-2,4,6,8-tetraenal (**261d**)—show a decreasing tendency toward self-condensation with increasing chain length. Indeed, chloroformylations of **261** have been successfully performed (144). To improve yields and avoid side reactions, Nikolajewski et al. (144) have found it desirable first to deactivate the **10a** by addition of 1 equivalent of methanol to the DMF solution, and then to run the reaction at 70–80°C for 3–4 hr. Compound **16a** hydrolyzes too quickly in aqueous media to be isolated and therefore was immediately transformed to **221a** (28%); **16b** could be obtained only in 18% yield as its perchlorate salt, whereas direct conversion to **221b** gives a 62% yield. The higher chloroiminium salts **16c** (74%) and **16d** (58%) are quite stable, in contrast to their derived cyanine salts **221c** and **221d**.

Crotonophenone (**262**), the benzylideneacetones **266a–c**, and cinnamylideneacetone (**266d**) are very smoothly substituted by either **3a** or **10a** in DMF. Compound **262** gives the 5-chloro-5-phenylpenta-2,4-dienylideniminium perchlorate (65%). This yields, by displacement of chlorine with dimethylamine, the pentamethinium perchlorate **264**, which was transformed into 2-phenylpyridine (**265**) (24).

From the reaction mixtures of **266a–d**, the colored, crystalline salts (**267a–d**) separate after a short time (143). Salt **267a** has also been isolated as its perchlorate (85%) and was then converted into the corresponding 2-styrylpentamethinium salt (145). Hydrolysis with aqueous sodium acetate transforms **267a–d** into the 3-chloro-3-styrylacroleins

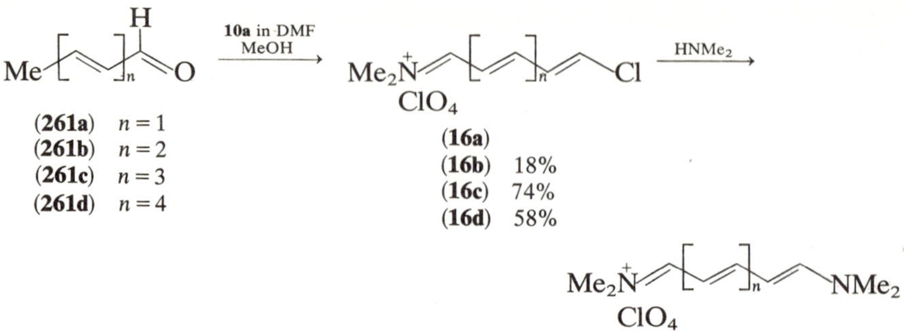

$$\text{Me}\underbrace{}_{n}\overset{\text{H}}{\underset{\text{O}}{\diagdown}} \xrightarrow{\text{10a in DMF}\atop\text{MeOH}} \text{Me}_2\overset{+}{\text{N}}\underbrace{}_{n}\diagdown\text{Cl}\quad \xrightarrow{\text{HNMe}_2}$$

(261a) n = 1
(261b) n = 2
(261c) n = 3
(261d) n = 4

ClO₄

(16a)
(16b) 18%
(16c) 74%
(16d) 58%

$$\text{Me}_2\overset{+}{\text{N}}\underbrace{}_{n}\diagdown\text{NMe}_2$$

ClO₄

(221a) 28%
(221b) 62%
(221c) 80%
(221d) 58%

268a–c and 3-chloro-3-(4-phenyl-1,3-butadienyl)acrolein (**268d**). Boiling these four compounds with sodium hydroxide–dioxane effected fragmentation to the 1-phenyl-1-buten-3-ynes (**269a–c**) and 1-phenyl-1,3-hexadien-5-yne (**269d**) in excellent yield (143).

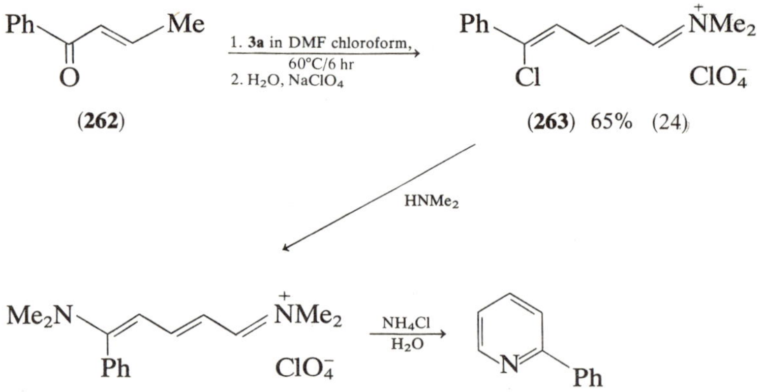

Ph–CH=CH–C(O)–Me (262)
$\xrightarrow[\text{2. H}_2\text{O, NaClO}_4]{\text{1. 3a in DMF chloroform,}\atop 60°\text{C/6 hr}}$
Ph–C(Cl)=CH–CH=CH–$\overset{+}{\text{N}}$Me₂ ClO₄⁻ (263) 65% (24)

↓ HNMe₂

Me₂N–C(Ph)=CH–CH=CH–$\overset{+}{\text{N}}$Me₂ ClO₄⁻ (264) 44%

$\xrightarrow{\text{NH}_4\text{Cl}\atop\text{H}_2\text{O}}$ 2-phenylpyridine (265) 72%

Chloroformylation of Steroidal Ene- and Dieneones. Three compounds, 17β-acetoxy-3-chloro-19-norandrosta-3,5-diene (**271a**), 17β-acetoxy-3-chloro-4-formyl-19-norandrosta-3,5-diene (**271b**), and 17β-acetoxy-3-chloro-6-formyl-19-norandrosta-3,5-diene (**271c**), were isolated from a reaction mixture of 19-nortestosterone acetate (**270**) and **10a** in TCE at 60°C after 3 hr and work-up with aqueous sodium acetate (149). This indicates that in a conjugated system electrophilic attack does not take place exclusively at the terminal carbon. A double substitution as in the related dienamine **190** (Section II-C-2) was not observed. The

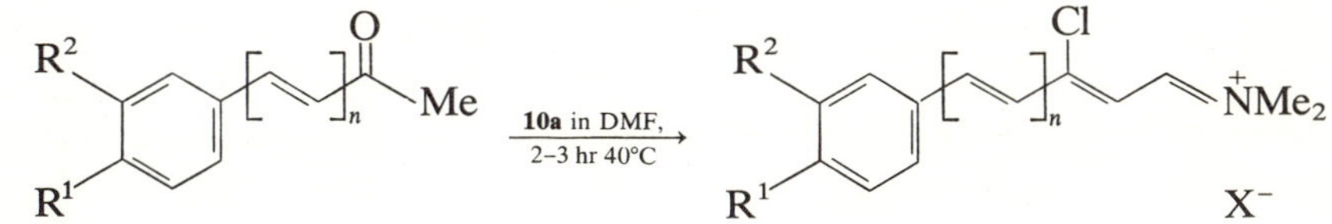

II. C—C BOND-FORMING REACTIONS

(266a) n = 1 R¹, R² = H
(266b) n = 1 R¹ = OMe, R² = H
(266c) n = 1 R¹, R² = OMe
(266d) n = 2 R¹, R² = H

10a in DMF,
2–3 hr 40°C

(267a) X = PO$_2$Cl$_2$ 63%
(267b) X = Cl 75%
(267c) X = Cl 81%
(267d) X = PO$_2$Cl$_2$ 64%

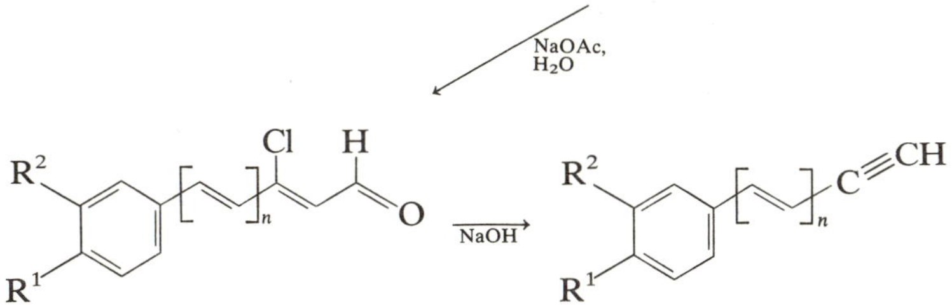

NaOAc, H$_2$O

(268a) 95%
(268b) 90%
(268c) 90%
(268d) 92%

NaOH →

(269a) 64%
(269b) 81%
(269c) 98%
(269d) 88%

structure of **271b** was determined from spectral evidence and by preparing the pyrazole **272** by treatment of **271b** with hydrazine in acetic acid at 110°C (149).

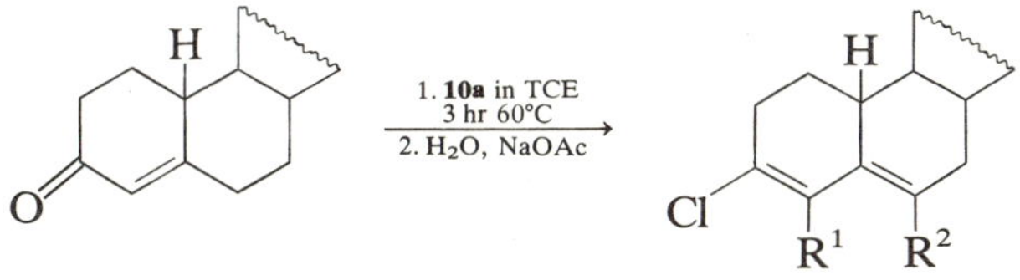

(270)

1. 10a in TCE 3 hr 60°C
2. H$_2$O, NaOAc

(271a) R¹, R² = H
(271b) R¹ = CHO, R² = H
(271c) R¹ = H, R² = CHO

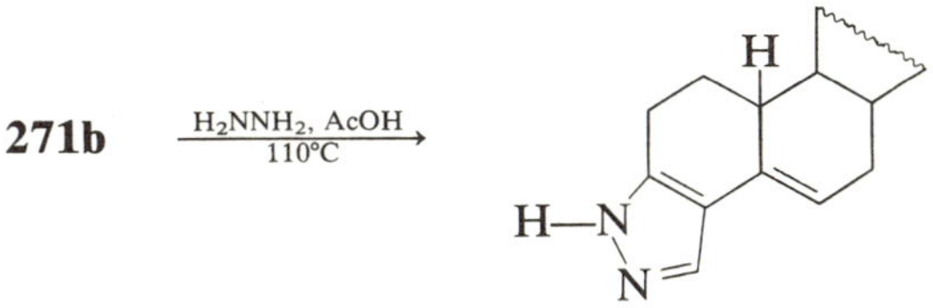

271b $\xrightarrow{\text{H}_2\text{NNH}_2,\ \text{AcOH}}_{110°C}$

(272)

3β-Acetoxy-5-en-7-ketosteroids **273** have been used as precursors of the 3,5-dien-7-ones **274**. Under the action of **10a** in DMF, **273** eliminates acetic acid to give intermediately **274**, which then undergoes the expected substitution by **10a** to yield the 7-chloro-2,4,6-trienes **275a** and the 7-chloro-2-formyl-2,4,6-trienes **275b** (127):

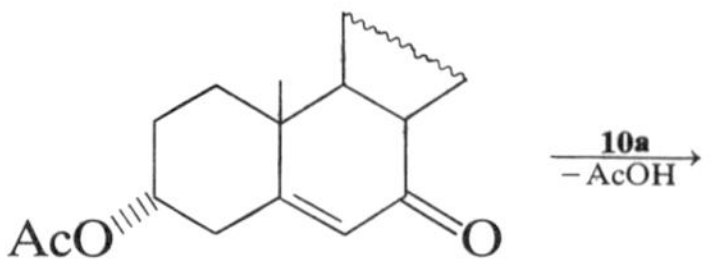

(**273**)

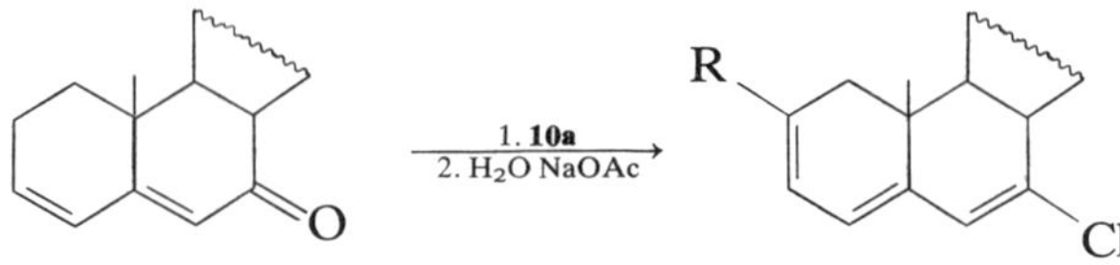

(**274**)

(**275a**) R=H
(**275b**) R=CHO

In an analogous manner a series of steroidal 4,6-dien-3-ones **276** reacts with **10a** in DMF at 70°C (2 hr) to yield a mixture of 3-chloro-2-formyl-2,4,6-trienes **277b**, 3-chloro-2,4,6-trienes **277a**, and 3-chloro-3,5,7-trienes **278** (128). A 3-chloro-3,5,7-triene (**278**) has also been isolated in the chloroformylation reaction of 17β-propionoxyandrosta-4,6-dien-3-one (130). The convertibility of chlorotrienes **275a** and **277a** to their formyl compounds **275b** and **277b** led to the conclusion that these are intermediates in the reaction (127,128). As a proof of structure, two examples of **277b** were transformed to their pyrazole derivatives **279** (128).

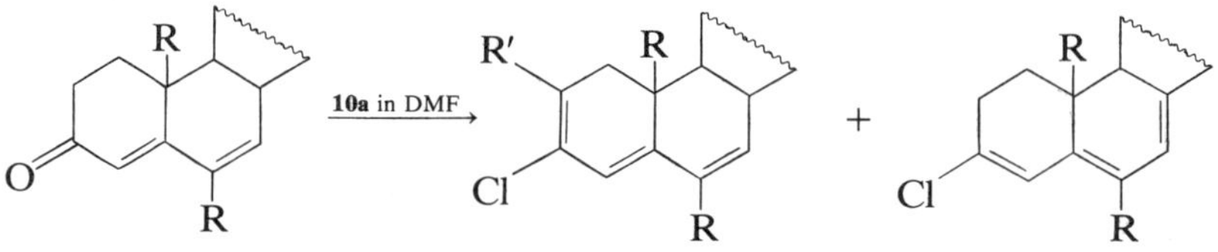

(**276**) R=H, Me (**277a**) R′=H (**278**)
(**277b**) R′=CHO

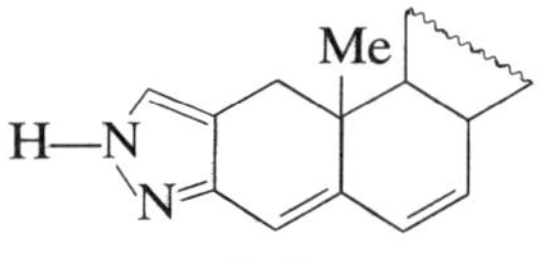

(**279**)

3. *Polysubstitutions of Ketones by Chloromethyleniminium Salts*

Aryl methyl ketones generally undergo only monosubstitution, except when strong electron-donating substituents are present in the benzene ring. *p*-Dimethylaminoacetophenone first reacts with **3a** to give **181**; subsequent formylation affords **182** and **183** (21) (see Section II-C-1). Similarly, monoformylation of 3,4-dimethoxyacetophenone (**240x**) yields **241x** while diformylation affords derivatives of acylmalondialdehyde (**282**) (133).

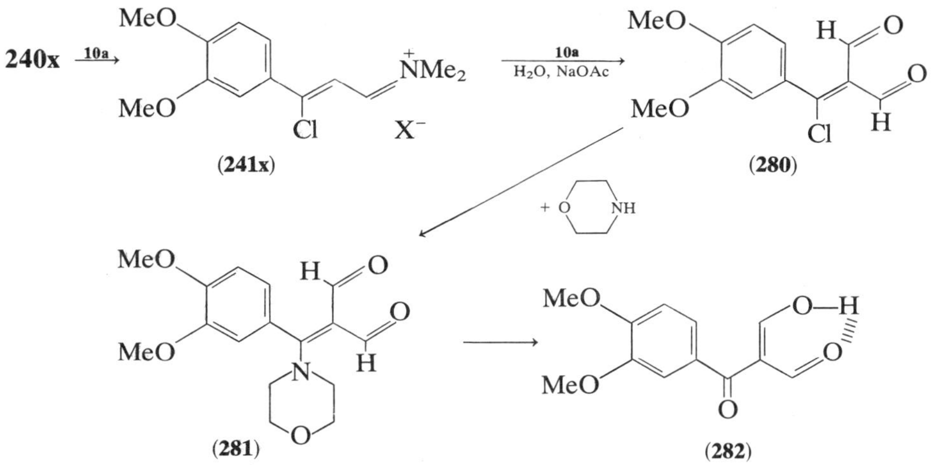

In the formylation of acetone (**240a**) and ethyl methyl ketone (**240b**) a much more complex, multistep reaction occurs (14) by the action of an excess of **3a** or **10a** (molar ratio 1:5) at 65°C. The moderate yield of 3-chlorocrotonaldehyde (**242a**), obtained by formylating acetone at low temperature, results from competing polysubstitution. The course of this reaction, reminiscent of the formylation of isobutene **99–100** (see Section II-B-2), may be initiated by proton abstraction of the iminium salt **241** to give the corresponding dienamine **241′**. The latter undergoes a rapid substitution by **3a** or **10a**, passing through a 3-chloropentamethinium intermediate (**283**), to give the bisiminium salt **284**, the threefold formylation product (see also Section II-C-2, **188–185**).

Acetone (**240a**) gives **284a** in 90% yield as its perchlorate salt (**3a**, 3 hr, 65°C or **10a**, 6 hr, 50–55°C; quenching with ice, NaClO$_4$). The same reaction with ethyl methyl ketone (**240b**) gives the perchlorate **285b** in 74% yield by hydrolysis of an aldiminium group in the work-up (12).

A series of reactions has also been performed with **284a** and **285b** (14). The chlorine in **284a** and **285b** is readily displaced by nucleophiles and thus triggers a remarkable fragmentation reaction that yields, after hydrolysis with aqueous potassium carbonate, 2-ethynyl-3-dimethylamino-acroleins **287a** and **287b**, as well as the other compounds depicted in the scheme:

The structures of **284a** and its derivative compounds **286** and **288a** were determined by ring closure reactions to the corresponding pyridine derivatives **291** and **292**:

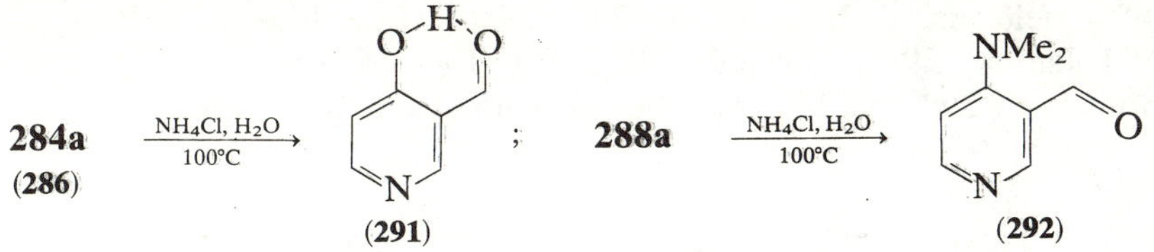

284a $\xrightarrow{\text{NH}_4\text{Cl, H}_2\text{O}}_{100°\text{C}}$ (291) ;   **288a** $\xrightarrow{\text{NH}_4\text{Cl, H}_2\text{O}}_{100°\text{C}}$ (292)
(286)

Double substitution in diethyl ketone (**240c**), cyclopentanone (**240k**), cyclohexanone (**240l**), and even isopropyl methyl ketone (**240d**) can be best achieved by using a large excess of **3a** at high temperature. Cyclobutanone is, however, quite unreactive and gives under vigorous conditions only the monoformylated 2-chlorocyclobutenecarboxaldehyde. The reaction path resembles the formylation course of acetone and terminates at the stage of the 3-chloropentamethinium salt (**293**). A third formylation is impossible since all available positions are occupied by alkyl groups. Starting from cyclopentanone (**240k**), the perchlorate **293k** was isolated in nearly quantitative yield after quenching the reaction mixture by ice. The homologue **293l** undergoes a partial hydrolysis under the same conditions of work-up to give **294l** in 69% yield. Work-up of the cyclopentanone reaction mixture with potassium carbonate affords **295k**, which might be better prepared by treatment of **294k** with dimethylamine. Treating **295k** with hot aqueous potassium carbonate eliminates one formyl group and forms a mixture of the two isomeric enaminocarbonyl compounds **296** and **297** (15):

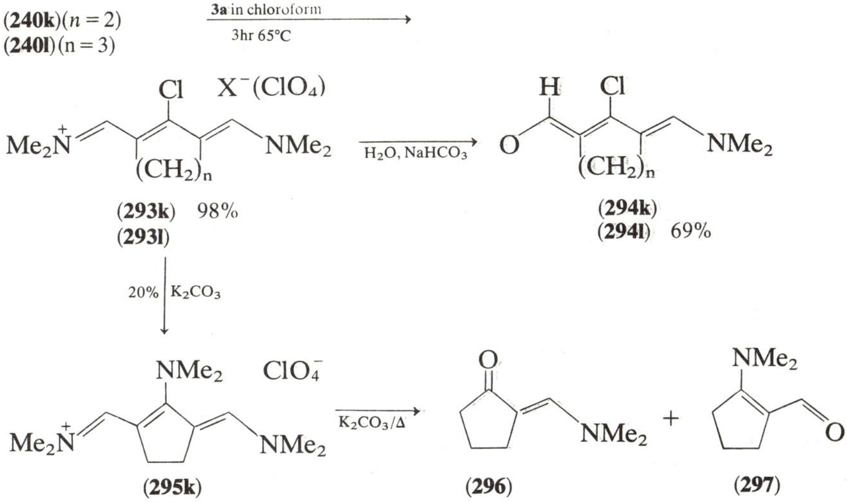

(**240k**)(n = 2)
(**240l**)(n = 3) $\xrightarrow[\text{3hr 65°C}]{\text{3a in chloroform}}$

(**293k**) 98%
(**293l**)
$\xrightarrow{\text{H}_2\text{O, NaHCO}_3}$
(**294k**)
(**294l**) 69%

20% | K$_2$CO$_3$

(**295k**) $\xrightarrow{\text{K}_2\text{CO}_3/\Delta}$ (**296**) + (**297**)

The dienamine **241′d** arising during the chloroformylation of isopropyl methyl ketone (**240d**) has only the β position available for further substitution. Thus one obtains the trimethinium perchlorate salt (**298**) in 16% yield using **3a** (15), or 24% yield with **10a** (21). With dimethylamine, **298** is converted to **299** (50%); sodium hydroxide transforms both **298** and **299** to isobutyrylmalondialdehyde (**300**) (21):

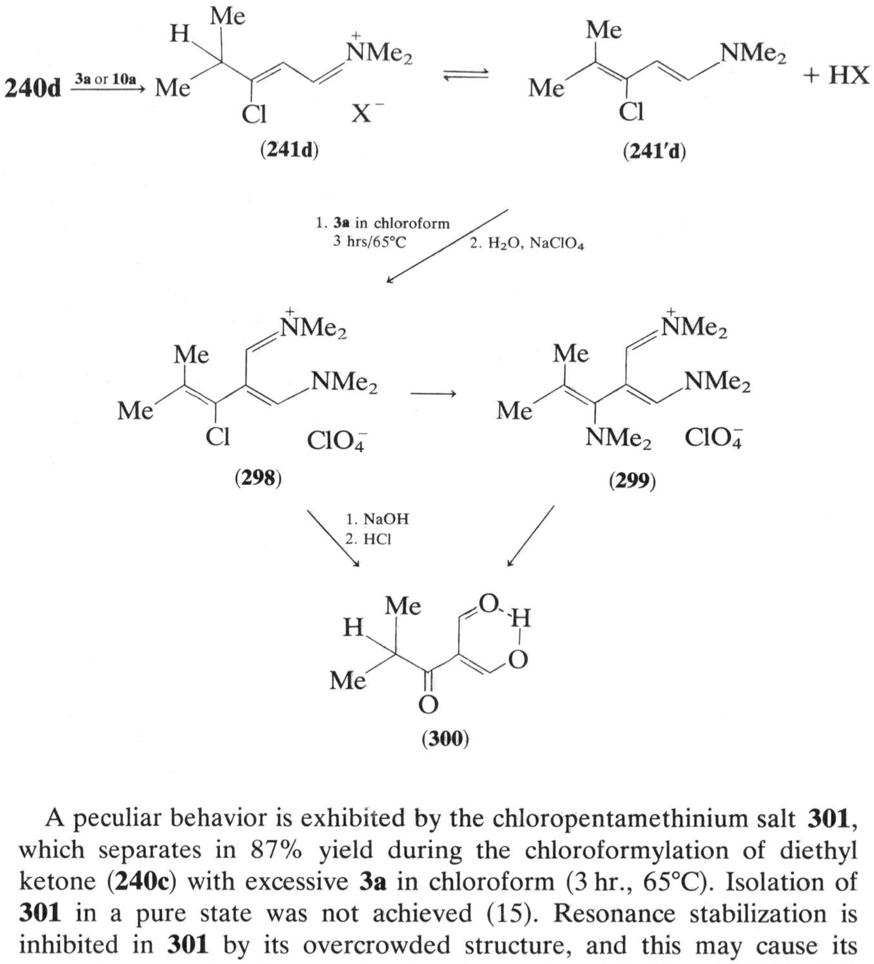

A peculiar behavior is exhibited by the chloropentamethinium salt **301**, which separates in 87% yield during the chloroformylation of diethyl ketone (**240c**) with excessive **3a** in chloroform (3 hr., 65°C). Isolation of **301** in a pure state was not achieved (15). Resonance stabilization is inhibited in **301** by its overcrowded structure, and this may cause its unusual properties. On heating in nitromethane (3.5 hr, 100°C), **301** cyclizes to the dihydrochloride of 2-chloro-1-methyl-3-methylene-4,5-bis(dimethylamino)cyclopentene (**303**) in 64% yield; this resembles reaction sequence **226–229** (see Section II-D-3). The same base (**303**) has been also isolated by treating **301** with aqueous sodium acetate or potassium carbonate. In acidic work-up, the hydrated derivative **304** is formed as a by-product (15).

Treatment of **301** with silver oxide in acetonitrile below 60°C yields 1,5-bis(dimethylamino)-2,4-dimethylpenta-1,4-dien-3-one (**302**); this may serve as evidence for structure **301** (15):

(**241c**) (**241′c**) (**301**) (**301**) (**303**) (**302**) (**304**)

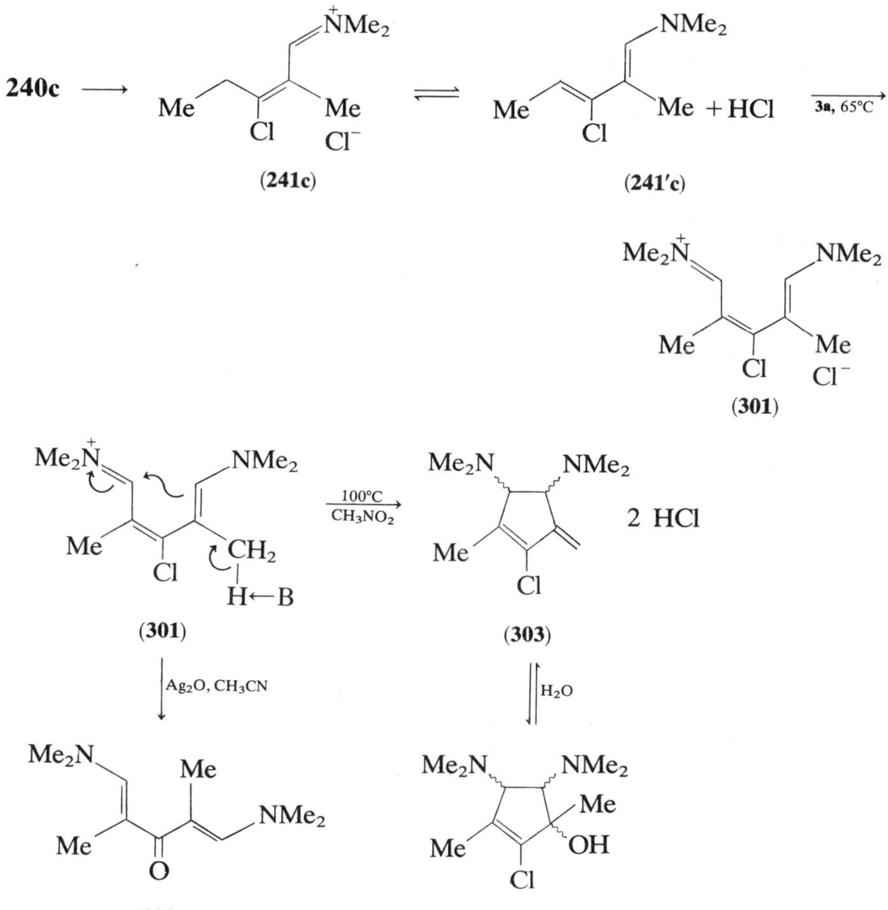

Disubstitution of Steroidal Ketones by Chloromethyleniminium Salts. In only one instance has a double formylation of a simple steroidal ketone been described: 17β-acetoxy-17α-methyl-5α-androstan-3-one (**305**) gives, on heating with **10a** in DMF (18 hr, 50–55°C) and subsequent hydrolysis of the mixture by sodium acetate in water, the diformyl compound **306** in moderate yield (129):

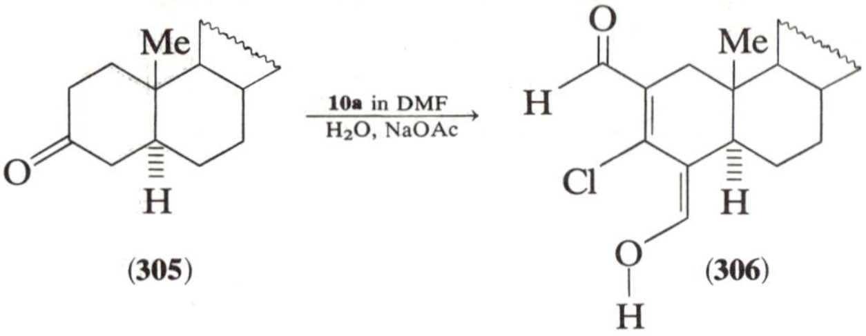

(305) → (306)

More complex is the chloroformylation and simultaneous dehydrogenation of 17β-acetoxyoestra-4,6-dien-3-one (**307a**) and 17-acetoxy-19-norpregna-4,6-dien-3,20-dione (**307b**) with **10a** in DMF (1 hr, 40°C). One isolates, along with the expected main products, 3-chloro-2,4,6-trien-2-carboxaldehydes **308a** and **308b**, the 1,3,5(10),6-tetraendialdehydes **309a** and **309b** in moderate yields (128):

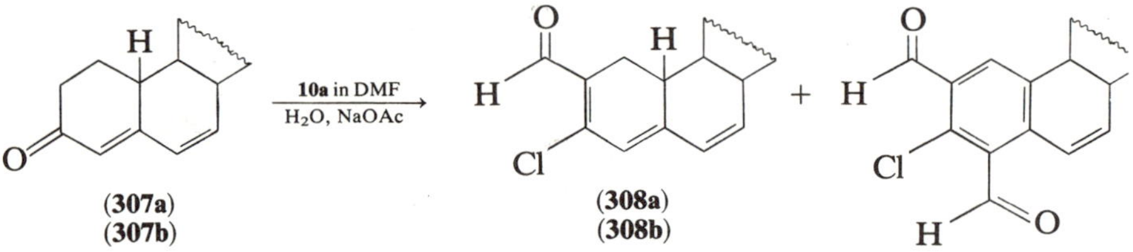

(307a)
(307b)
→
(308a)
(308b)
+
(309a)
(309b)

4. Cyclizations by Action of Chloromethyleniminium Salts on Ketones

The indene **310** was obtained in 75% yield by a smooth cyclization of the iminium salt (**241y**), formed in the chloroformylation of propioveratrone (**240y**) by **10a** at 60°C. The cyclization occurred when the reaction mixture was heated for 1 hr at 100°C (132). The closely related ring closure of **74** to the indene (**76**) (Section II-B-1) may be recalled here.

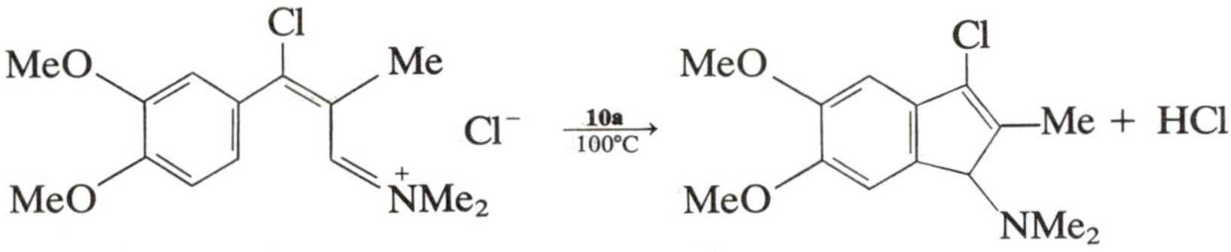

(241y) → (310) + HCl

The β-chloro-3,4-dimethoxy-α-methylcinnamaldehyde (**242y**) cyclizes in ethereal hydrogen chloride to give 3-chloro-5,6-dimethoxy-2-methylindene (132).

II. C—C BOND-FORMING REACTIONS

The isoflavone **313** is obtained in the reaction of **10a** in DMF with 2,4-dihydroxydeoxybenzoin (**311**) (150). Attack by **10a** on the activated (enolized) methylene group to give **312**, followed by cyclization onto oxygen and loss of dimethylamine, is the presumed reaction course; however, initial attack on oxygen cannot be excluded.

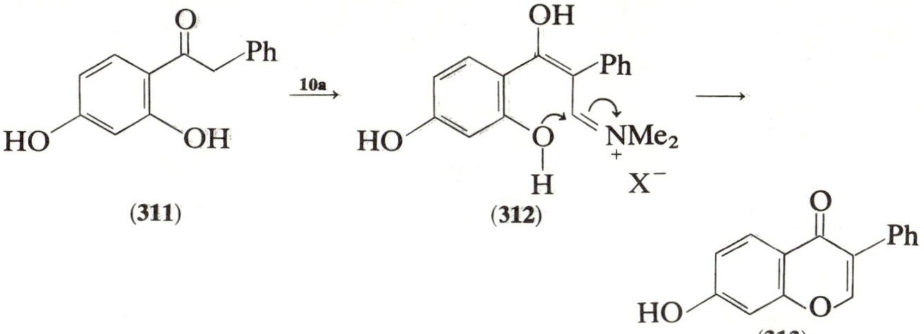

Chromenone-3-carboxaldehydes (**316**) have been obtained by treating *o*-hydroxyacetophenone (**314**) or related *o*-hydroxyarylmethyl ketones with an excess of **10a** in DMF at 45–60°C (142). The diformylated ketone (**315**), assuming a similar mechanism, was thought to be an intermediate. However, double substitution by **10a** under mild conditions without displacement of oxygen by chlorine is not the normal reaction course of acetophenones. It seems more plausible to assume that **10a** reacts first at the hydroxyl group; this is followed by a ring closure to the chromone by elimination of dimethylamine, and, finally, a rapid, second formylation occurs in position 3 of the chromenone (this is a β position of an enol ether).

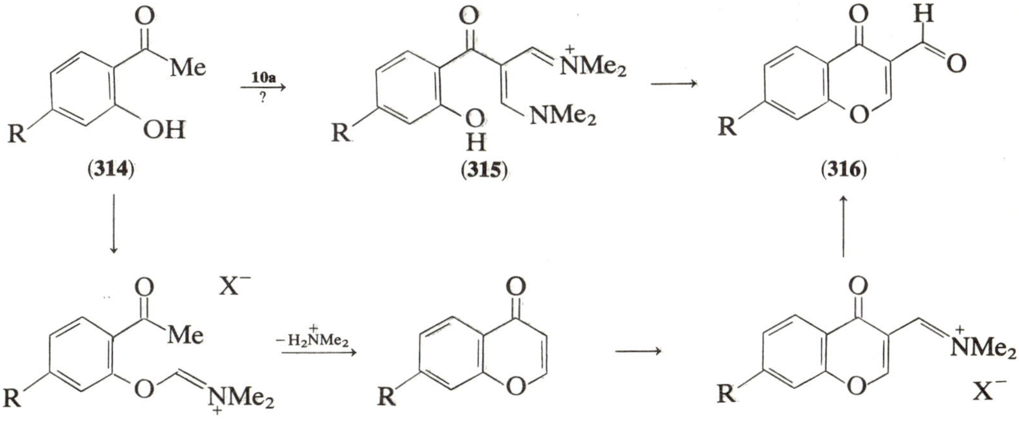

TABLE XIII
Chromenone-3-carboxaldehydes and Related Compounds by Formylation

Starting compound	Product	Reaction time, hr	Reaction temp., °C	Yield, %
314, R = H	**316**, R = H	1	45	81
314, R = Me	**316**, R = Me	1	45	86
2-Acetyl-1-hydroxy-naphthalene	Benzo(h)chromone-3-carboxaldehyde	3	60	68
1-Acetyl-2-hydroxy-naphthalene	Benzo(f)chromone-3-carboxaldehyde	3	60	63
4-Methyl-6-acetyl-7-hydroxycoumarin	6-Methyl-4,8-dioxo-4H,8H-pyrano(3.2-g)-chromene-3-carboxaldehyde	2	60	52
4-Methyl-8-acetyl-7-hydroxycoumarin	6-Methyl-7,8-dioxo-7H,10H-pyrano(3.4-f)-chromene-9-carboxaldehyde	2	60	57

Attempts to prepare 3-benzyl-3-chloro-2-phenylacrolein by chloroformylation of dibenzyl ketone (**320**) with **10a** produced, along with recovered starting ketone (60%), 3,5-diphenylpyrone (**323**) in 35% yield (138). Since **323** contains two additional carbon atoms, a double formylation of **320**, facilitated by the highly reactive benzylic methylene groups, must have taken place:

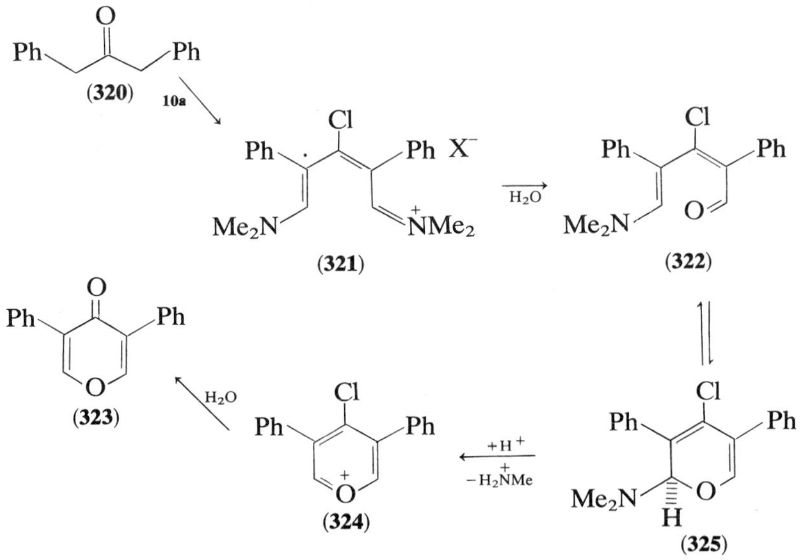

The most important step in the proposed mechanism is the electrocyclic 6-π-dienone-α-pyran ring closure (151) of the 5-dimethylaminopenta-2,4-dienal **322**. The sterically overcrowded iminium salt **321** (Section II-E-3) lacks resonance stabilization and should therefore hydrolyze readily, even in acidic media, to **322**. By acid-catalyzed loss of dimethylamine, the pyran **325** is transformed irreversibly to the 4-chloropyrylium ion **324**, whose hydrolysis leads finally to **323**.

5. *Bromo- and Iodoformylations of Ketones*

A simultaneous replacement of the carbonyl oxygen by the chlorine of **3a** or **10a** takes place, as a rule, when a formiminium group is introduced into ketones. This reaction, however, opens up the possibility of using bromo and iodo analogues of **3a** and **10a** to prepare the otherwise difficultly accessible 3-bromo- or 3-iodoacroleins (17,18,148). Since DMF does not react with carbonyl bromide, and carbonyl iodide does not exist, the desired bromomethylenedimethyliminium bromide (**3b**) and iodomethylenedimethyliminium iodide (**3c**) were obtained by treating **3a** in chloroform with gaseous hydrogen bromide or hydrogen iodide. The former compound exhibits the same properties as the adducts of DMF with either phosphorus oxybromide or phosphorus tribromide (17,148). For the latter adduct, structure **10b** has been suggested (148):

$$Me_2N{=}\overset{+}{C}H{-}Br\ \bar{O}PBr_2$$
(10b)

Double substitution has not been observed in reactions of **3b** or **10b** except with cyclohexanone. A small amount of the dialdehyde **326** (17) could be isolated:

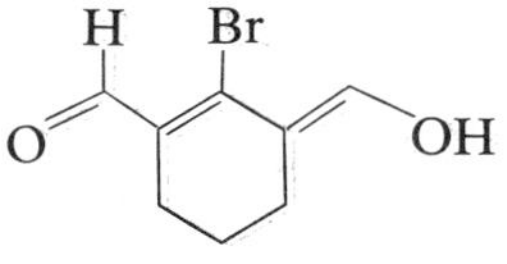

(326)

In contrast to expectation, **3b** and **10b** are less reactive than **3a** and **10a**. The donation of unshared electron pairs from the halogens to the electron-deficient carbon atom should decreasingly stabilize the iminium cation in the sequence $F > Cl > Br > I$, and therefore the electrophilicity of the corresponding halomethyleniminium ion should increase in the same order. However, steric demand of the various cations and *d*-orbital conjugation should not be neglected.

TABLE XIV

3-Bromoacroleins $R^1-\underset{\underset{Br}{|}}{C}=\underset{\underset{}{|}}{\overset{\overset{R^2}{|}}{C}}-CH=O$

No.	Starting ketone	Aldehyde R^1	R^2	Reaction temp., °C	Reaction time, hr	Agent	Solvent	Yield, %
a	Acetone	Me	H	20	12	10b	Chloroform	20
				60	1	3b	Chloroform	27
b	Pinacone	CMe$_3$	H	60	4	10b	Chloroform	75
c	Cyclopentanone	—(CH$_2$)$_3$—		20	12	10b	Chloroform	45
				65	1.5	3b	Chloroform	31
d	Cyclohexanone	—(CH$_2$)$_4$—		20	12	10b	Chloroform	54
e	Cycloheptanone	—(CH$_2$)$_5$—		70	3	10b	Chloroform	45
				65	1.5	3b	Chloroform	67
f	Cyclooctanone	—(CH$_2$)$_6$—		60	12	10b	Chloroform	37
				65	1.5	3b	Chloroform	63
g	Acetophenone	Ph	H	60	2	10b	Chloroform	45
				60	1.5	3b	Chloroform	68
h	Propiophenone	Ph	Me	60	5	10b	Chloroform	71
				70	3	3b	Chloroform	85
i	Ethyl methyl ketone	Me	Me	60	7	10b	Chloroform	36
j	Benzyl methyl ketone	Me	Ph	60	2	10b	Chloroform	25
				60	1	3b	Chloroform	56
k	Deoxybenzoin	Ph	Ph	60	3	10b	Chloroform	75
l	17β-Acetoxy-5α-androstan-3-one (148)	17β-Acetoxy-3-bromo-2-formyl-5α-androst-2-ene		75	1	10b	TCE	40
	3-Iodoacrolein							
	Acetophenone (18)	β-Iodocinnamaldehyde		80	1.5	3c	DMF	80

The 3-bromoacroleins and the β-iodocinnamaldehyde are unstable compounds and have been characterized as their semicarbazones or oximes.

6. Chloroacroleins and Chlorovinyl Ketones by Other Methods

The 3-dimethylaminoacroleins **203** and, in a wider sense, 3-dimethyl aminovinyl ketones **211** may be considered as vinylogues of DMF itself. They exhibit, in general, greater nucleophilicity than DMF and react with carbonyl chloride (A), phosphorus oxychloride (B), and thionyl chloride and *p*-toluenesulfonyl chloride (C) in the same way as DMF to give

II. C—C BOND-FORMING REACTIONS

TABLE XV
Prepared 3-Chloroacroleins **242** (7)

Starting compound	Product residue **242**: R	Method	Yield, %
204 Na salt	H	D	73
203b	Me	A	56
203c	Et	A	84
		B	76
203g	n-C$_5$H$_{11}$	B	79
203h	Ph—CH$_2$	A	85
203i	Ph	A	86

resonance-stabilized 3-chloroallylideniminium salts **241**, which are vinylogues of **3a** or **10a**, identical to those formed by chloroformylation of ketones. Hydrolysis of **241** with ice water leads, as mentioned in Section II-E-1, to 3-chloroacroleins **242** (7). Moreover, sodium salts of malondialdehydes **204** afford, on treatment with thionyl chloride, 3-chloro-

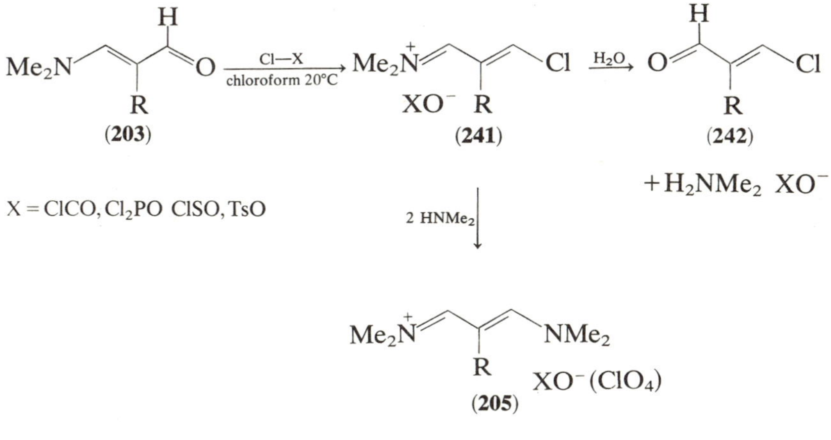

acroleins **242** (method D). The following sequence of alternate transformations of α-dimethylaminomethylenepropiophenone (**211b**), leading finally to the α-chloromethylenepropiophenone (**242′**), demonstrates the synthetic value of compounds prepared by chloroformylation with **3a** and

10a (7):

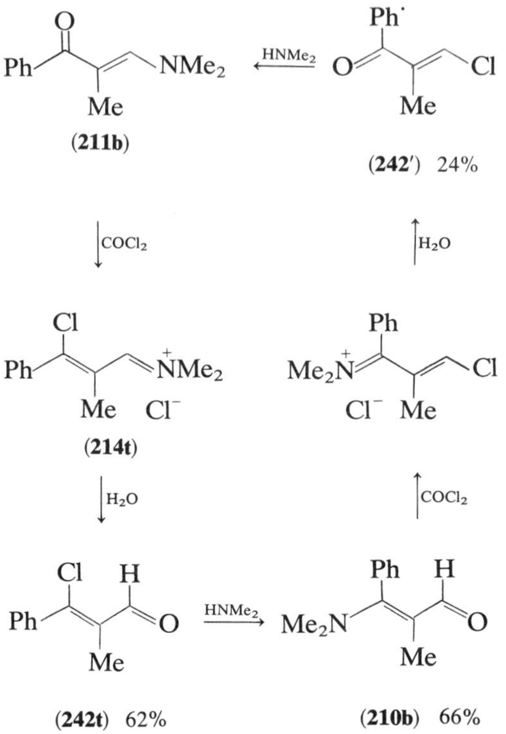

Compound **3a** transforms **203** and **211** to the corresponding 3-chloro-allylideniminium chlorides **241** (see Section II-A, **15** and **16**), and, by applying the bromo analogue **3b** (17), the 3-bromoacroleins, via the intermediate 3-bromoallylideniminium bromides, can be obtained. This fact is of synthetic importance, since both **203** and **211** fail to react with carbonyl bromide or phosphorus oxybromide in the expected manner.

TABLE XVI

Prepared 3-bromoacroleins: Br—CH=C(R)—CH=O
(17)

Starting compound	Product R	Yield, %
203a	H	23 (50 as semicarbazone)
203b	Me	82
203i	Ph	75

F. REACTIONS OF CHLOROMETHYLENIMINIUM SALTS WITH CARBONAMIDES POSSESSING AN α-METHYLENE GROUP

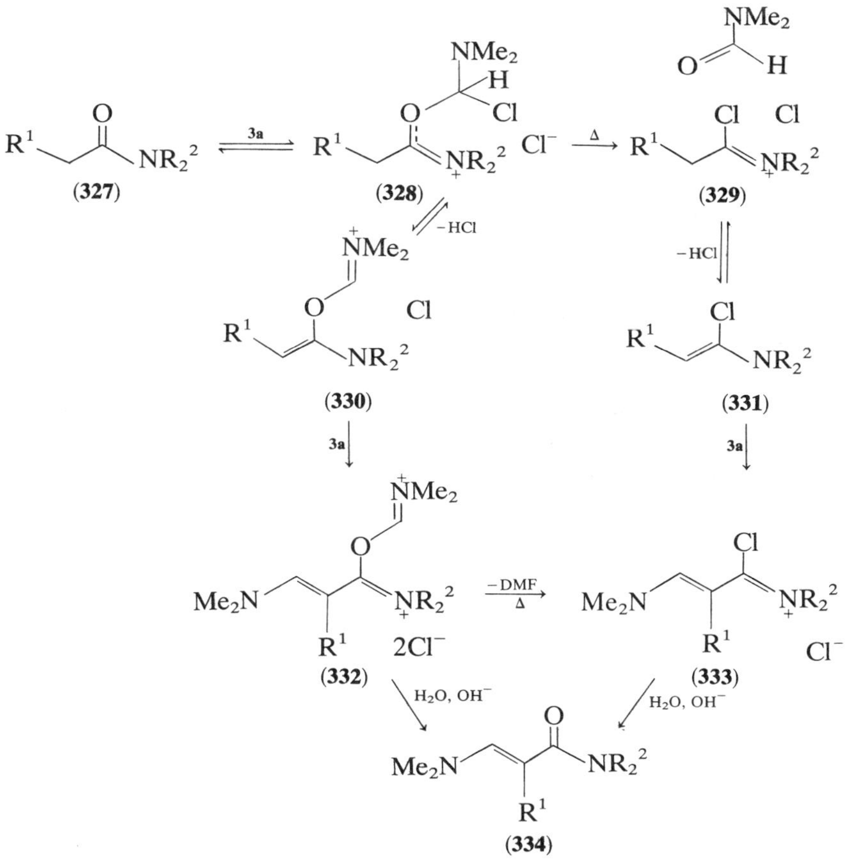

"Chloroformylations" are not confined to reactions of **3a** or **10a** with ketones; they have been extended to include substitutions of carbonamides with these reagents. In carbonamides the carbonyl oxygen is distinctly more basic than in ketones. Therefore it seems reasonable, supposing a reaction mechanism similar to that for ketones (Section II-E), to assume that the first step is an attack of the chloromethyleniminium ion on the carbonyl oxygen of the amide (**327** to **328**). The reaction path may branch off at **328** to the iminochloride (**329**), which gives rise to the formation of an α-chloroenamine (**331**) by loss of a proton. This may be the usual path of the reaction of N,N-dimethylacetamide and related open-chain amides with **3a** or **10a**. Some amides, especially cyclic and vinylogous ones, afford at low temperatures some chlorine-free formylation products; this indicates a formylation at the α-carbon atom of **327** before the displacement of oxygen by chlorine (158,161,163,168,169). In our scheme this means that a proton abstraction occurs, transforming **328** into the enamine **330**. The nucleophilic α-chloroenamines **331**, which are also intermediates in the reaction of carbonyl chloride with **327** (155), and **330** undergo fast substitution by **3a** or **10a** to give, respectively, **333** and **332**.

N,N-dimethylacetamide (**327a**) undergoes substitution by **3a** or **10a** in a vigorous, exothermic reaction. Cooling below 0°C is needed to stop the reaction at monoformylation (**333a**). Further substitution to **335** occurs readily, and after hydrolytic work-up the acrylic acid derivative **336** is obtained in 76% yield (16):

$$(327a) \xrightarrow[3\ hr\ 70°C]{10a} \left[Me_2N\text{–CH=C}(\overset{+}{N}Me_2Cl)(=\overset{+}{N}Me_2)\ 2X^- \right] \xrightarrow{K_2CO_3, H_2O} Me_2N\text{–CH=C(CHO)–C(=O)NMe_2}$$

(**335**) (**336**) 76%

With primary carbonamides the strongly dehydrating properties of **3a**, **10a**, and related reagents give rise to the formation of the corresponding nitriles **328** and **329** ($R^2 = H$), representing possible intermediates (170–172). An amino group in the substrate, as in anthranilic amide (171) or 3-aminopyrazine-2-carbonamide (172), is simultaneously transformed to the dimethylaminomethylenamino residue.

The action of **10a** on isatin-β-oximes effects a second-order Beckmann rearrangement (fragmentation), leading finally to 2-dimethylaminomethylenaminobenzonitriles (173) which are identical to the products formed in the reaction of anthranilic amide with **10a** (171).

Obviously, tertiary acetyl amides react with **10a** in analogy to the

TABLE XVII
Formylation Products of N,N-Dimethylcarbonamides

333	R^1	R^2	Yield, %	Ref.	
a	H	Me	86	16	as
b	Ph	Me	71	16	perchlorates
c	CN	Me	43	63	
	CN	Me	35[a]	30	
d	CO$_2$Et	Me	40[a]	30	

[a] After exchange of chlorine by the dimethylamino group.

scheme given for chloroformylation. Thus 10-acetylphenothiazine affords (4 hr at 60°C) the corresponding chlorotrimethinium salt **333** (as the perchlorate) in 83% yield (164). Moreover, a series of 1-acetyl-Δ3-pyrrolin-2-ones has been transformed by **10a** and its bromo analogue to the acroleins **337** (159):

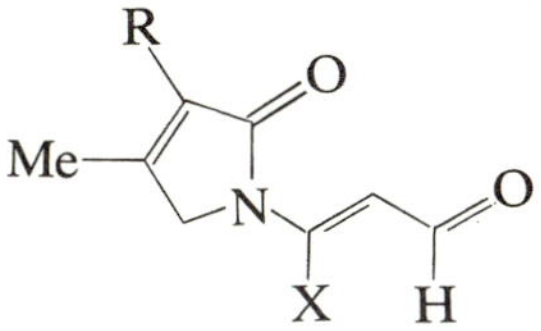

(**337a**) R = Me, X = Cl   72%
(**337b**) R = Me, X = Br   60%
(**337c**) R = Et, X = Cl   69%
(**337d**) R = Et, X = Br   56%

1. Lactams

More importantly, various lactams—3,4,5,6-tetrahydro-2*H*-1,4-thiazin-3-ones (153), 2*H*-1,4-benzoxazin-3-ones (156), 2*H*-1,4-benzthiazin-3-ones (153,157), and 1,3,4,5-tetrahydro-2*H*-benzazepinone (153)—have been chloroformylated by **10a**. The cyclic chlorovinylaldehydes thus obtained were used in heterocyclizations (e.g., with hydrazines to condensed pyrazoles) and other transformations.

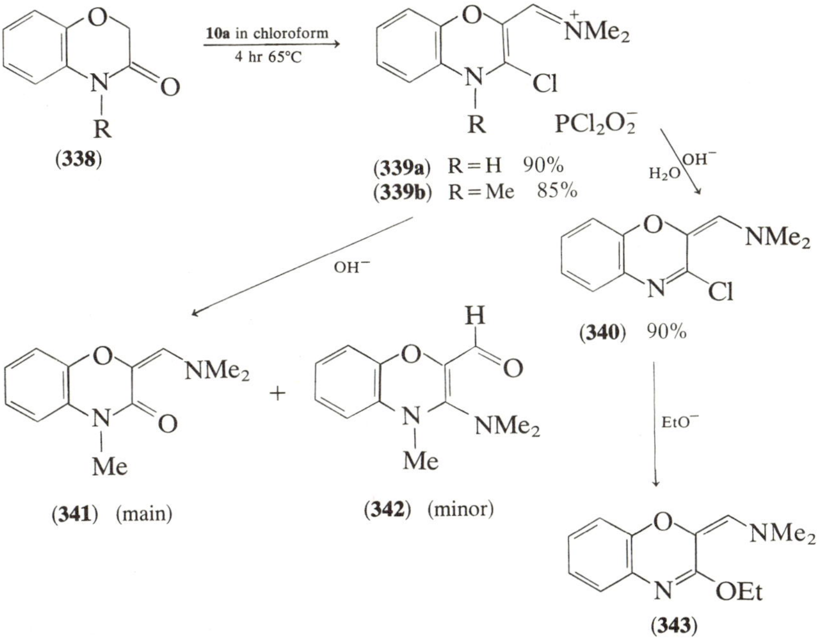

(**338**)

10a in chloroform, 4 hr 65°C →

(**339a**) R = H   90%
(**339b**) R = Me 85%

$PCl_2O_2^-$

(**340**) 90%

(**341**) (main)   +   (**342**) (minor)

(**343**)

(344) (345) (346)

a R, R' = H
b R = Me, R' = H
c R, R' = Me

As a representative example, the chloroformylation of 2H-1,4-benzoxazin-3-one (**338**) and some of its transformations (156) may be shown.

In the chloroenamine **340** the chlorine can be displaced by alkoxides. With secondary amines (e.g., morpholine) both the chlorine and the dimethylamino group are displaced. Chloroenamine **340** also reacts with hydrazine to give in 95% yield a tautomeric mixture of the possible pyrazolo-1,4-benzoxazines (**344a**, **345a**, and **346a**). Analogously **339b** gives, with hydrazine, the pyrazoles **344b** and **345b**. The reaction of **339b** with methylhydrazine affords, in a 4:1 ratio, a mixture of **344c** and **345c**, which can be separated chromatographically.

2. Pyrrolinones and Phthalimidines

As in the preceding example, 2-chloropyrrole-3-carboxaldehydes and 2-chloroindole-3-carboxaldehyde (**347**) have also been obtained in 50–60% yield by treating Δ4-pyrrolin-2-ones and 2-oxindole with **10a** in chloroform (6 hr, 65°C), followed by work-up with ice water (158). These new, synthetically valuable compounds are easily transformed into condensed heterocycles, as shown for **347** (158).

The formylation of 4-ethoxycarbonyl-5-methyl-Δ4-pyrrolin-2-one by **10a** (158) or its bromo analogue (161) at low temperatures (−20°C) takes place without any exchange of oxygen by halogen, producing the corresponding 3-dimethylaminomethylene compound and elucidating the true reaction course of the chloroformylation of such lactams.

The Δ3-pyrrolin-2-ones **348a–c** and, in a wider sense, the related phthalimidines **348d–f** can be considered to be lactams with a vinylogous methylene-carbonamide grouping. They all undergo smooth haloformylations with **10a**, **10b**, and **3a** (from DMF and oxalyl chloride), yielding the otherwise inaccessible, functionalized derivatives of pyrrole (**351a–c**) and isoindole (**351d–f**) (159–163). A successful extension of this method employs as substrates cyclic lactams of pyridine and pyrimidine (**248g–k**)

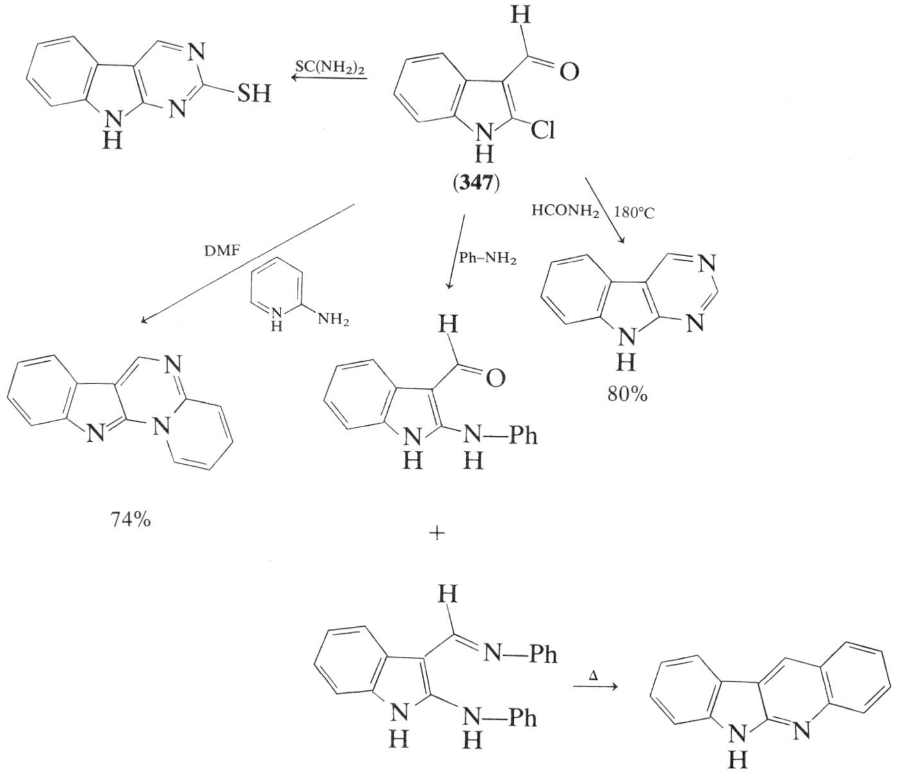

(163). The chloroformylation was generally performed by heating **348** with **10a** or DMF–POBr$_3$ (molar ratio 1:2.5) in chloroform solution for 6–18 hr under reflux. When a 1:1 ratio of the formylating agent **3a** and **348a–c** was used, the halogen-free 2,1′-dipyrromethenes were isolated instead of the expected products (159). The usual method of work-up with ice water–sodium hydroxide may be improved considerably by first neutralizing the reaction mixture with dimethylamine, and then isolating the enamine **350**, which is finally hydrolyzed by aqueous acetic acid. The preparation of the halogen-free pyrrolealdehydes and isoindolealdehydes **352** succeeds by selective hydrogenolysis of **351** with hydrogen–palladium on barium sulfate in the presence of alkoxide (yields >90%). Under identical conditions, **350** leads to the corresponding 2-dimethylamino-methylpyrroles (Mannich bases) (160).

3. Pyrazolones

The reaction of 5-pyrazolones with **10a** proceeds to afford the (originally protonated) 4-dimethylaminomethylene derivatives **353**. No

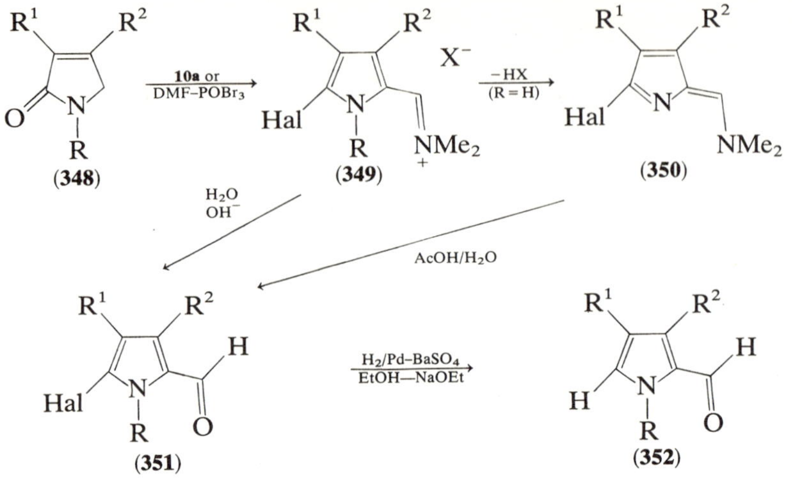

TABLE XVIII

Formylation Products of Pyrrolinones, Phthalimidines, and Related Compounds

No.	Halogen	R	R^1	R^2	Yield, % 349	350	351	Ref.
a	Cl	H	Me	Me		62	23	159
a'	Br	H	Me	Me	85[a]	81	85	159
b	Cl	H	Et	Me			21	159
b'	Br	H	Et	Me	50		45	159, 160
c'	Br	H	CH$_2$CH$_2$CO$_2$Me	Me			28	159
						80	95	160
d	Cl	H	—CH=CH—CH=CH—		81	64	86	162
d'	Br	H	—CH=CH—CH=CH—			72	84	162
e	Cl	Me	—CH=CH—CH=CH—		92			162
e'	Br	Me	—CH=CH—CH=CH—				87	162
f	Cl	PhCH$_2$CH$_2$	—CH=CH—CH=CH—		80		80	161
g	Cl	H	—CH=CH—CH=N—			50		163
g'	Br	H	—CH=CH—CH=N—			25		163
h	Cl	Me	—CH=CH—CH=N—				80	163
h'	Br	Me	—CH=CH—CH=N—		85[a]		85	163
i'	Br	PhCH$_2$CH$_2$	—C=N—C=CH— Me Me				65	163
j	Cl	Ph	—N=C—N=C— OMe OMe				20	163
k	Cl	Ph	—N=C—N=C— NHPh Pip				70	163

[a] X = Br, otherwise ClO$_4$; Pip = N-piperidinyl.

oxygen–chlorine exchange has been observed (165,166). Treatment of **353** first with $POCl_3$ and then with water introduces a chlorine, yielding the corresponding 5-chloropyrazole-4-carboxaldehyde, in which the reactive halogen can easily be displaced by amines, alkoxides, and mercaptides. With thioglycolic acid and an excess of alkali, cyclization to a thieno(3,2-*d*)pyrazole takes place (165).

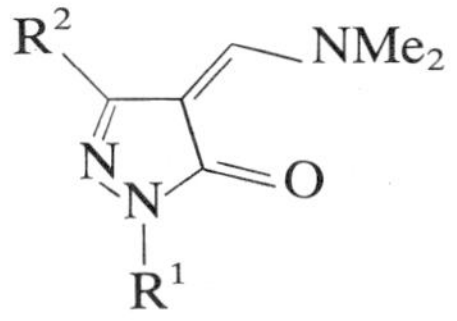

(**353**) $R^1 = Ph, Me$; $R^2 = Ph, Me, H, Cl, CO_2Et$

Hydrolysis of **353** by aqueous base leads to the 4-formyl-5-pyrazolones. Formation of a 4,4′-di-(5-pyrazolonyl)methine by the action of aqueous acid on **353** has been observed (165,166).

4. *Hydroxypyrimidines*

Unactivated pyrimidines (e.g. 4,6-dichloropyrimidine) do not react with **3a** or **10a**. However, the free 5 position of derivatives of barbituric acid (**355a–e**), uracils (**355f–i**) (167), and 4-hydroxy-6-oxodihydropyrimidines (**360**) corresponds to a β enamide position and can be smoothly formylated by **3a** or **10a**.

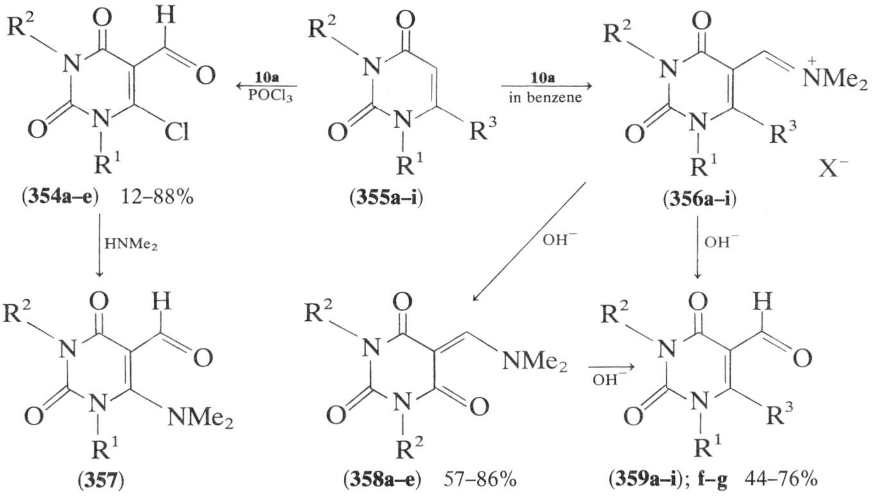
10a
POCl3
10a
in benzene
HNMe2
OH−
OH−
OH−
(354a–e) 12–88%
(355a–i)
(356a–i)
(357)
(358a–e) 57–86%
(359a–i); f–g 44–76%

TABLE XIX

Hydroxypyrimidines Subjected to Formylation

355	R¹	R²	R³		R¹	R²	R³
a	Me	Me	OH	f	Me	Me	Me
b	c-Hex[a]	c-Hex[a]	OH	g	Me	c-Hex[a]	Me
c	Ph	Ph	OH	h	Me	Ph	Me
d	H	n-Bu	OH	i	c-Hex[a]	Me	H
e	H	Ph	OH				

[a] c-Hex = cyclohexyl.

A chloroformylation, that is, a simultaneous replacement of the 4-hydroxy group by chlorine, takes place only if the barbituric acids **355a–e** are heated with neat **10a** and an excess of POCl₃ (167). Similarly, in **361** or **362**, the oxygen functions can only be replaced in a separate operation with POCl₃ in *N*-dimethylaniline to give **363**, which has been transformed by reductive dehalogenation to the pyrimidine-5-carboxaldehyde (168).

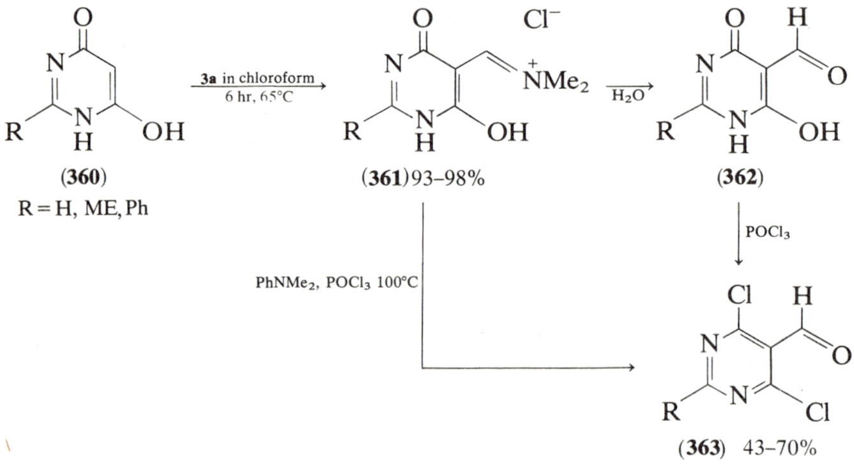

(360) R = H, ME, Ph

(361) 93–98%

(362)

(363) 43–70%

G. REACTIONS OF CHLOROMETHYLENIMINIUM SALTS WITH ACTIVATED METHYLENE AND METHYL GROUPS

Methyl groups attached to or conjugated with the C–N double bond of a tertiary iminium function are highly activated to electrophilic attack. Loss of a proton, of course, transforms such methyl groups into nucleophilic, enamine-like methylene groups that react very readily with **3a** or **10a**.

1. Methyl Groups in Polymethinium Salts

Only methyl groups in odd-numbered positions of a polymethinium chain fulfill the aforementioned condition. For example, the 3-dimethylamino-1-methylallylidenedimethyliminium perchlorate gives, by double formylation with **10a** in DMF at ambient temperature, **286** in 83% yield (21). Likewise the 5-dimethylamino-3-methylpenta-2,4-dienylidenedimethyliminium perchlorate **102** undergoes a threefold formylation to the trication **105** (22) (see also Section II-B-2).

2. Methyl Groups in Cycloiminium Salts and Related Compounds

The reactivity of the methyl group in 2- and 4-methylcycliminium salts (e.g., **364**) has been known for a long time. The corresponding free methylene bases (**365**), are, in general, very unstable, are isolable only in some instances, and are strong nucleophiles. Using the more conveniently prepared salts **364** normally results in a double substitution by **3a** in DMF (60°C) (177).

Formylation of the intermediate hemicyanine **367**, a potential enamine, appears to be a faster process than the attack of **3a** on **365** in its equilibrium concentration. Bosshard et al. have succeeded in preparing the monoformylated species **366b** and **367b** by reaction of equimolar amounts of the free base **365b** and **3a** in dichloromethane at 0°C (169) (see Section II-C, **150** and **151**).

Condensation of the intermediate **367** with **364** during the formylation reaction gives rise to colored tricarbocyanines, especially when chloroform is the solvent (177).

3. Methyl Groups in Pyrylium Salts

The methyl groups in 2-methyl-4,6-diphenylpyrylium perchlorate (171), 4-methyl-2,6-diphenylpyrylium perchlorate (171,173), 4-methylflavylium perchlorate, and 4-methylbenz(f)flavylium perchlorate (171,172), not surprisingly, have been found to be particularily reactive. DMF–acetic anhydride, a very mild formylating agent, converts the methyl group in these oxonium salts to the dimethylaminovinyl group, whereas the much more electrophilic **10a** leads by double substitution to a trimethinium salt (171). In 5,6,7,8-tetrahydro-4-phenylflavylium perchlorate and 6,7,8,9-tetrahydrobenz(a)xanthylium perchlorate only monoformylation is possible, the activated methylene group being attacked by either DMF–acetic anhydride or **10a** to yield a dimethylaminomethylene derivative (171,172). The remarkable conversion of 2-dimethylaminovinyl-4,6-diphenylpyrylium perchlorate (**371**) by dimethylamine to 4-dimethylamino-2-phenylbenzophenone (**375**) (171)

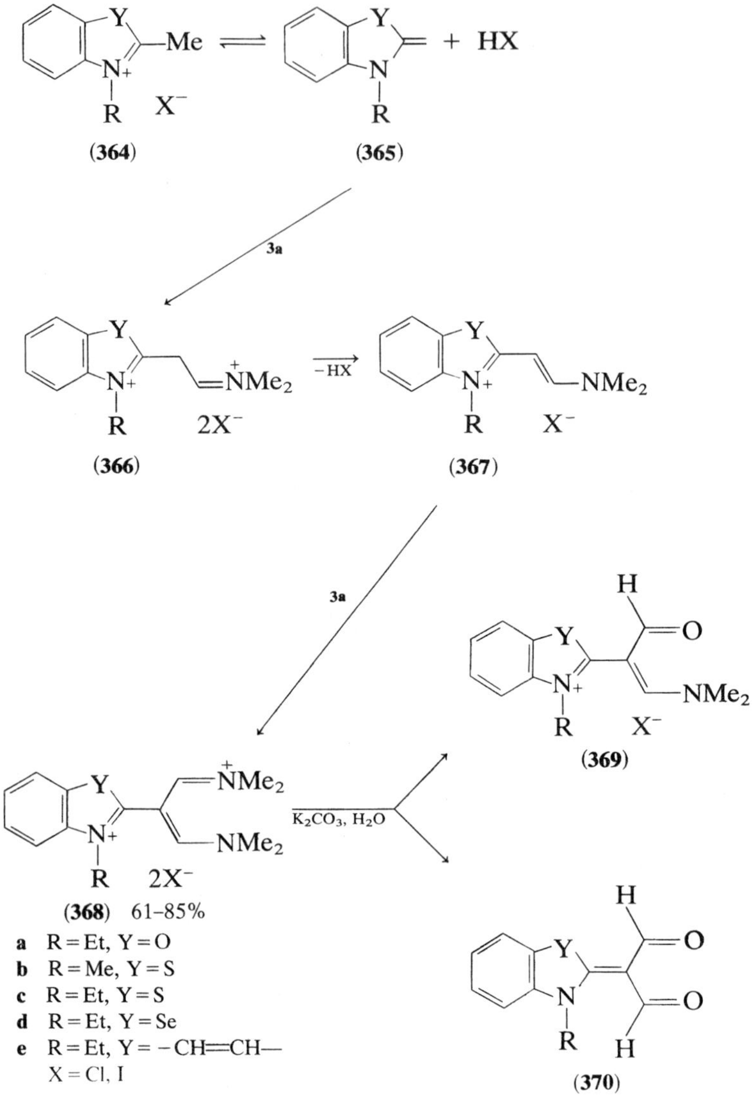
(364)
(365)
+ HX
3a
−HX
(366)
(367)
3a
(369)
K₂CO₃, H₂O
(368) 61–85%
a R = Et, Y = O
b R = Me, Y = S
c R = Et, Y = S
d R = Et, Y = Se
e R = Et, Y = −CH═CH—
X = Cl, I
(370)

may involve first an electrocyclic ring opening of the intermediate **372**, and then a recyclization of the aminoheptatrienone **373** to the cyclohexadiene **374**, followed by an elimination to **375**. Attempting to hydrolyze **371** by aqueous sodium hydroxide results in the formation of 2-formylmethylene-4,6-diphenyl-2*H*-pyran, which emerges as the main product (171):

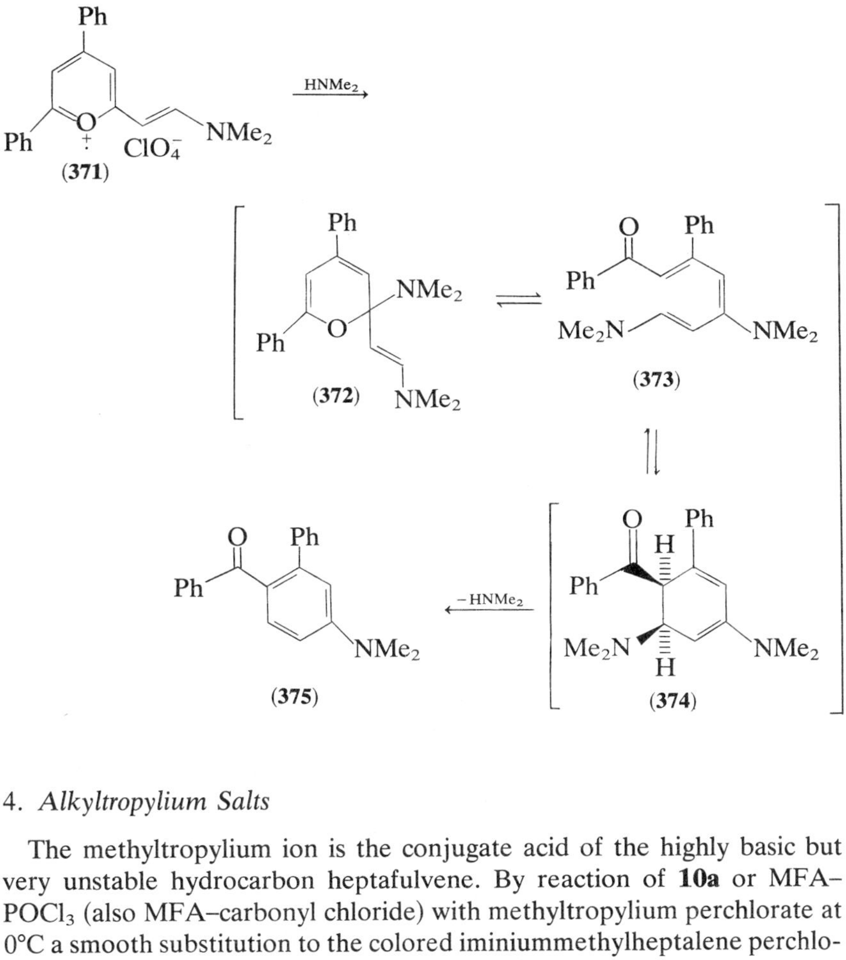

4. Alkyltropylium Salts

The methyltropylium ion is the conjugate acid of the highly basic but very unstable hydrocarbon heptafulvene. By reaction of **10a** or MFA–$POCl_3$ (also MFA–carbonyl chloride) with methyltropylium perchlorate at 0°C a smooth substitution to the colored iminiummethylheptalene perchlorates **376a** and **b** takes place. In the preparation of the substituted derivatives **376c–e**, the needed tropylium salts were easily generated *in situ* by hydride abstraction from the more conveniently prepared tropilidenes with phosphorus pentachloride (170). The 4,5-benzo derivative of **376a** has also been synthesized (170).

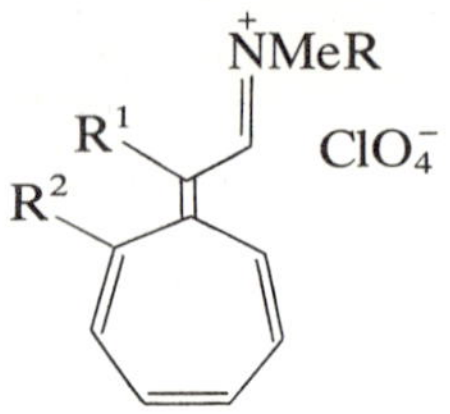

(376) R = Me, Ph

TABLE XX

Iminiummethylheptalene Derivatives

376	R^1	R^2	Yield, %
a	H	H	81–96
b	H	Me	91
c	Me	H	63–93
d	Ph	H	62–63
e	—CH$_2$	—CH$_2$—	—

5. Methyl Groups in Electron-Deficient Heteroaromatics

The methyl group in 4-picoline shows considerably reduced reactivity in comparison with N-alkyl-4-picolinium ions. Nevertheless, double formylation by **10a** at 70°C (6 hr) occurs smoothly, and after hydrolytic work-up the 3-dimethylamino-2-(4-pyridyl)acroleins and the pyridyl-4-malondialdehyde were isolated in high yield (20). Therefore Arnold (20) has proposed quaternization of the nitrogen of the picoline by **10a**, followed by proton abstraction before substitution. Throughout the reaction, the evolving hydrogen chloride may act as catalyst.

A more elaborate mechanism has been proposed by Bredereck et al. (179) for the related formylation of 4-methylpyrimidine (**377**) by **3a** in chloroform. It takes into account the double substitution of the methyl group by an excess of **3a** to yield **384** and pyrimidyl-4-malondialdehyde. It also explains the formation of 4-dimethylaminovinylpyrimidine (**385**), using 4-methylpyrimidine hydrochloride (**380**) and equimolar amounts of **3a**. A short version of this mechanism is given on p. 313.

Double formylation by **3a** or **10a** to a malondialdehyde derivative, according to the reaction course described, has also been observed with the following compounds: 4-methyl-2-phenyl- and 2-methyl-4-phenyl-pyrimidine (175), 2-methylbenzoxazoles (176,177), 2-methylbenzthiazoles (177,178), 2-methylselenazole (177), and also quinaldine and lepidine (177). 2-Hydroxyacetophenoneoximes can be employed as a

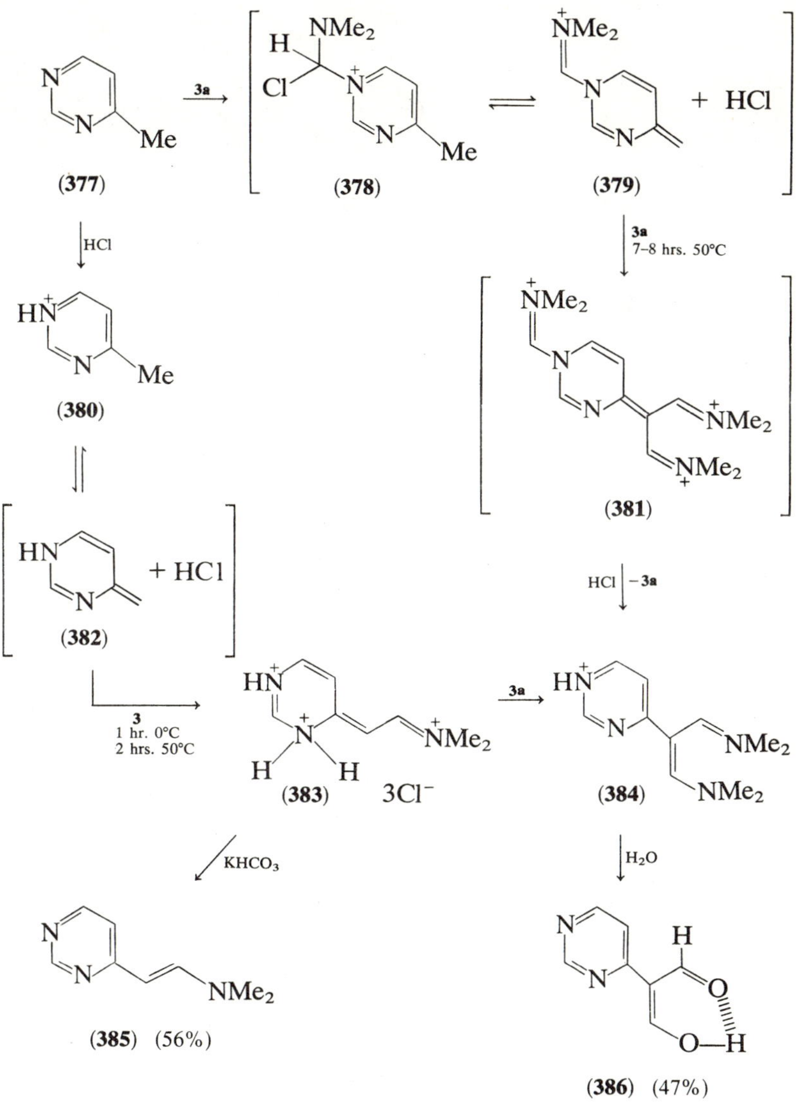

replacement for 2-methyloxazoles, since **10a** effects a Beckmann rearrangement and dehydrates the oximes to the necessary 2-methyloxazoles (178). 2,6-Dimethyl-4-pyrimidone and 4-chloro-2,6-dimethylpyrimidine both give mixtures of 4-dimethylamino-6-methylpyrimidin-2-malonaldehyde and 4-dimethylamino-2-methylpyrimidin-6-malonaldehyde. Replacement of the oxygen function by chlorine and nucleophilic displacement of the chlorine by dimethylamine occurs, perhaps in the work-up step (175). The oxygen of 2-methyl-3-phenyl-4-quinazolone, however, remains untouched, and the corresponding 4-quinazolone-2-malonaldehyde is obtained (180). Similarly, 6-methylpurine reacts with **10a** (1 hr at 120°C) by disubstitution to the 3-dimethylamino-2-(6-purinyl)allylidenedimethyliminium chloride, which is stable in aqueous solution at low temperature and pH 3 (183). From such a solution, many derivatives, including purine-6-malonaldehyde (82% yield), have been prepared by alkaline hydrolysis. Conversely, treatment of purine hypoxanthine, adenine, 8-methyladenine, guanine, and guanosine with **10a** gave no identifiable products (183). In many instances hetarylmalonaldehydes thus prepared have been transformed to their pyrazoles, oxazoles, or pyrimidines by the usual methods.

Notably, 2-picoline failed (in contrast to quinaldine) to give an isolable formylation product (20), whereas pyridine-2-ethyl acetate and pyridine-2-acetonitrile have been formylated to the expected dimethylaminovinyl derivative **387**. The latter compound has found use in a new approach to the quinolizinones **391** and **392** by reaction with ketene (174):

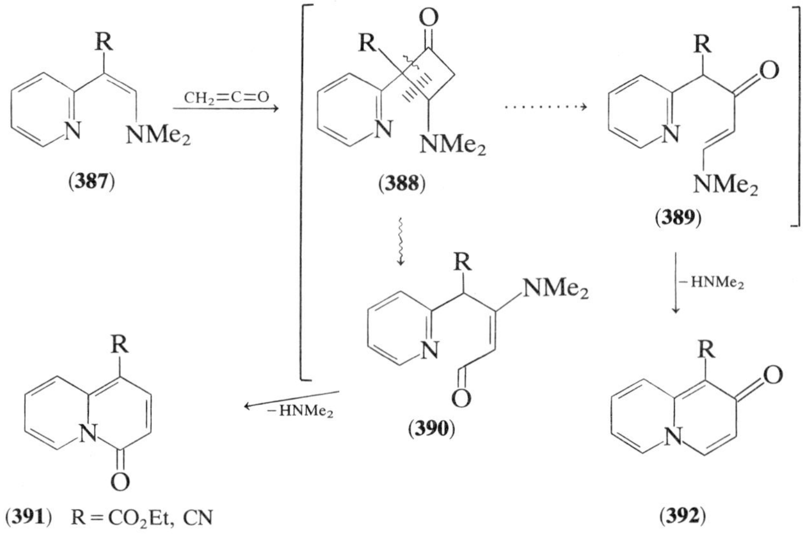

6. Cyclizations Involving Formylation of Activated Methyl Groups

The action of **10a** in DMF (3 hr at 100°C) on 2-amino-3-methylpyrazine and its N-methyl analogue furnishes the iminium salt **394** and, after hydrolysis, pyrrolo(2,3-*b*)pyrazine-3-carboxaldehyde (**395**) (181):

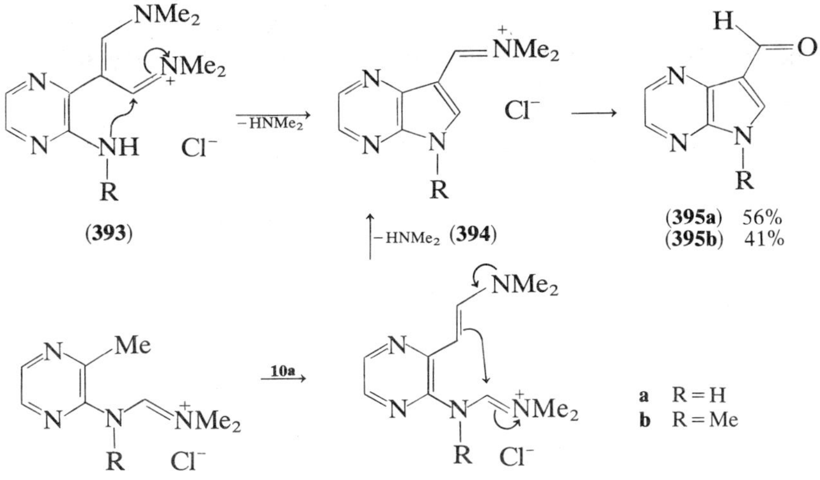

(**393**) → −HNMe₂ → (**394**) → (**395a**) 56%
(**395b**) 41%

(**396**) → 10a → (**397**)

a R = H
b R = Me

Double formylation to give **393**, followed by intramolecular nucleophilic attack by the vicinal amino group, has been proposed as a reasonable explanation of the reaction course (182). Isolation of the amidine **396** under milder reaction conditions (below 50°C), however, casts doubt on this proposal and implies that the first attack of **10a** takes place on the amino group. The next reaction step would certainly lead to **397**, which then cyclizes to **394**.

In a very similar manner, 3-amino-4-picoline and 3-aminoquinaldine, treated with **10a** in DMF (8 hr at 100°C), afford, respectively, pyrrolo(2,3-*c*)pyridine-3-carboxaldehyde (**398**) in 19% yield and pyrrolo(3,2-*b*)quinoline-3-carboxaldehyde (**399**) in 35% yield (182). The conclusions about the reaction course reached in the former case are also valid here. Similarly, an amidinium compound like **396** has been obtained from 3-amino-4-picoline under milder conditions.

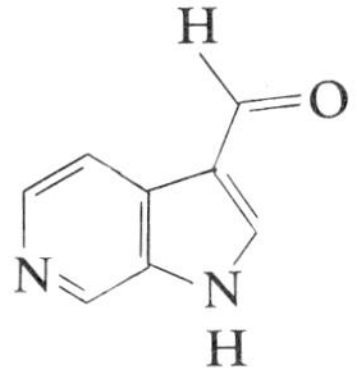

(**398**) 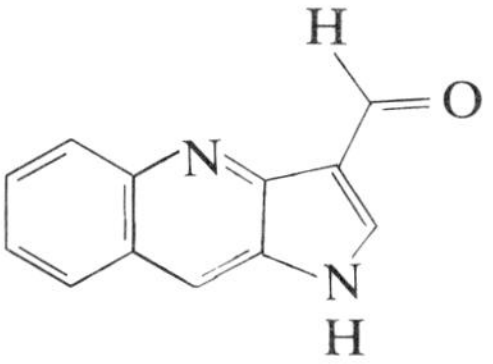   (**399**)

7. Methyl Groups in Nitro Compounds

The treatment of 2,4,6-trinitrotoluene with **10a** yields **400a**, which has been transformed into a malonaldehyde, an oxazole, and a pyrazole derivative. 2,4,6-Trinitroxylene was similarly converted to **400b**, which could be isolated as its dimethylaminoacrolein derivative. Analogously, 2,6-dinitro-4-toluic acid reacted to give 3-dimethylamino-2-(2,6-dinitro-4-carboxyphenyl)acrolein. The phenolic hydroxy group and, to some extent, one nitro group are exchanged by chlorine during the formylation of 2,4,6-trinitrocresol with **10a**, and a mixture of **400c** and **400d** results:

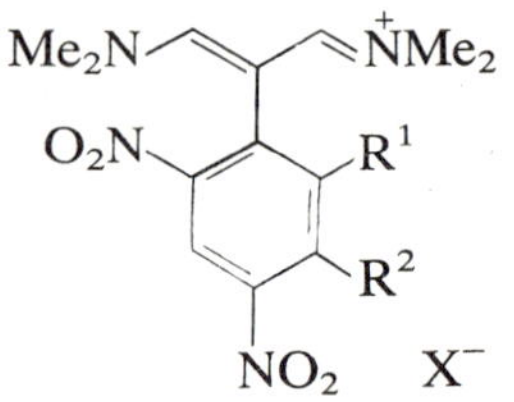

(**400a**) $R^1 = NO_2$, $R^2 = H$, $X = NO_3$
(**400b**) $R^1 = NO_2$, $R^2 = Me_2N-CH=C-CH=\overset{+}{N}Me_2 X^-$
  $X = ClO_4$
(**400c**) $R^1 = NO_2$, $R^2 = Cl$, $X = ClO_4$
(**400d**) $R^1 = R^2 = Cl$, $X = ClO_4$

Attempts to formylate 2,4-dinitrotoluene, 4-chloro-2,6-dinitrotoluene, 2-chloro-4,6-dinitrotoluene, and 2,4,6-trinitroethylbenzene were not successful (184).

8. Carboxylic Acid Derivatives, Nitriles

3-Methylindole-2-methyl acetate has been converted by formylation with **10a** (at 90°C) to its dimethylaminomethylene derivative, 3-dimethylamino-2-(3-methyl-2-indolyl)methyl acrylate, in 31% yield (185). As would be expected, acetonitrile, with its slightly reactive methyl group, undergoes double formylation by **10a** (20 hr at 100°C, acetonitrile in large excess) to **170**, and from the reaction mixture the 2-cyano-3-dimethylaminoacrolein is isolated in 32% yield (186).

The much more reactive malononitrile **401** is smoothly formylated by **3a** or **10a** in chloroform at 80°C (1 hr). Instead of monosubstitution to the expected dimethylaminomethylene derivative **402**, 2 moles of the reagent

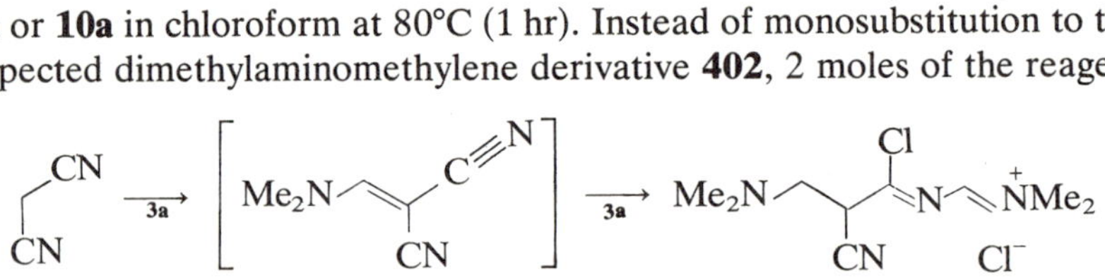

(**401**) (**402**) (**403**)

is consumed and the pale yellow 2-aza-3-chloro-4-cyano-5-dimethylaminopenta-2,4-dienylidenedimethyliminium chloride (**403**) is obtained as a perchlorate salt in 96% yield (187). In a Ritter reaction the chloromethyleniminium ion of **3a** or **10a** adds to the triple bond of one cyano group in **402**, in a reaction promoted by the electron-donating properties of the dimethylamino group. Independently synthesized **402** reacts with **3a** in the cold to yield **403**. Analogously, the lower vinylogue of **402**, the dimethylcyanamide, adds **3a** to afford the 2-aza-3-chloro-3-dimethylaminoallylideniminium chloride in high yield (187).

Smooth addition of **3a** or **10a** in the cold to the C–N triple bond of ordinary nitriles (**404**) has recently been accomplished by bubbling gaseous hydrogen chloride into the reaction mixture (189). This suggests an intermediate formation of imino chlorides from **404** and hydrogen chloride; these chlorides are then formylated to **405**, which are isolated as the crystalline acylamidinium perchlorates **406** (189).

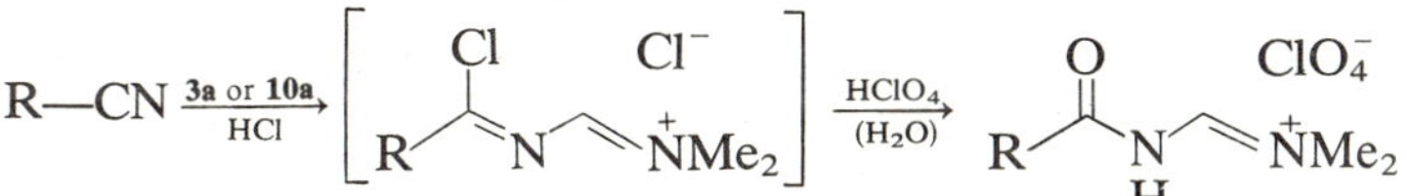

$$R-CN \xrightarrow[HCl]{3a\ or\ 10a,} \left[\begin{array}{c} Cl \quad Cl^- \\ R \diagdown N \diagup \overset{+}{N}Me_2 \end{array} \right] \xrightarrow[(H_2O)]{HClO_4} \begin{array}{c} O \quad ClO_4^- \\ R \diagdown N \diagup \overset{+}{N}Me_2 \\ H \end{array}$$

(**404**)   (**405**)   (**406**) 75–98%

R = Me, Et, Ph

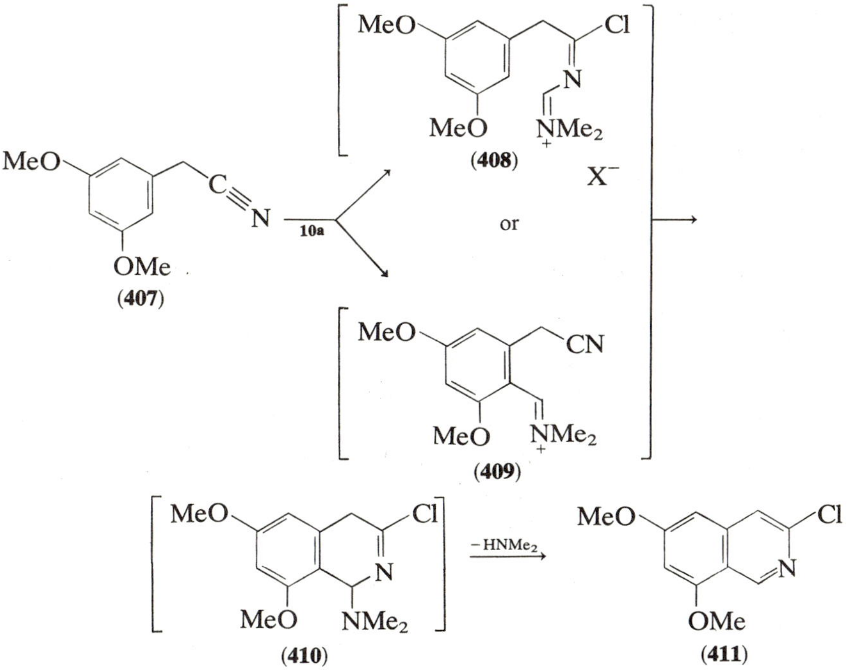

Cyclization of 3,5-dimethoxyphenylacetonitrile (**407**) by **10a** involves an intramolecular electrophilic ring closure, passing through either the chloromethyleneamidinium ion **408** or, by prior formylation of the aromatic nucleus, the iminium salt **409**, and leading to 3-chloro-6,8-dimethoxyisoquinoline (**411**) (188). The expected attack on the activated methylene group in **407** does not take place.

H. REACTIONS OF CHLOROMETHYLENIMINIUM SALTS WITH CARBOXYLIC ACIDS

1. *Substituted Acetic Acids*

Because of the work of Arnold, formylation of substituted acetic acids or their salts, where R is an aromatic or heteroaromatic ring, halogen, carboxyl, or some other substituent, has become a valuable technique for preparing many substituted trimethinium salts (**209**), 3-dimethylaminoacroleins, and malonaldehydes (16,28,31).

As shown in the preceding sections, C–C bond-forming reactions take place only when **3a** or **10a** react with sufficiently nucleophilic substrates, which may either be present as starting compounds or be formed from precursors during the course of the reaction. As is the case with enamines, enols, and enol ethers, most of the reactive compounds are olefinic in nature. What are the nucleophilic species in the formylation reactions of acetic acids **412**? Reichardt and Halbritter (190) have proposed a plausible mechanism. As is well known, **3a** or **10a** first attacks the carboxylic oxygen in **412**, forming an acid chloride (**413**) (47) and an equivalent of DMF. Reversible loss of hydrogen chloride may generate the aldoketene **414**, provided that the substituent R promotes the abstraction of a proton from **413**. Unsubstituted aliphatic carboxylic acids, therefore, fail to give formylation products in isolable amounts (16).

The highly nucleophilic ketene **414** immediately adds **3a**, thus forming the 3-dimethylaminoacrylyl chloride **415**. Electrophilic substitution of 3-dimethylaminoacrylic acid derivatives is quite possible at position 2 (194) (see **169** and **170**, Section II-C-1). The intermediate **416**, formed from **415** and DMF by analogy to dimethylcarbamyl chloride, its lower vinylogue (195), undergoes an intramolecular substitution with elimination of carbon dioxide, yielding **209**.

When chloroacetic acid was reacted with an excess of **10a** at temperatures above 70°C, instead of the expected 2-chloro-**209** the iminium salt **176a** (Section II-C-1), a derivative of the triformylmethane (**177a**), was obtained in 60% yield. The formylation of bromoacetic acid under

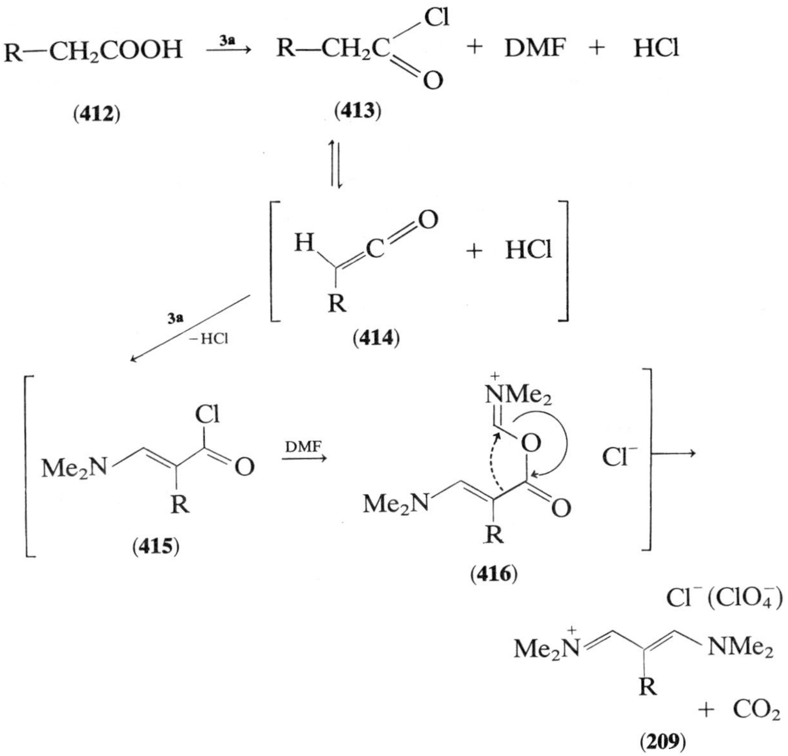

II. C—C BOND-FORMING REACTIONS

R—CH$_2$COOH $\xrightarrow{3a}$ R—CH$_2$C(Cl)=O + DMF + HCl

(412)   (413)

$\Updownarrow$

$\left[\begin{array}{c} H\\ \diagdown C=O \\ R \end{array} + HCl \right]$

(414)

$\xrightarrow[-HCl]{3a}$

$\left[\begin{array}{c} Me_2N\diagdown Cl \\ C=C\diagdown \\ R C=O \end{array} \right]$ $\xrightarrow{DMF}$ $\left[\begin{array}{c} \stackrel{+}{N}Me_2 \\ Me_2N \diagdown O \\ R O \end{array} Cl^- \right] \longrightarrow$

(415)   (416)

Cl$^-$ (ClO$_4^-$)

Me$_2\stackrel{+}{N}$=CH—C(R)=CH—NMe$_2$

+ CO$_2$

(209)

TABLE XXI
Trimethinium Salts (ClO$_4^-$) 209 Obtained by Formylation of Acetic Acids 412

R	Reaction time, hr	Reaction temp., °C	Yield, %	Ref.
Ph	3	70–90	92	16, 192
4-MeOC$_6$H$_4$	3	80–90	60	192
4-Cl—C$_6$H$_4$	3	80–90	83	192
4-Br—C$_6$H$_4$	3	80–90	65	192
4-NO$_2$C$_6$H$_4$	6	70	90	192
3,4-(MeO)$_2$C$_6$H$_3$	3	70	69	16
α-Naphthyl	3	70	39, 40	16, 192
β-Naphthyl	3	80–90	72	192
F—[a]	12	70	15	16
	0.5	80	37	190
Cl—	12	70	85, 70	16, 193, 194
β-Indolyl	3	90	90	16
COOEt[a]	4	90	58	16

[a] Isolated after hydrolysis as 3-dimethylaminoacrolein derivative.

various conditions afforded only the salt **176a**, partially as its diperbromide (191). It appears possible that in **209** the halogen is so strongly affected by the electron-deficient iminium nitrogen that a reductive dehalogenation or, in the case of bromine, a cleavage by liberation of free halogen occurs.

The formylation of both benzene-1,4-diacetic acid and benzene-1,3-diacetic acid proceeds analogously by tetrasubstitution to the corresponding bistrimethinium salts in 82% and 65% yields, respectively (26). The benzene-1,2-diacetic acid, however, leads by simultaneous cyclization to an indene derivative, the iminium salt **195** in low yield (26).

Glycine hydrochloride and its N-methyl, N-benzyl, and N-phenyl derivatives have been successfully formylated by **10a** in DMF (4 hr at 80°C, then 2 hr at 125°C) (31). A reaction of the amino group to give an amidinium derivative occurs simultaneously, and the diperchlorate **417** is isolated. Deprotonation by triethylamine transforms **417a** to **418**, a trimethinium salt with a protected but easily released amino group: a highly useful compound for synthetic purposes. Hydrolysis of **417a** by

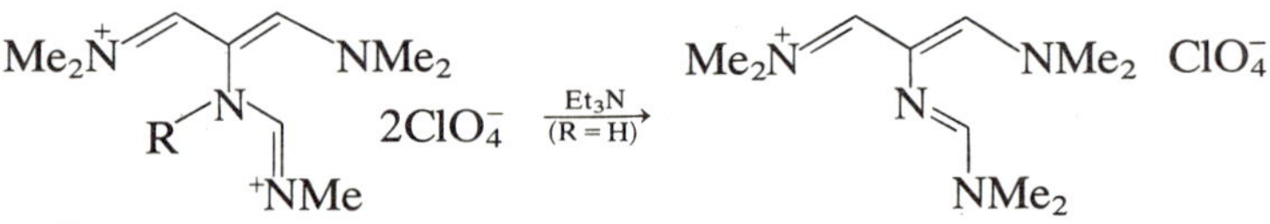

(**417a**) R = H	74%	
(**417b**) R = Me	56%	
(**417c**) R = CH$_2$Ph	62%	
(**417d**) R = Ph	62%	

(**418**) 80%

heating with 2N NaOH (3 hr at 60°C) affords aminomalonaldehyde, which is stable only as a salt, and from which (by treatment with nitrous acid) the extremely reactive diazomalonaldehyde (32) can be prepared.

2. Malonic Acids

The 2-alkyltrimethinium salts (**209**), inaccessible by the formylation of unsubstituted, aliphatic homologues of acetic acid, can be conveniently obtained by using the corresponding malonic acids (28). Adding a malonic acid to **10a** in the cold produces an immediate, vigorous evolution of carbon dioxide, accompanied by an exothermic reaction. The intermediate ketene **414** may be formed by a fragmentation of the

TABLE XXII

Trimethinium Salts (ClO$_4^-$) **209** Obtained by Formylation of Malonic Acids

R	Reaction time, hr	Reaction temp., °C	Yield, %	Ref.
Me	2	80–90	80	193
Et	2	80–90	80	193
n-Bu	2	80–90	90	193
n-Bu	6	90	31	28
PhCH$_2$	2	80–90	92	193
	6	90	41	28
Allyl[a]	6	90	50	28

[a] As 3-dimethylaminoacrolein.

initially formed half-acid chloride **419**:

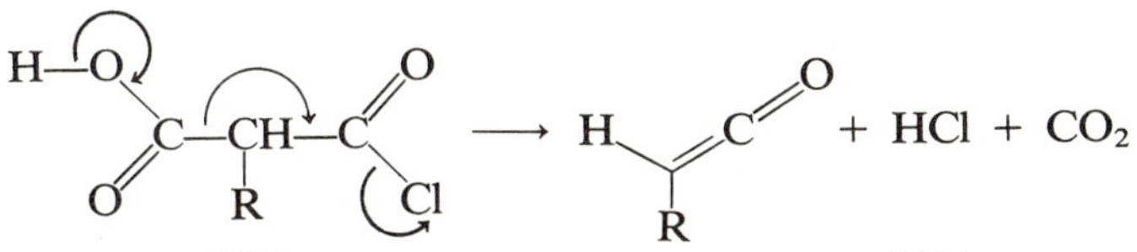

$$\underset{(419)}{\text{HO}\diagup\underset{\text{O}}{\overset{\text{O}}{\text{C}}}-\underset{\text{R}}{\text{CH}}-\underset{\text{Cl}}{\overset{\text{O}}{\text{C}}}} \longrightarrow \underset{(414)}{\text{H}\diagdown\underset{\text{R}}{\overset{\text{O}}{\text{C}}}\diagup} + \text{HCl} + \text{CO}_2$$

Not surprisingly, malonic acid itself is further substituted to **176a** (16). A 4:3:1.3 ratio of DMF to POCl$_3$ to malonic acid has been found to afford optimum yields (193). Malonic acids with bulky alkyl groups (e.g., isopropyl) failed to undergo this conversion.

I. REACTIONS OF CHLOROMETHYLENIMINIUM SALTS WITH HYDRAZONES AND AZINES

At first glance hydrazones should display behavior similar to that of their parent aldehydes and ketones with **10a**. Indeed, the phenylhydrazones of acetophenones and acetone (**420a–c**) undergo electrophilic attack on the methyl group, followed by subsequent cyclization of **420** to the iminium salts **422** (6 hr at 70–80°C). Upon hydrolysis these salts yield 1-phenylpyrazole-4-carboxaldehydes (**423a–c**) (197). The synthesis of 1-unsubstituted pyrazole-4-carboxaldehydes (**423d–g**, R^1 = H) has also been achieved by using the semicarbazones of acetophenones and of acetone (**420d–g**, R^1 = CONH$_2$) in the reaction with **10a** in DMF (4 hr at 60–70°C) (198).

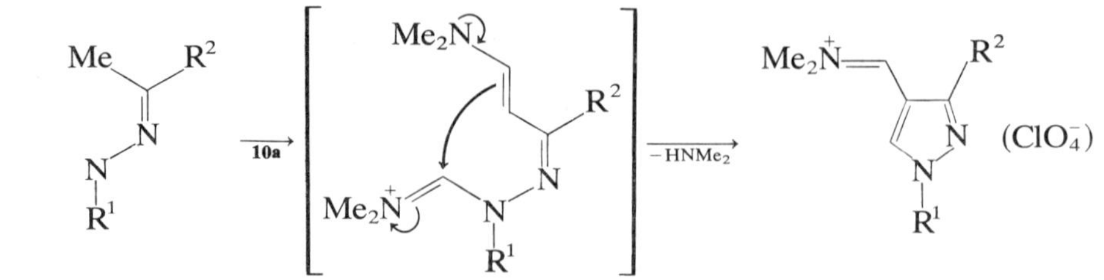

(420) R¹ = Ph, CONH₂   (421) R¹ = Ph, H   (422) OH⁻

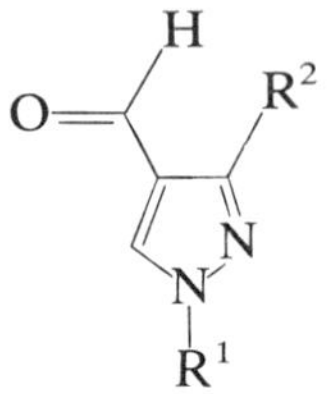

(423)

TABLE XXIII

(197, 198)

Pyrazole-4-carboxaldehydes by Formylation of Hydrazones

423	R¹	R²	Yield, %
a	Ph	Ph	96
b	Ph	4-NO₂C₆H₄	72
c	Ph	Me	77
d	H	Ph	85
e	H	2-MeOC₆H₄	95
f	H	2-Thienyl	83
g	H	Me	—

The carboxamido group was replaced by the dimethyliminiomethyl residue during the formylation process, as shown for various benzaldehyde semicarbazones (**424**), which yield the corresponding amidrazones **426**:

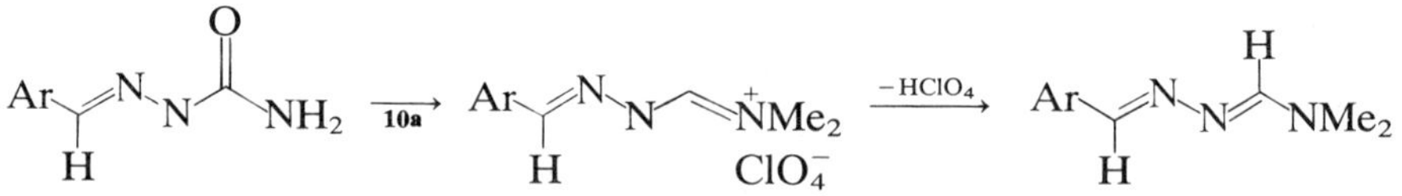

(424)   (425)   (426) (57–93%),

Ar = Ph, 2-MeOC₆H₄, 2-HOC₆H₄,   2-Cl—C₆H₄, 4-Cl—C₆H₄

This finding appears to be an argument favoring **421** as a key intermediate and precludes a proposed double formylation of the methyl group in **420** (197) before cyclization.

On the other hand, we must also consider hydrazones as "azaenamines," which, in analogy to simple enamines, should undergo electrophilic substitution at the position β to the secondary amino group (here the α-carbon atom). Such a reaction with **10a** has been observed with the N,N-tetramethylenehydrazones of benzaldehydes (**427**) (the phenylhydrazones failed to react in this way). Compound **427** affords the iminium salts **428**, which may be hydrolyzed to the hydrazones of phenylglyoxals **429** (196):

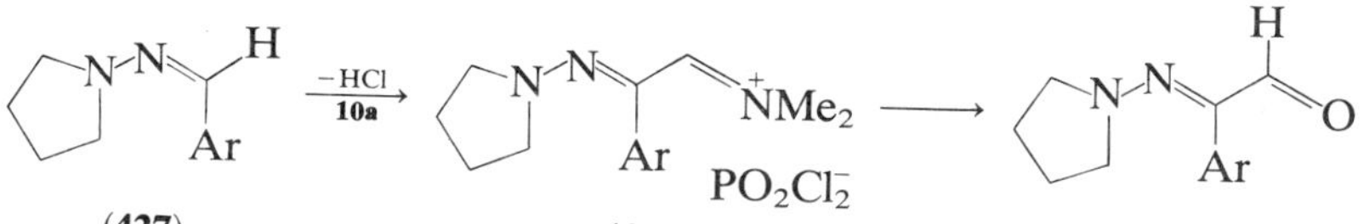

Ar = Ph, 4-MeOC$_6$H$_4$, 4-NO$_2$C$_6$H$_4$, 3-NO$_2$C$_6$H$_4$ (57–82%)

In the acetophenoneanil **430**, the imino nitrogen reacts with 1 mole of **10a** to form the isolable salt **431** (199). This reaction is analogous to the attack of **3a** on the carbonyl oxygen of acetophenone itself, affording **234** by the proposed mechanism (Section II-E).

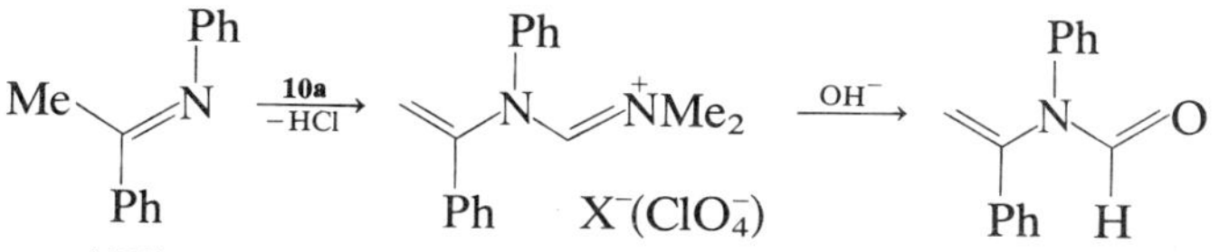

Treatment of **431** with sodium hydroxide leads to the stable N-formylenamine (**432**), whereas dilute acids hydrolyze **431** and **432** to the corresponding ketone. Acetophenoneazine, structurally related to **430**, has been converted by an excess of **10a** in nearly quantitative yield to the iminium salt **434**, hydrolysis of which gives the corresponding pyrazolealdehyde. Postulation of the intermediate **433** in this cyclizing formylation is substantiated by the observation that the acetophenone N,N-diphenylhydrazone, which cannot form such a formylation product, undergoes under the acidic conditions of the reaction a Fischer Indole-type ring closure with subsequent formylation to the 1,2-diphenylindole-3-carboxaldehyde.

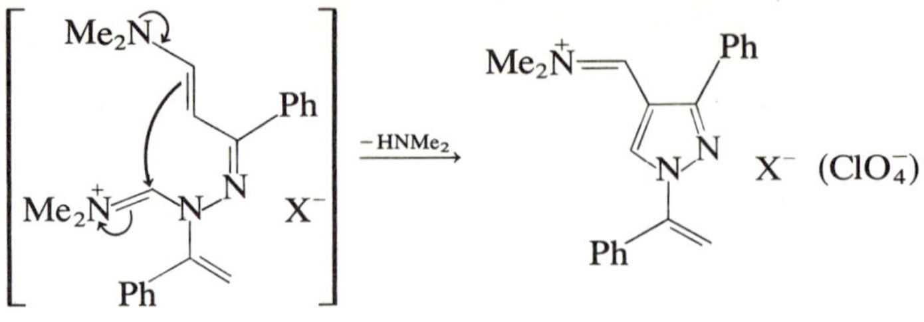

J. REACTIONS OF CHLOROMETHYLENIMINIUM SALTS WITH ALIPHATIC DIAZO COMPOUNDS

Not surprisingly, the strongly nucleophilic diazoacetophenone (**435a**) and ethyl diazoacetate (**435b**) are both very readily formylated by **3a** in chloroform (below 0°C) to the diazoiminium chlorides **436**. Half of the starting diazo compound is consumed by the released hydrogen chloride, forming the chloro compounds **437** and an equivalent of nitrogen. Careful hydrolysis with water transforms the crude **436** to the corresponding formyldiazo derivatives **438** (27):

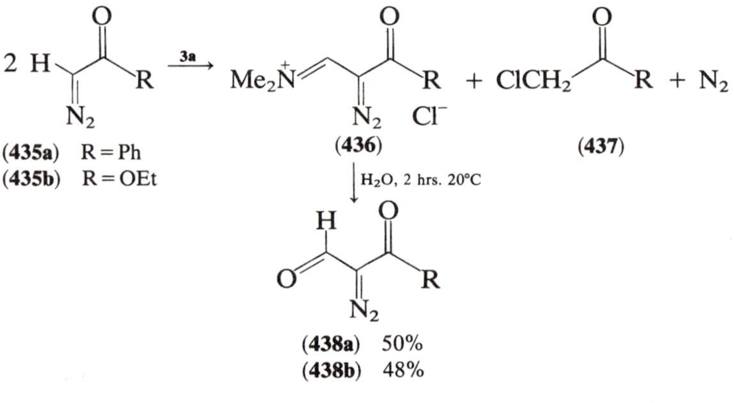

(435a) R = Ph
(435b) R = OEt

(438a) 50%
(438b) 48%

In contrast to the diazocarbonyl derivatives, treatment of the much more reactive diazomethane with **3a**, even at −65°C, does not lead to nitrogen-containing products. Rather, 1,3-dichloro-2-dimethylamino-propane is isolated in 45% yield (27).

III. Cyclizations with Chloromethyleniminium Salts

There are a considerable number of examples of reactions with **3a** or **10a** in which the formylation step of a properly substituted starting

TABLE XXIV
Survey of Cyclizations During Formylation Reactions

Derivative of	Formed from	Example	Section
Indene	Styrenes	**74–76**	II-B-1
Indene	Propioveratrone	**240–310**	II-E-4
Indene	Benzene-1,2-diacetic acid		II-H-1
Cyclopentadiene	Diphenylbutadiene	**83–86**	II-B-1
Cyclopentene	Unsat. acetal	**226–229**	II-D-3
Cyclopentene	Diethyl ketone	**301–303**	II-E-3
Chromenones	o-Hydroxyacetophenones	**311–313** **314–316**	II-E-4
4-Pyrone	Dibenzyl ketone	**320–323**	II-E-4
Diazaindoles and relatives	Aminomethylpyrazines	**393–395**	II-G-5
Isoquinoline	3,4-Dimethoxy-phenylacetonitrile	**407–411**	II-G-7
Pyrazoles	Acetophenonehydrazones	**420–423**	II-I

compound initiates a subsequent cyclization reaction. To provide a brief survey of the different classes of compounds that have been obtained by cyclization during formylation reactions, Table XXIV lists examples considered in detail in the sections cited. The examples that follow are concerned with reactions in which formylation takes place, as well as cyclization to an aromatic ring.

The cyclization step probably proceeds as the final stage of the reaction because formylation of the aromatic system that is not activated by substituents having strong electron-releasing effects is impossible. Treatment of the heptamethinium perchlorate **221b** with **3a** and DMF in chloroform (2 hr at 80°C) leads to the 1,3,5-triformylbenzene (**441**) (25). The acyclic intermediate **439**, initially formed by double formylation, undergoes a cyclization that might well be interpreted as an intramolecular, electrophilic attack on the enamine β position of **439** (25). There are indications, however, that this cyclization is, rather, an electrocyclic hexatriene-cyclohexadiene ring closure of **439'** to **440**, followed by elimination of dimethylamine (192,193,200).

With acetylacetone (**442**) a heptamethinium derivative **443** may be the intermediate; this cyclizes to 2,5-dichlorobenzaldehyde (**444**). Equivalently, 3-penten-2-one (**445a**), mesityl oxide (**445b**), and 4-dimethylamino-3-penten-2-one (**445c**) yield, by triformylation and subsequent cyclization, the 4-chloroisophthalic dialdehydes **446a–c**. With the quite similar methyl ether of acetylacetone (**445d**) the reaction takes a slightly

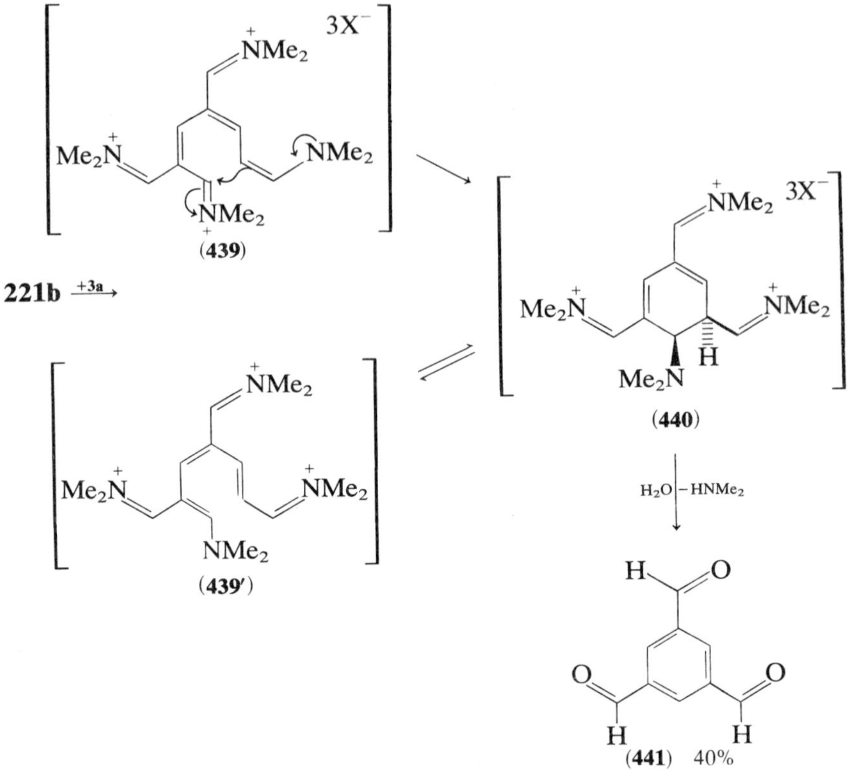

326 THE VILSMEIER-HAACK-ARNOLD ACYLATIONS

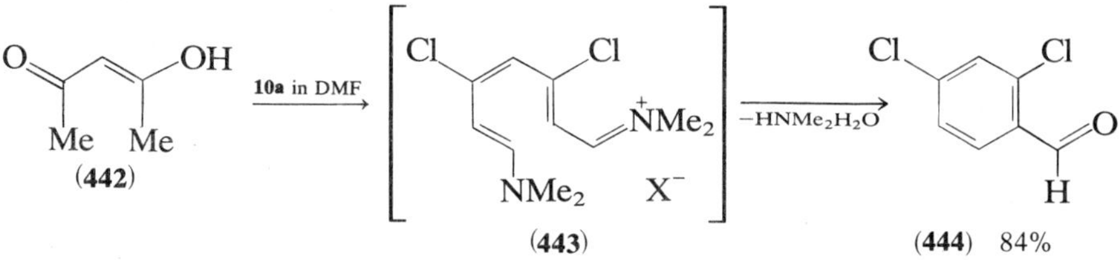

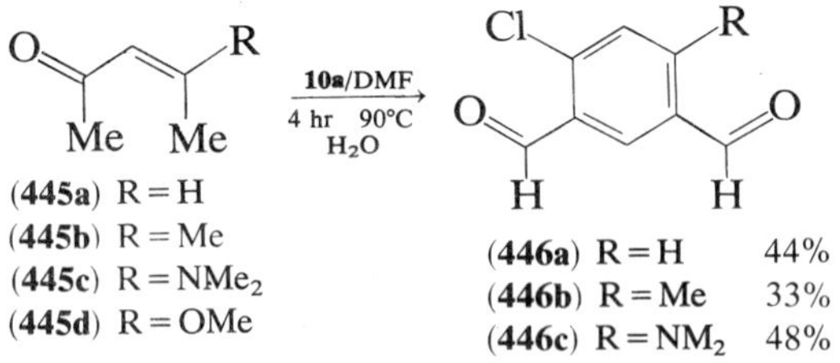

different course, and, in addition to **444** (23%), the 3-chloroanisole (**449**) has been isolated in 42% yield (25). Here the formation of **449** can be explained only by assuming an electrocyclic cyclization of the intermediary dimethylaminohexatriene (**448**).

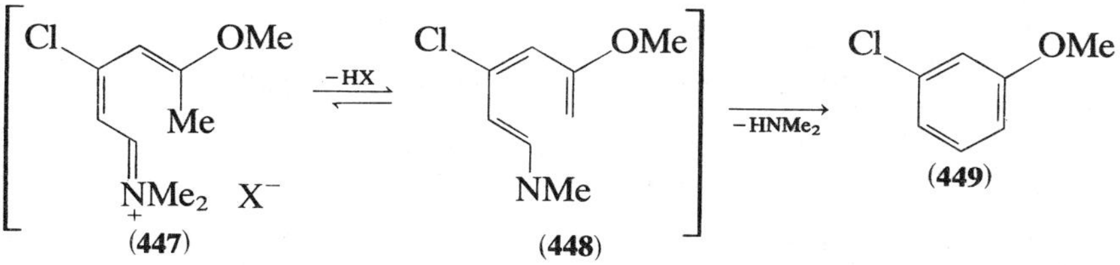

(**447**) (**448**) (**449**)

In a similar reaction, the 1-methylpentamethinium salt **450** affords 4-dimethylaminoisophthalic dialdehyde (**451**) in 64% yield. Presumably a double formylation of the methyl group occurs, and the heptamethinium salt thus formed then cyclizes in the described manner (25):

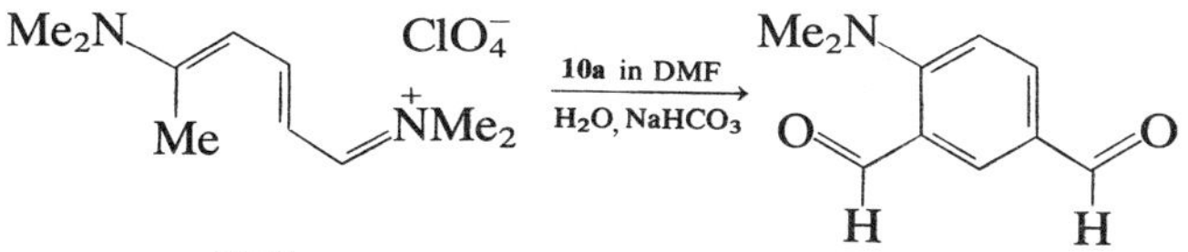

(**450**) (**451**)

Some benzylisoquinolines, such as the methylpapaverines **452** and their isomers **454**, give cyclization products (i.e., the dehydroberberinium salts **453** and **455**) on treatment with **10a** (201). An electrophilic substitution

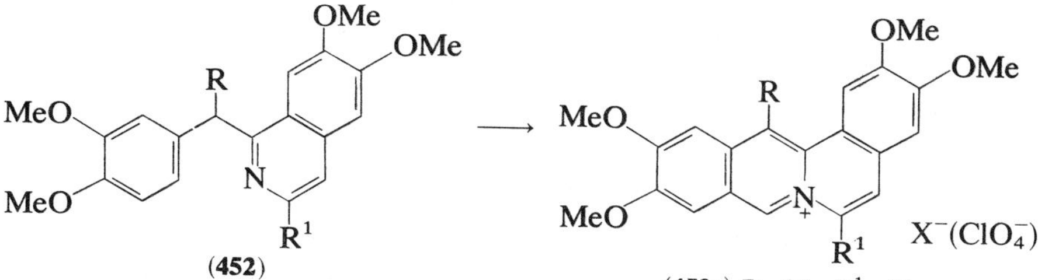

(**452**)

(**453a**) R = Me, R[1] = H
(**453b**) R = H, R[1] = Me

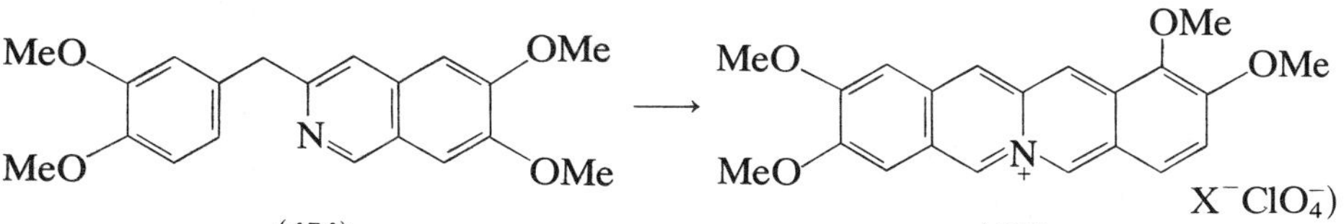

(**454**) (**455**)

of the benzene ring, introducing an aldiminium group, may occur, followed by ring closure and subsequent elimination of dimethylamine.

Similarly, treatment of 3,3'-dithienylmethane with **10a** gives a 33% yield of benzo(1,2-*b*;5,4-*b'*)dithiophene (206).

IV. Vilsmeier-Haack-Arnold Related Acylations

Since the Vilsmeier-Haack-Arnold formylations are among the most valuable and widely used reactions in organic chemistry, there have been many attempts to extend their scope to similar systems. The use of tertiary acyl amides (other than those of formic acid) as imino chlorides or complexes with phosphoryl chloride is limited to a few highly nucleophilic substrates which are then formylated in the usual manner (33–37). The amide chlorides of fatty acids (**456**), related to **3a** or **10a**, tend to lose a proton, forming chloroenamines (**457**), which then react with a second molecule of **456**, affording **458** and, after hydrolysis, **459**. This smooth self-condensation therefore parallels the reaction of **327** with **3a** or **10a** (Section II-F) (202,203).

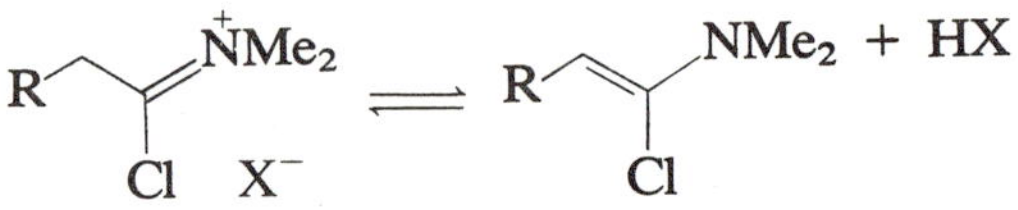

By analogy to the preceding reaction, N-alkyl-2-piperidone should condense in the presence of POCl$_3$ to yield **460**. Indeed, the β-ketoamide **461**, formed by conventional hydrolysis of **460**, and the tricyclic iminium salt **462** have been synthesized (207). The latter compound may arise from **460** by an intramolecular elimination of hydrogen chloride (See p. 329).

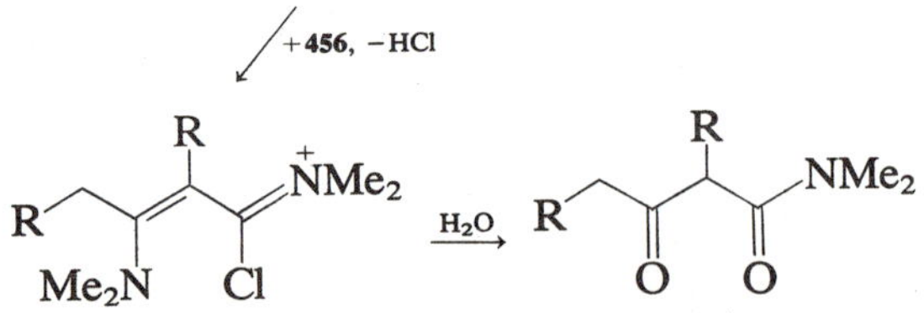

Thermolysis of N,N-dimethylcyanoacetamide (**333c**) with twice the amount of POCl$_3$ (2 hr at 100°C) should lead to the condensation product **463**. However, this intermediate then undergoes an intramolecular Ritter reaction, forming the 6-chloro-2,4-bis(dimethylamino)pyridine-3-carbonitrile **464** in 84% yield (204a):

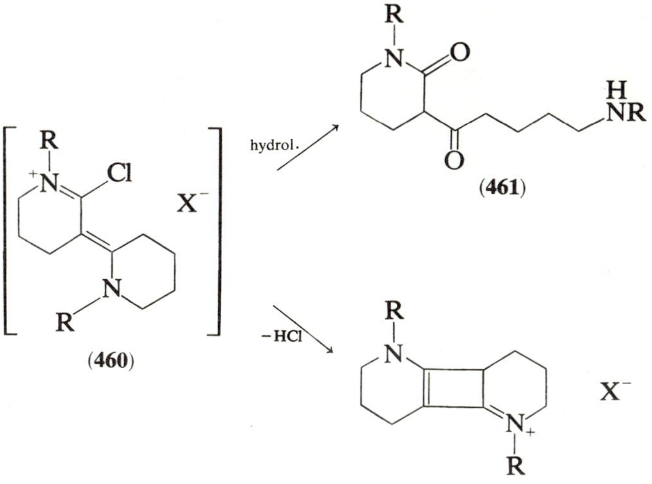

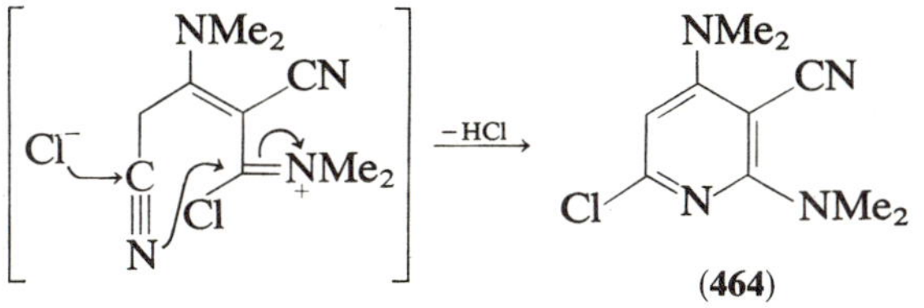

The corresponding C-alkylated N,N-dialkylcyanoacetamides do not react in this manner, except when the alkyl group is an isopropyl or benzyl residue, in which case an elimination of one group during the condensation-cyclization has been observed. A mixed condensation with N,N-dimethylphenylacetamide has been achieved, affording 5-alkyl-6-chloro-4-dialkylamino-2-dimethylamino-3-phenylpyridine (204c).

The mono N-methylcyanacetamide also condenses with POCl$_3$ but then cyclizes in a slightly different way to 6-amino-2-chloro-4-methylamino-1-methylpyridinium-3-carbonitrile chloride (204b). 3-Aminocrotononitrile and 3-aminocinnamonitrile react with **10a** β to the enamine nitrogen. However, treatment of 3-aminocinnamonitrile with N,N-dimethylbenzamide and POCl$_3$ yields 6-chloro-2,4-diphenyl-pyrimidine (**466**) by cyclization of the intermediate amidinium salt (**465**) (205):

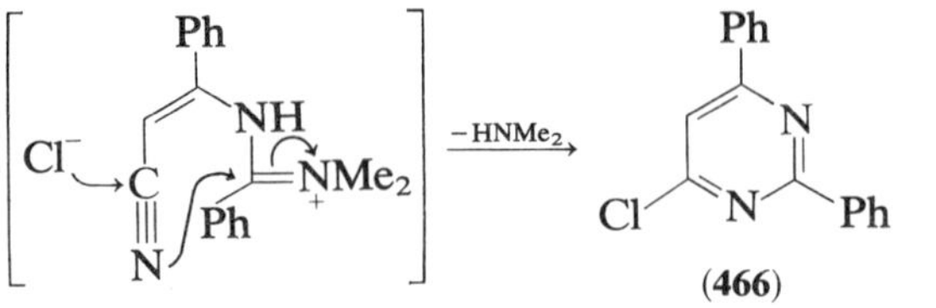

(**465**) →[−HNMe₂] (**466**)

The weakly electrophilic N,N-dialkylethoxycarbonylacetamide–POCl₃ complex transforms β-naphtholes into 3-dialkylamino-1-oxo-1H-naphtho(2,1-b)pyrans, which give, after hydrolysis, 1-hydroxy-3-oxo-3H-naphtho(2,1-b)pyrans (208).

A fascinating chemistry has been developed about the use of unsubstituted formamide with phosphoryl chloride (209). Formamide–POCl₃ alone gives a 50% yield of adenine, and, in mixture with N-methylformamide, a low yield of 7-methyladenine is realized. With formamide–POCl₃ and straight-chain fatty acid amides, 4-amino-5-alkylpyrimidines (16–32%) have been obtained. Cyclic amides such as pyrrolidone and piperidone react with the reagent to yield the corresponding bicyclic pyrimidines. Depending on the substitution in the benzene nucleus, desoxybenzoins and formamide–POCl₃ yield either 4,5-diphenylpyrimidines or 3-phenylquinolines, whereas benzoins give 4,5-diphenylimidazoles.

4-Chloropyrylium and 4-chloroflavylium salts, formed from the reaction of pyrones or flavones with POCl₃, represent highly electrophilic agents, comparable in reactivity to **3a** or **10a**. They react with N,N-dimethylacetamide through the intermediacy of chloroenamine (**331a**) to yield pyranylideneiminium salts (e.g., **467**) (210):

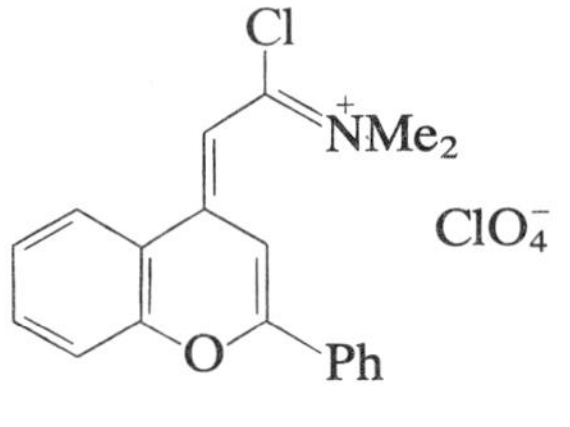

(**467**) 75%

A. VINYLOGOUS FORMYLATIONS

One of the most important extensions of the Vilsmeier-Haack-Arnold reaction may be the reaction of vinylogous formamides. Obviously the 3-chloroallylidenedimethyliminium ion **16**, formed from 3-dimethylamino-

IV. VILSMEIER-HAACK-ARNOLD RELATED ACYLATIONS 331

acrolein (DAA) and POCl$_3$ or carbonyl chloride, shows only weakly electrophilic properties because of high resonance stabilization and a basic nitrogen. The related POCl$_3$ complex, prepared from the corresponding 3-methylanilinoacrolein (MAA), appears to be more electrophilic. The POCl$_3$ complex with the next higher vinylogue, 5-methylanilinopenta-2,4-dienal (Zincke aldehyde) (MAP), has been used with some success (211). In these complexes phosphoryl chloride is replaced by hexachlorocyclotriphosphazatriene, N$_3$P$_3$Cl$_6$, which appears to form adducts similar to those of cyanuric chloride (**6** in Section I-B) (212).

Similarly, acetic anhydride (211,213–215) and acyl bromides (213) have been employed to activate MAA and MAP to electrophilic substitutions, by generating the corresponding, highly reactive *O*-acyliminium ions. In all these reactions, as in formylations with **3a** and **10a**, the first isolable products are deeply colored polyeniminium salts (211,213).

TABLE XXV

Starting compound	Product	Reagent	Yield, %	Ref.
N,N-Dimethylaniline	4-Me$_2$N-cinnamaldehyde	MAA POCl$_3$	70–80	211
N,N-Dimethylaniline	4-Me$_2$N-cinnamaldehyde	MAA N$_3$P$_3$Cl$_6$	54	212
N,N-Dimethylaniline	4-Me$_2$N-cinnamaldehyde	MAA PhCOBr	10–23	213
N,N-Diethylaniline	4-Et$_2$N-cinnamaldehyde	MAA POCl$_3$	84	211
1-Phenyl-1-(4'-dimethylaminophenyl)ethylene	1-Phenyl-1-(4'-dimethylaminophenyl)pentadienal	MAA POCl$_3$	80–95	211
Resorcin dimethyl ether	2,4-Dimethoxycinnamaldehyde	MAA POCl$_3$	90	211
N,N-Dimethylaniline	4-Me$_2$N-phenylpentadienal	MAP POCl$_3$	18–20	211
N,N-Dimethylaniline	4-Me$_2$N-phenylpentadienal	MAP N$_3$P$_3$Cl$_6$	—	212
Azulene	3-(Azulene-1')-acrolein	MAA POCl$_3$	97	211
Azulene	5-(Azulene-1')-2,4-pentadienal	MAP POCl$_3$	90–95	211
Other azulenes	Azulenepolyenals	MAA, MAP	ca. 90	211, 100
6-Dimethylaminofulvene	6-Dimethylamino-2-(3'-dimethyliminio-allyl)fulvene perchlorate	DAA (COCl)$_2$	—	100

Thus a synthesis of otherwise inaccessible substituted pentamethinium salts **469** has been achieved by the reaction of esters of the 3-dimethylaminoacrylic acids **468** with 3-dimethylaminoacroleins **203** in acetic anhydride–acetic acid in the presence of pyridine perchlorate (214). If one starts with the *tert*-butyl esters of **468**, an elimination of the *tert*-butyloxycarbonyl group in **469** may be effected, without saponification, by treatment with hydrogen bromide in glacial acetic acid (214):

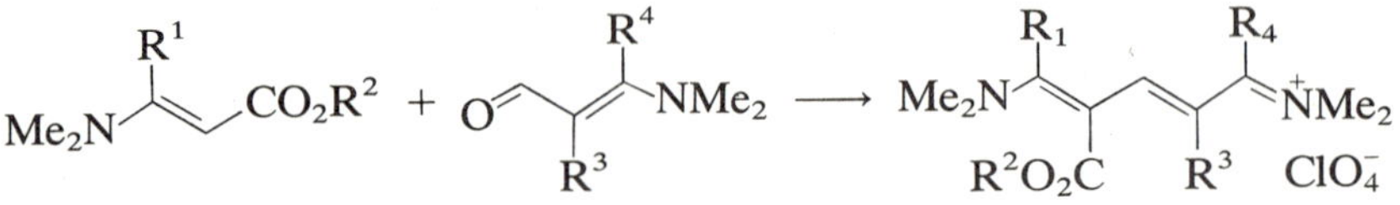

$$Me_2N-CR^1=CH-CO_2R^2 \;+\; OHC-CR^3=CR^4-NMe_2 \;\longrightarrow\; Me_2N-CR^1=C(CO_2R^2)-CH=CR^3-CR^4=\overset{+}{N}Me_2 \; ClO_4^-$$

(**468**)   (**203**)   (**469**) 71–99%

$R^1, R^3 = H$, Me; $R^2 =$ Me, Et, CMe$_3$; $R^4 = H$, Ph

The self-condensation of **203a** ($R^3, R^4 = H$) also occurs under the reaction conditions, and 4-formyl-5-dimethylaminopenta-2,4-dienal (**186**) (see Section II-C) may be isolated in 18% yield (214). Analogously, the pentamethinium salt **470** has been obtained by condensing the N,N-dimethylcyclopentanoneiminium perchlorate with MAA in acetic anhydride with a trace of pyridine (215,216).

Similarly, MAA and PAA condense, in the presence of acetic anhydride, with activated methylene compounds (e.g., phenacylpyridinium bromides). Thus the cyanine dyes **471** have been prepared in excellent yields (213).

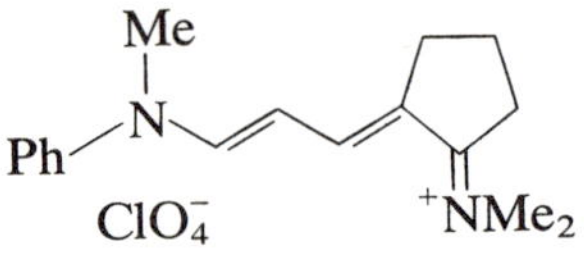

(**470**) 75%

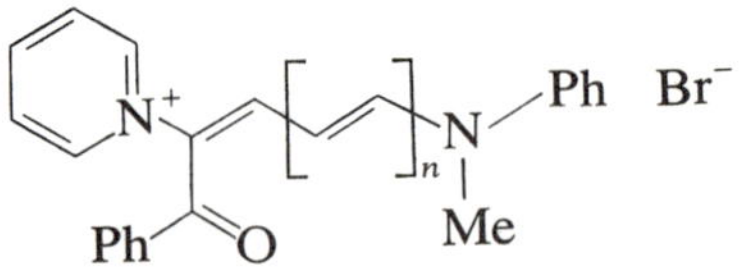

471 $n = 1, 2$

Many vinylogous amides might well be tried as potential candidates for reactions like the Vilsmeier-Haack-Arnold formylations. Thus *p*-dimethylaminobenzaldehyde, which may be considered the phenylogue of DMF, condenses with N,N-dimethylaniline in the presence of POCl$_3$ to give Michlers hydrol blue (211). It also reacts with **468** (214) and phenacylpyridinium bromide (213) in the presence of acetic anhydride.

With these last examples we reach the end of our discussion of the Vilsmeier-Haack-Arnold reaction. To terminate this chapter, the synthesis of bipyrroles (**472**) from the reaction of pyrrolin-2-ones and pyrroles with $POCl_3$ is depicted (217):

(**472**) 79%

REFERENCES

1. O. Dimroth and R. Zoeppritz, *Ber. Dtsch. Chem. Ges.*, **35,** 995 (1902).
2. A. Vilsmeier and A. Haack, *Ber. Dtsch. Chem. Ges.*, **60,** 119 (1927).
3. A. Vilsmeier, *Chem. Z.*, **75,** 133 (1951).
4. R. Witzinger, *J. Prakt. Chem.*, **154,** 25 (1939); H. Lorenz and R. Witzinger, *Helv. Chim. Acta*, **28,** 600 (1945).
5. Z. Arnold and F. Sorm, *Collect. Czech. Chem. Commun.*, **23,** 452 (1958).
6. Z. Arnold and J. Zemlicka, *Collect. Czech. Chem. Commun.*, **24,** 786 (1959).
7. Z. Arnold and J. Zemlicka, *Collect. Czech. Chem. Commun.*, **24,** 2378 (1959).
8. Z. Arnold and J. Zemlicka, *Collect. Czech. Chem. Commun.*, **24,** 2385 (1959).
9. Z. Arnold, *Collect. Czech. Chem. Commun.*, **25,** 1308 (1960).
10. Z. Arnold, *Collect. Czech. Chem. Commun.*, **25,** 1313 (1960).
10a. W. von E. Doering, M. Saunders, H. G. Boyton, H. W. Earhart, E. F. Wadley, W. R. Edwards, and G. Laber, *Tetrahedron*, **4,** 178 (1958).
11. Z. Arnold and J. Zemlicka, *Collect. Czech. Chem. Commun.*, **25,** 1318 (1960).
12. Z. Arnold, *Collect. Czech. Chem. Commun.*, **26,** 1113 (1961).
13. Z. Arnold, *Collect. Czech. Chem. Commun.*, **26,** 1723 (1961).
14. J. Zemlicka and Z. Arnold, *Collect. Czech. Chem. Commun.*, **26,** 2838 (1961).
15. J. Zemlicka and Z. Arnold, *Collect. Czech. Chem. Commun.*, **26,** 2852 (1961).
16. Z. Arnold, *Collect. Czech. Chem. Commun.*, **26,** 3051 (1961).
17. Z. Arnold and A. Holy, *Collect. Czech. Chem. Commun.*, **26,** 3059 (1961).
18. Z. Arnold and A. Holy, *Collect. Czech. Chem. Commun.*, **27,** 2886 (1962).
19. Z. Arnold, *Collect. Czech. Chem. Commun.*, **27,** 2993 (1962).
20. Z. Arnold, *Collect. Czech. Chem. Commun.*, **28,** 863 (1963).
21. Z. Arnold and A. Holy, *Collect. Czech. Chem. Commun.*, **28,** 869 (1963).
22. Z. Arnold and A. Holy, *Collect. Czech. Chem. Commun.*, **28,** 2040 (1963).
23. Z. Arnold, *Collect. Czech. Chem. Commun.*, **28,** 2047 (1963).
24. Z. Arnold, *Collect. Czech. Chem. Commun.*, **30,** 47 (1965).
25. A. Holy and Z. Arnold, *Collect. Czech. Chem. Commun.*, **30,** 53 (1965).
26. Z. Arnold, *Collect. Czech. Chem. Commun.*, **30,** 2783 (1965).
27. F. M. Stojanovic and Z. Arnold, *Collect. Czech. Chem. Commun.*, **32,** 2155 (1967).
28. J. Kucera and Z. Arnold, *Collect. Czech. Chem. Commun.*, **32,** 3792 (1967).
29. Z. Arnold, *Collect. Czech. Chem. Commun.*, **38,** 1168 (1973).
30. A. Holy and Z. Arnold, *Collect. Czech. Chem. Commun.*, **38,** 1371 (1973).
31. Z. Arnold, J. Sauliova, and V. Krchnak, *Collect. Czech. Chem. Commun.*, **38,** 2633 (1973).

32. Z. Arnold and J. Sauliova, *Collect. Czech. Chem. Commun.*, **38**, 2641 (1973).
33. O. Bayer, in *Methoden der Organischen Chemie* (Houben-Weyl), Vol. VII/1, Sauerstoff-Verbindungen II, Georg Thieme Verlag, Stuttgart, 1954, pp. 29–36.
34. V. I. Minkin and G. N. Dorofeenko, *Russ. Chem. Rev.*, **29**, 599–618 (1960).
35. M. R. de Maheas, *Bull. Soc. Chim. Fr.* **1962**, 1989–1999.
36. G. A. Olah and St. J. Kuhn, in *Friedel Crafts and Related Reactions*, G. A. Olah, Ed., Vol. III, Part 2, Interscience, New York, 1964, pp. 1211–1256.
37. G. Hazebroucq, *Ann. Pharm. Fr.*, **24**, 793–806 (1966).
38. H. Ulrich, in *The Chemistry of Imidoyl Halides*, Plenum Press, New York, 1968, pp. 87–97; C. Jutz, *Chem. Labor Betrieb*, **19**, 290–294, 337–343 (1968).
39. D. Burn, *Chem. Ind.*, **1973**, 870–873; J. S. Pizey, in *Synthetic Reagents*, Vol. 1, Ellis Horwood, Ed., John Wiley and Sons, New York, 1974, pp. 54–70.
40. H. A. Staab and A. P. Datta, *Angew. Chem.*, **75**, 1203 (1963); *Angew. Chem. Int. Ed.*, **3**, 132 (1964).
41. F. Wenzel and L. Bellak, *Monatsh. Chem.*, **35**, 965 (1914).
42. A. Rieche, H. Gross, and E. Höft, *Chem. Ber.*, **93**, 88 (1960).
43. H. Gold, *Angew. Chem.*, **72**, 956 (1960).
44. M. D. Scott and H. Spedding, *J. Chem. Soc. C*, **1968**, 1603.
45. R. Oda and K. Yamamoto, *Nippon Kagaku Zasshi*, **83**, 1202 (1962).
46. H. H. Bosshard and H. Zollinger, *Helv. Chim. Acta*, **42**, 1659 (1959).
47. H. H. Bosshard, R. Mory, M. Schmid, and H. Zollinger, *Helv. Chim. Acta*, **42**, 1653 (1959).
48. C. Jutz, *Chem. Ber.*, **91**, 850 (1958).
49. Z. Arnold, *Chem. Listy*, **52**, 2013 (1958).
50. G. J. Martin and M. Martin, *Bull. Soc. Chim. Fr.*, **1963**, 1637.
51. M. L. Filleux-Blanchard, M. T. Quemeneur, and G. J. Martin, *J. Chem. Soc. Chem. Commun.*, **1968**, 836.
52. G. J. Martin, S. Poignant, M. L. Filleux, and M. T. Quemeneur, *Tetrahedron Lett.*, **1970**, 5061.
53. G. J. Martin and S. Poignant, *J. Chem. Soc. Perkin II*, **1972**, 1964.
54. G. J. Martin and S. Poignant, *J. Chem. Soc. Perkin II*, **1974**, 642.
55. P. Linda, G. Marino, and S. Santini, *Tetrahedron Lett.*, **1970**, 4223.
56. S. Alunni, P. Linda, G. Marino, S. Santini, and G. Savelli, *J. Chem. Soc. Perkin II*, **1972**, 2070.
57. P. Linda, A. Lucarelli, G. Marino, and G. Savelli, *J. Chem. Soc. Perkin II*, **1974**, 1610.
58. H. Bredereck, R. Gompper, K. Klemm, and H. Rempfer, *Chem. Ber.*, **92**, 837 (1959).
59. H. Fritz and R. Oehl, *Liebigs Ann. Chem.*, **749**, 159 (1971).
60. G. Ferre and A.-L. Palomo, *Tetrahedron Lett.*, **1969**, 2161; K. Kugawa and T. Kawashima, *Chem. Pharm. Bull.*, **19**, 2629 (1971).
61. F. S. Fawcett, C. W. Tullock, and D. D. Coffman, *J. Amer. Chem. Soc.*, **84**, 4275 (1962).
62. C. Jutz, unpublished results.
63. R. M. Wagner, Dissertation, Tec. Univ., Munich, 1972.
64. F. Dallacker and F.-E. Eschelbach, *Liebigs Ann. Chem.*, **689**, 171 (1965).
65. J. G. Dingwall, D. H. Reid, and K. Wade, *J. Chem. Soc. C*, **1969**, 913.
66. A. Streitwieser, Jr., *Molecular Orbital Theory for Organic Chemists*, John Wiley and Sons, New York, 1961, p. 335.

67. C. Grundmann and J. M. Dean, *Angew. Chem.*, **77**, 966 (1965); C. Grundmann and H. Hooks, *Angew. Chem.*, **78**, 747 (1966).
68. K. Hafner and C. Bernhard, *Angew. Chem.*, **69**, 533 (1957); *Liebigs Ann. Chem.*, **625**, 108 (1959).
69. F. Dallacker, D. Bernabei, R. Katzke, and P.-H. Benders, *Chem. Ber.*, **104**, 2526 (1971).
70. S. Akabori, *Bull. Chem. Soc. Jap.*, **1**, 96 (1926).
71. G. F. Smith, *J. Chem. Soc.*, **1954**, 3842.
72. H. K. Hall, Jr., *J. Amer. Chem. Soc.*, **78**, 2717 (1956).
73. D. E. Horning and J. M. Muchowski, *Can. J. Chem.*, **48**, 192 (1972).
74. P. S. Clezy and A. J. Liepa, *Aust. J. Chem.*, **23**, 2461 (1970).
75. J.-P. Marquet, E. Bisagni, and J. A. Louisfert, *Bull. Soc. Chim. Fr.*, **1973**, 2323.
76. N. Gjøs and S. Gronowitz, *Acta Chem. Scand.*, **24**, 99 (1970).
77. A. N. Sukiasjan, V. P. Litvinov, and Y. L. Gol'dfarb, *Izv. Akad. Nauk SSSR, Ser. Khim.*, **1970**, 1345.
78. P. A. Konstantinov, N. M. Koloskova, R. I. Shupik, and M. N. Volkov, *Zh. Obshch. Khim.*, **43**, 872 (1973).
79. Y. L. Gol'dfarb and M. A. Kalik, *Khim. Geterotsikl. Soedin.*, **1971**, 171.
80. C. H. Tilford, W. J. Hudak, and R. E. Lewis, *J. Med. Chem.*, **14**, 328 (1971).
81. C. F. Candy, R. A. Jones, and P. H. Wright, *J. Chem. Soc. C*, **1970**, 2563.
82. B. Bonnett and G. F. Stephenson, *J. Org. Chem.*, **30**, 2891 (1965).
83. R. Royer and L. Rene, *Bull. Soc. Chim. Fr.*, **1970**, 1037.
84. R. Royer, P. Demerseman, J. F. Rossignol, and A. Chetin, *Bull. Soc. Chim. Fr.*, **1971**, 2072.
85. V. I. Shvedov, A. K. Chizhov, and A. N. Grinev, *Khim. Geterotsikl. Soedin.*, **1971**, 339.
86. S. McKenzie and D. H. Reid, *J. Chem. Soc. C*, **1970**, 145.
87. N. Y. Koshelev, A. V. Reznichenko, L. S. Efros, and I. Y. Kvitko, *Zh. Org. Khim.*, **1973**, 2201.
88. M. Y. Uritskaya, V. V. Vasil'eva, S. S. Liberman, and L. N. Yakhontov, *Khim. Farm. Zh.*, **7**, 8 (1973).
89. G. Duguay, D. H. Reid, K. O. Wade, and R. G. Webster, *J. Chem. Soc. C*, **1971**, 2829.
90. J. Bignebat and H. Quiniou, *C. R. Acad. Sci. Paris*, **269C**, 1129 (1969); J. Bignebat and H. Quiniou, *Bull. Soc. Chim. Fr.*, **1972**, 4181.
91. C. J. Schmidle and P. G. Barnett, *J. Amer. Chem. Soc.*, **78**, 3209 (1956).
92. C. Jutz and R. Heinicke, *Chem. Ber.*, **102**, 623 (1969).
93. C. Jutz, W. Müller, and E. Müller, *Chem. Ber.*, **99**, 2479 (1966).
94. H. Hartmann, *J. Prakt. Chem.*, **312**, 1194 (1970).
95. A. W. Nichol, *J. Chem. Soc. C*, **1970**, 903.
96. D. T. Witiak, D. R. Williams, S. V. Kakodkar, G. Hite, and M.-S. Shen, *J. Org. Chem.*, **39**, 1242 (1974).
97. E. J. Seus, *J. Org. Chem.*, **30**, 2818 (1965).
98. C. Jutz and W. Müller, *Chem. Ber.*, **100**, 1536 (1967).
99. M. J. Grimwade and M. G. Lester, *Tetrahedron*, **25**, 4535 (1969).
100. K. Hafner and K. H. Vöpel, *Angew. Chem.*, **71**, 672 (1959); K. Hafner, *Angew. Chem.*, **72**, 514 (1960); K. Hafner and M. Kreuder, *Angew. Chem.*, **73**, 657 (1961); K. Hafner, K. H. Vöpel, and C. König, *Liebigs Ann. Chem.*, **666**, 52 (1963); K. Hafner, K. H. Häfner, C. König, M. Kreuder, G. Ploss, G. Schulz, E. Sturm, and K. H. Vöpel, *Angew. Chem.*, **75**, 35 (1963).

101. H. Meerwein, W. Florian, N. Schön, and G. Stopp, *Liebigs Ann. Chem.*, **641,** 7 (1960); Z. Arnold and J. Zemlicka, *Collect. Czech. Chem. Commun.*, **25,** 1302 (1960).
102. C. Jutz, R. Kirchlechner, and W. Müller, *Angew. Chem.*, **77,** 1027 (1965); R. Kirchlechner, Thesis, Tec. Univ., Munich, 1965.
103. N. Roh and G. Kochendörfer, Ger. Pat. 677,207 (1937); Z., **11,** 3195 (1939).
104. French Pat. 839,359; Z., **11,** 1774 (1939).
105. W. Ziegenbein, *Angew. Chem.*, **77,** 380 (1965).
106. C. Reichardt and K. Schagerer, *Angew. Chem.*, **85,** 346 (1973).
107. J. Kucera and Z. Arnold, *Collect. Czech. Chem. Commun.*, **32,** 1704 (1967).
108. H. R. Müller and M. Seefelder, *Liebigs Ann. Chem.*, **728,** 88 (1969).
109. M. Seefelder, French Pat. 150, 1589 (BASF).
110. C. Jutz and H. Amschler, *Chem. Ber.*, **97,** 3331 (1964).
111. A. Holy, J. Krupicka, and Z. Arnold, *Collect. Czech. Chem. Commun.*, **30,** 4127 (1965).
112. Z. Arnold, *Experientia*, **15,** 415 (1959); Czech. Pat. 97,162; *C.A.*, **56,** 3466c.
113. R. Sciaky, U. Pallini, and A. Consonni, *Gazz. Chim. Ital.* **96,** 1284 (1966).
114. D. Burn, G. Cooley, M. T. Davies, J. W. Ducker, B. Ellis, P. Feather, A. K. Hiscock, D. N. Kirk, A. P. Leftwick, V. Petrow, and D. M. Williamson, *Tetrahedron*, **20,** 597 (1964).
115. S. M. Makin, O. Shavrygina, M. T. Berezhnaya, and T. P. Kolobova, *Zh. Org. Khim.*, **8,** 1394 (1972).
116. H. Bredereck, F. Effenberger, and G. Simchen, Ger. Pat. B65,348 IVb/120 (1961); *Chem. Ber.*, **96,** 1350 (1963); H. Bredereck, F. Effenberger, and D. Zeyfang, *Angew. Chem.*, **77,** 219 (1965).
117. A. Ferwanah, W. Pressler, and C. Reichardt, *Tetrahedron Lett.*, **1973,** 3979; C. Reichardt and E.-U. Würthwein, *Synthesis*, **1973,** 604.
118. B. Eistert and F. Haupter, *Chem. Ber.*, **92,** 1921 (1959).
119. M. G. Lester, V. Petrow, and O. Stephenson, *Tetrahedron*, **20,** 1407 (1964).
120. T. Sugasawa, K. Sasakura, and T. Toyoda, *Chem. Pharm. Bull.*, **22,** 763 (1974).
121. S. M. Makin, O. A. Shavrygina, M. I. Berezhnaya, and G. V. Kirillova, *Zh. Org. Khim.*, **8,** 674 (1972).
122. I. N. Nazarov, I. I. Nazarova, and I. V. Torgov, *Dokl. Akad. Nauk SSSR*, **122,** 82 (1958).
123. H. Normant and G. Martin, *Bull. Soc. Chim. Fr.*, **1963,** 1646.
124. Z. Arnold and J. Zemlicka, *Proc. Chem. Soc.*, **1958,** 227; Czech. Pat. 91,565; *C.A.*, **56,** 3358i.
125. G. Kalischer, A. Scheyer, and K. Keller, Ger. Pat. 514,415 (1927); *C. A.*, **25,** 1536.
126. A. E. Pohland and W. R. Benson, *Chem. Rev.*, **66,** 161 (1966).
127. H. Laurent, G. Schulz, and R. Wiechert, *Chem. Ber.*, **99,** 3057 (1966).
128. H. Laurent and R. Wiechert, *Chem. Ber.*, **101,** 2393 (1968).
129. G. W. Moersch and W. A. Neuklis, *J. Chem. Soc.*, **1965,** 788.
130. A. Consonni, F. Mancini, U. Pallini, B. Patelli, and R. Sciaky, *Gazz. Chim. Ital.*, **100,** 244 (1970).
131. K. Bodendorf and R. Mayer, *Chem. Ber.*, **98,** 3554 (1965); K. Bodendorf and P. Kloss, *Angew. Chem.*, **75,** 139 (1963).
132. K. Bodendorf and R. Mayer, *Chem. Ber.*, **98,** 3565 (1965).
133. V. Dressler and K. Bodendorf, *Arch. Pharm.*, **303,** 481 (1970).
134. W. Ziegenbein and W. Franke, *Angew. Chem.*, **71,** 573, 628 (1959); W. Ziegenbein and W. Lang, *Chem. Ber.*, **93,** 2743 (1960).

135. M. A. Volodina, A. P. Terent'ev, L. G. Roshcupkina, and V. G. Mishina, *Zh. Obshch. Khim.*, **34,** 469 (1964).
136. M. Weissenfels, H. Schurig, and G. Huehsam, *Z. Chem.*, **6,** 471 (1966).
137. H. Schellhorn, S. Hauptmann, and H. Frischleder, *Z. Chem.*, **13,** 97 (1973).
138. M. Weissenfels, M. Pulst, and P. Schneider, *Z. Chem.*, **13,** 175 (1973).
139. D. T. Drewry and R. M. Scrowsdon, *J. Chem. Soc. C*, **1969,** 2750.
140. C. Simionescu and M. Pastravanu, *Bull. Acad. Pol. Sci., Ser. Sci. Chim.*, **20,** 505 (1972).
141. T. S. Liksandru, *Zh. Obshch. Khim.*, **42,** 1991 (1972).
142. H. Harnisch, *Liebigs Ann. Chem.*, **765,** 8 (1972).
143. J. Lötzbeyer and K. Bodendorf, *Chem. Ber.*, **100,** 2620 (1967).
144. H. E. Nikolajewski, S. Dähne, and B. Hirsch, *Chem. Ber.*, **100,** 2616 (1967).
145. V. M. Vlasov and O. V. Zakharova, *Zh. Org. Khim.*, **10,** 66 (1974).
146. R. Sciaky and U. Pallini, *Tetrahedron Lett.*, **1964,** 1839.
147. J. Schmitt, J. J. Panhouse, A. Hallot, P.-J. Cornu, H. Pluchet, and P. Comoy, *Bull. Soc. Chim. Fr.*, **1964,** 2753.
148. J. Schmitt, J. J. Panhouse, P.-J. Cornu, H. Pluchet, A. Hallot, and P. Comoy, *Bull. Soc. Chim. Fr.*, **1964,** 2760.
149. R. Sciaky and F. Mancini, *Tetrahedron Lett.*, **1965,** 137.
150. S. A. Kagal, P. Madhavannair, and K. Venkataraman, *Tetrahedron Lett.*, **1962,** 593.
151. G. W. Fischer and W. Schroth, *Chem. Ber.*, **102,** 590 (1969); E. N. Marvell, G. Caple, T. A. Cosink, and G. Zimmer, *J. Amer. Chem. Soc.*, **88,** 619 (1966); P. Schiess and H. L. Chia, *Helv. Chim. Acta*, **53,** 485 (1970); P. Schiess, H. L. Chia, and C. Suter, *Tetrahedron Lett.*, **1968,** 5747.
152. J. M. F. Gagan, A. G. Lane and D. Lloyd, *J. Chem. Soc. C*, **1970,** 2484.
153. O. Aki and Y. Nakagawa, *Chem. Pharm. Bull.*, **20,** 1325 (1972).
154. N. Coniac, G. Hazebroucq, and J. Gardent, *C. R. Acad. Sci. Paris*, **268C,** 2031 (1969).
155. R. Buyle and H. G. Viehe, *Tetrahedron*, **24,** 4217 (1968).
156. M. Mazharuddin and G. Thyagarajan, *Tetrahedron Lett.*, **1971,** 307.
157. S. R. Shah and S. Seshadri, *Indian J. Chem.*, **10,** 820, 977 (1972).
158. K. E. Schulte, J. Reisch, and U. Stoess, *Angew. Chem.*, **77,** 1141 (1965); *Arch. Pharm.*, **305,** 523 (1972).
159. H. von Dobeneck and F. Schnierle, *Tetrahedron Lett.*, **1966,** 5327; F. Schnierle, H. Reinhard, N. Dieter, E. Lippacher, and H. von Dobeneck, *Liebigs Ann. Chem.*, **715,** 90 (1968).
160. H. von Dobeneck and T. Messerschmitt, *Liebigs Ann. Chem.*, **751,** 32 (1971).
161. T. Messerschmitt, U. von Specht, and H. von Dobeneck, *Liebigs Ann. Chem.*, **751,** 50 (1971).
162. H. von Dobeneck, H. Reinhard, H. Deubel, and D. Wolkenstein, *Chem. Ber.*, **102,** 1357 (1969).
163. B. Hansen and H. von Dobeneck, *Chem. Ber.*, **105,** 3630 (1972).
164. A. Kirschner, Dissertation, Tec. Univ., Munich, 1969.
165. N. Yu. Koshelev, I. Ya. Kvitko, and L. S. Efros, *Zh. Org. Khim.*, **8,** 1750 (1972); B. A. Porai-Koshits, I. Ya. Kvitko, and E. A. Shutkova, *Pharm. Chem. J. (N.Y.)*, **1970,** 138.
166. M. A. Kira and W. A. Bruckner, *Acta Chim. Acad. Sci. Hung.*, **56,** 47 (1968).
167. S. Senda, K. Hirota, G. N. Yang, and M. Shirahashi, *Yakugaku Zasshi (J. Pharm. Soc. Jap.)*, **91,** 1372 (1971).

168. H. Bredereck, G. Simchen, H. Wagner, and A. A. Santos, *Liebigs Ann. Chem.*, **766**, 73 (1972).
169. H. H. Bosshard, E. Jenny, and H. Zollinger, *Helv. Chim. Acta*, **44**, 1203 (1961).
170. C. Jutz, *Chem. Ber.*, **97**, 2050 (1964).
171. G. A. Reynolds and J. A. Van Allan, *J. Org. Chem.*, **34**, 2736 (1969).
172. N. E. Shelepin, N. S. Loseva, L. E. Nivorozhkin, and V. I. Minkin, *Khim. Geterotsikl. Soedin.*, **1971**, 733.
173. H. Khedija, H. Strzellecka, and M. Simalty, *Bull. Soc. Chim. Fr.*, **1973**, 218.
174. T. Kato and T. Chiba, *J. Pharm. Soc. Jap.*, **89**, 1464 (1969).
175. T. Kato, H. Yamanaka, and H. Hiranuma, *J. Pharm. Soc. Jap.*, **90**, 870 (1970).
176. M. R. Chandramohan and S. Seshadri, *Indian J. Chem.*, **10**, 573 (1972).
177. J. Ciernik, *Collect. Czech. Chem. Commun.*, **37**, 2273 (1972).
178. M. R. Jayanth, H. A. Naik, D. R. Tatke, and S. Seshadri, *Indian J. Chem.*, **11**, 1112 (1973).
179. H. Bredereck, G. Simchen, and P. Speh, *Liebigs Ann. Chem.*, **737**, 46 (1970).
180. R. S. Pandit and S. Seshadri, *Indian J. Chem.*, **11**, 532 (1973).
181. S. Klutchko, H. V. Hansen, and R. I. Meltzer, *J. Org. Chem.*, **30**, 3454 (1965).
182. B. A. Clark, J. Parrick, P. J. West, and A. H. Kelly, *J. Chem. Soc. C*, **1970**, 498.
183. D. M. Brown and A. Giner-Sorolla, *J. Chem. Soc. C*, **1971**, 128.
184. V. L. Zbarskij, A. A. Borisenko, and E. Orlova, *Z. Org. Khim.*, **6**, 520 (1970); V. L. Zbarskij, G. M. Sutov, V. F. Zilin, and E. Orlova, *Z. Org. Khim.*, **4**, 1970 (1968).
185. C. F. Jones, D. A. Taylor, and D. P. Bowyer, *Tetrahedron*, **30**, 957 (1974).
186. C. Reichardt and W.-D. Kermer, *Synthesis*, **1970**, 538.
187. C. Jutz and W. Müller, *Angew. Chem.*, **78**, 1059 (1966). W. Müller, Dissertation, Tec. Univ., Munich, 1967.
188. T. Koyama, T. Hirota, I. Ito, and M. Toda, *J. Pharm. Soc. Jap.*, **89**, 1492 (1969).
189. J. Liebscher and H. Hartmann, *Z. Chem.*, **14**, 358 (1974).
190. C. Reichardt and K. Halbritter, *Liebigs Ann. Chem.*, **737**, 99 (1970).
191. Z. Arnold, *Collect. Czech. Chem. Commun.*, **30**, 2125 (1965).
192. C. Jutz, R. Kirchlechner, and H.-J. Seidel, *Chem. Ber.*, **102**, 2301 (1969).
193. C. Jutz and E. Schweiger, *Chem. Ber.*, **107**, 2383 (1974).
194. Z. Arnold and A. Holy, *Collect. Czech. Chem. Commun.*, **30**, 40 (1965).
195. Z. Arnold, *Collect. Czech. Chem. Commun.*, **24**, 760 (1959).
196. R. Brehme and H. E. Nikolajewski, *Z. Chem.*, **8**, 226 (1968).
197. M. A. Kira, A. Bruckner-Wilhelm, F. Ruff, and J. Borsi, *Acta Chim. Hung.*, **56**, 189 (1968); M. A. Kira, M. O. Abdel-Rahman, and K. Z. Gadalla, *Tetrahedron Lett.*, **1969**, 109.
198. M. A. Kira, M. N. Aboul-Enein, and M. I. Korkor, *J. Heterocycl. Chem.*, **7**, 25 (1970).
199. M. A. Kira, Z. M. Nofal, and K. Z. Gadalla, *Tetrahedron Lett.*, **1970**, 4215.
200. C. Jutz and M. Wagner, *Angew. Chem.*, **84**, 299 (1972); C. Jutz, *Angew. Chem.*, **86**, 781 (1974).
201. W. Wiegrebe, D. Sasse, H. Reinhart, and L. Faber, *Z. Naturforsch.*, **B25**, 1408 (1970).
202. H. Eilingsfeld, M. Seefelder, and H. Weidinger, *Angew. Chem.*, **72**, 836 (1960).
203. H. Bredereck, R. Gompper, and K. Klemm, *Chem. Ber.*, **92**, 1456 (1959), and *Angew. Chem.*, **71**, 32 (1959); H. Bredereck and K. Bredereck, *Chem. Ber.*, **94**, 2278 (1961).

204a. A. L. Cossey, R. L. Harris, J. L. Huppatz, and J. N. Phillips, *Angew. Chem.*, **84,** 1183 (1972).
204b. A. L. Cossey, R. L. Harris, J. L. Huppatz, and J. N. Phillips, *Angew. Chem.*, **84,** 1184 (1972).
204c. A. L. Cossey, R. L. Harris, J. L. Huppatz, and J. N. Phillips, *Angew. Chem.*, **84,** 1185 (1972).
205. R. R. Crenshaw and R. A. Partyka, *J. Heterocycl. Chem.*, **7,** 871 (1970).
206. M. Ahmed, J. Ashby, and O. Meth-Cohn, *J. Chem. Soc. Chem. Commun.*, **1970,** 1094.
207. S. Akaboshi and T. Kutsuma, *J. Pharm. Soc. Jap.*, **89,** 1029, 1035, 1039, 1045 (1969).
208. A. Ermili and G. Roma, *Gazz. Chim. Ital.*, **101,** 269 (1971); A. Ermili, G. Roma, and A. Balbi, *Gazz. Chim. Ital.*, **101,** 651 (1971); A. Ermili, G. Roma, and F. Braguzzi, *Ann. Chim. (Roma)*, **62,** 458 (1972); A. Ermili, G. Roma, M. Mazzoi, A. Balbi, A. Cuttica, and N. Passerini, *Farmaco, Ed. Sci.*, **29,** 225 (1974).
209. K. Morita, S. Kobayashi, H. Shimadzu, and M. Ochiai, *Tetrahedron Lett.*, **1970,** 861; T. Koyama, M. Toda, T. Hirota, Y. Katsuse, and M. Yamato, *J. Pharm. Soc. Jap.*, **90,** 11 (1970); T. Koyama, M. Toda, T. Hirota, M. Hashimoto, and M. Yamato, *J. Pharm. Soc. Jap.*, **89,** 1688 (1969); T. Koyama, Y. Katsuse, M. Toda, T. Hirota, and M. Yamato, *Yakugaku Zasshi*, **90,** 1207 (1970).
210. J. A. Van Allen, C. C. Petropoulos, and G. A. Reynolds, *J. Heterocycl. Chem.*, **6,** 803 (1969).
211. C. Jutz, *Chem. Ber.*, **91,** 851 (1958); C. Jutz, *Angew. Chem.*, **70,** 270 (1958).
212. G. P. Stepanova and B. I. Stepanov, *Ž. Vses. Obšč. D. I. Mendeleeva*, **15,** 357 (1970), and *Zh. Org. Khim.*, **7,** 1013 (1971).
213. K. Dickore and F. Kröhnke, *Chem. Ber.*, **93,** 1068, 2479 (1960); H. G. Nordmann and F. Kröhnke, *Angew. Chem.*, **81,** 747 (1969), and *Liebigs Ann. Chem.*, **731,** 80 (1970).
214. Z. Arnold and A. Holy, *Collect. Czech. Chem. Commun.*, **30,** 40 (1965).
215. E. E. Nikolajewski, S. Dähne, D. Leupold, and B. Hirsch, *Tetrahedron*, **24,** 6685 (1968).
216. C. Jutz and E. Schweiger, *Synthesis*, **1974,** 193.
217. J. Bordner and H. Rapoport, *J. Org. Chem.*, **30,** 3825 (1965).

ADDENDUM

Some additional applications of the Vilsmeier-Haack-Arnold acylations were published since this chapter was written. However, no fundamentally new facts or reaction principles of the chloromethyleniminium salts could be revealed.

To Section II-A-2

Dibenzo(a,1)pyrene was substituted by MFA–POCl$_3$ only in the 10 position, dibenzo(a,e)fluoranthene gave a mixture of monoformylation products (218). An electrophilic displacement of bromine in 1,3-dibromo-azulene by **10a** at 150°C takes place affording 3-bromoazulene-1-aldehyde (92%), and in an analogous manner from the latter the dialdehyde (**27**) (23%) was obtained (219). Treatment of benzyl-1-azulylketone with **10a** causes both substitution of the azulene nucleus in position 3 (38%) and attack of the reagent on the keto-methylene group forming after hydrolytic work up 3-(azulyl-1)-3-chloro-2-phenylacrolein (54%) (220) (see also Section II-E-1).

To Section II-A-3

2,5-Dimethyl and 1,2,5-trimethylpyrrole-3,4-dialdehyde, obtained by formylation of the corresponding pyrroles with **10a** were used starting a new synthesis of 2-azaazulenes (221). Only monoformylation at a methin position of the ring was observed by action of **10a** on porphin metal complexes (metal = Cu, Mg, Zn, Ni, Co, Mn) (222) and on aethioporphyrin-I-Cu(II) (223). 5-Amino-3-methyl-1-phenylpyrazole was substituted, as expected, by treatment with **10a** at position 4 under simultaneous transformation of the amino group to the dimethylaminomethylenamino group (224).

In contrast to corresponding pyrazoles, 1,3,5-triphenyl and 1,5-diphenyl-3-styryl-Δ^4 pyrazolines are not attacked at 4 position but were substituted by **10a** in one benzene nucleus forming the 1-*p*-formylphenyl derivatives (225). In 2-(2'-thienyl)indole the more reactive indole nucleus was substituted by **10a** yielding the 3-formyl derivative (226). Some indoles: 1-*R*-5-acetoxy-2-methylindoles (*R* = Me, Ph, CH$_2$Ph) (227) and 2-*R*-benz(*e*)indoles and 2-*R*-benz(*g*)indoles (*R* = H, Me, Ph) were formylated by **10a** to their 3-carboxaldehydes (228).

Imidazo(1,5-*a*)pyridine, resembling the indolizine in its reactivity, was substituted by **10a** yielding mainly the 1-formyl besides the 3-formyl derivative (229).

1,3-Di(ethoxycarbonyl)pyrido(2,1,6-*de*)quinolizine (cycl[3,3,3]azine) with its antiaromatic, cyclic conjugated system could be subjected to formylation by **10a** yielding a mixture of the corresponding 4- and 6-monoformyl derivatives (230).

In a synthesis of the likewise antiaromatic cycl(4,3,2)azine ring, **10a** reacted with 3*a*-aza-4-azulenone, which behaves like a 1-acyl-pyrrole, at the only free α-pyrrole position forming the 3-carboxaldehyde (73%) (231).

To Section II-E-1

A series of 3-chloroacroleins (**242**), partly new, were prepared by treatment of the corresponding methylene ketones with **10a**. The obtained **242** were transformed by action of sulfide and alkylhalides to thiophenes (232).

To Section II-E-4

Cyclization to chromenone-3-carboxaldehydes by reaction of **10a** with substituted 2-hydroxyacetophenones in 14–80% yield was once more described (33).

To Section II-G-5

6-Chlorobenzoxazol-2-malonaldehyde (**370a**, R = H) as starting compound preparing benzoxazolodiazepines and other heterocycles was obtained in 60% by treatment of 4-chloro-2-hydroxyacetophenonoxime with **10a** (234). The intermediate 6-chloro-2-methylbenzoxazole must be formed in a Beckman rearrangement followed by double formylation of the reactive methyl group.

To Section II-G-8

The formation of *N*-chloromethin-formamidinium salts **405** by reaction of benzonitriles or phenylacetonitriles with **10a** in presence of hydrogen chloride were further investigated, **405** isolated in form of their perchlorates and used in a synthesis of thiopyrylium and 1,3-thiazinium salts (235). Some additional 3-chloroisoquinolines **411** by the cyclizing formylation of substituted phenylacetonitriles by **10a** were described (236). Treatment of **407** by **10a** at 90–95°C leads after hydrolytic workup not only to **411** (62%), but minor amounts of 3-chloro-6,8-dimethoxyisoquinoline-4-carboxaldehyde (1%) could also be isolated. The isomeric 3-chloro-6,8-dimethoxyisoquinoline-5-carboxaldehyde was formed in 40% yield by reaction of **411** itself with **10a**.

Perspectives

The great potentialities of the C-bond forming reactions of chloromethyleniminium salts for organic syntheses are by no means exploited. This is also true for the synthetic value of the various and numerous compounds accessible by such formylation reactions. In this chapter many mechanistic proposals are given for the first time that are more or less suggestive. A thorough investigation of the real reaction mechanisms seems to be desirable and useful. Such examinations not only may allow improvement or optimization of the reaction conditions in some cases, but also should lead to an extension of the scope of these formylation reactions by chloromethyleniminium salts.

REFERENCES

218. O. Perin-Roussel, P. Jacquignon, and F. Perin, *C.R. Acad. Sci.*, Serie C **280,** 1315 (1975).
219. Y. Porshnev, N. Porshnev, and E. M. Treshchenko, *Zh. Org. Khim.* **11,** 657 (1975).
220. Y. Porshnev, N. Porshnev, and M. I. Cherkashin, *Izv. Akad. Nauk. S.S.S.R., Ser. Khim.* **1975,** 2322.
221. R. Kreher, G. Vogt, and M.-L. Schulz, *Angew. Chem.* **87,** 840 (1975).
222. R. Schlözer and J.-H. Fuhrop, *Angew. Chem.* **87,** 388 (1975).
223. G. V. Ponomarev, B. V. Rozynov, and C. B. Maravin, *Khim. Geterotsikl. Soedin.* **1975,** 139.
224. J. Häufel and W. Breitmaier, *Angew. Chem.* **86,** 671 (1974).
225. L. A. Kutulya, A. E. Shevchenko, Y. Surov, and N. Surov, *Khim. Geterotsikl Soedin.* **1975,** 250.
226. B. S. Holla and S. Y. Ambekar, *J. Indian Chem. Soc.* **51,** 965 (1974).
227. A. N. Grinev, V. I. Shedov, N. K. Chizov and T. F. Vlasova, *Khim. Geterotsikl. Soedin.* **1975,** 1250.
228. V. N. Eraksina, L. B. Shagalov, and N. N. Surorov, *Khim. Geterotsikl. Soedin.* **1975,** 1257.
229. O. Fuentes and W. N. Paudler, *J. Heterocycl. Chem.* **12,** 379 (1975).
230. D. Farquhar, Th. T. Gough, and D. Leaver, *J. Chem. Soc. (Perkin Trans. I.* **1976,** 341.
231. W. Flitsch, A. Gurke, and B. Müter, *Chem. Ber.* **108,** 2969 (1975).
232. P. Cagniant and G. Kirsch, *C. R. Acad. Sci. Ser. C* **281,** 35 (1975).
233. A. Nohara, T. Umetani, and Y. Sanno, *Tetrahedron* **30,** 3553 (1974).
234. S. M. Jain and R. A. Pawar, *Indian J. Chem.* **13,** 304 (1975).
235. J. Liebscher and H. Hartmann, *Z. Chem.* **15,** 16 (1975).

CHEMISTRY OF DICHLOROMETHYLENIMINIUM SALTS (PHOSGENIMINIUM SALTS)

Z. JANOUSEK and H. G. VIEHE, *Laboratoire de Chimie Organique, Université de Louvain, Place L. Pasteur, 1, B-1348-Louvain-la-Neuve, Belgium.*

CONTENTS

I. Introduction	344
A. Structure of N,N-Disubstituted Dihalomethyleniminium Salts	344
B. Comparison of Dichloromethyleniminium Salts with Related Classes of Compounds	345
C. Previous Work	348
II. Syntheses of Phosgeniminium Salts	348
A. Chlorination of Thiocarbamoyl Chlorides	349
B. Chlorination of Thiurame Disulfides	351
C. Chlorination of Thioformamides	352
D. Chlorination of Dithiocarbamates and of C-Sulfonylthioformamides	352
E. Syntheses of Dichloromethyleniminium Chlorides from Dichloromethylenimines and from Cyanogen Chloride	354
F. C–C Double-Bond Cleavage Reactions Leading to **1**	355
G. High-Temperature Chlorination of Tertiary Amines	356
H. Synthesis of Dibromomethyleniminium Bromides	357
I. Synthesis of Trifluoromethylamines	357
III. The Reaction Potential of Dichloromethyleniminium Salts	358
A. Chlorine Substitution in Dichloromethyleniminium Chlorides Leading to C–C Bond Formation	358
1. Reaction with Grignard and Other Metallorganic Compounds	358
2. Reactions with Activated Methylene Groups	359
3. Reactions with Ketones	363
4. Reactions of **1** with Tertiary Acetamides to Yield 1,3-Dichlorotrimethinecyanines and Their Use in Synthesis	371
5. Electrophilic Substitution on Aromatic Compounds	378
B. C–N Bond Formation with Phosgeniminium Salts	379
1. Reaction of **1** with Amines, Sulfonamides, and Hydrazines	379
2. Reactions with Secondary Acetamides	384
C. Chlorine Substitution with Azide, Isocyanate, and Phosphinyl Groups	388
D. Chlorine Substitution in **1** by Oxygen and Sulfur	391
1. Reaction with Phenols and Thiophenols	391
2. Reaction with Alcohols, Glycols, and Their Thio Analogues	392
3. Ring Cleavage of Cyclic Ethers with **1**	394
4. Reaction with Aldehydes, Carboxylic Acids, Primary Amides, and Oximes	395

E. Addition Reactions of Phosgeniminium Salts to C–C Multiple Bonds 396
  1. Addition to Activated Acetylenes 396
  2. Addition to Enamines 397
  3. Reaction of Phosgeniminium Salts with Vinyl Ethers 399
F. Addition to C–N Multiple Bonds 400
  1. Reactions with Activated Nitriles 400
  2. Hydrogen Chloride-Initiated Additions of **1** to Nitriles 403
  3. Reactions with Imines 403
  4. Addition to Cyanates 406
  5. Addition to Heterocumulenes 406
G. Heterocyclization Reactions with Phosgeniminium Salts 409
IV. Conclusions and Outlook 412
References 416

I. Introduction

A. STRUCTURE OF N,N-DISUBSTITUTED DIHALOMETHYLEN-IMINIUM SALTS

Because of their inherent reactivity and easy availability dichloromethyleniminium salts (**1**) are the most important dihalomethyleniminium salts. They are formally derived from trichloromethylamines (**1a**) by dissociation. In practice, this dissociation is complete for systems with alkyl or aryl substituents on nitrogen. The covalent form (**1a**) may expectedly become predominant for **1** where R and R′ are electron-withdrawing substituents:

$$\underset{R}{\overset{R'}{>}}\overset{+}{N}=C\underset{Cl}{\overset{Cl}{<}} \quad Cl^{-} \rightleftharpoons \underset{R}{\overset{R'}{>}}N-C\underset{Cl}{\overset{Cl}{<}}Cl$$

  (**1**)   (**1a**)

Similarly, the tribromomethylamines (**2a**) exist predominantly in the iminium form (**2**), whereas the iodo derivatives (**3** and **3a**) are still unknown. In contrast to the chloro and bromo derivatives, trifluoromethylamines (**4a**) show no tendency to dissociate, but difluoromethyleniminium salts (**4**) are probably the actual reactive intermediates in reactions involving trifluoromethylamines. Accordingly, **4a** are rather volatile and distillable liquids, soluble in nonpolar solvents.

$$\underset{R}{\overset{R'}{>}}\overset{+}{N}=C\underset{X}{\overset{X}{<}} \quad X^{-} \rightleftharpoons \underset{R}{\overset{R'}{>}}N-C\underset{X}{\overset{X}{<}}X$$

  (**2**)   X = Br   (**2a**)
  (**3**)   X = I   (**3a**)
  (**4**)   X = F   (**4a**)

I. INTRODUCTION

The structure of dichloromethyleniminium salts (**1**) has been unequivocally established both by spectroscopy and by their chemical behavior. To emphasize their relationship with phosgene, these compounds have frequently been referred to as phosgeniminium or PI salts.

Dichloromethyleniminium salts (**1**) represent colorless salts, which are mostly hygroscopic, but they can be stored indefinitely in a dry atmosphere. Hydrolysis occurs also in the presence of other compounds that can be dehydrated (e.g., alcohols, nitromethane, DMSO, and trifluoroacetic acid). The hydrolysis products—the tertiary carbamoyl chlorides (**5**)—can be isolated:

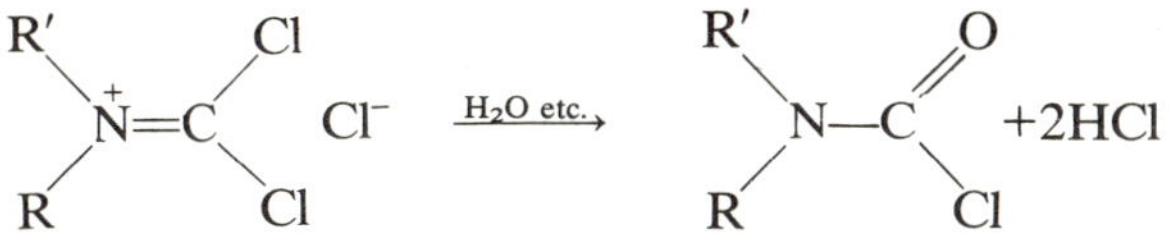

$$\underset{(\mathbf{1})}{\overset{R'}{\underset{R}{\diagup}}\overset{+}{N}=C\overset{Cl}{\underset{Cl}{\diagdown}}\quad Cl^-} \xrightarrow{H_2O \text{ etc.,}} \underset{(\mathbf{5})}{\overset{R'}{\underset{R}{\diagup}}N-C\overset{\diagup\!\!\!\diagup O}{\underset{Cl}{\diagdown}}\quad +2HCl}$$

Compounds **1** are generally soluble only in liquid sulfur dioxide, although thionyl chloride may be used as a suitable solvent for NMR measurements. Chloroform, methylene chloride, acetonitrile, and nitrobenzene are occasionally good solvents for PI salts.

The majority of iminium salts show a tendency to N-dealkylation at elevated temperatures. This is also true of **1**, which in the molten state decompose rapidly to alkyl chlorides and dichloromethylenimines (**11**):

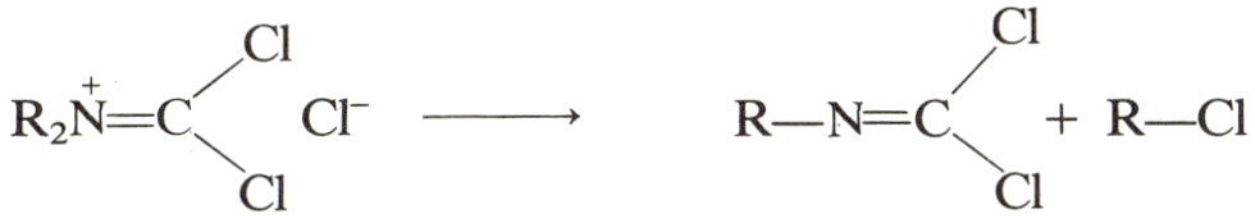

$$\underset{(\mathbf{1})}{R_2\overset{+}{N}=C\overset{Cl}{\underset{Cl}{\diagdown}}\quad Cl^-} \longrightarrow \underset{(\mathbf{11})}{R-N=C\overset{Cl}{\underset{Cl}{\diagdown}}} + R-Cl$$

Dichloromethyleniminium salt **1**, R = CH$_3$, begins to decompose perceptibly above 130°, and this reaction becomes very fast at the melting point (~190°). Because of this decomposition and sensitivity to moisture, the melting points of **1** may vary considerably and are less characteristic for this class of compounds. When suspended in inert solvents, **1** may decompose at temperatures even lower than 100°. The above reaction can be inversed, using strong alkylating agents and dichloromethylenimines (see Section II-E).

B. COMPARISON OF DICHLOROMETHYLENIMINIUM SALTS WITH RELATED CLASSES OF COMPOUNDS

Even if one restricts the comparison of dichloromethyleniminium chloride (**1**) to the simplest related systems with C—N and C—O double bonds, the relationship between the iminium salts (**1, 6, 7**), the carbonyl compounds (**8–10**), and the imines (**11–13**) is instructive.

In Scheme 1 the arrows indicate the increasing ease of nucleophilic addition to the C–X double bond:

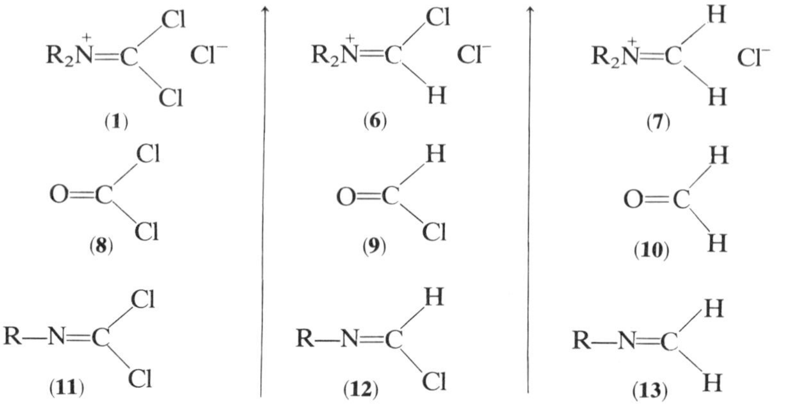

$R_2\overset{+}{N}=C(Cl)(Cl)\ Cl^-$ $R_2\overset{+}{N}=C(Cl)(H)\ Cl^-$ $R_2\overset{+}{N}=C(H)(H)\ Cl^-$
(**1**) (**6**) (**7**)

$O=C(Cl)(Cl)$ $O=C(H)(Cl)$ $O=C(H)(H)$
(**8**) (**9**) (**10**)

$R-N=C(Cl)(Cl)$ $R-N=C(H)(Cl)$ $R-N=C(H)(H)$
(**11**) (**12**) (**13**)

Carbonic acid derivatives   Formic acid derivatives   Formaldehyde derivatives

Scheme 1

In each column the iminium salts show higher electrophilicity than the corresponding carbonyl derivatives, which in turn are more electrophilic in nature than the imino compounds. Thus it is evident that in the "phosgene group" the phosgeniminium salts (**1**) are the most electrophilic (**1** > **8** > **11**), and in the second column the chloromethyleniminium salts (Vilsmeier-Haack-Arnold reagents) (**6**) are again the most reactive members (**6** > **9** > **12**). Analogously, methyleniminium salts (**7**) are stronger electrophiles than formaldehyde (**10**) and its imine (**13**).

The difference in "horizontal" reactivity toward nucleophilic addition is much less obvious, and the chemical evidence clearly shows that **6** is the most reactive member of the first-row iminium salts. Steric hindrance due to the presence of two bulky chlorine atoms at the iminium carbon atoms should be partly responsible for the drop in reactivity in going from **6** to **1**.

From a synthetic point of view, **1** may be expected to have the greatest scope, since successive chlorine substitutions in **1** by hydrogens or alkyl and aryl groups lead to iminium chlorides (**6, 7**) or their derivatives (Scheme 2). In fact, **1** are the most reactive trichloromethane derivatives hitherto isolated.

Although many reactions of **1** remain unexplored, appreciable progress has been achieved since systematic study of them began in 1969.

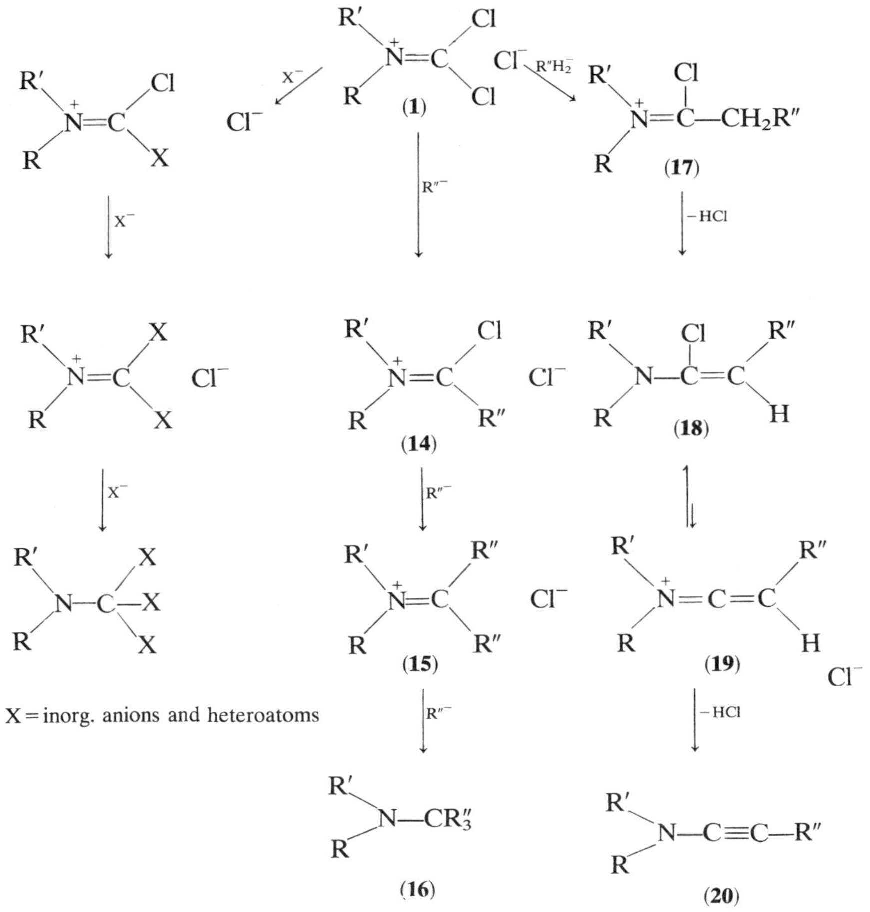

I. INTRODUCTION

Scheme 2

Reaction scheme 2 shows that, in addition to heterosubstitution leading to a great number of carbonic acid derivatives (see Sections III-B and III-D), homosubstitutions with carbanions are equally useful. Tertiary amines (**16**), for example, having a *tert*-alkyl group as substituent, are obtained in high yield via the chloromethyleniminium (**14**) and methyleniminium (**15**) salts.

Active methylene compounds yield the versatile α-chloroenamines (**18**), the precursors of the very reactive keteniminium salts (**19**) (see the chapter by Ghosez and Marchand-Brynaert).

C. PREVIOUS WORK

Knowledge of the chemistry of N,N-dialkyldichloromethyleniminium salts (**1**) is much more recent than that of methylene- and chloromethyleniminium salts (**7**) and (**6**). The first report concerning **1** was published as late as 1959 (1) and went almost unnoticed for another decade (2). In 1969 we became interested in these reagents (3), which we named phosgeniminium (PI) salts. At first their potential as starting materials in ynamine synthesis (Scheme 2) was their attractive feature, but their importance as general building blocks in organic chemistry soon became obvious. Independent work, mainly by Kukhar and his group (4), has contributed to this new and rapidly developing chemistry.

II. Syntheses of Phosgeniminium Salts (1)

Phosgeniminium salts (**1**) are derivatives of carbon dioxide, which is formed on extended hydrolysis:

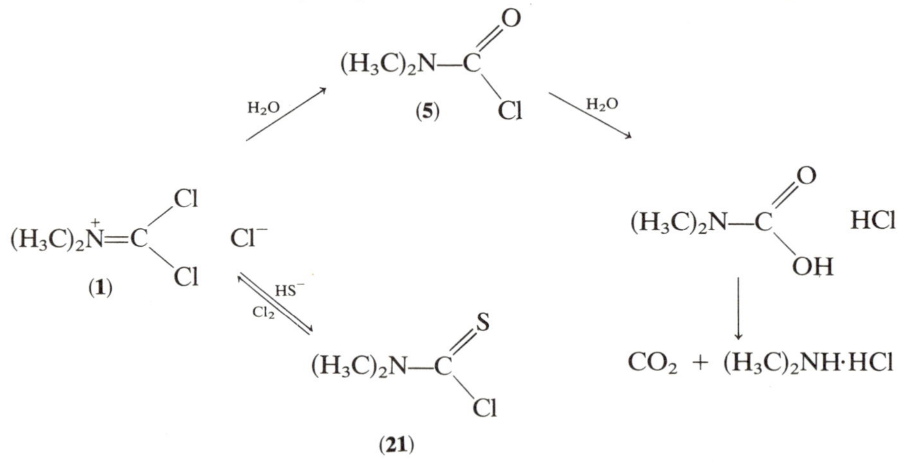

Mild hydrolysis, however, stops at the carbamoyl chloride stage.

Thiolysis of **1** has been reported (5) to proceed exothermally when hexamethyldisilylthiane is used, to give very good yields of thiocarbamoyl chlorides (**21**).

Because the conversion of **5** back to **1** has not yet been realized, the chlorination of **21** and other compounds containing the thiocarbonyl group remains the only practical approach to **1** if one does not take into account the still exceptional preparations starting with either cyanogen chloride or phosgenimines (**11**), which can be protonated to the parent iminium salts (**1**, R = H). Similarly, alkylation of **11** succeeds only with strong reagents (see Section II-E). Thus, so far all syntheses but one start with compounds that already contain the C–N bond and the carbon atom at the oxidation level of the final PI products.

Carbon tetrachloride and tetrabromide react with secondary amines via both ionic and radical pathways (6). Nevertheless, there is no evidence that **1** are intermediates in these reactions, which frequently give rise to complex mixtures.

The synthetic methods leading to **1** are summarized in Scheme 3.

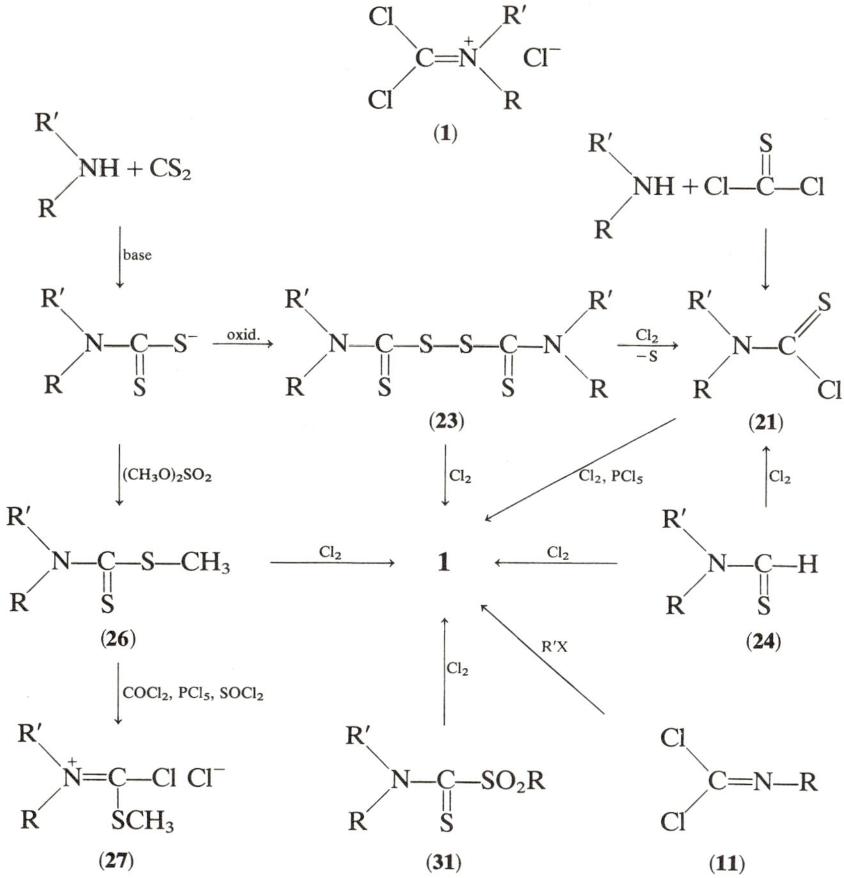

A. CHLORINATION OF THIOCARBAMOYL CHLORIDES (21)

Thiocarbamoyl chlorides (**21**) are ideal starting compounds for the synthesis of phosgeniminium salts, since the thione sulfur is easily substituted by two chlorine atoms.

The chlorination agent generally employed is either elementary

TABLE I

Synthesis of **1** by Chlorination of Thiocarbamoyl Chlorides

R	R'	Yield, %	Melting point, °C	Ref.
CH_3	CH_3	80	194–196	3,4
C_2H_5	C_2H_5	94	130	1,8,9
CH_3	$CH_2C_6H_5$	70	85	8
CH_3	Cyclohexyl	76	144	8
$-(CH_2)_5-$		69	120–122	9
$-(H_2C)_2-O-(CH_2)_2-$		58	150–153	9
CH_3	$p\text{-}ClC_6H_4$	90	111–113	2
CH_3	$p\text{-}FC_6H_4$	91	105–110	2
CH_3	$p\text{-}BrC_6H_4$	94	115–116	2
$n\text{-}C_3H_7$	$n\text{-}C_3H_7$	90	125	10
$n\text{-}C_4H_9$	$n\text{-}C_4H_9$	95	—	10

chlorine or phosphorus pentachloride. Chlorinations are usually run in methylene chloride or chloroform solutions, in which **1** are insoluble and precipitate. With an excess of chlorine, a complex of **1** with 1-mole of chlorine is formed. This complex dissociates to **1** and chlorine on heating under vacuum or dissolution in acetonitrile (4,7).

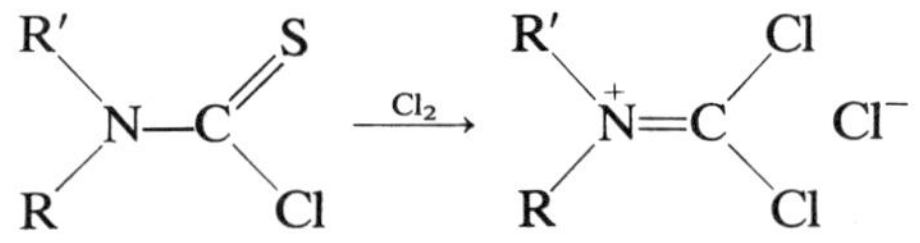

$$\underset{\mathbf{(21)}}{\overset{R'}{\underset{R}{>}}N-\overset{S}{\underset{Cl}{\overset{\|}{C}}}} \xrightarrow{Cl_2} \underset{\mathbf{(1)}}{\overset{R'}{\underset{R}{>}}\overset{+}{N}=\overset{Cl}{\underset{Cl}{\overset{}{C}}}\ Cl^-}$$

The first reaction step probably involves the formation of adduct **22**, which ionizes to the chlorosulfenylchloromethyleniminium salt (**22a**):

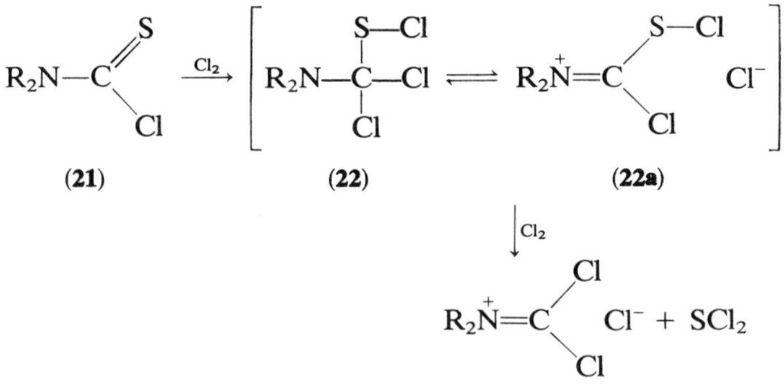

$$R_2N-\overset{S}{\underset{Cl}{\overset{\|}{C}}} \xrightarrow{Cl_2} \left[R_2N-\overset{S-Cl}{\underset{Cl}{\overset{|}{C}}}-Cl \rightleftharpoons R_2\overset{+}{N}=\overset{S-Cl}{\underset{Cl}{\overset{}{C}}}\ Cl^- \right]$$

$$\mathbf{(21)} \qquad \mathbf{(22)} \qquad \mathbf{(22a)}$$

$$\downarrow Cl_2$$

$$R_2\overset{+}{N}=\overset{Cl}{\underset{Cl}{\overset{}{C}}}\ Cl^- + SCl_2$$

$$\mathbf{(1)}$$

To our knowledge, **22** has never been isolated, but an analogous compound (**22b**) was trapped by intramolecular cyclization, as shown in the following example using bromine as halogenating agent (11):

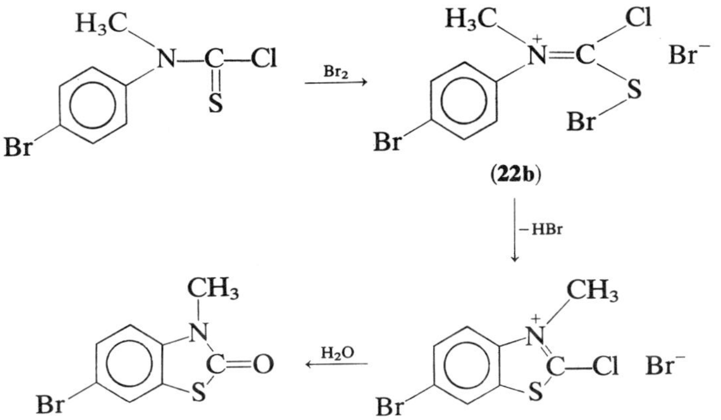

The scope of this facile synthesis of **1** is limited only by the accessibility of thiocarbamoyl chlorides (**21**). Early claims of the preparation of "thiocarbamoyl chloride perchlorides" were apparently incorrect, and these compounds were probably PI salts or their complex with chlorine (12,13).

B. CHLORINATION OF THIURAME DISULFIDES (23)

Since the controlled chlorination of thiurame disulfide (**23**) is a well-known (13–15) synthetic method for obtaining thiocarbamoyl chlorides; it is obvious that exhaustive chlorination leads directly to phosgeniminium salts (**1**). Compounds **23**, which are unexpensive, commercially available products used as fungicides and vulcanization accelerators, are obtained in many cases simply from carbon disulfide and secondary amines, followed by oxidation (Scheme 3).

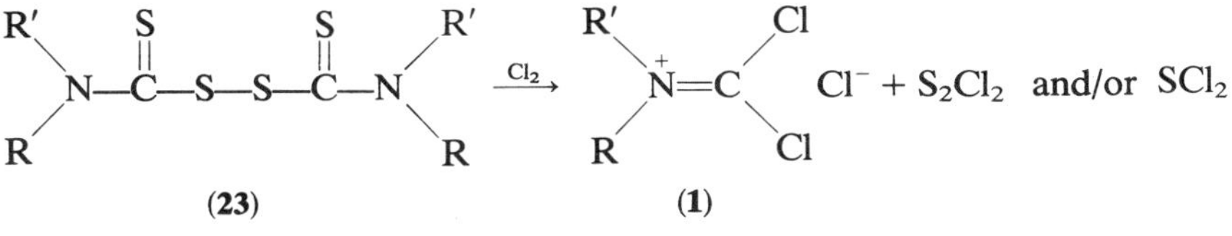

TABLE II

Synthesis of **1** by Chlorination of **23**

R	R'	Yield, %	Melting point, °C	Ref.
CH_3	CH_3	85	190	7,10
C_2H_5	C_2H_5	82	130	8
i-C_3H_7	i-C_3H_7	60	207–208	8
—$(CH_2)_4$—		86	127–129	8
—$(CH_2)_5$—		70	137–139	8,9
—$(CH_2)_6$—		73	111	10
—$(CH_2)_2$—O—$(CH_2)_2$—		54	156–159	8,9
$C_6H_5CH_2$	$C_6H_5CH_2$	80	—	10
C_2H_5	$C_6H_5CH_2$	77	Oil	10
i-C_4H_9	i-C_4H_9	81	Oil	10

C. CHLORINATION OF THIOFORMAMIDES (24)

Tertiary thioformamides (**24**) can be chlorinated to thiocarbamoyl chlorides (**21**). This transformation can be effected by a number of chlorinating agents, for example, elementary chlorine, sulfuryl chloride, or sulfur dichloride (11,16). Again, an excess of chlorine affords **1** as final products.

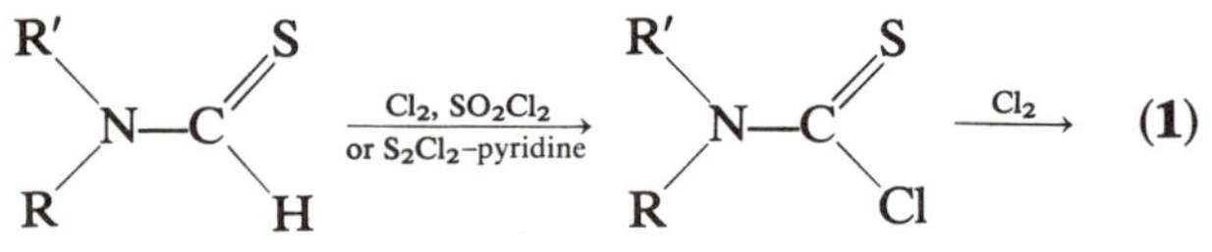

$$\underset{(24)}{\overset{R'}{\underset{R}{>}}N-\overset{S}{\underset{H}{C}}} \xrightarrow{Cl_2,\ SO_2Cl_2 \atop or\ S_2Cl_2-pyridine} \underset{(21)}{\overset{R'}{\underset{R}{>}}N-\overset{S}{\underset{Cl}{C}}} \xrightarrow{Cl_2} \quad (1)$$

For example, **1**, R, R' = CH_3, was obtained in 83% yield from N,N-dimethylthioformamide. By the same method **1**, R, R' = CH_3, C_6H_5, was prepared in 70% yield (7); this is interesting since the other methods give unsatisfactory results, and chlorination of the aromatic ring is sometimes observed (2).

D. CHLORINATION OF DITHIOCARBAMATES (26) AND OF C-SULFONYLTHIOFORMAMIDES (31)

Dithiocarbamic esters (**26**) have been reported to form mercaptoformamide chlorides (chloromercaptomethyleniminium salts, **27** with reactive inorganic acid halides such as phosgene, thionyl chloride, and phosphorus pentachloride (17,18). It was later found (7,8) that both sulfur atoms in **26** are cleaved by elementary chlorine, and once again **1** are formed as the final products:

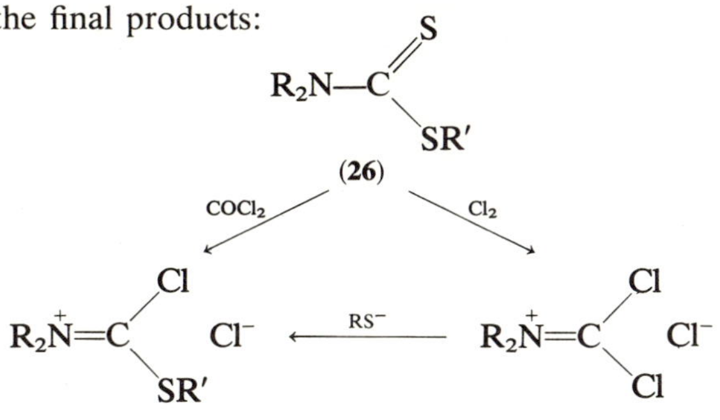

$$R_2N-C\begin{smallmatrix}S\\\parallel\\SR'\end{smallmatrix}$$
(26)

$$R_2\overset{+}{N}=C\begin{smallmatrix}Cl\\ \\SR'\end{smallmatrix} Cl^- \xleftarrow{RS^-} R_2\overset{+}{N}=C\begin{smallmatrix}Cl\\ \\Cl\end{smallmatrix} Cl^-$$
(27) (1)

Since dithiocarbamic esters are more easily available than thioformamides, dithiuram disulfides, and thiocarbamyl chlorides, this is the method of choice in many cases (Table III).

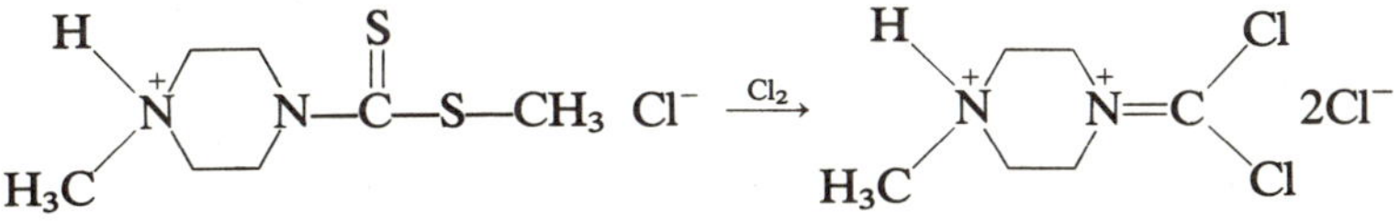

[Structure showing chlorination of piperazine dithiocarbamate to give (28) with 2Cl⁻]

Even bis dichloromethyleniminium chlorides (**29,30**) could be synthesized by this procedure:

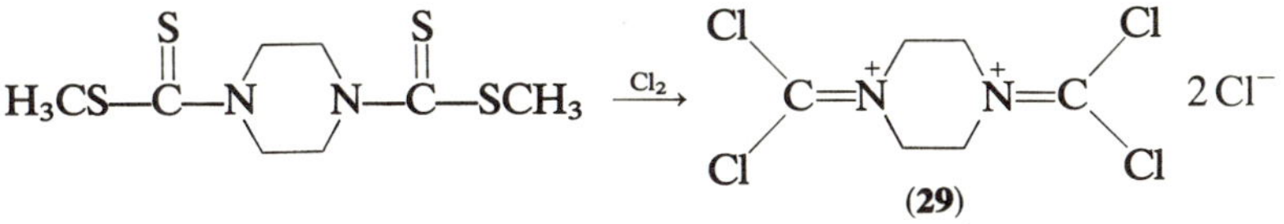

[Structure showing piperazine bis-dithiocarbamate → (29) with 2 Cl⁻]

[Structure showing acyclic bis-dithiocarbamate → Cl₂C=N⁺(CH₃)—CH₂CH₂—N⁺(CH₃)=CCl₂ 2Cl⁻ (30)]

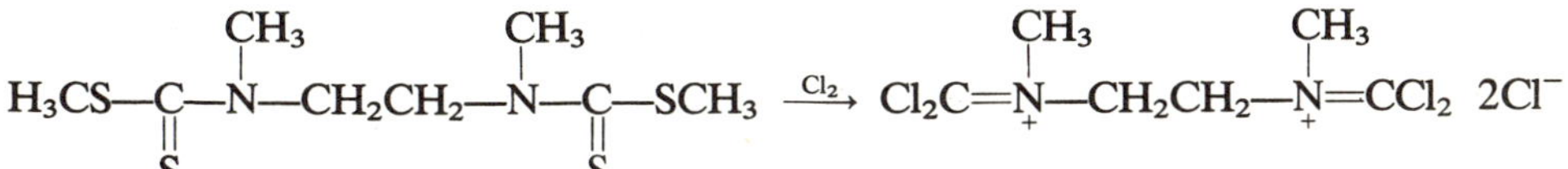

TABLE III
Synthesis of **1** by Chlorination of Dithiocarbamates (**26**) (7,8)

R	R'	Yield, %	Melting point, °C
CH$_3$	CH$_3$	75	190
—(CH$_2$)$_4$—		90	127–129
—(CH$_2$)$_5$—		94	137–139
—(CH$_2$)$_2$—O—(CH$_2$)$_2$—		89	156–159
—(CH$_2$)$_2$—N(CH$_3$)—(CH$_2$)$_2$—		81	185
29		91	190
30		98	175

C-sulfonylthioformamides (**31**) are oxidation products of dithiocarbamic esters and react analogously with chlorine (19):

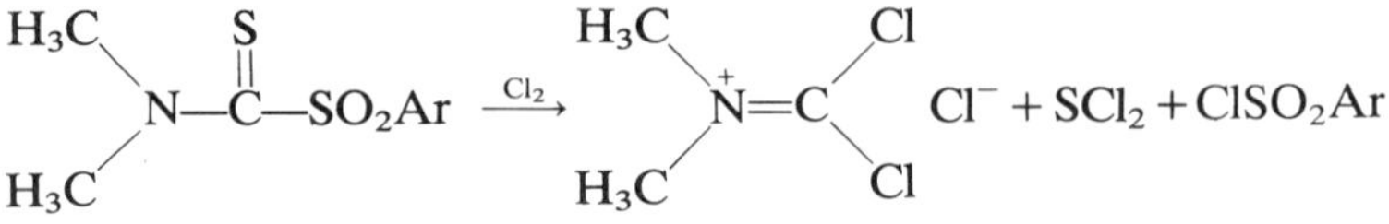

$$\underset{\underset{H_3C}{H_3C}}{>}N-\overset{S}{\underset{\parallel}{C}}-SO_2Ar \xrightarrow{Cl_2} \underset{\underset{H_3C}{H_3C}}{>}\overset{+}{N}=C\underset{Cl}{\overset{Cl}{<}} \quad Cl^- + SCl_2 + ClSO_2Ar$$

(**31**)   (**1**)   81%

E. SYNTHESIS OF DICHLOROMETHYLENIMINIUM CHLORIDES FROM DICHLOROMETHYLENIMINES AND FROM CYANOGEN CHLORIDE

Dichloromethylenimines (isocyanide dichlorides) (20) (**11**) are very weak bases and therefore can be transformed into the iminium salts (**32,33**) only by strong alkylating agents or strong acids. Thus this alkylation is the reversal of the thermal N-dealkylation of **1**.

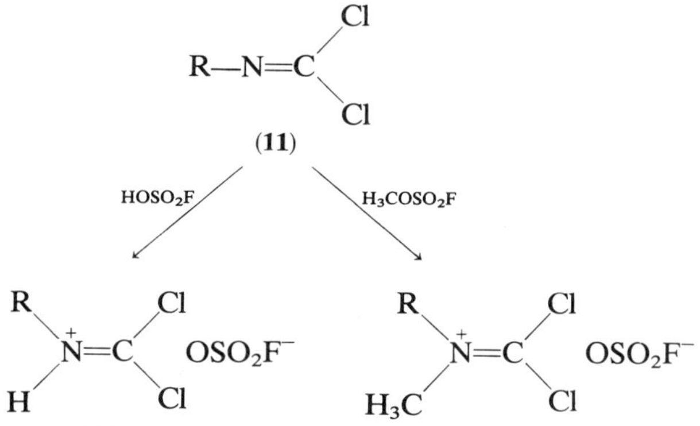

$$R-N=C\underset{Cl}{\overset{Cl}{<}}$$
(**11**)

HOSO₂F ↙   ↘ H₃COSO₂F

$$\underset{H}{\overset{R}{>}}\overset{+}{N}=C\underset{Cl}{\overset{Cl}{<}} \quad OSO_2F^- \qquad \underset{H_3C}{\overset{R}{>}}\overset{+}{N}=C\underset{Cl}{\overset{Cl}{<}} \quad OSO_2F^-$$

(**32**)   (**33**)

TABLE IV

Alkylation and Protonation of Isocyanide Dichlorides (**11**)

Compounds	R	Yield, %	Melting point, °C	Ref.
32a	C_6H_5	96	90–92	8
32b	CH_3	93	—	8
33a	CH_3	95	221	8
33b	Cyclohexyl	92	120–125	8
33c	Phenyl	81	91	8

Cyanogen chloride reacts with alcohols in the presence of hydrogen chloride via the unstable intermediate **34** to a mixture of products (21), but these unsubstituted phosgeniminium salts can be isolated in the presence of inorganic chlorides capable of forming complex anions, for example, with ferric chloride (**35**) (22) or with antimony pentachloride (**36**) (22,23):

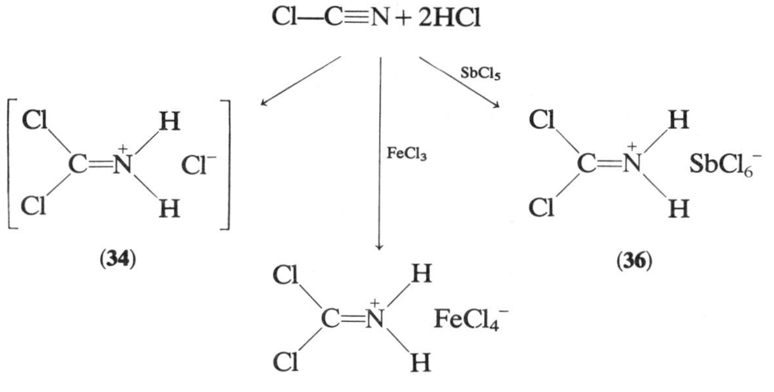

$$Cl-C\equiv N + 2HCl$$

$$\left[\begin{array}{c} Cl \\ Cl \end{array} \!\!\!\! C=\overset{+}{N} \!\!\!\! \begin{array}{c} H \\ H \end{array} \right] Cl^- \qquad \begin{array}{c} Cl \\ Cl \end{array} \!\!\!\! C=\overset{+}{N} \!\!\!\! \begin{array}{c} H \\ H \end{array} SbCl_6^-$$

(**34**)   (**36**)

$$\begin{array}{c} Cl \\ Cl \end{array} \!\!\!\! C=\overset{+}{N} \!\!\!\! \begin{array}{c} H \\ H \end{array} FeCl_4^-$$

(**35**)

In this context it should be stressed that a change in counterion largely alters the reactivity of **1**. Up to now, nearly all the chemistry has been done with **1**, where $X^- = Cl$, although **1**, R, R' = CH_3, $X^- = SbCl_6^-$, prepared from **1**, $X^- = Cl$, and $SbCl_5$ in 95% yield, is reported to react violently with ether, acetone, and nitromethane (4,10,111). Likewise, the tetrachloroaluminate was synthesized in 65% yield, using aluminum chloride in methylene chloride (10).

It was found that **1** react with triethyloxonium fluoroborate by anion exchange to **1**, $X = BF_4^-$; dimethyl sulfate is reported (24) to form compound **37**:

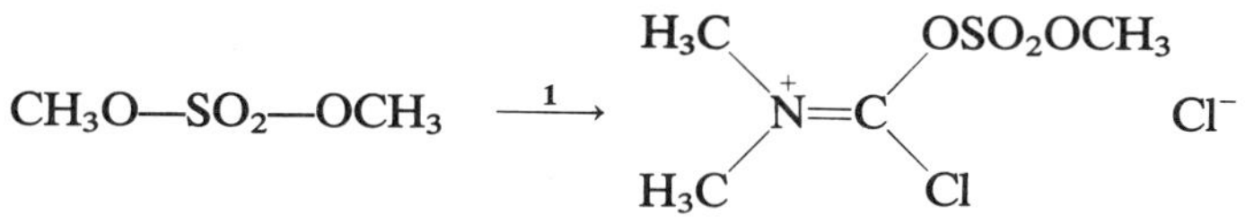

$$CH_3O-SO_2-OCH_3 \xrightarrow{\quad 1 \quad} \begin{array}{c} H_3C \\ H_3C \end{array} \!\!\!\! \overset{+}{N}=C \!\!\!\! \begin{array}{c} OSO_2OCH_3 \\ Cl \end{array} \quad Cl^-$$

(**37**)

F. C–C DOUBLE-BOND CLEAVAGE REACTIONS LEADING TO 1

Malononitrile derivative **38** undergoes cleavage to **1** with chlorine (4). In view of the finding that **39** undergoes the same fragmentation, such reactions are probably quite general (25).

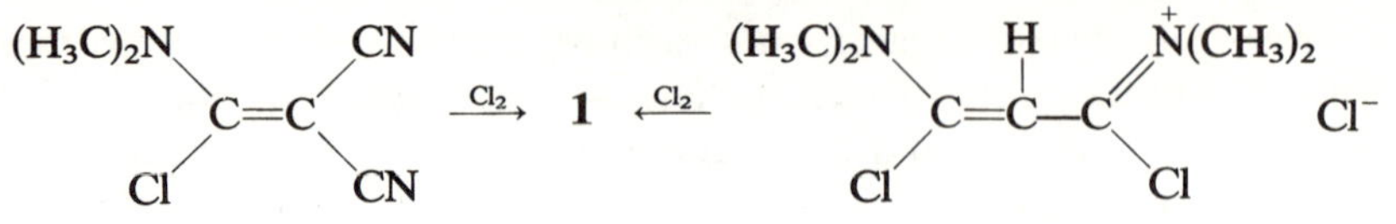

356   CHEMISTRY OF DICHLOROMETHYLENIMINIUM SALTS

$$(H_3C)_2N\underset{Cl}{\diagdown}C=C\underset{CN}{\overset{CN}{\diagup}} \quad \xrightarrow{Cl_2} \quad \mathbf{1} \quad \xleftarrow{Cl_2} \quad (H_3C)_2N\underset{Cl}{\diagdown}C=C\underset{Cl}{\overset{H}{-}}C\underset{Cl}{\overset{\overset{+}{N}(CH_3)_2}{\diagup}} \quad Cl^-$$

(38)   (39)

G. HIGH-TEMPERATURE CHLORINATION OF TERTIARY AMINES

High-temperature chlorination of *N,N*-dimethylaniline affords pentachlorophenylisocyanide dichloride as the final product, but under controlled conditions **40** can be isolated in 70% yield (26). Aliphatic tertiary amines can also be chlorinated, and the intermediate polychloroamines have been isolated in certain cases (27).

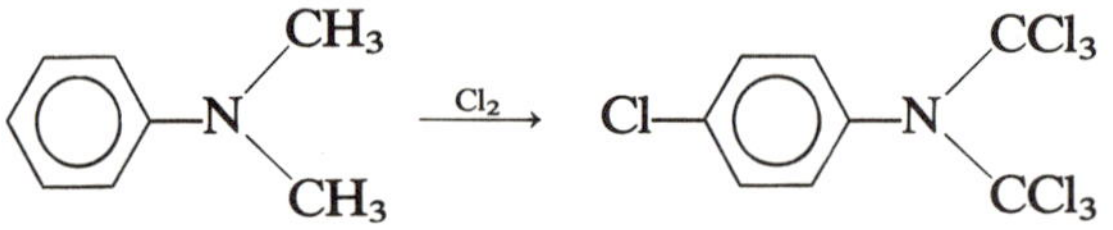

$$Ph-N(CH_3)_2 \xrightarrow{Cl_2} Cl-C_6H_4-N(CCl_3)_2$$

(40)

$$ClCH_2CH_2-N(CH_3)_2 \xrightarrow{Cl_2} Cl_3C-CHCl-N(CCl_3)(CHCl_2)$$

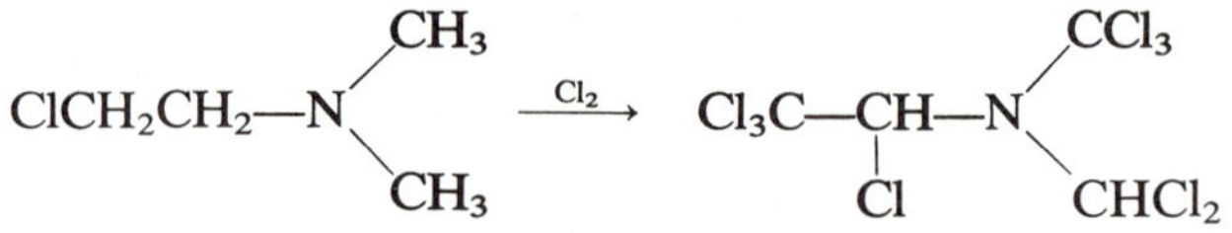

(41)

Polychlorinated amines (**40,41**) may be considered as covalent compounds since they are distillable and soluble in nonpolar organic solvents (e.g., in tetrachloromethane or petroleum ether). Unlike their iminium counterparts, they are comparatively inert and react slowly with water. Their hydrolysis requires heating with formic acid (27):

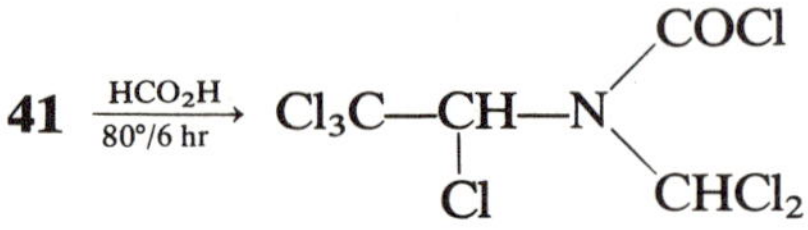

$$\mathbf{41} \xrightarrow[80°/6\ hr]{HCO_2H} Cl_3C-CHCl-N(COCl)(CHCl_2)$$

Chlorination of chloromethyleniminium chloride does not give **1**; rather, covalent tris(dichloromethyl)amine (**42**) is formed:

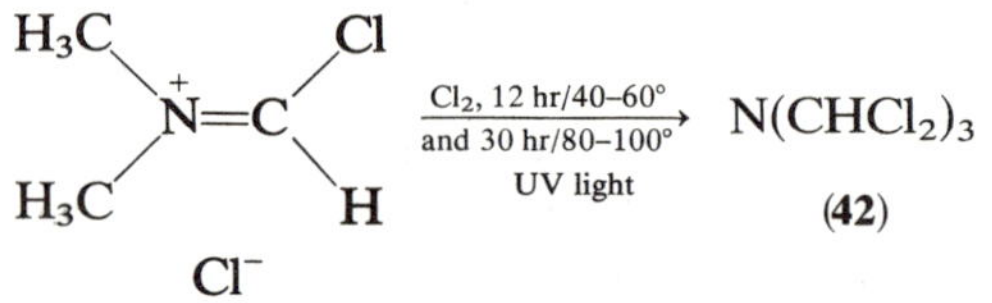

$$\underset{H_3C}{\overset{H_3C}{\diagdown}}\overset{+}{N}=C\underset{H}{\overset{Cl}{\diagup}} \quad Cl^- \quad \xrightarrow[UV\ light]{\substack{Cl_2,\ 12\ hr/40-60° \\ and\ 30\ hr/80-100°}} \quad N(CHCl_2)_3 \quad (42)$$

H. SYNTHESIS OF DIBROMOMETHYLENIMINIUM BROMIDES

Bromination of dithiuram disulfides (**23**) with elementary bromine has been used for the synthesis of bromophosgeniminium bromide (**2**):

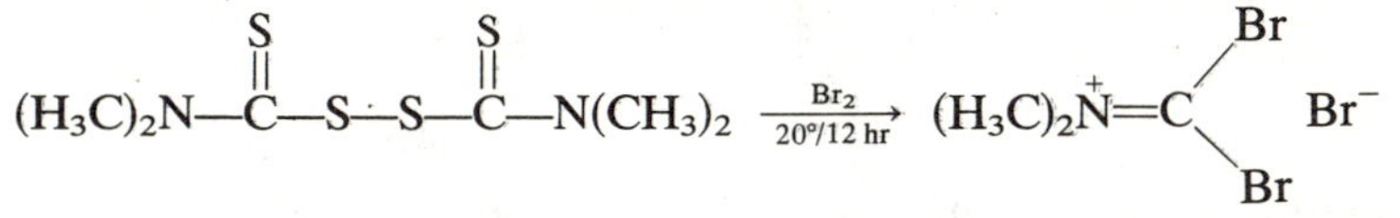

$$(H_3C)_2N-\underset{\underset{(23)}{}}{\overset{\overset{S}{\|}}{C}}-S-S-\overset{\overset{S}{\|}}{C}-N(CH_3)_2 \xrightarrow[20°/12\,hr]{Br_2} (H_3C)_2\overset{+}{N}=C\underset{Br}{\overset{Br}{\diagup}} \quad Br^- \quad (2)$$

Compound **2** represents a white salt that melts at 176° upon recrystallization from acetonitrile (24,28). In analogy to **1**, **2** hydrolyzes to dimethylcarbamoyl bromide and ammonolysis yields dimethylcyanamide (28) but preliminary results indicate that **2** is less reactive than the corresponding chloro derivative (**1**). This decrease in reactivity can be attributed to increased steric hindrance at the carbonium atom.

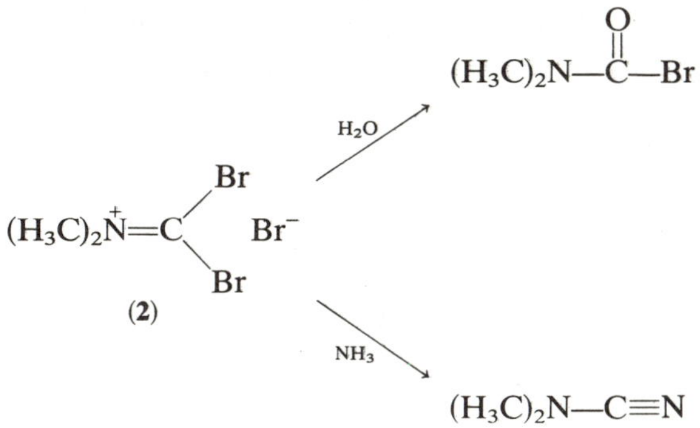

$$(H_3C)_2\overset{+}{N}=C\underset{Br}{\overset{Br}{\diagup}} \, Br^- \quad \underset{NH_3}{\overset{H_2O}{\rightrightarrows}} \quad \begin{array}{l}(H_3C)_2N-\overset{\overset{O}{\|}}{C}-Br \\ \\ (H_3C)_2N-C\equiv N\end{array}$$

(2)

I. SYNTHESIS OF TRIFLUOROMETHYLAMINES

The phosgeniminium fluorides (**4**) are probably the reactive dissociation products of the apparently more stable and covalent trifluoromethylamines (**4a**):

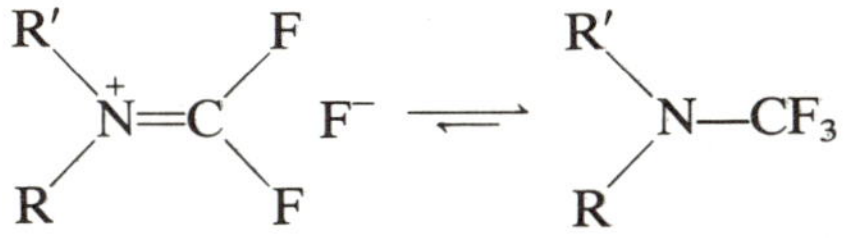

$$\underset{R}{\overset{R'}{\diagdown}}\overset{+}{N}=C\underset{F}{\overset{F}{\diagup}} \, F^- \rightleftharpoons \underset{R}{\overset{R'}{\diagdown}}N-CF_3$$

(4)   (4a)

A similar approach could be used for the synthesis of **4a**, because dithiuram disulfides (**23**) are cleaved by a number of fluorinating agents, for example, sulfur tetrafluoride (29), dimethylaminosulfur trifluoride

(30,112,113), and carbonyl fluoride (31):

$$\left[\begin{array}{c} R' \\ \diagdown \\ R \diagup N-\underset{\underset{S}{\|}}{C}-S \end{array} \right]_2 \xrightarrow{COF_2 \text{ etc.}} \begin{array}{c} R' \\ \diagdown \\ R \diagup N-CF_3 \end{array}$$

(23)   (4a)

An alternative method starts with **1**, and chlorine is replaced by fluorine simply on dissolving in hydrofluoric acid (32) or by the action of antimony trifluoride (2,33).

Because of their covalent structure, the reactivity of **4a** is lower than that of **1**, but these fluoro derivatives are promising reagents (e.g., for replacing hydroxyl groups with fluorine atoms).

III. The Reaction Potential of Dichloromethyleniminium Salts

Phosgeniminium salts are versatile building blocks for organic synthesis, comparable to Vilsmeier-Haack-Arnold (hereafter abbreviated as VHA) reagents (**6**) and to Mannich reagents (**7**):

$$R_2\overset{+}{N}=C\begin{array}{c}Cl\\ \diagdown \\ \diagup \\ Cl\end{array} Cl^- \qquad R_2\overset{+}{N}=C\begin{array}{c}H\\ \diagdown \\ \diagup \\ Cl\end{array} Cl^- \qquad R_2\overset{+}{N}=C\begin{array}{c}H\\ \diagdown \\ \diagup \\ H\end{array} Cl^-$$

(1)   (6)   (7)

Although **1** are somewhat less reactive than **6**, experience indicates that they are more versatile than **6** and **7** because of their higher oxidation level. The lower reactivity of **1**, $R = CH_3$, can be partly explained by its insolubility in organic solvents. Accordingly, the more soluble **1**, $R = C_2H_5$, reacts much faster. The influences of substituents and of counterions have yet to be investigated.

The reactions of **1** will be classified according to the nature of the arising bonds (C–C or C–heteroatom) and to the reaction type, for example chlorine substitution reactions in **1** or addition reactions to multiple bonds. Although this classification may sometimes be arbitrary, (e.g., with ambident partners), it has the advantage of simplicity.

A. CHLORINE SUBSTITUTION IN DICHLOROMETHYLEN-IMINIUM CHLORIDES LEADING TO C–C BOND FORMATION

1. *Reaction with Grignard and Other Metallorganic Compounds*

Grignard reagents substitute all three chlorine atoms, forming tertiary amines having an α_N tertiary carbon atom (see Table V). These amines are difficult to synthesize by other methods.

III. POTENTIAL OF DICHLOROMETHYLENIMINIUM SALTS

TABLE V
Tertiary Amines (43) from 1 and Grignard Reagents

R	R'	Me X	Yield, %	Boiling or melting point, °C/torr	Ref.
CH_3	C_2H_5	MgI	60	166–167/760	34
CH_3	C_2H_5	MgBr	54	32/0.1	35
CH_3	n-C_3H_7	MgBr	40	95-100/15	34
—$(CH_2)_5$—	n-C_3H_7	MgBr	31	62–64/0.1	34
CH_3	n-C_4H_9	MgBr	55	125–130/12	34
C_2H_5	n-C_4H_9	MgBr	37	130–135/12	34
CH_3	n-C_5H_{11}	MgI	45	85–90/0.05	34
CH_3	—$CH_2C_6H_5$	MgCl	84	64	35
C_3H_7	CH_3	Li—	35	161–164/760	10

$$\begin{array}{c} R \\ \diagdown \\ \end{array} \overset{+}{N}=C \begin{array}{c} Cl \\ \diagup \\ \\ \diagdown \\ Cl \end{array} \quad Cl^- \xrightarrow{3R'MeX} \begin{array}{c} R \\ \diagdown \\ \end{array} N - \underset{R'}{\overset{R'}{\underset{|}{C}}} - R'$$

(43)

All attempts to stop the reaction at mono- or disubstitution have failed up to now.

2. Reactions with Activated Methylene Groups

Malononitrile ($pK_a = 10.38$) condenses very easily with **1**, even in the absence of triethylamine. Cyanoacetates and especially malonic diesters, being weaker carbon-acids, require the presence of a base:

$$(H_3C)_2\overset{+}{N}=C\begin{array}{c}Cl\\ \\Cl\end{array} \; Cl^- + H_2C\begin{array}{c}X\\ \\Y\end{array} \xrightarrow{base} (H_3C)_2N\begin{array}{c} \\ \\Cl\end{array}C=C\begin{array}{c}X\\ \\Y\end{array}$$

(44)   (45)

A triphenylphosphonium group combined with a carboethoxy group activates sufficiently, and no base is required (Table VI). Nevertheless, better yields are obtained by using the corresponding phosphorane (37). Cyanoacetic acid reacts smoothly with 2 equivalents of **1** via the corresponding acid chloride to **45d**.

The presence of a base is useful also for another reason: it prevents hydrogen chloride from adding to the cyano group activated by the amino

TABLE VI

Condensation of **44** with **1**

44	X	Y	45	Yield, %	Boiling or melting point, °C/torr	Ref.
a	CN	CN	a	77	100/0.5, 37–39	7,9
b	CN	CO_2Et	b	85	135/0.02, 48	9,38
c	CO_2Et	CO_2Et	c	70	174/0.15,[a] 40–43	7,9
d	COCl	CN	d	90	86–90	7
e	C_6H_5	CN	e	33	145/0.6[b]	36
f	CO_2Et	$(H_5C_6)_5P^+$	f	62	—	37

[a] Ref. 9: b.p. 115–119/0.1.
[b] Mixture of cis+trans.

group in β position. Acetonitrile, **1**, and triethylamine do not react properly, and the presence of organometallic bases is apparently required. In the presence of hydrogen chloride, acetonitrile reacts via the amide chloride by another mechanism to azapentamethinecyanines (Section III-F-2). Phenylacetonitrile affords a moderate yield of **45e** only after previous treatment with 2 equivalents of butyllithium (136).

It may be interesting to compare **1** to dimethylformide chloride (**6**), which also requires the presence of triethylamine in the reaction with malon diester (39,40). In contrast, **6** reacts with malononitrile (41a) to the azapentamethinecyanine (**47**), apparently by N-acylation of the intermediate enamine dinitrile (**46**):

$$(H_3C)_2\overset{+}{N}=C\begin{matrix}H\\Cl\end{matrix}\ Cl^- + CH_2\begin{matrix}CN\\CN\end{matrix} \longrightarrow \left[(H_3C)_2N-C=C\begin{matrix}CN\\CN\end{matrix}\right]$$

(6)   (46)

$$46 \xrightarrow{(6)} \left[(H_3C)_2N-\underset{H}{C}-\underset{Cl}{\overset{CN}{C}}-\underset{H}{C}-N-C-N(CH_3)_2\right]^+ Cl^-$$

(47)

Tetrachlorocyclopentadiene reacts with **1** in boiling tetrahydrofuran to form 31% of 6-chloro-6-dimethylaminotetrachlorofulvene (41b).

The α-chloroenaminonitriles and esters (**45**) are stable and distillable compounds. They can be considered as derivatives of methanetricarboxylic acid. Although the chlorine in **45** is less reactive than that in ordinary α-chloroenamines (see the chapter by Ghosez and Marchard-Brynaert), it can be readily displaced by a number of nucleophiles (9,38).

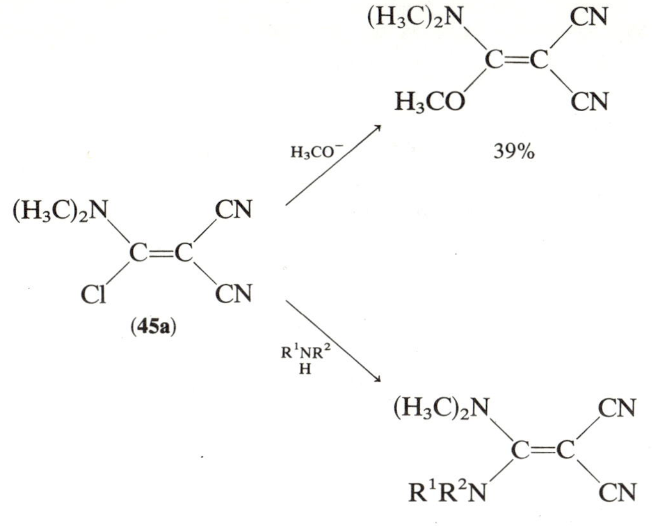

$$(H_3C)_2N\diagdown_{C=C}\diagup^{CN}_{CN}$$
$$H_3CO\diagup \qquad\qquad CN$$

H$_3$CO$^-$ ↗ 39%

$$(H_3C)_2N\diagdown_{C=C}\diagup^{CN}$$
$$Cl\diagup \qquad CN$$
(45a)

↘ R^{1}NR^2
   H

$$(H_3C)_2N\diagdown_{C=C}\diagup^{CN}$$
$$R^1R^2N\diagup \qquad CN$$

(45g) R^1, R^2 = C$_6$H$_5$, H 95%; H, H 45%; O⌐ 63%

Although **45g**, where both R^1 and R^2 are alkyl substituents, can be prepared from **44** and tetrasubstituted urea dichlorides (39), alcoholysis or aminolysis of **45** with primary amines (ammonia) can become synthetically useful.

Furthermore, **45** are suitable starting materials for heterocyclizations with hydrazines, hydroxylamines, and amidines (38,42). Thus phenylhydrazine and methylhydrazine cyclize with **45a** to diaminopyrazoles **48** or **49**:

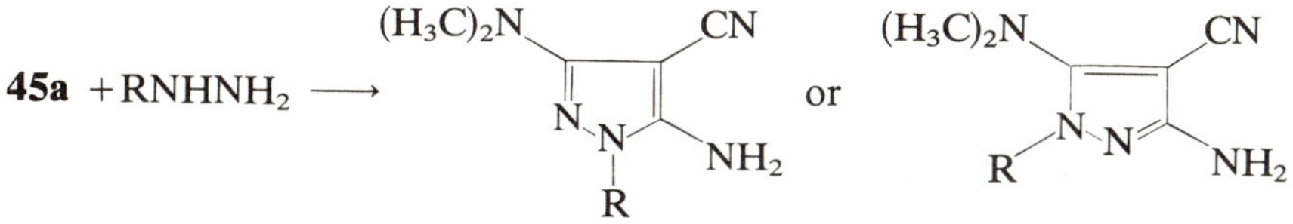

45a + RNHNH$_2$ ⟶ [pyrazole with (H$_3$C)$_2$N, CN, NH$_2$, N-R] or [pyrazole isomer]

(48) R = CH$_3$, C$_6$H$_5$ (49)

Both cyclizations are selective, that is, only one isomer, either **48** or **49**, is formed, but the course of the reaction remains to be investigated (114).

The same problem is encountered in the cyclization of the diester **45c** with phenylhydrazine:

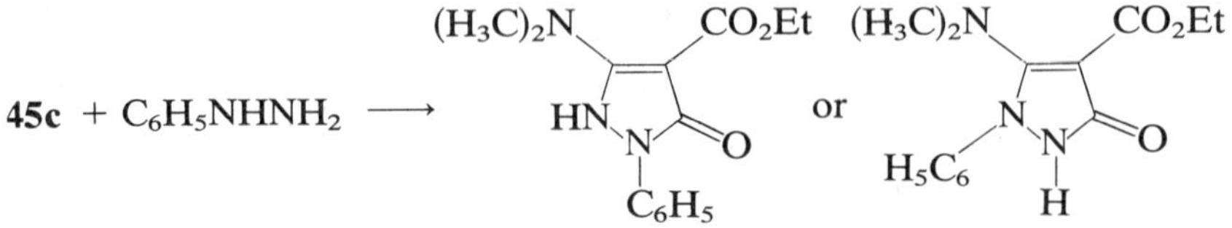

45c + C$_6$H$_5$NHNH$_2$ ⟶ [pyrazolone with (H$_3$C)$_2$N, CO$_2$Et, HN-N-C$_6$H$_5$] or [isomer with H$_5$C$_6$-N, N-H]

In **45b** the cyclization can take place on both the cyano and the ester group. Interestingly, phenylhydrazine adds to the cyano group, forming one of the isomers (**50,51**), whereas methylhydrazine displaces the alcoxy group to yield **52** or **53**:

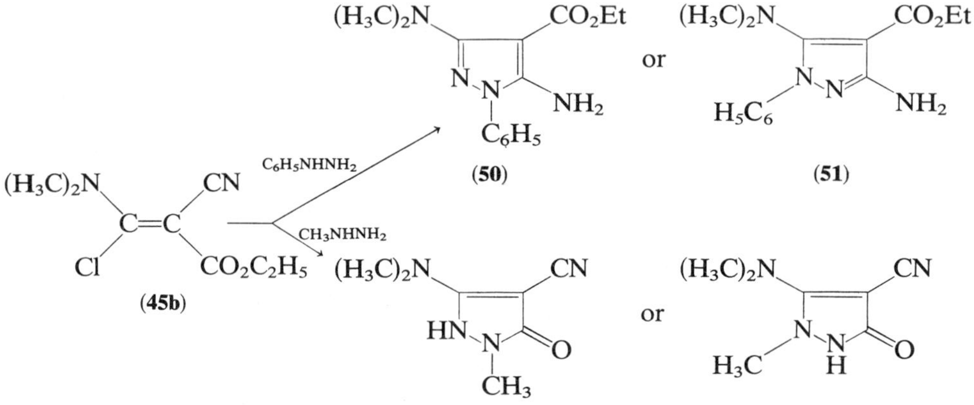

[Scheme: (45b) (H$_3$C)$_2$N-C(Cl)=C(CN)(CO$_2$C$_2$H$_5$) reacts with C$_6$H$_5$NHNH$_2$ to give pyrazoles **(50)** or **(51)** bearing CO$_2$Et and NH$_2$ groups; reacts with CH$_3$NHNH$_2$ to give pyrazolones **(52)** or **(53)** bearing CN group]

Nonambiguous chemistry is observed with symmetrical bisnucleophiles, for example, benzamidine:

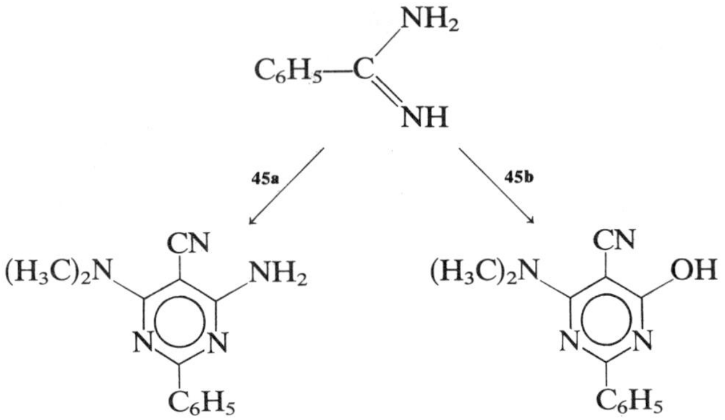

$$C_6H_5-C(NH_2)=NH$$

with **45a** → pyrimidine: (H$_3$C)$_2$N, CN, NH$_2$, C$_6$H$_5$

with **45b** → pyrimidine: (H$_3$C)$_2$N, CN, OH, C$_6$H$_5$

Tricyanomethane anion reacts with **1** in the isomeric ketenimine form under C–N bond formation and will be mentioned in Section III-B-1.

3. Reactions with Ketones

Compounds **1** readily replace nucleophilic oxygen atoms in alcohols, aldehydes, ketones, carboxylic acids, and epoxides by chlorine (see Section III-D). Methyl and methylene ketones undergo, an addition to this oxygen–chlorine exchange, a carbon–carbon condensation, leading to β-chloroacrylic amide chlorides. Scheme 4 exemplifies these interesting reactions.

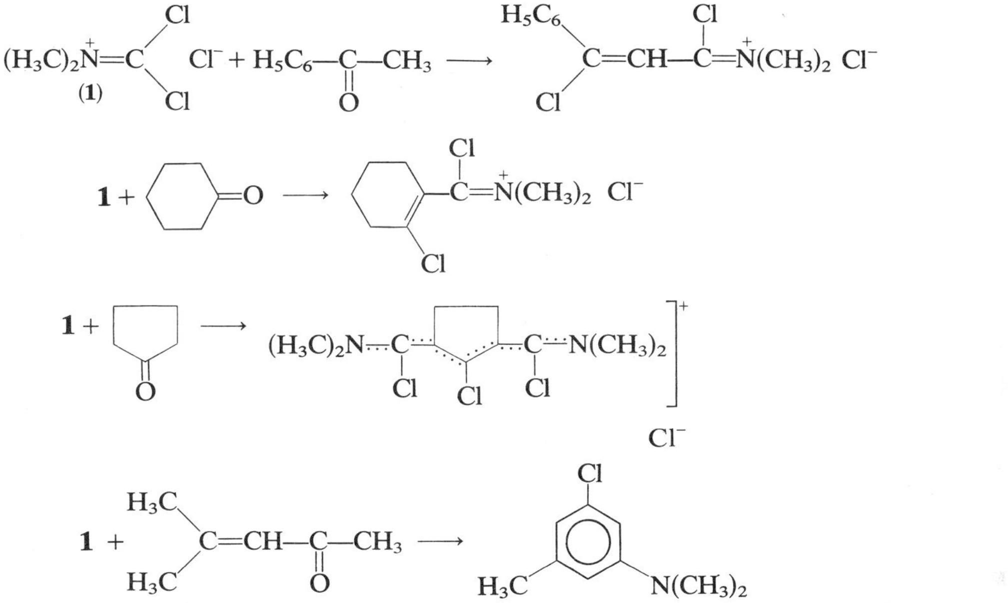

$$(H_3C)_2\overset{+}{N}=C\begin{smallmatrix}Cl\\Cl\end{smallmatrix}\ Cl^- + H_5C_6-\underset{O}{\overset{}{C}}-CH_3 \longrightarrow \begin{smallmatrix}H_5C_6\\Cl\end{smallmatrix}C=CH-\underset{}{\overset{Cl}{C}}=\overset{+}{N}(CH_3)_2\ Cl^-$$

$$\mathbf{1} + \text{cyclohexanone} \longrightarrow \text{2-chloro-1-(dichloro/chloro)cyclohexenyl iminium } Cl^-$$

$$\mathbf{1} + \text{cyclopentanone} \longrightarrow [(H_3C)_2N\cdots\underset{Cl}{\overset{}{C}}\cdots\text{cyclopentane}\cdots\underset{Cl}{\overset{}{C}}\cdots N(CH_3)_2]^+\ Cl^-$$

$$\mathbf{1} + \begin{smallmatrix}H_3C\\H_3C\end{smallmatrix}C=CH-\underset{O}{\overset{}{C}}-CH_3 \longrightarrow \text{3-chloro-5-methyl-N,N-dimethylaniline derivative}$$

These examples illustrate the value of these reactions. For simplicity different kinds of ketones will be dealt with, as follows:
(a) Methyl ketones.
(b) Methylene ketones.
(c) Methine ketones.
(d) Methyl vinyl ketones.
(e) Synthetic use of condensation products.

(a) *Reaction of* **1** *with Methyl Ketones.* Acetophenone (**54**) might be expected to condense with **1** at the α_{CO} methyl group, forming the α-chloro-β-benzoylenamine (**55**). In practice, however, 2 equivalents of **1** are consumed and β-chlorocinnamide chloride (**56**); together with dimethylcarbamoyl chloride, is the final product (7,45). This can be accounted for in two ways. In the first, **55** as vinylogous amide reacts with

1 at the nucleophilic carbonyl group by oxygen–chlorine exchange, yielding **56**:

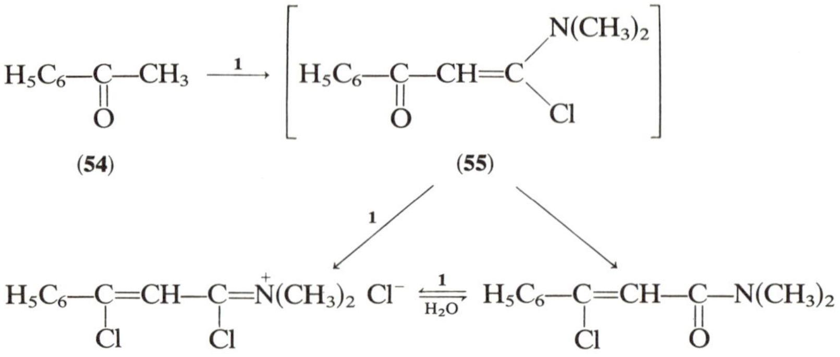

$$H_5C_6-\underset{O}{\underset{\|}{C}}-CH_3 \xrightarrow{\mathbf{1}} \left[H_5C_6-\underset{O}{\underset{\|}{C}}-CH=C\underset{Cl}{\overset{N(CH_3)_2}{\diagup}} \right]$$

(54)   (55)

$$H_5C_6-\underset{Cl}{\underset{|}{C}}=CH-\underset{Cl}{\underset{|}{C}}=\overset{+}{N}(CH_3)_2 \; Cl^- \underset{H_2O}{\overset{\mathbf{1}}{\rightleftarrows}} H_5C_6-\underset{Cl}{\underset{|}{C}}=CH-\underset{O}{\underset{\|}{C}}-N(CH_3)_2$$

(56)   (57)

Alternatively, **55** rearranges first to β-chlorocinnamide (**57**)—such rearrangements are well documented (43,44)—and the latter amide is converted to **56** with the phosgeniminium salt present. It has not yet been established which mechanism is operating. The reaction is quite general, and ring-substituted acetophenones (*p*-Cl, *m*-NO$_2$, *p*-NO$_2$, *p*-OCH$_3$) react equally well (7).

The hydrolysis of **56** stops at the β-chloroacrylamide stage (**57**), indicating the great difference in reactivity between the two chlorine atoms. Of the two possible geometrical isomers only one is formed, probably that with a phenyl and a carbamoyl group in the trans arrangement (7).

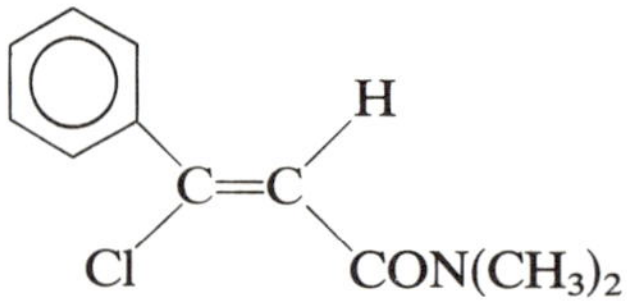

$$\underset{Cl}{\overset{C_6H_5}{\diagdown}}C=C\underset{CON(CH_3)_2}{\overset{H}{\diagup}}$$

(57)

Compounds **57** eliminate hydrogen chloride in the presence of a base, producing arylpropiolic acid amides (**58**): (115):

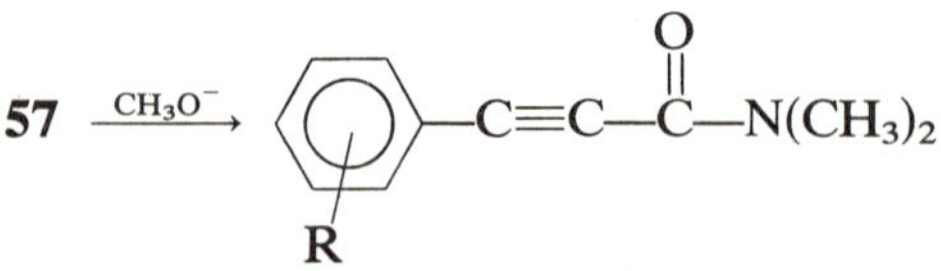

$$\mathbf{57} \xrightarrow{CH_3O^-} \underset{R}{C_6H_4}-C\equiv C-\underset{O}{\underset{\|}{C}}-N(CH_3)_2$$

(58)

Aminolysis of **56** with primary amines yields phenylpropiolamidines (**59a**). In two cases, R = *tert*-butyl and isopropyl, the β-chlorocinnamic amidines (**59b**) could be isolated (46):

$$\mathbf{56} + \text{RNH}_2 \longrightarrow \text{H}_5\text{C}_6\text{—C} \equiv \text{C—C} \begin{smallmatrix} \diagup \text{N(CH}_3)_2 \\ \diagdown \text{N—R} \end{smallmatrix} \quad \text{and/or}$$

(**59a**)

$$\text{H}_5\text{C}_6\text{—}\underset{\underset{\text{Cl}}{|}}{\text{C}}\text{=CH—C} \begin{smallmatrix} \diagup \text{N(CH}_3)_2 \\ \diagdown \text{NR} \end{smallmatrix}$$

(**59b**)

The condensation of methyl ketones with **1** is very sensitive to steric hindrance. This is probably the reason for the sluggish and unproductive reaction of **1** with *o*-methylacetophenone (7). Another example is the failure of propiophenone to react with **1**, whereas phenylacetone reacts expectedly at the methylene group. Open-chain aliphatic ketones such as acetone, diethyl ketone, chloroacetone, and methyl ethyl ketone have given, so far, only mixtures of products, resulting probably from polycondensation and aldolization, but these reactions remain to be studied. Significant is the failure of *tert*-butyl methyl ketone to react with **1**, whereas the Vilsmeier reagent reacts well (47). The same holds true for methyl isopropyl ketone; sterical factors seem to be intervening.

(*b*) *Reactions at* α_{CO} *Methylene Groups.* A smooth reaction occurs with cyclanones (45,48) in refluxing chloroform. Cyclohexanone and higher ketones ($n = 3, 4, 5, 9$) react in the same fashion as acetophenone to form monocondensation products (**60a–d**, yields 80–95%), whereas cyclopentanone and cyclobutanone afford the interesting 1,3,5-trichloropentamethinecyanines (**61a,61b**, yield 77 and 47%, respectively).

The difference in reactivity of cyclanones, to produce either **60** or **61** depending on the ring size, may derive from the higher carbon–hydrogen acidity of the smaller cycles. Furthermore, the latter form more stable exocyclic double bonds and thus favor formation of the cyanines (**61**) resulting from twofold condensation. Compounds **60** and **61** hydrolyze to the corresponding cyclic β-chlorovinylamides (**62,63**). Again, the vinylic chlorine atoms are quite inert to substitution.

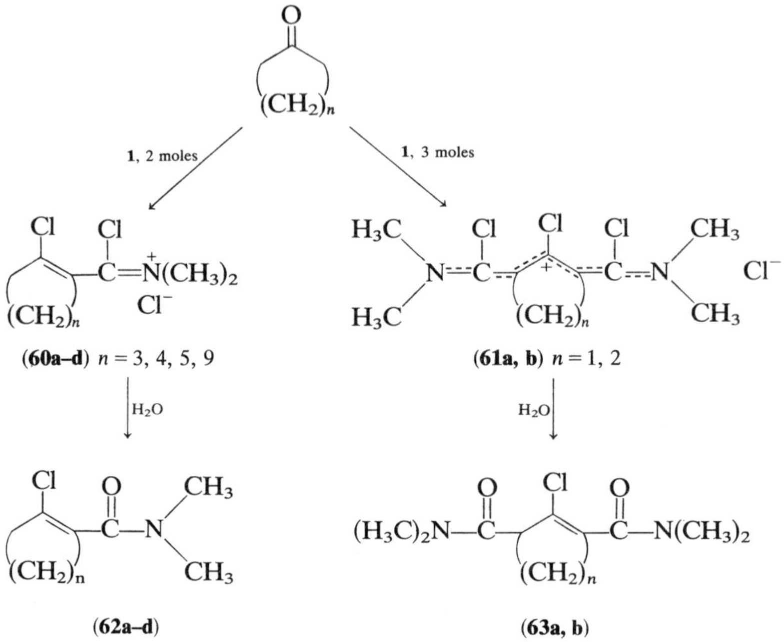

α-Tetralone reacts (48) as readily as cyclohexanone to give **64** in 81% yield, which in turn can be hydrolyzed to the amide (**65**):

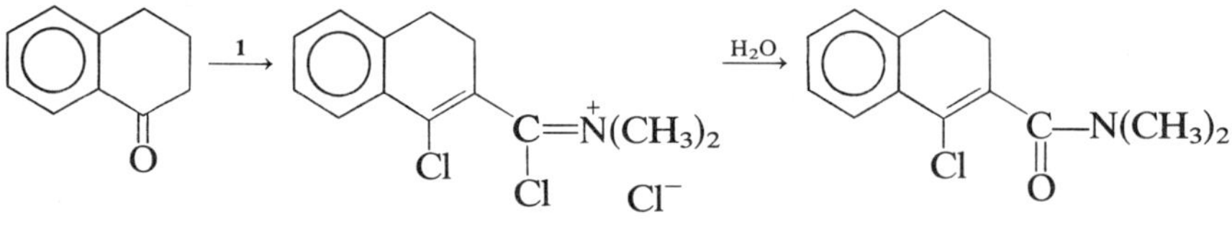

TABLE VII

Chloroacrylamides (**62**) and (**63**) from **1** and Cyclanones

Compound	n	Yield, %	Melting or boiling point, °C/torr	Ref.
62a	3	72	99–102/0.5	7,45
62b	4	87	130/0.6	7,45
62c	5	75	140/0.6	7,45
62d	9	84	140/0.02	48
63a	1	85	200/0.1	48
63b	2	78	107–108	48

Likewise, β-tetralone yields **66** in 57% yield:

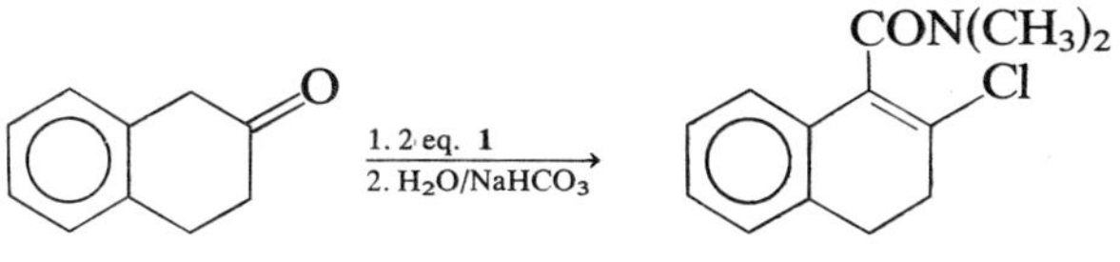

In the case of α-methylcyclohexanone both the methine and the methylene group react slowly, producing a mixture of the β-ketoamide (**67**) and the β-chlorocyclohexenylcarboxamide (**68**) in 40% yield after hydrolysis (48):

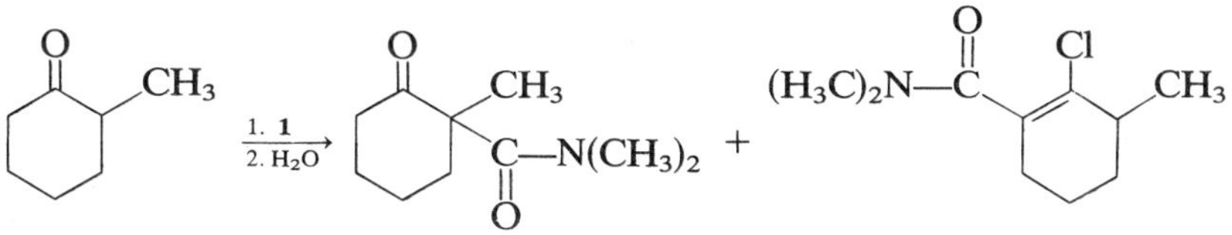

As expected, ketones carrying both a methyl and a methylene group, such as phenylacetone (**69**) or acetoacetates (**70**), react exclusively at the methylene group. Phenylacetone forms, in almost quantitative yield, the substituted crotonamide chloride (**71**), which then affords 75% of the parent crotonamide (**72**) on hydrolysis:

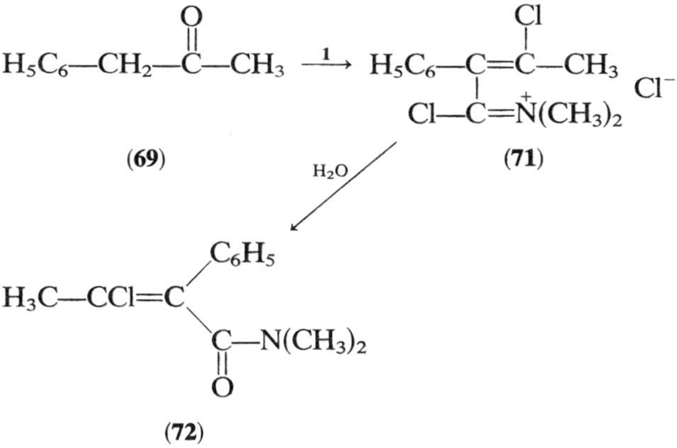

Acetoacetic acid esters condense with **1** first to alkylidenemalonic acid derivatives (**73**), as shown by their hydrolysis to **74**. Since the methyl group in **73** is activated by vinylogy, the reaction does not stop here and **73** slowly reacts further to afford a 1,5-dichloropentamethinecyanine (**75**), which cyclizes spontaneously to the more stable α-pyrone system

(**76**). In practice both **74** and **76** are obtained in moderate yields upon hydrolysis of the reaction mixture, but their separation is easy (7,45).

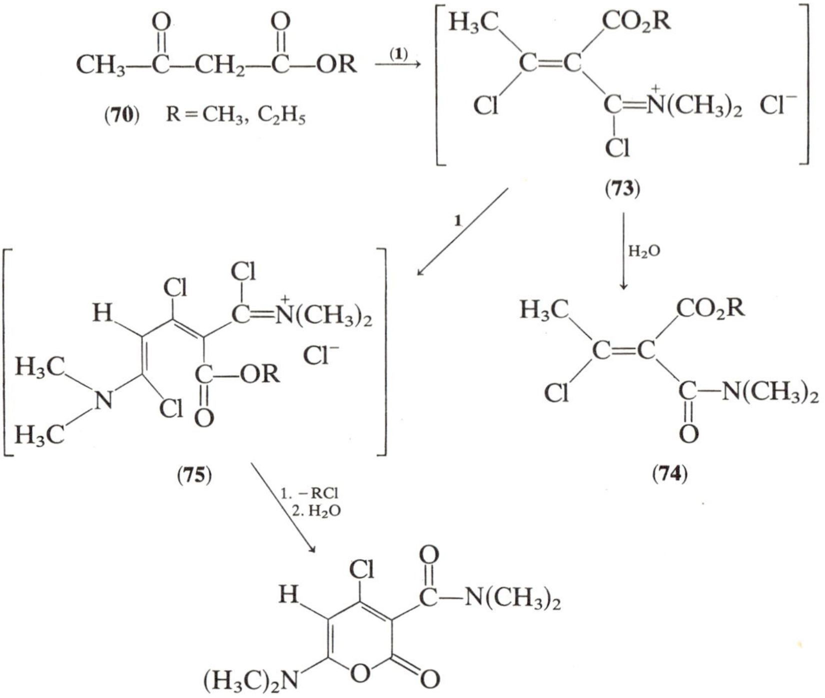

$$CH_3-\overset{O}{\underset{\|}{C}}-CH_2-\overset{O}{\underset{\|}{C}}-OR \xrightarrow{(1)}$$

(**70**) R = CH$_3$, C$_2$H$_5$

$$\left[\begin{array}{c} H_3C \diagdown \diagup CO_2R \\ C=C \\ Cl \diagup \diagdown C=\overset{+}{N}(CH_3)_2 \ Cl^- \\ Cl \end{array} \right]$$

(**73**)

↓ H$_2$O

$$\begin{array}{c} H_3C \diagdown \diagup CO_2R \\ C=C \\ Cl \diagup \diagdown \underset{\|}{C}-N(CH_3)_2 \\ O \end{array}$$

(**74**)

$$\left[\begin{array}{c} H \diagdown Cl \overset{Cl}{\underset{|}{C}}=\overset{+}{N}(CH_3)_2 \\ H_3C \diagdown Cl^- \\ N \underset{\|}{C}-OR \\ H_3C \diagup Cl O \end{array} \right]$$

(**75**)

↓ 1. –RCl
  2. H$_2$O

(H$_3$C)$_2$N pyranone ring with H, Cl, C–N(CH$_3$)$_2$, O substituents

(**76**)

(c) *Reactions of* **1** *with Methine Ketones.* Because of the steric hindrance, methine ketones are expected to react very slowly or not at all. In addition to the already mentioned 2-methylcyclohexanone, only 2,5-dimethylcyclohexanone has been studied. In this case the carbamylated product (**77**) is obtained in 20% yield (48):

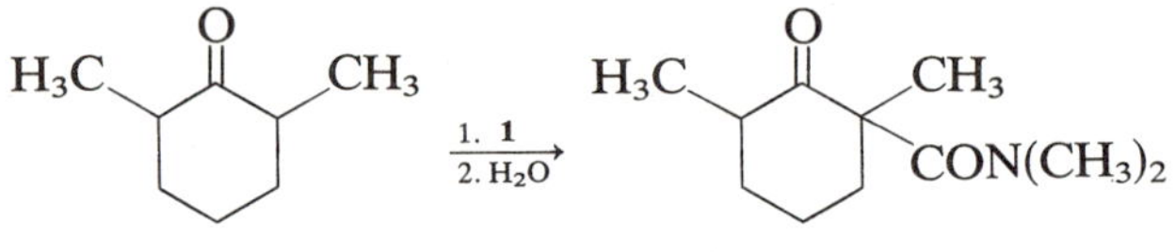

2,5-dimethylcyclohexanone $\xrightarrow[2.\ H_2O]{1.\ \mathbf{1}}$ 2,5-dimethyl-2-(N,N-dimethylcarbamoyl)cyclohexanone

(**77**)

(d) *Reactions with Vinyl Ketones.* Benzalacetone (**78**) reacts at the terminal methyl group, as shown by subsequent hydrolysis to the 5-phenyl-3-chloro-2,4-pentadienoic acid amide (**79**), but the overall yield is only 18%. Subsequent treatment with potassium *tert*-butylate affords the vinylogue of phenylpropiolamide (**80**):

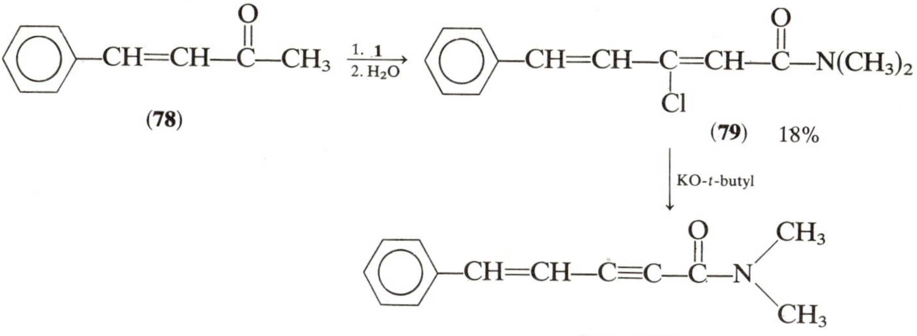

$$\text{Ph-CH=CH-C(=O)-CH}_3 \xrightarrow[2.\,H_2O]{1.\,\mathbf{1}} \text{Ph-CH=CH-C(Cl)=CH-C(=O)-N(CH}_3)_2$$

(**78**)   (**79**)   18%

$$\downarrow \text{KO-}t\text{-butyl}$$

$$\text{Ph-CH=CH-C≡C-C(=O)-N(CH}_3)_2$$

(**80**)   75%

As already mentioned (Scheme 4), the reaction of mesityl oxide (**81**) with **1** leads, via condensation followed by cyclization, to the benzene derivative (**82**):

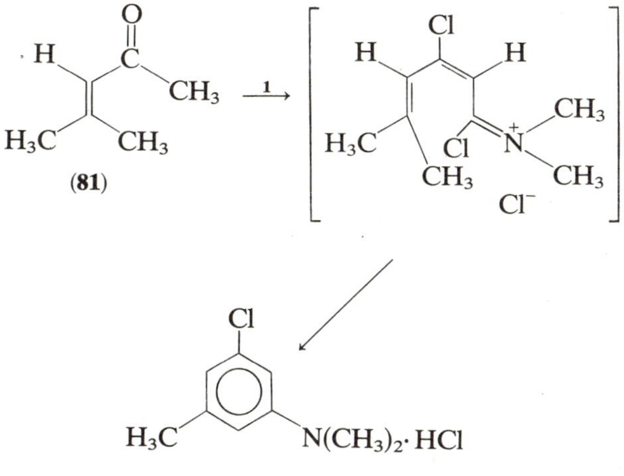

(**82**)   30%

Here an α_{CO} and a vinylogous methyl group are involved, evidently in this sequence, since purely vinylogous methyl groups either do not seem to react (e.g., crotonophenone) or react very sluggishly (isophorone) (7).

Isophorone reacts both on the α-methylene and the vinylogous methyl group to the bisamide (**83**) in poor yield (7). The reaction sequence is unknown, but it is probable that the methylene reacts first and the methyl group is then activated by vinylogy, in the same manner as with **73**:

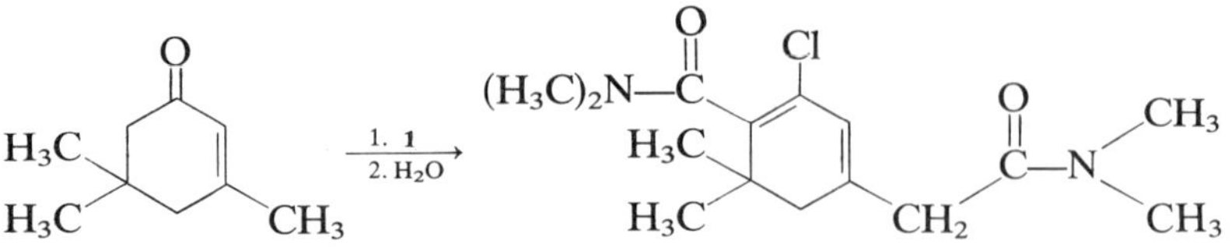

(**83**)

(e) *Synthetic Use of Condensation Products from Ketones and* **1**. The use of β-chlorocinnamamide chlorides in the synthesis of phenylpropiolic acid amides and amidines has already been illustrated. Furthermore, heterocyclizations using the cyclic precursors (**60**) and substituted hydrazines are the subject of a study (48) which reveals that phenylhydrazine affords usually only one isomer (**84**), whereas methylhydrazine, being less discriminative, sometimes gives rise to mixtures of both **84** and **85**. This is true especially for the cyclohexanone product (**60**, $n = 1$), where a mixture of equal amounts of isomers is obtained in 40% yield. As shown in Section III-H, **85** are also formed from the corresponding ketonehydrazones and **1** in excellent yields (Table VIII).

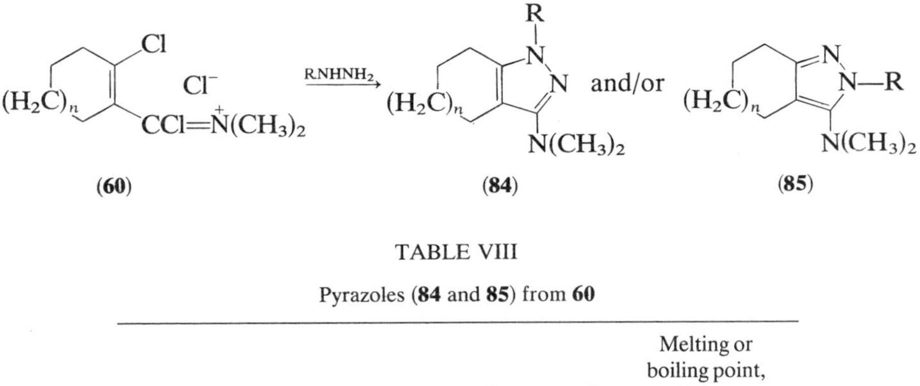

TABLE VIII

Pyrazoles (**84** and **85**) from **60**

Compound	n	R	Method[a]	Yield, %	Melting or boiling point, °C/torr
84a	1	C_6H_5	A	40	195
84b	1	CH_3	B	65	100/0.3
84c	2	C_6H_5	A	90	150/0.1
84d	7	C_6H_5	A	56	160/0.07
85a	2	CH_3	A	76	105/0.05
85b	3	CH_3	A	80	115/0.05

[a] A: direct method using the monosubstituted hydrazine; B: using dimethylhydrazine.

The two-isomer problem was circumvented by using N,N-dimethyl-hydrazine. In this case the isolable amidrazones (**86**) may be thermally cyclized to **84**, $R = CH_3$:

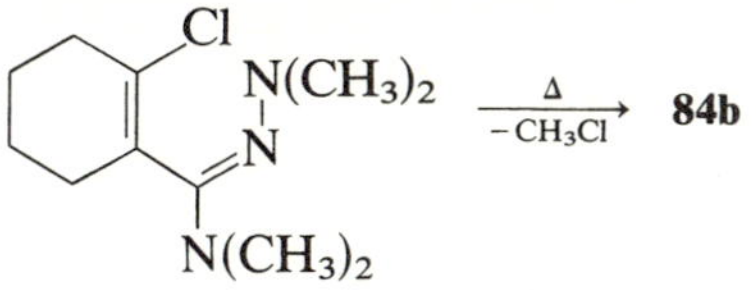

$$\text{(86)} \xrightarrow[-CH_3Cl]{\Delta} \textbf{84b}$$

(**86**)

As vinylogues of **1**, **60** are very reactive amide chlorides and condense with activated methylene groups (48):

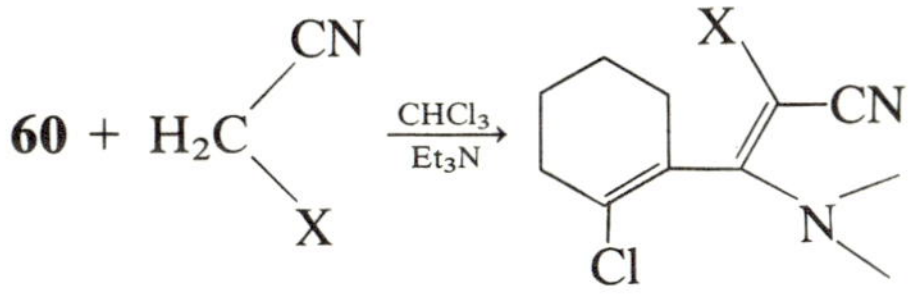

$$\textbf{60} + H_2C\begin{smallmatrix}CN\\X\end{smallmatrix} \xrightarrow[Et_3N]{CHCl_3}$$

(**87a**) X = CN 47%
(**87b**) X = CO$_2$Et 90%

4. Reactions of **1** with Tertiary Acetamides to Yield 1,3-Dichlorotrimethinecyanines and Their Use in Synthesis

The nucleophilic carbonyl oxygen in tertiary amides is easily attacked with phosgeniminium salts (**1**) to form the corresponding amide chlorides. Substituted tertiary acetamides possessing an α_{CO} methylene group (7,49,55) react further with **1** to give stable, delocalized 1,3-dichlorotrimethinecyanines (**88**). This is an elegant method of synthesis for these long-sought activated malondiamide derivatives (39,50–52).

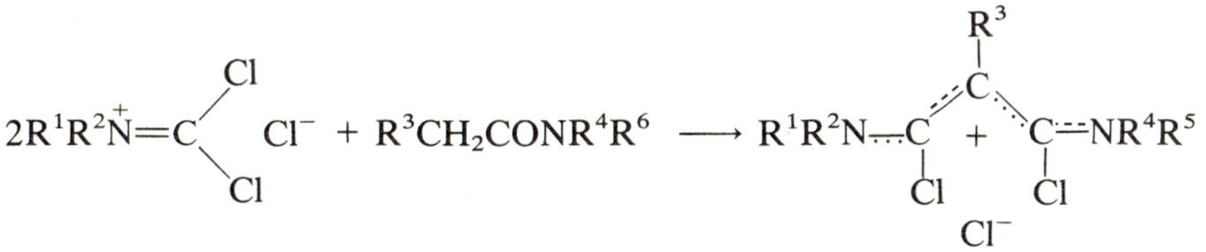

$$2R^1R^2\overset{+}{N}{=}C\begin{smallmatrix}Cl\\Cl\end{smallmatrix}\ Cl^- + R^3CH_2CONR^4R^6 \longrightarrow R^1R^2N\overset{R^3}{\underset{Cl}{\cdots}}\overset{|}{C}\cdots\overset{|}{\underset{Cl}{C}}{=}NR^4R^5\ Cl^-$$

(**1**) (**88**)

Addition of **1** to ynamines gives the same products (**88**). The high yields of **88** demonstrate that phosgeniminium salts (**1**) are much more reactive than phosgene itself, which produces α-chloro-β-chlorocarbonyl-enamines (**89**) in only moderate yields even when activating substituents such as chlorine and phenyl are present (53):

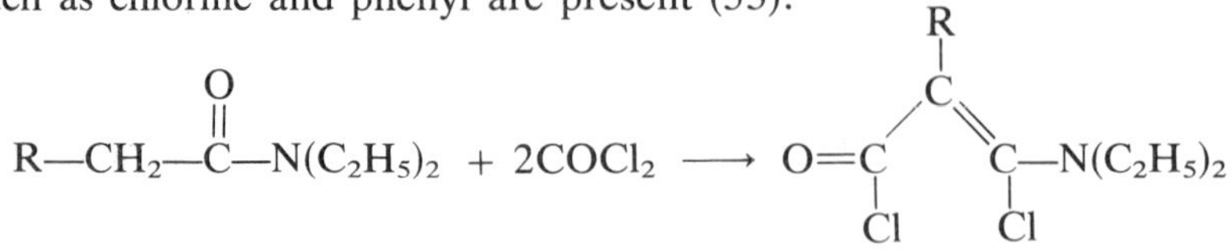

$$R-CH_2-\overset{O}{\underset{\|}{C}}-N(C_2H_5)_2 + 2COCl_2 \longrightarrow O=\underset{Cl}{\overset{}{C}}-\overset{\overset{R}{|}}{\underset{}{C}}=\underset{Cl}{\overset{}{C}}-N(C_2H_5)_2$$

(89a) R = Cl 65%
(89b) R = C$_6$H$_5$ 30%

Vilsmeier-Haack-Arnold reagents react similarly to give cyanines (90), which are malonic aldehyde amide (91) derivatives, (54):

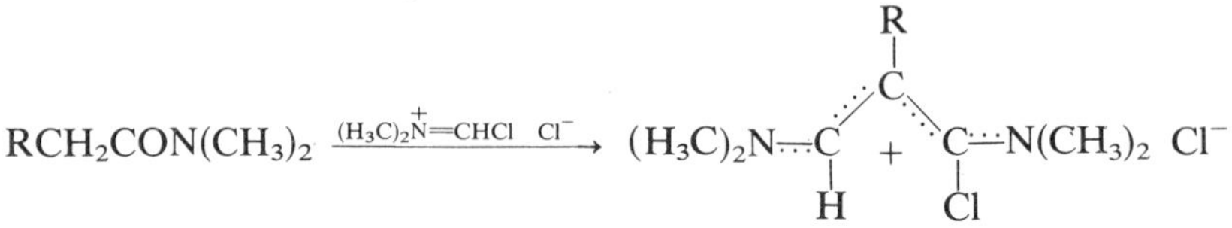

$$RCH_2CON(CH_3)_2 \xrightarrow{(H_3C)_2\overset{+}{N}=CHCl\ Cl^-} (H_3C)_2N\cdots\underset{H}{\overset{}{C}}\cdots\overset{\overset{R}{|}}{\underset{}{C}}\cdots\underset{Cl}{\overset{}{C}}\cdots N(CH_3)_2\ Cl^-$$

(90)

$$90 \xrightarrow{H_2O} \underset{H}{\overset{R}{\underset{}{\underset{}{C}}}}\underset{CHO}{\overset{CON(CH_3)_2}{}}$$

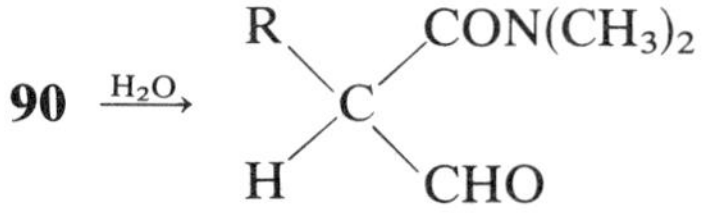

(91)

The cyanines (88) so far reported are listed in Table IX.

TABLE IX
1,3-Dichlorotrimethinecyanines

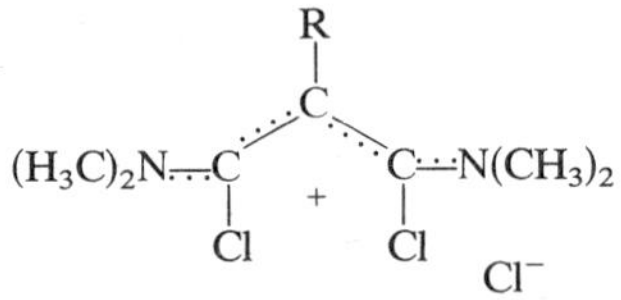

$$(H_3C)_2N\cdots\underset{Cl}{\overset{}{C}}\cdots\overset{\overset{R}{|}}{\underset{}{C}}\cdots\underset{Cl}{\overset{}{C}}\cdots N(CH_3)_2\ Cl^-$$

88	R	Yield, %	UV (CH$_2$Cl$_2$) λ_{max}, nm	Ref.
a	H	91	346	49
b	C$_2$H$_5$	88	388	49
c	C$_6$H$_5$	90	397, 278	49
d	Cl	88	410	49
e	CH$_3$	90	393	55
f	F	71	400	7
g	CH(CH$_3$)$_2$	75	408	7
h	OCH$_3$	95	406	55
i	OC$_2$H$_5$	98	409	55
j	OCH(CH$_3$)$_2$	92	407	55
k	OC$_6$H$_5$	99	405	55
l	OCOCH$_3$	—	392	56

III. POTENTIAL OF DICHLOROMETHYLENIMINIUM SALTS 373

Moreover, cyclic derivatives of **88** are obtained from tertiary lactams, as shown in the example with *N*-methylpyrrolidone (7):

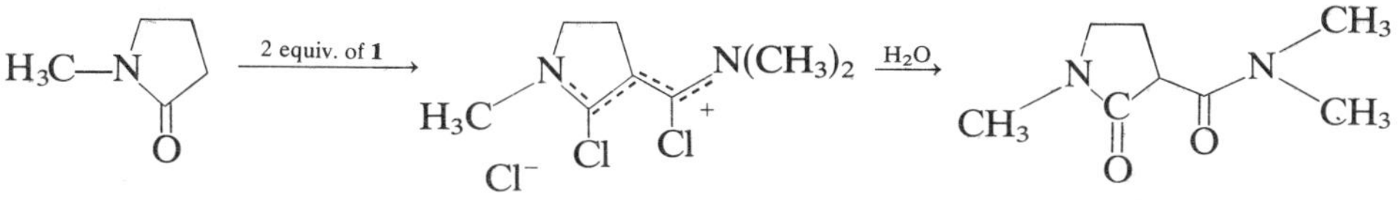

Recently, the reaction of **1** with 2-methylbenzoxazole and thiazole was reported to give **88m** and **88n**, where R are heterocyclic groups. These cyanines were hydrolyzed to the corresponding malondiamides (**92m,92n**) without isolation (57).

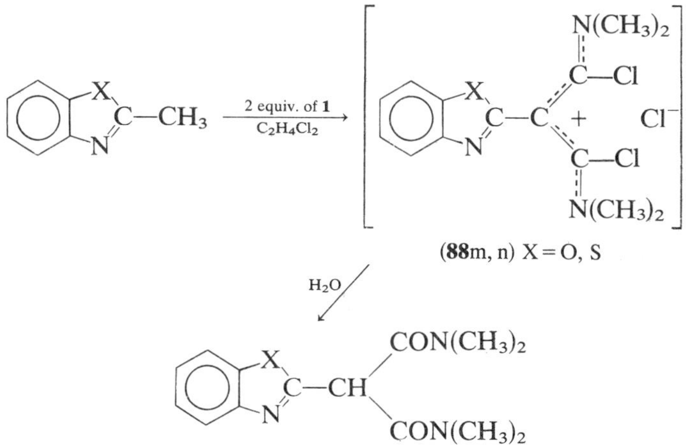

(**88m, n**) X = O, S

(**92m**) X = O, 18%
(**92n**) X = S, 35%

In the case where the heterocyclic residue bears a positive charge the intermediary α-chloroenamine is too deactivated and the reaction stops there (57). Another reason is that **93** is already a trimethinecyanine (**88**), in which one chlorine is replaced by a sulfur atom:

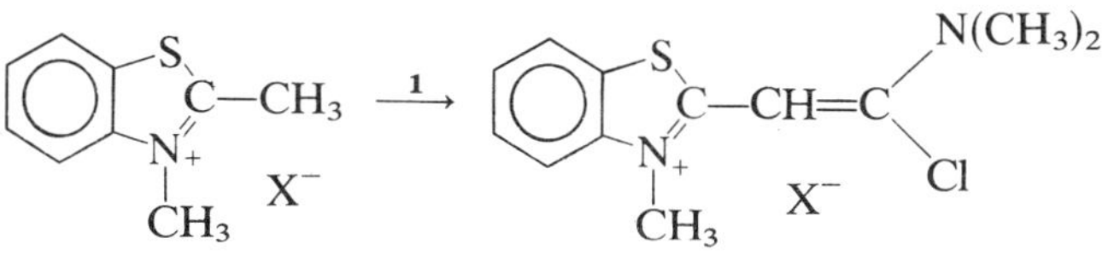

(**93**)

1,4-Phenylene and α, ω-alkylene bisacetamides react with 4 equivalents of **1** to yield biscyanines (**88o,88p**) (58):

$$\begin{array}{c}\text{CH}_2\text{CON(CH}_3)_2\\|\\\text{A}\\|\\\text{CH}_2\text{CON(CH}_3)_2\end{array} \xrightarrow{\text{4 equiv. of }\mathbf{1}} \begin{array}{c}(\text{H}_3\text{C})_2\text{N}\diagdown \quad \diagup \text{N(CH}_3)_2\\\text{C}\!=\!\overset{+}{\text{C}}\!=\!\text{C}\\\diagup \qquad \diagdown\\\text{Cl} \qquad\quad \text{Cl}\\|\\\text{A}\\|\\(\text{H}_3\text{C})_2\text{N}\diagdown \quad \diagup \text{N(CH}_3)_2\\\text{C}\!=\!\overset{+}{\text{C}}\!=\!\text{C}\\\diagup \qquad \diagdown\\\text{Cl} \qquad\quad \text{Cl}\end{array}\ 2\ \text{Cl}^-$$

(**88o**) A = —⟨C₆H₄⟩—   90%

(**88p**) A = —(CH$_2$)$_4$—   50%

In contrast, tetramethyladipamide affords the cyclic cyanine (**88q**). Prior self-condensation (59,60) of the intermediary bisamide chloride to the chloride of β-aminocyclopentenecarboxamide should be ruled out since the latter does not react further with **1**.

$$\begin{array}{c}\text{CH}_2\text{CON(CH}_3)_2\\|\\(\text{CH}_2)_2\\|\\\text{CH}_2\text{CON(CH}_3)_2\end{array} \xrightarrow{\mathbf{1}} \begin{array}{c}(\text{H}_3\text{C})_2\text{N}\diagdown\qquad\qquad\text{Cl}\\\text{C}\!=\!\!\!\langle\text{cyclopentene}\rangle\!\!\!-\!\text{C}\!=\!\overset{+}{\text{N}}\text{(CH}_3)_2\ \text{Cl}^-\\\diagup\qquad\\\text{Cl}\qquad \text{N(CH}_3)_2\end{array}$$

(**88q**)

TABLE X
Malonic Acid Derivatives (**94,95,96**)

Compound	R	R'	Yield, %	Melting or boiling point, °C/torr	Ref.
94a	H	—	70	104/0.6	49,55
94b	C$_2$H$_5$	—	75	76	49,55
94c	C$_6$H$_5$	—	78	149	49,55
94d	Cl	—	73	92	49,55
94e	F	—	51	120/0.3	7
94f	OCH$_3$	—	95	63	55
94g	OC$_2$H$_5$	—	98	76	55
94h	O-i-propyl	—	92	53	55
94i	OC$_6$H$_5$	—	99	116	55
94j	OCOCH$_3$	—	60	82/0.04	55
95a	H	—	70	107	49
95b	C$_2$H$_5$	—	83	178	49
95c	C$_6$H$_5$	—	55	193	49
95d	OCOCH$_3$	—	—	121–122	56
96a	H	CH$_3$	61	75/0.5	49
96b	H	C$_6$H$_5$	75	229	61,68

III. POTENTIAL OF DICHLOROMETHYLENIMINIUM SALTS

The dichlorotrimethinecyanines (**88**) have three mobile chlorine atoms that are readily substituted with OH, SH, OR, SR, NR$_2$, and NHR groups to malonamides (**94**), dithiomalonamides (**95**), and malonamidines (**96**). These reactions represent the most facile synthesis of these compounds (Table X).

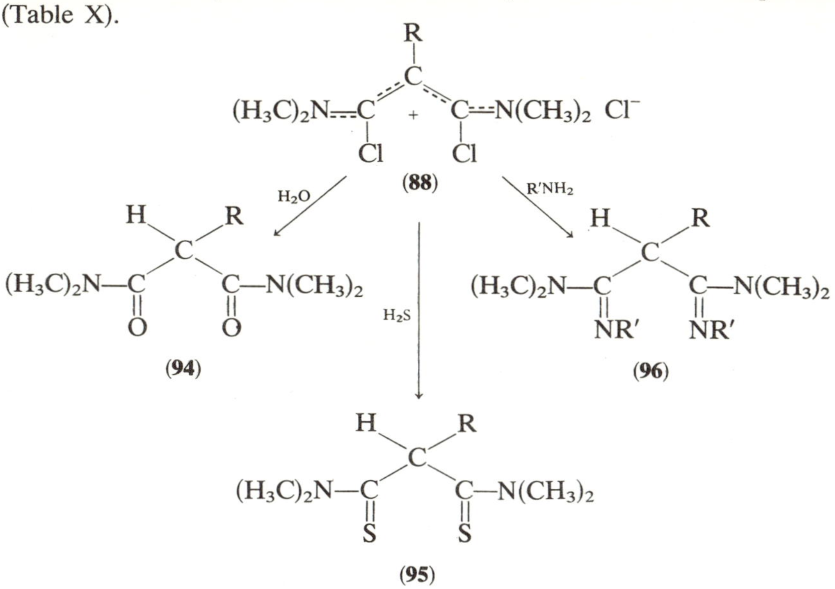

$$(H_3C)_2N{=}\overset{\displaystyle R}{\underset{\displaystyle Cl}{C}}{=}\overset{+}{\underset{\displaystyle Cl}{C}}{=}N(CH_3)_2 \quad Cl^-$$
(**88**)

With H$_2$O →

$$(H_3C)_2N-\underset{\displaystyle \underset{O}{\|}}{C}\overset{\displaystyle H \quad R}{\underset{}{\diagdown C \diagup}}\underset{\displaystyle \underset{O}{\|}}{C}-N(CH_3)_2$$
(**94**)

With R'NH$_2$ →

$$(H_3C)_2N-\underset{\displaystyle \underset{NR'}{\|}}{C}\overset{\displaystyle H \quad R}{\underset{}{\diagdown C \diagup}}\underset{\displaystyle \underset{NR'}{\|}}{C}-N(CH_3)_2$$
(**96**)

With H$_2$S ↓

$$(H_3C)_2N-\underset{\displaystyle \underset{S}{\|}}{C}\overset{\displaystyle H \quad R}{\underset{}{\diagdown C \diagup}}\underset{\displaystyle \underset{S}{\|}}{C}-N(CH_3)_2$$
(**95**)

Moreover, cyanine **88a** has proved to be a versatile source of ynamines and tetraaminoallenes. Thus triethylamine eliminates (7) hydrogen chloride to give the unstable ynamine amide chloride (**97**). Analogously, ynamine amide (**98**) is formed in 70% yield with aqueous alkali at low temperature (7,62):

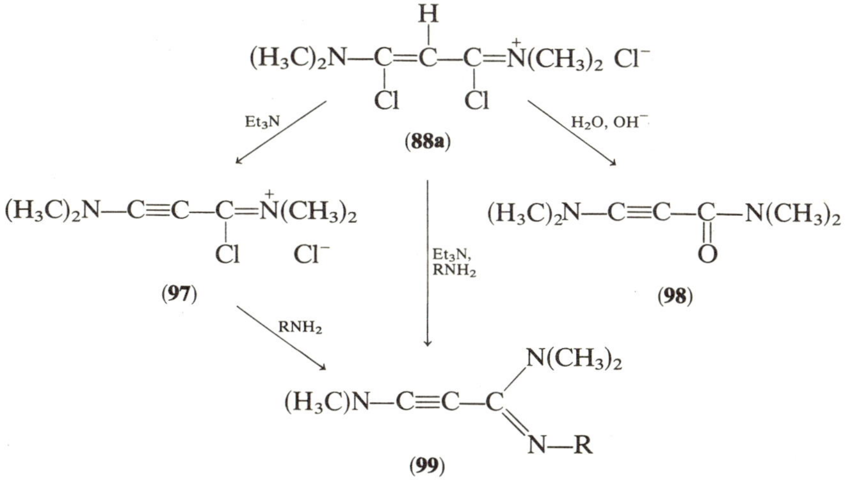

$$(H_3C)_2N-\underset{\displaystyle Cl}{C}{=}\overset{\displaystyle H}{C}-\underset{\displaystyle Cl}{C}{=}\overset{+}{N}(CH_3)_2 \quad Cl^-$$
(**88a**)

Et$_3$N ↙   H$_2$O, OH$^-$ ↘

$$(H_3C)_2N-\underset{\displaystyle Cl}{C}{\equiv}C-\overset{+}{C}{=}N(CH_3)_2 \quad Cl^-$$
(**97**)

$$(H_3C)_2N-C{\equiv}C-\underset{\displaystyle \underset{O}{\|}}{C}-N(CH_3)_2$$
(**98**)

↓ Et$_3$N, RNH$_2$

↘ RNH$_2$

$$(H_3C)N-C{\equiv}C-C\overset{\displaystyle \diagup N(CH_3)_2}{\diagdown N-R}$$
(**99**)

Amide chloride (**97**) formed *in situ* is transformed to relatively stable ynamine amidines (**99**) with primary amines (63) (Table XI). Ynamine amidines (**99**), where R are substituted aromatics, are unstable since they undergo electrocyclizations to condensed systems having a 2,4-bis(dialkyl-amino)quinoline moiety (64):

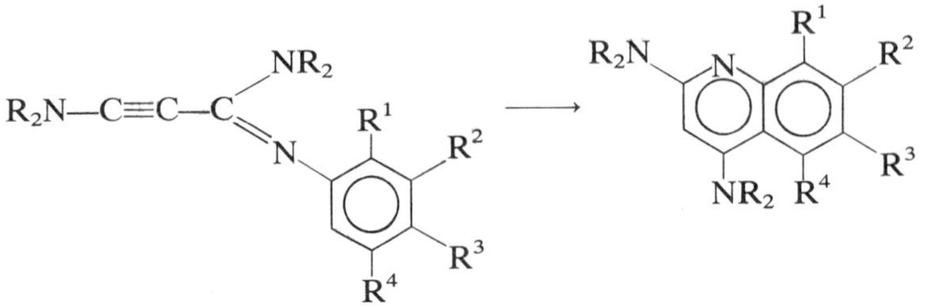

Ynamine amidine (**99a**) reacts smoothly with carbon dioxide and phenylisocyanate to enaminoisoxazolones (**99f,99g**) (63):

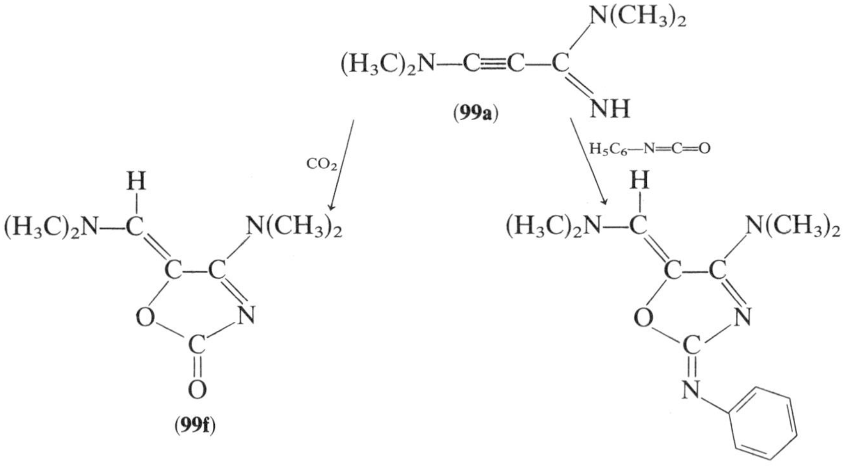

TABLE XI

Ynamine Amidines (**99**)

99	R	Yield, %
a	H	65
b	CH_3	50
c	$i\text{-}C_3H_7$	63
d	$t\text{-}C_4H_9$	49
e	C_6H_{11}	54

TABLE XII

Cyanines (**100**) and Allene (**101**)

Compound	R	X⁻	Yield, %	Melting or boiling point, °C/torr
100a	CH_3	ClO_4	85	174
100b	C_2H_5	Cl	91	Liquid
101a	CH_3	—	—	—
101b	C_2H_5	—	70	120/0.5

Secondary amines replace both chlorine atoms in (**88a**), thus forming the very stable 1,1,3,3-tetrakis(disalkylamino)alkyl cations (**100**). These cations may be regarded as protonation products of allenetetramine (**101**); in fact, they can be deprotonated with strong bases such as n-butyllithium or sodium amide (7,65) (Table XII).

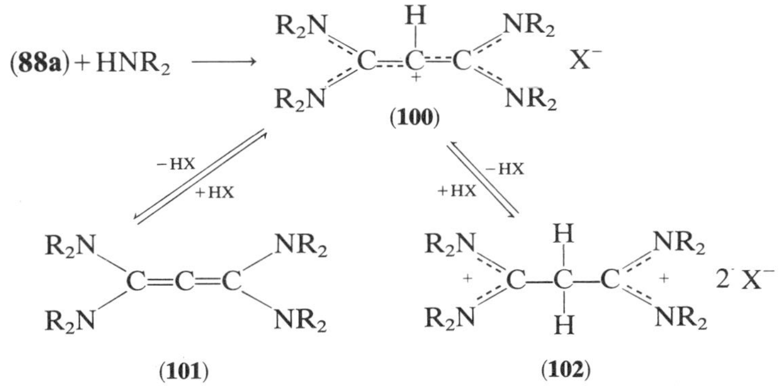

$$(88a) + HNR_2 \longrightarrow \underset{R_2N}{\overset{R_2N}{>}}C = \underset{|}{\overset{H}{C}} = C\underset{NR_2}{\overset{NR_2}{<}} \quad X^-$$

(**100**)

$$\underset{R_2N}{\overset{R_2N}{>}}C = C = C\underset{NR_2}{\overset{NR_2}{<}}$$

(**101**)

$$\underset{R_2N}{\overset{R_2N}{>}}C - \underset{H}{\overset{H}{C}} - C\underset{NR_2}{\overset{NR_2}{<}} \quad 2\,X^-$$

(**102**)

On the other hand, the conjugation in **100** leads to a high negative charge at C_2, which is confirmed by NMR measurements (66,67). The ^{13}C signal is found at a very high field (-72.4 ppm), and the same holds true for the chemical shift of the methine hydrogen ($\delta = 3.6$ ppm). Consequently, **100** can also be protonated (7) with strong acids, for example, fluorosulfonic acid, to a malonbisamidinium salt (**102**). Tetraaminoallenes react with carbon dioxide and disulfide, sulfur dioxide, and sulfur to give dipolar adducts, and phenyl cyanate effects cyanation at C_2 (66).

Furthermore, the biselectrophilic system in **88** is of general applicability to the synthesis of aminated heterocyclic compounds. A few examples of cyclizations with hydrazines, hydroxylamines, 1,2-diamines, and amidines leading to five to seven membered rings are listed below (55,68,116).

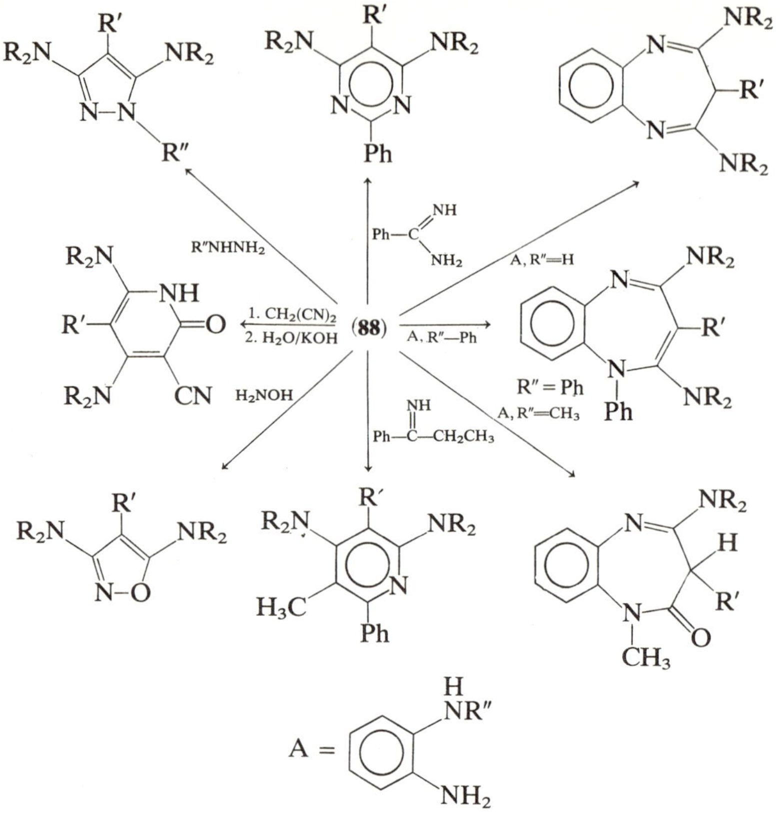

Likewise, the biscyanines (**88o,88p**) have been cyclized to the corresponding bispyrazoles (58).

5. Electrophilic Substitution on Aromatic Compounds

So far, only strongly activated aromatics have been reported to react with **1**:

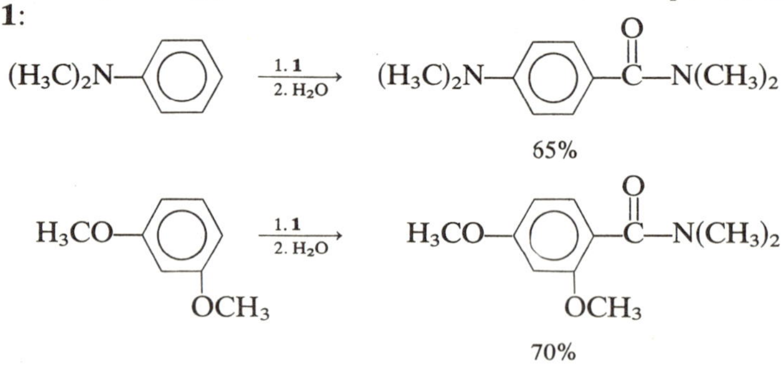

Anisol does not react with **1**, even in the presence of aluminum chloride (7,69). In contrast to chloromethyleniminium salts (70), **1** reacts with phenol by O-acylation (see Section III-D-1).

So far, only a limited number of electron-rich heterocycles have been studied. Pyrrole and indole form the expected amide chlorides with **1**, but N-methylpyrrole gives a 2:1 mixture of both α- and β-substituted products (69).

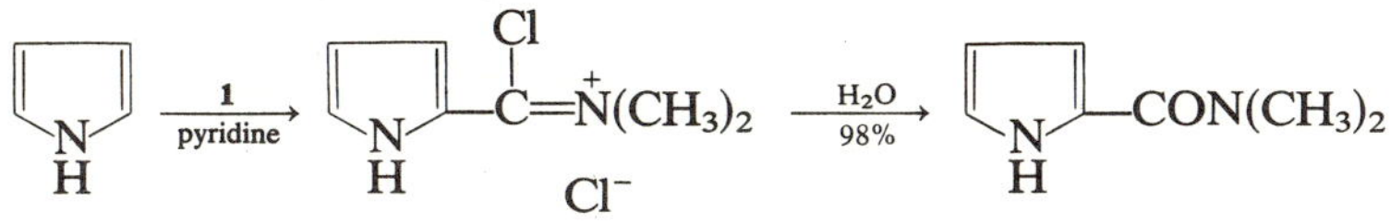

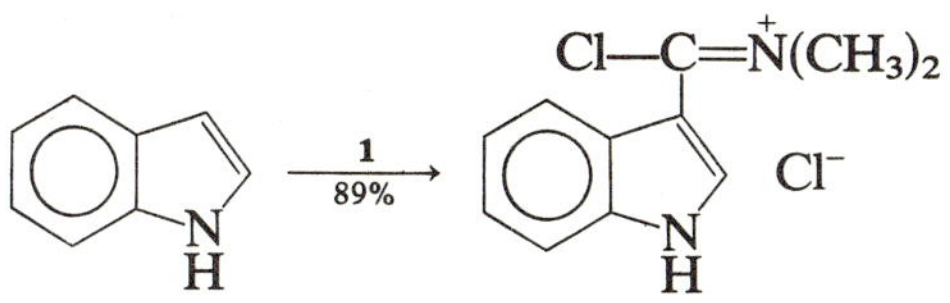

The less nucleophilic furan reacts much more slowly, with only 40% yield of the corresponding amide chloride (69).

The reactive **102a** affords interesting cyclization products:

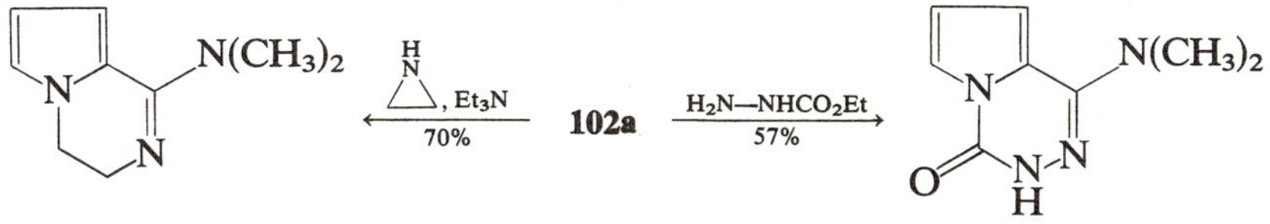

B. C–N BOND FORMATION WITH PHOSGENIMINIUM SALTS

1. *Reaction of **1** with Amines, Sulfonamides, and Hydrazines*

The reaction of **1** with amines represents a facile synthesis of cyanamides (**103**), chloroformamidines (**104,107**), guanidines (**105**), and urea dichlorides (**106**):

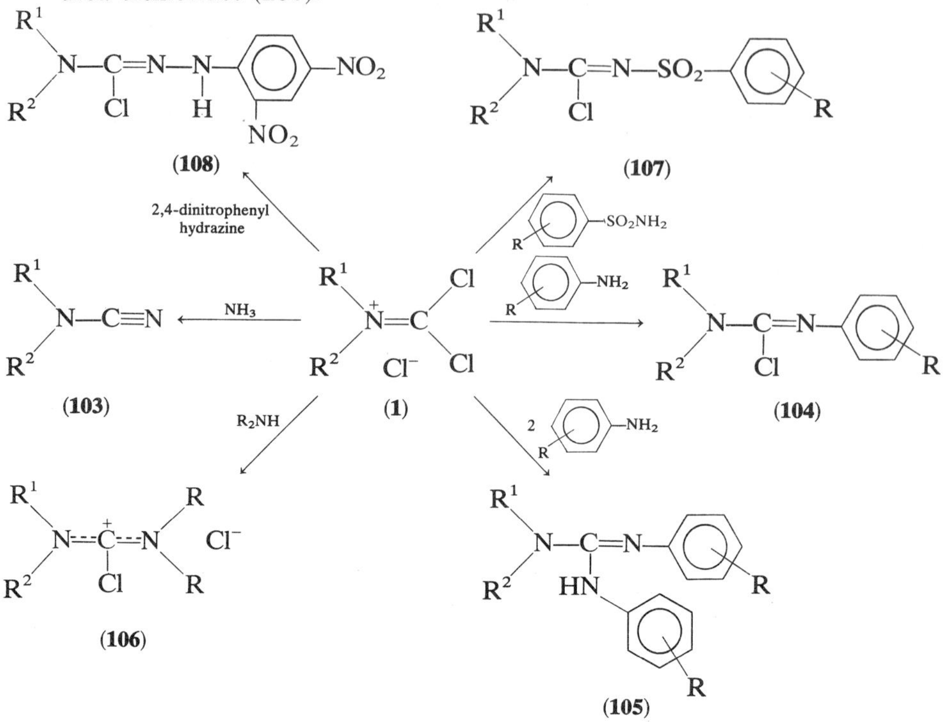

TABLE XIII

Reaction of **1**, $R^1 = R^2 = CH_3$, with Amines

Compound	R	Yield, %	Melting or boiling point, °C/torr	Ref.
104a	H	67	81/0.5	7
104b	4-$CO_2C_2H_5$	96	65	3
104c	4-NO_2	80	102	3
104d	2,4-Dinitro	87	97	3
105	4-NO_2	58	270[a]	4
106	C_6H_5	97	—	7
107a	H	94	66	7,71
107b	4-CH_3	76	152–164	4,69,71
107c	4-Br	95	126–127	69,71
107d	4-Cl	85	143	4,69,71
107e	4-NO_2	90	180–181	69,71
108	2,4-Dinitro	87	198	7

[a] Hydrochloride.

As shown in Table XIII, in addition to hydrazines and other particular amines even very weak aromatic amines, such as dinitroaniline and sulfonamides, give high yields of the corresponding chloroformamidines.

It is interesting that N-chloro compounds react with **1** in the same manner as do the corresponding primary amines or sulfonamides except that elementary chlorine is evolved (71,119,120):

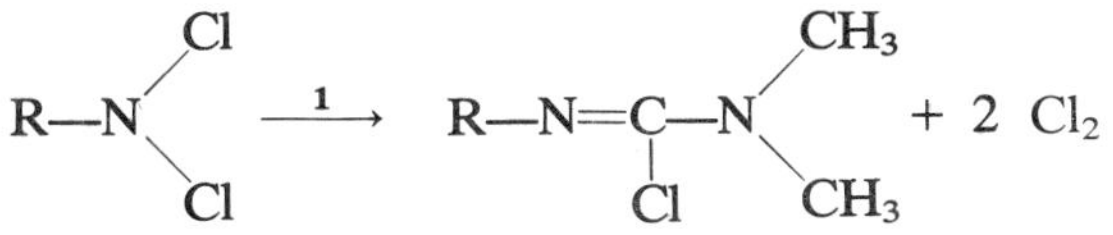

$$\text{R-NCl}_2 \xrightarrow{\mathbf{1}} \text{R-N=C(Cl)-N(CH}_3)_2 + 2\text{ Cl}_2$$

N,N',N"-trichloroisocyanuric acid reacts in the same manner with **1**, but the intermediate undergoes subsequent fragmentation to the isocyanate (**109a**), which is in equilibrium with the isomeric carbamoyl chloride (**109b**) (72):

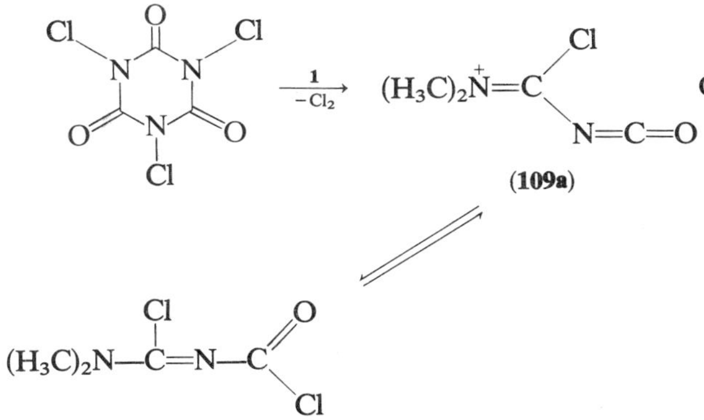

[Structure of trichloroisocyanuric acid] $\xrightarrow[-\text{Cl}_2]{\mathbf{1}}$ $(H_3C)_2\overset{+}{N}=C(Cl)(N=C=O) \;\; Cl^-$

(**109a**)

$(H_3C)_2N-\underset{Cl}{C}=N-C(=O)Cl$

(**109b**)

O-Substituted hydroxylamines condense in the same manner to N-alkoxychloroformamidines (**110**), as shown in the following example (73):

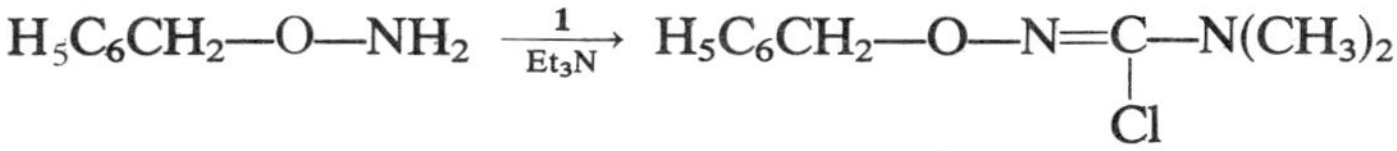

$$H_5C_6CH_2-O-NH_2 \xrightarrow[\text{Et}_3\text{N}]{\mathbf{1}} H_5C_6CH_2-O-N=\underset{Cl}{C}-N(CH_3)_2$$

(**110**) 91%

Hydroxylamine hydrochloride and **1** afford 35% of the N,N-dimethyl-N'-dimethylcarbamoylchloroformamidine at 85° and 64% of the known azacyanine (**166a**) when heated to 110° (117). Whereas 2,4-dinitrophenylhydrazine reacts only at the primary amino group (7) to **108**, the more basic phenyl- and methylhydrazine react preferentially with 2 equivalents of **1**, forming the reactive hydrazine derivatives (**111**)

(74, 118):

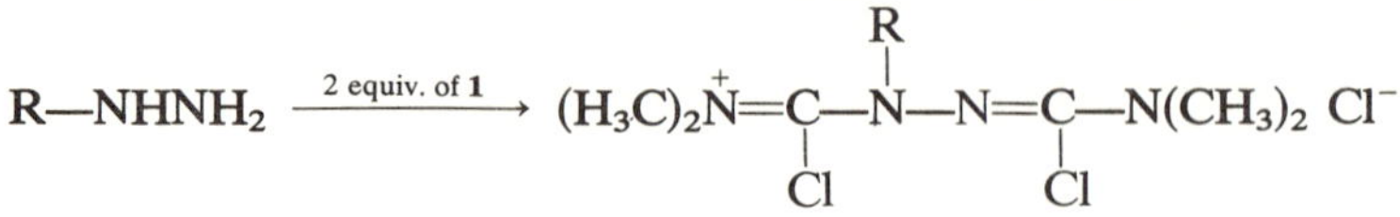

$$R-NHNH_2 \xrightarrow{\text{2 equiv. of 1}} (H_3C)_2\overset{+}{N}=\underset{Cl}{C}-\underset{|}{\overset{R}{N}}-N=\underset{Cl}{C}-N(CH_3)_2 \; Cl^-$$

(111) R = CH₃, C₆H₅

(111) R = CH₃ + [2-aminophenol] $\xrightarrow{90\%}$ [imidazolium product] Cl⁻

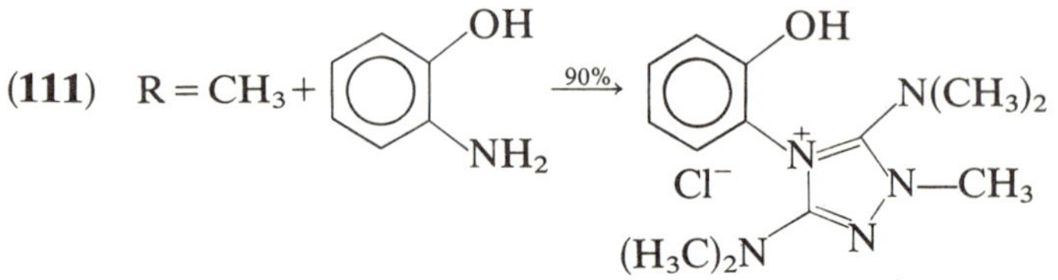

Enamines having free NH₂ or NCl₂ groups such as **112** condense with **1**, as do ordinary primary amines (75, 120):

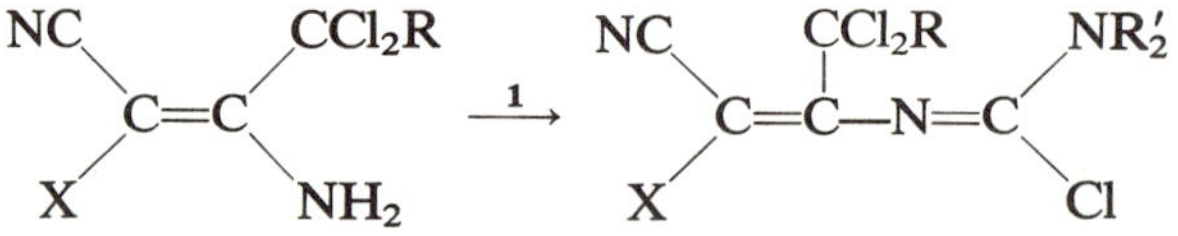

$$\underset{X}{\overset{NC}{\diagdown}}C=C\underset{NH_2}{\overset{CCl_2R}{\diagup}} \xrightarrow{1} \underset{X}{\overset{NC}{\diagdown}}C=C\underset{N=C}{\overset{CCl_2R}{|}}\underset{Cl}{\overset{NR'_2}{\diagup}}$$

(112) (113) R = H, Cl; R' = CH₃, Et; X = CN, CO₂R

Compounds **113**, where X = CO₂R, cyclize spontaneously to the corresponding 4,5-disubstituted 2-dimethylamino-1,3-oxazine-6-ones by the loss of alkyl chloride (75).

The tricyanomethane anion (76) reacts in its tautomeric ketenimine form, yielding N-(1-chloro-2,2-dicyanovinyl)chloroformamidines (**114**):

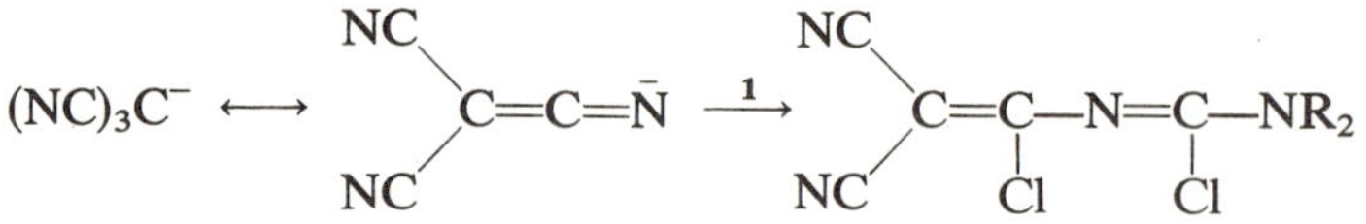

$$(NC)_3C^- \longleftrightarrow \underset{NC}{\overset{NC}{\diagdown}}C=C=\bar{N} \xrightarrow{1} \underset{NC}{\overset{NC}{\diagdown}}C=\underset{Cl}{\overset{|}{C}}-N=\underset{Cl}{\overset{|}{C}}-NR_2$$

(114) R =	
CH₃	70%
C₂H₅	76%
—(CH₂)₄—	74%
—(CH₂)₂O(CH₂)₂—	72%

The VHA reagent reacts in the same fashion to the corresponding formamidines (76). Tetracyanopropene anion reacts with **1** also at a terminal cyano group (121).

Carbamates behave like amines since they form N-carboalkoxychloroformamidines (**115**) (35). This method is complementary to the well-known addition of chloroformates to cyanamides, which works satisfactorily only with thermally stable aryl and certain alkyl (e.g., trichloroethyl) chloroformates (77).

III. POTENTIAL OF DICHLOROMETHYLENIMINIUM SALTS 383

$$H_2N-CO_2C_2H_5 \xrightarrow{1} (H_3C)_2N-\underset{\underset{Cl}{|}}{C}=N-CO_2C_2H_5$$

(115a) 72%

$$H_3C-\underset{\underset{H}{|}}{N}-CO_2C_2H_5 \xrightarrow{1} (H_3C)_2N\overset{+}{=\!=\!=}\underset{\underset{Cl}{|}\;\underset{CH_3}{|}}{C}\!-\!-\!-N-CO_2C_2H_5 \quad Cl^-$$

(115b) 95%

Ethylglycine hydrochloride reacts smoothly with **1** to the corresponding chloroformamidinium chloride (**116**) in almost quantitative yield (35,78):

$$H_3\overset{+}{N}CH_2-CO_2Et \;\; Cl^- \xrightarrow{1} (H_3C)_2N\overset{+}{=\!=\!=}\underset{\underset{Cl}{|}}{C}\!-\!-\!-\overset{\overset{H}{|}}{N}-CH_2CO_2Et \quad Cl^-$$

(116)

Sulfamide (**69**) reacts on both amino groups to give the bischloroformamidine (**117**):

$$H_2N-SO_2-NH_2 \xrightarrow{1} (H_3C)_2N-\underset{\underset{Cl}{|}}{C}=N-SO_2-N=\underset{\underset{Cl}{|}}{C}-N(CH_3)_2$$

(**117**) 97%, m.p. 128°

Among the secondary amines leading to urea dichlorides, ethylenimine merits attention because of the rearrangement (35) of the primary substitution product (**118**) to the β-chloroethylchloroformamidine (**119**):

$$\mathbf{1} \xrightarrow{\overset{H}{\underset{\triangle}{N}}} \left[\underset{\underset{Cl}{|}}{\triangleright N}\overset{+}{=\!=\!=}C\!-\!-\!-N\overset{CH_3}{\underset{CH_3}{\diagdown}} \;\; Cl^- \right] \longrightarrow Cl-CH_2-CH_2-N=\underset{\underset{Cl}{|}}{C}-N(CH_3)_2$$

(118)   (119)   84%

Likewise, isocyanide dichlorides form β-chloroethylcarbodiimides with ethylenimine (79). N-Methyltoluenesulfonamide (**7**) reacts smoothly, but the tosylchloroformamidinium salt (**120**) could not be isolated since it undergoes loss of tosyl chloride:

$$Tos-N\overset{CH_3}{\underset{H}{\diagup}} \xrightarrow{1} \left[Tos-\underset{\underset{CH_3}{|}}{\overset{+}{N}}\!=\!=\!=\underset{\underset{Cl^-}{}}{C\!-\!-\!-N(CH_3)_2} \right] \longrightarrow$$

(120)

$$(H_3C)_2N-\underset{\underset{Cl}{|}}{C}=NCH_3 + Tos-Cl$$

Phosgene readily acylates tertiary aliphatic amines (80), but von Braun degradation (i.e., loss of alkyl chloride) takes place below 0°C. The same is true for isocyanide dichlorides (81), which of course require higher temperatures.

$$R_3N + COCl_2 \xrightarrow{-20°} [R_3N\text{---}COCl^+ \; Cl^-] \xrightarrow{0°} R_2N\text{---}COCl + RCl$$

$$(H_5C_2)_3N + H_5C_6N=C\begin{array}{c}Cl\\ \diagdown \\ Cl\end{array} \xrightarrow{100°} (H_5C_2)_2N\text{---}\underset{\underset{Cl}{|}}{C}=NC_6H_5 + C_2H_5Cl$$

Tertiary aliphatic amines react exothermally with **1**, forming complexes that have not yet been studied but appear to be thermally stable. Triethylamine is widely used as a basic catalyst and hydrogen chloride scavenger.

2. Reactions with Secondary Acetamides

In analogy to phosgene (82), phosgeniminium salts (**1**) attack the amide nitrogen atoms of secondary acetamides, yielding N-(α-chlorovinyl)-chloroformamidinium salts (**121**) (7,83,84):

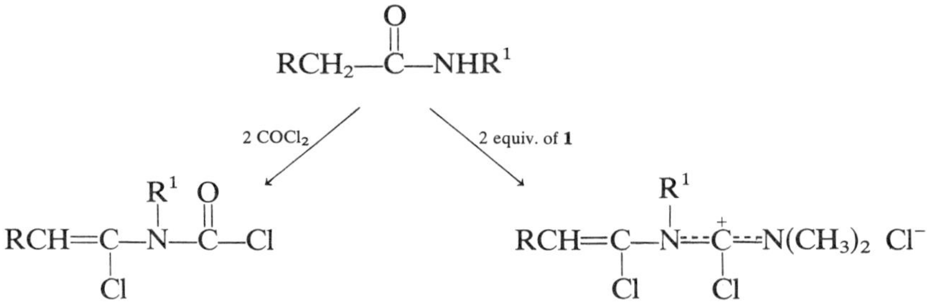

$$RCH_2\text{---}\overset{O}{\overset{\|}{C}}\text{---}NHR^1$$

2 COCl$_2$ ↙   ↘ 2 equiv. of **1**

$$RCH=\underset{\underset{Cl}{|}}{\overset{\overset{R^1}{|}}{C}}\text{---}\overset{O}{\overset{\|}{N}}\text{---}\overset{}{C}\text{---}Cl \qquad\qquad RCH=\underset{\underset{Cl}{|}}{\overset{\overset{R^1}{|}}{C}}\text{---}\overset{+}{N}\text{=}\underset{\underset{Cl}{|}}{C}\text{=}N(CH_3)_2 \; Cl^-$$

(**121**)

Compounds **121** that have so far been synthesized are listed in Table XIV.

TABLE XIV

Reaction of Secondary Amides with **1**

121	R	R^1	Melting point, °C	Yield, %	Ref.
a	H	CH_3	Oil	90	83
b	CH_3	CH_3	Oil	82	84
c	CH_3	C_2H_5	Oil	80	84
d	$t\text{-}C_4H_9$	CH_3	110	71	84
e	C_6H_5	CH_3	82–85	97	83
f	—$(CH_2)_4$—		Oil	86	83
g	Cl	CH_3	154	85	83
h	SC_2H_5	CH_3	Oil	79	84
i	OC_6H_5	CH_3	Oil	84	84
j	OCH_3	CH_3	Oil	82	84
k	F	CH_3	Oil	87	32,122

The α_{CO} carbon atoms in the starting amides may bear two additional groups since N-methylisobutyrylamide and N-methyldiphenylacetamide react smoothly, as well.

$$\begin{array}{c}H_3C\\\diagdown\\CH-\overset{O}{\underset{H}{C}}-NCH_3\end{array}\xrightarrow{\text{2 equiv. of }\mathbf{1}}\begin{array}{c}H_3CCH_3\\\diagdown|\\C=C-N\overset{+}{=\!\!=}C\cdots N(CH_3)_2\ Cl^-\\\diagup||\\H_3CClCl\end{array}$$

$$\begin{array}{c}H_5C_6\\\diagdown\\CH-\overset{O}{\underset{H}{C}}-NCH_3\end{array}\xrightarrow[-CH_3Cl]{\text{2 equiv. of }\mathbf{1}}\begin{array}{c}H_5C_6\\\diagdown\\C=C-N=C-N(CH_3)_2\\\diagup||\\H_5C_6ClCl\end{array}$$

(**122d**)

As shown, in this case the loss of methyl chloride occurs in refluxing chloroform during the reaction, and the aminodichloroazabutadiene (**122d**) is the final product (85).

The reactivity of the vinylic chlorine in (**121**) depends on the substituent R but generally is lower than that of the chloroformamidinium group. This was demonstrated by selective hydrolysis of **121e** to the α-chlorovinylurea (**123d**):

$$\mathbf{121}\xrightarrow[NaHCO_3]{H_2O}H_5C_6CH=\underset{Cl}{\overset{\underset{\displaystyle CH_3}{|}}{C}}-\underset{}{N}-\underset{O}{\overset{\displaystyle \|}{C}}-N(CH_3)_2$$

(**123d**)

TABLE XV

Partial Hydrolysis of Chlorovinylurea Dichlorides (121) to N-(α-chlorovinyl)ureas (123)

123	R	R^1	Yield, % Method A[a]	Yield, % Method B[a]	Melting or boiling point, °C/torr	Ref.
a	H	C_2H_5	—	77	118/0.5	84
b	CH_3	CH_3	50	85	120/0.1	84
c	t-C_4H_9	CH_3	—	60	115/0.7	84
d	C_6H_5	CH_3	88	91	140/0.01	83
e	Cl	CH_3	65	—	68–71/0.5	7
f	SC_2H_5	CH_3	—	94	140/0.1	84
g	OC_6H_5	CH_3	85	—	140/0.1	84
h	OCH_3	CH_3	70	—	80/0.04	84

[a] A: using H_2O–$NaHCO_3$; B: using propylene oxide–chloroform.

Later, it was found that the partial hydrolysis can be achieved more simply and with greater selectivity (84) by using propylene oxide:

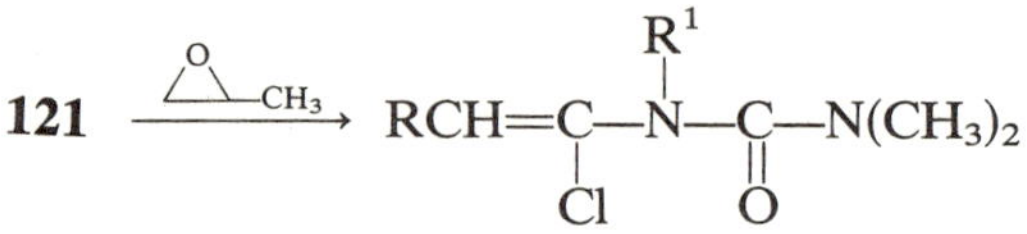

$$121 \xrightarrow{\triangle\text{—}CH_3 \atop O} RCH=\underset{Cl}{C}-\underset{R^1}{N}-\underset{O}{\overset{\|}{C}}-N(CH_3)_2$$

$$(123)$$

It is obvious that both **121** and **123** may occur in either cis or trans configurations and their proportions can be assessed from the NMR spectra. The reaction leading to the α-chlorovinylurea dichlorides is regioselective, since in the cases studied only one isomer, either cis or trans, was present. On the other hand, the vinylureas (**123**) obtained are frequently mixtures of both isomers.

TABLE XVI

N-Acyl-N,N',N'-Trialkyl Ureas (124)

117	R	R^1	Yield, %	Boiling point, °C/torr	Ref.
a	H	CH_3	90	—	7
b	CH_3	C_2H_5	87	—	84
c	t-C_4H_9	CH_3	85	—	84
d	C_6H_5	CH_3	71	135/0.2	83
e	—$(CH_2)_4$—		75	120/0.5	7
f	Cl	CH_3	—	—	7
j	—$(CH_2)_2$—		31	120/0.7	7
n	OCH_3	CH_3	90	110–120/0.5	84

Extended hydrolysis of **121** is a facile route to various tetrasubstituted acyl ureas (**124**). Several examples are summarized in Table XVI.

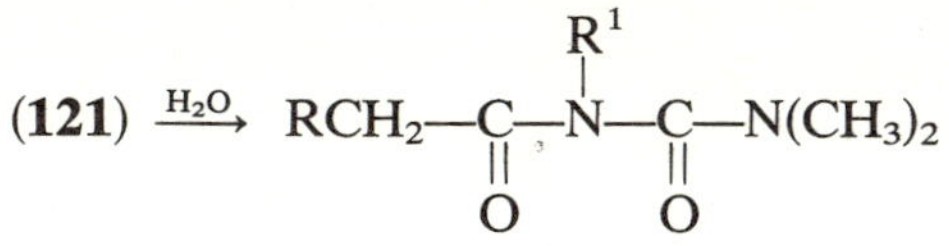

$$(121) \xrightarrow{H_2O} RCH_2-\underset{\underset{O}{\|}}{C}-\underset{R^1}{N}-\underset{\underset{O}{\|}}{C}-N(CH_3)_2$$
$$(124)$$

Next, the α-chlorovinylureas (**123**) may undergo hydrogen chloride elimination, the ease of which depends on the nature of substituents R and on the geometry of the starting vinyl compound. So far, two N-acylated ynamines (yne-ureas) (**125a,125b**) have been prepared by this method in 70 and 40% yields, respectively, (7,84):

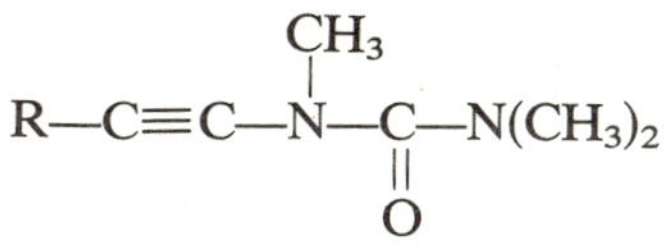

$$R-C\equiv C-\underset{CH_3}{N}-\underset{\underset{O}{\|}}{C}-N(CH_3)_2$$

(**125a**) R = C$_6$H$_5$
(**125b**) R = t-butyl

In analogy to **1**, the iminium salts (**121**) undergo thermic degradation accompanied by a loss of methyl chloride to the conjugated 1,3-dichloro-1-aminoazabutadienes (**122**):

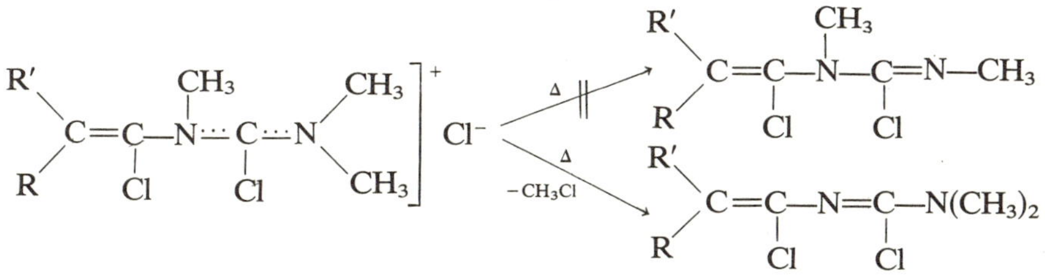

$$\underset{(121)}{\left[\underset{R}{\overset{R'}{\diagdown}}C=\underset{Cl}{\overset{CH_3}{\underset{|}{C}}}-N\cdots\underset{Cl}{\overset{}{C}}\cdots N\underset{CH_3}{\overset{CH_3}{\diagup}}\right]^+ Cl^-} \xrightarrow[-CH_3Cl]{\Delta} \begin{array}{c} \underset{R}{\overset{R'}{\diagdown}}C=\underset{Cl}{\overset{CH_3}{C}}-N-\underset{Cl}{\overset{}{C}}=N-CH_3 \\ \underset{R}{\overset{R'}{\diagdown}}C=\underset{Cl}{\overset{}{C}}-N=\underset{Cl}{\overset{}{C}}-N(CH_3)_2 \end{array}$$
(**122**)

TABLE XVII

Azabutadienes (**122**) via Pyrolysis of **121**

R	R'	Yield, %	Boiling point, °C/torr
H	CH$_3$	86	150/0.15
H	C$_6$H$_5$	80	135–136/0.08
H	t-C$_4$H$_9$	71	71–72/0.5
C$_6$H$_5$	C$_6$H$_5$	60	180/0.04
H	F	84	13010.1

Compound **122**, R = C₆H₅, afforded moderate yield of *N*-phenylethinylchloroformamidine (**126**) on treatment with lithium diethylamide (84):

$$H_5C_6-CH=\underset{Cl}{C}-N=\underset{Cl}{C}-N(CH_3)_2 \longrightarrow H_5C_6-C\equiv C-N=\underset{Cl}{C}-N(CH_3)_2$$
$$(126)$$

As 1,3-biselectrophiles, **121** cyclize with hydrazines to derivatives of 1,2,4-triazole (68).

C. CHLORINE SUBSTITUTION WITH AZIDE, ISOCYANATE, AND PHOSPHINYL GROUPS

The chlorine substitution in **1** with azide anion has been studied in some detail. Although the monosubstitution can be achieved with sodium or lithium azide in methylene chloride, the use of trimethylsilylazide in chloroform is more practical (35,86a).

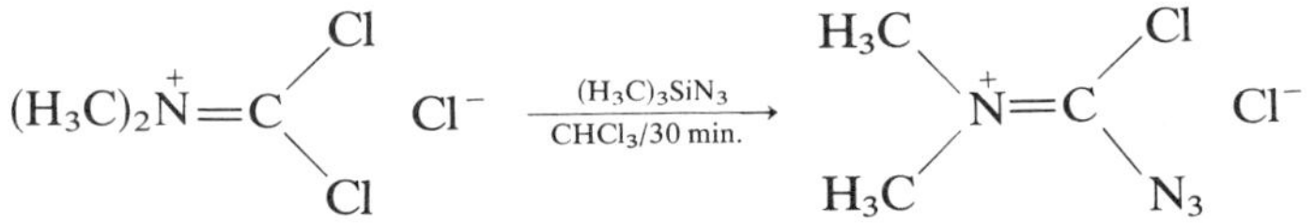

$$(H_3C)_2\overset{+}{N}=C\begin{pmatrix}Cl\\ \\Cl\end{pmatrix} Cl^- \xrightarrow[CHCl_3/30\ min.]{(H_3C)_3SiN_3} \begin{matrix}H_3C\\ \\H_3C\end{matrix}\overset{+}{N}=C\begin{pmatrix}Cl\\ \\N_3\end{pmatrix} Cl^-$$

(**1**) (**127**) 80–90%

Compound **127** is a very hygroscopic solid, hydrolyzing to dimethylcarbamoylazide. Primary amines react with **127** to the corresponding azidoformamidines (**128**), which exist as aminotetrazoles (**129**) (35). The latter are also formed in the reaction of the corresponding chloroformamidines with sodium azide (35,86a):

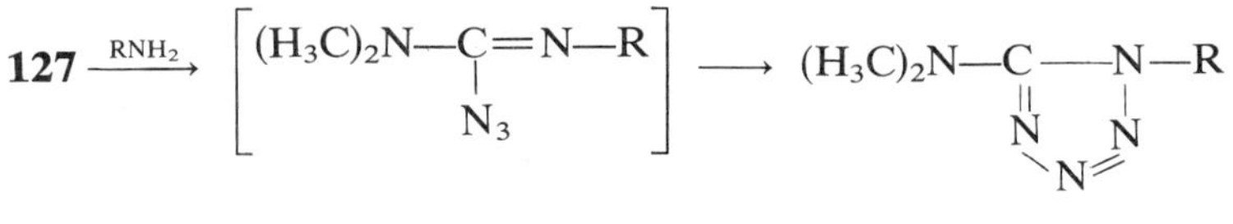

127 $\xrightarrow{RNH_2}$ $\left[(H_3C)_2N-\underset{N_3}{C}=N-R\right]$ $\longrightarrow$ (H₃C)₂N—C—N—R (tetrazole ring)

(**128**)

(**129a**), R = C₆H₅CH₂,   80%
(**129b**), R = C₆H₅, 71%   71%
(**129c**), R = p-C₆H₄CO₂Et, 91%

N-alkoxychloroformamidines (**110**) proved to be inert toward substitution with azide anion, and the only method of preparation of 1-alkoxy-5-dimethylaminotetrazoles (**130**) consists in condensation of **127** with

Q-substituted hydroxylamines (73):

127 $\xrightarrow{\text{RONH}_2}_{\text{Et}_3\text{N}}$ RO—N—C—N(CH$_3$)$_2$ $\xleftarrow{\text{N}_3^-}$ RO—N=C—N(CH$_3$)$_2$
         │ ‖   │
         N N   Cl
          ＼N／
                                                          (110)

(**130a**), R = CH$_3$, 50%
(**130b**), R = C$_6$H$_5$CH$_2$, 43%

Although primary diazidomethyleniminium salts (**131**) are known (86b), no tertiary diazidophosgeniminium salt (**132**) has been reported as yet:

$$\text{H}_2\overset{+}{\text{N}}=\text{C}\begin{matrix}\diagup\text{N}_3\\ \diagdown\text{N}_3\end{matrix}\quad \text{X}^- \qquad \begin{matrix}\text{R}\diagdown\\ \text{R}\diagup\end{matrix}\overset{+}{\text{N}}=\text{C}\begin{matrix}\diagup\text{N}_3\\ \diagdown\text{N}_3\end{matrix}\quad \text{X}^-$$

(**131**)   (**132**)

Reportedly, silver isocyanate reacts with **1** to the monosubstitution product (**109**) in low yield (24), and the indirect route using chlorine isocyanate trimer (N,N',N''-trichloroisocyanuric acid) is more feasible (72). The third method involves reacting **115a** with **1** (119):

                Cl
                │
1 $\xrightarrow{\text{AgNCO}}_{27\%}$ (H$_3$C)$_2$N—C—N=C=O $\rightleftharpoons$ (H$_3$C)$_2$$\overset{+}{\text{N}}$=C—N=C=O
                │   │
                Cl   Cl   Cl$^-$
              (**109a**)   (**109b**)   ⇅

1 + (**115a**) $\xrightarrow{62\%}$ (H$_3$C)$_2$N—C=N—COCl
                                        │
                                        Cl
                                    (**109c**) b.p. 84–85°/0.05

The IR spectrum of **109** shows both an isocyanate band at $\nu = 2270\ \text{cm}^{-1}$ and a carbonyl group at $\nu = 1730\ \text{cm}^{-1}$, indicating a tautomeric equilibrium as formulated above. Compound **109** is a colorless mobile liquid that distills under vacuum without decomposition. Aniline, methanol, and triethyl phosphite gave the expected products (**133–135**):

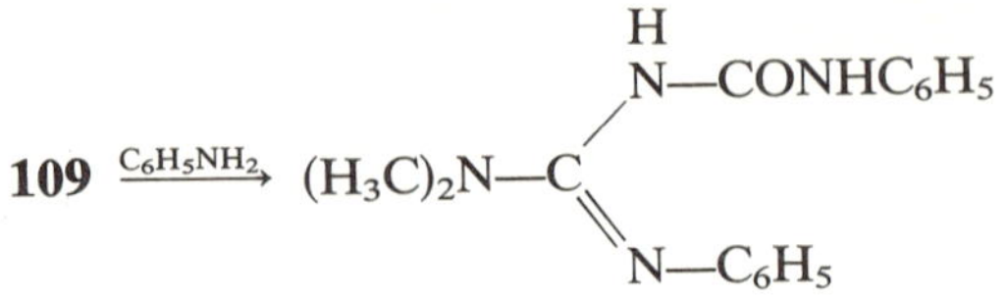

$$109 \xrightarrow{C_6H_5NH_2} (H_3C)_2N-C \begin{array}{c} \nearrow N-CONHC_6H_5 \\ \searrow N-C_6H_5 \end{array}$$
$$\text{H}$$

(**133**)

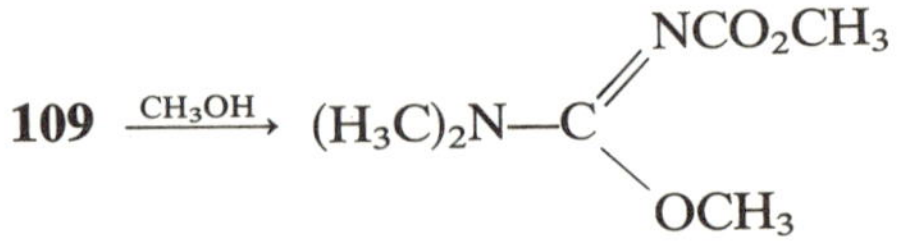

$$109 \xrightarrow{CH_3OH} (H_3C)_2N-C \begin{array}{c} \nearrow NCO_2CH_3 \\ \searrow OCH_3 \end{array}$$

(**134**)

$$109 \xrightarrow{(EtO)_3P} (H_3C)_2N-C \begin{array}{c} \nearrow N-COP(O)(OEt)_2 \\ \searrow P(O)(OEt)_2 \end{array}$$

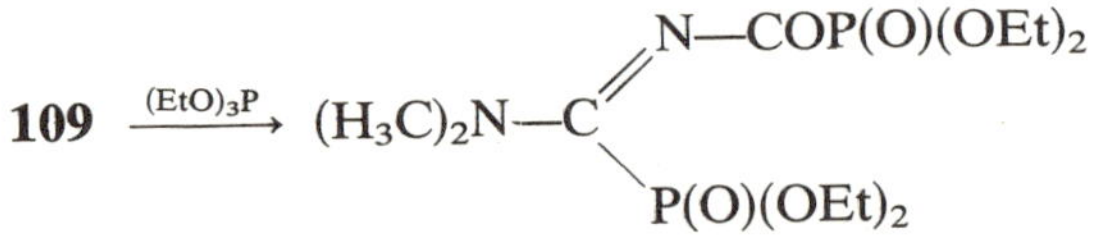

(**135**)

In the last example it is clearly the tautomer (**109c**) which reacts with phosphite, and the primary product undergoes the Arbuzov rearrangement to *N,N*-dimethyl-*N'*-(diethoxyphosphinylcarbonyl)-*C*-diethoxyphosphinylformamidine (**135**).

It is also reported (87) that **1** react very vigorously with 3 equivalents of trialkyl phophites to form dialkylaminomethanetriphosphonic acid esters (**136**):

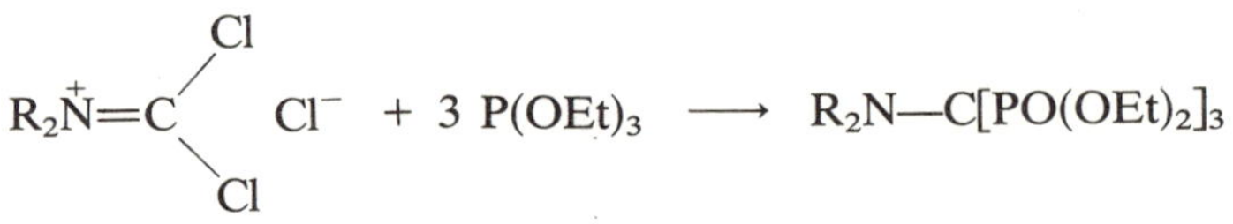

$$R_2\overset{+}{N}=C \begin{array}{c} \nearrow Cl \\ \searrow Cl \end{array} Cl^- + 3\ P(OEt)_3 \longrightarrow R_2N-C[PO(OEt)_2]_3$$

(**136a**), R = CH$_3$,   70%
(**136b**), R = —(CH$_2$)$_2$O(CH$_2$)$_2$—,   77%

A recent patent (88) claims that **136** have a flameproofing action and can be used as stabilizers for polymers.

N-Sulfinyltrimethylsilylamine and **1** afford quantitative yields of chloroformamidines (**109d**) that are formally analogous to **109c** (123):

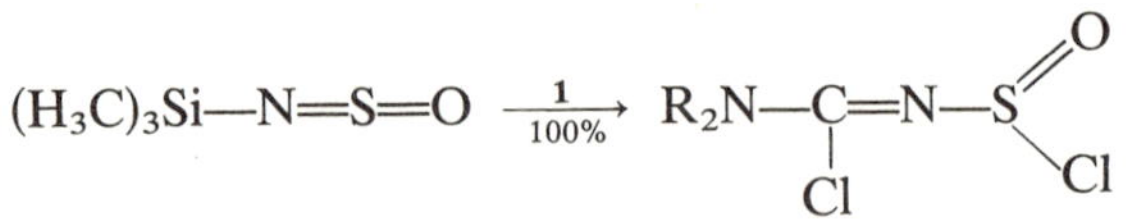

$$(H_3C)_3Si-N{=}S{=}O \xrightarrow[100\%]{1} R_2N-\underset{Cl}{\underset{|}{C}}=N-S\begin{array}{c}\nearrow O \\ \searrow Cl\end{array}$$

(**109d**), R = CH$_3$, C$_2$H$_5$

D. CHLORINE SUBSTITUTION IN **1** BY OXYGEN AND SULFUR

1. *Reaction with Phenols and Thiophenols*

Phenol (3,7) reacts successively to a phenoxychloromethyleniminium salt (**137**) and a diphenoxymethyleniminium salt (**138**) (124,125):

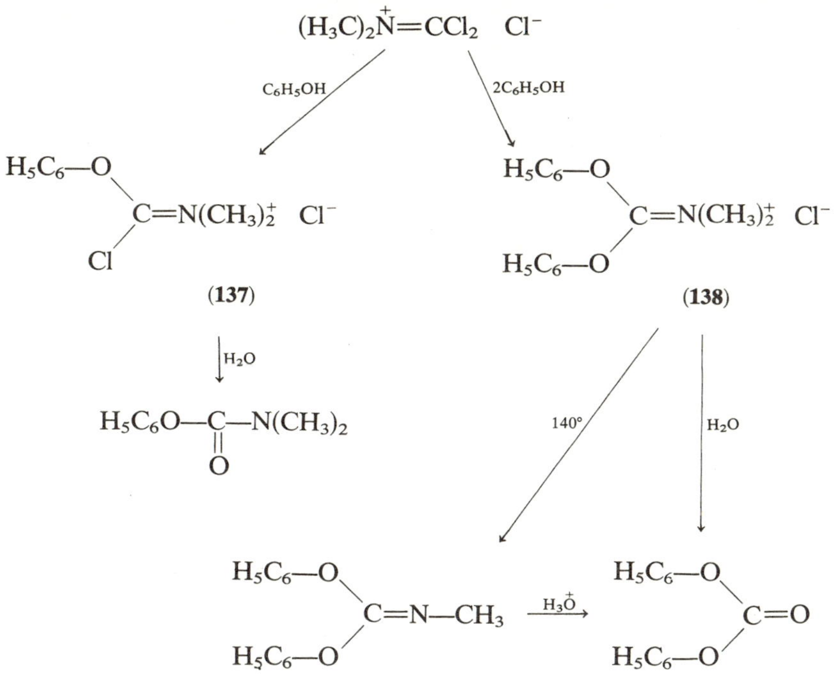

$$(H_3C)_2\overset{+}{N}=CCl_2 \quad Cl^-$$

With C_6H_5OH:

$$H_5C_6-O\diagdown C=N(CH_3)_2^+ \quad Cl^-$$
$$Cl\diagup$$
(**137**)

With $2 C_6H_5OH$:

$$H_5C_6-O\diagdown C=N(CH_3)_2^+ \quad Cl^-$$
$$H_5C_6-O\diagup$$
(**138**)

(**137**) $\xrightarrow{H_2O}$ $H_5C_6O-\underset{\underset{O}{\|}}{C}-N(CH_3)_2$

(**138**) $\xrightarrow{140°}$ $H_5C_6-O\diagdown C=N-CH_3$ $\xrightarrow{H_3\overset{+}{O}}$ $H_5C_6-O\diagdown C=O$
$\qquad\qquad H_5C_6-O\diagup$ $\qquad\qquad\qquad\qquad H_5C_6-O\diagup$
$\qquad\qquad\qquad$(**139**)

(**138**) $\xrightarrow{H_2O}$ $H_5C_6-O\diagdown C=O$
$\qquad\qquad H_5C_6-O\diagup$

The salt (**137**) hydrolyzes to the parent amide, *N,N*-dimethylphenyl carbamate, whereas the iminium carbonate (**138**) forms diphenyl carbonate. More interesting is the clean transformation of (**138**) into the iminocarbonate (**139**) at 140°. Similarly, *o*-dihydroxybenzene forms the stable cyclic iminium carbonate (**140**):

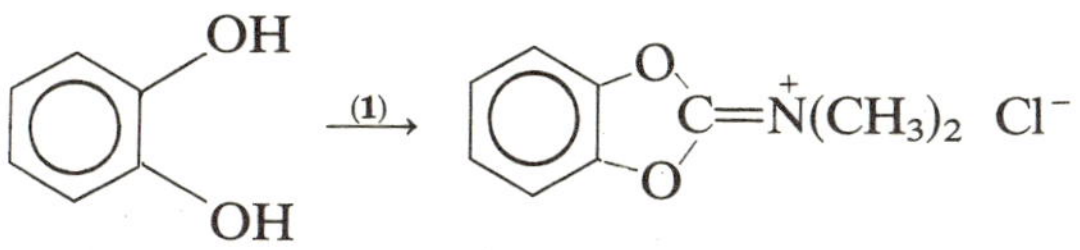

o-C₆H₄(OH)₂ $\xrightarrow{(1)}$ benzo[1,3]dioxol-2-ylidene-$\overset{+}{N}(CH_3)_2$ Cl^-
(**140**)

Resorcine and hydroquinone form insoluble polymeric iminium carbonates. Thiophenol affords a phenylthiochloromethyleniminium salt in almost quantitative yield (35).

2. Reaction with Alcohols, Glycols, and Their Thio Analogues

Primary, secondary, and tertiary aliphatic alcohols are transformed to the corresponding chlorides apparently through the intermediacy of unstable alkoxychloromethyleniminium salts. It has been found that 1,2-, 1,3-, and even some 1,4-glycols (89) react with **1** to mostly unstable cyclic iminium carbonates (**142**), which cleave instantaneously to chlorocarbamates (**143,144**):

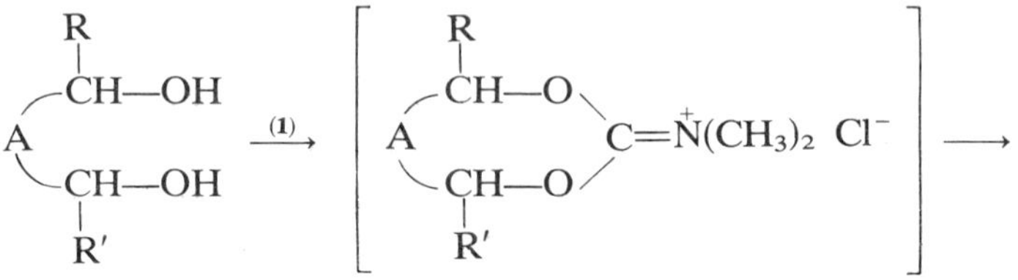

$$A\begin{matrix}-CH-OH\\ -CH-OH\\ R'\end{matrix} \quad \xrightarrow{(1)} \quad \left[A\begin{matrix}-CH-O\\ \\ -CH-O\\ R'\end{matrix}C=\overset{+}{N}(CH_3)_2 \; Cl^- \right] \longrightarrow$$

(141)   (142)

$$\begin{matrix}-CH-\overset{O}{\overset{\|}{C}}-N(CH_3)_2\\ -CH-Cl\\ R'\end{matrix} \quad + \quad A\begin{matrix}-CH-Cl\\ -CH-\underset{O}{\underset{\|}{C}}-N(CH_3)_2\\ R'\end{matrix}$$

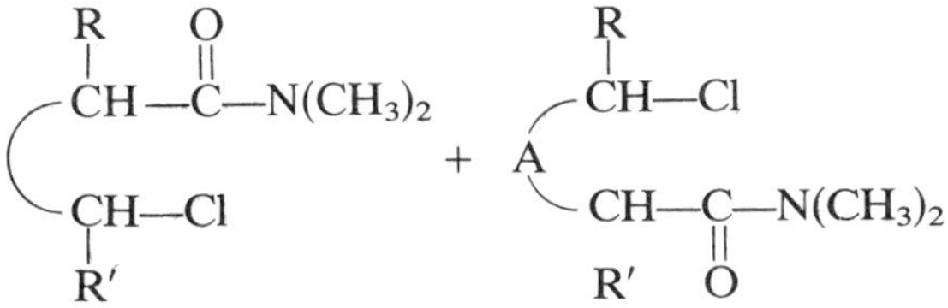

(143)   (144)

The ring opening that represents nucleophilic substitution of oxygen by chlorine is stereospecific with inversion of configuration. Thus *cis*-1,2-cyclohexandiol leads in high yield to *trans*-chlorocarbamate (**145**):

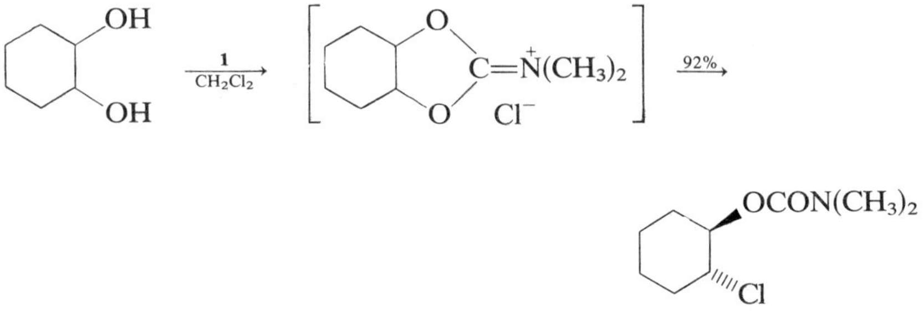

cis-cyclohexane-1,2-diol $\xrightarrow[CH_2Cl_2]{1}$ [cyclic iminium carbonate with $C=\overset{+}{N}(CH_3)_2$, Cl^-] $\xrightarrow{92\%}$

trans-cyclohexyl-OCON(CH$_3$)$_2$ / Cl (**145**)

In contrast, 1,2-*trans*-cyclohexanediol forms a stable iminium carbonate that does not undergo ring opening to the corresponding chlorocarbonate (78). Another stable iminium carbonate (**146**) arises from pinacol and **1**. In these cases steric hindrance inhibits the S_N2 attack by the chloride ion which would lead to ring opening.

TABLE XVIII

Chlorocarbamates (**143,144**) from **1** and Glycols

A	R	R'	Yield, %	Boiling point, °C/torr	Ratio (**143/144**)
—	H	H	100	90/13	—
—CH$_2$—	H	H	96	100–105/13	—
H$_3$C—C—CH$_3$	H	H	95	102–110/13	—
—CH$_2$—	CH$_3$	CH$_3$	90	100/13	—
—CH$_2$—CH$_2$—	H	H	40	100–102/13	—
—	H	C$_2$H$_5$	86	107/17	10/90
—CH$_2$—	H	CH$_3$	90	115/20	20/80

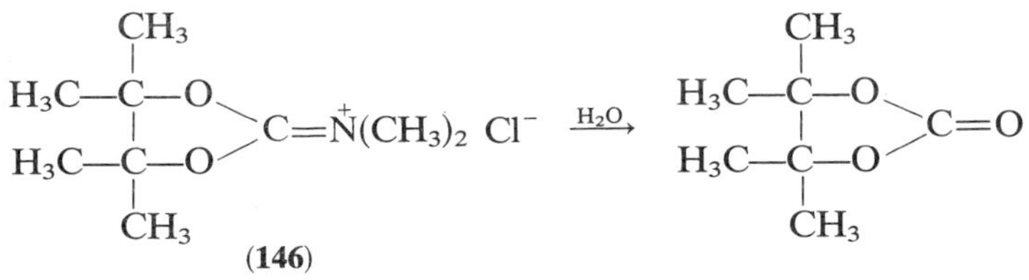

$$\begin{array}{c} CH_3 \\ | \\ H_3C-C-O \\ | \\ H_3C-C-O \\ | \\ CH_3 \end{array} C=\overset{+}{N}(CH_3)_2 \; Cl^- \xrightarrow{H_2O} \begin{array}{c} CH_3 \\ | \\ H_3C-C-O \\ | \\ H_3C-C-O \\ | \\ CH_3 \end{array} C=O$$

(**146**)

The hypothesis of a S_N2 mechanism is also supported by the directiochemistry of chlorocarbamate formation from asymmetric diols, where chloride attacks preferentially the sterically less hindered side (see Table XVIII). It is interesting that cyanogen chloride and phosgenimines react with glycols in the presence of HCl as unsubstituted and monosubstituted iminium salts (**32,34**) leading to primary and secondary carbamates (21,89).

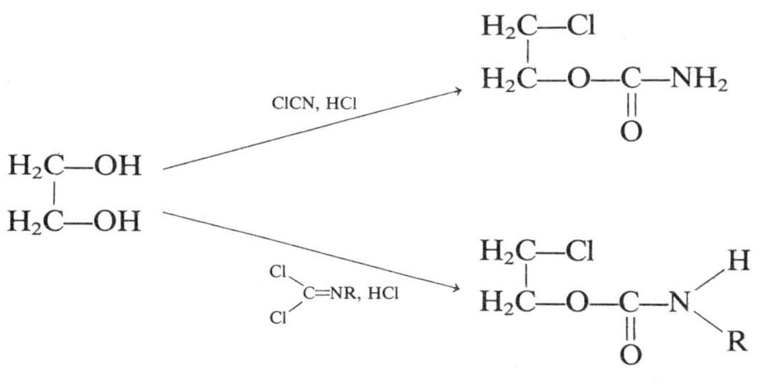

$$\begin{array}{c} H_2C-OH \\ | \\ H_2C-OH \end{array} \xrightarrow{ClCN,\, HCl} \begin{array}{c} H_2C-Cl \\ | \\ H_2C-O-C-NH_2 \\ \parallel \\ O \end{array}$$

$$\xrightarrow[Cl]{Cl\diagdown C=NR,\, HCl} \begin{array}{c} H_2C-Cl \\ | \\ H_2C-O-C-N\diagup^H_{\diagdown R} \\ \parallel \\ O \end{array}$$

The stable iminium carbonates are reactive intermediates that can be used further in synthesis. For example, **147** was transformed to carbonate tosylhydrazone (**148**), which afforded the olefin (**149**) on thermolysis (90):

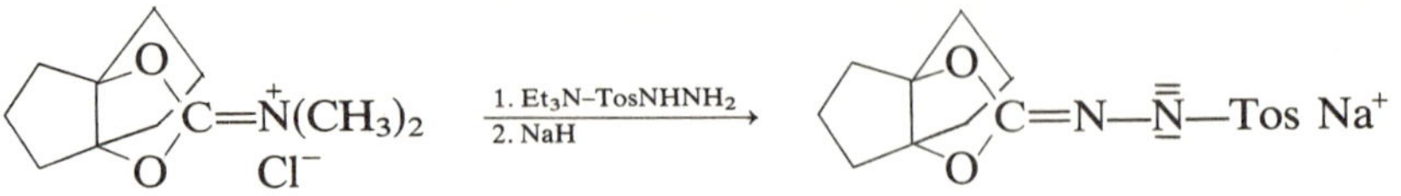

$$\text{(147)} \xrightarrow[\text{2. NaH}]{\text{1. Et}_3\text{N-TosNHNH}_2} \text{(148)}$$

$$\textbf{148} \xrightarrow{\Delta} \text{(149)}$$

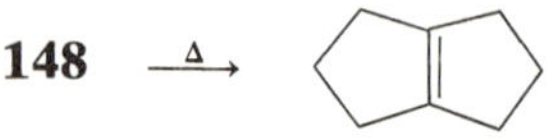

As would be expected, C–S bonds are not cleaved by chloride anions (78), and mercaptoethanol forms the S-(2-chloroethyl)thiocarbamate (**150**) and dithiols give stable iminium dithiocarbonates (**151**). The latter hydrolyze to dithiocarbonates (**152**) and dealkylate on heating to iminodithiocarbonates (**153**).

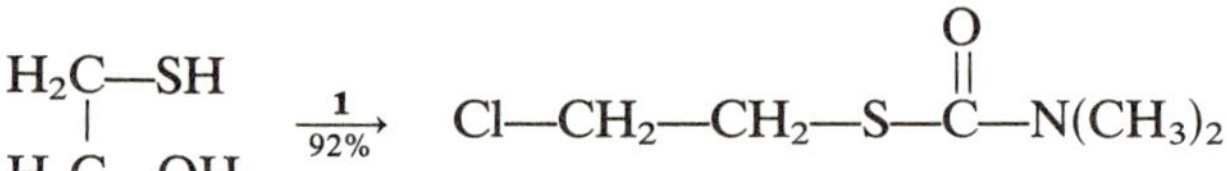

$$\begin{array}{c} \text{H}_2\text{C--SH} \\ | \\ \text{H}_2\text{C--OH} \end{array} \xrightarrow[92\%]{1} \text{Cl--CH}_2\text{--CH}_2\text{--S--C(=O)--N(CH}_3)_2 \quad \textbf{(150)}$$

$$\begin{array}{c} \text{H}_2\text{C--SH} \\ | \\ \text{H}_2\text{C--SH} \end{array} \xrightarrow[90\%]{1} \begin{array}{c} \text{H}_2\text{C--S} \\ | \quad \quad \text{C=}\overset{+}{\text{N}}(\text{CH}_3)_2 \text{ Cl}^- \\ \text{H}_2\text{C--S} \end{array} \quad \textbf{(151)}$$

$$\begin{array}{c} \text{H}_2\text{C--S} \\ | \quad \quad \text{C=O} \\ \text{H}_2\text{C--S} \end{array} \xleftarrow[75\%]{\text{H}_2\text{O}} \textbf{151} \xrightarrow[40\%]{200°} \begin{array}{c} \text{H}_2\text{C--S} \\ | \quad \quad \text{C=N--CH}_3 \\ \text{H}_2\text{C--S} \end{array}$$

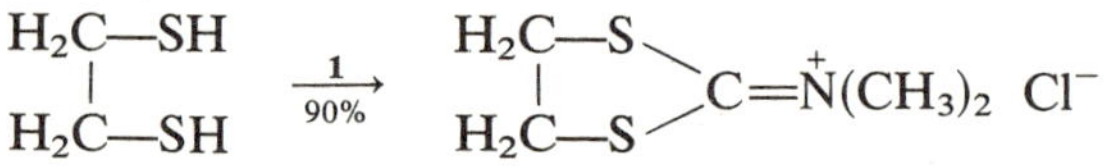

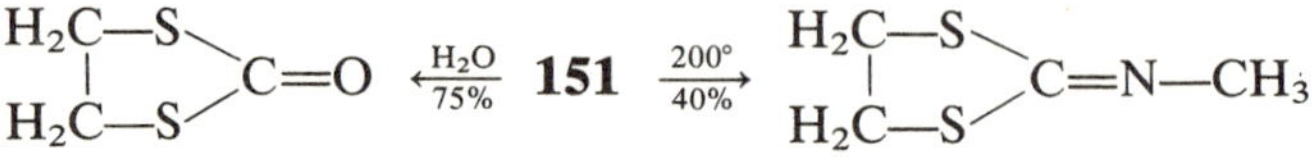

(**152**)   (**153**)

3. Ring Cleavage of Cyclic Ethers with **1**

As anticipated, epoxides react exothermally with **1** to form 1,2-dichlorides (**154**) (Table XIX):

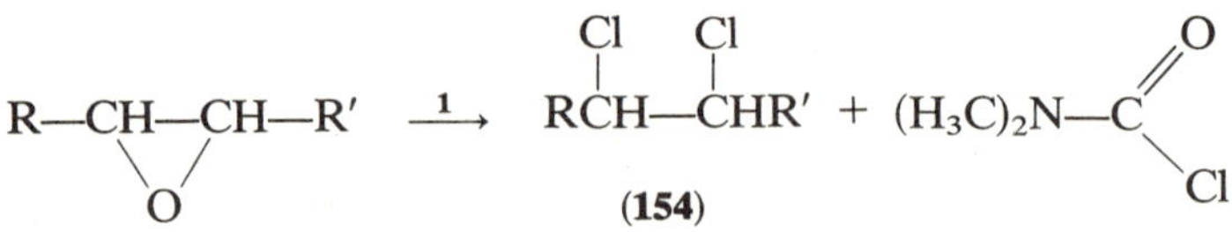

$$\text{R--CH--CH--R'} \xrightarrow{\textbf{1}} \underset{(\textbf{154})}{\text{RCH(Cl)--CHR'(Cl)}} + (\text{H}_3\text{C})_2\text{N--C(=O)Cl}$$

TABLE XIX
Dichlorides (**154**) from **1** and Epoxides

154	R	R'	Yield, %	Ref.
a	H	H	77	91
b	C$_6$H$_5$	H	93	91
c	CH$_2$Cl	H	70	91
d	—(CH$_2$)$_4$—		60^a	78,92

^{a} Mixture of cis and trans forms.

Tetrahydrofuran is sometimes used as solvent for reactions with **1**, but it should be taken into account that this compound is cleaved slowly to 1,4-dichlorobutane. The same is true for dioxane, dibutyl, and diisopropyl ether (91).

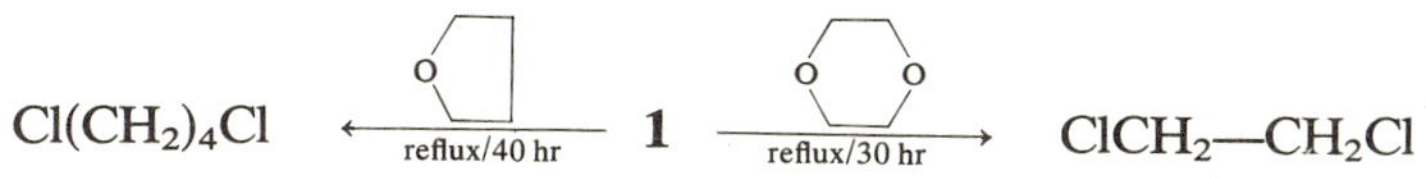

$$\text{Cl(CH}_2)_4\text{Cl} \xleftarrow[\text{reflux/40 hr}]{} \text{[THF]} \quad \mathbf{1} \quad \text{[dioxane]} \xrightarrow[\text{reflux/30 hr}]{} \text{ClCH}_2\text{—CH}_2\text{Cl}$$

Surprisingly, ethyl acetate requires only 10 hr at 50° to cleave to ethyl chloride and acetyl chloride. Butyrolactone afforded γ-chlorobutyric chloride in 52% yield (91).

4. Reaction with Aldehydes, Carboxylic Acids, Primary Amides, and Oximes

It is not surprising that carboxylic acids form acid chlorides under very mild conditions. Tertiary amides devoid of an α_{CO} methylene group are transformed to amide chlorides (**155**), and **1** is superior to phosgene because of its higher reactivity and easy handling.

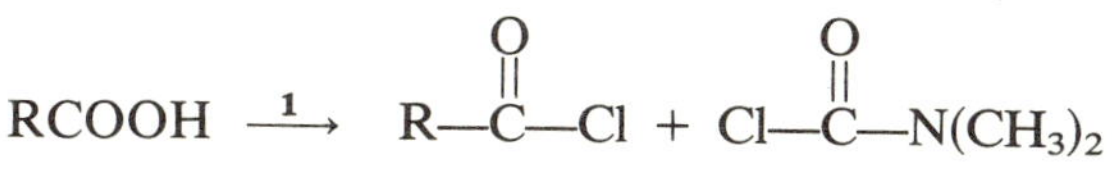

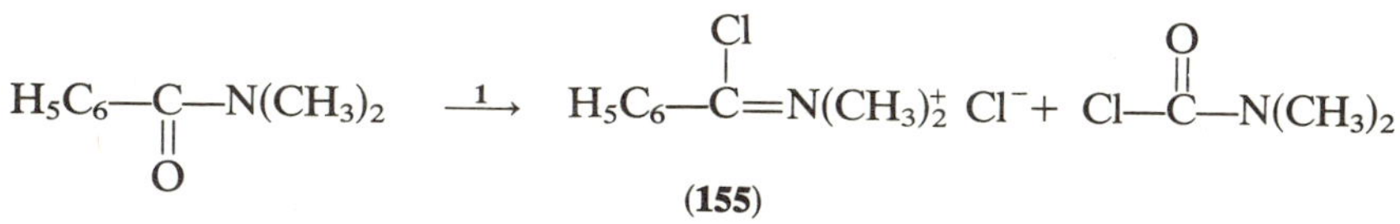

$$\text{RCOOH} \xrightarrow{\mathbf{1}} \text{R—}\underset{\underset{\text{O}}{\|}}{\text{C}}\text{—Cl} + \text{Cl—}\underset{\underset{\text{O}}{\|}}{\text{C}}\text{—N(CH}_3)_2$$

$$\text{H}_5\text{C}_6\text{—}\underset{\underset{\text{O}}{\|}}{\text{C}}\text{—N(CH}_3)_2 \xrightarrow{\mathbf{1}} \text{H}_5\text{C}_6\text{—}\underset{\underset{\text{Cl}}{|}}{\text{C}}\text{=N(CH}_3)_2^+ \text{ Cl}^- + \text{Cl—}\underset{\underset{\text{O}}{\|}}{\text{C}}\text{—N(CH}_3)_2$$

(**155**)

Dimethylformamide might be expected to undergo a similar exchange leading to the chloromethyleniminium chloride but tetramethylchloroformamidinium chloride was isolated instead (24):

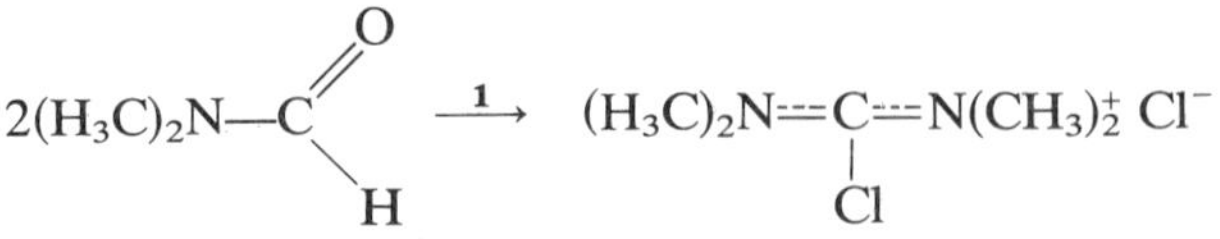

$$2(H_3C)_2N-\underset{H}{\overset{O}{\overset{\|}{C}}} \xrightarrow{1} (H_3C)_2N=\underset{Cl}{C}=N(CH_3)_2^+ \ Cl^-$$

20%

Aldoximes are dehydrated to nitriles; for example, acetaldoxime gives 92% of acetonitrile. Similarly, para-substituted aromatic aldoximes (p-chloro, methoxy, dimethylamino, and nitro) have been converted to the nitriles in ~90% yields (93). The reaction between **1** and acetophenone oxime gave the expected Beckmann rearrangement:

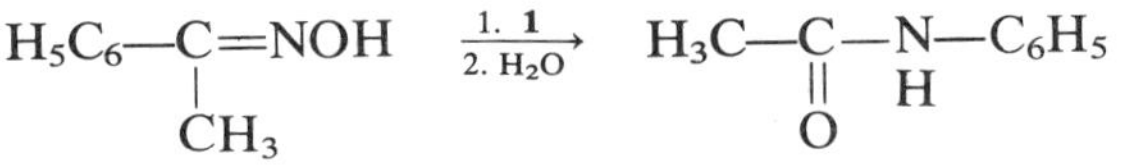

$$H_5C_6-\underset{CH_3}{\overset{}{C}}=NOH \xrightarrow[2.\ H_2O]{1.\ \mathbf{1}} H_3C-\underset{O}{\overset{\|}{C}}-\underset{H}{\overset{}{N}}-C_6H_5$$

86%

In contrast to ketones, aldehydes undergo only oxygen–chlorine exchange, and no C–C bond formation has as yet been detected.

$$RCH_2CHO \xrightarrow{1} RCH_2-CHCl_2 + (H_3C)_2N-COCl$$

$$CH_3-CH=CH-CHO \xrightarrow{1} CH_3-CH=CH-CHCl_2 + (H_3C)_2N-COC$$

In marked contrast, crotonaldehyde and its vinylogues react with the VHA reagent, giving not only an oxygen–chlorine exchange but also a carbon–carbon condensation (94), leading to conjugated iminium salts (**156**):

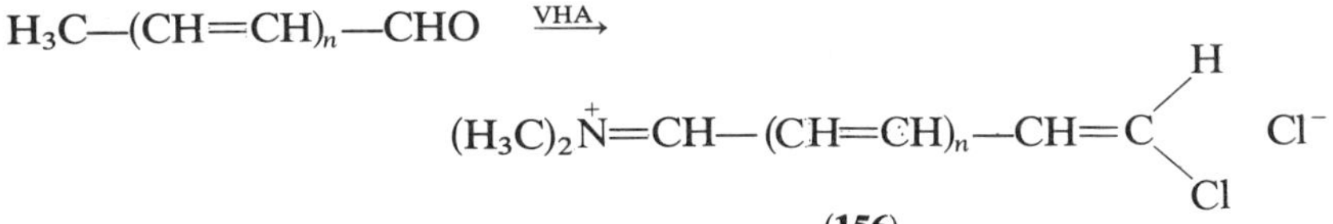

$$H_3C-(CH=CH)_n-CHO \xrightarrow{VHA}$$

$$(H_3C)_2\overset{+}{N}=CH-(CH=CH)_n-CH=C\overset{H}{\underset{Cl}{\diagdown}} \ Cl^-$$

(**156**)

E. ADDITION REACTIONS OF PHOSGENIMINIUM SALTS TO C–C MULTIPLE BONDS

1. *Addition to Activated Acetylenes*

Phosgeniminium salts (**1**) add to ynamines (95) in practically quantitative yield to form dichlorotrimethinecyanines (**88**), which can be obtained also from **1** and tertiary acetamides:

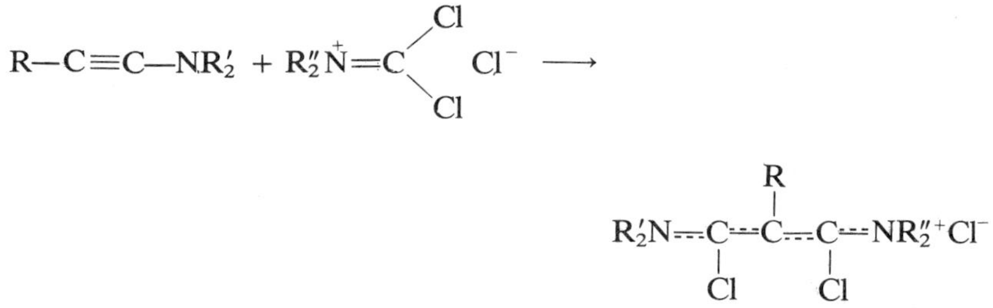

$$R-C\equiv C-NR'_2 + R''_2\overset{+}{N}=C\begin{smallmatrix}Cl\\Cl\end{smallmatrix}\;Cl^- \longrightarrow$$

$$R'_2N=\!\!=\!\!\underset{Cl}{C}=\!\!=\underset{|}{\overset{R}{C}}=\!\!=\underset{Cl}{C}=\!\!=NR''^+_2\,Cl^-$$
(88)

It is obvious that this reaction is practical only in particular cases when the tertiary acetamides do not react satisfactorily with **1** and when the corresponding ynamines are available.

2. Addition to Enamines

The electrophilic reagents **1** add readily to enamines in the same way as do phosgene and VHA reagents (96). Thus piperidino-, morpholino-, and pyrrolidinoisobutenes add **1** instantaneously to form nonconjugated bisiminium salts (**157**) in high yield (75,97):

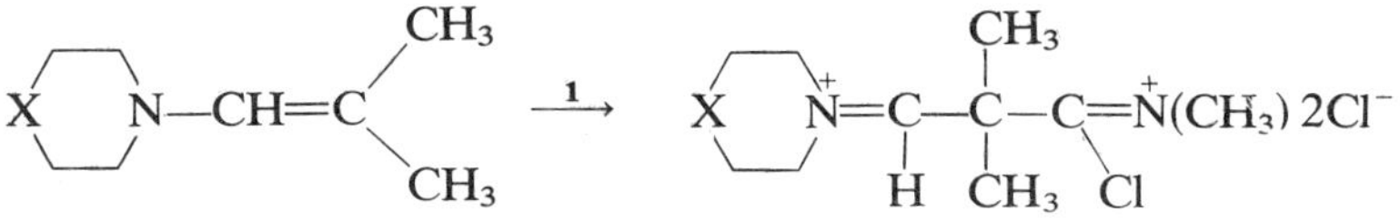

$$X\diagdown N-CH=C\begin{smallmatrix}CH_3\\CH_3\end{smallmatrix} \xrightarrow{\;1\;} X\diagdown \overset{+}{N}=\underset{H}{C}-\underset{CH_3}{\overset{CH_3}{C}}-\underset{Cl}{C}=\overset{+}{N}(CH_3)\;2Cl^-$$
(157)

$$\mathbf{157} \xrightarrow{H_2O} \begin{smallmatrix}H_3C\\H_3C\end{smallmatrix}\diagdown C\diagup\begin{smallmatrix}CON(CH_3)_2\\CHO\end{smallmatrix}$$

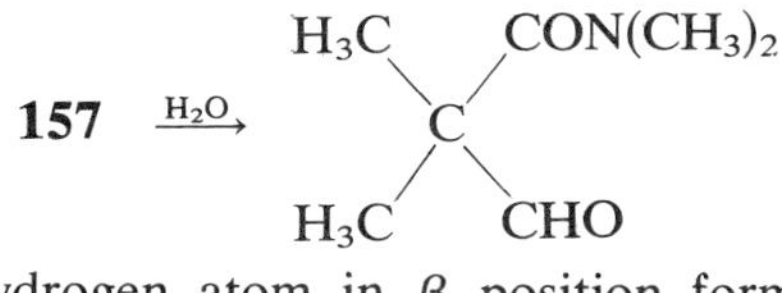

Enamines bearing a hydrogen atom in β position form 1-chloro-2,3-dialkyltrimethinecyanines (**158**):

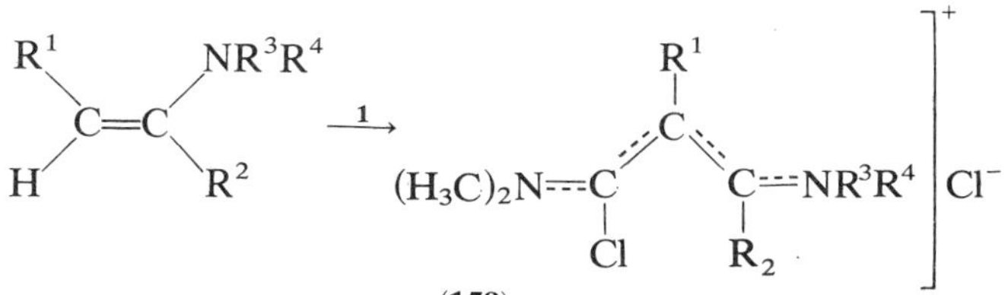

$$\begin{smallmatrix}R^1\\H\end{smallmatrix}\diagdown C=C\diagup\begin{smallmatrix}NR^3R^4\\R^2\end{smallmatrix} \xrightarrow{\;1\;} \left[(H_3C)_2N=\!\!=\underset{Cl}{C}\diagdown\overset{R^1}{\underset{R_2}{C}}\diagup C=\!\!=NR^3R^4\right]^+ Cl^-$$
(158)

TABLE XX
Cyanines (**158**) from **1** and Enamines

158	R^1	R^2	R^3	R^4	Yield, %	Ref.
a	—(CH$_2$)$_4$—		—(CH$_2$)$_4$—		95	97,98
b	—(CH$_2$)$_3$—		—(CH$_2$)$_4$—		90	97,98
c	H	C$_6$H$_5$	—(CH$_2$)$_4$—		92	97
d	—(CH$_2$)$_4$—		—(CH$_2$)$_2$—O—(CH$_2$)$_2$—		96	75,97
e	o-C$_6$H$_4$—CH$_2$—		—(CH$_2$)$_4$—		94	97,98
f	t-C$_4$H$_9$	H	CH$_3$	CH$_3$	95	97
g	H	H	—(CH$_2$)$_3$—C=O		94	97
h	H	N(CH$_3$)$_2$	CH$_3$	CH$_3$	95	99
j	CN	H	CH$_3$	CH$_3$	40	36
k	CO$_2$C$_2$H$_5$	CH$_3$	C$_2$H$_5$	C$_2$H$_5$	98	75

These barely known compounds arise also from self-condensation of suitable amide chlorides, but they have usually been worked up without previous isolation (60).

Aminolysis to the β-aminoacryl amidines (**159**) and hydrolysis to the β-aminoacrylamides (**160**) or further to β-ketoamides (**161**) shows the ease of substitution reactions of **158**:

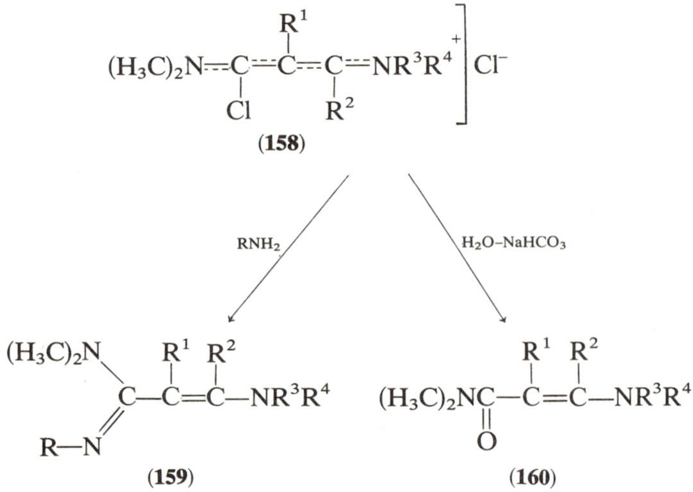

$$[(H_3C)_2N=C(Cl)-C(R^1)=C(R^2)-NR^3R^4]^+ Cl^-$$
(**158**)

↙ RNH$_2$ ↘ H$_2$O–NaHCO$_3$

(H$_3$C)$_2$N\C(=N-R)-C(R^1)=C(R^2)-NR^{3}R^4
(**159**)

(H$_3$C)$_2$NC(=O)-C(R^1)=C(R^2)-NR^{3}R^4
(**160**)

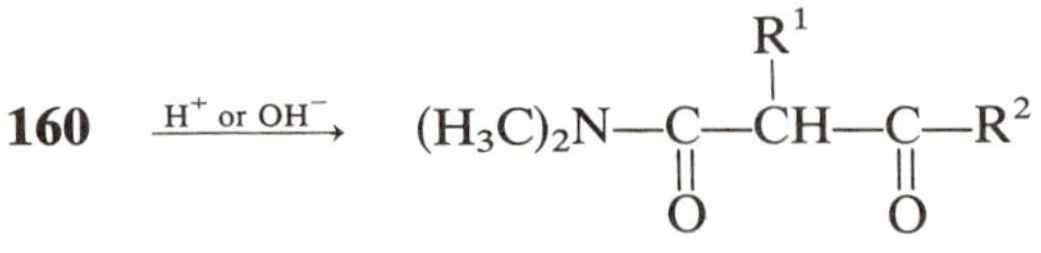

160 $\xrightarrow{H^+ \text{ or } OH^-}$ (H$_3$C)$_2$N-C(=O)-CH(R^1)-C(=O)-R^2
(**161**)

III. POTENTIAL OF DICHLOROMETHYLENIMINIUM SALTS

Enamines and phosgeniminium salts have been varied in particular because of the interest in **158** as 1,3-biselectrophiles for heterocyclizations leading to aminopyrazoles (**162**) and pyrimidines (**163**) (7,98):

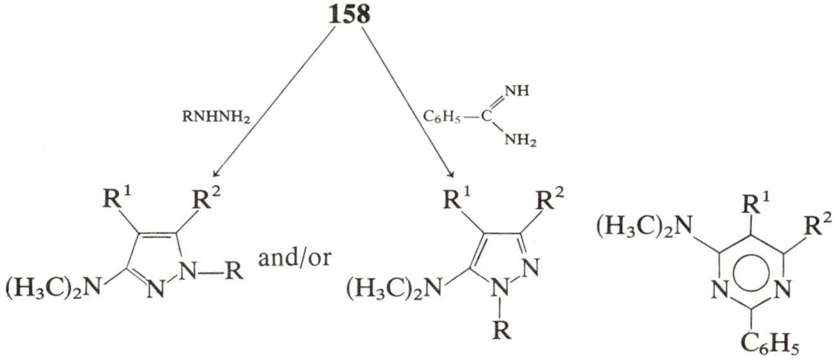

To our knowledge N-vinylpyrrolidone is the only enamide that has been reacted with **1**. The result indicates that enamides react in the same way as enamines (97).

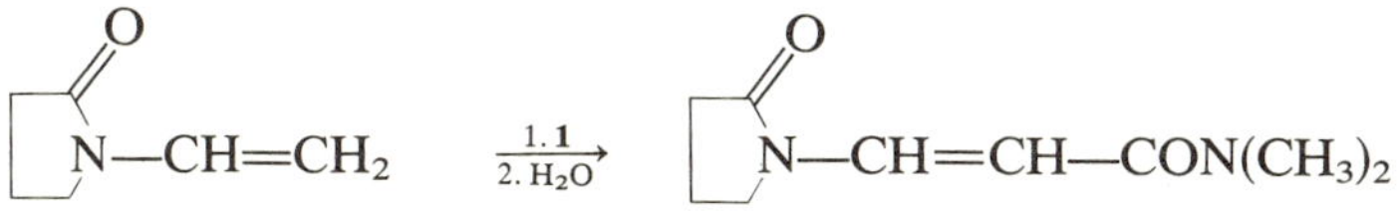

3. Reaction of Phosgeniminium Salts with Vinyl Ethers

Ethyl vinyl ether and 5,6-dihydropyrane add **1** to yield β-alkoxyamide chlorides, but the fact that the reaction times are much longer than with enamines reflects the weaker nucleophilic nature of vinyl ethers (7,100):

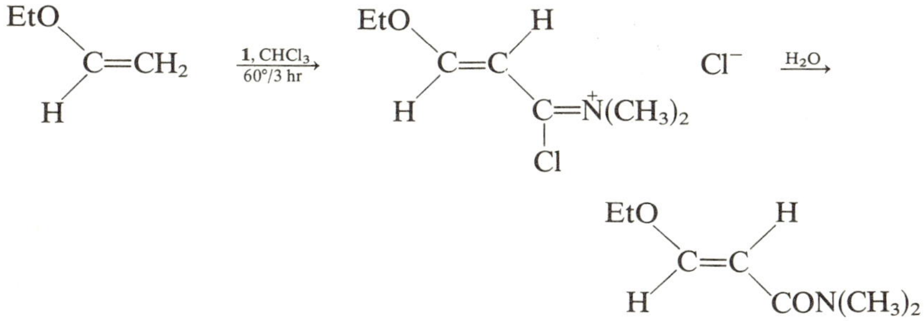

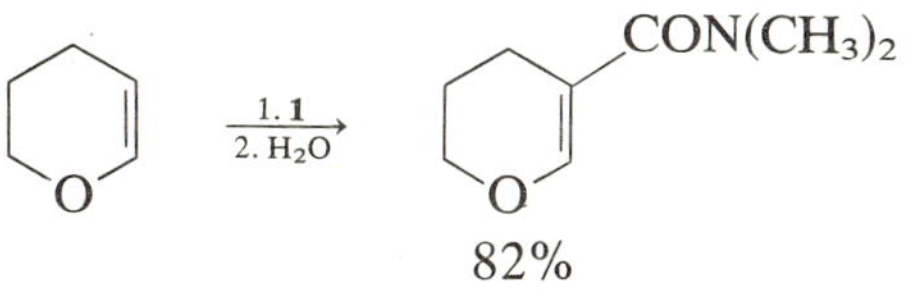

These amide chlorides, too, are promising starting materials for derivatization and, in particular, for heterocyclizations.

Keten acetals (**164**) undergo a more complex reaction with **1** in which 2-carboalkoxy-1,3-dichlorocyanines (**165**) are formed. The latter could not be made from malonamid esters and **1**. It is likely that α-chloro-β-carboalkoxyamines are intermediates in this reaction (78).

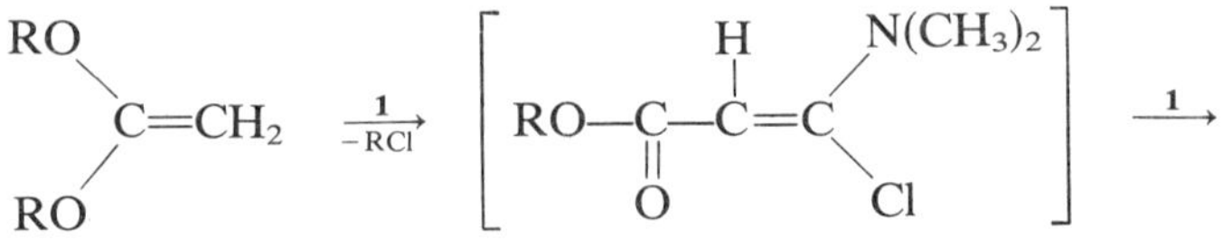

$$\underset{RO}{\overset{RO}{>}}C=CH_2 \xrightarrow[-RCl]{1} \left[RO-\underset{\underset{O}{\|}}{C}-\underset{H}{\overset{}{C}}=C\underset{Cl}{\overset{N(CH_3)_2}{<}} \right] \xrightarrow{1}$$

(**164**)

$$(H_3C)_2N\!=\!\!\overset{}{\underset{Cl}{C}}\!=\!\!\overset{CO_2R}{\underset{}{C}}\!=\!\!\overset{}{\underset{Cl}{C}}\!=\!\!N(CH_3)_2^+ \ Cl^-$$

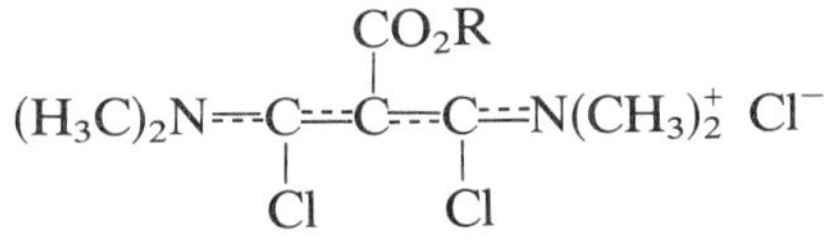

(**165**) R = Et   50%

F. ADDITION TO C–N MULTIPLE BONDS

1. *Reactions with Activated Nitriles*

Acetonitrile and benzonitrile do not react with phosgeniminium salts (**1**) unless they are first transformed into the corresponding imidoyl chlorides by means of hydrogen chloride. On the other hand, tertiary cyanamides add **1** with the same ease as do ynamines, yielding 1,3-dichloroazatrimethinecyanines (**166**). These interesting compounds are as versatile as 1,3-dichlorotrimethines (**68**), discussed in Section III-A-4 (7,101,102).

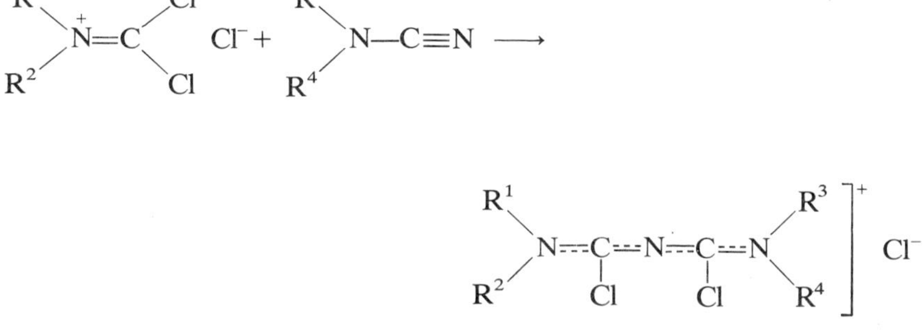

$$\underset{R^2}{\overset{R^1}{>}}\overset{+}{N}=C\underset{Cl}{\overset{Cl}{<}} \ Cl^- + \underset{R^4}{\overset{R^3}{>}}N-C\equiv N \longrightarrow$$

$$\left[\underset{R^2}{\overset{R^1}{>}}N\!=\!\!\overset{}{\underset{Cl}{C}}\!=\!\!N\!=\!\!\overset{}{\underset{Cl}{C}}\!=\!\!N\underset{R^4}{\overset{R^3}{<}} \right]^+ Cl^-$$

(**166**)

TABLE XXI
1,3-Dichloroazacyanines (166)

166	R^1	R^2	R^3	R^4	Yield, %
a	CH_3	CH_3	CH_3	CH_3	93
b	CH_3	CH_3	CH_3	C_6H_5	93
c	CH_3	$CH_2C_6H_5$	CH_3	CH_3	96
d	CH_3	$CH_2C_6H_5$	CH_3	$CH_2C_6H_5$	92

Alternatively, the cyanamides can be generated *in situ* by using N,N-disubstituted ureas and an excess of **1**.

The UV spectrum of **166a** shows a marked hypsochromic shift of 64 nm as compared to that of **88a** ($\lambda_{max} = 346$ nm), which is due to the aza substitution. Compounds **166** may be considered as biuret trichlorides since their hydrolysis and thiolysis afford 1,1,5,5-tetrasubstituted biurets in low yield:

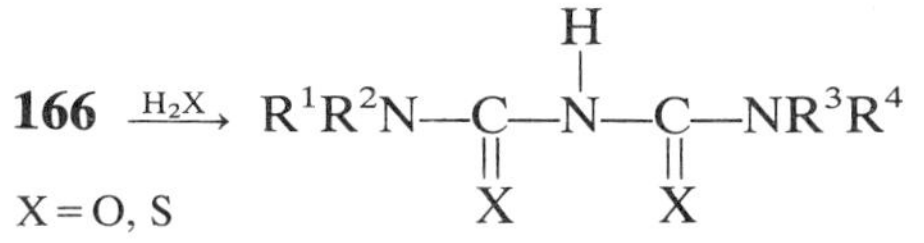

$$\mathbf{166} \xrightarrow{H_2X} R^1R^2N-\underset{\underset{X}{\|}}{C}-\underset{\underset{H}{|}}{N}-\underset{\underset{X}{\|}}{C}-NR^3R^4$$

X = O, S

The reaction of **166a** with sodium cyclopentadienide provides an elegant one-step synthesis of 1,3-bis(dimethylamino)-2-azapentalene (103a):

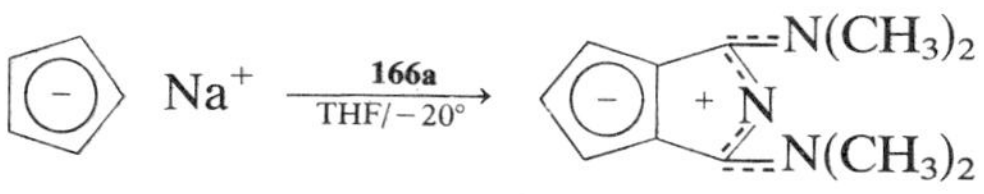

81%

Azacyanines (**166**) are also valuable reagents in heterocyclic synthesis (68,101). For example, cyclizations with monosubstituted hydrazines result in 3,5-bis(dialkylamino)-1,2,4-triazoles (**167**), and hydroxylamine gives the oxadiazole (**168a**):

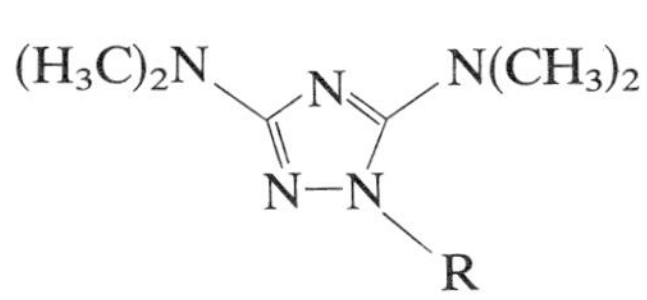

(**167**) R = H, CH_3, C_6H_5, 70–90%

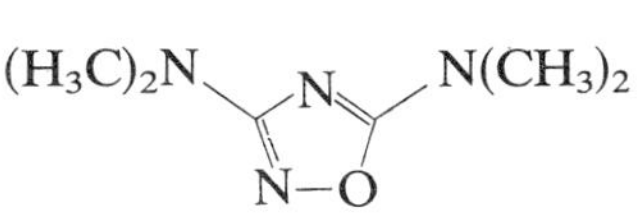

(**168a**) 95%

Similarly, **166a** undergoes cyclizations with benzamidine and propiophenonimine to diaminotriazine and pyrimidine (**166b,168c**):

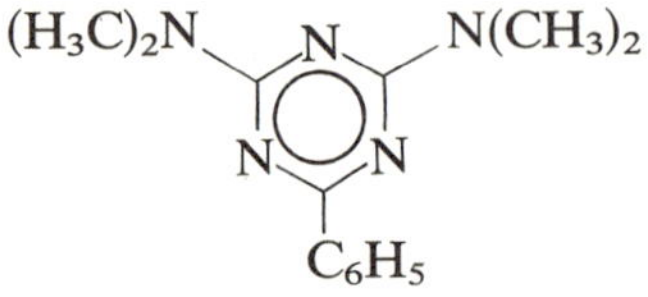

(**168b**) 99%

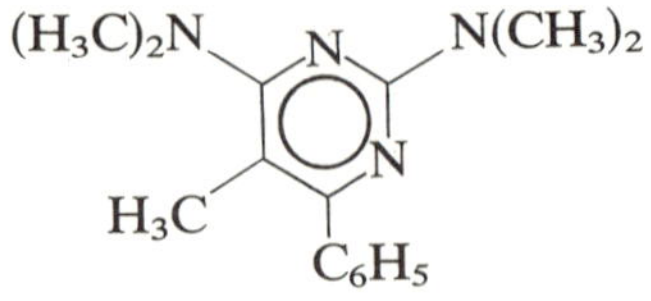

(**168c**) 89%

In these reactions azacyanines are more reactive and give better yields than dithiolium salts (103b), which have been used for the same purpose. Aminolysis of **166** results in the interesting tetrakis(dimethylamino)-azaallyl cation (**169**), isolated as perchlorate (35):

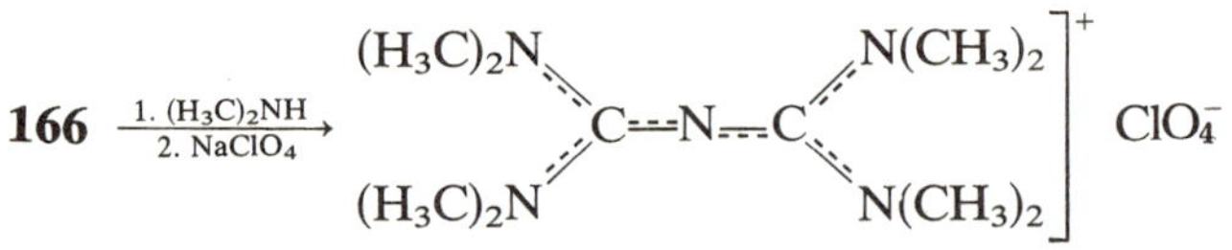

166 $\xrightarrow[\text{2. NaClO}_4]{\text{1. (H}_3\text{C)}_2\text{NH}}$ [(H$_3$C)$_2$N)$_2$C---N---C(N(CH$_3$)$_2$)$_2$]$^+$ ClO$_4^-$

(**169**) 60%, m.p. 190°

The vinylogue of dimethylcyanamide, 2-dimethylaminoacrylonitrile, behaves as an enamine toward **1**, giving 1-chloro-2-cyanotrimethinecyanine (Table XX), but α-substituted β-aminoacylonitriles (**170**) undergo addition to the activated cyano group (36):

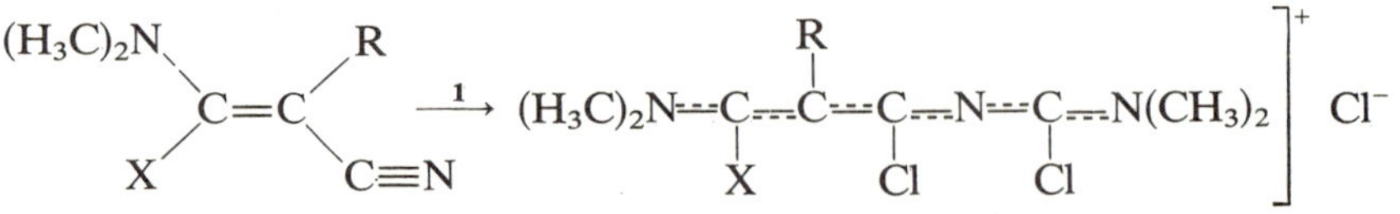

(**170**)

(**171**) R = C$_6$H$_5$, X = Cl 97%

Cyanamide and urea react with an excess of **1** to yield the known 1,3,5-trichloro-2,4-diazapentamethinecyanines (**172**), which are more commonly prepared from phosgene and tertiary cyanamides (36,104,105):

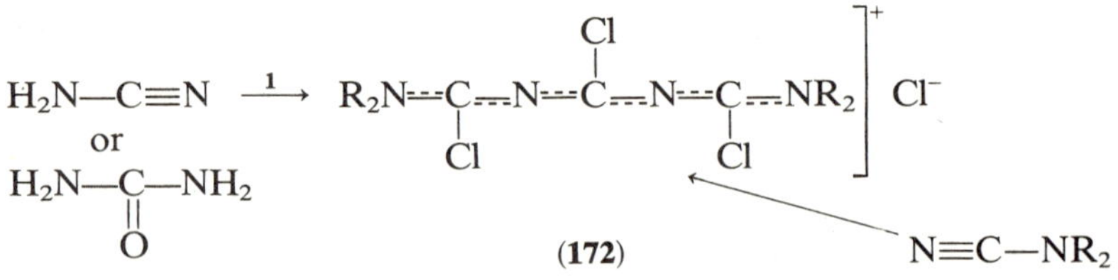

H$_2$N—C≡N
or
H$_2$N—C(=O)—NH$_2$

$\xrightarrow{1}$ [R$_2$N---C(Cl)---N---C(Cl)---N---C(Cl)---NR$_2$]$^+$ Cl$^-$

(**172**)

N≡C—NR$_2$

2. Hydrogen Chloride-Initiated Additions of 1 to Nitriles

The reactions of nitriles with hydrogen chloride can lead to a variety of products (see the chapter by S. Yahagida et al.) which arise from the intermediary chloroalkylideniminium salts (**173**). In the presence of **1**, these intermediates are trapped before they dimerize, and the stable, delocalized azapentamethines (**174**) are final products. The mechanisms can be depicted as follows:

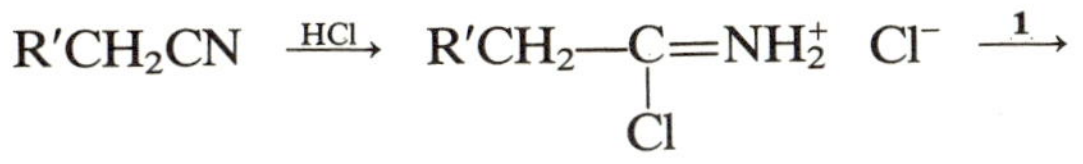

$$R'CH_2CN \xrightarrow{HCl} R'CH_2-\underset{Cl}{C}=NH_2^+ \; Cl^- \xrightarrow{1}$$

(173)

$$R'CH_2-\underset{Cl}{C}=N-\underset{Cl}{\overset{+}{C}}=NR_2 \; Cl^- \underset{+HCl}{\overset{-HCl}{\rightleftharpoons}} R'CH=\underset{Cl}{C}-N=\underset{Cl}{C}-NR_2 \xrightarrow{1}$$

$$\left[R_2N \cdots \underset{Cl}{C} \cdots \overset{R'}{\underset{Cl}{C}} \cdots \underset{Cl}{C} \cdots N \cdots C \cdots NR_2 \right]^+ Cl^-$$

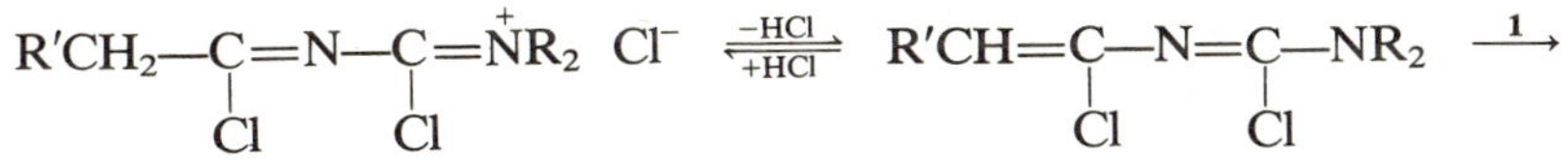

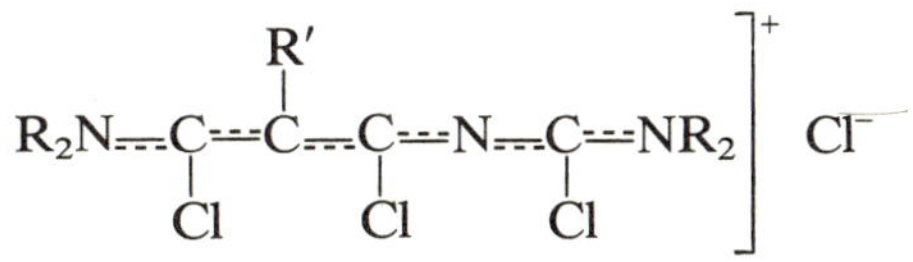

(174)

3. Reactions with Imines

So far, only imines nonsubstituted at the nitrogen atom have been studied. Different products are obtained, depending on the number of hydrogen atoms in α position to the imine group.

Nonenolizable benzophenonimine forms the chloroformamidine (**175a**), which in analogy to N-chlorocarbonylbenzophenonimine (106) may be supposed to exist in tautomeric equilibrium with the covalent

TABLE XXII

Trichloroazapentamethines (**174**) (36)

R:	CH_3	CH_3	CH_3	C_2H_5	C_2H_5
R':	H	CH_3	C_6H_5	CH_3	C_6H_5
Yield, %	50	72	60	56	82

form (**175b**):

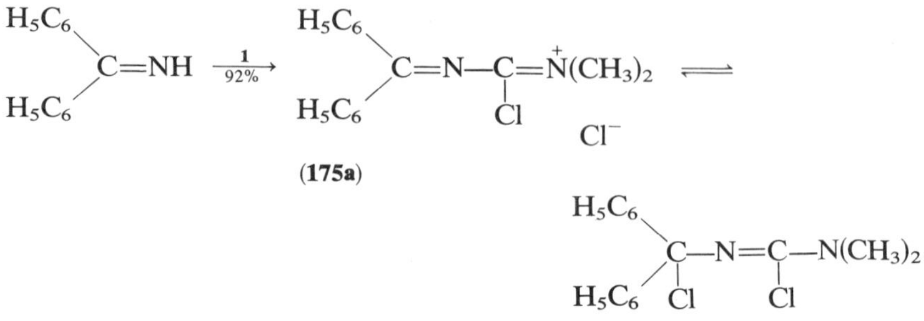

$$(H_5C_6)_2C=NH \xrightarrow[92\%]{1} (H_5C_6)_2C=N-\overset{+}{\underset{Cl}{C}}=N(CH_3)_2 \; Cl^- \rightleftharpoons$$

(**175a**)

$$(H_5C_6)_2\underset{Cl}{C}-N=\underset{Cl}{C}-N(CH_3)_2$$

(**175b**)

The reactions of **175** with nucleophiles are accompanied by fragmentation to form benzophenone (or its imine) and dimethylcyanamide (97):

$$(H_5C_6)_2C=O \xleftarrow{H_2O} \mathbf{175} \xrightarrow{RNH_2} (H_5C_6)_2C=NR + (H_3C)_2N-C\equiv N$$

N-Benzalamines smoothly form addition products **176**, which give back benzaldehyde upon hydrolysis (75,97):

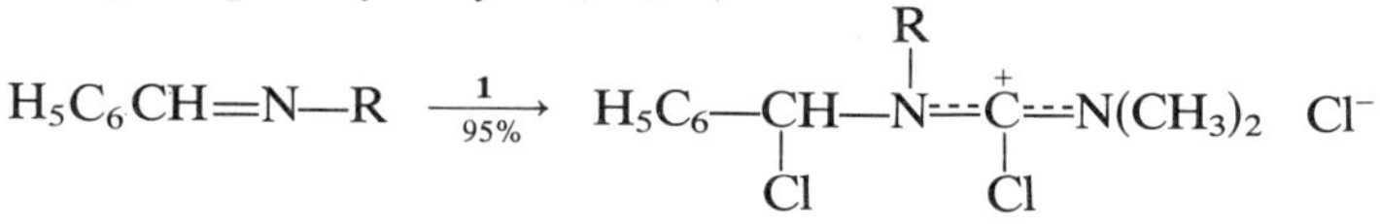

$$H_5C_6CH=N-R \xrightarrow[95\%]{1} H_5C_6-\underset{Cl}{\underset{|}{CH}}-\overset{R}{\underset{|}{N}}\cdots\overset{+}{C}\cdots N(CH_3)_2 \; Cl^-$$
$$ Cl$$

(**176**) R = *n*-butyl, C$_6$H$_5$

Synthetically more useful are products from **1** and enolizable ketimines. Thus phenylisopropylketimine yields the interesting *N*-vinylchloroformamidine (**177**):

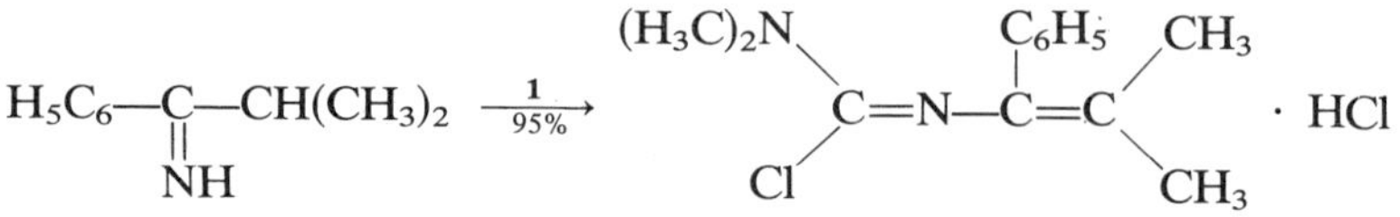

$$H_5C_6-\underset{NH}{\overset{\|}{C}}-CH(CH_3)_2 \xrightarrow[95\%]{1} \underset{Cl}{\overset{(H_3C)_2N}{\diagdown}}C=N-\underset{CH_3}{\overset{C_6H_5}{C}}=C\diagdown_{CH_3} \cdot HCl$$

(**177**)

Hydrolysis of **177** gives rise to the expected *N*-vinylurea (**178**), but the fragmentation still competes to a certain extent (97):

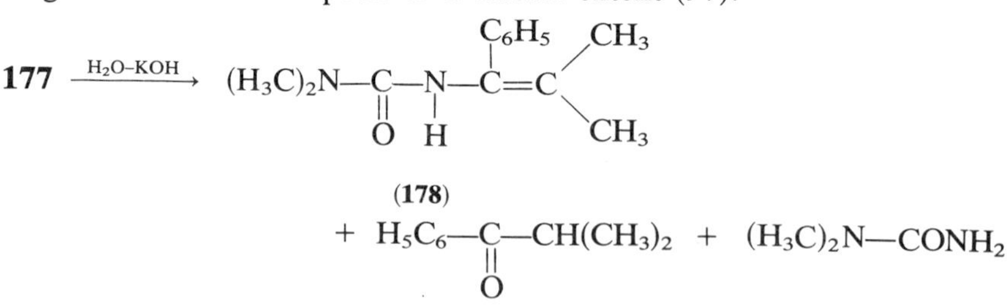

$$\mathbf{177} \xrightarrow{H_2O-KOH} (H_3C)_2N-\underset{O}{\overset{\|}{C}}-\underset{H}{\overset{|}{N}}-\underset{CH_3}{\overset{C_6H_5}{C}}=C\diagdown_{CH_3}$$

(**178**)

$$+ \; H_5C_6-\underset{O}{\overset{\|}{C}}-CH(CH_3)_2 \; + \; (H_3C)_2N-CONH_2$$

The reaction with dimethylamine affords the *N*-vinylguanidine (**179**); *sym*-dimethylhydrazine and ethyl aminoacetate yield the cyclization products **180** and **181**, respectively (97):

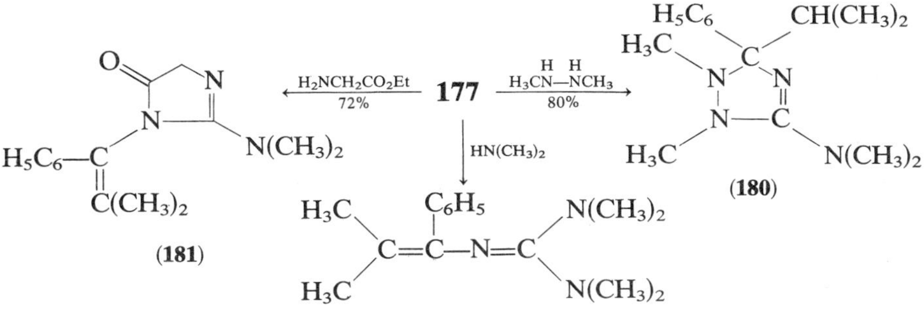

As noted in Section III-B-2, azabutadienes containing a chlorine atom instead of an alkyl or aryl group in position 3 arise from thermolysis of **121**, produced in turn from **1** and secondary amides.

Imines displaying at least two hydrogen atoms in α position to the imine group react with 2 moles of **1** to afford 1,5-dichloro-2-azapentamethinecyanines (**182**):

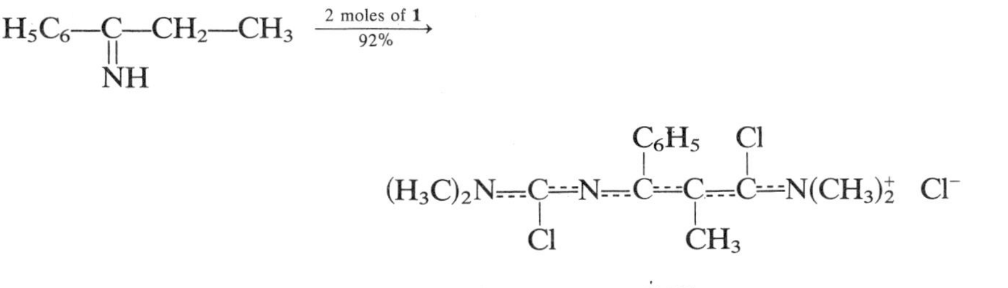

Compounds **182** bear close resemblance to azapentamethinecyanines (**174**), which arise in the reactions among nitriles, **1**, and hydrogen chloride. Under methyl chloride elimination **182** cyclizes thermally to one of two possible aminochloropyrimidines, and ammonia leads to 2,4-bis-(dimethylamino)-5-methyl-6-phenylpyrimidine (97). Hydrolysis produces a cleavage of the cyanine chain between carbon and nitrogen:

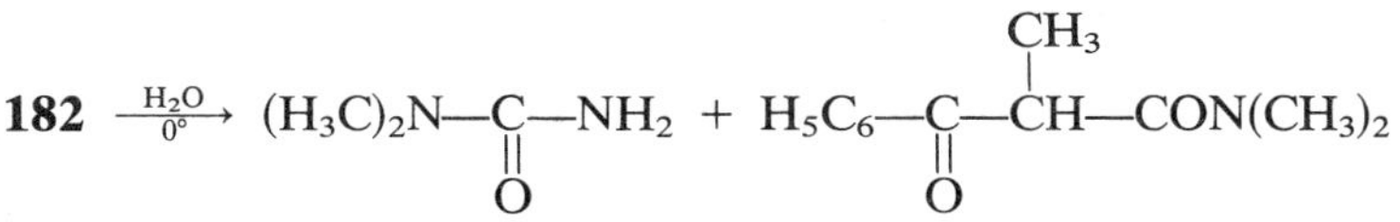

4. Addition to Cyanates

Aryl cyanates react in a 2:1 ratio with **1**, but it has not yet been established which of the two possible adducts (**183a,183b**) is formed. Ammonolysis forms the triazine (**184**) in either case (102).

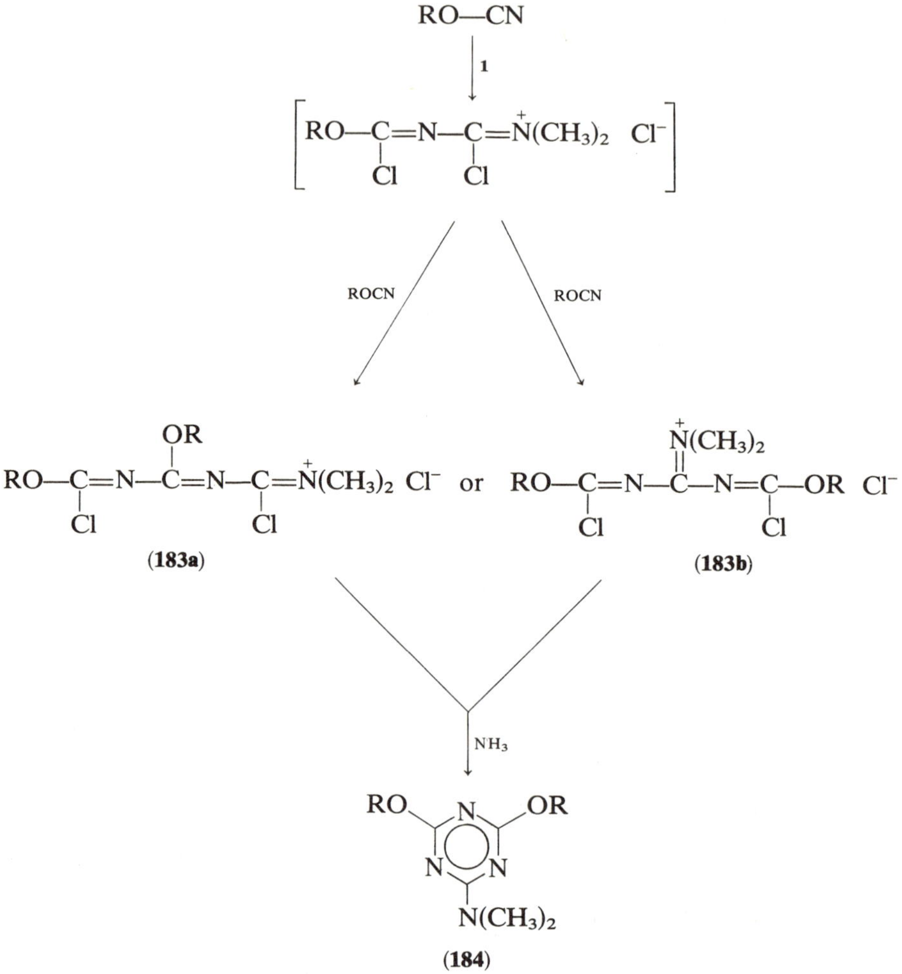

$$\text{RO—CN} \xrightarrow{\mathbf{1}} \left[\text{RO—}\underset{\text{Cl}}{\text{C}}\text{=N—}\underset{\text{Cl}}{\text{C}}\text{=}\overset{+}{\text{N}}(\text{CH}_3)_2 \ \text{Cl}^- \right]$$

Branches with ROCN lead to:

$$\text{RO—}\underset{\text{Cl}}{\text{C}}\text{=N—}\underset{\overset{|}{\text{OR}}}{\text{C}}\text{=N—}\underset{\text{Cl}}{\text{C}}\text{=}\overset{+}{\text{N}}(\text{CH}_3)_2 \ \text{Cl}^- \quad \text{or} \quad \text{RO—}\underset{\text{Cl}}{\text{C}}\text{=N—}\underset{\overset{\|}{\overset{+}{\text{N}}(\text{CH}_3)_2}}{\text{C}}\text{—N=}\underset{\text{Cl}}{\text{C}}\text{—OR} \ \text{Cl}^-$$

(**183a**)   (**183b**)

$\xrightarrow{\text{NH}_3}$ triazine ring with RO, OR, N(CH$_3$)$_2$ substituents

(**184**)

5. Addition to Heterocumulenes

Additions to allenes, isocyanates, and isothiocyanates have not yet been mentioned in the literature. On the other hand, ketenes, carbodimides, and ketenimines add **1** with varying ease, depending on their nucleophilic nature.

III. POTENTIAL OF DICHLOROMETHYLENIMINIUM SALTS 407

Ketenes having a hydrogen atom form α-chloro-β-chlorocarbonyl-enamines (78) (**185**), which are also known to arise from amides and phosgene (53):

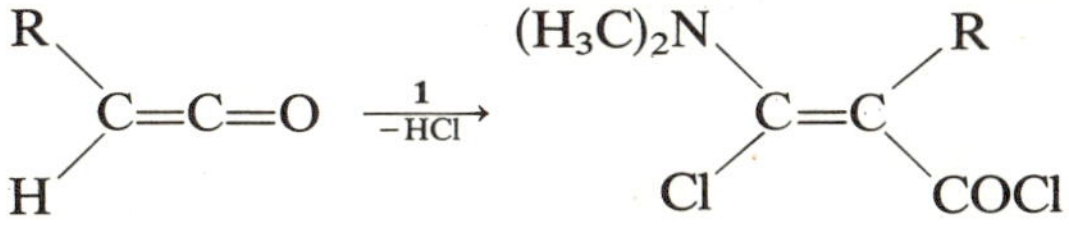

$$\underset{H}{\overset{R}{>}}C=C=O \xrightarrow[-HCl]{1} \underset{Cl}{\overset{(H_3C)_2N}{>}}C=C\underset{COCl}{\overset{R}{<}}$$

R = C$_6$H$_5$, Cl

(**185a**) R = C$_6$H$_5$ 70%
(**185b**) R = Cl   75%
(**185c**) R = F   86%
(**185d**) R = CH$_3$ 70%
(**185e**) R = C$_2$H$_5$ 68%

Dichloroketene gives the same product as chloroketene since the primary adduct loses elementary chlorine. Diphenylketene, however, behaves in a different manner, forming a dichloro-β-lactam (**186**), the structure of which is based on spectral data (78):

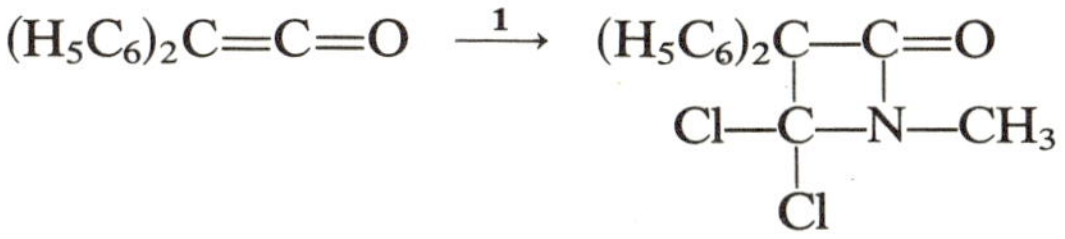

$$(H_5C_6)_2C=C=O \xrightarrow{1} \begin{array}{c} (H_5C_6)_2C-C=O \\ | \quad\quad | \\ Cl-C-N-CH_3 \\ | \\ Cl \end{array}$$

(**186**) 75%, m.p. 152–154°

As the study shows, the loss of methyl chloride seems to occur after the cycloaddition step because isocyanide dichlorides do not react with diphenylketene under the same conditions.

The more nucleophilic carbodiimides give almost quantitative yields of biuret dichlorides (**187**), which are isomers of the azacyanines (**166**) obtained from **1** and tertiary cyanamides:

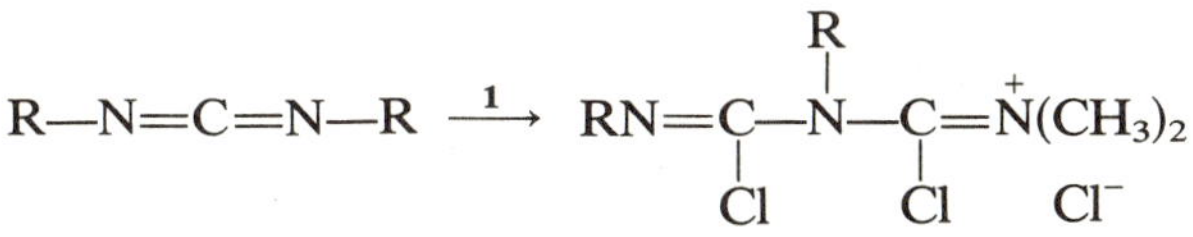

$$R-N=C=N-R \xrightarrow{1} \begin{array}{c} R \\ | \\ RN=C-N-C=\overset{+}{N}(CH_3)_2 \\ | \quad\quad | \\ Cl \quad\quad Cl \quad Cl^- \end{array}$$

(**187a**) R = CH$_3$ 95%
(**187b**) R = C$_6$H$_{11}$ 95%

The same study (107) shows also that N,N'-disubstituted ureas give the same products with an excess of **1**, thus forming the carbodiimides *in situ*. Biuret trichlorides (**187**) hydrolyze to the corresponding biurets, and aminolyze to biguanides (**188a,188b**):

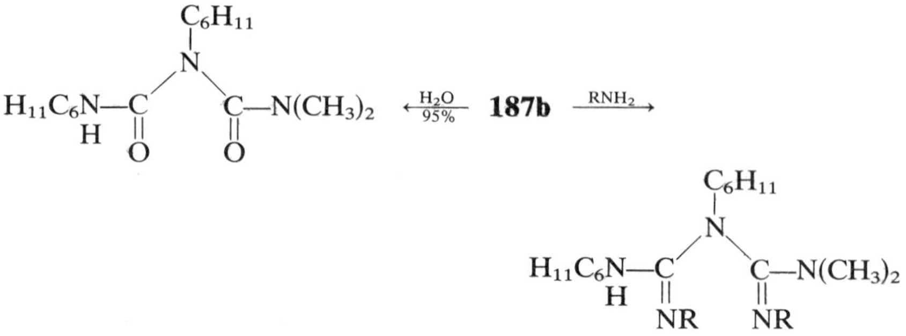

$$H_{11}C_6N-\underset{H}{\overset{\underset{|}{N-C_6H_{11}}}{C}}\underset{O}{\overset{O}{C}}-N(CH_3)_2 \xleftarrow{\underset{95\%}{H_2O}} \mathbf{187b} \xrightarrow{RNH_2}$$

$$H_{11}C_6N-\underset{H}{\overset{\underset{|}{N-C_6H_{11}}}{C}}\underset{NR}{\overset{NR}{C}}-N(CH_3)_2$$

(**188a**) R = CH$_3$ 83%
(**188b**) R = H   80%

Cyclizations between **187b** and phenylhydrazine, benzamidine, and *o*-phenylenediamine have also been effected:

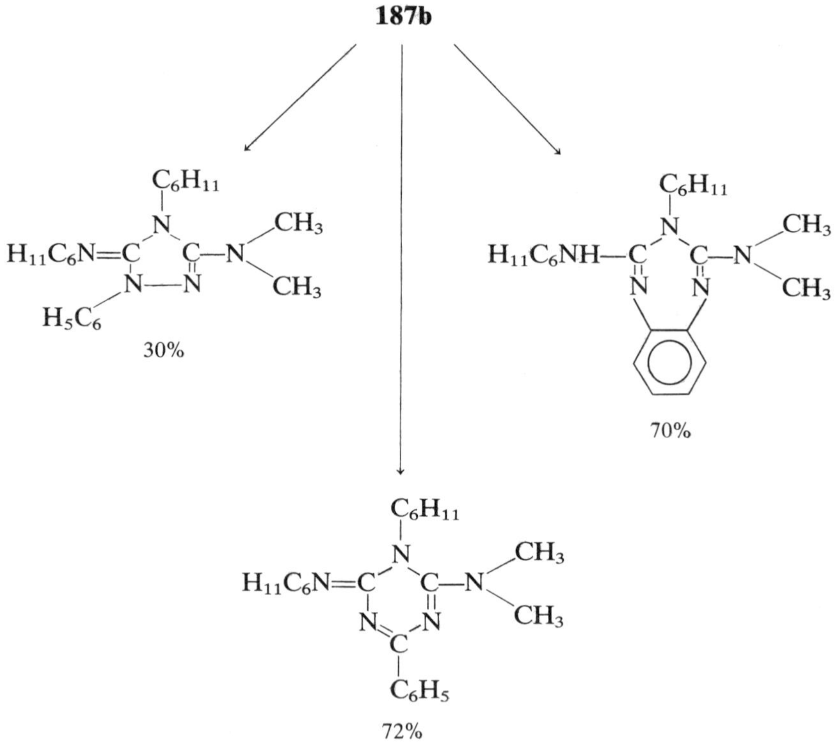

187b

30%

70%

72%

Ketenimine (**189**) adds **1** quantitatively to form **190**, which can also be prepared from *N*-ethylphenylacetamide and **1** (85):

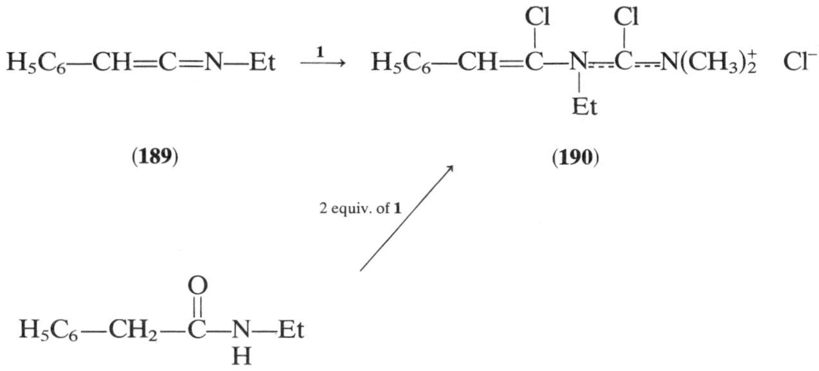
H5C6—CH=C=N—Et
1
H5C6—CH=C—N—C—N(CH3)2+ Cl−
Cl
Cl
Et
(189)
(190)
2 equiv. of 1
O
H5C6—CH2—C—N—Et
H

G. HETEROCYCLIZATION REACTIONS WITH PHOSGENIMINIUM SALTS

As shown in Section III-D, pyrocatechol, pinacol, and ethanedithiol react with **1** to form the corresponding iminium carbonates and thiocarbonates. Only the latter compounds have been studied since they are easily synthesized by alkylation of dithiocarbamates (108).

A great number of ortho-disubstituted aromatic and heterocyclic systems offer themselves for heterocyclizations with **1** (Table XXIII):

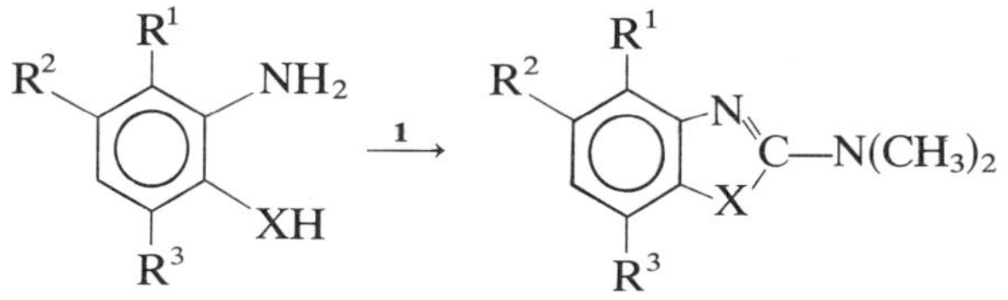
R1
R2
NH2
XH
R3
1
R1
R2
N
C—N(CH3)2
X
R3

(**191**)

TABLE XXIII

Cyclization of Ortho-disubstituted Benzenes with **1**

X	R^1	R^2	R^3	Yield, %	Melting point, °C	Ref.
O	H	H	H	90	90–91	9,109
O	H	H	NO_2	74	164	109
O	H	NO_2	Cl	71	163	109
O	Cl	Cl	Cl	60	139–140	109
S	H	H	H	70	87	109
NH	H	H	H	95	300	9,74
NC_6H_5	H	H	H	97	77–79	74

Likewise, naphthalene-1,8-diamine (9) was cyclized with **1** to the corresponding dimethylaminopyrimidine (**192**):

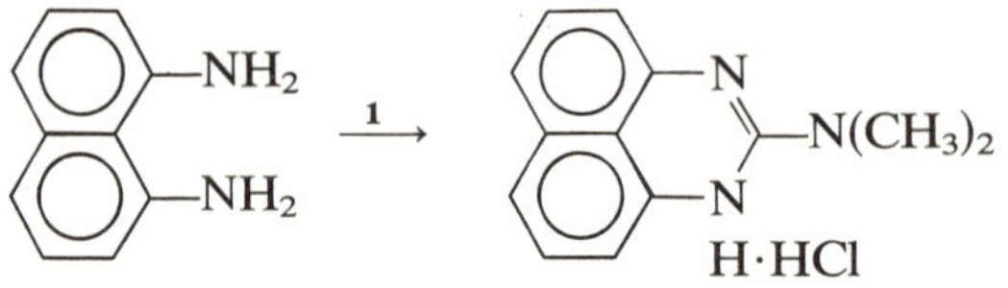

(**192**) 98%, m.p. 255–260°

Various benzimidazoles, benzoxazoles, and benzothiazoles have also been prepared using chloromercaptomethyleniminium salts; the result depends markedly on the basicity of the solvent used (17).

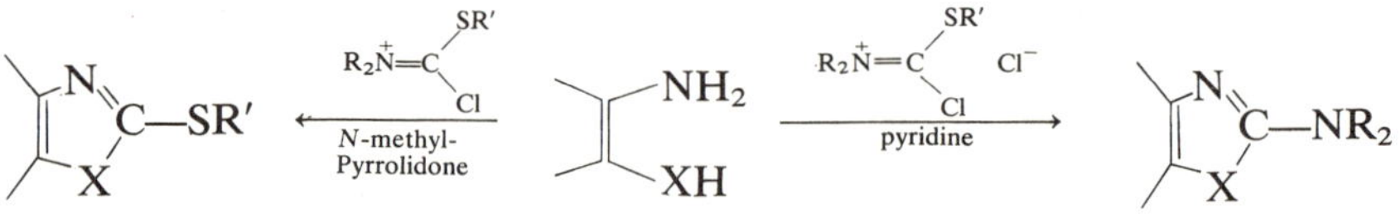

Carbohydrazides cyclize with **1** to 2-dialkylamino-1,3,4-oxadiazoles (**193**) (Table XXIV):

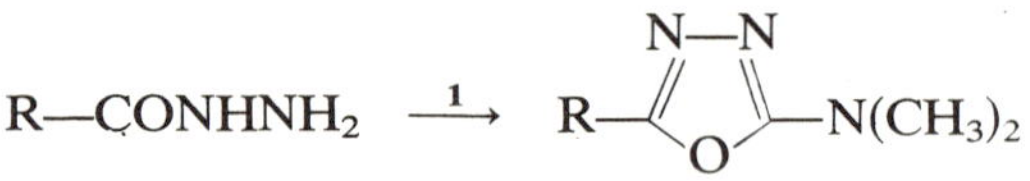

(**193**)

TABLE XXIV

Oxadiazoles (**193**)

R	Melting or boiling point, °C/torr	Yield, %	Ref.
CH_3	100/0.1	90	109
$C_6H_5CH_2$	75–76	93	109
4-Pyridyl	118–120	95	109
C_6H_5	185–187	92	9,74
3-Nitrophenyl	136	91	74
2-Nitrophenyl	148	89	74

5-Chloro-3-amino-1,2,4-triazoles (**194**) can be obtained from semicarbazides and **1** (74):

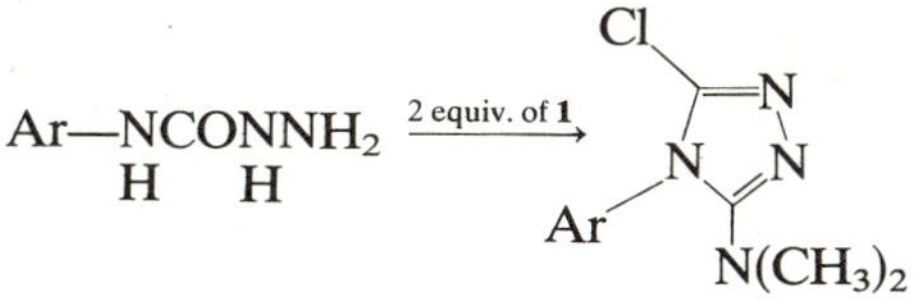

(**194a**) Ar = C$_6$H$_5$ 73%

(**194b**) Ar = O$_2$N—⟨⟩—NO$_2$ 50%

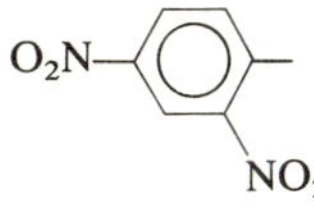

Both open-chain and cyclic amidrazones cyclize readily to 1,2,4-triazoles:

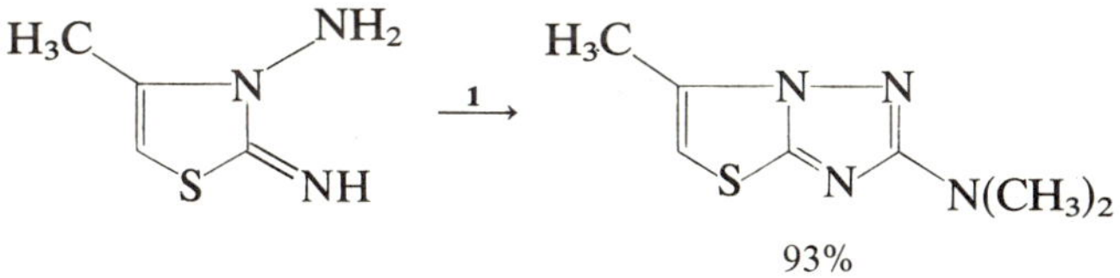

93%

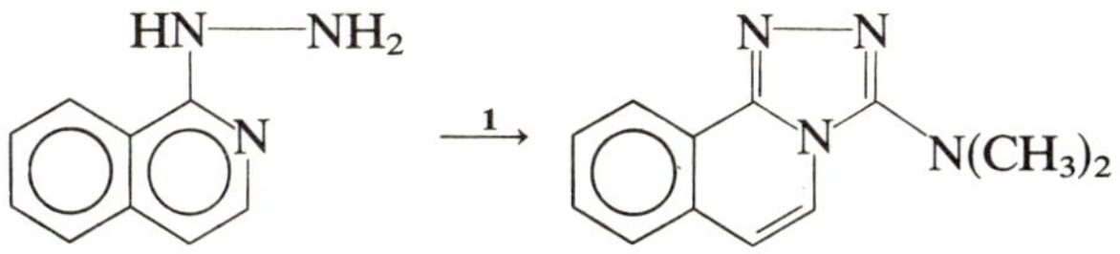

89%

Also important are reactions in which chloroformamidinium salts formed in the first step undergo an internal S_E on the adjacent aromatic or heterocyclic nucleus (74,109):

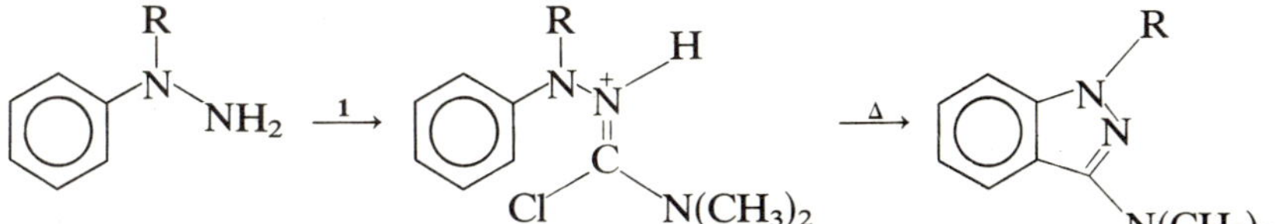

(**195a**) R = CH$_3$ 82%
(**195b**) R = C$_6$H$_5$ 83%

As shown above, 1-substituted phenylhydrazines form semicarbazide dichlorides, which cyclize readily to 3-dimethylaminoindazoles. A synthetically very useful reaction using this principle has been described (110). Hydrazones of both aliphatic and cyclic ketones form high yields of

TABLE XXV

Aminopyrazoles (**196**) from Hydrazones and **1** (97)

R^1	R^2	R^3	R^4	Yield, %
CH_3	C_2H_5	CH_3	CH_3	81
C_6H_5	C_2H_5	CH_3	CH_3	87
CH_3	CH_3	C_2H_5	CH_3	70
CH_3	C_6H_5	H	CH_3	24
CH_3	—$(CH_2)_3$—		CH_3	66
CH_3	—$(CH_2)_4$—		CH_3	82
CH_3	—$(CH_2)_5$—		CH_3	84
CH_3	—$(CH_2)_6$—		CH_3	84
CH_3	—$(CH_2)_4$—		C_2H_5	98
CH_3	—$(CH_2)_4$—		—$(CH_2)_2$—O—$(CH_2)_2$—	—
CH_3	—$(CH_2)_4$—		—$(CH_2)_2$—N(CH_3)—$(CH_2)_2$—	37

aminopyrazoles and indazoles with **1** on simple refluxing in chloroform:

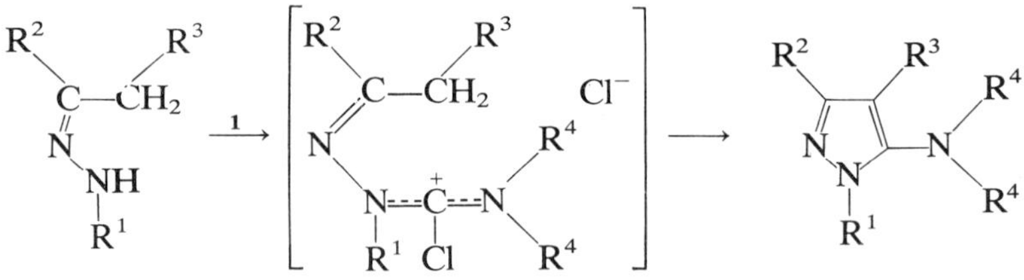

In the last compound in Table XXV the intermediate semicarbazone dichloride is stable and could be cyclized only upon prolonged refluxing in phosphorus oxychloride.

IV. Conclusions and Outlook

Phosgeniminium salts (**1**) represent a highly reactive and easily available synthon. Being derivatives of carbonic or carbamic acid (126) these salts have a higher oxidation level and a greater scope than the VHA and Mannich-Böhme reagents.

Correspondingly, even weak nucleophiles condense with **1**, replacing one, two, or all three chlorine atoms. Monocondensation leads to highly

IV. CONCLUSIONS AND OUTLOOK

versatile amide chlorides:

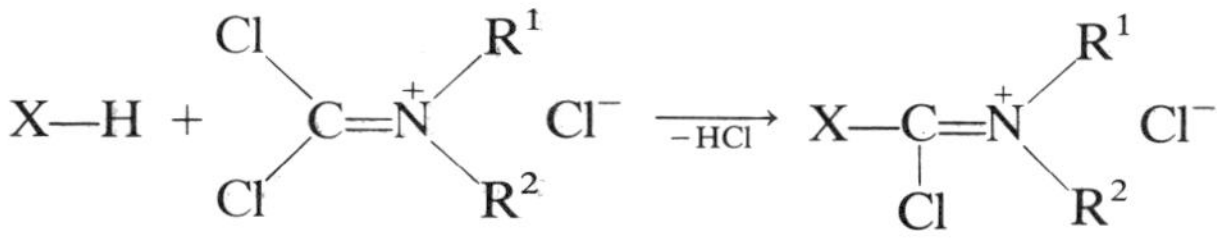

$$X-H + \underset{Cl}{\overset{Cl}{C}}=\overset{+}{N}\underset{R^2}{\overset{R^1}{\diagdown}} \quad Cl^- \xrightarrow{-HCl} X-\underset{Cl}{\overset{}{C}}=\overset{+}{N}\underset{R^2}{\overset{R^1}{\diagdown}} \quad Cl^-$$

$X = R_3C, R_2N, ArO, ArS, N_3, -N=C=O$ etc.

Compounds containing two reactive hydrogen atoms give rise to still reactive α-chloroenamines, chloroformamidines, iminium carbonates and thiocarbonates, etc.:

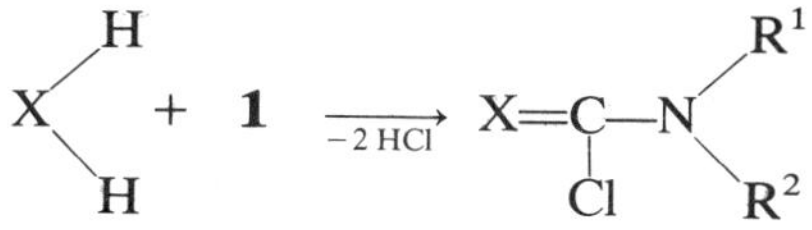

$$X\diagup^{H}_{\diagdown H} + \mathbf{1} \xrightarrow{-2\,HCl} X=\underset{Cl}{C}-N\underset{R^2}{\overset{R^1}{\diagdown}}$$

$X = R_2C=, RN=$

$$\left(\begin{array}{c}XH\\XH\end{array}\right. + \mathbf{1} \xrightarrow{-2\,HCl} \left(\begin{array}{c}X\\X\end{array}\right.\!\!C=\overset{+}{N}\underset{R^2}{\overset{R^1}{\diagdown}} \quad Cl^-$$

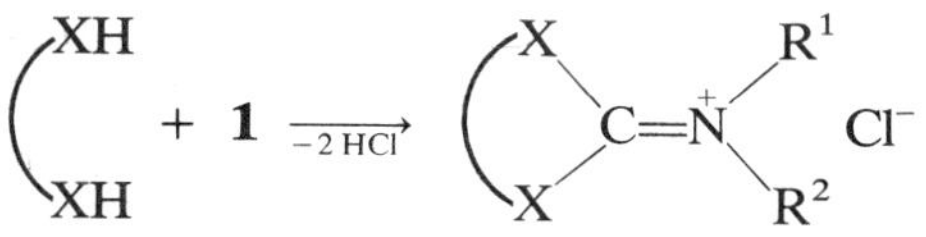

$X = O, O; S, S; R-N, R-N; O, S; R-N, O; R-N, S$

Tricondensations either afford compounds containing triple bonds or lead to various heterocyclic systems:

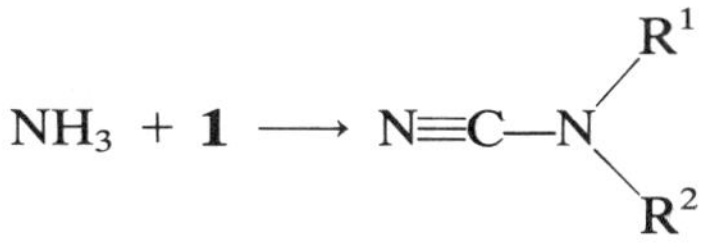

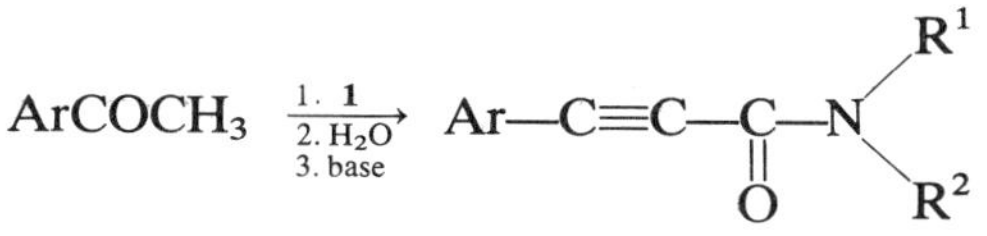

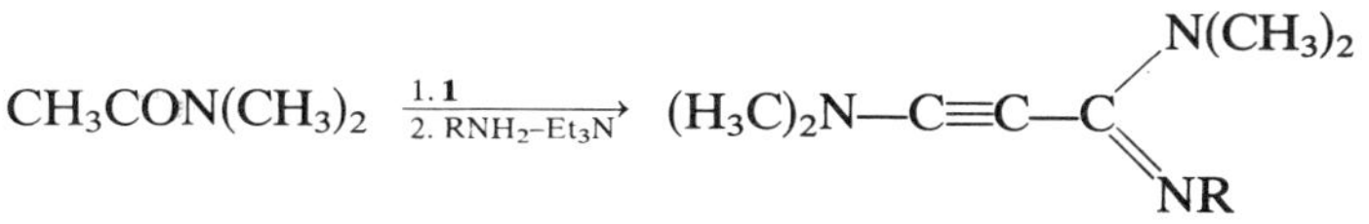

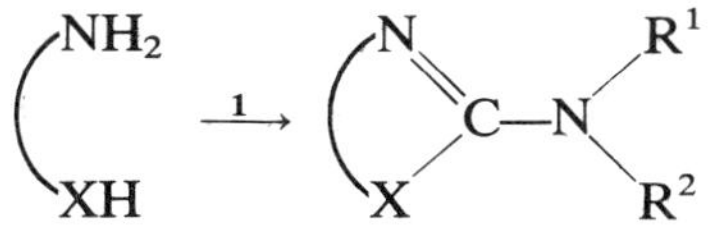

$$NH_3 + \mathbf{1} \longrightarrow N\equiv C-N\underset{R^2}{\overset{R^1}{\diagdown}}$$

$$ArCOCH_3 \xrightarrow[\substack{2.\,H_2O\\3.\,\text{base}}]{1.\,\mathbf{1}} Ar-C\equiv C-\underset{\underset{O}{\|}}{C}-N\underset{R^2}{\overset{R^1}{\diagdown}}$$

$$CH_3CON(CH_3)_2 \xrightarrow[2.\,RNH_2-Et_3N]{1.\,\mathbf{1}} (H_3C)_2N-C\equiv C-C\underset{NR}{\overset{N(CH_3)_2}{\diagdown}}$$

$$\left(\begin{array}{c}NH_2\\XH\end{array}\right. \xrightarrow{\mathbf{1}} \left(\begin{array}{c}N\\X\end{array}\right.\!\!C-N\underset{R^2}{\overset{R^1}{\diagdown}}$$

$X = O, S, NR^3$

Moreover, di- and tricondensations can also lead to polymerizations.

Phosgeniminium salts are precursors of a great number of 1,3-bis- or 1,3-triselectrophilic compounds, some of which also are valuable synthons:

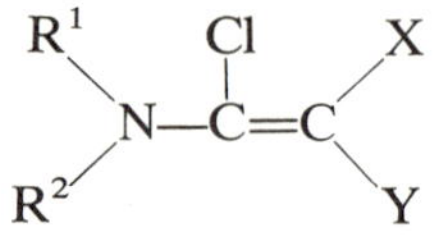

(45) $R^1R^2N-CCl=CXY$

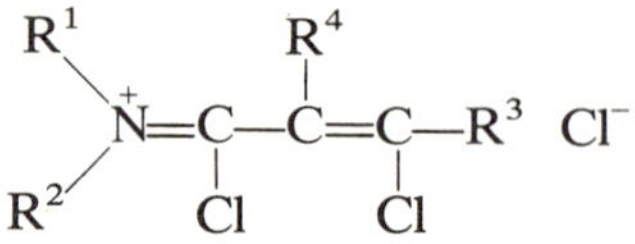

(60) $[R^1R^2N=C(Cl)-C(R^4)=C(Cl)-R^3]^+ Cl^-$

(88) $[R^1R^2N\cdots C(Cl)\cdots C(R^5)\cdots C(Cl)\cdots NR^3R^4]^+ Cl^-$

(109) $R^1R^2N-C(Cl)=N-C(=O)Cl$

(115) $[R^1R^2N\cdots C(Cl)\cdots N(R^3)-CO_2R^4]^+ Cl^-$

(121) $[R^1R^2N\cdots C(Cl)\cdots N(R^3)-C(Cl)=C(R^4)(R^5)]^+ Cl^-$

(122) $R^1R^2N-C(Cl)=N-C(Cl)=C(R^3)(R^4)$

(158) $[R^1R^2N\cdots C(Cl)\cdots C(R^5)\cdots C(R^6)\cdots NR^3R^4]^+ Cl^-$

(166) $[R^1R^2N\cdots C(Cl)\cdots N\cdots C(Cl)\cdots NR^3R^4]^+ Cl^-$

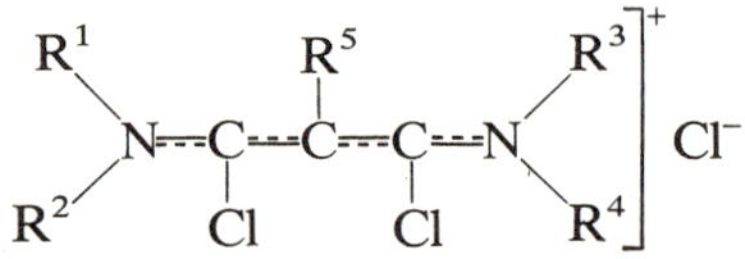

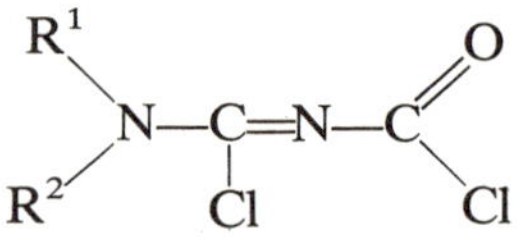

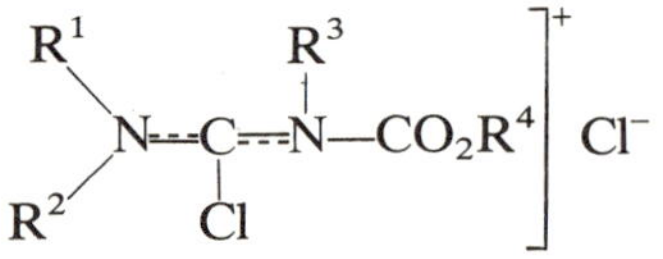

(185) $R^1R^2N-C(Cl)=C(R)-C(=O)Cl$

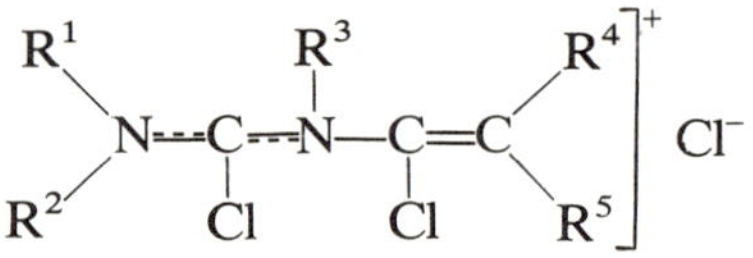

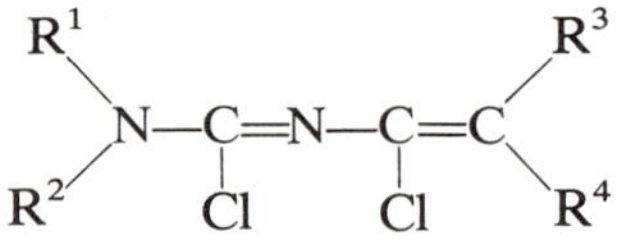

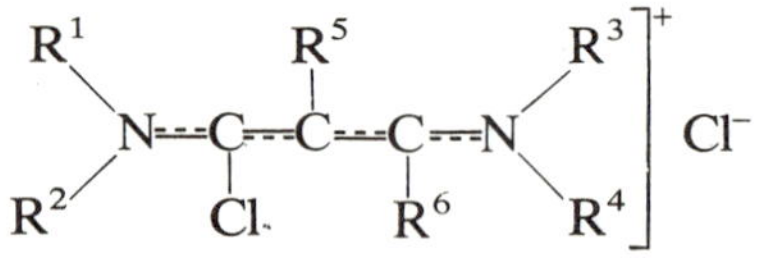

(187) $[R^1R^2N=C(Cl)-N(R)-C(Cl)=NR]^+ Cl^-$

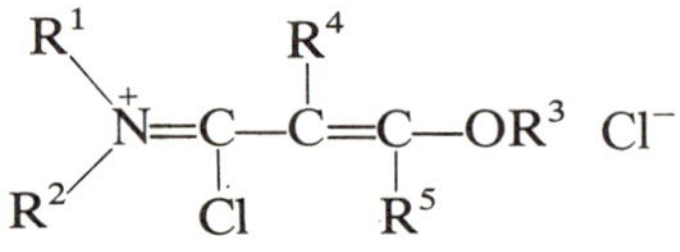

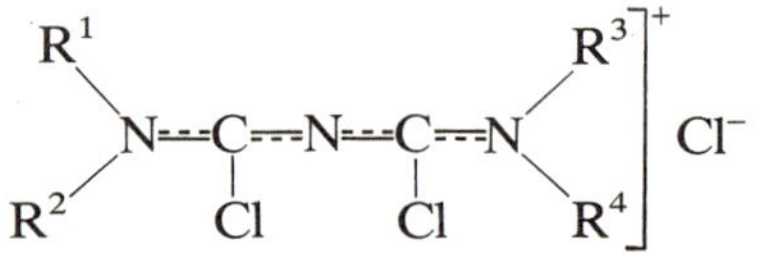

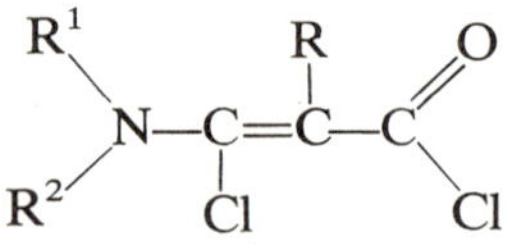

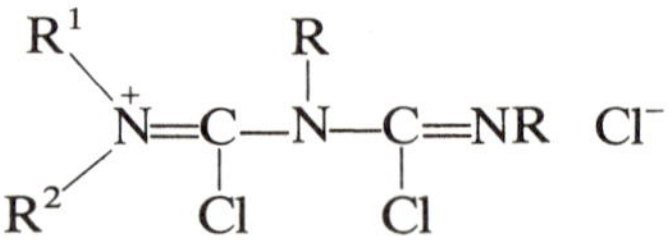

Stable 1,4-bis- or 1,4-triselectrophiles are less numerous because of frequent intramolecular reactions:

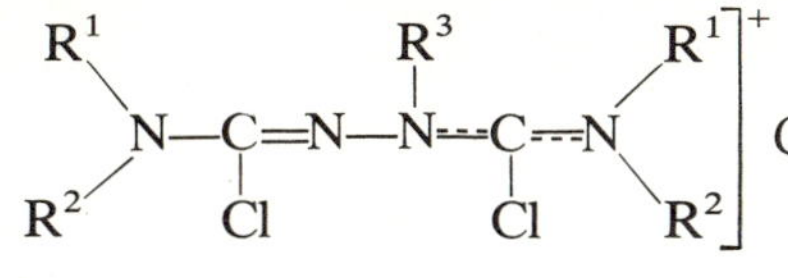

$$\begin{bmatrix} R^1 & R^3 & R^1 \\ \diagdown & | & \diagup \\ N-C=N-N=C=N \\ \diagup & | & | & \diagdown \\ R^2 & Cl & Cl & R^2 \end{bmatrix}^+ Cl^-$$

(111)

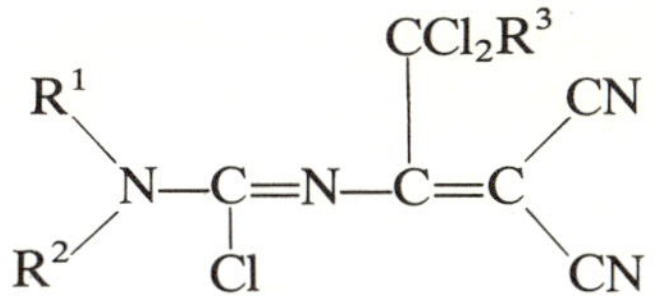

$$\begin{array}{c} CCl_2R^3 \\ R^1 & | & CN \\ \diagdown & | & \diagup \\ N-C=N-C=C \\ \diagup & | & \diagdown \\ R^2 & Cl & CN \end{array}$$

(113)

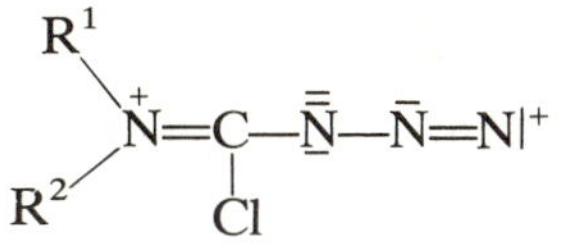

$$\begin{array}{c} R^1 \\ \diagdown + & \bar{=} \\ N=C-\bar{N}-\bar{N}=N|^+ \\ \diagup & | \\ R^2 & Cl \end{array}$$

(127)

$$\begin{bmatrix} CH_3 \\ | \\ \diagup=N-N=C=N \diagdown\diagup N-CH_3 \\ | \\ Cl \end{bmatrix}^+ Cl^-$$

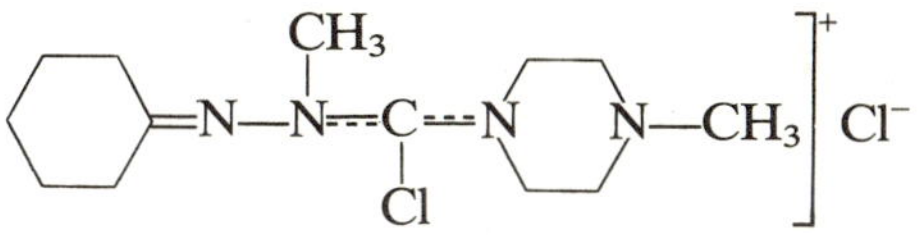

(195)

Bischloroformamidines (**117**) and pentamethine cyanines (**182**) are 1,5-biselectrophiles. Correspondingly, pentamethines (**61,172,174**) are 1,3,5-triselectrophiles:

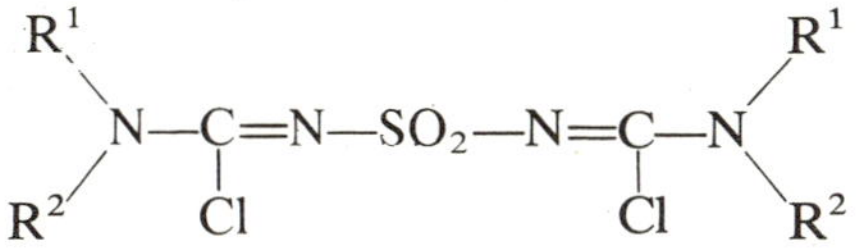

$$\begin{array}{c} R^1 & R^1 \\ \diagdown & \diagup \\ N-C=N-SO_2-N=C-N \\ \diagup & | & | & \diagdown \\ R^2 & Cl & Cl & R^2 \end{array}$$

(117)

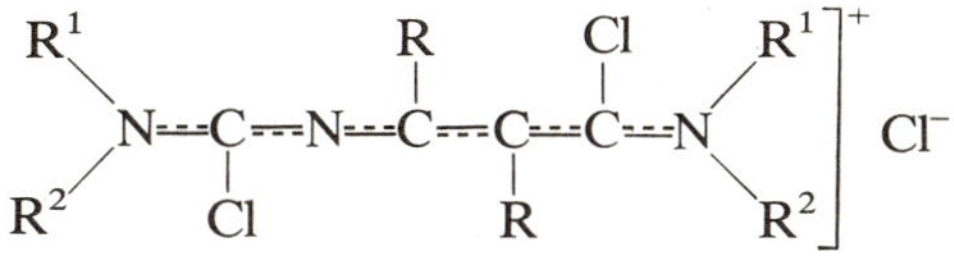

$$\begin{bmatrix} R^1 & R & Cl & R^1 \\ \diagdown & | & | & \diagup \\ N=C=N=C=C=C=N \\ \diagup & | & | & \diagdown \\ R^2 & Cl & R & R^2 \end{bmatrix}^+ Cl^-$$

(182)

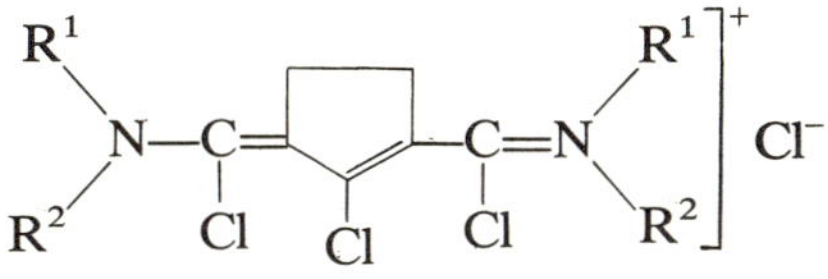

$$\begin{bmatrix} R^1 & R^1 \\ \diagdown & \diagup \\ N-C=\underset{Cl}{\underset{|}{\bigcirc}}=C=N \\ \diagup & | & | & \diagdown \\ R^2 & Cl & Cl & R^2 \end{bmatrix}^+ Cl^-$$

(61)

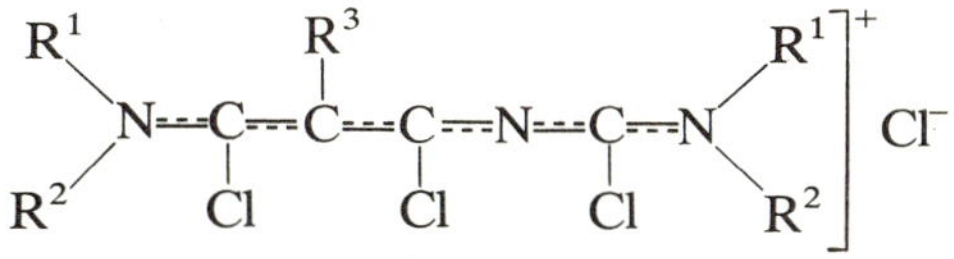

$$\begin{bmatrix} R^1 & R^3 & R^1 \\ \diagdown & | & \diagup \\ N=C=C=C=N=C=N \\ \diagup & | & | & | & \diagdown \\ R^2 & Cl & Cl & Cl & R^2 \end{bmatrix}^+ Cl^-$$

(174)

$$\begin{bmatrix} R^1 & Cl & R^1 \\ \diagdown & | & \diagup \\ N=C=N=C=N=C=N \\ \diagup & | & | & \diagdown \\ R^2 & Cl & Cl & R^2 \end{bmatrix}^+ Cl^-$$

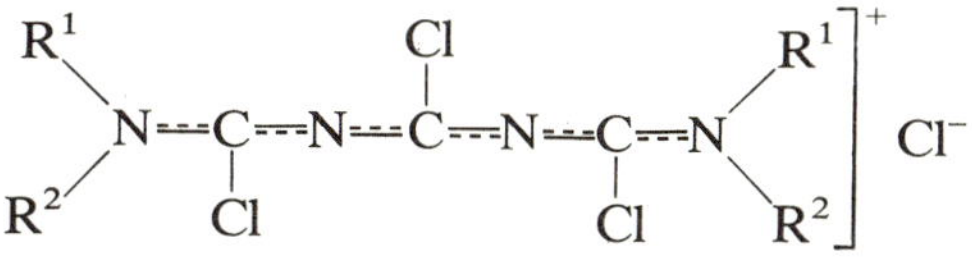

(172)

Although incomplete, this list demonstrates clearly the scope of PI salts as stable but reactive synthons that will continue to find widespread use in synthetic organic chemistry.

REFERENCES

1. N. N. Jarovenko and A. S. Vasileva, *Zh. Obshch. Khim.*, **29,** 3786 (1959); *C. A.*, **54,** 19466 (1960).
2. L. M. Yagupolskij and M. I. Dronkina, *Zh. Obshch. Khim.*, **36,** 1309 (1966); *C. A.*, **65,** 16885 (1966).
3. H. G. Viehe and Z. Janousek, *Angew. Chem.*, **83,** 614 (1971); *Angew. Chem. Int. Ed.*, **10,** 573 (1971).
4. V. P. Kukhar, V. I. Pasternak, and A. V. Kirsanov, *Zh. Org. Khim.*, **7,** 2084 (1971); *C. A.*, **76,** 13713 (1972).
5. L. N. Markovskij, T. N. Dubinina, E. S. Levchenko, V. P. Kukhar, and A. V. Kirsanov, *Zh. Org. Khim.*, **8,** 1822 (1972).
6. K. G. Hancock and D. A. Dickinson, *J. Org. Chem.*, **39,** 331 (1974), and the references cited therein.
7. Z. Janousek, Dissertation, Louvain, 1972.
8. Z. Janousek, J. Gorissen, B. Marin, C. Humblet, G. Duchêne, J. Motte-Collard, B. Stelander, R. Mérényi, and H. G. Viehe, unpublished results.
9. V. P. Kukhar, B. I. Pasternak, and G. V. Pesotzkaja, *Zh. Org. Khim.*, **9,** 39 (1973).
10. U. Schlottmann, Dissertation, Marburg/Lahn, 1972.
11. W. Walter and R. F. Becker, *Liebigs Ann. Chem.*, **755,** 145 (1971).
12. E. Jaul and W. W. Lewis, (Sharples Chem. Inc.), Fr. Pat. 1,085,980 (1954); *C. A.*, **51,** 4418 (1957).
13. Houben-Weyl, Vol. IX, Thieme Verlag, Stuttgart, 1955, p. 830.
14. R. H. Goshorn, W. W. Levis, Jr., E. Jaul, and E. J. Ritter, *Organic Syntheses*, Coll. Vol. IV, N. Rabjohn, Ed., John Wiley and Sons, New York, 1963, p. 307.
15. G. Zumach and E. Kühle, *Angew. Chem.*, **82,** 63 (1970); *Angew. Chem. Int. Ed.*, **9,** 54 (1970).
16. U. Hasserodt, *Chem. Ber.*, **101,** 113 (1968).
17. H. Eilingsfeld and L. Möbius, *Chem. Ber.*, **98,** 1293 (1965); cf. Belg. Pat. 660,941; *C. A.*, **64,** 3354 (1966).
18. R. L. Harris, *Tetrahedron Lett.*, **1970,** 5217.
19. N. H. Nilson, C. Jacobsen, O. N. Sørensen, N. K. Haunste, and A. Senning, *Chem. Ber.*, **105,** 2854 (1972).
20. E. Kühle, *Neuere Methoden der Präparativen Organischen Chemie*, Vol. VI, W. Foerst, Ed., Verlag Chemie, Weinheim, 1970, p. 119; E. Kühle, B. Anders, and E. Zumach, *Angew. Chem.*, **79,** 663 (1967); *Angew. Chem. Int. Ed.*, **6,** 649 (1967).
21. R. W. Addor, *J. Org. Chem.*, **29,** 738 (1964).
22. R. Fuks and W. A. Harteminck, *Bull. Soc. Chim. Belg.*, **82,** 23 (1973); *Tetrahedron*, **29,** 297 (1973), and the references cited therein.
23. E. Allenstein and A. Schmidt, *Chem. Ber.*, **97,** 1286 (1964).
24. V. P. Kukhar, V. I. Pasternak, M. I. Povolotskij, and N. G. Pavlenko, *Zh. Org. Khim.*, **10,** 449 (1974).
25. F. Huys, Mémoire de Licence, Louvain-la-Neuve, 1975.
26. H. Holtschmidt, E. Degener, H. G. Schmelzer, H. Tarnow, and W. Zecher, *Angew Chem.*, **80,** 942 (1968); *Angew. Chem. Int. Ed.*, **7,** 856 (1968).

27. K. Grohe, E. Degener, and H. Holtschmidt, *Liebigs Ann. Chem.*, **730,** 133 (1969).
28. Z. Janousek and B. Le Clef, unpublished results.
29. R. J. Harder and W. C. Smith, *J. Am. Chem. Soc.*, **83,** 3422 (1961).
30. L. N. Markovskij, V. E. Pashinnik, and A. V. Kirsanov, *Synthesis*, **1973,** 787.
31. F. S. Fawcett, C. W. Tullock, and D. D. Coffman, *J. Am. Chem. Soc.*, **84,** 4275 (1962).
32. J. Gorissen, Dissertation, Louvain-la-Neuve, 1977.
33. L. M. Yagupolskij and M. I. Dronkina, *Zh. Obshch. Khim.*, **36,** 1343 (1966).
34. V. P. Kukhar and V. I. Pasternak, *Synthesis*, **1972,** 611.
35. P. George, Dissertation, Louvain-la-Neuve, 1976.
36. B. Stelander, Dissertation, Louvain-la-Neuve, 1976.
37. B. Fontaine, Mémoire de Licence, Louvain-la-Neuve, 1974.
38. A. Bettencourt, Mémoire de Licence, Louvain-la-Neuve, 1973.
39. H. Bredereck and K. Bredereck, *Chem. Ber.*, **94,** 2278 (1961).
40. N. D. Harris, *Synthesis*, **1971,** 220.
41a. C. Jutz and W. Miller, *Angew. Chem.*, **78,** 1059 (1966); *Angew. Chem. Int. Ed.*, **5,** 1042 (1966).
41b. F. G. Fick and K. Hartke, *Tetrahedron Lett.*, **36,** 3105 (1974).
42. A. Bettencourt, Z. Janousek, and H. G. Viehe, unpublished results.
43. H. J. Gais, K. Hafner, and M. Neuenschwander, *Helv. Chim. Acta*, **52,** 2641 (1969).
44. A. Niederhauser and M. Neuenschwander, *Helv. Chim. Acta*, **56,** 1331, 2427 (1973).
45. H. G. Viehe, Z. Janousek, and M. A. Deffrenne, *Angew. Chem.*, **83,** 616 (1971); *Angew. Chem. Int. Ed.*, **10,** 575 (1971).
46. J. Baum, unpublished results.
47. Z. Arnold and J. Zemlicka, *Proc. Chem. Soc.*, **1956,** 227; *Collect. Czech. Chem. Commun.*, **24,** 2385 (1959).
48. M. A. Deffrenne, Dissertation, Louvain-la-Neuve, 1976.
49. Z. Janousek and H. G. Viehe, *Angew. Chem.*, **83,** 615 (1971); *Angew. Chem. Int. Ed.*, **8,** 574 (1971).
50. H. Bredereck, F. Effenberger, and H. P. Bayerlin, *Chem. Ber.*, **97,** 3076 (1964); H. Bredereck, F. Effenberger, and G. Simchen, *Angew. Chem.*, **73,** 493 (1961).
51. K. Hafner, K. F. Bangert, and V. Orfanos, *Angew. Chem.*, **79,** 414 (1967); *Angew. Chem. Int. Ed.*, **6,** 451 (1967).
52. H. Weingarten and W. A. White, *J. Org. Chem.*, **31,** 2874 (1966).
53. R. Buyle and H. G. Viehe, *Tetrahedron*, **24,** 4217 (1968).
54. Z. Arnold, *Collect. Czech. Chem. Commun.*, **26,** 3851 (1961).
55. G. J. de Voghel, T. L. Eggericks, Z. Janousek, and H. G. Viehe, *J. Org. Chem.*, **39,** 1233 (1974).
56. B. Gowenko, Mémoire de Licence, Louvain-la-Neuve, 1973.
57. L. A. Lazukina and V. P. Kukhar, *Khim. Geterosikl. Soedin.*, **6,** 771 (1974); *Chem. Inform.*, **40,** 430 (1974).
58. M. Huys, Mémoire de Licence, Louvain-la-Neuve, 1974.
59. M. Seefelder, *Chem. Zentralbl.*, **1961,** 5620.
60. H. Eilingsfeld, M. Seefelder, and H. Weidinger, *Chem. Ber.*, **96,** 2899 (1963).
61. C. Fripiat, Mémoire de Licence, Louvain-la-Neuve, 1974.
62. G. J. de Voghel, Mémoire de Licence, Louvain, 1971.
63. B. Caillaux, P. George, F. Tataruch, Z. Janousek, and H. G. Viehe, *Chimia* (in press).
64. H. G. Viehe, G. J. de Voghel, and F. Smets, *Chimia*, **30,** 189 (1976).
65. H. G. Viehe, Z. Janousek, R. Gompper, and D. Lach, *Angew. Chem.*, **85,** 581 (1973); *Angew. Chem. Int. Ed.*, **12,** 566 (1973).
66. H.-U. Wagner, *Chem. Ber.*, **107,** 634 (1974).

67. E. Oeser, *Chem. Ber.*, **107,** 627 (1974).
68. G. J. De Voghel, Dissertation, Louvain-la-Neuve, 1976.
69. G. Duchêne, Dissertation, Louvain-la-Neuve, 1976.
70. B. C. Challis and J. A. Challis, in *The Chemistry of Amides*, J. Zabicky, Ed., Interscience, London, 1970, p. 805.
71. V. P. Kukhar, A. M. Pinchuk, and M. V. Shevchenko, *Zh. Org. Khim.*, **9,** 43 (1973).
72. V. P. Kukhar, M. V. Shevchenko, and N. A. Kirsanova, *Zh. Org. Chem.*, **9,** 1815 (1973).
73. Z. Janousek and W. Lwowski, unpublished results.
74. F. Hervens, Dissertation, Louvain-la-Neuve, 1976.
75. N. D. Bondarchuk, V. V. Momot, L. A. Lazukina, G. V. Pesotskaja, and V. P. Kukhar, *Zh. Org. Khim.*, **10,** 735 (1974).
76. V. K. Kukhar, N. G. Pavlenko, and A. V. Kirsanov, *Zh. Org. Khim.*, **9,** 305 (1973).
77. E. Grigat, Ger. Pat. 2,042,255; *C. A.*, **76,** 140257 (1972).
78. B. Le Clef, Dissertation, Louvain-la-Neuve, 1976.
79. D. A. Tomalia, T. J. Giacobbe, and W. A. Springer, *J. Org. Chem.*, **36,** 2142 (1971).
80. J. A. Strepikheev, T. G. Perlova, and L. A. Zhivekova, *Zh. Org. Khim.*, **4,** 1891 (1968); *C. A.*, **70,** 28207 (1969).
81. E. Kühle and R. Wegler, Fr. Pat. 81,999; *C. A.*, **60,** 11955 (1964).
82. J. H. Ottenheym and J. W. Garritsen, Brit. Pat. 901,169 (1962); *C. A.*, **58,** 6810 (1963); Ger. Pat. 1,157,210 (1963); *C. A.*, **60,** 6756 (1964). For a review on phosgene see H. Babad and A. G. Zeiler, *Chem. Rev.*, **73,** 75 (1973).
83. Z. Janousek, J. Collard, and H. G. Viehe, *Angew. Chem.*, **84,** 993 (1972); *Angew. Chem. Int. Ed.*, **11,** 917 (1972).
84. E. Goffin, Dissertation, Louvain-la-Neuve, 1974.
85. Y. Legrand, Dissertation, Louvain-la-Neuve, 1977.
86a. P. George and H. G. Viehe, *Chimia* **29,** 209 (1975).
86b. A. Schmidt, *Chem. Ber.*, **100,** 3725 (1967).
87. V. P. Kukhar, V. I. Pasternak, and A. V. Kirsanov, *Zh. Obshch. Khim.*, **42,** 1169 (1972).
88. DOS 2,237,879 (Henkel Co); *Angew. Chem.*, **86,** 821 (1974); *Angew. Chem. Int. Ed.*, **13,** 747 (1974).
89. B. Le Clef, J. Mommaerts, B. Stelander, and H. G. Viehe, *Angew. Chem.*, **85,** 445 (1973); *Angew. Chem. Int. Ed.*, **12,** 404 (1973).
90. W. T. Borden, P. W. Concannon, and D. I. Phillips, *Tetrahedron Lett.*, **1973,** 3161.
91. V. P. Kukhar, L. A. Lazukina, and A. V. Kirsanov, *Zh. Org. Khim.*, **9,** 304 (1973).
92. Cf. S. Masson and A. Thuillier, *C. R. Acad. Sci. Paris*, **273,** 251 (1971).
93. V. P. Kukhar and V. I. Pasternak, *Synthesis*, **1974,** 563.
94. H. E. Nikolajewski, S. Dähne, and B. Hirsch, *Chem. Ber.*, **100,** 2616 (1967).
95. For a review see *Chemistry of Acetylenes*, H. G. Viehe, Ed., Marcel Dekker, New York, 1969, p. 861.
96. Y. Ito, S. Katsuragawa, M. Okano, and R. Oda, *Tetrahedron*, **23,** 1961, 2159 (1967).
97. Th. Michel-Van Vyve, unpublished results.
98. H. G. Viehe, Th. Van Vyve, and Z. Janousek, *Angew. Chem.*, **84,** 991 (1972); *Angew. Chem. Int. Ed.*, **11,** 916 (1972).
99. J. Weber, Mémoire de Licence, Louvain-la-Neuve, 1973.
100. J. Mommaerts, Mémoire de Licence, Louvain-la-Neuve, 1970.
101. Z. Janousek and H. G. Viehe, *Angew. Chem.*, **85,** 90 (1973); *Angew. Chem. Int. Ed.*, **10,** 74 (1973).
102. C. Humblet, Mémoire de Licence, Louvain, 1971.

103a. H. J. Gais and K. Hafner, *Tetrahedron Lett.*, **1974,** 771.
103b. J. E. Oliver, S. C. Chang, R. T. Brown, J. B. Stokes, and A. B. Borkovec, *J. Med. Chem.*, **15,** 315 (1972).
104. K. Bredereck and R. Richter, *Chem. Ber.*, **99,** 2465 (1966).
105. Z. Csürös, R. Soos, A. Anfus-Ercsényi, I. Bitter, and J. Tamas, *Acta Chim. Hung.*, **76,** 81 (1973); *ibid.*, **77,** 443 (1973); *ibid.*, **78,** 409, 419 (1973).
106. L. I. Samaraj, O. W. Wishnewskij, and G. I. Derkatsch, *Angew. Chem.*, **80,** 620 (1968); *Angew. Chem. Int. Ed.*, **7,** 621 (1968).
107. A. Elgavi and H. G. Viehe, *Angew. Chem.* (in press).
108. T. Nakai, Y. Ueno, and M. Okawara, *Bull. Soc. Chem. Jap.*, **43,** 3175 (1970).
109. F. Hervens and H. G. Viehe, *Angew. Chem.*, **85,** 446 (1973); *Angew. Chem. Int. Ed.*, **12,** 405 (1973).
110. Th. Van Vyve and H. G. Viehe, *Angew. Chem.*, **86,** 45 (1974); *Angew. Chem. Int. Ed.*, **13,** 79 (1974).
111. This increase in reactivity could not be confirmed by the present authors. On the contrary, all the PI salts where X^- are complex anions are very unreactive.
112. L. N. Markovski and V. E. Pashinnik, *Synthesis*, **1975,** 801.
113. W. J. Middleton, *J. Org. Chem.*, **40,** 574 (1975).
114. The X-ray study shows that phenylhydrazine forms the expected isomer (**48**).
115. J. S. Baum and H. G. Viehe, *J. Org. Chem.*, **41,** 183 (1976).
116. G. J. de Voghel, T. L. Eggericks, B. Clamot, and H. G. Viehe, *Chimia*, **30,** 191 (1976).
117. V. P. Kukhar and V. I. Pasternak, *Zh. Org. Khim.*, **11,** 2223 (1975).
118. G. Germain, J. P. Declerq, M. van Meersche, F. Hervens, and H. G. Viehe, *Bull. Soc. Chim. Belg.*, **84,** 1005 (1975).
119. V. P. Kukhar and M. V. Shevchenko, *Zh. Org. Khim.*, **11,** 71 (1975).
120. N. D. Bondarchuk, V. V. Momot, and B. B. Gavrilenko, *Zh. Obshch. Khim.*, **45,** 873 (1975).
121. V. P. Kukhar and N. G. Pavlenko, *Zh. Org. Khim.*, **10,** 36 (1974).
122. J. Gorissen, Z. Janousek, and H. G. Viehe, *Isr. J. Chem.* (in press).
123. L. N. Markovskij, J. G. Shermolovich, V. I. Gorbatenko, and V. I. Shevchenko, *Zh. Org. Khim.*, **11,** 751 (1975).
124. A. A. Svishchuk, N. K. Machnovskij, N. V. Sukharenko, and E. D. Basalkevitch, *Ukr. Khim. Zh.*, **41,** 510 (1975).
125. F. G. Fick and K. Hartke, *Tetrahedron Lett.*, **1975,** 1133.
126. For a review on Reactive Carbonic Acid Derivatives see E. Kühle, *Angew. Chem. Int. Ed.*, **12,** 630 (1973).

α-HALOENAMINES AND KETENIMINIUM SALTS

By L. GHOSEZ and J. MARCHAND-BRYNAERT, *Laboratoire de Chimie Organique de Synthèse, Université de Louvain, Place Louis Pasteur, 1, B-1348-Louvain-la-Neuve, Belgium*

CONTENTS

- I. Introduction 422
- II. General Summary of Structure and Reactivity 424
- III. Synthesis of α-Haloenamines 427
    - A. General Aspects 427
    - B. α-Chloroenamines by Elimination Reactions 428
        - 1. From Amide Chlorides Derived from α-Disubstituted Acetamides and α-Substituted Lactams 428
        - 2. From Amide Chlorides Derived from α-Monosubstituted Acetamides and Lactams 432
        - 3. From Amide Chlorides Derived from Phosgeniminium Salts and Activated C—H Bonds 435
        - 4. From α,β-Dihaloamines 436
    - C. α-Haloenamines from Electrophilic Addition to Ynamines 437
        - 1. H—X, Alk—X and X_2 Additions to Ynamines 437
        - 2. Acylation of Ynamines 440
        - 3. Addition of Sulfenyl Chlorides to Ynamines 441
    - D. α-Haloenamines by Nucleophilic Additions to Haloacetylenes 442
    - E. α-Haloenamines by Substitution Reactions 442
        - 1. Halovinylation of Metal Amides 442
        - 2. Substitution on α-Heterosubstituted Enamines 444
    - F. Miscellaneous Methods 447
        - 1. Reaction of Trivalent Phosphorus Compounds with α-Substituted Dichloroacetamides 447
        - 2. Reactions of Phenyl(trichloromethyl)mercury with Secondary Amines and Some Allyl Amines 449
        - 3. Pyrolysis of Perfluorocyclobutane Derivatives 450
        - 4. Phosphorylation of Tricyanomethane Derivatives 450
- IV. Structural Data and General Properties of α-Haloenamines 450
    - A. Stability and Solubility Properties 450
    - B. Structural Data on α-Haloenamines 465
    - C. Geometrical Isomerism in α-Haloenamines 467
    - D. Structure of Keteniminium Salts 468
- V. Nucleophilic Substitutions on α-Haloenamines 470
    - A. Mechanistic Aspects 470

B. Nucleophilic Substitutions with Formation of C–Heteroatom Bonds 473
  1. Reactions with Hydroxide, Alkoxide, and Thiolates 473
  2. Reactions with Carboxylate Anions 475
  3. Halide Exchange 477
  4. Reactions with Amines 478
  5. Reactions with Azide Ions 482
C. Nucleophilic Substitutions with Formation of C–C Bonds: the Aminoalkenylation Reactions 484
  1. Reactions with Cyanide Ion 484
  2. Reactions with Ambident Anions 486
  3. Reactions with Grignard Reagents and Organolithium Compounds 489
  4. Reactions with Aromatic Compounds 490
D. Oxidation Reactions 495
VI. Reactions of α-Haloenamines with Electrophiles 496
A. General Aspects 496
B. Hydration 496
C. Addition of Hydrogen Halides 497
D. Halogenation 497
E. Reaction with Alcohols and Carboxylic Acids 498
F. Acylation of α-Haloenamines 500
  1. Reaction with Acid Chloride 501
  2. Reaction with Amide Chlorides 503
  3. Reactions with Heterocumulenes 504
VII. Elimination to Ynamines 507
VIII. Cycloaddition Reactions of Keteniminium Ions 508
A. General Aspects 508
B. Cycloadditions to Olefins 510
C. Cycloadditions to Conjugated Dienes 514
D. Cycloadditions to Allenes 519
E. Cycloadditions to Acetylenes 519
F. Cycloadditions to Imines 522
G. Diels-Alder Reactions of Alkenylketeniminium Ions 522
IX. Transition-Metal Complexes from α-Chloroenamines 524
X. Entertainment! 525
References 528

I. Introduction

Keteniminium salts form a class of heterocumulenes that are characterized by the cumulative arrangement of iminium and olefin functional groups:

$$\text{>C=C=}\overset{+}{\text{N}}\text{<} \quad X^- \qquad \text{>C=C=}\bar{\text{N}}\text{<}$$

Keteniminium salt   Ketenimine

According to the IUPAC rules for the nomenclature of organic compounds, these heterocumulenes should be termed "1-alkylideneammonium ions." However, throughout this chapter, we shall

I. INTRODUCTION

use the name "keteniminium" in accordance with the designation adopted by *Chemical Abstracts* for the related ketenimines.

Establishment of the chemistry of keteniminium salts is of fairly recent origin. Until a few years ago they were considered only as transient entities, and their existence had been inferred mainly from studies on the course of reactions of ketenimines (1) and ynamines (2) with electrophilic reagents but also from the products of N-alkylation of alkylidene bis-dimethylamine (3) (Scheme 1). In these reactions, however, either an external nucleophile or the electron-rich starting materials immediately consume the highly electrophilic keteniminium ions, thus precluding their isolation or even their identification.

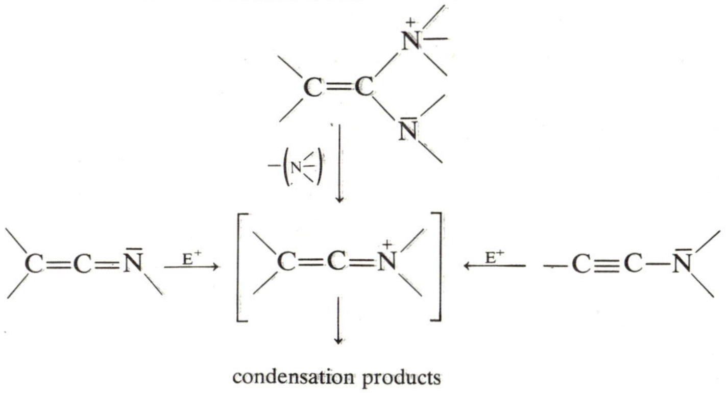

Scheme 1

In 1967 we became interested in these new heterocumulenes because of their potential to react in [2+2] cycloadditions as "activated ketenes." It occurred to us that α-haloenamines, formally the enamine derivatives of carboxylic acid halides, might react as keteniminium halides and thus become practical sources of keteniminium salts:

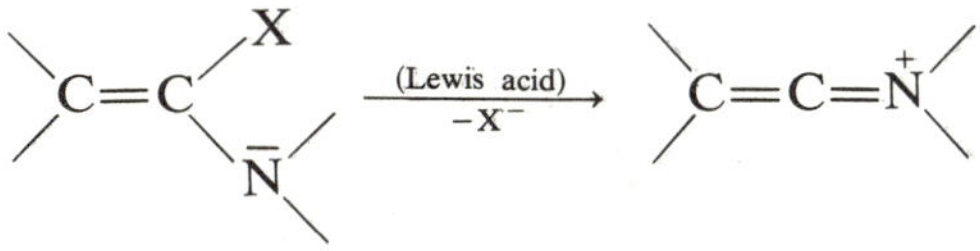

X = F, Cl, Br, I

It is quite remarkable that this reaction principle had not been explored before in spite of the fact that the first α-chloroenamines were prepared 46 years ago by von Braun and Heymans (4). Further examples of α-haloenamines were described later, in particular by the Speziale group (5) in the United States and by Yakubovich et al. (6) in the U.S.S.R.

There is ample evidence that at the present time α-chloroenamines are the most practical sources of keteniminium salts. Even in the absence of Lewis acids, they display high reactivity toward nucleophilic reagents and should thus be regarded as keteniminium chlorides. Other α-haloenamines have been studied much less but seem to exhibit similar chemical behavior.

A full account is given here of the chemistry of α-haloenamines and the keteniminium salts derived therefrom. The substantial progress already made demonstrates the theoretical and synthetic interest of these reactive substances, although the details of many reactions must still be investigated.

We realize that a review of keteniminium chemistry should include a survey of the reactions of ketenimines and ynamines. The breadth of the subject, however, forces us to be arbitrary. Since comprehensive reviews (1,2) already published on these fascinating reagents provide a full account of the experimental data, these aspects will not be reviewed here; only reactions of comparative interest will be mentioned.

The literature has been reviewed up to early 1975.

II. General Summary of Structure and Reactivity

The cumulative arrangement of iminium and olefinic double bonds results in a unique set of properties that are interesting to compare with those of simpler iminium ions, as well as of related heterocumulenes.

Keteniminium ions can be regarded as alkenyl carbocations, which are strongly stabilized by electron donation from the nonbonded electron pair of a nitrogen atom:

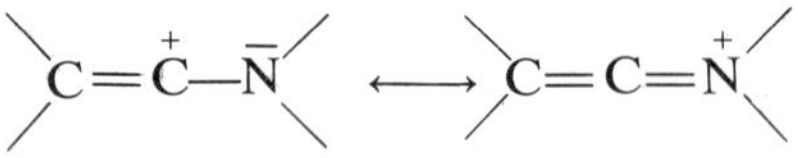

$$\diagup\hspace{-0.2em}C=\overset{+}{\underset{}{C}}-\bar{N}\diagdown \longleftrightarrow \diagup\hspace{-0.2em}C=C=\overset{+}{N}\diagdown$$

It is now evident that they are not only the most reactive electrophiles among related heterocumulenes but also by far the most electrophilic of all iminium ions (Scheme 2).

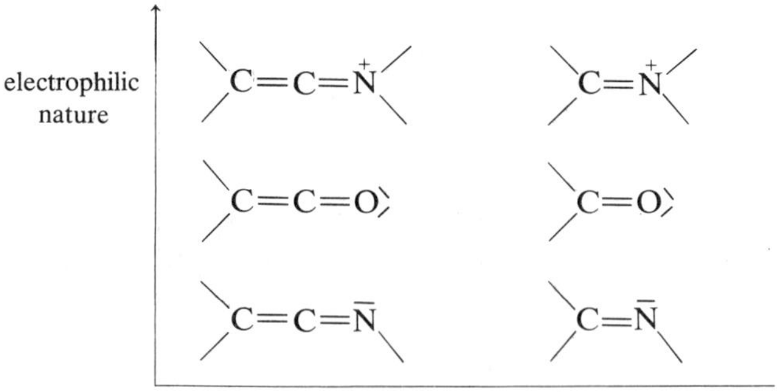

Scheme 2

II. GENERAL SUMMARY OF STRUCTURE AND REACTIVITY

Several factors are responsible for this high reactivity of keteniminium ions toward nucleophilic reagents:

1. Carbenium centers that are sp hybridized are more electrophilic than the corresponding sp^2 centers.
2. The approach of a nucleophilic reagent toward the sp carbon atom in the keteniminium ion is less hindered than the approach toward a trigonal carbon atom in other iminium ions.
3. Nucleophilic addition on a keteniminium ion brings a nitrogen lone pair and eventually a lone pair of the nucleophilic reagent into conjugation with the C–C π bond:

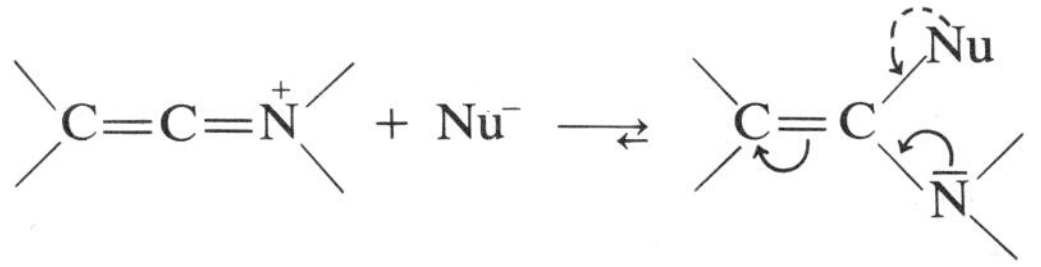

$$\mathrm{\searrow\!\!C\!\!=\!\!C\!\!=\!\!\overset{+}{N}\!\!\nearrow} + \mathrm{Nu^-} \rightleftarrows \mathrm{\searrow\!\!C\!\!=\!\!C\!\!\nearrow^{Nu}_{\searrow N\nearrow}}$$

This enhanced electrophilic nature of keteniminium cations as compared with other iminium ions has some very important consequences. Direct observation of stable long-lived keteniminium salts is possible only by the use of very weakly nucleophilic counterions (BF_4^-, $ZnCl_3^-$, PF_6^-, ...) in inert solvents (7–9). In addition, keteniminium chlorides can be formulated as reactive intermediates whose covalent tautomers, α-chloroenamines, predominate at equilibrium. This observation is in contrast with the behavior of methylene-, alkylidene-, chloromethylene-, and dichloromethyleneammonium chlorides, where the ionic forms are usually thermodynamically more stable (10,11) (Scheme 3). The equilibrium is also displaced in favor of the covalent structure in the case of keteniminium fluorides that exist as α-fluoroenamines. This again seems to be the case for the bromides and iodides.

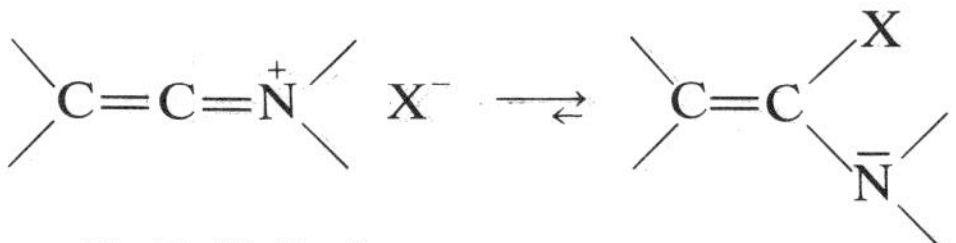

$$\mathrm{\searrow\!\!C\!\!=\!\!C\!\!=\!\!\overset{+}{N}\!\!\nearrow}\ X^- \rightleftarrows \mathrm{\searrow\!\!C\!\!=\!\!C\!\!\nearrow^{X}_{\searrow N\nearrow}}$$

X = F, Cl, Br, I

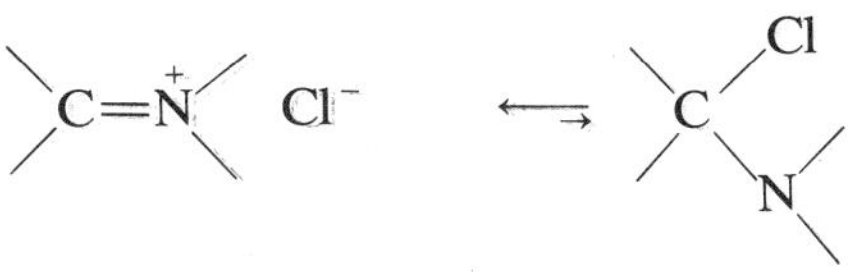

$$\mathrm{\searrow\!\!C\!\!=\!\!\overset{+}{N}\!\!\nearrow}\ Cl^- \longleftarrow \mathrm{\searrow\!\!C\!\!\nearrow^{Cl}_{\searrow N\nearrow}}$$

Scheme 3

Even if the covalent structure is more stable, the possibility of an equilibrium with a keteniminium halide confers on α-haloenamines a versatile chemical behavior and increases considerably their reaction potential (Scheme 4). As enamines derived from a carboxylic acid halide, α-haloenamines are expected to react with an electrophilic reagent

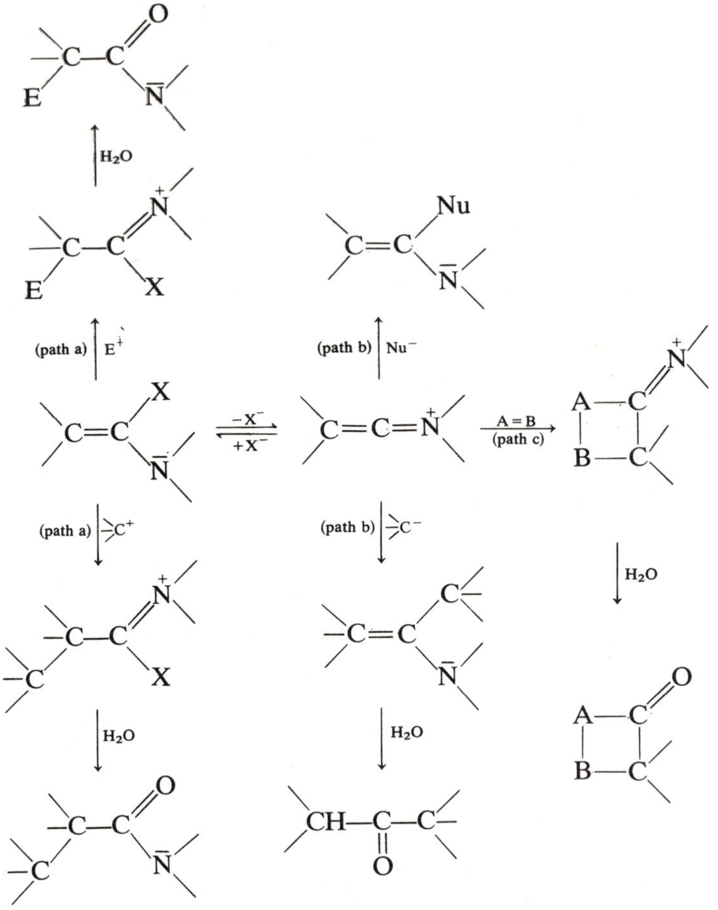

Scheme 4

(E^+, $>C^+$) on C_2 as well as on nitrogen. Further hydrolysis then yields a carboxamide substituted at the α position (path a). On the other hand, ionization to a keteniminium salt generates an electrophilic center at C_1 which might react with various nucleophilic reagents (Nu^-, $-C^-$) to yield 1-substituted enamines (path b). This synthetic process has been called "aminoalkenylation." When a carbon nucleophile is used, the sequence "aminoalkenylation–hydrolysis" yields ketones. Finally, as reactive heterocumulenes, keteniminium salts undergo extremely facile [2+2] cycloadditions to various types of unsaturated substrates (A═B), permitting easy preparation of four-membered rings (path c).

Although many reactions remain to be investigated, it is already obvious that α-haloenamines and their keteniminium tautomers are versatile "building blocks" that will be extremely useful in organic synthesis.

III. Synthesis of α-Haloenamines

A. GENERAL ASPECTS

A necessary requirement for a practical use of a new class of reagents is availability in good yields and large quantities from cheap and readily obtainable materials.

Methods that meet these requirements are now available for the preparation of α-haloenamines, but none of them is entirely general. Most of these methods are based on three different synthetic principles:

1. *Elimination* reactions from amide halides.
2. *Addition* to acetylenic triple bonds.
2. *Substitution* on olefinic double bonds.

The selection of the synthetic route will depend on the desired structure and, in particular, on the substitution at the C_2 atom of the enamine bond.

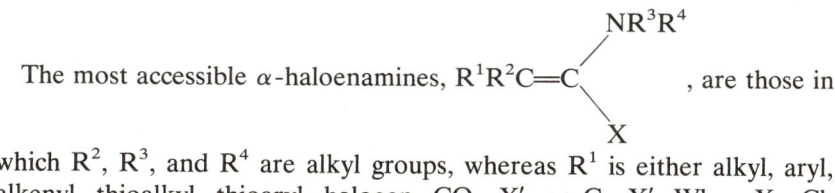

The most accessible α-haloenamines, R^1R^2C═$C\begin{smallmatrix}NR^3R^4\\X\end{smallmatrix}$, are those in which R^2, R^3, and R^4 are alkyl groups, whereas R^1 is either alkyl, aryl, alkenyl, thioalkyl, thioaryl, halogen, CO—X′, or C—X′. When X = Cl, they can be prepared by *HCl elimination* from amide chlorides, which are readily available from tertiary amines. In most cases these α-chloroenamines can easily be converted to the corresponding α-fluoro-, α-bromo-, or α-iodoenamines. Unfortunately, this simple and effective method does not apply as well to the preparation of α-chloroenamines bearing a hydrogen substituent at C_2 (R^1 = H), while R_3 and R_4 are alkyl groups. The latter are thermally less stable and often condense under the experimental conditions. Therefore, in practice, it is often advisable to prepare them *in situ* or, preferably, to replace them by equivalent synthons, such as α-chloroenamines in which R_3 is an aryl group or the hydrogen at C_2 is temporarily replaced by halogen, thioalkyl, or carbonyl substituents.

α-Haloenamines bearing electron-attracting substituents at C_2 (—CO—, —SO_2—, —CS—, R—S—) can be prepared by electrophilic *additions* to ynamines. This method is sometimes handicapped, however, by the cost and difficulty encountered in preparing the starting materials. Many of these stabilized α-chloroenamines are more accessible from the reaction of dichloromethyleneammonium salts or from phosgene on substituted acetamides or malonic derivatives.

The potentially useful synthesis by direct *substitution* of halogen on 1,1-dihaloalkenes is rather limited in scope. Nevertheless it permits the preparation of the reactive α,β-dichlorovinylamines and several α-fluoro-β-dihalovinylamines.

Trichlorovinylamines, which were the first α-haloenamines thoroughly studied, are obtained from the reaction of trichloroacetamides with trialkyl phosphites or tertiary phosphines or from dichloroacetamides with pyrocatechol phosphortrichloride.

In the next sections we shall review in greater detail the methods of synthesis of α-haloenamines outlined above. Structurally related compounds such as α-haloenamides (12), α-haloenureas (13,14), α-haloenisocyanates (14,15), 3-chloroisoquinoline derivatives (16) and others (17,18), which show a different chemical behavior, will not be treated here.

B. α-CHLOROENAMINES BY ELIMINATION REACTIONS

1. *From Amide Chlorides Derived from α-Disubstituted Acetamides and α-Substituted Lactams*

A brief experimental report by von Braun and Heymans (4) in 1929 provides the basis of the presently most useful method of synthesis of alkyl- and aryl-substituted α-chloroenamines (19). This method involves first the formation of an amide chloride from a tertiary amide and a "chlorinating" reagent ($COCl_2$, PCl_5), followed by thermal- or base-induced dehydrochlorination.

The first representative α-chloroenamine (**2**) was obtained (4) from the reaction of PCl_5 on N,N-diethyl-1-methylbutyramide (**1**), followed by slow distillation of the amide chloride to effect the elimination of hydrogen chloride:

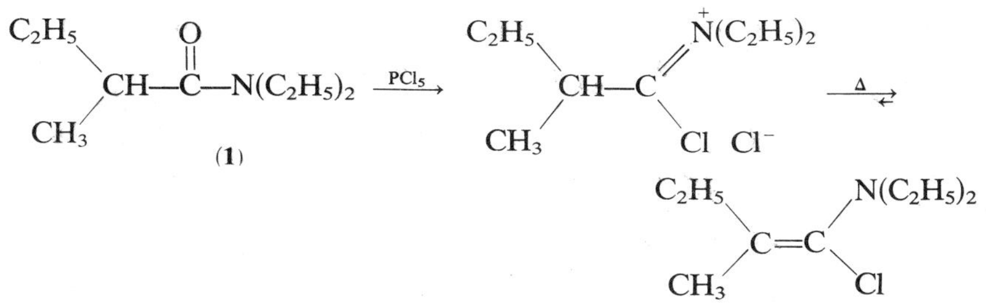

III. SYNTHESIS OF α-HALOENAMINES

$$\underset{(1)}{\underset{CH_3}{\overset{C_2H_5}{>}}CH-\overset{O}{\overset{\|}{C}}-N(C_2H_5)_2} \xrightarrow{PCl_5} \underset{CH_3}{\overset{C_2H_5}{>}}CH-\overset{\overset{+}{N}(C_2H_5)_2}{\underset{Cl}{\overset{\|}{C}}} Cl^- \xrightarrow{\Delta} \underset{(2)}{\underset{CH_3}{\overset{C_2H_5}{>}}C=C\underset{Cl}{\overset{N(C_2H_5)_2}{<}}}$$

However, prolonged heating of the reaction mixture leads to partial formation of the α-chloroimide chloride (**3**), resulting from the chlorination of **2** followed by the loss of ethyl chloride (20):

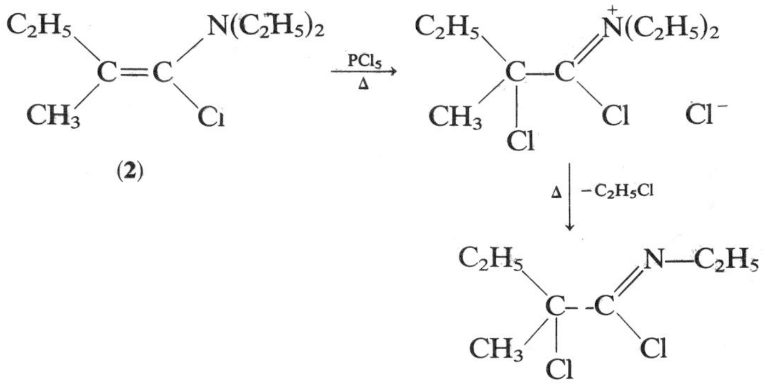

$$\underset{(2)}{\underset{CH_3}{\overset{C_2H_5}{>}}C=C\underset{Cl}{\overset{N(C_2H_5)_2}{<}}} \xrightarrow[\Delta]{PCl_5} \underset{CH_3}{\overset{C_2H_5}{>}}\underset{Cl}{\overset{|}{C}}-\overset{\overset{+}{N}(C_2H_5)_2}{\underset{Cl}{\overset{|}{C}}} Cl^-$$

$$\Delta \downarrow -C_2H_5Cl$$

$$\underset{(3)}{\underset{CH_3}{\overset{C_2H_5}{>}}\underset{Cl}{\overset{|}{C}}-\overset{N-C_2H_5}{\underset{Cl}{\overset{\|}{C}}}}$$

Further chlorination may be avoided by replacing phosphorus pentachloride by pyrocatechol phosphortrichloride, a reagent introduced by Gross and Gloede (21) in 1963. Trichlorovinyl-*N*-diethylamine (**5**) was obtained (22) in this manner from the corresponding dichloroacetamide (**4**):

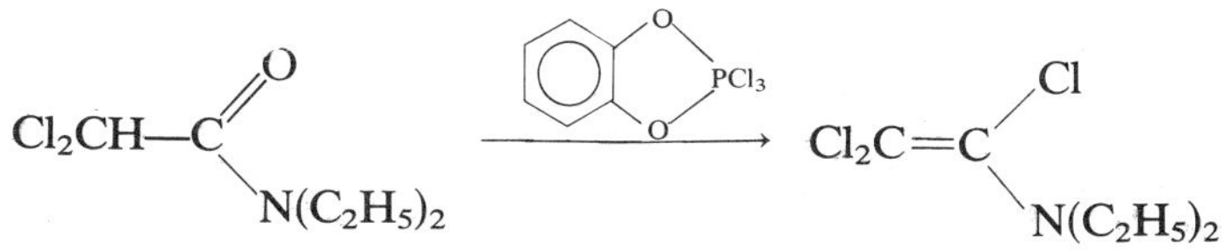

$$\underset{(4)}{Cl_2CH-C\underset{N(C_2H_5)_2}{\overset{O}{<}}} \xrightarrow{\text{(catechol)}PCl_3} \underset{(5)}{Cl_2C=C\underset{N(C_2H_5)_2}{\overset{Cl}{<}}}$$

A mixture of *N,N*-dimethylisobutyramide, triethylamine, and phosphorus oxychloride also results (20) in the formation of the corresponding α-chloroenamine (**6**):

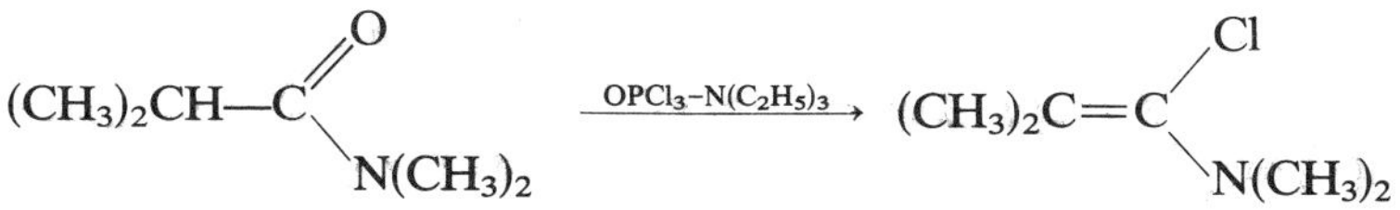

$$(CH_3)_2CH-C\underset{N(CH_3)_2}{\overset{O}{<}} \xrightarrow{OPCl_3-N(C_2H_5)_3} \underset{(6)}{(CH_3)_2C=C\underset{N(CH_3)_2}{\overset{Cl}{<}}}$$

Amide chlorides are even more easily obtained by the reaction of phosgene with tertiary amides at room temperature, a method first described by Hallmann (23a) in 1876 and well documented (23b) since, specially by Eilingsfeld and Seefelder (24). These amide chlorides are obtained in excellent yields from phosgene and methylene chloride solutions of amides, $R^1R^2CHCONR^3R^4$, where R^2, R^3, and R^4 are alkyl groups and R^1 is alkyl, alkenyl, aryl, or chlorine. When R^1 is a thioalkyl or thioaryl group, the amide solution is first saturated with HCl and the phosgenation is conducted in the presence of 0.1 equivalent of dimethylformamide (59). The amide chlorides split off HCl readily to give the corresponding α-chloroenamines.

The thermal dehydrochlorination is usually conducted in a Soxhlet apparatus by refluxing a suspension of the amide chlorides in toluene or xylene and passing the condensed vapor through calcium oxide. However, this elimination step is effected more practically and usually in better yields by treatment of a methylene chloride solution of the amide chloride with a tertiary base such as trimethyl- or triethylamine, followed by precipitation of the amine hydrochloride with petroleum ether (25). This

(**6**) 75–85%

(2 isomers)

(**7** path a) 66%
(path b) 91%

(**8**) 46%

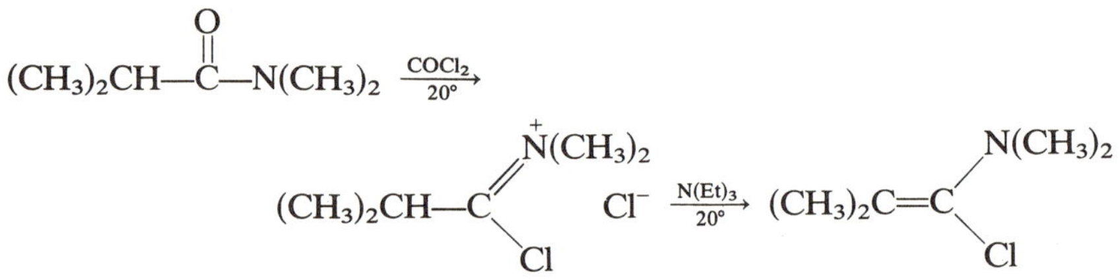

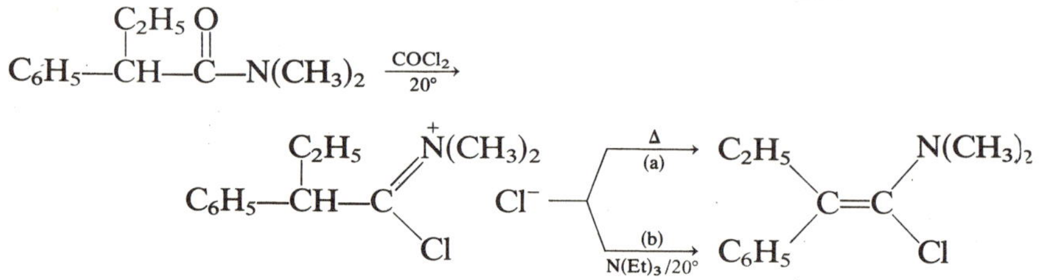

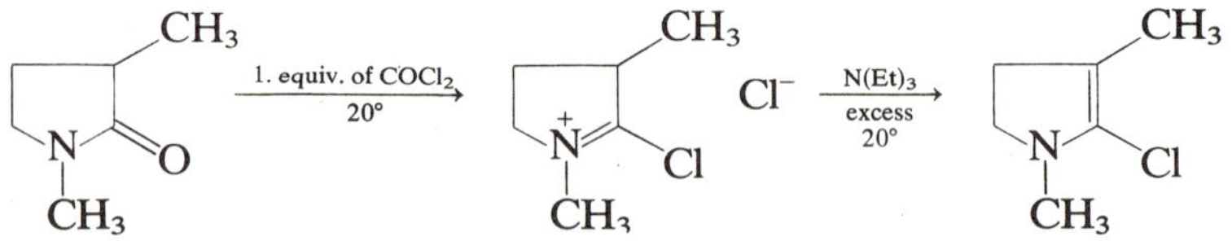

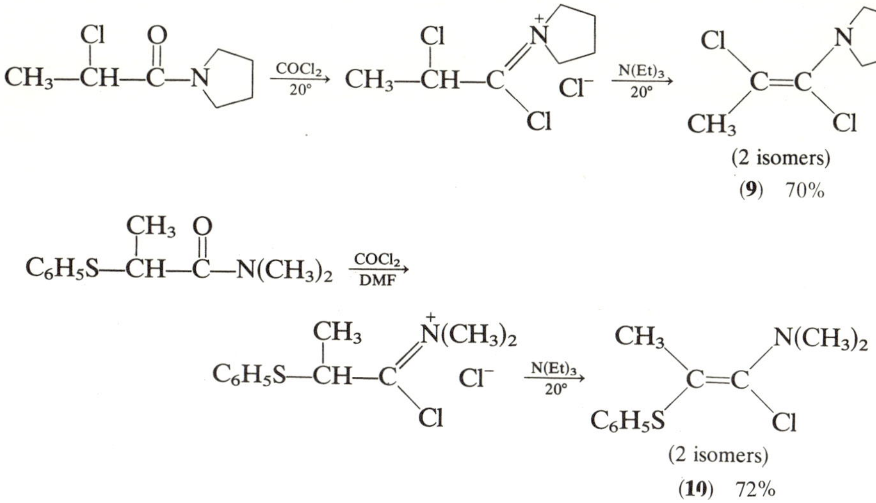

III. SYNTHESIS OF α-HALOENAMINES

[Scheme showing reaction of CH₃—CHCl—CO—N(pyrrolidine) with COCl₂ at 20° to give CH₃—CHCl—C(=N⁺pyrrolidine)Cl₂⁻ then with N(Et)₃ at 20° to give (CH₃)(Cl)C=C(pyrrolidine)(Cl)]

(2 isomers)
(**9**) 70%

[Scheme showing C₆H₅S—CH(CH₃)—CO—N(CH₃)₂ with COCl₂/DMF to give C₆H₅S—CH(CH₃)—C(=N⁺(CH₃)₂)Cl · Cl⁻ then with N(Et)₃ at 20° to give (CH₃)(C₆H₅S)C=C(N(CH₃)₂)(Cl)]

(2 isomers)
(**10**) 72%

Scheme 5

method permits the convenient preparation of large quantities of α-chloroenamines, usually in excellent yields (26) (Scheme 5). A vinyl-α-chloroenamine (**11**), a reactive derivative of isoprene, was prepared in the same way (27,28):

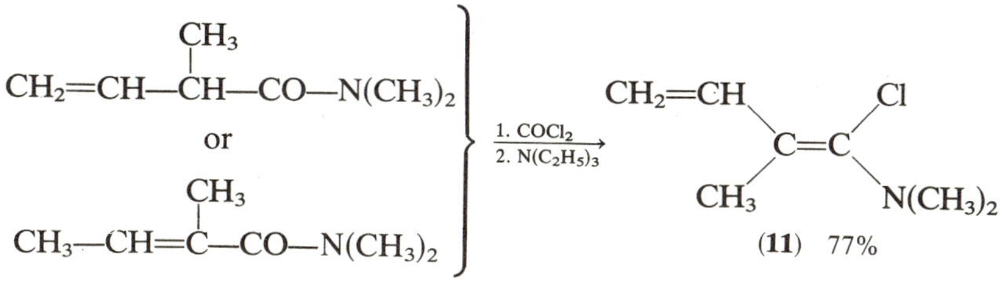

$$\left. \begin{array}{c} CH_2=CH-CH(CH_3)-CO-N(CH_3)_2 \\ \text{or} \\ CH_3-CH=C(CH_3)-CO-N(CH_3)_2 \end{array} \right\} \xrightarrow[2.\ N(C_2H_5)_3]{1.\ COCl_2} \begin{array}{c} (CH_2=CH)(CH_3)C=C(Cl)(N(CH_3)_2) \\ (\mathbf{11})\ 77\% \end{array}$$

The method was also successfully applied (29a) to the synthesis of the crystalline bis-α-chloroenamine (**12**), whose structure was determined accurately by X-ray diffraction (29b).

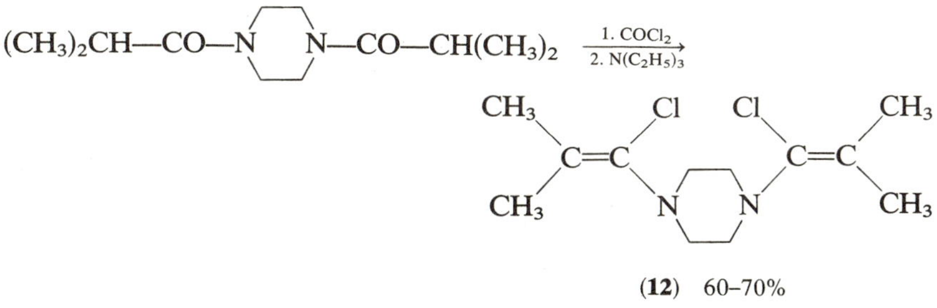

(CH₃)₂CH—CO—N⌒N—CO—CH(CH₃)₂ $\xrightarrow[2.\ N(C_2H_5)_3]{1.\ COCl_2}$

[structure: (CH₃)₂C=C(Cl)—N⌒N—C(Cl)=C(CH₃)₂ piperazine-bridged]

(**12**) 60–70%

The less nucleophilic anilides (R^3 = phenyl, R^4 = alkyl) are less reactive under the same conditions, but phosgenation of the thioanilides (**13**), followed by HCl elimination (30), gives good yields of the corresponding α-chloroenamine (**14**):

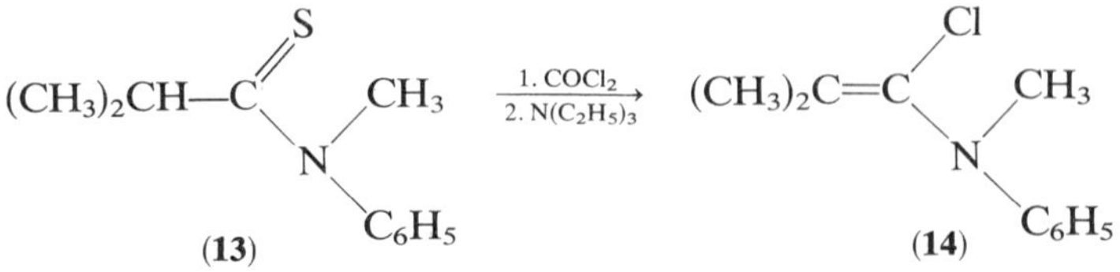

$$(CH_3)_2CH-C(=S)(N(C_6H_5)CH_3) \xrightarrow[2.\ N(C_2H_5)_3]{1.\ COCl_2} (CH_3)_2C=C(Cl)(N(C_6H_5)CH_3)$$

(**13**)   (**14**)

2. From Amide Chlorides Derived from α-Monosubstituted Acetamides and Lactams

Several complications occur when the preceding reaction sequence is applied to C_1-monosubstituted acetamides or lactams (Scheme 6) (20,31).

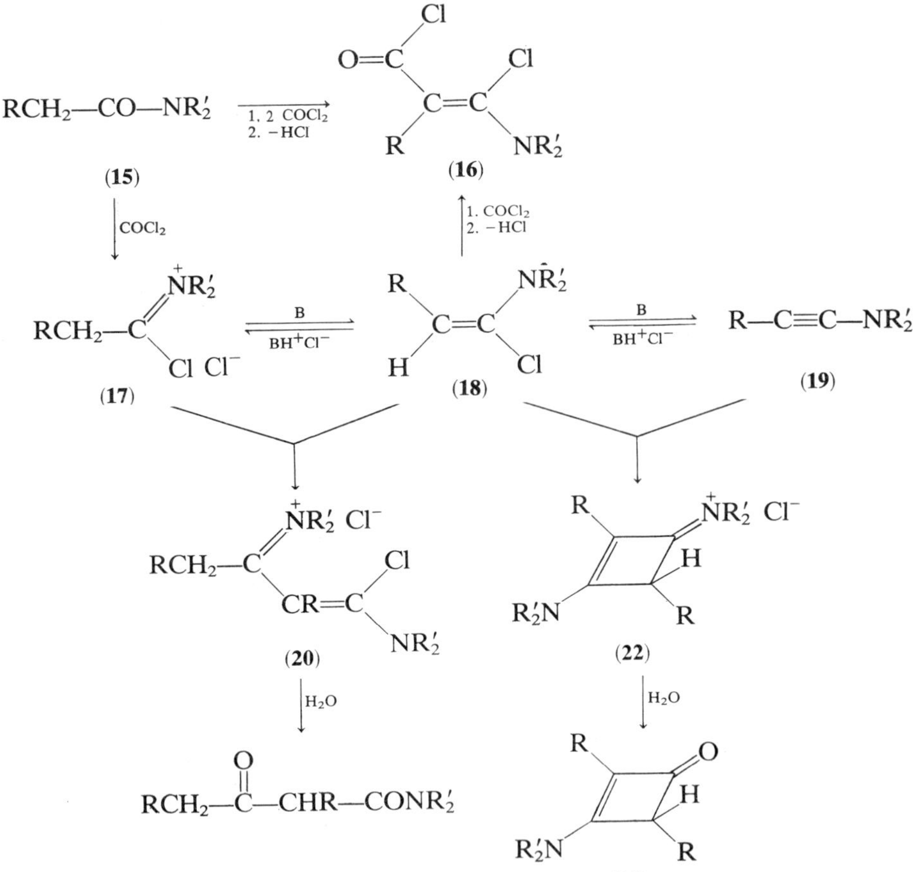

$$RCH_2-CO-NR_2' \xrightarrow[2.\ -HCl]{1.\ 2\ COCl_2} (16)$$

Scheme 6

III. SYNTHESIS OF α-HALOENAMINES

1. The phosgenation of α-monosubstituted acetamides (**15**) and butyrolactams gives mixtures of amide chlorides (**17**) and β-chlorocarbonyl-α-chloroenamines (**16**) (31b,c). These reactive derivatives of malonic acids are formed in increasing amounts in accordance with the anion-stabilizing nature of R. A common precursor for **16** and **17** has been postulated because their relative yields were not affected when the amide/phosgene molar ratio, temperature, concentration, or reaction time was modified (31c). However, in chlorinated solvents that dissolve the amide chlorides, the acylation of the α-chloroenamine (**18**) in equilibrium with the amide chloride has been shown to occur to a certain extent (31c). Actually this acylation reaction is of preparative value since it allows for simple conversion of the acetamides or lactams into β-chlorocarbonyl-α-chloroenamines, an interesting class of bifunctional reagents that have been used for heterocyclic synthesis (31d). This method certainly supersedes the ynamine method (31a) (see Section III-3) because of the ready availibility of the starting materials (Scheme 7). It is possible to limit or even suppress the formation of these acylated α-chloroenamines by addition of gaseous HCl to the amide before adding phosgene (19,20).

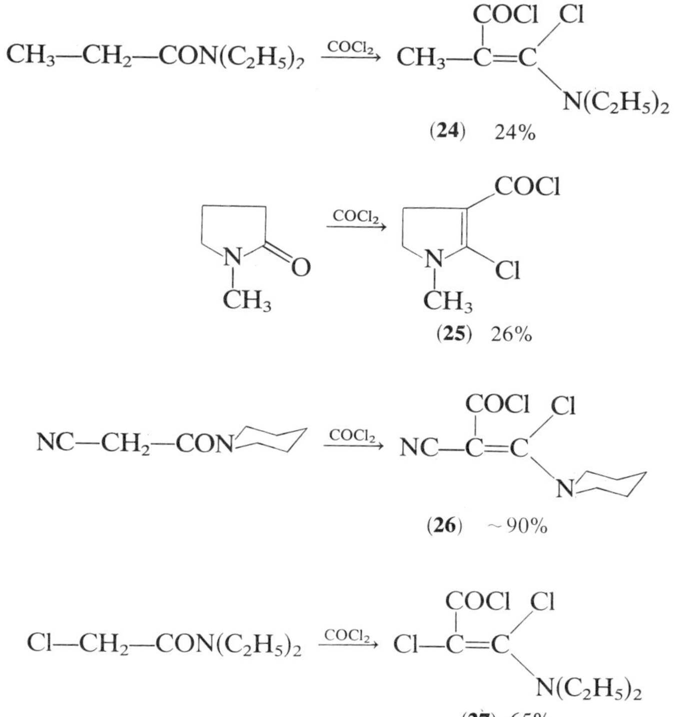

$CH_3-CH_2-CON(C_2H_5)_2 \xrightarrow{COCl_2} CH_3-C(COCl)=C(Cl)(N(C_2H_5)_2)$

(**24**) 24%

[N-methylpyrrolidinone] $\xrightarrow{COCl_2}$ [3-chloro-2-acyl chloride N-methyl dihydropyrrole]

(**25**) 26%

$NC-CH_2-CON\diagup \xrightarrow{COCl_2} NC-C(COCl)=C(Cl)(N\diagup)$

(**26**) ~90%

$Cl-CH_2-CON(C_2H_5)_2 \xrightarrow{COCl_2} Cl-C(COCl)=C(Cl)(N(C_2H_5)_2)$

(**27**) 65%

Scheme 7

2. When the elimination is effected by adding the tertiary base to a dichloromethane solution of amide chloride derived from a monosubstituted acetamide or a lactam, the α-chloroenamine (**18**) that is primarily formed can be partially acylated by the amide chloride to give, after hydrolysis, a β-ketoamide (**21**) (19,20) (Scheme 6). This reaction, which is of preparative interest, also occurs when these amide chlorides are heated (20,32) or when monosubstituted acetamides are reacted with PCl_5, PCl_3, or $OPCl_3$ (32). However, the formation of these condensation products can be suppressed by conducting the elimination step in an apolar solvent, like CCl_4, that does not dissolve the amide chloride (20) or by using the less reactive anilides. In this way the β-unsubstituted α-chloroenamine (**28**) has ben prepared in moderate yields (20):

$$CH_3-\overset{O}{\underset{\underset{CH_3}{N}}{C}}-C_6H_5 \quad \xrightarrow[\text{2. N(Et)}_3/\text{CCl}_4]{\text{1. HCl-COCl}_2} \quad CH_2=\overset{Cl}{\underset{\underset{CH_3}{N}}{C}}-C_6H_5$$

(**28**) 44%

3. With α-chloroenamines derived from monosubstituted acetamides, a second elimination of HCl may take place. It has been found that this dehydrochlorination occurs with exceptional ease for a vinylic chloride since a weak base such as triethylamine is capable of abstracting HCl from these α-chloroenamines even at room temperature (20,25) (Scheme 6). Therefore careful control of the conditions for the elimination of HCl from acetamide chloride is required in order to avoid further elimination to the ynamine (**19**), which, under the experimental conditions, cycloadds rapidly to the α-chloroenamine to yield (19,20,25) a cyclobutenylcyanine (**22**) (Scheme 6).

It is thus not entirely surprising that, up to now, all attempts to prepare the pure α-chloroenamines derived from N,N-dimethylacetamide, propionamides, and butyramides have been unsuccessful. However when the C_2 carbon atom bears a bulky group such as *tert*-butyl, less acylation is observed and the α-chloroenamine (**29**) predominates in the reaction mixture (33).

$$(CH_3)_3C-CH_2CON(CH_3)_2 \quad \xrightarrow[\text{2. N(Et)}_3]{\text{1. COCl}_2}$$

$$\underset{H}{(CH_3)_3C}\diagdown C=C \diagup^{Cl}_{N(CH_3)_2} \quad + \quad \underset{O=C-Cl}{(CH_3)_3C}\diagdown C=C\diagup^{Cl}_{N(CH_3)_2}$$

(**29**) (**30**)

III. SYNTHESIS OF α-HALOENAMINES

It must be pointed out that these monosubstituted α-chloroenamines, despite the difficulties encountered in their preparation, can be reacted *in situ*.

3. *From Amide Chlorides Derived from Phosgeniminium Salts and Activated C–H Bonds*

These reactions were treated in detail in the chapter by Janousek and Viehe.

The reactive phosgeniminium salts produce with monosubstituted tertiary acetamides (34,35) the synthetically interesting dichlorotrimethinecyanines (**31**). The reaction is analogous to the previously described phosgenation of the same amides, which yields β-chlorocarbonyl-α-chloroenamines. However, with phosgeniminium salts the yields are higher (Scheme 8).

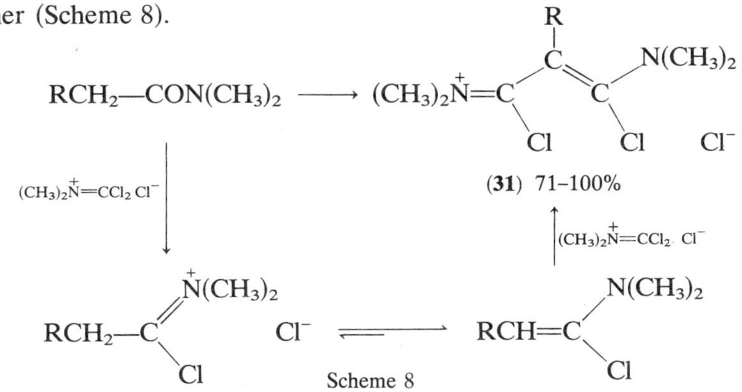

$$RCH_2-CON(CH_3)_2 \longrightarrow (CH_3)_2\overset{+}{N}=C\diagdown\overset{\displaystyle R}{\underset{Cl}{C}}=C\diagup\overset{N(CH_3)_2}{\underset{Cl}{}} \quad Cl^-$$

(**31**) 71–100%

Scheme 8

$$RCH_2-C\diagup\overset{\overset{+}{N}(CH_3)_2}{\underset{Cl}{}} Cl^- \rightleftharpoons RCH=C\diagup\overset{N(CH_3)_2}{\underset{Cl}{}}$$

Similarly malondinitrile, ethyl cyanoacetate, ethyl malonate, cyanoacetic acid, and carbethoxymethylenetriphenylphosphonium chloride condense with dichloromethyleneammonium chloride to give β-functionalized α-chloroenamines (34b,35,36), for example:

$$CH_2\diagup\overset{CN}{\underset{CN}{}} + (CH_3)_2\overset{+}{N}=CCl_2 \; Cl^- \xrightarrow{N(Et)_3} \overset{NC}{\underset{NC}{}}\diagdown C=C\diagup\overset{Cl}{\underset{N(CH_3)_2}{}}$$

(**32**) 77%

In the same way, the β-chlorocarbonyl-α-chloroenamines can be obtained in good yields from monosubstituted acetyl chlorides and phosgeniminium (37):

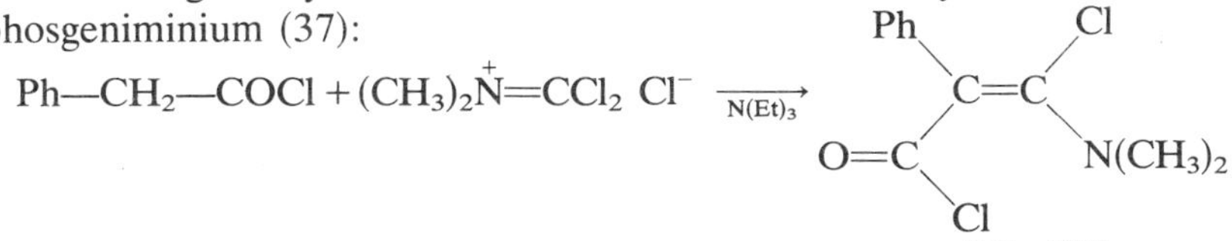

$$Ph-CH_2-COCl + (CH_3)_2\overset{+}{N}=CCl_2 \; Cl^- \xrightarrow{N(Et)_3} \overset{Ph}{\underset{O=C\diagdown Cl}{}}\diagdown C=C\diagup\overset{Cl}{\underset{N(CH_3)_2}{}}$$

(**33**) 70%

4. From α,β-Dihaloamines

From the preceding discussions it is clear that the removal of a β-hydrogen from amide chlorides (α-chloroalkylideneammonium chloride) to give an α-chloroenamine is an extremely facile reaction (Scheme 9).

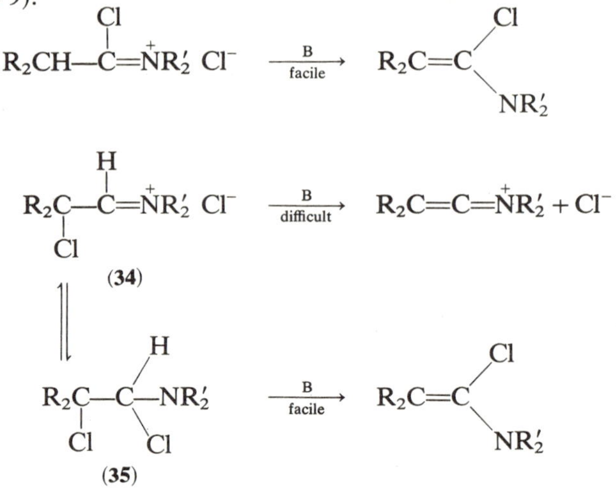

$$R_2CH-\overset{Cl}{\underset{}{C}}=\overset{+}{N}R'_2 \; Cl^- \xrightarrow[\text{facile}]{B} R_2C=C\overset{Cl}{\underset{NR'_2}{\diagdown}}$$

$$R_2C-\overset{H}{\underset{Cl}{C}}=\overset{+}{N}R'_2 \; Cl^- \xrightarrow[\text{difficult}]{B} R_2C=C=\overset{+}{N}R'_2 + Cl^-$$

(34)

$$R_2C-\overset{H}{\underset{Cl}{C}}-NR'_2 \xrightarrow[\text{facile}]{B} R_2C=C\overset{Cl}{\underset{NR'_2}{\diagdown}}$$

(35)

Scheme 9

The isomeric β-chloroalkylideneammonium salts (**34**) could also lead, in principle, to the α-haloenamines or their keteniminium tautomers by dehydrochlorination. However, in general, this elimination process is expected to be much more difficult or perhaps impossible, since it involves the abstraction of a proton attached to the carbon atom of the iminium functional group and the primary formation of the highly energetic keteniminium structure. However an elimination reaction from the covalent tautomers α,β-dihaloamines (**35**) should be far easier, and therefore any structural variation increasing the amount of this covalent form should facilitate the base-catalyzed elimination reaction to an α-chloroenamine (Scheme 9). One such structural variation would be reduction of the basicity of the amino group. Indeed, several 1-halo-*N,N*-bis-(trifluoromethyl)alkenylamines have been prepared according to this principle (38,39):

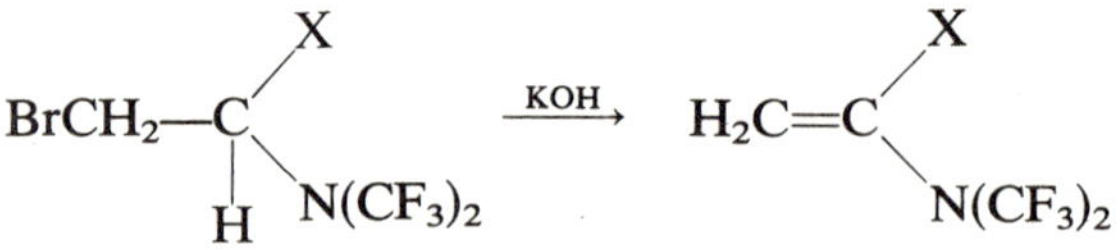

$$BrCH_2-\overset{X}{\underset{H}{\overset{|}{C}}}\diagdown N(CF_3)_2 \xrightarrow{KOH} H_2C=C\overset{X}{\underset{N(CF_3)_2}{\diagdown}}$$

(**36**) X = Br
(**37**) X = F

The dehalogenation of **38** with zinc in ethanol has also been shown (39) to give an α-haloenamine (**39**):

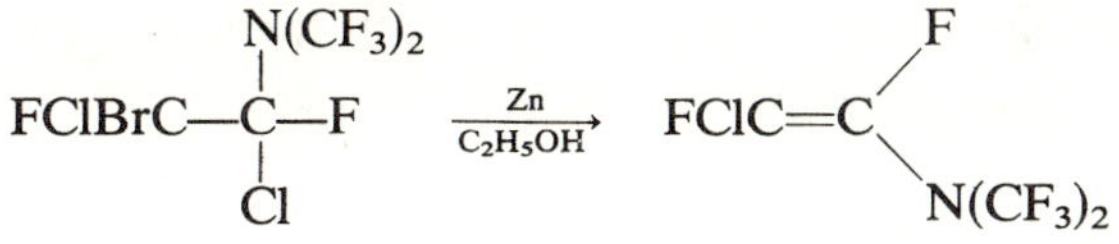

$$\text{FClBrC}-\underset{\underset{\text{Cl}}{|}}{\overset{\overset{\text{N(CF}_3)_2}{|}}{\text{C}}}-\text{F} \xrightarrow[\text{C}_2\text{H}_5\text{OH}]{\text{Zn}} \text{FClC}=\text{C}\underset{\text{N(CF}_3)_2}{\overset{\text{F}}{\diagup}}$$

(**38**)   (**39**)

C. α-HALOENAMINES FROM ELECTROPHILIC ADDITION TO YNAMINES

Addition of reagents R^+X^- to ynamines provides a potentially interesting route toward α-haloenamines as long as the starting ynamines are readily available. As will be seen, however, the course of these reactions is strongly dependent on the nature of the electrophilic reagent.

1. H—X, Alk—X, and X_2 Additions to Ynamines

The addition of hydrogen or alkyl chlorides and bromides to alkyl- or aryl-substituted ynamines leads to α-chloro- or α-bromoenamines sufficiently electrophilic to react further with ynamines to give reactive intermediates that undergo intramolecular N- or, more often, C-alkylation, yielding (2,40) stable alleneamidinium salts (**40**) or cyclobutenylcyanines (**41**) (Scheme 10).

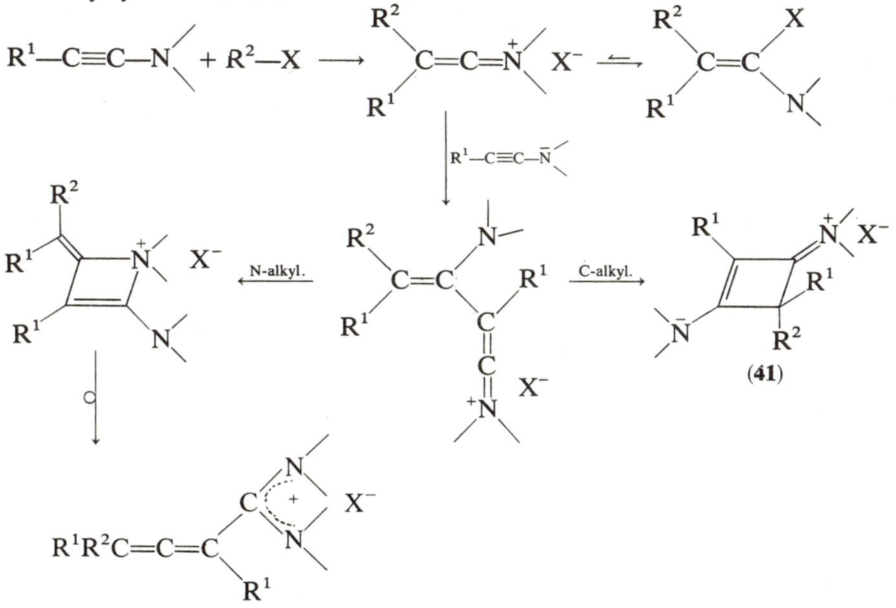

(**40**)   Scheme 10

Use of an excess of HCl at low temperature (0°C) converts the ynamines into amide chlorides (25), whereas triethylamine hydrochloride leads to the formation of cyclobutenylcyanines (2,40) (Scheme 11). The reaction with methyl iodide (40) follows the same course.

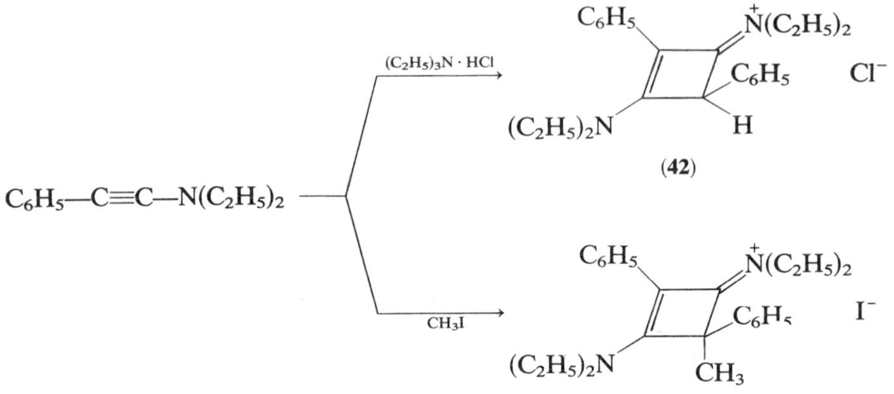

$C_6H_5-C\equiv C-N(C_2H_5)_2$

$(C_2H_5)_3N\cdot HCl$ → (**42**) cyclobutenyl product with C_6H_5, $\overset{+}{N}(C_2H_5)_2$, $(C_2H_5)_2N$, C_6H_5, H, Cl^-

CH_3I → (**43**) cyclobutenyl product with C_6H_5, $\overset{+}{N}(C_2H_5)_2$, $(C_2H_5)_2N$, C_6H_5, CH_3, I^-

Scheme 11

Polar bromination of alkyl- and aryl-substituted ynamines with bromine–dioxane complexes (2,40) also gives the cyclobutenecyanines or alleneamidinium salts (Scheme 12).

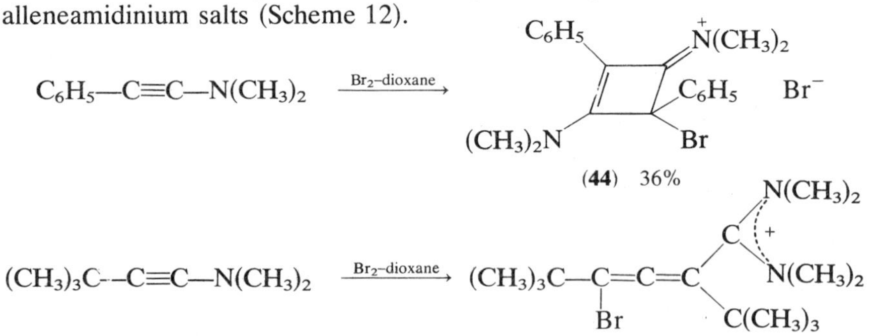

$C_6H_5-C\equiv C-N(CH_3)_2$ $\xrightarrow{Br_2\text{-dioxane}}$ (**44**) 36% cyclobutenyl with C_6H_5, $\overset{+}{N}(CH_3)_2$, $(CH_3)_2N$, C_6H_5, Br, Br^-

$(CH_3)_3C-C\equiv C-N(CH_3)_2$ $\xrightarrow{Br_2\text{-dioxane}}$ $(CH_3)_3C-\underset{Br}{C}=C=C\overset{N(CH_3)_2}{\underset{C(CH_3)_3}{\overset{+}{\cdots N(CH_3)_2}}}$ (**45**) 87%

Scheme 12

In contrast with these results, the addition of HF (41) or HCl (42,43) at 0° in THF to dimethylaminopropynal was found to yield the β-formyl-α-fluoro- or β-formyl-α-chloroenamines, which are incapable of further reaction with ynamine and can be isolated in good yields:

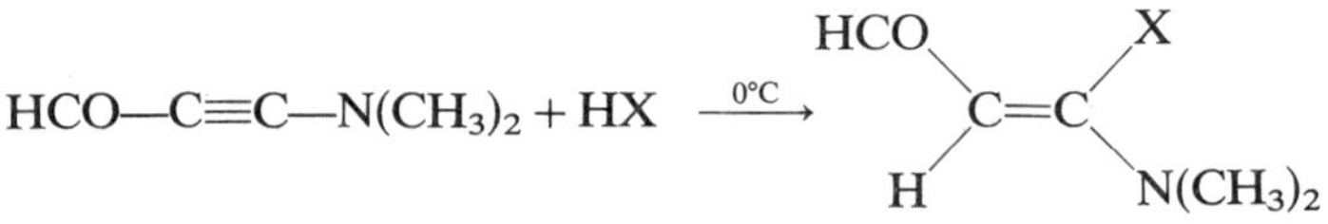

$HCO-C\equiv C-N(CH_3)_2 + HX \xrightarrow{0°C}$ $\underset{H}{\overset{HCO}{>}}C=C\underset{N(CH_3)_2}{\overset{X}{<}}$

(**46**) X = F 65%
(**47**) X = Cl 97%

III. SYNTHESIS OF α-HALOENAMINES

β-Alkyl-α-fluoroenamines can also be obtained in good yields from the corresponding ynamines and KHF_2 (29a):

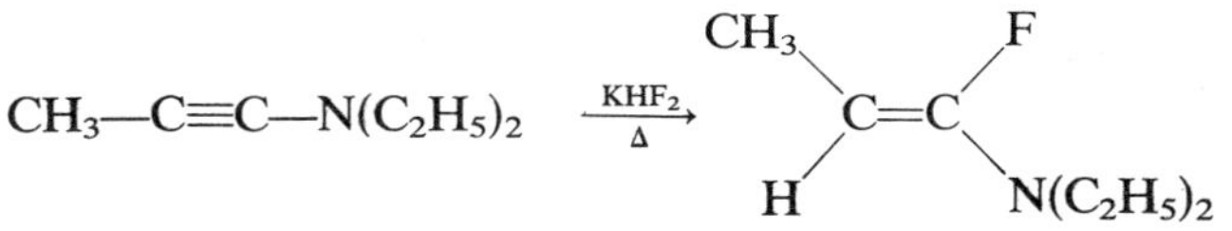

$$CH_3-C\equiv C-N(C_2H_5)_2 \xrightarrow[\Delta]{KHF_2} \underset{H}{\overset{CH_3}{>}}C=C\underset{N(C_2H_5)_2}{\overset{F}{<}}$$

(**48**)
(2 isomers)

Diethylaminopropyne reacts with various fluoroalkenes (44) to give either a cycloadduct and/or a derivative of 1-diethylamino-1-fluoro-isoprene (**49**):

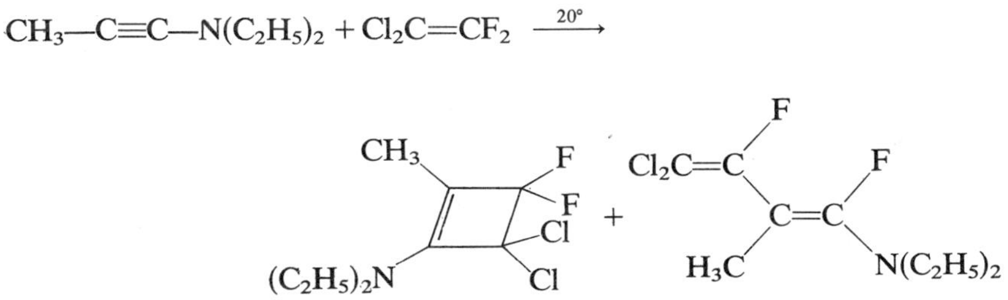

$$CH_3-C\equiv C-N(C_2H_5)_2 + Cl_2C=CF_2 \xrightarrow{20°}$$

[cyclobutene with CH_3, $(C_2H_5)_2N$ on one double bond carbon and F, Cl, F, Cl substituents] 20%

+

$Cl_2C=C(F)-C(CH_3)=C(F)N(C_2H_5)_2$ represented as $Cl_2C=C\overset{F}{<}$ attached to $\underset{H_3C}{>}C=C\underset{N(C_2H_5)_2}{\overset{F}{<}}$ (**49**)

$$CH_3-C\equiv C-N(C_2H_5)_2 + \underset{Cl}{\overset{F}{>}}C=CF_2 \xrightarrow{20°}$$

[cyclobutene with CH_3, $(C_2H_5)_2N$ on double bond carbons and F, F, F, Cl substituents] 60%

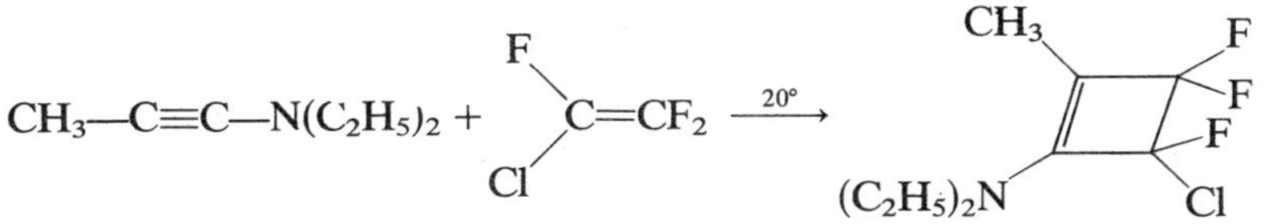

Several 1-halo-*N*,*N*-bis(trifluoromethyl)alkenylamines (**36,50**) have been obtained by the addition of HBr or Br_2 to *N*,*N*-bis(trifluoromethyl)-ynamine under photochemical conditions (45) or in the presence of aluminum bromide (46,38b):

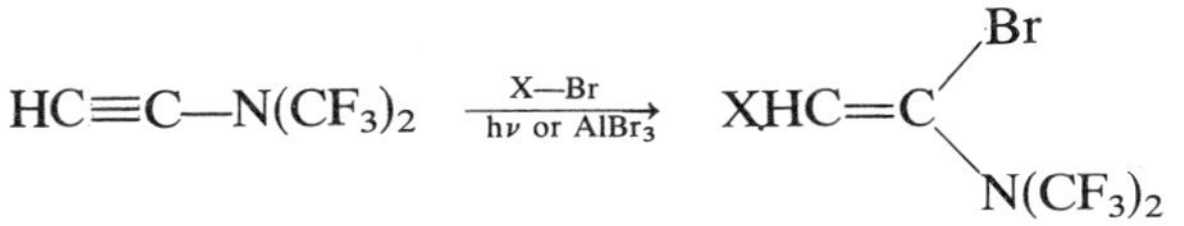

$$HC\equiv C-N(CF_3)_2 \xrightarrow[h\nu \text{ or } AlBr_3]{X-Br} XHC=C\underset{N(CF_3)_2}{\overset{Br}{<}}$$

(**36**) X = H
(**50**) X = Br

Finally (38a,46,47), photochemical addition of perfluoro-N-haloamines to N,N-bis(trifluoromethyl)ynamine gave **51** and **52**:

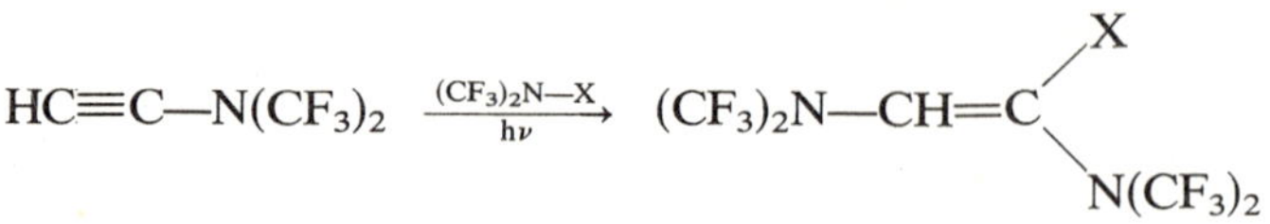

$$HC\equiv C-N(CF_3)_2 \xrightarrow[h\nu]{(CF_3)_2N-X} (CF_3)_2N-CH=C\begin{smallmatrix}X\\ \\N(CF_3)_2\end{smallmatrix}$$

(**51**) X = Br
(**52**) X = Cl

2. Acylation of Ynamines

As seen in previous examples, when the addition step leads to α-fluoroenamines or to α-chloro- and bromoenamines bearing anion-stabilizing substituents or weakly basic amino groups, no further reaction is observed and α-haloenamines are obtained in good yields from ynamines. This condition is found in the acylation of ynamines with acyl halides such as phosgene, thiophosgene, acetyl or benzoyl chlorides, and thionyl chloride (2,31a,40). Several β-acyl-α-chloroenamines (**53–56**) have been obtained by this route. Yields of crude products are reported to be quantitative (Scheme 13). These α-chloroenamines are bifunctional reagents that have been used for the synthesis of various heterocyclic compounds (31d).

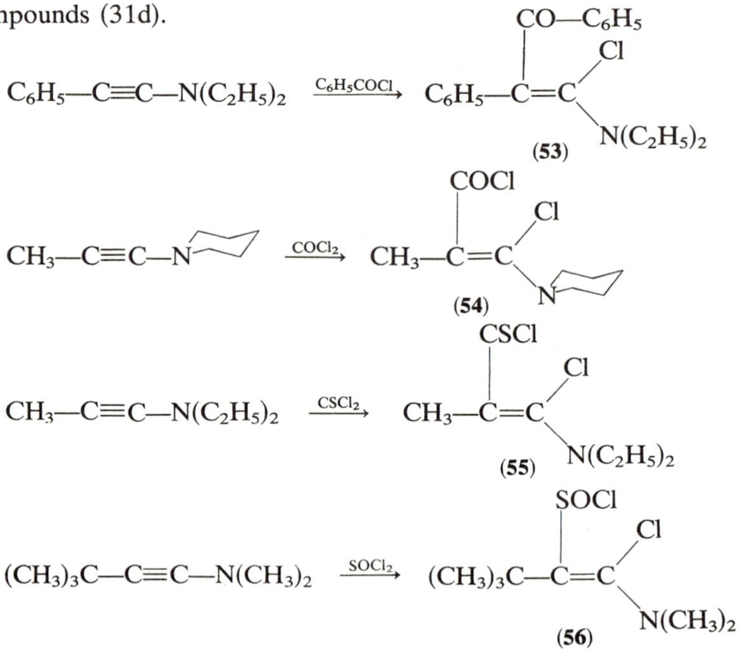

Scheme 13

In the same way, β-acyl-α-fluoroenamines (**57**) can be obtained by treatment of ynamines with acyl fluorides (29a):

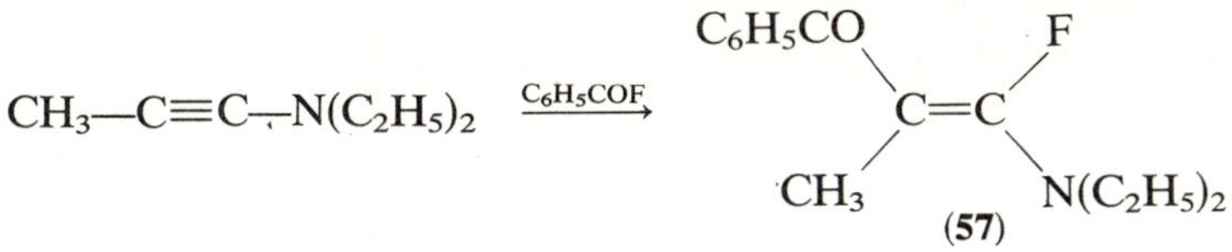

$$CH_3—C≡C—N(C_2H_5)_2 \xrightarrow{C_6H_5COF} \underset{(57)}{\underset{CH_3 \quad\quad N(C_2H_5)_2}{\overset{C_6H_5CO \quad\quad F}{\searrow C=C \swarrow}}}$$

3. Addition of Sulfenyl Chlorides to Ynamines

α-Chloroenamines bearing a thioether group at C_2 also exhibit a reduced electrophilic nature that permits their preparation by addition of sulfenyl chlorides to ynamines (48–50). The first example was described by Senning (48), who reacted N,N-diethylaminopropyne with trichloromethanesulfenyl chloride.

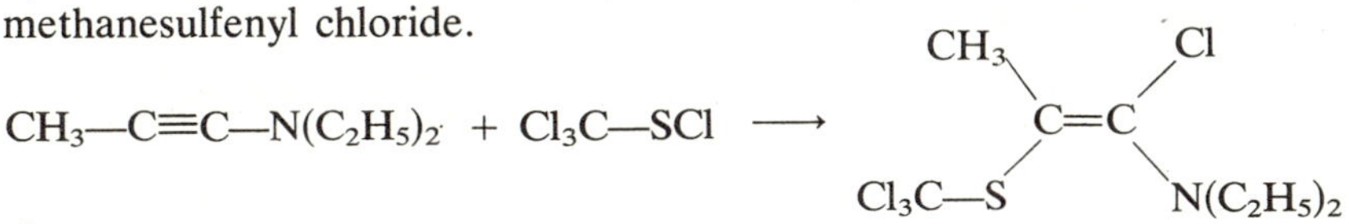

$$CH_3—C≡C—N(C_2H_5)_2 + Cl_3C—SCl \longrightarrow \underset{Cl_3C—S \quad\quad N(C_2H_5)_2}{\overset{CH_3 \quad\quad Cl}{C=C}}$$

(2 isomers)
(**58**) 76%

Similarly the addition of methylsulfenyl chloride (50) to the ynamine thioether (**59**) gives quantitatively a bisthioalkyl-α-chloroenamine (**60**):

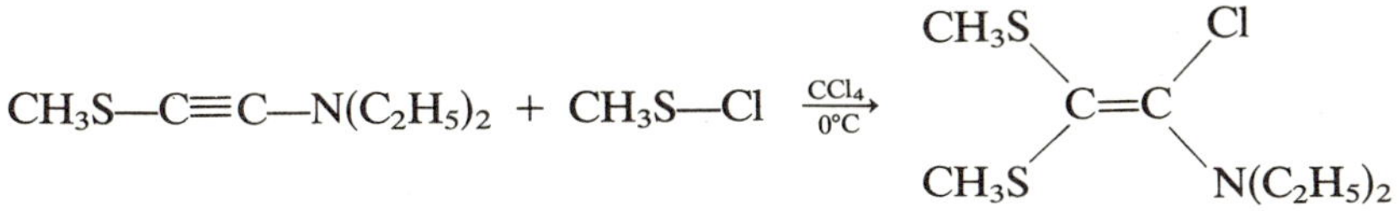

$$\underset{(59)}{CH_3S—C≡C—N(C_2H_5)_2} + CH_3S—Cl \xrightarrow[0°C]{CCl_4} \underset{(60)}{\underset{CH_3S \quad\quad N(C_2H_5)_2}{\overset{CH_3S \quad\quad Cl}{C=C}}}$$

With sulfenyl chloride two successive additions were observed (49a):

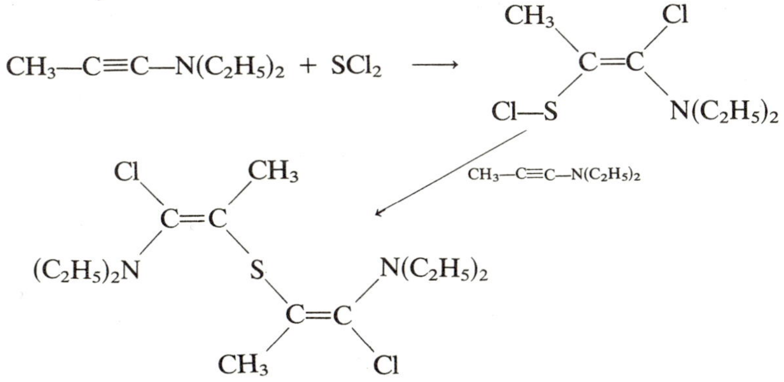

$$CH_3—C≡C—N(C_2H_5)_2 + SCl_2 \longrightarrow \underset{Cl—S \quad\quad N(C_2H_5)_2}{\overset{CH_3 \quad\quad Cl}{C=C}}$$

$$\xleftarrow{CH_3—C≡C—N(C_2H_5)_2}$$

$$\underset{(C_2H_5)_2N \quad\quad S}{\overset{Cl \quad\quad CH_3}{C=C}} \underset{CH_3 \quad\quad Cl}{\overset{\quad\quad N(C_2H_5)_2}{C=C}}$$

(2 isomers)
(**61**)

α-Chloroenamines bearing thioether groups at C_2 are of synthetic interest since the thioether group can be easily replaced by hydrogen after reaction. Hence, they can be considered as appropriate substitutes for the less readily obtainable β-monosubstituted α-chloroenamines.

D. α-HALOENAMINES BY NUCLEOPHILIC ADDITIONS TO HALOACETYLENES

This method was used by Ott et al. (51) to prepare one of the earliest representatives of the α-chloroenamines class; the unstable 1,2-dichloro-1-diethylaminoethylene (**62**) was obtained, but in unreported yields, from the exothermic addition of diethylamine to dichloroacetylene:

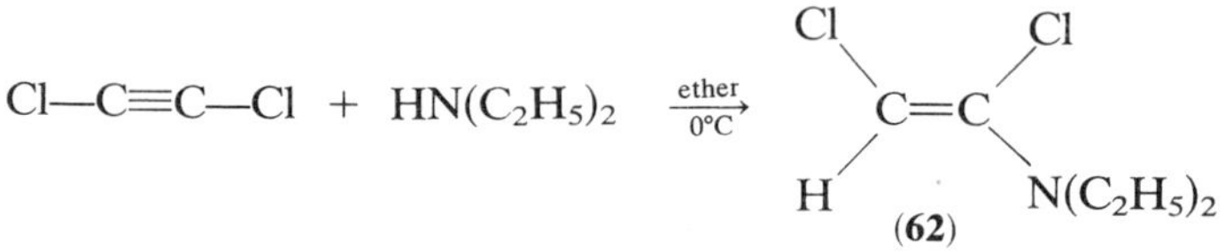

$$Cl-C{\equiv}C-Cl + HN(C_2H_5)_2 \xrightarrow[0°C]{ether} \underset{(62)}{\underset{H \quad\quad N(C_2H_5)_2}{\overset{Cl \quad\quad Cl}{C=C}}}$$

Because of the instability of dihaloacetylenes, the reaction is of no practical value per se. However, the same product can be obtained by generating the dichloroacetylene *in situ* (2b,52,53a). Unfortunately this method is of very limited scope since monochloroacetylenes react with nucleophilic reagents to give adducts with the wrong regiochemistry (53b):

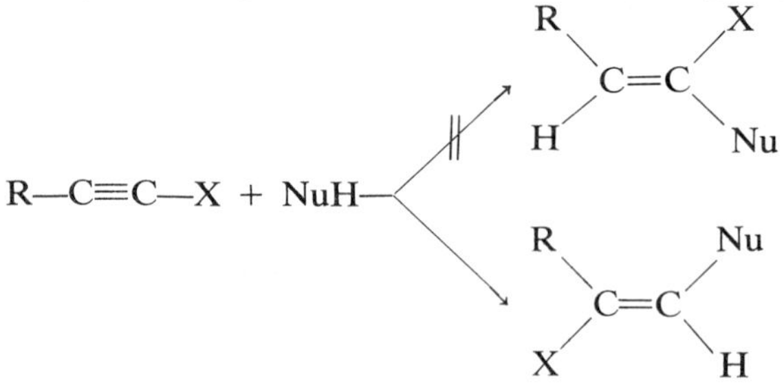

$$R-C{\equiv}C-X + NuH \nrightarrow \underset{H \quad Nu}{\overset{R \quad X}{C=C}}$$
$$\searrow \underset{X \quad H}{\overset{R \quad Nu}{C=C}}$$

E. α-HALOENAMINES BY SUBSTITUTION REACTIONS

1. *Halovinylation of Metal Amides*

α-Haloenamines can be obtained by direct introduction of an amine group on an olefinic double bond by nucleophilic substitution of a suitable nucleofuge. The method is often handicapped, however, by the enhanced electrophilic reactivity of the α-haloenamines obtained, so that disubstitution may be difficult to avoid. Therefore this route is usually restricted to the preparation of less reactive α-haloenamines and, in particular, α-fluoroenamines. The fluorovinylation of alkyl- and aryl-substituted lithium or potassium amides was studied by England et al.

(54) and later, more extensively, by Yakubovich (6). With such fluoroolefins as tetrafluoroethylene, chlorotrifluoroethylene, trifluoroethylene, vinylidene fluoride, and hexafluoropropylene, good yields of the corresponding α-fluoroenamines were obtained (6):

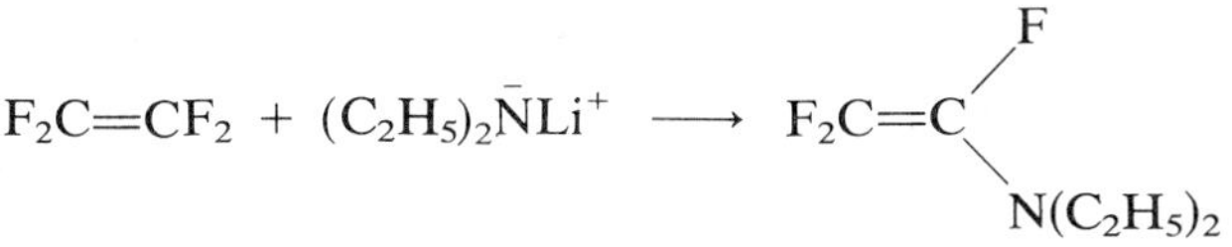

$$F_2C{=}CF_2 + (C_2H_5)_2\bar{N}Li^+ \longrightarrow F_2C{=}C\begin{smallmatrix}F\\N(C_2H_5)_2\end{smallmatrix}$$

(63) 60–80%

$$H_2C{=}CF_2 + \text{(carbazole-N-Li}^+\text{)} \longrightarrow H_2C{=}C\begin{smallmatrix}F\\N\text{(carbazolyl)}\end{smallmatrix}$$

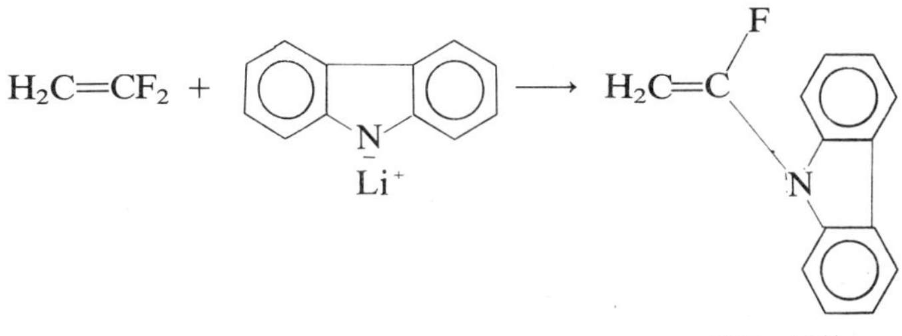

(64) 51%

In these reactions the initial aliphatic lithium amides were obtained by the action of butyllithium (from dibutylmercury and metallic lithium) or naphthyllithium on a secondary amine. When lithium amides are prepared by the reaction of lithium with an alkyl bromide or iodide followed by reaction with a secondary amine, the initially formed α-fluoroenamines may react further with the lithium bromide or iodide present to give α-bromo- or α-iodoenamines in addition to or to the exclusion of the desired product (6,55):

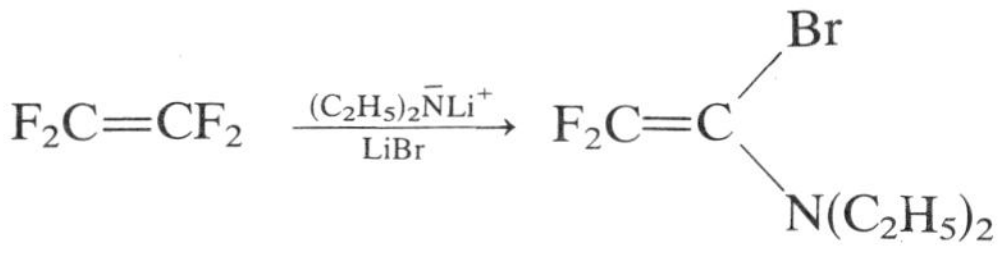

$$F_2C{=}CF_2 \xrightarrow[\text{LiBr}]{(C_2H_5)_2\bar{N}Li^+} F_2C{=}C\begin{smallmatrix}Br\\N(C_2H_5)_2\end{smallmatrix}$$

(65) 31%

$$ClFC{=}CF_2 \xrightarrow[\text{LiI}]{(C_2H_5)_2\bar{N}Li^+} \begin{smallmatrix}F\\Cl\end{smallmatrix}C{=}C\begin{smallmatrix}F\\N(C_2H_5)_2\end{smallmatrix} + \begin{smallmatrix}F\\Cl\end{smallmatrix}C{=}C\begin{smallmatrix}I\\N(C_2H_5)_2\end{smallmatrix}$$

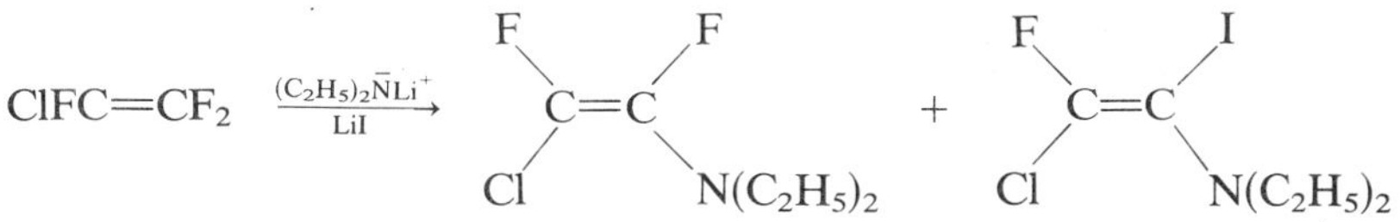

(66) 27%   (67) 13%

In one case (56) a trialkyltin amide has been used for the fluorovinylation reaction: dimethylaminotrimethyl stannate and chlorotrifluoroethylene react smoothly to give (56) the α-fluoroenamine (**68**).

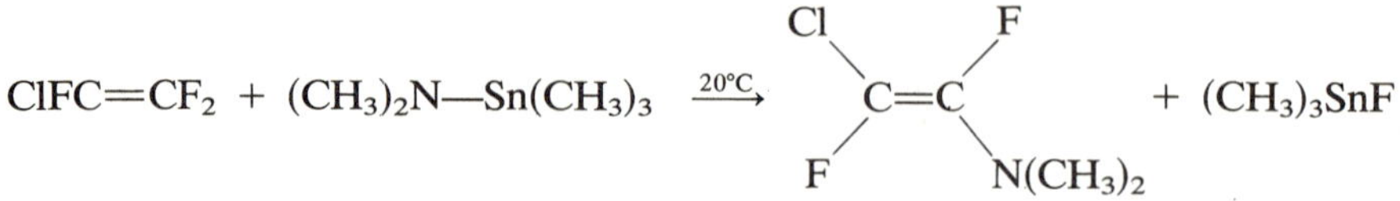

$$ClFC=CF_2 + (CH_3)_2N-Sn(CH_3)_3 \xrightarrow{20°C} \underset{F}{\overset{Cl}{>}}C=C\underset{N(CH_3)_2}{\overset{F}{<}} + (CH_3)_3SnF$$

(**68**) 69%

The reaction of trichloroethylene with lithium amides (2b,52,53a) is a good preparative method for dichlorovinylamines (**69**):

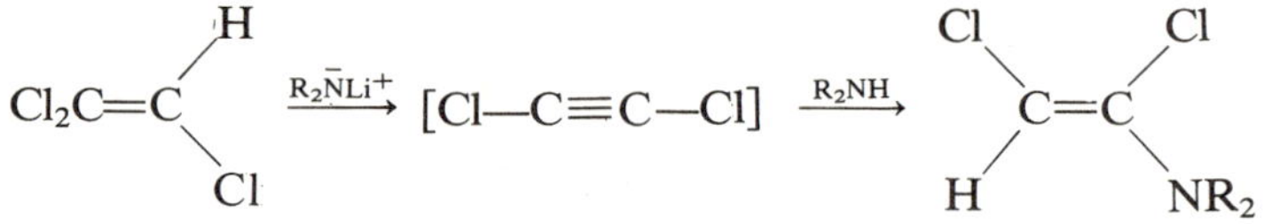

$$Cl_2C=C\underset{Cl}{\overset{H}{<}} \xrightarrow{R_2\overset{-}{N}Li^+} [Cl-C\equiv C-Cl] \xrightarrow{R_2NH} \underset{H}{\overset{Cl}{>}}C=C\underset{NR_2}{\overset{Cl}{<}}$$

(**69**)

When the nitrogen substituents are alkyl groups, the dichlorovinylamine is unstable and is usually reacted *in situ*. The corresponding diarylamino analogues, however, are stable. As mentioned above, dichloroacetylene is probably the reactive intermediate in these substitutions.

The reaction of cyclic secondary amines with hexachlorobutadiene in a 2:1 molar ratio yields (57) the first perchlorodienamines (**70,71**):

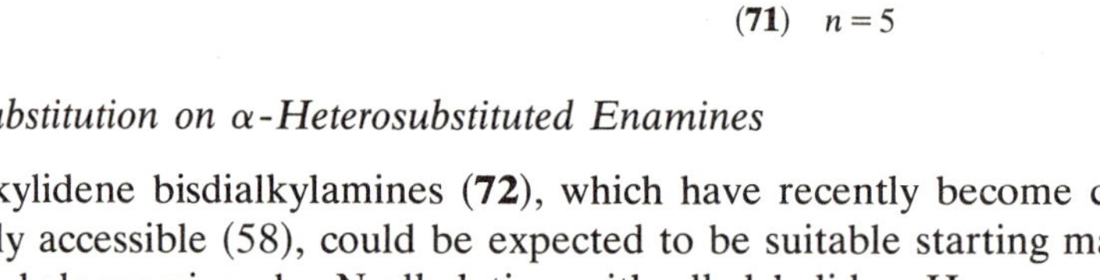

$$Cl_2C=\underset{Cl}{\overset{Cl}{C}}-\underset{Cl}{\overset{}{C}}=CCl_2 \xrightarrow[\Delta]{(CH_2)_n\ NH} Cl_2C=\underset{Cl}{\overset{Cl}{C}}-\underset{Cl}{\overset{}{C}}=C\overset{}{\underset{Cl}{<}}N\underbrace{(CH_2)_n}$$

(**70**) n = 4
(**71**) n = 5

2. Substitution on α-Heterosubstituted Enamines

Alkylidene bisdialkylamines (**72**), which have recently become conveniently accessible (58), could be expected to be suitable starting material for α-haloenamines by N-alkylation with alkyl halides. However, these enediamines show (3) a much stronger tendency to alkylate at carbon with the formation of the charge-stabalizing amidinium compound (**73**). Furthermore the less abundant initial products of N-alkylation (**74**) are unstable and react (3) with the starting enediamine to yield a condensation product (**75**) (Scheme 14).

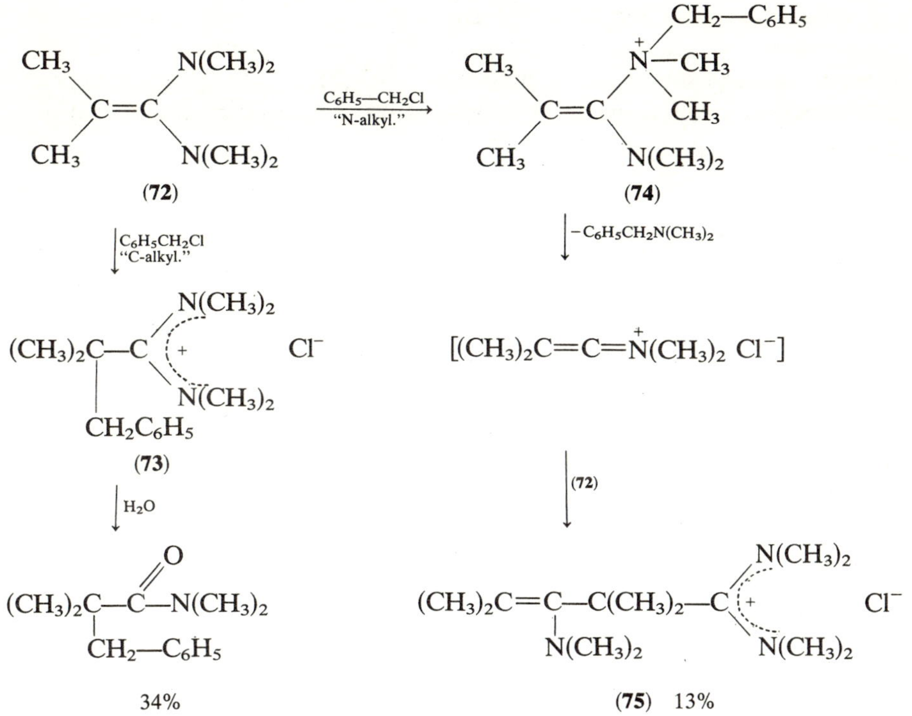

III. SYNTHESIS OF α-HALOENAMINES

[Scheme 14 structures]

(72) → (via C₆H₅—CH₂Cl, "N-alkyl.") → (74)

(72): (CH₃)(CH₃)C=C(N(CH₃)₂)(N(CH₃)₂)

(74): (CH₃)(CH₃)C=C(N⁺(CH₂C₆H₅)(CH₃)CH₃)(N(CH₃)₂)

C₆H₅CH₂Cl "C-alkyl." ↓ (from 72)

(73): (CH₃)₂C(CH₂C₆H₅)—C⁺(N(CH₃)₂)(N(CH₃)₂) Cl⁻

From (74): −C₆H₅CH₂N(CH₃)₂ ↓

[(CH₃)₂C=C=N⁺(CH₃)₂ Cl⁻]

(73) ↓ H₂O

(CH₃)₂C(CH₂—C₆H₅)—C(=O)—N(CH₃)₂   34%

↓ (72)

(75): (CH₃)₂C=C(N(CH₃)₂)—C(CH₃)₂—C⁺(N(CH₃)₂)(N(CH₃)₂) Cl⁻   13%

Scheme 14

It is interesting that the conversion of an enediamine into an α-chloro-enamine was observed when 2-methylpropenylidene bisdimethylamine (72) was reacted with phosphorus trichloride or dichlorophenylphosphine (9). These reagents are believed to form first a one-to-one complex (76), which slowly disappears to yield 1-chloro-N,N-2-trimethylpropenylamine (6) (Scheme 15). It remains to be demonstrated whether this or analogous routes can be used for the synthesis of β-monosubstituted α-chloro-enamines.

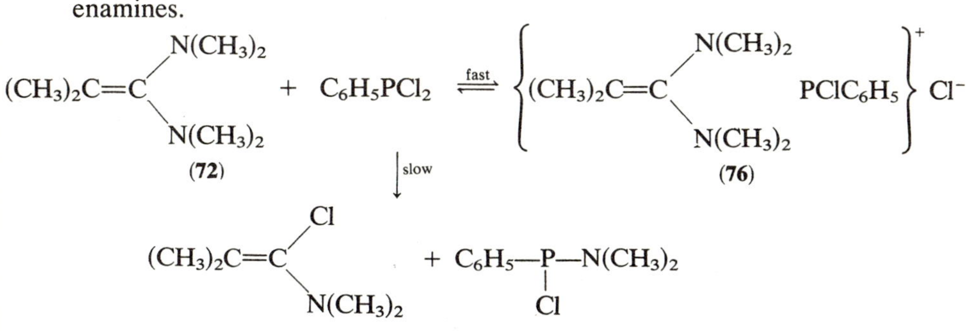

(CH₃)₂C=C(N(CH₃)₂)(N(CH₃)₂) + C₆H₅PCl₂ ⇌ (fast) {(CH₃)₂C=C(N(CH₃)₂)(N(CH₃)₂)—PClC₆H₅}⁺ Cl⁻

(72)   (76)

↓ slow

(CH₃)₂C=C(Cl)(N(CH₃)₂) + C₆H₅—P(Cl)—N(CH₃)₂

(6)

Scheme 15

The interconversion of α-haloenamines through their keteniminium tautomers has been successfully used (29a) for the synthesis of the first alkyl- and aryl-substituted α-fluoroenamines (**77,78**). These are obtained in excellent yields from the readily available β,β-disubstituted α-chloroenamines and potassium or cesium fluoride (29a):

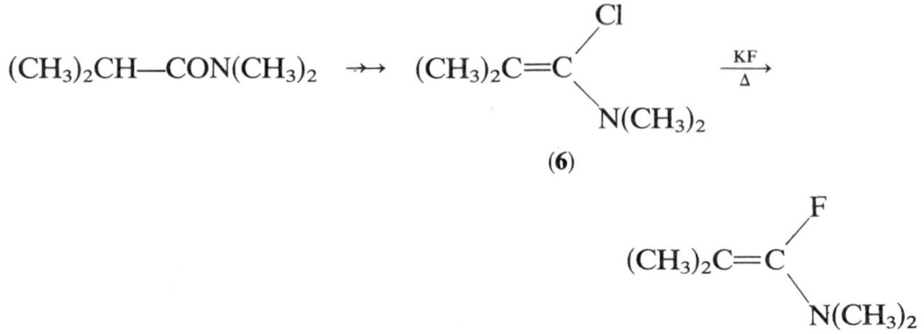

$$(CH_3)_2CH-CON(CH_3)_2 \longrightarrow (CH_3)_2C=C\begin{matrix}Cl\\N(CH_3)_2\end{matrix} \xrightarrow{KF,\Delta}$$

(**6**)

$$(CH_3)_2C=C\begin{matrix}F\\N(CH_3)_2\end{matrix}$$

(**77**)

$$C_6H_5(C_2H_5)CH-CON(CH_3)_2 \longrightarrow \begin{matrix}C_6H_5\\C_2H_5\end{matrix}C=C\begin{matrix}Cl\\N(CH_3)_2\end{matrix} \xrightarrow{KF,\Delta}$$

(**7**)

$$\begin{matrix}C_6H_5\\C_2H_5\end{matrix}C=C\begin{matrix}F\\N(CH_3)_2\end{matrix}$$

(**78**)

This simple method was also successfully applied to the synthesis of some C_2-heterosubstituted α-fluoroenamines (59,60):

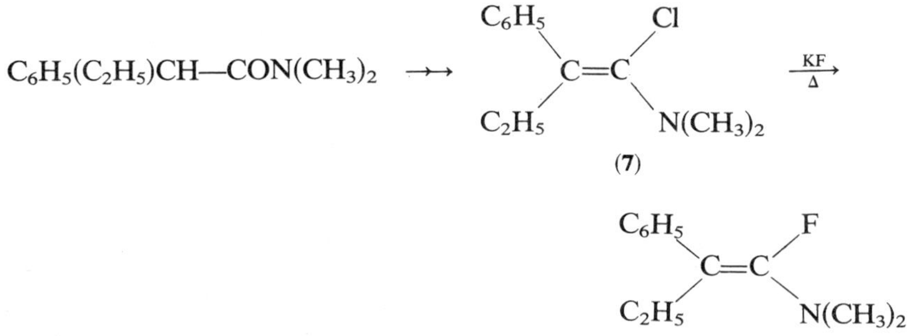

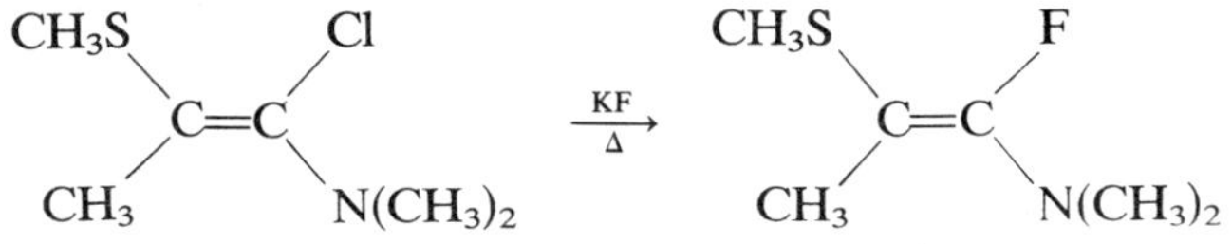

$$\begin{matrix}CH_3S\\CH_3\end{matrix}C=C\begin{matrix}Cl\\N(CH_3)_2\end{matrix} \xrightarrow{KF,\Delta} \begin{matrix}CH_3S\\CH_3\end{matrix}C=C\begin{matrix}F\\N(CH_3)_2\end{matrix}$$

(**79**)

$$\begin{matrix}Cl\\CH_3\end{matrix}C=C\begin{matrix}Cl\\N(CH_3)_2\end{matrix} \xrightarrow{KF,\Delta} \begin{matrix}Cl\\CH_3\end{matrix}C=C\begin{matrix}F\\N(CH_3)_2\end{matrix}$$

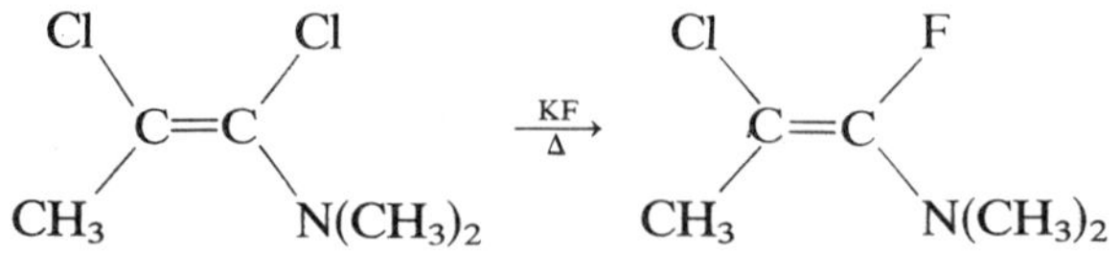

(**80**)

III. SYNTHESIS OF α-HALOENAMINES

The ability of α-fluoroenamines to undergo fluorine substitution in the presence of lithium bromide or iodide was mentioned earlier (6,55). In this manner 1-fluoro-N,N-2-trimethylpropenylamine (**77**) has been converted (29a) into bromo- or iodoenamines (**81** and **82**):

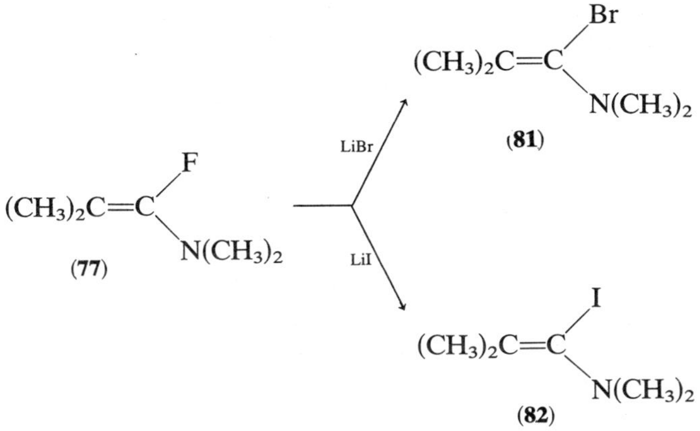

$$(CH_3)_2C=C\begin{smallmatrix}F\\N(CH_3)_2\end{smallmatrix} \quad (77)$$

$$\xrightarrow{LiBr} (CH_3)_2C=C\begin{smallmatrix}Br\\N(CH_3)_2\end{smallmatrix} \quad (81)$$

$$\xrightarrow{LiI} (CH_3)_2C=C\begin{smallmatrix}I\\N(CH_3)_2\end{smallmatrix} \quad (82)$$

The α-iodoenamine (**82**) has been obtained directly from the α-chloroenamine (**6**) by treatment with methyl iodide (61).

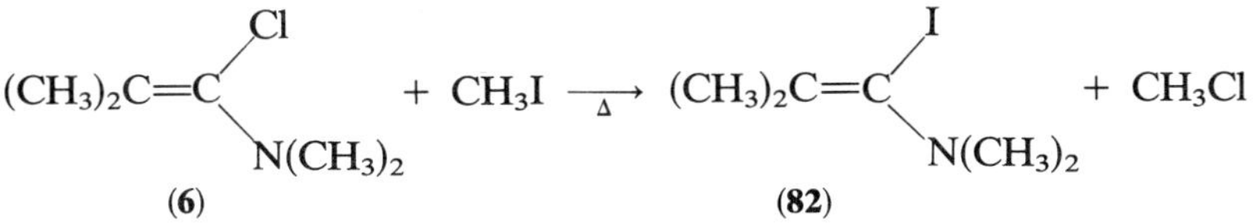

$$(CH_3)_2C=C\begin{smallmatrix}Cl\\N(CH_3)_2\end{smallmatrix} + CH_3I \xrightarrow{\Delta} (CH_3)_2C=C\begin{smallmatrix}I\\N(CH_3)_2\end{smallmatrix} + CH_3Cl$$
$$\quad (6) \qquad\qquad\qquad\qquad\qquad (82)$$

F. MISCELLANEOUS METHODS

1. *Reaction of Trivalent Phosphorus Compounds with α-Substituted Dichloroacetamides*

Several di- or trichloroenamines have been prepared from the reaction of phosphines or phosphites with α-substituted dichloroacetamides (5). Tertiary trichloroacetamides give the corresponding trichlorovinylamines in 23–83% yields (5a,b). In general, trialkylphosphines react faster and give higher yields and purer products than the phosphorus esters.

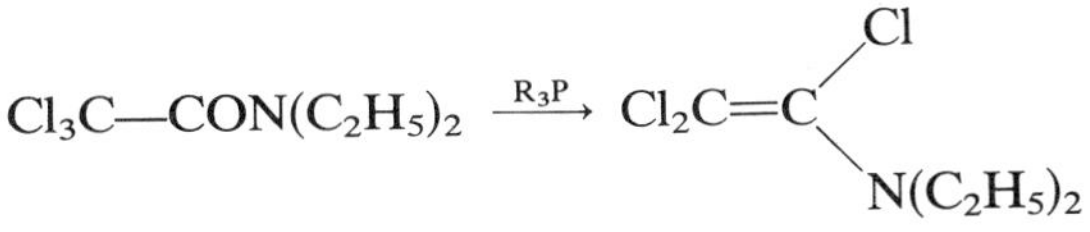

$$Cl_3C-CON(C_2H_5)_2 \xrightarrow{R_3P} Cl_2C=C\begin{smallmatrix}Cl\\N(C_2H_5)_2\end{smallmatrix}$$

R = C_2H_5O 73% at 145–155°C
R = C_4H_9 83% at 20°C

Triphenylphosphine reacts more sluggishly than either phosphites or alkyl phosphines.

In contrast to the foregoing case, the corresponding 2,2-dichloropropionamides and 2,2-dichloroacetamides do not produce any enamine and give only partial recovery of starting material (5c):

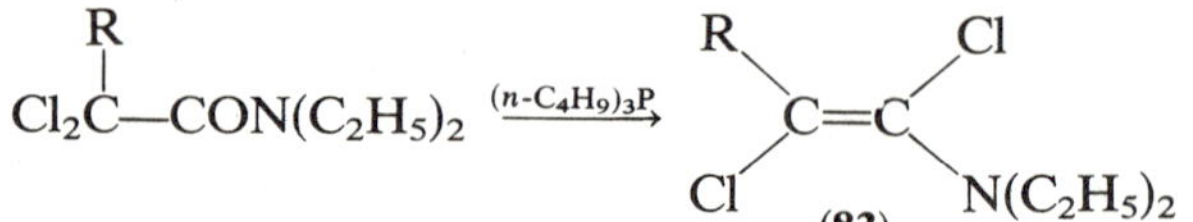

$$Cl_2\overset{\overset{R}{|}}{C}-CON(C_2H_5)_2 \xrightarrow{(n-C_4H_9)_3P} \underset{Cl}{\overset{R}{>}}C=C\underset{N(C_2H_5)_2}{\overset{Cl}{<}}$$

(83)

R = H   0%
R = CH$_3$ 0%
R = C$_6$H$_5$ 70%

However, the 2-phenyl-substituted dichloroacetamides lead (5c) to a high yield of α-chloroenamines. It appears that the anion-stabilizing group (Cl, C$_6$H$_5$) at the α-carbon atom of the amide facilitates the chlorine migration reaction. Furthermore electron-withdrawing groups bonded at the amide nitrogen atom also favor the reaction, as shown by the facile reaction of N,N-diphenyl-2,2-dichloroacetamide with triphenylphosphine, in contrast to the inertness of N,N-diethyl-2,2-dichloroacetamide (5c).

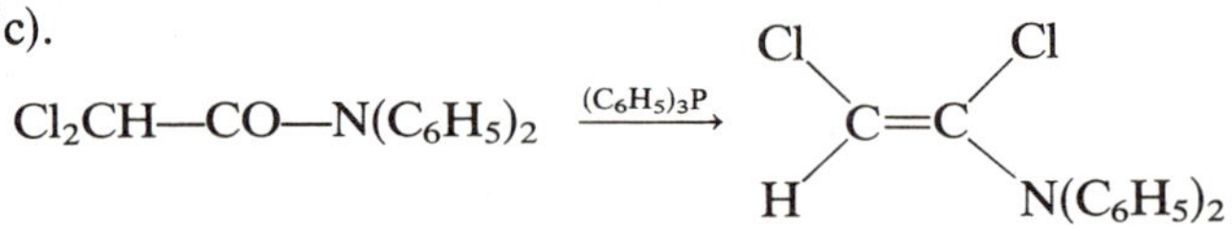

$$Cl_2CH-CO-N(C_6H_5)_2 \xrightarrow{(C_6H_5)_3P} \underset{H}{\overset{Cl}{>}}C=C\underset{N(C_6H_5)_2}{\overset{Cl}{<}}$$

(84) 84%

A kinetic study (5d) has shown that the reaction of triphenylphosphine and α,α-dichloro-α-phenyl-N-methylacetanilide is a second-order polar reaction that is strongly accelerated by electron-attracting substituents attached to the α-phenyl group ($\rho = +2.6$). The results are consistent with a mechanism involving initial attack of the phosphorus atom on the α-chlorine atom to give a phosphonium enolate ion pair (Scheme 16).

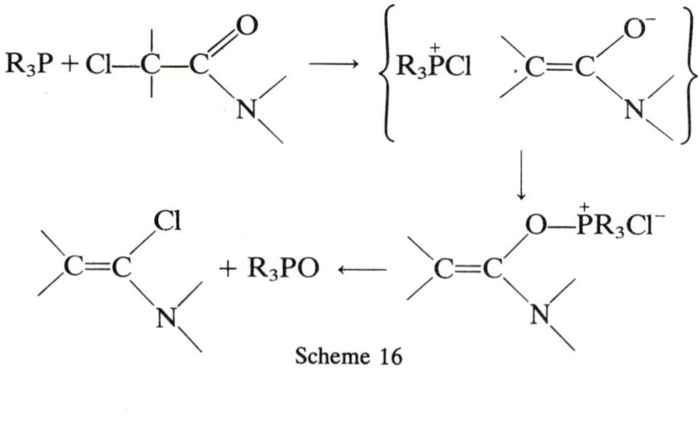

Scheme 16

As anticipated, when the reaction was extended to secondary or primary trichloroacetamides, the trichlorovinylamines were unstable and isomerized to the imidoyl chloride (**85**) or to dichloroacetonitrile (**86**), respectively (5c):

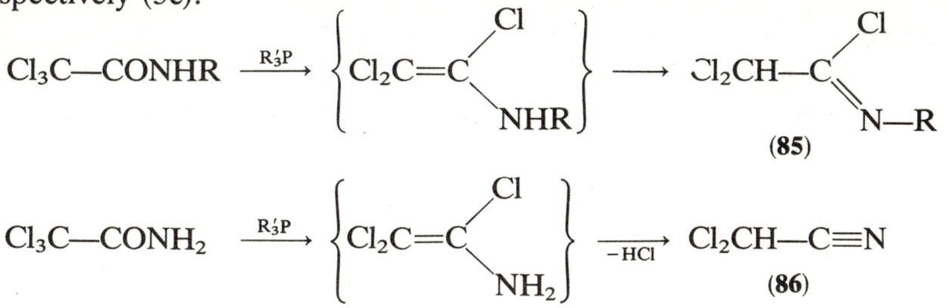

$$Cl_3C-CONHR \xrightarrow{R_3'P} \left\{ Cl_2C=C\begin{matrix}Cl\\NHR\end{matrix} \right\} \longrightarrow Cl_2CH-C\begin{matrix}Cl\\\|\\N-R\end{matrix}$$
$$(\mathbf{85})$$

$$Cl_3C-CONH_2 \xrightarrow{R_3'P} \left\{ Cl_2C=C\begin{matrix}Cl\\NH_2\end{matrix} \right\} \xrightarrow{-HCl} Cl_2CH-C\equiv N$$
$$(\mathbf{86})$$

2. Reactions of Phenyl(trichloromethyl)mercury with Secondary Amines and Some Allyl Amines

The reaction of secondary amines with dichlorocarbene generated from chloroform and potassium *tert*-butoxide yields formamides as products (62). In contrast, when phenyl(trichloromethyl)mercury is used as the carbene generator, the reaction proceeds in a different manner, yielding mixtures of trichlorovinylamines and the corresponding dichloroketene-*N,N*-acetals (63a):

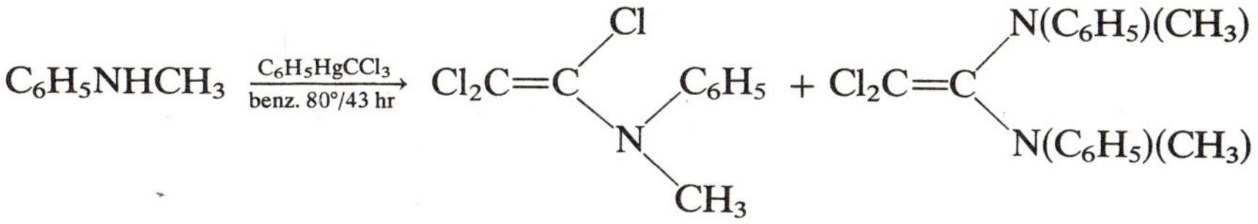

$$C_6H_5NHCH_3 \xrightarrow[\text{benz. } 80°/43\text{ hr}]{C_6H_5HgCCl_3} Cl_2C=C\begin{matrix}Cl\\N\\CH_3\end{matrix}C_6H_5 + Cl_2C=C\begin{matrix}N(C_6H_5)(CH_3)\\N(C_6H_5)(CH_3)\end{matrix}$$

(**87**) 15%   (**88**) 84%

When a series of acyclic butenylamines was reacted with phenyl-(trichloromethyl)mercury, two other reaction paths were observed: (a) a cleavage reaction leading to trichlorovinylamines, and (b) cyclopropane formation (63b). The yields of trichlorovinylamines increase as the basicity of the nitrogen atom decreases.

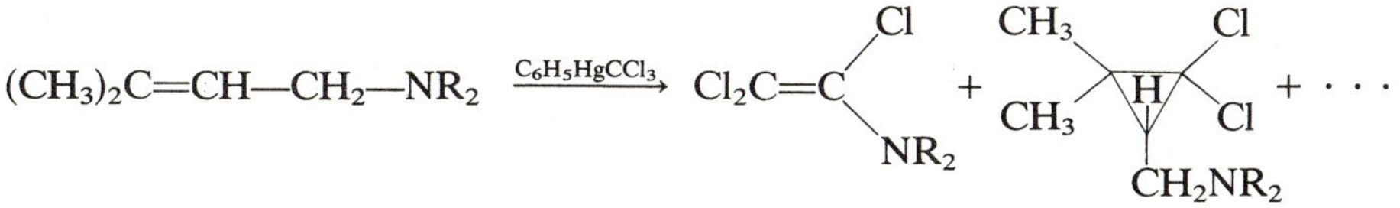

$$(CH_3)_2C=CH-CH_2-NR_2 \xrightarrow{C_6H_5HgCCl_3} Cl_2C=C\begin{matrix}Cl\\NR_2\end{matrix} + \begin{matrix}CH_3\\CH_3\end{matrix}\triangle\begin{matrix}Cl\\H\\Cl\\CH_2NR_2\end{matrix} + \cdots$$

(**5**) R = C$_2$H$_5$,   44%   0%
(**89**) R = C$_6$H$_5$,   <1%   56%

It is suggested that participation of the mercury reagent rather than a carbene intermediate is involved in the cleavage step.

3. Pyrolysis of Perfluorocyclobutane Derivatives

The flow pyrolysis of perfluoro-1-dimethylamino-2-methoxycyclobutane or perfluoro-1,2-bis(dimethylamino)cyclobutane has been reported to give perfluoro-N,N-dimethylvinylamine (**90**) among other products (39):

$$\begin{array}{c} F_2C-CF-N(CF_3)_2 \\ |\quad\quad| \\ F_2C-CF-X \end{array} \xrightarrow{600°} F_2C=C\begin{array}{c} F \\ \diagdown \\ N(CF_3)_2 \end{array} + \cdots$$

X = OCF$_3$, N(CF$_3$)$_2$ \quad\quad (**90**)

4. Phosphorylation of Tricyanomethane Derivatives

An interesting example of an α-chloroenamine with a primary amino group (**91**) has been obtained by Kukhar et al. (64):

$$(NC)_3\bar{C}\,Na^+ + PCl_5 \xrightarrow[20°]{benz.} \begin{array}{c} NC \\ \diagdown \\ NC \end{array} C=C \begin{array}{c} N=PCl_3 \\ \diagup \\ Cl \end{array}$$

$$\downarrow H_2O$$

$$\begin{array}{c} NC \\ \diagdown \\ NC \end{array} C=C \begin{array}{c} NH_2 \\ \diagup \\ Cl \end{array}$$

(**91**)

IV. Structural Data and General Properties of α-Haloenamines

A. STABILITY AND SOLUBILITY PROPERTIES

Some data on the α-haloenamines known to date are reported in Tables I–IV. These compounds appear to be stable only when the amino group is tertiary. Secondary α-haloenamines tautomerize to the more stable imidoyl halides (5c), whereas primary α-haloenamines give the corresponding nitriles (5c):

$$\begin{array}{c} \diagdown \\ \diagup \end{array}C=C\begin{array}{c} X \\ \diagup \\ \diagdown \\ N \\ | \\ H \end{array} \rightleftharpoons \begin{array}{c} \diagdown \\ \diagup \end{array}CH-C\begin{array}{c} X \\ \diagdown\diagdown \\ N-R \end{array}$$

$$\begin{array}{c} \diagdown \\ \diagup \end{array}C=C\begin{array}{c} X \\ \diagup \\ \diagdown \\ NH_2 \end{array} \xrightarrow{R} \begin{array}{c} \diagdown \\ \diagup \end{array}CH-C\equiv N + HX$$

TABLE I
α-Chloroenamines

C_n	Compound	No.	Method of synthesis (section)	Yield, %	Boiling point, °C/mm Hg	Ref.
	2-Alkyl-substituted					
C_5	$CH_3-CH=C(Cl)N(CH_3)_2$	18	III-B-2	~10	—	20
C_6	$(CH_3)_2C=C(Cl)N(CH_3)_2$	6	III-B-1	78–82	129–130/760	26
			III-E-2	75	40/25	9
	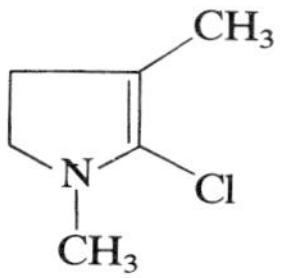 ![structure with CH3, Cl, N-CH3 ring]	8	III-B-1	46	82/80	65
C_7	$CH_3-CH=C(Cl)N$ 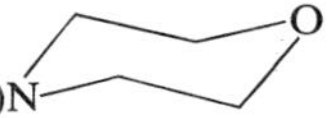 (morpholine)	92	III-B-2	~10	—	20
C_8	$(CH_3)_2C=C(Cl)N(C_2H_5)_2$	93	III-B-1	85	78/13	25
	$(CH_3)_2C=C(Cl)N$ 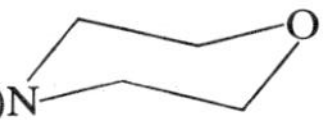 (morpholine)	94	III-B-1	72	56/3	25
	$(CH_3)_2C=C(Cl)N$  (pyrrolidine)	95	III-B-1	76	110/14	60
	$(CH_3)_3C-CH=C(Cl)N(CH_3)_2$	29	III-B-2	60	59–62/16	33,60
	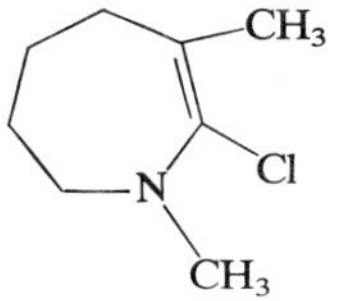 ![azepine structure with CH3, Cl, N-CH3]	96	III-B-1	66	82/15	65

TABLE I (continued)

C_n	Compound	No.	Method of synthesis (section)	Yield, %	Boiling point, °C/mm Hg	Ref.
C_9	 (cyclohexyl)CH=C(Cl)N(...)	97	III-B-1	85	95–100/15	25
	$(CH_3)_2C=C(Cl)N(...)$	2	III-B-1	—	76–85/13	4
	$C_2H_5(CH_3)C=C(Cl)N(C_2H_5)_2$	28	III-B-2	44	44/0.5	20
	$CH_2=C(Cl)N(CH_3)(C_6H_5)$	98	III-B-2	13	32–34/0.4	33
C_{10}	$(CH_3)_2CH—CH=C(Cl)N(C_2H_5)_2$	99	III-B-1	75	96/2	60
	$(CH_3)_2C=C(Cl)N(CH(CH_3)_2)$	100	III-B-2	18	—	33
	$(CH_3)_3C—CH=C(Cl)N(C_2H_5)_2$	101	III-B-2	49	65/1.5	20
C_{11}	$CH_3—CH=C(Cl)N(CH_3)(C_6H_5)$	14	III-B-1	55	103/2	30
	$(CH_3)_2C=C(Cl)N(CH_3)(C_6H_5)$					
	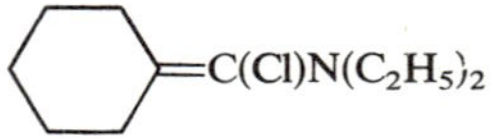 cyclohexylidene=C(Cl)N(C_2H_5)_2	102	III-B-1	40	55/0.2	33
C_{12}	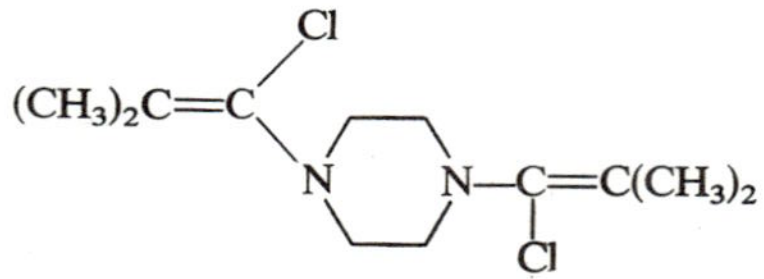 piperazine with two $(CH_3)_2C=C(Cl)$— groups	12	III-B-1	50–60	Solid	29a
C_{13}	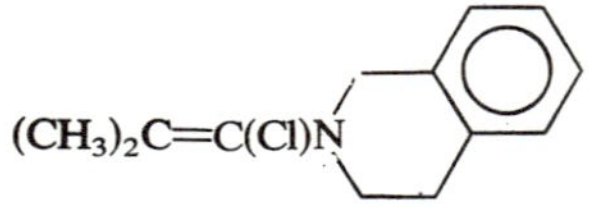 tetrahydroisoquinoline-N-C(Cl)=C(CH_3)_2	103	III-B-1	40	140/2	29a

	Compound	No.	Method	Yield (%)	b.p. (°C/torr)	Ref.
C_{15}	$C_6H_5-CH=C(Cl)N(CH_3)(C_6H_5)$	16	III-B-2	45	180/1	20
C_{16}	$(CH_3)_2C=C(Cl)N(C_6H_{11})_2$	104	III-B-1	80	115–120/0.8 (m.p. ~40°C)	29a
C_{18}	$(CH_3)_2C=C(Cl)N(CH_2-C_6H_5)_2$	105	III-B-1	60	—	29a
	$(C_6H_5-CH_2)_2C=C(Cl)N(CH_3)_2$	106	III-B-1	60	80–90/0.05	29a
C_{19}	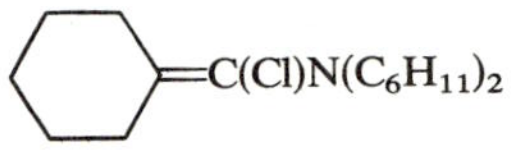 ⌬=C(Cl)N(C_6H_{11})_2	107	III-B-1	40	150/0.6 (m.p. ~80°C)	29a

2-Aryl- and alkenyl-substituted

	Compound	No.	Method	Yield (%)	b.p. (°C/torr)	Ref.
C_7	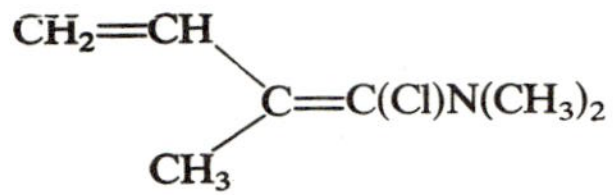 $CH_2=CH-C(CH_3)=C(Cl)N(CH_3)_2$	11	III-B-1	77	50/10	27,28
C_{10}	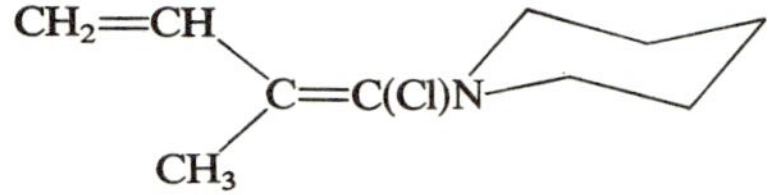 $CH_2=CH-C(CH_3)=C(Cl)N\text{(cyclohexyl)}$	108	III-B-1	34	55/0.6	27
C_{11}	$C_6H_5(CH_3)C=C(Cl)N(CH_3)_2$ (2 isomers)	109	III-B-1	76	86–87/1	33,60
C_{12}	$C_6H_5(C_2H_5)C=C(Cl)N(CH_3)_2$ (2 isomers)	7	III-B-1	91	76–78/0.6	60,66
	 $C_6H_5-CH=C(Cl)N\text{(morpholinyl)}$	110	III-B-2	—	—	30,25

2-Acyl-substituted

	Compound	No.	Method	Yield (%)	b.p. (°C/torr)	Ref.
C_4	$(CN)_2C=C(Cl)NH_2$	91	III-F-4	—	—	64
C_5	$HCO-CH=C(Cl)N(CH_3)_2$	47	III-C-1	97	0/12	42,43
	$(ClCO)(Cl)C=C(Cl)N(CH_3)_2$	31	III-B-3	75	70–75/0.1	37

TABLE I (continued)

C_n	Compound	No.	Method of synthesis (section)	Yield, %	Boiling point, °C/mm Hg	Ref.
C_6	$(CN)_2C=C(Cl)N(CH_3)_2$	**32**	III-B-3	77	180/0.5	35
	$CH_3CO-CH=C(Cl)N(CH_3)_2$	**111**	III-C-1	—	—	43
	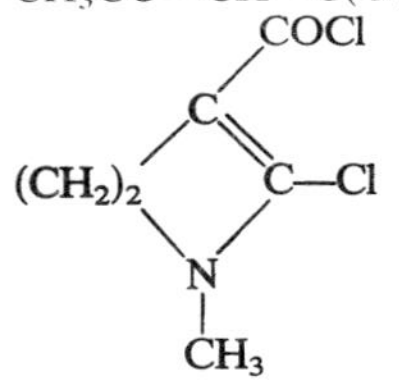 ![structure with COCl, Cl, (CH2)2, N-CH3]	**25**	III-B-2	26	Solid	31b
	$(COCl)(CH_3)C=C(Cl)N(CH_3)_2$	**112**	III-B-2	30	—	20
			III-B-3	70	70–74/0.1	37
C_7	$(ClCO)(C_2H_5)C=C(Cl)N(CH_3)_2$	**113**	III-B-2	32	74–76/0.5	20
			III-B-3	68	—	37
C_8	$(NC)(COOC_2H_5)C=C(Cl)N(CH_3)_2$	**114**	III-B-3	85	135/0.02	35,36a,b
	$(ClCO)(CH_3)C=C(Cl)N(C_2H_5)_2$	**24**	III-B-2	24	78–80/0.5	31b,20
	$(ClCS)(CH_3)C=C(Cl)N(C_2H_5)_2$	**55**	III-C-2	60–70	—	31a
	$(ClSO)((CH_3)_3C)C=C(Cl)N(CH_3)_2$	**56**	III-C-2	60–70	—	31a
	$(COCl)(i-C_3H_7)C=C(Cl)N(CH_3)_2$	**115**	III-B-2	30	—	20
	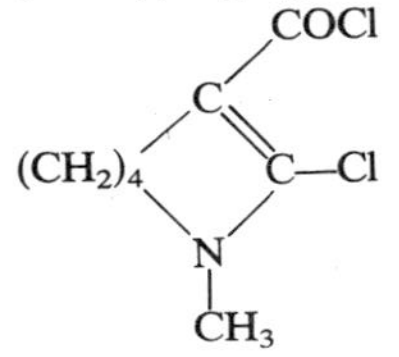 ![structure with COCl, Cl, (CH2)4, N-CH3]	**116**	III-B-2	21	Solid	31b

	Structure	Method	Yield (%)	b.p. (°C/torr)	Ref.
C_9	 **26** (ClCO)(CN)C=C(Cl)N⟩	III-B-2	12	89–90/2.75	67
	30 (ClCO)((CH$_3$)$_3$C)C=C(Cl)N(CH$_3$)$_2$	III-B-2	30	—	20
			15	60–62/2	31b
	54 (ClCO)(CH$_3$)C=C(Cl)N⟩	III-C-2	60–70	—	31a,40
		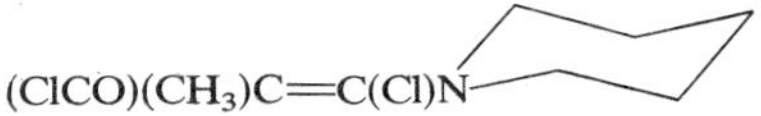 III-B-2	36	118–120/0.5	31b
	117 (ClCO)(C$_2$H$_5$)C=C(Cl)N(C$_2$H$_5$)$_2$	III-C-2	60–70	—	31a,c,2a
		III-B-2	34	76–78/0.5	31b
	118 (CH$_3$CO)(CH$_3$)C=C(Cl)N(C$_2$H$_5$)$_2$	III-C-2	60–70	—	31a,d
		III-C-2	60–70	—	31a,40
C_{10}	**119** (COOC$_2$H$_5$)$_2$C=C(Cl)N(CH$_3$)$_2$	III-B-3	70	—	35
	120 HCO—CH=C(Cl)N(CH$_3$)(C$_6$H$_5$)	III-C-1	—	—	43
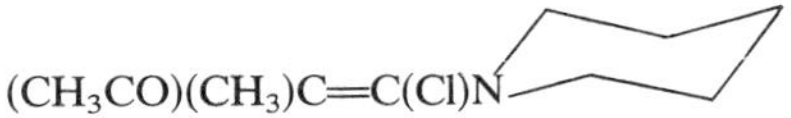	**121** (CH$_3$CO)(CH$_3$)C=C(Cl)N⟩	III-C-2	91	—	31a,40
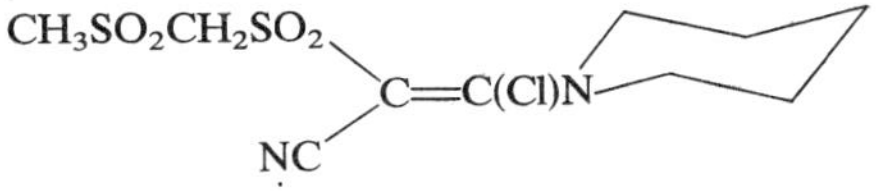	**122** CH$_3$SO$_2$CH$_2$SO$_2$\C=C(Cl)N⟩ /NC	III-C-2	23	Solid (m.p. = 147–148°C)	68
C_{11}	**33** (COCl)(C$_6$H$_5$)C=C(Cl)N(CH$_3$)$_2$	III-B-2	30	—	20
		III-B-3	70	155–160/0.5	37
C_{13}	**123** (ClCO)(C$_6$H$_5$)C=C(Cl)N(C$_2$H$_5$)$_2$	III-B-2	30	76–78/0.5 (solid)	31b
		III-C-2	60–70		31a,d
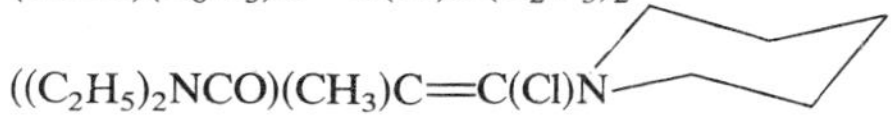	**124** ((C$_2$H$_5$)$_2$NCO)(CH$_3$)C=C(Cl)N⟩	III-F	59	112–114/0.05	31c
C_{14}	 **125** (CH$_3$–⟨C$_6$H$_4$⟩–SO$_2$)(CH$_3$)C=C(Cl)N(C$_2$H$_5$)$_2$	III-C-2	60–70	—	31a

TABLE I (continued)

C_n	Compound	No.	Method of synthesis (section)	Yield, %	Boiling point, °C/mm Hg	Ref.
C_{15}	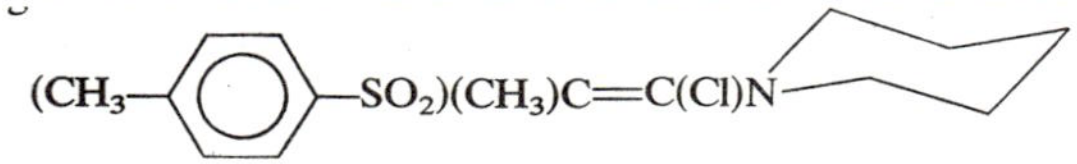 (CH₃)⟨C₆H₄⟩-SO₂)(CH₃)C=C(Cl)N⟨cyclohexyl⟩	126	III-C-2	60–70	—	31a
	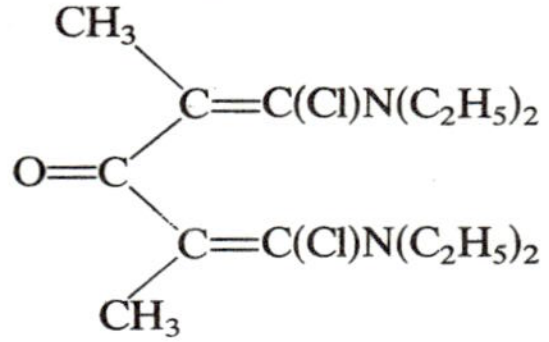 CH₃-C(=C(Cl)N(C₂H₅)₂)-C(O)-C(CH₃)=C(Cl)N(C₂H₅)₂	127	III-C-2	60–70	—	31a
C_{19}	(C₆H₅CO)(C₆H₅)C=C(Cl)N(C₂H₅)₂	53	III-C-2	88	—	31a,40
	2-Heterosubstituted					
C_4	Cl—CH=C(Cl)N(CH₃)₂	128	III-E-1	—	—	52
	Cl₂C=C(Cl)N(CH₃)₂	129	III-F-1	60	65–66/24	5a
C_5	CH₃(Cl)C=C(Cl)N(CH₃)₂	130	III-B-1	30–40	60–65/12	66
C_6	ClCH=C(Cl)N(C₂H₅)₂	62	III-D	77	79–82/13	51
	Cl₂C=C(Cl)N(C₂H₅)₂	5	III-F-1	82	67/6.2	5a,b,c
			III-B-1	—	—	69
			III-F-2	44	—	17a,63
	FClC=C(Cl)N(C₂H₅)₂	131	III-F-1	13	34–35/4.3	5c
	(CF₃)₂N—CH=C(Cl)N(CF₃)₂	52	III-C-1	—	—	46,47
	(CH₃S)(CH₃)C=C(Cl)N(CH₃)₂	132	III-B-1	50	83.5/10	59
	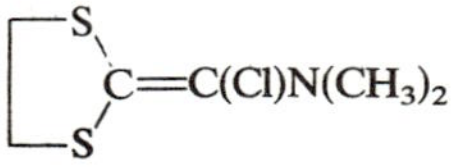 ⟨S-CH₂-CH₂-S⟩C=C(Cl)N(CH₃)₂	133	III-B-1	10–20	—	59

	Compound	No.	Method	Yield (%)	bp (°C/torr)	Ref.
C_7	$Cl(CH_3)C=C(Cl)N(C_2H_5)_2$	134	III-F-1	—	—	5c
	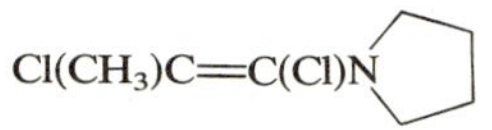 $Cl(CH_3)C=C(Cl)N\bigcirc$	9	III-B-1	70	55/0.2	60,66
C_8	$(ClCO)(Cl)C=C(Cl)N(C_2H_5)_2$	27	III-B-2	65	102-104/6	31b,2a
	$(CH_3OCO)(Cl)C=C(Cl)N(C_2H_5)_2$	135	III-F	72	64-66/0.01	31c
	$Cl-CH=C(Cl)N(CH(CH_3)_2)_2$	136	III-E-1	—	—	52
	$Cl_2C=C(Cl)N(nC_3H_7)_2$	137	III-F-1	52	—	5a
	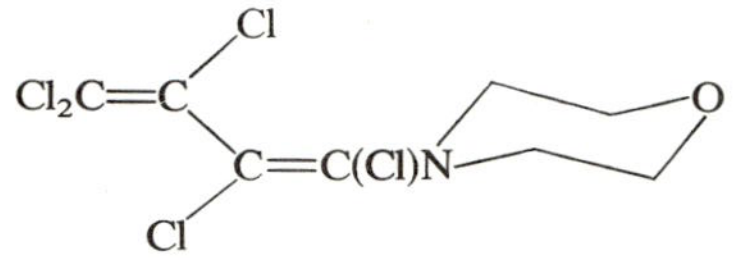 $Cl_2C=C(Cl)N\bigcirc O$ (morpholine)	138	III-E-1	64	125-129/0.01	57
	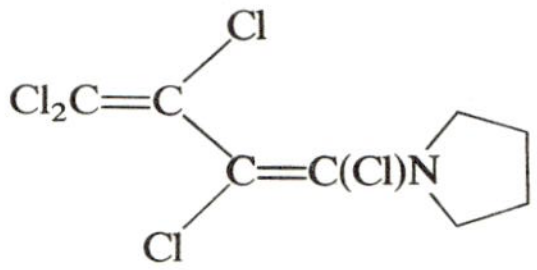 $Cl_2C=C(Cl)N\bigcirc$ (pyrrolidine)	70	III-E-1	76	—	57
C_9	$(ClCOS)(CH_3)C=C(Cl)N(C_2H_5)_2$	139	III-C-3	100	—	49
	$(Cl_3C-S)(CH_3)C=C(Cl)N(C_2H_5)_2$	58	III-C-3	76	112/1.5	48
	$(CH_3S)_2C=C(Cl)N(C_2H_5)_2$	60	III-C-3	100	98-100/0.3	50,59
	$Cl_2C=C(Cl)N(CH_3)(C_6H_5)$	87	III-F-1	11	94-98/0.4-0.7	5b
			III-B-1	—	—	69
			III-F-2	36	—	63
	$ClCH=C(Cl)N(CH_3)(C_6H_5)$	140	III-F-1	17	84-89/1.1	5c
	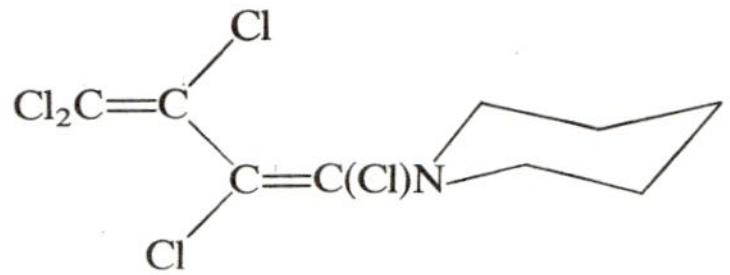 $Cl_2C=C(Cl)N\bigcirc$ (piperidine)	71	III-E-1	66	110-140/0.7-2	57

TABLE I (continued)

C_n	Compound	No.	Method of synthesis (section)	Yield, %	Boiling point, °C/mm Hg	Ref.
C_{11}	$((C_2H_5)_2NCO)(Cl)C=C(Cl)N(C_2H_5)_2$	**141**	III-F	68	99–101/0.1	31c
	$(Cl_3C-S)(C_6H_5)C=C(Cl)N(CH_3)_2$	**142**	III-C-3	48	170–180/0.2	49
	$(C_6H_5S)(CH_3)C=C(Cl)N(CH_3)_2$	**10**	III-B-1	72	102/0.2	59
	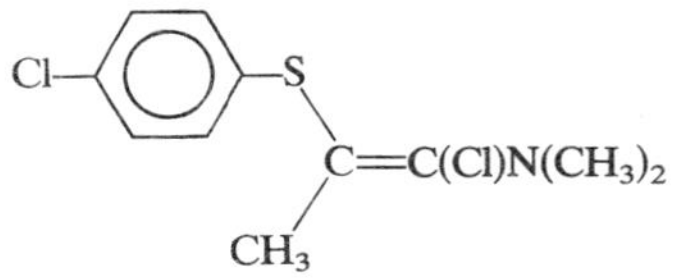 4-Cl-C$_6$H$_4$-S-C(CH$_3$)=C(Cl)N(CH$_3$)$_2$	**143**	III-B-1	70	—	59
C_{12}	$(C_6H_5)(Cl)C=C(Cl)N(C_2H_5)_2$	**83**	III-F-1	70	110–119/1.1	5c, 17a
C_{13}	$(C_6H_5S)(CH_3)C=C(Cl)N(C_2H_5)_2$	**144**	III-C-3	100	—	50, 59
	$(C_6H_5S)(CH_3S)C=C(Cl)N(C_2H_5)_2$	**145**	III-C-3	74	120/10.4	50, 59
C_{14}	$Cl_2C=C(Cl)N(C_6H_5)_2$	**89**	III-F-1	55	Solid	5a
			III-F-2	45		63
			III-B-1	—		69
	$ClCH=C(Cl)N(C_6H_5)_2$	**84**	III-F-1	84	154–158/0.5	5c
	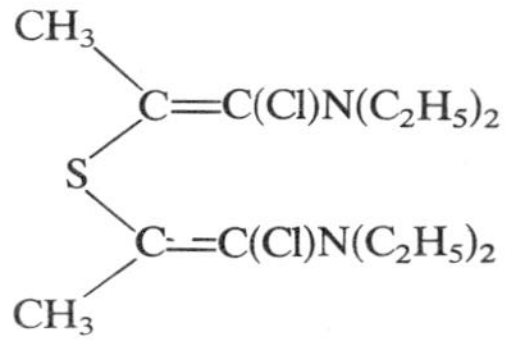 CH$_3$-C(=C(Cl)N(C$_2$H$_5$)$_2$)-S-C(=C(Cl)N(C$_2$H$_5$)$_2$)-CH$_3$ (2 isomers)	**61**	III-C-3	15–20	Solid (m.p. = 16–20°C)	49a

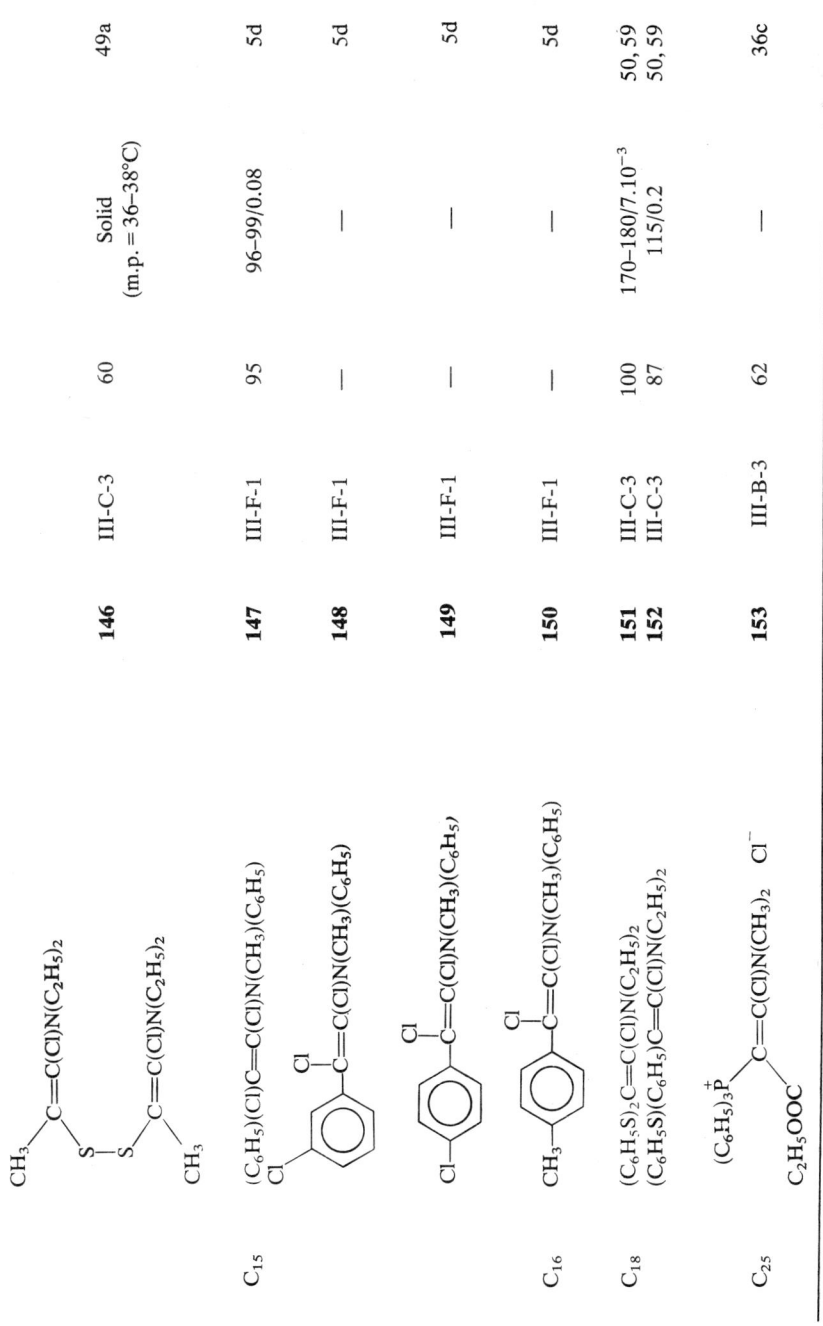

C$_{15}$	**146**	III-C-3	60	Solid (m.p. = 36–38°C)	49a
	147	III-F-1	95	96–99/0.08	5d
	148	III-F-1	—	—	5d
	149	III-F-1	—	—	5d
C$_{16}$	**150**	III-F-1	—	—	5d
C$_{18}$	**151**	III-C-3	100	170–180/7.10^{-3}	50, 59
	152	III-C-3	87	115/0.2	50, 59
C$_{25}$	**153**	III-B-3	62	—	36c

TABLE II
α-Fluoroenamines

C_n	Compound	No.	Method of synthesis (section)	Yield, %	Boiling point, °C/mm Hg	Ref.
	2-Alkyl-substituted					
C_4	$CH_2=C(F)N(CF_3)_2$	37	III-B-4	100	—	38b
C_6	$(CH_3)_2C=C(F)N(CH_3)_2$	77	III-E-2	70	91/760	29a
C_7	$CH_3-CH=C(F)N(C_2H_5)_2$ (2 isomers)	48	III-C-1	100	—	29a
C_{12}	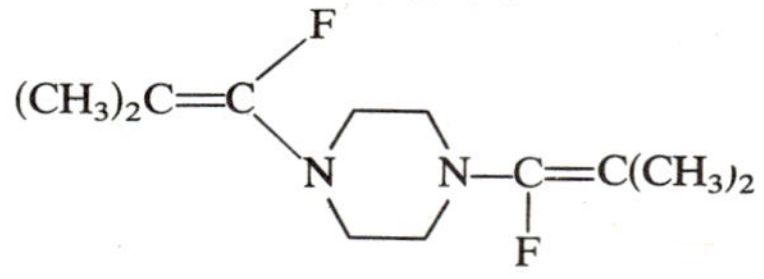 $(CH_3)_2C=C(F)-N\text{-piperazinyl-}C(F)=C(CH_3)_2$	155	III-E-2	55–60	120/12	29a
C_{14}	$H_2C=C(F)N(C_{12}H_8)$	64	III-E-1	51	126–128/2	6a
	2-Aryl- and alkenyl-substituted					
C_9	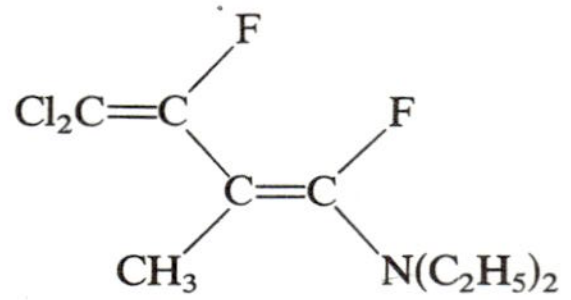 $Cl_2C=C(F)-C(CH_3)=C(F)N(C_2H_5)_2$	49	III-C-1	—	—	44
C_{10}	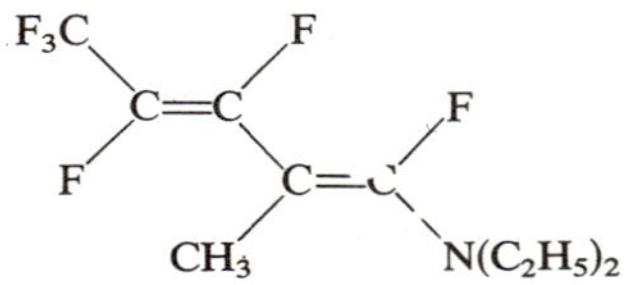 $F_3C-C(F)=C(F)-C(CH_3)=C(F)N(C_2H_5)_2$	156	III-C-1	30	70/12	44

C_{11}	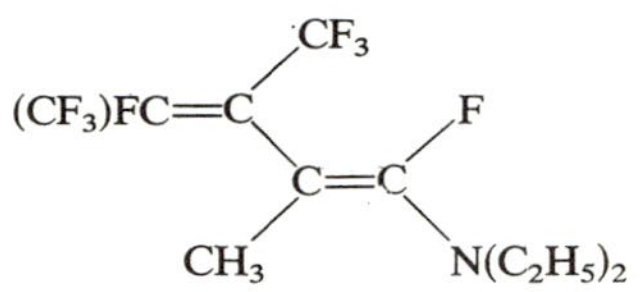 (CF$_3$)FC=C(CF$_3$)—C(F)=C(CH$_3$)N(C$_2$H$_5$)$_2$ (cis + trans)	**157**	III-C-1	95	65–66/15	44
C_{12}	(C$_6$H$_5$)(C$_2$H$_5$)C=C(F)N(CH$_3$)$_2$	**78**	III-E-2	50–60	90/13	29a
C_{14}	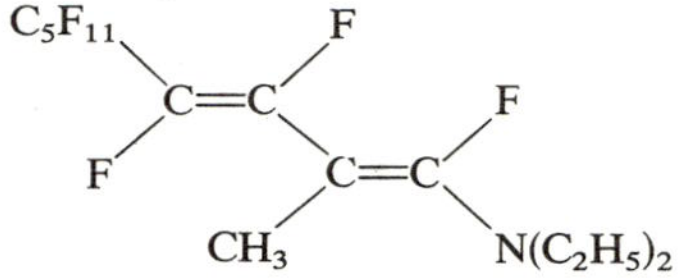 C$_5$F$_{11}$—C(F)=C(F)—C(F)=C(CH$_3$)N(C$_2$H$_5$)$_2$	**158**	III-C-1	—	—	44

2-Acyl-substituted

C_5	HCO—CH=C(F)N(CH$_3$)$_2$	**46**	III-C-1	65	—	41
C_6	(FCO)$_2$C=C(F)N(CH$_3$)$_2$	**159**	III-B-2	15	143–145/1 (m.p. = 75–77°)	70
C_{14}	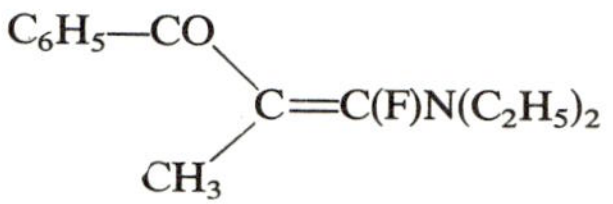 C$_6$H$_5$—CO—C(CH$_3$)=C(F)N(C$_2$H$_5$)$_2$	**57**	III-C-2	100	—	29a

2-Heterosubstituted

C_4	FClC=C(F)N(CH$_3$)$_2$	**68**	III-E-1	69	—	56
	F$_2$C=C(F)N(CF$_3$)$_2$	**90**	III-B-4	99	11.1/760	39
	ClFC=C(F)N(CF$_3$)$_2$		III-F-3	—	11.1/760	39
	ClFC=C(F)N(CF$_3$)$_2$	**39**	III-B-4	87	41.2/760	39
	FHC=C(F)N(CF$_3$)$_2$	**160**	III-B-4	98	24.9/760	39
C_5	(CH$_3$)(Cl)C=C(F)N(CH$_3$)$_2$	**80**	III-E-2	50	—	60
C_6	F$_2$C=C(F)N(C$_2$H$_5$)$_2$	**63**	III-E-1	60	87–89/755	6a
	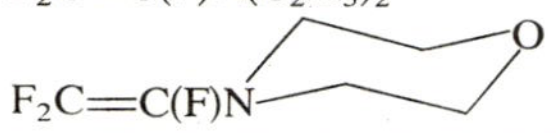 F$_2$C=C(F)N⟨(CH$_2$)$_4$O⟩	**161**	III-E-1	—	34–37/15	6a

TABLE II (continued)

C_n	Compound	No.	Method of synthesis (section)	Yield, %	Boiling point, °C/mm Hg	Ref.
	$FClC=C(F)N(C_2H_5)_2$	66	III-E-1	27–57	42/30	55
	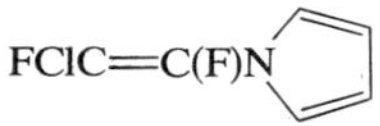 $FClC=C(F)N\text{(pyrrole)}$ (cis + trans)	162	III-E-1	—	70/100	54
	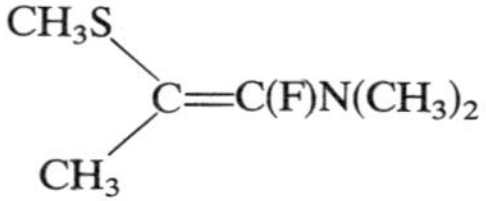 $\begin{array}{c}CH_3S\\ \diagdown\\ C=C(F)N(CH_3)_2\\ \diagup\\ CH_3\end{array}$	79	III-E-2	~100	85–86/55	59
C_7	 $F_2C=C(F)N\text{(cyclo)}$	163	III-E-1	70	42–44/35	6a
	$(CF_3)FC=C(F)N(C_2H_5)_2$	164	III-E-1	25	58–59/49	6a
C_8	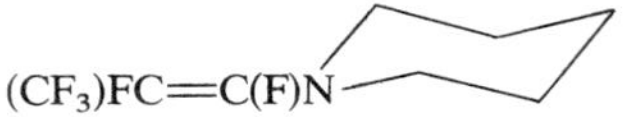 $(CF_3)FC=C(F)N\text{(cyclo)}$	165	III-E-1	43	52–53/8	6a
	$(CH_3S)_2C=C(F)N(C_2H_5)_2$	166	III-E-2	~100	—	59
	$(CF_3S)_2C=C(F)N(C_2H_5)_2$	167	III-F	62	52–54/0.9	71
C_9	$F_2C=C(F)N(CH_3)(C_6H_5)$	168	III-E-1	73	45–46/1	6a
	$FClC=C(F)N(CH_3)(C_6H_5)$	169	III-E-1	71.5	84–85/3	6a
C_{14}	$F_2C=C(F)N(C_6H_5)_2$	170	III-E-1	70	80–82/1.4	6a
	$F_2C=C(F)N(C_{12}H_8)$	171	III-E-1	34	108/1	6a
	$FCH=C(F)N(C_{12}H_8)$	172	III-E-1	48.5	148–152/2	6a
	$FClC=C(F)N(C_{12}H_8)$	173	III-E-1	49	145–146/2	6a
C_{15}	$(CF_3)FC=C(F)N(C_{12}H_8)$	174	III-E-1	—	110–117/1.5	6a

TABLE III
α-Bromoenamines

C_n	Compound	No.	Method of synthesis (section)	Yield, %	Boiling point, °C/mm HG	Ref.
	2-Alkyl-substituted					
C_4	$CH_2=C(Br)N(CF_3)_2$	36	III-C-1	95	60.7/760	46
			III-B-4	99	60.7/760	38a
C_5	$CF_3-CH=C(Br)N(CF_3)_2$	175	III-B-4	90	—	38a
	$\underset{CH_3}{\overset{Br}{}}C=C\underset{H}{\overset{N(CF_3)_2}{}}$ (trans)	176	III-C-1	95	94/755	38b
C_6	$(CH_3)_2C=C(Br)N(CH_3)_2$	81	III-E-2	~90	50/12	29a
	2-Heterosubstituted					
C_4	$BrCH=C(Br)N(CF_3)_2$ (trans)	50	III-C-1	97	114/750	46
	$HFC=C(Br)N(CF_3)_2$ (cis + trans)	177	III-B-4	23	—	38b
				39	70/751	39
	$F_2C=C(Br)N(CF_3)_2$	178	III-B-4	99	50.7/760	39
C_5	$CF_3(Br)C=C(Br)N(CF_3)_2$ (trans)	179	III-C-1	95	117/745	45
C_6	$F_2C=C(Br)N(C_2H_5)_2$	65	III-E-1	76	39/27	6a
	$ClFC=C(Br)N(C_2H_5)_2$	180	III-E-1	31	36/3	6a
	$(CF_3)_2N-CH=C(Br)N(CF_3)_2$ (trans)	51	III-C-1	97	100.1/760	38a
			III-C-1	94	100.1/760	45
	$(CF_3)_2N-C(Br)=C(Br)N(CF_3)_2$ (trans)	181	III-C-1	-97	132/748	45
C_7	$F_2C=C(Br)N\diagdown\diagup$ 	182	III-E-1	41	49/6	6a
	$ClFC=C(Br)N\diagdown\diagup$ 	183	III-E-1	—	69–75/5.5	6a

TABLE IV
α-Iodoenamines

C_n	Compound	No.	Method of synthesis (section)	Yield %	Boiling point, °C/mm Hg	Ref.
	2-Alkyl-substituted					
C_5	CF_3—CH=C(I)N(CF_3)_2 (cis + trans)	184	III-C-1	99	95.1/760 98.1/760	38a
C_6	$(CH_3)_2C$=C(I)N(CH_3)_2	82	III-E-2	85	61-63/9	29a
C_8	$(CH_3)_2C$=C(I)N⌐⌐O 	185	III-E-2	85	70/0.3	29a
	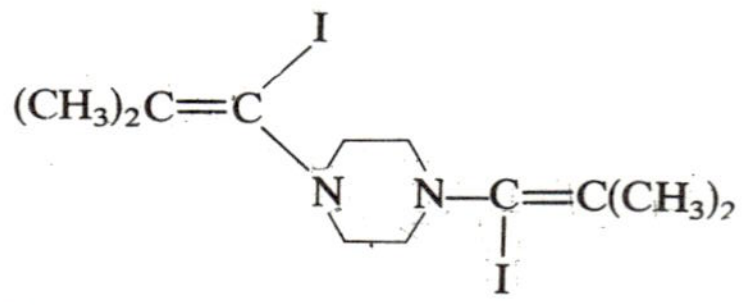 (CH_3)_2C=C(I)—N⌐⌐N—C(I)=C(CH_3)_2	186	III-E-2	60	Solid	29a
	2-Heterosubstituted					
C_6	FClC=C(I)N(C_2H_5)_2	67	III-E-1	18–43	75–78/12	55

IV. STRUCTURAL DATA AND GENERAL PROPERTIES OF α-HALOENAMINES

However, the presence of substituents that enter into conjugation with the enamine function stabilizes the enamine form (64):

$$\underset{NC}{\overset{NC}{>}}C=C\underset{NH_2}{\overset{Cl}{<}} \longleftrightarrow (CN)_2CH-C\equiv N + HCl$$

(91)

The substituents on the nitrogen may be alkyl, aryl, trifluoromethyl, or part of a heterocyclic ring; the substituents on the C_2 atom may be hydrogen, alkyl, alkenyl, aryl, halogen, thioalkyl, acyl, cyano, etc.

α-Haloenamines are liquids or crystalline solids that are soluble in nonpolar sovents such as benzene, ether, or dichloromethane, as well as in chloroform, acetonitrile, or dimethylformamide. They are usually very sensitive to moisture and react rapidly with water, aqueous acids, or bases to give the corresponding amides (25,31a,57,72):

$$\underset{R^2}{\overset{R^1}{>}}C=C\underset{NR^3R^4}{\overset{X}{<}} \xrightarrow{H_2O} \underset{R^2}{\overset{R^1}{>}}CH-CO-NR^3R^4$$

They are also unstable in protic solvents (31a,c,72,73).

The thermal stability of α-haloenamines varies considerably with the substitution. Thus tetramethyl-α-chloroenamine (6) is a stable colorless liquid that distills at 129–130°C at atmospheric pressure without decomposition (26). On the other hand, 2-monosubstituted 1-chloro-N,N-dialkylenamines are very unstable and are difficult to obtain in the pure state (20,33). The simplest α-chloroenamine, 1-chloro-1-dimethylamino-ethylene, is still unknown. Usually the thermal stability also decreases with increasing basicity of the amino group (20,60).

B. STRUCTURAL DATA ON α-HALOENAMINES

The physical properties mentioned above indicate that keteniminium halides exist predominantly in the covalent α-haloenamine structures, in contrast to all other iminium halides (10,11,24). This view is completely supported by all their spectroscopic properties as well as by X-ray analysis (29b).

The IR spectra of α-haloenamines show no characteristic absorption for cumulenes around 2000 cm^{-1} (19,20,29a,33,59,60). α-Chloroenamines show an absorption band at 1635–1645 cm^{-1} for the C=C stretch (19,20). Conjugation of the enamine function with a carbonyl group lowers (31) the frequency to 1530–40 cm^{-1}. Replacement of chlorine by fluorine increases (29a) the frequency to 1730–1805 cm^{-1}.

The proton (33) and ^{13}C magnetic resonance spectra of α-haloenamines (74) resemble those of the corresponding enamines, thus supporting the covalent structure.

The crystalline α-chloroenamine (**12**) has been analyzed (29b) by X-ray diffraction (Scheme 17). The length of the olefinic C–C double bond is normal (1.33 Å) (75), indicating that the free pair on the nitrogen does

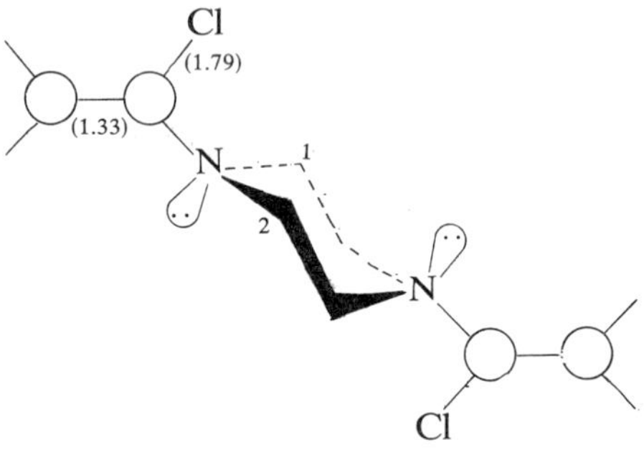

Scheme 17

not interact significantly with the π electrons of the double bond. This is confirmed by the striking observation that the plane of the olefinic system is almost perpendicular to the plane formed by nitrogen and carbon atoms 1 and 2 (Scheme 17) of the six-membered ring. Furthermore the carbon–chlorine distance is exceptionally long: its value (1.79 Å) is even greater then would be expected for a normal C_{sp^3}–Cl bond (75). This suggests a hyperconjugative interaction between the C–Cl bond and the lone-pair electrons of the nitrogen:

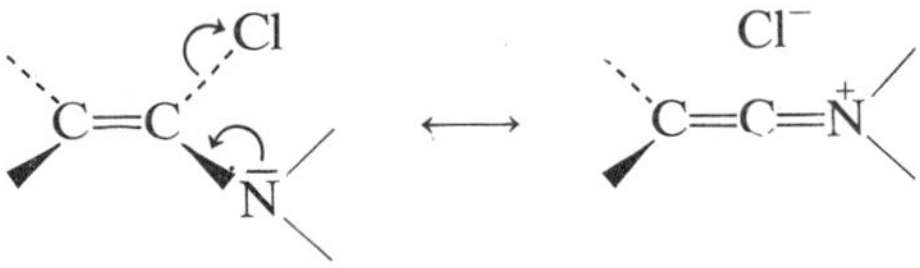

The existence of this effect, which, in the cases studied, overcomes the conjugation of the lone pair with the π olefinic system, is also clearly demonstrated (76) by nuclear quadrupole resonance frequencies for the chlorine atom (Table V). If the sole factor influencing the NQR frequency of the chlorine were the electronegativity of the nitrogen substituent, the NQR frequency would be expected to be increased in relation to that of vinyl chloride. Experimentally, however, the opposite situation is found,

IV. STRUCTURAL DATA AND GENERAL PROPERTIES OF α-HALOENAMINES

TABLE V
^{35}Cl Nuclear Quadrupole Resonance Absorption Frequencies

Compound	q, MHz	Ref.
CH$_3$Cl	34.029	77
H$_2$C=CH—Cl	33.414	78
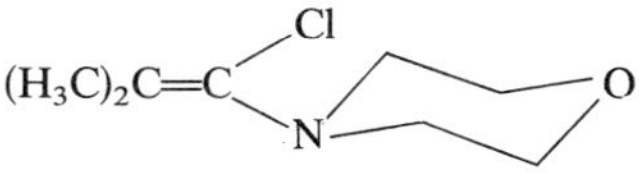 (H$_3$C)$_2$C=C(Cl)—N(morpholino)	31.15 (±0.09)	76
(H$_3$C)$_2$C=C(Cl)—N(C$_2$H$_5$)$_2$	30.50 (±0.05)	76
CH$_3$—O—CH$_2$—Cl	30.181	79

showing that nitrogen has not withdrawn electrons but rather injected them, making the chlorine sbustituent more negative.

C. GEOMETRICAL ISOMERISM IN α-HALOENAMINES

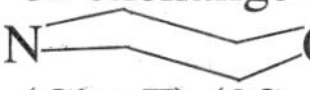

α-Haloenamines bearing two different substituents at C$_2$ may exist in the E or Z configuration. The methods of synthesis presently available usually give mixture of isomers (Tables I-IV) which have been detected and sometimes identified by PMR spectroscopy. The ratio of the two isomers is determined primarily by thermodynamic control. The fact that the geometrical isomers are often readily interconvertible in solution or as neat liquids at room temperature precludes their separation. Interconversion is also easily observable in 2-dimethyl-1-chloroenamines, where the two methyl groups are in a different magnetic environment and should normally give two different signals (33,74). This is indeed observed when, for instance, the spectrum of tetramethyl-α-chloroenamine (**6**) is taken at 25°C in CCl$_4$ or C$_2$Cl$_4$. However, at higher temperature or in CDCl$_3$, CD$_3$CN, CD$_3$NO$_2$, etc., these methyl protons give a single resonance line at a frequency that is the average of those of the two separate signals (33,74). The rate of exchange increases with the basicity of the nitrogen atom [N(CH$_3$)$_2$ > N(morpholino)] (33,74) and the nucleofugal property of the halogen atom (Cl ≫ F) (29a,74). The available data favor an exchange mechanism involving an intermediate (or transition-state) keteniminium halide:

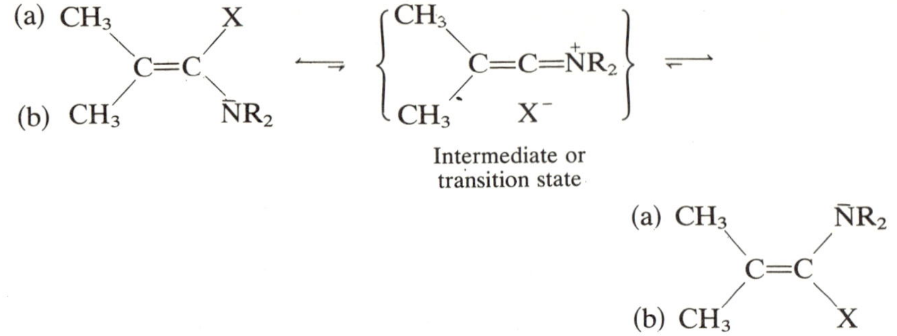

(a) CH$_3$, (b) CH$_3$ \C=C/ X, NR$_2$ ⟶ { CH$_3$, CH$_3$ \C=C=N$^+$R$_2$ X$^-$ } ⇌

Intermediate or
transition state

(a) CH$_3$, (b) CH$_3$ \C=C/ NR$_2$, X

D. STRUCTURE OF KETENIMINIUM SALTS

Direct observation of a keteniminium ion is possible when the nucleophilic halide ions are replaced by such counteranions as BF_4^- and PF_6^- (9,80). These salts have been obtained in solutions when α-chloroenamines were treated with the appropriate silver salts in inert solvents such as methylene chloride or chloroform (7,9,80).

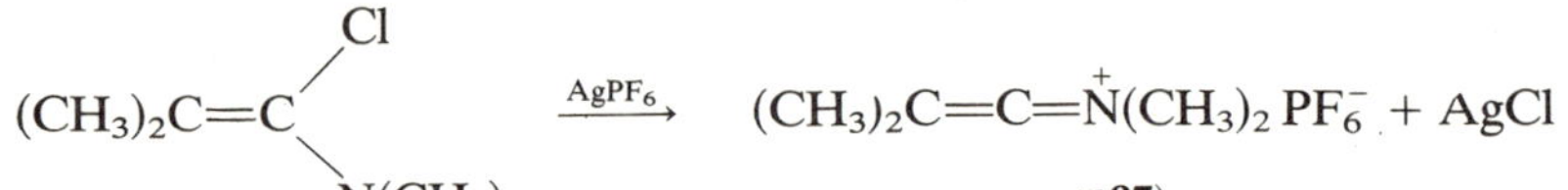

$(CH_3)_2C=C\genfrac{}{}{0pt}{}{Cl}{N(CH_3)_2}$ $\xrightarrow{AgPF_6}$ $(CH_3)_2C=C=\overset{+}{N}(CH_3)_2$ PF_6^- + AgCl

(6)   (187)

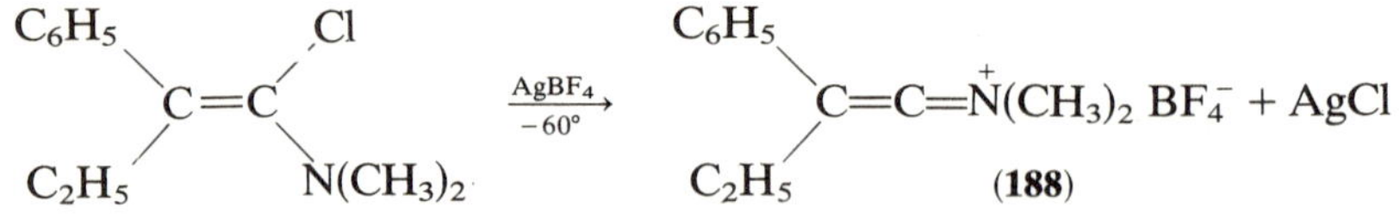

$\begin{matrix}C_6H_5\\C_2H_5\end{matrix}C=C\begin{matrix}Cl\\N(CH_3)_2\end{matrix}$ $\xrightarrow[-60°]{AgBF_4}$ $\begin{matrix}C_6H_5\\C_2H_5\end{matrix}C=C=\overset{+}{N}(CH_3)_2$ BF_4^- + AgCl

(7)   (188)

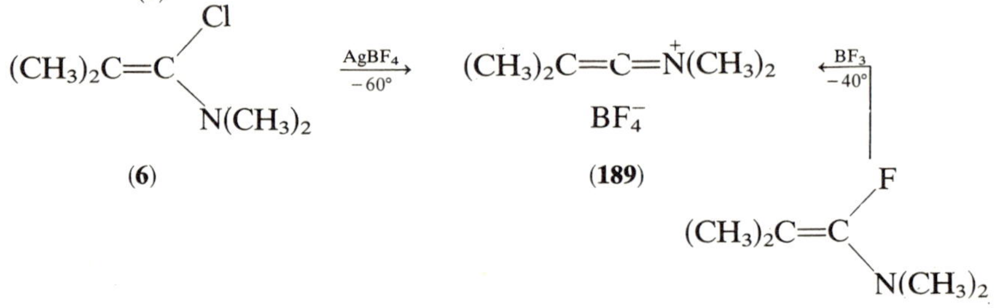

$(CH_3)_2C=C\genfrac{}{}{0pt}{}{Cl}{N(CH_3)_2}$ $\xrightarrow[-60°]{AgBF_4}$ $(CH_3)_2C=C=\overset{+}{N}(CH_3)_2$ $\xleftarrow[-40°]{BF_3}$

(6) BF_4^-

                      (189)

$(CH_3)_2C=C\genfrac{}{}{0pt}{}{F}{N(CH_3)_2}$

(77)

$(CH_3)_2C=C\genfrac{}{}{0pt}{}{Cl}{N(CH_3)_2}$ $\xrightarrow[20-40°]{ZnCl_2}$ $(CH_3)_2C=C=\overset{+}{N}(CH_3)_2$ $ZnCl_3^-$

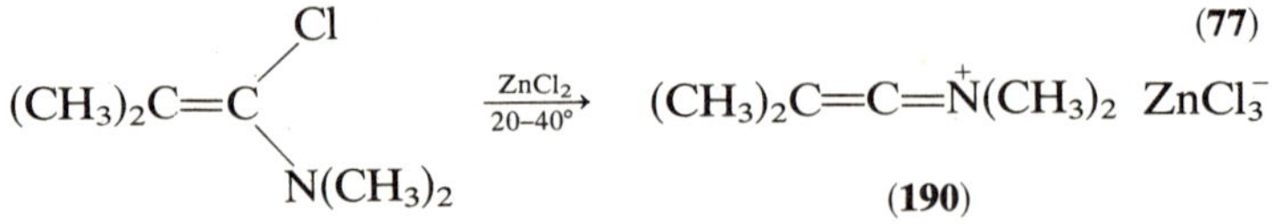

(6)   (190)

IV. STRUCTURAL DATA AND GENERAL PROPERTIES OF α-HALOENAMINES   469

The reaction of tetramethyl-α-fluoroenamine (**77**) with boron trifluoride at −40°C also gives (29a) the keteniminium salt (**189**). The corresponding zinc salt (**190**) is obtained on reaction of tetramethyl-α-chloroenamine with zinc chloride at room temperature (8,60).

These solutions show an IR absorption bond at 2020–30 cm^{-1}, which is typical for the cumulative arrangement of the C–C and C–$\overset{+}{N}$ double bonds (60). The PMR spectra confirm the structures: for instance, for tetramethylketeniminium ion the NMR spectra show two sharp singlets which, as expected, appear at lower fields than the corresponding signals in the α-haloenamines (9,74,80).

The "cumulene" structure of keteniminium salts is clearly indicated (74) by ^{13}C NMR. Thus the central sp carbon gives rise to a signal at a very low field (215.0 ppm from TMS) and is even more deshielded than the corresponding carbon atom in ketenes [194.0 ppm from TMS (81)]. However, the C_2 atom gives a signal at 88.6 ppm, a much lower field than is found for ketenes [2.5 ppm from TMS (81)].

Comparison of the structure of keteniminium ions with the structures of related ketenes is quite instructive. An important feature of the electronic structures of these molecules is the number of electrons involved in the two orthogonal π systems containing two electrons each. On the other hand, ketenes and ketenimines are characterized by the presence of a π system containing two electrons, the C–O and C–N double bonds, respectively, and a four-π-electron system involving the two π electrons of the olefinic double bond and the lone pair of the heteroatom (Scheme 18).

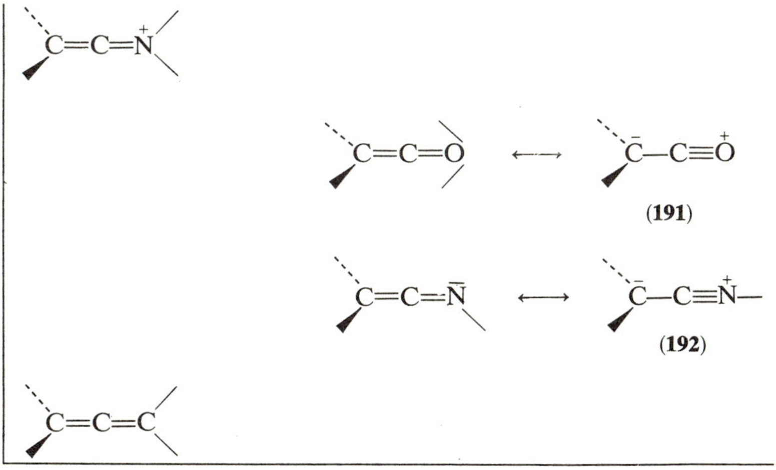

$$\diagdown\!\!\!\!\!C\!\!=\!\!C\!\!=\!\!\overset{+}{N}\!\diagup$$

$$\diagdown\!\!\!\!\!C\!\!=\!\!C\!\!=\!\!O \quad \longleftrightarrow \quad \diagdown\!\!\!\!\!\overset{-}{C}\!\!-\!\!C\!\!\equiv\!\!\overset{+}{O}$$
(**191**)

$$\diagdown\!\!\!\!\!C\!\!=\!\!C\!\!=\!\!\bar{N} \quad \longleftrightarrow \quad \diagdown\!\!\!\!\!\overset{-}{C}\!\!-\!\!C\!\!\equiv\!\!\overset{+}{N}\!\!-$$
(**192**)

$$\diagdown\!\!\!\!\!C\!\!=\!\!C\!\!=\!\!C\!\diagup$$

Scheme 18

TABLE VI

Electronic Populations of the $p\pi$ Orbitals in $H_2C=C=X$

	Allene	Ketenimine	Ketene	Keteniminium
$C_{2(p_y)}$	1.09	1.18	1.29	0.85
$C_{1(p_y)}$	0.93	0.95	0.96	1.16
$C_{1(p_z)}$	0.93	0.85	0.77	0.55
$N_{(p_z)}$		1.17		1.52
$O_{(p_z)}$			1.27	
$O_{(p_y)}$			1.74	

The contribution of valence bond structures such as **191** and **192** is significant, as indicated by the net population of the different $p\pi$ orbitals (82) (Table VI).

A significant excess of electronic charge at the p_y orbital on the C_2 atom in ketenes and ketenimines results from the interaction with the lone pair of the heteroatom. This may also be partly responsible for the strong shielding of the ^{13}C signal for the C_2 atom in ketenes (81). On the other hand, the same p_y orbital at C_2 of keteniminium shows an electron deficiency induced by the positively charged iminium group.

Experimentally the ambident character of ketenes reveals itself in their strong tendency to oligomerize or polymerize (83). Ketenimines have been shown to undergo cycloaddition reactions involving either of the two orthogonal π systems, depending on the reaction partner (84,85). In contrast keteniminium salts do not show any tendency to dimerize (80): a solution of tetramethylketeniminium tetrafluoroborate can be kept at room temperature for several months without any observable change in the NMR spectra (28). This behavior is in marked contrast with the high thermal instability of dimethylketene (83) under comparable conditions.

V. Nucleophilic Substitutions on α-Haloenamines

A. MECHANISTIC ASPECTS

A large variety of nucleophilic reagents has been used to displace the halide ion on α-haloenamines to form substitution products. These substitution reactions are remarkably fast with alkyl-substituted α-chloroenamines, in contrast to what is normally encountered for simple alkenyl chlorides (86). With these α-chloroenamines the reaction usually takes place at room temperature and yields are excellent (19,25).

The displacement of the 1-haloatom could proceed by the following

mechanisms:
1. Direct substitution

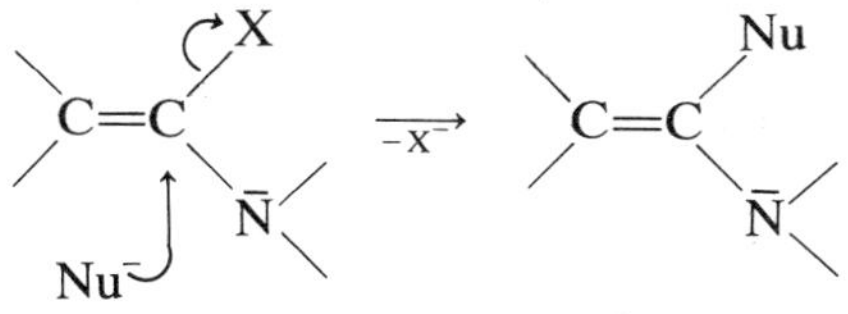

2. α-Elimination and α-addition

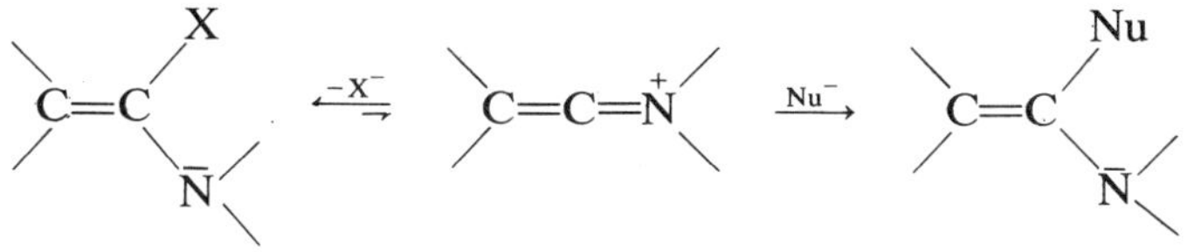

3. α-Addition followed by α-elimination

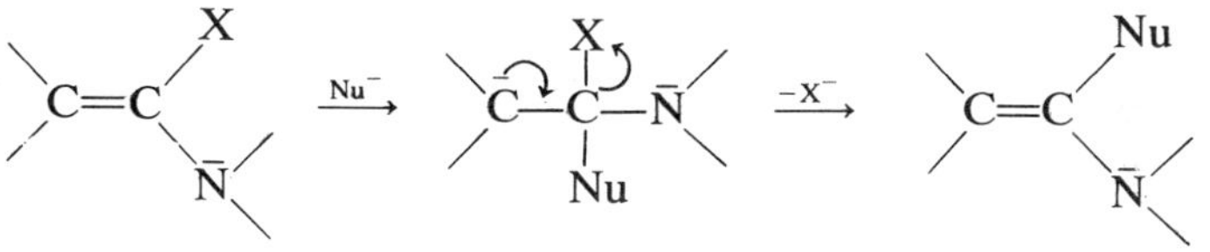

4. Elimination to an ynamine, followed by addition

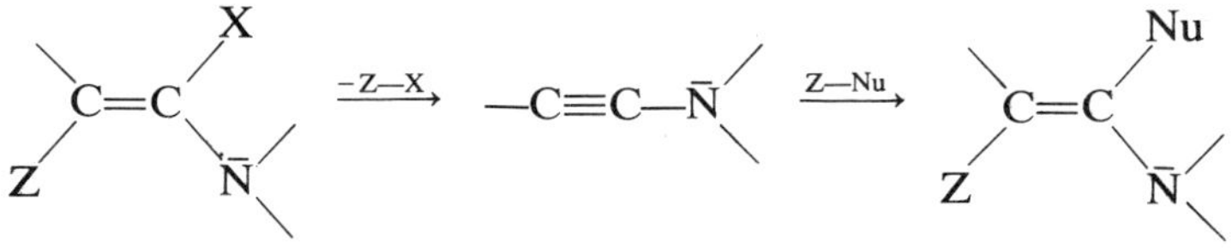

The available experimental data allow rejection of mechanisms 3 and 4 for the following reasons:

1. An addition–elimination mechanism implying the formation of a carbanionic center at C_2 as the rate-determining step requires that the rate of substitution under a given set of conditions be enhanced by the presence of electron-withdrawing groups at C_2, as found for simpler ethylenic substrates (86). Experimentally, however, the reverse situation is observed. Thus for the reaction of α-chloroenamines with sodium methoxide (72,87) the rate of substitution decreases in the following order:

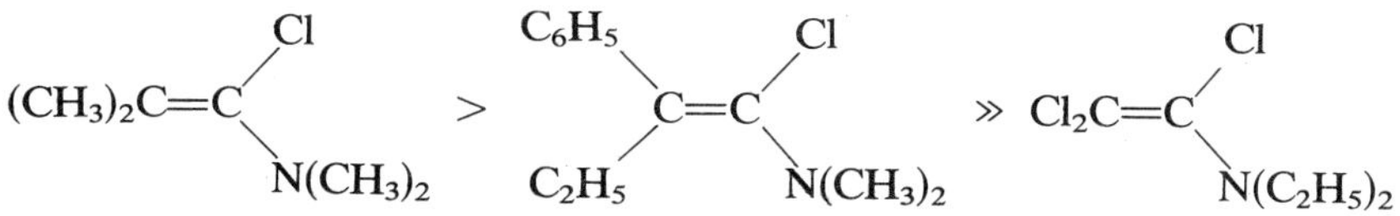

$(CH_3)_2C=C(Cl)(N(CH_3)_2)$ > $(C_6H_5)(C_2H_5)C=C(Cl)(N(CH_3)_2)$ ≫ $Cl_2C=C(Cl)(N(C_2H_5)_2)$

For these types of substrates at least, mechanism 3 should be rejected. This type of mechanism cannot definitely be ruled out, however, for α-haloenamines bearing at C_2 such functional groups as CO or CN.

2. Elimination to an ynamine (mechanism 4) is clearly impossible for α-haloenamines bearing no hydrogen (or eventually halogen) at C_2. Moreover, even when the elimination step is possible, the second step is feasible only under protic conditions (2). Therefore mechanism 4 is very limited in scope.

No mechanistic studies are yet available that would allow differentiation between mechanisms 1 and 2. However, several features suggest the formation of a transient keteniminium chloride.

1. In aprotic medium the α-chloroenamines react faster than the α-fluoroenamines in agreement with the order of increasing bond energies (29a). Thus the reaction of tetramethyl-α-chloroenamine and sodium methoxide takes place readily in ether at room temperature (87), whereas the corresponding α-fluoroenamine is inert under these conditions (29a).

2. The reaction is catalyzed by Lewis acids. The α-chloroenamines react instantaneously with silver salts [AgCN (66), AgN_3 (33,88)] to give the corresponding substitution products and silver chloride. The reactions of α-fluoroenamines are catalyzed by the lithium cation (29a). Thus, whereas tetramethyl-α-fluoroenamine is completely inert toward sodium methoxide in ether, it reacts rapidly with lithium methoxide (29a).

3. The rate of substitution under a given set of conditions increases with the basic strength of the amino group (33,89):

$$-N\begin{matrix}CH_3\\C_6H_5\end{matrix} \ll -N\frown O < -N(CH_3)_2$$

This corresponds to the order of increasing stability of the keteniminium ions:

$$\begin{matrix}\\ \\ \end{matrix}C=C=\overset{+}{N}\begin{matrix}C_6H_5\\CH_3\end{matrix} \ll \begin{matrix}\\ \\ \end{matrix}C=C=\overset{+}{N}\frown O < \begin{matrix}\\ \\ \end{matrix}C=C=\overset{+}{N}(CH_3)_2$$

4. The rate of substitution is faster in solvents like dimethylformamide or acetonitrile than in ether or benzene: 1-chloro-1 dimethylamino-2 phenylprop-1-ene (**109**) and sodium azide do not react in ether at room temperature, but in DMF at room temperature they give the aminoazirine in less than 2 hr (33).

5. Strongly nucleophilic aromatic compounds such as pyrrole and furan (Section V-C-4) react with α-chloroenamines, often without added catalysts, to give substitution products (90,91). This is clearly reminiscent of the behavior of typical iminium salts (10,11,24a,92).

Although mechanism 2 appears more feasible from the data presently available, no definitive mechanistic picture can be drawn as yet, and one should bear in mind the possibility of a continuous spectrum of mechanisms between limiting mechanisms 1 and 2, depending on the experimental parameters. The stereochemistry of these substitution reactions is still unknown although it seems that in most cases mixtures of stereoisomers are obtained (66,87).

B. NUCLEOPHILIC SUBSTITUTIONS WITH FORMATION OF C–HETEROATOM BONDS

1. Reactions with Hydroxide, Alkoxide, and Thiolates

α-Haloenamines react (19) readily with aqueous sodium hydroxide to yield the corresponding substitution products, which tautomerize to the stable tertiary amides (**193**):

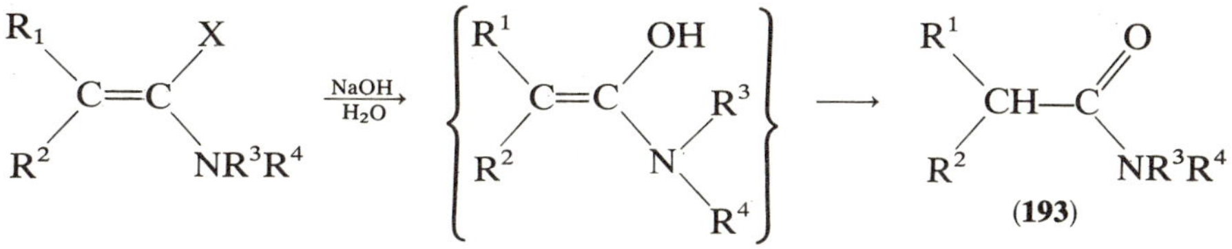

The reactions of α-chloroenamines with alkoxides or thiolates also take place under mild conditions, usually at room temperature, to give the stable ketene O,N-ketal (**194**) or S,N-ketal (**195**) in high yields (25,87) (Table VII):

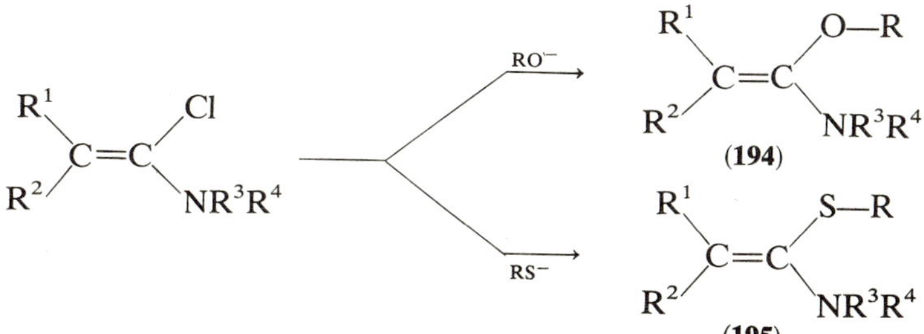

TABLE VII

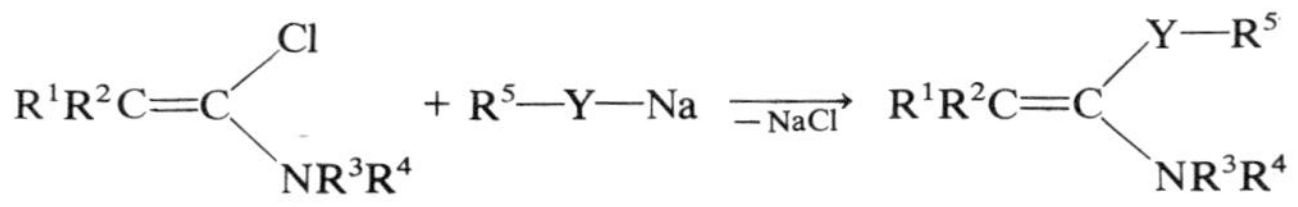

$$R^1R^2C=C\underset{NR^3R^4}{\overset{Cl}{\diagdown}} + R^5-Y-Na \xrightarrow{-NaCl} R^1R^2C=C\underset{NR^3R^4}{\overset{Y-R^5}{\diagdown}}$$

α-Chloroenamines				R^5-Y-Na, n equiv.	Experimental conditions	Yield, %	Boiling point, °C/mm Hg	Ref.
R^1	R^2	R^3	R^4					
CH_3	CH_3	$-(CH_2)_5-$		C_2H_5ONa, 1	THF/20°C/1–3 hr	80	83/0.37	25
CH_3	CH_3	CH_3	CH_3	C_2H_5ONa, 1	Ether/20°C/3 hr	86	56/29	87
CH_3	CH_3	CH_3	CH_3	CH_3ONa, 1	Ether/20°C/3 hr	82	78/108	87
C_6H_5	C_2H_5	CH_3	CH_3	CH_3ONa, 1	Ether/20°C/3 hr	83	69/0.2	87
Cl	Cl	C_2H_5	C_2H_5	C_2H_5ONa, 1	–/20°C/22 hr	83	—	72
CH_3	CH_3	$-(CH_2)_5-$		C_2H_5SNa, 1	THF/20°C/1–3 hr	90	76/0.3	25
CH_3	CH_3	CH_3	CH_3	C_2H_5SNa, 1	Ether/25°C/3 hr	87	84/28	87
COCl	CH_3	$-(CH_2)_5-$		CH_3ONa, 2	Ether/25°C/10 hr	74	104–106/0.5	31c
COCl	Cl	C_2H_5	C_2H_5	CH_3ONa, 2	Ether/25°C/10 hr	63	82–84/0.01	31c
COCl	CH_3	$-(CH_2)_5-$		C_2H_5SNa, 2	THF/25°C/10 hr	68	170–171/4	31c

V. NUCLEOPHILIC SUBSTITUTIONS ON α-HALOENAMINES

With β-chlorocarbonyl-α-chloroenamines, the reaction first takes place at the acyl chloride functional group to give an isolable substitution product, which further reacts with the nucleophilic reagent to yield the ketene O,N-acetal (**196**) or S,N-acetal (**197**) (31c):

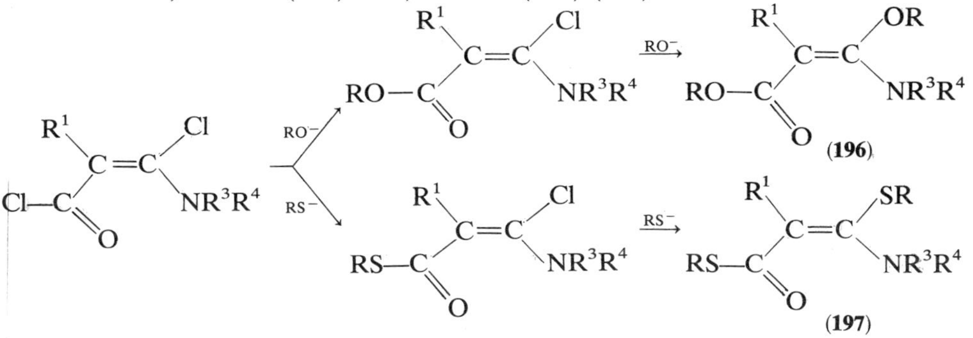

2. Reactions with Carboxylate Anions

The sodium salts of carboxylic acids do not react with α-chloroenamines at room temperature. The reaction must be conducted at 80°C and gives (93) the α-acyloxyenamine (**198**), contaminated by the acid anhydride (**199**) and the isobutyramide (**200**):

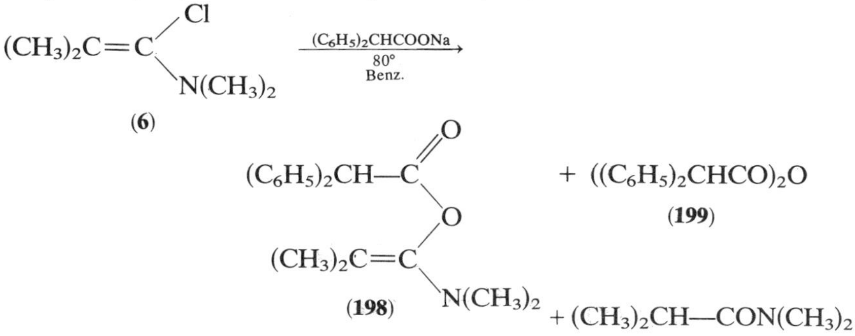

These α-acyloxyenamines (**201**) are obtained (93) in quantitative yields at lower temperature (−10°, −20°C) by reaction of the α-chloroenamines with silver carboxylates (Table VIII):

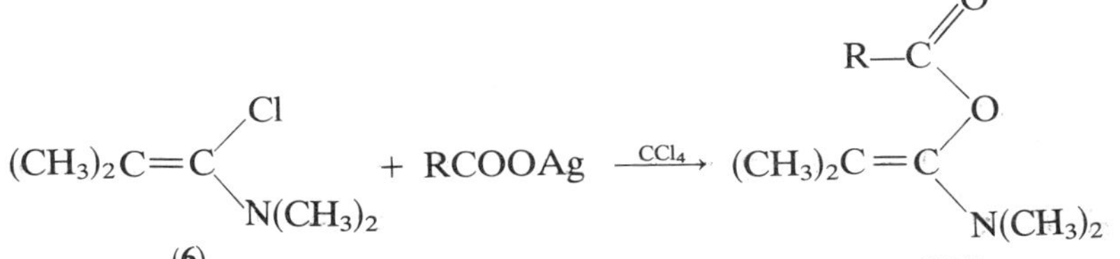

TABLE VIII

$$R^1R^2C=C\begin{matrix}Cl\\NR^3R^4\end{matrix} + R^5COO^-M^+ \longrightarrow R^1R^2C=C\begin{matrix}O-C(=O)-R^5\\NR^3R^4\end{matrix}$$

α-Chloroenamines				Carboxylic compounds	Experimental conditions	Yield, %	Ref.
R^1	R^2	R^3	R^4				
CH_3	CH_3	CH_3	CH_3	PhCOOAg	$CCl_4/20°C/\frac{1}{2}$ hr	80	93, 96
CH_3	CH_3	CH_3	CH_3	PhCOOH	$N(Et)_3/CH_2Cl_2/20°C/1$ hr	74	93
CH_3	CH_3	CH_3	CH_3	CH_3COOH	$N(Et)_3/CH_2Cl_2/20°C/1$ hr	80	93
CH_3	CH_3	CH_3	CH_3	tert-C_4H_9COOH	$N(Et)_3/CCl_4/20°C/1$ hr	84	93
CH_3	CH_3	CH_3	CH_3	$Cl_3C-COOH$	$N(Et)_3/CCl_4/-10°C/1$ hr	81	93
CH_3	CH_3	CH_3	CH_3	CF_3-COOH	$N(Et)_3/CCl_4/-10°C/\frac{1}{2}$ hr	[a]	93
CH_3	CH_3	$-(CH_2)_5-$		$Ph_2CHCOONa$	Benzene/20°C	No reaction	93
					THF/20°C	No reaction	93
					Benzene/80°C/1 hr	70	93
CH_3	CH_3	CH_3	CH_3	$Ph_2CHCOOAg$	$CCl_4/20°C/1$ hr	77	93
CH_3	CH_3	CH_3	CH_3	$Ph_2CHCOOH$	$N(Et)_3/CH_2Cl_2/20°C/1$ hr	81	95
CH_3	CH_3	$-(CH_2)_5-$		$Ph_2CHCOOAg$	$CCl_4/20°C/1$ hr	73	93
CH_3	CH_3	CH_3	CH_3	$(CH_3)_2CH-COOAg$	$CCl_4/20°C/1$ hr	81	93
CH_3	CH_3	CH_3	CH_3	$(CH_3)_2CH-COOAg$	$CCl_4/20°C/1$ hr	75	93
CH_3	CH_3	$-(CH_2)_2-O-(CH_2)_2-$		$(CH_3)_2CH-COOAg$	$CCl_4/20°C/1$ hr	88	93
C_6H_5	C_2H_5	CH_3	CH_3	$(CH_3)_2CH-COOAg$	$CCl_4/20°C/1$ hr	85	93
CH_3	CH_3	CH_3	CH_3	$Cl_2CH-COOAg$	$CCl_4/20°C/\frac{1}{2}$ hr	83	97
CH_3	CH_3	CH_3	CH_3	$Cl_2CH-COOH$	$N(Et)_3/CH_2Cl_2/20°C/1$ hr	86	93
CH_3	CH_3	$-(CH_2)_2-O-(CH_2)_2-$		$Cl_2CH-COOAg$	$CCl_4/20°C/1$ hr	86	93
CH_3	CH_3	CH_3	CH_3	$ClCH_2-COOAg$	$CCl_4/20°C/\frac{1}{2}$ hr	~100	97
CH_3	CH_3	CH_3	CH_3	S–CH–COOH (1,3-dithiane)	$N(Et)_3/CH_2Cl_2/20°C/1$ hr	92	94, 95

[a] Unstable at room temperature.

V. NUCLEOPHILIC SUBSTITUTIONS ON α-HALOENAMINES

They are also readily prepared (93) from the reaction of α-chloroenamines with carboxylic acids in the presence of triethylamine (Table VIII):

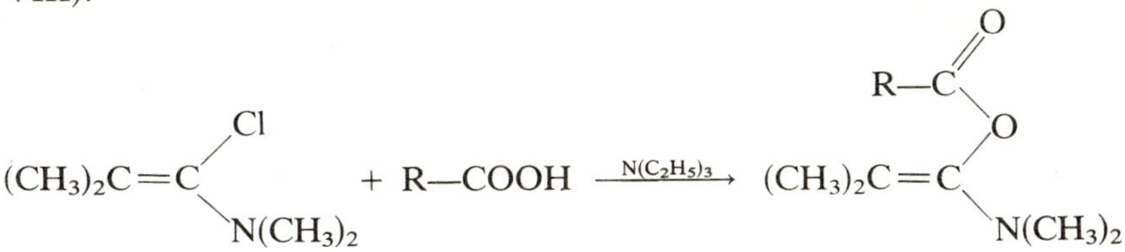

$$(CH_3)_2C=C(Cl)(N(CH_3)_2) + R-COOH \xrightarrow{N(C_2H_5)_3} (CH_3)_2C=C(OC(=O)R)(N(CH_3)_2)$$

(6) → (201)

This reaction does not involve a keteniminium intermediate but goes through the amide chloride (see Section VI).

The α-acyloxyenamines are useful for the preparation of ketenes or β-ketoamides (93–95), for example:

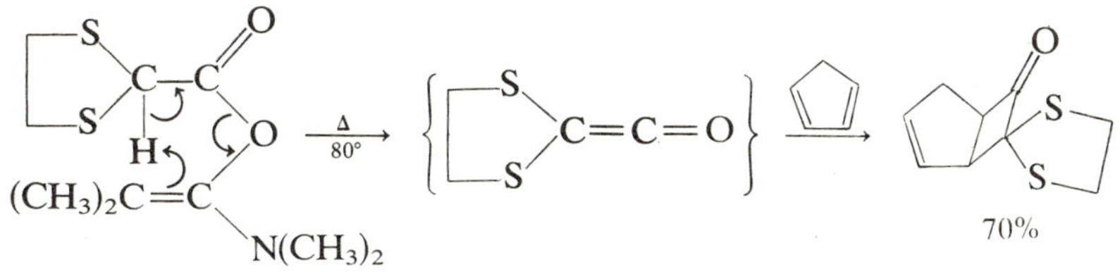

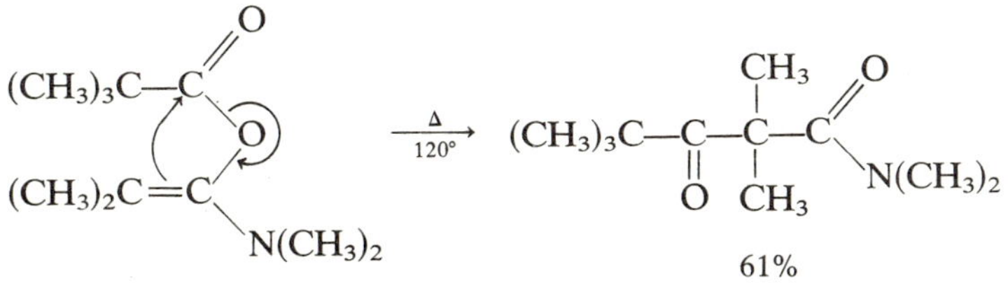

[reaction scheme: dithiolane-substituted α-acyloxyenamine heated at 80° → ketene intermediate {S_2C=C=O} + cyclopentadiene → bicyclic adduct, 70%]

[reaction scheme: (CH_3)_3C—C(=O)—O—C(=C(CH_3)_2)—N(CH_3)_2 $\xrightarrow{\Delta, 120°}$ (CH_3)_3C—C(=O)—C(CH_3)_2—C(=O)—N(CH_3)_2, 61%]

3. Halide Exchange

Halide exchange in α-haloenamines is an equilibrium process, but, as already discussed in connection with the methods of preparation of α-haloenamines (see Section III-E-2), it is often possible to shift the equilibrium toward the desired α-haloenamine. Thus it may be recalled here that several α-fluoroenamines have been obtained from the corresponding α-chloroenamines by heating with potassium fluoride neat or in a solvent such as chlorobenzene (29a):

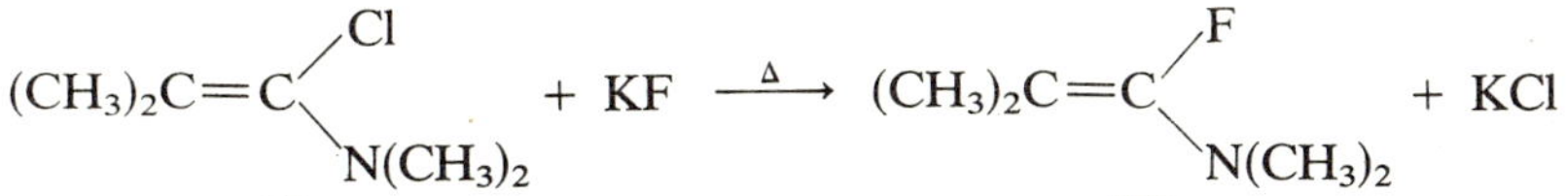

$$(CH_3)_2C=C(Cl)(N(CH_3)_2) + KF \xrightarrow{\Delta} (CH_3)_2C=C(F)(N(CH_3)_2) + KCl$$

(6) → (77)

Conversely, α-fluoroenamines can be converted to α-chloro-, α-bromo-, or α-iodoenamines by reaction with lithium chloride, bromide, or iodide, at room temperature, in methylene chloride (29a):

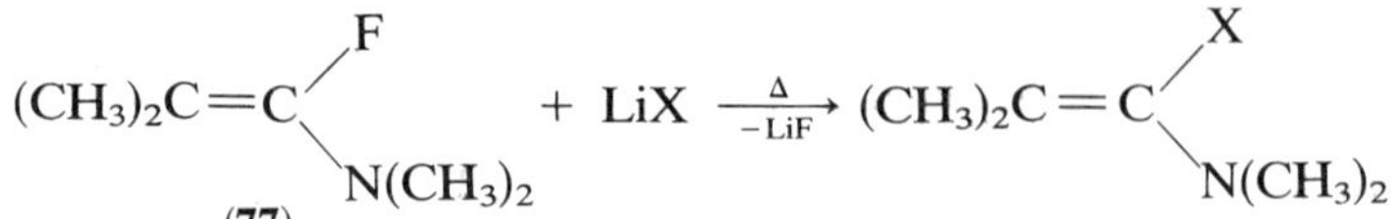

$$(CH_3)_2C=C\begin{smallmatrix}F\\N(CH_3)_2\end{smallmatrix} \quad + \text{ LiX} \xrightarrow[-\text{LiF}]{\Delta} (CH_3)_2C=C\begin{smallmatrix}X\\N(CH_3)_2\end{smallmatrix}$$
(77)   X = Cl, Br, I

4. Reaction with Amines

Amidines (**202**) are formed from the reaction of α-chloroenamines with primary aliphatic or aromatic amines (31a,72,98) as well as with cyanamide (99). With the strongly electrophilic β-alkyl or aryl-α-chloroenamines, the reaction is fast even at room temperature (Table IX).

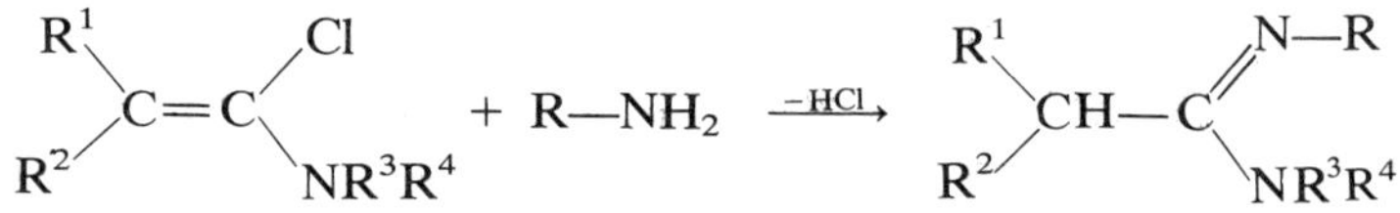

$$R^1\!\!\!\!\diagdown\!\!\!\!\!\!\diagup Cl \\ C=C \\ R^2\!\!\!\!\diagup\!\!\!\!\!\!\diagdown NR^3R^4 \quad + \text{ R—NH}_2 \xrightarrow{-\text{HCl}} \quad R^1\!\!\!\!\diagdown\!\!\!\!\!\!\diagup N-R \\ CH-C \\ R^2\!\!\!\!\diagup\!\!\!\!\!\!\diagdown NR^3R^4$$

R = alk., aryl, CN   (**202**)

As with alkoxides or thiolates, β-chlorocarbonyl-α-chloroenamines are attacked first at the acyl chloride group, and a further molecule of amine then gives (31a) the corresponding amidines (**203**):

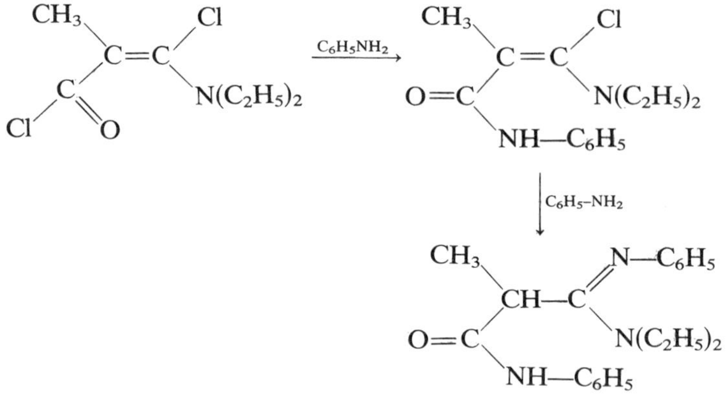

$$\begin{array}{c}CH_3\\C=C\\Cl\!\!-\!\!C\!\!=\!\!O\end{array}\!\!\begin{array}{c}Cl\\N(C_2H_5)_2\end{array} \xrightarrow{C_6H_5NH_2} \begin{array}{c}CH_3\\C=C\\O=C\\NH\!-\!C_6H_5\end{array}\!\!\begin{array}{c}Cl\\N(C_2H_5)_2\end{array}$$

$$\downarrow C_6H_5\text{-}NH_2$$

$$\begin{array}{c}CH_3\\CH\!-\!C\\O=C\\NH\!-\!C_6H_5\end{array}\!\!\begin{array}{c}N\!-\!C_6H_5\\N(C_2H_5)_2\end{array}$$

(**203**)

With trichlorovinylamines, the amines react in the order of increasing pK_a values and the reaction is catalyzed by traces of hydrogen chloride (72). It is probable that trichlorovinylamines react with amines by a mechanism involving the initial addition of a proton to give an amide

TABLE IX

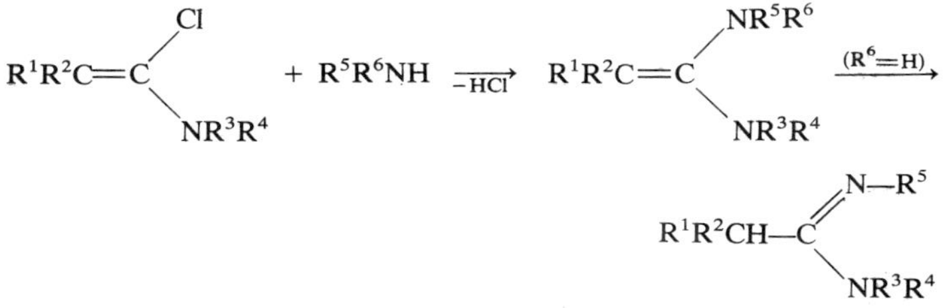

$$R^1R^2C(Cl)(NR^3R^4) + R^5R^6NH \xrightarrow{-HCl} R^1R^2C(NR^5R^6)(NR^3R^4) \xrightarrow{(R^6=H)} R^1R^2CH-C(=NR^5)(NR^3R^4)$$

α-Chloronamine				Amine, n equiv.	Experimental conditions	Yield, %	Boiling point, °C/mm, or melting point, °C	Ref.
R^1	R^2	R^3	R^4					
Cl	Cl	C_2H_5	C_2H_5	C_6H_5—NH_2, 1	Benzene/reflux/16 hr	78	88/0.02	72
Cl	Cl	C_2H_5	C_2H_5	C_6H_5—$NH_3^+Cl^-$, 1	Benzene/reflux/22 hr	82	88/0.02	72
Cl	Cl	C_2H_5	C_2H_5	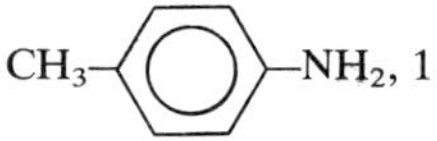 CH_3—C$_6$H$_4$—NH_2, 1	Benzene/reflux/4.5 hr	92	87/0.02	72
Cl	Cl	C_2H_5	C_2H_5	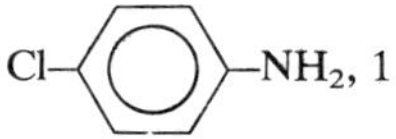 Cl—C$_6$H$_4$—NH_2, 1	Benzene/reflux/22 hr	66	95/0.02	72
Cl	Cl	C_2H_5	C_2H_5	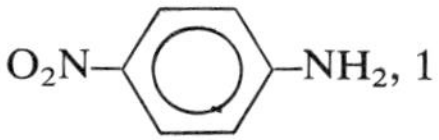 O_2N—C$_6$H$_4$—NH_2, 1	Benzene/reflux/2.5 hr	87	m.p. = 104	72
Cl	Cl	C_2H_5	C_2H_5	 C_2H_5O—C$_6$H$_4$—NH_2, 1	Benzene/reflux/4.5 hr	19	118/0.06	72

TABLE IX (continued)

α-Chlorenamine				Amine, n equiv.	Experimental conditions	Yield, %	Boiling point, °C/mm, or melting point, °C	Ref.
R^1	R^2	R^3	R^4					
Cl	Cl	C_2H_5	C_2H_5	n-C_4H_9—NH_2, 1	Benzene/reflux/20 hr	23	102/8.1	72
Cl	Cl	C_2H_5	C_2H_5	n-C_4H_9—$NH_3^+Cl^-$, 2	Benzene/reflux/48 hr	44	102/8.1	72
Cl	Cl	C_2H_5	C_2H_5	$(C_2H_5)_2NH$, 1	Benzene/reflux/24 hr	68	102/8.1	72
Cl	Cl	C_2H_5	C_2H_5	$(C_2H_5)_3N$, 1	Benzene/reflux/24 hr	No reaction		72
Cl	Cl	C_2H_5	C_2H_5		Benzene/reflux/4 hr	No reaction		72
CH_3CO	CH_3	C_2H_5	C_2H_5	C_6H_5—NH_2, 2	CH_2Cl_2/20°C/3 hr	60–70	m.p. = 47	31a
COCl	CH_3	—$(CH_2)_5$—		C_6H_5—NH_2, 4	CH_2Cl_2/20°C/3 hr	60–70	m.p. = 131	31a
COCl	CH_3	C_2H_5	C_2H_5	C_6H_5—NH_2, 4	CH_2Cl_2/20°C/3 hr	60–70	m.p. = 84	31a
COCl	C_2H_5	C_2H_5	C_2H_5	C_6H_5—CH_2—NH_2, 4	CH_2Cl_2/20°C/3 hr	60–70	m.p. = 164	31a
CSCl	CH_3	C_2H_5	C_2H_5	C_6H_5—NH_2, 4	CH_2Cl_2/20°C/3 hr	60–70	m.p. = 161	31a
COCl	CH_3	—$(CH_2)_5$—		⟨NH⟩, 4	Ether/20°C/1 hr	79[a]	m.p. = 94[a]	31a
COCl	CH_3	C_2H_5	C_2H_5	C_6H_5—NH_2, 2	CH_2Cl_2/20°C/3 hr	60–70	m.p. = 80	31a
CH_3—C$_6$H$_4$—SO_2	CH_3	—$(CH_2)_5$—		C_6H_5—NH_2, 2	CH_2Cl_2/20°C/3 hr	60–70	m.p. = 94	31a
CH_3	CH_3	CH_3	CH_3	$(CH_3)_2NH$, 10	Petroleum ether/20°C	90–100	65/30	98
CH_3	CH_3	CH_3	CH_3	H_2N—CN, 1	CH_2Cl_2/20°C	~100	—	99
CH_3	Cl	—$(CH_2)_4$—		H_2N—CN, 1	CH_2Cl_2/20°C	~100	—	99
CH_3	CH_3	—$(CH_2)_5$—		$(C_6H_{11})_2N^-Li^+$, 1	Ether/0°C/3 hr	80	155/0.9	25, 19
CH_2=CH	CH_3	CH_3	CH_3	$(CH_3)_2N^-Li^+$, 1	0°C	70	—	27

[a] Hydrolysis product: CH_3—CH(CON⟨⟩)$_2$.

chloride, which is attacked by the amine. Accordingly, when triethylamine is added initially, the trichlorovinylamine is recovered unchanged (72).

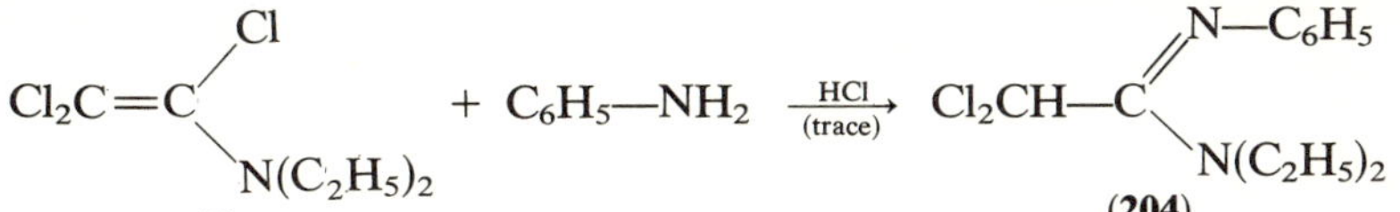

$$Cl_2C=C\begin{matrix}Cl\\N(C_2H_5)_2\end{matrix} + C_6H_5-NH_2 \xrightarrow[\text{(trace)}]{HCl} Cl_2CH-C\begin{matrix}N-C_6H_5\\N(C_2H_5)_2\end{matrix}$$
(5)   (204)

Secondary amines were found to be unreactive toward trichlorovinylamines (72), but tetramethyl-α-chloroenamine (**6**) (98) and even β-chlorocarbonyl-α-chloroenamine (**54**) (31a) reacted at room temperature to give the ketene *N,N*-acetals (**205,206**). These are also the reaction products (19,25) of lithium dicyclohexylamide with the α-chloroenamine (**97**):

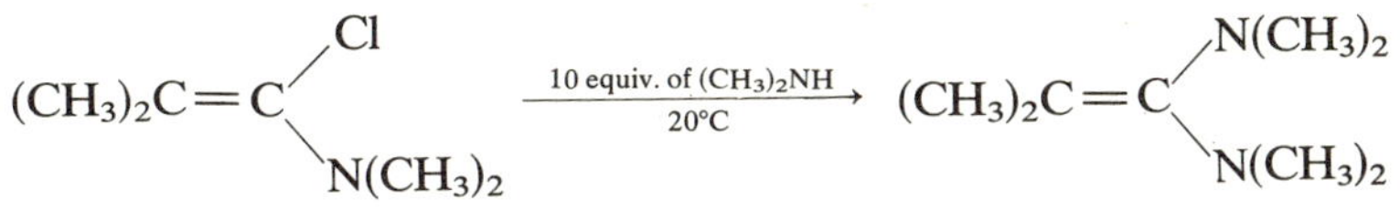

$$(CH_3)_2C=C\begin{matrix}Cl\\N(CH_3)_2\end{matrix} \xrightarrow[20°C]{10\text{ equiv. of }(CH_3)_2NH} (CH_3)_2C=C\begin{matrix}N(CH_3)_2\\N(CH_3)_2\end{matrix}$$
(6)   (205)

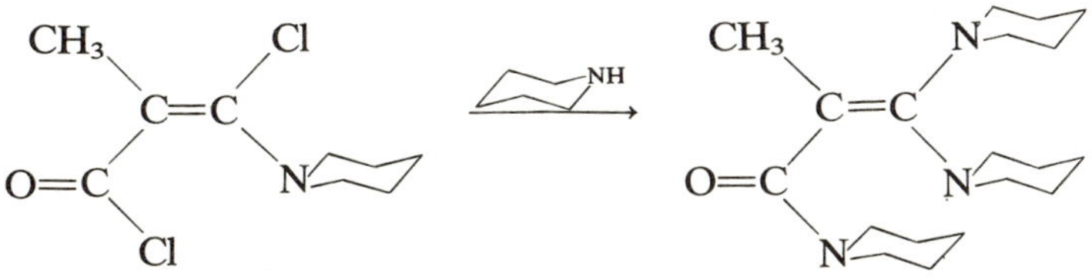

(54) → (206) [with piperidine NH]

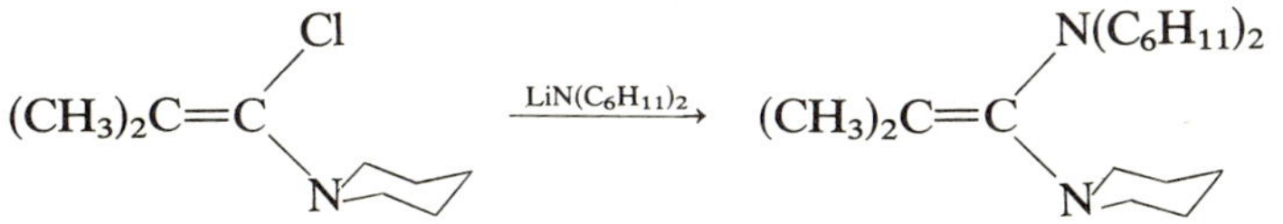

(97) → (207) [with LiN(C₆H₁₁)₂ giving (CH₃)₂C=C(N(C₆H₁₁)₂)(N-piperidyl)]

The reaction of the vinyl-α-chloroenamine (**11**) with lithium dimethylamide gives the strongly nucleophilic diene (**208**), which combines readily with electrophilic olefins to form in high yields adducts that can be hydrolyzed to functionalized cyclohexenones (27):

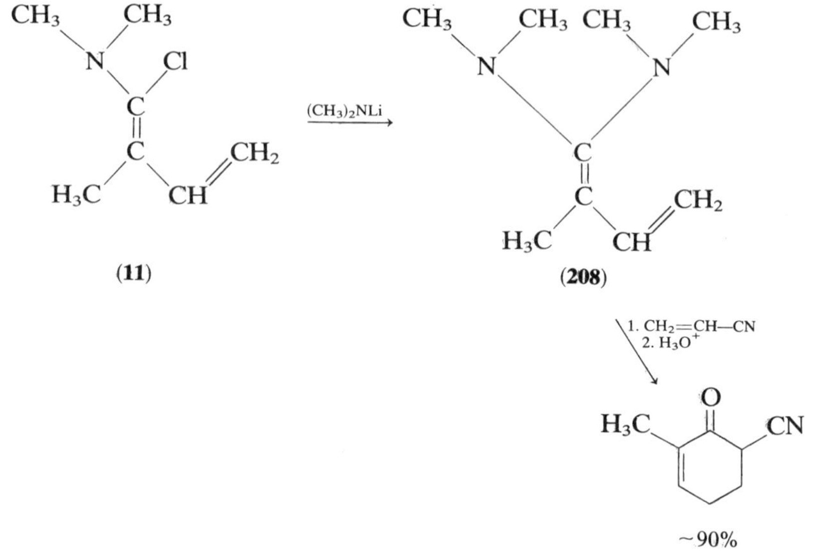

[Scheme: compound (11) with (CH₃)₂NLi → compound (208); then 1. CH₂=CH—CN, 2. H₃O⁺ → cyclohexenone product, ~90%]

5. Reactions with Azide Ions

The reaction of sodium azide with α-chloroenamines (33,88) provides a convenient approach to the chemistry of 2-amino-1-azirines, a new class of amidines (Table X).

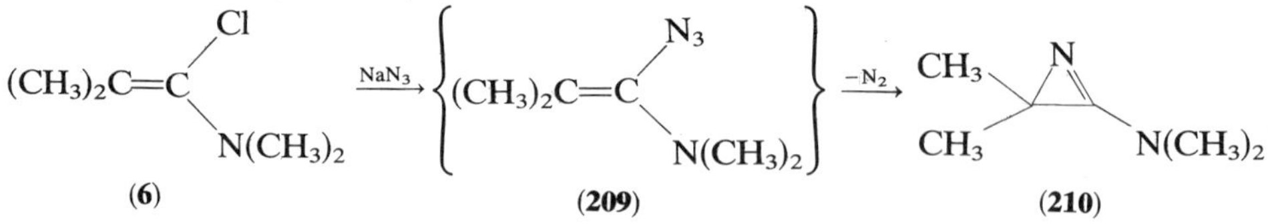

[Scheme: (6) —NaN₃→ {(209)} —−N₂→ (210)]

An intermediate α-azidoenamine (209) is probably formed in the first step by substitution of chlorine by the azide ion. These α-azidoenamines are very unstable and lose nitrogen so quickly that they cannot be isolated, even at low temperature (−40°C). The assumption of this intermediate is nevertheless justified: it has been detected (33) spectroscopically in the reaction of sodium azide with 2-isopropyl- or tert-butyl-1-chloroenamines, and trapped by a molecule of ynamine (211) formed in situ by dehydrochlorination of the α-chloroenamine (29) (Scheme 19). With the highly electrophilic α-chloroenamines (Table X) the reaction takes place readily at room temperature. Less reactive α-chloroenamines (Table X) require the presence of a catalyst such as Ag⁺ or, more conveniently, a polar solvent such as dimethylformamide.

TABLE X

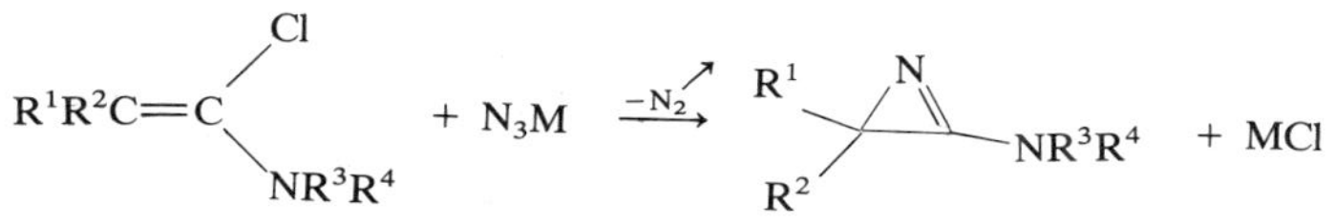

$$R^1R^2C=C\begin{matrix}Cl\\NR^3R^4\end{matrix} + N_3M \xrightarrow{-N_2} \begin{matrix}R^1\\R^2\end{matrix}\begin{matrix}N\\C-C\end{matrix}NR^3R^4 + MCl$$

α-Chloroenamine				N_3M	Experimental conditions	Yield, %	Boiling point, °C/mm Hg	Ref.
R^1	R^2	R^3	R^4					
CH_3	CH_3	CH_3	CH_3	NaN_3	Ether, CCl_4, or $CH_3CN/20°C/2$ hr	95–99	97–98/76	33, 88
CH_3	CH_3	C_2H_5	C_2H_5	AgN_3	Ether/−5°C/2 hr	97		33
CH_3	CH_3	—$(CH_2)_5$—		NaN_3	Ether/20°C/6 hr	94	42/1	33, 88
CH_3	CH_3			NaN_3	Ether/20°C/5 hr	94	48–49/0.3	33, 88
CH_3	CH_3	—$(CH_2)_2$—O—$(CH_2)_2$—		NaN_3	$CH_3CN/20°C/1$ hr	95		33
					Ether/40°C		No reaction	33
					$CHCl_3/65°C$		No reaction	33
					$DMF/20°C/2$ hr	91	72–74/0.3	33, 88
C_6H_5	CH_3	CH_3	CH_3	AgN_3	Ether/20°C/1 hr	96		33, 88
C_6H_5	C_2H_5	CH_3	CH_3	AgN_3	Ether/−20°C/$1\frac{1}{2}$ hr	95	84/3	33, 88
				NaN_3	Ether/20°C		No reaction	33
				AgN_3	$DMF/20°C/1\frac{1}{2}$ hr	73	100–101/0.8	33, 88
—$(CH_2)_5$—		C_2H_5	C_2H_5	NaN_3	Ether/20°C/$1\frac{1}{2}$ hr	98		33, 88
i-C_3H_7	H	C_2H_5	C_2H_5	AgN_3	Ether/60°C/6 hr	92	62–64/0.2	33
tert-C_4H_9	H	CH_3	CH_3	AgN_3	$CCl_4/20°C/1\frac{1}{2}$ hr	45	86–88/12	33
				AgN_3	$CCl_4/20°C/1\frac{1}{2}$ hr	18	64–66/0.5	33
$CH_2=CH$—	CH_3	CH_3	CH_3	NaN_3	Ether/20°C/3 hr	70	30/1	27

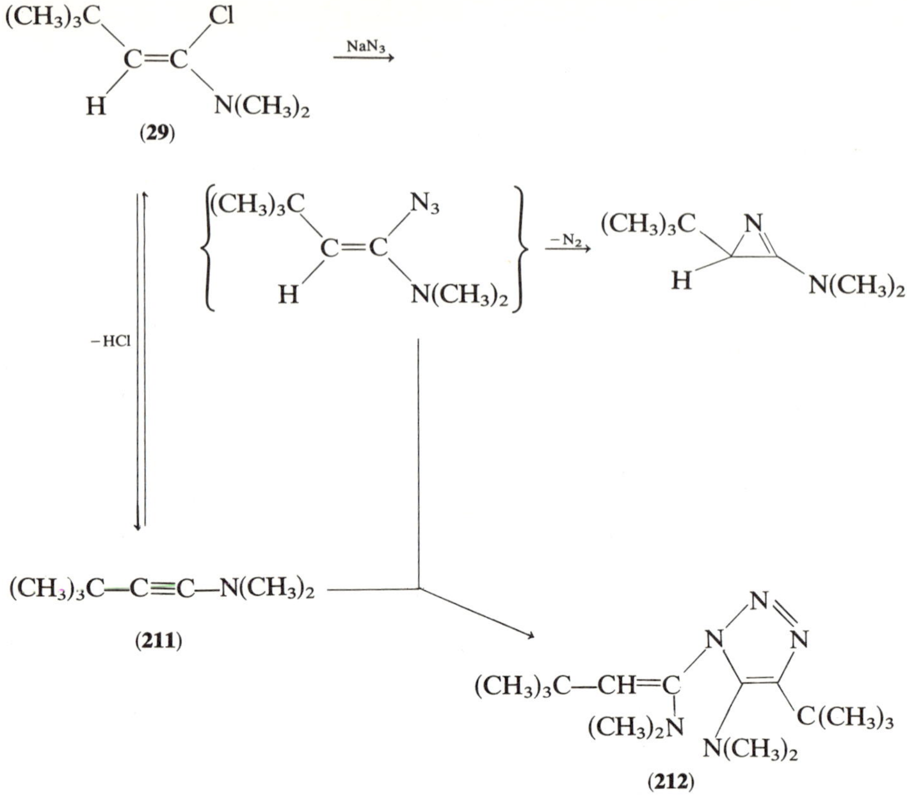

Scheme 19

By this procedure 2-aminoazirines have become available in large quantities (33). They are stable liquids, soluble in water and in organic solvents. It is interesting that tetramethyl-2-aminoazirine (210), when subjected to gas-phase pyrolysis (100) at 250–300°C, rearranges in high yields to an activated heterodiene (214), which has been used for the construction of pyridine rings (215) (Scheme 20). The stabilized carbene (213) has been postulated as an intermediate in this rearrangement, which is analogous to the ring opening of aziridines (101).

C. NUCLEOPHILIC SUBSTITUTIONS WITH FORMATION OF C–C BONDS: THE AMINOALKENYLATION REACTIONS

1. Reactions with Cyanide Ion

The reaction between potassium cyanide and α-chloroenamines represents the most convenient preparation (66) of α-cyanoenamines (216)

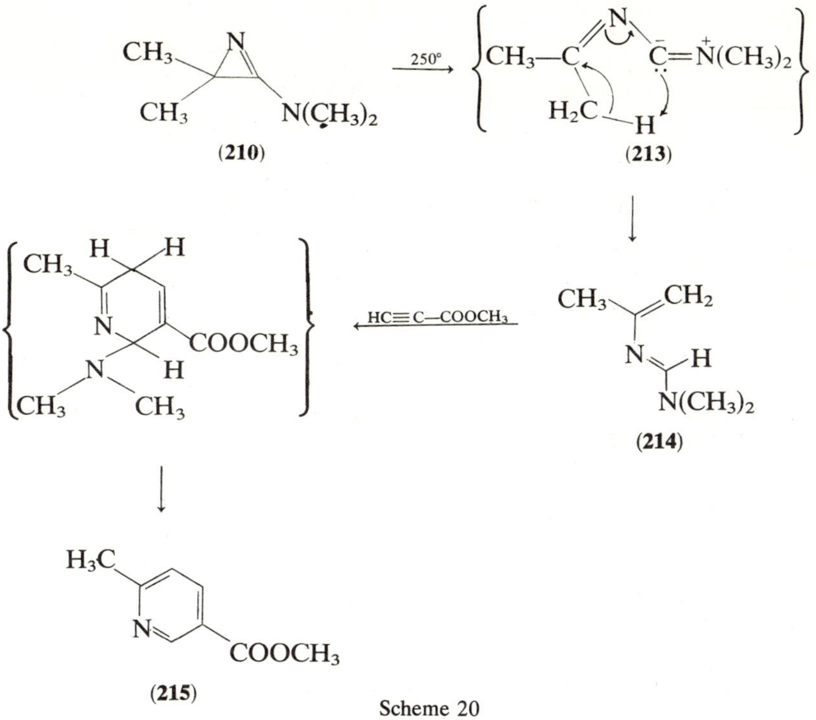

V. NUCLEOPHILIC SUBSTITUTIONS ON α-HALOENAMINES

Scheme 20

(Table XI). The reaction is usually effected in refluxing acetonitrile. Zinc cyanide in refluxing chloroform has also been used successfully (66) (Scheme 21). The use of silver ion instead of potassium ion accelerates

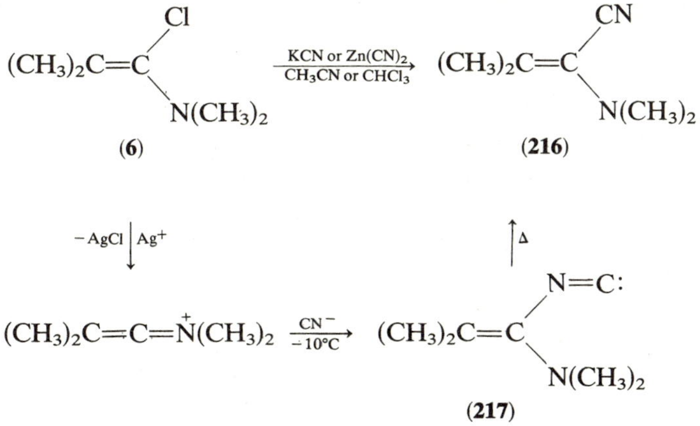

Scheme 21

TABLE XI

$$R^1R^2C=C\begin{matrix}Cl\\NR^3R^4\end{matrix} \xrightarrow{CN^-} R^1R^2C=C\begin{matrix}CN\\NR^3R^4\end{matrix}$$

R^1	R^2	R^3	R^4	Cyanide	Experimental conditions	Yield, %	Ref.
CH_3	CH_3	CH_3	CH_3	KCN	$CH_3CN/\triangle/35$ hr	90	66
CH_3	CH_3	CH_3	CH_3	AgCN	$CCl_4/-10°/1$ hr	100[a]	66
CH_3	CH_3	CH_3	CH_3	$Zn(CN)_2$	$CHCl_3/\triangle/4$ hr	80	66
CH_3	CH_3	CH_3	CH_3	$(CH_3)_4N^+CN^-$	$CH_3CN/20°/1$ min	100	66
C_6H_5	C_2H_5	CH_3	CH_3	KCN	$CH_3CN/\triangle/40$ hr	90	66
C_6H_5	C_2H_5	CH_3	CH_3	AgCN	$CCl_4/-10°/1$ hr	100[b]	66
Cl	CH_3	CH_3	CH_3	$Zn(CN)_2$	$CHCl_3/\triangle/8$ hr	80	66
Cl	CH_3	CH_3	CH_3	AgCN	$CCl_4/20°/60$ hr	95[c]	66
Cl	CH_3	—$(CH_2)_4$—		$Zn(CN)_2$	$CHCl_3/\triangle/8$ hr	80	66
Cl	CH_3	—$(CH_2)_4$—		AgCN	$CCl_4/20°/60$ hr	95[c]	66
CH_3S	CH_3S	C_2H_5	C_2H_5	$(CH_3)_4N^+CN^-$	$CH_3CN/20°/10$ min	~60	59

[a] Isonitrile isomerizes at room temperature.
[b] Isonitrile isomerizes at refluxing $CHCl_3$.
[c] Isonitrile does not isomerize to nitrile.

the substitution reaction by favoring the formation of the keteniminium ion but also promotes attack at the more electronegative atom (66). Therefore, at −10°C, with silver cyanide, α-isocyanoenamines (**217**), a new class of isonitriles, are formed in high yields. At higher temperatures they rearrange quantitatively to the corresponding α-cyanoenamines (66).

These α-cyanoenamines are key intermediates in a new general method of synthesis of α-diketones from carboxylic acid derivatives (66) (Scheme 22).

2. Reactions with Ambident Anions

The reaction of tetramethyl-α-chloroenamine (**6**) with acetylacetone, at room temperature, in the presence of triethylamine, yields a mixture of the C- and the O-aminoalkenylation products (**218,219**); the latter is not stable and rearranges (89) to the conjugated amide (**220**) (Scheme 23). The isomer ratio is concentration and solvent dependent (89).

Under the same experimental conditions (102), nitromethane gives a mixture of products, which probably results from a primary O-aminoalkenylation product (**221**) (Scheme 24). The methacrylamide (**222**) is the major component of the mixture when the sodium salt of nitromethane is used (102).

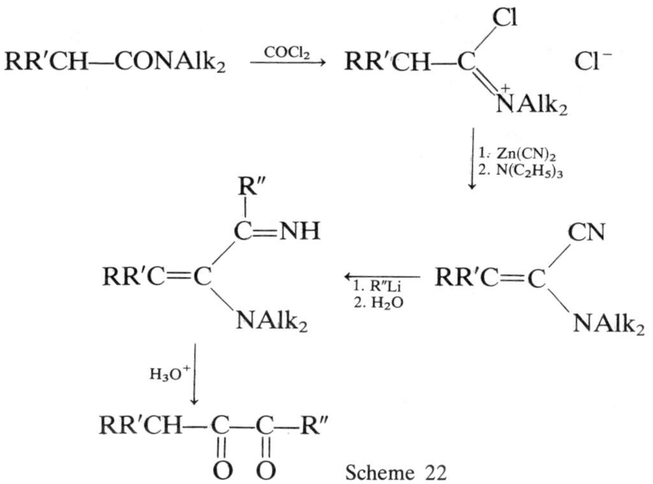

V. NUCLEOPHILIC SUBSTITUTIONS ON α-HALOENAMINES

Scheme 22

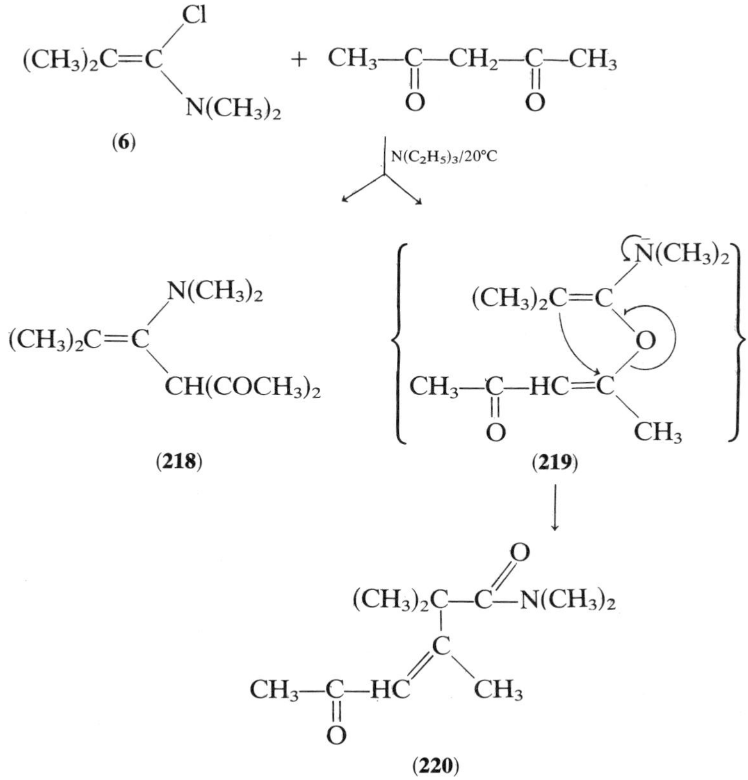

Scheme 23

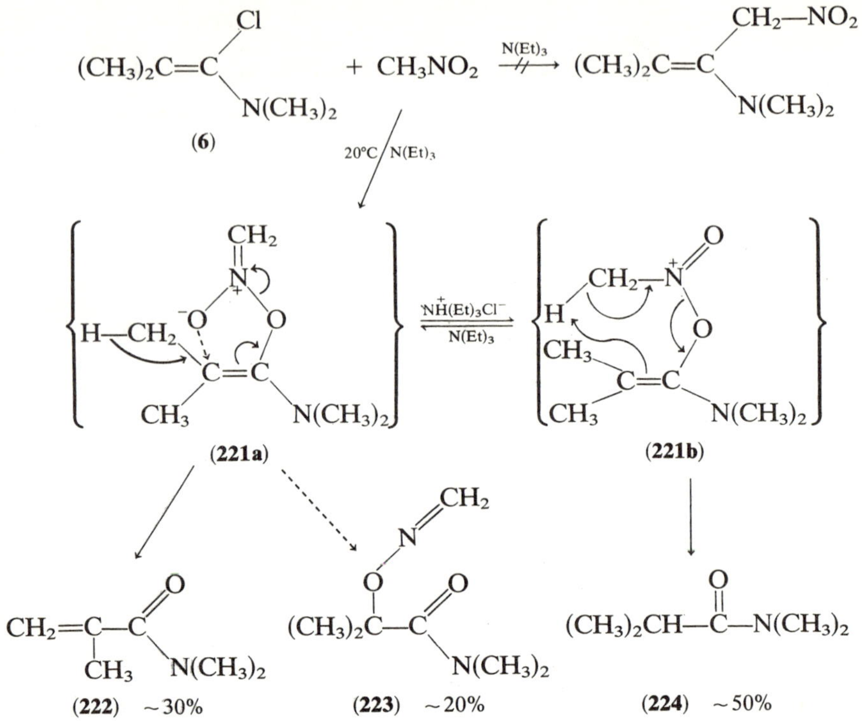

Scheme 24

The sodium salt of malonitrile reacts with **6** to give (102) the C-aminoalkenylation product (**225**), which tautomerizes to the more stable **226**. The corresponding α-fluoroenamine (**77**) reacts (29a) with malonitrile itself to give **226**:

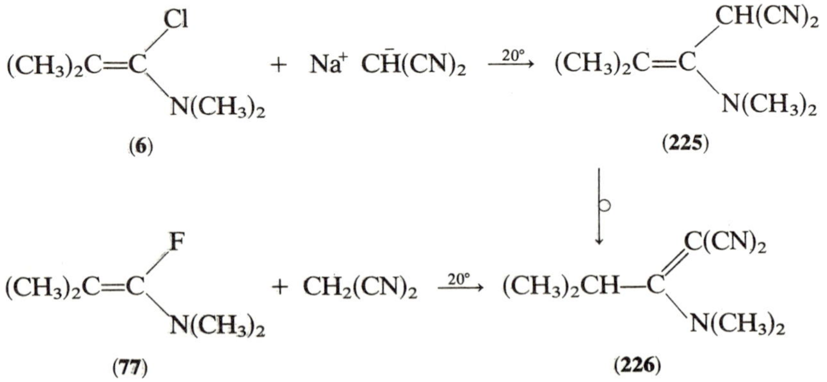

Benzamide was found (99) to react as an ambident reagent with **6**. When the reaction was conducted in the presence of triethylamine, N-aminoalkenylation occurred and an acyl amidine (**227**) was formed (Scheme 25). In the absence of triethylamine, the products were *N,N*-dimethylisobutyramide and benzonitrile, which probably result from the fragmentation of an O-aminoalkenylation product (**228**). In this case the α-chloroenamine (**6**) behaved as a dehydrating agent.

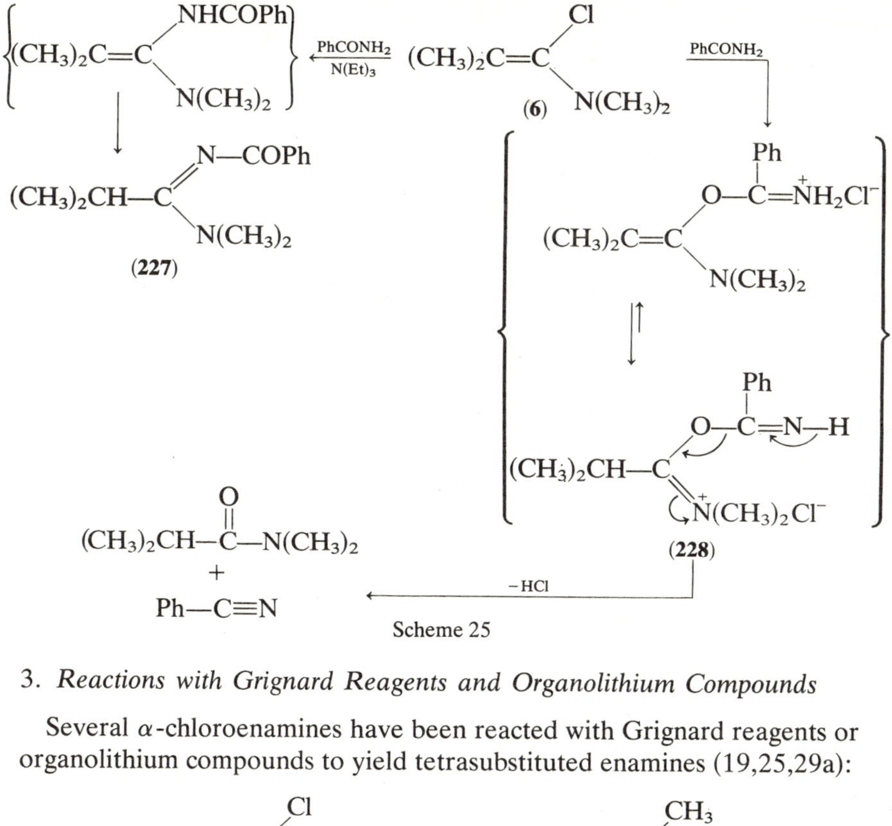

Scheme 25

3. Reactions with Grignard Reagents and Organolithium Compounds

Several α-chloroenamines have been reacted with Grignard reagents or organolithium compounds to yield tetrasubstituted enamines (19,25,29a):

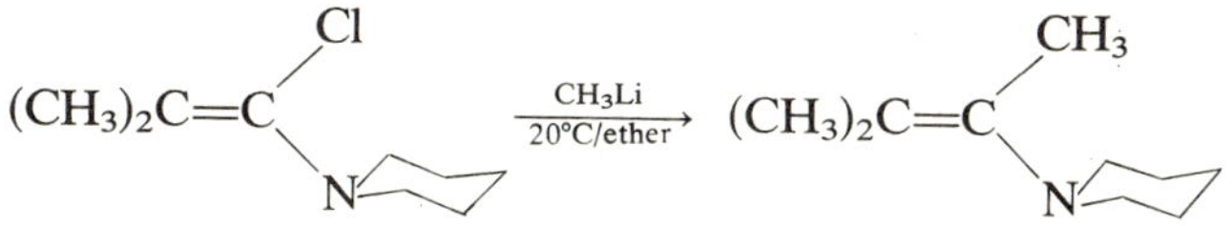

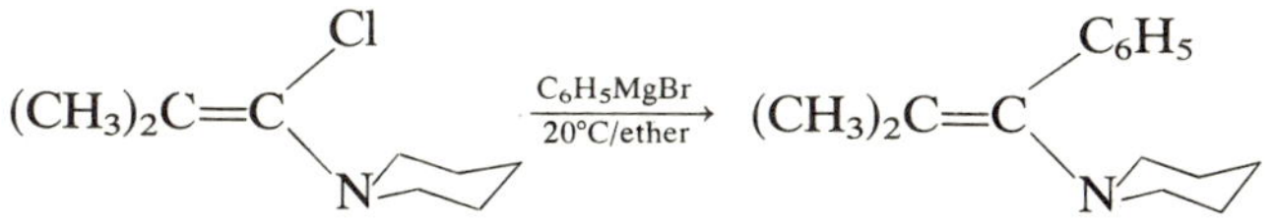

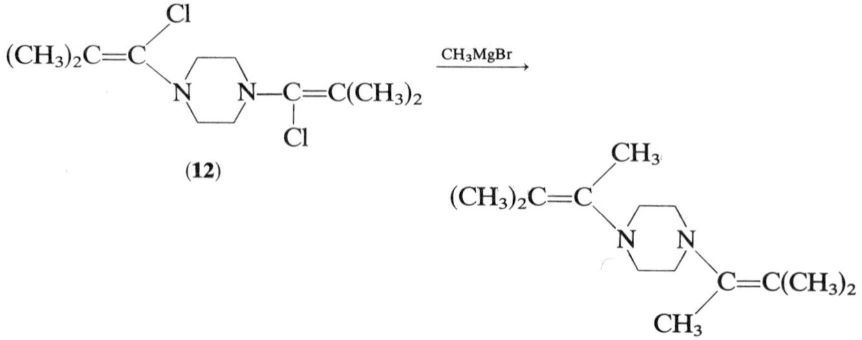

$$(CH_3)_2C=C\begin{smallmatrix}Cl\\ \diagdown\end{smallmatrix}N\overset{\frown}{}N-C=C(CH_3)_2 \xrightarrow{CH_3MgBr}$$
$$\underset{Cl}{}$$
(12)

$$(CH_3)_2C=C\begin{smallmatrix}CH_3\\ \diagdown\end{smallmatrix}N\overset{\frown}{}N\begin{smallmatrix}\diagup\\ CH_3\end{smallmatrix}C=C(CH_3)_2$$
(231)

Tetramethyl-α-fluoroenamine (**77**) and methyllithium react (29a) in a similar way to give the pentamethylenamine (**232**):

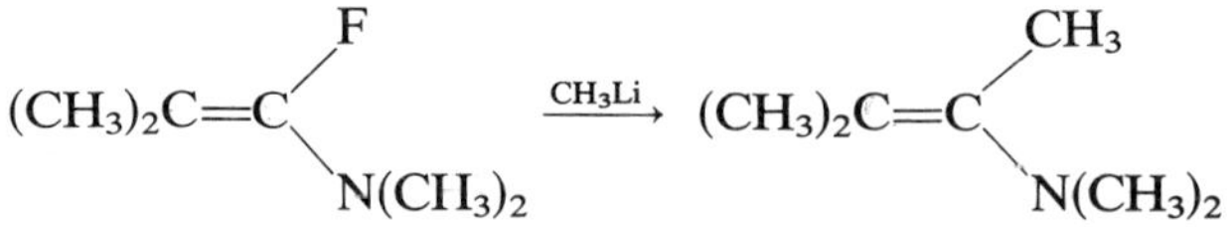

$$(CH_3)_2C=C\begin{smallmatrix}F\\ \diagdown\\ N(CH_3)_2\end{smallmatrix} \xrightarrow{CH_3Li} (CH_3)_2C=C\begin{smallmatrix}CH_3\\ \diagdown\\ N(CH_3)_2\end{smallmatrix}$$
(77) (232)

4. Reactions with Aromatic Compounds

The reactions of α-chloroenamines with aromatic compounds (90,91) in the presence of triethylamine represent one of the most interesting applications of the aminoalkenylation principle. It makes possible the direct introduction of an enamine functional group on an aromatic nucleus under nonacidic conditions (Table XII). The reaction of **6** with furan (90) is a typical example of the process:

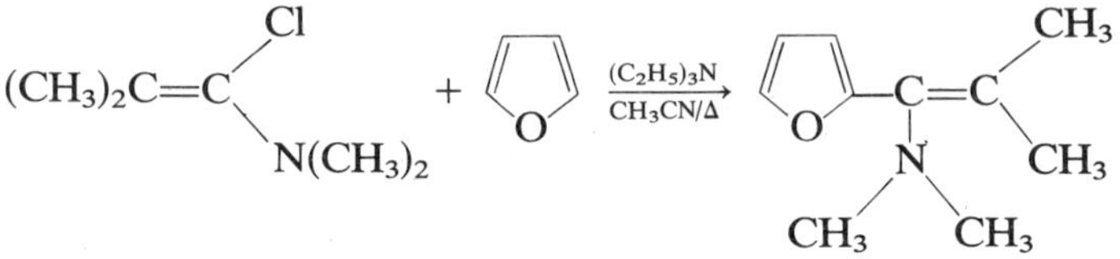

$$(CH_3)_2C=C\begin{smallmatrix}Cl\\ \diagdown\\ N(CH_3)_2\end{smallmatrix} + \underset{O}{\bigcirc} \xrightarrow[CH_3CN/\Delta]{(C_2H_5)_3N} \underset{O}{\bigcirc}-C=C\begin{smallmatrix}CH_3\\ \diagdown\\ CH_3\end{smallmatrix}$$
(6)   (233)

This behavior is reminiscent of that of Vilsmeier reagents or other iminium derivatives (10,11,92) and is best explained by a predissociation of the α-chloroenamine into a keteniminium chloride. It is thus not surprising that with aromatics like thiophene and anisole the keteniminium salt must first be formed by adding a Lewis acid to the α-chloroenamine. The solution is then reacted with the aromatic compounds (91) (Scheme 26).

TABLE XII

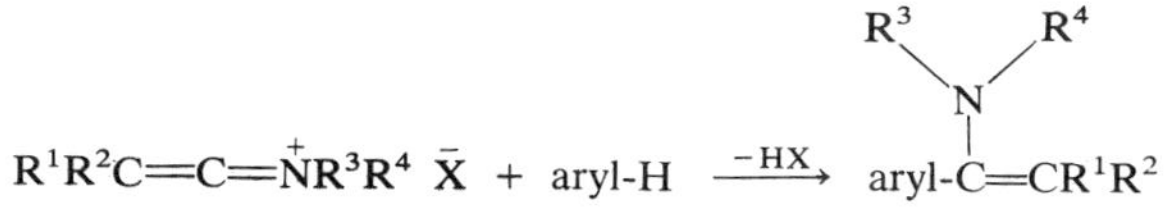

$$R^1R^2C=C=\overset{+}{N}R^3R^4 \;\bar{X} + \text{aryl-H} \xrightarrow{-HX} \begin{matrix} R^3 \;\; R^4 \\ \diagdown N \diagup \\ | \\ \text{aryl-C}=CR^1R^2 \end{matrix}$$

R^1	R^2	R^3	R^4	X^-	Aryl-H	Experimental conditions	Yield, %	Ref.
CH_3	CH_3	CH_3	CH_3	Cl^-	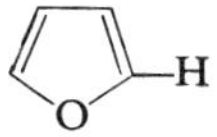 furan	$N(Et)_3/CH_3CN/\triangle/24$ hr	83	90
CH_3	CH_3	C_2H_5	C_2H_5	Cl^-	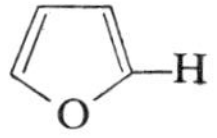 furan	$N(Et)_3/CH_3CN/\triangle/24$ hr	85	90
CH_3	CH_3	—$(CH_2)_5$—		Cl^-	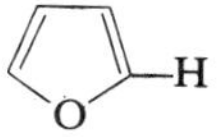 furan	$N(Et)_3/CH_3CN/\triangle/24$ hr	85	90
CH_3	CH_3	CH_3	CH_3	Cl^-	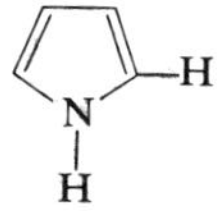 pyrrole	$N(Et)_3/\text{ether}/20°C/6$ hr	94	90
CH_3	CH_3	C_2H_5	C_2H_5	Cl^-	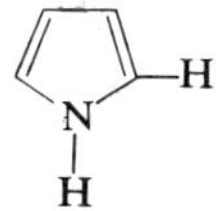 pyrrole	$N(Et)_3/\text{ether}/20°C/6$ hr	95	90

TABLE XII (continued)

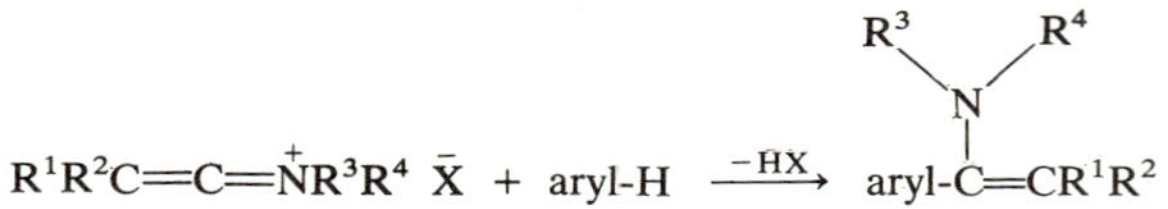

$$R^1R^2C=C=\overset{+}{N}R^3R^4 \ \bar{X} + \text{aryl-H} \xrightarrow{-HX} \text{aryl-C}=CR^1R^2 \text{ with } N(R^3)(R^4)$$

R^1	R^2	R^3	R^4	X^-	Aryl-H	Experimental conditions	Yield, %	Ref.
CH_3	CH_3	—$(CH_2)_5$—		Cl^-	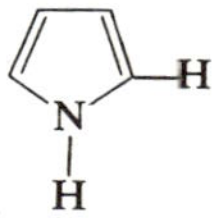 pyrrole (N–H)	$N(Et)_3$/ether/20°C/6 hr	94	90
CH_3	CH_3	CH_3	CH_3	BF_4^-	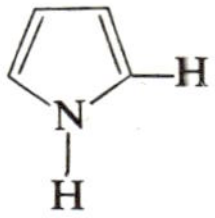 pyrrole (N–H)	CH_2Cl_2/20°C/1 hr	100[a]	89
CH_3	CH_3	CH_3	CH_3	Cl^-	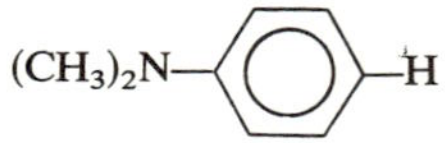 $(CH_3)_2N$–C$_6$H$_4$–H	$N(Et)_3$/CH_3CN/Δ/4 hr	89[b]	90
CH_3	CH_3	CH_3	CH_3	Cl^-	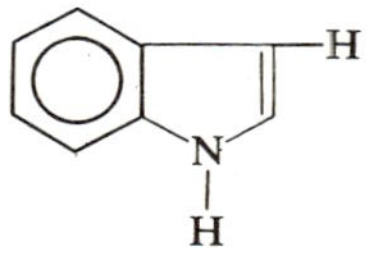 indole	$N(Et)_3$/ether/20°C	60	91

CH$_3$	CH$_3$	CH$_3$	ZnCl$_3^-$	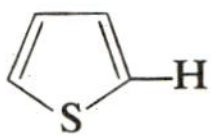 thiophene (2-H)	CH$_2$Cl$_2$/40°C/N(Et)$_3$	50	91
CH$_3$	CH$_3$	CH$_3$ 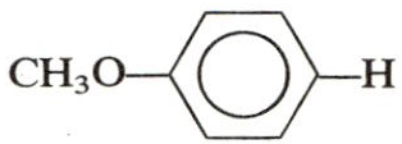	ZnCl$_3^-$	4-methoxyphenyl	CH$_2$Cl$_2$/40°C/N(Et)$_3$	50	91
CH$_2$=CH	CH$_3$	CH$_3$ 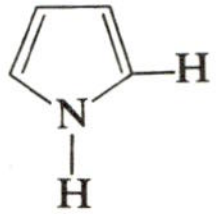	BF$_4^-$	pyrrole (2-H)	CH$_2$Cl$_2$/−40°C	70[a]	91
CH$_2$=CH	CH$_3$	CH$_3$ 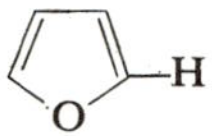	BF$_4^-$	furan (2-H)	CH$_2$Cl$_2$/−40°C	40[a]	
CH$_3$	Cl	—(CH$_2$)$_4$—		ZnCl$_3^-$	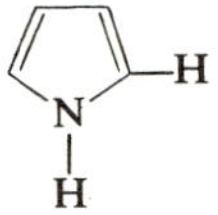 pyrrole (2-H)	CH$_2$Cl$_2$/20°C	50[b]	91

[a] Salt.
[b] Hydrolysis product.

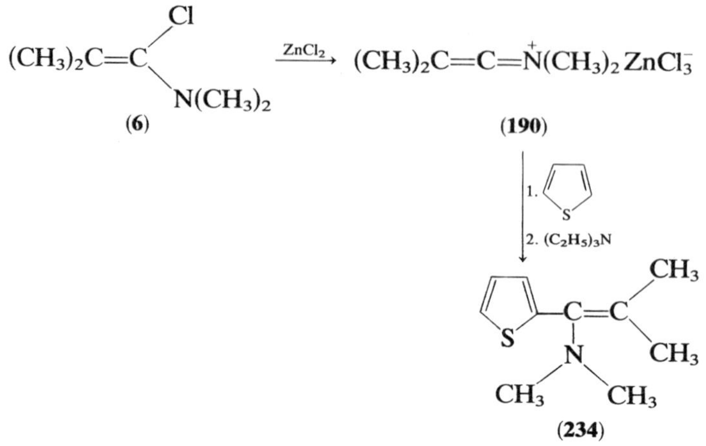

Scheme 26

Extension of the aminoalkenylation principle to the reaction of vinyl-α-chloroenamine (**11**) with pyrrole gives an interesting iminium salt (**235**), which undergoes (91) an electrocyclic ring closure at 180°C (Scheme (27)):

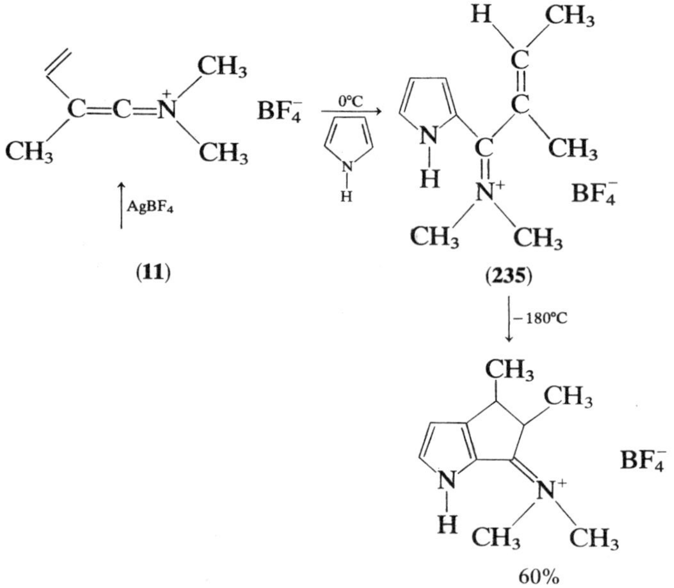

Scheme 27

The intramolecular counterpart of the aminoalkenylation process is found in the case of 2-thioaryl-α-chloroenamines. When exposed to zinc chloride or silver tetrafluoroborate, they undergo (50,59) a smooth cyclization to the 3-diethylaminobenzothiophenes (**236**) (Scheme 28).

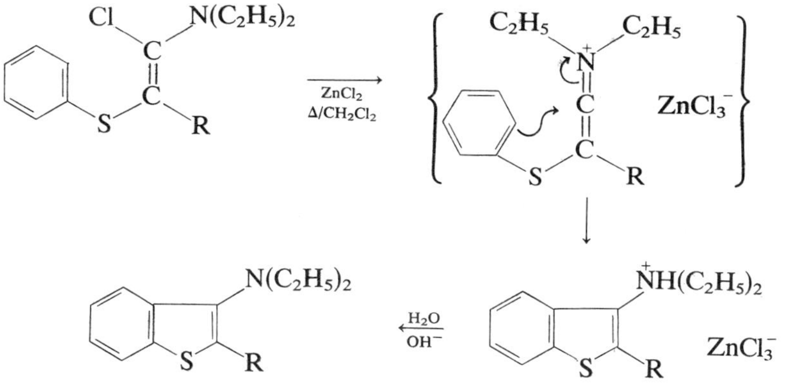

(**236**) 70%
R = PhS, Ph, CH$_3$S, CH$_3$.

Scheme 28

D. OXIDATION REACTIONS

α-Chloroenamine (**6**) reacts (103) with dimethyl sulfoxide, in the presence of triethylamine, to give the methacrylamide (**222**), contamined by N-dimethylisobutyramide (Scheme 29).

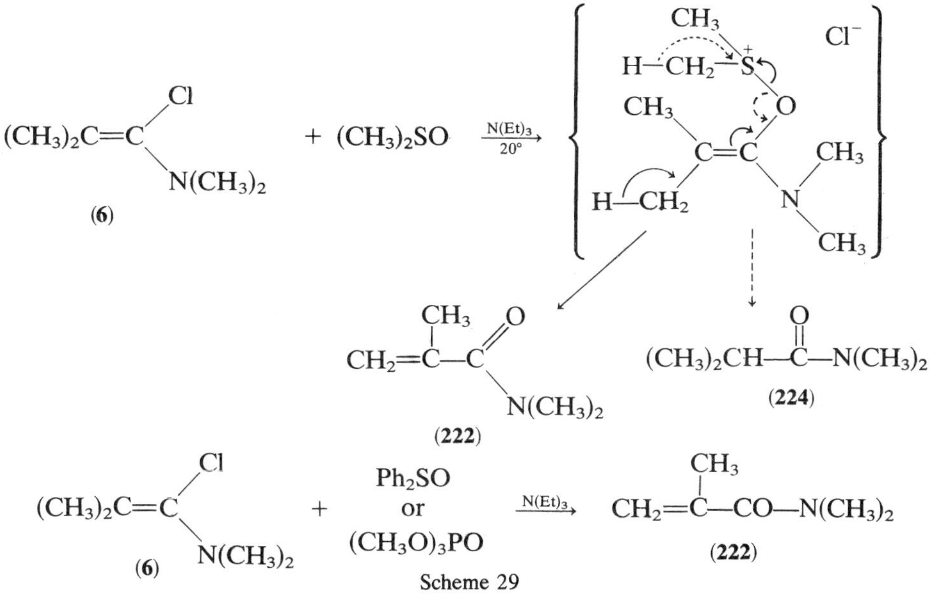

Scheme 29

The formation of side product **224** can be suppressed by using diphenyl sulfoxide or trimethyl phosphite as an oxidizing agent (104).

VI. Reactions of α-Haloenamines with Electrophiles

A. GENERAL ASPECTS

Section V illustrated the electrophilic nature of α-haloenamines, which results from the interaction of the lone pair of the nitrogen atom with the C–Cl bond. However, in these polyfunctional molecules the lone pair can also enter into conjugation with the π olefinic system and thus develop another reactive center at the C_2 atom. Indeed, in the presence of electrophilic reagents, α-haloenamines undergo electrophilic additions to give derivatives of amide halides, which may further react with nucleophiles present in the reaction mixture:

$$\underset{N}{\overset{X}{C=C}} \xrightarrow{E^+} \underset{E\quad N^+}{\overset{X}{C-C}} \longrightarrow \text{products}$$

No quantitative data are yet available that would allow comparison of the reactivities of different α-haloenamines with various electrophiles. Qualitatively, however, it can be said that α-chloroenamines bearing one hydrogen at C_2 are more reactive than the corresponding C_2 disubstituted systems (105). Furthermore α-chloroenamines appear less reactive toward electrophiles than do the corresponding enamines. For instance, tetramethyl-α-chloroenamine cannot be acylated (105), in contrast to the corresponding enamine (106). α-Fluoroenamines are generally more nucleophilic than α-chloroenamines (29a).

B. HYDRATION

It was shown in Section V-B-1 that α-haloenamines are readily hydrolyzable under alkaline conditions. Under these conditions a keteniminium chloride appears to be the most plausible reactive intermediate. Hydrolysis of α-haloenamines under acidic conditions also occurs very rapidly and exothermically (19). Here, however, the reactive intermediate is the iminium ion resulting from the protonation at C_2, as in the case of simple enamines (106).

$$\underset{N}{\overset{X}{C=C}} \xrightarrow{H^+} \underset{H\quad N^+}{\overset{X}{C-C}} \xrightarrow{H_2O} \underset{H}{\overset{O}{C-C-N}}$$

C. ADDITION OF HYDROGEN HALIDES

α-Chloroenamines react (25,72,107) vigorously and instantaneously with dry hydrogen chloride to form quantitatively the amide chloride salts (**237**). Similarly (29a) α-fluoroenamines and hydrogen fluoride give the covalent, volatile amide fluorides (**238**):

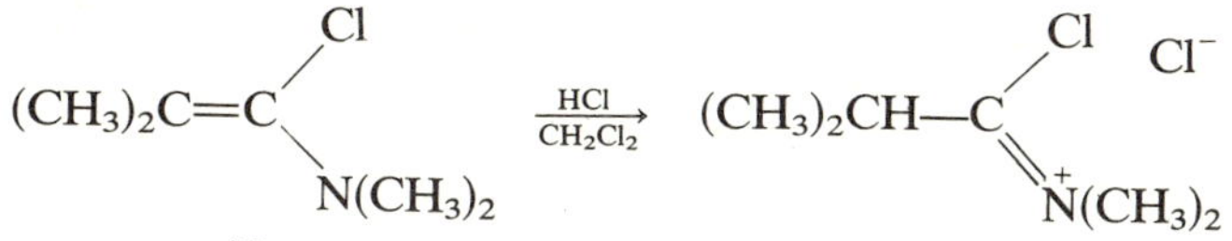

$$(CH_3)_2C=C\begin{smallmatrix}Cl\\N(CH_3)_2\end{smallmatrix} \xrightarrow[CH_2Cl_2]{HCl} (CH_3)_2CH-C\begin{smallmatrix}Cl\\ \\ \overset{+}{N}(CH_3)_2\end{smallmatrix} \; Cl^-$$
(**6**)   (**237**) ~100%

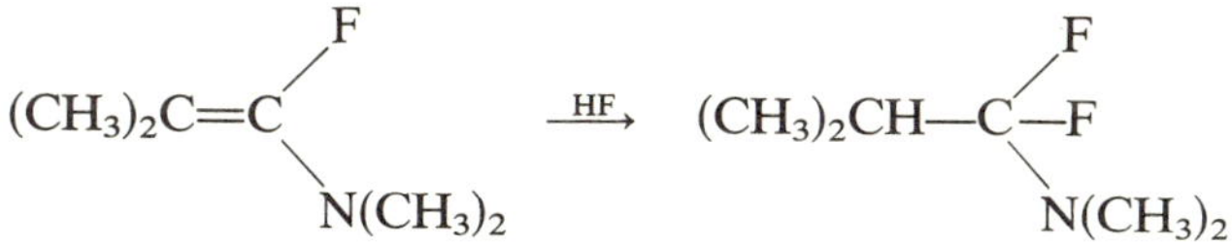

$$(CH_3)_2C=C\begin{smallmatrix}F\\N(CH_3)_2\end{smallmatrix} \xrightarrow{HF} (CH_3)_2CH-C\begin{smallmatrix}F\\|\\N(CH_3)_2\end{smallmatrix}-F$$
(**77**)   (**238**) 80%

D. HALOGENATION

The polar chlorination or bromination of α-haloenamines proceeds in complete analogy with the protonation and gives α,β-dihaloammonium halides, which are readily hydrolyzed to α-halocarboxamides (25,57,105,108). Thus tetramethyl-α-chloroenamine (**6**) has been brominated to **239**, which, on hydrolysis, gives (108) N,N-dimethyl-α-bromoisobutyramide (**240**):

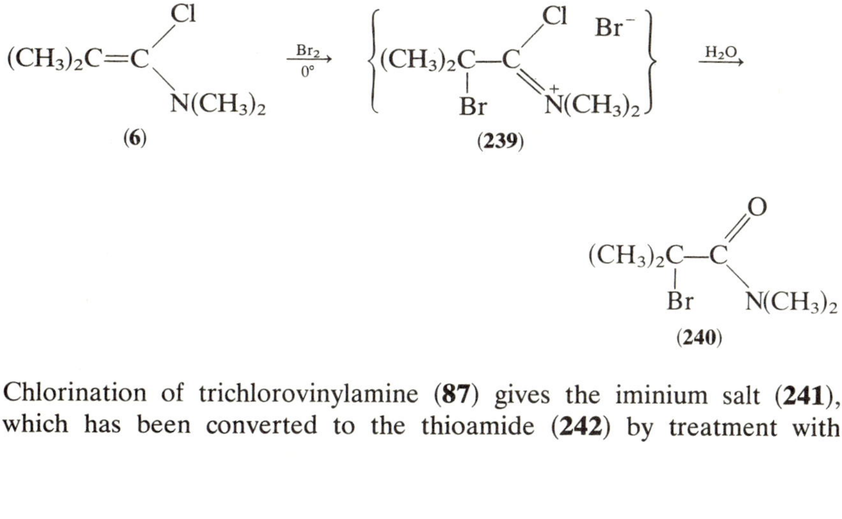

$$(CH_3)_2C=C\begin{smallmatrix}Cl\\N(CH_3)_2\end{smallmatrix} \xrightarrow[0°]{Br_2} \left\{(CH_3)_2\underset{Br}{\overset{Cl}{C}}-C\begin{smallmatrix}\\ \\ \overset{+}{N}(CH_3)_2\end{smallmatrix} Br^-\right\} \xrightarrow{H_2O}$$
(**6**)   (**239**)

$$(CH_3)_2\underset{Br}{\overset{|}{C}}-C\begin{smallmatrix}\nearrow O\\ \searrow N(CH_3)_2\end{smallmatrix}$$
(**240**)

Chlorination of trichlorovinylamine (**87**) gives the iminium salt (**241**), which has been converted to the thioamide (**242**) by treatment with hydrogen sulfide (107):

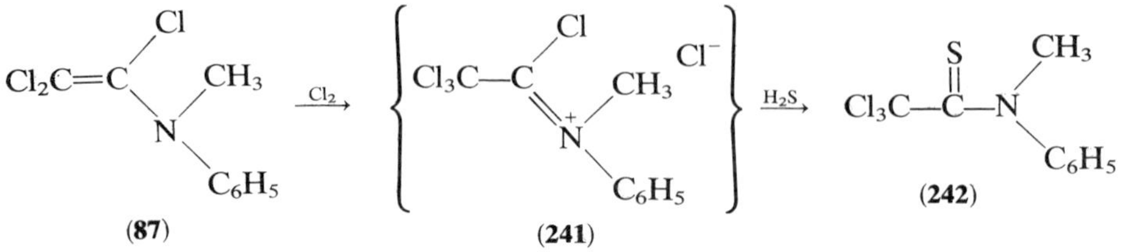

$$Cl_2C=C(Cl)(CH_3)(N(C_6H_5)) \xrightarrow{Cl_2} \left\{ Cl_3C-C(Cl)(CH_3)=\overset{+}{N}(C_6H_5)\ Cl^- \right\} \xrightarrow{H_2S} Cl_3C-C(=S)-N(CH_3)(C_6H_5)$$

(87)   (241)   (242)

α-Chloroenamines are certainly the reactive intermediates in the halogenation of amide chlorides (32,108) (Scheme 30).

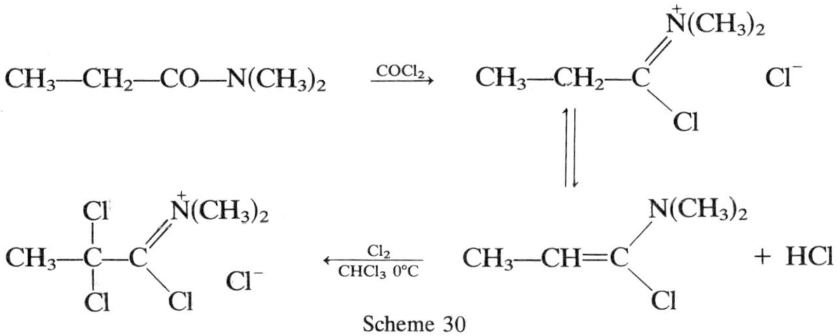

$$CH_3-CH_2-CO-N(CH_3)_2 \xrightarrow{COCl_2} CH_3-CH_2-C(=\overset{+}{N}(CH_3)_2)(Cl)\ Cl^-$$

$$CH_3-C(Cl)(Cl)-C(=\overset{+}{N}(CH_3)_2)(Cl)\ Cl^- \xleftarrow[CHCl_3\ 0°C]{Cl_2} CH_3-CH=C(N(CH_3)_2)(Cl) + HCl$$

Scheme 30

These halogenation reactions are of preparative value as a possible variation of the classical Hell-Volhard-Zelinsky method for the halogenation of carboxylic derivatives.

E. REACTION WITH ALCOHOLS AND CARBOXYLIC ACIDS

α-Chloroenamines are effective reagents for the replacement of hydroxyl groups by chlorine. These reactions, which were discovered by Speziale and Freeman (72), usually occur in the absence of HCl and under very mild conditions. They have been used successfully for the chlorination of hydroxyl-containing molecules, which are destroyed under the conditions normally used for this transformation. Tetramethyl-α-chloroenamine (6) has been found (73,93,109) to be much more reactive toward alcohols and carboxylic acids than are the trichlorovinylamines used by Speziale (Table XIII).

α-Chloroenamines have also been used for the conversion of optically active alcohols to the chlorides with inversion of configuration (72–74). The ease of preparation and large yields of product with high optical purity make this method an attractive one for the preparation of active alkyl halides.

TABLE XIII

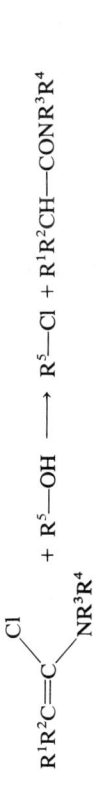

$$R^1R^2C=C(Cl)(NR^3R^4) + R^5-OH \longrightarrow R^5-Cl + R^1R^2CH-CONR^3R^4$$

R^1	R^2	R^3	R^4	R^5—OH	Experimental conditions	Yield, %	Ref.
Cl	Cl	C_2H_5	C_2H_5	C_2H_5—OH	60°C/1 hr/pure alcohol (trace HCl)	86	72
Cl	Cl	C_2H_5	C_2H_5	sec-C_4H_9—OH	70°C/1 hr/pure alcohol (trace HCl)	81[a]	72
Cl	Cl	C_2H_5	C_2H_5	tert-C_4H_9—OH	60°C/1 hr/pure alcohol (trace HCl)	60	72
C_6H_5	Cl	CH_3	C_6H_5	n-C_3H_7—OH	60°C/1 hr	59	5d
CH_3	CH_3	CH_3	CH_3	$PhCH_2$—OH	−30°C/CH_2Cl_2	100	103
CH_3	CH_3	CH_3	CH_3	n-C_4H_9—OH	−30°C/CH_2Cl_2	89	73
CH_3	CH_3	CH_3	CH_3	sec-C_4H_9—OH	−30°C/CH_2Cl_2	100	73
CH_3	CH_3	CH_3	CH_3	tert-C_4H_9—OH	−30°C/CH_2Cl_2	100	73
Cl	Cl	C_2H_5	C_2H_5	CH_3—COOH	50°C/2 hr/pure acid	72	72
Cl	Cl	C_2H_5	C_2H_5	Ph—COOH	85°C/3 hr/benzene	72	72
CH_3	CH_3	CH_3	CH_3	H—COOH	−60°C/CS_2/1 hr	90[b]	109
CH_3	CH_3	CH_3	CH_3	Cl_3C—COOH	20°C/CCl_4/instantaneous	94	109
CH_3	CH_3	CH_3	CH_3	$(C_2H_5S)_2CH$—COOH	20°C/CCl_4/instantaneous	95	109

[a] Inversion of configuration.
[b] Amide isolated after addition of pyrrolidine at −60°C.

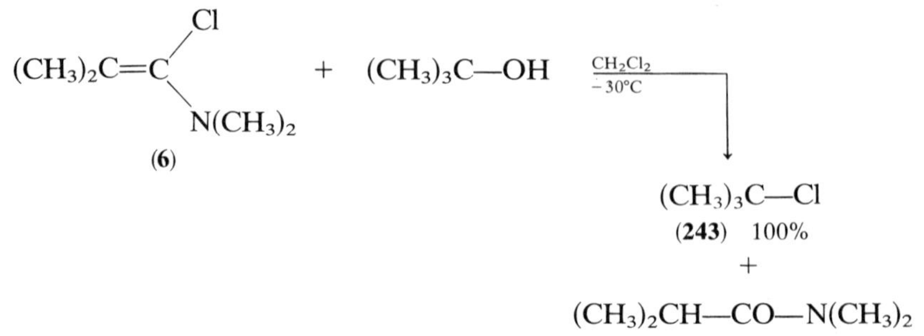

$(CH_3)_2C=C\overset{Cl}{\underset{N(CH_3)_2}{}}$ + $(CH_3)_3C-OH$ $\xrightarrow[-30°C]{CH_2Cl_2}$

(6)

$(CH_3)_3C-Cl$
(243) 100%
+
$(CH_3)_2CH-CO-N(CH_3)_2$

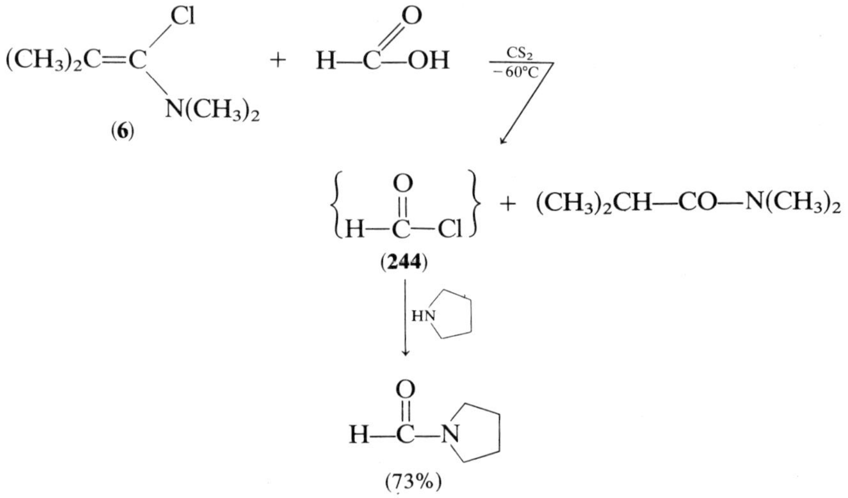

$(CH_3)_2C=C\overset{Cl}{\underset{N(CH_3)_2}{}}$ + $H-\overset{O}{\underset{}{C}}-OH$ $\xrightarrow[-60°C]{CS_2}$

(6)

$\left\{H-\overset{O}{\underset{}{C}}-Cl\right\}$ + $(CH_3)_2CH-CO-N(CH_3)_2$

(244)

$\downarrow HN\overset{}{\diagup}$

$H-\overset{O}{\underset{}{C}}-N\diagup$

(73%)

In the same way (29a) acid fluorides can be obtained easily from the tetramethyl-α-fluoroenamine (**77**) and carboxylic acids:

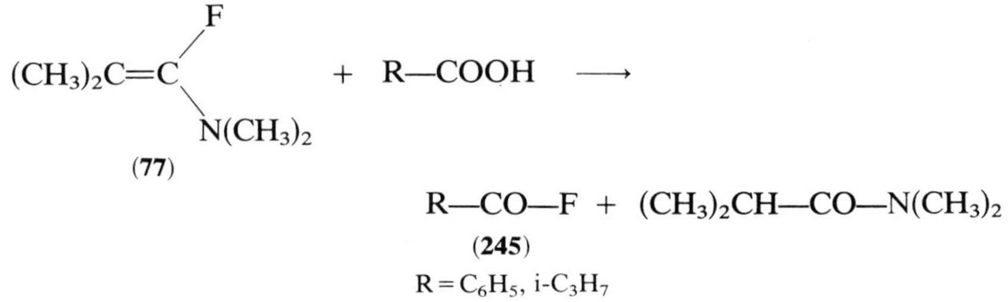

$(CH_3)_2C=C\overset{F}{\underset{N(CH_3)_2}{}}$ + R—COOH ⟶

(77)

R—CO—F + $(CH_3)_2CH-CO-N(CH_3)_2$
(245)
R = C_6H_5, i-C_3H_7

F. ACYLATION OF α-HALOENAMINES

The acylation of α-haloenamines has not been studied as thoroughly as that of enamines. The results presently available, however, indicate that, in general, α-chloroenamines are acylated less readily than are the corresponding enamines (105). 2-Disubstituted α-chloroenamines react only with strong electrophiles such as chlorosulfonyl isocyanate (105). On the other hand, 2-monosubstituted α-chloroenamines can be acylated by various reagents, such as phosgene (19,20,25), ketenes (105), or phenyl isocyanate (105). α-Fluoroenamines are more nucleophilic and usually react faster than the corresponding α-chloroenamines (29a).

1. *Reaction with Acid Chlorides*

It was mentioned in Section III-B-2 that α-chloroenamines derived from α-monosubstituted acetamides react with phosgene to give β-chlorocarbonyl-α-chloroenamines (19,20,25):

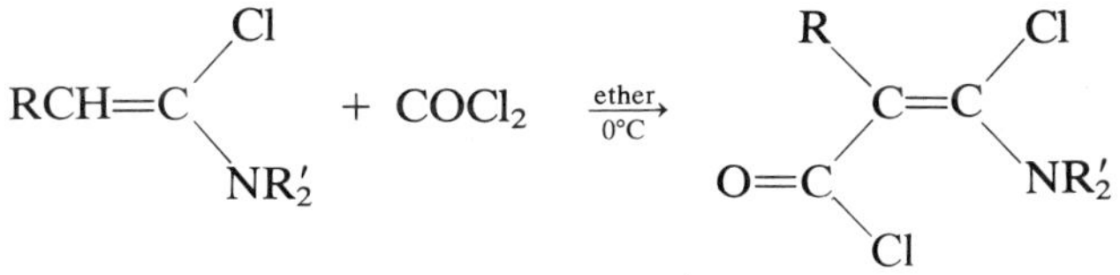

$$RCH=C\begin{matrix}Cl\\NR'_2\end{matrix} + COCl_2 \xrightarrow[0°C]{ether} \begin{matrix}R\\O=C\\Cl\end{matrix}C=C\begin{matrix}Cl\\NR'_2\end{matrix}$$

(**123**) R = C$_6$H$_5$, R' = C$_2$H$_5$ 60%
(**54**) R = CH$_3$, R'$_2$ = —(CH$_2$)$_5$—   70%

Tetramethyl-α-chloroenamine does not react with phosgene under the same experimental conditions (105). The explanation for this unreactivity could be the lack of coplanarity of the nitrogen lone pair with the π system (see Section IV-B: X-ray analysis). It is interesting, however, that the five-membered α-chloroenamine (**8**) is rapidly acylated (110):

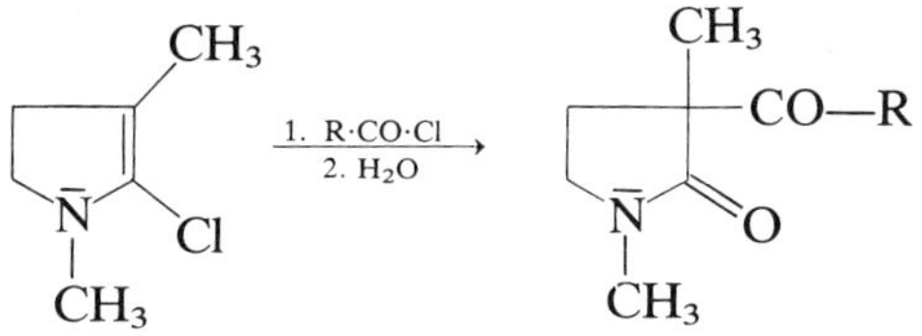

(**8**)   R = Cl   100%   (**246**)
         R = CH$_3$ 87%
         R = Ph   83%

This undoubtedly arises from the geometry of the ring, which forces the free pair of the nitrogen atom into conjugation with the double bond. Similarly tetramethylfluoroenamine (**77**) is smoothly acylated by phosgene (29a). The primary acylation product (**247**) has been converted (29a) into various malonic acid derivatives by direct hydrolysis or by thermal demethylation followed by hydrolysis (Scheme 31).

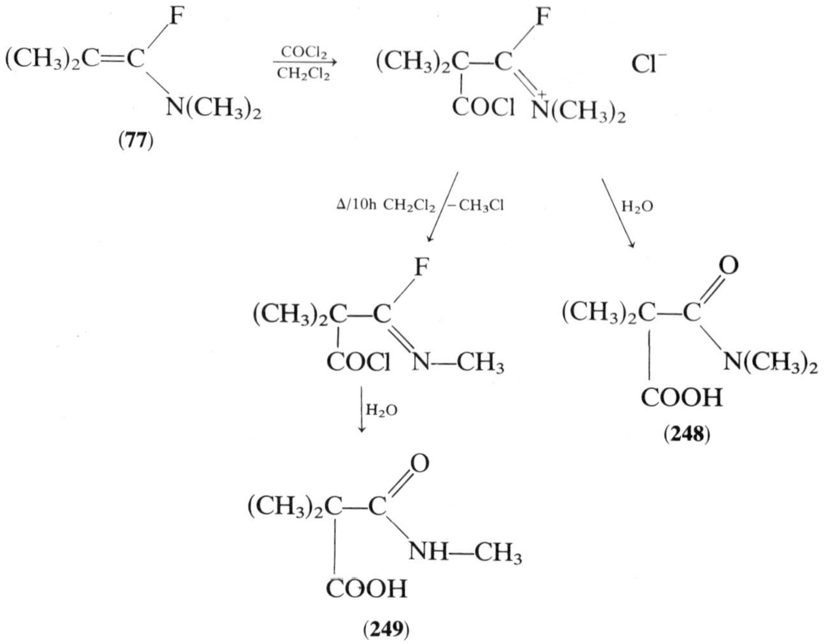

Scheme 31

β-Chlorocarbonyl-α-chloroenamines and analogous β-acyl-α-chloroenamines are 1,3-biselectrophilic reagents that are exceedingly useful for heterocyclizations (31d), for example:

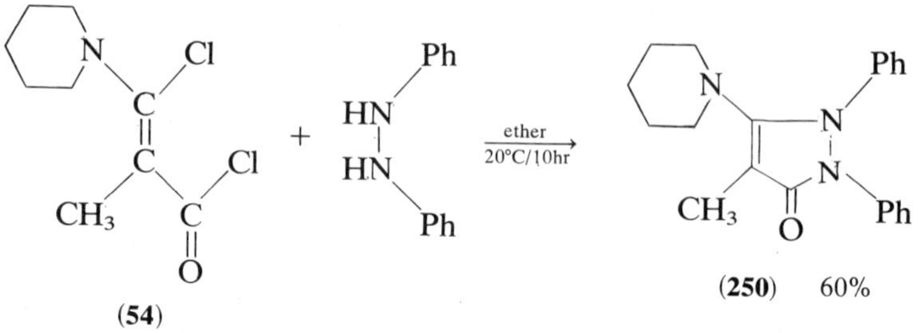

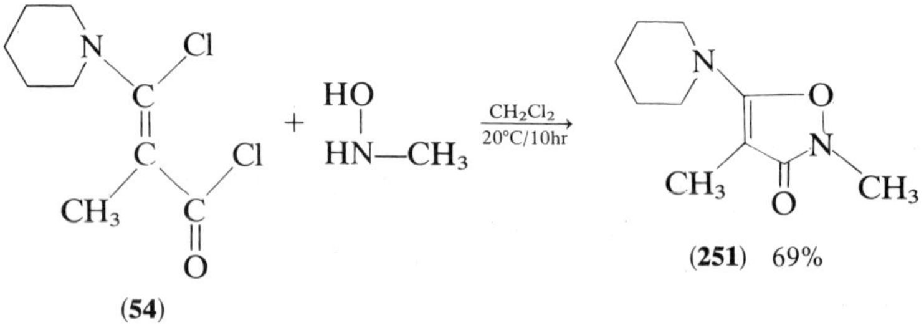

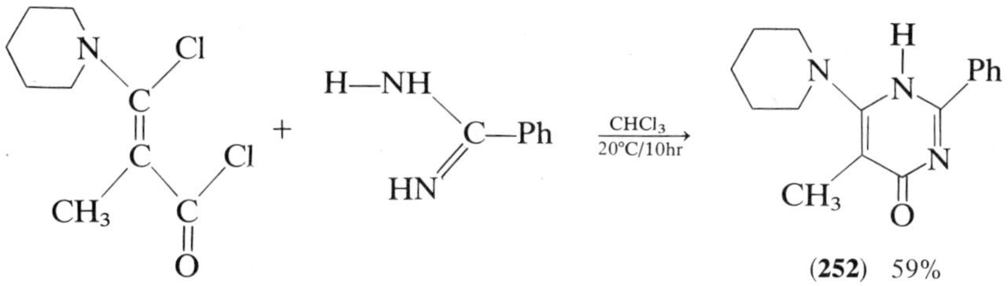

VI. REACTIONS OF α-HALOENAMINES WITH ELECTROPHILES

[Structures: compound (54) reacting with H-NH/HN=C-Ph amidine in CHCl₃ at 20°C/10hr to give compound (252) 59%]

α-Chloroenamines have been shown (5b,111) to be the reactive intermediates in the reaction (Scheme 32) of acetamide derivatives with oxalyl chloride to give derivatives of furanones (**253**) (Table XIV).

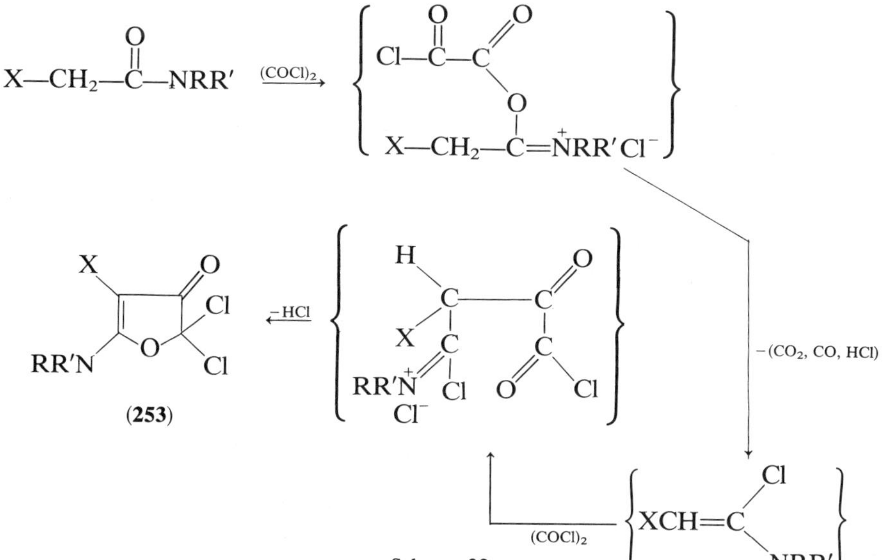

[Scheme 32: X—CH₂—C(O)—NRR′ with (COCl)₂ gives intermediates leading to furanone (253) with loss of HCl, and via (COCl)₂ to XCH=C(Cl)(NRR′), losing CO₂, CO, HCl]

Scheme 32

2. Reaction with Amide Chlorides

The condensation of β-monosubstituted α-chloroenamines with amide chlorides was mentioned in Section III-B-2. The resulting products are bisiminium salts (**254**) formally derived from β-ketoamides. In practice the isolation of the α-chloroenamine is not required, and the condensation can be effected directly from the tertiary amide by treatment with reagents such as PCl_5, $OPCl_3$ and $COCl_2$, followed by thermal dehydrochlorination (32,112). However, the best results are obtained when a monosubstituted N,N-dialkyl acetamide is treated with phosgene at room

TABLE XIV

$$X-CH_2-CONRR' \xrightarrow[CH_2Cl_2/2\ hr/20°C]{2\ ClCOCOCl}$$

(product structure shown with X, R, R', and Cl, Cl, O substituents)

X	R	R'	Yield, %	Ref.
PhO	CH_3	Ph	63	111
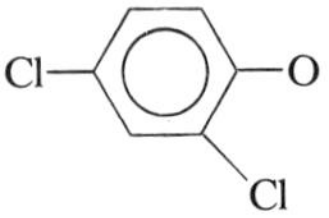 Cl-C₆H₃(Cl)-O	CH_3	Ph	70	111
CH_3O	CH_3	Ph	76	111
PhO	C_2H_5	C_2H_5	82	111
CH_3	C_2H_5	C_2H_5	34	111
CN	CH_3	Ph	32	111
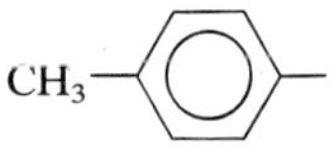 CH_3-C₆H₄-	CH_3	Ph	56	111
PhS	C_2H_5	C_2H_5	34	111
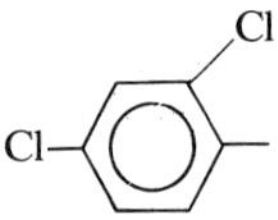 Cl-C₆H₃(Cl)-	C_2H_5	C_2H_5	81	111
Cl	C_2H_5	C_2H_5	94	5b
Cl	CH_3	CH_3	87	5b
Cl	CH_3	Ph	91	5b
Cl	Ph	Ph	84	5b,111
Ph	C_2H_5	C_2H_5	57	5b

temperature, and the resulting amide chloride is treated with 0.5 equivalent of triethylamine (20) (Scheme 33).

The bisiminium salts can be identified by hydrolysis to the β-ketoamides (21). The overall yield from 15 to 21 is excellent [R, R' = CH_3: 83% (19)]. The bisiminium salts (254) are useful reagents for heterocyclizations (20).

3. *Reactions with Heterocumulenes*

2-Monosubstituted-α-chloroenamines (18), generated *in situ* by base-induced dehydrochlorination of amide chlorides (17), react readily with

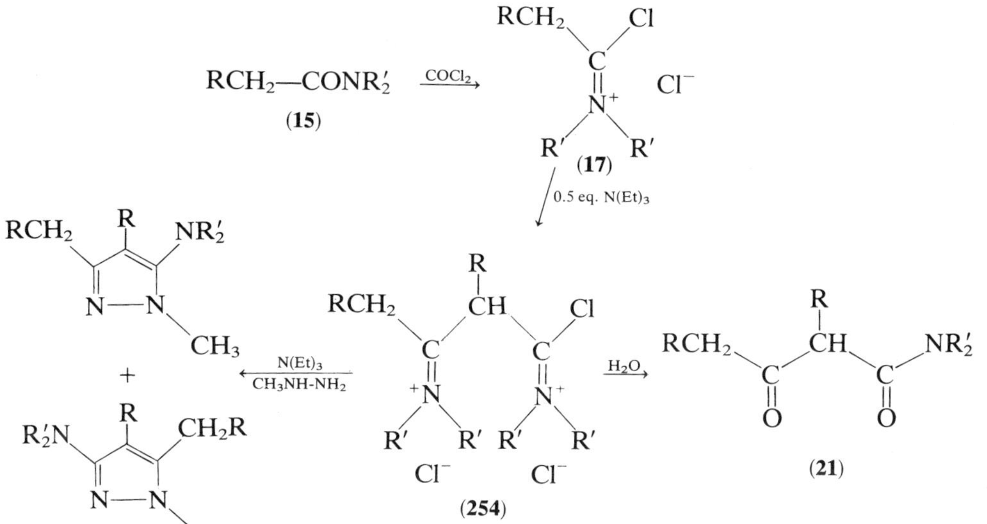

VI. REACTIONS OF α-HALOENAMINES WITH ELECTROPHILES

Scheme 33

diphenyl- or dimethylketene to yield cyclobutanone adducts (105). The reaction probably occurs by way of a zwitterionic intermediate (**255**). The primary adducts (**256**) can be detected spectroscopically but are thermally unstable. After treatment with triethylamine or diluted sodium hydroxide they give (105) the stable aminocyclobutenones (**257**) (Scheme 34).

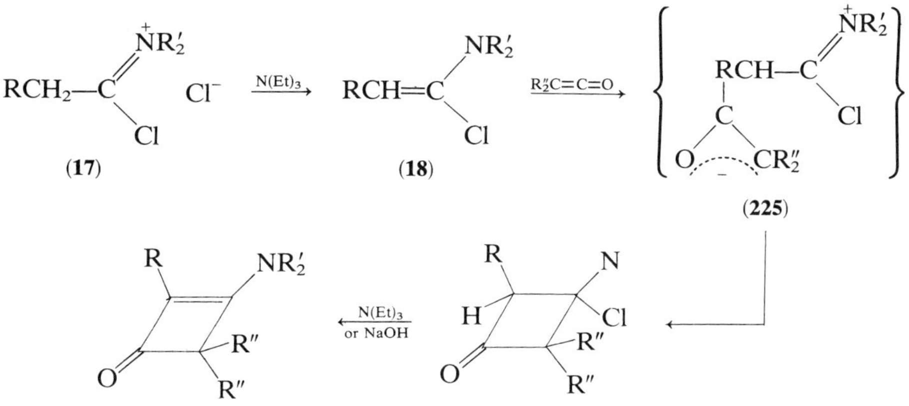

Scheme 34

The α-chloroenamines derived from isobutyramides are totally inert toward diphenylketene, even under forced conditions (105). However, they react with the highly electrophilic chlorosulfonyl isocyanate to give (105), after hydrolysis, derivatives of malondiamides (**259**) (Scheme 35).

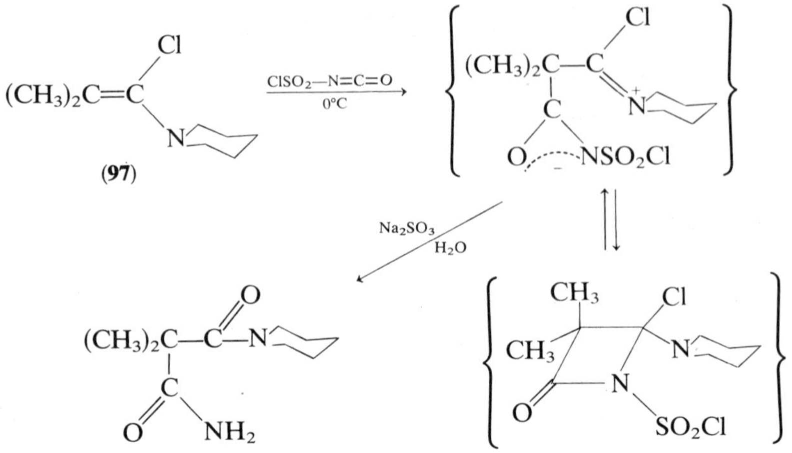

(97) (259) (258)

Scheme 35

The higher nucleophilicity of tetramethyl-α-fluoroenamine (**77**), as compared with the corresponding α-chloroenamine (**6**), reveals itself in the ability of **77** to react (29a) with keteniminium salts at −15°C (Scheme 36). The course of the reaction depends on the nature of the anion (29a):

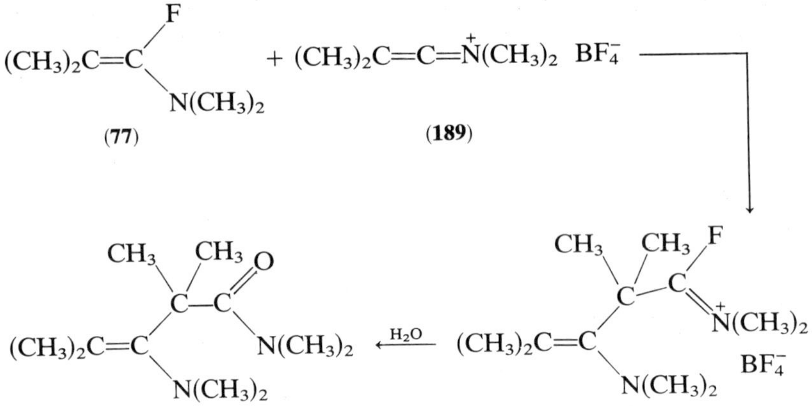

(77) (189) (261) (260)

Scheme 36

with the tetrafluoroborate, the primary addition product **260** is stable and, after hydrolysis, yields the amide (**261**). With the covalent keteniminium iodide (**82**), however, a β-lactam (**265**) is obtained after hydrolysis (29a). The formation of **265** can be explained by an equilibration between the open-chain addition product **262** and the cyclic form **263**, which is demethylated by the nucleophilic iodide ion (Scheme 37). In this

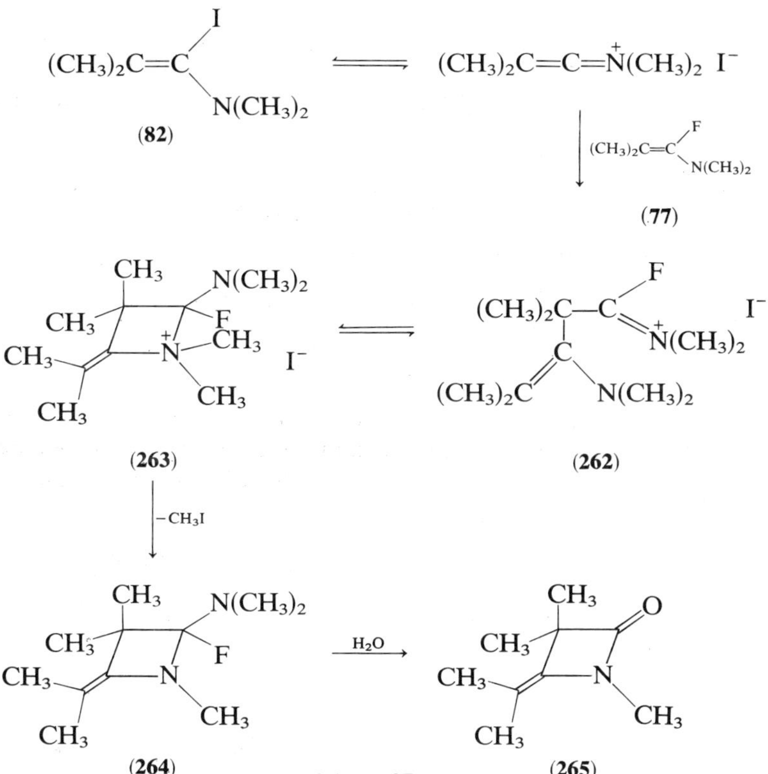

Scheme 37

reaction, which is a good illustration of the bifunctional character of α-haloenamines, the α-fluoroenamine behaves like an enamine derived from a carboxylic acid halide and the α-iodoenamine like a keteniminium iodide, a synthon equivalent to an acyl cation (R—$\overset{+}{C}$=O).

VII. Elimination to Ynamines

Elimination of HCl from an α-chloroenamine occurs much more readily than the dehydrochlorination of simple alkenyl chlorides (86). In most cases it can be promoted by a weak base such as triethylamine at room temperature (19,25). These elimination reactions deserve more detailed investigations since they offer a potentially useful method (2) of preparing ynamines. For the present, at least, however, the method is limited by secondary reactions that are not readily controlled. When N,N-disubstituted lithium amides are used as bases, the elimination often proceeds smoothly (2) but is accompanied by the formation of alkylidene diamines (**267**) (Scheme 38).

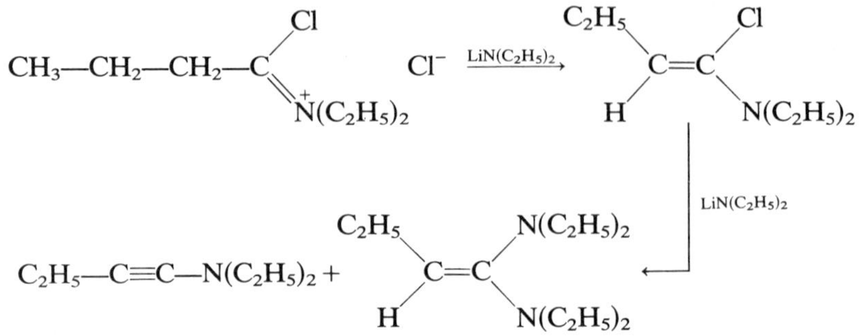

Scheme 38

(266) 38%   (267) 32%

With weaker tertiary bases such as triethylamine, the reaction is often followed by cycloaddition of the ynamine and the α-chloroenamine to yield (2,19,20,25) a cyclobutenecyanine (268) (cf. Section III and VIII) (Scheme 39).

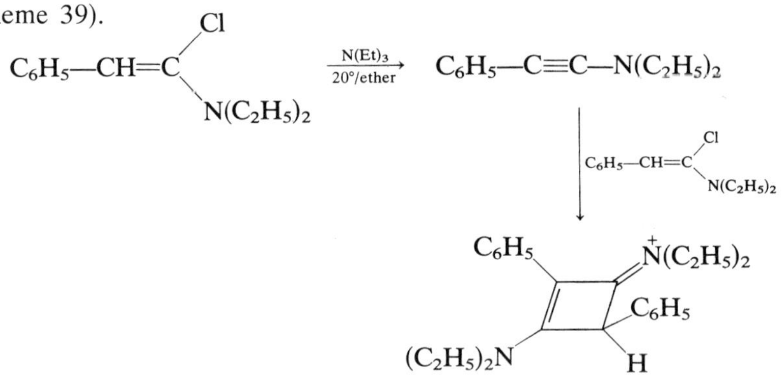

Scheme 39

VIII. Cycloaddition Reactions of Keteniminium Ions

A. GENERAL ASPECTS

As *electron-rich olefins* intermediate between olefins and enamines in polarizability, α-haloenamines are suitable for use in cycloadditions to electrophilic substrates such as carbenes, nitrenes, and electrophilic π systems. The reactions of 2-monosubstituted α-chloroenamines with ketenes (105) mentioned in Section VI are apparently the only examples known at present. Clearly, these reactions offer great synthetic potential and merit further attention. It is worth noting that, in this case, the α-haloenamine is a synthon equivalent to a ketene or a ketenimine reacting as the *nucleophilic partner* in a cycloaddition (Scheme 40).

VIII. CYCLOADDITION REACTIONS OF KETENIMINIUM IONS

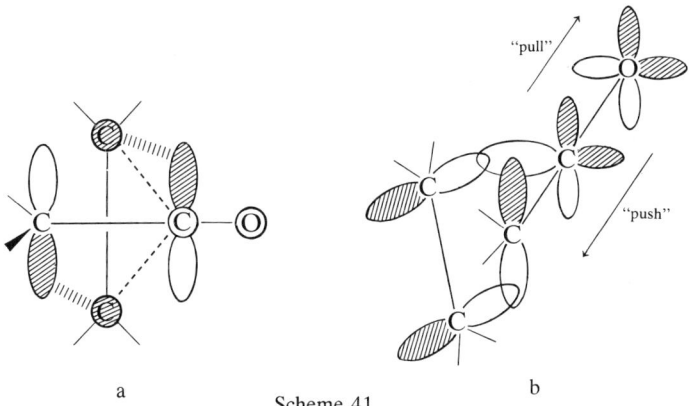

Scheme 40

On the other hand, α-haloenamines are sources of keteniminium salts, which, as the "iminium derivatives" of ketenes, are expected to react with special ease as the electrophilic partner in (2+2) cycloaddition reactions. Thermal cycloadditions of ketenes to weakly polar olefins have been studied extensively (83,113), and the results analyzed in terms of a $(\pi^2 s + \pi^2 a)$ concerted reaction (114) in which the ketene is the antarafacial partner (Scheme 41a). This description emphasizes the crucial role of the reactive carbonyl function of a ketene molecule, which contributes strong bonding interactions (115) normally absent in the transition state $(\pi^2 s + \pi^2 a)$ reactions for simple olfins.

An alternative description of the transition state is shown in Scheme 41b. The reaction is viewed as involving electrophilic attack of the carbonyl carbon of the ketene on one end of the olefinic component, accompanied by nucleophilic attack of the orthogonal enolate function at the other end of the olefinic bond (116). The transition state is of the Hückel type and, as it contains six electrons, is aromatic. The reaction is therefore thermally allowed. In our view (117) this description stresses very clearly the important structural features that render ketenes especially reactive in (2+2) cycloadditions: a strongly electron-deficient π system (the "pull" system) orthogonal to a nucleophilic π component (the "push" system). In this respect ketenes resemble singlet carbenes, which are also composed of orthogonal "pull" and "push" systems and are very reactive in (2+2) cycloadditions.

It is expected that replacement of the carbonyl group by the more electrophilic iminium function should increase the reactivity of the cumulene in (2+2) cycloadditions. However, the keteniminium ion lacks a strong "push" component (82). This reveals itself in the electronic population of the $p\pi$ orbital (Table VI) (82), as well as, chemically, in the absence of dimerization (7,80). Therefore keteniminium ions should not simply be considered as "superketenes." Indeed, whereas they show a higher reactivity than ketenes in (2+2) cycloadditions to simple olefins (7,8,80) or acetylenes (80,118), they behave like allenes toward cisoid dienes and give (4+2) cycloadducts (80,119).

In spite of their recent discovery, cycloadditions of keteniminium ions to unsaturated substrates such as olefins, acetylenes, or imines now serve as valuable and effective methods for the formation of four-membered rings. It will be apparent that one cannot speak of a single mechanism for the cycloaddition of keteniminium ions to unsaturated molecules. Although there is probably a spectrum of mechanisms ranging from the stepwise with a discrete intermediate carbenium ion to a multicenter process, it is expected that highly nucleophilic or crowded substrates will usually follow the stepwise pathway.

B. CYCLOADDITIONS TO OLEFINS (TABLE XV)

The exceptional reactivity of keteniminium ions in (2+2) cycloadditions is well illustrated by the fast addition of tetramethylketeniminium tetrafluoroborate (or zincate) to ethylene at room temperature and atmospheric pressure (8,80):

$(CH_3)_2C{=}C{=}\overset{+}{N}(CH_3)_2\ X^- + CH_2{=}CH_2 \xrightarrow[20°C]{CH_2Cl_2}$

(189) $X = BF_4^-$
(190) $X = ZnCl_3^-$

(269)

TABLE XV

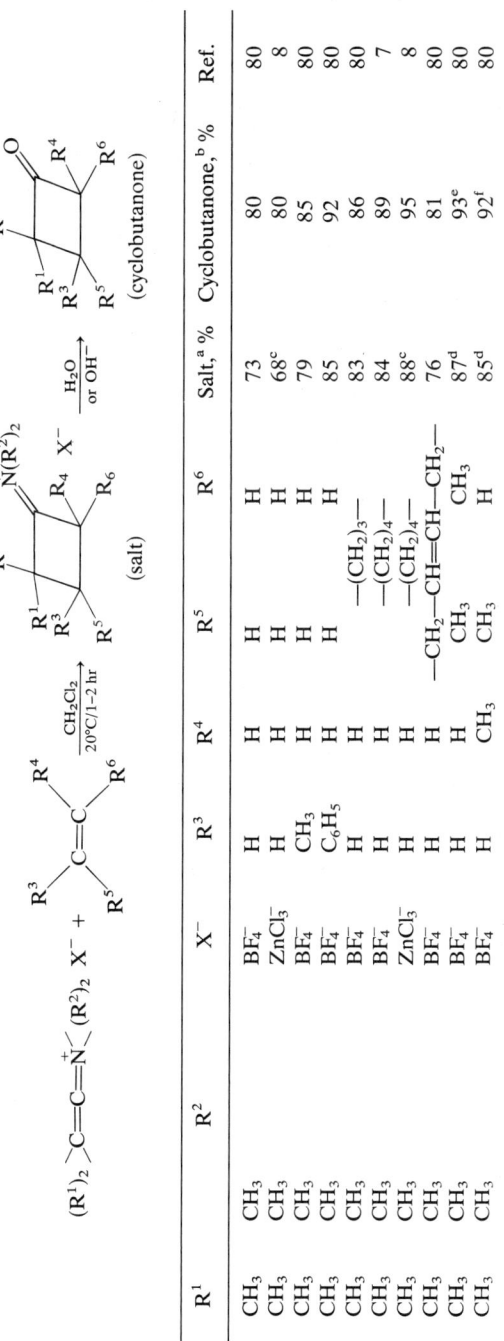

R^1	R^2	X^-	R^3	R^4	R^5	R^6	Salt,[a] %	Cyclobutanone,[b] %	Ref.
CH_3	CH_3	BF_4^-	H	H	H	H	73	80	80
CH_3	CH_3	$ZnCl_3^-$	H	H	H	H	68[c]	80	8
CH_3	CH_3	BF_4^-	CH_3	H	H	H	79	85	80
CH_3	CH_3	BF_4^-	C_6H_5	H	H	H	85	92	80
CH_3	CH_3	BF_4^-	H	H	—$(CH_2)_3$—		83	86	80
CH_3	CH_3	BF_4^-	H	H	—$(CH_2)_4$—		84	89	7
CH_3	CH_3	$ZnCl_3^-$	H	H	—$(CH_2)_4$—		88[c]	95	8
CH_3	CH_3	BF_4^-	H	H	—CH_2—CH=CH—CH_2—		76	81	80
CH_3	CH_3	BF_4^-	H	H	CH_3	CH_3	87[d]	93[e]	80
CH_3	CH_3	BF_4^-	H	CH_3	CH_3	H	85[d]	92[f]	80

TABLE XV (continued)

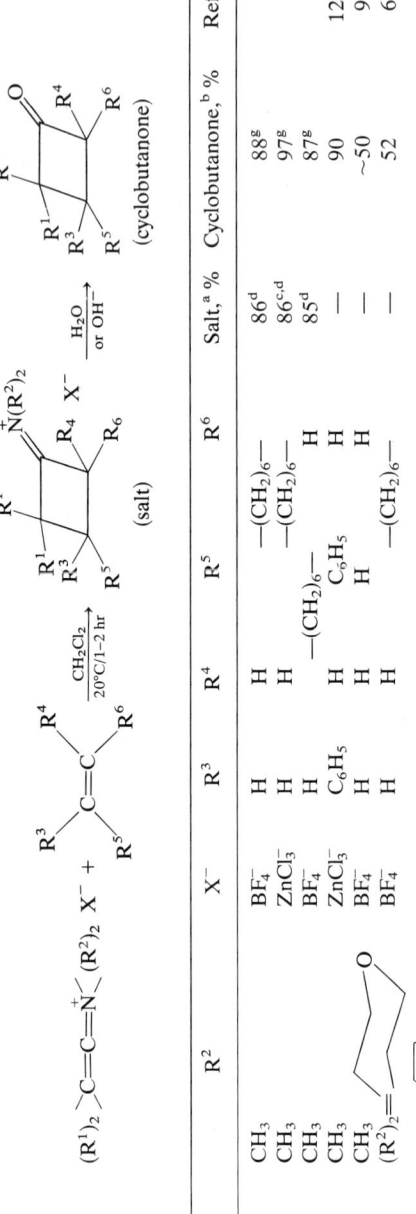

R^1	R^2	R^3	R^4	R^5	R^6	X^-	Salt,[a] %	Cyclobutanone,[b] %	Ref.
CH$_3$	CH$_3$	H	H	—(CH$_2$)$_6$—		BF$_4^-$	86[d]	88[g]	7
CH$_3$	CH$_3$	H	H	—(CH$_2$)$_6$—		ZnCl$_3^-$	86[c,d]	97[g]	8
CH$_3$	CH$_3$	H	H	—(CH$_2$)$_5$—	H	BF$_4^-$	85[d]	87[g]	7
CH$_3$	CH$_3$	C$_6$H$_5$	H	C$_6$H$_5$	H	ZnCl$_3^-$	—	90	120
Cl	CH$_3$	H	H	H	H	BF$_4^-$	—	~50	98
CH$_3$	(R^2)$_2$=	H	H	—(CH$_2$)$_6$—		BF$_4^-$	—	52	60
CH$_3$	(R^2)$_2$=	H	H	—(CH$_2$)$_4$—		ZnCl$_3^-$	—	65	60
CH$_3$	C$_2$H$_5$	H	H	—(CH$_2$)$_4$—		ZnCl$_3^-$	—	50	60

[a] Pure product crystallized from chloroform–ether.
[b] Hydrolysis of crude reaction mixtures. (The yields are corrected by VPC: the corresponding amides are the sole by-products.)
[c] Salt crystallized as perchlorate.
[d] No contamination by the stereoisomer (IR and NMR analysis).
[e] 17% contamination after hydrolysis (VPC analysis).
[f] 4% contamination after hydrolysis (VPC analysis).
[g] ≤5% contamination after hydrolysis (NMR and VPC analysis).

VIII. CYCLOADDITION REACTIONS OF KETENIMINIUM IONS 513

With propylene or styrene the cycloaddition occurs regiospecifically (80):

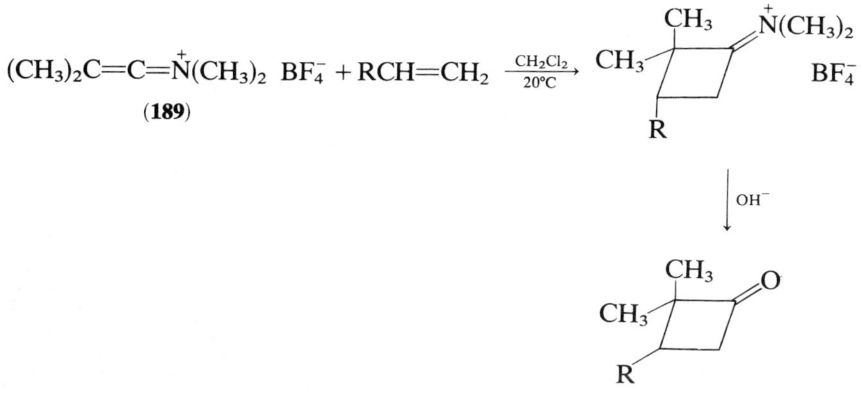

$(CH_3)_2C=C=\overset{+}{N}(CH_3)_2$ BF_4^- + $RCH=CH_2$ $\xrightarrow[20°C]{CH_2Cl_2}$ [cyclobutane with CH_3, CH_3, R, and $=\overset{+}{N}(CH_3)_2$] BF_4^-

(189)

$\downarrow$ OH$^-$

[cyclobutanone with CH_3, CH_3, R]

(270) R = CH$_3$
(271) R = C$_6$H$_5$

The reactions are cis stereospecific (7,80): thus *cis*-cyclooctene and tetramethylketeniminium tetrafluoroborate give only the cis adduct (272):

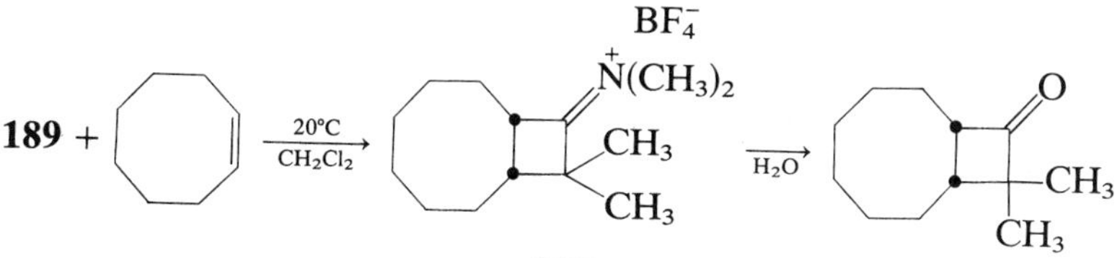

189 + [cyclooctene] $\xrightarrow[CH_2Cl_2]{20°C}$ [bicyclic adduct with $\overset{+}{N}(CH_3)_2$ BF_4^-, CH_3, CH_3] (272) $\xrightarrow{H_2O}$ [bicyclic ketone with CH_3, CH_3, =O]

The addition of **189** to 2,2-dimethyl-1-methylenecyclopropane proceeds (120) without any of the skeletal rearrangements that might be expected from a stepwise reaction involving carbenium ion intermediate:

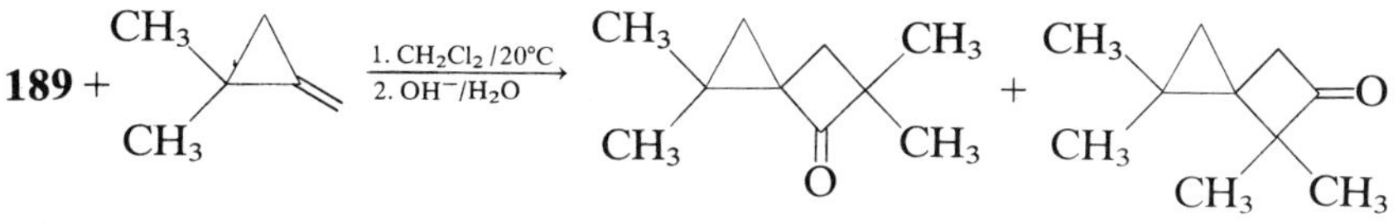

189 + [2,2-dimethyl-1-methylenecyclopropane] $\xrightarrow[\text{2. OH}^-/\text{H}_2\text{O}]{\text{1. CH}_2\text{Cl}_2/20°C}$ (273) + (274)

Furthermore, with dicyclopentadiene, tetramethylketeniminium reacts

(80) like ketenes (113b,121) or dihalocarbenes (122) and adds preferentially to the less-strained double bond:

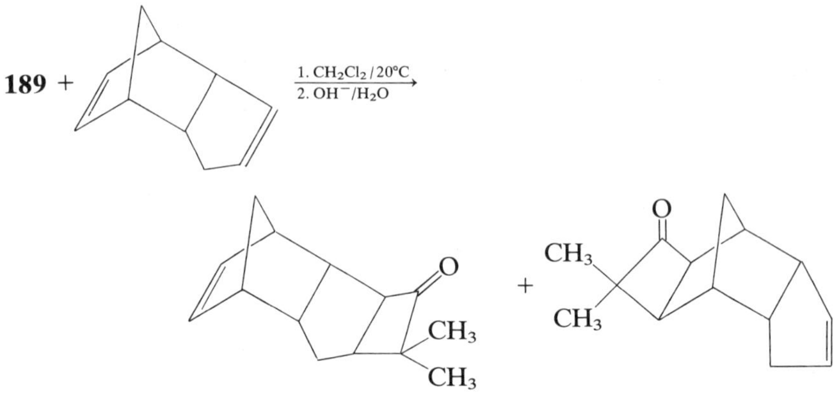

(**275**) 90%   (**276**) 10%

These results are fully consistent with the hypothesis of a concerted mechanism, implying an orthogonal approach of the reactants. Such a transition state is quite sensitive to steric effects. Thus N,N-diisopropyl-dimethylketeniminium ion (**277**) does not give any adduct with cyclohexene even after 24 hr at 100°C (60), whereas **189** reacts completely in less than 1 hr at room temperature (80):

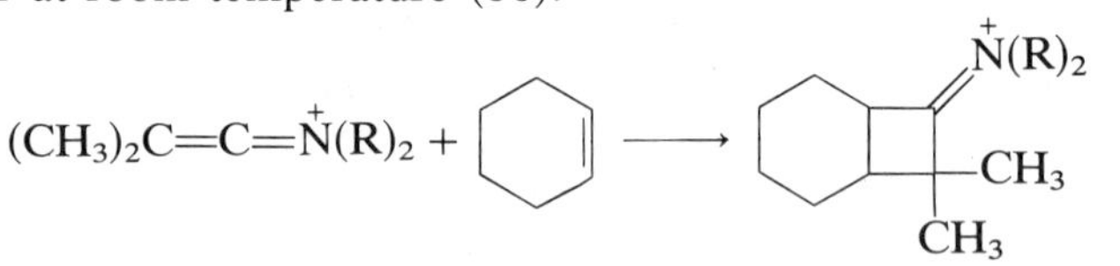

$(CH_3)_2C=C=\overset{+}{N}(R)_2$ +

(**189**) R = CH_3
(**277**) R = i-C_3H_7

(**278**) R = CH_3 89%
       R = i-C_3H_7 0%

With 1,1-disubstituted olefins the geometry required for the transition state is also less readily attained. Thus isobutene and **189** give (120) only a low yield of the four-membered ring adduct (**279**). Hydrolysis of the reaction mixture leads to the isolation of a linear ketone (**283**) in 60% yield and a cyclobutanone (**280**) in 21% yield (120) (Scheme 42). Both products may arise from a common intermediate (**281**) with a tertiary carbeñium ion center (eventually in equilibrium with **282**), or they can be formed by competing (2+2) concerted addition and (4+2) ene reaction.

C. CYCLOADDITIONS TO CONJUGATED DIENES

The behavior of ketenes and that of allenes toward conjugated dienes are strikingly different. Ketenes give exclusively cycloadducts resulting

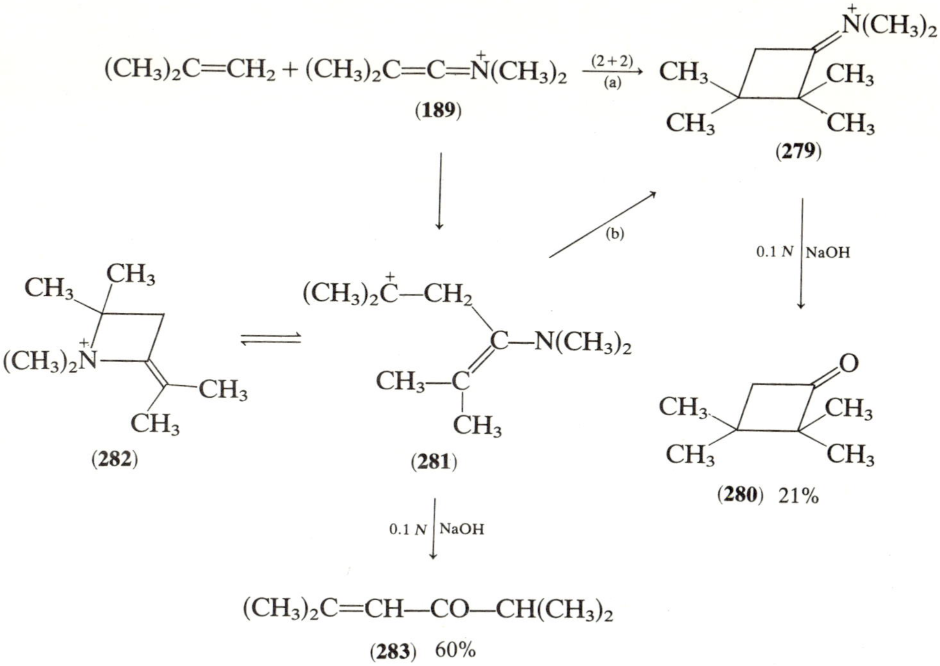

VIII. CYCLOADDITION REACTIONS OF KETENIMINIUM IONS

$$(CH_3)_2C=CH_2 + (CH_3)_2C=C=\overset{+}{N}(CH_3)_2 \xrightarrow[(a)]{(2+2)}$$
(189)

[Structure **(279)**: cyclobutane with =N⁺(CH₃)₂, three CH₃ groups]

[Structure **(282)**: (CH₃)₂N⁺ cyclobutane with CH₃ groups] ⇌ (CH₃)₂Ċ—CH₂ / CH₃—C / C—N(CH₃)₂ / CH₃ **(281)**

(b) →

(279) $\xrightarrow{0.1\,N\;\text{NaOH}}$ **(280)** 21% [cyclobutanone with CH₃ groups]

(281) $\xrightarrow{0.1\,N\;\text{NaOH}}$ $(CH_3)_2C=CH—CO—CH(CH_3)_2$
(283) 60%

Scheme 42

from a (2+2) addition across the C–C double bond of the cumulene (113), whereas allenes give (4+2) cycloadducts (123). It is interesting that tetramethylketeniminium (**189**) shows an intermediate behavior. With transoid dienes it reacts like ketenes and gives (2+2) cycloadducts (7,80) in high yield (Table XVI).

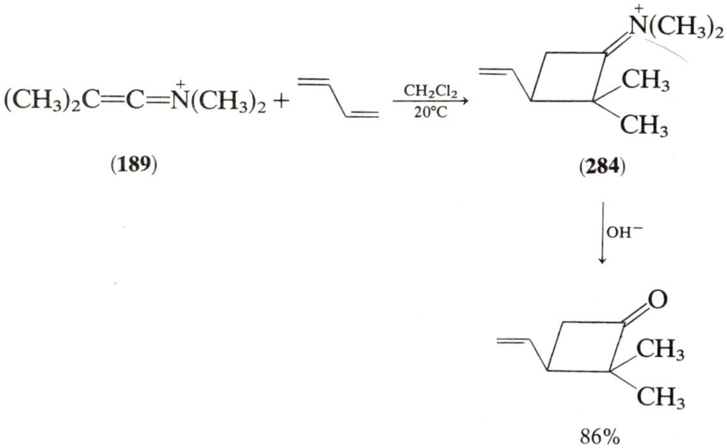

$$(CH_3)_2C=C=\overset{+}{N}(CH_3)_2 + \text{[butadiene]} \xrightarrow[20°C]{CH_2Cl_2}$$
(189)

(284) [vinyl-substituted cyclobutane with =N⁺(CH₃)₂ and two CH₃]

$\xrightarrow{OH^-}$ [vinylcyclobutanone with two CH₃] 86%

TABLE XVI

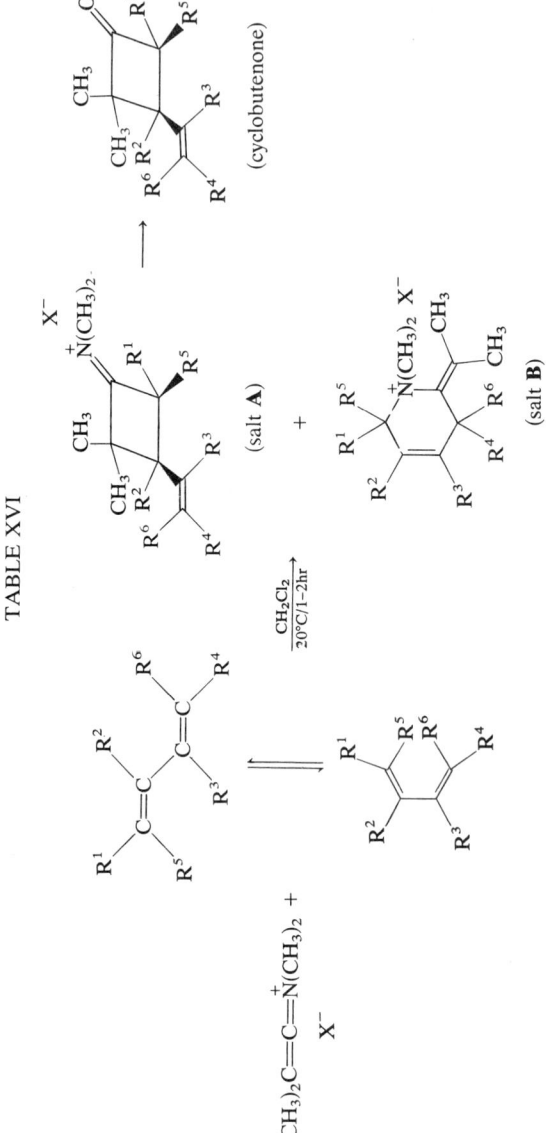

X^-	R^1	R^2	R^3	R^4	R^5	R^6	Salt,[a] %		Cyclobutanone,[b] %	Ref.
							A	B		
BF_4^-	H	H	H	H	H	H	84	0	90	7
$ZnCl_3^-$	H	H	H	H	H	H	48[c]	0	90–100	98
BF_4^-	H	H	H	H	H	CH_3	85[d]	0	93	7
$ZnCl_3^-$	H	H	H	H	H	CH_3	50[c,d]	0	90–100	98
BF_4^-	H	H	H	CH_3	H	H	82[d]	0	90	7
$ZnCl_3^-$	H	H	H	CH_3	H	H	53[c,e]	0	90–100	98
BF_4^-	CH_3	H	H	CH_3	H	H	80[e]	0	84[f]	80
BF_4^-	H	H	H	CH_3	CH_3	H	87[d,e]	0	91[g]	80
BF_4^-	H	CH_3	H	H	H	H	—	≤7	71[h]	80
BF_4^-	H	CH_3	CH_3	H	H	H	—	50	18–20	80
BF_4^-	H	H	H	H	—CH_2—		0	88	0	119
BF_4^-	H	H	H	H	—$(CH_2)_2$—		0	75	0	119

[a] Pure products crystallized from chloroform-ether.
[b] Corrected yields by VPC after hydrolysis of the crude reaction mixtures.
[c] Crystallized as perchlorate.
[d] No contamination by the regioisomer.
[e] No contamination by the stereoisomer (NMR analysis).
[f] 3% contamination by the stereoisomer after hydrolysis (VPC analysis).
[g] 11% contamination by the stereoisomer after hydrolysis (VPC analysis).
[h] Mixture of the two regioisomers.

The cycloaddition of **189** to transoid dienes is also regio- and stereo-specific, as illustrated by the reaction with *cis,trans*-hexadiene (80):

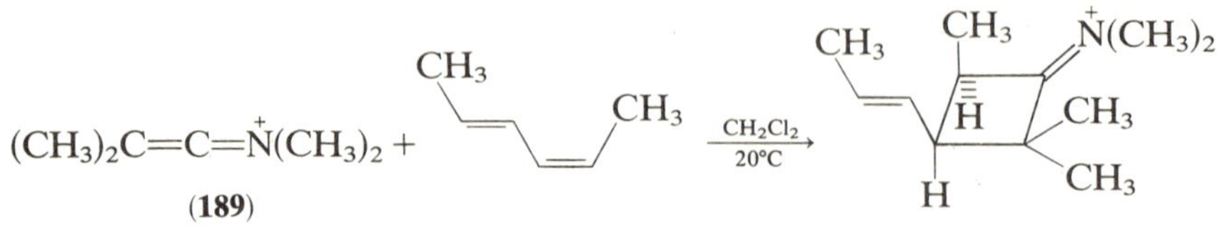

$$(CH_3)_2C=C=\overset{+}{N}(CH_3)_2 + \text{CH}_3\text{-CH=CH-CH=CH-CH}_3 \xrightarrow[20°C]{CH_2Cl_2} \textbf{(285)} \quad 91\%$$

This example also demonstrates the higher reactivity of cis, as compared to trans, 1,2-disubstituted double bonds—a result in agreement with the known relative ketenophilic activity of olefins (113,121). Similarly, *cis*- and *trans*-piperylenes yield only adducts resulting from an addition of **189** on the unsubstituted double bond of the diene (7,80):

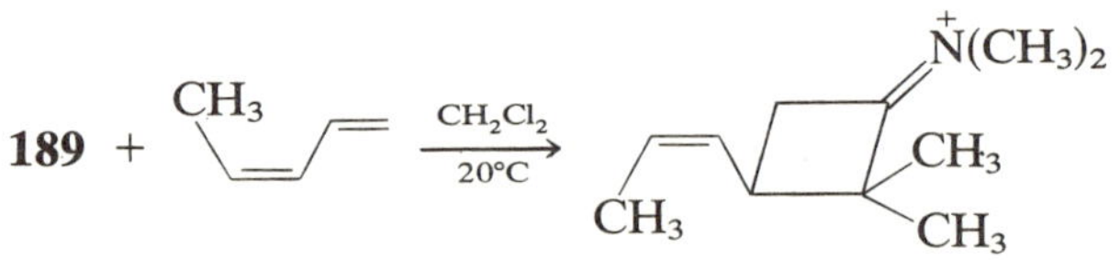

$$\textbf{189} + \text{CH}_3\text{-CH=CH-CH=CH}_2 \xrightarrow[20°C]{CH_2Cl_2} \textbf{(286)} \quad 93\%$$

$$\textbf{189} + \text{CH}_2\text{=CH-CH=CH-CH}_3 \xrightarrow[20°C]{CH_2Cl_2} \textbf{(287)} \quad 90\%$$

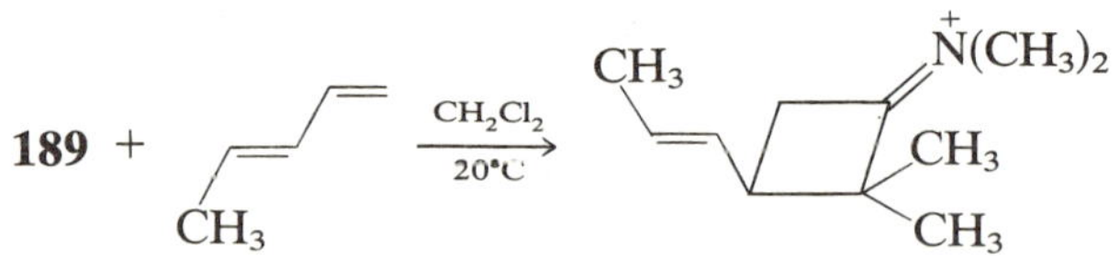

These reactions may also be satisfactorily interpreted by a multicenter mechanism.

Unexpectedly, toward cis-fixed dienes, tetramethylketeniminium ion **189** reacts as an allene. Thus cyclopentadiene and cyclohexadiene add across the C–$\overset{+}{N}$ double bond of the cumulene (80,119):

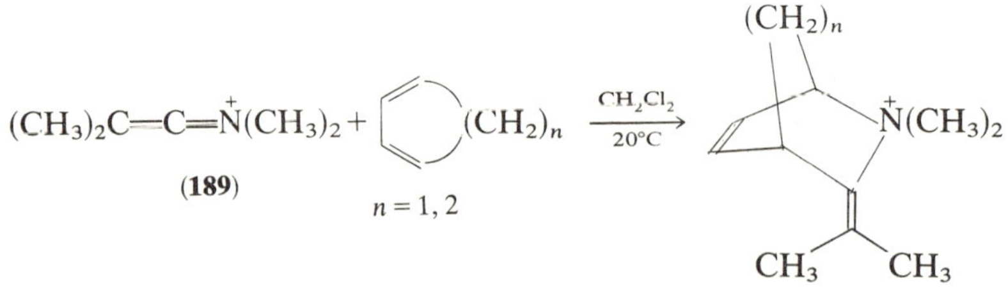

$$(CH_3)_2C=C=\overset{+}{N}(CH_3)_2 + \text{cyclic diene}(CH_2)_n \xrightarrow[20°C]{CH_2Cl_2} \textbf{(288)}$$

$n = 1, 2$

VIII. CYCLOADDITION REACTIONS OF KETENIMINIUM IONS

An important factor governing the mode of cycloaddition is the population of transoid and skew conformations of the dienes. The skew (cisoid) conformation is expected to favor the (4+2) over the (2+2) process (124). These results show that keteniminium ion should be regarded, not simply as a "superketene," but also as an "activated allene."

D. CYCLOADDITIONS TO ALLENES

Few data are available on the reactions of keteniminium ions with allenes, but the known results are striking. Allene itself gives (120) the "normal" adduct (**289**) with **189**, whereas tetramethylallene gives (120) an "abnormal" product (**290**) arising from a cycloaddition across the iminium bond of **189**. 1,1-Dimethylallene gives two products (120): the unsubstituted double bond of the allene adds across the C–C double bond of **189**, whereas the disubstituted double bond reacts with the iminium bond (Scheme 43 and Table XVII).

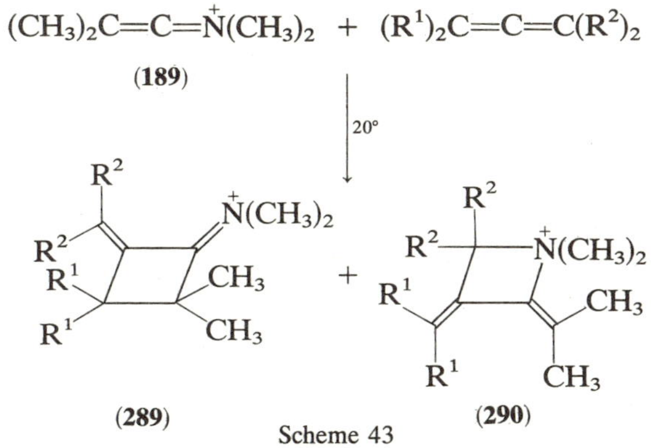

$$(CH_3)_2C=C=\overset{+}{N}(CH_3)_2 \;+\; (R^1)_2C=C=C(R^2)_2$$

(**189**)

Scheme 43

TABLE XVII

R^1	R^2	**289**, %	**290**, %
H	H	70	0
CH$_3$	CH$_3$	0	80
H	CH$_3$	40	40

E. CYCLOADDITIONS TO ACETYLENES

Simple acetylenic compounds are usually poor ketenophiles (83). However, they combine, even at room temperature, with tetramethylketeniminium salt to form (80,118) cyclobuteniminium salts in good yields (Table XVIII). The most striking example of these cycloadditions is the

TABLE XVIII

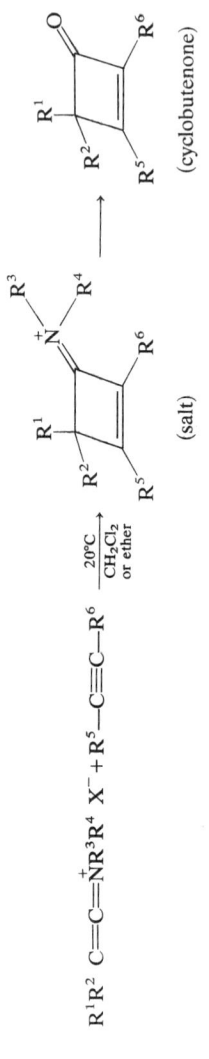

R^1	R^2	R^3	R^4	X^-	R^5	R^6	Salt,[a] %	Cyclobutenone,[b] %	Ref.
CH_3	CH_3	CH_3	CH_3	BF_4^-	C_6H_5	C_6H_5	82	90	80
CH_3	CH_3	CH_3	CH_3	BF_4^-	C_2H_5	C_2H_5	80	80	118
CH_3	CH_3	CH_3	CH_3	$ZnCl_3^-$	C_2H_5	C_2H_5	—	95	118
CH_3	CH_3	CH_3	CH_3	$ZnCl_3^-$	CH_3	CH_3	56	60	118
CH_3	CH_3	CH_3	CH_3	BF_4^-	H	H	77	—	118
CH_3	CH_3	CH_3	CH_3	BF_4^-	CH_3	H	80[c]	—	118
CH_3	CH_3	CH_3	CH_3	BF_4^-	tert-C_4H_9	H	70[c]	~55[c]	118
CH_3	CH_3	CH_3	CH_3	$ZnCl_3^-$	tert-C_4H_9	H	94[c]	~49[c]	118
CH_3	CH_3	—$(CH_2)_5$—		Cl^-	$N(C_2H_5)_2$	C_6H_5	70	—	25
C_6H_5	H	C_2H_5	C_2H_5	Cl^-	$N(C_2H_5)_2$	C_6H_5	80	—	19

[a] Pure products crystallized from chloroform-ether.
[b] Corrected yields by VPC after hydrolysis of the crude reaction mixtures.
[c] Mixture of the two regioisomers.

reaction of **189** with acetylene itself to give *N,N*-4,4-tetramethylcyclobut-2-en-1-iminium tetrafluoroborate (**291**) in 77% yield (118). The adduct does not give the corresponding cyclobutenone on hydrolysis but undergoes a ring-opening reaction. However, it combines (118) readily with cyclopentadiene to give a Diels-Alder adduct (**292**) (Scheme 44).

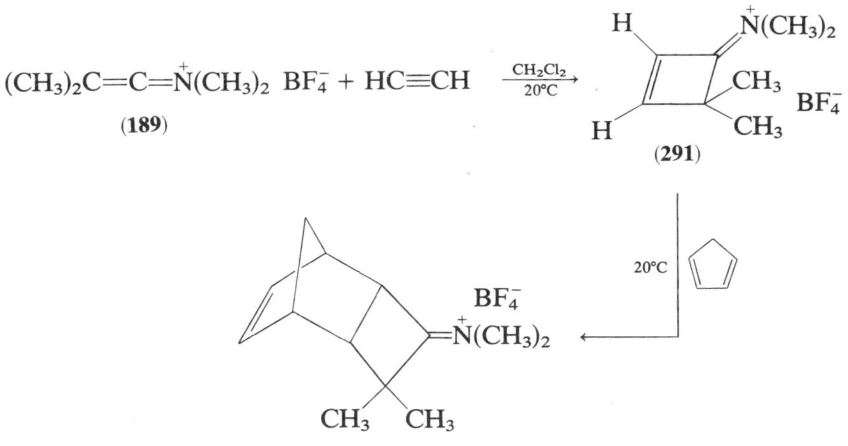
(CH3)2C=C=N+(CH3)2 BF4− + HC≡CH
CH2Cl2
20°C
H
N+(CH3)2
CH3
CH3
BF4−
H
20°C
BF4−
=N+(CH3)2
CH3 CH3

(**189**) (**291**) (**292**)

Scheme 44

Monosubstituted acetylenes and **189** give (118) mixtures of regioisomers (Table XVIII). The adducts from **189** and diphenyl- or diethylacetylene hydrolyze readily to the corresponding cyclobutanones (80,118). As expected, the strongly nucleophilic ynamines react exothermally even with α-chloroenamines (keteniminium chloride) to give (19,25), the cyclobutenecyanines (**293**):

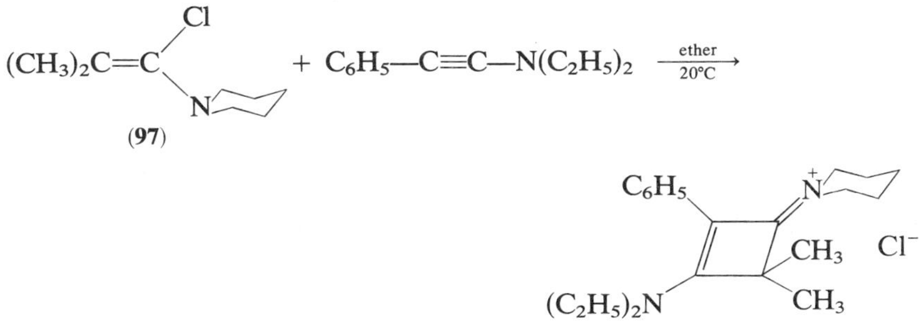
Cl
(CH3)2C=C
N
+ C6H5—C≡C—N(C2H5)2
ether
20°C
C6H5
N+
CH3
Cl−
(C2H5)2N
CH3

(**97**) (**293**) 70%

These cycloadditions are of synthetic value since they offer a simple method for the preparation of functionalized cyclobutenes.

F. CYCLOADDITIONS TO IMINES

It could be safely predicted that the strongly ketenophilic C–N double bond (83) would react very readily with keteniminium ions. This is indeed the case, and even α-chloroenamines cycloadd (125) readily to Schiff bases at room temperature to form the new α-azetidinylideneammonium salts (**294**), the iminium derivatives of β-lactams (Scheme 45 and Table XIX).

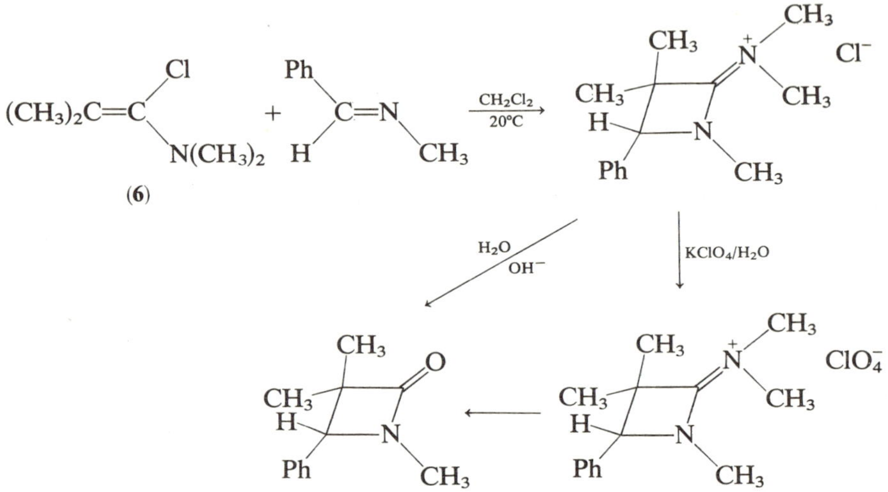

Scheme 45

It is worth noting that these salts are readily hydrolyzable with 0.1 N potassium hydroxide. Except in one case, these hydrolyses do *not* lead to cleavage of the intracyclic C–N bond and release of the internal strain but rather give (125) the azetidinone (**295**)! Therefore the cycloaddition represents a new method of synthesis of these four-membered heterocycles.

G. DIELS-ALDER REACTIONS OF ALKENYLKETENIMINIUM IONS

2-Alkenyl-1-chloroenamines are sources of 2-alkenylketeniminium ions, which, as reactive dienes, should be useful for the synthesis of carbo- and heterocyclic six-membered rings by cycloadditions. At present only one type of such reactions has been described. 1-Chloro-1-dimethylaminoisoprene (**11**) reacts with acetonitrile (27) or benzonitrile (99) at 100°C in the presence of triethylamine to give, after loss of HCl, the

TABLE XIX

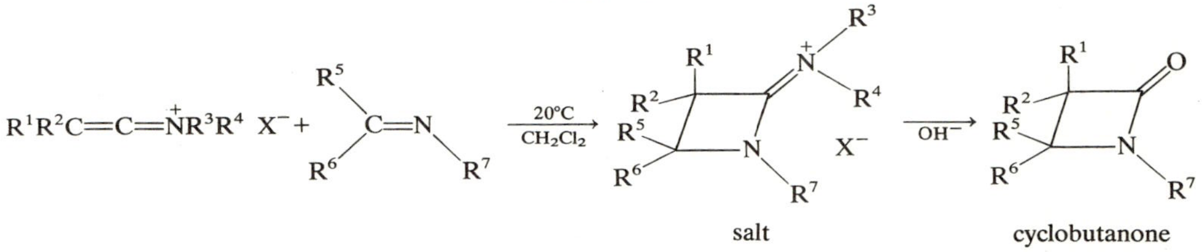

$$R^1R^2C=C=\overset{+}{N}R^3R^4 \; X^- + \underset{R^6}{\overset{R^5}{>}}C=N-R^7 \xrightarrow[CH_2Cl_2]{20°C} \text{salt} \xrightarrow{OH^-} \text{cyclobutanone}$$

R^1	R^2	R^3	R^4	X^-	R^5	R^6	R^7	Salt,[a] %	β-Lactam,[b] %	Ref.
CH_3	CH_3	CH_3	CH_3	Cl^-	C_6H_5	H	CH_3	73[c]	82	125
CH_3	CH_3	CH_3	CH_3	Cl^-	C_6H_5	H	C_6H_5	47[c]	68	125
CH_3	CH_3	CH_3	CH_3	Cl^-	$S-CH_2-C_6H_5$	H	tert-C_4H_9	60[c]	42	125
H	Cl	CH_3	CH_3	Cl^-	C_6H_5	H	C_6H_5	58[c,d]	—	52
CH_3S	CH_3S	C_2H_5	C_2H_5	Cl^-	C_6H_5	H	C_6H_5	65[c]	—	59
CH_3	CH_3	CH_3	CH_3	BF_4^-	C_6H_5	C_6H_5	CH_3	—	72	125
CH_3	CH_3	CH_3	CH_3	BF_4^-	C_6H_5	H	C_6H_{11}	—	69	89
CH_3	CH_3	—$(CH_2)_5$—		BF_4^-	C_6H_5	C_6H_5	C_6H_5	—	60	89
Cl	Cl	CH_3	CH_3	BF_4^-	C_6H_5	H	C_6H_5	100	100	98
Cl	Cl	CH_3	CH_3	BF_4^-	C_6H_5	H	C_6H_5	100	—	98

[a] Pure products crystallized from chloroform-ether.
[b] Corrected yields after hydrolysis of the crude reaction mixtures.
[c] Isolated as perchlorate.
[d] Trans isomer only.

pyridine derivatives (**296**):

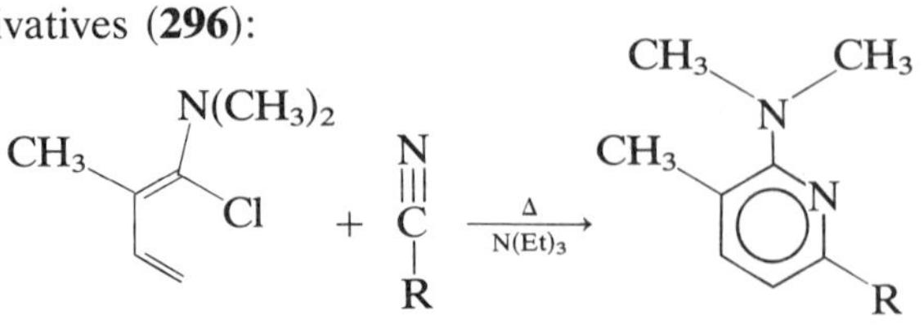

$$CH_3\text{-}C(N(CH_3)_2)=CH\text{-}CH=CH_2 \text{ (with Cl)} \quad (\mathbf{11})$$

$$+ \quad N\equiv C\text{-}R \quad \xrightarrow{\Delta, N(Et)_3} \quad$$

pyridine product (**296**) with substituents $N(CH_3)_2$, CH_3, and R

$R = CH_3, C_6H_5, \text{i-}C_3H_7S$

With thioisopropylnitrile (**99**) the reaction is very fast, even at 50°C.

IX. Transition-Metal Complexes from α-Chloroenamines

New transition-metal organometallic compounds, some with novel structural features, have been prepared by King and Hodges (126) from tetramethyl-α-chloroenamine and metal carbonyl anions at ambient temperature. Reaction of **6** with $NaRe(CO)_3$ or $NaFe(CO)_2C_5H_5$ gives complexes **297** and **298**, where the neutral $(CH_3)_2C=C\ N(CH_3)_2$ unit acts as a one-electron donor:

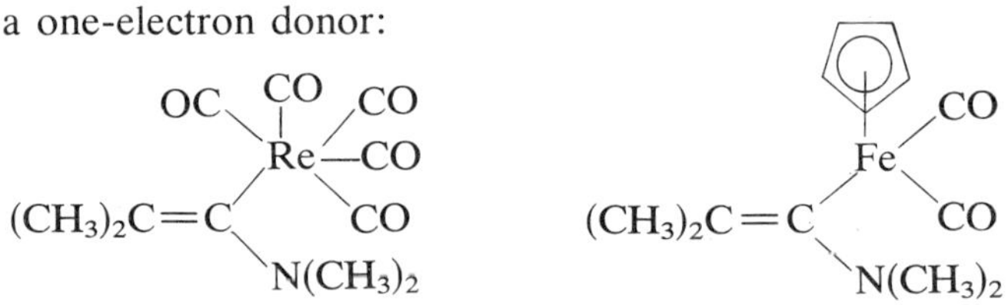

$$(CH_3)_2C=C\begin{pmatrix}Re(CO)_4\\N(CH_3)_2\end{pmatrix} \quad (\mathbf{297}) \qquad (CH_3)_2C=C\begin{pmatrix}Fe(CO)_2C_5H_5\\N(CH_3)_2\end{pmatrix} \quad (\mathbf{298})$$

Complexes in which the neutral $(CH_3)_2C=C\ N(CH_3)_2$ is a three-electron donor have also been obtained. With $NaCo(CO)_4$ **6** gives a complex (**299**) in which the keteniminium ligand is bonded through the C–C bond, whereas with $NaMo(CO)_3C_5H_5$ it forms a complex in which the keteniminium ligand binds through the C–N double bond (126):

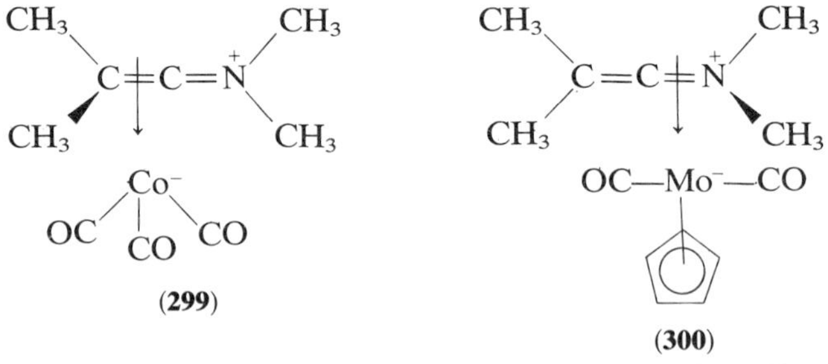

$$(CH_3)_2C=C=N^+(CH_3)_2 \text{ with } Co(CO)_3 \quad (\mathbf{299}) \qquad (CH_3)_2C=C=N^+(CH_3)_2 \text{ with } Mo(CO)_2C_5H_5 \quad (\mathbf{300})$$

Cyclic acyl derivatives **301** and **303**, in which a carbonylated neutral $[(CH_3)_2C=C\ N(CH_3)_2]$ CO unit acts as a three-electron donor, have also been prepared from **6** and NaMn(CO)$_5$ and NaW(CO)$_3$C$_5$H$_5$, respectively (126):

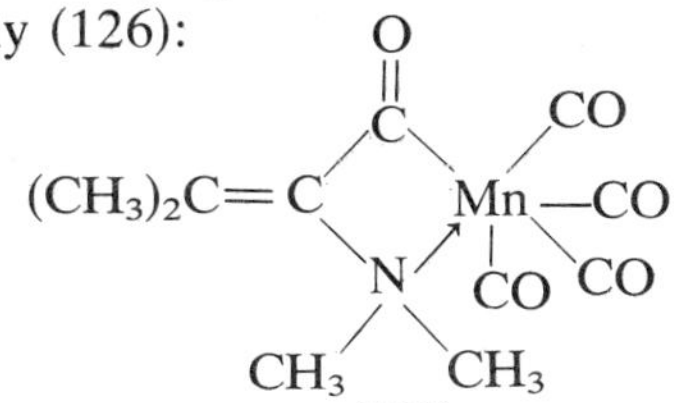

$$(CH_3)_2C=C \begin{array}{c} \overset{O}{\overset{\parallel}{C}} \\ N \\ | \\ CH_3 \end{array} \begin{array}{c} CO \\ Mn-CO \\ CO \\ CH_3 \end{array}$$
(**301**)

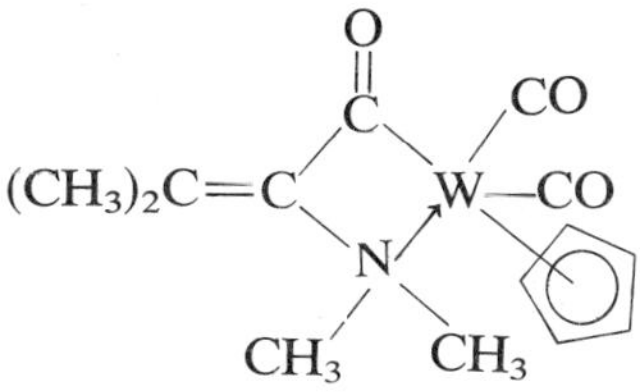

$$(CH_3)_2C=C \begin{array}{c} \overset{O}{\overset{\parallel}{C}} \\ N \\ | \\ CH_3 \end{array} \begin{array}{c} CO \\ W-CO \\ (C_5H_5) \\ CH_3 \end{array}$$
(**303**)

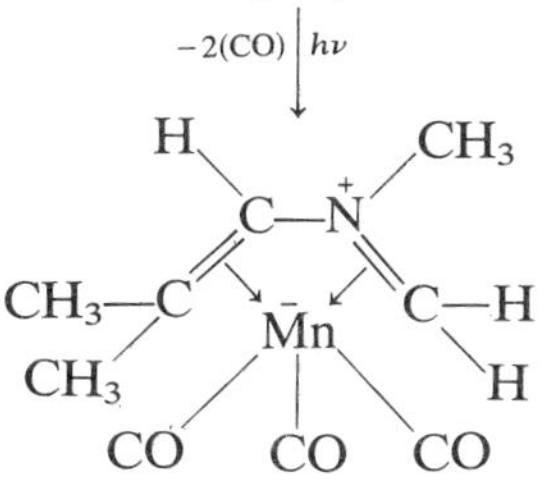

$\downarrow$ −2(CO) | $h\nu$

$$\begin{array}{c} H \qquad CH_3 \\ C-\overset{+}{N} \\ CH_3-C \diagup \quad \diagdown C-H \\ CH_3 \qquad Mn \qquad H \\ | \\ CO \quad CO \quad CO \end{array}$$
(**302**)

Ultraviolet irradiation of **301** in hexane results in the loss of two carbonyl groups and migration of one of the *N*-methyl protons to the vinylic carbon next to the amino group. Complex **302** is formed in 28% yield with the five-electron donor 2-azabutadiene ligand. Analogous transition-metal complexes have been obtained from 1-*N*-piperidinoisobutene (126).

X. Entertainment!

In the preceding sections the versatile chemical behavior of α-haloenamines has been illustrated by several examples of additions, substitutions, and cycloadditions with either electron-poor or electron-rich partners. In these reactions the α-haloenamines reacted either through its nucleophilic C$_2$ atom, being thus synthetically equivalent to an enolate of an acid halide (Scheme 46a), or through its electrophilic C$_1$ atom, being then equivalent to an acyl cation (Scheme 46b):

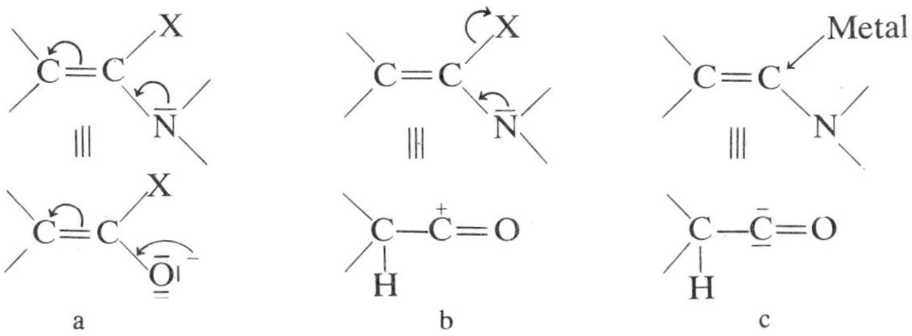

Scheme 46

We turn now to a discussion of an attractive possibility that consists in inverting the polarity of the C_1 atom by replacing the C–halogen bond by a C–metal bond. This transformation should allow for a new reaction called "nucleophilic aminoalkenylation," in which the "1-enaminocarbanion," which is synthetically equivalent to an acyl anion, is transferred onto an electrophilic reagent (Scheme 46c).

One obvious way of achieving this goal is the transformation of an α-chloroenamine into the corresponding Grignard reagent. α-Chloroenamines **6**, **94**, and **14** react with magnesium in boiling tetrahydrofuran (30). When the reaction mixtures are quenched with D_2O, three kinds of products are obtained (30), depending on the structure of the starting material (Scheme 47). In addition to the enamines (**304**), which result from hydrogen abstraction, and the interesting 2,3-diamino-1,3-butadiene derivatives (**305**), which are coupling products, a 2-aminoalkenyl Grignard reagent (**306**) can be formed. This product has been titrated by the method of Watson and Eastham (127) and converted to the corresponding 1-deuterioenamines (**307**) with D_2O (30).

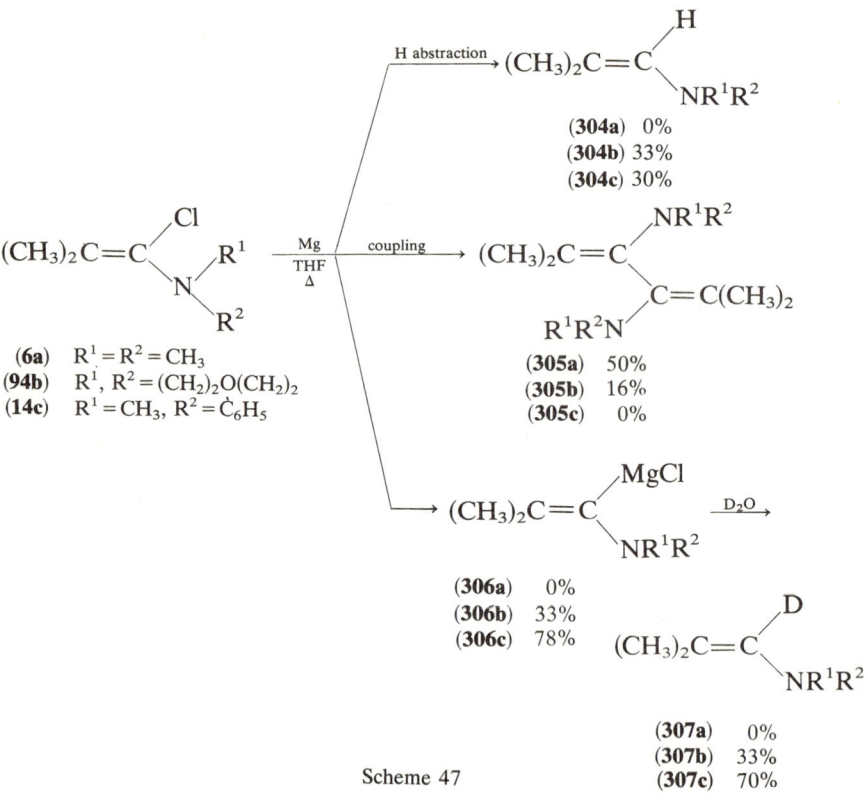

Scheme 47

It can be seen in Scheme 47 that the amount of Grignard reagent formed in the reaction decreases with increasing basicity of the amino function. The best amino substituent that has been found to date for the Grignard formation is —N(CH$_3$)C$_6$H$_5$. The Grignard reagent (**306c**) formed from the α-chloroenamine (**14**) has been used (30) in several nucleophilic aminoalkenylations (Scheme 48). It is quite obvious that

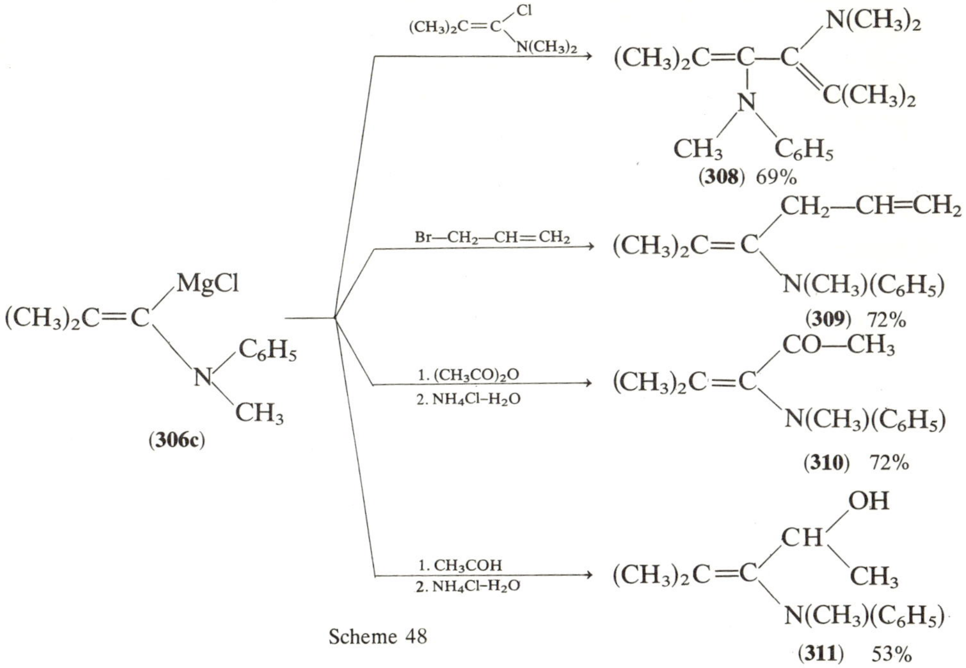

Scheme 48

additional synthetic applications have to be awaited of these new organometallic reagents, which add another dimension to the chemistry of α-haloenamines.

ACKNOWLEDGMENTS

The incentive for our work was provided by the pioneering studies of Dr. B. Haveaux in Louvain, combined with a fruitful collaboration with Professors H. G. Viehe and R. Fuks and Dr. R. Buyle, formerly at Union Carbide European Research Associates in Brussels. The results realized at the University of Louvain are due to the enthusiasm, creativity, and skillful work of M. X. H. Bui, Dr. A. Colens, M. Demuylder, Dr. M. De Poortere, Professor M. L. Herr, Dr. A. M. Hesbain-Frisque, C. Hoornaert, M. Houtekie, Dr. A. M. Léonard, Dr. T. F. Lin-Ho, Dr. R.

Maurin, Dr. P. Notté, Dr. J. Rémion, Dr. M. Rens, Dr. A. Sidani, Dr. E. Sonveaux, M. Staelens-Vanhorenbeeck, C. T'Kint, J. Toye, and Dr. C. Zamar-Wiaux.

We wish to acknowledge also the helpful collaboration of several young and able students and the technical assistance of A. Malengraux-Dekoker and H. Vanlierde. They all deserve our gratitude.

We are grateful for the financial support of l'Institut pour l'Encouragement de la Recherche Scientifique dans l'Industrie et l'Agriculture, Union Carbide European Research Associates, Fonds de la Recherche Fondamentale Collective, and Fonds National de la Recherche Scientifique, without whose help this work would not have been possible. Finally, we thank Dr. M. O'Donnell, who is responsible for making our English comprehensible to readers.

REFERENCES

1. G. R. Krow, *Angew. Chem. Int. Ed.*, **10**, 435 (1971) (a review).
2a. H. G. Viehe, *Angew. Chem. Int. Ed.*, **6**, 767 (1967) (a review).
2b. H. G. Viehe, in *Chemistry of Acetylenes*, Marcel Dekker, New York, 1969, Chap. XII: Ynamines (a review).
3a. C. F. Hobbs and H. Weingarten, *J. Org. Chem.*, **33**, 2385 (1968).
3b. C. F. Hobbs and H. Weingarten, *J. Org. Chem.*, **39**, 918 (1974).
4. J. von Braun and A. Heymans, *Ber. Dtsch. Chem. Ges.*, **62**, 409 (1929).
5a. A. J. Speziale and R. C. Freeman, *J. Amer. Chem. Soc.*, **82**, 903 (1960).
5b. A. J. Speziale and L. R. Smith, *J. Org. Chem.*, **27**, 4361 (1962).
5c. A. J. Speziale and L. R. Smith, *J. Amer. Chem. Soc.*, **84**, 1868 (1962).
5d. A. J. Speziale and L. J. Taylor, *J. Org. Chem.*, **31**, 2450 (1966).
6a. Y. Yakubovich, A. P. Sergeev, and I. N. Belyaeva, translated from *Dokl. Akad. Nauk. SSSR*, **161** (6), 1362 (1965).
6b. Y. Yakubovich, A. P. Sergeev, and T. I. Novozhilova, *Zh. Org. Khim.*, **6** (11), 2192 (1970).
7. J. Marchand-Brynaert and L. Ghosez, *J. Amer. Chem. Soc.*, **94**, 2870 (1972).
8. A. Sidani, J. Marchand-Brynaert, and L. Ghosez, *Angew. Chem. Int. Ed.*, **13**, 267 (1974).
9. H. Weingarten, *J. Org. Chem.*, **35**, 3970 (1970).
10. H. Böhme and M. Haake, the chapter entitled "Methyleniminium Salts" in this book.
11. Z. Janousek and H. G. Viehe, the chapter entitled "Chemistry of Dichloromethyleniminium Salts" in this book.
12a. Y. Legrand and H. G. Viehe, unpublished results.
12b. E. Goffin, Dissertation, Louvain, 1974.
13. Z. Janousek, J. Collard, and H. G. Viehe, *Angew. Chem. Int. Ed.*, **11**, 917 (1972).
14a. M. Ohoka, S. Yanagida, and S. Komari, *J. Org. Chem.*, **36**, 3542 (1971).
14b. S. Yanagida, *Bull. Soc. Chim. Jap.*, **46**, 1275 (1973).
15. S. Yanagida and S. Komari, *Synthesis*, **1973**, 189.
16a. G. Simchen, *Angew. Chem. Int. Ed.*, **5**, 663 (1966).
16b. G. Simchen and W. Krämer, *Chem. Ber.*, **102**, 3656–3666 (1969).
17a. R. D. Partos and A. J. Speziale, *J. Amer. Chem. Soc.*, **87**, 5068 (1965).

17b. R. D. Partos and K. W. Ratts, *J. Amer. Chem. Soc.*, **88,** 4996 (1966).
18a. J. H. Ottenheym and J. W. Garritsen, Brit. Pat. 901,169 (1962); *C.A.,* **58,** 6810 (1963).
18b. B. Prajsnar, *C.A.,* **59,** 5136 (1963).
19. B. Haveaux, Dissertation, Louvain, 1968.
20. M. Houtekie, Dissertation, Louvain, 1976.
21. H. Gross and J. Gloede, *Chem. Ber.,* **96,** 1387 (1963).
22. V. Jäger and H. G. Viehe, unpublished results.
23a. F. Hallmann, *Ber. Dtsch. Chem. Ges.,* **9,** 846 (1876).
23b. J. von Braun, *Angew. Chem.,* **47,** 611 (1934).
24a. H. Eilingsfeld, M. Seefelder, and H. Weidinger, *Angew. Chem.,* **72,** 836 (1960); *Chem. Ber.,* **96,** 2671 (1963).
24b. H. Eilingsfeld, G. Neubauer, M. Seefelder, and H. Weidinger, *Chem. Ber.,* **97,** 1232 (1964).
25. L. Ghosez, B. Haveaux and H. G. Viehe, *Angew. Chem. Int. Ed.,* **8,** 454 (1969).
26. M. Rens, J. Toye, and L. Ghosez, report submitted to *Organic Synthesis.*
27. E. Sonveaux, Dissertation, Louvain, 1975.
28. C. T'Kint and L. Ghosez, unpublished results.
29a. A. Colens, Dissertation, Louvain, 1976.
29b. A. Colens, L. Ghosez, J. P. Declercq, G. Germain, N. Molhant, and M. Van Meerssche, unpublished results.
30. C. Zamar-Wiaux, Dissertation, Louvain, 1975.
31a. R. Buyle and H. G. Viehe, *Tetrahedron,* **24,** 3987 (1968).
31b. R. Buyle and H. G. Viehe, *Tetrahedron,* **24,** 4217 (1968).
31c. R. Buyle and H. G. Viehe, *Tetrahedron,* **25,** 3447 (1969).
31d. R. Buyle and H. G. Viehe, *Tetrahedron,* **25,** 3453 (1969).
32. H. Eilingsfeld, M. Seefelder, and H. Weidinger, *Chem. Ber.,* **96,** 2899 (1963).
33. M. Rens, Dissertation, Louvain, 1973.
34a. Z. Janousek and H. G. Viehe, *Angew. Chem. Int. Ed.,* **10,** 574 (1971).
34b. Z. Janousek, Dissertation, Louvain, 1972.
34c. G. S. de Voghel, T. L. Eggerichs, and H. G. Viehe, *J. Org. Chem.,* **39,** 1233 (1974).
35. H. G. Viehe and Z. Janousek, *Angew. Chem. Int. Ed.,* **12,** 806 (1973).
36a. H. G. Viehe, Z. Janousek, and M. A. Deffrenne, *Angew. Chem. Int. Ed.,* **10,** 575 (1971).
36b. A. Bettercourt, Mémoire, Louvain, 1973.
36c. B. Fontaine, Mémoire, Louvain, 1974.
37. H. G. Viehe, B. Le Clef, and A. Elgavi, unpublished results.
38a. J. Freear and A. E. Tipping, *J. Chem. Soc. C,* **1968,** 1096.
38b. J. Freear and A. E. Tipping, *J. Chem. Soc. C,* **1969,** 1955.
39. R. N. Haszeldine and A. E. Tipping, *J. Chem. Soc. C,* **1968,** 398.
40. H. G. Viehe, R. Buyle, R. Fuks, R. Mérényi, and J. M. F. Oth, *Angew. Chem. Int. Ed.,* **6,** 77 (1967).
41. A. Niederhauser and M. Neuenschwander, *Helv. Chim. Acta,* **56,** 1331 (1973).
42. K. Hafner and M. Neuenschwander, *Angew. Chem. Int. Ed.,* **7,** 460 (1968).
43. H. J. Gais, K. Hafner, and M. Neuenschwander, *Helv. Chim. Acta,* **52,** 2641 (1969).
44. J. C. Blazejwski, D. Cantacuzène, and C. Wakselman, *Tetrahedron Lett.,* **1974,** 2055.
45. J. Freear and A. E. Tipping, *J. Chem. Soc. C,* **1969,** 1848.
46. J. Freear and A. E. Tipping, *J. Chem. Soc. C,* **1969,** 411.
47. F. S. Fawcett (to E. I. du Pont de Nemours and Co.), U.S. Pat. 3,311,599 (1967).
48. A. Senning, *Acta Chem. Scand.,* **22,** 1370 (1968).

49a. N. Schindler, *Synthesis*, **1971**, 656.
49b. N. Schindler and W. Plöger, *Chem. Ber.*, **104**, 2021 (1971).
50. P. Notté, R. Maurin, and L. Ghosez, unpublished results.
51. E. Ott, G. Dittus, and H. Weissenburger, *Chem. Ber.*, **76B**, 80-88 (1943).
52. H. Vanlierde, A. Van Camp, and L. Ghosez, unpublished results.
53a. S. Y. Delavarenne and H. G. Viehe, *Chem. Ber.*, **103**, 1198–1209 (1970).
53b. S. Y. Delavarenne and H. G. Viehe, in *Chemistry of Acelylenes*, Marcel Dekker, New York, 1969, Chap. X.
54. D. C. England, L. R. Melby, M. A. Dietrich, and R. V. Lindsey, Jr., *J. Amer. Chem. Soc.*, **82**, 5116 (1960).
55. R. Sauvetre and J. F. Normant, *Bull. Soc. Chim. Fr.*, **1972**, 3202.
56. T. A. George and M. F. Lappert, *Chem. Commun.*, **1966**, 463.
57. P. Hegenberg and G. Maahs, *Angew. Chem. Int. Ed.*, **5**, 895 (1966).
58a. H. Weingarten and W. A. White, *J. Org. Chem.*, **31**, 2874 (1966).
58b. H. Weingarten and W. A. White, *J. Amer. Chem. Soc.*, **88**, 850 (1966).
59. P. Notté, Dissertation, Louvain, 1976.
60. M. Demuylder and L. Ghosez, unpublished results.
61. B. Techy, A. Colens, and L. Ghosez, unpublished results.
62. M. Saunders and R. W. Murray, *Tetrahedron*, **11**, 1 (1960).
63a. W. E. Parham and J. R. Potoski, *Tetrahedron Lett.*, **1966**, 2311.
63b. W. E. Parham and J. R. Potoski, *J. Org. Chem.*, **32**, 278 (1967).
64. P. V. Kukhar, N. G. Pavlenko, and A. V. Kirsanov, *Zh. Obshch. Khim.*, **43** (105), 9, 1896 (1973).
65. M. Staelens and L. Ghosez, unpublished results.
66. J. Toye and L. Ghosez, *J. Amer. Chem. Soc.*, **97**, 2276 (1975).
67. M. Rens, Mémoire, Louvain, 1968.
68. T. Sasaki and A. Kojima, *J. Chem. Soc. C*, **1970**, 476.
69. A. J. Speziale and R. C. Freeman, U.S. Pat. 3,145,230; *C.A.*, **61**, 11891f.
70. F. S. Fawcett, C. W. Tullock, and D. Coffman, *J. Amer. Chem. Soc.*, **84**, 4275 (1962).
71. J. F. Harris, Jr., *J. Org. Chem.*, **32**, 2063 (1967).
72. A. J. Speziale and R. C. Freeman, *J. Amer. Chem. Soc.*, **82**, 909 (1960).
73a. A. M. Hesbain-Frisque, J. Verbelen, and L. Ghosez, unpublished results.
73b. J. M. Damien, Mémoire, Louvain, 1975.
74. A. M. Hesbain-Frisque and L. Ghosez, unpublished results.
75. "Tables of Interatomic Distances and Configurations of Molecules and Ions," Supplement 1956–1959, Special Publication No. 18, London, 1965.
76. M. L. Herr and L. Ghosez, unpublished results.
77. R. Livingston, *J. Phys. Chem.*, **57**, 496 (1953).
78. J. H. Goldstein and R. Livingston, *J. Chem. Phys.*, **19**, 1613 (1951).
79. E. A. C. Lucken, *J. Chem. Soc.*, **1959**, 2954.
80. J. Marchand-Brynaert, Dissertation, Louvain, 1973.
81. J. Firl and W. Runge, *Angew. Chem. Int. Ed.*, **12**, 668 (1973).
82. J. M. André, L. Ghosez, E. Sonveaux, and J. Delhalle, unpublished results.
83. H. Ulrich, *Cycloaddition Reactions of Heterocumulenes*, Academic Press, New York, 1967.
84. C. de Perez and L. Ghosez, *Angew. Chem. Int. Ed.*, **10**, 184 (1971).
85. E. Sonveaux and L. Ghosez, *J. Amer. Chem. Soc.*, **95**, 5417 (1973).
86a. S. Patai, *The Chemistry of Alkenes*, Interscience, New York, 1964, Chap. VII, p. 525 (a review).
86b. Z. Rappoport, *Adv. Phys. Org. Chem.*, **7**, 1–144 (1969).

86c. G. Modena, *Acc. Chem. Res.*, **4**, 73 (1971).
87. M. Demuylder, Mémoire, Louvain, 1972.
88. M. Rens and L. Ghosez, *Tetrahedron Lett.*, **1970**, 3765.
89. J. Marchand-Brynaert and L. Ghosez, unpublished results.
90. J. Marchand-Brynaert and L. Ghosez, *J. Amer. Chem. Soc.*, **94**, 2869 (1972).
91. M. Xuan-Hoa Bui, Dissertation, Louvain, 1976.
92a. H. Gross and E. Höft, *Angew. Chem. Int. Ed.*, **6**, 335 (1967).
92b. H. Ulrich, *Chemistry of Imidoyl Halides*, Plenum Press, New York, 1968.
92c. H. E. Zaugg, *Synthesis*, **1970**, 49.
93. J. Remion, Dissertation, Louvain, 1974.
94. E. Cossement, Dissertation, Louvain, 1973.
95. J. P. Dejonghe and L. Ghosez, unpublished results.
96. R. Van Binst, Mémoire, Louvain, 1972.
97. A. Van Camp, Mémoire, Louvain, 1971.
98. A. Sidani and L. Ghosez, unpublished results.
99. C. T'Kint and L. Ghosez, unpublished results.
100. A. Demoulin, H. Gorissen, A. M. Hesbain-Frisque, and L. Ghosez, *J. Am. Chem. Soc.*, **97**, 4409 (1975).
101. R. Huisgen, W. Scheer, and H. Huber, *J. Amer. Chem. Soc.*, **89**, 1753 (1967).
102. D. Goossens, Mémoire, Louvain, 1974.
103. A. M. Hesbain-Frisque, F. Saintes, and L. Ghosez, unpublished results.
104. M. Gillard, Mémoire, Louvain, 1975.
105. T. F. Lin-Ho, Dissertation, Louvain, 1972.
106. A. G. Cook, *Enamines: Synthesis, Structure and Reactions*, Marcel Dekker, New York, 1969.
107. A. J. Speziale and L. R. Smith, *J. Org. Chem.*, **28**, 3492 (1963).
108. C. T'Kint, Mémoire, Louvain, 1973.
109. J. Van Uytbergen-Van Hamme, Mémoire, Louvain, 1974.
110. M. Staelens-Vanhorenbeeck and L. Ghosez, unpublished results.
111. A. J. Speziale, L. R. Smith, and J. E. Fedder, *J. Org. Chem.*, **30**, 4303 (1965).
112. H. Bredereck, R. Gompper, and K. Klemm, *Chem. Ber.*, **92**, 1456 (1959).
113a. R. Huisgen et al., *Angew. Chem. Int. Ed.*, **3**, 753 (1964); *Tetrahedron Lett.*, **1968**, 4485–4497; *Chem. Ber.*, **102**, 3444–3460 (1969).
113b. L. Ghosez et al., *Angew. Chem. Int. Ed.*, **7**, 643 (1968); *ibid.*, **8**, 72 (1969); *Tetrahedron*, **27**, 615 (1971).
113c. A. S. Dreiding et al., *Helv. Chim. Acta*, **53**, 417 (1970).
113d. W. T. Brady et al., *J. Amer. Chem. Soc.*, **86**, 616 (1964); *ibid.*, **91**, 5679 (1969); *Tetrahedron Lett.*, **1970**, 819; *Synthesis*, **1971**, 415.
113e. T. Do Minh and O. P. Strausz, *J. Amer. Chem. Soc.*, **92**, 1766 (1970).
114a. R. B. Woodward and R. Hoffmann, *Acc. Chem. Res.*, **1**, 17 (1968).
114b. R. B. Woodward and R. Hoffmann, *Angew. Chem. Int. Ed.*, **8**, 781 (1969).
114c. R. B. Woodward and R. Hoffmann, *The Conservation of Orbital Symmetry*, Verlag Chemie, Academic Press, New York, 1970, p. 163.
115a. R. Sustmann, A. Ansmann, and F. Vahrenholt, *J. Amer. Chem. Soc.*, **94**, 8099 (1972).
115b. H. V. Wagner and R. Gompper, *Tetrahedron Lett.*, **1970**, 2819; *ibid.*, **1971**, 4061.
116. H. E. Zimmerman, *Acc. Chem. Res.*, **4**, 272 (1971).
117. L. Ghosez and M. O'Donnell, in *Pericyclic Reactions*, R. Lehr and A. P. Marchand, Eds., Academic Press, New York (in preparation).

118. C. Hoornaert, A. M. Hesbain-Frisque, and L. Ghosez, *Angew. Chem. Int. Ed.*, **14,** 569 (1975).
119. J. Marchand-Brynaert and L. Ghosez, *Tetrahedron Lett.*, **1974,** 377–380.
120. A. M. Leonard, Dissertation, Louvain, 1976.
121a. R. Montaigne, Dissertation, Louvain, 1968.
121b. W. Dumont, Dissertation, Louvain, 1972.
122. P. Laroche, Dissertation, Louvain, 1965.
123. S. Patai, *The Chemistry of Alkenes*, Interscience, New York, 1964, p. 1076.
124a. D. Craig, J. J. Shipman, and R. B. Fowler, *J. Amer. Chem. Soc.*, **83,** 2885 (1961).
124b. P. D. Bartlett et al., *J. Amer. Chem. Soc.*, **90,** 2049–2056 (1968).
124c. R. Sustmann et al., *Tetrahedron Lett.*, **1971,** 2721; *Angew. Chem. Int. Ed.*, **11,** 840 (1972).
125. M. De Poortere, J. Marchand-Brynaert, and L. Ghosez, *Angew. Chem. Int. Ed.*, **13,** 268 (1974).
126. R. B. King and K. C. Hodges, *J. Amer. Chem. Soc.*, **96,** 1263 (1974).
127. S. C. Watson and J. F. Eastham, *J. Organomet. Chem.*, **9,** 165 (1967).

N-HETEROIMINIUM SALTS

By J. ELGUERO and C. MARZIN, *Laboratoire de Synthèse et d'Etude Physicochimique d'Hétérocycles Azotés, Place Eugène Bataillon, F-34060-Montpellier Cedex, France*

CONTENTS

I. Introduction	534
A. Equilibria Occurring in *N*-Heteroiminium Salts	535
II. Hydrazonium Salts	538
A. Synthesis	538
1. Protonation of Enehydrazines **V** (Condensation of Carbonyl Compounds with 1,2-Disubstituted Hydrazines)	538
2. Hydrazone Quaternization	540
3. Hydrazone Protonation	544
4. Diaziridine Opening in Acidic Medium	546
B. Reactivity	546
1. Chemical Reactivity	546
2. Spectroscopic Properties	548
C. Hydrazonium Salts as Intermediates	548
1. Intermediates $\diagdown\!\!\text{N}\!-\!\overset{+}{\text{N}}\text{H}\!=\!\text{C}\!\diagup$	548
2. Intermediates $\diagdown\!\!\text{N}\!-\!\overset{+}{\text{N}}\text{R}\!=\!\text{C}\!\diagup$	552
III. Azinium Salts	556
A. Synthesis	556
1. Condensation of Carbonyl Compounds with Monosubstituted Hydrazines	556
2. Alkylation and Acylation of Azines	559
3. Azine Protonation	560
B. Reactivity	561
1. Hydrolysis	561
2. Cyclization	562
3. Reduction	562
4. Spectroscopic Properties	563
C. Azinium Salts as Intermediates	563
1. Intermediates $\diagdown\!\!\text{C}\!=\!\overset{+}{\text{N}}\text{H}\!-\!\text{N}\!=\!\text{C}\!\diagup$	563
2. Intermediates $\diagdown\!\!\text{C}\!=\!\overset{+}{\text{N}}\text{R}\!-\!\text{N}\!=\!\text{C}\!\diagup$	564

IV. Azomethine Imines 566
   A. Introduction 566
   B. Synthesis and Reactivity 567
   C. Stabilized Azomethine Imines 568
V. Diazenes and Diazenium Ions 570
VI. Oximinium Salts 571
   A. Synthesis 571
      1. Alkylation of Oximes, Nitroso Derivatives, and Nitrones 572
      2. Protonation of Oximes, Nitroso Derivatives, Nitrones, and Oxaziridines. 574
   B. Reactivity 576
      1. Chemical Reactivity 576
      2. Spectroscopic Properties 577
   C. Oximinium Salts as Intermediates 577
VII. Nitrones 579
   A. Introduction 579
   B. Synthesis 579
   C. Reactivity 580
   D. Heterosubstituted Nitrones 580
VIII. Sulfur-Containing N-Heteroiminium Salts 581
   References 582

I. Introduction

N-Heteroiminium salts (**2**) and (**3**) are closely related to the iminium salts (**1**), as they derive from them by substituting the groups on nitrogen by a heteroatom, Y or Z (oxygen, sulfur, nitrogen).

$$
\begin{array}{ccc}
\underset{R}{\overset{R}{\diagdown}}\overset{+}{N}=C\underset{R''}{\overset{R'}{\diagup}}\ X^- & \underset{R}{\overset{Y}{\diagdown}}\overset{+}{N}=C\underset{R''}{\overset{R'}{\diagup}}\ X^- & \underset{Z}{\overset{Y}{\diagdown}}\overset{+}{N}=C\underset{R''}{\overset{R'}{\diagup}}\ X^- \\
(\mathbf{1}) & (\mathbf{2}) & (\mathbf{3})
\end{array}
$$

Two classes of compounds must be distinguished according to whether or not the Y (or Z) heteroatom bears a negative charge. In the first case (dipolar ions) synthetically interesting and thus well-studied compounds are obtained, but they are out of the framework of this review; we shall mention them incidentally, however, insofar as they are related to the cationic structures. In the second class the products obtained are unstable, so that only a few studies have been undertaken on them and in many cases they have been considered only as unisolated intermediates.

Another way to classify the N-heteroiminium salts takes into account the possibility that such functions are part of a heterocyclic ring: the interest of such a classification is to show that some of them are known only as part of a heterocycle. We have excluded heteroaromatic nuclei such as **4** and **5**, although formally they have an N-heteroiminium structure. For the same reason anilines are not classified among enamines: the

I. INTRODUCTION

existence of the aromatic sextet provides a special reactivity to these molecules.

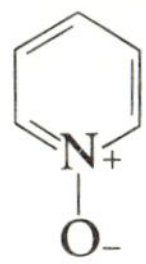

(4) pyridine N-oxide

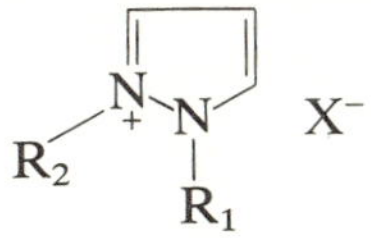

(5) N,N'-disubstituted pyrazolium salt with R₂ and R₁ substituents, X⁻ counterion

A. EQUILIBRIA OCCURRING IN N-HETEROIMINIUM SALTS

It is interesting to compare the acid-base equilibria of iminium salts (**1**) with those of N-heteroiminium ones (**2**). For the first ones it is known (1, Section II-B-1) that, if one of the N-substituents is a hydrogen atom, the imino tautomer **II** is generally more stable than the enamine **I** (see Scheme 1), but the difference in stability is small and depends on the

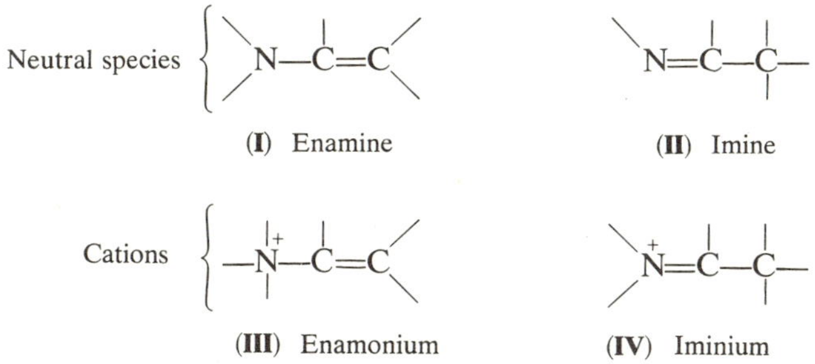

Neutral species:
(**I**) Enamine: $\rangle$N—C=C$\langle$
(**II**) Imine: $\rangle$N=C—C$\langle$

Cations:
(**III**) Enamonium: —N⁺—C=C$\langle$
(**IV**) Iminium: $\rangle$N⁺=C—C$\langle$

Scheme 1

nature of the substituents. Protonation of the enamines **I** leads to the enamonium ions **III** (kinetic products), which isomerize rapidly into the more stable iminium salts **IV** (thermodynamical products). Therefore it is possible to reach the iminium ions **IV** from a protonation of enamines or imines. According to the experimental conditions and to the substrate, alkylation of the enamines **I** (1, Section II-B-2) leads to the enamonium or iminium ions **III** or **IV**.

If analogous schemes are built for the N-heteroiminium ions (**2**), the complexity introduced by the second heteroatom is immediately striking: twelve species must be considered instead of four. The hydrazonium cation **XV** (see Scheme 2) could be theoretically reached from the six structures **V–X** by a protonation followed by a prototropism: we shall show (Section II-A-1) that the only general method is to protonate the N,N'-disubstituted enehydrazines **V**. Scheme 3 has been drawn similarly

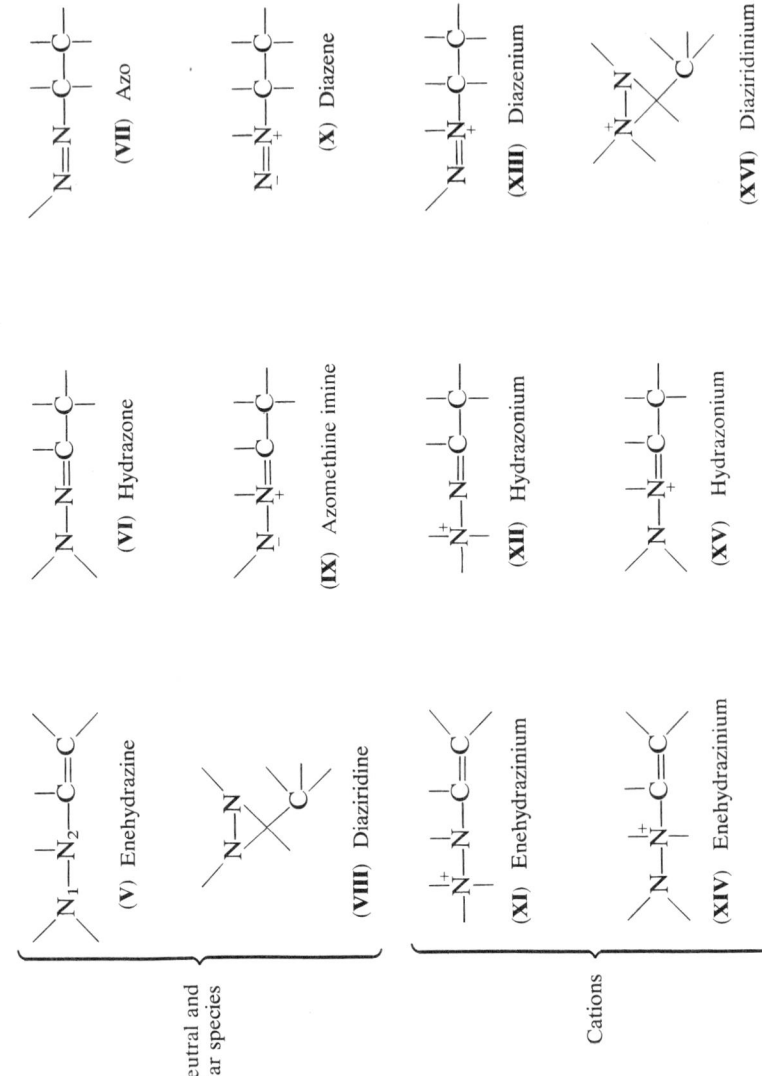

Scheme 2

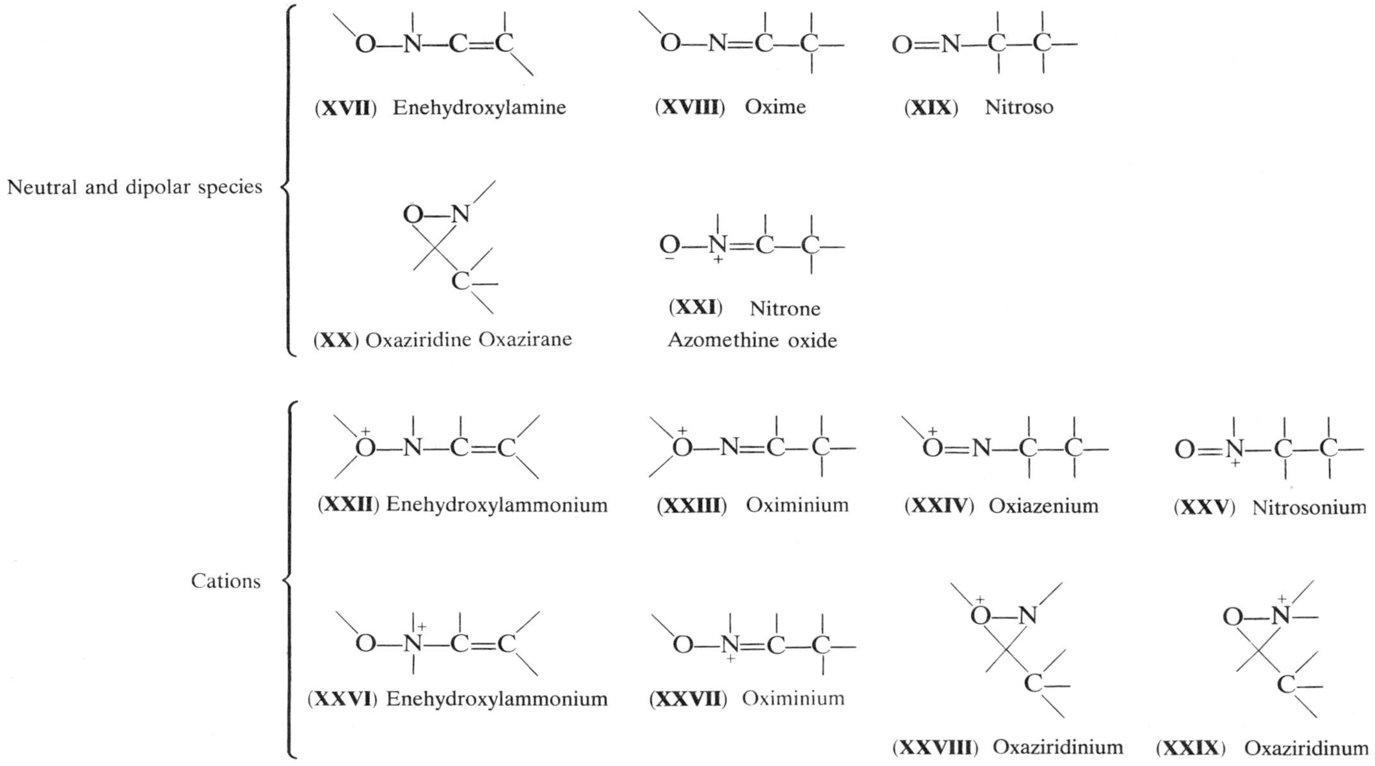

Neutral and dipolar species

(XVII) Enehydroxylamine
(XVIII) Oxime
(XIX) Nitroso
(XX) Oaziridine Oxazirane
(XXI) Nitrone Azomethine oxide

Cations

(XXII) Enehydroxylammonium
(XXIII) Oximinium
(XXIV) Oxiazenium
(XXV) Nitrosonium
(XXVI) Enehydroxylammonium
(XXVII) Oximinium
(XXVIII) Oxaziridinium
(XXIX) Oxaziridinium

Scheme 3

537

for the corresponding oxygen heteroiminium (**2**, Y = O); there is no dipolar structure corresponding to the diazene X, but two oxaziridinium cations can exist which are O- or N-protonated **XXVIII** or **XXIX**.

The main difference between these two series is the greater stability of the hydrazonium **XII** over **XIII**, whereas the oximinium **XXVII** is more stable than **XXIII** because of the lower basicity of oxygen-containing derivatives as compared to nitrogen ones. Thus it is possible to obtain a N-heteroiminium structure (**XXVII**) from all the neutral or dipolar species described in Scheme 3.

Finally, we have not been able to make an analogous scheme for the corresponding sulfur-containing derivatives because the only structure that has been described in this case is the oxime analogue, the alkylidene-sulfenamide **XXX** (Section VIII):

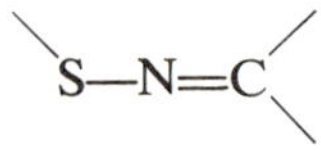

$$S-N=C$$

(XXX)

II. Hydrazonium Salts

A. SYNTHESIS

1. *Protonation of Enehydrazines* **V** (*Condensation of Carbonyl Compounds with 1,2-Disubstituted Hydrazines*)

In the noncyclic series the only hydrazonium salt synthesis has been performed by Schiess and Grieder (2) from condensation of a trisubstituted hydrazine with ketones. The orientation of the reaction depends on the ketone and on the acid. A hydrazonium salt (**6**) has been obtained in the case of R′, R″ = $(CH_2)_3$ or R′, R″ = $(CH_2)_4$ and only with perchloric acid. The structure of the perchlorates **6** has been established by IR(KBr), UV(CH_3CN), and NMR($CDCl_3$) spectroscopy.

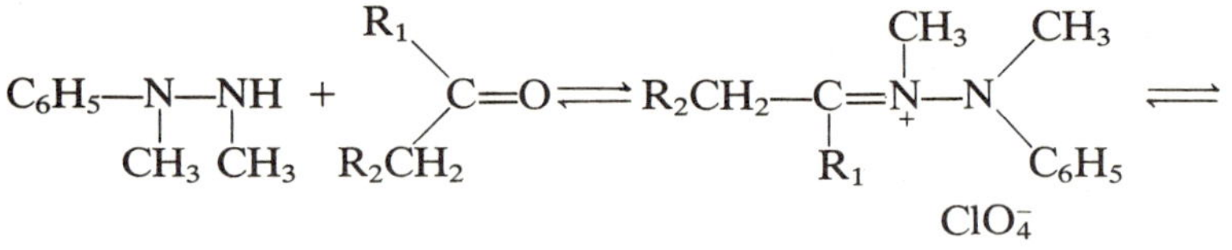

$$C_6H_5-\underset{\underset{CH_3}{|}}{N}-\underset{\underset{CH_3}{|}}{NH} + \underset{\underset{R_2CH_2}{}}{\overset{R_1}{C}}=O \rightleftharpoons R_2CH_2-\underset{\underset{R_1}{|}}{\overset{CH_3}{\underset{+}{C}}}=\overset{CH_3}{\underset{\underset{C_6H_5}{}}{N}}-N \rightleftharpoons$$

$$ClO_4^-$$

(6)

$$R_2CH=\underset{\underset{R_1}{|}}{\overset{}{C}}-\underset{}{N}-\overset{CH_3}{\underset{\underset{C_6H_5}{}}{N}}$$

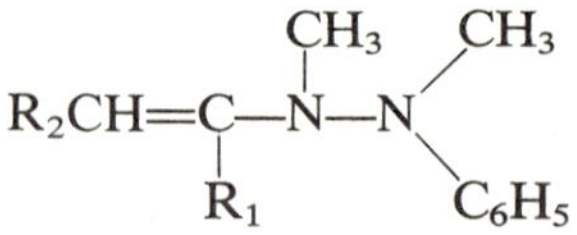

(7)

Unlike the acyclic series, for which structures **XV** are difficult to isolate, cyclic hydrazonium salts are easy to prepare as pyrazolinium salts (**8**):

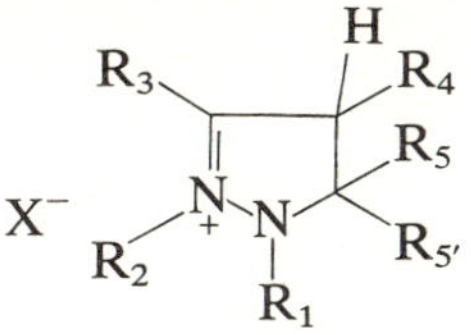

(**8**) 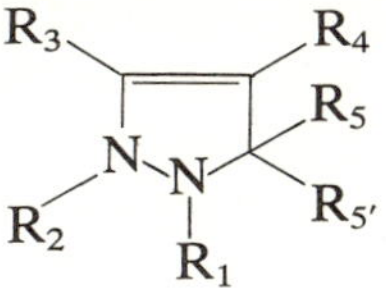 (**9**)

These pyrazolinium salts (**8**) may be obtained either directly or by protonation of 3-pyrazolines (**9**). The direct synthesis is a condensation of 1,2-disubstituted hydrazines (especially of 1,2-dimethylhydrazine, $R_1 = R_2 = CH_3$) with a carbonyl compound ($R_3COCH_2R_4$) and formol ($R_5 = R'_5 = H$) (Hinman's reaction) (3–8), with two molecules of carbonyl compound ($R_5 = R_3$, $R'_5 = CH_2R_4$) (14,15), or with an α,β-unsaturated carbonyl compound [$R_3COC(R_4)=C(R_5R'_5)$] (5,9–13). Exceptionally, the action of the 1,3-dibromopropane on a heterocycle with an azo group leads also to a pyrazolinium salt (**10**) (16), the dication (**11**) could be an intermediate of this reaction*:

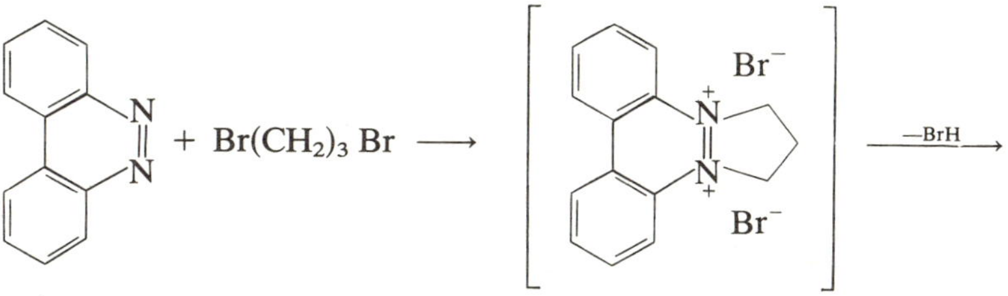

$+ Br(CH_2)_3 Br \longrightarrow$

(**11**)

$\xrightarrow{-BrH}$

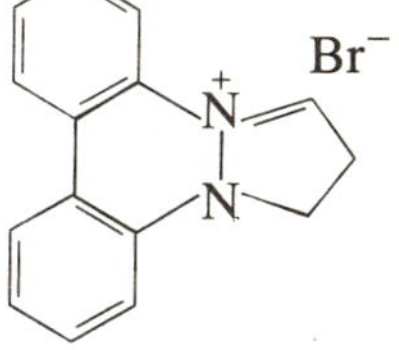

(**10**)

3-Pyrazolines (**9**) can be prepared in various ways: action of organomagnesium compounds on pyrazolidones (17); reduction with lithium aluminum hydride of pyrazolidones (18,19) [thiopyrazolidones yield only pyrazolidines (20,21)], of pyrazolones (18,20,22), and of pyrazolium salts

---

* The opposite reaction **10** + $H^+ \rightarrow$ **11** has never been reported.

(7,8,23–25); and 1,3-dipolar cycloaddition of azomethine imines with triple bonds (26,27).

The second way to obtain pyrazolinium salts (**8**) (4,5,28,31) is to protonate 3-pyrazolines (**9**) [the stereochemistry of this protonation has been discussed (29,30)]; in some cases it has been possible to observe the formation of a kinetic product (**12**) protonated on the sp^2 N_1 (28,32)*. Finally, the 3-methylenepyrazolidines (**13**) in equilibrium with the 3-pyrazolines (if R_3 is a CHRR' group) are protonated to the same pyrazolinium cation (**8**, R_3 = CHRR') (15).

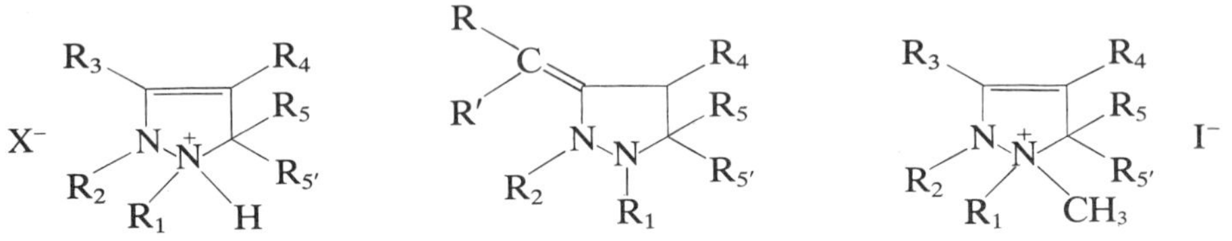

(**12**)   (**13**)   (**14**)

Quaternization of 3-pyrazolines occurs on N_1 with formation of the cation (**14**) (28,32) with the same structure (**XI**) as the kinetic protonation product (**12**). A C-methylation or a transposition of N—CH_3 toward C—CH_3 has never been observed (as is the case with cyclic enamines (1, Section II-B-2)). It seems that hydrazonium salts cannot be obtained in this way.

2. Hydrazone Quaternization

Generally quaternization of hydrazones (**VI**) occurs on the sp^3 nitrogen, which is more strongly nucleophilic, leading to salts **XII**: this appears to be the case in the acyclic series as well as in the heterocyclic one. For early studies on noncyclic hydrazones, see the two general reviews in Refs. 34 and 35a. Amidrazones such as **15** react in a similar way (36,37) to afford the quaternary salts (**16**):

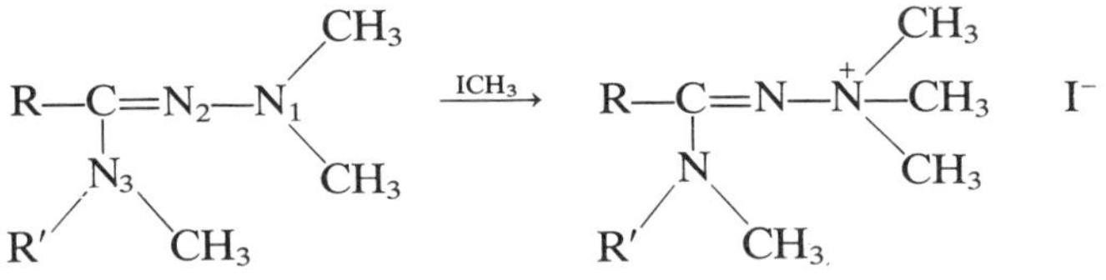

(**15**) R = C_6H_5, C_2H_5; R' = C_6H_5, CH_3 (**16**)

In the case of a cyclic hydrazone of type **17** methyl iodide does not react on the sp^2 nitrogen atom (**18**) as was proposed earlier, but reacts

* This N-protonated compound rearranges rapidly to the C-protonated form unless N_2 is a bridgehead nitrogen, in which case the cation **12** is stable (33).

like the other hydrazones, leading to **19** (39):

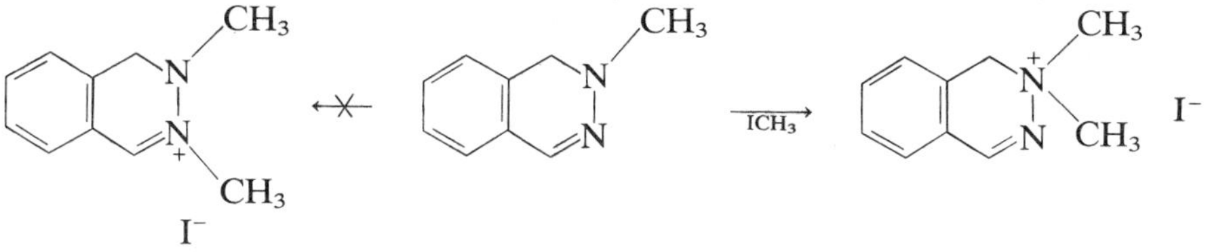

(**18**)　　　　(**17**)　　　　(**19**)

The 1,4,5,6-tetrahydropyridazines (**20**) give rise to the same type of quaternization; chemical (40,41) and spectroscopic (42) (UV and NMR) evidence has been provided.

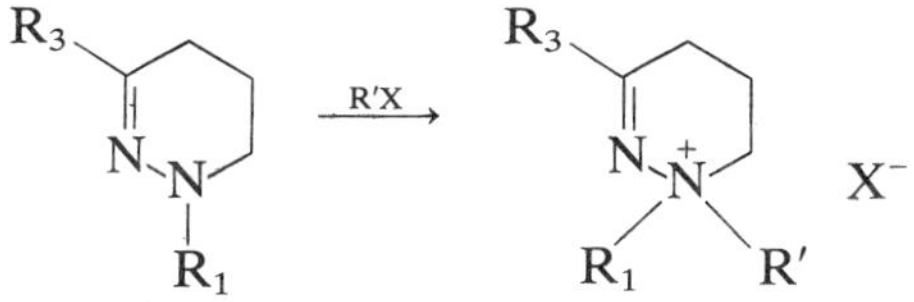

(**20**)　　　　(**21**)

The five-membered rings containing the hydrazone skeleton—the 2-pyrazolines (**22**)—behave similarly toward quaternization that occurs on the sp^3 nitrogen (**23**) (17,43–45):

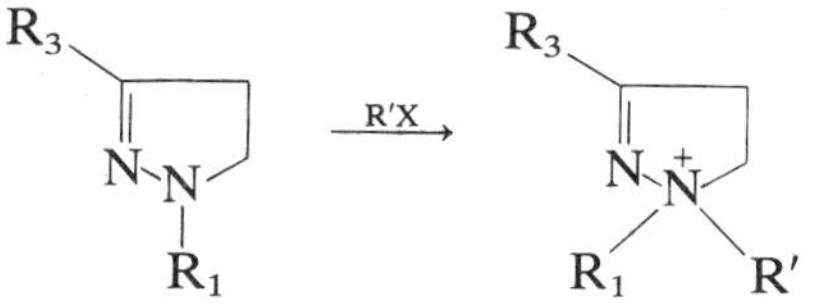

(**22**)　　　　(**23**)

In the case of the pyrazoline (**22**, $R^1 = CH_3$, $R^3 = CH=NOH$) quaternization occurs also on nitrogen N_1, which is thus more nucleophilic than the oxime one (45a, Section VI-A-1).

However, some exceptions to the quaternization of hydrazones on the sp^3 nitrogen exist if alkylation on the sp^2 nitrogen leads to a salt with its charge delocalized between two heteroatoms. Thus amidrazones of type **24** are quaternized on N_2 and afford cations of structure **25** (37):

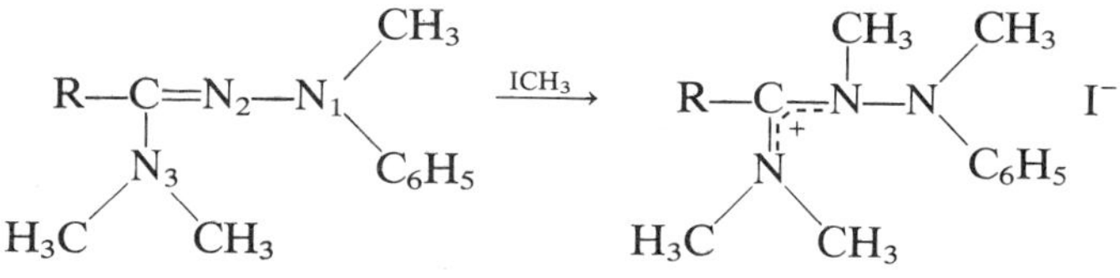

$R = H, C_2H_5$
(**24**)　　　　(**25**)

The authors explain the different reactivity of **24** as compared to **15** (methylation on N_1) by a decrease in the nucleophilicity of N_1 due to the electron-withdrawing effect of the phenyl group, which overcomes the greater steric hindrance existing in a structure such as **25**. A resonance-stabilized cation has also been observed (37) in the case of the amidrazones (**26**), even with $R' = C_6H_5$, the stabilizing factor being the occurrence of a hydrogen bond as in the chelated charge delocalized salt **27**:

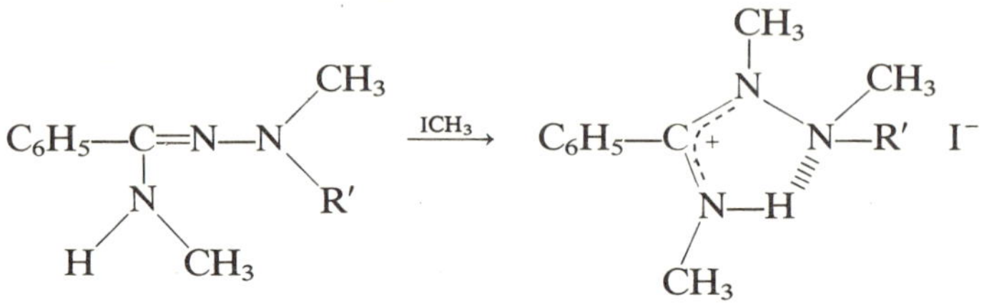

$$C_6H_5-C=N-N\begin{array}{c}CH_3\\R'\end{array} \quad\xrightarrow{ICH_3}\quad $$

$R' = CH_3, C_6H_5$

(**26**) (**27**)

Similar stabilized structures are obtained by quaternization of hydrazide imide (**28**), (**29**), (37), and thiohydrazides (**30**) (46,47):

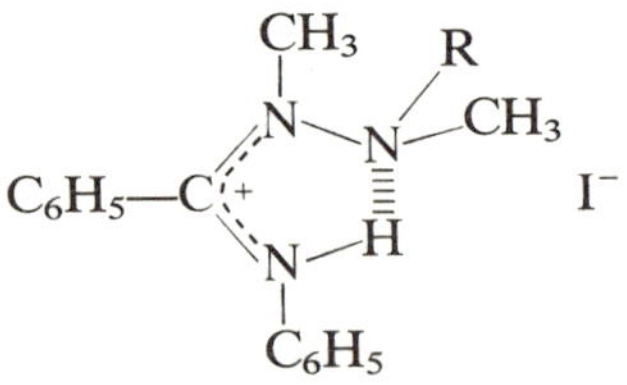

(**28**) 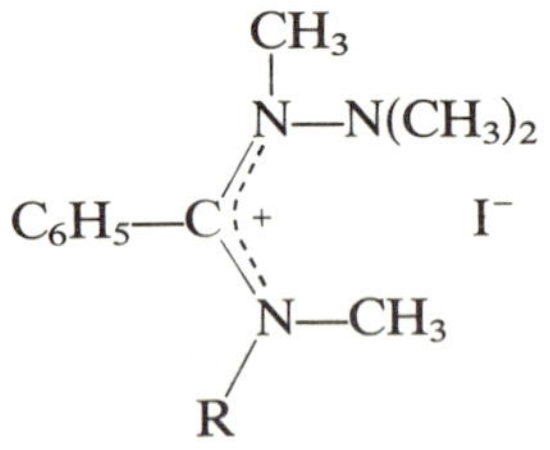 (**29**)

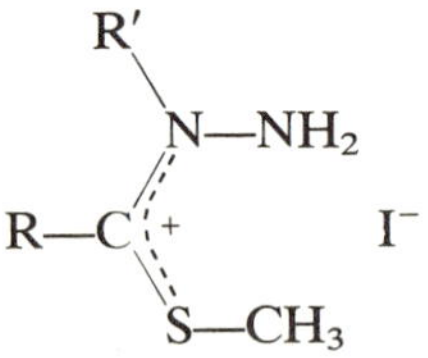

(**30**)

Although methylation of cyclic amidrazones such as the 3-amino-2-pyrazolines (**31**) occurs on the sp^3 nitrogen (**32**) (48,49), it is possible to isolate resonance-stabilized salts (**34**) by protonation of 1-substituted 2-methyl-3-iminopyrazolidine (**33**) (50):

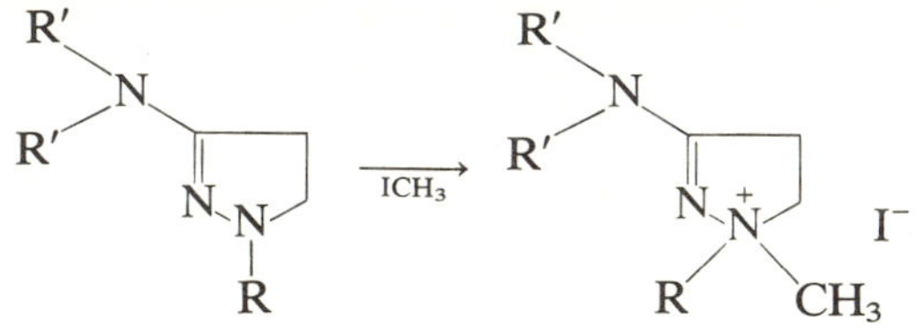

(**31**) R = CH$_3$, C$_6$H$_5$;   (**32**)
R' = H, CH$_3$

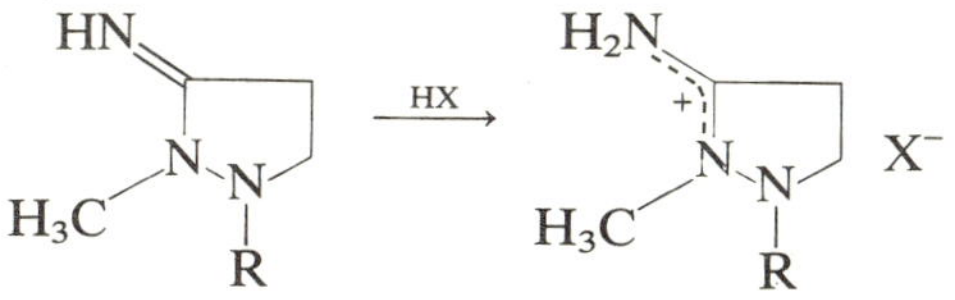

(**33**) R = CH$_3$, C$_6$H$_5$ (**34**)

Such a structure has been proved by UV spectroscopy: the absorption of **34** is different from that of **31** and **32**, but identical to that of the 3-amino-2-pyrazoline in HCl, which has been shown to be protonated on N$_2$ (49) (see Section II-A-3).

Two examples of hydrazone quaternization leading to the formation of a classical hydrazonium salt (**XV**) have been described. Houlihan and Theuer (51) found that the carbon chain of N-chloroalkylpyrazolines derived from **35** can react either on N$_1$ (**36**) or on N$_2$ (**37**), depending on its length.

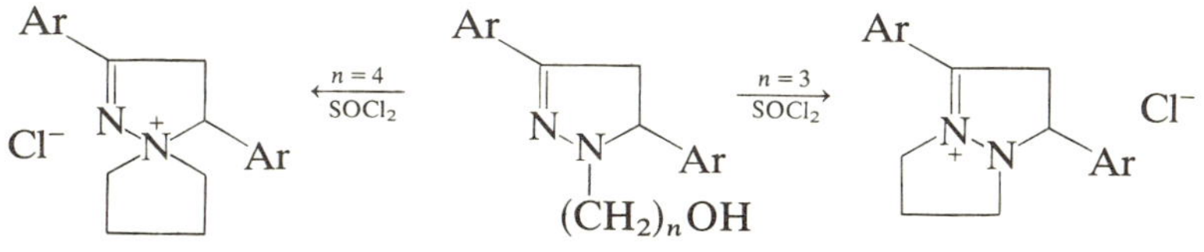

(**36**)   (**35**)   (**37**)

Another intramolecular alkylation has been reported by Schmitz (51a) on heating the 2,4-dinitrophenylhydrazone (**38**); alkylation on the sp^2 nitrogen (**39**) may be favored either because of the formation of a six-membered ring or because of the dinitrophenyl substituent on the sp^3

nitrogen (Section II-C-1):

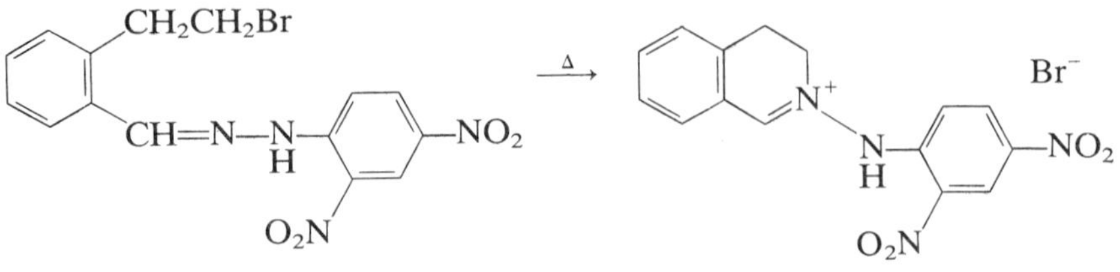

(38) → (39)

3. Hydrazone Protonation

Although several stable aliphatic and aromatic hydrazonium salts have been isolated as halogenoantimonites, bismuthites, and platinates (52,53) or perchlorates (54), no structural study of them has ever been carried out, so that the site of protonation is still unknown. The three possibilities for hydrazones to add a proton (cations **XII**, **XIII**, and **XV**) have been considered, but only as intermediates, to explain their reactions or their E–Z isomerization in acidic medium; therefore they will be considered in Section II-C, which deals with hydrazonium salts as intermediates.

However, some studies have been undertaken on the protonation of cyclic hydrazones. A first claim for a protonation of 3,6-dimethyl-6-phenyl-1,4,5,6-tetrahydropyridazine (55) on the sp^2 nitrogen was later ruled out; in fact, UV and NMR spectroscopy (42) shows that protonation occurs on the sp^3 nitrogen (**40**), which is thus more basic, as it is more nucleophilic (see Section II-A-2).

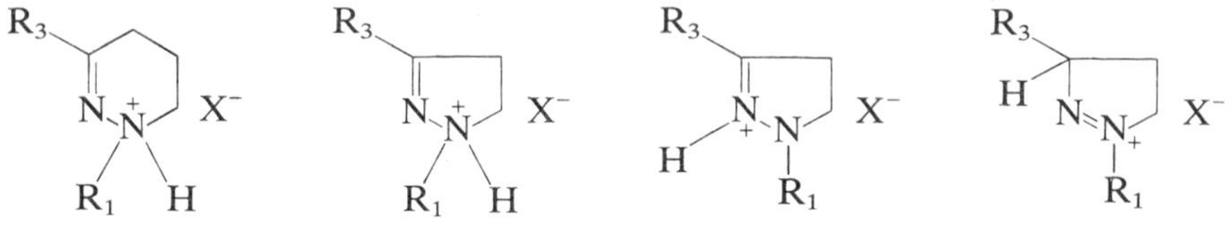

(**40**) $R_1 = H, R_3 = C_6H_5$;   (**41**)   (**42**)   (**43**)
    $R_1 = CH_3, R_3 = C_6H_5$

In the five-membered ring hydrazones, the 2-pyrazolines, the normal site of protonation is also the sp^3 nitrogen, giving rise to the salt (**41**). This result has been established by UV and NMR spectroscopy (4,17) and X-ray diffraction (56). However, new data show that if a salt such as **41** is thermodynamically favored, the two other possible conjugate acids **42** and **43** occur (57) in acidic medium (see Section II-C).

Luth and Trotter (58)* described a very peculiar result, which corresponds (Scheme 2) to the transformation **VII → XV**.† The salt obtained from the action of hydrobromic acid on the 1-pyrazoline (**44**) has been studied by X-ray diffraction; Luth and Trotter proposed structure **45** (a 2-pyrazoline of type **42** protonated on the sp^2 nitrogen), but as the hydrogen atoms have not been localized, structure **45** cannot be considered as definitively established, especially in view of the very strange geometry the authors obtained:

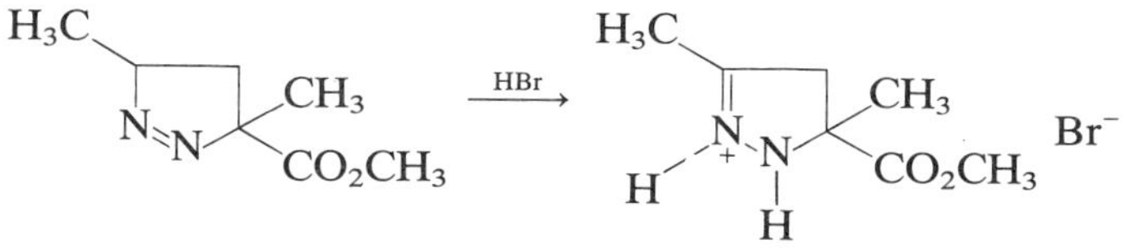

(**44**)   (**45**)

Like quaternization, protonation of amidrazones (**46**) may occur on the sp^2 nitrogen to give rise to a resonance-stabilized structure (**47**) (59):

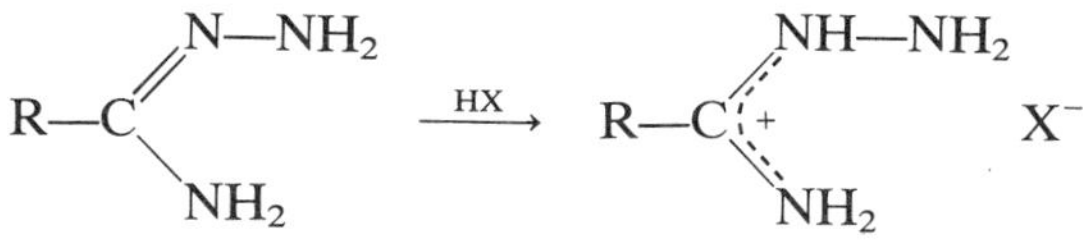

(**46**)   (**47**)

Ultraviolet spectroscopy shows (49) that a similar electron delocalized cation (**49**) is obtained by protonation of 3-amino-2-pyrazolines (**48**):

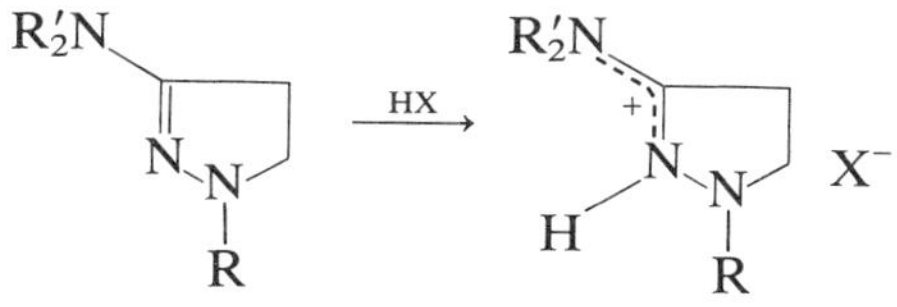

(**48**)   (**49**)

---

* These authors seem to have been unaware of the work of Nardelli and Fava (56) on the structure of the pyrazoline (**41**, $R_1 = R_3 = H$) hydrochloride.

† The diazenium cation (**XIII**) has not been obtained by protonation or quaternization of the azo derivative (**VII**), but it can be prepared from the 1,1-disubstituted hydrazines (Section V).

4. Diaziridine Opening in Acidic Medium

Schmitz (60) described the acid hydrolysis of the diaziridine (50) into the hydrazonium salt (51) (Scheme 2: VIII → XVI → XV):

This reaction has been thoroughly studied (61) kinetically: the first step is a protonation to form a diaziridinium ion (52), which then undergoes an ionic opening, the latter step determining the rate of hydrolysis. These diaziridines are much more difficult to hydrolyze than the open-chain compounds containing a $>$N—C—N$<$ group. Schmitz (60) described also a formation of 51 by an acidic opening of hexahydrotetrazines (for a discussion see Section IV-C).

B. REACTIVITY

1. Chemical Reactivity

In view of the scarcity of acyclic hydrazonium structures (2), it is not surprising that all the reactivity studies but one (the oxidation of hydrazone perchlorates; see Section II-C-2) deal with heterocyclic compounds—more precisely, with pyrazolinium salts (8). Nucleophiles can attack these compounds at three possible sites, which are shown in Scheme 4.

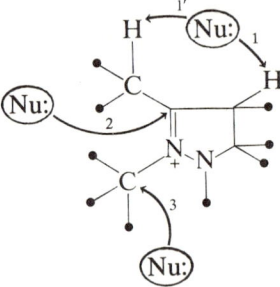

Scheme 4

(a) *Attacks 1 and 1'* (*Scheme 4*). This attack is analogous to the formation of enamines from the action of bases on iminium salts (1, Section IV-I). Pyrazolinium salts (8) afford 3-pyrazolines (9) (62) and methylenepyrazolidines (**13**) (15). This acidity of the hydrogen atoms in β position to the C–N⁺ double bond explains their exchange in refluxing D$_2$O (5,62). The mechanism of these attacks and the stereoelectronic factors that favor 1 over 1' have been discussed (62); a ψ-E$_2$ mechanism and an orthosynclinal attack have been proposed.

(b) *Attack 2 (Scheme 4).* Because of this type of attack, pyrazolium salts can be classified as *N*-heteroiminium ions (**2**) and can be included in this review.

For most of the nucleophiles, the equilibrium is totally shifted toward the pyrazolidine form, although it is possible to demonstrate their iminium character by the action of a second nucleophile. Thus, adding Grignard reagents to 3-hydroxy- (**54b**) or 3-cyano- (**54c**) pyrazolidines affords 3-R pyrazolidines (**54e**) via the iminium hydroxide (**53b**) and cyanide (**53c**) (63):

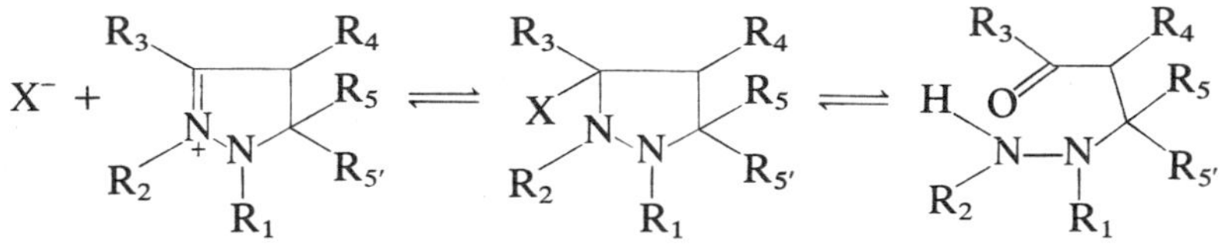

(**53a**) X = Cl, Br, ClO$_4$ (6, 10) (**54b**) X = OH (11, 12, 64)   (**54'**)
(**53b**) X = OH   (**54c**) X = CN (64)
(**53c**) X = CN   (**54d**) X = H (8, 51, 64)
                                        (**54e**) X = R, Ar (63, 64)

Similarly to the obtaining of enamines from iminium salts (1), carbinolhydrazines (**54b**) are not intermediates in the synthesis of 3-pyrazolines (**9**) (62), contrary to the statements of some authors (12). Rather, they exist in equilibrium with the open-chain hydrazinoketone (**18**) or Mannich bases (**54'**) (62). The formation of **54'** from **53** corresponds to the hydrolysis of iminium salts (1, Section IV-B).

(c) *Attack 3 (Scheme 4).* It is well known that in ammonium salts a secondary reaction occurs, often together with the Hofmann degradation, resulting in an attack on the carbon in a position to the positively charged nitrogen (65). The peculiarity of attack 3 in this case is that the charged nitrogen has sp^2 hybridization: the products obtained have a 2-pyrazoline structure (**22**) (62).

A fourth process (i.e., **XV** to **XVI**) can occur in the pyrazolinium salts, which, as shown in Scheme 2, is characteristic of these structures because it requires the presence of a >C–N⁺< double bond and of a heteroatom. It is difficult to confirm, however, because the diaziridinium ion is unstable, and its formation can be inferred only from the isolated products. For this reason this type of reaction will be described in Section II-C-1.

2. Spectroscopic Properties

In addition to the properties described in Section II-A-1 for the hydrazonium salt **6**, information about pyrazolinium salts (**53a**) obtained by UV (5) NMR (5,10–12,30), and X-ray diffraction (66,67) is available.

$$\text{\textbackslash N—\overset{+}{N}=C\textbackslash} \quad \text{with H on middle N}$$

C. HYDRAZONIUM SALTS AS INTERMEDIATES

Here two cases must be distinguished, depending on whether or not the substituent on the sp^2 nitrogen is a hydrogen atom. This distinction is important; it is difficult to demonstrate without ambiguity the occurrence of an intermediate hydrazonium salt, but if this salt is of type $>\overset{+}{N}-NH=C<$, one must consider a mixture of prototropic cations, one of them, not necessarily the major one, having hydrazonium structure **XV**.

1. Intermediates $>N-\overset{H}{\underset{+}{N}}=C<$

As pointed out in Section II-A-3, hydrazones, at least the cyclic ones, are protonated on the sp^3 nitrogen atom to give a cation of type **XII**. The problem of the relative basicities of the two nitrogens of hydrazones **VI** [or (what means the same thing) of the relative stabilities of the two cations **55** and **56**] has been qualitatively discussed in connection with the basicity of 2-pyrazolines (68). In theory the presence of the three cations **41**, **42**, and **43** should be considered in acidic medium, but cation **41** is sufficient by itself to explain the pK_a's of 2-pyrazolines.*

$$\underset{R}{\overset{R}{\diagdown}} H-\overset{+}{N}-N=C \underset{R''}{\overset{R'}{\diagup}} \quad X^- \qquad \underset{R}{\overset{R}{\diagdown}} N-\overset{H}{\underset{+}{N}}=C \underset{R''}{\overset{R'}{\diagup}} \quad X^-$$

$$\qquad\qquad (55) \qquad\qquad\qquad\qquad (56)$$

*In ref. 68 only N—H and N—CH$_3$ 2-pyrazolines (**41**, $R_1 = H$ or CH_3) are under consideration. If the conclusions remain valid for $R_1 = C_6H_5$ (17), it is not certain that they are so if R_1 is strongly electron withdrawing, for instance, if $R_1 = 2,4$-$C_6H_3(NO_2)_2$ (69).

It is well known that alkyl (70a) and aryl (70c) hydrazones can be hydrolyzed in acidic medium. The mechanism generally accepted (70c) starts with protonation on the sp^2 nitrogen to form cation **57**, followed by a nucleophilic attack of type 2 (Section II-B-1):

$$\begin{array}{c}R\\ \diagdown \\ H\end{array}\!\!N\!-\!N\!=\!C\!\begin{array}{c}R'\\ \diagup \\ R''\end{array} \xrightarrow{H^+} \begin{array}{c}R\\ \diagdown \\ H\end{array}\!\!N\!-\!\overset{H}{\underset{+}{N}}\!=\!C\!\begin{array}{c}R'\\ \diagup \\ R''\end{array} \xrightarrow[-H^+]{H_2O} \begin{array}{c}R\\ \diagdown \\ H\end{array}\!\!N\!-\!\overset{H}{\underset{}{N}}\!-\!\overset{R'}{\underset{OH}{C}}\!-\!R''$$

(57)

$\downarrow H^+$

$$\begin{array}{c}R\\ \diagdown \\ H\end{array}\!\!N\!-\!\overset{+}{N}H_3 \;+\; O\!=\!C\!\begin{array}{c}R'\\ \diagup \\ R''\end{array} \longleftarrow \begin{array}{c}R\\ \diagdown \\ H\end{array}\!\!N\!-\!\overset{H}{\underset{H}{\overset{+}{N}}}\!-\!\overset{R'}{\underset{OH}{C}}\!-\!R''$$

Grandberg et al. (71) used Hückel's method to calculate the π-electron density of the nitrogen atoms of phenylhydrazones (**58**):

$$\underset{H}{\overset{C_6H_5}{\diagdown}}\overset{+0.1358}{N_\alpha}\!-\!\underset{-0.2732}{N_\beta}\!=\!C\!\begin{array}{c}C\\ \\ C\end{array} \qquad \underset{R}{\overset{R_1}{\diagdown}}\overset{+}{N}\!-\!N\!=\!C\!\begin{array}{c}R'\\ \\ R''\end{array}$$

(58) (59)

They reached the conclusion that the β-nitrogen is much more basic and nucleophilic than the α-nitrogen; this allows them to explain the formation of tryptamines (Section II-C-2). This view, totally inconsistent with the formation of salts (**59**) by quaternization of hydrazones (Section II-A-2), is a consequence of an incorrect use of the results obtained by theoretical calculations. The basicity and the π-electron density of two atoms with different hybridizations are in no way related. [It would be better to calculate the relative stabilities of the two cations **55** and **56** (71a).]

The observation that E-Z isomerization of hydrazones is easier in an acidic medium also raises the problem of the site of protonation. Two problems are related to this process: first, its mechanism, which can be a lateral shift or an internal rotation around the C-N double bond (72-74), and, second, the energy necessary for the isomerization to occur. Up to now no definite conclusion has been drawn about the mechanism. As to the isomerization barrier, it is, as could be expected, very high; the direct equilibration of isomers E and Z has been studied (75–77), and it has even been possible to isolate the isomers (78); usually (in NMR spectroscopy) an increase in temperature does not provoke any phenomenon (79–81). The only successful results obtained from temperature NMR experiments (82,83) occur in chlorinated or protic solvents: the first effect is still unknown, but the influence of acids has been more extensively studied. Kinetic studies (84) show also an acid catalysis of the E–Z isomerization. Such an effect is difficult to explain if the isomerization process occurs only through an inversion mechanism in acidic solution; the inversion would be made difficult by protonation on the sp^3 nitrogen (**55**) because of the greater electronegativity of $-\overset{+}{N}\!\!\lneq$. Protonation on the sp^2 carbon (**60**) or on the sp^2 nitrogen (**56**) decreases the C–N double-bond nature and thus the rotation mechanism is favored. Another possible explanation of the phenomenon would be protonation on the imino nitrogen with an enehydrazine-type tautomerism (**61**), which would facilitate rotation about the C–N bond.

$$
\begin{array}{cc}
\underset{R}{\overset{R}{>}}\!\overset{+}{N}\!\!=\!\!N\!-\!\underset{H}{\overset{R'}{\underset{|}{C}}}\!\!\underset{R''}{\overset{}{\,}} \quad X^- & \underset{R}{\overset{R}{>}}\!N\!-\!\underset{H}{\overset{H}{\underset{|}{\overset{+}{N}}}}\!-\!\underset{}{\overset{R'}{\underset{}{C}}}\!=\!\underset{R'''}{\overset{R''}{\underset{}{C}}} \quad X^- \\
(\mathbf{60}) & (\mathbf{61})
\end{array}
$$

Thus it seems that isomerization by a rotation mechanism would be more favored in acidic solution (this does not say anything about the isomerization process in neutral solutions), but this acid-catalyzed isomerization does not solve the problem of the site of protonation.

The occurrence of an iminium salt (**62**) has been postulated by Hammerum (85,86) to explain the acid-catalyzed dimerization of the formolalkylhydrazones:

$$
\begin{array}{c}
R\!-\!NH\!-\!\overset{+}{N}H\!=\!CH_2 \\
(\mathbf{62}) \\
+ \\
R\!-\!NH\!-\!N\!=\!CH_2
\end{array}
\longrightarrow
\left[
\begin{array}{c}
\underset{HN}{\overset{R}{>}}\!\!\underset{\underset{CH_2}{|}}{\overset{H_2C}{>}}\!\!\underset{\overset{+}{N}H}{\overset{N}{<}}\!\!R \\
(\mathbf{63})
\end{array}
\right]
\xrightarrow{-H^+}
\begin{array}{c}
\underset{HN}{\overset{R}{>}}\!\!\underset{\underset{}{N}}{\overset{N}{>}}\!\!\underset{}{\overset{NH}{<}}\!\!R \\
(\mathbf{64})
\end{array}
$$

Other authors propose analogous mechanisms for the dimerization of various hydrazones (87–89).

Szmant and McGinnis (90) suggest protonation on the sp^2 nitrogen of hydrazones as the first step in the formation of azines from hydrazones.

A hydrazonium intermediate has also been postulated to react with olefins to yield a pyrazolidine (**65**) (91):

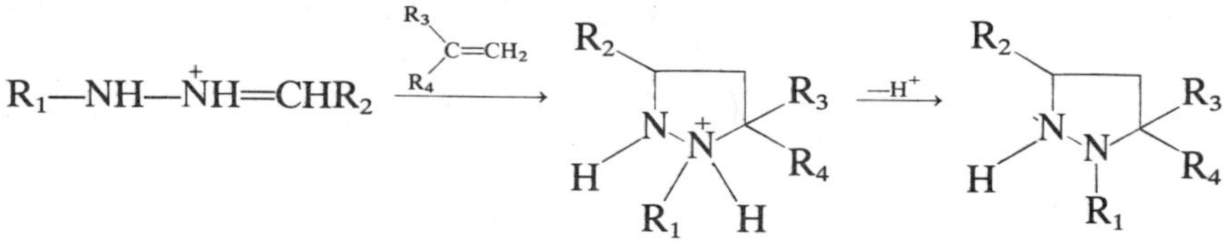

We have already pointed out that in the heterocyclic series the stable cation of the 2-pyrazolines has structure **41**; however, the complete behavior of these cyclic hydrazonium salts in acidic medium can be explained only if cations **42** and **43** (57) occur. A hydrazonium cation such as **42** explains the acid-catalyzed epimerization of the carbon in position 4 [through a 3-pyrazoline (**9**)], as well as the H–D exchange of the hydrogen atoms in position 4.

Another process can occur only after protonation on the sp^2 nitrogen: the attack of the sp^3 nitrogen lone pair on the electrophilic iminium carbon. (This type of attack was reported in Section II-B-1). Such a mechanism has been proposed in two cases. One has been described by Baumes et al. (92,93) from the pyrazolinium salts (**66**), which, like all the N_1-unsubstituted pyrazolium salts, have not been isolated; the authors extended their mechanism to salt **68**. In the case of 1,2-disubstituted pyrazolinium salts (**53a**), such diaziridinium intermediates have never been invoked.

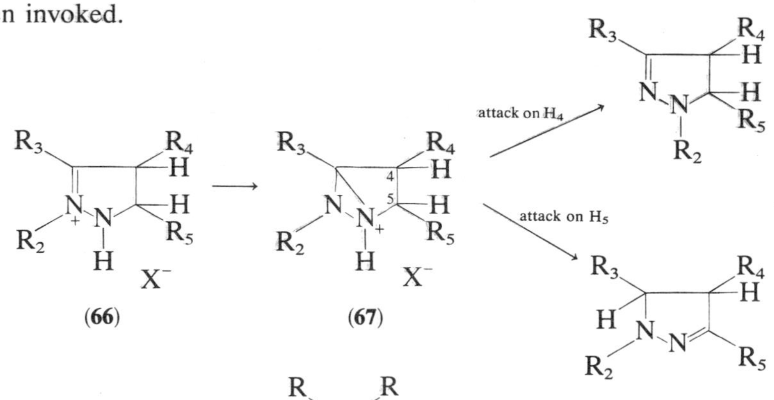

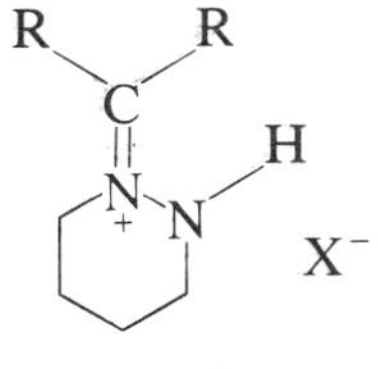

A similar process has been proposed by Moore and Binkert (94) to explain the formation of the N-aminopyridinium salt (**72**) from the diazepinone (**69**):

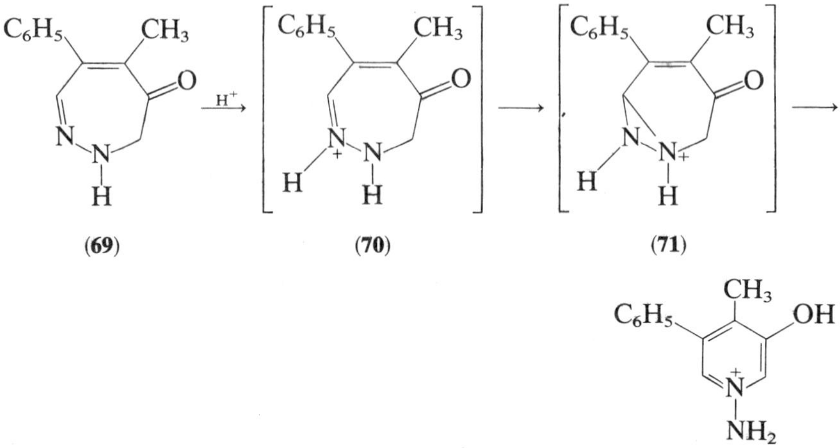

[structures (**69**) → (**70**) → (**71**) → (**72**)]

2. Intermediates $\ \diagdown\!\!\!N\!-\!\overset{+}{N}\!=\!C\!\diagup\ $ (with R on central N)

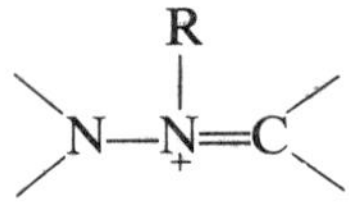

This type of intermediate has often been postulated because it is essential to explain several reactions important in heterocyclic chemistry.

(a) Condensation of Hydrazines with Carbonyl Compounds. The first step in this type of condensation in acidic medium is the formation of the hydrazonium ion (**74**), which is usually not isolated (for an exception see **6**, Section II-A-1) because it undergoes various transformations due to its high reactivity.

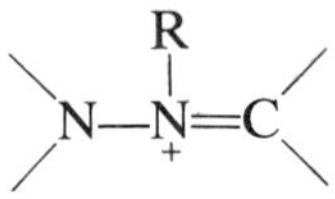

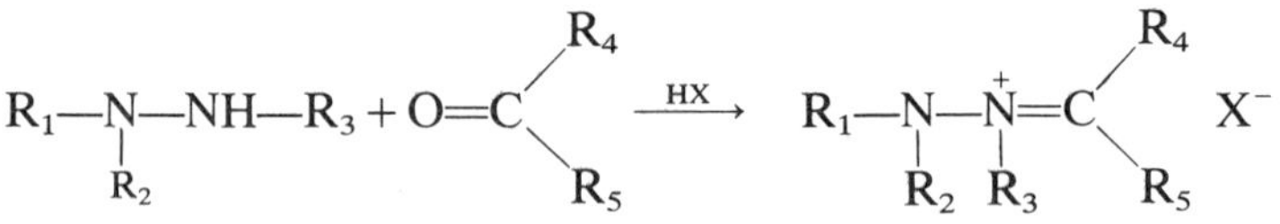

$$R_1\!-\!\underset{R_2}{N}\!-\!NH\!-\!R_3 + O\!=\!C\!\underset{R_5}{\overset{R_4}{\diagup}} \xrightarrow{HX} R_1\!-\!\underset{R_2}{N}\!-\!\overset{+}{\underset{R_3}{N}}\!=\!C\!\underset{R_5}{\overset{R_4}{\diagup}}\ \ X^-$$

$R_2 = H$ or $\neq H$ (**73**)   (**74**)

Kornet and Thio (95) wrote a resonance-stabilixed structure for this type of cation; it seems excessive to consider a diaziridinium cation (**75**)

II. HYDRAZONIUM SALTS

as a resonance form of the hydrazonium cation:

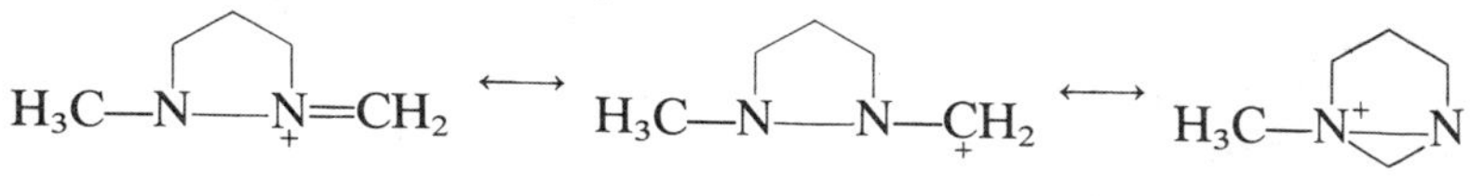

$$H_3C-\overset{+}{N}-N=CH_2 \longleftrightarrow H_3C-N-\overset{+}{N}-CH_2 \longleftrightarrow H_3C-\overset{+}{N}\overset{\frown}{=\!=\!=}N$$

(75)

If $R_2 = H$ and if R_4 (or R_5) has a functional group capable of reacting with N_1, a heterocyclization occurs; pyrazole derivatives are obtained in this way from a condensation of hydrazines with 1,3-difunctional compounds (13).

If these conditions are not fulfilled, the hydrazonium salt (**74**) can undergo reactions of the type described in Section II-B-1: attack 1 on hydrazonium salt **74** which would give rise to an enehydrazine intermediate like the one proposed in the Fischer synthesis (14), or attack 3 with formation of a hydrazone (**76**) (96):

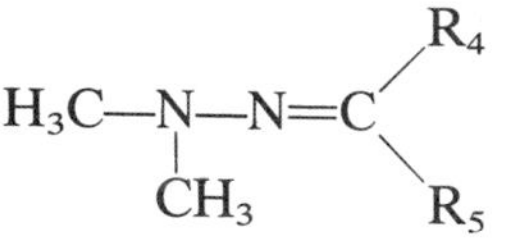

$$H_3C-\underset{CH_3}{N}-N=C\underset{R_5}{\overset{R_4}{<}}$$

(76)

But the hydrazonium (**74**) has two reactive sites: N_1, which is nucleophilic, and C_3, which is electrophilic. These two sites can react on each other in an intramolecular process leading to a diaziridinium ion (**XVI**) (Section II-B-1) or in a bimolecular process affording a hexahydrotetrazine (**64**) (Section II-C-1). If the carbonyl compound (**73**) is the formol ($R_4 = R_5 = H$), the salt (**74**) can undergo a Mannich reaction (condensation with a carbonyl compound, R_6COCH_3, in its enol form) or a Ugi condensation (with an isocyanide, ArNC). These are analogous to reactions of iminium salts (1, Section IV-H). In the first case the Mannich base (**77**) is isolated if $R_2 \neq H$, and the 3-pyrazoline (**78**) if $R_2 = H$ (3,6). In the second reaction the amide (**79**) is obtained (95):

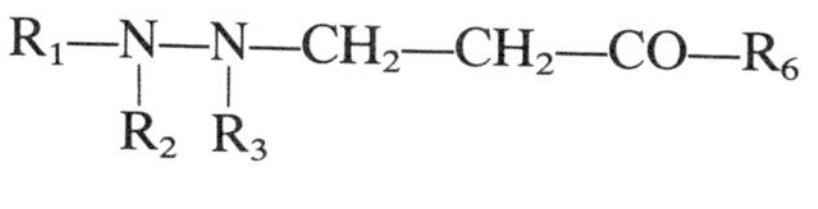

$$R_1-\underset{R_2}{N}-\underset{R_3}{N}-CH_2-CH_2-CO-R_6$$

(77)

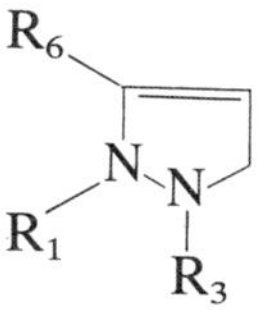

(78)

$$R_1-\underset{R_2}{N}-\underset{R_3}{N}-CH_2-CO-NH-Ar$$

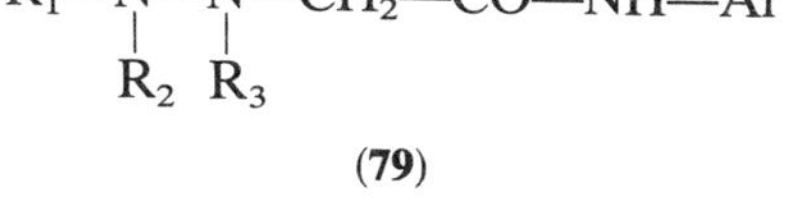

(79)

If R_4 and R_5 are not hydrogen atoms and if $R_2 = H$, the hydrazonium salt (**74**) reacts with carbonyl compounds that have at least one hydrogen atom in α position to the carbonyl group, leading to another nonisolated intermediate (**80**) with a hydrazonium–enehydrazine structure (15):*

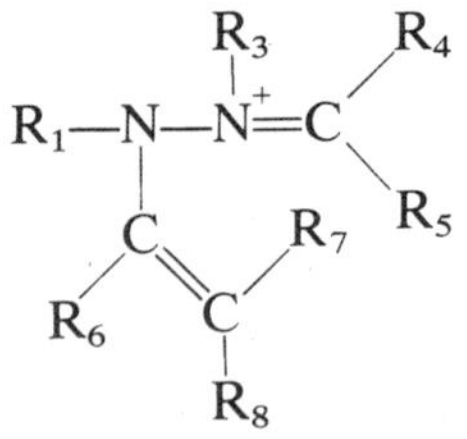

(**80**)

From the salt (**80**) it is possible to get pyrroles (**82**) (Piloty's reaction) through the dienehydrazine (**81**) (15,98,99), or salts of 3-pyrazolines (**83**) (14,15,98) by direct cyclization; if $R_1 = H$, 2-pyrazolines are obtained (100) by a cyclization of azines:

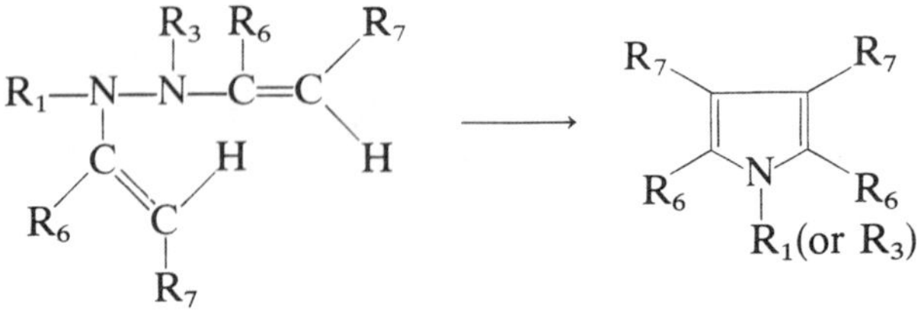

(**81**)   (**82**)

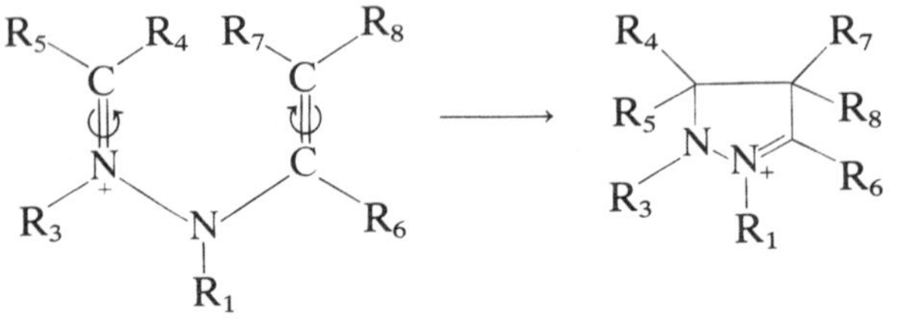

(**80**)   (**83**)

The cyclization of **80** to **83** has been described as occurring via a concerted, disrotatory process (15,100,101).

The oxidation by Theilacker and Leichtle (54) of the hydrazone perchlorate (**84**) by nitrobenzene into either the pyrazoline (**86**) or the

---

* Because of its hydrazonium structure, **80**, like **74**, can undergo an attack of type 2 from nucleophilic reagents like hydrazines (14,92) on the carbon.

pyrazole (**87**) [the authors (54) could not determine which] could as well occur through a hydrazonium–enehydrazine intermediate (**85**):

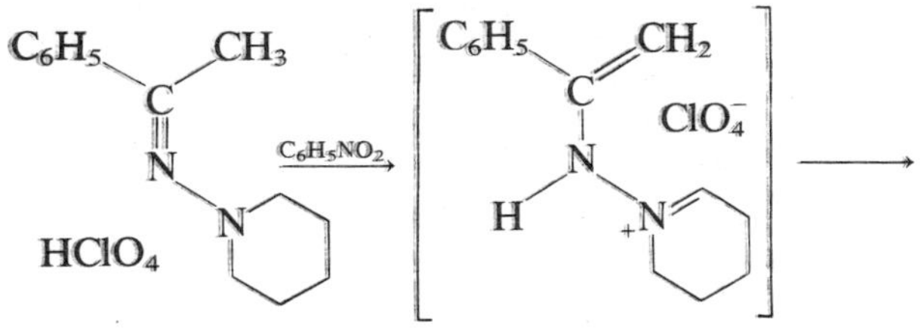

(**84**) → (**85**) 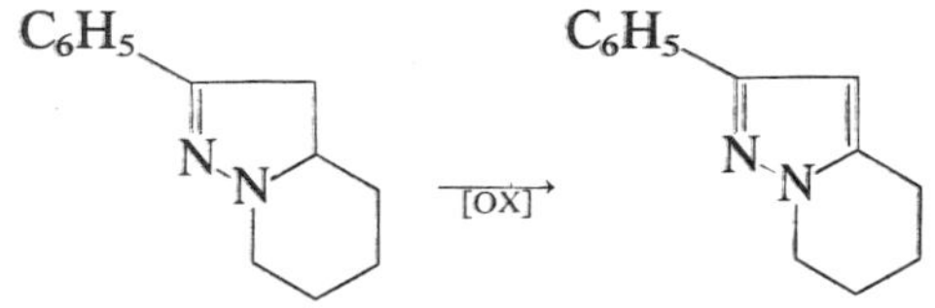 → (**86**) —[OX]→ (**87**)

(*b*) *Hydrazone Quaternization.* Two types of reaction can be reported in which a nonisolated hydrazonium intermediate has been postulated in order to explain the compounds obtained by quaternization of hydrazones. The first example deals with an intermolecular process (102):

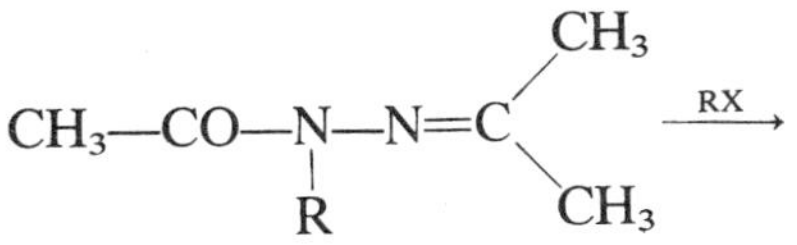

$$CH_3—CO—N(R)—N=C(CH_3)(CH_3) \xrightarrow{RX}$$

(**88**)

$$[CH_3—CO—N(R)—\overset{+}{N}(R)=C(CH_3)(CH_3)\ X^-] \xrightarrow{HCl} R—NH—NH—R$$

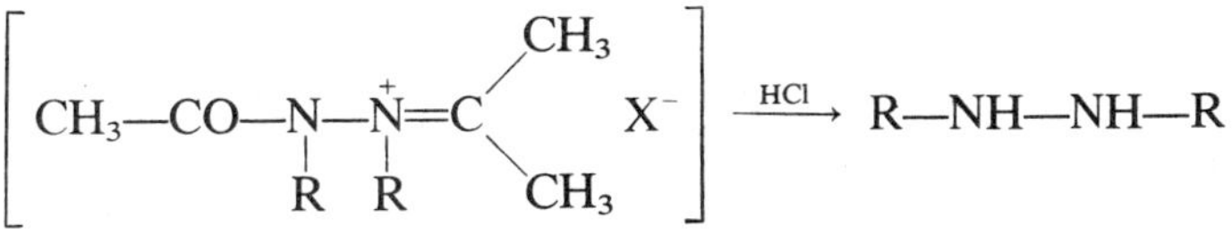

(**89**)　　　　　　　　　　　(**90**)

The quaternization of the hydrazone (**88**) on the sp^2 nitrogen (**89**) finds substantiation in the symmetrical structure of the isolated hydrazine (**90**); it appears that a strongly electron-attracting group on the sp^3 nitrogen of hydrazone **88** orients the quaternization to the sp^2 nitrogen, unlike what is usually observed (Section II-A-2).

The second process is intramolecular: the reaction of aryl hydrazines with γ-chloroketones yields tetrahydropyridazines (**91**) and tryptamines (**93**), according to Grandberg and co-workers (71,71b); the authors assume the existence of a hydrazonium intermediate (**92**) to explain the formation of tryptamines (**93**):

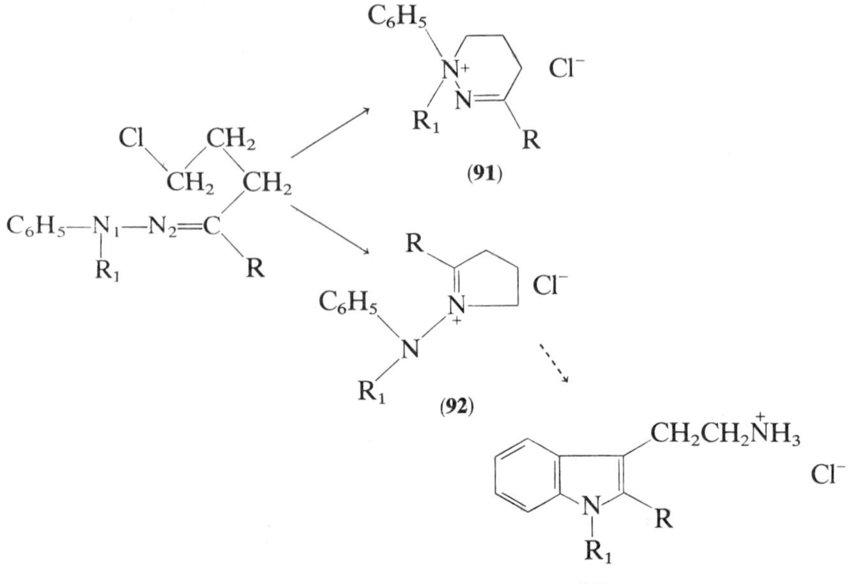

$$\text{C}_6\text{H}_5\text{—N}_1\text{—N}_2\text{=C}$$

(91), (92), (93)

The transformation **92 → 93** occurs by a mechanism analogous to that of the Fischer synthesis. The authors believe that N_2 is more nucleophilic (see discussion in Section II-C-1) and that the formation of **92** is favored unless $R_1 = C_6H_5$.

(*c*) *Reduction of Heterocycles with Metal Hydrides.* The reduction of pyrazolones (18), pyrazolidones (18,22), thiopyrazolones (20,21) or pyrazolium salts (24,103) with lithium aluminium hydride or of 3-pyrazolines with sodium borohydride in acetic acid (64) gives rise to totally reduced products (pyrazolidines). This reaction can be explained by the formation of pyrazolinium salts or of their complexes with the metallic ions present in the medium.

III. Azinium Salts

A. SYNTHESIS

1. *Condensation of Carbonyl Compounds with Monosubstituted Hydrazines*

Most of the work on this type of condensation has been undertaken by Lamchen et al. In the acyclic series, acetone condenses with methylhydrazine in the presence of hydrochloric acid and stannic chloride to afford the N-methyldimethylketazine chlorostannate (**94**) (104):

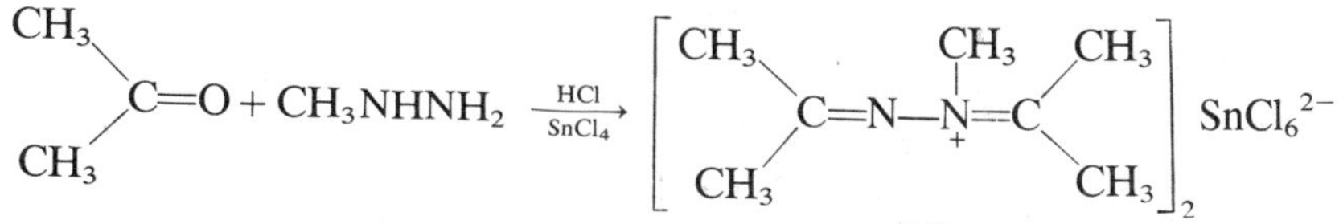

$$CH_3\text{-}C(CH_3)=O + CH_3NHNH_2 \xrightarrow[SnCl_4]{HCl} \left[(CH_3)_2C=N\text{-}\overset{+}{N}(CH_3)=C(CH_3)_2 \right]_2 SnCl_6^{2-}$$

(**94**)

With the same procedure other N-methylaldazinium and ketazinium complex salts have been obtained (105). However, in the case of the salicylaldazine the methochloride was unexpectedly obtained (105). It has been suggested (34) that the first step of this reaction is condensation of the aldehyde on the NH_2 group of the methylhydrazine, leading to the N-methylhydrazone.

If phenylhydrazine is used in place of methylhydrazine, the N-phenyl-ketazinium salt is not obtained (104,105). If the same kind of reaction is performed with hydrazine, a pyrazolinium derivative (**95**) is isolated (106):

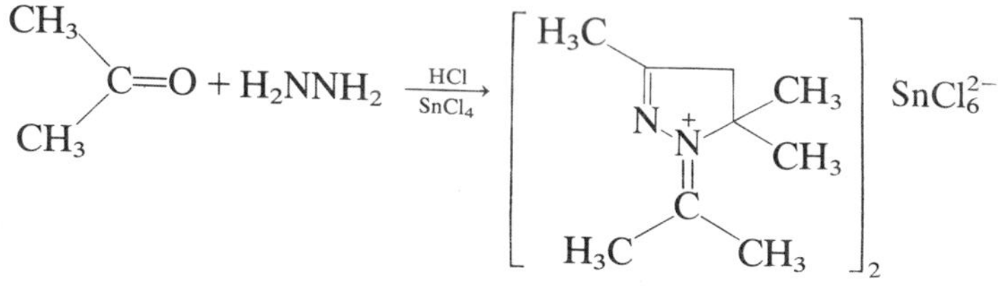

$$CH_3\text{-}C(CH_3)=O + H_2NNH_2 \xrightarrow[SnCl_4]{HCl} \left[\text{pyrazolinium} \right]_2 SnCl_6^{2-}$$

(**95**)

The same type of semicyclic azinium salt (**98**) can be obtained from the action of a carbonyl compound, aldehyde or ketone, on a pyrazolinium hydrochloride (**96**) or on a dipyrazolinium hexachlorostannate (**97**):

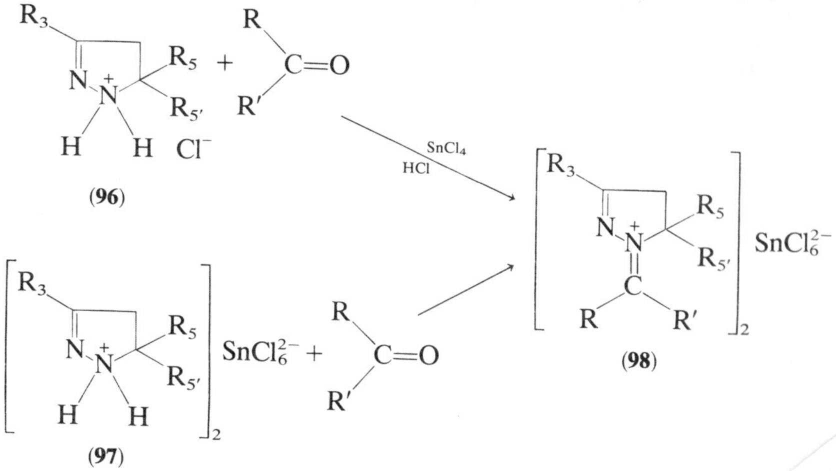

(**96**) + R-CO-R' $\xrightarrow[HCl]{SnCl_4}$ (**98**)

(**97**) + R-CO-R' → (**98**)

Most of these reactions have been performed with 3,5,5-trimethyl- (**96**, $R_3 = R_5 = R'_5 = CH_3$) and 5-methyl- (**96**, $R_3 = R_5 = H$, $R'_5 = CH_3$) 2-pyrazolines, and spectroscopic evidence demonstrates the structure of the pyrazolinium salts (**98**) (107,108).

No reaction has been observed in attempts to condense 1-substituted 2-pyrazolines with carbonyl compounds (104). In a similar way 3-imino-pyrazolidine sulfates (**99**) condense with carbonyl compounds, as reported by Dorn and co-workers (109,110), the salt (**100**) being stabilized by the 3-amino group:

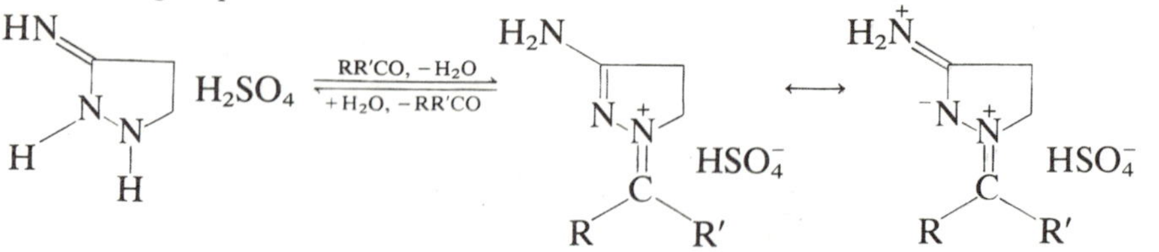

(**99**) (**100**)

Cyclic azinium salts may be obtained as isopyrazolium salts (**102**) through direct condensation of the β-diketone (**101**) with a monosubstituted hydrazine in acidic medium (111):

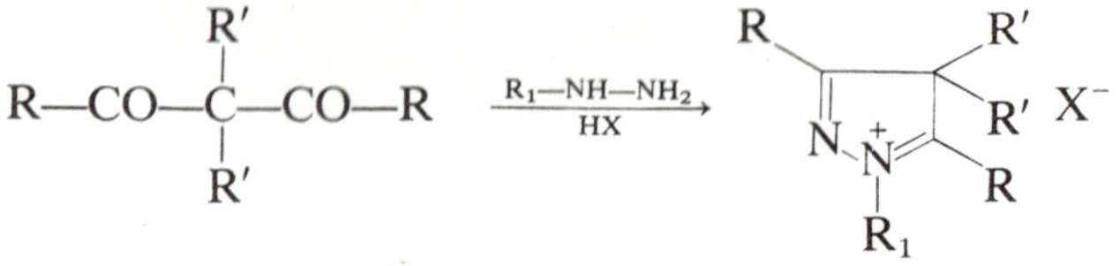

(**101**) (**102**)

The 2-pyrazoline (**104**) has been isolated (112) from the reaction of methylhydrazine with the β-diketone (**103**); the action of picric acid on this enamine-like compound (**104**) leads to the isopyrazolium salt (**105**), identified by NMR and IR spectroscopy:

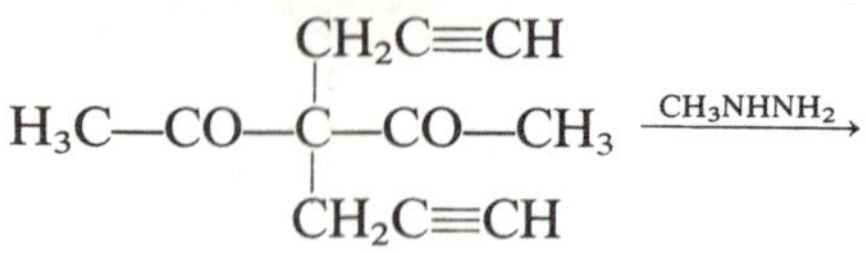

(**103**)

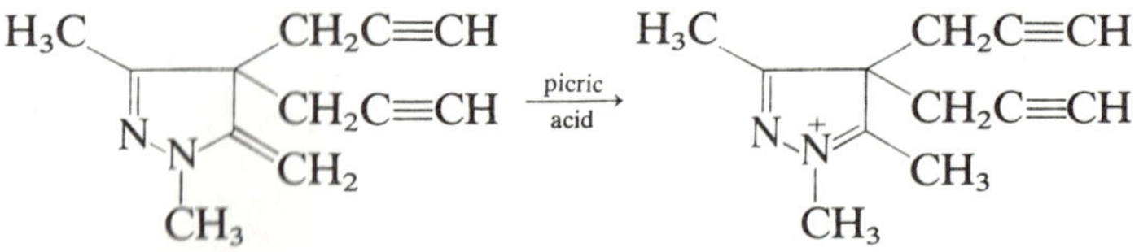

(**104**) (**105**)

2. Alkylation and Acylation of Azines

Direct methylation of azines has been shown to afford an azinium salt in only one case: the action of dimethyl sulfate on benzaldazine, first described in 1910 (113) and since then used as a pathway to N-methylhydrazine (114,115) by hydrolysis of the quaternary salt (**106**):

$$C_6H_5-CH=N-N=CH-C_6H_5 + (CH_3)_2SO_4 \longrightarrow$$

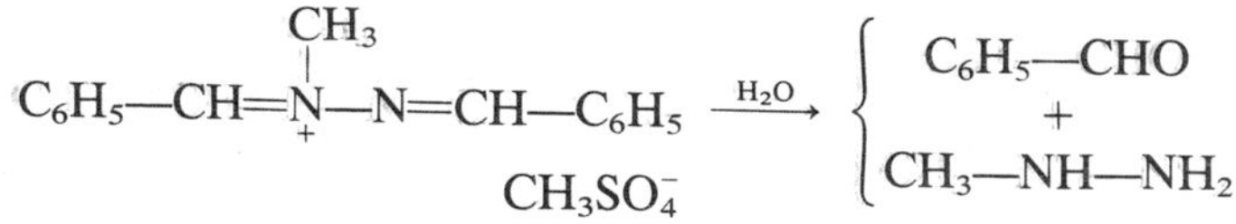

$$\underset{(106)}{C_6H_5-CH=\overset{+}{\underset{|}{N}}(CH_3)-N=CH-C_6H_5 \cdot CH_3SO_4^-} \xrightarrow{H_2O} \begin{cases} C_6H_5-CHO \\ + \\ CH_3-NH-NH_2 \end{cases}$$

Recently (115a) a convenient synthesis of **106** has been described, using triethyloxonium tetrafluoroborate ($\overset{+}{N}-C_2H_5$) and methyl fluorosulfonate.

The action of alkyl halides on azines (**107**) with a proton in α position to the C–N double bond does not stop at the alkylated azines but leads after cyclization to the N—H- and N-alkylated pyrazolines (**108** and **109**) (116,117). The authors assume an intermediate analogous to **106**. Recently (**117a**) the intermediate was isolated; its cyclization leads to a pyrrole derivative (Piloty's reaction, Section II-C-2).

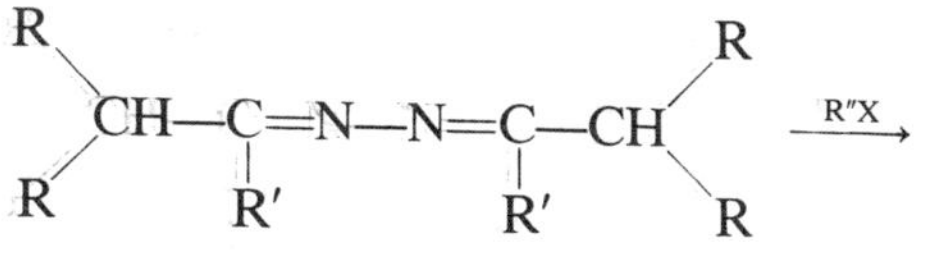

$$\underset{(107)}{\underset{R}{\overset{R}{>}}CH-\underset{R'}{\overset{|}{C}}=N-N=\underset{R'}{\overset{|}{C}}-CH\underset{R}{\overset{R}{<}}} \xrightarrow{R''X}$$

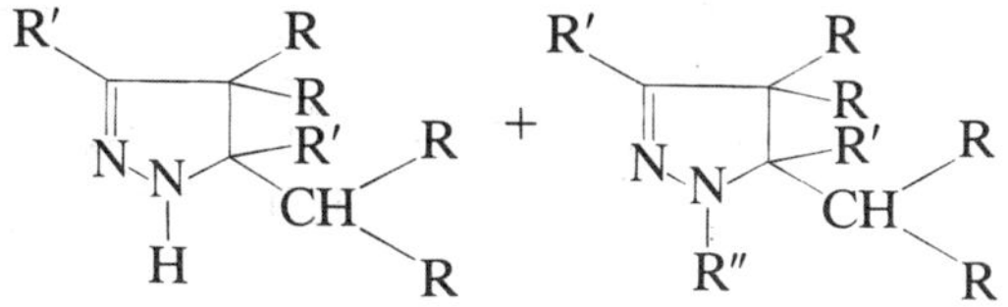

(108)   (109)

An hydrazonium–enehydrazine intermediate of type **80** may be assumed to occur in this reaction (see Section II-C-2). In a similar way benzoyl chloride leads to cyclization to give a benzoylpyrazoline (118,119).*

In the case of ethyl chloroformate, Böhme and Ebel (119) were able to isolate the azinium salt (**110**), characterized only by its analysis and its

---

* This reaction must be carried out in anhydrous conditions. Otherwise the cyclization does not occur; instead there is formation of the dibenzoylhydrazine.

hydrolysis into hydrazine. It appears that this compound would rather exist as a covalent structure (**110a**) (1, Section II-A-3):

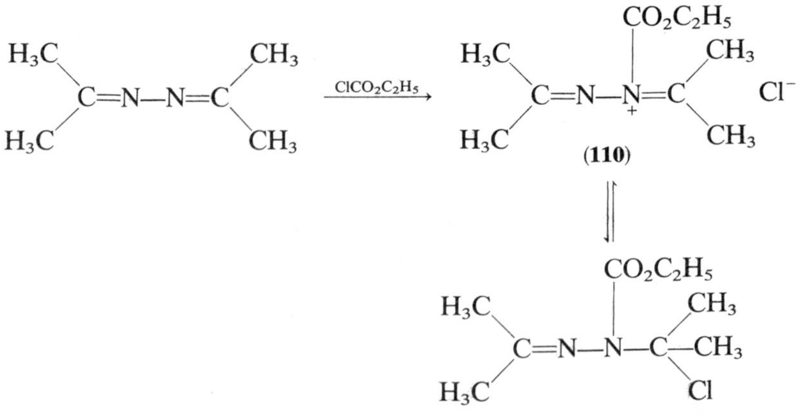

$$\underset{H_3C}{\overset{H_3C}{>}}C=N-N=C\underset{CH_3}{\overset{CH_3}{<}} \xrightarrow{ClCO_2C_2H_5} \underset{H_3C}{\overset{H_3C}{>}}C=N-\overset{+}{\underset{(110)}{N}}=C\underset{CH_3}{\overset{\overset{CO_2C_2H_5}{|}}{CH_3}} \quad Cl^-$$

$$\Updownarrow$$

$$\underset{H_3C}{\overset{H_3C}{>}}C=N-N-\underset{Cl}{\overset{\overset{CO_2C_2H_5}{|}}{C}}-CH_3$$
$$(\mathbf{110a})$$

It has been shown that salicylaldazine can form chelated metal salts (**111**) (120) and that dimethylketazine can form an adduct, $(CH_3)_2C=N-N=C(CH_3)_2, 2Al(CH_3)_3$, with trimethylaluminum (121):

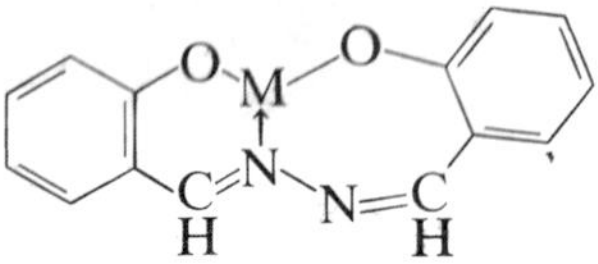

(**111**)

Cyclic azines such as the isopyrazoles (**112**) undergo a quaternization to the azinium salts (**102**) (122):

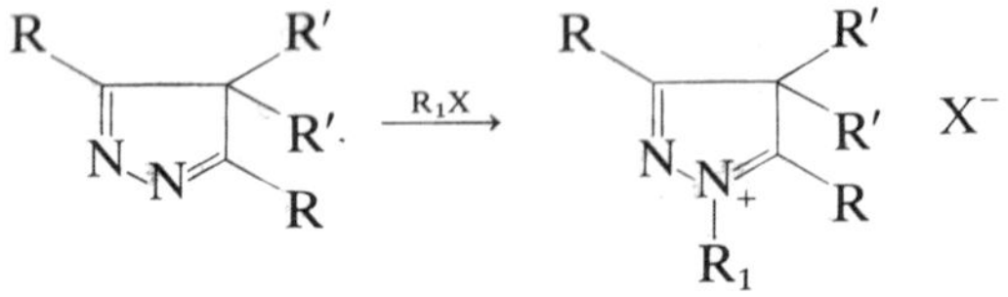

(**112**) $\xrightarrow{R_1X}$ (**102**)

3. *Azine Protonation*

Aromatic aldazines or ketazines (**113**) are very easily protonated with halogen acids or sulfuric acid and the salts can be isolated (**114**, R = H or Ar, X = Br, Cl, I, SO_4H); more complex salts have also been easily obtained (123). For $X^- = Br_3^-$ or I_3^- see ref. 123a.

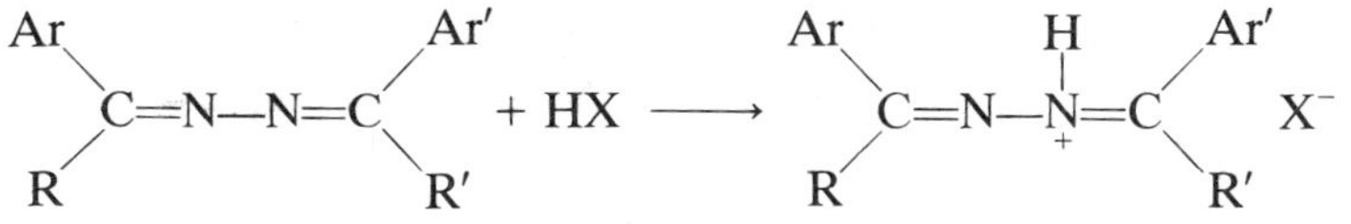

III. AZINIUM SALTS 561

$$\underset{R}{\overset{Ar}{>}}C=N-N=C\underset{R'}{\overset{Ar'}{<}} + HX \longrightarrow \underset{R}{\overset{Ar}{>}}C=N-\overset{H}{\underset{+}{N}}=C\underset{R'}{\overset{Ar'}{<}} \ X^-$$

(113)   (114)

A comparative study of the colors and UV spectra of numerous azines (**113**, $R = R' = H$) and of their hydrochlorides (**114**, $R = R' = H$) has shown (124) the loss of conjugation in the azinium salts.

With regard to the aliphatic ketazines and aldazines, salts are more difficult to prepare (129) and only complex salts of dimethylketazine have been obtained (**115**, $MX_6 = PtBr_6, PtCl_6$):

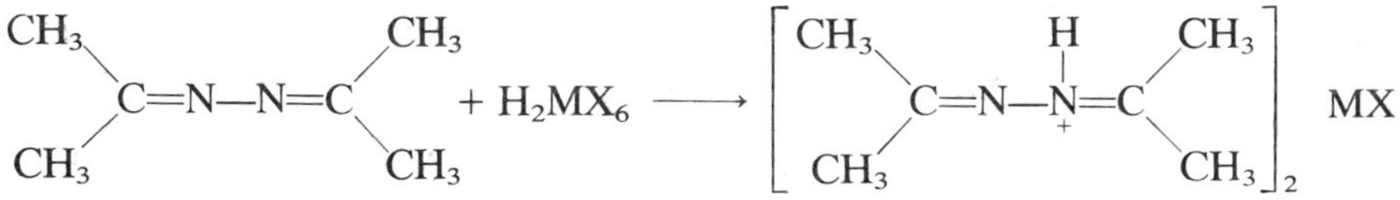

$$\underset{CH_3}{\overset{CH_3}{>}}C=N-N=C\underset{CH_3}{\overset{CH_3}{<}} + H_2MX_6 \longrightarrow \left[\underset{CH_3}{\overset{CH_3}{>}}C=N-\overset{H}{\underset{+}{N}}=C\underset{CH_3}{\overset{CH_3}{<}} \right]_2 MX$$

(115)

Some simple and complex salts of mixed azines have also been isolated (123).

In all cases concerning aliphatic azines (symmetrical or mixed), results must be taken with caution because the possibility of cyclization to pyrazolines cannot be ruled out (100).

B. REACTIVITY

Although several azinium salts have been isolated, only a few studies have been undertaken on their reactivity, so that no general conclusions can be drawn.

1. Hydrolysis

Acyclic azinium salts (**116**) are very easily hydrolyzed in acidic medium (34,105,125) and give rise to the carbonyl compound and the hydrazinium salt, certainly starting with an attack on the sp^2 carbon, $>C=\overset{+}{N}<$. This type of reaction can be related to the behavior of hydrazonium salts toward $OH^{\ominus}$ (see Section II-B-1).

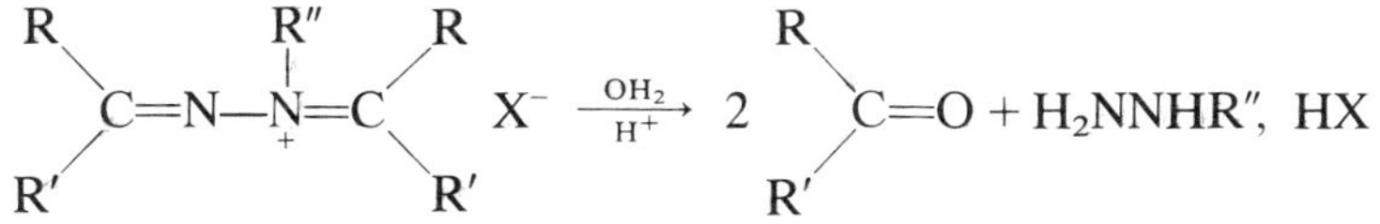

$$\underset{R'}{\overset{R}{>}}C=N-\overset{R''}{\underset{+}{N}}=C\underset{R'}{\overset{R}{<}} \ X^- \ \xrightarrow[H^+]{OH_2} \ 2 \ \underset{R'}{\overset{R}{>}}C=O + H_2NNHR'', \ HX$$

(116)

In the case of methylbenzaldazinium methyl sulfate (**106**), water is sufficient to induce hydrolysis (113):

$$C_6H_5CH=N-\overset{\underset{\mid}{CH_3}}{\overset{+}{N}}=CHC_6H_5 + 3\ OH_2 \longrightarrow$$

(**106**) $CH_3SO_4^-\ 2\ C_6H_5CHO + CH_3NHNH_2, H_2SO_4 + CH_3OH$

Cyclic azinium salts such as **95** are also hydrolyzed with acids (106) and even with D_2O in DMSO-d_6 (108).

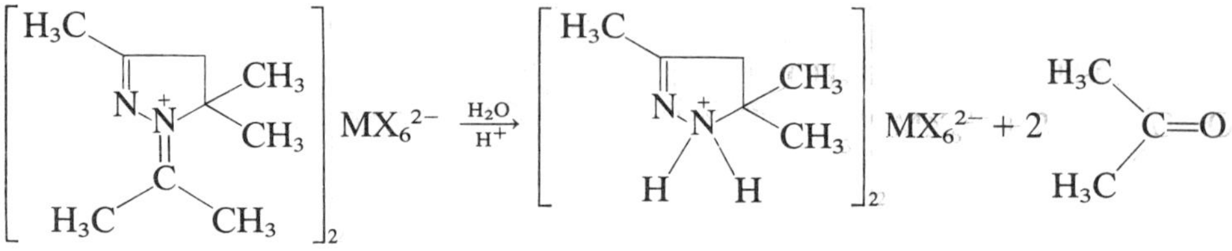

[structure **(95)**] MX_6^{2-} $\xrightarrow[H^+]{H_2O}$ [structure **(117)**] MX_6^{2-} + 2 $\begin{array}{c}H_3C\\H_3C\end{array}C=O$

The easy hydrolysis can be one of the reasons why many kinds of azinium salts cannot be isolated. On the other hand, hydrolysis can serve as an analytical method for the salt, or as proof of the formation of an azinium salt in the medium.

2. Cyclization

Cyclization of the *N*-carbethoxyazinium salt (1, Section IV-H) (**110**) into pyrazoline (**118**) with triethylamine has been described (119), the intermediary salt certainly having a hydrazonium–enehydrazine structure (**80**), as discussed in Section II-C-2:

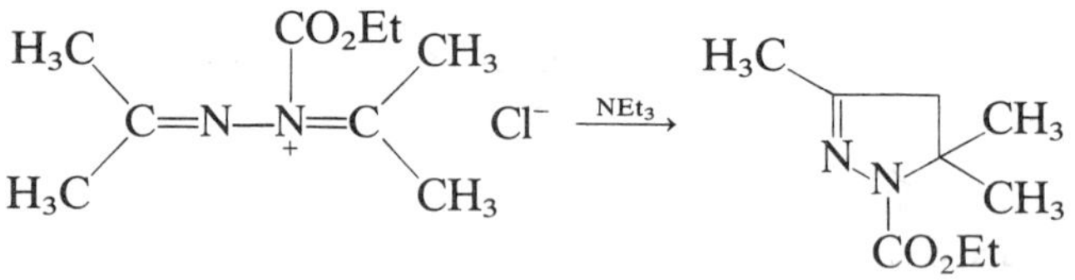

[structure **(110)**] $\xrightarrow{NEt_3}$ [structure **(118)**]

3. Reduction

Lithium aluminum hydride reduces iminium salts (1, Section IV-C) to tertiary amines; in a similar process it also reduces the isopyrazolium salts

(**119**, R = C$_2$H$_5$, C$_6$H$_5$, CH$_2$C$_6$H$_5$) to 2-pyrazolines (**111**):

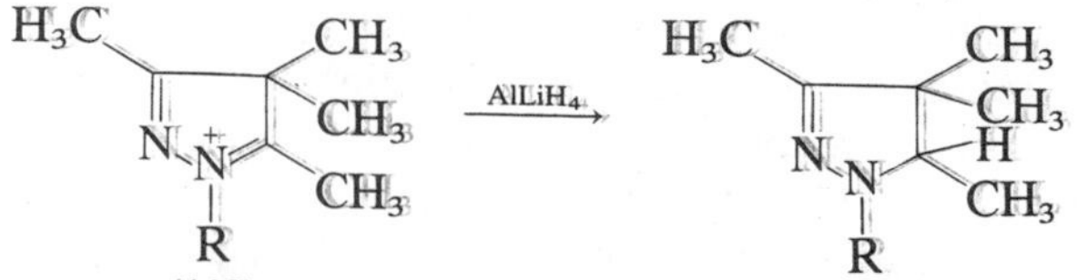

$$\underset{(119)}{\text{pyrazolinium}} \xrightarrow{\text{AlLiH}_4} \text{pyrazoline}$$

The iminium bond of the pyrazolinium salts (**100**) can be hydrogenated to give 2-iminopyrazolidines (**120**), as reported by Dorn and co-workers (109,110):

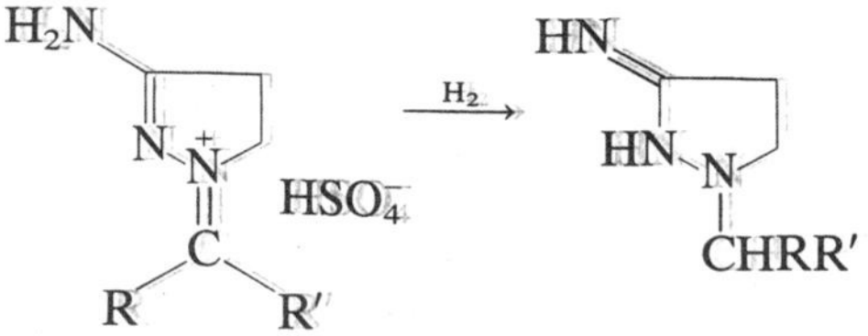

$$(\mathbf{100}) \xrightarrow{H_2} (\mathbf{120})$$

4. Spectroscopic Properties

Most of the azinium salts were isolated at a time when spectroscopic methods did not exist or were scarcely used. However, some UV information can be found about the hydrochlorides of benzaldazine and of its derivatives (124). Semicyclic azinium salts (**98**) have received more attention by IR and NMR spectroscopy (107,108); UV and NMR data have been given for the isopyrazolium (**102**) (111,112).

C. AZINIUM SALTS AS INTERMEDIATES

1. Intermediates

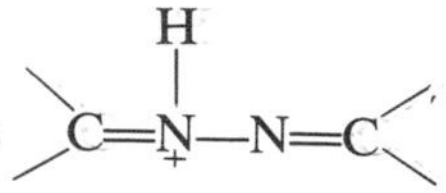

$$\overset{H}{\underset{+}{C=N-N=C}}$$

Azines (**121**) are readily hydrolyzed in acidic medium (70d) to the carbonyl compound and the hydrazone, which in turn are hydrolyzed to the carbonyl compound and hydrazine (see Section II-C-1). The first step is protonation of the azine, which then undergoes a nucleophilic attack of OH$^\ominus$ (Section III-B-1):

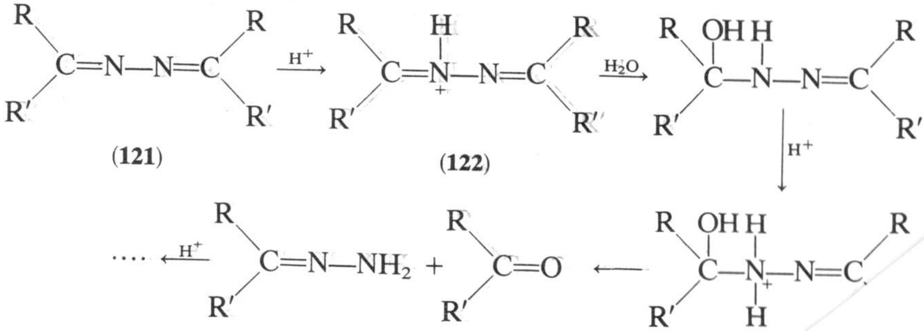

$$(\mathbf{121}) \xrightarrow{H^+} (\mathbf{122}) \xrightarrow{H_2O} \cdots$$

$$\cdots \xleftarrow{H^+} C=N-NH_2 + C=O \leftarrow \cdots$$

In early work Russian authors (118) proposed a nonconcerted mechanism for the cyclization of azines, the first step being protonation of the azine. More recent results described in Section II-C-2 explain this reaction more clearly. Zirngibl and Tam (126) consider a hydrazonium–enehydrazine structure (**123**) as an intermediate in the cyclization of the butyraldazine, as do Elguero et al. (100), but they assume that protonation of the azine occurs before enamine formation:

$$PrCH=N-N=CH\ Pr \xrightarrow{H^+}$$

$$\begin{array}{c} Pr \\ \diagdown \\ C \\ \parallel \\ N \\ | \\ \overset{+}{N} \\ | \\ H \end{array} \begin{array}{c} H \\ | \\ CH_2Et \\ | \\ C \\ | \\ H \end{array} \longrightarrow \begin{array}{c} Pr \\ \diagdown \\ C \\ \parallel \\ \overset{+}{N} \\ | \\ H \end{array} \begin{array}{c} H\ H \\ \diagdown / \\ C \\ | \\ C \\ \diagup \\ N \\ | \\ H \end{array} \begin{array}{c} Et \\ \diagup \\ \\ \\ \\ H \end{array}$$

(**123**)

Comrie (127) proposed that the formation of benzonitrile, benzaldehyde, and desylamine in the reaction of benzoin (**124**) with hydrazine hydrochloride proceeds via the benzoin azine, which is protonated (**125**) and then undergoes an abnormal Beckmann-type rearrangement:

$$C_6H_5-\underset{\underset{OH}{|}}{CH}-\underset{\underset{O}{\parallel}}{C}-C_6H_5 \xrightarrow{N_2H_4,\ HCl}$$

(**124**)

$$\left[C_6H_5-\underset{\underset{OH}{|}}{CH}-\underset{\underset{}{\overset{C_6H_5}{|}}}{C}=N-\overset{H}{\underset{\underset{C_6H_5}{|}}{\overset{+}{N}}}=C-\underset{\underset{Cl^-}{}}{\overset{OH}{\overset{|}{CH}}}-C_6H_5 \right] \longrightarrow \begin{cases} C_6H_5-CHO \\ C_6H_5-CN \\ C_6H_5-\underset{\underset{NH_2}{|}}{CH}-CO-C_6H_5 \end{cases}$$

(**125**)

2. *Intermediates* $\quad \diagdown C=\overset{\overset{R}{|}}{\underset{+}{N}}-N=C \diagup$

As with the cyclization of azines by acids, Russian authors (116,117) considered the alkylation of azines to be the first step in their cyclization by alkyl halides (see Section III-A-2).

The oxidation of benzaldehyde N,N-disubstituted hydrazones (**126**) with lead tetraacetate yields monosubstituted hydrazones (128) whose formation can be explained by the occurrence of an intermediary azinium

III. AZINIUM SALTS

salt (127):

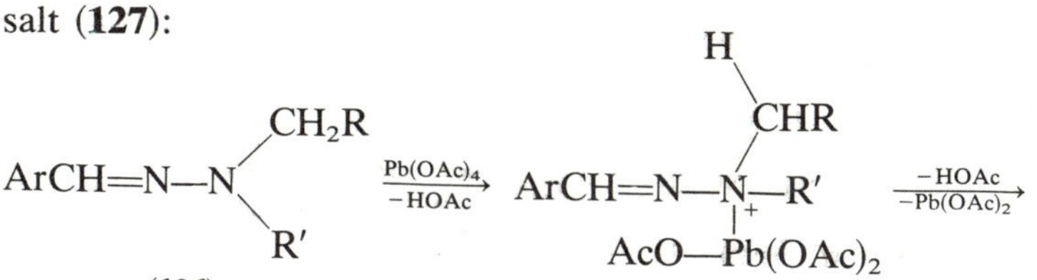

$$ArCH=N-N(CH_2R)(R') \xrightarrow[-HOAc]{Pb(OAc)_4} ArCH=N-\overset{+}{N}(CHRH)(R')(OPb(OAc)_2·AcO) \xrightarrow[-Pb(OAc)_2]{-HOAc}$$

(126)

$$ArCH=N-\overset{+}{N}(R')=CHR \longrightarrow \begin{cases} ArCH=N-NHR' \\ RCHO \end{cases}$$

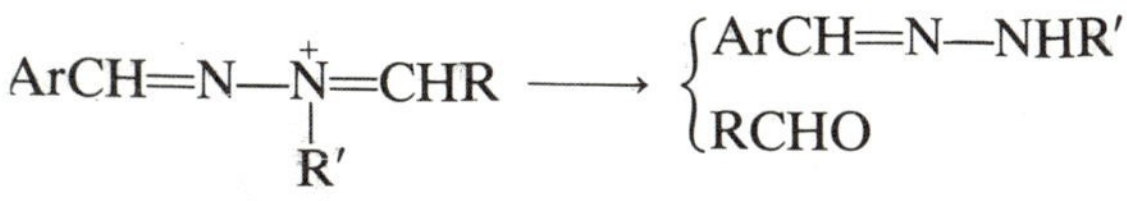

(127)

Cyclic azinium salts have been considered as intermediates of some reactions: for instance, an isopyrazolium (128) must be involved in the halogenation of pyrazoles in nitric acid (129):

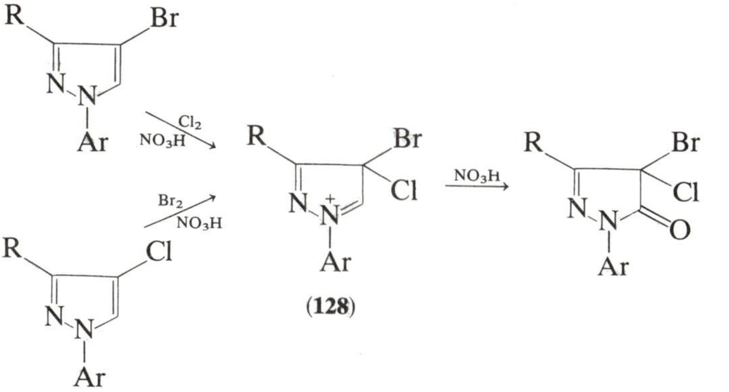

[Scheme showing pyrazole halogenation via isopyrazolium salt (128)]

Reduction of 2-methyl-4,4a,5,6,7,8-hexahydro-3-cinnolone (129) by lithium aluminum hydride to 131 and 132 has been described as proceeding via the azinium salt (130) (130):

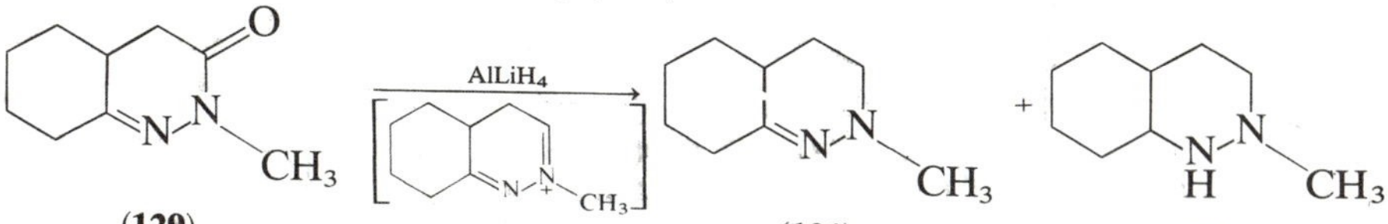

[Scheme showing reduction of (129) via azinium salt (130) to (131) and (132)]

An azinium salt (133) has also been considered to be involved in the reduction of 1,4,5,6-tetrahydropyridazines with lithium aluminum hydride:

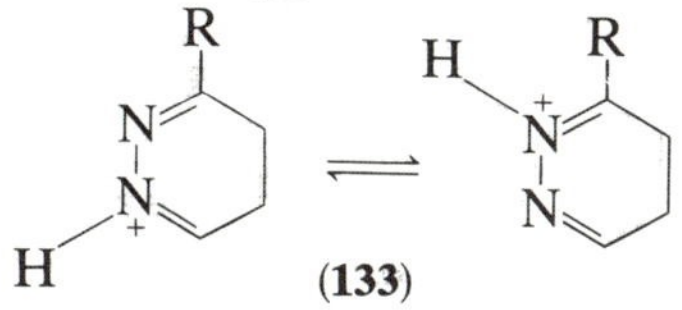

[Equilibrium structures for (133)]

IV. Azomethine Imines

A. INTRODUCTION

We pointed out in the introduction to this chapter that our purpose here is not a complete study of this type of compounds, which are better classified among 1,3-dipoles (132,133,133a). Rather we shall give a brief resumé of their properties, especially those connecting them with hydrazonium salts (**XV**) and diaziridines (**VIII**).

The opening of diaziridines (**VIII**) into hydrazonium salts (**XV**) was discussed in Section II-A-4. As far as the valence isomerism **VIII ⇌ IX** is concerned, the position of the equilibrium depends on the nature of the substituents; in some cases the stable form is the diaziridine (60,70b,134), while in others it is the dipolar ion (135–138). The two isomers differ in their reactions with certain reagents such as phenyl isocyanate, which behaves as an electrophile toward the diaziridines (**134**) and as a dipolarophile toward the azomethine imines (**136**):

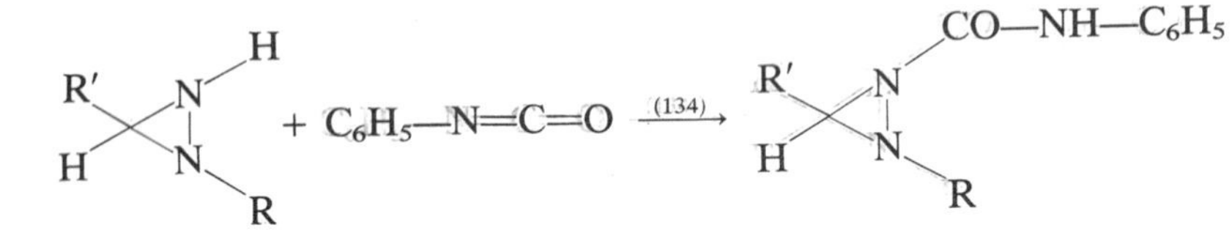

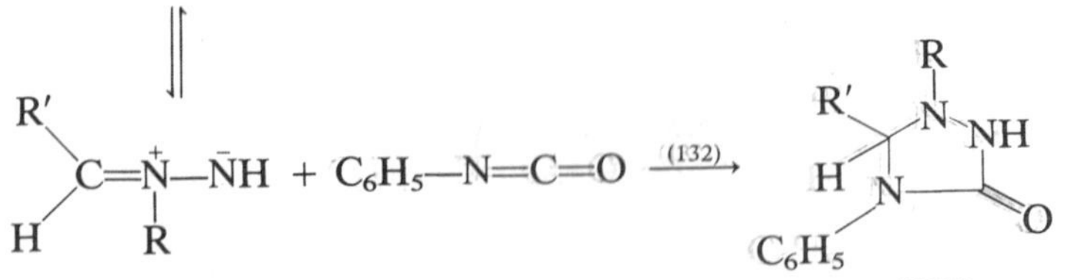

To explain the reactivity of some stable azomethine imines (**138**) it is sometimes necessary to take into account the diaziridine form (**139**) (139):

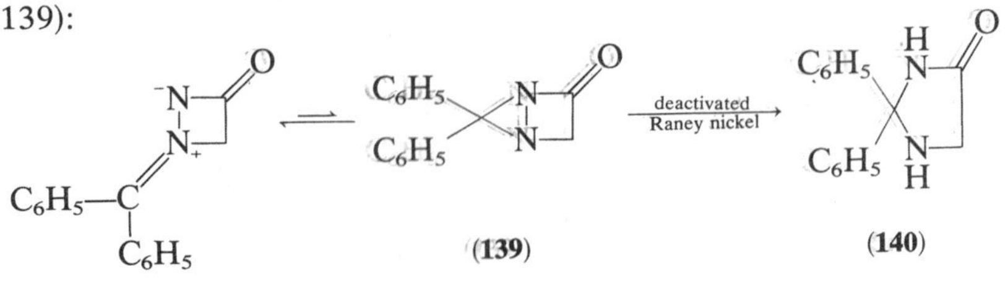

On the contrary, one can consider that the reaction between a 1,3,3-trialkyldiaziridine and phenyl isocyanate occurs via the 1,3-dipolar ion (**142**) and not directly from the diaziridine (**141**) as written by Schmitz (134):

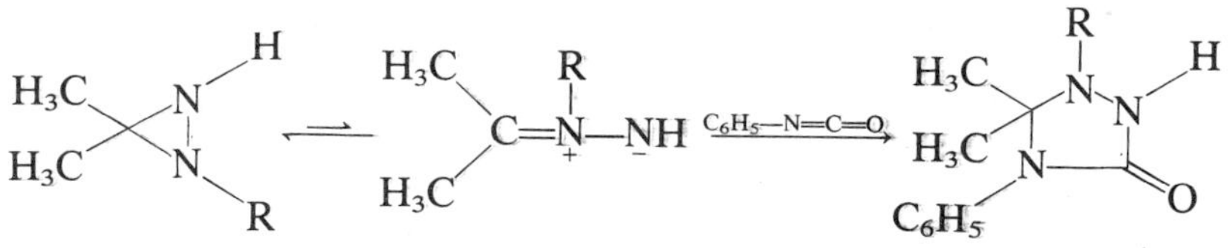

(**141**) (**142**) (**143**)

As a matter of fact the relationships connecting the two valence isomers **VIII** and **IX** have not been entirely resolved; thus Schmitz (60) demonstrated that the product of the action of chloramine with dihydroisoquinoline had a diaziridine (**144**) and not an azomethine-imine (**146**) structure. Protonation affords the hydrazonium salt (**145**) (Section II-A-4) (60), which regenerates with bases, not the diaziridine (**144**), but the dipole (**146**), as shown by Huisgen (132,133); this dipole undergoes 1,3-dipolar cycloadditions or dimerizes but does not cyclize to a diaziridine.

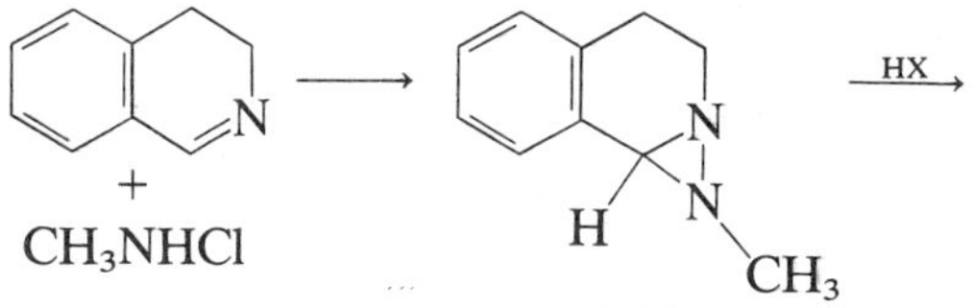

(**144**)

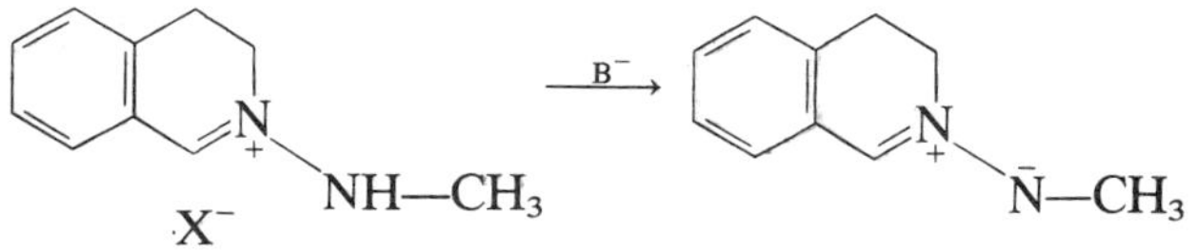

(**145**) (**146**)

B. SYNTHESIS AND REACTIVITY

Scheme 5 summarizes the chemistry of azomethine imines (**IX**), excluding the problem of diaziridines.

These compounds can be obtained from a condensation of carbonyl compounds (usually aldehydes) with 1,2-disubstituted hydrazines, via the carbinolhydrazines (**147**) (133,140), or on heating the hexahydrotetrazines (**148**) (133,135,141), on heating the addition product (**149**) (132), or by a retrocycloaddition reaction (135). We shall stress their accessibility by the action of bases on hydrazonium salts (**150**) (Section IV-1: **145** → **146**).

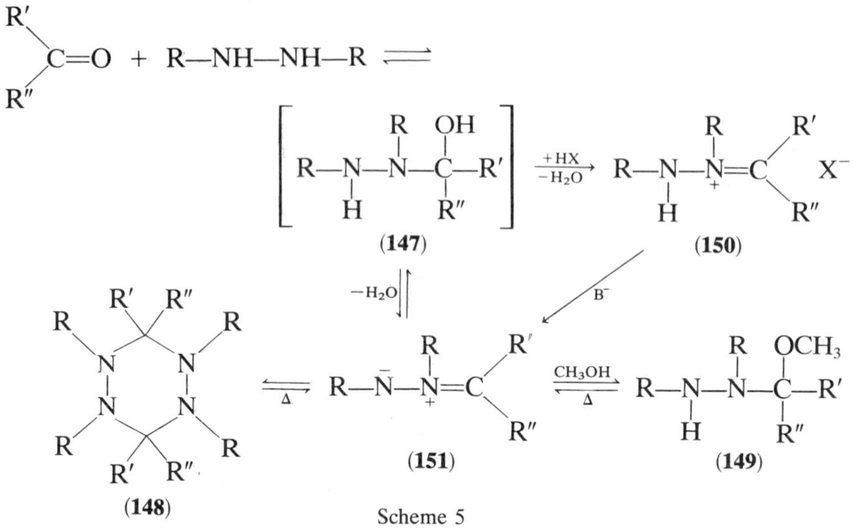

Scheme 5

$$R'R''C=O + R-NH-NH-R \rightleftharpoons$$

$$\left[\begin{array}{c} R\ OH \\ | \ \ | \\ R-N-N-C-R' \\ | \ \ \ \ \ | \\ H \ \ \ R'' \end{array} \right] \xrightarrow[-H_2O]{+HX} \begin{array}{c} R \ \ \ \ R' \\ | \ \ \ \ \ \diagup \\ R-N-N=C \\ | \ \ + \ \ \diagdown \\ H \ \ \ \ \ R'' \end{array} X^-$$

(147)   (150)

(148) hexahydrotetrazine structure

$$\rightleftharpoons_\Delta R-\bar{N}-\overset{+}{N}=C\diagup^{R'}_{R''} \xrightleftharpoons[\Delta]{CH_3OH} \begin{array}{c} R \ \ \ OCH_3 \\ | \ \ \ \ | \\ R-N-N-C-R' \\ | \ \ \ \ \ \ \ | \\ H \ \ \ \ \ R'' \end{array}$$

(151)   (149)

The reactivity of azomethine imines (**151**) is due essentially to their 1,3-dipolar character (132,133), but the literature reports the ease with which they dimerize into hexahydrotetrazines (**148**) (132,140,141,141a) and the possibility that they have to add methanol to yield product **149** (132) [it is difficult to know whether the attack of methanol on the sp^2 carbon of the iminium occurs on the dipolar ion (**151**) or on the protonated form (**150**)]. The reaction between two 1,3-dipoles [e.g., **151** + **151** → **148**, **170** + **170** → **208** (Section VII-C), **152** + **170** → 1,2,4,5-oxatriazine (141b)], if concerted and suprafacial ($\pi_4^s + \pi_4^s$), is forbidden in the ground state. This fact and the stereochemical consequences derived have not received much attention, except in ref. (141c).

We pointed out in Section II-C-1 that the tetrazine (**148**) can be formed from the hydrazonium salt (**150**). Grashey et al. (142) report that some compounds arising formally from a 1,3-dipolar cycloaddition can as well be formed from the hydrazonium salt. As the intermediates are not isolated, it is difficult to know whether the carbinolhydrazine (**147**) leads to the final products via the hydrazonium salt or via the azomethine imine.

C. STABILIZED AZOMETHINE IMINES

Two stabilized forms of azomethine-imines are known (**152** and **153**), but only the first type has been systematically investigated. Compound **154**, the first stable azomethine imine isolated by Huisgen et al. (136,137) must be also classified in this series of compounds.

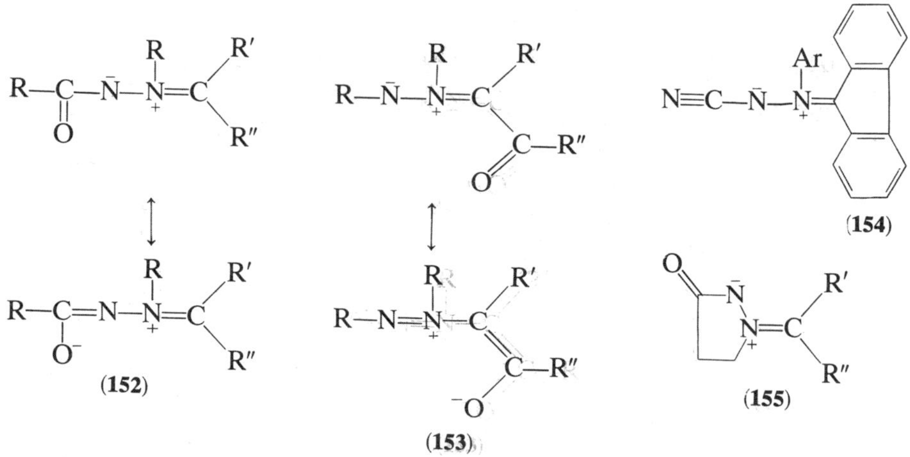

IV. AZOMETHINE IMINES

Scheme 6 summarizes the properties of amide imides (**152**) (143); in the heterocyclic series (**155**) Dorn's approach is mainly of interest, and in

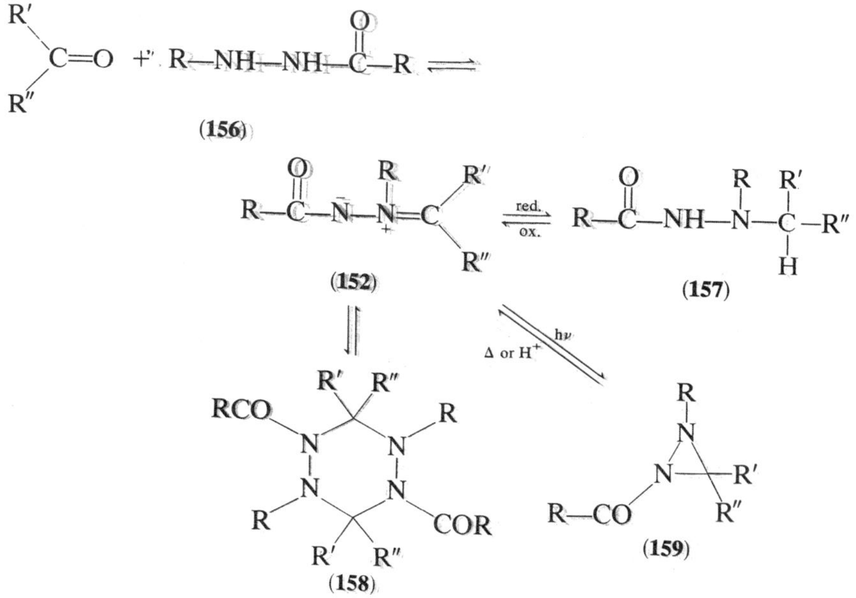

Scheme 6

the acyclic series, that of Oppolzer. These compounds can be prepared by condensation of carbonyl compounds with hydrazides (**156**) (27,144–148) or by oxidation of hydrazides (**157**) (149,150); in the heterocyclic series intramolecular quaternization of the hydrazone (**160**) can be used (139).

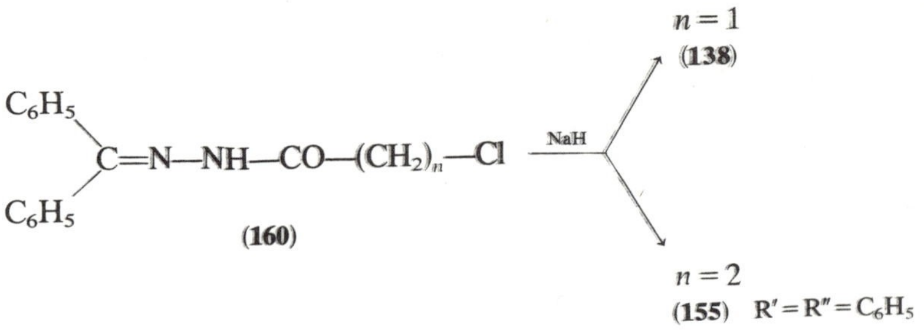

$$\underset{C_6H_5}{\overset{C_6H_5}{\diagdown}}C=N-NH-CO-(CH_2)_n-Cl \xrightarrow{NaH}$$

(160)

$n = 1$ (138)

$n = 2$ (155) R' = R'' = C₆H₅

Here again most of the studies (26,27,143,144,151) of these compounds deal with their 1,3-dipole character; however, it has been reported that they can be catalytically reduced by hydrogen (26,146,152,153) or by sodium borohydride in methanol (139). They can also be hydrolyzed to the acyl hydrazine (156) and the carbonyl compound (139,145). Finally, they can dimerize to N,N'-diacylated hexahydrotetrazine (158) (152). Structures of type 153, such as compound 161, have been used as 1,3-dipoles (151,154); protonation occurs on the oxygen atom to afford the salt (162), but as this cation is aromatic, one cannot deduce that O-protonation prevails generally over N-protonation.

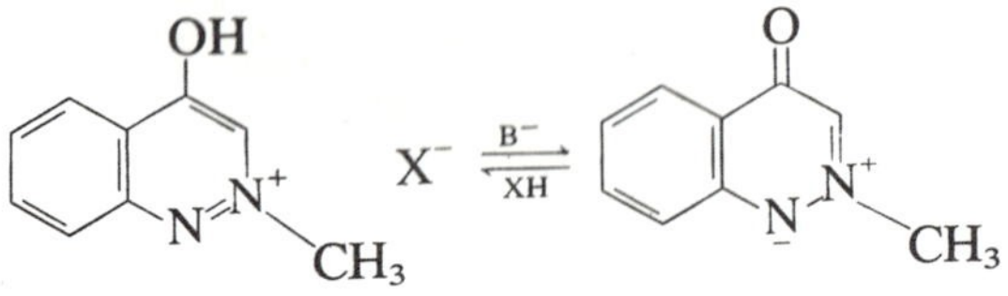

(162)   (161)

It has been shown (155) that heterocycles such as 155 cyclize to N-acyl diaziridines by irradiation (transformation 152 → 159) and that the 1,3-dipole is regenerated on heating or acidification.

V. Diazenes and Diazenium Ions

No one seems to have connected the diazenium ion (**XIII**) such as (**163**) and the hydrazonium ions (**XV**) such as (**164**), although a tautomerism can be considered between both structures:

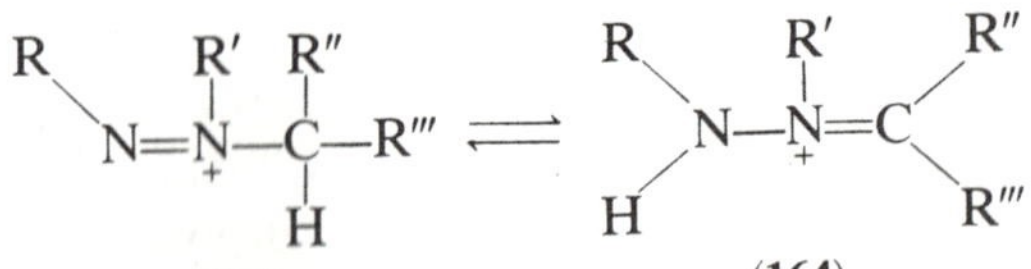

(163)   (164)

However, in the important reviews published by Lemal (156) and Hünig (157,157a) an equilibrium is reported between the diazene (**165**) and the azomethine imine (**167**) (Section IV) via the diazenium cation (**166**) (see also TRef. 141a):

$$\underset{(\mathbf{165})}{\underset{-}{N}=\underset{+}{N}-\underset{H}{\overset{R'}{\underset{|}{C}}}-R'''} \quad \overset{+H^+}{\underset{-H^+}{\rightleftharpoons}} \quad \underset{(\mathbf{166})}{\underset{H}{\overset{H}{\diagdown}}\underset{+}{N}=\underset{}{N}-\underset{H}{\overset{R'}{\underset{|}{C}}}-R'''} \quad \overset{-H^+}{\longrightarrow} \quad \underset{(\mathbf{167})}{\underset{-}{\overset{H}{\diagdown}}\underset{+}{N}-N=C\diagup^{R'}_{R'''}}$$

In fact, **164**, R = H, is the N-protonated form of the azomethine imine (**167**).

Like the azomethine imines (**167**) (Section IV-B) and the hydrazonium salts (**164**) (Section II-C-1), the diazenes (**165**) form cyclic dimers (**168**); prototropy (**165** → **167**) leads us back to a case discussed in Section IV-B. Particularly interesting is the diazene–hydrazone rearrangement (**165** → **169**). Lemal (156) carefully investigated this problem and examined various mechanisms [with occurrence of an azo (**VII**), an azomethine imine (**IX**), a diaziridine (**VIII**), and a tetrazine (**168**)] but did not reach any final conclusion.

$$(\mathbf{168}) \quad \longleftarrow (\mathbf{165}) \longrightarrow \quad R'-NH-N=C\diagup^{R''}_{R'''} \quad (\mathbf{169})$$

VI. Oximinium Salts

A. SYNTHESIS

The chemistry of oximinium salts (**XXVII**) has not been widely investigated, especially that of acyclic compounds. Usually direct syntheses such as condensation of an N-substituted hydroxylamine with a carbonyl compound afford, not oximinium salts, but nitrones (**170**) (158,159):

$$\underset{R''}{\overset{R'}{\diagdown}}C=O + R-NHOH \longrightarrow \underset{R''}{\overset{R'}{\diagdown}}C=\underset{+}{N}-R \quad \underset{O^-}{\overset{}{|}}$$
$$(\mathbf{170})$$

This difference in behavior between 1,2-disubstituted hydrazines [formation of hydrazonium salts (**XV**) (Section II-A-1)] and N-substituted hydroxylamines certainly arises from the lower basicity of the nitrones (**XXI**) as compared to that of the azomethine imines (**IX**).

Similarly, α,β-unsaturated carbonyl compounds do not condense with N-methylhydroxylamine to yield a 2-isoxazolinium salt but rather give either a nitrone (160) or a 5-hydroxyisoxazolidine (161).

The only exception concerns the synthesis of 2-methoxy-1,1,2,3,3-pentamethylguanidine perchlorate (**172**) from tetramethylchloroformamidine chloride (**171**) and O,N-dimethylhydroxylamine (162). In this case, however, the charge is delocalized because of the two dimethylamino groups (this is proved by the nonrestricted rotation about the C–N double bond).

$$(CH_3)_2\overset{+}{N}=CH-N(CH_3)_2 + CH_3NHOCH_3 \longrightarrow$$
$$\underset{Cl}{} \quad Cl^-$$

$$(CH_3)_2N\diagdown \atop (CH_3)_2N\diagup C=\overset{+}{N}\diagup CH_3 \atop \diagdown OCH_3 \quad ClO_4^-$$

(**171**)   (**172**)

1. Alkylation of Oximes, Nitroso Derivatives, and Nitrones

(*a*) *Acyclic Alkoximinium Salts.* Alkylation of acyclic oximes leads to nitrones (**170**) and/or to O-alkyl oximes (**173**), depending on the oxime itself, on its E–Z isomerism (163), on the alkylating reagent, and on the experimental conditions (35b,164,165). O-Alkyl oximes do not undergo further alkylation (35b):

$$\underset{R''}{\overset{R'}{\diagdown}}C=N-OH \xrightarrow{RX} \underset{R''}{\overset{R'}{\diagdown}}C=\overset{O^-}{\underset{|}{\overset{+}{N}}}-R + \underset{R''}{\overset{R'}{\diagdown}}C=N-OR$$

(**170**)   (**173**)

However, it is possible to alkylate oxime acetates (**174**) on the sp^2 nitrogen and thus obtain the salt (**175**) (166):

$$\underset{R''}{\overset{R'}{\diagdown}}C=N-OCOCH_3 \xrightarrow{RX} \underset{R''}{\overset{R'}{\diagdown}}C=\overset{+}{\underset{R}{N}}-OCOCH_3 \quad X^-$$

(**174**)   (**175**)

With methyl iodide, methyl tosylate, and dimethyl sulfate the reaction is very slow even at high temperature, but it occurs faster with triethyloxonium or trimethyloxonium tetrafluoroborate (166). Other authors

(115a) reacting triethyloxonium tetrafluoroborate with benzaldehyde oxime obtained the O-alkylated salt.

Nitroso compounds (**176**) are alkylated to oximinium salts (**177**) with triethyloxonium fluoroborate (167); the intermediate that Baldwin et al. propose could have structure **XXV** (Scheme 3).

$$R-N=O + BF_4O(C_2H_5)_3 \longrightarrow \left[\begin{array}{c} R \\ \diagdown \overset{+}{N}=O \\ \diagup \\ C_2H_5 \end{array} \right] BF_4^- \longrightarrow \begin{array}{c} CH_3 \\ \diagdown \\ C=\overset{+}{N}-OH \\ \diagup \quad | \\ H \quad R \end{array} BF_4^-$$

(**176**)   (**177**)

The same reagent has been reported to alkylate nitrones (**178**) (35b). Such a powerful alkylating reagent, however, does not seem to be necessary to obtain oximinium salts (**179**) from nitrones; usual alkylating agents such as sulfates, iodides, and bromides react easily (168).

$$\begin{array}{c} R' \\ \diagdown \\ C=\overset{+}{N}-CH_3 \\ \diagup \quad | \\ R'' \quad O^- \end{array} \xrightarrow{RX} \begin{array}{c} R' \\ \diagdown \\ C=\overset{+}{N}-CH_3 \\ \diagup \quad | \\ R'' \quad OR \end{array} X^-$$

(**178**)   (**179**)

(*b*) *Cyclic Oximinium Salts: 2-Oxazolinium Salts.* Methylation of 2-oxazolines (**180**) with dimethyl sulfate leads to the expected 2-oxazolinium salts (**181**), isolated as chloroferrates (169,170) and perchlorates (160); the latter derivatives have been thoroughly studied by UV and NMR spectroscopy (160).

[Structure **180** with substituents R_3, R_4, R'_4, R_5, R'_5 on oxazoline ring]
$$\xrightarrow[\text{2. FeCl}_3, \text{HCl or ClO}_4\text{Na}]{\text{1. SO}_4(\text{CH}_3)_2}$$
[Structure **181** with H_3C on N] $FeCl_4^-$ or ClO_4^-

Similar methylation occurs in the case of the 3-amino-2-isoxazolines (**182**) (171), and the salt (**183**) is obtained as shown from IR and NMR evidence.

[Structure **182**: R—N(R)— on isoxazoline]
$$\xrightarrow[R=CH_3, SO_4(CH_3)_2]{R=H, ICH_3}$$
[Structure **183** with H_3C] X^-

From the results described here, it follows that oximinium salts (**XXVII**) can be obtained by alkylation of some O-alkyl oximes (**XVIII**, O—R), nitroso compounds (**XIX**), and nitrones (**XXI**), but not from oximes (**XVIII**, O—H) (see Scheme 3).

2. Protonation of Oximes, Nitroso Derivatives, Nitrones, and Oxaziridines

(a) *Oximes*. Oximes readily form stable salts (35b) but usually not O-alkyl oxime [see reference (171a) for stable benzophenone O-methyl oxime salts]. In both cases, however, the presence of an acid can induce isomerization (164,172); the mechanism of this acid-catalyzed isomerization could be discussed on the basis of the same arguments as are used for hydrazone isomerization (Section II-C-1), but here the problem is simplified because, as we shall see later, the site of protonation has been nearly established.

Thus Olah and Kiousky (173) have shown that acetone and acetophenone oximes undergo protonation on the nitrogen in neat fluorosulfonic acid or in the mixture FSO_3H–SbF_5–SO_2: the oximinium salt (**184**) was detected by NMR spectroscopy (presence of two low field peaks corresponding to the OH and NH protons and coupling of the methyl protons with NH).

$$\underset{R}{\overset{CH_3}{>}}C=N-OH \quad \xrightarrow[\text{or } FSO_3H-SbF_5-SO_2]{FSO_3H} \quad \underset{R}{\overset{CH_3}{>}}C=\overset{+}{\underset{|}{N}}-OH \quad (184)$$

This site of protonation was postulated earlier by Saitô and Nukada (174,175) to explain the IR spectra of oxime hydrochlorides, the cancellation of the lone-pair anisotropy (175) in these salts, and the change of conformation in 2-substituted cyclohexanone oximinium salts (176).*

The pK_a's of some dioximes have been determined spectrophotometrically (177a).

(185) [cyclohexane-1,3-dione dioxime, =NOH groups] $\xrightarrow{HCl}$ (186) [protonated form with =NHOH and =$\overset{+}{N}$HOH, Cl⁻]

---

* Cyclohexanone oximinium hydrochloride can undergo a second protonation on oxygen, leading to an unstable dihydrochloride (174).

In spite of all these data favoring this type of protonation some authors do not eliminate O-protonation. For instance, both models like **187** and **188** explain the effect in NMR spectroscopy of the addition of hydrochloric acid vapor in benzenic solutions of oximes: an upfield shift of the protons syn to the hydroxyl group and a downfield shift of the protons anti (178). However, these results must be considered with caution because of the possibility of an E–Z isomerization in acidic medium.

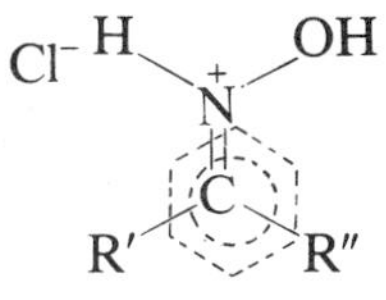

(187) 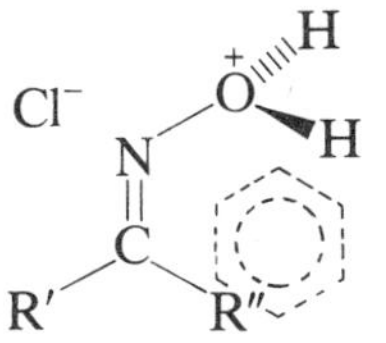   (188)

Both types of protonation can also explain the upfield shift observed for the ^{14}N resonance of acetoxime hydrochloride, compared to that of the oxime itself (179). The same problem arises if one wants to determine the site of coordination of oximes with paramagnetic shift reagents that dramatically separate syn and anti protons: Eu(dpm)$_3$ can coordinate with nitrogen (180) or with oxygen (181), but the authors themselves indicate that this represents only a predominant tendency.

(*b*) *Nitroso Derivatives.* It has been shown (182) by NMR spectroscopy that nitrosocyclohexane (**189**) in SO$_2$ gives rise to an N-protonated salt (**190**) if a mixture of FSO$_3$H and SbF$_5$ is employed:

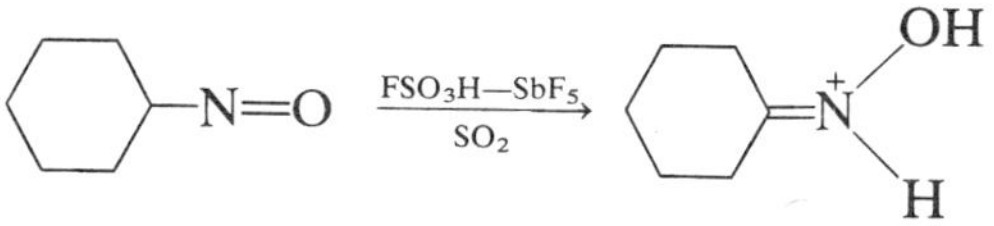

(189)   (190)

(*c*) *Nitrones.* A study of the pK_a's of nitrones (**191**) has been undertaken (183).

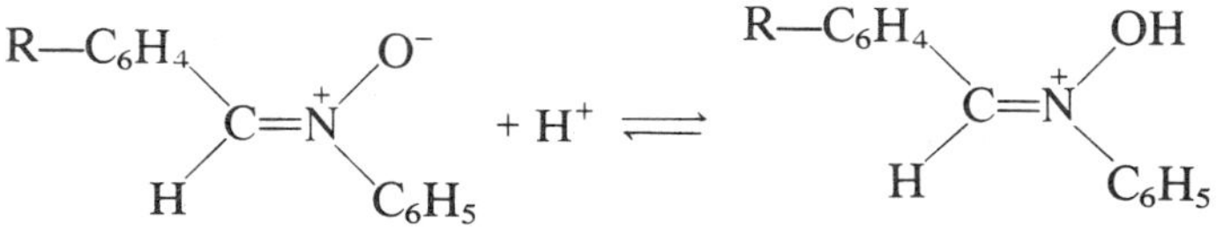

(191)   (192)

(d) *Oxaziridines.* The opening of an oxaziridine (**193**) with a strong acid such as methanesulfonic acid has been reported (184). The product is a hygroscopic salt that has not been characterized by Emmons. By analogy to the conversion of diaziridines to hydrazonium salts (Section II-A-4), an oximinium salt (**194**) might be formed by protonation on oxygen of the oxadiaziridine (salt **XXVIII** in Scheme 3); however, the N-protonated species is favored by theoretical calculations (185).

$$\underset{(193)}{\underset{C_6H_5}{\overset{H}{\diagup}}\overset{O}{\underset{\diagdown}{N}}\underset{C_4H_9\text{-}t}{\diagdown}} \xrightarrow{CH_3SO_3H} \underset{(194)}{C_6H_5-CH=\overset{+}{\underset{|}{N}}-C_4H_9\text{-}t} \quad CH_3SO_3^-$$

From all the data just reported, it can be seen that only cations of type **XXVII** (see Scheme 3) have actually been shown to exist from protonation of oximes (**XVIII**), nitroso derivatives (**XIX**), nitrones (**XXI**), or oxaziridines (**XX**); however, the possibility of the existence of cation **XXIII** is not completely ruled out by NMR studies.

These results, as well as those on alkylation (Section VI-A-1), clearly show the difference between hydrazonium and oximinium salts: in the first case salt **XII** is more stable than salt **XV**, whereas in the second case salt **XXIII** is less stable than salt **XXVII**. Such behavior results from the lower basicity and nucleophilicity of oxygen as compared to nitrogen with the same hybridization.

B. REACTIVITY

1. *Chemical Reactivity*

Only a few studies have been undertaken on the reactivity of oximinium salts, mainly in the acyclic series.

Hydrolysis in acidic medium of acyclic oximinium salts (**179**) has been reported (166,168) mainly because of interest in the synthesis of O,N-dialkylated hydroxylamines. This is the result of normal attack of nucleophiles on the sp^2 carbon of iminium salts, as already reported for hydrazonium salts (see Section II-B-1).

$$\underset{(179)}{\underset{R''}{\overset{R'}{\diagdown}}C=\overset{+}{\underset{|}{N}}-CH_3} \xrightarrow[H_2SO_4]{OH_2} \underset{R''}{\overset{R'}{\diagdown}}C=O + CH_3NHOR$$

In a similar way cyclic oximinium salts (**181**) undergo a nucleophilic attack (186) on the carbon in position 3:

$$RO^- + \underset{(181)}{\begin{array}{c}R_3\\\\H_3C\end{array}\!\!\!\overset{R_4}{\underset{\overset{+}{N}}{=}}\!\!\!\overset{H}{\underset{O}{\diagup}}\!\!\overset{R_5}{\underset{R'_5}{\diagdown}}} \rightleftharpoons \underset{(195)^*}{\begin{array}{c}R_3\\\\H_3C\end{array}\!\!\!\overset{R_4}{\underset{N}{-}}\!\!\!\overset{H}{\underset{O}{\diagup}}\!\!\overset{R_5}{\underset{R'_5}{\diagdown}}}$$

Derivatives (**195**) with $R = CH_3$ and $R_3 = Ar$ are stable. 2-Oxazolinium salts such as **197** exchange with D_2O the hydrogen atoms in β position to the $C\!-\!\overset{+}{N}$ double bond (160)† because of their acidity (for a similar reactivity of pyrazolinium salts see Section II-B-1):

(**197**) ClO_4^- $\xrightarrow{D_2O}$ (**198**)

Pyrolysis (173) of the acyclic oximinium salts (**184**) of acetone and acetophenone at about 100° affords N-alkylnitrilinium ions (**199**):

$$\underset{(184)}{\begin{array}{c}R\\\diagdown\\\\H_3C\diagup\end{array}\!\!\!C\!=\!\overset{H}{\underset{+}{N}}\!-\!OH} \xrightarrow{\Delta} \underset{(199)}{H_3C\!-\!C\!\equiv\!\overset{+}{N}\!-\!R}$$

2. Spectroscopic Properties

For some acyclic oximinium salts IR and NMR data have been reported (172–175), including ^{14}N resonance studies (179). The UV and NMR spectra of 2-oxazolinium salts have been described (160).

C. OXIMINIUM SALTS AS INTERMEDIATES

We reported in Section V-A-2 the E–Z isomerization undergone by oximes when the oximinium salt is formed; this isomerization can occur in the course of a reaction in acidic medium even if the salt is not isolated, as it was in the case of hydrazones (Section II-C-1).

Numerous investigations have been undertaken at the acid-catalyzed formation and hydrolysis of oximes (187–189). The protonated oxime is

---

* If $R = H$, this compound is in equilibrium with the open-chain form, (**196**) $R_3COC\text{-}HR_4CR_5R'_5ONHCH_3$.

† In some cases hydrolysis into β-oxyamino ketone can occur (160).

the reactive species, and the rate-determining step depends on the pH.

Oximinium salts are usually not isolated from the action of acids on oxaziridines (**200**) (for an exception see Section V-A-2), but they can be considered as intermediates in their hydrolysis (184,190):

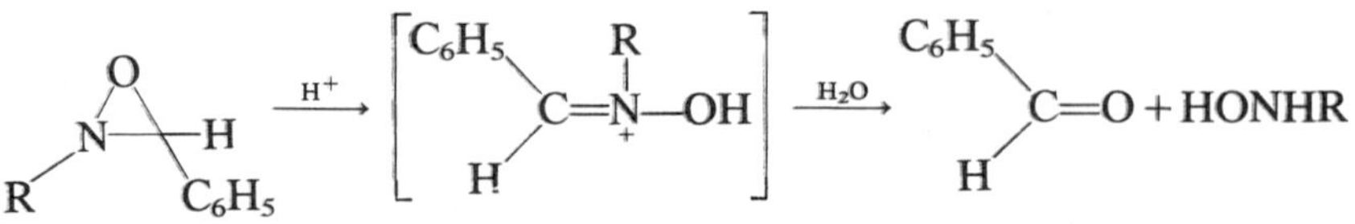

$$\underset{(\mathbf{200})}{\underset{R}{\overset{O}{\diagdown}}\underset{C_6H_5}{\overset{}{N}}\hspace{-0.5em}-\hspace{-0.5em}H} \xrightarrow{H^+} \left[\underset{(\mathbf{201})}{\underset{H}{\overset{C_6H_5}{\diagdown}}C\hspace{-0.3em}=\hspace{-0.3em}\underset{+}{N}\hspace{-0.3em}-\hspace{-0.3em}OH}\right] \xrightarrow{H_2O} \underset{H}{\overset{C_6H_5}{\diagdown}}C\hspace{-0.3em}=\hspace{-0.3em}O + HONHR$$

Acid-catalyzed cyclization of oximes leads to isoxazolines (**203**) via the oximinium salt (**202**) (191):

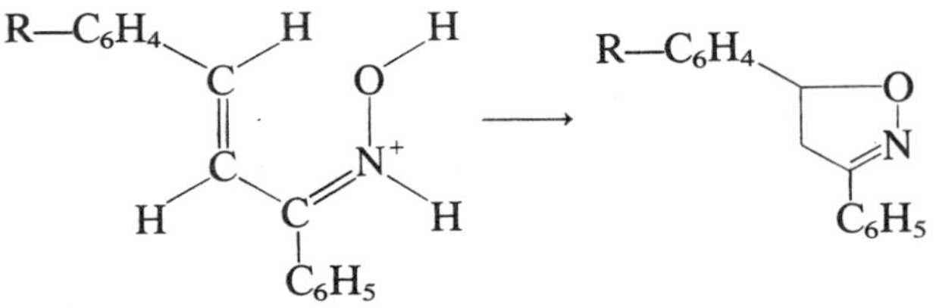

(**202**) → (**203**)

Oximes of cyclohexenones can aromatize in the presence of acylating reagents and strong acids (35b,192); this Semmler-Wolf reaction usually competes with the Beckmann rearrangement and is supposed to occur through the N-protonated salt of the O-acylated oxime (**204**):

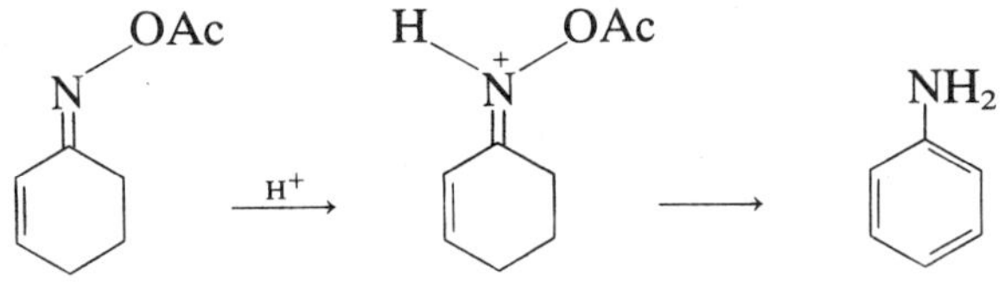

(**204**)

The reaction of the *O*-alkylketoxime (**205**) with monosubstituted malonyl chlorides is postulated to proceed via the oximinium salt (**206**) to afford the 1-alkoxy-4-hydroxypyrid-2-one (**207**) (193):

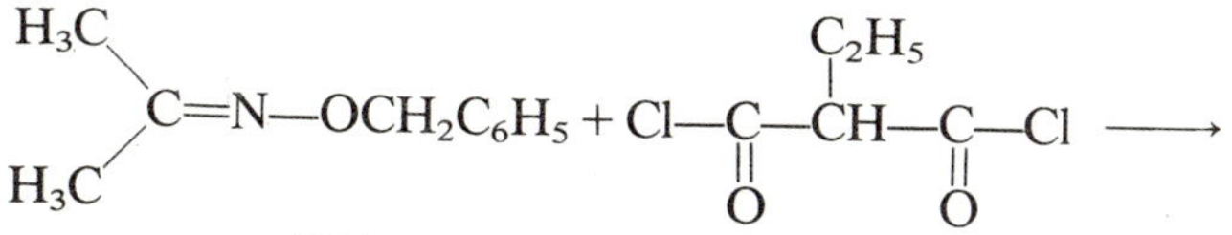

$$\underset{H_3C}{\overset{H_3C}{\diagup}}C=N-OCH_2C_6H_5 + Cl-\underset{\underset{O}{\|}}{C}-\underset{\overset{C_2H_5}{|}}{CH}-\underset{\underset{O}{\|}}{C}-Cl \longrightarrow$$

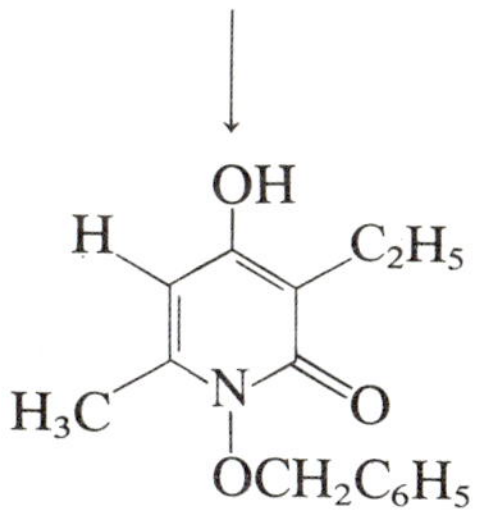

(**205**)

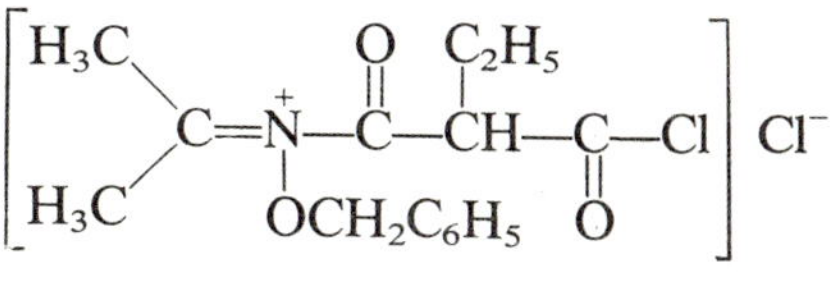

$$\left[\underset{H_3C}{\overset{H_3C}{\diagup}}C=\overset{+}{\underset{\underset{OCH_2C_6H_5}{|}}{N}}-\underset{\underset{O}{\|}}{C}-\underset{\overset{C_2H_5}{|}}{CH}-\underset{\underset{O}{\|}}{C}-Cl\right]Cl^-$$

(**206**)

↓

(**207**) [1-benzyloxy-3-ethyl-4-hydroxy-6-methyl-pyrid-2-one structure with OH, H, C₂H₅, H₃C, N-OCH₂C₆H₅, =O substituents]

Finally, we can point out that the oxyazenium ions (**XXIV**) (194) (Scheme 3) are much more elusive than the diazenium ions (Section 5) (**XIII**) (Scheme 2).

VII. Nitrones

A. INTRODUCTION

As in the case of azomethine imines (Section IV-A), we will not examine the 1,3-dipolar behavior of nitrones [known also as azomethine oxides (**XXI**) (132,133,151)]. We shall briefly discuss only the reactions included in Scheme 7, omitting the reactions more closely related to the subject of this review, such as the protonation or alkylation of nitriones into oximinium salts (**XXVII**), because they have already been reported in Sections V-A-1 and V-A-2.

B. SYNTHESIS

We indicated in Section V-A that the condensation of a carbonyl compound with an N-substituted hydroxylamine leads, not to an oximinium salt, but to a nitrone, which shows the stability of these compounds. This explains why the dimer (**208**) is scarcely observed, the

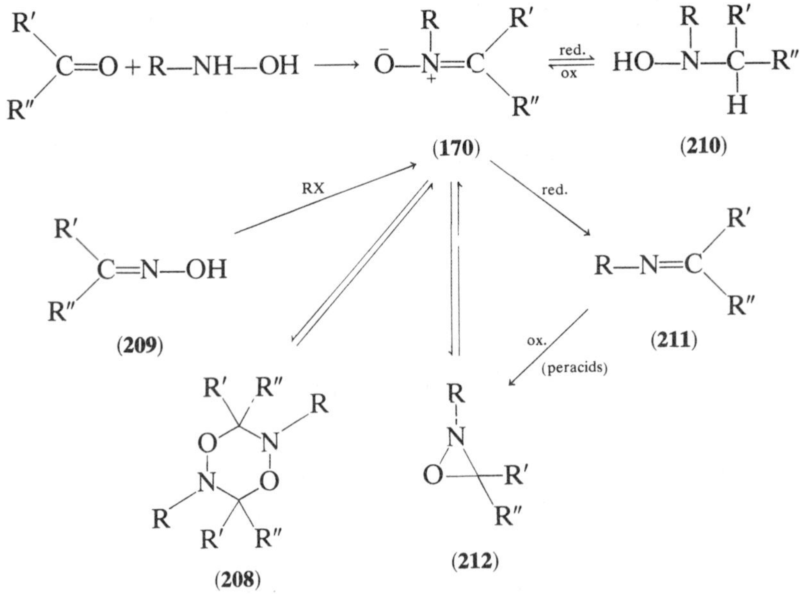

Scheme 7

equilibrium being shifted toward the nitrone form (195,196) (for an example of a stable dimer see Ref. 197).

Nitrones are also obtained (195,196) by alkylation of oximes (**209**), by oxidation of hydroxylamines (**210**), and by opening of oxaziridines. The oxidation of imines (**211**) with peracids yields oxaziridines (**212**) (134a, 199) but not nitrones (**170**). For heterocyclic nitrones, even nonaromatic ones, consult Ref. 198.

C. REACTIVITY

We are concerned here with reactions that are the inverse of those we have just reported: reduction to hydroxylamines (**210**) or to imines (**211**), (195,196), dimerization (**208**), and cyclization to oxaziridines (**212**) 134a,195,196,199). The oxidation of nitrones with lead tetraacetate has been reviewed recently (200). The photocyclization of nitrones to oxaziridines (Scheme 3: **XXI → XX**) is a very important reaction from a fundamental (201) as well as preparative (198,202,203) standpoint.

D. HETEROSUBSTITUTED NITRONES

The synthesis and reactivity of the following systems have been investigated in the acyclic series [case of the azine N-oxides (70d)] but more especially in the heterocyclic series:

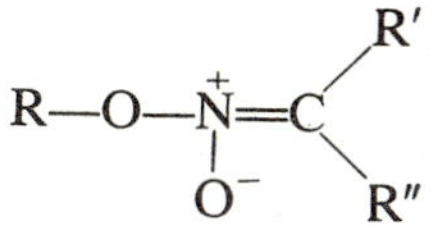

$$\text{R—O—}\overset{+}{\underset{O^-}{N}}\text{=C}\genfrac{}{}{0pt}{}{R'}{R''} \quad (204)$$

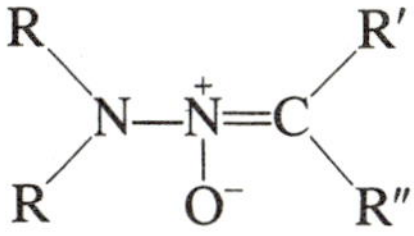

$$\genfrac{}{}{0pt}{}{R}{R}\text{N—}\overset{+}{\underset{O^-}{N}}\text{=C}\genfrac{}{}{0pt}{}{R'}{R''} \quad (205), \qquad \genfrac{}{}{0pt}{}{R}{R'}\text{C=N—}\overset{+}{\underset{O^-}{N}}\text{=C}\genfrac{}{}{0pt}{}{R''}{R'''} \quad (206,207)$$

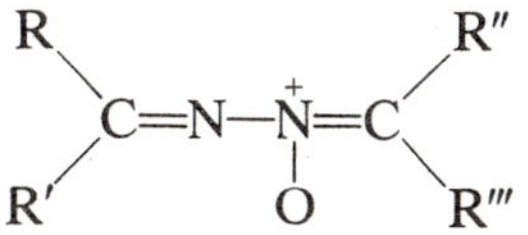

VIII. Sulfur-Containing N-Heteroiminium Salts

It is striking to notice the large discrepancy between nitrogen and oxygen derivatives, on the one hand, and sulfur derivatives, on the other hand: no cations RS—$\overset{+}{N}$=C$\genfrac{}{}{0pt}{}{}{}$ and no dipolar compounds $^-$S—$\overset{+}{N}$=C$\genfrac{}{}{0pt}{}{}{}$ are known.

In this particular case, the classical similarity of functions $>$NR, $>$O, and $>$S is exaggerated: compounds R—S—NH$_2$ are better compared to amides R—CO—NH$_2$ than to O-substituted oximes and thus are called "sulfenamides." They are synthetized similarly to amides (action of amines or ammonia on sulfenic acids), but their reactivity is quite different.

Particularly interesting here is the reaction of N-unsubstituted sulfenamides (**213**) with carbonyl compounds, aldehydes, or ketones (208,209):

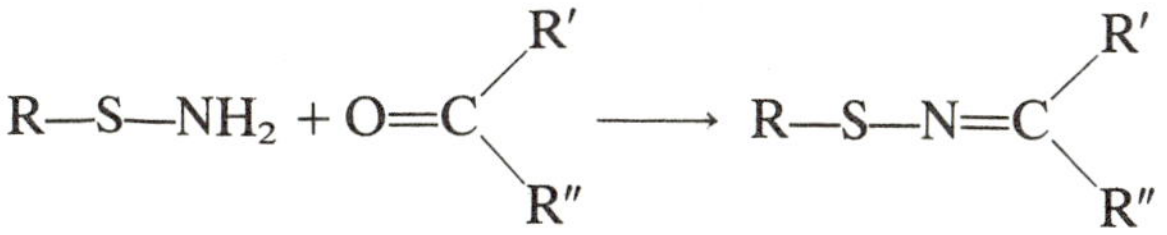

$$\text{R—S—NH}_2 + \text{O=C}\genfrac{}{}{0pt}{}{R'}{R''} \longrightarrow \text{R—S—N=C}\genfrac{}{}{0pt}{}{R'}{R''}$$

(**213**)   (**214**)

(**215**) 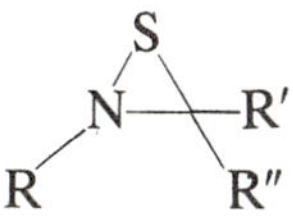   (**216**)

Since this reaction is analogous to that undergone by hydroxylamines, some authors (210) named compounds (**214**) "thiooximes." The E–Z isomerization of these compounds has been investigated and compared to that of oximes (211–213), but aside from their acid hydrolysis into carbonyl compound, ammonia, and "disulfide" (213), little is known about their behavior. For instance, their quaternization, which would lead to thiooximinium, has not been studied.

Furthermore, five-membered heterocyclic compounds containing such a functional group, the 2-isothiazolines (**215**), are not yet known.

Thiaziridines (**216**) have never been isolated, but they are very often postulated as reaction intermediates (214).

REFERENCES

1. H. Böhme and M. Haake, "Methyleniminium Salts," chapter in this book.
2. P. Schiess and A. Grieder, *Tetrahedron Lett.*, **1969**, 2097.
3. R. L. Hinman, R. D. Ellefson, and R. D. Campbell, *J. Amer. Chem. Soc.*, **82**, 3988 (1960).
4. J.-L. Aubagnac, J. Elguero, and R. Jacquier, *Tetrahedron Lett.*, **1965**, 1171.
5. J.-L. Aubagnac, J. Elguero, and R. Jacquier, *Bull. Soc. Chim. Fr.*, **1970**, 3316.
6. J.-L. Aubagnac, J. Elguero, and R. Jacquier, *Bull. Soc. Chim. Fr.*, **1969**, 3300.
7. N. M. Omar and A. V. El'tsov, *J. Org. Chem. (USSR)*, **4**, 709 (1968).
8. N. M. Omar and A. V. El'tsov, *J. Org. Chem. (USSR)*, **4**, 1246 (1968).
9. J.-L. Aubagnac, J. Elguero, and R. Jacquier, *Bull. Soc. Chim. Fr.*, **1968**, 3869.
10. J.-L. Aubagnac, J. Elguero, and R. Jacquier, *Bull. Soc. Chim. Fr.*, **1969**, 3292.
11. J.-L. Aubagnac, J. Elguero, and R. Jacquier, *Bull. Soc. Chim. Fr.*, **1971**, 3758.
12. R. Jacquier, C. Pellier, C. Petrus, and F. Petrus, *Bull. Soc. Chim. Fr.*, **1971**, 646.
13. G. Coispeau and J. Elguero, *Bull. Soc. Chim. Fr.*, **1970**, 2717.
14. J.-P. Chapelle, J. Elguero, R. Jacquier, and G. Tarrago, *Bull. Soc. Chim. Fr.*, **1970**, 240.
15. J.-P. Chapelle, J. Elguero, R. Jacquier, and G. Tarrago, *Bull. Soc. Chim. Fr.*, **1970**, 3147.
16. D. G. Farnum, R. J. Alaimo, and J. M. Dunston, *J. Org. Chem.*, **32**, 1130 (1967).
17. P. Bouchet, J. Elguero, and R. Jacquier, *Tetrahedron Lett.*, **1966**, 6409.
18. P. Bouchet, J. Elguero, and R. Jacquier, *Tetrahedron*, **22**, 2461 (1966).
19. C. Dittli, J. Elguero, and R. Jacquier, *Bull. Soc. Chim. Fr.*, **1969**, 4469.
20. C. Dittli, J. Elguero, and R. Jacquier, *Bull. Soc. Chim. Fr.*, **1969**, 4466.
21. C. Dittli, J. Elguero, and R. Jacquier, *Bull. Soc. Chim. Fr.*, **1969**, 4474.
22. J. Elguero, R. Jacquier, and D. Tizané, *Tetrahedron*, **27**, 133 (1971).
23. J.-L. Aubagnac, J. Elguero, R. Jacquier, and D. Tizané, *Tetrahedron Lett.*, **1967**, 3705.
24. J. Elguero, R. Jacquier, and D. Tizané, *Bull. Soc. Chim. Fr.*, **1970**, 1121.
25. A. V. El'tsov and N. M. Omar, *J. Org. Chem. (USSR)*, **4**, 692 (1968).
26. H. Dorn and A. Otto, *Angew. Chem. Int. Ed.*, **7**, 214 (1968).
27. W. Oppolzer, *Tetrahedron Lett.*, **1970**, 2199.
28. J.-L. Aubagnac, J. Elguero, and R. Jacquier, *Bull. Soc. Chim. Fr.*, **1969**, 3316.
29. G. Coispeau, J. Elguero, R. Jacquier, and D. Tizané, *Bull. Soc. Chim. Fr.*, **1970**, 1581.
30. J. Elguero, R. Jacquier, and D. Tizané, *Tetrahedron*, **27**, 123 (1971).
31. M. J. Kornet and H. S. I. Tan, *J. Pharm. Sci.*, **61**, 235 (1972).
32. J.-L. Aubagnac, J. Elguero, R. Jacquier, and D. Tizané, *Tetrahedron Lett.*, **1967**, 3709.

33. J. Elguero and M.-C. Pardo, *Anal. Quim.*, **70,** 991 (1974).
34. H. J. Sisler, G. M. Omietanski, and B. Rudner, *Chem. Rev.*, **57,** 1021 (1957).
35. P. A. S. Smith, *The Chemistry of Open-Chain Organic Nitrogen Compounds*, Vol. II, W. A. Benjamin, New York, 1966: (a) p. 119; (b) p. 29.
36. R. F. Smith, D. S. Johnson, C. L. Hyde, T. C. Rosenthal, and A. C. Bates, *J. Org. Chem.*, **36,** 1155 (1971).
37. R. F. Smith, D. S. Johnson, R. A. Abgott, and M. J. Madden, *J. Org. Chem.*, **38,** 1344 (1973).
38. R. C. Elderfield and S. L. Wythe, in *Heterocyclic Compounds*, Vol. 6, R. C. Elderfield, Ed., John Wiley and Sons, New York, 1957, p. 190.
39. R. F. Smith and E. D. Otremba, *J. Org. Chem.*, **27,** 879 (1962).
40. K. N. Zelenin and V. G. Kamerdinerov, *J. Org. Chem. (USSR)*, **1,** 1935 (1965).
41. K. N. Zelenin and V. G. Kamerdinerov, *Iz. Vyssh. Ucheb. Zaved., Khim. Teknol.*, **12,** 911 (1962); *C. A.*, **72,** 12670 (1970).
42. J.-L. Aubagnac, J. Elguero, R. Jacquier, and R. Robert, *C.R. Acad. Sci. Paris*, **270,** 1829 (1970).
43. B. V. Ioffe and K. N. Zelenin, *Tetrahedron Lett.*, **1962,** 481.
44. B. V. Ioffe and K. N. Zelenin, *J. Gen. Chem. (USSR)*, **33,** 3521 (1963).
45. L. Pappalardo, Thesis, Montpellier, 1971.
45a. N. H. Shapranova, I. N. Somin and S. G. Kuznetsov. *Chem. Heterocycl. Compd. (USSR)*, **1973,** 1093.
46. R. Grashey and M. Baumann, *Angew. Chem.*, **81,** 115 (1969).
47. R. Grashey, M. Baumann, and H. Bauer, *Chem. Ztg.*, **96,** 224 (1972).
48. G. F. Duffin and J. D. Kendall, *J. Chem. Soc.*, **1954,** 408.
49. J. Elguero, R. Jacquier, and S. Mondon, *Bull. Soc. Chim. Fr.*, **1970,** 1576.
50. G. Coispeau, Thesis, Montpellier, 1970.
51. W. J. Houlihan and W. J. Theuer, *J. Org. Chem.*, **33,** 3941 (1965).
51a. E. Schmitz, *Chem. Ber.*, **91,** 1495 (1958).
52. W. Pugh, *J. Chem. Soc.*, **1953,** 3445.
53. E. G. Sohn, R. H. Marks, and W. Pugh, *J. Chem. Soc.*, **1955,** 1753.
54. W. Theilacker and O. R. Leichtle, *Liebegs Ann. Chem.*, **572,** 121 (1951).
55. C. G. Overberger and G. Kesslin, *J. Org. Chem.*, **27,** 3898 (1962).
56. M. Nardelli and G. Fava, *Acta Cryst.*, **15,** 214 (1962).
57. J. Elguero, R. Jacquier, and C. Marzin, *Tetrahedron Lett.*, **1970,** 3099.
58. H. Luth and J. Trotter, *Acta Cryst.*, **19,** 614 (1965).
59. D. G. Neilson, R. Roger, J. W. M. Heatlie, and L. R. Newlands, *Chem. Rev.*, **69,** 151 (1969).
60. E. Schmitz, *Chem. Ber.*, **95,** 676 (1962).
61. Cs. Szantay and E. Schmitz, *Chem. Ber.*, **95,** 1759 (1962).
62. J.-L. Aubagnac, J. Elguero, and R. Jacquier, *Bull. Soc. Chim. Fr.*, **1969,** 3306.
63. J.-L. Gilles, Thesis, Montpellier, 1973.
64. J.-L. Aubagnac, J. Elguero, and R. Jacquier, *Bull. Soc. Chim. Fr.*, **1969,** 3302.
65. E. H. White and D. J. Woodcoock, in *The Chemistry of the Amino Group*, S. Patai, Ed., Interscience, London, 1968, p. 416.
66. G. Lepicard, D. Saint-Giniez, R. Jacquier, and C. Rérat, *C.R. Acad. Sci. Paris*, **267C,** 1786 (1968).
67. J.-L. Aubagnac, J. Elguero, B. Rérat, C. Rérat, and Y. Uesu, *C.R. Acad. Sci. Paris*, **274C,** 1192 (1972).
68. J. Elguero, E. Gonzalez, and R. Jacquier, *Bull. Soc. Chim. Fr.*, **1969,** 2054.
69. P. Bouchet, J. Elguero, and E. Gonzalez, unpublished results.

70. E. Müller, *Methoden der Organischen Chemie* (Houben-Weyl), Vol. X/2, Stickstoffverbindungen, I.G. Tieme Verlag, Stuttgart, 1967: (a) p. 56; (b) p. 75; (c) p. 310; (d) p. 118.
71. I. I. Grandberg, T. Y. Zuyanova, N. M. Przheval'skii and V. I. Minkin, *Chem. Heterocycl. Compd.* (*USSR*), **1970**, 693.
71a. J. Arriau, A. Dargelos, and J. Elguero, work in progress.
71b. I. I. Grandberg, *Chem. Heterocycl. Compd.* (*USSR*), **1974**, 579.
72. H. Kessler, *Tetrahedron Lett.*, **1968**, 2041.
73. N. P. Marullo and E. H. Wagener, *Tetrahedron Lett.*, **1969**, 2555.
74. J. C. Tobin, A. F. Hegarty, and F. L. Scott, *J. Chem. Soc. B*, **1971**, 2198.
75. G. J. Karabatsos and R. A. Taller, *J. Amer. Chem. Soc.*, **85**, 3624 (1963).
76. G. J. Karabatsos and R. A. Taller, *Tetrahedron Lett.*, **1968**, 3557.
77. G. J. Karabatsos and R. A. Taller, *Tetrahedron Lett.*, **1968**, 3923.
78. A. Krebs and H. Kimling, *Liebigs Ann. Chem.*, **740**, 126 (1970).
79. D. Wurmb-Gerlich, F. Vögtle, A. Mannschreck, and H. A. Staab, *Liebigs Ann. Chem.*, **708**, 36 (1967).
80. A. Mannschrek and U. Koelle, *Tetrahedron Lett.*, **1967**, 863.
81. C. I. Stassinopoulo, C. Zioudrou, and G. J. Karabatsos, *Tetrahedron Lett.*, **1972**, 3671.
82. Y. Shvo and A. Nahlieli, *Tetrahedron Lett.*, **1970**, 4273.
83. H. O. Kalinowski, H. Kessler, D. Leibfritz, and A. Pfeffer, *Chem. Ber.*, **106**, 1023 (1973).
84. J. P. Idoux and J. A. Sikorski, *J. Chem. Soc. Perkin II*, **1972**, 921.
85. Hammerum, *Tetrahedron Lett.*, **1972**, 949.
86. Hammerum, *Acta Chem. Scand.*, **27**, 779 (1973).
87. W. Skorianetz and E. Kovats, *Helv. Chim. Acta*, **53**, 251 (1970).
88. F. E. Condon and D. Forcasiu, *J. Amer. Chem. Soc.*, **92**, 6625 (1970).
89. A. N. Gafarov, *J. Org. Chem.* (*USSR*), **6**, 1552 (1970).
90. H. Szmant and C. McGinnis, *J. Amer. Chem. Soc.*, **72**, 2890 (1950).
91. K.-D. Hesse, *Liebegs Ann. Chem.*, **743**, 50 (1971).
92. R. Baumes, Thesis, Montpellier, 1973.
93. R. Baumes, R. Jacquier, and G. Tarrago, *Bull. Chim. Fr.* **1976**, 260.
94. J. A. Moore and J. Binkert, *J. Amer. Chem. Soc.*, **81**, 6029 (1959).
95. M. J. Kornet and P. A. Thio, *J. Pharm. Sci.*, **58**, 724 (1969).
96. J.-P. Chapelle, J. Elguero, R. Jacquier, and G. Tarrago, *Bull. Soc. Chim. Fr.*, **1970**, 3145.
97. J. E. Baldwin and H. H. Basson, *Chem. Commun.*, **1969**, 795.
98. H. Fritz and P. Uhrhan, *Liebigs Ann. Chem.*, **744**, 81 (1971).
99. R. Baumes, R. Jacquier, and G. Tarrago, *Bull. Soc. Chim. Fr.*, **1973**, 317.
100. J. Elguero, R. Jacquier, and C. Marzin, *Bull. Soc. Chim. Fr.*, **1970**, 4119.
101. J. Elguero, *Bull. Soc. Chim. Fr.*, **1971**, 1925.
102. A. N. Kost and R. S. Sagitullin, *J. Gen. Chem.* (*USSR*), **33**, 855 (1963).
103. J. Elguero, R. Jacquier, and S. Mignonac-Mondon, *Bull. Soc. Chim. Fr.*, **1972**, 2807.
104. M. Lamchen, W. Pugh, and A. M. Stephen, *J. Chem. Soc.*, **1954**, 2429.
105. M. Lamchen and A. M. Stephen, *J. Chem. Soc.*, **1955**, 2044.
106. W. Pugh, *J. Chem. Soc.*, **1954**, 2423.
107. J. Elguero and R. Jacquier, *Bull. Soc. Chim. Fr.*, **1965**, 2961.
108. E. B. Rathbone and A. M. Stephen, *J. S. Afr. Chem. Inst.*, **24**, 155 (1971).
109. H. Dorn and D. Arndt, *Z. Chem.*, **9**, 336 (1969).
110. H. Dorn and A. Otto, *Chem. Ber.*, **103**, 2505 (1970).

111. J. Elguero, R. Jacquier, and D. Tizané, *Bull. Soc. Chim. Fr.*, **1968**, 3866.
112. D. T. Manning, H. A. Coleman, and R. A. Langdale-Smith, *J. Org. Chem.*, **33**, 4413 (1968).
113. J. Thiele, *Liebigs Ann. Chem.*, **376**, 239 (1910).
114. L. Maaskant, *Rec. Trav. Chim.*, **56**, 211 (1937).
115. A. H. Blatt, *Org. Synth., Coll.* **II**, 395 (1943).
115a. S. S. Mathur and H. Suschitzky, *Tetrahedron Lett.*, **1975**, 785.
116. A. N. Kost, G. A. Golubeva, and I. I. Grandberg, *J. Gen. Chem. (USSR)*, **26**, 2201 (1956).
117. A. N. Kost and I. I. Grandberg, *J. Gen. Chem. (USSR)*, **26**, 2905 (1956).
117a. H. Posvic, R. Dombro, H. Ito, and T. Telinski, *J. Org. Chem.*, **39**, 2575 (1974).
118. A. N. Kost, I. I. Grandberg, and E. B. Evreinova, *J. Gen. Chem. (USSR)*, **28**, 503 (1958).
119. H. Böhme and S. Ebel, *Pharmazie*, **5**, 296 (1965).
120. C. S. Marvel and P. V. Bonsignore, *J. Amer. Chem. Soc.*, **81**, 2668 (1959).
121. M. Bender, Z. Buczkowski, and J. Plenkiewicz, *Bull. Acad. Pol. Sci.*, **17**, 637 (1969).
122. K. V. Auwers and F. Bergmann, *Ann. Chem.*, **472**, 287 (1929).
123. E. G. Sohn, R. H. Marks, and W. Pugh, *J. Chem. Soc.*, **1955**, 1753.
123a. F. L. Scott and P. A. Cashell, *J. Chem. Soc. C*, 2674 (1970).
124. B. M. Bogoslovsky and T. I. Jakovenko, *J. Gen. Chem. (USSR)*, **27**, 159 (1957).
125. E. C. Gilbert, *J. Amer. Chem. Soc.*, **51**, 3394 (1929).
126. L. Zirngibl and S. W. Tam, *Helv. Chim. Acta*, **53**, 1927 (1970).
127. A. M. Comrie, *J. Chem. Soc. C*, **1971**, 2807.
128. J. B. Aylward, *J. Chem. Soc. C*, **1970**, 1494.
129. J. Elguero, R. Jacquier, and H. C. N. Tien Duc, *Bull. Soc. Chim. Fr.*, **1967**, 2617.
130. J. Daunis, M. Guerret-Rigail, and R. Jacquier, *Bull. Soc. Chim. Fr.*, **1972**, 1994.
131. J.-L. Aubagnac, J. Elguero, R. Jacquier, and R. Robert, *Bull. Soc. Chim. Fr.*, **1972**, 2859.
132. R. Huisgen, *Angew. Chem. Int. Ed.*, **2**, 565 (1963).
133. R. Huisgen, R. Grashey, and J. Saver, in *The Chemistry of Alkenes*, Vol. 1, S. Patai, Ed., Interscience, London, 1964, p. 739.
133a. C. G. Stuckwisch, *Synthesis*, **1973**, 469.
134. E. Schmitz, in *Advances in Heterocyclic Chemistry*, Vol. 2, A. R. Katritzky, Ed., Academic Press, New York, 1963: (a) p. 85; (b) p. 104.
135. R. Grashey and K. Adelsberger, *Angew. Chem. Int. Ed.*, **1**, 267 (1962).
136. R. Huisgen, R. Fleichmann, and A. Eckell, *Tetrahedron Lett.*, **1960**, 1.
137. R. Huisgen and A. Eckell, *Tetrahedron Lett.*, **1960**, 5.
138. R. B. Greenwald and E. C. Taylor, *J. Amer. Chem. Soc.*, **90**, 5272 (1968).
139. R. B. Greenwald and E. C. Taylor, *J. Amer. Chem. Soc.*, **90**, 5273 (1968).
140. G. Zinner, W. Kliegel, W. Ritter, and H. Böhlke, *Chem. Ber.*, **99**, 1678 (1966).
141. H. Dorn and H. Dilcher, *Liebigs Ann. Chem.*, **717**, 104 (1968).
141a. P. R. Farina and H. Tieckelmann, *J. Org. Chem.*, **38**, 4259 (1973).
141b. H. W. Heine and L. Heitz, *J. Org. Chem.*, **39**, 3192 (1974).
141c. A. Guingant and J. Renault, *Bull. Soc. Chim. Fr.* **1976**, 291.
142. R. Grashey, R. Huisgen, K. K. Sun, and R. M. Moriarty, *J. Org. Chem.*, **30**, 74 (1965).
143. W. J. McKillip, E. A. Sedor, B. M. Culbertson, and S. Wawzonek, *Chem. Rev.*, **73**, 255 (1973).
144. W. Oppolzer, *Tetrahedron Lett.*, **1970**, 3091.
145. H. Dorn and A. Otto, *Chem. Ber.*, **101**, 3287 (1968).

146. H. Dorn and A. Otto, *Z. Chem.*, **8**, 273 (1968).
147. J. C. Howard, G. Gever, and P. H. L. Wei, *J. Org. Chem.*, **28**, 868 (1963).
148. W. O. Godtfredsen and S. Vangedal, *Acta Chem. Scand.*, **9**, 1498 (1955).
149. W. Oppolzer and H. P. Weber, *Tetrahedron Lett.*, **1972**, 1711.
150. H. Dorn and A. Otto, *Tetrahedron*, **24**, 6809 (1968).
151. J. Bastide, J. Hamelin, F. Texier, and Y. Vo Quang, *Bull. Soc. Chim. Fr.*, **1973**, 2871.
152. H. Dorn and A. Zubek, *Z. Chem.*, **8**, 270 (1968).
153. P. De Mayo and J. J. Ryan, *Chem. Commun.*, **1967**, 88.
154. D. E. Ames and B. Novitt, *J. Chem. Soc. C*, **1969**, 2355.
155. M. Schulz and G. West, *J. Prakt. Chem.*, **312**, 161 (1970).
156. D. M. Lemal, in *Nitrenes*, W. Lwowski, Ed., Interscience, New York, 1970, p. 345.
157. S. Hünig, *Helv. Chim. Acta*, **54**, 1721 (1971).
157a. J. Cramer, H. Hansen, and S. Hünig. *Chem. Commun.*, **1974**, 264.
158. J. Hamer and A. Macaluso, *Chem. Rev.*, **64**, 473 (1964).
159. G. R. Delpierre and M. Lamchen, *Quart. Rev.*, **19**, 329 (1965).
160. A. Belly, C. Petrus and F. Petrus, *Bull. Soc. Chim. Fr.*, **1973**, 1390.
161. A. Belly, R. Jacquier, F. Petrus, and J. Verducci, *Bull. Soc. Chim. Fr.*, **1972**, 395.
162. V. J. Bauer, W. Fulmor, G. O. Morton, and S. R. Safir, *J. Amer. Chem. Soc.*, **90**, 6846 (1968).
163. E. Buehler, *J. Org. Chem.*, **32**, 26 (1967).
164. O. L. Brady, F. P. Dunn, and R. F. Goldstein, *J. Chem. Soc.*, **1926**, 2386.
165. P. A. S. Smith and J. E. Robertson, *J. Amer. Chem. Soc.*, **84**, 1197 (1962).
166. H. O. House and F. A. Richey, Jr., *J. Org. Chem.*, **34**, 1430 (1969).
167. J. E. Baldwin, R. G. Pudussery, B. Sklarz, and M. K. Sultan, *Chem. Commun.*, **22**, 1361 (1968).
168. E. I. du Pont de Nemours and Co., Belg. Pat., 614,730; *C. A.*, **57**, 16484 (1962).
169. A. H. Blatt and N. Gross, *J. Amer. Chem. Soc.*, **77**, 5424 (1955).
170. R. P. Barnes and F. L. McMillian, *J. Chem. Eng. Data*, **8**, 280 (1963).
171. R. Jacquier, J. L. Olive, C. Petrus, and F. Petrus, unpublished results.
171a. W. Theilacker, I. Gerstenkorn, and F. Gruner, *Liebigs Ann. Chem.*, **563**, 104 (1949).
172. H. Saitô and K. Nukada, *J. Mol. Spectrosc.*, **18**, 1 (1965).
173. G. A. Olah and T. E. Kiousky, *J. Amer. Chem. Soc.*, **90**, 4666 (1968).
174. H. Saitô, K. Nakada, and M. Ohno, *Tetrahedron Lett.*, **1964**, 2124.
175. H. Saitô and K. Nukada, *J. Mol. Spectrosc.*, **18**, 1 (1965).
176. H. Saitô, I. Terasawa, M. Ohno, and K. Nukada, *J. Amer. Chem. Soc.*, **91**, 6696 (1969).
177. Y. Iwakura, U. Keikichi, and H. Kazuo, *Bull. Chem. Soc. Jap.*, **43**, 873 (1970).
177a. V. I. Shlenskaya, T. I. Tikhivinskaya, and A. A. Biryukov, *Vest. Mosk. Univ. Khim.*, **11**, 337 (1970).
178. B. L. Fox, J. E. Reboulet, R. E. Rondeau, and H. M. Rosenberg, *J. Org. Chem.*, **35**, 4234 (1970).
179. M. Witanowski, L. Stefaniak, H. Januszewski, S. Szymanski, and G. A. Webb, *Tetrahedron Lett.*, **1973**, 2833.
180. K. D. Berlin and S. Rengaraju, *J. Org. Chem.*, **36**, 2912 (1971).
181. Z. W. Wolkowski, *Tetrahedron Lett.*, **1971**, 825.
182. G. A. Olah and R. H. Schlosberg, *J. Amer. Chem. Soc.*, **90**, 6461 (1968).
183. V. A. Bren, E. A. Medyantseva, and V. I. Minkin, *Reakt. Sposobnost Org. Soedin.*, **5** (4), 988 (1968).
184. W. D. Emmons, *J. Amer. Chem. Soc.*, **79**, 5742 (1957).
185. R. Bonaccorsi, E. Scrocco, and J. Tomasi, *Theor. Chim. Acta*, **21**, 17 (1971).

Author Index

Numbers in parentheses are reference numbers and show that an author's work is referred to although his name is not mentioned in the text. Numbers in *italics* indicate the pages on which the full references appear.

Abdel-Rahman, M. O. 321(197), 323(197), *338*
Abduganiev, E. G., 196(399), *216*
Abgott, R. A., 540(37), 541(37), 542(37), *583*
Aboul-Enein, M. N., 321(198), *338*
Adam, A., 173(342), 200(406), *215, 216*
Adam, D., 200(406), *216*
Adameik, J. A., 57(157), *99*
Addor, R. W., 47(111), 69(111), 72(111), 85(111), *98,* 355(21), 393(21), *416*
Adelsberger, K., 566(135), 567(135), *585*
Adrian, G., 75(217), 78(217), *101,* 134 (180), *211*
Aguiar, A. M., 175(351), *215*
Ahlbrecht, H., 75(214), 78(214), *101,* 116(70), *208*
Ahmad, V. U., 111(20d), 162(20d), *207*
Ahmed, M., 328(206), *339*
Ahond, A., 28(13), *96,* 144(207), *211*
Ahrens, K. H., 75(235), 78(235), 89(235), *101,* 196(396), *216*
Akabori, S., 236(70), *335*
Akaboshi, S., 328(207), *339*
Aki, O., 303(153), *337*
Alaimo, R. J., 539(16), *582*
Alais, L., 78(246, 395), *102, 105,* 112 (42, 45), 114(45), 212(42), *207*
Alkalay, D., (301), *214*
Allen, L. C., 2(3), *19*
Allendörfer, H., 142(199), *211*
Allenstein, E., 28(11), 29(27, 29), 30(29, 31), 39(60), 42(80), 44(99), 47(80), 48(118), 49(123, 125), 74(11, 31), 75(11, 27, 29, 31, 80, 99, 118, 123, 125), *96, 97, 98, 99,* 146(219), *212,* 355(23), *416*
Alonso, G., 200(406), *216*
Al-Showiman, S., 34(373), 84(373), *105*

Alt, G. H., 73(193), 85(193, 295), 86(295), *100, 103,* 112(32), 114(32, 58), 115(32, 58), 117(82), *207, 208*
Altreuther, (N. I.), 43(92), *98*
Alumni, S., 95(362), *104,* 229(56), 234(56), *334*
Ambekar, S. Y., 340(226), *342*
American Cyanamid Company, 138(187), *211*
Ames, D. E., 570(154), *586*
Ammon, H. L., 34(49), 93(49), *97*
Amschler, H., 40(67), 42(67), *97,* 263(110), *336*
Anders, B., 354(20), *416*
André, J. M., 470(82), 510(82), *530*
Andrieux, C. P., 95(357, 393), *104,* 206(421), *217*
Anfus-Ercsényi, A., 402(105), *419*
Angibeaud, P., 78(246, 395), *102, 105,* 112(42), 218(42), *207*
Anselme, J.-P., 108(3), *206*
Ansmann, A., 509(115a), *531*
Aono, K., 63(171), *100*
Arndt, D., 558(109), 563(109), *584*
Arnold, Z., 26(9), 48(117), 49(121), 69(117, 121), 74(121), 75(9), 78(241), 86(184, 300, 307), 95(184, 358), *96, 98, 99, 100, 101, 102, 104,* 151(274), *213,* 226(5, 6, 7, 8. 9, 10, 11, 12, 13, 14, 15, 16, 17, 18, 19, 20, 21, 22, 23, 24, 25, 26, 27, 28, 29, 30, 31, 32), 229(18, 49), 230(18, 23), 231(17, 23, 49, 81), 250(26, 29), 251(23), 255(10, 101), 259(29), 260(19, 29), 261(107), 262(11, 16, 21, 107), 263(21, 111), 265(9, 26, 112), 267(5), 268(22), 269(6), 272(24), 274(8, 15, 124), 275(8), 276(15), 278(8, 15), 285(24), 289(12, 14, 21), 291(15), 292(15, 21) 293(15), 294(10), 297(17, 18) 299(7),

300(17), 302(30, 76), 309(21, 22), 312 (20), 314(20), 318(16, 28, 31, 194, 195), 319(16, 194), 320(26, 28, 31, 32, 191), 321(16, 28), 324(27), 325(25), 327(25), 331(214), 332(214), *333, 334, 336, 338, 339,* 365(47), 372(54), *417*
Arriau, J., 549(71a), *584*
Ashby, J., 328(206), *339*
Aubagnac, J. L., 53(145, 146), 69(145), *99,* 92(344), *104,* 158(300), 160(300, 302), *214,* 539(4, 5, 6, 9, 10, 11), 540 (4, 5, 23, 28, 32), 541(42), 544(4, 42), 546(62), 547(5, 6, 10, 11, 62, 64), 548 (5, 10, 11, 67), 553(6), 556(64), 558(6) *582, 583, 585*
Aufderhaar, E., 192(388), *216*
Auterhoff, G., 69(182), 71(182), 73(182), 75(212), 78(212), 95(102), *100, 101,* 116(73), 126(73), *208,* 126(147, 148, 150), 152(148), 157(147, 148), 177 (150), 183(147), 184(147), 186(147), *208, 210*
Auwers, K. V., 560(122), *585*
Axenrod, T., 63(170), *100*
Aylward, J. B., 564(128), *585*

Babad, H., 384(82), *418*
Babayan, A. T., 194(395), *216*
Babcock, J. C., 75(201), 86(201), 94(201), *100*
Backhaus, P., 128(161), 144(206, 211), 163(206), 164(206), 181(211), *210, 211*
Backstez, M., 148(231), *212*
Badische Anilin-& Soda Fabrik, 140(189), *211*
Bahner, C. T., 116(76), *208*
Baker, A. D., 144(212), *211*
Balaban, A. T., 78(268), 93(347), *102, 104*
Balaban, J. E., 167(326), *214*
Balbi, A., 330(208), *339*
Baldeschwieler, J. P., 61(163), *99*
Baldwin, J. E., (97), *584,* 573(167), *586*
Bangert, K. F., 371(51), *417*
Baranav, S. N., 203(412), *216*
Barefoot, R. D., 82(275), *102*
Barili, P. L., 63(170), *100*
Barltrop, J. A., 581(210), *587*
Barnes, R. P., 573(170), *586*
Barnett, P. G., 244(91), *335*
Barr, J. T., 116(76), *208*
Bartell, L. S., (13), *19*
Bartfeld, H.-D., 52(138), 69(138), *99,* 119(125), *209*
Bartlett, P. D., 519(124b), *532*
Bartulin, J., 111(28), *207*
Basalkevitch, E. D., 391(124), *419*

Basha, A., 111(20d), 162(20d), *207*
Basson, H. H., (97), *584*
Bastide, J., 570(151), 579(151), *586*
Bates, A. C., 549(36), *583*
Bather, P. A., 33(46), *97,* 144(210), *211*
Bauer, H., 542(47), *583*
Bauer, L., 54(149), 57(149), *99*
Bauer, V. J., 572(162), *586*
Bauld, N. L., 37(57), 69(57), *97,* 143(202), *211*
Baum, J., 365(46), *417*
Baum, J. S., 40(76), 43(76), 49(76), 65(76), 67(76), 89(319), *97, 103,* 364(115), *419*
Baum, M. W., 67(181), 77(181), *100*
Baumann, M., 542(46, 47), *583*
Baumes, R., 551(92, 93), 554(92, 99), *584*
Bayer, O., 227(33), 237(33), 258(33), 328 (33), 341(33), *334*
Bayerlein, F., 164(314), *214*
Bayerlin, H. P., 371(50), *417*
Beasley, J. G., 118(88), 149(88), *208*
Beauchamp, J. L., 3(9), *19*
Becker, E. D., 78(252), *102*
Becker, P., 109(12), *207*
Becker, R. F., 351(11), 352(11), *416*
Beekman, P., 142(199), *211*
Beierl, D., 75(205), *101*
Beilmann, J.-F., 54(153), 63(153), 67(153), *99*
Belegratis, K., 579(193), *587*
Bell, C. L., 54(149), 57(149), *99*
Bellak, L., 228(41), *334*
Belloc, A.-M., 75(204), *101*
Belly, A., 572(160, 161), 573(160), 577 (160, 186), *586, 587*
Belyaeva, I. N., 423(6a), 433(6a), 447(6a), 460(6a), 461(6a), 462(6a), 463(6a), *528*
Bender, M., 560(121), *585*
Benders, P.-H., 236(69), *335*
Ben-Ishai, D., 122(119), *209*
Benson, W. R., 126(275), *336*
Bent, H. L., 4(18), *19*
Bentz, L. O., 166(324), *214*
Berezhnaya, M. I., 272(121), *336*
Berezhnaya, M. T., 267(115), *336*
Berg, G., 116(80), *208*
Berger, A., 78(369), *105*
Bergmann, F., 560(122), *585*
Berlin, K. D., 575(180), *586*
Bernabei, D., 236(69), *335*
Bernhard, C., 236(68), *335*
Berti, G., 63(170), *100*
Bestmann, H. J., 200(403, 404), *216*
Bettencourt, A., 360(38), 361(38, 42), *417,* 435(36b), 454(36b), *529*
Beveridge, D. L., 2(1), *19*

Beyerlin, H. P., 75(225), 101
Beyl, V., 39(60), 75(60), *97*
Bignebat, J., 242(90), *335*
Binde, L., 173(342), *215*
Bingham, R. C., 2(2b), *19*
Binkert, J., 552(94), *584*
Binsch, G., 63(167, 167a), 77(167), 78 (167a), *100*
Birchall, T., 42(81, 82), 43(81, 82), 63(81), *98*
Birdsell, B., 49(126), *99*
Biryukov, A. A., 574(177a), *586*
Bisagni, E., 237(75), *335*
Bitter, I., 75(219, 220), *101*, 402(105), *419*
Black, D. St. C., 582(214), *587*
Blatt, A. H., 559(115), *585*, 573(169), *586*
Blazejwski, J. C., 439(44), 460(44), 461(44), *529*
Block, B. P., 177(360), *215*
Bochow, H., 219(431), *222*
Bock, E., 54(151), *99*
Bodendorf, K., 276(131), 278(131, 133), 279(131), 280(131), 281(131), 282(131), 285(143), 286(143), 289(133), 294(132), *336, 337*
Bodor, N., 2(2a), *19*
Boer, Th. J. de, 32(38), 38(58), 78(370), *96, 97, 105,* 122(128), *209*
Bogoslovsky, B. M., 561(124), 563(124), *585*
Böhlke, H., 125(142), 127(142), *210,* 567(140), 568(140), *585*
Bohlmann, F., 89(323), *103,* 150(272), *213*
Bohme, H., 6(22), 16(22), 17(22), *19,* 25 (3), 28(16, 17, 18), 69(182), 71(182), 72 (182, 192), 75(212, 235), 78(212, 235), 95(182), *96, 100, 101, 102,* 108(9), 109 (15), 111(20a, 21, 24), 112(35, 36, 38), 116(73, 75, 80), 119(98, 99, 103, 104, 106, 107, 108, 115), 121(127), 122(36, 132, 133, 134), 123(137), 124(140), 125(143), 126(73, 137, 146, 147, 148, 149, 150, 152), 127(156, 157), 128(38, 161, 162), 129(9), 130(166), 131(132, 162, 172, 173), 132(175), 133(134, 146, 178), 134(9), 135(38), 136(137, 184), 137(166), 138(140, 186), 139(162), 140(38, 157, 191), 141(194, 195, 196), 142(186), 144(38, 206), 145(132, 133, 172, 173, 214, 215), 147(221, 228), 148(232), 150(127, 221), 151(146), 152(146, 148, 149), 153(191), 155(9, 127), 156(166), 157(147, 148, 149, 295), 159(15), 160(305b), 161(295), 162(20a, 196, 310, 311), 163(206, 313), 164(99, 103, 196, 206), 165(319, 320, 321), 166(319), 167(99, 103, 106, 186, 311), 168(106, 321, 328), 169(140, 329, 333), 170(132, 172, 173, 174, 336, 337), 171(338, 339, 341), 172(340), 173(175, 228, 310), 175(103, 175, 328), 176(337), 177(150, 162), 179(36, 127, 157), 180(140, 363, 364, 365), 181(98, 99, 103, 363, 365, 367), 182(24), 183(127, 147), 184(15, 75, 147, 149, 295), 185(15), 186(15, 147), 187(15, 36, 152, 178), 188(374), 189(378, 379), 190(381, 382, 383), 191(385), 192(386), 194(394), 195(313, 396), 196(400), 198(75, 400), 200(174, 196, 385, 405), 202(405, 409), 203(173, 413, 414), 205(174), 206(147), *207, 208, 210, 211, 212, 213, 215, 216,* 425(10), 465(10), 473(10), 490(10), *528, 529,* 535(1), 540(1), 547(1), 553(1), 559(119), 560(1), 562(1, 119), *582, 585*
Bohn, H. J., 112(38), 128(38), 135(38), 140(38), 144(38), *207*
Boll, E., 145(214), *212*
Bomke, U., 152(281), 155(281), 188(374), 202(281), *213, 215*
Bonaccorsi, R., 576(185), *586*
Bondarchuk, N. D., 381(120), 382(75, 120), 397(75), 398(75), 400(75), 404(75), *418, 419*
Bonner, O. D., 75(366), *104*
Bonnett, B., 240(82), *335*
Bonsignore, P. V., 560(120), *585*
Borden, W. T., 394(90), *418*
Bordner, J., 333(217), *339*
Borisenko, A. A., 316(184), *338*
Borkovec, A. B., 402(103b), *419*
Borsetti, A. P., 205(419), *217*
Borsi, J., 321(197), 323(197), *338*
Bose, A. K., 182(370, 371), *215*
Bosshard, H. H., 29(28), 30(28), 32(28), 35(28), 75(207), 86(207), *96, 101,* 229 (46, 47), 232(47), 302(169), 309(169), 318(47), *334, 338*
Botlan, D. Le, 48(114), 49(114), 78(262), 80(114), 81(262), 86(114), *98, 102*
Bott, K., 147(220), 194(393), *212, 216*
Bouchet, P., 539(17, 18), 541(17), 544(17), 547(18), 548(69), 556(18), *582, 583*
Bourn, A. J. R., 63(167), 77(167), 78(167b), *100*
Bowyer, D. P., 316(185), *338*
Boyd, G. V., 75(227), 78(227, 254), *101, 102*
Boyton, H. G., 226(10a), *333*
Bradley, R. B., 78(252), *102*
Bradley, W. E., 134(179), *211*
Bradsher, C. K., 193(392), *216*

Brady, O. L., 572(164), 574(164), *586*
Brady, W. T., 509(113d), 515(113d), 518(113d), *531*
Braguzzi, F., 330(208), *339*
Bränden, C.-I., 93(348), *104*
Braun, J. von, 122(129), *209*, 423(4), 428(4), 430(23b), 452(4), *528, 529*
Breckwoldt, J., 79(279), 83(279), *102*, 118(95), 153(95), *209*
Bredereck, H., 26(8), 75(8, 224, 225), *96, 101*, 229(58), 237(58), 268(116), 302 (168), 308(168), 312(179), 328(203), *334, 336, 338*, 360(39), 361(39), 371(39, 50), *417*, 503(112), *531*
Bredereck, K., 328(203), *338*, 360(39), 361(39), 371(39), 402(104), *417, 419*
Brehme, R., 323(196), *338*
Breitmaier, E., 66(180), *100*
Breitmaier, W., 340(224), *342*
Bremanis, E., 156(289), *213*
Bren, V. A., 575(183), *586*
Breslow, R., 118(94), *209*
Bridson, J. N., 196(398), *216*
Briggs, P. R., (317), *103*
Brizzolara, A., 114(53), *208*
Broadeus, C. D., 29(21), 75(21), *96*, 142 (200), *211*
Broese, R., 162(310), 173(310), 181(367), *214, 215*
Brown, C., 581(211), *587*
Brown, D. M., 314(183), *338*
Brown, R. T., 402(103b), *419*
Bruckner, W. A., 307(166), *337*
Bruckner-Wilhelm, A., 321(197), 323(197), *338*
Brun, L., 93(348), *104*
Bruylants, A., 78(384), *105*
Buchler, J. W., 39(64), *97*, 116(77), *208*
Buczkowski, Z., 560(121), *585*
Buder, W., 43(94), 75(94), *98*
Buehler, E., 572(163), *586*
Buenker, R. J., 6(23), *19*
Bui, Xuan-Hoa, M., 473(91), 491(91), 492 (91), 493(91), 494(91), *531*
Buijle, R., 41(77), 56(77), 69(77), *97*
Bunzl, K. W., 75(366), *104*
Burg, A. B., 144(208), *211*
Bürger, W., 129(165), 177(353), 180 (165), 181(165), 200(165), 202(165), 203(165), *210, 215*
Burke, S. S., 221(437), *223*
Burn, D., 227(39), 266(39, 114), 274(114), *334, 336*
Busing, W. R., 91(337), *104*
Butler, R. N., 580(200), *587*
Buyle, R., 302(155), *337*, 372(53), 407(53), *417*, 432(31a, 31b, 31c), 433(31a, 31b, 31c, 31d), 437(40), 438(40), 440(31a, 31d, 40), 454(31a, 31b), 455(31a, 31b, 31c, 40), 456(31a, 40), 457(31b, 31c), 458(31c), 465(31a, 31c), 466(31a, 31b, 31c, 31d), 474(31c), 475(31c), 478(31a), 480(31a), 481(31a), 502(31d), *529*

Cabaret, D., 157(293, 294, 296), 186(294), *213*
Cabell, M., Jr., 150(263), *213*
Cabré, F. R. M., 219(432), *222*
Cagniant, P., 341(232), *342*
Caillaux, B., 376(63), *417*
Calin, M., 47(108), *98*
Callot, H., 54(153), 63(153), 67(153), *99*
Camerman, A., 93(347), *104*
Campbell, R. D., 539(3), 553(3), *582*
Candy, C. F., 240(81), *335*
Cantacuzéne, D., 439(44), 460(44), 461(44), *529*
Caple, G., 297(151), *337*
Carlson, R., 116(69c), *208*
Carr, R. L. K., 42(86), *98*
Carrié, R., 34(48), *97*, 150(260, 261), 220 (436), *213, 223*
Carvey, F. A., 579(194), *587*
Cashell, P. A., 560(123a), *585*
Cavé, A., 28(13), *96*, 144(207), *211*
Cerutti, P., 31(36), 32(36), *96*, 206(422), *217*
Cervinka, O., 52(135), 75(208), *99, 101*, 157(292), 158(299), *213, 214*
Challis, B. C., 379(70), *418*
Challis, J. A., 379(70), *418*
Chandramohan, M. R., 312(176), *338*
Chang, H. S., 86(308), *103*
Chang, S. C., 402(103b), *419*
Chapelle, J.-P., 539(14, 15), 540(15), 547 (15), 553(14, 96), 554(14, 15), *582, 584*
Chauvière, G., 157(293, 294, 296), 186 (294), *213*
Cheng-fan, I. J., 221(437), *223*
Chentli-Benchikha, F. J., 90(329, 332), 91(335), 92(340), *104*
Cherbuliez, E., 119(100, 101), *209*
Cherkashin, M. I., 340(220), *342*
Chetin, A., 241(84), *335*
Chia, H. L., 297(151), *337*
Chiang, Y., 55(155), 86(155, 299), *99, 103*
Chiba, T., 314(174), *338*
Childs, A. F., 162(307), *214*
Chizhov, A. K., 241(85), *335*
Chizov, N. K., 340(227), *342*
Chupp, J. P., 111(20b, 20c), 162(20b, 20c), *207*

AUTHOR INDEX

Ciabattoni, J., 150(263), *213*
Ciernik, J., 309(177), 312(177), *338*
Ciganek, E., 149(238), *212*
Clamot, B., 377(116), *419*
Clark, B. A., 315(182), *338*
Cleghorn, H. P., 95(359), *104*
Clemens, D. H., 69(183), *100*
Clementi, R., 165(317), *214*
Clezy, P. S., 237(74), *335*
Coffman, D., 461(70), *530*
Coffman, D. D., 89(320), *103,* 230(61), *334,* 358(31), *417*
Coispeau, G., 539(13), 540(29), 543(50), 553(13), *582, 583*
Coleman, H. A., 558(112), 563(112), *585*
Colens, A., 30(33), 68(33), *96,* 431(29a, 29b), 439(29a), 441(29a), 446(29a), 447(29a, 61), 452(29a), 453(29a), 460 (29a), 461(29a), 463(29a), 464(29a), 465(29a, 29b), 466(29a, 29b), 467(29a), 469(29a), 472(29a), 477(29a), 478(29a), 479(29a), 488(29a), 489(29a), 490(29a), 496(29a), 497(29a), 500(29a), 501(29a), 506(29a), *529, 530*
Collard, J., 78(256), *102,* 384(83), 385(83), 386(83), *418,* 428(13), *528*
Collet, C., 163(312), 169(312), *214*
Colman, P. M., 91(333), *104*
Colson, J. G., 42(86), *98*
Comay, P., 284(147, 148), 285(148), 297 (148), *337*
Combelas, P., 75(229, 231, 232), *101*
Comrie, A. M., 564(127), *585*
Concannon, P. W., 394(90), *418*
Condon, F. E., 550(88), *584*
Coniac, N., (154), *337*
Consonni, A., 265(113), 275(130), 288 (130), *336*
Conti, F., 79(287, 288), *103*
Cook, A. G., 496(106), *531*
Cook, D., 43(89, 91), 54(91), *98*
Cooley, G., 266(114), 274(114), *336*
Cope, A. C., 149(238), *212*
Copeland, R. F., 4(15), *19,* 75(197), 89 (197), 90(197), *100,* 150(273), *213*
Cornu, P.-J., 284(147, 148), 285(148), 297 (148), *337*
Corral, R. A., 148(234, 235), *212*
Cosink, T. A., 151(297), *337*
Cossar, B. C., 123(136), 165(136), 201(136), 201(407, 408), *216*
Cossement, E., 476(94), 477(94), *531*
Cossey, A. L., 329(204a, 204b, 204c), *339*
Costisella, B., 29(25), *96,* 136(185), *211,* 162(309), 165(309), 177(353), *214, 215*

Courseille, C., 89(390), *105*
Coustumer, G. le, 78(262), 81(262), 84(285), *102, 103*
Cox, L. E., 94(351), *104*
Craig, D., 519(124a), *532*
Cramer, F., 94(356), *104*
Cramer, J., 571(157a), *586*
Crenshaw, R. R., (205), *339*
Crepaux, D., 51(134), 63(134), 77(239), *99, 101*
Crist, D. R., 205(419), *217*
Cross, H., 47(110), *98*
Cruege, F., 75(229), *101*
Csizmadia, I. G., 580(201), 587
Csürös, Z., 75(219, 220), *101,* 402(105), *419*
Culbertson, B. M., 569(143), 570(143), *585*
Curl, R. F., 4(12), *19*
Curtis, R. M., 4(16), *19*
Cuttica, A., 330(208), *339*

Dabard, R., (381), *105*
Dähler, G., 165(319, 320), 166(319), *214*
Dahms, G., 89(322), *103*
Dähne, M., 122(133), 145(133), 171(339), *210, 215*
Dähne, S., 24(2), 39(65), 40(65), 48(113), 49(2, 113), 65(174), 69(65, 186), 78(259), *96, 97, 98, 100, 102,* 118(89), *208,* 160 (303), *214,* 285(144), *337,* 331(215), 332(215), *339,* 396(94), *418*
Dalberga, I. E., 218(430), *222*
Dallacker, F., 232(64), *334,* 236(69), *335*
Daltrozzo, E., 48(112), 56(112), 85(112, 294), 86(112, 294), *98, 103*
Damico, R., 29(21), 75(21), *96,* 142(200), *211*
Damien, J. M., 465(73b), 498(73b), 499 (73b), *530*
Danion-Bougot, R., 34(48), *97,* 150(260), 220(436), *213, 223*
Danion, D., 34(48), *97,* 150(260), 220(436), *213, 223*
Danyluk, S. S., 94(355), *104*
Dargelos, A., 549(71a), *584*
Datta, A. P., 228(40), *334*
Dauben, H. J., Jr., 37(56), 69(56), 71(56), *97,* 143(203), *211*
Daunis, J., 565(130), *585*
Davies, M., 40(68), 97
Davies, M. T., 266(114), 274(114), *336*
Davis, F. A., 581(212, 213), *587*
Davis, G. T., 150(255), *213*
Dayagi, S., 108(3), 206
Dayal, B., 182(370), *215*

AUTHOR INDEX

Dean, J. M., 235(67), *335*
Dean, R. R., 77(239), *101*
Decker, H., 109(12), *207*
Declercq, J.-P., 89(392), 90(329, 330, 332), 91(335), *104, 105,* (118), *419,* 431(29b), 465(29b), 466(29b), *529*
Deffrenne, M. A., 35(53), *97,* 363(45), 365 (45, 48), 366(45, 48), 367(48), 368(45, 48), 370(48), 371(48), *417,* 435(36a), 454(36a), *529*
Degener, E., 147(227), *212,* 356(26, 27), *416, 417*
Degnani, Y., 108(3), *206*
Dehmel, H., 148(232), 162(311), 167(311), *212, 214*
Deikalo, A. A., (415), *217*
Dejonghe, J. P., 476(95), 477(95), *531*
Delavarenne, S. Y., 442(53a, 53b), 444(53a), *530*
Delhalle, J., 470(82), 510(82), *530*
Delpierre, G. R., 571(159), 580(196), *586, 587*
DeMayo, P., 570(153), *586*
Demerseman, P., 241(84), *335*
Demoulin, A., 484(100), *531*
Demuylder, M., 446(60), 451(60), 452(60), 453(60), 457(60), 461(60), 465(60), 469 (60), 471(87), 472(87), 473(87), 474(87), 512(60, 514(60), *530, 531*
Denkstein, J., 135(181), 138(181), *211,* 169(331), *214*
Dennis, W. H., Jr., 150(256), *213*
DePoortere, M., 522(125), 523(125), *532*
Derkatsch, G. I., 403(106), *419*
Desai, P., 94(356), *104*
Deubel, H., 304(162), 306(162), *337*
Dewar, M. J. S., 2(2a, 2b, 3), *19*
Deyrup, J. A., 78(382), 89(382), *105,* 217 (423), 219(423), *222*
Dick, A., 162(310), 173(310), *214*
Dickinson, D. A., 349(6), *416*
Dickore, K., 331(213), 332(213), *339*
Diepers, W., 192(387, 388), 193(387), *216*
Dieter, N., 303(159), 304(159), 305(159), 306(159), *337*
Dietrich, M. A., 443(54), 462(54), *530*
Dilcher, H., 567(141), 568(141), *585*
Dillard, J. G., 3(10), 7(10), *19*
Dimroth, O., 226(1), *333*
Dingwall, J. G., 233(65), 235(65), 242(65), *334*
Ditchfield, R., 2(4b), 3(4b), *19*
Dittli, C., 539(19, 20, 21), 556(20, 21), *582*
Dittus, G., 442(51), 456(51), *530*
Dixneuf, P., (381), *105*

Dobeneck, H. von, 302(161, 163), 303(159), 304(159, 160, 161, 162, 163), 305(159, 160, 163), 306(159, 160, 161, 162, 163), *337*
Doering, W. von E., 226(10a), *333*
Dolbier, W. R., 581(207), *587*
Dolgushina, I. Y., 119(119, 120), 168(120), 175(120), *209*
Dombro, R., (117a), *585*
Donia, R. A., 166(324), *214*
Donohue, F., 93(345), *104*
Dorn, H., 540(26), 558(109, 110), 563(109, 110), 567(141), 568(141), 569(145, 146, 150), 570(26, 145, 146, 152), *582, 584, 585, 586*
Dornseiften, J. W., 38(58), *97*
Dorofeenko, G. N., 227(34), 235(34), 237 (34), 258(34), 328(34), *334*
Dörr, F., 85(293), 86(293), *103*
Dörscheln, W., 31(36), 32(36), *96,* 206(422), *217*
Dos 2, 237, 879 (Henkel Cie), 390(88),*418*
Dougherty, T. J., 40(73), 84(73), *97*
Drach, B. S., 119(115a, 119, 120), 168(120), 175(120, 350), 177(356, 359, 359a), *209, 215*
Dreiding, A. S., 509(113c), 515(113c), 518(113c), *531*
Dressler, V., 278(133), 289(133), *336*
Drewry, D. T., 283(139), *337*
Driesen, G., 119(98), 181(98), *209*
Dronkina, M. I., 348(2), 350(2), 352(2), 358(2, 33), *416, 417*
Droxyl, K., 3(10), 7(10), *19*
Dubinina, T. N., 348(5), *416*
Duchêne, G., 25(5), 30(5), 31(5), 32(5), 56(156), 65(5), 67(5), 77(5), 88(5), *96, 99,* 350(8), 352(8), 353(8), 354(8), 379 (69), 380(69), 383(69), *416, 418*
Ducker, J. W., 266(114), 274(114), *336*
Duffin, G. F., 542(48), *583*
Duguay, G., 242(89), 243(89), *335*
Duhamel, L., 29(24), *96,* 112(49), 115(67), 116(72), 116(74), 126(74, 151), 127 (151), 160(151), 163(312), 169(312), 189(376), 189(377), 219(429), *208, 210, 214, 216, 222*
Duhamel, P., 29(24), *96,* 112(49), 115(67), 116(72), 116(74), 126(74, 151), 127 (151), 160(151), 163(312), 169(312), 189 (376), 189(377), *208, 210, 214, 216*
Dumont, W., 514(121b), 518(121b), *532*
Dunn, F. P., 572(164), 574(164), *586*
Dunning, K. W., 135(183), *211*
Dunning, W. J., 135(183), *211*
Dunston, J. M., 539(16), *582*

duPont de Nemours, E. I., and Co., 573(168), 576(168), *586*
Duptin, J.-F., 580(199), *587*
Dusseau, Ch. H. V., 32(38), *96,* 122(128), *209*
Dybowski, V., 169(332), *214*
Dyer, J. R., 5(14), 6(14), *19*
Dygos, D. K., 150(262), *213*

Earhart, H. W., 226(10a), *333*
Eastham, J. F., 526(127), *532*
Ebel, S., 111(21, 24), 182(24), *207,* 559 (119), 562(119), *585*
Eckell, A., 566(136, 137), 568(136, 137), *585*
Edlund, U., 112(41), *207*
Edwards, F. T., 86(308), *103*
Edwards, W. R., 226(10a), *333*
Effenberger, F., 218(428), *222,* 268(116), *336,* 371(50), *417*
Efros, L. S., 242(87), *335,* 307(165), *337*
Ege, G., 78(257), *102*
Eggerichs, T. L., 40(72), 48(72), 65(72), *97,* 69(185), *100,* 371(55), 372(55), 374 (55), 377(55, 116), *417, 419,* 435(34c), *529*
Ehrenberg, A., 166(323), *214*
Eicher, T., 118(94), *209*
Eichler, D., 28(18), *96,* 124(140), 138 (140), 169(140), 180(140), 200(405), 202(405, 410), *210, 216*
Eiden, F., 119(108, 115), 147(228), 162 (310), 173(228, 310), 181(367), *209, 212, 214, 215*
Eilingsfeld, H., 75(226), *101,* 328(202), *338,* 352(17), 410(17), *416,* 374(60), 398(60), *417,* 430(24a, 24b), 434(32), 465(24a, 24b), 473(24a), 498(32), 503(32), *529*
Eisner, V., 86(302), *103*
Eistert, B., 269(118), *336*
Elderfield, R. C., (38), *583*
Elgavi, A., 407(107), *419,* 435(37), 453 (37), 454(37), 455(37), *529*
Elguero, J., 51(129), 53(144, 145, 146, 147), 69(145, 147), 70(147), 78(242), 86(315), 75(367), 92(344), *99, 102, 103, 104, 105,* 158(300), 160(300, 302), *214,* 539(4, 5, 6, 9, 10, 11, 13, 14, 15, 17, 18, 19, 20, 21, 22), 540(4, 5, 15, 23, 24, 28, 29, 30, 32, 33), 541(17, 42), 542(49), 543(49), 544(4, 17, 42, 57), 545(49), 546(62), 547(5, 6, 10, 11, 15, 18, 62, 64), 548(5, 10, 11, 30, 67, 68, 69), 549(71a), 551(57), 553(6, 13, 14, 96), 554(14, 15, 100, 101), 556(18, 20, 21, 22, 24, 64, 101), 558(107, 111), 561(100, 129), 563(107, 111), 564(100), 565(129), *582, 583, 584, 585*
Elkik, E., 78(243, 247), *102,* 114(55), 116 (69a), *208*
Ellefson, R. D., 539(3), 553(3), *582*
Ellenberg, H., 112(36), 122(36), 133(178), 141(193), 179(36), 187(36, 178), *207, 211*
Ellis, B., 266(114), 274(114), *336*
El'tsov, A. V., 539(7, 8), 540(7, 8, 25), 547(8), *582*
Emmons, W. D., 69(183), *100,* 576(184), 578(184), *586*
Engel, R., 144(212), *211*
Engelhardt, G., 65(174), *100,* 177(353), *215*
England, D. C., 443(54), 462(54), *530*
Epishina, L. V., 66(177), 76(177), *100*
Epsztajn, J., 78(271), *102*
Eraksina, V. N., 340(228), *342*
Eremenko, L. T., 141(192), *211*
Ermili, A., 330(208), *339*
Eschelbach, F.-E., 232(64), *334*
Eschenmoser, A., 28(14), *96,* 145(216), 151(216), 152(216), 204(216), *212*
Ettenberger, F., 75(224, 225), *101*
Evans, S., 581(213), *587*
Evreinova, E. B., 559(118), 564(118), *585*

Faber, L., 327(201), *338*
Farbenfabriken Bayer, A. G., 119(112), 127(158), 131(171), 167(112), 175(348), *209, 210, 215*
Farbwerko, Hoechst, 121(126), 122(126, 135), 126(153), 177(355), *209, 210, 214*
Farina, P. R., 568(141a), 571(141a), *585*
Farnum, D. G., 539(16), *582*
Farquhar, D., 340(230), *342*
Fava, G., 544(56), 545(56), *583*
Fawcett, F. S., 89(320), *103,* 230(61), *334,* 358(31), *417,* 440(47), 456(47), *529,* 461(70), *530*
Feather, P., 266(114), 274(114), *336*
Fedder, J. E., 503(111), 504(111), *531*
Federolf, E., 150(266), *213*
Feeney, F., 49(126), *99*
Feer, G., 119(100), *209*
Feil, D., 91(337), *104*
Feinauer, R., 192(388), *216*
Feist, F., 119(114), *209*
Feldmann, K., 48(112), 56(112), 85(112, 292, 294), 86(112, 292, 294), *98, 103*
Felson-Reingold, D., 144(212), *211*

Felton, H. R., 165(317), *214*
Ferre, G., 230(60), *334*
Fersht, A. R., 86(312), *103*
Ferwanah, A., 117(269), *336*
Fick, F.-G., 181(366), *215,* 360(41b), *417,* 391(125), *419*
Field, F. H., 3(10), 7(10), *19*
Fields, E. K., 177(354), *215*
Fields, T. L., 75(236), 89(236), *101,* 118(93), *209*
Filleux, M. L., 83(277), 95(277), *102,* 229(52), 230(52), *334*
Filleux-Blanchard, M. L., 48(114), 49(114), 78(262), 80(114), 81(262), 83(276), 84(285), 86(114), *98, 102, 103,* 229(51), *334*
Fink, W., 162(308), 165(308), 189(308), *214*
Firl, J., 469(81), 470(81), *530*
Firrel, N. F., 115(62), *208*
Fischer, E. O., 33(374), 65(374), 78(374), *105*
Fischer, G. W., 86(314), *103,* 297(151), *337*
Fleckenstein, L. J., 149(238), *212*
Fleichmann, R., 566(135), 568(135), *585*
Flitsch, W., 52(138), 69(138), *99,* 119(125), *209,* 340(231), *342*
Florian, W., 255(101), *336*
Fluck, E., 40(71), 78(248), *97, 102,* 131(102, 168, 169), 155(169), 162(169), 164(169), 165(169), 183(169), *210*
Flurry, R. L., Jr., 4(15), *19,* 75(197), 89(197), 90(197), *100,* 150(273), *213*
Fodor, G., 34(52), 35(52), 78(371), *97, 105,* 149(250), *212*
Foglizzo, R., 75(203), *100*
Fonken, A. E., 75(201), 86(201), 94(201), *100*
Fontaine, B., 359(37), 360(37), *417,* 435(36c), 459(36c), *529*
Forcasiu, D., 550(88), *584*
Forster, M. O., 109(11), *207*
Fowler, R. B., 519(124a), *532*
Fox, B. L., 575(1; 8), *586*
Frankel, G., 45(102), 78(368), 84(102), *98, 105*
Franke, W., 278(134), 282(134), *336*
Franconi, C., 45(102), 79(287), 84(102), *98, 103*
Franklin, J. L., 3(10), 7(10), *19*
Franzen, V., 149(239), *212*
Fraser, J. E., 95(365), *104*
Fraser, R. R., 149(251), *212*
Fraunberger, F., 179(361, 362), *215*
Freear, J., 436(38a, 38b), 439(38b, 45, 46), 440(38a, 46), 456(46), 460(38b), 463(38a, 38b, 45, 46), 464(38a), *529*
Freedman, M. H., 42(83), *98*
Freeman, J. P., 577(189), 581(205), *587*
Freeman, R. E., 423(5a), 447(5a), 456(5a, 69), 457(5a, 69), 458(5a, 69), 465(72), 471(72), 474(72), 478(72), 479(72), 480(72), 481(72), 497(72), 498(72), 499(72), *528, 530*
Freiberg, J., 75(209), *101,* 129(164, 165), 136(164), 177(353), 180(165), 181(165), 200(165), 202(165), 203(165), *210, 215*
Freil, D., 91(338), *104*
French, Pat, 258(104), *336,* 359, 839
Fresenius, W., 191(385), 192(386), 200(385), *216*
Frey, H. O., 78(257), *102*
Fridman, A. L., 217(426), *222*
Friedrich, H. J., 85(377), *105*
Fripiat, C., 374(61), *417*
Frischleder, H., 278(137), 279(137), *337*
Fritz, H., 86(304), *103,* 229(59), 229(231), *334,* 554(98), *584*
Fryer, C. W., 79(287), *103*
Fuentes, O., 340(229), *342*
Fuhrop, J.-H., 340(222), *342*
Fujita, K., 95(360), *104,* 154(285), *213*
Fujita, T., 75(221, 222), 78(221, 222), *101*
Fuks, R., 41(77, 79), 55(79), 56(77), 69(77), *97, 98,* 199(401), *216,* 355(22), *416,* 437(40), 438(40), 440(40), 455(40), 456(40), *529*
Fulmer, R. W., 53(141), *99,* 112(40), 143(40), 150(40), *207*
Fulmor, W., 572(162), *586*

Gabriel, S., 116(79), *208,* 119(102), 148(233), *209, 212*
Gadalla, K. Z., 321(197), 323(197, 199), *338*
Gafarov, A. N., 550(89), *584*
Gafurov, R. G., 141(192), *211*
Gagan, J. M. F., 279(152), *337*
Gais, H. J., 364(43), 401(103a), *417, 419,* 438(43), 453(43), 454(43), 455(43), *529*
Gal, J., 34(52), 35(52), *97,* 88(318), *103*
Galloy, J., 28(376), 63(376), 90(330), *104, 105*
Gardent, J., (154), *337*
Garret, B. R., 165(317), *214*
Garrigou-Lagrange, C., 75(204, 231, 232), *101*
Garritsen, J. W., 384(82), *418,* 428(18a), *529*
Gase, R. A., 219(432), *222*

AUTHOR INDEX

Gash, V. W., 52(136), 53(141), 74(136), *99*, 112(40), 143(40), 150(40), *207*
Gaskin, J. E., 95(359), *104*
Gassman, P. G., 150(262), *213*
Gati, Gy., 94(352), *104*
Gavrilenko, B. B., 381(120), 382(120), *419*
Geita, L. S., 218(430), *222*
Geoffre, S., 89(390), *105*
George, P., 44(100), 47(100), *98*, 359(35), 376(63), 382(35), 383(35), 388(35, 86a), 391(35), 402(35), *417, 418*
George, T. A., 444(51), 461(51), *530*
Gerland, H., 111(25), *207*
Germain, G., 89(392), 90(329, 330, 332), 91(335), 92(340), *104, 105,* (118), *419,* 431(29b), 465(29b), 466(29b), *529*
Gerstenkorn, I., 574(171a), *586*
Gever, G., 569(147), *586*
Gey, E., 24(2), *96*
Ghosez, L., 30(32, 33), 36(55),37(55), 48 (120),56(32), 65(32, 55), 67(32), 68(33), 75,(216),*96, 97, 98, 101,* 80(25, 98, 99), 81(25, 98), 82(88), 83(88), 84(66, 100), 85(66), 86(66, 89), 89(25, 99), 90(90), 91(90), 92(86, 90), 95(50, 103), 97(25), 98(73a, 74), 99(73a, 103), 101(25, 110), 107(25), 108(25), 109(113b, 117), 110 (7, 8, 82, 118, 119), 111(7, 8), 112(7, 8, 60, 98), 113(7), 114(60, 113b), 115(7), 116(7, 98), 117(7, 98, 119), 118(114b, 119), 119(118), 120(25, 118), 121(25, 118), 122(99, 125), 123(52, 89, 98, 125), 124(99), 425(7, 8), 430(25), 431(26, 28, 29b), 434(25), 438(25), 441(50), 442, (52), 444(52),446(60), 451(25, 26, 60, 65), 452, (25, 60), 453(25, 28, 60, 66), 456(52, 66), 457(50, 52, 60, 66), 458(50), 459(50), 461(60, 61), 465(25, 26, 29b, 60, 73a), 466(29b, 74, 76), 467(74, 76), 468(7), 469(8, 60, 74), 470(25, 28, 82, 84, 85), 472(66, 88, 89), 473(25, 66, 90), 474 (25), 476(95), 477(95), 478(98, 99), *528, 529, 530, 531, 532*
Giacobbe, T. J., 383(79), *418*
Giam, C. S., 78(273), *102*
Gibson, J. D., 116(76), *208*
Giger, W., 54(152), 57(152), 61(152), 67 (152), *99,* 78(389), *105*
Gil, R., 75(367), *105*
Gil, V. M. S., 66(179), *100*
Gilbert, E. C., 561(125), *585*
Gillard, 496(104), *531*
Gilles, J.-E., 547(63), *583*
Gillespie, R. J., 42(81, 82), 43(81, 82), 63(81), *98*
Gillies, D. G., 77(238), *101*
Giner-Sorolla, A., 314(183), *338*
Gjøs, N., 238(76), *335*
Gloede, J., 75(209), *101,* 129(164, 165), 136(164), 157(298), 180(165), 181(165), 200(165), 202(165), 203(165), *210, 214,* 429(21), *529*
Gnichtel, H., 581(206), *587*
Godtfredsen, W. O., 569(148), *586*
Goff, D. L., 581(213), *587*
Goffin, E., 41(78), *98*, 384(84), 385(84), 386(84), 387(84), 388(84), *418*, 428(12b), *528*
Gold, H., 49(122), 69(122), *99, * 229(43), *334*
Gol'dfarb, Y. L., 238(77, 79), *335*
Goldman, N. L., 144(212), *211*
Goldstein, J. H., 467(78), *530*
Goldstein, R. F., 572(164), 574(164), *586*
Goldstein, S., 144(212), *211*
Goldsworthy, L. J., 162(307), *214*
Golubeva, G. A., 559(116), 564(116), *585*
Gompper, R., 26(8), 43(92), 50(127), 75(8), 84(284), *96, 98, 99, 105,* 177(355), *215,* 229(58), 237(58), 328(203), *334, 338,* 377(65), *417,* 503(112), 509(115b), *531*
Gonzalez, E., 548(68, 69), *583*
Goossens, D., 486(102), 488(102), *531*
Gorbatenko, V. I., 390(125), *419*
Gorissen, H., 484(100), *531*
Gorissen, J., 25(5), 30(5), 31(5, 35), 32(5), 65(5), 67(5), 77(5), 88(5), 350(8), 352 (8), 353(8), 354(8), 358(32), 385(32, 122), *96, 416, 417, 419*
Goshorn, R. H., 351(14), *416*
Göth, H., 31(36), 32(36), *96,* 206(422), *217*
Goto, T., 78(269), *102*
Goubeau, J., 28(11), 74(11), 75(11), *96,* 146(219), *212*
Gough, Th. T., 340(230), *342*
Goulden, J. D. S., 44(98), *98*
Gowenko, B., 372(56), 374(56), *417*
Grahn, W., 78(387), *105*
Grandberg, I. I., 549(71), 555(71), *584,* 555(71b), *584,* 559(116, 117, 118), 564 (116, 117, 118), *585*
Grant, D. M., 66(176), 76(176), *100*
Grashey, R., 542(46, 47), 566(133, 135), 567(133, 135), 568(133, 142), 579(133), *583, 585*
Grätz, J. Grätzel von, 112(39), 160(39), *207*
Grayson, B. T., 581(211), *587*
Grdinic, M., 34(51), 35(51), *97*
Green, R., 165(317), *214*

Greenberg, H., 165(317), *214*
Greenwald, R. B., 566(138, 139), 569(139), 570(139), *585*
Gribble, G. W., 150(257), *213*
Grieder, A., 86(311), *103,* 538(2), 546(2), *582*
Griesinger, A., 112(43), 184(372), 187(373), 205(43), *207, 215*
Griffin, G. E., 54(154), 68(154), 77(154), *99*
Grigat, E., 382(77), *418*
Grigoryan, D. V., 194(395), *216*
Grill, W., 94(353), 95(353), *104*
Grillot, G. F., 142(198), *211,* 165(317), *214*
Grimwade, M. J., 254(99), 255(99), 256(99), *335*
Grinev, A. N., 241(85), *335,* 340(227), *342*
Grinvalde, A. K., 218(430), *222*
Grivas, J. C., 40(70), 75(70), *97*
Grob, C. A., 149(247), *212*
Grohe, K., 147(226, 227), 157(226), *212,* 356(27), *417*
Gronowitz, S., 238(76), *335*
Gross, H., 29(25), 75(209), *96, 101,* 129 (164, 165), 136(164, 185), 145(218), 157(298), 162(309), 165(309), 177(353), 180(165), 181(165), 190(384), 200(165), 202(165), 203(165), *210, 211, 212, 214, 215, 216,* 228(42), *334,* 429(21), 473 (92a), 491(92a), *529, 531*
Gross, N., 573(169), *586*
Grosse, D., 48(115), 80(115), 81(115), *98*
Grundmann, C., 235(67), *335*
Gruner, F., 574(171a), *586*
Guerret-Rigail, M., 565(130), *585*
Guibé-Jampel, E., 86(306), *103*
Guingant, A., 568(141c), *585*
Gund, P., 4(17), 13(17), 14(17), *19*
Gurke, A., 340(231), *342*

Haack, A., 226(2), 229(2), 233(2), *333*
Haage, K., 190(384), *216*
Haake, M., 6(22), 16(22), 17(22), *19,* 73 (192), *100,* 126(147, 149, 150), 131(173), 145(173), 152(149), 157(147, 149, 297), 169(334), 170(173, 336, 339), 176(337), 177(150), 183(147), 184(147, 149), 186 (147), 187(297), 203(173), 206(147), 220 (435), *210, 211, 213, 214, 223,* 425(10), 465(10), 473(10), 490(10), *528,* 535(1), 540(1), 547(1), 553(1), 560(1), 562(1), *582*
Haas, A., 89(322), *103*
Haas, D. J., 91(336), *104*
Hafner, K., 236(68), 255(100), 257(100), 266(100), 331(100), *335,* 364(43),
371(51), 401(103a), *417, 419,* 438(42, 43), 453(42, 43), 454(43), 455(43), *529*
Häfner, K. H., 255(100), 257(100), 266 (100), 331(100), *335*
Häfner, L., 180(364), *215*
Hagemann, H., 111(23), *207*
Hahn, V., 34(51), 35(51), *97*
Haider, A., 163(312), 169(312), *214*
Hain, K., 119(121), *209*
Halbritter, K., 318(190), *338*
Hall, H. K., Jr., 237(72), *335*
Halleux, A., 115(65), 116(71), *208*
Hallmann, F., 430(23a), *529*
Hallot, A., 284(147, 148), 285(148), 297 (148), *337*
Hamelin, J., 34(48), *97,* 150(260), 220(436), *213, 223,* 570(151), 579(151), *586*
Hamer, J., 571(158), 580(195), *586, 587*
Hamilton, E. E. P., 112(34), *207*
Hamilton, W. C., 93(349), *104*
Hammerum, 550(85, 86), *584*
Hammond, G. S., 40(73, 74, 75), 82(74, 75), 84(73), *97*
Hancock, K. G., 349(6), *416*
Handa, R., 92(341), *104*
Hanlon, S., 75(199, 200), *100*
Hansen, B., 302(163), 304(163), 305(163), 306(163), *337*
Hansen, H., 571(157a), *586*
Hansen, H. V., 315(181), *338*
Hansen, K. C., 175(351), *215*
Hansson, C., 150(268), *213*
Harada, K., 108(3), *206*
Harano, K., 111(27), *207*
Harder, R. J., 78(240), 89(240), *101,* 151 (275), *213,* 357(29), *417*
Harding, G. F., 162(307), *214*
Hardy, W. B., 144(209), *211*
Hargreaves, J. R., 72(190), 85(190), *100,* 115(61), *208*
Harkern, S., 91(337), *104*
Harnisch, H., 295(142), *337*
Harris, D. L., 60(160), *99*
Harris, D. R., 91(336), *104*
Harris, J. F., Jr., 462(71), *530*
Harris, N. D., 360(40), *417*
Harris, R. L., 329(204a, 204b, 204c), *339,* 352(18), *416*
Harris, R. L. N., 78(255), *102*
Harrison, A. C., 3(8), *19*
Harteminck, W. A., 355(22), *416*
Hartke, K., 45(104), 46(104), 75(216), 78 (218), *98, 101,* 109(17a), 111(20a, 21, 28), 123(137), 125(143), 126(137), 136 (137), 141(196), 151(277), 162(20a, 196),

AUTHOR INDEX

164(196), 181(366), 194(394), 200(196), 212(409), *207, 210, 211, 213, 215, 216,* 360(41b), *417,* 391(125), *419*
Hartmann, H., 72(188), *100,* 245(94), *335,* 317(189), *338,* 341(235), *342*
Hashimoto, N., 28(14), *96,* 145(216), 151 (216), 152(216), 204(216), *212*
Hasserodt, V., 352(16), *416*
Hassner, A., 221(437), *223*
Haszeldine, R. N., 436(39), 450(39), 461 (39), 463(39), *529*
Hata, Y., 205(420), *217*
Hauck, F. P. Jr., 52(139), 74(139), *99,* 150(253), 185(253), *212*
Häufel, J., 340(224), *342*
Haug, W., 150(266), *213*
Haunste, N. K., 354(19), *416*
Haupter, F., 269(118), *336*
Hauptmann, S., 278(137), 279(137), *337*
Hauser, C. R., 109(13), 179(13), 185(13), *207*
Haveaux, B., 428(19), 430(25), 433(19), 434(19, 25), 438(25), 451(25), 452(25), 453(25), 465(19, 25), 470(19, 25), 473(19, 25), 474(25), 480(19, 25), 481(19, 25), 489(19, 25), 496(19), 497(25), 501(19, 25), 504(19), 507(19, 25), 508(19, 25), 520(19, 25), 521(19, 25), *529*
Hay, A. S., 53(141, 142), *99,* 112(40), 143(40), 150(40), 164(315), 179(315), 185(315), 187(315), *207, 214*
Hayes, L. J., 579(194), *587*
Hazebroucq, G., 227(37), 235(37), 237(37), 328(37), *334,* (154), *337*
Heatlie, J. W. M., 545(59), *583*
Hegarty, A. F., 549(74), *584*
Hegenberg, P., 444(57), 457(57), 465(57), 497(57), *530*
Hehre, W. J., 2(46), 3(4a, 4b), 7(24), *19*
Heine, H. W., 568(141b), *585*
Heinicke, R., 86(301), *103,* 245(92), 249 (92), *335*
Heiss, J., 85(292), 86(292), *103*
Heitz, L., 568(141b), *585*
Heitzer, H., 147(226, 227), 157(226), *212*
Hellmann, H., 5(20), *19,* 28(15), 29(15), 32(15), 69(15), 71(15), 74(15), 85(15), *96,* 108(4), 111(4), 112(37), 119(4), 149(240), 150(4), 151(37), 182(4), 192 (388), 202(4), *206, 207, 212, 216*
Herbig, H., 110(119), *209*
Herbison-Evans, D., 61(162), 62(162), 78 (250), 89(250), *99, 102*
Herboth, O.-E., 112(36), 121(127), 122(36), 132(174), 150(127), 155(127), 170(174), 179(36, 127), 183(127), 187(36), 200 (174), 205(174), *207, 209, 211*
Hercules, D. M., 94(350, 351), *104*
Herlem, D., 150(271), *213*
Herr, M. E., 75(201), 86(201), 94(201), 100
Herr, M. L., 466(76), 467(76), *530*
Herron, J. T., 3(10), 7(10), *19*
Hervens, F., 89(392), *105,* 382(74), 409(74, 109), 410(74, 109), 411(74, 109), *418, 419*
Hesbain-Frisque, A.-M., 30(32), 36(55), 37(55), 56(32), 65(32, 55), 67(32), 465 (73a), 466(74), 467(74), 469(74), 484 (100), 495(100), 498(73a, 74), 499(73a, 103), 510(118), 519(118), 520(118), 521(118), *96, 97, 530, 531, 532*
Hesse, K.-D., 551(91), *584*
Hesse, M., 87(316), *103*
Heyl, F. W., 75(201), 86(201), 94(201), *100*
Heymans, A., 423(4), 428(4), 452(4), *528*
Hickmott, P. W., 72(190), 75(210), 78(210), 85(190), *100, 101,* 114(59), 115(59, 61, 62, 63, 64), *208*
Hilgetagi, G., 116(69b), *208*
Hilp, M., 25(3), *96, 102,* 126(146), 127 (156), 133(146), 145(215, 217), 151(146), 152(146), 153(217), *210, 212*
Hinman, R. L., 74(195), 75(195), 86(195, 299), *100, 103,* 539(3), 553(3), *582*
Hiranuma, H., 312(175), 314(175), *338*
Hirota, K., 307(167), 308(167), *337*
Hirota, T., 318(188), 330(209), *338, 339*
Hirsch, B., 69(186), *100,* 118(89), *208,* 160(303), *214,* 285(144), *337,* 331(215), 332(215), *339,* 396(94), *418*
Hiscock, A. K., 266(114), 274(114), *336*
Hite, G., 246(96), *335*
Hitzel, V., 189(379), *216*
Hobbs, Ch. F., 39(63), 85(63), *97,* 46(106), *98,* 423(3a, 3b), 444(3a, 3b), *528*
Hoch, H., 6(21), *19,* 112(33), 114(33), 115(33), *207*
Hock, H., 75(211), *101*
Hodges, K. C., 524(126), 525(126), *532*
Hoffmann, A. R., 193(391), *216*
Hoffmann, R., 509(114a, 114b, 114c), *531*
Höft, E., 228(42), *334,* 473(92a), 491(92a), *531*
Hohlneicher, G., 85(377), *105*
Holla, B. S., 340(226), *342*
Hollowell, C. D., (13), *19*
Holtschmidt, H., 147(225, 226, 227), 157 (225, 226), *212,* 356(26, 27), *416, 417*
Holý, A., 26(9), 75(9), 95(358), *96, 104,* 226(17, 18, 21, 22, 25, 30), 229(18), 230(18), 231(17, 18), 262(21), 263(21), 263(111), 268(22), 289(21), 292(21),

297(17, 18), 300(17), 302(30), 309(21, 22), 318(194), 319(194), 325(25), 327(25), 331(214), 332(214), *333, 336, 338, 339*
Holysz, R. P., 75(234), *101,* 116(68), 186(68), *208*
Honig, B., 18(29), *20*
Hooks, H., 235(67), *335*
Hoornaert, C., 510(118), 519(118), 520(118), 521(118), *532*
Hoover, F. W., 119(109), 132(109), *209*
Hooz, J., 196(398), *216*
Hopkins, B. J., 72(190), 75(210), 78(210), 85(190), *100, 101,* 115(61, 62, 63), *208*
Horie, M., 116(78), *208*
Hornaert, C., 36(55), 37(55), 65(55), *97*
Horner, L., 578(190), *587*
Horning, D. E., 237(73), *335*
Hospital, M., 89(390), *105*
Houben-Weyl, 351(13), *416*
Hougardy, M.-L., 33(43), 61(43), 65(43), 67(43), *97*
Houlihan, W. J., 543(51), 547(51), *583*
House, H. I., 113(50), 115(30), *208*
House, H. O., 149(241), 196(241), *212,* 572(166), 576(166), *586*
Houtekie, M., 48(120), 56(120), *98,* 429(20), 432(20), 433(20), 434(20), 451(20), 452(20), 453(20), 454(20), 455(20), 465(20), 501(20), 504(20), 508(20), *529*
Höver, W., 75(212), 78(212), *101,* 116(73, 75), 126(73), 132(175), 173(175), 184(75), 198(75), *208, 211*
Howard, J. C., 569(147), *586*
Hub, L., 75(208), *101*
Huber, H., 484(101), *531*
Huber, L. K., 177(360), *215*
Hubert-Brierre, Y., 150(271), *213*
Hudak, W. J., 240(80), *335*
Hudson, R. F., 581(211), *587*
Huehsam, G., 278(136), 281(136), 282(136), *337*
Huisgen, R., 149(243), 164(243), *212,* 164(314), *214,* 484(101), 509(113a), 515(113a), 518(113a), *531,* 566(132, 133, 136, 137), 567(132, 133), 568(132, 133, 136, 137, 142), 579(132, 133), *585*
Hull, L. A., 150(255, 256), *213*
Humblet, C., 25(5), 30(5), 31(5), 32(5), 65(5), 67(5), 77(5), 88(5), *96,* 350(8), 352(8), 353(8), 354(8), 400(102), 406(102), *416, 419*
Hünig, S., 6(21), *19,* 75(211), *101,* 86(298), *103,* 109(16), 112(33), 114(33), 152(33), *207,* 571(157, 157a), *586*
Hunkapiller, M. W., 66(178), 67(178), 77(178), 78(178), *100*
Hunt, W. C., 122(131), *210*
Hupe, D. J., 150(269), *213*
Huppatz, J. L., 329(204a, 204b, 204c), *339*
Husson, H.-P., 28(13), 96, 144(207), *211*
Hyde, C. L., 540(36), *583*
Huys, F., 355(25), *416*
Huys, M., 374(58), 378(58), *417*

Idoux, J. P., 550(84), *584*
Inbal, Z., 119(122), *209*
Inukai, T., 115(60), *208*
Ioffe, B. V., 541(43, 44), *583*
Ioko, A., 3(8), *19*
Isobe, M., 78(269), *102*
Issleib, K., 176(352a), *215*
Ito, H., (117a), *585*
Ito, I., 318(188), *338*
Ito, Y., 33(44), *97,* 110(19), 114(19), *207,* 396(96), *418*
Ivanov, B. E., 139(188), 177(188), *211*
Iwakura, Y., 574(177), *586*
Iwamura, H., 219(433), 221(433), *222*

Jack, J. J., 94(350, 351), *104*
Jacobsen, C., 354(19), *416*
Jacobson, R. A., 89(326), *103*
Jacquier, R., 51(129), 53(143, 144, 145, 146, 147), 69(145, 147), 70(147), 80(242), 86(315), 92(343), 75(367), *99, 102, 103, 104, 105,* 158(300), 160(300, 302), *214,* 539(4, 5, 6, 9, 10, 11, 12, 14, 15, 17, 18, 19, 20, 21, 22), 540(4, 5, 15, 23, 24, 28, 29, 30), 541(17, 42), 542(49), 543(49), 544(4, 17, 42, 57), 545(49), 546(62), 547(5, 6, 10, 11, 12, 15, 18, 62, 64), 548(5, 10, 11, 12, 30, 66, 68), 551(57, 93), 553(6, 14, 96), 554(14, 15, 99, 100), 556(18, 20, 21, 22, 24, 64, 103), 558(103, 111), 561(100, 129), 563(103, 111), 564(100), 565(129, 130), 572(161), 573(171), 574(171), *582, 583, 584, 585, 586*
Jacquignon, P., 340(218), *342*
Jäger, V., 429(22), *529*
Jain, S. M., 341(234), *342*
Jakovenka, T. I., 561(124), 563(124), *585*
Jann, K., 31(37), 42(37), *96,* 205(417), *217*
Janousek, Z., 25(5), 30(5, 34), 31(5), 32(5), 35(53, 54), 36(54), 41(54), 47(34, 54), 48(54, 119), 49(54, 119, 124), 69(185), 73(191), *96, 97, 98, 99, 100,* 348(3), 350(3, 7, 8), 352(7, 8), 353(7, 8), 354(8), 357(28), 360(7), 361(42), 363(7, 45), 364(7), 365(7, 45), 366(7, 45), 368(7, 45), 369(7), 371(7, 49, 55), 372(7, 49, 55),

373(7), 374(7, 49, 55), 375(7), 376(63), 377(7, 55, 65), 379(7), 380(3, 7), 381(7, 73), 383(7), 384(7, 83), 385(83, 122), 386(7, 83), 387(7), 389(73), 391(3, 7), 398(98), 399(7, 98), 400(7, 101), *416, 417, 418, 419,* 425(11), 428(13), 435(34a, 34b, 35, 36a), 454(35, 36a), 455(35), 465(11), 473(11), 490(11), *528, 529*
Janssen, M. J., 44(97), 45(97), 69(97), 72(189), 86(97, 189), *98, 100,* 157(291), *213*
Januszewski, H., 575(179), 577(179), *586*
Janz, G. J., 94(355), *104*
Jarovenko, N. N., 348(1), 350(1), *416*
Jaul, E., 351(12, 14), *416*
Jayanth, M. R., 312(178), 313(178), *338*
Jencks, W. P., 86(312), *103,* 577(188), *587*
Jennings, W. B., 34(373), 84(373), *105*
Jenny, E., 75(207), 86(207), *101,* 302(169), 309(169), *338*
Jensen, L. H., 93(347), *104*
Jentzsch, W., 39(62), 75(62), *97*
Johnson, D. S., 540(36, 37), 541(37), 542 (37), *583*
Johnson, J. L., 75(201, 234), 86(201), 94 (201), *100,* 116(68), 186(68), *208*
Johnson, S. L., 78(266), *102*
Johnson, T. E., 168(327), *214*
Jonas, V., 75(198), 79(283), 84(283), *100, 102*
Jones, C. F., 316(185), *338*
Jones, R. A., 240(81), *335*
Jones, R. A. Y., 74(196), 78(264), 86(196), *100, 102*
Jongejan, E., 32(38), 78(370), *96, 105,* 122(128), *209*
Jugie, G., 95(394), *105*
Junga, M., 180(365), 181(365), *215*
Jürgens, E., 578(190), *587*
Jutz, C., 40(66, 67), 42(67), 48(66, 115, 116), 49(66), 71(187), 78(260), 80(115), 81(115), 85(116), 86(116, 301), *97, 98, 100, 102, 103,* 143(204), 160(304), *211, 214,* 227(38), 229(48), 231(62), 245(92), 249(92), 250(93), 251(93), 251(98), 252(98), 254(98), 257(98), 258(102), 263(110), 302(170), 311(170), 316(187), 317(187), 319(192), 325(192), 319(193), 321(193), 325(193), 325(200), 331(211), 332(211), 332(216), *334, 335, 336, 338, 339,* 360(41a), *417*
Jutz, Ch., 33(42), 69(42), 74(42), 86(42), *97*

Kaderábek, V., 169(331), *214*
Kagal, S. A., 150(295), *337*

Kakodkar, S. V., 246(96), *335*
Kalbfus, W., 33(374), 65(374), 78(374), *105*
Kalik, M. A., 238(79), *335*
Kalinowski, H.-O., 65(173), *100,* 78(385), *105,* 550(83), *584*
Kalischer, G., 274(125), *336*
Kamerdinerov, V. G., 541(40, 41), *583*
Kaminski, J. M., 581(212), *587*
Kan-Fan, Ch., 28(13), *96,* 144(207), *211*
Kapur, J. C., 182(370, 371), *215*
Karabatsos, G. J., 550(75, 76, 77), *584,* 550(81), *584*
Kárpáti-Adám, E., 75(219), *101*
Karplus, M., 18(29), *20*
Kass, M. B., 205(419), *217*
Kata, K., 112(47), *207*
Katagiri, I., 75(221, 222), 78(221, 222), *101*
Kato, T., 312(175), 314(174, 175), *338*
Katritzky, A. R., 52(135), 74(196), 78(264, 265, 271, 388), 86(196), *99, 100, 102, 105,* 580(198), *587*
Katsuragawa, K., 110(19), 114(19), *207*
Katsuragawa, S., 33(44), *97,* 396(96), *418*
Katsuse, Y., 330(209), *339*
Katzke, R., 69(236), *335*
Kausen, H., 85(293), 86(293), *103*
Kawashima, T., 230(60), *334*
Kawazoe, Y., 63(171), *100*
Kazuo, H., 574(177), *586*
Keenan, T. R., 33(45), *97,* 92(118), 205 (418), *209, 217*
Keikichi, U., 574(177), *586*
Keitzer, G., 122(134), 133(134), *210,* 180(363), 181(363), *215*
Keller, K., 274(125), *336*
Kellner, K., 177(357), *215*
Kelly, A. H., 315(182), *338*
Kelly, P., 147(222), 150(222), *212*
Kendall, J. D., 542(48), *583*
Kendall, M. C. R., 150(269), *213*
Kenyon, G., 18(28), *20*
Kermer, W.-D., 316(186), *338*
Kessler, H., 45(103), 46(103, 105), 65(173), 78(385), 80(103, 105), *98, 100, 105,* 549(72), *584,* 550(83), *584*
Kesslin, G., 544(55), *583*
Khafizov, Kh., 196(399), *216*
Khedija, H., 57(158), *99,* 302(173), 309 (173), *338*
Khmelnitski, L. I., 66(177), 76(177), *100*
Khuong-Huu, F., 150(271), *213*
Kiefer, H., 111(22), *207*
Kiesel, R., 147(223), *212*
Kiltz, H-H., 28(19), 67(19), 69(19), *96,*

142(201), 151(201), 157(201), 183(201), 211
Kilwing, W., 124(141), 210
Kimling, H., 78(550), 584
King, F. E., 162(307), 214
King, G. S. D., 91(335), 104, 199(401), 216
King, R. B., 524(126), 525(126), 532
Kintzinger, J. P., 51(133), 63(133), 99
Kinus, M., 78(388), 105
Kiousky, T. E., 574(173), 577(173), 586
Kiovsky, T. E., 51(131), 99
Kipping, F. B., 175(344), 215
Kira, M. A., 307(166), 337, 321(197, 198), 323(197, 199), 338
Kirchlechner, R., 258(102), 336, 319(192), 325(192), 338
Kirillova, G. V., 272(121), 336
Kirk, D. N., 266(114), 274(114), 336
Kirrmann, A., 116(69a), 208
Kirsanov, A. V., 119(119), 175(350), 177 (356, 359a), 209, 215, 348(4, 5), 350(4), 355(4), 358(30), 380(4), 382(76), 390 (87), 395(91), 416, 417, 418, 450(64), 453(64), 465(64), 530
Kirsanova, N. A., 381(72), 389(72), 418
Kirsch, G., 341(232), 342
Kirschner, A., 303(164), 337
Klages, F., 94(353, 354), 95(353), 104
Klainer, J. A., 29(23), 33(23), 34(23), 43(23), 57(23), 69(23), 96
Klauke, E., 147(226), 157(226), 212
Klemm, K., 26(8), 75(8), 96, 229(58), 237(58), 328(203), 334, 338, 503(112), 531
Kliegel, W., 119(97), 124(138, 139), 125 (142), 127(97, 142), 209, 210, 567(140), 568(140), 585
Kloetzel, M. C., 52(137), 75(137), 99
Kloss, P., 276(131), 278(131), 279(131), 280(131), 281(131), 282(131), 336
Klotz, I. M., 75(199, 200), 100, 75(230), 78(230), 90(331), 101, 104
Klug, W., 89(322), 103
Klutchko, S., 315(181), 338
Knoll, F., 25(4), 28(4), 63(4), 94(4), 96, 152(282), 154(282), 155(282), 157(282), 176(282), 213
Knotz, F., 119(105), 175(346, 349), 209, 215
Kobayashi, S., 330(209), 339
Köbrich, G., 190(380), 216
Koch, L., 128(162), 131(162), 139(162), 145(215), 177(162), 210, 212
Kochendörfer, G., 258(103), 336
Koelle, U., 550(80), 584
Köhler, E., 108(9), 112(38), 128(38, 162), 129(9), 131(162), 134(9), 135(38), 139 (162), 140(38), 144(38), 155(9), 177 (162), 207, 210
Köhler, P., 122(130), 209
Kojima, A., 455(68), 530
Kolbeck, W., 149(243), 164(243), 212
Kollman, P. A., 3(7), 4(17, 19), 7(7), 8(7), 13(17), 14(7, 17), 15(7), 18(28), 19, 20, 24(1), 96
Kolobova, T. P., 267(115), 336
Koloskova, N. M., 238(78), 335
Komari, S., 75(221, 222), 78(221, 222), 101, 428(14a, 15), 528
König, C., 255(100), 257(100), 266(100), 331(100), 335
Konshina, L. O., 217(426), 222
Konstantinov, P. A., 238(78), 335
Korepin, A. G., 141(192), 211
Korkor, M. I., 321(198), 338
Kornet, M. J., 540(31), 582, 552(95), 553 (95), 584
Korte, A. M., 95(365), 104
Koshelev, N. Y., 242(87), 335
Koshelev, N. Yu., 307(165), 337
Kost, A. N., 203(411), 216, 555(102), 559 (116, 117, 118), 564(116, 117, 118), 584, 585
Kostjanovskij, R. G., 133(177), 211, 196 (399), 216
Kotowyez, G., 54(151), 99
Kotschy, J., 85(293), 86(293), 103
Kovats, E., 550(87), 584
Koyama, T., 318(188), 330(209), 338, 339
Krack, W., 165(320), 214
Krakover, E., 78(168), 100
Kramer, H. E. A., 50(127), 84(127, 284), 99, 103, 112(44), 207
Krämer, W., 428(16b), 528
Krause, W., 147(221), 150(221), 212, 156(288), 213
Krchňák, V., 86(307), 103, 226(31), 318 (31), 320(31), 333
Krebs, A., 32(40), 36(40), 57(40), 69(40), 71(40), 79(279, 280), 83(40, 279, 280), 86(280), 97, 102, 118(94, 95), 153(95), 209, 550(78), 584
Kreher, R., 340(221), 342
Kreienbühl, P., 29(26), 38(26), 96, 43(95), 68(95), 98, 108(7), 153(7), 206
Kreiter, C. G., 33(374), 65(374), 78(374), 105
Krekel, A., 86(304), 103
Kreuder, M., 255(100), 257(100), 266(100), 331(100), 335
Kreuzkamp, N., 122(132), 131(132), 145 (132), 170(132), 210, 166(322), 214

Kröhnke, F., 42(85), *98,* 331(213), 332 (213), *339*
Krokhina, S. S., 139(188), 177(188), *211*
Krow, G. R., 75(249), *102,* 108(8), 153(8), 206, 423(1), 424(1), *528*
Krumm, U., 25(4), 28(4), 63(4), 94(4), *96,* 152(282), 154(282), 155(282), 157(282), 176(282), *213*
Krupička, J., 95(358), *104,* 263(111), *336*
Ku, A. T., 92(342), *104*
Kučera, J., 69(184), 86(184), 95(184), *100,* 226(28), 261(107), 262(107), 318(28), 320(28), 321(28), *333, 336*
Kuchitsu, K., (13), *19*
Kucherenko, A. P., 204(416), 205(416), *217*
Kugawa, K., 230(60), *334*
Kühle, E., 351(15), 354(20), 384(81), *416, 418*
Kuhn, M., 39(59), 40(59), 41(59), 45(59), 46(59), 75(59), *97*
Kuhn, R., 75(206), *101,* 118(86), *208*
Kuhn, S. J., 78(251), *102*
Kuhn, St. J., 227(36), 235(36), 237(36), 258(36), 328(36), *334*
Kukhar, V. K., 382(76), *418*
Kukhar, V. P., 31(378), 41(378), 95(378), *105,* 348(4, 5), 350(4, 9), 352(9), 355(4, 24), 357(24), 359(34), 360(9), 361(9), 373(57), 380(4, 71), 381(71, 72, 117, 119), 382(75, 121), 389(24, 72, 119), 390(87), 395(91), 396(24, 93), 397(75), 398(75), 400(75), 404(75), 409(9), 410 (9), *416, 417, 418, 419,* 450(64), 453(64), 465(64), *530*
Kulow, E., 128(160), 131(160), *210*
Kulpe, S., 91(334), *104*
Kunath, D., 157(298), *214*
Kunert, F., 142(199), *211*
Kunovskaya, D. M., 116(81), *208*
Kuthan, J., 86(302), *103*
Kutsuma, T., 328(207), *339*
Kutulya, L. A., 340(225), *342*
Kutzelnigg, W., 40(69), 42(87), 44(96), 45(96), 46(96), 75(61, 87, 96), *97, 98*
Kuznetsov, S. G., 541(45a), *583*
Kvitko, I. Y., 242(87), *335*
Kvitko, I. Ya., 307(165), *337*
Kyziol, J., (69d), *208*

Laber, G., 226(10a), *333*
Lach, D., 377(65), *417*
Lachmann, A., 108(10), *207*
Lafferty, R. H., 116(76), *208*
Lagowski, J. M., 580(198), *587*
Lambert, J. B., 63(167, 167a), 77(167), 78(167a, 267), 86(267), *100, 102*
Lamchen, M., 118(85), *208,* 557(104, 105), 561(105), 571(159), 580(196), *584, 586, 587*
Landesman, H., 114(53), *208*
Landesman, H. K., 114(52), *208*
Landwehr, H. K., 111(20b), 162(20b), *207*
Lane, A. G., 279(152), *337*
Lang, W., 278(134), 282(134), *336*
Langdale-Smith, R. A., 558(112), 563(112), *585*
Lappert, M. F., 444(56), 461(56), *530*
Laroche, P., 514(122), *532*
Larsen, J. W., 85(296), *103*
Lascomb, J., 75(229), *101,* 75(231), *101*
Laurent, H., 256(127), 275(127, 128), 288(127, 128), 294(128), *336*
Lavine, L. R., 93(345), *104*
Lazukina, L. A., 373(57), 382(75), 395(91), 397(75), 398(75), 400(75), 404(75), *417, 418*
Leaver, D., 340(230), *342*
Lebedev, O. V., 66(177), 76(177), *100*
Lebedev, Yu. A., 581(204), *587*
Lecher, H. Z., 144(209), *211*
LeClef, B., 47(109), *98,* 357(28), 392(89), 393(89), *417, 418,* 383(78), 392(78), 394 (78), 395(78), 407(78), *418,* 435(37), 453(37), 454(37), 455(37), *529*
Lednicer, D., 109(13), 179(13), 185(13), *207*
Leftwick, A. P., 266(114), 274(114), *336*
Legrand, Y., 384(85), 409(85), *418,* 428 (12a), *528*
Lehn, J. M., 4(11), 6(11), *19,* 51(133, 134), 63(133, 134), 77(239), *99, 101*
Lehners, W., 112(36), 122(36, 133), 131 (172), 132(174), 133(134), 145(133, 172), 170(172, 174), 179(36), 187(36), 200 (174), 205(174), *207, 210, 211*
Leibfritz, D., 45(103), 46(103, 105), 80 (103, 105), *98,* 550(83), *584*
Leichtle, O. R., 544(54), 554(54), 555(54), *583*
Leitch, L. C., 144(205), 164(205), *211*
Leitz, C., 78(249), *102,* 108(8), 153(8), *206*
Lemal, D. M., 571(156), *586*
Lengyel, I., 179(362), *215*
Leonard, A. M., 511(120), 513(120), 514 (120), 519(120), *532*
Leonard, N. J., 6(22), 16(22), 17(22), *19,* 29(22, 23), 31(22, 37), 32(22), 33(23, 45), 34(22, 23), 37(22), 38(22), 42(37), 43(23), 52(136, 139), 53(141, 142), 57 (23, 157), 69(23), 75(233), *96, 97, 99, 101,* 112(40), 118(87, 92), 143(40),

150(40, 252, 253), 151(87), 152(87), 164(315), 179(315), 185(253, 315), 187 (315), 205(417, 418), *207, 209, 212, 214, 217*
Lepicard, G., 92(343), *104,* 548(66), *583*
Lerche, G., 28(16), *96,* 141(194), *211*
Leroy, F., 89(390), *105*
Lesinka, J., 114(57), *208*
Lester, M. G., 254(99), 255(99), 256(99), *335,* 271(119), *336*
Letourneau, F., 34(52), 35(52), 78(371), *97, 105,* 149(250), *212*
Leubner, I. H., 78(263), 86(263), *102*
Leuchs, H., 111(25, 26), *207*
Leupold, D., 160(303), *214,* 331(215), 332(215), *339*
Levchenko, E. S., 348(5), *416*
Levitan, L. P., 125(155), *210*
Lewis, R. E., 240(80), *335*
Lewis, W. W., 351(12), *416*
Lewis, W. W., Jr., 351(14), *416*
Ley, K., 111(23), *207*
Liberman, S. S., 242(88), *335*
Libman, N. M., 118(91), *209*
Lichter, R. L., 61(165), 63(165), 64(165), *99*
Liebscher, J., 317(189), *338,* 341(235), *342*
Liepa, A. J., 237(74), *335*
Liksandru, T. S., 281(141), 282(141), *337*
Liler, M., 63(93), 69(93), 86(93), *98*
Lin-Ho, T. F., 496(105), 497(105), 501(105), 505(105), 508(105), *531*
Linda, P., 95(361, 362), *104,* 229(55, 56, 57), 234(55, 56, 57), *334*
Lindblad, C.-G., 75(228), 78(228), *101*
Linde, R. van der, 38(58), *97*
Lindsay-Smith, J. R., 33(46), *97,* 144(210), *211*
Lindsey, R. V., Jr., 443(54), 462(54), *530*
Linnett, J. W., 4(18), *19*
Liotta, D., 144(212), *211*
Lippacher, E., 303(159), 304(159), 305 (159), 306(159), *337*
Lippmaa, E., 65(174), 66(177), 76(177), *100*
Lischewski, M., 176(352a), *215*
Litterscheid, F. M., 145(213), *211*
Litvinov, V. P., 238(77), *335*
Livingston, R., 467(77, 78), *530*
Lloyd, D., 78(261), 95(359), *102, 104,* 279(152), *337*
Lo, C. P., 167(325), *214*
Lo, D. H., 2(2a, 2b), *19*
Lockerente, S. R. de, 78(384), *105*
Loeppky, R. A., 150(258), *213*
Loeppky, R. N., 33(47), *97*
Loewenstein, A., 78(369), *105*
Lorenz, H., 226(4), 229(4), 243(4), *333*
Loseva, N. S., 302(172), 309(172), *338*
Lötzbeyer, J., 285(143), 286(143), *337*
Louisfert, J. A., 237(75), *335*
Love, A. L., 112(48), *207*
Lowe, J. U., Jr., 82(275), *102*
Luca, D. C. de, 150(255), *213*
Lucarelli, A., 229(57), 234(57), *334*
Lucken, E. A. C., 467(79), *530*
Lukasczyk, G., 94(354), *104*
Lukasiewiz, A., 114(57), *208*
Lundberg, K. L., 175(352), *215*
Lunt, E., 78(271, 388), *102*
Lüpke, N.-P., 169(332), *214*
Luth, H., 545(58), *583*
Luther, H., 195(397), *216*
Lwowski, W., 381(73), 389(73), *418*
Lyerla, J. R., 42(83), *98*
Lyle, J. L., 78(275), *102*

Maag, H., 28(14), *96,* 145(216), 151(216), 152(216), 204(216), *212*
Maahs, G., 444(57), 457(57), 465(57), 497(57), *530*
Maaskant, L., 559(114), *585*
Macaluso, A., 571(158), 580(195), *586, 587*
McCullagh, L. N., 78(380), *105*
McDonagh, A. F., 78(245), *102,* 152(280), *213*
McFarlane, H. C. E., 63(169), *100*
McFarlane, W., 63(169), *100*
McGinnis, C., 550(90), *584*
Machnovskij, N. K., 391(124), *419*
McIntyre, J. S., 78(251), *102*
McKelvey, J., 4(17), 13(17), 14(17), *19*
McKennis, J. S., 78(383), 79(286), *103, 105*
McKenzie, S., 241(86), *335*
Mackie, D., 78(261), *102*
McKillip, W. J., 569(143), 570(143), *585*
McLeod, C. M., 162(306), 165(306), *214*
McManus, S. P., 85(296), *103*
McMillian, F. L., 573(170), *586*
McNab, H., 78(261), *102*
Madden, M. J., 540(37), 541(37), 542(37), *583*
Madhavannair, P., 295(150), *337*
Mägi, M., 66(177), 76(177), *100*
Mague, J. T., 175(351), *215*
Maheas, M. R., de, 227(35), 235(35), 237 (35), 258(35), 328(35), *334*
Majer, J., 135(181), 138(181), *211*
Majeste, R., 4(15), *19,* 75(197), 89(197), 90(197), *100,* 150(273), *213*
Makin, S. M., 267(115), 272(121), *336*
Maksimov, V. I., 149(249), *212*

AUTHOR INDEX

Malhorta, S. S., 85(291), 86(291), *103*
Malii, V. A., 127(155), *210*
Mancelle, N., 189(376, 377), *216*
Mancini, F., 275(130), 286(149), 287(149), 288(130), *336, 337*
Manhas, M. S., 182(370, 371), *215*
Mann, F. G., 175(344), *215*
Manning, D. T., 558(112), 563(112), *585*
Mannschreck, A., 78(253), *102,* 550(79, 80), *584*
Marakowski, J., 78(249), *102,* 108(8), 108(153), *206*
Maravin, C. B., 340(223), *342*
Marchand-Brynaert, J., 32(39), 50(39), *97,* 75(216), *101,* 425(7, 8), 468(7, 80), 469(8, 80), 470(80), 472(89), 473(90), 486(89), 490(90), 491(90), 492(89, 90), 510(7, 8, 80, 119), 511(7, 8, 80), 512(7, 8), 513(7, 80), 514(80), 515(7, 80), 516(7), 517(7, 80, 119), 518(7, 80, 119), 519(80), 520(80), 521(80), 522(125), 523(89, 125), *528, 530, 531, 532*
Marin, B., 25(5), 30(5), 31(5), 32(5), 65(5), 77(5), 88(5), *96,* 350(8), 352(8), 353(8), 354(8), *416*
Marino, G., 95(361, 362), *104,* 229(55, 56, 57), 234(55, 56, 57), *334*
Markovskij, L. N., 348(5), 358(30), 390(123), *417, 416, 419,* 358(112), *419*
Marks, R. H., 544(53), *583,* 560(123), 561(123), *585*
Marmar, R. S., 177(358), *215*
Marquet, J-P., 237(75), *335*
Marshall, D. R., 78(261), *102*
Martin, D., 140(190), *211*
Martin, F., 160(305b), *214,* 169(333), 172(340), *214, 215*
Martin, G., 26(10), 29(10), 50(10), *96,* 152(279), *213,* 273(123), *336*
Martin, G. J., 26(7), 35(375), 67(7), 78(386), 83(7, 276, 277, 278), 84(375), 95(7, 277, 278, 394), *96, 102, 105,* 229(50, 51, 52, 53, 54), 230(52, 53, 54), 231(53, 54), 267(50), *334*
Martin, M., 26(10), 29(10), 50(10), *96,* 152(279), *213,* 229(50), 267(50), *334*
Martin, M. L., 35(375), 78(375), 84(375), *105*
Martin, R. B., 93(363), *104*
Martin, S. F., 177(355), *215*
Martin, W. B., 108(5), 111(5), 119(5), 150(5), 179(5), 182(5), 202(5), *206*
Martirosyan, G. T., 194(395), *216*
Marullo, N. P., 549(73), *584*
Marvel, C. S., 560(120), *585*
Marvell, E. N., 297(151), *337*

Marzin, C., 544(57), 551(57), *583,* 554(100), 561(100), 564(100), *584*
Mason, R., 75(213), *101,* 155(286), *213*
Mason, S. F., 86(297), *103*
Mass, W., 157(291), *213*
Masson, Cf.S., 395(92), *418*
Masui, M., 95(360), *104,* 154(285), *213*
Mathias, A., 66(179), *100*
Mathison, I. W., 118(88), 149(88), *208*
Mathur, S. S., 109(17b), *207,* 559(115a), 573(115a), *585*
Matthews, B. W., 91(333), *104*
Matthies, D., 119(116, 117, 118, 121), 162(116), 164(116), 167(116), 175(116, 117), 181(117), 182(368), *209, 215*
Maurin, R., 441(50), 457(50), 458(50), 459(50), 495(50), *530*
Mayer, H., 580(197), *587*
Mayer, R., 276(131), 278(131), 279(131), 280(131), 281(131), 282(131), 294(132), *336*
Mayer, S., 150(267), *213*
Mazharuddin, M., 304(156), *337*
Mazzoi, M., 330(208), *339*
Mecke, R., 39(59), 40(59, 69), 41(59), 42(87), 44(96), 45(59, 96), 46(59, 96), 75(59, 69, 87, 96), *97, 98*
Medyantseva, E. A., 575(183), *586*
Meerssche, M. van, 89(392), 90(329, 330, 332), 91(335), 92(340), *104, 105,* (118), *419,* 431(29b), 465(29b), 466(29b), *529*
Meerwein, H., 142(199), *211,* 255(101), *336*
Meiboom, S., 78(369), *105*
Meierhofer, A., 179(362), *215*
Meiser, P., 40(71), *97,* 78(248), *102,* 131(167, 168, 169), 155(169), 162(169), 164(169), 165(169), 183(169), *210*
Meisinger, M. A. P., 149(238), *212*
Melby, L. R., 443(54), 462(54), *530*
Mellish, C. E., 4(18), *19*
Meltzer, R. I., 315(181), *338*
Mérényi, R., 25(5), 28(12), 29(12), 30(5), 31(5, 12), 32(5), 33(43), 41(12, 77, 79), 42(12), 54(12), 55(12, 79), 56(12, 77), 58(12), 61(12, 43), 63(376), 65(5, 12, 43), 67(5, 43, 181), 69(77), 70(77), 77(5, 181), 88(5), *96, 97, 98, 100, 105,* 350(8), 352(8), 353(8), 354(8), *416,* 437(40), 438(40), 440(40), 455(40), 456(40), *529*
Merz, W., 33(41), 69(41), 74(41), 86(41), *97*
Messerschmitt, T., 302(161), 304(160, 161), 305(160), 306(160, 161), *337*
Meth-Cohn, O., 328(206), *339*
Methoden der Organische Chemie(Houben-

Weyl), 156(290), *213*
Metzner, E. K., 78(258), *102*
Meyer, G., 119(99, 107), 164(99), 167(99), 181(99), *209*
Meyer, H., 86(304), *103*
Meyer-Dulheuer, K.-H., 128(159), 130(166), 137(166), 156(166), *210*
Meyers, E. A., 4(15), *19,* 75(197), 89(197), 90(197), *100,* 150(273), *213*
Michel, A., 90(332), 91(335), *104*
Michelot, R., 57(158), *99,* 78(246, 395), *102, 105,* 112(42, 45), 114(45), 218(42), *207*
Michel-Van Vyve, Th., 397(97), 398(97), 399(97), 404(97), 405(97), *418*
Middleton, W. J., 358(113), *419*
Mignonac-Mondon, S., 556(103), *584*
Miki, S., 118(96), *209*
Miles, H. T., 78(252), *102*
Milgrom, A. E., 127(155), *210*
Miller, L. A., 75(233), *101,* 150(252), *212*
Miller, N. E., 175(352), *214*
Miller, W., 360(41a), *417*
Mills, H. H., 91(336), *104*
Minh, T. Do., 509(113e), 515(113e), 518 (113e), *531*
Minkin, V. I., 227(34), 235(34), 237(34), 258(34), 328(34), *334,* 302(172), 309 (172), *338,* 549(71), 555(71), *584,* 575 (183), *586*
Miroshnichenko, E. A., 581(204), *587*
Mishina, V. G., 283(135), *337*
Miskevich, G. N., 119(115a), *209*
Miýake, M., 75(222), 78(222), *101*
Miyamoto, T., 118(96), *209*
Möbius, L., 75(226), *101,* 352(17), 410(17), *416*
Modena, G., 470(86c), 471(86c), 507(86c), *531*
Moder, T. I., 18(28), *20*
Moeller, T., 78(244), *102,* 131(170), *210*
Moersch, G. W., 275(129), 283(129), 284 (129), 294(129), *336*
Möhrle, H., 150(265, 266, 267), *213*
Mole, M. L., Jr., 78(258), *102*
Molhant, N., 431(29b), 465(29b), 466(29b), *529*
Mollier, Y., 78(262), 81(262), 84(285), *102*
Mommaerts, J., 392(89), 393(89), 399(100), *418*
Momot, V. V., 381(120), 382(75, 120), 397 (75), 398(75), 400(75), 404(75), *418, 419*
Mondon, S., 542(49), 543(49), 545(49), *583*
Montaigne, R., 514(121a), 518(121a), *532*
Moore, J. A., 552(94), *584*

Morf, D., 173(341), *215*
Morgan, K. J., 581(210), *587*
Moriarty, R. M., 568(142), *585*
Morita, K., 330(209), *339*
Morrison, A. L., 109(14), *207*
Morschel, H., 142(199), *211*
Morton, G. O., 572(162), *586*
Morton, P. D., 177(360), *215*
Mory, R., 229(47), 232(47), 318(47), *334*
Moskowitz, M., 165(317), *214*
Motoyama, Yoshiaki, 95(364), *104,* 141 (197), 150(197), *211*
Motte-Collard, J., 25(5), 30(5), 31(5), 32(5), 65(5), 67(5), 77(5), 88(5), *96,* 350(8), 352(8), 353(8), 354(8), *416*
Mourot, D., 149(23b), *212*
Muchowski, J. M., 237(73), *335*
Muenster, L. J., 45(10), *98*
Mühlstädt, M., 195(397), *216*
Müller, A., 125(143), 138(186), 142(186), 167(186), 175(345), 194(394), 203(413, 414), *210, 211, 215, 216*
Müller, E., 33(42), 69(42), 74(42), 86(42), *97,* 250(93), 251(93), *335,* 549(70a, 70c), 563(70d), 566(70b), 580(70d), *584*
Müller, H.-J., 89(323), *103,* 150(272), *213*
Müller, H.-R., 86(305), *103,* 261(108), *336*
Müller, W., 33(42), 48(116), 69(42), 74(42), 78, (260), 85(116), 86(42), *97,* 86(116), *98, 102,* 250(93), 251(93, 98), 252(98), 254(98), 257(98), 258(102), 316(187), 317(187), *335, 336, 338*
Mundlos, E., 121(127), 132(174), 150(127), 155(127), 170(174), 179(127), 180(363), 181(363), 183(127), 200(174), 205(174), *209, 211, 215*
Murray, R. W., 449(62), *530*
Müter, B., 340(231), *342*
Myers, G. R., 132(176), *211*

Nagata, S., 62(166), *99*
Nagy, D. B., 78(384), *105*
Nahlieli, A., 550(82), *584*
Naik, H. A., 312(178), 313(178), *338*
Nakada, K., 574(174), 577(174), *586*
Nakagawa, Y., 303(153), *337*
Nakai, T., 409(108), *419*
Nardelli, M., 544(56), 545(56), *583*
Naulet, N., 78(386), *105*
Nazarov, I. N., 273(122), *336*
Nazarova, I. I., 273(122), *336*
Nefedov, V. I., 94(352), *104*
Nehring, R., 192(388), *216*
Neilson, D. G., 545(59), *583*
Nelson, R. B., 150(257), *213*
Net, R. W. de, 95(365), *104*

Neubauer, G., 430(24b), 465(24b), *529*
Neuenschwander, M., 150(259), *213,* 438
  (41, 42, 43), 453(42, 43), 454(43), 455
  (43), 461(41), *529*
Neuklis, W. A., 275(129), 283(129), 284
  (129), 294(129), *336,* 364(43, 44), *417*
Neuman, R. C., Jr., 40(73, 74, 75), 75(198),
  100, 79(283), 82(74, 75), 84(73, 282,
  283), *97, 102*
Newlands, L. R., 545(59), *583*
Newman, H., 75(236), 89(236), *101,* 118
  (93), *209*
Nichol, A. W., 246(95), *335*
Niederhauser, A., 150(259), *213,* 364(44),
  417, 438(41), 46(41), *529*
Nieman, C., 78(368), *105*
Nie-Sarink, M. J. de, 219(432), *222*
Nikolajewski, E. E., 331(215), 332(215),
  339
Nikolajewski, H. E., 69(186), *100,* 118(89),
  208, 160(303), *214,* 285(144), *337,* 323
  (196), *338,* 396(94), *418*
Nilson, N. H., 354(19), *416*
Nineham, A. W., 162(307), *214*
Nishimura, J., 43(95), 68(95), *98*
Nishimura, S., 219(433), 221(433), *222*
Nivorozhkin, L. E., 302(172), 309(172),
  338
Nofal, Z. M., 323(199), *338*
Nohara, A., (233), *342*
Nolte, K.-D., 24(2), 96, 65(174), *100*
Nordmann, H. G., 42(85), *98,* 331(213),
  332(213), *339*
Norman, R. O, C., 33(46), *97,* 144(210),
  211
Normant, H., 273(123), *336*
Normant, J. F., 443(55), 447(55), 462(55),
  464(55), *530*
Norris, W. L., 162(307), *214*
Notté, P., 430(59), 441(50), 446(59), 456
  (59), 457(50, 59), 458(50, 59), 459(50,
  59), 462(59), 465(59), 495(50, 59), *530*
Novak, A., 75(203), *100*
Novikov, S. S., 66(177), 76(177), *100*
Novitt, B., 570(154), *586*
Novozhilova, T. I., 423(66), 443(66), 447
  (6b), *528*
Nukada, K., 51(130, 132), 75(130, 132),
  99, 574(172, 175, 176), 577(172, 175),
  586

Ochiai, M., 330(209), *339*
Oda, R., 33(44), *97,* 110(19), 114(19), *207,*
  229(45), *334,* 396(96), *418*
O'Donnell, M., 509(117), *531*
Oehl, R., 229(59), 231(59), *334*

Oel, H. Y., 166(322), *214*
Oeser, E., 89(328), *104,* 377(67), *418*
Oettmeier, W., 179(362), *215*
Ogloblin, K. O., 116(81), *208*
Ohmori, H., 95(360), *104,* 154(285), *213*
Ohnishi, M., 63(171), *100*
Ohno, M., 51(132), 75(132), *99,* 574(174,
  176), 577(174), *586*
Ohoka, M., 75(221, 222), 78(221, 222),
  101, 428(14a), *528*
Ohtsuru, M., 63(171), *100*
Ojha, N. D., 78(270), *102*
Okano, M., 33(44), *97,* 110(19), 114(19),
  207, 396(96), *418,* 409(108), *419*
Olah, G. A., 29(26), 38(26), 45(90, 95), 47
  (107), 48(108), 51(131), 54(90, 150), 76
  (237), *96, 98, 99, 101,* 108(7), 153(7),
  206, 227(36), 235(36), 237(36), 258(36),
  328(36), *334,* 574(173), 575(182), 577
  (173), *586*
Olah, J. A., 54(150), *99*
Olin, J. F., 111(20b), 162(20b), *207*
Olive, J. L., 573(171), 574(171), *586*
Oliver, J. E., 402(103b), *419*
Ollmann, G., 129(165), 180(165), 181(165),
  200(165), 202(165), 203(165)
Olsen, R. K., 112(48)
O'Mahony, T. A. F., 580(200), 587
Omar, N. M., 539(7, 8), 540(7, 8, 25), 547
  (8), *582*
Omietanski, G. M., 540(34), 557(34), 561
  (34), *583*
Opitz, G., 5(20), *19,* 28(15), 29(15), 33(41),
  69(15, 41), 71(15), 74(15, 41), 85(15),
  86(41), 112(37, 43), 149(240), 151(37),
  184(372), 187(373), 205(43), *207, 212,
  215*
Oppolzer, W., 540(27), 569(27, 144, 149),
  570(27, 144), *582, 586*
Orazi, O. O., 148(234, 235), *212*
Orfanos, V., 371(51), *417*
Orlova, E., 316(184), *338*
Orth, H., 136(184), *211,* 171(338), *215*
Osmers, K., 127(157), 140(157), 179(157),
  210, 114(56), 115(56), 139(56), *208,*
  196(400), 198(400), *216*
Oth, J. M. F., 41(77), 56(77), 69(77), *97,*
  437(40), 438(40), 440(40), 455(40),
  456(40), *529*
Otremba, E. D., 541(39), *583*
Ott, A. C., 75(234), *101,* 116(68), 186(68),
  208
Ott, E., 442(51), 456(51), *530*
Ottenbrite, R. M., 132(176), *211*
Ottenheym, J. H., 384(82), *418,* 428(18a),
  529

Otto, A., 540(26), 558(110), 563(110), 569(145, 146, 150), 570(26, 145, 146), *582, 584, 585, 586*
Otto, H.-H., 119(104), *209,* 166(321), 168 (321), *214*
Overberger, C. G., 544(55), *583*
Overchuk, N. A., 54(150), *99*

Pagani, G., 78(272), 86(272), *102*
Pálinkás, J., 75(220), *101*
Pallini, U., 265(113), 275(130), 283(146), 284(146), 285(146), 288(130), *336, 337*
Palmer, M. H., 60(161), *99*
Palmer, R., 581(213), *587*
Paloma, A.-L., 230(60), *334*
Pandit, R. S., 314(180), *338*
Pandit, U. K., 219(432), *222*
Panhouse, J. J., 284(142, 148), 285(148), 297(148), *337*
Pappalardo, L., 541(45), *583*
Pardo, M.-C., 540(33), *583*
Parham, W. E., 449(63a, 63b), 456(63a, 63b), 457(63a, 63b), 458(63a, 63b), *530*
Parrick, J., 315(182), *338*
Parsons, A. E., 40(68), *97*
Partos, R. D., 428(17a, 17b), 456(17a), 458(17a), *528, 529*
Partyka, R. A., (205), *339*
Pasche, W., 125(144), 128(144), 131(144), *210*
Pashinnik, V. E., 358(30), *417,* 358(112), *419*
Passerini, N., 330(208), *339*
Pasternak, B. I., 350(9), 352(9), 360(9), 361(9), 409(9), 410(9), *416*
Pasternak, R. A., 4(16), *19*
Pasternak, V. I., 31(378), 41(378), 95(378), *105,* 348(4), 350(4), 355(4, 24), 357(24), 358(34), 380(4), 381(117), 389(24), 390(87), 396(24, 93), *416, 417, 418, 419*
Pastravanu, M., 281(140), 282(140), *337*
Patai, S., 108(3), *206,* 470(86a), 471(86a), 507(86a), 515(123), *530, 532*
Patelli, B., 275(130), 288(130), *336*
Patin, H., 149(236), *212*
Pattison, V. A., 42(86), *98*
Patrick, J. B., 74(194), *100*
Paudler, W. N., 340(229), *342*
Paukstelís, J. V., 6(22), 16(22), 17(22), *19,* 29(22), 31(22), 32(22), 34(22), 37(22), 38(22), *96,* 108(1), 114(1), 118(87), 151 (1, 87), 152(87), 157(1), *206, 208*
Paul, L., 116(69b), *208*
Pavlenko, N. G., 31(378), 41(378), 95(378), *105,* 355(24), 357(24), 389(24), 396(24), 382(76, 121), *416, 418, 419,* 450(64), 453(64), 465(64), *530*
Pawar, R. A., 341(234), *342*
Pawellek, F., 142(199), *211*
Peat, I. R., 42(83), *98*
Pederson, R. L., 75(234), *101,* 116(68), 186(68), *208*
Pehk, T., 65(174), *100*
Pellier, C., 53(143), *99,* 539(12), 547(12), 548(12), *582*
Perez, C., de, 470(84), *530*
Pergosin, P. S., 64(172), *100*
Perin, F., 340(218), *342*
Perin-Roussel, O., 340(218), *342*
Perlova, T. G., 384(80), *418*
Pesotskaja, G. V., 350(9), 352(9), 361(9), 382(75), 397(75), 398(75), 400(75), 404(75), *418,* 409(9), 410(9), *416*
Petersen, H., 119(124), *209*
Peterson, R. A., 118(94), *209*
Peterson, S. W., 89(325), *103*
Petropoulas, C. C., 86(309), *103,* 330 (210), *339*
Petrow, V., 119(271), *336,* 266(114), 274(114), *336*
Petrus, C., 53(143), *99,* 539(12), 547(12), 548(12), 572(160), 573(160, 171), 574(171), 577(160), *582, 586*
Petrus, F., 53(143), *99,* 539(12), 547(12), 548(12), 572(160, 161), 573(160, 171), 574(171), 577(160), *582, 586*
Petukhov, S. A., 217(426), *222*
Pfeffer, A., 550(83), *584*
Philipsborn, W. von, 79(288), *103*
Phillips, B. A., 34(52), 35(52), *97,* 88 (318), *103*
Phillips, D. I., 394(90), *418*
Phillips, J. N., 329(204a, 204b, 204c), *339*
Pielichowski, J., 112(46), *207, 208*
Pilotti, A., 75(228), 78(228), *101*
Pinchuk, A. M., 380(71), 381(71), *418*
Pinkus, J. L., 52(137), 75(137), *99*
Pittman, C. V., Jr., 85(296), *103*
Pizey, J. S., 227(39), 266(39), *334*
Plant, S. G. P., 162(307), *214*
Plappert, P., 109(15), 157(295), 159(15), 161(295), 184(15, 295), 186(15), 187 (15), *207, 213*
Plé, G., 115(67), 116(72, 74), 126(74), 217(425), 218(429), *208, 222*
Plenkiewicz, J., 560(121), *585*
Pletcher, J., 89(327), *103*
Plöger, W., 441(49b), 457(49b), 458(49b), *530*
Ploss, G., 255(100), 257(100), 266(100), 331(100), *335*
Pluchet, H., 284(147, 148), 285(148),

297(148), *337*
Podszun, W., 218(428), *222*
Pohland, A. E., 275(126), *336*
Poignant, S., 26(7), 35(375), 67(7), 83(7, 277, 278), 95(7, 277, 278), *96, 102,* 229(52, 53, 54), 230(52, 53, 54), 231 (53, 54), *334*
Poirer, J.-M., 112(49), 163(312), 169(312), *208, 214*
Pollak, I. E., 142(198), *211*
Polyakov, S. A., 154(284), *213*
Ponomarev, G. V., 340(223), *342*
Pople, J. A., 2(1, 4b), 3(4a, 4b), 7(24), 8(1, 4a), *19*
Popoff, I. C., 177(360)
Porai-Koshits, A., 307(165), *337*
Porshnev, N., 340(219, 220), *342*
Porshnev, Y., 340(219, 220), *342*
Porter, S. K., 89(326), *103*
Posner, J., 118(94), *209*
Post, H. W., 135(182), *211*
Posvic, H., (117a), *585*
Potier, P., 28(13), *96,* 144(207), *211*
Potoski, J. R., 449(53a, 63b), 456(63a, 63b), 457(63a, 63b), 458(63a, 63b), *530*
Povolotskij, M. I., 31(378), 41(378), 95 (378), *105,* 355(24), 357(24), 389(24), 396(24), *416*
Prajsnar, B., 428(18b), *529*
Pressler, W., 269(117), *336*
Prihadko, A. S., 66(177), 76(177), *100*
Prilepskaya, A. N., 203(411), *216*
Profft, E., 175(347), *215*
Pruett, R. L., 116(76), *208*
Przheval'skii, N. M., 549(71), 555(71), *584*
Pucher, G. W., 168(327), *214*
Pudussery, R. G., 573(167), *586*
Pugh, W., 118(85), *208,* 544(52, 53), 557 (104, 106), 560(123), 561(123), 562 (106), *583, 584, 585*
Pugmire, R. J., 66(176), 76(176), *100*
Pulst, M., 278(138), 279(138), 296(138), *337*
Putzeys, J.-P., 90(330), 92(340), *104*
Pyun, C., 78(249), *102,* 108(8), 153(8), *206*

Quemeneur, M. T., 83(276, 277), 95(277), *102,* 229(51, 52), 230(52), *334*
Quiniou, H., 242(90), *335*
Quis, P., 44(99), 48(118), 49(125), 75(99, 118, 125), *98, 99*
Quivoron, C., 75(229), *101*

Radau, M., 75(218), 78(218), *101,* 109(17a), *207*
Radeglia, R., 24(2), *96,* 65(174), *100*
Radom, L., 7(24), *19*
Rahman, A. W., 111(20d), 162(20d), *207*
Ramey, K., 78(249), *102*
Randall, E. W., 61(163), 63(167), 64(172), 77(167, 238), 78(167b), *99, 100, 101*
Ranft, J., 48(113), 49(113), 78(259), 86 (259), *102*
Ranftaand, J., 39(65), 40(65), 69(65), *97*
Rapoport, H., 333(217), *339*
Rapp, C., 116(69c), *208*
Rapp, K. E., 116(76), *208*
Rappoport, Z., 470(86b), 471(86b), 507 (86b), *530*
Raschig, F., 147(224), *212*
Rathbone, E. B., 558(108), 562(108), 563 (108), *584*
Ratts, K. W., 111(20c), 162(20c), *207,* 428(17b), *529*
Reavill, R. E., 78(265), *102*
Reboulet, J. E., 575(178), *586*
Redpath, C. R., 75(223), *101*
Reewes, L. W., 78(168), *100*
Reiber, H. G., 117(83), *208*
Reichardt, C., 259(106), 269(117), *336,* 316(186), 318(190), *336, 338*
Reichardt, Ch., 78(387), *105*
Reid, D. H., 233(65), 235(65), 242(65), *334,* 241(86), *335,* 242(89), 243(89), *335*
Reid, E. E., 581(208), *587*
Reid, S. T., 580(203), *587*
Reiner, M. T., 116(70), *208*
Reiner, M. Th., 75(214), 78(214), *101*
Reinhard, H., 303(159), 304(159, 162), 305(159), 306(159, 162) *337*
Reinhard, W., 94(356), *104*
Reinhart, H., 327(201), *338*
Reinheckel, H., 190(384), *216*
Reisch, J., 302(158), 304(158), *337*
Reliquet, A., 48(114), 49(114), 80(114), 86(114), *98,* 109(18), *207*
Reliquet-Clesse, F., 48(114), 49(114), 80 (114), 86(114), *98,* 109(18), *207*
Remion, J., 475(93), 476(93), 477(93), 498(93), *531*
Rempfer, H., 28(8), 75(8), *96,* 229(58), 237(58), *334*
Renaud, R. N., 144(205), 164(205), *211*
Renault, J., 568(141c), *585*
Rene, L., 241(83), *335*
Rengaraju, S., 575(180), *586*
Rens, M., 431(26), 434(33), 451(26, 33), 452(33), 453(33), 455(67), 465(26, 33),

466(33), 467(33), 472(33, 88), 482(33, 88), 483(33, 88), 484(33), *529, 530, 531*
Renshaw, R. R., 165(316), *214*
Rérat, B., 92(344), *104,* 548(67), *583*
Rérat, C., 92(343, 344), 93(346), *104,* 548(66, 67), *583*
Reuterhäll, A., 75(228), 78(228), *101*
Rey-Lafon, M., 75(229), *101*
Reynolds, D. D., 123(136), 165(136), 201(136), *210*
Reynolds, D. R., 201(407, 408), *216*
Reynolds, G. A., 86(309, 310), *103,* 302(171), 309(171), 311(171), 330(210), *338, 339*
Reynolds, W. F., 42(83), *98*
Reznichenko, A. V., 242(87), *335*
Rhoades, D. F., 37(56), 69(56), 71(56), *97,* 143(203), *211*
Richards, J. H., 66(178), 67(178), 77(178), 78(178), *100*
Richards, R. E., 61(162), 62(162), 78(250), 89(250), *99, 102*
Richey, F. A., Jr., 572(166), 576(165), *586*
Richter, R., 128(163), 131(163), 147(163), *210,* 402(104), *419*
Ricoleau, G., 35(375), 78(375), 84(375), *105*
Rieche, A., 228(42), *334*
Riesz, E., 581(209), *587*
Rim, T. S., 143(202), *211*
Rim, Yong Sung, 37(57), 69(57), *97*
Rinderknecht, H., 109(14), *207*
Rindone, B., 150(254), *212*
Riordan, R. P., 177(360), *215*
Ritter, E., 122(133), 145(133, 215), *210, 212*
Ritter, E. J., 351(14), *416*
Ritter, W., 119(97, 111, 113), 125(142), 127(97, 142), 167(111, 113), *209, 210,* 567(140), 568(140), *585*
Rittner, S., 94(356), *184*
Robb, M. A., 580(210), *587*
Robert, R., 541(42), 544(42), *583, 585*
Roberts, B. W., 63(167, 167a), 77(167), 78(167a, 267), 86(267), *100, 102*
Roberts, J. D., 61(165), 63(165, 167, 167a), 64(165), 77(107), 78(167a, 207), 86(267), *99, 102*
Robertson, J. E., 572(165), *586*
Robinson, G. M., 162(306), 165(306), *214*
Robinson, R., 112(34), *207*
Roehr, J., 112(38), 128(38), 135(38), 140(38), 141(195), *211,* 144(38), *207*
Roger, R., 545(59), *583*

Roh, N., 258(103), *336*
Röhr, A., 89(391), *105,* 222(440), *223*
Rokhlim, E. M., 196(399), *216*
Rollett, F. S., 93(345), *104*
Roma, G., 330(208), *339*
Rondeau, R. E., 575(178), *586*
Rosenberg, H. M., 575(178), *586*
Rosenblatt, D. H., 150(255, 256), *213*
Rosenstock, H. M., 3(10), 7(10), *19*
Rosenthal, T. C., 540(36), *583*
Roshcupkina, L. G., 283(135), *337*
Ross, S. D., 150(270), *213*
Rossbach, F., 75(218), 78(218), *101,* 109(17a), *207*
Rossignol, J. F., 241(84), *335*
Rostolan, J. de, 28(13), *96,* 144(207), *211*
Roth, E. A., (13), *19*
Roth, H. J., 150(264), *213*
Rothenberg, S., 3(7), 7(7), 8(7), 14(7), 15(7), *19,* 24(1), *96*
Rothrock, H. S., 119(109), 132(109), *209*
Röver, E., 122(129), *209*
Rowatt, R. J., 175(352), *215*
Royer, R., 241(83, 84), *335*
Rozynov, B. V., 340(223), *342*
Rucci, G., 75(213), *101,* 155(286), *213*
Ruchti, L., 149(244), *212*
Rudner, B., 540(34), 557(34), 561(34), *583*
Ruff, F., 321(197), 323(197), *338*
Rumon, K. A., 78(266), *102*
Runge, W., 469(81), 470(81), *530*
Rusche, J., 47(110), *98,* 145(218), *212*
Russmann, H., 180(365), 181(365), *215*
Russo, S. F., 75(199), *100*
Ruthe, V., 219(434), *223*
Ryan, J. J., 34(52), 35(52), 78(371), *97, 105,* 149(250), *212,* 570(153), *586*

Sachs, F., 147(229), 162(229), *212,* 175(343), *215*
Sadanandam, Y. S., 126(152), 187(152), *210*
Safir, S. R., 572(162), *586*
Sagitullin, R. S., 555(102), *584*
Saha, N. N., 92(341), *104*
Saigh, A. A. R., 127(154), *210*
Saintes, F., 495(103), 499(103), *531*
Saint-Giniez, D. de, 92(343), *104,* 548(66), *583*
Saitô, H., 51(130, 132), 62(166), 75(130, 132), *99,* 574(172, 174, 175, 176), 577(172, 174, 175), *586*
Sakharov, Y. K., 127(155), *210*
Saladin, E., 75(215), *101*
Salamon, G., 45(104), 46(104), *98*
Salomaa, P., 577(187), *587*

Samaraj, L. I., 403(106), *419*
Samoilova, Z. E., 196(399), *216*
Samojlova, Z. E., 133(177), *211*
Sanno, Y., (233), *342*
Santini, S., 95(361, 362), *104,* 229(55, 56), 234(55, 56), *334*
Santos, A. A., 302(168), 308(168), *338*
Sasaki, T., 455(68), *530*
Sasakura, K., 271(120), *336*
Sasse, D., 327(201), *338*
Sastry, K. V. L. N., 4(12), *19*
Sauliová, J., 86(307), *103,* 226(31, 32), 318(31), 320(31, 32), *333, 334*
Saunders, M., 226(10a), *333,* 449(62), *530*
Sauvetre, R., 443(55), 447(55), 462(55), 464(55), *530*
Saveant, J.-M., 95(357, 393), *104,* 206(421), *217*
Savelli, G., 93(362), *104,* 229(56, 57), 234(56, 57), *334*
Saver, J., 566(133), 567(133), 568(133), 579(133), *585*
Sax, M., 89(327), *103*
Sayigh, A. A. R., 111(30), 128(163), 131(163), 147(163), *207, 210*
Schaefer, T., 54(151), *99*
Schaeffer, H. F., 3(6), 4(6), 5(6), 8(6), *19*
Schagerer, K., 259(106), *336*
Schaumann, E., 89(391), *105,* 222(438, 439, 440), *223*
Scheer, W., 484(101), *531*
Scheffold, R., 75(215), *101*
Scheibe, G., 40(66), 48(66, 115), 49(66), 80(115), 81(115), 85(292, 294, 377), 86(292, 294), *97, 98, 103, 105*
Schellhorn, H., 278(137), 279(137), *337*
Scheyer, A., 274(125), *336*
Schiess, P., 86(311), *103,* 297(151), *337,* 538(2), 546(2), *582*
Schindler, N., 441(49a, 49b), 457(49a, 49b), 458(49a, 49b), 459(49a), *530*
Schittenhelm, D., 150(265), *213*
Schleyer, P. von, 2(3), *19*
Schlipf, E., 86(313), *103,* 119(123), 156(123), 175(123), *209*
Schlosberg, R. H., 575(182), *586*
Schlottmann, B. U., 25(6), 88(6), 94(6), *96,* 151(278), 152(283), *213,* 154(278), *213*
Schlottmann, U., 350(10), 352(10), 355(10), 359(10), *416*
Schlötzer, A., 111(26), *207*
Schlözer, R., 340(222), *342*
Schmelzer, H. G., 356(26), *416*
Schmid, H., 31(36), 32(36), *96,* 206(422), *217*
Schmid, K. H., 28(20), *96,* 118(90), 146(90), *208,* 147(224), *212*
Schmid, K. R., 33(374), 65(374), 78(374), *105*
Schmid, M., 229(47), 232(47), 318(47), *334*
Schmidle, C. J., 244(91), *335*
Schmidt, A., 28(11), 29(27, 29), 30(29, 30, 31), 39(60), 42(80), 43(94), 47(80), 50(30), 74(11, 31), 75(11, 27, 29, 30, 31, 80, 94, 202, 205), *96, 97, 98, 100, 101,* 218(427), *222,* 146(219), *212,* 355(23), 389(86b), *416, 418*
Schmidt, E., 122(130), 145(213), *209, 211*
Schmidt, R. R., 86(313), *103,* 119(123), 156(123, 287), 175(123), 193(390, 391), *209, 213, 216*
Schmitt, J., 284(147, 148), 285(148), 297(148), *337*
Schmitz, E., 543(51a), 546(60, 61), 566(60, 134), 567(60, 134), 580(134a), *583, 585*
Schneider, P., 278(138), 279(138), 296(138), *337*
Schneider, W., 219(431), *222*
Schneider, W. C., 54(148), 63(148), *99*
Schnierle, F., 303(159), 304(159), 305(159), 306(159), *357*
Schoeller, W. W., 218(428), *222*
Schoffner, J., 54(149), 57(149), *99*
Schön, N., 255(101), *336*
Schönberg, A., 580(202), *587*
Schönenberger, H., 173(342), 200(406), *215, 216*
Schönherr, H. J., 581(206), *587*
Schram, E. P., 147(223), *212*
Schreiber, J., 28(14), *96,* 145(216), 151(216), 152(216), 204(216), *212*
Schretzmann, H., 75(206), *101,* 118(86), *208*
Schroth, W., 297(151), *337*
Schubert, H. W., 5(20), *19,* 28(15), 29(15), 32(15), 69(15), 71(15), 74(15), 85(15), *96,* 112(37), 151(37), 187(373), *207, 215*
Schubert, W. M., 95(364), *104,* 141(197), 150(197), *211*
Schulte, K. E., 302(158), 304(158), *337*
Schulz, G., 255(100), 256(127), 257(100), 266(100), 275(127), 288(127), *336,* 331(100), *335*
Schulz, M., 570(155), *586*
Schulz, M.-L., 340(221), *342*
Schumann, D., 89(323), *103,* 150(272), *213*
Schunemann, D., 119(98, 115), 162(310), 173(310), *214,* 181(98), *209*
Schurig, H., 278(136), 281(136), 282(136), *337*

AUTHOR INDEX

Schuster, E., 116(69b), *208*
Schuttenberg, H., 148(235), *212*
Schwartz, A., 581(213), *587*
Schwartz, H., 119(106), 167(106), 168(106), 175(106), *209*, 168(328), 175(328), *214*
Schweiger, E., 160(304), *214*, 319(193), 321(193), 325(193), 332(216), *338, 339*
Sciaky, R., 265(113), 275(130), 283(146), 284(146), 285(146), 286(149), 287(149), 288(130), *336, 337*
Scolastico, C., 150(254), *212*
Scott, F. L., 549(74), *584*, 560(123a), *585*, 580(200), *587*
Scott, M. D., 39(61), 40(61), 49(61), 67(61), *97*, 229(44), *334*
Scrocco, E., 576(185), *586*
Scrowsdon, R. M., 283(139), *337*
Searle, D. E., 165(316), *214*
Sedor, E. A., 569(143), 570(143), *585*
Seebach, D., 189(375), *216*
Seefelder, M., 86(305), *103*, 261(108), 262(109), 328(202), *336, 338*, 374(59, 60), 398(60), *417*, 430(24a, 24b), 434(32), 465(24a, 24b), 473(24a), 498(32), 503(32), *529*
Seeliger, W., 192(387, 388), 193(387), *216*
Seidel, H. J., 319(192), 325(192), *338*
Seiffert, W., 40(66), 48(66, 115), 49(66), 80(115), 81(115), *97, 98*
Seitz, G., 119(103), 164(103), 167(103), 175(103), 181(103), *209*
Sellers, C. F., 149(248), *212*
Selton, B., 162(307), *214*
Semple, B., 60(161), *99*
Senda, S., 307(167), 308(167), *337*
Senning, A., 142(222), 150(222), *212*, 354(19), *416*, 441(48), 457(48), *529*
Sergeev, A. P., 423(6a, 6b), 443(6a, 6b), 447(6a, 6b), 460(6a), 461(6a), 462(68), 463(6a), *528*
Seshadri, S., 303(157), *337*, 312(176), *338*, 312(178), 313(178), *338*, 314(180), *338*
Seufert, W. G., 218(428), *222*
Seus, E. J., 248(97), *335*
Seyferth, D., 177(358), *215*
Shagalov, L. B., 340(228), *342*
Shah, S. R., 303(157), *337*
Shannon, T. W., (317), *103*
Shapranova, N. H., 541(45a), *583*
Shavrygina, O. A., 267(115), 272(121), *336*
Shedov, V. I., 340(227), *342*
Sheinkmann, A. K., 203(411, 412), 204(416), 205(416), *127, 216*
Shelepin, N. E., 302(172), 309(172), *338*
Shen, M.-S., 246(96), *335*

Shermolovich, J. G., 390(123), *419*
Shestaeva, M. M., 160(305a), *214*
Shevchenko, A. E., 340(225), *342*
Shevchenko, M. V., 380(71), 381(71, 72, 119), 389(72, 119), *418*
Shevchenko, V. I., 390(123), *419*
Shimadzu, H., 330(209), *339*
Shipman, J. J., 519(124a), *532*
Shirahashi, M., 307(167), 308(167), *337*
Shlenskaya, V. I., 574(177a), *586*
Shotton, J. A., 166(324), *214*
Shropshire, E. Y., 69(183), *100*
Shupik, R. I., 238(78), *335*
Shutkova, E. A., 307(165), *337*
Shvedov, V. I., 241(85), *335*
Shvo, Y., 550(82), *584*
Sickmuller, A., 140(191), 153(191), *211*
Sidani, A., 425(8), 469(8), 478(98), 480(98), 481(98), 510(8), 511(8), 512(8, 98), 516(98), 517(98), 528(98), *528, 531*
Siddall, T. H., III, 84(281), *102*
Sieveking, S., 89(391), *105*, 222(438, 439, 440), *223*
Sikorski, J. A., 550(84), *584*
Simalty, M., 302(173), 309(173), *338*
Simchen, G., 75(224), *101*, 268(116), 302(168), 308(168), 312(179), *336, 338*, 371(50), *417*, 428(16a, 16b), *528*
Simionescu, C., 281(140), 282(140), *337*
Simon, W., 54(152), 57(152), 61(152), 67(152), *99*, 78(389), *105*
Singh Walia, J., 34(50), 69(50), *97*
Singh Walia, P., 34(50), 69(50), *97*
Sinitsa, A. D., 119(120), 168(120), 175(120), *209*
Siret, P., 126(151), 127(151), 160(151), *210*
Sisler, H. J., 540(34), 557(34), 561(34), *583*
Sklarz, B., 573(167), *586*
Skorianetz, W., 550(87), *584*
Slegeir, W. A. R., 581(212, 213), *587*
Smallcombe, S. H., 66(178), 67(178), 77(178), 78(178), *100*
Smets, F., 376(64), *417*
Smith, G. E. P., Jr., 166(324), *214*
Smith, G. F., 86(303), *103*, 237(71), *335*
Smith, H. E., 78(245), *102*, 152(280), *213*
Smith, I. C., 54(143), 63(148), *99*
Smith, J. A. S., 75(223), 95(394), *101, 105*
Smith, L. R., 423(5b, 5c), 447(5b, 5c), 448(5c), 449(5c), 450(5c), 456(5b, 5c), 457(5b, 5c), 458(5c), 497(107), 498(107), 503(5b, 111), 504(5b, 111), *528, 531*
Smith, P. A. S., 33(47), *97*, 79(286), *103*, 108(2), 150(258), 157(2), *206, 213,*

540(35a), 572(35b, 165), 573(35b), 574 (35b), 578(35b), *583*
Smith, R., 88(318), *103*
Smith, R. F., 540(36, 37), 541(37, 39), 542(37), *583*
Smith, T. D., 42(84), 43(84), *98*
Smith, W. C., 78(240), 89(240), *101,* 151 (275), *213,* 357(29), *417*
Snyder, L., 7(25), *19*
Sobell, H. M., 89(324), *103*
Sohn, E. G., 544(53), *583,* 560(123), 561 (123), *585*
Sollenberger, P. Y., 95(363), *104*
Somin, I. N., 541(45a), *583*
Song Loong, W., 91(338), *104*
Sonnek, G., 190(384), *216*
Sonveaux, E., 431(27), 453(27), 470(82, 85), 480(27), 482(27), 483(27), 510(82), 522(27), *529, 530*
Soos, R., 75(219, 220), *101,* 402(105), *419*
Sorensen, O. N., 354(19), *416*
Sorum, F., 226(5), 267(5), *333*
Specht, U. von, 302(161), 304(161), 306 (161), *337*
Spedding, H., 39(61), 40(61), 49(61), 67 (61), *97,* 229(44), *334*
Speh, P., 312(179), *338*
Spencer, T. A., 150(269), *213*
Speziale, A. J., 73(193), 85(193), 85(295), 86(295), *100, 103,* 114(58), 115(58), 117(82), 423(5a, 5b, 5c, 5d), 428(17a), 447(5a, 5b, 5c, 5d), 448(5c, 5d), 449(5c), 450(5c), 456(5a, 5b, 5c, 17a, 69), 457 (5a, 5b, 5c, 69), 458(5a, 17a, 69), 459(5d), 465(72), 471(72), 474(72), 478(72), 479 (72), 480(72), 481(72), 497(72, 107), 498(72, 107), 499(5d, 72), 503(5b, 111), 504(5b, 111), *528, 530, 531*
Spinner, E., 43(88), 45(88), *98*
Spohn, K. H., 66(180), *100*
Springer, W. A., 383(79), *418*
Staab, H. A., 78(253), *102,* 228(40), *334,* 550(79), *584*
Stachel, H. D., 111(29), *207*
Staelens, M., 451(65), *530*
Staelens-Vanhorenbeeck, M., 501(110), *531*
Stafford, J. E., 75(201), 86(201), 94(201), *100*
Stam, C. H., 91(339), *104*
Stamhuis, E. J., 157(291), *213*
Stammberger, W., 28(17), *96,* 169(329), 190(381, 382, 383), 199(402), *214, 216*
Stanley, J. W., 118(88), 149(88), *208*
Stassinopoulo, C. I., 550(81), *584*
Stefaniak, L., 575(179), 577(179), *586*
Steglich, W., 179(361, 362), *215*

Steinberg, H., 32(38), 78(370), *96, 105,* 122(128), *209*
Steinbrückner, C., 149(245), 199(245), *212*
Stelender, B., 25(5), 30(5), 31(5), 32(5), 65(5), 67(5), 77(5), 88(5), *96,* 350(8), 352(8), 353(8), 354(8), 360(36), 392(89), 393(89), 398(36), 402(36), *416, 417, 418*
Stencamp, R. S., 91(333), *104*
Stepanov, B. I., 331(212), *339*
Stepanova, G. P., 331(212), *339*
Stephen, A. M., 118(85), *208,* 557(104, 105), 558(108), 561(105), 562(108), 563(108), *584*
Stephenson, G. F., 240(82), *335*
Stephenson, O., 271(119), *336*
Stevenson, H. B., 119(109), 132(109), *209*
Stewart, J. M., 89(327), *103*
Stewart, R., 45(101), *98,* 86(308), *103*
Stewart, R. F., 3(4a), *19*
Stewart, T. D., 117(83), *208,* 134(179), *211*
Stewart, W. E., 84(281), *102*
Stoess, U., 302(158), 304(158), *337*
Stohrer, W.-D., 218(428), *222*
Stojanovic, F. M., 226(27), 324(27), *333*
Stokes, J. B., 402(103b), *419*
Stopp, G., 255(101), *336*
Stork, G., 114(51, 52, 53), *208*
Strausz, O. P., 509(113e), 515(113e), 518 (113e), *531*
Streitwieser, A., Jr., 235(66), 244(66), *334*
Strepikheev, J. A., 384(80), *418*
Struve, G. E., 18(28), *20*
Strzellecka, H., 302(173), 309(173), *338*
Stuckwisch, C. G., 566(133a), *585*
Sturm, E., 255(100), 257(100), 266(100), 331(100), *335*
Sugasawa, T., 271(120), *336*
Sukharenko, N. V., 391(124), *419*
Sukiasjan, A. N., 238(77), *335*
Sultan, M. K., 573(167), *586*
Sulzer, G., 119(101), *209*
Sun, K. K., 568(142), *585*
Sundaralingem, M., 92(342), *104*
Surorov, N. N., 340(228), *342*
Surov, N., 340(225), *342*
Surov, Y., 340(225), *342*
Suschitzky, H., 109(17b), *207,* 115(64), 149(248), *208, 212,* 559(115a), 573 (115a), *585*
Sustmann, R., 2(3), *19,* 509(115a), *531,* 519(124c), *532*
Suter, C., 297(151), *337*
Sutov, G. M., 316(184), *338*
Sviridov, E. P., 175(350), 177(356, 359, 359a), *215*

AUTHOR INDEX

Svishchuk, A. A., 391(124), *419*
Swinbourne, F. J., 52(135), *99*
Swingle, R. B., 149(251), *212*
Szabo, W. A., 78(382), 89(382), *105*, 217(423), 219(423), *222*
Szalin, J. W., 94(352), *104*
Szantay, Cs., 546(61), *583*
Szilágyi, P., 43(90), 54(90), *98*
Szmant, H., 550(90), *584*
Szmuszkovicz, J., 114(51, 53, 54), *208*
Szymanski, S., 575(179), 577(179), *586*

Taguchi, T., 111(27), *207*
Takayama, K., 11(27), *207*
Takhistov, V. V., 154(284), *213*
Taller, R. A., 550(75, 76, 77), *584*
Tam, J. W. O., 75(230), 78(230), 90(331), *101, 104*
Tam, S. W., 564(126), *585*
Tamas, J., 402(105), *419*
Tan, H. S. I., 540(31), *582*
Tanaka, Y., 62(166), *99*
Tarnow, H., 356(26), *416*
Tarrago, G., 78(242), *102*, 539(14, 15), 540(15), 547(15), 551(93), 553(14, 96), 554(14, 15, 99), *582, 584*
Tartakovskii, V. A., 581(204), *587*
Tataruch, F., 376(63), *417*
Tatke, D. R., 312(178), 313(178), *338*
Tattershall, B. M., 151(276), *213*
Tattershall, B. W., 89(321), *103*
Taurins, A., 40(70), 75(70), *97*
Tawara, Y., 58(159), 70(159), *99*
Taylor, D. A., 316(185), *338*
Taylor, E. C., 566(138, 139), 569(139), 570(139), *585*
Taylor, L. J., 423(5d), 447(5d), 448(5d), 459(5d), 499(5d), *528*
Tchoubar, B., 78(395), *105*
Techy, B., 447(61), *530*
Telinski, T., (117a), *585*
Terasawa, I., 574(176), *586*
Terent'ev, A. P., 283(135), *337*
Terrel, R., 174(51, 53), *208*
Teubner, H., 175(347), *215*
Texier, F., 150(261), *213*, 570(151), 579 (151), *586*
Thamer, D., *214*, 125(145), 162(145), 165(318), 169(330), *214*, 175(145), *210*
Theilacker, W., 544(54), 554(54), 555(54), *583*, 574(171a), *586*
Thesing, J., 580(197), *587*
Thiele, J., 559(113), 562(113), *585*
Thier, W., 192(388), *216*
Theuer, W. J., 543(51), 547(51), *583*
Thill, B. P., 78(270), *102*

Thio, P. A., 552(95), 553(95), *584*
Thomas, J. L., 75(379), *105*
Thomas, P. D., 75(233), *101*, 150(252), *212*
Thomas, W. A., 54(154), 68(154), 77(154), *99*
Thuillier, A., 395(92), *418*
Thyagarajan, G., 304(156), *337*
Tieckelmann, H., 568(141a), 571(141a), *585*
Tiefenthaler, H., 31(36), 32(36), *96, 206* (422), *217*
Tien Duc, H. C. N., 561(129), 565(129), *585*
Tikhivinskaya, T. I., 574(177a), *586*
Tilborg, W. J. M., van, 32(38), *96*, 122(128), *209*
Tilford, C. H., 240(80), *355*
Tipping, A. E., 436(38a, 38b, 39), 439(38b, 45, 46), 440(38a, 46), 450(39), 456(46), 460(38b), 461(39), 463(38a, 38b, 39, 45, 46), 464(36a), *529*
Tizané, D., 53(144, 147), 69(147), 70(147), 86(315), *99, 103*, 539(22), 540(23, 24, 29, 30, 32), 548(30), 556(22, 24), 558 (11), 563(111), *582, 585*
T'Kint, C., 431(28), 453(28), 470(28), 478 (99), 480(99), 489(99), 497(108), 498 (108), 522(99), 524(99), *529, 531*
Tobin, J. C., 549(74), *584*
Toda, M., 318(188), 330(209), *338, 339*
Tokarev, A. K., 203(412), *216*
Tollary, M. S., 34(373), 84(373), *105*
Toma, C., 78(268), *102*
Tomalia, D. A., 78(270), *102*, 383(79), *418*
Tomasi, R., 576(185), *586*
Tomita, K., 89(324), *103*
Tompa, A. S., 82(275), *102*
Tompsett, A. L. L., 162(307), *214*
Torgov, I. V., 273(122), *336*
Tori, K., 63(171), *100*
Torsell, K., 75(228), 78(228), *101*
Toye, J., 431(26), 451(26), 453(66), 456 (66), 457(66), 465(26), 472(66), 473(66), 484(66), 485(66), 486(66), *529, 530*
Toyoda, T., 271(120), *336*
Trager, W. F., 3(7), 7(7), 8(7), 14(7), 15(7), *19*, 24(1), *96*
Tramontini, M., 149(242), *212*
Trefonas, L. M., 4(15), *19*, 75(197), 89(197), 90(197), *100*, 150(273), *213*
Treshchenko, E. M., 340(219), *342*
Trifunac, A. D., 142(198), *211*
Trotter, J., 545(58), *583*
Tsai, M., 182(371), *215*
Tschoubar, B., 112(45), 114(45), *207*
Tsuchimoto, M., 219(433), 221(433), *222*

Tsuge, O., 116(78), *208*
Tullock, C. W., 89(320), *103*, 230(61), *334, 358*(31), *417,* 461(70), *530*
Tzschach, A., 177(357), *215*

Udalova, V. S., 127(155), *210*
Ueno, Y., 409(108), *419*
Uesu, Y., 92(344), *104,* 548(67), *583*
Ugi, I., 149(245, 246), 199(245, 246), *212*
Uhrhan, P., 554(98), *584*
Ulrich, H., 111(30), 127(154), 128(163), 131(163), 147(163), *207, 210,* 227(38), *334,* 470(83), 473(92b), 491(92b), 509 (83), 519(83), 522(83), *530, 531*
Umetani, T., (233), *342*
Union Oil Company of California, 192(389), *216*
Unterhalt, B., 125(145), 162(145), 165 (318), 175(145), *210, 214,* 169(330), *214,* 578(191), *587*
Uray, G., 111(19a), *207*
Urbain, R., 115(64), *208*
Uritskaya, M. Y., 242(88), *335*
Urushadze, M. V., 196(399), *216*
Utermann, J., 86(298), *103,* 109(16), *207*

Vahrenholt, F., 509(115a), *531*
Van Allen, J. A., 86(309, 310), *103,* 302 (171), 309(171), 311(171), 330(210), *338, 339*
Van Binst, R., 476(96), *531*
Van Camp, A., 476(96), *531,* 442(52), 444(52), 456(52), 457(52), 523(52), *530*
Van Dormael, A., 85(290), 86(290), *103*
Vangedal, S., 569(148), *586*
Vanlierde, H., 442(52), 444(52), 456(52), 457(52), 523(52), *530*
Van Raalte, D., 3(8), *19*
Van Uytbergen-Van Hamme, 498(109), 499 (109), *531*
Van Vyve, T., 73(191), *100*
Van Vyve, Th., 398(98), 399(98), 411(110), *418, 419*
Vasileva, A. S., 348(1), 350(1), *416*
Vasil'eva, V. V., 242(88), *335*
Vattina, L. P., 154(284), *213*
Vaudescal, P., 78(243), *102,* 114(55), 116 (69a), *208*
Vaultier, M., 34(48), *97,* 150(260), 220 (436), *213, 223*
Veenlandt, J. V., 38(58), *97*
Veith, H. J., 87(316), *103*
Venkataraman, K., 295(150), *337*
Venot, A., 75(217), 78(217), *101,* 134 (180), *211*
Verbelen, J., 465(73a), 498(73a), 499 (73a), *530*
Verbruggen, R., 67(181), 77(181), *100*
Verducci, J., 572(161), *586*
Vida, J. A., 138(186a), *211*
Viehe, G., 25(5), 30(5, 34), 31(5, 35), 32(5), 35(53), 40(72, 76), 41(77, 79), 43(76), 44(100), 47(34, 100, 109), 48(72, 119), 49(76, 119, 124), 55(79), 56(77, 156), 65(5), 67(5, 76), 69(77, 124, 185), 70(77), 73(101), 77(5), 78(256), 88(5), 89(392), *96, 97, 98, 99, 100, 102, 105*
Viehe, H. G., 9(27), *20,* 115(65), 116(71), 199(401), *208, 216,* 302(155), *337,* 348 (3), 350(3, 8), 352(8), 353(8), 354(8), 361(42), 363(45), 364(115), 365(45, 47), 366(45), 368(45), 371(55), 372(53, 55), 374(55), 376(63, 64), 377(55, 65, 116), 380(3), 384(83), 385(83, 122), 386(86a), 388(86a), 391(8), 392(89), 393(89), 398 (98), 399(98), 400(101), 401(101), 407 (53, 107), 409(109), 410(109), 411(109, 110), *416, 417, 418, 419,* 423(2a, 2b), 424(2a, 2b), 428(12a), 429(22), 430(25), 432(31a, 31b, 31c), 433(31a, 31b, 31c, 31d, 35, 37), 434(25), 435(34a, 34c), 437(2a, 2b, 40), 438(2a, 2b, 25, 40), 440 (2a, 2b, 31a, 31d, 40), 442(2b, 53a), 444 (2b, 53a), 451(25), 452(25), 453(25, 37), 454(31a, 31b, 31c, 31d, 35, 37, 40), 455 (2a, 31a, 31b, 31c, 31d, 35, 37, 40), 456 (31a, 40), 457(2a, 31b, 31c), 458(31c), 465(11, 25, 31a, 31c), 466(31a, 31b, 31c, 31d), 470(25), 472(2a, 2b), 473(11, 25), 474(25, 31c), 475(31c), 478(31a), 480 (25, 31a), 481(25, 31a), 489(25), 490(11), 499(25), 501(25), 502(31d), 507(2a, 2b, 25), 508(2a, 2b, 25), 520(25), 521(25), *528, 529, 530*
Vilsmeier, A., 226(2, 3), 229(23), 233(2, 3), *333*
Vlasov, V. M., 285(145), *337*
Vlasova, T. F., 340(227), *342*
Vlasvo, V. M., 154(284), *213*
Voghel, G. de, 69(185), 100
Voghel, G. J. de, 371(55), 372(55), 374(55, 68), 375(62), 376(64), 377(55, 68, 116), 388(68), 401(68), *417, 418, 419*
Voghel, G. S. de, 435(34c), *529*
Vogt, G., 340(221), *342*
Vögtle, F., 78(253), *102,* 550(79), *584*
Volgnandt, P., 218(427), *222*
Volkov, M. N., 238(78), *335*
Volodina, M. A., 283(135), *337*
Volz, H., 28(19), 67(19), 69(19), 78(372), 142(201), 149(244), 151(201), 157(201), 183(201), *96, 105, 211, 212*

Vopel, K. H., 255(100), 257(100), 266(100), 331(100), *335*
VoQuang, Y., 570(151), 579(151), *586*
Vouros, P., (317), *103*

Wade, K., 233(65), 235(65), 242(65), *334*
Wade, K. O., 242(89), 243(89), *335*
Wadley, E. F., 226(10a), *333*
Wagener, E. H., 549(73), *584*
Wager, J. S., 115(66), 117(66)
Wagner, E. C., 112(131), *210*
Wagner, H., 302(168), 308(168), *338*
Wagner, H.-U., 65(175), *100,* 377(66), *418*
Wagner, H. V., 509(115b), *531*
Wagner, M., 325(200), *338*
Wagner, P., 163(313), 195(313), *214,* 196(400), 198(400), *216*
Wagner, R. M., 232(63), 302(63), *334*
Wakselman, C., 439(44), 460(44), 461(44), *529*
Wakselman, M., 86(306), *103*
Walia, J. S., 112(31), *207*
Walia, P. S., 112(31), *207*
Walker, G. N., (301), *214*
Walter, W., 89(391), *105,* 222(438, 439, 440), *223,* 351(11), 352(11), *416*
Wanatabe, M., 205(420), *217*
Ward, J. J., 78(380), *105*
Warshel, A., 18(29), *20*
Washburn, R. M., 52(137), 75(137), *99*
Watnick, C. M., 578(192), *587*
Watson, K. G., 582(214), *587*
Watson, S. C., 526(127), *532*
Wawzonek, S., 569(143), 570(143), *585*
Webb, G. A., 575(179), 577(179), *586*
Weber, H. P., 569(149), *586*
Weber, J., 398(99), *418*
Webster, R. G., 242(89), 243(89), *335*
Weglein, R. C., 150(255), *213*
Wegler, R., 384(81), *418*
Wehrli, F. W., 54(152), 57(152), 61(152), 67(152), *99*
Wei, P. H. L., 569(147), *586*
Weidinger, H., 328(202), *338,* 374(60), 398(60), *417,* 430(24a, 24b), 434(32), 465(24a, 24b), 473(24a), 498(32), 503(32), *529*
Weingarten, H., 39(63), 85(63), *97,* 46(106), 50(128), *98, 99,* 115(66), 117(66), 371(52), *417,* 423(3a, 3b), 425(9), 444(3a, 3b, 58a, 58b), 445(9), 451(9), 468(9), 469(9), *528, 530*
Weinheimer, A. J., 78(258), *102*
Weinstein-Lanse, F., 144(212), *211*
Weise, W., 140(190), *211*
Weissenburger, H., 442(51), 456(51), *530*

Weissenfels, M., 278(136, 138), 279(138), 281(136), 282(136), 296(138), *337*
Wellman, K. M., 60(160), *99*
Welvart, Z., 157(293, 294, 296), 186(294), *213*
Wengenmayr, H., 40(66), 48(66), 49(66), *97*
Wenzel, F., 228(41), *334*
West, G., 570(155), *586*
West, P. J., 315(182), *338*
Westlake, A. H., 78(244), *102,* 131(170), *210*
Weutten, W., 175(347), *215*
Weygand, F., 149(237), 179(361, 362), *212, 215*
Wheland, G. W., 85(289), *103*
Whipple, E. B., 55(155), 74(195), 75(195), 86(155, 195, 299), *99, 100, 103*
Whitaker, D. R., 66(178), 67(178), 77(178), 78(178), *100*
White, A. I., 64(172), *100*
White, A. M., 46(107), *98*
White, E. H., 547(65), *583*
White, W. A., 371(52), *417,* 444(58a, 58b), *530*
Whiting, M. C., 85(291), 86(291), *103*
Whitman, P. J., 128(163), 131(163), 147(163), *210*
Wiberg, N., 28(20), 39(64), *96, 97,* 116(77), 118(90), 146(90), *208,* 147(224), *212*
Wickberg, B., 150(268), *213*
Wiechert, R., 256(127), 275(127, 128), 288(127, 128), 294(128), *336*
Wieder, M. J., 63(170), *100*
Wiegrebe, W., 327(201), *338*
Wiessler, M., 217(423), *222*
Wigert, H., 119(110), *209*
Wilber, W. R., 138(186b), *211*
Williams, D. R., 246(96), *335*
Williams, H. K. R., 150(255), *212*
Williams, J. E., 2(3), 3(7), 7(7), 8(7), 14(7), 15(7), *19,* 24(1), *96*
Williams, J. M., 89(325), *103*
Williams, W. M., 581(207), *587*
Williamson, D. M., 266(114), 274(114), *336*
Wilson, J. D., 39(63), 85(63), *97*
Wimmer, T., 182(369), *215*
Wishnewskij, O. W., 403(106), *419*
Witanowski, M., 61(164), 62(164), *99,* 575(179), 577(179), *586*
Witiak, D. T., 246(96), *335*
Witkop, B., 53(140), 74(140, 194), 75(140), *99*
Witzinger, R., 226(4), 229(4), 243(4), *333*
Wolkenstein, D., 304(162), 306(162), *337*
Wolkowski, Z. W., 575(181), *586*

AUTHOR INDEX

Woodcoock, D. J., 547(65), *583*
Woodward, R. B., 509(114a, 114b, 114c), *531*
Woolsey, G. B., 75(366), *104*
Worsham, J. E., 91(337), *104*
Wright, P. H., 240(81), *335*
Wulfman, D. S., 78(380), *105*
Wulkow, G., 111(25), *207*
Wunderlich, K., 142(199), *211*
Wurker, W., 118(84), *208*
Wurmb-Gerlich, D., 550(79), *584*
Würthwein, E.-U., 269(117), *336*
Wynberg, H., 157(291), *213*
Wythe, S. L., (38), *583*

Yagupolskij, L. M., 348(2), 350(2), 352(2), 358(2, 33), *416, 417*
Yakhontov, L. N., 242(88), *335*
Yakubovich, Y., 423(6a, 6b), 443(6a, 6b), 447(6a, 6b), 460(6a), 461(6a), 462(6a), 463(6a), *528*
Yamamoto, K., 229(45), *334*
Yamanaka, H., 312(175), 314(175), *338*
Yanagi, K., 116(78), *208*
Yanagida, S., 75(221, 222), 78(221, 222), *101*, 428(14a, 14b, 15), *528*
Yang, G. N., 307(167), 308(167), *337*
Yates, K., 86(308), *103*
Yoneda, S., 118(96), *209*
Yoo, C. S., 89(327), *103*
Yoshida, Z., 58(159), 70(159), *99*, 118(96), *209*
Yoshizawa, R., 115(60), *208*
Young, J. E., Jr., (13), *19*
Young, L. B., 84(282), *103*

Zagorevskii, V. A., 160(305a), *214*
Zaitsev, I. A., 160(305a), *214*
Zakharova, O. V., 285(145), *337*
Zamar-Wiaux, C., 432(30), 452(30), 526(30), 527(30), *529*
Zaugg, H. E., 95(365), *104*, 108(5, 6), 111 (5, 6), 150(5, 6), 179(5), 182(5, 6), 187 (6), 192(6), 193(6), 198(6), 202(5, 6), 147(230), *206, 212,* 473(92c), 491(92c), *531*
Zbarskij, V. L., 316(184), *338*
Zecher, W., 356(26), *416*
Zedler, A., 91(334), *104*
Zeiler, A. G., 384(82), *418*
Zelenin, K. N., 541(40, 41, 43, 44), *583*
Zemlicka, J., 226(6, 7, 8, 11, 14, 15), 225(101), 262(11), 269(6), 274(8, 15, 124), 275(8), 276(15),278(8, 15), 289(14), 291(15), 292(15), 293(15), 299(7), *333, 336,* 365(47), *417*
Zeyfang, D., 268(116), *336*
Zhivekova, L. A., 384(80), *418*
Ziegenbein, W., 259(105), 278(134), 282 (134), *336*
Ziegler, E., 111(19a), *207,* 182(369), *215,* 579(193), *587*
Ziegler, F., 189(378), *216*
Zilin, V. F., 316(184), *338*
Zimmer, G., 297(151), *337*
Zimmerman, H. E., 509(116), *531*
Zincke, T., 118(84), *208*
Zinner, G., 119(97), 124(138, 139, 141), 125(142), 127(97, 142), *209, 210,* 169(332), *214,* 219(434), *223,* 567 (140), 568(140), *585*
Zinner, H., 119(110), *209*
Zioudrou, C., 550(81), *584*
Zirngibl, L., 564(126), *585*
Zoeppritz, R., 226(1), *333*
Zollinger, H., 29(28), 30(28), 32(28), 35(28), *96,* 75(207), 86(207), *101,* 229(46, 47), 232(47), 318(47), *334,* 302(169), 309(169), *338*
Zschunke, A., 176(352a), *215*
Zubek, A., 570(152), *586*
Zumach, G., 351(15), 354(20), *416*
Zuyanova, T. Y., 549(71), 555(71), *584*

Subject Index

Acetals, cyclization during formylation of, 274
Acetamide derivatives, reaction with oxalyl chloride, 503
Acetoacetic esters, 367
Acetonitrile, double formylation of, 316
N-Acetoxymethylnitramines, 135
N-Acetyliminium salts, ring closure of, 221
Acid chlorides, from α-chloroenamines and carboxylic acids, 500
Acid fluorides, from α-fluoroenamines and carboxylic acids, 500
Activated methyl groups, cyclization during formylation of, 314
Acyclic iminium compounds, ^{13}C nmr spectra of, 65
N-Acylated enamines, 114
Acylimide dichlorides, 147
N-Acyliminium salts, 110, 119
  cycloaddition reactions of, 192
  deprotonation of, 219
α-Acyloxyenamines, from α-chloroenamines and carboxylate anions, 475
  from α-chloroenamines and carboxylic acids, 477
  thermal decomposition of, 477
N-Acylpiperidine, chlorination of, 148
Adenine, 330
Aldehydes, condensation with salts of secondary amines, 118
Aldimines, 217
  acylation of, 110
  alkylation of, 109
  halogenation of, 112
  protonation of, 108
  reactions of, 108
2-Alkenyl keteniminium salts, Diels-Alder reaction of, 522, 524

Alkyl chlorides, from α-chloroenamines and alcohols, 498–501
Alkylaryliminium salts, 138
α-Alkylated carbonyl compounds, preparation of, 114
Alkylidene bis-dialkylamines, alkylation of, 423, 444
  from α-chloroenamines and lithium amides, 507
  reaction with phosphorus trichloride and dichlorophenylphosphine, 445
  see also Ketene N,N-ketal
Alkylidene phosphoranes, 199
N-Alkylmercaptomethylamides, 142
3-Alkyloxazolidines, cleavage of, 141
Alkyltropylium salts, formylation of, 311
Alleneamidinium salts, 437
  from alkylation of ynamines, 437
  from bromination of ynamines, 438
Allenes, cycloaddition to conjugated dienes, 514
  electronic structure of, 470
Amadori rearrangement, 149
Amide chlorides, addition reactions of, 110
  derived from α-disubstituted acetamides and α-substituted lactams, 428–432
  preparation of, 428–432
  thermal dehydrochlorination of, 428, 430
  treatment with a tertiary base, 430
  derived from α-monosubstituted acetamides and lactams, 432–435
  HCl elimination from, 434
  preparation of, 433
  derived from phosgeniminium salts and activated C-H bonds, 435
  from α-chloroenamines and HCl, 497
  from ynamines and HCl, 438
  halogenation of, 498

619

reaction with, α-chloroenamines, 434
triethylamine, 503–504
Amide fluorides, from α-fluoroenamines
and HF, 497
Amide imides, 569
Amides, and α-Ketoamides, 84
charge distribution of, 24
from hydrolysis of α-haloenamines, 465, 473, 496
protonated, 74, 85
protonated and alkylated, spectra of, 42–43
reaction with phosgene/PCl$_5$, 428–430, 434, 503
see also Dichloroacetamides; α-Halo-carboxamides
Amidines, 84
acylamidines, 489
charge distribution of, 24
from α-chloroenamines and primary amines, 478
proton affinities for, 14
protonated, 85
Amidinium salts, 75, 83, 85
organometallic, 117
spectra of, 39–41
spectra of iminologues, 48, 49
spectra of vinylogues, 48–50, 56–58
stabilization energy of, 14
vinylogues of, 50, 85
Amidium salts, 75, 83, 85
spectra of, 42–43
vinylogues of, 85
β-Amidoaldehydes, 194
Amidomethylation, of alcohols, 162
of amines, 173
of arenes, 202
of arsines, 177
of carboxylic acids, 164
of CH-acidic compounds, 181
of cyanates, 175
of cyanides, 179
of dialkylanilines, 202
of dienes, 192
of enamines, 198
of heterocycles, 204
of imides, 175
of olefins, 192
of phenols, 162
of phosphinates, 177

of phosphine oxides, 177
of phosphines, 177
of phosphites, 177
of sulfodiimides, 220
of thiocyanates, 168, 175
of thiols, 167
Aminals, aminoalkylated, 169
Aminals, cleavage with, acyl halides, 123
alkyl or aryl halides, 131
alkyl chlorocarbonates, 123
carbonic anhydrides, 128
halogens or cyanogen bromide, 121
hydrogen halides, 122
inorganic acid halides and anhydrides, 128
monoquaternary salts of, 131, 145, 170
N-hydroxy, 169
reaction with bis(trifluoromethyl)ketene, 196
sulfonyl, sulfinyl or sulfenyl chlorides, 123
unsymmetrical, 169
Amidrazones, protonation of, 545
quaternization of, 541
Amine oxides, reactions of, 143
Amines, tertiary, action of halogens on, 147
formation from N,S-acetals, 142
from methyleniminium salts, 157
high-temperature chlorination of, 147
hydride abstraction from, 142
oxidation of, 150
photochemical chlorination of, 147
photochemical demethylation of, 150
preparation of, 183, 187
α-Aminoacetals, 163
α-Aminoaldehydes, 189
Aminoalkenylation, 427, 484–495
intramolecular, 494–495
nucleophilic, 526–527
Aminocyclobutenones, from α-chloro-enamines and ketones, 505. *See also* Cyclobutenones
Aminofulyenes, charge distribution of, 24
Aminoindenes, 246
Aminomercaptals, 189
α-Aminonitriles, cyanide elimination from, 117
3-Amino-2-pyrazoline, 543
Aminomethylation, 154, 164
of alcohols, 162

SUBJECT INDEX 621

of amides, 169
of aminals, 173
of amines, 169, 170
of azides, 173
of CH-active compounds, 180
of dialkylanilines, 202
of enamines, 196
of enol borinates, 196
of enol ethers, 195
of heterocycles, 203
of hydrazines, 172
of imides, 169
of olefins, 190
of oxetanes, 164
of oxiranes, 163
of phenols, 162, 200
of phenol ethers, 188, 200
of phosphines, 175, 220
of phosphine alkylenes, 199
of phosphites, 177
of thiocarbonates, 166
of thiols, 164
of tetracycline, 169
twofold, 191, 202
Anilides, reaction with phosgene, 432, 434
Anilines, o-substituted, 410
Arsines, 177
Arylpropiolic acid amides, 364
vinylogues of, 369
5-Arylpent-2,4-dien-1-als, preparation of, 245
2-Azaazulenes, preparation of, 340
Azabutadiene, aminodichloro, 385, 387
Azacyanines, spectra of, 48, 49
Azadichlorotrimethinecyanines, 400
Azadichloropentamethinecyanines, 405
Azaenamines (hydrazones), formylation of, 323
7-Azaindoles, formylation of, 242
Azatetraamino trimethinecyanines, 402
Azatrichloropentamethinecyanines, 403
Azetidinones, see β-Lactams
α-Azetidinylideneammonium salts, from α-chloroenamines and imines, 522–523
Azines, acylation of, 559
alkylation of, 559
hydrolysis of, 563
protonation of, 560
Azinium salts, 561

Aziridinium salts, deprotonation of, 205
synthesis of, 205
Aziridinones, decomposition of, 146
Azirines, acylation of, 221
Azirines, 2-amino, from α-chloroenamines and azide ions, 472, 482–484
gas-phase pyrolysis of, 484–485
Azomethine imines, 566, 571
dimerization of, 568
Azomethine ylides, 219
Azulene-1,3-dialdehyde, 236

Benzalacetone, 368
Benzo(b)furans, formylation of, 241
Benzothiophenes, 3-amino, from intramolecular aminoalkenylation, 495
Benzylisoquinolines, cyclization by formylation, 327
Bisamidinium salts, 116
1,2-Bisiminium bromides, 116
Biuret dichlorides, 407
Biurets, 1,1,5,5-tetrasubstituted, 401
Biuret trichlorides, 401
3-Bromoacroleins, preparation of, 297, 301
3-Bromoallylideniminium bromides, as intermediates, 300
β-Bromoalkylamines, 186
α-Bromoenamines, preparation of, from α-fluoroenamines, 433, 447
reaction with ynamines, 437
see also α-Haloenamines
Bromoformylation, of ketones, 297
N-Bromoiminium bromides, preparation of, 112
Bromomethylenedimethyliminium bromide, 297

Carbodiimides, 407
alkylation of, 109
Carbohydrazides, 410
Carbonium ions, 76
stability of, 8
Chemical shifts, 76
Chemical shifts, ^{19}F, 77
Chemical shifts, ^{14}N (^{15}N), 77
3-Chloroacroleins, fragmentation of, 280
preparation of, 271, 278, 280, 299, 340
stereoisomerism of, 280
use in heterocyclic synthesis, 275
α-Chloroamines, 183

α-Chloro-β-chlorocarbonylenamines, 371
α-Chloroenamines, acidic hydrolysis of, 496
  acylation of, 433, 434
  β-Acyl substituted, heterocyclizations with, 502–503
    preparation of, 440
  β-alkyl and β-aryl substituted preparation of, 428–430, 445
  bis α-chloroenamines, preparation of, 431
    X-ray diffraction of, 466
  bromination of, 497–498
  chlorination of, 429, 497–498
  β-chloro substituted, preparation of, 442, 444, 447–449
  β-chlorocarbonyl substituted, preparation of, 433, 435, 501
  dehydration with, 489
  β-Dichloro substituted, see Trichlorovinylamine
  elimination of HCl from, 434, 507
  β-Formyl substituted, preparation of, 438
  β-Functionalized, preparation of, 435
  Grignard reagents of, 526
  β-Monosubstituted, preparation of, 434, 435
  nuclear quadrupole resonance of, 466–467
  with a primary amino group, preparation of, 450
  reaction with, acetylacetone, 496–487
    alcohols, 498–501
    alkoxides, 471, 472, 473–475
    amide chlorides, 434, 503
    aromatic compounds, 473, 490–495
    azide ions, 472, 482–484
    benzamide, 489
    calcium chloride, 469
    carboxylic acids, 477, 498–501
    chlorosulfonyl isocyanate, 505–506
    cyanamide, 478
    cyanide ions, 484–486
    dimethylsulfoxide, 495
    diphenylsulfoxide, 496
    dry HCl, 497
    Grignard reagents, 489–490
    hydroxide, 473
    imines, 522–523
    ketenes, 505
    magnesium, 526
    metal amides, 481
    methyl iodide, 447

    nitromethane, 486–488
    organolithium compounds, 489–490
    oxalyl chloride, 503
    phosgene, 433, 501
    potassium or cesium fluoride, 446, 477
    primary amines, 478–481
    secondary amines, 481
    silver salts, 468, 472
    sodium salt of malononitrile, 488
    thiolates, 473–475
    trimethylphosphite, 496
    ynamines, 437, 508, 521
  β-Thioalkyl substituted, preparation of, 441–442
  transition metal complexes from, 524–525
  β-unsubstituted, preparation of, 434
  β-vinyl substituted, preparation of, 431, 444
  reaction with nitriles, 522, 524
  see also α-Haloenamines
α-Chloroenaminonitriles and esters, cyclization with, 360
Chloroformamidine, phenylethynyl, 388
Chloroformamidines, 379, 403
  vinyl, 404
Chloroformylation, of carboxamides, 301, 302
  of lactams, 303
  mechanism of, 275
  of pyrrolinones, 304
  of steroidal ene- and dienones, 286
  of steroidal ketones, 283
Chloroiminium salts, 117
N-Chloromethylalkylnitramines, 125, 135, 138, 141, 217
N-Chloromethylamides, 192
Chloromethylene dibenzoate, 228
Chloromethylenedimethyliminium chloride, 226
  formation of, 231
  reaction with, aliphatic olefins, 250
    3,3-dimethylaminoacrolein, 231
    DMF, 231
    styrene, 243
Chloromethyliminium chloride, addition of nitriles, 317
Chloromethyliminium chlorides, 228
Chloromethyleniminium salts, cyclization with, 294, 324
  formylating potential of, 232

SUBJECT INDEX

reaction with, acetic acids, 318
  activated methylene groups, 308
  aliphatic diazo compounds, 324
  aliphatic olefins, 250
  aromatic hydrocarbons, 234
  azines, 321
  carboxamides posessing an -methylene group, 301
  dienamines, 258, 263
  enamines, 258
  enol ethers and ketals, 271
  hydrazones, 321
  ketones, 288
  malonic acids, 320
  methyl ketones, 274
  methylene ketones, 274
  olefins, 243
  polyenals, 285
  polyenones, 285
  stereoidal ketones, 293
  trienamines, 258, 266
  vinyl ethers and acetals, 267
  vinylogous enol ethers and acetals, 271
N-Chloromethylimides, 192
α-Chloromethylmethylnitrosamine, 217
Chlorotrimethinecyanines, 397
N-(β-Chlorovinyl)chloroformamidinium salts, 384
Chlorovinyl ketones, preparation of, 298
Chromemane-3-carboxaldehydes, by formylation, 296
Conductivity measurements, 94
Conductometric studies, 154
Conformational properties, of creatine phosphate, 18
  of substituted creatines, 18
Coupling constants, carbon-nitrogen, 64, 76, 78
  proton-carbon, 67, 76, 77, 84
  proton-fluorine, 68
  proton-nitrogen, 63, 78
  proton-proton, 59–60, 77
Creatine phosphate, conformational properties of, 18
Creatines, conformational properties of, 18
Cyanoacetates, 359
Cyanoacetic acid, 359
Cyanamides, 379
Cyanide elimination, from α-aminonitriles, 117

Cyanines, 83, 85
  charge distribution of, 25
  protonated, 74
  spectra of, 48, 49
2-Cyanoenamines, from α-chloroenamines and cyanide ion, 484–486
  synthesis of β-diketones with, 486–487
Cyanogen chloride, reaction with alcohols, 355
Cyanuric chloride, reaction with DMF, 229
Cyclic iminium salts, 75
  spectra of, 52–55, 56–58
Cycloadditions, of α-haloenamines, 508
  of keteniminium ions, 508–524
Cyclobutanones, from α-chloroenamines and ketenes, 505
  from keteniminium salts, and dienes, 516–517
  and olefins, 511–512
Cyclobutenones, from keteniminium salts and acetylenes, 519–521. *See also* Aminocyclobutenones
Cyclobuteniminium salts, Diels–Alder reaction of, 521
  from keteniminium salts, and acetylenes, 519–520
  and dienes, 516–517
  and olefins, 511–512
Cyclobuten(yl)cyanines, 437
  from alkylation of ynamimes, 438
  from bromination of ynamines, 438
  from ynamines and α-chloroenamines, 434, 508, 521
Cyclobutylmalondialdehyde, 269
Cyclohexenones, from β-vinyl-α-chloroenamines, 481–482
Cyclohexylmalondialdehyde, 269
Cyclopentylmalondialdehyde, 369
Cyclopropylideniminium salts, 118

Deuteroporphyrin-IX-Cu(II) dimethyl ester, formylation of, 240
Dialkylacyloxymethylamines, cleavage with hydrogen halides and other acids, 133
Dialkylalkoxymethylamines, cleavage with acid halides, acid anhydrides and cyanogen bromide, 135
  cleavage with hydrogen halides and other acids, 133
Dialkylalkylmercaptomethylamines, 164

cleavage with acyl halides and halogens, 141
Dialkylaminoacetonitriles, 179
α-Dialkylaminoalkanols, esters of, 128
Dialkylaminomethyldialkyl phosphonates, 177
Dialkylaminomethyldiethyl phosphonates, 130
Dialkylaminomethyl esters, 164
  nmr data of, 153
Dialkylaminomethyl ethers, reaction with bis(trifluoromethyl)ketene, 196
Dialkylaryloxymethylamines, cleavage with acid halides, acid anhydrides, and cyanogen bromide, 135
Dialkylazidomethylamines, 173
Diaryl aldazines, monoquaternization of, 109
Diazatrichloropentamethine cyanines, 402
Diazene, 571
Diazene-hydrazone rearrangement, 571
Diazenium ion, 570
Diaziridines, 566
  acid hydrolysis of, 546
Diaziridinium ion, 547
Dibromomethyleniminium bromides, synthesis of, 357
1,3-Dicarbonyl derivatives, by formylation, 267
Dichloroacetamide (α-substituted), reaction with trivalent phosphorus compounds, 447–449
Dichloromethyl alkyl ethers, 228
Dichloromethylenimines, alkylation of, 354
  protonation of, 354
Dichloromethyleniminium salts, 344
  addition to, t-cyanamides, 400
    cyanates, 406
    heterocumulenes, 406
    imines, 403
    vinyl ethers, 399
    ynamines, 371
  alkylation of, 348
  comparison with related compounds, 345
  hydrolysis of, 345
  properties of, 345
  reaction with, t-acetamides, 371
    activated methylene groups, 359
    aldehydes, 395
    alcohols and thiols, 392
    amidrazones, 411
    arylamines, 381
    carbohydrazides, 411
    carboxylic acids, 395
    complex counterions, 355
    cyclic ethers, 394
    DMF, 396
    dimethyl sulfate, 355
    enamines, 382, 397
    ethyl glycine hydrochloride, 383
    Grignard reagents, 358
    hydrazines, 381
    hydrazones, 411
    hydroxylamines, 381
    inorganic anions, 388
    ketones, 363, 368
    ketene acetals, 400
    2-methylbenzoxazole, 373
    nitriles and HCl, 403
    nitrogen compounds, 379, 397, 411
    oximes, 395
    phenols and thiophenols, 391
    phenylhydrazines, 411
    phosphites, 390
    primary amides, 395
    secondary acetamides, 384
    semicarbazides, 411
    sulfamide, 383
    tetrachlorocyclopentadiene, 360
    trichloroisocyanuric acid, 381
    tricyanomethane anion, 362
    triethyloxonium tetrafluoroborate, 355
  structure of, 345
  S_E on aromatic compounds, 377
  synthesis of, 348
  thermolysis of, 345
  thiolysis of, 348
  use in heterocyclizations, 409
1,5-Dichloropentamethinecyanine, 367
1,3-Dichlorotrimethinecyanines, 371, 396
  aminolysis of, 375
  heterocyclization of, 377
  hydrolysis of, 375
  reaction with arylamines, 376
  thiolysis of, 375
Diels-Alder reaction, 194
α,β-Dihaloamines, base-catalyzed elimination reaction of, 436
1,4-Dihydropyridines, 219
α-Diketones, synthesis from

SUBJECT INDEX 625

α-cyanoenamines, 486–487
3,5-Dimethoxyphenylacetonitrile, cyclization by formation of, 317
2,4-Dimethoxystilbene, 248
4,6-Dimethoxy-3-stilbenecarboxaldehyde, 248
β-Dimethylaminoacrolein, 267, 268
   preparation of, 267, 268
   reaction with POCl$_3$, 330
4-Dimethylamino-6-hydroxy-3,5-bis(p-methoxyphenyl)-fulvene-2-carboxaldehyde, 249
Dimethylaminomethyl methyl ethers nmr spectrum of, 153
4-Dimethylamino-α-phenylcinnamaldehyde, 248
Dimethylaminomethylphosphines, 175
4-Dimethylaminostilbene, 248
Dimethylaminosulfur trifluoride, 357
N,N-Dimethylcarbonamides, formylation of, 302
Dimethylchloromethylamine, 154
Dimethylformamide chloride, reaction with, malonic diesters, 360
   malononitrile, 360
Dimethylformamide, reaction with POCl$_3$, 230
Dimethylformamide/POCl$_3$, as formylating agent, 226
Dimethylthioformamide, reaction with POCl$_3$, 233
α,ω-Diphenylpolyenes, formylation of, 248
Dithiocarbamates, chlorination of, 352
Dithiuram disulfides, reaction with fluorinating agents, 357

Electron density, 76
Electron spin resonance, 94
Electrophilic substitution, intramolecular, 236
Enamides, bromination of, 116
Enamines, 84
   acylation of, 114
   alkylation of, 112
   α-chloro-β-chlorocarbonyl, 407
   formylation of, 258
   from methyleniminium salts, 158
   hydrolysis of, 157
   oxidation of, 218
   protonation of, 74, 112, 217
   reactions of, 112, 160
   reaction with, halogens and inorganic halides, 116
      sulfonyl isocyanates, 221
      sulfonyl isothiocyanates, 221
   synthesis of, 177, 187
   vicinal, 115
Enamino ketones, reactions of, 117
Endiamines, 217
   acylated derivatives of, 115
   oxidation of, 218
Enehydrazines, protonation of, 538
Enol borinates, 196
Enol ethers, 195
ESCA, 94
Eschweiler-Clarke methylation, 149
Ethylenimine, 383
Ethylideniminium hexachloroantimonate, 218
5-Ethyl-5-phenylbarbituric acid, 138
5-Ethyl-5-phenylbarbituric acid, bis(halomethyl) derivatives, 138
Exchange processes, 25, 82

Fischer synthesis, 553, 556
α-Fluoroenamines, β-acyl substituted, preparation of, 441. *See also* α-Haloenamines
   β-alkyl substituted, preparation of, 439, 446
   β-dihalosubstituted, preparation of, 442–444, 450
   β-formyl substituted, preparation of, 438
   β-heterosubstituted, preparation of, 446
   preparation of, from α-chloroenamines, 446
   reaction with, boron trifluoride, 469
      carboxylic acids, 500
      HF, 497
      keteniminium salts. 506
      α-iodoenamines, 506–507
      lithium bromide, 443, 447, 478
      lithium chloride, 478
      lithium iodide, 443, 447, 478
      lithium methoxide, 472
      methyl lithium, 490
      malonitrile, 488
      phosgene, 501
   β-unsubstituted, preparation of, 443
Fluoromethyldialkylamines, 126
   nmr studies of, 153

Formamide/POCl$_3$, 330
Formanilide/POCl$_3$, as formylating agent, 226
3-Formylmethylenephthalimidines, 262
Formylpiperidine, reaction with POCl$_3$, 234
Formylpyrrolidine, reaction with POCl$_3$, 234
Furanones, from acetamides and oxalyl chloride, 503
Furans, formylation of, 237

Guanidine, vinyl, 405
Guanidines, proton affinities for, 14
Guanidinium ions, stabilization energy of, 14
Guanidinium salts, 82, 84
  spectra of, 45-46

N-α-Haloalkylamides, 110, 119, 155, 218
N-α-Haloalkylimides, 119
α-Haloamines, 150, 154
  bicyclic, 218
α-Halocarboxamides, from halogenation of α-chloroenamines, 497
α-bifunctional character of, 426, 507, 525
  equilibrium with keteniminium halides, 425, 465
  geometrical isomerism, 467-468
  halide exchange, 477-478
  l-halo-N,N-bis(trifluoromethyl)alkenyl-amines, preparation of, 436-437, 439
  hydrolysis of, 496
  IR spectra of, 465-466
  nucleophilic substitution on, mechanistic aspects, 470-473
  proton and ^{13}C nmr of, 466, 467
  reaction with electrophiles, general aspects, 496, 500-501
  solubility properties, 465
  stability of, 449, 450-465
  synthesis of, general aspects, 427-428
    by electrophilic additions to ynamines, 437-442
    by halovinylation of metal amides, 442-443
    by nucleophilic additions to haloacetylenes, 442
    by substitution on -heterosubstituted enamines, 444-447
    from amide chlorides, 428-435
    from α,β-dihaloamines, 436
  thermal stability of, 465

X-ray diffraction of, 466
  see also α-Bromo-; α-Chloro-; α-Fluoro-; and α-Iodoenamines
β-Haloenamines, 112
β-Haloiminium salts, 112
N-Halomethyl amides, 148
Halomethylammonium halides, electrophilic potential of, 232
  fragmentation of, 145
N-Halomethyl imides, 148
Heterocyclization, with β-acyl-α-chloroenamines, 502-530
  with bisiminium salts, 504-505
Heterogeminals, 121
N-Heterosubstituted iminium salts, spectra of, 51
Hexahydrotetrazines, 553, 568, 570
Hinman's reaction, 539
Hydrazones, dimerization of, 550
  intramolecular quaternization of, 569
  E-Z isomerization of, 544, 549
  protonation of, 544
  quaternization of, 540, 555
Hydrazonium salts, 538, 552
Hydride abstraction, from tertiary amines, 142
Hydrogen bonding, 75
N-α-Hydroxyalkylamides, halogenation of, 119
N-α-Hydroxyalkylimides, halogenation of, 119
Hydroxypyrimidines, formylation of, 307

Imidazo(1,5-a)pyridine, formylation of, 340
Imide chlorides, from α-chloroenamines, 429
Imines, basicity of, 3
  proton affinities for, 14
  protonated, 74, 76
Iminium carbonates, 84, 391
  ring opening of, 392
  spectra of, 47
  use in synthesis, 393
Iminium compounds, cumulated, 50
Iminium ions, ability of double bond to conjugate with, 10
  effects of substituents on electronic structure, 8
  electronic structure theory, 2
  ground-state geometry of, 3
  orbital energies and atomic populations, 5
  rotational barriers, 6, 13

stability of, 7
stabilization energy of, 14
stretching frequencies of, 5
structural and spectral properties, 2
Iminium methylheptalene derivatives, formation of, 311
Iminium pentabromocarbonates, 218
Iminium salts, alkoxychloro, 392
  as intermediates, 149
  bicyclic, 118
  chloromercapto, 410
  chlorophenoxy, 391
  chlorophenylthio, 391
  cycloaddition with ynamines, 199
  dealkylation of, 345, 385
  diazido, 389
  dichloro, 345. See also Dichloromethyleniminium salts
  diphenoxy, 391
  electrophilic character of, 346
  formation from aziridines, 220
  hydride transfer to, 186
  reaction with Grignard reagents, 184, 185
  secondary, 126, 142
  ternary, 117, 142
  vinylogues of, 118
Iminium thiocarbonates, 84
  spectra of, 47
Iminologues, of amidinium salts, 48, 49
Indanones, 247
Indene, formylation of, 250
Indoles, formylation of, 241, 316
Infrared spectroscopy, of iminium salts, 26, 74
α-Iodoenamines, preparation of, from α-chloroenamines, 447
  from α-fluoroenamines, 443, 447
  reaction with α-fluoroenamines, 506–507
  see also α-Haloenamines
Iodoformylation, of ketones, 297
Iodomethylenedimethyliminium iodide, 297
Iron(III)hemin, formylation of, 246
Isocyanide dichlorides (see Dichloromethylenimines, 147, 354
α-Isocyanoenamines, from α-chloroenamines and silver cyanide, 486
  rearrangement of, 486
E-Z Isomerization, of hydrazones, 449, 544
  of oximes, 575

of thiooximes, 581
Isopropylideniminium hexachloroantimonate, 218
Isopyrazoles, quaternization of, 560
Isopyrazolium salts, 558
  reduction of, 562

Ketene N-N ketal, from α-chloroenamines and amines, 481. See also Alkylidene bis-dialkylamines
Ketene O-N ketal, from α-chloroenamines and alkoxides, 473–475
Ketene S-N ketal, from α-chloroenamines and thiolates, 473–475
Ketenes, cycloaddition to conjugated dienes, 514
  ^{13}C nmr spectra of, 469
  electronic structure of, 469–470
  preparation of, from α-acyloxyenamines, 477
  reaction with α-chloroenamines, 505, 508
  thermal (2+2) cycloadditions, 509–510
Ketenimines, 409
  acylation of, 110
  alkylation of, 109
  electronic structure of, 469–470
  halogenation of, 112
  protonation of, 108
  reactions of, 108, 423
Keteniminium halides, see α-Haloenamines
Keteniminium salts, with acetylenes, 519–521
  2-alkenyl substituted, Diels-Alder reaction of, 522, 524
  with allenes, 519
  ^{13}C nmr spectra of, 469
  with cis-fixed dienes, 518–519
  counterions, 425, 468
  covalent tautomers, 425
  cycloaddition reactions of, 427
    regiospecificity of, 513, 518, 521
  with imines, 522–523
  with olefins, 510–514
  selectivity of, 514, 518
  stereospecificity of, 513, 518
  steric effects in, 514
  theoretical aspects, 423, 508–510
  with transoid dienes, 515–516
β-Ketoamides, from amide chlorides, 434, 504–505
  preparation of, from α-acyloxyenamines, 477

Ketones, condensation with salts of secondary amines, 118
Kinetics, 83, 95

β-Lactams, from α-chloroenamines and imines, 522–523
  dichloro, 407
  from α-fluoroenamines and α-iodoenamines, 506–507
Leuckart-Wallach reaction, 149
Lithium azide, 388

Malondialdehydes, preparation of, 259
Malondiamides, 373, 505
  activated derivatives of, 371
  dithio derivatives of, 375
Malonic acid derivatives, 433, 501–502
Malonic diesters, 359
Malonitrile, double formylation of, 316, 359
Mannich condensation, 149, 154, 181, 553
Mannich-Böhme reagents, see Chloromethyleniminium salts
Mass spectroscopy, of iminium salts, 86
Mercaptoformamide chlorides, see Chloromercaptomethyleniminium salts
Mercuric acid dehydrogenations, 143
Merocyanines, alkylated, 74
Methyl groups, in cycloiminium salts, formylation of, 309
  in electron deficient heteroaromatics, formylation of, 312
  in polymethinium salts, formylation of, 309
  in pyrylium salts, formylation of, 309
  on nitro compounds, formylation of, 316
7-Methyladenine, 330
2-Methylbenzothiazole, 373
2-Methylbenzoxazole, 373
Methylbis(chloromethyl)amine, 131, 155, 183
3,4-Methylenedioxystyrenes, conversion to aminoindenes, 248
  cyclization by formylation of, 246
Methyleniminium hexachloroantimonate, 146
Methyleniminium salts, aldol condensation reactions of, 160
  analysis of, 156
  anion exchange with, 155

C–C bond formation with, 179
C–N bond formation with, 169, 219
C–O bond formation with, 162, 219
C–P bond formation with, 175
C–S bond formation with, 164
conductometric studies of, 154
cycloaddition reactions with, 194
deprotonation of, 157, 219
handling of, 156
H–D exchange in, 160
hydrolysis of, 156
infrared absorption of, 151
mass spectra of, 154
nuclear magnetic resonance of, 152
photochemical addition of methanol, 206
physical properties of, 150, 154
polarographic studies of, 154
reactions, 154, 155, 156, 162, 164, 179, 219
reaction with, aromatic compounds, 200
  CH-acidic compounds, 180
  cyanides, 179
  diazoalkanes, 205
  1,3-didnes, 194
  enamines, 196
  enol ethers, 195
  Grignard reagents, 183
  heterocycles, 203
  magnesium powder, 206
  multiple-bond systems, 190
  phosphine alkylenes, 199
reduction with, 1,4-dihydropyridines, 219
  hydrides, 157
stability, structure and solubility of, 150
synthesis of, 180
ultraviolet absorption spectra of, 151
N-Methylformanilides, $POCl_3$ complexes of, 226, 233
Methyl p-methoxyphenylformamide, reaction with $POCl_3$, 233
N-Methyl methyleniminium hexachloroantimonate, 218
Methyl p-nitrophenylformamide, reaction with $POCl_3$, 233
N-Methylpyrrolidone, 373
Michaelis-Arbosov rearrangement, 177

Naphthalene-1,8-diamine, 410
Nicotinic aldehyde, from crotonaldehyde, 265

Nitrogen heteroaromatic compounds, charge
    distribution of salts of, 25
Nitrones, 571, 579
    alkylation of, 572
    dimerization of, 580
    photocyclization of, 580
    protonation of, 574
    reactions of, 143
Nitroso derivatives, alkylation of, 572
    protonation of, 574
Nitrosyl chloride, addition to imines, 217
Nuclear magnetic resonance, of iminium
    salts, 26, 76
Nuclear magnetic resonance, ^{13}C, 65, 76
    ^{19}F, 68

Olefins, 190
Oxalyl chloride, reaction with DMF, 231
Oxazenium ions, 579
Oxaziridines, 580
Oxaziridines, protonation of, 574
2-Oxazolinium salts, 573, 577
Oximes, alkylation of, 572
    E-Z isomerization of, 575
    protonation of, 75, 574
Oximinium salts, 538, 571, 576, 578

Pentalene, azabisamino, 401
Pentamethinium salts, formylation of, 262
    preparation of, 332
Perchlorodienamines, synthesis of, 444
Perfluorocyclobutane derivatives, pyrolysis
    of, 450
Phenylacetone, 367
Phenylacetonitrile, 360
Phenylpropiolamidines, 365
Phosgeniminium salts, 346
    reaction with, activated C−H bonds, 435
        amides, 435
        monosubstituted acetyl chlorides, 435
    see also Dichloromethyleniminium salts
Phosphine oxides, 177
Phosphines, 177
Phosphites, 177
Phosphoranes, 359
Phthalimidines, formylation of, 305
Piloty's reaction, 554, 559
Polarimetry, 94
Polarographic methods, 154
Polarography, of iminium salts, 94

Polonovski reaction, 149, 164
Polyeniminium salts, 331
Positive charge delocalization, 89
Positive charge localization, 24, 76, 83
Pressure of decomposition, 95
Protonated N-heterocycles, ^{13}C nmr spectra
    of, 66
Pyrazoles, 553
Pyrazole-3-carboxaldehydes, from
    hydrazones, 323
Pyrazolidines, 547
Pyrazolidones, reduction of, 539, 556
2-Pyrazolines, 544, 547, 558
    basicity of, 548
    quaternization of, 541
3-Pyrazolines, 539, 546, 553
    protonation of, 540
    quarternization of, 540
4-Pyrazolines, formylation of, 340
Pyrazolinium salts, 539, 546, 551
Pyrazolium salts, 539
Pyrazolones, 539
    formylation of, 305
    reduction of, 556
Pyridinium salts, 76, 94
    spectra of, 52−55
Pyrimidines, formylation of, 312
Pyrocatechol phosphortrichloride, 429
Pyrroles, 554, 559
    formylation of, 237, 240
    reaction with POCl$_3$, 333
Pyrrolinones, formylation of, 305

Quaternary enammonium salts, 112

Rearrangement, diazene-hydrazone, 571
Resonance and inductive effects, on
    carbonium ion stability, 8
Resorcinol, 226
Ritter reaction, 316, 328
Rotation barriers, 78, 79−81

Seleno(3,2-d)pyrazoles, formylation of, 242
Semmler-Wolff reaction, 578
Sodium azide, 388
Strecker synthesis, 149
Styrenes, formylation of, 245
Substitution reactions, nucleophilic, of
    α-haloenamines, 470−496
Sulfenamides, 581

C-Sulfonylthioformamides, chlorination of, 352
Sulfur tetrafluoride, 357

Tautomeric structures, 25, 26, 87
Tertiary amines, from dichloromethyleniminium salts, 359
  high-temperature chlorination of, 356
Tetraaminoallenes, 375
Tetraaminoethenes, bromination of, 116
Tetrahydropyridazines, 555, 565
  quaternization of, 541
α-Tetralone, 366
β-Tetralone, 367
Tetramethyladipamide, 374
Tetramethylketeniminium tetrafluoroborate, see Keteniminium salts
Tetrazoles, amino, 388
  aminoalkoxy, 388
6a-Thiathiophthenes, formylation of, 242
1,3-Thiazinium salts, preparation of, 341
Thieno(3,2-d)pyrazoles, formylation of, 242
Thioamides, charge distribution of, 24
  protonated and alkylated, spectra of, 44
  phosgenation of, 432
Thioamidium salts, 75
  spectra of, 44
Thiocarbamoy chlorides, 348, 349
  alkylation of, 348
  chlorination of, 349
  protonation of, 348
Thiocyanates, 175
Thioformamides, chlorination of, 352
3-Thioformylindolizines, 241
Thiooximes, E-Z isomerization of, 581
Thiophenes, formylation of, 237, 238
  preparation of, 341
Thiopyrazolones, reduction of, 556
Thiopyrylium salts, preparation of, 341
Thiouronium salts, spectra of, 45–46
Thiurame disulfides, chlorination of, 351
Trialkylfluoromethylhydrazines, 127
Triaminoethenes, bromination of, 116
Tribromomethylamines, 344
Trichloromethylamines, 344
Trichloromethylmercury, reaction with, allylamines, 449–450
  secondary amines, 449–450
Trichlorovinylamine, chlorination of, 497

reaction with, (optically active) alcohols, 498
  amines, 478–481
  synthesis of, 428, 429, 447–449
Tricyanomethane derivatives, phosphorylation of, 450
Trifluoromethylamines, 344
  synthesis of, 357
Trimethinium salts, from acetic acids by formylation, 319
  formylation of, 262, 263
  from malonic acids by formylation, 321
Triose reductone, ethers of, 269
Tris(chloromethyl)amine, 131, 155, 162, 183, 189
Tscherniac-Einhorn synthesis, 193

Ugi's four-component condensation, 149, 553
Ultraviolet and visible spectroscopy, of iminium salts, 26, 69–73, 74, 84
Urea dichlorides, 361, 380, 383
Ureas, ethinyl, 387
  tetrasubstituted, 387
  vinyl, 386, 404
Uronium salts, 75, 84
  spectra of, 45–46

Vilsmeier-Haack-Arnold reagents, 358
  basicity of the anions, 231
  formation of, 229
  related formylations, 328
  structure of, 229
  see also Dichloromethyleniminium salts
Vinylogous amides, reactions of, 332
Vinlogous formylations, 330
Vinylogues, of amidinium salts, 48, 49, 50, 56–58, 85
  of amidium salts, 56–58, 85
  of iminium salts, 33–38, 84
N-Vinylpyrrolidone, 399
von Braun amide degradation, 149

X-ray diffraction structures, of iminium salts, 89
X-ray photoelectron spectra, of pyridinium salts, 94

Ynamine amides, 375
Ynamine amidines, 376

Ynamines, 375, 396
  acylation of, 440–441
  addition of HX, alkyl-X, and X_2, 437–440
  addition of sulfenylchlorides, 441–442
  cycloaddition with iminium salts, 199
  from monosubstituted α-chloroenamines, 434, 507
  photochemical addition of perfluoro N-haloamines, 440
  reaction with, α-chloroenamines, 434, 508, 521
    electrophiles, 423, 437–442
    fluoroalkenes, 439

Zincke aldehyde, 331